Trigonometric Formulas

$$\sin\theta = \frac{y}{r} \qquad \cos\theta = \frac{x}{r} \qquad \tan\theta = \frac{y}{x}$$

$$\csc\theta = \frac{r}{y} \qquad \sec\theta = \frac{r}{x} \qquad \cot\theta = \frac{x}{y}$$

$$\sin^2\theta + \cos^2\theta = 1$$
$$1 + \tan^2\theta = \sec^2\theta$$
$$1 + \cot^2\theta = \csc^2\theta$$

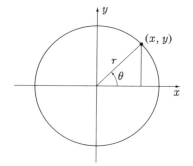

Cosine law- $c^2 = a^2 + b^2 - 2\,ab\cos\theta$

Sine law- $\dfrac{\sin A}{a} = \dfrac{\sin B}{b} = \dfrac{\sin C}{c}$

$$\sin(A + B) = \sin A \cos B + \cos A \sin B$$
$$\sin(A - B) = \sin A \cos B - \cos A \sin B$$
$$\cos(A + B) = \cos A \cos B - \sin A \sin B$$
$$\cos(A - B) = \cos A \cos B + \sin A \sin B$$

$$\tan(A + B) = \frac{\tan A + \tan B}{1 - \tan A \tan B} \qquad \tan(A - B) = \frac{\tan A - \tan B}{1 + \tan A \tan B}$$

$$\sin 2A = 2\sin A \cos A$$
$$\cos 2A = \cos^2 A - \sin^2 A = 1 - 2\sin^2 A = 2\cos^2 A - 1$$
$$\tan 2A = \frac{2\tan A}{1 - \tan^2 A}$$

$$\sin A \sin B = \frac{1}{2}[-\cos(A + B) + \cos(A - B)]$$

$$\sin A \cos B = \frac{1}{2}[\sin(A + B) + \sin(A - B)]$$

$$\cos A \cos B = \frac{1}{2}[\cos(A + B) + \cos(A - B)]$$

$$\sin A + \sin B = 2\sin\left(\frac{A + B}{2}\right)\cos\left(\frac{A - B}{2}\right)$$

$$\sin A - \sin B = 2\cos\left(\frac{A + B}{2}\right)\sin\left(\frac{A - B}{2}\right)$$

$$\cos A + \cos B = 2\cos\left(\frac{A + B}{2}\right)\cos\left(\frac{A - B}{2}\right)$$

$$\cos A - \cos B = -2\sin\left(\frac{A + B}{2}\right)\sin\left(\frac{A - B}{2}\right)$$

Inverse Trigonometric Function	Principal Values
$\mathrm{Sin}^{-1} x$	$-\dfrac{\pi}{2} \le y \le \dfrac{\pi}{2}$
$\mathrm{Tan}^{-1} x$	$-\dfrac{\pi}{2} < y < \dfrac{\pi}{2}$
$\mathrm{Cos}^{-1} x$	$0 \le y \le \pi$

Inverse Trigonometric Function	Principal Values
$\mathrm{Cot}^{-1} x$	$0 < y < \pi$
$\mathrm{Csc}^{-1} x$	$-\pi < y \le -\dfrac{\pi}{2}, \quad 0 < y \le \dfrac{\pi}{2}$
$\mathrm{Sec}^{-1} x$	$-\pi \le y < -\dfrac{\pi}{2}, \quad 0 \le y < \dfrac{\pi}{2}$

Canadian Cataloging in Publication Data

Trim, Donald W.
    Calculus for engineers
ISBN 0-13-785494-3
1. Calculus. 2. Engineering mathematics I. Title
QA303.T735 1998 515 C97-931476-3

 © 1998 Prentice-Hall Canada, Inc., Scarborough, Ontario
A Division of Simon & Schuster/A Viacom Company

Prentice-Hall, Inc., Upper Saddle River, New Jersey
Prentice-Hall International (UK) Limited, London
Prentice-Hall of Australia, Pty. Limited, Sydney
Prentice-Hall of Hispanoamericana, S.A., Mexico City
Prentice-Hall of India Private Limited, New Delhi
Prentice-Hall of Japan, Inc., Tokyo
Simon & Schuster Southeast Asia Private Limited, Singapore
Editora Prentice-Hall do Brasil, Ltda., Rio de Janeiro

ISBN 0-13-785494-3

Publisher: *Patrick Ferrier*
Acquisitions Editor: *Sarah Kimball*
Development Editor: *Maurice Esses*
Marketing Manager: *Catherine Koop*
Copy Editor: *Edie Franks*
Production Editor: *Mary Ann McCutcheon*
Production Coordinator: *Deborah Starks*
Art Director: *Mary Opper*
Cover Design: *Gary Beelik*
Cover Image: *Tony Stone/Van Scoy Photography*
Page Layout: *PreT$_E$X, Inc.*
Technical Checkers: *Ishuwa Sikaneta*
                    Bernie Brooks
                    Jon-Paul Voroney

1 2 3 4 5 C 02 01 00 99 98

Printed and bound in United States

Visit the Prentice Hall Canada Web site! Send us your comments,
browse our catalogues, and more. **www.phcanada.com** Or reach us
through e-mail at **phcinfo_pubcanada@prenhall.com**

In memory of my father, who,
like his son, was a man of few words.
Our love was expressed in other ways.

Brief Table of Contents

Contents

Preface

A Special Focus

This text has been written especially for students in engineering and the physical sciences. At the same time, students from other disciplines might well find this book appealing. Like every instructor, I have developed certain views on the best pedagogical approach to calculus. These views are often independent of the area of study of the student.

It is the choice of examples and exercises that help to focus a text for students in a specific discipline. Throughout this book, I have chosen to emphasize the physical applications of calculus. To this end, I have consulted engineers and physicists, and I have searched the literature carefully for physically meaningful examples and exercises.

Three categories of applications recur throughout the book to demonstrate the indispensibility of calculus in engineering and the physical sciences:

- Velocity, speed, and acceleration play a major role in many physical systems. Velocity and speed are introduced graphically in Chapter 1, defined as derivatives of displacement in Chapter 3, and then given a fuller discussion in Chapter 4. They are treated again from an integration point of view in Chapter 5. Instead of expecting the use of standard physics formulas, we encourage students to state time conventions, choose coordinate systems, and develop whatever equations are necessary to solve a particular problem.

- Resistors, capacitors, and inductors and how they relate to charge and current are introduced in an intuitive way in Chapter 1. As calculus unfolds, intuitive ideas are formalized mathematically as derivatives and antiderivatives.

- Differential equations provide mathematical models for a wide variety of applications. They are introduced relatively early in the book (very briefly in Chapter 3 and a bit more fully in Chapter 5) so that students become familiar with the ideas surrounding this important branch of calculus. Only the simplest of equations are solved in the early chapters. Differential equations are discussed in detail in Chapter 16.

Other Features

A careful effort has been made to incorporate features that will enhance applications and facilitate learning.

- Chapter 1 provides a useful precalculus review and anticipates some of the important applications that will be discussed later throughout the book. This chapter includes a brief review of plane analytical geometry (straight lines and conic sections) and functions (including trigonometric, exponential, and logarithm functions).

- Electronic calculators or software packages such as Maple, Mathematica, MathCad, and MatLab are now commonplace in many engineering and science courses. And the use of these devices is becoming more and more prevalent in the workplace. Accordingly, we have taken advantage of the plotting facilities of calculators and computers in this book. At the same time, a great disparity exists among calculators and computers with respect to taking limits, differentiating, and integrating, and we have not therefore written a totally integrated text that uses these devices to do these operations.

- Chapter 4 contains a section on Problem Solving (Section 4.13) with hints on how to manage word problems. It discusses how to think about a problem, how to key on certain words, and how to develop a series of questions and answers that will lead to a solution.

- Numerous Examples with worked-out solutions are given throughout the book.

- Key terms are boldfaced where they are defined in the text, and they are listed near the end of each chapter.

- Important Definitions, Theorems, and Corollaries are highlighted in Boxes where they are introduced in the text.

- A special icon, consisting of an exclamation mark in a triangular sign, is placed in the margin to alert the reader to pitfalls that commonly entrap students. Such pitfalls may be inappropriate calculations or misinterpretations of ideas.

- A concise Summary is given near the end of each chapter.

- A list of the main Derivative and Integration Formulas is given on the inside of the back cover for easy reference.

- The Exercises at the ends of the sections and the Review Exercises at the ends of the chapters have been graded into 3 difficulty levels. Exercises with a single asterisk ($*$) are the easiest. They are routine problems designed to reinforce fundamentals. Exercises with two asterisks ($**$) demand more thought. This grade of exercise contains most of the applications. Fundamentals should be mastered before tackling applications to problems from other fields. Exercises with 3 asterisks ($***$) are the most challenging. They should be attempted only after most of the double-asterisk problems have been successfully solved. Answers to all the even-numbered exercises, except the challenging three-asterisk ones, are provided near the back of the book.

- A set of Comprehensive Review Problems and Applications, divided into 2 large sections, is given near the end of the book for additional practice and review. It can be used to help prepare for exams in single and multivariable calculus. This set provides more than 350 problems on Chapters 1–11 and more than 230 problems for Chapters 12–16. Answers to all the questions are provided at the end of each section.

- Three Appendices are included near the back of the book: Appendix A on Mathematical Induction, Appendix B on Determinants, and Appendix C on Complex Numbers.

Approach

For the most part, the approach in this book is intuitive, making free use of geometry and familiar physical settings to motivate and illustrate concepts. For instance, limits are introduced intuitively, but an optional section is included to allow exposure to the mathematical definition of a limit. Derivatives are introduced and defined algebraically as instantaneous rates of change, and interpreted geometrically as slopes of tangent lines. The definite integral is introduced through two geometric and two physical problems, but its definition is algebraic.

This book begins with an introduction to the fundamentals of plane analytic geometry in order to provide a way of visualizing problems. The most difficult part of a solution to a problem is frequently the initial step. Once started, the solution usually unfolds smoothly and easily, but that first step sometimes seems impossible. One of the best ways to start a problem is with a diagram. A picture, no matter how rough, is invaluable in giving a "feeling" for what is going on. It displays the known facts surrounding the problem; it permits you to see what the problem really is and how it relates to the known facts; and it often suggests that all-important first step. We encourage students to develop the habit of making diagrams at every opportunity — not just to solve problems, but to understand what calculus is all about.

The main goal of this book is to develop a way of thinking. Above all else, this book is designed to help students think logically. We do not provide an exhaustive list of formulas to memorize so that students can solve problems by rote. Instead, we give a few formulas and, we hope, a great deal of insight into the ideas surrounding these formulas. By working on the examples and problems in this book, students learn how to organize facts and interpret them mathematically. They learn how to decide exactly what a particular problem is, and how to produce a step-by-step procedure by which to solve it.

Order of Topics

The order of topics in the book is fairly standard. The main exceptions are the following:

- Conic sections are introduced in Chapter 1 as illustrations of how the form of the equation of a curve dictates its shape and conversely how the shape of a curve influences its equation. Detailed discussions of properties of conic sections are delayed until Chapter 10.

- From the outset, we address what it means for x_0 to approximate a solution of the equation $f(x) = 0$ to a given number of decimal places or with error less than some specified amount. Approximations are introduced in a geometric way in Chapter 1 through the Intermediate Value Theorem. Newton's iterative procedure for approximating solutions to equations is the first application of differentiation in Chapter 4. Numbers in subsequent applications need not then be contrived in order to lead to equations with elementary solutions. Throughout, we demand that the accuracy of approximations be clearly identified.

- Trigonometric, exponential, and logarithm functions are reviewed in Chapter 1. Derivatives of these functions are developed in Chapter 3 making them available for applications of differentiation in Chapter 4 and integration in Chapters 5–7.

- Limits at infinity and infinite limits are presented in Chapter 2, along with general discussions on limits.

- In-depth discussions of inverse trigonometric functions and hyperbolic functions, and their derivatives, are delayed until Chapter 8. To discuss these functions in Chapter 3 would bring the topic of differentiation to an early conclusion, but this is overshadowed by what we call the "one-at-a-time" principle. Do not do more than one new thing at a time; two or more new ideas presented simultaneously can lead to confusion. By restricting the domain of functions that can be differentiated to relatively simple ones in Chapter 3, we can focus on the principles of differentiation.

- Applications of the definite integral (Chapter 7) are presented before techniques of integration (Chapter 9). In this way the student sees the diversity of applications of integration and can appreciate why it is necessary to antidifferentiate more difficult functions. Applications are covered again in the examples and exercises of Chapter 9.

- Taylor polynomials are introduced in an optional section in Chapter 4 as a way to approximate a function with a polynomial. The higher the degree of the polynomial, the better the approximation.

- In Chapter 11, Taylor series and power series are discussed before series of numbers. In this way, we can introduce important applications more quickly.

Treatment of Particular Topics

A subtle difference in how a topic is treated can significantly affect whether a student fully understands that topic. From my experiences in teaching calculus, I have developed certain preferences which I have incorporated in this book.

- In the review of logarithm functions in Chapter 1, we treat logarithms as powers. When we study definite integrals with variable upper limits in Section 6.5, we give the integral definition of the natural logarithm function.

- Relative extrema are discussed in Section 4.3, whereas absolute extrema are discussed in Section 4.6. Because they are different and are used differently, they are treated in separate sections. Curve sketching uses relative extrema, while most applied extrema problems involve absolute extrema.

- Antiderivatives or indefinite integrals are treated in a chapter unto themselves, namely, Chapter 5. Relegating them to a section in a chapter on differentiation would not do them justice. Conceptually integration is simple — backwards differentiation. But this operation is so important, and it is so much more difficult to perform, only a separate chapter does it justice. Integration techniques are further explored in Chapter 9.

- Antidifferentiation is approached from an organizational viewpoint. The emphasis is on learning to organize an integrand into a form in which integration is obvious. Needless memorization of formulas is discouraged.

- Antiderivatives are used to evaluate definite integrals. This is the fundamental theorem of integral calculus (Section 6.4). The fact that a definite integral with a variable upper limit (Section 6.5) can be regarded as an antiderivative is given secondary importance. At first, we emphasize the importance of the antiderivative as a calculational tool for definite integrals (one-thing-at-a-time).

- In Chapter 6, the definite integral is motivated through four types of problems: area, volume, work, and fluid flow. At the same time, definite integrals are defined algebraically as limits of Riemann sums. This approach permits us to use the definition of a definite integral in a wide variety of applications without fear that it has been associated with any one application in particular. Definitions of double, triple, line, and surface integrals as limit-summations in multivariable calculus follow quite naturally.

- We prefer the notation Sin^{-1} for the inverse sine function. It is so natural once the notation $f^{-1}(x)$ has been introduced for inverse functions in general. We capitalize the first letter to remind ourselves that this is indeed an inverse trigonometric function and that the -1 superscript should not be interpreted as a power. In Section 8.2, where inverse trigonometric functions are introduced for the first time, we also present the arcsin notation so that students become familiar with it.

- There is no agreement in the mathematical community on principal values for the inverse secant and inverse cosecant functions. Our choice is based on practical reasons for engineers and physical scientists. For values outside the first quadrant, we use third-quadrant angles stated between $-\pi$ and $-\pi/2$. Third-quadrant angles simplify derivatives of these functions. Choosing angles between $-\pi$ and $-\pi/2$ matches the usual methods when working in polar coordinates and with complex variables. When polar angle θ is restricted to an interval of length 2π, it is most often chosen to be $-\pi < \theta < \pi$. This is especially true in complex analysis where principal values of the argument of a complex number are always chosen in this interval. Principal values of all multivalued functions (logarithm, inverse trigonometric, and inverse hyperbolic) use these principal values.

- We have chosen to keep the polar coordinate r nonnegative. Nothing is lost by this convention. Curves may have slightly different equations depending on which convention is used. But by choosing $r > 0$, considerable simplification is achieved when equations of curves in polar coordinates are rewritten in Cartesian coordinates.

- In Chapter 11, we apply the limit ratio test for convergence of infinite series only to series with positive terms. It is not used directly as a test for convergence of series with positive and negative terms. We discuss convergence one step at a time. First we discuss convergence of series of positive terms using the comparison, limit comparison, integral, limit ratio, and limit root tests. Next we introduce series with positive and negative terms, and use the above tests as tests for absolute convergence.

- Three-dimensional analytic geometry and vectors in Chapter 12 provide the tools for multivariable calculus. We stress the value of drawing curves and surfaces in space, using the curve sketching tools learned in Chapters 1–4. Such diagrams are essential to the evaluation of double, triple, line, and surface integrals, and to an appreciation of many of the ideas of differential calculus. Vectors are handled algebraically and geometrically. Every algebraic definition is interpreted geometrically, and every geometric definition is followed by an algebraic equivalent. Differentiation and integration of vectors dependent on a single parameter lead to discussions on tangent and normal vectors to curves, curvature and arc length, and three-dimensional kinematics.

- Gradients are useful in many areas of applied mathematics. We introduce gradients in Section 13.4, and apply them to directional derivatives and normal vectors to curves and surfaces in Sections 13.8 and 13.9.

- Chain rules for composite multivariable functions are endless in variations and applications. In Chapter 13, we show how to appreciate each term in a chain rule as a contribution of particular variables to the overall rate of change of a function, and then to create a schematic diagram to handle the most complicated functional situations.

- In Chapter 14, our definitions of double and triple integrals as limit-summations are analogous to the definition of the definite integral presented earlier. The evaluation by double and triple iterated integrals is geometric. Through representative boxes, rectangles, strips, and columns, we show how to visualize the summation process and affix appropriate limits to integrals. There is no algebraic manipulation of inequalities; a thoughtfully prepared diagram does it all. The geometric approach also helps us visualize integrations in polar, cylindrical, and spherical coordinates. We demonstrate that many of the applications of the definite integral in Chapter 7 are handled much more easily with double integrals.

- For simplicity, we keep definitions to a minimum in Chapter 15. We present one kind of line integral $f(x, y, z)\,ds$ and then treat integrals of the form $P\,dx + Q\,dy + R\,dz$ as a special case when $f(x, y, z)$ is the tangential component of some vector function defined along C. They can all be evaluated by substituting from parametric equations of the curve. Alternatively, it may be expedient to use Green's or Stokes's theorems or determine whether the integral is independent of path. Likewise, we present one surface integral $f(x, y, z)\,dS$, and then in many applications, we treat $f(x, y, z)$ as the normal component of some vector function defined on S.

- Chapter 16 provides a sound introduction to differential equations. It contains sections on separable equations, linear first-order equations, and second-order equations easily reduced to first-order equations. The exercises introduce (first-order) homogeneous and Bernoulli equations. Considerable space (four sections) is devoted to the important topic of linear differential equations. Applications to Newtonian mechanics, vibrating mass-spring systems, and electric circuits are discussed in detail. Other applications are introduced through examples and exercises.

Supplements

Four supplements have been carefully prepared to aid instructors and students:

- An Instructor's Solutions Manual, providing complete solutions to all the exercises in the text.

- A Test Item File, organized by chapter, with the level of difficulty (i.e., easy, moderate, and difficult) indicated for each question.
- PH Custom Test (for Windows), a special computerized version of the Test Item File, which enables instructors to edit existing questions, add questions, and generate tests.
- A Student's Solutions Manual, providing complete solutions to all the even-numbered exercises, except those challenging questions marked with three asterisks.

Acknowledgments

I am grateful to the many people who offered such helpful suggestions for the preparation of this book. I would particularly like to thank the following instructors who provided formal reviews:

Thomas G. Berry (University of Manitoba)
Dietrich Burbulla (University of Toronto)
W. A. Cebuhar (Queen's University)
Jack Macki (University of Alberta)
Angelo Mingarelli (Carleton University)

Any errors in the text or in the supplements are the responsibility of the author. I would appreciate having them brought to my attention. Please send them to me care of Acquisitions Editor, Mathematics, College Division, Prentice Hall Canada Inc., 1870 Birchmount Road, Scarborough, Ontario, M1P 2J7.

About the Author

Donald W. Trim received his honours degree in mathematics and physics in 1965. After proceeding to his Masters Degree, he obtained his Doctorate degree with a dissertation in general relativity. As a graduate student at the University of Waterloo, he discovered that teaching was his life's ambition. He quickly became well known for his teaching. In 1971 he was invited to become the first member of the Department of Applied Mathematics in the Faculty of Science at the University of Manitoba. In 1976, after only five years at the university, he received one of the university's highest awards for teaching excellence based on submissions by faculty and graduating students. At the University of Manitoba, no individual can receive this award twice within a ten year span. Professor Trim again won the award in 1988. In 1997, he was awarded a 3M teaching fellowship. In addition, the Faculty of Engineering presented him with a gold replica of the engineer's iron ring in appreciation for his service.

Professor Trim's skills in teaching are reflected in the notes and books he has written. These include an Introduction to Applied Mathematics, an Introduction to Complex Analysis and its Applications, Applied Partial Differential Equations, and this calculus text. All have received most favourable reviews from students. To bridge the gap between calculus and partial differential equations, he is currently writing a book on ordinary differential equations.

Professor Trim is a member of the Canadian Applied Mathematics Society and the Mathematical Association of America. His present interests are in partial differential equations and mathematical education.

Plane Analytic Geometry and Functions

A sound knowledge of analytic geometry, a union of algebra and geometry, is essential to the study of calculus. On the one hand, it provides a way to describe geometric objects algebraically; on the other hand, it permits a geometric visualization of algebraic statements. Our approach to calculus is visual; we draw pictures at every possible opportunity, by hand, by graphing calculator, and by computer. We draw pictures to introduce ideas, to illustrate concepts, to reinforce principles, and to solve problems. Analytic geometry is the basis for these pictures. In Section 1.1 we give a brief review of plane analytic geometry. We discuss curves and their representations by use of equations, paying particular attention to straight lines. In Section 1.2 we review circles, parabolas, ellipses, and hyperbolas. Students who have previously studied analytic geometry could skip these sections and begin with functions in Section 1.3, but we strongly recommend against this. Even the best-prepared student can benefit by a cursory reading of Sections 1.1 and 1.2. Not only is it helpful to refresh one's memory on concepts learned some time ago, it is also wise to become familiar with the terminology, notation, and conventions set forth in these sections. Section 1.4 is essential reading; it introduces some major applications that we deal with later and indicates why calculus is so necessary. Trigonometric, exponential, and logarithm functions are reviewed in Sections 1.5 and 1.6 with special attention to properties important to calculus. Finally, in Section 1.7, we discuss some vital issues that arise when it becomes necessary to approximate solutions to equations.

FIGURE 1.1

1.1 Plane Analytic Geometry and Straight Lines

Points in the plane are identified by an ordered pair (x, y) of real numbers called their Cartesian coordinates (Figure 1.1). The x-coordinate of a point P is its perpendicular distance from the y-axis, and the y-coordinate is its perpendicular distance from the x-axis.

The axes divide the plane into four parts which, beginning at the upper right and proceeding counterclockwise, are called respectively the first, second, third, and fourth **quadrants**. The axes themselves are not considered part of any quadrant. Points in the first quadrant have x- and y-coordinates that are both positive; points in the second quadrant have a negative x-coordinate and a positive y-coordinate; and so on (Figure 1.2).

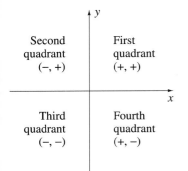

FIGURE 1.2

If P and Q are any two points with coordinates (x_1, y_1) and (x_2, y_2) respectively (Figure 1.3), then by the Pythagorean relation for the right-angled triangle PQS,

$$\|PQ\|^2 = \|PS\|^2 + \|QS\|^2,$$

where $\|PQ\|$ denotes the length of the line segment joining P and Q, and $\|PS\|$ and $\|QS\|$ denote the lengths of the related line segments. But[†]

$$\|PS\| = |x_2 - x_1| \qquad \text{and} \qquad \|QS\| = |y_2 - y_1|,$$

[†] The vertical lines around $|x_2 - x_1|$ and $|y_2 - y_1|$ denote absolute values. They operate on whatever is between them to produce a nonnegative result according to $|x| = x$ if $x \geq 0$ and $|x| = -x$ if $x < 0$. We discuss absolute values from a functional point of view in Section 1.3.

FIGURE 1.3

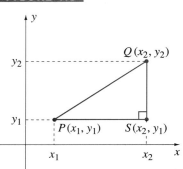

and therefore[†]

$$\|PQ\|^2 = |x_2 - x_1|^2 + |y_2 - y_1|^2 \quad \text{or}$$

$$\|PQ\| = \sqrt{(x_2 - x_1)^2 + (y_2 - y_1)^2}. \tag{1.1}$$

Equation 1.1 expresses the length of a line segment joining points P and Q in terms of their coordinates (x_1, y_1) and (x_2, y_2). For example, if the coordinates of P and Q in Figure 1.4 are $(-3, 4)$ and $(2, -1)$, then

$$\|PQ\| = \sqrt{(2 + 3)^2 + (-1 - 4)^2} = 5\sqrt{2}.$$

Suppose $R(x, y)$ in Figure 1.5 is the midpoint of the line segment joining $P(x_1, y_1)$ and $Q(x_2, y_2)$. Since triangles PTR and RUQ are congruent, it follows that $\|RU\| = \|PT\|$, and for P and Q as shown in Figure 1.5, this condition becomes $x_2 - x = x - x_1$. Consequently,

FIGURE 1.4

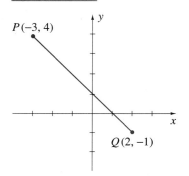

$$x = \frac{x_1 + x_2}{2}. \tag{1.2a}$$

Similarly, equality of $\|RT\|$ and $\|QU\|$ gives

$$y = \frac{y_1 + y_2}{2}. \tag{1.2b}$$

Derivation of these results assumed that $x_2 > x_1$ and $y_2 > y_1$, but the same formulas are valid for any x_1, x_2, y_1, and y_2 whatsoever. Coordinates of the midpoint of a line segment, therefore, are averages of coordinates of the ends of the line segment.

FIGURE 1.5

FIGURE 1.6

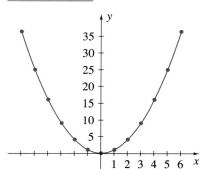

An equation such as $y = x^2$ specifies a relationship between numbers represented by the letter x and those represented by the letter y: y must always be the square of x. Algebraically, we speak of pairs of values x and y that satisfy this equation. If we write these pairs in the form (x, y), and interpret each pair as coordinates of a point, we find that all such points lie on the curve in Figure 1.6. This curve, then, is a geometric visualization of solution pairs of the equation $y = x^2$. Every pair of values x and y that satisfies $y = x^2$ is represented by a point on the curve. Conversely, the coordinates (x, y) of every point on the curve provide a pair of numbers that satisfies $y = x^2$. Algebraic solutions, then, are represented geometrically as points on a curve; points on the curve provide algebraic solutions.

[†] $\sqrt{x}$ always denotes the positive square root of x, so that $\sqrt{4} = 2$.

If x and y are the coordinates of a point in the plane, then the point is on the curve in Figure 1.6 if and only if x and y satisfy the equation $y = x^2$. In other words, this equation completely characterizes all points on the curve. We therefore call $y = x^2$ the **equation of the curve**. Thus the equation of a curve is an equation that the coordinates of every point on the curve satisfy and at the same time is an equation that no point off the curve satisfies. In the remainder of this section we discuss straight lines and their equations. In Section 1.2, we show that parabolas, circles, ellipses, and hyperbolas can be recognized by the distinctive forms of their equations. Once again note the algebraic-geometric interplay; the form of an equation dictates the shape of the curve and, conversely, the shape of a curve determines the form of its equation.

The Straight Line

The **slope** of the line through the points (x_1, y_1) and (x_2, y_2) in Figure 1.7a is defined as the quotient m of

$$\frac{y_2 - y_1}{x_2 - x_1}. \tag{1.3}$$

It is customary to call the difference $y_2 - y_1$ the **rise** because it represents the vertical distance between the points, and $x_2 - x_1$ the **run**, the horizontal distance between the points. It is easy to show using similar triangles that m is independent of the two points chosen on the line; that is, no matter which two points we choose on the line to evaluate m, the result is always the same. The four numbers x_1, x_2, y_1, and y_2 vary, but the ratio $(y_2 - y_1)/(x_2 - x_1)$ remains unchanged. A horizontal line (Figure 1.7b) has slope zero (since $y_2 - y_1 = 0$), whereas the slope of a vertical line is undefined (since $x_2 - x_1 = 0$). In Figure 1.7c, line l_1, which leans to the right, has positive slope and line l_2, which leans to the left, has negative slope.

FIGURE 1.7a

FIGURE 1.7b

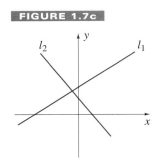

FIGURE 1.7c

In Figure 1.8 we have shown the line through the points $(1, 2)$ and $(3, 5)$. Its slope is $(5 - 2)/(3 - 1) = 3/2$. To find the equation for this line, we let (x, y) be the coordinates of any other point on the line. Since the slope of the line must also be given by $(y - 2)/(x - 1)$, it follows that

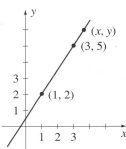

FIGURE 1.8

$$\frac{y - 2}{x - 1} = \frac{3}{2},$$

and, when this equation is simplified,

$$2y = 3x + 1.$$

If (x, y) are the coordinates of any point not on this line, then they do not satisfy this equation because the slope $(y - 2)/(x - 1)$ joining (x, y) to $(1, 2)$ is not equal to $3/2$. Thus, $2y = 3x + 1$ is the equation for the straight line through $(1, 2)$ and $(3, 5)$.

FIGURE 1.9

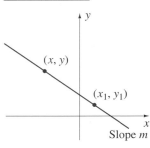

Slope m

We can use this procedure to find the equation for the straight line through any point (x_1, y_1) with any slope m (Figure 1.9). If (x, y) are the coordinates of any other point on the line, then the slope of the line is $(y - y_1)/(x - x_1)$; therefore

$$\frac{y - y_1}{x - x_1} = m \quad \text{or} \quad y - y_1 = m(x - x_1). \tag{1.4}$$

This is called the **point-slope** formula for the equation of a straight line; it uses the slope m of the line and a point (x_1, y_1) on the line to determine the equation of the line.

Other formulas for the equation of a line are used when different characteristics of the line are given. They are listed below:

Form of Equation	Name of Equation	Characteristics Determining the Line
$\dfrac{y - y_1}{y_2 - y_1} = \dfrac{x - x_1}{x_2 - x_1}$	Two-point formula	Two points (x_1, y_1) and (x_2, y_2) on the line
$y = mx + b$	Slope y-intercept formula	m is the slope; b is the y-intercept
$y = m(x - a)$	Slope x-intercept formula	m is the slope; a is the x-intercept
$\dfrac{x}{a} + \dfrac{y}{b} = 1$	Two-intercept formula	a and b are x- and y-intercepts

Point-slope formula 1.4 encompasses all of these. The characteristics in the right column determine the slope of a line and a point on it; therefore, with minimal calculations, the point-slope formula can be used in all situations. For example, if we know that the slope of a line is m and the y-intercept is b, then a point on the line is $(0, b)$. Hence, 1.4 gives $y - b = m(x - 0)$, or $y = mx + b$, the slope y-intercept formula.

There is one other equation that is worth mentioning. Vertical lines, which do not have slopes, cannot be represented in form 1.4; they have form $x = k$, for some constant k. On the other hand, all lines can, for various values of the constants A, B, and C, be represented in the form

$$Ax + By + C = 0. \tag{1.5}$$

This is often called the **general equation** of a line.

EXAMPLE 1.1

Find the equation of the line through the points $(-1, 1)$ and $(2, 3)$.

SOLUTION Since the slope of the line is $(3 - 1)/(2 + 1) = 2/3$, we can use the point $(-1, 1)$ and the slope $2/3$ in point-slope formula 1.4 to give

$$y - 1 = \frac{2}{3}(x + 1) \quad \text{or} \quad 3y = 2x + 5.$$

The same result is obtained if the point $(2, 3)$ is used in place of $(-1, 1)$. The two-point formula could also be used.

EXAMPLE 1.2

FIGURE 1.10

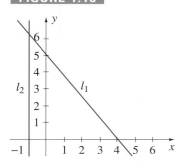

Find equations for the lines in Figure 1.10.

SOLUTION   Since the slope of l_1 is $-5/4$ and its y-intercept is 5, we use formula 1.4 with the point $(0, 5)$,

$$y - 5 = -\frac{5}{4}(x - 0) \quad \text{or} \quad 4y = 20 - 5x.$$

The equation of l_2 is clearly $x = -1$.

To find the point of intersection of two straight lines—say, $3y = 2x + 5$ and $4y + x = 14$ (Figure 1.11)—we find the point whose coordinates (x, y) satisfy both equations. If we solve each equation for x and equate the expressions, we obtain $(3y - 5)/2 = 14 - 4y$, which immediately yields $y = 3$. Either of the original equations then gives $x = 2$, and the point of intersection of the lines is $(2, 3)$. Note once again the algebraic-geometric interplay. Geometrically, $(2, 3)$ is the point of intersection of two straight lines. Algebraically, the two numbers constitute the solution of the equations $3y = 2x + 5$ and $4y + x = 14$.

Of particular importance to the study of lines are the concepts of parallelism and perpendicularity.

FIGURE 1.11

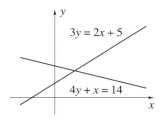

DEFINITION 1.1
Two lines are said to be **parallel** if they have no point of intersection.

EXAMPLE 1.3

Verify that the lines with equations $2x - y = 4$ and $4x - 2y = 7$ are parallel.

SOLUTION   If we solve each of these equations for y and equate the resulting expressions, we obtain

$$2x - 4 = 2x - \frac{7}{2},$$

an obvious impossibility. Consequently, the lines do not intersect.

Geometrically, the following is clear.

THEOREM 1.1
Two lines are parallel if and only if they have the same slope.

For example, if we write the equations of the lines in Example 1.3 in the form $y = 2x - 4$ and $y = 2x - 7/2$, coefficients of the x-terms identify slopes of the lines. Since each has slope 2, the lines are parallel.

DEFINITION 1.2
Two lines are said to be **perpendicular** if they intersect at right angles.

The following theorem gives a test for perpendicularity of straight lines in terms of slopes.

THEOREM 1.2

Two lines with nonzero slopes m_1 and m_2 are perpendicular if and only if

$$m_1 = -\frac{1}{m_2} \qquad \text{or} \qquad m_1 m_2 = -1. \qquad (1.6)$$

PROOF Suppose nonparallel lines l_1 and l_2 with nonzero slopes m_1 and m_2 intersect at a point Q (Figure 1.12).

FIGURE 1.12

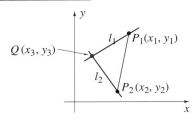

Suppose the coordinates of Q are x_3 and y_3; let $P_1(x_1, y_1)$ and $P_2(x_2, y_2)$ be any two points on l_1 and l_2 respectively. Lines l_1 and l_2 are perpendicular if and only if triangle $Q P_1 P_2$ is right-angled at Q; that is, if and only if,

$$\|P_1 P_2\|^2 = \|Q P_1\|^2 + \|Q P_2\|^2.$$

Using distance formula 1.1, this condition can be expressed as

$$(x_1 - x_2)^2 + (y_1 - y_2)^2 = (x_1 - x_3)^2 + (y_1 - y_3)^2 + (x_2 - x_3)^2 + (y_2 - y_3)^2.$$

When terms are expanded, and the equation is simplified, the result is

$$x_3^2 - x_1 x_3 + x_1 x_2 - x_2 x_3 = -y_3^2 + y_1 y_3 - y_1 y_2 + y_2 y_3.$$

After factoring each side, this equation can be rearranged into the form

$$\frac{y_3 - y_2}{x_3 - x_2} = -\frac{x_3 - x_1}{y_3 - y_1} = -\frac{1}{\dfrac{y_3 - y_1}{x_3 - x_1}}.$$

Since $m_1 = (y_3 - y_1)/(x_3 - x_1)$ and $m_2 = (y_3 - y_2)/(x_3 - x_2)$, we have shown that lines l_1 and l_2 are perpendicular if and only if

$$m_2 = -\frac{1}{m_1}. \qquad \blacksquare$$

EXAMPLE 1.4

Find the equation of the straight line that passes through the point $(2, 4)$ and is perpendicular to the line $3x + y = 5$.

SOLUTION   Since the slope of the given line $y = -3x + 5$ is -3, the required line has slope $1/3$. Using point-slope formula 1.4, we find that the equation of the required line is

$$y - 4 = \frac{1}{3}(x - 2) \quad \text{or} \quad x - 3y = -10.$$

These ideas are put to the test in the following situation; what comes out is a very useful formula. Figure 1.13 shows a spring attached on one end to a fixed point $P(x_1, y_1)$. The other end Q is attached to a sleeve that slides along the line $Ax + By + C = 0$. We assume for this discussion that the spring is stretched for any position Q on the line.

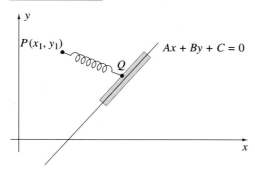

FIGURE 1.13

We know that the greater the stretch of the spring, the greater the force that it exerts on the sleeve. Furthermore, the spring force is smallest when the length of the spring is smallest, and we propose to find this length. This clearly occurs when PQ is perpendicular to the line $Ax + By + C = 0$, and the slope of line PQ in this case is B/A, the negative reciprocal of the slope $-A/B$ of $Ax + By + C = 0$. The equation of line PQ (using slope B/A and point P) is $y - y_1 = (B/A)(x - x_1)$. We can find the coordinates of Q by solving the following equations for x and y,

$$y - y_1 = \left(\frac{B}{A}\right)(x - x_1), \quad Ax + By + C = 0.$$

If we solve each of these for y and equate the results,

$$y_1 + \left(\frac{B}{A}\right)(x - x_1) = -\frac{C}{B} - \frac{A}{B}x.$$

Gathering terms in x on the left and transposing the rest,

$$\left(\frac{B}{A} + \frac{A}{B}\right)x = \frac{Bx_1}{A} - \frac{C}{B} - y_1,$$

from which

$$x = \left(\frac{Bx_1}{A} - \frac{C}{B} - y_1\right)\frac{AB}{A^2 + B^2} = \frac{B^2x_1 - ABy_1 - AC}{A^2 + B^2}.$$

We could substitute this into $Ax + By + C = 0$ to find the y-coordinate of Q, but we can be more efficient if we look ahead. When we find coordinates (x, y) for Q, the minimum length of PQ is $\sqrt{(x - x_1)^2 + (y - y_1)^2}$. What we really need then are expressions for $x - x_1$ and $y - y_1$. We have x so that

$$x - x_1 = \frac{B^2 x_1 - AB y_1 - AC}{A^2 + B^2} - x_1$$

$$= \frac{-A^2 x_1 - AB y_1 - AC}{A^2 + B^2} = \frac{-A(Ax_1 + By_1 + C)}{A^2 + B^2}.$$

The equation $y - y_1 = (B/A)(x - x_1)$ now gives

$$y - y_1 = \frac{B}{A}\left[\frac{-A(Ax_1 + By_1 + C)}{A^2 + B^2}\right] = \frac{-B(Ax_1 + By_1 + C)}{A^2 + B^2}.$$

It follows that the shortest length for PQ is

$$\sqrt{\frac{A^2(Ax_1 + By_1 + C)^2}{(A^2 + B^2)^2} + \frac{B^2(Ax_1 + By_1 + C)^2}{(A^2 + B^2)^2}} = \frac{|Ax_1 + By_1 + C|}{\sqrt{A^2 + B^2}}.$$

As a result of this discussion we have the formula

$$\frac{|Ax_1 + By_1 + C|}{\sqrt{A^2 + B^2}} \tag{1.7}$$

for the shortest (or perpendicular) distance from a point (x_1, y_1) to a line $Ax + By + C = 0$. For example, the distance from $(-2, -1)$ to the line $3x + 2y = 2$ in Figure 1.14 is

$$\frac{|3(-2) + 2(-1) - 2|}{\sqrt{3^2 + 2^2}} = \frac{10}{\sqrt{13}}.$$

FIGURE 1.14

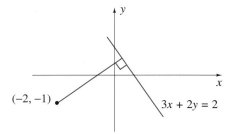

$(-2, -1)$

$3x + 2y = 2$

In Exercises 1–4 find the distance between the points.

* **1.** $(1, 3)$, $(3, 4)$ * **2.** $(-2, 1)$, $(4, -2)$

* **3.** $(-1, -2)$, $(-3, -8)$ * **4.** $(3, 2)$, $(-4, -1)$

* **5.–8.** Find the midpoint of the line segment joining the points in Exercises 1–4.

In Exercises 9–18 draw the line described and find its equation.

* **9.** Through the points $(1, 2)$ and $(-3, 4)$

* **10.** Through the points $(3, -6)$ and $(5, -6)$

* **11.** Through the point $(-2, -3)$ with slope 3

* **12.** Through the point $(1, -3/2)$ with slope $-1/2$

* **13.** The y-axis

* **14.** The x-axis

* **15.** Through the point $(4, 3)$ and crossing the x-axis at -2

* **16.** Through the point $(-1, -2)$ and crossing the y-axis at 4

* **17.** Crossing the x- and y-axes at 1 and -3, respectively

* **18.** Through the origin and the midpoint of the line segment joining $(3, 4)$ and $(-7, 8)$

In Exercises 19–26 determine whether the lines are perpendicular, parallel, or neither.

* **19.** $y = -x + 4$, $y = x + 6$ * **20.** $x + 3y = 4$, $2x + 6y = 7$

* **21.** $x = 3y + 4$, $y = x/3 - 2$ * **22.** $2x + 3y = 1$, $3x - 2y = 5$

* **23.** $y = 3x + 2$, $y = -x/2 + 1$ * **24.** $x - y = 5$, $2x + 3y = 4$

* **25.** $x = 0$, $y = 5$ * **26.** $x + y + 2 = 0$, $3x - y = 4$

In Exercises 27–32 find the point of intersection of the lines.

* **27.** $x + y = 0$, $x - 2y = -3$ * **28.** $x = 1$, $y = 2$

* **29.** $3x + 4y = 6$, $x - 6y = 3$ * **30.** $y = 2x + 6$, $x = y + 4$

* **31.** $x/2 + y/3 = 1$, $2x - y/4 = 15$ * **32.** $14x - 2y = 5$, $3x + 10y = 12$

In Exercises 33–38 find the (shortest) distance from the point to the line.

* **33.** $(3, 4)$ to $x + y = 1$ * **34.** $(1, -3)$ to $x + 2y = 3$

* **35.** $(5, 1)$ to $x - y = 4$ * **36.** $(3, 1)$ to $y = -x$

* **37.** $(6, -2)$ to $x = -1$ * **38.** $(4, -2)$ to $15x - 2y + 3 = 0$

In Exercises 39–46 find the equation of the line described.

** **39.** Parallel to $x + 2y = 15$, and through the point of intersection of $2x - y = 5$ and $x + y = 4$

** **40.** Perpendicular to $x - y = 4$, and through the point of intersection of $2x + 3y = 3$ and $x - y = 4$

** **41.** Parallel to the line through $(1, 2)$ and $(-3, 0)$, and through the point $(5, 6)$

** **42.** Perpendicular to the line through $(-3, 4)$ and $(1, -2)$, and through the point $(-3, -2)$

** **43.** Crossing through the first quadrant to form an isosceles triangle with area 8 square units

** **44.** Through $(3, 5)$ and crossing through the first quadrant to form a triangle with area 30 square units

** **45.** Has slope 2, and that part in the second quadrant has length 3

** **46.** Parallel to the x-axis, below the point of intersection of the lines $x = y$ and $x + y = 4$, and forms with these lines a triangle with area 9 square units

** **47.** A median of a triangle is a line segment drawn from a vertex to the midpoint of the opposite side. Find equations for the three medians of the triangle with vertices $(1, 1)$, $(3, 5)$, and $(0, 4)$. Show that all three medians intersect in a point called the centroid of the triangle.

** **48.** If (x_1, y_1), (x_2, y_2), (x_3, y_3), and (x_4, y_4) are vertices of any quadrilateral, show that the line segments joining the midpoints of adjacent sides form a parallelogram.

** **49.** Find the equation of the perpendicular bisector of the line segment joining $(-1, 2)$ and $(3, -4)$. (The perpendicular bisector is the line that cuts the line segment in half and is perpendicular to it.)

** **50.** Find coordinates of the point that is equidistant from the three points $(1, 2)$, $(-1, 4)$, and $(-3, 1)$.

** **51.** Prove Theorem 1.1.

** **52.** Generalize the result of equations 1.2 to prove that if a point R divides the length PQ so that

$$\frac{\text{length } PR}{\text{length } RQ} = \frac{r_1}{r_2},$$

where r_1 and r_2 are positive integers, then the coordinates of R are

$$x = \frac{r_1 x_2 + r_2 x_1}{r_1 + r_2} \quad \text{and} \quad y = \frac{r_1 y_2 + r_2 y_1}{r_1 + r_2}.$$

*** **53.** Prove that in any triangle the sum of the squares of the lengths of the medians is equal to three-quarters of the sum of the squares of the lengths of the sides. (A median of a triangle is a line segment drawn from one vertex to the midpoint of the opposite side.)

*** **54.** Let P be any point inside an equilateral triangle (see figure below). Show that the sum of the distances of P from the three sides is always equal to the height h of the triangle.

1.2 Conic Sections

Parabolas, circles, ellipses, and hyperbolas arise in a multitude of applications. As we are interested only in illustrating the algebraic-geometric interplay of analytic geometry, we do not give a complete development of these curves together with their many properties. We show only how the form of the equation for each conic section relates to its shape. Detailed discussions of conic sections in terms of focus and directrix can be found in Sections 10.5 and 10.6.

The Parabola

When the y-coordinate of a point (x, y) on a curve is related to its x-coordinate by an equation of the form

$$y = ax^2 + bx + c, \tag{1.8}$$

where a, b, and c are constants (with $a \neq 0$), the curve is called a **parabola**. Figure 1.6 (page 2) shows the simplest of all parabolas, $y = x^2$. For every point (x, y) to the right of the y-axis on this curve, there is a point equidistant to the left of the y-axis which has the same y-coordinate; that is, the point $(-x, y)$ is also on the curve. Putting it another way, that part of the parabola to the left of the y-axis is the image in the y-axis (thought of as a mirror) of that part to the right of the y-axis. Such a curve is said to be symmetric about the y-axis. It happens whenever the equation of a curve is unchanged when each x therein is replaced by $-x$. In other words, we have the test:

A curve is **symmetric** *about the y-axis if its equation remains unchanged when x is replaced by* $-x$.

The parabola $y = 9 - x^2$ is shown in Figure 1.15. It is symmetric about the y-axis; replacing x by $-x$ leaves the equation unchanged. This parabola is said to "open downward", whereas the parabola $y = x^2$ in Figure 1.6 "opens upward". The sign in front of the x^2-term dictates which way a parabola opens.

FIGURE 1.15

FIGURE 1.16

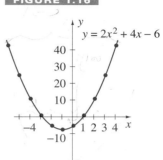

The parabola $y = 2x^2 + 4x - 6$ is plotted in Figure 1.16. It is not symmetric about the y-axis; its equation changes when x is replaced by $-x$. The parabola appears to be symmetric about the line $x = -1$ with lowest point $(-1, -8)$. To prove that this is indeed the case, we rewrite the equation as

$$y = 2(x^2 + 2x) - 6 = 2(x + 1)^2 - 8.$$

This form clearly indicates that the smallest value for y is -8, and it occurs when $x = -1$. The parabola is symmetric about $x = -1$.

The technique used above to rewrite $y = 2x^2 + 4x - 6$ as $2(x + 1)^2 - 8$ is called *completing the square*. If we apply the same technique to the general parabola $y = ax^2 + bx + c$, we obtain

$$y = a\left(x^2 + \frac{b}{a}x\right) + c = a\left(x + \frac{b}{2a}\right)^2 + \left(c - \frac{b^2}{4a}\right). \tag{1.9}$$

This form for the equation of the parabola shows the following:

1. When $a > 0$, the parabola opens upward and has a minimum at the point

$$\left(-\frac{b}{2a}, c - \frac{b^2}{4a}\right).$$

FIGURE 1.17

2. When $a < 0$, the parabola opens downward and has a maximum at the point

$$\left(-\frac{b}{2a}, c - \frac{b^2}{4a}\right).$$

3. The parabola is symmetric about the line $x = -b/(2a)$ (Figure 1.17); that is, for every point P on one side of the line, there is a point Q on the other side that is the mirror image of P in the line.

4. The parabola crosses the x-axis when

$$0 = y = a \left(x + \frac{b}{2a} \right)^2 + \left(c - \frac{b^2}{4a} \right).$$

To solve this equation for x, we first write

$$\left(x + \frac{b}{2a} \right)^2 = \frac{b^2}{4a^2} - \frac{c}{a},$$

and then take square roots of each side,

$$x + \frac{b}{2a} = \pm \sqrt{ \frac{b^2}{4a^2} - \frac{c}{a} }.$$

Finally, then,

$$x = -\frac{b}{2a} \pm \sqrt{ \frac{b^2}{4a^2} - \frac{c}{a} } = \frac{-b \pm \sqrt{b^2 - 4ac}}{2a}. \tag{1.10}$$

This is called the **quadratic formula**. It determines points where the parabola $y = ax^2 + bx + c$ crosses the x-axis. Equivalently, it defines roots of the quadratic equation $ax^2 + bx + c = 0$. When $b^2 - 4ac > 0$, there are two solutions (Figure 1.18a); when $b^2 - 4ac = 0$, there is only one solution (Figure 1.18b); and when $b^2 - 4ac < 0$, there is no solution (Figure 1.18c), that is, no real solution. The quantity $b^2 - 4ac$ is called the **discriminant**.

FIGURE 1.18a

FIGURE 1.18b

FIGURE 1.18c

When x and y in equation 1.8 are interchanged, the resulting equation

$$x = ay^2 + by + c \tag{1.11}$$

still defines a parabola; it opens to the right or left, rather than up or down. Figures 1.19 and 1.20 show the parabolas $x = y^2 + 1$ and $x = -y^2 + 4y - 4$, respectively. The parabola $x = y^2 + 1$ is symmetric about the x-axis; any point (x, y) on the parabola is symmetric with the corresponding point $(x, -y)$. In general, *a curve is symmetric about the x-axis if its equation remains unchanged when y is replaced by $-y$.*

FIGURE 1.19

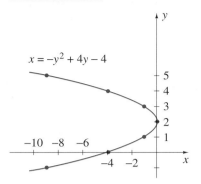

FIGURE 1.20

EXAMPLE 1.5

Find equations for the parabolas in Figures 1.21 and 1.22.

FIGURE 1.21

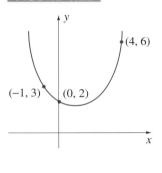

FIGURE 1.22

SOLUTION   The fact that the parabola in Figure 1.21 is symmetric about the x-axis means that b in equation 1.11 must vanish; that is, its equation must be of the form $x = ay^2 + c$. Since the points $(2, 0)$ and $(0, 3)$ are on the parabola, their coordinates must satisfy the equation of the parabola,

$$2 = a(0)^2 + c, \qquad 0 = a(3)^2 + c.$$

These imply that $c = 2$ and $a = -c/9 = -2/9$. Thus, the equation of the parabola is $x = -2y^2/9 + 2$.

The parabola in Figure 1.22 has no special attributes that we can utilize (such as the position of the line of symmetry in Figure 1.21). We therefore use the facts that its equation must be of the form $y = ax^2 + bx + c$, and the three points $(-1, 3)$, $(0, 2)$, and $(4, 6)$ are on the parabola. Substitution of these coordinates into the equation gives

$$3 = a(-1)^2 + b(-1) + c,$$

$$2 = a(0)^2 + b(0) + c,$$

$$6 = a(4)^2 + b(4) + c.$$

The second equation yields $c = 2$, and when this is substituted into the other two equations,

$$a - b = 1,$$

$$16a + 4b = 4.$$

These can be solved for $a = 2/5$ and $b = -3/5$; therefore, the required equation is $y = 2x^2/5 - 3x/5 + 2$.

EXAMPLE 1.6

When a shell is fired from the artillery gun in Figure 1.23, it follows a parabolic path

$$y = -\frac{4.905}{v^2 \cos^2 \theta} x^2 + x \tan \theta,$$

where v is the muzzle velocity of the shell and θ is the angle at which the shell is fired. Find the range R of the shell and the maximum height attained by the shell.

FIGURE 1.23

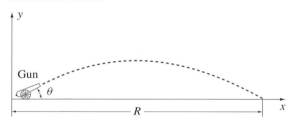

SOLUTION   We can find R by setting $y = 0$ and solving for x:

$$0 = \left(\frac{-4.905}{v^2 \cos^2 \theta} x + \tan \theta \right) x.$$

One solution is $x = 0$ corresponding to the firing position of the shell. The other solution gives the range of the shell,

$$x = R = \frac{\tan \theta \; v^2 \; \cos^2 \theta}{4.905} = \frac{v^2 \; \sin \theta \; \cos \theta}{4.905}.$$

Maximum height of the shell is attained when $x = R/2$ in which case

$$y = \frac{-4.905}{v^2 \cos^2 \theta} \left(\frac{v^2 \sin \theta \; \cos \theta}{2(4.905)} \right)^2 + \tan \theta \left(\frac{v^2 \sin \theta \; \cos \theta}{2(4.905)} \right)$$

$$= \frac{-v^2 \sin^2 \theta}{19.62} + \frac{v^2 \sin^2 \theta}{9.81} = \frac{v^2 \sin^2 \theta}{19.62}.$$

EXERCISES 1.2A

In Exercises 1–12 draw the parabola. Use a calculator or computer to plot the parabola as a check.

∗ **1.** $y = 2x^2 - 1$

∗ **2.** $y = -x^2 + 4x - 3$

∗ **3.** $y = x^2 - 2x + 1$

∗ **4.** $3x = 4y^2 - 1$

∗ **5.** $x = y^2 + 2y$

∗ **6.** $2y = -x^2 + 3x + 4$

∗ **7.** $x + y^2 = 1$

∗ **8.** $2y^2 + x = 3y + 5$

∗ **9.** $y = 4x^2 + 5x + 10$

∗ **10.** $x = 10y^2$

∗ **11.** $y = -x^2 + 6x - 9$

∗ **12.** $x = -(4 + y)^2$

∗ **13.** Find x- and y-intercepts for the following parabolas:

(a) $y = x^2 - 2x - 5$

(b) $x = 4y^2 - 8y + 4$

In Exercises 14–17 determine the equation for each parabola shown.

∗ **14.**

∗ **15.**

∗ **16.**

∗ **17.**

In Exercises 18–23 find all points of intersection for the curves. In each case draw or plot the curves.

∗ **18.** $y = 1 - x^2,\ y = x + 1$

∗ **19.** $y + 2x = 0,\ y = 1 + x^2$

∗ **20.** $y = 2x - x^2 - 6,\ 25 + x = 5y$

∗ **21.** $x = y(y - 1),\ 2y = 2x + 1$

∗ **22.** $x = -y^2 + 1,\ x = y^2 + 2y - 3$

∗ **23.** $y = 6x^2 - 2,\ y = x^2 + x + 1$

∗∗ **24.** For what angle θ is the range of the artillery shell in Example 1.6 largest?

∗∗ **25.** The cable of the suspension bridge in the figure below hangs in the shape of a parabola. The towers are 200 m apart and extend 50 m above the roadway. If the cable is 10 m above the roadway at its lowest point, find the length of the supporting rods 30 m from the towers.

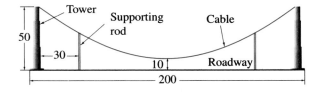

** **26.** Find points of intersection for the parabolas

$$y = (x - 2)^2, \qquad 5x = y^2 + 4.$$

** **27.** Find the height of the parabolic arch in the figure below.

** **28.** Determine the equation of a parabola of type 1.8 passing through the points $(1, 2)$, $(-3, 10)$, and $(3, 4)$.

** **29.** Resistance R in ohms in a platinum resistance thermometer is related to temperature T in degrees Celsius by the equation

$$R = R_0(1 + aT + bT^2)$$

where R_0, a, and b are constants.

    (a) Use the facts that resistances at temperatures $0°C$, $100°C$, and $700°C$ are respectively, 10.000, 13.946, and 24.172 to determine R_0, a, and b.

    (b) Plot a graph of the function on the interval $0 \leq T \leq 700$.

    (c) At what temperature is the resistance 20 ohms?

*** **30.** The parabola $y = x^2 - 1$ represents a wall (x and y measured in metres). A rope is attached to a stake at position $(3, 4)$, pulled tight, and wrapped around that part of the wall containing the vertex of the parabola. At what point does it meet the parabola?

FIGURE 1.24

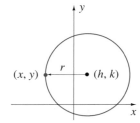

FIGURE 1.25

The Circle

When the x- and y-coordinates of points on a curve are related by an equation of the form

$$(x - h)^2 + (y - k)^2 = r^2, \tag{1.12}$$

where h, k, and $r > 0$ are constants, the curve is called a **circle**. It takes but a quick recollection of distance formula 1.1 to convince ourselves that this definition of a circle coincides with our intuitive idea of a circle. If we write equation 1.12 in the form

$$\sqrt{(x - h)^2 + (y - k)^2} = r,$$

the left side is the distance from the point (x, y) to the point (h, k). Equation 1.12 therefore describes all points (x, y) at a fixed distance r from (h, k), a circle centred at (h, k) with radius r (Figure 1.24). For example, the radius of the circle in Figure 1.25 is equal to 2, and its equation is therefore

$$(x + 1)^2 + (y - 2)^2 = 4.$$

When equation 1.12 is expanded, we have

$$x^2 - 2hx + h^2 + y^2 - 2ky + k^2 = r^2 \qquad \text{or}$$

$$x^2 + y^2 - 2hx - 2ky + h^2 + k^2 - r^2 = 0.$$

This shows that the equation of a circle may be given in another form, namely,

$$x^2 + y^2 + fx + gy + e = 0. \qquad (1.13)$$

Given this equation, the centre and radius can be identified by reversing the expansion and completing the squares of $x^2 + fx$ and $y^2 + gy$. For instance, if $x^2 + y^2 + 2x - 3y - 5 = 0$, then

$$0 = (x + 1)^2 + (y - 3/2)^2 - 5 - 1 - 9/4 = (x + 1)^2 + (y - 3/2)^2 - 33/4.$$

The centre of the circle is therefore $(-1, 3/2)$ and its radius is $\sqrt{33}/2$.

When the centre of a circle is the origin $(0, 0)$, equation 1.12 simplifies to

$$x^2 + y^2 = r^2. \qquad (1.14)$$

 Be careful to use equation 1.12, not 1.14, when the centre of the circle is not the origin. It is a common error to use 1.14.

EXAMPLE 1.7

Figure 1.26 shows an arc of a circle. Find the equation for the circle.

FIGURE 1.26

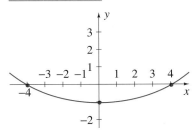

SOLUTION   From the symmetry of the figure, we see that the centre of the circle is on the y-axis. Its equation must be of the form

$$x^2 + (y - k)^2 = r^2.$$

Because $(4, 0)$ is a point on the circle, these coordinates must satisfy its equation; that is,

$$16 + (0 - k)^2 = r^2 \qquad \text{or} \qquad 16 + k^2 = r^2.$$

Similarly, since $(0, -1)$ is on the circle,

$$(-1 - k)^2 = r^2 \qquad \text{or} \qquad 1 + 2k + k^2 = r^2.$$

If we subtract these two equations, we obtain $2k - 15 = 0$, from which we see that $k = 15/2$. Consequently, $r^2 = 16 + k^2 = 16 + 225/4 = 289/4$, and the equation of the circle is

$$x^2 + (y - 15/2)^2 = 289/4.$$

EXAMPLE 1.8

The straight lines in Figure 1.27a represent a welded frame where AOB is a right angle. The circle represents a wheel rotating on a pin through O perpendicular to the plane of the frame. If the minimum clearance between circle and side AB must be 10 cm, find the maximum radius of the wheel.

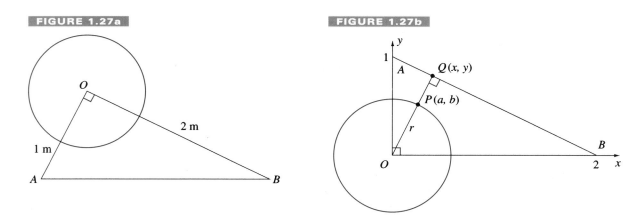

FIGURE 1.27a

FIGURE 1.27b

SOLUTION   Suppose we rotate and flip the figure and establish the coordinate system in Figure 1.27b. If r is the radius of the circle, then its equation is $x^2 + y^2 = r^2$. The shortest distance $\|PQ\|$ between wheel and AB occurs when line OPQ is perpendicular to AB. Since the slope of the line AB is $-1/2$, it follows that the slope of OP is 2, and therefore $b/a = 2$. Because P is on the circle, we also have $a^2 + b^2 = r^2$. When we substitute $b = 2a$ into this equation,

$$a^2 + 4a^2 = r^2 \qquad \Longrightarrow \qquad a = \frac{r}{\sqrt{5}}, \quad b = \frac{2r}{\sqrt{5}}.$$

According to formula 1.7, the distance from the origin to line AB is

$$\|OQ\| = \frac{|(0) + 2(0) - 2|}{\sqrt{1^2 + 2^2}} = \frac{2}{\sqrt{5}}.$$

For the length of PQ to be 1/10 m, it follows that $r + 1/10 = 2/\sqrt{5}$, and the radius of the wheel is $r = (2/\sqrt{5} - 1/10)$ m.

EXERCISES 1.2B

In Exercises 1–10 draw the circle. Use a calculator or computer to plot the circle as a check.

∗ **1.** $x^2 + y^2 = 50$

∗ **2.** $(x + 5)^2 + (y - 2)^2 = 6$

∗ **3.** $x^2 + 2x + y^2 = 15$

∗ **4.** $x^2 + y^2 - 4y + 1 = 0$

∗ **5.** $x^2 - 2x + y^2 - 2y + 1 = 0$

∗ **6.** $2x^2 + 2y^2 + 6x = 25$

∗ **7.** $3x^2 + 3y^2 + 4x - 2y = 6$

∗ **8.** $x^2 + 4x + y^2 - 2y = 5$

∗ **9.** $x^2 + y^2 - 2x - 4y + 5 = 0$

∗ **10.** $x^2 + y^2 + 6x + 3y + 20 = 0$

In Exercises 11–16 find an equation for the circle.

∗ **11.**

∗ **12.**

∗ **13.**

∗ **14.**

∗∗ **15.**

∗∗ **16.**

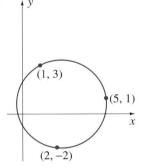

∗∗ **17.** Find the equation of a circle that passes through the points $(3, 4)$ and $(1, -10)$, and has its centre:

    (a) on the line $2x + 3y + 16 = 0$;

    (b) on the line $x + 7y + 19 = 0$.

In Exercises 18–21 find points of intersection for the curves.

∗ **18.** $x^2 + 2x + y^2 = 4$, $y = 3x + 2$

∗ **19.** $x^2 + y^2 - 4y + 1 = 0$, $2x + y = 1$

∗∗ **20.** $x^2 + y^2 = 9$, $y = 3x^2 + 4$

∗∗ **21.** $(x + 3)^2 + y^2 = 25$, $y^2 = 16(x + 1)$

∗∗ **22.** Show that every equation of the form 1.13 represents a circle, a point, or nothing at all.

∗∗ **23.** Prove that the perpendicular bisector of a chord of a circle always passes through the centre of the circle.

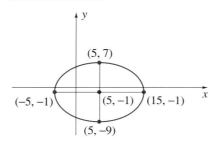

FIGURE 1.31

EXERCISES 1.2C

In Exercises 1–8 draw the ellipse. Use a calculator or computer to plot the ellipse as a check.

∗ **1.** $x^2/25 + y^2/36 = 1$ ∗ **2.** $7x^2 + 3y^2 = 16$

∗ **3.** $9x^2 + 289y^2 = 2601$ ∗ **4.** $3x^2 + 6y^2 = 21$

∗ **5.** $x^2 + 16y^2 = 2$ ∗ **6.** $x^2 + 2x + 4y^2 - 16y + 13 = 0$

∗ **7.** $9x^2 + y^2 - 18x - 6y = 26$ ∗ **8.** $x^2 + 4x + 2y^2 + 16y + 32 = 0$

∗ **9.** Find the equation of an ellipse that passes through the points $(-2, 4)$ and $(3, 1)$.

∗∗ **10.** Find the width of the elliptic arch in the figure below.

In Exercises 11–16 find all points of intersection for the curves. In each case draw the curves.

∗ **11.** $x^2 + 4y^2 = 4$, $y = x$ ∗ **12.** $16x^2 + 9y^2 = 144$, $y = x + 3$

∗∗ **13.** $9x^2 - 18x + 4y^2 = 27$, $2y = \sqrt{3}x + \sqrt{3}$

∗∗ **14.** $9x^2 - 18x + 4y^2 = 27$, $2y = -\sqrt{3}x + 5\sqrt{3}$

∗∗ **15.** $x^2 + 4y^2 - 8y = 0$, $y = x^2$ ∗∗ **16.** $x^2 + 4y^2 = 4$, $y = x^2 - 4$

The Hyperbola

Changing one sign in the equation of an ellipse leads to a curve with totally different characteristics. The set of points whose coordinates (x, y) satisfy an equation of the form

$$\frac{x^2}{a^2} - \frac{y^2}{b^2} = 1 \qquad \text{or} \qquad (1.17a)$$

$$\frac{y^2}{b^2} - \frac{x^2}{a^2} = 1, \qquad (1.17b)$$

where a and b are positive constants, is called a **hyperbola**. Hyperbola 1.17a crosses the x-axis at $x = \pm a$, but does not cross the y-axis. Since the equation remains unchanged when x is replaced by $-x$ and y is replaced by $-y$, the hyperbola is symmetric about both the x-axis and the y-axis. By taking positive square roots of

$$\frac{y^2}{b^2} = \frac{x^2}{a^2} - 1,$$

we obtain

$$\frac{y}{b} = \sqrt{\frac{x^2}{a^2} - 1},$$

from which

$$y = \frac{b}{a}\sqrt{x^2 - a^2}.$$

This equation describes the top half of the hyperbola. Because a is a fixed constant, we can say that, for large values of x, values of a^2 are insignificant compared to values of x^2, and values of y are approximately equal to bx/a. This means that for large values of x, the hyperbola is very close to the line $y = bx/a$. We have shown these facts in Figure 1.32a. The complete hyperbola is shown in Figure 1.32b. The lines $y = \pm bx/a$ which the hyperbola approaches for large positive and negative values of x are called **asymptotes** (straight lines that the curve approaches for large positive or negative values of x) of the hyperbola.

FIGURE 1.32a

FIGURE 1.32b

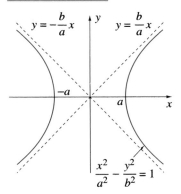

Hyperbola 1.17b is shown in Figure 1.33.

FIGURE 1.33

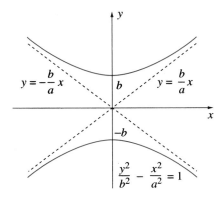

EXAMPLE 1.11

Find the equation of a hyperbola that cuts the y-axis at $y = \pm 5$ and has lines $y = \pm x/\sqrt{3}$ as asymptotes.

SOLUTION   Since the hyperbola crosses the y-axis at $y = \pm 5$, we write its equation in form 1.17b with $b = 5$,

$$\frac{y^2}{25} - \frac{x^2}{a^2} = 1.$$

When we solve this equation for y in terms of x, the result is

$$y = \pm \frac{5}{a}\sqrt{x^2 + a^2}.$$

Since asymptotes of this hyperbola are $\pm 5x/a$, it follows that $5/a = 1/\sqrt{3}$, or, $a = 5\sqrt{3}$. The equation of the hyperbola is therefore

$$\frac{y^2}{25} - \frac{x^2}{75} = 1.$$

When x and y in equations 1.17a and b are replaced by $x - h$ and $y - k$, the resulting equations

$$\frac{(x-h)^2}{a^2} - \frac{(y-k)^2}{b^2} = 1 \qquad \text{and} \qquad (1.18\text{a})$$

$$\frac{(y-k)^2}{b^2} - \frac{(x-h)^2}{a^2} = 1 \qquad (1.18\text{b})$$

still describe hyperbolas. They are shown in Figures 1.34 and 1.35. These are the hyperbolas of Figures 1.32b and 1.33 shifted so that the asymptotes intersect at the point (h, k). Equations of the asymptotes are $y = k \pm b(x - h)/a$, and the lines $x = h$ and $y = k$ are now axes of symmetry.

FIGURE 1.34

FIGURE 1.35

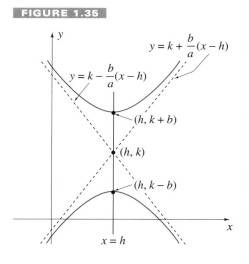

EXAMPLE 1.12

Find asymptotes for the hyperbola $x^2 - y^2 + 4x + 10y = 5$.

SOLUTION   If we complete squares on x- and y-terms, we obtain

$$(x + 2)^2 - (y - 5)^2 = -16 \quad \text{or} \quad \frac{(y - 5)^2}{16} - \frac{(x + 2)^2}{16} = 1.$$

The axes of symmetry of the hyperbola are $x = -2$ and $y = 5$, intersecting at the point $(-2, 5)$. When we solve for y in terms of x, the result is

$$y = 5 \pm \sqrt{(x + 2)^2 + 16}.$$

Therefore, the asymptotes are $y = 5 \pm (x + 2)$. The hyperbola is shown in Figure 1.36.

FIGURE 1.36

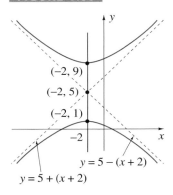

Hyperbolas are but one of many curves that have asymptotes. For instance, in Figure 1.37 the curve $y = (x^3 - 3x^2 + 1)/(x^2 + 1)$ is asymptotic to the line $y = x - 3$. The curve $y = 2e^{-x^2/100}$ is asymptotic to the x-axis (Figure 1.38). "Limits" in Chapter 2 provides a unifying structure for all types of asymptotes.

FIGURE 1.37

FIGURE 1.38

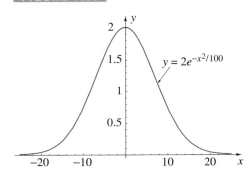

EXAMPLE 1.13

To the right of the right branch of the hyperbola $x^2 - 4y^2 = 5$ in Figure 1.39 is a dense forest. A person P_1 stands at position $(15, 10)$. A second person P_2 walks along the line $x = -1$ in the negative y direction. At what point does P_2 disappear from P_1's view?

FIGURE 1.39

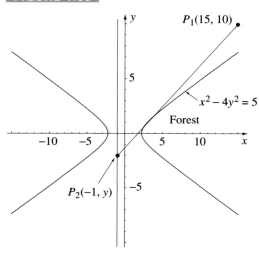

SOLUTION P_2 disappears from view when the line joining P_1 and P_2 just touches the right half of the hyperbola. If we let m be the slope of $P_1 P_2$, then, using point-slope formula 1.4, the equation of line $P_1 P_2$ is

$$y - 10 = m(x - 15).$$

To show the required position of P_2, line $P_1 P_2$ must intersect the hyperbola in exactly one point (most lines intersect in two points, or not at all). Points of intersection are found by solving

$$y - 10 = m(x - 15) \qquad \text{and} \qquad x^2 - 4y^2 = 5.$$

Substituting from the first equation into the second gives

$$x^2 - 4[m(x - 15) + 10]^2 = 5.$$

If we expand the second term on the left,

$$x^2 - 4[m^2(x - 15)^2 + 20m(x - 15) + 100] = 5.$$

This can be rearranged into the form

$$(1 - 4m^2)x^2 + (120m^2 - 80m)x + (-900m^2 + 1200m - 405) = 0.$$

Given a value for m, solutions for x of this quadratic equation are x-coordinates of points of intersection of the line with slope m and the hyperbola. We want only one solution. Consequently, the discriminant must be equal to zero,

$$(120m^2 - 80m)^2 - 4(1 - 4m^2)(-900m^2 + 1200m - 405) = 0.$$

When terms are expanded, this equation reduces to

$$176m^2 - 240m + 81 = 0.$$

The quadratic factors

$$(4m - 3)(44m - 27) = 0,$$

and therefore solutions are $m = 3/4$ and $m = 27/44$. Thus, lines through $(15, 10)$ with slopes $3/4$ and $27/44$ touch the hyperbola at only one point. The line with smaller slope $27/44$ touches the left branch of the hyperbola. The line with larger slope $3/4$ touches the right branch; this is the line we want. Its equation is $y - 10 = (3/4)(x - 15)$, and it cuts the line $x = -1$ when $y = 10 + (3/4)(-1 - 15) = -2$. Thus, P_2 disappears from view at the point $(-1, -2)$.

We shall find a much easier solution to this problem when we have studied some calculus, but the above solution shows that it can be done without calculus, albeit not easily.

◤

EXERCISES 1.2D

In Exercises 1–10 draw the hyperbola. Use a calculator or computer to plot the hyperbola as a check.

* **1.** $y^2 - x^2 = 1$ * **2.** $x^2 - y^2 = 1$

* **3.** $x^2 - y^2/16 = 1$ * **4.** $25y^2 - 4x^2 = 100$

* **5.** $y^2 = 10(2 + x^2)$ * **6.** $3x^2 - 4y^2 = 25$

* **7.** $x^2 - 6x - 4y^2 - 24y = 11$ * **8.** $9x^2 - 16y^2 - 18x - 64y = 91$

* **9.** $4y^2 - 5x^2 + 8y - 10x = 21$ * **10.** $x^2 + 2x - 16y^2 + 64y = 79$

* **11.** Find the equation of a hyperbola that passes through the point $(1, 2)$ and has asymptotes $y = \pm 4x$.

In Exercises 12–18 find all points of intersection for the curves. In each case draw the curves.

* **12.** $x^2 - 2y^2 = 1, x = 2y$ * **13.** $9y^2 - 4x^2 = 36, y = x$

* **14.** $9y^2 - 4x^2 = 36, x = 3y$ * **15.** $3x^2 - y^2 = 3, 2x + y = 1$

** **16.** $x^2 - 2x - y^2 = 0, x = y^2$ ** **17.** $x^2 - 2x - y^2 = 0, x = -y^2$

** **18.** $9(x - 1)^2 - 4(y - 1)^2 = 36, 27x = 5(y - 1)^2$

1.3 Functions and Their Graphs

Most quantities that we encounter in everyday life are dependent on many, many other quantities. For example, think of what might be affecting room temperature as you read this sentence—thermostat setting, outside temperature, wind conditions, insulation of the walls, ceiling, and floors, and perhaps other factors that you can think of. Functional notation allows interdependences of such quantities to be represented in a very simple way.

When one quantity depends on a second quantity, we say that the first quantity is a function of the second. For example, the volume V of a sphere depends on its radius r; in particular, $V = 4\pi r^3/3$. We say that V is a function of r. When an object is dropped, the distance d (metres) that it falls in time t (seconds) is given by the formula $d = 4.905t^2$. We say that d is a function of t. Mathematically, we have the following definition.

> **DEFINITION 1.3**
>
> A quantity y is said to be a function of a quantity x if there exists a rule by which we can associate exactly one value of y with each value of x. The rule that associates the value of y with each value of x is called the **function**.

If we denote the rule or function in this definition by the letter f, then the value that f assigns to x is denoted by $f(x)$, and we write

$$y = f(x). \tag{1.19}$$

In our first example above, we write $V = f(r) = 4\pi r^3/3$, and the function f is the operation of cubing a number and then multiplying the result by $4\pi/3$. For $d = f(t) = 4.905\, t^2$, the function f is the operation of squaring a number and multiplying the result by 4.905.

We call x in equation 1.19 the **independent variable** because values of x are substituted into the function, and y the **dependent variable** because its values depend on the assigned values of x. The **domain** of a function is the set of all specified (real) values for the independent variable. It is an essential part of a function and should always be specified. Whenever the domain of a function is not mentioned, we assume that it consists of all possible values for which $f(x)$ is a real number.

As the independent variable x takes on values in the domain, a set of values of the dependent variable is obtained. This set is called the **range** of the function. For the function $V = f(r) = 4\pi r^3/3$, which represents the volume of a sphere, the largest possible domain is $r > 0$, and the corresponding range is $V > 0$. Note that mathematically the function $f(r)$ is defined for negative as well as positive values of r, and $r = 0$; it is because of our interpretation of r as the radius of a sphere that we restrict $r > 0$. The function $d = f(t) = 4.905\, t^2$ represents the distance fallen in time t by an object that is dropped at time $t = 0$. If it is dropped from a height of 20 m, then it is clear that the range of this function is $0 \le d \le 20$. The domain that gives rise to this range is $0 \le t \le \sqrt{20/4.905}$.

EXAMPLE 1.14

Find the largest possible domain for the function

$$f(x) = \sqrt{\frac{8 + 2x - x^2}{x + 1}}.$$

SOLUTION   We begin by factoring the quadratic in the numerator

$$f(x) = \sqrt{\frac{(2 + x)(4 - x)}{x + 1}},$$

and note that $f(x)$ is defined whenever $(2 + x)(4 - x)/(x + 1) \ge 0$. This occurs when $(2 + x)(4 - x) = 0$, and when $(2 + x)(4 - x)$ and $x + 1$ are both positive or both negative. Figure 1.40 contains the parabola $y = (2 + x)(4 - x)$ and the line $y = x + 1$. They are both below the x-axis when $x < -2$, and both above the x-axis for $-1 < x < 4$. Since $(2 + x)(4 - x) = 0$ when $x = -2$ and $x = 4$, it follows that the largest domain of the function consists of all values of x in the intervals $x \le -2$ and $-1 < x \le 4$.

FIGURE 1.40

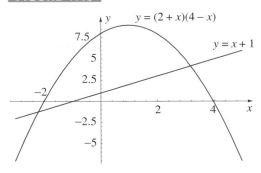

EXAMPLE 1.15

Does the equation $x - 4 - y^2 = 0$ define y as a function of x for $x \geq 4$?

SOLUTION   For any $x > 4$, the equation has two solutions for y:

$$y = \pm\sqrt{x - 4}.$$

Since the equation does not define exactly one value of y for each value of x, it does not define y as a function of x.

If we add to the equation in Example 1.15 an additional restriction such as $y \geq 0$, then y is defined as a function of x, namely,

$$y = \sqrt{x - 4}.$$

Note that $x - 4 - y^2 = 0$ does, however, define x as a function of y:

$$x = y^2 + 4.$$

In this case, y is the independent variable and x is the dependent variable. In other words, whenever an equation (such as $x - 4 - y^2 = 0$) is to be regarded as defining a function, it must be made clear which variable is to be considered as independent and which as dependent. For example, the equation $x - 4 - y = 0$ defines y as a function of x and x as a function of y.

In the study of calculus and its applications we are interested in the behaviour of functions; that is, for certain values of the independent variable, what can we say about the dependent variable? The simplest and most revealing method for displaying characteristics of a function is a graph. To obtain the graph of a function $f(x)$ we use a plane coordinatized with Cartesian coordinates x and y (in short, the Cartesian xy-plane). The **graph** of the function $f(x)$ is defined to be the curve with equation $y = f(x)$, where x is limited to the domain of the function. For example, the graph of the function $f(x) = x^3 - 27x + 1$ is shown in Figure 1.41.

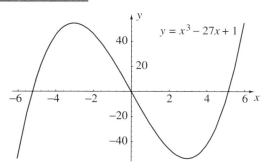

$y = x^3 - 27x + 1$

This curve is a pictorial or geometric representation of the function $f(x) = x^3 - 27x + 1$. The value of the function for a given x is visually displayed as the y-coordinate of that point on the curve with x-coordinate equal to the given x. This graph clearly illustrates properties of the function that may not be so readily obvious from the algebraic definition $f(x) = x^3 - 27x + 1$. For instance,

1. For negative values of x, the largest value of $f(x)$ is $f(-3) = 55$; for positive values of x, the smallest value of $f(x)$ is $f(3) = -53$.

2. As x increases, values of $f(x)$ increase for $x < -3$ and $x > 3$, and values of $f(x)$ decrease for $-3 < x < 3$.

3. $f(x)$ is equal to zero for three values of x—one a little less than -5, one a little larger than 0, and one a little larger than 5.

Graphing calculators and computers have become indispensable tools for modern scientists; one reason for this is the ability of these electronic devices to graph functions so quickly. We hasten to point out, however, that machine-generated graphs may sometimes be misleading; care must be taken not to make rash assumptions based on machine output. We illustrate with some examples below. In addition, because calculators and computers have finite screen resolution, it may not always be perfectly clear how to interpret every aspect of a machine-generated graph. Mathematical analysis may be required to confirm or deny what is being suggested by the graph. This is very important! We use machine-generated graphs extensively, but we are always prepared to corroborate what we see, or don't see, with rigorous mathematical analysis.

When we ask you to "plot" a graph, we intend you to use a graphing calculator or computer. When we ask you to "draw" a graph, we expect you to do so without these devices. Sometimes a drawing is more informative than a plot.

EXAMPLE 1.16

A very powerful software package gives the graph for the function $f(x) = x + \sin(2\pi x)$ on the interval $0 \le x \le 24$ in Figure 1.42. How do you feel about what you see?

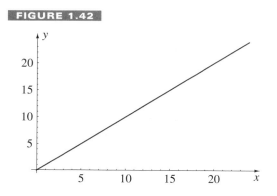

SOLUTION   The graph appears to be the straight line $y = x$, and therefore $\sin(2\pi x)$ is always zero. This is ridiculous; it is a fluke, resulting from the choice of sampling points taken by the software package in plotting the graph. Choosing the plot interval to be $0 \le x \le 25$ gives Figure 1.43a. This is still misleading. The plot in Figure 1.43b gives a true indication of the nature of the function. Examples like this are rare, but they can occur. Use your common sense.

FIGURE 1.43a

FIGURE 1.43b

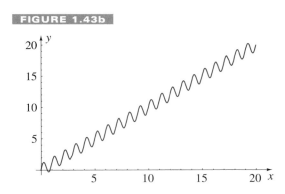

We encountered absolute values in Section 1.1. The absolute value function, denoted by $|x|$, is defined as

$$|x| = \begin{cases} -x, & x < 0 \\ x, & x \ge 0. \end{cases} \tag{1.20}$$

It defines the size or "magnitude" of its argument without regard for sign. The graph of the function is composed of two straight lines with slopes ± 1 meeting at right angles at the origin. This does not appear to be the case in Figure 1.44. Why?

FIGURE 1.44

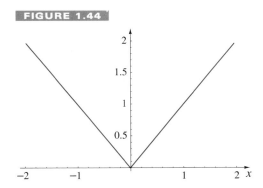

Quite often the argument of the absolute value function is a function of x rather than x itself. Such is the case in the following example.

EXAMPLE 1.17

A computer-generated graph of the function $f(x) = |x^3 + 5|$ is shown in Figure 1.45. It appears to touch the x-axis somewhere between $x = -2$ and $x = -1$, but we cannot be positive of this graphically. In addition, we cannot be sure whether the graph is rounded or whether it comes to a sharp point at this same value of x. Perform some elementary graphing to answer these questions.

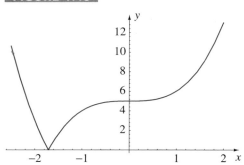

SOLUTION   To draw the curve $y = |x^3 + 5|$, we begin with $y = x^3$ in Figure 1.46a. Next we draw the curve $y = x^3 + 5$ by adding 5 to every ordinate in Figure 1.46a. The result in Figure 1.46b is the curve in Figure 1.46a shifted upward 5 units. It crosses the x-axis at $-5^{1/3}$. The last step is to take the absolute value of every ordinate on the curve $y = x^3 + 5$. This changes no ordinate that is already positive, but changes the sign of any ordinate that is negative. The resulting final drawing is shown in 1.46c. Clearly then, the graph touches the x-axis at $x = -5^{1/3}$, and there is a sharp point on the graph at $(-5^{1/3}, 0)$.

FIGURE 1.46a

FIGURE 1.46b

FIGURE 1.46c

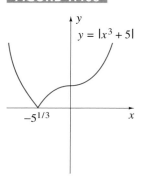

We might get the impression from Figures 1.44 and 1.46c that absolute values lead to curves with sharp points. This is not always the case. When absolute values are placed around the function x^3 in Figure 1.47a, the resulting function $|x^3|$ is shown in Figure 1.47b. There is no sharp point at the origin.

FIGURE 1.47a

FIGURE 1.47b

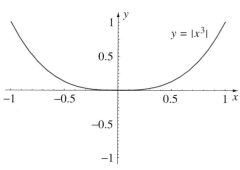

The absolute value of the function $2x^3 - 11x^2 + 16x - 7$ in Figure 1.48a leads to the graph in Figure 1.48b for $|2x^3 - 11x^2 + 16x - 7|$. It has a sharp point at the position between $x = 3$ and $x = 4$ where it touches the x-axis, but not at $x = 1$.

FIGURE 1.48a

FIGURE 1.48b

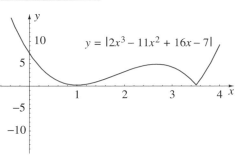

EXAMPLE 1.18

Plot a graph of the function $f(x) = 2\sqrt{4 - x^2}$ (using a calculator or computer). Does the graph look familiar? Confirm what it suggests.

SOLUTION   The domain of the function is $-2 \leq x \leq 2$. Its graph in Figure 1.49 appears to be the top half of an ellipse. To confirm this, we square $y = 2\sqrt{4 - x^2}$, obtaining $y^2 = 4(4 - x^2)$, or $4x^2 + y^2 = 16$. This is indeed the equation of an ellipse. Only the top half of the ellipse is defined by $y = 2\sqrt{4 - x^2}$ since the square root requires y to be nonnegative.

FIGURE 1.49

FIGURE 1.50

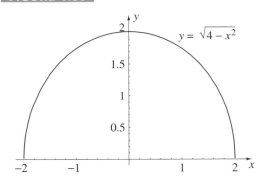

EXAMPLE 1.19

Repeat Example 1.18 with $f(x) = \sqrt{4 - x^2}$.

SOLUTION   The graph in Figure 1.50 once again suggests the top half of an ellipse. Squaring $y = \sqrt{4 - x^2}$ leads to $x^2 + y^2 = 4$, a circle, not an ellipse. The reason the graph does not appear to be semicircular is that the computer has chosen different scales for the x- and y-axes.

EXAMPLE 1.20

Plot a graph of the function $f(x) = x + \sqrt{3x^2 - 4}$, $x \geq 2/\sqrt{3}$. Does it appear to have an asymptote? Confirm this mathematically.

SOLUTION   The plot in Figure 1.51 appears to be fairly straight, suggesting an asymptote. To confirm this, we note that for large x, the 4 is insignificant compared to $3x^2$, and therefore

$$f(x) \approx x + \sqrt{3x^2} = x + \sqrt{3}x = (1 + \sqrt{3})x.$$

Thus, the line $y = (1 + \sqrt{3})x$ is an asymptote for the graph.

FIGURE 1.51

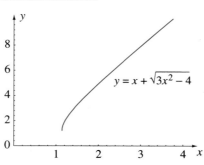

$y = x + \sqrt{3x^2 - 4}$

EXAMPLE 1.21

When a ball is dropped from the top of a building 20 m high at time $t = 0$, the distance d (in metres) that it falls in time t is given by the function $d = f(t) = 4.905t^2$. Plot its graph.

SOLUTION   In this example independent and dependent variables are denoted by letters t and d which suggest their physical meaning—time and distance—rather than the generic labels x and y. With the axes labelled correspondingly as the t-axis and d-axis, we draw part of the parabola $d = 4.905t^2$ in Figure 1.52. The remainder of the parabola has no physical significance in the context of this problem.

FIGURE 1.52

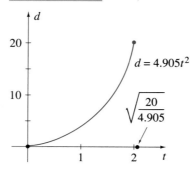

$d = 4.905t^2$

$\sqrt{\dfrac{20}{4.905}}$

Even and Odd Functions

The parabola $y = x^2 + 1$ in Figure 1.53 is the graph of the function $x^2 + 1$; it is symmetric about the y-axis. The curve in Figure 1.54 is the graph of the function $(x^4 - 2x^2 + 5)/(x^2 + 6)$; it is also symmetric about the y-axis. These are examples of a special class of functions identified in the following definition.

FIGURE 1.53

FIGURE 1.54

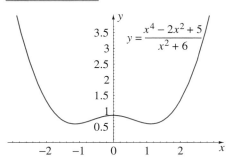

DEFINITION 1.4

A function $f(x)$ is said to be **even** if for each x in its domain

$$f(-x) = f(x); \tag{1.21a}$$

it is said to be **odd** if

$$f(-x) = -f(x). \tag{1.21b}$$

The first of these implies that the equation $y = f(x)$ for the graph of an even function is unchanged when x is replaced by $-x$; therefore, the graph of an even function is symmetric about the y-axis. As a result, $f(x) = x^2 + 1$ and $f(x) = (x^4 - 2x^2 + 5)/(x^2 + 6)$ are even functions.

Equation 1.21b implies that if (x, y) is any point on the graph of an odd function, so too is the point $(-x, -y)$. This is illustrated by the graph of the odd function $f(x) = 2x^5 - 3x^3 + x$ in Figure 1.55. Another way to describe the graph of an odd function is to note that either half $(x < 0$ or $x > 0)$ is the result of two reflections of the other half, first in the y-axis and then in the x-axis. Alternatively, either half of the graph is a result of rotating the other half by π radians (one-half a revolution) around the origin.

FIGURE 1.55

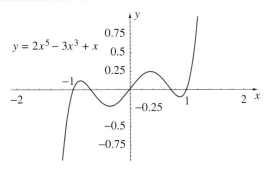

FIGURE 1.62a **FIGURE 1.62b** **FIGURE 1.62c**

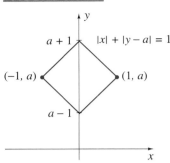

Polynomials and Rational Functions

Two very important classes of functions are polynomials and rational functions. A **polynomial of degree n** is a function of the form

$$f(x) = a_n x^n + a_{n-1} x^{n-1} + \cdots + a_2 x^2 + a_1 x + a_0 \tag{1.22}$$

where n is a nonnegative integer, and $a_0, a_1, \ldots, a_n$ are real numbers ($a_n \neq 0$). Special names have become associated with some of the lower-degree polynomials because of the frequency with which they arise:

$f(x) = a_1 x + a_0$	polynomial of degree 1, linear polynomial, linear function
$f(x) = a_2 x^2 + a_1 x + a_0$	polynomial of degree 2, quadratic polynomial, quadratic function
$f(x) = a_3 x^3 + a_2 x^2 + a_1 x + a_0$	polynomial of degree 3, cubic polynomial, cubic function

Polynomials are defined for all values of x. Graphs of linear and quadratic polynomials are straight lines and parabolas. The cubic polynomial $x^3 + 12x^2 - 40x + 6$ and the quartic $x^4 + 10x^3 + 6x^2 - 64x + 5$ are plotted in Figures 1.63 and 1.64. Figure 1.63 has a low point just to the right of $x = 1$ and a high point just to the left of $x = -9$. We could approximate these points more and more closely by reducing the plot interval. Calculus provides an analytic way to determine them.

FIGURE 1.63 **FIGURE 1.64**

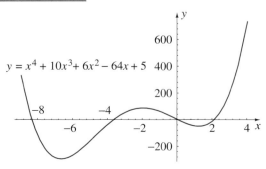

A **rational function** $R(x)$ is defined as the quotient of two polynomials $P(x)$ and $Q(x)$:

$$R(x) = \frac{P(x)}{Q(x)}. \tag{1.23}$$

Rational functions are undefined at points where $Q(x) = 0$. For example, the rational function

$$f(x) = \frac{x^3 + 3x^2 - 12x + 1}{x^2 - 2x - 3}$$

is undefined at $x = -1$ and $x = 3$. The computer-generated graph on the interval $-2 \le x \le 4$ in Figure 1.65a indicates that function values become very large positively and negatively near $x = -1$ and $x = 3$. We investigate this behaviour in detail in Chapter 2. The graph may or may not be accompanied by messages indicating that division by 0 is encountered at $x = -1$ and $x = 3$, depending on the plot interval specified to the plotting device. A plot on the larger interval $-10 \le x \le 10$ in Figure 1.65b suggests that the graph has an asymptote. We shall see why in Chapter 2.

FIGURE 1.65a

FIGURE 1.65b

Every function $f(x)$ can be represented pictorially by its graph, the curve with equation $y = f(x)$. But what about the reverse situation? Does every curve in the xy-plane represent a function $f(x)$? The curves in Figure 1.66, which both extend between $x = a$ and $x = b$, illustrate that the answer is no. The curve in Figure 1.66a represents a function, whereas the curve in Figure 1.66b does not, because for values of x between a and c there are two possible values of y. In other words, *a curve represents a function f(x) if every vertical line that intersects the curve does so at exactly one point.*

FIGURE 1.66a

FIGURE 1.66b

EXERCISES 1.3

In Exercises 1–6 find the largest possible domain of the function.

* **1.** $f(x) = \sqrt{9 - x^2}$ * **2.** $f(x) = 1/(x - 2)$

* **3.** $f(x) = \dfrac{1}{x\sqrt{x^2 + 4}}$ * **4.** $f(x) = \dfrac{1}{x\sqrt{x^2 - 4}}$

** **5.** $f(x) = \dfrac{1}{x\sqrt{4 - x^2}}$ ** **6.** $f(x) = \sqrt{\dfrac{x^2 - 4}{9 - x^2}}$

In Exercises 7–16 determine algebraically whether the function is even, odd, or neither even nor odd. Plot each function to confirm your conclusion geometrically.

* **7.** $f(x) = 1 + x^2 + 2x^4$ * **8.** $f(x) = x^5 - x$

* **9.** $f(x) = 12x^2 + 2x$ * **10.** $f(x) = x^{1/5}$

* **11.** $f(x) = (x - 1)/(x + 1)$ * **12.** $f(x) = x(x^2 + x)$

* **13.** $f(x) = \dfrac{x^2 + 1}{1 - 2x^4}$ * **14.** $f(x) = x^3 + \sqrt{x}$

* **15.** $f(x) = \dfrac{x|x|}{3 + x^2}$ * **16.** $f(x) = \dfrac{x^2 + \sqrt{x^4 + 1}}{x^6 + 3x^2}$

* **17.** Verify that when an odd function $f(x)$ is defined at $x = 0$, its value must be $f(0) = 0$.

* **18.** Prove each of the following:

    (a) The product of two even functions or two odd functions is an even function.

    (b) The product of an even and an odd function is an odd function.

In Exercises 19–48 first draw, then plot a graph of the function.

* **19.** $f(x) = \sqrt{5 - x^2}$ * **20.** $f(x) = -\sqrt{5 - x^2}$

* **21.** $f(x) = -\sqrt{x^2 - 5}$ * **22.** $f(x) = \sqrt{x^2 - 5}$

* **23.** $f(x) = \sqrt{5 - 4x^2}$ * **24.** $f(x) = -\sqrt{5 - 4x^2}$

* **25.** $f(x) = |x| + 2x$ * **26.** $f(x) = |x| - 2x$

* **27.** $f(x) = -\sqrt{-x}$ * **28.** $f(x) = x^{1/3}$

* **29.** $f(x) = x^{2/3}$ * **30.** $f(x) = 3x^{3/2}$

** **31.** $f(x) = 3|x|^{3/2}$ ** **32.** $f(x) = x\sqrt{x + 1}$

** **33.** $f(x) = x\sqrt{x^2 - 1}$ ** **34.** $f(x) = -x\sqrt{4 - 9x^2}$

** **35.** $f(x) = x^2\sqrt{4 - x}$ ** **36.** $f(x) = x^2\sqrt{x^2 - 4}$

** **37.** $f(x) = x^2\sqrt{4 - 9x^2}$ ** **38.** $f(x) = |x^2 - x - 12|$

** **39.** $f(x) = \sqrt{2x - x^2}$ ** **40.** $f(x) = \sqrt{2x - 4x^2}$

** **41.** $f(x) = \sqrt{4x^2 - 2x}$ ** **42.** $f(x) = \sqrt{2x - x^2 - 4}$

** **43.** $f(x) = x + 2 + \sqrt{x}$ ** **44.** $f(x) = \sqrt{9 - 4x^2} + \sqrt{4x^2 - 9}$

** **45.** $f(x) = x^2 + |x| - 2$ ** **46.** $f(x) = \sqrt{(x^2 - 4)^2}$

** **47.** $f(x) = \sqrt{2 - \sqrt{1 + x}}$ ** **48.** $f(x) = \sqrt{(x^2 - 1)^2 - (x^2 - 1)}$

* **49.** What condition ensures that a curve in the xy-plane represents a function $x = f(y)$?

** **50.** The *greatest integer function* is denoted by $f(x) = [\![x]\!] = $ greatest integer that does not exceed x.

    (a) Draw a graph of the greatest integer function.

    (b) If first-class postage is 40¢ for each 50 gm, or fraction thereof, up to and including 500 gm, draw a graph of this cost function.

    (c) Express the cost function in (b) in terms of the greatest integer function.

In Exercises 51–62 plot the curve. Indicate whether the curve defines y as a function of x.

* **51.** $y = x^2 - x^6$ * **52.** $y = (x + 1)^2 - 2(x + 1)$

* **53.** $(x - 2)^2 + y^2 = 4$ * **54.** $y = \sqrt{4 - (x - 2)^2}$

* **55.** $x^2 - (y + 5)^2 = 1$ * **56.** $x = -\sqrt{64 + 9(y + 5)^2}$

* **57.** $y = (x + 1)^4 - (x + 1)^2$ * **58.** $|x| + |y + 2| = 1$

* **59.** $y = \left|3 - |x + 2|\right| - 1$ * **60.** $|x| - |y| = 1$

* **61.** $x = 4 - |y|$ * **62.** $(y - 1)^2 = (x - 1)^2[1 - (x - 1)^2]$

** **63.** A cherry orchard has 255 trees, each of which produces an average of 25 baskets of cherries. For each additional tree planted, the yield per tree decreases by one-twelfth of a basket. If x represents the number of extra trees (beyond 255) and Y the total yield, find Y as a function of x, and draw its graph. How many more trees should be planted for maximum yield?

** **64.** A rectangle with sides parallel to the axes is inscribed inside the ellipse $b^2x^2 + a^2y^2 = a^2b^2$ below. Find a formula for the area A of the rectangle in terms of x. Draw a graph of this function.

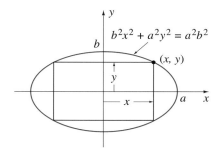

** **65.** A man 2 m tall walks along the edge of a straight road that is 5 m wide. On the other edge of the road stands a street light 10 m high. Find a functional relationship for the length of the man's shadow in terms of his distance from the point on his side of the road directly across from the light. Plot a graph of this function.

** **66.** When two substances A and B are brought together, a chemical reaction takes place to form a new substance C. It requires 2 L of A for each litre of B to produce 3 L of C. The rate R at which A and B react to form C is proportional to the product of the amounts of A and B present at that instant. If the original amounts of A and B are 20 L and 40 L, respectively, and if x represents the amount of C present in the reaction at any given time, find a formula for R as a function of x. Draw a graph of this function, and determine when the reaction rate is highest.

** **67.** In classical physics and engineering, the mass m of an object is constant, independent of how fast it is moving. In special relativity, however, m is given by the formula

$$m = \frac{m_0}{\sqrt{1 - (v^2/c^2)}},$$

where m_0 is the mass of the object when it is not moving, v is the speed of the mass, and c is a constant (the speed of light). Draw a graph of this function, and draw any conclusions that you feel are suggested.

✸✸ **68.** A square plate 4 m on each side is slowly submerged in a large tank of water. One diagonal is kept vertical and lowered at a rate of 0.5 m/s, entering the water at time $t = 0$. If A is the area of the submerged portion of the surface (one side only) at time t until complete submersion occurs, find A as a function of t and plot its graph.

✸✸ **69.** Because of construction, no passing is permitted on a 10 km stretch of highway. If cars travel at v km/h along this stretch, a safe distance between them must be maintained, and this distance increases as v increases. In particular, the highway traffic commission has determined that for speeds over 50 km/h, the distance in metres between cars should be at least

$$d = \frac{3v^2}{500}.$$

If it is supposed that everyone maintains the safe distance and the same constant speed v through the stretch, find the number q of cars leaving the "bottleneck" per hour as a function of speed v. Plot a graph of this function for $50 \le v \le 100$, and determine the speed that maximizes q.

✸✸ **70.** A box measuring 1 m on each side is attached to a rope as shown in the figure below. The rope passes over a pulley 10 m from the ground and a truck pulls on the other end in a horizontal direction along a line 1 m above the ground. We denote positions of the truck and the bottom of the box by x and y, respectively. Find y as a function of x if the truck starts at position $x = 5$ m and stops when the top of the box touches the pulley. Assume that the length of rope between truck and box is 25 m. Plot a graph of the function.

In Exercises 71–74 draw a graph of the function where $[\![x]\!]$ is the greatest integer function of Exercise 50.

✸✸ **71.** $f(x) = [\![2x]\!]$ ✸✸ **72.** $f(x) = x + [\![x]\!]$

✸✸ **73.** $f(x) = x[\![x]\!]$ ✸✸ **74.** $f(x) = [\![x + [\![x+1]\!]]\!]$

✸✸ **75.** Is $[\![f(x) + g(x)]\!] = [\![f(x)]\!] + [\![g(x)]\!]$?

✸✸ **76.** If a thermal nuclear reactor is built in the shape of a right circular cylinder of radius r and height h, then neutron diffusion theory requires r and h to satisfy an equation of the form

$$\frac{a^2}{r^2} + \frac{b^2}{h^2} = 1,$$

where a and b are positive constants. Draw a graph of the function $r = f(h)$ defined by this equation for $h \ge 2b$.

✸✸ **77.** Is there a difference between the graphs of the functions $f(x) = \ln(x^2)$ and $g(x) = 2\ln x$?

✸✸ **78.** Prove that every function can be expressed as the sum of an even function and an odd function, and that there is only one way to do this.

1.4 A Glimpse of Things to Come

Calculus is an indispensable tool for problems in a multitude of fields—engineering, physics, chemistry, medecine, and economics, to name a few. Because we have not yet introduced calculus, we cannot solve these problems here, but we can at least introduce them, and these initial discussions form the basis for future analysis. Pay careful attention to the ideas; it will pay dividends.

Displacement and Speed

Calculus plays a central role in discussions that trace the motion of particles and objects moving under the influence of forces. Without calculus, analysis of a simple system such as a mass on the end of a spring would be exceedingly difficult; planning lift-off, space travel, and reentry of spacecraft would be impossible. We introduce basic ideas connected with motion here, and with each new development of calculus we take the analysis a little further.

Suppose a particle is at some position on the x-axis at time $t = 0$, and for $t > 0$, various forces act on it resulting in motion that is confined to the x-axis (Figure 1.67).

FIGURE 1.67

The position of the particle can be identified by its x-coordinate, and x can always be regarded as a function of time t. Suppose, for example, that

$$x = f(t) = t^3 - 27t^2 + 168t + 20, \quad t \geq 0,$$

where x is measured in metres and t in seconds. This is called the **displacement function**; it gives a complete history of the particle's motion. It tells us not only where the particle is at any given time, but also contains information about the velocity, speed, and acceleration of the particle; for in-depth discussions of these, however, we need calculus. Even without calculus, much can be visualized about the motion of the particle by plotting a graph of the displacement function (Figure 1.68). Ordinates of this curve represent horizontal displacements of the particle relative to $x = 0$. For example, the height of the displacement graph at time $t = 0$ is 20. This means that the particle begins motion 20 m to the right of the origin. At $t = 4$ s, it is 324 m to the right of the origin, and at $t = 14$ s, it is 176 m to the left of the origin. Whenever ordinates are positive, the particle is to the right of the origin. Whenever ordinates are negative, the particle is to the left of $x = 0$. Figure 1.68 indicates that the particle is to the right of $x = 0$ for the first ten seconds and then again after $t \approx 17$ s. Between these times, it is to the left of the origin.

FIGURE 1.68

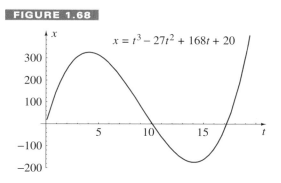

For $0 < t < 4$, values of x are getting larger. This means that the particle is moving to the right. For $4 < t < 14$, the particle moves to the left (values of x are getting smaller), and for $t > 14$, it moves to the right once again. This means that the particle stops and reverses direction at $t = 4$ s and at $t = 14$ s.

Everyone has an intuitive idea about speed, when something moves quickly and when it moves slowly. To discuss speed of the above particle quantitatively, we need calculus, but even now we can derive qualitative information about how fast the particle is moving. The more distance the particle covers in a given time interval, the faster it is moving. In Table 1.1, we have tabulated the distance that the particle travels each second for the first twenty seconds. The first two entries indicate that the particle travels 142 m in the first second, but only 94 m from $t = 1$ s to $t = 2$ s. The particle is therefore travelling faster duing the first second of its motion than during the next second. The remainder of the table suggests that the particle slows down until, as we know, it comes to a stop at $t = 4$ s. It then speeds up until somewhere around $t = 9$ s when it starts to slow down again, and comes to a stop at $t = 14$ s. It then picks up speed thereafter.

TABLE 1.1

second	1	2	3	4	5	6	7	8	9	10
distance	142	94	52	16	14	38	56	68	74	74

second	11	12	13	14	15	16	17	18	19	20
distance	68	56	38	32	30	52	94	142	196	256

We can see this graphically as well as in tabular form. The steeper the displacement graph, the larger the speed, the flatter the graph, the smaller the speed. This means that speed is greatest when the graph rises or falls quickly. At the beginning of the particle's motion (for small $t > 0$), the graph is relatively steep, and therefore the particle is moving quickly. Steepness decreases as we head toward $t = 4$ s, and therefore the particle is slowing down. At $t = 4$ s, the particle's speed is zero (it reverses direction). The particle then picks up speed, moving now to the left until steepness is greatest somewhere around $t = 9$ s. It then starts to slow down, coming to a stop at $t = 14$ s. It then moves to the right thereafter picking up speed (as steepness increases). Again, all of this is quite qualitative, but when we learn a little *differential* calculus, we can be much more specific and we can add information about the velocity and acceleration of the particle.

Electric Circuits

Modern electronic equipment contains a vast array of devices; most find their origin in three fundamental elements—capacitors, inductors, and resistors. How these elements relate to one another and how they affect voltages, charges, and currents in electric circuits can be fully understood with calculus. Concepts that we introduce here form the basis for many discussions to follow. We begin with capacitors.

FIGURE 1.69

Capacitor

Voltmeter

A **capacitor** is a device that stores equal amounts of positive charge and negative charge in such a way that the charges cannot neutralize one another. Suppose $Q > 0$ is the amount of positive charge, and therefore $-Q$ is the negative charge. Separation of these charges creates a potential difference V at the terminals of the capacitor. It can be measured by placing a voltmeter across the terminals as shown in Figure 1.69. The size of V depends on how charges are stored in the capacitor; different configurations lead to different potential differences. When we divide Q by the potential difference V that it produces, we obtain what is called the capacitance C of the capacitor,

$$C = \frac{Q}{V}.$$

$$(1.24)$$

It has units of coulombs per volt, called farads (F). The higher the capacitance, the greater the charge that can be stored per volt of potential difference.

If a 9 V battery is connected to a 10^{-6} F capacitor (Figure 1.70) and the switch is closed, the battery creates a flow of charge in the circuit until the capacitor is charged with $Q = 9(10^{-6})$ coulombs. This happens very quickly, so quickly that we cannot measure how fast the charge flows. The rate at which charge flows is called **current**. It is measured in amperes (A); one ampere is a flow of one coulomb of charge per second. The reason that we cannot measure the current as the battery in Figure 1.70 charges the capacitor is because we have not accounted for a resistance in the circuit. We do this now.

A **resistor** is an electronic device that retards the flow of charge in a circuit. Suppose potential difference V is maintained between the terminals A and B of the resistor in Figure 1.71, where the $+$ and $-$ signs indicate that A is at higher potential than B. This creates a flow of positive charge from B to A. If the rate of flow (current) is i A, the ratio

$$ R = \frac{V}{i} \tag{1.25} $$

is called the **resistance** of the resistor. It is measured in volts per ampere, called ohms (Ω). The higher the resistance, the smaller the current generated by a given voltage, or the larger the voltage required to produce a given current. For example, to maintain a current of 2 A through a 3 Ω resistor requires 6 V, but to maintain the same current through a 30 Ω resistor requires 60 V.

FIGURE 1.70

$C = 10^{-6}$ F

S

9 V

FIGURE 1.71

Potential
$+$ ——— difference ——— $-$
= V

B Resistor A

Current

FIGURE 1.72

C

i

V

R

S

Charge flows through a resistor or any other electric device when a potential difference is created between its ends. A device that creates and maintains potential difference is an **emf** device. Examples are batteries, electric generators, solar cells, and thermopiles. An emf device is shown in Figure 1.72 in a circuit including a capacitor with capacitance C, a resistor with resistance R, and a switch S. This is called an RC-circuit. The emf device keeps its upper terminal postively charged and its lower terminal negatively charged as indicated by the $+$ and $-$ signs. The result is that charge flows in the circuit creating a current i. *Kirchhoff's loop rule* for electric circuits states that the algebraic sum of the changes in potential encountered in a complete traversal of any circuit must be zero. Let us apply the rule to the circuit in Figure 1.72, when the switch is in the closed position, starting at the lower terminal of the emf device and proceeding clockwise. If the emf device creates voltage V, there is an increase of potential equal to V across the device. With current in the direction shown, positive charge builds up on the left terminal of the capacitor and negative charge on the right terminal. The left terminal is therefore at higher potential than the right one. The potential difference when the charge on the capacitor has reached Q is $V = Q/C$. Since potential drops across the capacitor, we regard the change as $-Q/C$. With current flowing down through the resistor, the potential drop across it is $-iR$ when the current is i. Taking these three changes into account, Kirchhoff's loop rule gives

$$ V - \frac{Q}{C} - iR = 0 \qquad \text{or} \qquad V = \frac{Q}{C} + iR. \tag{1.26} $$

To analyze this equation and determine the charge Q on the capacitor and current i in the circuit at any given time, we must express i in terms of Q. We know that i is the rate at which Q flows but in order to express this relationship precisely, we need the *derivative* from calculus. We can, however, discuss what happens in the circuit in a general way. Suppose, for example, that the charge on the capacitor is zero when the switch is closed at time $t = 0$. Just after $t = 0$, the charge on the capacitor is small with the result that so also is the voltage Q/C across the capacitor. Because V in 1.26 is constant, and Q/C is small, voltage iR across the resistor must constitute most of V. Consequently, current is essentially V/R when the switch is closed. As time progresses, charge builds up on the capacitor so that voltage across it increases. Voltage across the resistor must decrease and so also therefore must the current through it. As charge on the capacitor approaches CV, voltage across it approaches V, and voltage across the resistor approaches zero. Current in the circuit becomes less and less, eventually becoming zero when the capacitor is fully charged. Graphs of Q and i as functions of time are shown in Figures 1.73. Equations describing these curves will be discussed in Section 1.6.

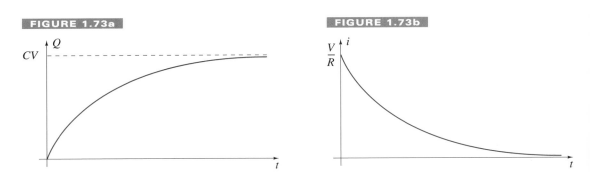

FIGURE 1.73a

FIGURE 1.73b

The third fundamental circuit element is the inductor. It is really impossible to discuss inductors in a meaningful way without calculus, but the following remarks give a general idea of how they fit in with capacitors and resistors. Voltage across a capacitor is related to charge; voltage across a resistor is related to current (the rate of change of charge); voltage across an inductor is related to the rate of change of current. In other words, an inductor reacts to changes in current. If V is the voltage across the terminals of an inductor and current is changing in the inductor at some rate, then the inductance of the inductor is defined as

$$L = \frac{V}{\text{Rate of change of current}}. \tag{1.27}$$

It has units of volts per ampere per second, called henries (H). When inductance is large, a small rate of change of current produces a large voltage across the terminals of the inductor.

With calculus we will be able to analyze circuits that contain all three fundamental elements, so-called LCR-circuits (Figure 1.74). Inductors will be incorporated into Kirchhoff's loop

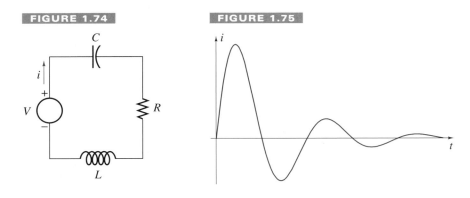

FIGURE 1.74

FIGURE 1.75

rule, and charge on the capacitor and current in the circuit will be determined at any given time. Typically, the current in an LCR-circuit looks like that in Figure 1.75. It is a combination of trigonometric functions that we review in Section 1.5 and an exponential function from Section 1.6.

Optimization Problems

Differential calculus can be used to solve optimization problems—problems wherein it is required to maximize or minimize some quantity. The following example is typical of such problems. A company wishes to construct a storage tank in the form of a box with a square horizontal cross-section. The volume of the tank must be 100 m³. The bottom must be stronger than the sides and top and, as a result, material for the bottom costs $4.75/m², whereas for the sides and top the cost is $1.25/m². The twelve edges of the tank are to be welded at a cost of $7.50/m. The company naturally wishes to minimize the total cost of materials for the tank. Let us determine these costs. The tank in Figure 1.76 has length and width x m and height y m.

The cost for materials for the top, sides, and bottom is

$$125(x^2 + 4xy) + 475x^2,$$

and that for the twelve welds is

$$750(8x + 4y).$$

Thus, total material costs are

$$125(x^2 + 4xy) + 475x^2 + 750(8x + 4y).$$

Since the tank must hold 100 m³, it follows that

$$100 = x^2 y.$$

When we solve this for y in terms of x, we obtain

$$y = \frac{100}{x^2}.$$

If we substitute this into the cost expression, we obtain a function of x that we denote by

$$C = f(x) = 125\left[x^2 + 4x\left(\frac{100}{x^2}\right)\right] + 475x^2 + 750\left[8x + 4\left(\frac{100}{x^2}\right)\right]$$

$$= 600x^2 + 6000x + \frac{50\,000}{x} + \frac{300\,000}{x^2}.$$

This function represents material costs for a 100 m³ tank when its horizontal cross-section is x m by x m. To minimize material costs it is necessary to find that value of x that makes $f(x)$ as small as possible. The graph of $f(x)$ in Figure 1.77 indicates that the smallest value for $f(x)$ is around $650.00 for x near 4 m. We could pinpoint this value of x more closely by plotting $f(x)$ on smaller and smaller intervals each of which captures the lowest point on the curve. Differential calculus provides a very simple way to find this point analytically.

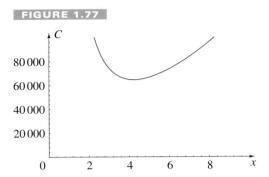

FIGURE 1.77

Area

Here is a completely different kind of problem. We have no formula for finding the area A in Figure 1.78a bounded by the parabola $y = x^2$, the x-axis, and the line $x = 1$.

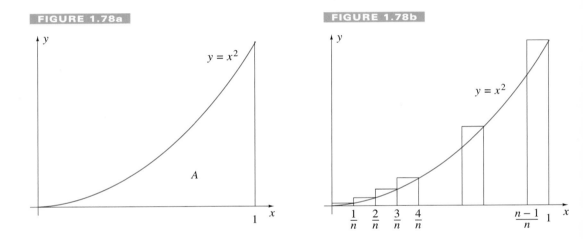

FIGURE 1.78a

FIGURE 1.78b

We could, however, find an approximation to the area as follows. Divide the interval $0 \le x \le 1$ into n subintervals of equal lengths by the points

$$\frac{1}{n}, \ \frac{2}{n}, \ \ldots, \ \frac{n-1}{n}.$$

On each subinterval draw a rectangle with height corresponding to the height of the parabola (Figure 1.78b). The area of each rectangle is its height multiplied by its width, and the width of each rectangle is $1/n$. Hence the area of the first rectangle is $(1/n^2)(1/n)$, that of the second is $(2/n)^2(1/n)$, and so on. By adding areas of all such rectangles we obtain an approximation to A,

$$A \approx \left(\frac{1}{n}\right)^2 \left(\frac{1}{n}\right) + \left(\frac{2}{n}\right)^2 \left(\frac{1}{n}\right) + \left(\frac{3}{n}\right)^2 \left(\frac{1}{n}\right) + \cdots + \left(\frac{n-1}{n}\right)^2 \left(\frac{1}{n}\right) + (1)^2 \left(\frac{1}{n}\right)$$

$$= \left(\frac{1}{n}\right) \left[\frac{1}{n^2} + \frac{2^2}{n^2} + \frac{3^2}{n^2} + \cdots + \frac{(n-1)^2}{n^2} + \frac{n^2}{n^2} \right]$$

$$= \frac{1}{n^3}[1^2 + 2^2 + 3^2 + \cdots + (n-1)^2 + n^2].$$

There is an algebraic formula for the sum of the squares of the first n positive integers, namely,

$$1^2 + 2^2 + 3^2 + \cdots + n^2 = \frac{n(n+1)(2n+1)}{6}.$$

Try substituting $n = 1, 2, 3, 4$ to convince yourself of the validity of this formula if you have never seen it before. Accepting the formula, it follows that area A in Figure 1.78a can be approximated by

$$A \approx \frac{1}{n^3}\left[\frac{n(n+1)(2n+1)}{6}\right] = \frac{2n^2 + 3n + 1}{6n^2}.$$

We believe that as the number of rectangles gets larger and larger, and this can be accomplished by making n larger and larger, the approximation gets better and better. To see what happens to the approximation for large n, we divide numerator and denominator by n^2,

$$A \approx \frac{2 + \dfrac{3}{n} + \dfrac{1}{n^2}}{6}.$$

When n gets very large, $3/n$ and $1/n^2$ both approach 0. In other words, the larger the value of n, the closer the approximation gets to $1/3$. It is our conclusion, then, that the area in Figure 1.78a is $1/3$. This looks reasonable on the basis that the area would seem to occupy about $1/3$ of the unit rectangle with vertices $(0, 0)$, $(1, 0)$, $(1, 1)$, and $(0, 1)$. That part of calculus called *integral* calculus gives a quick and easy way to find area bounded by much more complicated curves than those in Figure 1.78a, such as those in Figure 1.79.

FIGURE 1.79

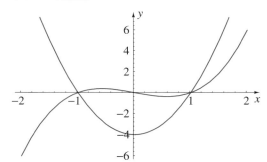

EXERCISES 1.4

∗ **1.** The first figure on page 50 shows the graph of the displacement function for a particle moving along the x-axis during the time interval $0 \le t \le 10$. Use the graph to answer the following questions, giving estimates when necessary.

    (a) Is the particle to the left or right of the origin at $t = 1$, $t = 5$, and $t = 9$?

    (b) How many times does the particle pass through the points $x = 10$, $x = -50$, and $x = 175$?

    (c) Is the particle moving to the left or right at $t = 1$, $t = 4$, and $t = 7$?

    (d) Is the speed greater at $t = 1/2$ or $t = 3$?

    (e) At what times is speed greatest and smallest?

    (f) At what times is the particle not moving?

(g) For how long is the particle moving to the left?

(h) For how long is the particle to the left of the origin?

(i) Is the particle speeding up or slowing down at $t = 5/2$ and $t = 8$?

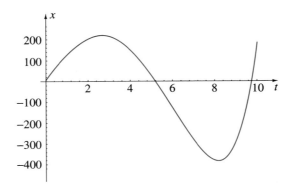

∗ **2.** The figure below shows the graph of the displacement function of a particle moving along the x-axis during the time interval $0 \le t \le 6$. Answer the following questions, estimating values from the graph when necessary.

(a) Is the particle to the left or right of the origin at times $t = 1$ and $t = 4$?

(b) Is the particle more often to the right or to the left of the point $x = 2$?

(c) What is the maximum distance from the origin achieved by the particle?

(d) What is the maximum distance from the point $x = 3$ achieved by the particle?

(e) Is the particle moving to the right or to the left at times $t = 1/2$ and $t = 3$?

(f) How many times does the particle change direction?

(g) Is the speed greater at $t = 7/2$ or at $t = 9/2$?

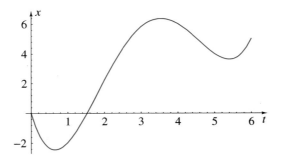

∗ **3.** Repeat Exercise 2 for the graph in the figure below.

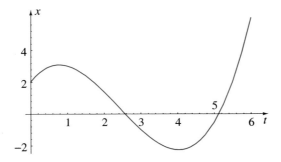

✶✶ **4.** Use the technique for finding the area in Figure 1.78 (page 48) to find that bounded by the x-axis, the vertical lines $x = 0$ and $x = 1$, and the parabola $y = 2x^2 + 1$.

✶✶ **5.** Repeat Exercise 4 for the area bounded by $y = 0$, $x = 1$, and $y = x^3$. You will need the fact that the sum of the cubes of the first n positive integers is $n^2(n + 1)^2/4$.

✶✶ **6.** Repeat Exercise 4 for the area bounded by the curves $y = x^2 + 1$ and $y = -2x^2$ and the vertical lines $x = -1$ and $x = 2$.

✶✶ **7.** An agronomist wishes to fence eight rectangular plots for experimentation as shown in the figure below. Each plot must contain 9000 m^2. If x represents the length of each plot, find a function for the amount of fencing in terms of x. Plot a graph of the function and estimate the minimum amount of fencing that can be used.

✶✶ **8.** A wall of a building is to be braced by a beam that must pass over a parallel wall 5 m high and 2 m from the building (figure below). Show that the length L of the beam in terms of the distance x from the building to the foot of the beam is

$$L = \sqrt{x^2 + \frac{25x^2}{(x - 2)^2}}.$$

(Ignore the dimensions of the beam and wall.) Plot the function and estimate the length of the shortest beam.

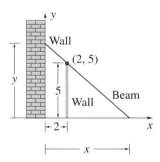

✶✶ **9.** A rectangular beam is to be cut from a circular log of radius 25 cm (see the first figure on page 52). In structural engineering, it is shown that the strength of a beam is proportional to the product of the width and the square of the depth. Show that the strength S of a beam with width w can be expressed in the form

$$S = kw(2500 - w^2),$$

where $k > 0$ is a constant. Plot a graph of the function and estimate the width of the beam with greatest strength.

** **10.** Show that

$$D = \sqrt{(x-5)^2 + [(3/4)\sqrt{16-x^2} - 8]^2}$$

represents the distance from the point $(5, 8)$ to any point (x, y) on the ellipse $9x^2 + 16y^2 = 144$. Plot a graph of the function and estimate coordinates of the point on the ellipse closest to $(5, 8)$.

** **11.** Triangle ABC in the figure below is isosceles, and AD is an altitude with length 6. Show that the sum of the lengths AP, BP, and CP when P is a point on AD a distance y from D is

$$L = 6 - y + 2\sqrt{9 + y^2}.$$

Plot the function and estimate the value of y that minimizes L.

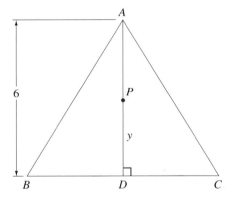

12. The graph in the figure on the right at the top of page 53 represents the charge on the (left half of the) capacitor in the figure on the left for $0 \le t \le 10$, the switch being closed at $t = 0$. Answer the following questions, making estimates when necessary.

    (a) What was the charge on the capacitor when the switch was closed?

    (b) Is the charge on the capacitor positive or negative at $t = 1$, $t = 4$, and $t = 9$?

    (c) For how long does the capacitor have positive charge?

    (d) At what times is there no charge on the capacitor?

    (e) For what times is positive charge being added to the capacitor? Rephrase this question in terms of current.

    (f) Is current clockwise or counterclockwise at $t = 1$ and $t = 4$?

    (g) For what times is the voltage across the capacitor negative?

    (h) Is current increasing or decreasing at $t = 7$?

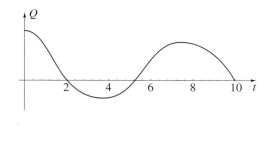

1.5 Trigonometry Review

Many physical systems exhibit an oscillatory nature: vibrations of a plucked guitar string, motion of a pendulum, alternating electric currents, fluctuations in room temperature as the thermostat continually engages and disengages the furnace on a cold day, and the rise and fall of tides and waves. Magnitudes of these oscillations are best described by the sine and cosine functions of trigonometry. In addition, rates at which these oscillations occur can be represented by *derivatives* of these functions from calculus. In this section, we briefly review the trigonometric functions and their properties, placing special emphasis on those aspects that are most useful in calculus.

In trigonometry, angles are measured in either degrees or radians; in calculus, angles are always measured in radians. In preparation, then, for the calculus of trigonometric functions, we work completely in radian measure. By definition, a **radian** is that angle subtended at the centre of a circle of radius r by an arc of equal length r (Figure 1.80). When an arc has length s (Figure 1.81), the number of "units of radius" in this arc is s/r; therefore, the angle θ subtended at the centre of the circle is

$$\theta = \frac{s}{r}. \tag{1.28}$$

In particular, if s contains π units of radius ($s = \pi r$), then s represents one-half the circumference of the complete circle, and $\theta = \pi r/r = \pi$ radians. In degree measure, this angle is $180°$; hence we can state that π radians is equivalent to $180°$. This statement enables us to convert angles expressed in degrees to radian measure and vice versa. If the degree measure of an angle is ϕ, then it is $\pi\phi/180$ radians; conversely, if an angle measures θ radians, then its degree measure is $180\theta/\pi$. For example, the radian measure for $\phi = 45°$ is $45\pi/180 = \pi/4$ radians.

Elementary trigonometry is concerned with relationships among angles and lengths and, in particular, angles and sides of triangles. This naturally restricts angles to the range $0 \leq \theta \leq \pi$. For the purposes of calculus, however, we need to talk about angle θ, where θ is any real number—positive, negative, or zero. To do this we first define what we mean by the **standard position of an angle**. If $\theta > 0$, we draw a line segment OP through the origin O of the xy-plane (Figure 1.82) in such a way that the positive x-axis must rotate counterclockwise through an angle θ to coincide with OP. In other words, we now regard an angle as rotation, rotation of the positive x-axis to some terminal position. If $0 < \theta < \pi/2$, then OP lies in the first quadrant; and if $\pi/2 < \theta < \pi$, then OP is in the second quadrant. But if $\theta > \pi$, we have a geometric representation of θ also. For example, angles $7\pi/4$ and $9\pi/4$ are shown in Figure 1.83. When $\theta < 0$, we regard θ as a clockwise rotation (Figure 1.84).

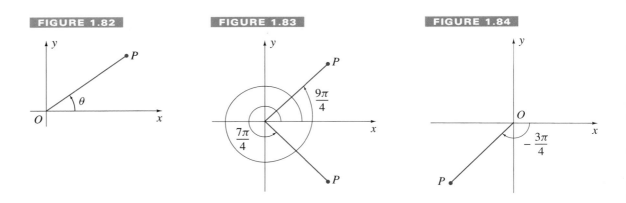

FIGURE 1.82 **FIGURE 1.83** **FIGURE 1.84**

For any angle θ, the angles $\theta + 2n\pi$, where n is an integer, have the same terminal position of OP as θ. They are different angles, however, because the positive x-axis must encircle the origin one or more times before reaching the terminal position in the case of $\theta + 2n\pi$.

With angles represented as rotations, it is easy to define the six trigonometric functions. If θ is an angle in standard position (Figure 1.85) and x and y are the coordinates of P, we define

$$\sin \theta = \frac{y}{r}, \quad \cos \theta = \frac{x}{r}, \quad \tan \theta = \frac{y}{x},$$
$$\csc \theta = \frac{r}{y}, \quad \sec \theta = \frac{r}{x}, \quad \cot \theta = \frac{x}{y}, \tag{1.29}$$

wherever these ratios are defined, and where $r = \sqrt{x^2 + y^2}$ is always assumed positive. Since x is positive in the first and fourth quadrants, and y is positive in the first and second, signs of the trigonometric functions in the various quadrants are as shown in Figure 1.86. Furthermore, since $r \neq 0$, $\sin \theta$ and $\cos \theta$ are defined for all θ, whereas $\tan \theta$ and $\sec \theta$ are not defined for $x = 0$, and $\csc \theta$ and $\cot \theta$ do not exist when $y = 0$.

FIGURE 1.85 **FIGURE 1.86**

Definitions 1.29 indicate that

$$\csc \theta = \frac{1}{\sin \theta}, \qquad \sec \theta = \frac{1}{\cos \theta}, \qquad \cot \theta = \frac{1}{\tan \theta}. \qquad (1.30)$$

In addition, the fact that $r^2 = x^2 + y^2$ leads to the identity

$$\sin^2 \theta + \cos^2 \theta = 1, \qquad (1.31\text{a})$$

which, in turn, implies that

$$1 + \tan^2 \theta = \sec^2 \theta, \qquad (1.31\text{b})$$

$$1 + \cot^2 \theta = \csc^2 \theta. \qquad (1.31\text{c})$$

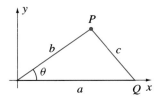

FIGURE 1.87

Identity 1.31a is simply a restatement of the Pythagorean relation ($r^2 = x^2 + y^2$), which allows us to express the hypotenuse of a right-angled triangle in terms of the other two sides. If the triangle is not right-angled (Figure 1.87), it is not possible to express one side, c, in terms of the other two sides, a and b alone, but it is possible to express c in terms of a and b and the angle θ between them. The coordinates of P and Q in Figure 1.87 are $(b \cos \theta, b \sin \theta)$ and $(a, 0)$; therefore, by distance formula 1.1,

$$c^2 = \| PQ \|^2 = (b \cos \theta - a)^2 + (b \sin \theta)^2$$
$$= b^2 (\cos^2 \theta + \sin^2 \theta) + a^2 - 2ab \cos \theta.$$

Since $\sin^2 \theta + \cos^2 \theta = 1$, we have

$$c^2 = a^2 + b^2 - 2ab \cos \theta. \qquad (1.32)$$

This result, called the **cosine law**, generalizes the Pythagorean relation to triangles that are not right-angled. It reduces to the Pythagorean relation $c^2 = a^2 + b^2$ when $\theta = \pi/2$.

If A, B, and C are the angles in the triangle of Figure 1.88, and a, b, and c are the lengths of the opposite sides, then by drawing altitudes of the triangle we can show that

FIGURE 1.88

$$\frac{\sin A}{a} = \frac{\sin B}{b} = \frac{\sin C}{c}. \qquad (1.33)$$

This result, known as the **sine law**, is also useful in many problems.

There is a large number of identities satisfied by the trigonometric functions. They can all be derived from the following **compound-angle formulas** for sines and cosines:

$$\sin (A + B) = \sin A \cos B + \cos A \sin B, \qquad (1.34\text{a})$$

$$\sin (A - B) = \sin A \cos B - \cos A \sin B, \qquad (1.34\text{b})$$

$$\cos (A + B) = \cos A \cos B - \sin A \sin B, \qquad (1.34\text{c})$$

$$\cos (A - B) = \cos A \cos B + \sin A \sin B. \qquad (1.34\text{d})$$

FIGURE 1.89

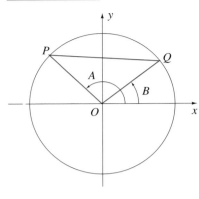

Let us prove one of these, say identity 1.34d, where A and B are the angles in Figure 1.89. If P and Q lie on the circle $x^2 + y^2 = r^2$, then their coordinates are $(r \cos A, r \sin A)$ and $(r \cos B, r \sin B)$ respectively. According to formula 1.1, the length of PQ is given by

$$\|PQ\|^2 = (r \cos A - r \cos B)^2 + (r \sin A - r \sin B)^2$$

$$= r^2(\cos^2 A + \sin^2 A) + r^2(\cos^2 B + \sin^2 B)$$

$$- 2r^2(\cos A \cos B + \sin A \sin B)$$

$$= 2r^2 - 2r^2(\cos A \cos B + \sin A \sin B).$$

But, according to cosine law 1.32,

$$\|PQ\|^2 = \|OP\|^2 + \|OQ\|^2 - 2\|OP\|\|OQ\| \cos(A - B)$$

$$= r^2 + r^2 - 2r^2 \cos(A - B)$$

$$= 2r^2 - 2r^2 \cos(A - B).$$

Comparison of these two expressions for $\|PQ\|^2$ immediately implies identity 1.34d. With equations 1.34, it is easy to derive the following identities:

$$\sin(-\theta) = -\sin\theta, \tag{1.35a}$$

$$\cos(-\theta) = \cos\theta, \tag{1.35b}$$

$$\tan(-\theta) = -\tan\theta, \tag{1.35c}$$

$$\csc(-\theta) = -\csc\theta, \tag{1.35d}$$

$$\sec(-\theta) = \sec\theta, \tag{1.35e}$$

$$\cot(-\theta) = -\cot\theta; \tag{1.35f}$$

$$\sin(\pi/2 - \theta) = \cos\theta, \tag{1.36a}$$

$$\cos(\pi/2 - \theta) = \sin\theta, \tag{1.36b}$$

$$\tan(\pi/2 - \theta) = \cot\theta, \tag{1.36c}$$

$$\csc(\pi/2 - \theta) = \sec\theta, \tag{1.36d}$$

$$\sec(\pi/2 - \theta) = \csc\theta, \tag{1.36e}$$

$$\cot(\pi/2 - \theta) = \tan\theta; \tag{1.36f}$$

$$\sin(\pi - \theta) = \sin\theta, \tag{1.37a}$$

$$\cos(\pi - \theta) = -\cos\theta, \tag{1.37b}$$

$$\tan(\pi - \theta) = -\tan\theta, \tag{1.37c}$$

$$\csc(\pi - \theta) = \csc\theta, \tag{1.37d}$$

$$\sec(\pi - \theta) = -\sec\theta, \tag{1.37e}$$

$$\cot(\pi - \theta) = -\cot\theta. \tag{1.37f}$$

By expressing $\tan{(A + B)}$ as $\sin{(A + B)}/\cos{(A + B)}$ and using identities 1.34a and 1.34c, we find a compound-angle formula for the tangent function,

$$\tan{(A + B)} = \frac{\tan A + \tan B}{1 - \tan A \tan B}, \tag{1.38a}$$

and, similarly,

$$\tan{(A - B)} = \frac{\tan A - \tan B}{1 + \tan A \tan B}. \tag{1.38b}$$

By setting $A = B$ in 1.34a and c and 1.38a, we obtain the **double-angle formulas**,

$$\sin 2A = 2 \sin A \cos A, \tag{1.39a}$$

$$\cos 2A = \cos^2 A - \sin^2 A, \tag{1.39b}$$

$$= 2 \cos^2 A - 1, \tag{1.39c}$$

$$= 1 - 2 \sin^2 A, \tag{1.39d}$$

$$\tan 2A = \frac{2 \tan A}{1 - \tan^2 A}. \tag{1.39e}$$

When pairs of compound-angle formulas 1.34 are added or subtracted, the **product formulas** result. For example, subtracting 1.34c from 1.34d gives

$$\sin A \sin B = \frac{1}{2}[-\cos{(A + B)} + \cos{(A - B)}]. \tag{1.40a}$$

The other product formulas are

$$\sin A \cos B = \frac{1}{2}[\sin{(A + B)} + \sin{(A - B)}], \tag{1.40b}$$

$$\cos A \cos B = \frac{1}{2}[\cos{(A + B)} + \cos{(A - B)}]. \tag{1.40c}$$

By setting $X = A + B$ and $Y = A - B$ in 1.40, we obtain the **sum and difference formulas**,

$$\sin X + \sin Y = 2 \sin\left(\frac{X + Y}{2}\right) \cos\left(\frac{X - Y}{2}\right), \tag{1.41a}$$

$$\sin X - \sin Y = 2 \cos\left(\frac{X + Y}{2}\right) \sin\left(\frac{X - Y}{2}\right), \tag{1.41b}$$

$$\cos X + \cos Y = 2 \cos\left(\frac{X + Y}{2}\right) \cos\left(\frac{X - Y}{2}\right), \tag{1.41c}$$

$$\cos X - \cos Y = -2 \sin\left(\frac{X + Y}{2}\right) \sin\left(\frac{X - Y}{2}\right). \tag{1.41d}$$

The following example is typical of problems in calculus when trigonometric functions are involved.

EXAMPLE 1.24

Write the expression $\cos^4 \theta$ in terms of $\cos 2\theta$ and $\cos 4\theta$.

SOLUTION If we replace A by θ in double-angle formula 1.39c,

$$\cos 2\theta = 2\cos^2 \theta - 1.$$

It follows that

$$\cos^2 \theta = \frac{1 + \cos 2\theta}{2}.$$

Consequently,

$$\cos^4 \theta = (\cos^2 \theta)^2 = \frac{1}{4}(1 + \cos 2\theta)^2 = \frac{1}{4}(1 + 2\cos 2\theta + \cos^2 2\theta).$$

But, if we now replace θ by 2θ in the identity $\cos^2 \theta = (1 + \cos 2\theta)/2$, we obtain

$$\cos^2 2\theta = \frac{1 + \cos 4\theta}{2}.$$

Thus,

$$\cos^4 \theta = \frac{1}{4}\left(1 + 2\cos 2\theta + \frac{1 + \cos 4\theta}{2}\right) = \frac{1}{8}(3 + 4\cos 2\theta + \cos 4\theta).$$

By regarding arguments of the trigonometric functions as angles, we have been stressing geometric properties of these functions. In particular, all the identities 1.34–1.41 have been based on definitions of the trigonometric functions as functions of angles. What is important about a function—be it a trigonometric function or any other kind of function—is that there is a number associated with each value of the independent variable. How we arrive at this number is irrelevant. As far as properties of the function are concerned, only its values are taken into account. Thus, when we discuss properties of a trigonometric function, what is important is not that its argument can be regarded as an angle or that its values can be defined as ratios of sides of a triangle, but that we know its values. Indeed, it is sometimes unwise to regard arguments of trigonometric functions as angles. Consider, for example, the motion of a mass m suspended from a spring with spring constant k (Figure 1.90). If $x = 0$ is the position at which the mass would hang motionless, then, when vertical oscillations are initiated, the position of m as a function of time t is always of the form

$$x = A \cos \left(\sqrt{\frac{k}{m}}t \right) + B \sin \left(\sqrt{\frac{k}{m}}t \right),$$

where A and B are constants. Clearly there are no angles associated with the motion of m, and it is therefore unnatural to attempt to interpret the argument $(\sqrt{k/m})t$ as a physical angle.

Henceforth, we consider the trigonometric functions as those of a real variable. If it is convenient to regard the argument as an angle, then we do so, but only if it is convenient. To emphasize this, we replace θ with our usual generic label for the independent variable of a function, namely, x. With this change, the trigonometric functions are $\sin x$, $\cos x$, $\tan x$, $\csc x$, $\sec x$, and $\cot x$. Their graphs are shown in Figures 1.91.

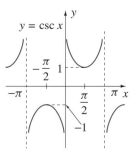

These graphs illustrate that $\sin x$, $\csc x$, $\tan x$, and $\cot x$ are odd functions, and $\cos x$ and $\sec x$ are even. Identities 1.35 confirm this. Trigonometric functions are *periodic*. A function $f(x)$ is said to be periodic if there exists a number T such that for all x,

$$f(x + T) = f(x). \tag{1.42}$$

The smallest such positive number T is called the **period** of $f(x)$. Clearly, then, $\sin x$, $\cos x$, $\csc x$, and $\sec x$ are periodic with period 2π, whereas $\tan x$ and $\cot x$ have period π.

When a periodic function is added to a periodic function, the resulting function may be periodic. For example, the function $f(x) = 3\sin 2x + 3\sqrt{3}\cos 2x$ is the addition of two π-periodic functions $3\sin 2x$ and $3\sqrt{3}\cos 2x$. It also has period π.

Function $f(x) = 2\sin 3x$ has period $2\pi/3$; function $g(x) = -3\cos 2x$ has period π. Their sum $2\sin 3x - 3\cos 2x$ in Figure 1.92 has period 2π; it is the smallest interval in which both functions reproduce themselves.

FIGURE 1.92

Seldom is the argument of a trigonometric function in applications just x; it is usually a function of x.

EXAMPLE 1.25

Draw the graph of the function $f(x) = \cos(x - \pi/3)$.

SOLUTION   The graph of this function is that of $\cos x$ translated $\pi/3$ units to the right (Figure 1.93).

FIGURE 1.93

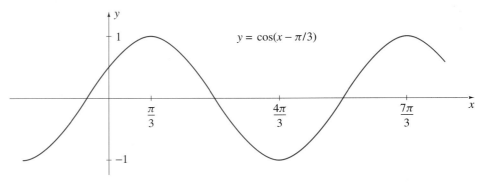

EXAMPLE 1.26

When a mass vibrates on the end of a spring as in Figure 1.94, and there is no friction with the surface, or air resistance, the position of the mass relative to its position when the spring is unstretched takes the form

$$x(t) = 3 \sin 2t + 3\sqrt{3} \cos 2t.$$

Numbers would be different, but $x(t)$ would be a combination of a sine function and a cosine function with the same argument. We have chosen simple numbers here so that unnecessarily complicated calculations do not obscure the significance of the discussion. Plot a graph of $x(t)$, and then draw a graph. What advantages are derived from the drawing as opposed to the plot?

FIGURE 1.94

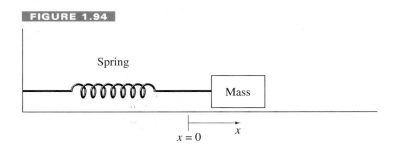

Spring

Mass

$x = 0$ x

SOLUTION   A plot of the function is shown in Figure 1.95.

FIGURE 1.95

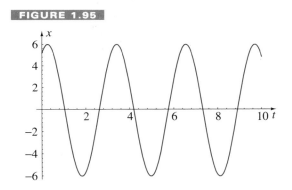

To draw a graph of $x(t)$, we first write it in the form $x(t) = R \sin(2t + \phi)$, where R and ϕ are constants. When we equate this to the given expression for $x(t)$, and expand $\sin(2t + \phi)$ with compound angle formula 1.34a, we obtain

$$3 \sin 2t + 3\sqrt{3} \cos 2t = R \sin(2t + \phi) = R[\sin 2t \cos \phi + \cos 2t \sin \phi].$$

For this equation to be valid for all t, it must be true for $t = 0$ and $t = \pi/4$. Substitution of these gives the two equations

$$3\sqrt{3} = R \sin \phi, \qquad 3 = R \cos \phi.$$

To solve these for R and ϕ, we square each equation and add the results,

$$27 + 9 = R^2 \sin^2 \phi + R^2 \cos^2 \phi = R^2.$$

This implies that $R = \pm 6$. If we choose $R = 6$ ($R = -6$ works equally well), then

$$3\sqrt{3} = 6\sin\phi, \qquad 3 = 6\cos\phi.$$

These equations are satisfied by $\phi = \pi/3$ (there are other angles also), and therefore $x(t)$ can be expressed in the form

$$x(t) = 6\sin(2t + \pi/3) = 6\sin[2(t + \pi/6)].$$

The function is most easily graphed by shifting the graph of $f(t) = \sin 2t$ in Figure 1.96a to the left by $\pi/6$ units and modifying the scale on the x-axis to change the amplitude from 1 to 6. The result is shown in Figure 1.96b.

FIGURE 1.96a

FIGURE 1.96b

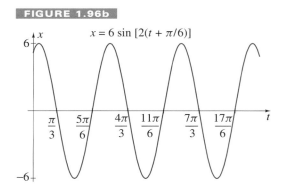

With $x(t)$ expressed in the form $6\sin(2t + \pi/3)$, it is clear that the function has period π and that the amplitude of the oscillations is 6. These are facts that we could surmise from the plot in Figure 1.95, but evidence would not be conclusive. Were we to need values of t at which the function has value 3, say, it is definitely advantageous to have $x(t)$ in the form $6\sin(2t + \pi/3)$. To solve $3 = 6\sin(2t + \pi/3)$, we write $\sin(2t + \pi/3) = 1/2$. All angles that have a sine equal to $1/2$ can be expressed as follows

$$\frac{\pi}{6} + 2n\pi \qquad \text{or} \qquad \frac{5\pi}{6} + 2n\pi,$$

where n is an integer. In other words, we can set

$$2t + \frac{\pi}{3} = \begin{cases} \dfrac{\pi}{6} + 2n\pi \\ \dfrac{5\pi}{6} + 2n\pi \end{cases}.$$

Consequently,

$$2t = \begin{cases} -\dfrac{\pi}{6} + 2n\pi \\ \dfrac{\pi}{2} + 2n\pi \end{cases} \qquad \Longrightarrow \qquad t = \begin{cases} -\dfrac{\pi}{12} + n\pi \\ \dfrac{\pi}{4} + n\pi \end{cases}.$$

You begin to appreciate the elegance of this solution and the simplicity of the result when you compare the magnitude of the problem were you to attempt to find, using a calculator or computer, all solutions of $3\sin 2t + 3\sqrt{3}\cos 2t = 3$. Think about it.

The curve in Figure 1.96b is an example of a **general sine function**,

$$f(x) = A \sin(\omega x + \phi), \tag{1.43}$$

where A and ω are positive constants and ϕ is also constant. The graph of the general sine function is shown in Figure 1.97. The number A, which represents half the range of the function, is called the **amplitude** of the oscillations. The period is $2\pi/\omega$, and $-\phi/\omega$ is called the **phase shift**.

FIGURE 1.97

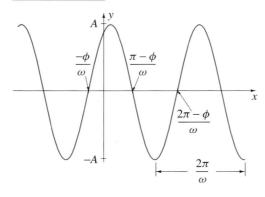

EXAMPLE 1.27

The emf device in the LC-circuit of Figure 1.98 produces a constant voltage of V volts. If the switch is closed at time $t = 0$, and there is no initial charge on the capacitor, the charge thereafter is given by

FIGURE 1.98

$$Q(t) = \frac{V}{C}\left(1 - \cos\frac{t}{\sqrt{LC}}\right), \quad t \geq 0.$$

Draw a graph of this function.

SOLUTION   We begin by drawing a graph of $\cos\left(t/\sqrt{LC}\right)$. It is a standard cosine function with period $2\pi\sqrt{LC}$ (Figure 1.99a). The graph of $Q(t)$ in Figure 1.99b is then obtained by turning Figure 1.99a upside down, shifting it upward 1 unit, and changing the scale on the Q-axis by a factor V/C.

FIGURE 1.99a

FIGURE 1.99b

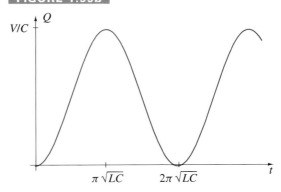

It is worthwhile pointing out here, as we are sure you realize, that electronic devices cannot plot curves containing unspecified parameters. This is another reason why we must develop our graphing skills and not rely totally on graphing calculators and computers.

Perpendicularity and parallelism deal with lines that make a right angle at their point of intersection or that make no angle since parallel lines do not intersect. Lines that intersect usually do so at angles other that $\pi/2$ radians. In order to determine the angle at which two lines intersect, we first define the inclination of a line.

> **DEFINITION 1.5**
> The **inclination** of a line l is the angle of rotation ϕ ($0 \leq \phi < \pi$) from the positive x-direction to the line.

Line $y = x + 3$ in Figure 1.100 has inclination $\pi/4$ radians, and line $y = -x + 4$ has inclination $3\pi/4$ radians.

FIGURE 1.100

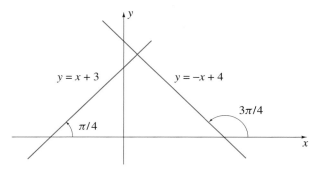

When the slope m of a line is positive, as shown in Figure 1.101a, it is clear that ϕ must be in the interval $0 < \phi < \pi/2$, and m and ϕ are related by $\tan \phi = m$. When the slope of l is negative as in Figure 1.101b, we use identity 1.38b to write

$$m = -\tan(\pi - \phi) = -\frac{\tan \pi - \tan \phi}{1 + \tan \pi \tan \phi} = \tan \phi.$$

Thus, whenever the slope m of a line is defined, the angle of inclination is related to m by the equation

$$\tan \phi = m. \tag{1.44}$$

In some sense this equation is true even when m is not defined. This occurs for vertical lines, which have no slope. For a vertical line, $\phi = \pi/2$, and $\tan \phi$ is undefined. Thus, equation 1.44 is also valid for vertical lines from the point of view that neither side of the equation is defined.

FIGURE 1.101a   FIGURE 1.101b

EXAMPLE 1.28

What is the angle of inclination of the line $2x - 3y = 4$?

SOLUTION   From $y = 2x/3 - 4/3$, the slope of the line is $2/3$. The angle whose tangent is equal to $2/3$ is $\phi = 0.588$ radians.

FIGURE 1.102

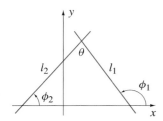

When two lines l_1 and l_2 with nonzero slopes m_1 and m_2 intersect (Figure 1.102), the angle θ $(0 < \theta < \pi)$ between the lines is given by the equation

$$\phi_1 = \theta + \phi_2 \quad \text{or} \quad \theta = \phi_1 - \phi_2.$$

By applying the tangent function to both sides of this equation and using identity 1.38b, we can express θ in terms of m_1 and m_2,

$$\tan\theta = \tan(\phi_1 - \phi_2) = \frac{\tan\phi_1 - \tan\phi_2}{1 + \tan\phi_1 \tan\phi_2} = \frac{m_1 - m_2}{1 + m_1 m_2}.$$

This equation determines θ when $\phi_1 > \phi_2$. When $\phi_1 < \phi_2$, the equation is replaced by

$$\tan\theta = \frac{m_2 - m_1}{1 + m_1 m_2}.$$

In both cases we may write

$$\tan\theta = \left| \frac{m_1 - m_2}{1 + m_1 m_2} \right| \tag{1.45}$$

for the acute angle between the lines.

EXAMPLE 1.29

Find the angle between the lines $2x - 3y = 4$ and $x + 4y = 6$.

SOLUTION   Since slopes of these lines are $2/3$ and $-1/4$, it follows that

$$\tan\theta = \left| \frac{2/3 - (-1/4)}{1 + (2/3)(-1/4)} \right| = \frac{11}{10}.$$

The acute angle whose tangent is equal to $11/10$ is $\theta = 0.833$ radians.

In Exercises 1–10, express the angle in radians.

∗ **1.** 30°　　∗ **2.** 60°　　∗ **3.** 135°　　∗ **4.** −90°　　∗ **5.** −300°

∗ **6.** 765°　　∗ **7.** 72°　　∗ **8.** −128°　　∗ **9.** 321°　　∗ **10.** −213°

In Exercises 11–20, express the angle in degrees.

∗ **11.** $\pi/3$　　∗ **12.** $-5\pi/4$　　∗ **13.** $3\pi/2$　　∗ **14.** 8π　　∗ **15.** $-5\pi/6$

∗ **16.** 1　　∗ **17.** −3　　∗ **18.** 2.5　　∗ **19.** −3.6　　∗ **20.** 11

∗ **21.** What angle is subtended at the centre of a circle of radius 4 by an arc of length (a) 2; (b) 7; (c) 3.2?

∗ **22.** Fill in entries for the following table.

	sin x	cos x	tan x	csc x	sec x	cot x
0						
$\pi/6$						
$\pi/4$						
$\pi/3$						
$\pi/2$						

For each of the triangles in Exercises 23–26 use the cosine law and/or the sine law to find the lengths of all three sides and the sizes of all the interior angles.

∗ **23.**

∗ **24.**

∗ **25.**

∗ **26.**

∗ **27.** Use compound-angle formulas 1.34 to prove identities 1.38.

∗ **28.** Verify double-angle formulas 1.39.

∗ **29.** Use compound-angle formulas 1.39 to prove the product formulas 1.40.

∗ **30.** Show that the sum and difference formulas 1.41 can be obtained from 1.40.

In Exercises 31–36 find the angle of inclination of the line.

∗ **31.** $x - y + 1 = 0$　　　　　　∗ **32.** $x + 2y = 3$

∗ **33.** $3x - 2y = 1$　　　　　　∗ **34.** $y - 3x = 4$

∗ **35.** $x = 4$　　　　　　　　　∗ **36.** $y = 2$

In Exercises 37–44 determine whether the lines are perpendicular, parallel, or neither. In the last case determine the angle between the lines.

* **37.** $y = -x + 4$, $y = x + 6$ * **38.** $x + 3y = 4$, $2x + 6y = 7$

* **39.** $x = 3y + 4$, $y = x/3 - 2$ * **40.** $2x + 3y = 1$, $3x - 2y = 5$

* **41.** $y = 3x + 2$, $y = -x/2 + 1$ * **42.** $x - y = 5$, $2x + 3y = 4$

* **43.** $x = 0$, $y = 5$ * **44.** $x + y + 2 = 0$, $3x - y = 4$

In Exercises 45–66 draw a graph of the function.

* **45.** $f(x) = 3 \sin x$ * **46.** $f(x) = \sin 2x$

* **47.** $f(x) = 3 \sin 2x$ * **48.** $f(x) = \sin(x + \pi/4)$

* **49.** $f(x) = 3 \sin(x + \pi/4)$ * **50.** $f(x) = \sin(2x + \pi/4)$

* **51.** $f(x) = 3 \sin(2x + \pi/4)$ * **52.** $f(x) = 4 \cos(x/3)$

* **53.** $f(x) = 2 \sin(x/2 - \pi)$ * **54.** $f(x) = 5 \cos(\pi/2 - 3x)$

* **55.** $f(x) = \sec 2x$ * **56.** $f(x) = \tan 3x$

* **57.** $f(x) = \csc(x - \pi/3)$ * **58.** $f(x) = \cot(x + \pi/4)$

* **59.** $f(x) = \tan^2 x$ ** **60.** $f(x) = \sqrt{1 - \cos^2 x}$

** **61.** $f(x) = \sqrt{1 + \tan^2 x}$ ** **62.** $f(x) = 5 - 2 \sec x$

** **63.** $f(x) = 4 + 2 \tan x$ ** **64.** $f(x) = \tan|x|$

** **65.** $f(x) = -|\cot 2x|$ ** **66.** $f(x) = 3 \csc(x/2)$

In Exercises 67–74 find all solutions of the equation.

* **67.** $\sin x = \sqrt{3}/2$ * **68.** $\cos x = 2$

* **69.** $3 \sin x = \cos x$ * **70.** $3 \cos^2 x = 1$

** **71.** $\sin 2x = \sin x$ ** **72.** $\sin^2 x + 3 \sin x + 1 = 0$

** **73.** $\sin^2 x + \sin x - 1 = 0$ ** **74.** $\sin x + \cos x = 1$

In Exercises 75–78 express each function as a general sine function, identifying its amplitude, period, and phase shift. Draw a graph of each function.

** **75.** $f(x) = \sin x - 2 \cos x$ ** **76.** $f(x) = 4 \sin 2x + \cos 2x$

** **77.** $f(x) = -2 \sin 3x - 4 \cos 3x$ ** **78.** $f(x) = \sin x \cos x$

In Exercises 79–83 verify the identity.

** **79.** $\cos 3x = 4 \cos^3 x - 3 \cos x$ ** **80.** $\sin 4x = 8 \cos^3 x \sin x - 4 \cos x \sin x$

** **81.** $\tan 3x = \dfrac{3 \tan x - \tan^3 x}{1 - 3 \tan^2 x}$ ** **82.** $\tan\left(\dfrac{x}{2}\right) = \dfrac{\sin x}{1 + \cos x}$

** **83.** $\dfrac{1 + \tan x}{1 - \tan x} = \tan\left(x + \dfrac{\pi}{4}\right)$

** **84.** Show that a function $f(x) = A \cos \omega x + B \sin \omega x$ can always be written in the form

$$f(x) = \sqrt{A^2 + B^2} \sin(\omega x + \phi),$$

where ϕ is defined by the equations

$$\sin \phi = \frac{A}{\sqrt{A^2 + B^2}} \quad \text{and} \quad \cos \phi = \frac{B}{\sqrt{A^2 + B^2}}.$$

SOLUTION   Graphs of these functions are shown in Figure 1.106, but no attempt has been made to use a scale on either the x- or y-axis. Notice here that for $x > 16$, $2^x > x^4$. This is always the situation for power and exponential functions. Given any exponential function a^x ($a > 1$), and any power function x^n ($n > 1$), there always exists a value of x, say X, such that when $x > X$, we have $a^x > x^n$. In short, exponential functions grow more rapidly than power functions for large values of x.

In terms of the exponential function, rules 1.46 take the form:

$$a^{x_1} a^{x_2} = a^{x_1 + x_2}, \tag{1.48a}$$

$$\frac{a^{x_1}}{a^{x_2}} = a^{x_1 - x_2}, \tag{1.48b}$$

$$\left(a^{x_1}\right)^{x_2} = a^{x_1 x_2}. \tag{1.48c}$$

With the exponential function in place, it is natural to define the logarithm function as the "inverse" of the exponential function. A general discussion of inverse functions is delayed until Chapter 8, however, and therefore an appreciation of exponentials and logarithms from this point of view will have to wait until that time. In this section, we give a discussion that contains the essence of inverse functions without the formality.

In the following discussion, a is a fixed number greater than 1. Given a value of y, we know what it means to calculate $x = a^y$. Suppose the value of x is given and we want to find the value of y. As a simple example, consider finding the value of y that satisfies the equation $100 = 10^y$. Clearly, the answer is 2. Likewise, the value of y satisfying the equation $1\,000\,000 = 10^y$ is 6, and that satisfying $1/1000 = 10^y$ is -3. In each case we are given a number and asked to find the exponent to which 10 must be raised to produce that number. This power is called the **logarithm** of the number to base 10. In other words, the logarithm of a number x to base 10 is the exponent to which we must raise 10 to get x. We denote the logarithm of x to base 10 by $\log_{10} x$. With this notation,

$$\log_{10} 100 = 2, \qquad \log_{10} 1\,000\,000 = 6, \qquad \log_{10}\left(\frac{1}{1000}\right) = -3.$$

Electronic calculators can find the logarithm of any number. For instance,

$$\log_{10} 4.67 = 0.669 \qquad \text{and} \qquad \log_{10} 0.359 = -0.445.$$

What these mean is that

$$10^{0.669} = 4.67 \qquad \text{and} \qquad 10^{-0.445} = 0.359.$$

Since we can find the logarithm of any number x, we have in $\log_{10} x$ a function; we call it the logarithm function to base 10. Given a value of x, $y = \log_{10} x$ is the exponent to which 10 must be raised to yield x.

Using 10 as the base for logarithms is natural since our everyday number system is based on 10. Mathematically, we can associate logarithms with any exponential function a^x, but it is customary to restrict a to be larger than 1 (see Exercise 32). We define the logarithm of x to base a, denoted by $\log_a x$, as the exponent to which to raise a to get x. For example,

$$\log_2 16 = 4, \qquad \log_9 27 = 3/2, \qquad \log_5 1/25 = -2.$$

Logarithms to base 10 are called **common** logarithms. All scientific calculators are programmed for them. These calculators are also programmed to give logarithms to another base denoted by the letter e. We will discuss them shortly.

A plot of the logarithm function $f(x) = \log_a x$ is shown in Figure 1.107. It passes through the point (1,0) for any a. In other words, $\log_a 1 = 0$. As values of x get closer and closer to 0, their logarithms become very large negative numbers. We cannot take the logarithms of 0 or negative numbers.

FIGURE 1.107

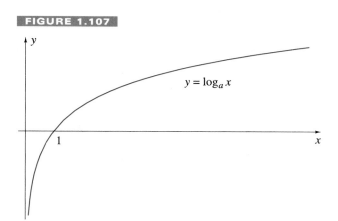

Since the logarithm $\log_a x$ is the exponent to which to raise a to get x, we can write that

$$y = \log_a x \qquad \text{only if} \qquad x = a^y. \tag{1.49}$$

When the first of these is substituted into the second, the result is

$$x = a^{\log_a x} \qquad (x > 0). \tag{1.50a}$$

When the second is substituted into the first,

$$y = \log_a (a^y),$$

or, since y is arbitrary, we may replace it with x and write

$$x = \log_a (a^x). \tag{1.50b}$$

These equations are valuable properties of the exponential and logarithm functions. They are used to simplify many expressions involving exponentials and logarithms in calculus.

Corresponding to rules 1.48 for the exponential function are the following rules for logarithms:

$$\log_a (x_1 x_2) = \log_a x_1 + \log_a x_2, \tag{1.51a}$$

$$\log_a \left(\frac{x_1}{x_2} \right) = \log_a x_1 - \log_a x_2, \tag{1.51b}$$

$$\log_a \left(x_1^{x_2} \right) = x_2 \log_a x_1. \tag{1.51c}$$

To prove 1.51a, say, we set $z = \log_a(x_1 x_2)$, in which case

$$a^z = x_1 x_2$$

$$= \left(a^{\log_a x_1}\right)\left(a^{\log_a x_2}\right) \qquad \text{(using 1.50a)}$$

$$= a^{\log_a x_1 + \log_a x_2}. \qquad \text{(using 1.48a)}$$

Thus,

$$\log_a x_1 + \log_a x_2 = z = \log_a(x_1 x_2).$$

We leave proofs of 1.51b and 1.51c to the exercises.

EXAMPLE 1.32

Simplify the following expressions:

(a) $3^{\log_3(x^2)}$ (b) $10^{-4\log_{10} x}$ (c) $\log_a\left(a^{-x+3}\right)$ (d) $\log_2 8 + \log_3(1/27)$

SOLUTION

(a) Identity 1.50a implies that

$$3^{\log_3(x^2)} = x^2.$$

(b) Since -4 intervenes between the logarithm and exponential operations, we cannot use 1.50a immediately. The -4 can be relocated, however, with 1.51c:

$$10^{-4\log_{10} x} = 10^{\log_{10}(x^{-4})} = \frac{1}{x^4} \quad \text{(if } x > 0\text{)}.$$

(c) Identity 1.50b gives

$$\log_a\left(a^{-x+3}\right) = -x + 3.$$

(d) Since $\log_2 8 = 3$ and $\log_3(1/27) = -3$,

$$\log_2 8 + \log_3(1/27) = 3 - 3 = 0.$$

EXAMPLE 1.33

Solve the following equations:

(a) $\log_5 x = -3$ (b) $\log_{10} x + \log_{10}(x + 1) = 0$ (c) $10^x - 12 + 10^{-x} = 0$

SOLUTION

(a) By means of 1.49,

$$x = 5^{-3} = \frac{1}{125}.$$

(b) Since $\log_{10} x + \log_{10}(x+1) = \log_{10}[x(x+1)]$, we can write

$$0 = \log_{10}[x(x+1)].$$

If we now take exponentials to base 10,

$$10^0 = 10^{\log_{10}[x(x+1)]} \qquad \text{or} \qquad 1 = x(x+1).$$

This quadratic equation has solutions

$$x = \frac{-1 \pm \sqrt{1+4}}{2} = \frac{-1 \pm \sqrt{5}}{2}.$$

Since x must be positive (the original equation demands this), the only solution is $x = (\sqrt{5}-1)/2$.

(c) If we multiply the equation by 10^x, the result is

$$0 = 10^{2x} - 12(10^x) + 1 = (10^x)^2 - 12(10^x) + 1.$$

But this is a quadratic equation in 10^x so that

$$10^x = \frac{12 \pm \sqrt{144-4}}{2} = 6 \pm \sqrt{35}.$$

Finally, we have

$$x = \log_{10}\left(6 \pm \sqrt{35}\right).$$

EXAMPLE 1.34

In acoustics, the minimum sound level detectable by the normal human ear is taken as the "tick of a watch at 20 feet under otherwise quiet conditions." This sound level, called the audible sound threshold, is said to have a decibel reading of 0. All other sounds are given decibel readings relative to this audible sound threshold. For example, a normal voice is approximately 70 decibels, the sound of a car is 100 decibels, and a jet engine 160 decibels. Mathematically, the loudness of a sound is said to be L decibels if

$$L = 10\log_{10}\left(\frac{I}{I_0}\right),$$

where I is the intensity of the sound and I_0 is the intensity of sound at the audible threshold. Use this definition to answer the following questions.

(a) Express the intensity I of a sound in terms of I_0 and its decibel reading L.

(b) Intensity is a measure of energy per unit area per unit time transmitted by sound waves. Find the intensity of the normal voice, a car, and a jet engine relative to the intensity I_0.

(c) If the pain threshold for sound has an intensity 10^{14} times I_0, what is its decibel reading?

(d) If the intensity I_1 of one sound is 10 times the intensity I_2 of a second sound, how do their decibel readings compare?

SOLUTION

(a) When we take both sides as exponents of powers of 10, and use properties 1.51c and 1.50a,

$$10^L = 10^{10 \log_{10}(I/I_0)} = 10^{\log_{10}(I/I_0)^{10}} = \left(\frac{I}{I_0}\right)^{10}.$$

If we take 10 roots of both sides,

$$\left(10^L\right)^{1/10} = \frac{I}{I_0} \qquad \text{or} \qquad I = I_0 10^{L/10}.$$

(b) Since the decibel level of the normal voice is 70, its intensity is

$$I = I_0 10^{70/10} = 10^7 I_0.$$

Similarly, the intensities of a car and a jet are 10^{10} and 10^{16} times I_0.

(c) For an intensity of $10^{14} I_0$, the decibel reading is

$$L = 10 \log_{10}(10^{14}) = 10(14) = 140.$$

(d) If L_1 and L_2 are the decibel readings for sounds with intensities I_1 and I_2, then

$$L_1 = 10 \log_{10}(I_1/I_0) \qquad \text{and} \qquad L_2 = 10 \log_{10}(I_2/I_0).$$

When we subtract these readings,

$$
\begin{aligned}
L_1 - L_2 &= 10 \log_{10}(I_1/I_0) - 10 \log_{10}(I_2/I_0) \\
&= 10[\log_{10}(I_1/I_0) - \log_{10}(I_2/I_0)] \\
&= 10 \log_{10}\left(\frac{I_1/I_0}{I_2/I_0}\right) \qquad \text{(using 1.51b)} \\
&= 10 \log_{10}\left(\frac{I_1}{I_2}\right) \\
&= 10 \log_{10}\left(\frac{10 I_2}{I_2}\right) \qquad \text{(since } I_1 = 10 I_2) \\
&= 10 \log_{10} 10 \\
&= 10.
\end{aligned}
$$

Thus, when the intensity of one sound is 10 times that of another, their decibel readings differ by 10.

It is sometimes necessary to change from one base of logarithms to another. If we take logarithms to base b on both sides of identity 1.50a, we immediately obtain

$$\log_b x = (\log_a x)(\log_b a). \qquad (1.52)$$

This equation defines $\log_b a$ as the conversion factor from logarithms to base a to logarithms to base b.

Before the discovery of calculus, the base of logarithms was invariably chosen to be 10. Such logarithms are called *common logarithms*; they correspond to the exponential function 10^x. Another base for exponentials and logarithms, however, that is much more convenient in most applications is a number, denoted by the letter e, and defined in a variety of ways. One way is to consider the numbers

$$\left(1 + \frac{1}{n}\right)^n \tag{1.53}$$

for ever increasing values of n. The numbers in Table 1.2 are steadily increasing but getting closer together. They suggest that for larger and larger values of n, the function $(1 + 1/n)^n$ is indeed getting closer to some number—to twelve decimal places this number is

$$e = 2.718281828459.$$

TABLE 1.2

n	$\left(1 + \frac{1}{n}\right)^n$
1	2.000 000
3	2.370 370
5	2.488 320
10	2.593 742
100	2.704 814
1 000	2.716 924
10 000	2.718 146
100 000	2.718 255
1 000 000	2.718 282

This number e is irrational with a nonterminating, nonrepeating decimal expansion. Why this number is so convenient as a base for logarithms and exponentials is shown in Section 3.10. For now, let us rewrite some of the more important formulas of this section with a set equal to e. The exponential function to base e is e^x, and equations 1.48 in terms of e^x read

$$e^{x_1} e^{x_2} = e^{x_1 + x_2}, \tag{1.54a}$$

$$\frac{e^{x_1}}{e^{x_2}} = e^{x_1 - x_2}, \tag{1.54b}$$

$$\left(e^{x_1}\right)^{x_2} = e^{x_1 x_2}. \tag{1.54c}$$

Logarithms to base e are usually given the notation $\ln x$ rather than $\log_e x$, and are called **natural**, or **Naperian logarithms** after the Scottish mathematician John Napier (1550–1617), who invented logarithms.

$$\ln x = \log_e x. \tag{1.55}$$

In terms of $\ln x$, rules 1.51 are

$$\ln (x_1 x_2) = \ln x_1 + \ln x_2, \tag{1.56a}$$

$$\ln \left(\frac{x_1}{x_2}\right) = \ln x_1 - \ln x_2, \tag{1.56b}$$

$$\ln \left(x_1^{x_2}\right) = x_2 \ln x_1. \tag{1.56c}$$

Identities 1.50 become

$$x = e^{\ln x}, \qquad x > 0, \tag{1.57a}$$

$$x = \ln(e^x). \tag{1.57b}$$

Graphs of e^x and $\ln x$ in Figures 1.108 and 1.109 have the same shape as those of a^x in Figure 1.103 and $\log_a x$ in Figure 1.107. Only "steepness" of the curves is affected by a change of base.

FIGURE 1.108

FIGURE 1.109

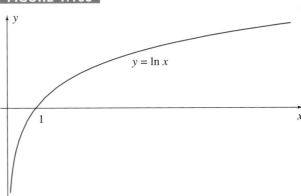

Neither of the functions e^x nor $\ln x$ is even or odd, but functions derived from them may be even or odd. This is illustrated in the following example.

EXAMPLE 1.35

Is the function $f(x) = e^{-ax^2}$, where $a > 0$ is a constant, even or odd? Draw its graph.

SOLUTION   Since

$$f(-x) = e^{-a(-x)^2} = e^{-ax^2} = f(x),$$

the function e^{-ax^2} is even. Graphs for $a = 1, 2,$ and 3 are plotted in Figures 1.110a–c. The value of a controls the spread of the curve. For any value of a, the graph passes through the point $(0, 1)$ and is asymptotic to the x-axis (Figure 1.110d). This curve is very important in statistics. It is called the *bell curve* or *normal distribution*.

FIGURE 1.110a

FIGURE 1.110b

FIGURE 1.110c

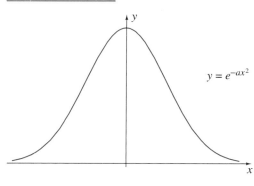

FIGURE 1.110d

You will be making graphs of functions at every turn in this book. We remind you that when we ask you to "plot" a graph, you are to use a graphing calculator or computer. When we ask you to "draw" a graph, you are to do so without these devices. In the previous example, we plotted graphs of $f(x) = e^{-ax^2}$ for $a = 1, 2$, and 3, and then drew $y = e^{-ax^2}$ by hand.

EXAMPLE 1.36

Draw a graph of the function $f(x) = \ln(x + a)$, where $a > 0$ is a constant.

SOLUTION   By translating the graph of $f(x) = \ln x$ in Figure 1.111a to the left by a units, we obtain the graph of $f(x) = \ln(x + a)$ in Figure 1.111b.

FIGURE 1.111a

FIGURE 1.111b

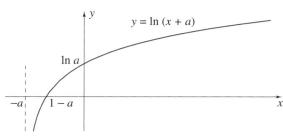

EXAMPLE 1.37

The emf device in the RC-circuit of Figure 1.112 produces a constant voltage of V volts. If the switch is closed at time $t = 0$, and there is no initial charge on the capacitor, the charge thereafter is given by

$$Q(t) = CV[1 - e^{-t/(RC)}], \quad t \geq 0.$$

Draw a graph of this function.

FIGURE 1.112

SOLUTION   A graph of $e^{t/(RC)}$ (Figure 1.113a) has the same shape as that in Figure 1.108; the constant RC only affects "steepness" of the curve. The graph of $e^{-t/(RC)}$ in Figure 1.113b is that of Figure 1.113a reflected in the vertical axis. The graph of $Q(t)$ in Figure 1.113c is then obtained by turning Figure 1.113b upside down, shifting it upward 1 unit, changing the scale on the vertical axis by a factor CV, and retaining only that part of the curve $t \geq 0$. It is asymptotic to the line $Q = CV$.

FIGURE 1.113a

FIGURE 1.113b

FIGURE 1.113c

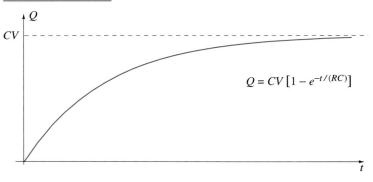

$$Q = CV \left[1 - e^{-t/(RC)} \right]$$

EXERCISES 1.6

In Exercises 1–13 find all values of x satisfying the equation.

* **1.** $\log_{10}(2 + x) = -1$ * **2.** $10^{3x} = 5$

* **3.** $\log_{10}(x^2 + 2x + 1) = 1$ * **4.** $\ln(x^2 + 2x + 10) = 1$

* **5.** $10^{5-x^2} = 100$ * **6.** $10^{1-x^2} = 100$

** **7.** $\log_{10}(x - 3) + \log_{10} x = 1$ ** **8.** $\log_{10}(3 - x) + \log_{10} x = 1$

** **9.** $\log_{10}[x(x - 3)] = 1$ ** **10.** $2\log_{10} x + \log_{10}(x - 1) = 2$

** **11.** $\log_a x + \log_a (x + 2) = 2$ ** **12.** $\log_a [x(x + 2)] = 2$

** **13.** $\log_{10}\left[\log_{10}\left(\dfrac{x + 3}{200x} \right) + 4 \right] = -1$

In Exercises 14–27 plot or draw a graph of the function.

** **14.** $f(x) = e^{-x^2}$ ** **15.** $f(x) = \ln(\cos x)$

** **16.** $f(x) = \log_a |x^2 - 1|$ ** **17.** $f(x) = a^{\log_a(2x+1)}$

** **18.** $f(x) = \log_{10}(4x)$ ** **19.** $f(x) = \ln(1 - x)$

** **20.** $f(x) = \ln(1 - x^2)$ ** **21.** $f(x) = \ln(x^2 - 1)$

** **22.** $f(x) = 10^{x+2}$ ** **23.** $f(x) = e^{2-x}$

** **24.** $f(x) = e^{x^2}$ ** **25.** $f(x) = e^{-x^2} \sin x$

** **26.** $f(x) = e^{-x^2} \cos x$ ** **27.** $f(x) = x^2 e^x, \quad -1 \le x \le 1$

** **28.** Is there a difference between the graphs of the functions $f(x) = \log_a(x^2)$ and $g(x) = 2\log_a x$?

** **29.**   (a) In the early period of reforestation, the percentage increase of timber each year is almost constant. If the original amount A_0 of a certain timber increases 3.5% the first year and 3.5% each year thereafter, find an expression for the amount of timber after t years.

    (b) How long does it take for timber of this type to double?

** **30.** A new car costs $20 000. In any year it depreciates to 75% of its value at the beginning of that year. What is the value of the car after t years?

** **31.** If the effective height of the earth's atmosphere (in metres) is the solution of the equation $10^{-6} = (1 - 2.08 \times 10^{-6} y)^{56}$, find y.

** **32.** Show that if y is the logarithm of x to base a, then $-y$ is the logarithm of x to base $1/a$.

** **33.** Prove 1.51b and 1.51c.

** **34.** Is identity 1.51a valid for any x_1 and x_2?

** **35.** The magnitude of an earthquake is measured in much the same way as noise level. An earthquake of minimal size is taken as having value 0 on the Richter scale. Any other earthquake of intensity I is said to have magnitude R on the Richter scale if

$$R = \log_{10}\left(\frac{I}{I_0}\right),$$

where I_0 is the intensity of the minimal earthquake being used as reference.

    (a) Express the intensity of an earthquake in terms of I_0 and its reading on the Richter scale.

    (b) What are readings on the Richter scale of earthquakes that have intensities 1.20×10^6 and 6.20×10^4, times I_0?

** **36.**   (a) If P dollars is invested at i % compounded n times per year, show that the accumulated value after t years is

$$A = P\left(1 + \frac{i}{100n}\right)^{nt}.$$

    (b) How long does it take to double an investment if interest is 8% compounded semiannually?

    (c) Calculate the maximum possible value of A if i is fixed but the number of times that interest is compounded is unlimited, that is, calculate $\lim_{n \to \infty} A$. This method of calculating interest is called *continuously compounded interest*.

    (d) What is the accumulated value of a $1000 investment after 10 years at 6% compounded continuously? Compare this with the accumulated value if interest is calculated only once each year.

In Exercises 37–40 find all values of x satisfying the equation.

** 37. $3a^{2x} + 3a^{-2x} = 10$

** 38. $2^x + 4^x = 8^x$

** 39. $3^{x+4} = 7^{x-1}$

** 40. $\log_x 2 = \log_{2x} 8$

** 41. Show that $f(x) = \ln\left(x + \sqrt{x^2+1}\right)$ is an odd function. Plot its graph to confirm this geometrically.

** 42. Repair costs on the car in Exercise 30 are estimated at \$50 the first year, increasing by 20% each year thereafter. Set up a function $C(t)$ that represents the average yearly cost associated with owning the car for t years.

** 43. A straight wire conductor has length $2L$ and circular cross-section of radius R. If the wire carries current $i > 0$, then the magnitude of the "vector" potential at a distance r from the centre of the wire is

$$f(r) = \begin{cases} \dfrac{\mu_0 i}{4\pi}\left[\ln\left(1 + \dfrac{4L^2}{R^2}\right) - 1 + \dfrac{r^2}{R^2}\right], & 0 \le r \le R \\[4mm] \dfrac{\mu_0 i}{4\pi}\ln\left(1 + \dfrac{4L^2}{r^2}\right), & r > R \end{cases}$$

where μ_0 is a positive constant.

    (a) Draw a graph of this function.

    (b) Find the radius $r > R$ for which $f(r) = f(0)$.

In Exercises 44–46 solve the given equation for x in terms of y.

** 44. $y = \dfrac{e^{2x} - e^{-2x}}{2}$

** 45. $y = \dfrac{e^x + e^{-x}}{2}$

** 46. $y = \dfrac{e^x - e^{-x}}{e^x + e^{-x}}$

1.7 Solving Equations

In writing this book we have assumed that you have access to a graphing calculator and/or a computer with a mathematical software package. Every software package has one or more equation-solving commands; most graphing calculators have an equation-solving routine. In other words, you have a device that solves equations either exactly or approximately. In this section we discuss some useful concepts related to accuracies of approximations. Suppose, for example, that we are to find all solutions of the equation

$$2x - 4 = \cos(x^2 - 7x + 10). \tag{1.58a}$$

We can express the equation in the equivalent form

$$f(x) = 2x - 4 - \cos(x^2 - 7x + 10) = 0, \tag{1.58b}$$

FIGURE 1.114

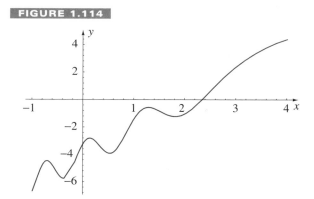

where solutions are now visualized as x-intercepts of the graph of the function $f(x)$. The graph in Figure 1.114 indicates that the only solution of the equation is between $x = 2$ and $x = 3$.

Suppose we use our calculator or computer to find this root and the result is $x = 2.323\,762\,277$. The number of digits depends on the device used and/or how it is programmed. For the purposes of our discussion here, the number of digits is immaterial. Naturally, if $2.323\,762\,277$ constitutes the full display of a calculator, we might be skeptical of the accuracy of the last digit, perhaps even the second last digit. If this is computer output, we might be unsure of how the machine arrived at the last digit. Did it round and, if so, what are its rules for rounding, or did it simply truncate after the 10th digit? There is a very simple way to verify the accuracy of this, or any other, approximate root of an equation. The following theorem provides it.

THEOREM 1.3   (Intermediate Value Theorem)

If $f(a)f(b) < 0$ for a function $f(x)$ continuous on $a \le x \le b$, then there exists at least one number c between a and b for which $f(c) = 0$.

A function $f(x)$ is **continuous** on an interval $a \le x \le b$ if its graph can be traced between a and b without lifting pencil from paper. A more mathematical definition of continuous will be given in Section 2.4. For our present discussion, the geometric version suffices.

Figure 1.115 illustrates the situation for Theorem 1.3. Condition $f(a)f(b) < 0$ requires that one of $f(a)$ and $f(b)$ be positive and the other be negative. (We have shown $f(a) < 0$ and $f(b) > 0$.) Because the graph can be traced from $(a, f(a))$ to $(b, f(b))$ without lifting pencil from paper, it follows that the graph must cross the x-axis at least once, and this is at point c.

FIGURE 1.115

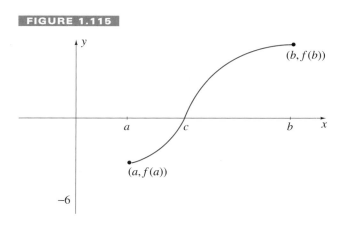

There are two common ways to discuss the accuracy of an approximation to the solution of an equation and, simple as the Intermediate Value Theorem is, it handles both situations.

Approximations Rounded to a Specified Number of Decimal Places

We say that x is an approximation to the root α of an equation

$$f(x) = 0, \tag{1.59}$$

correctly rounded to k decimal places, if x has k decimals, and α rounds to the same k decimals. For example, the approximation $x = 2.323\,762\,277$ rounded to four decimals is $x = 2.3238$. How can we verify that the root of equation 1.58b rounds to these same four decimals? We evaluate $f(x)$ at $x = 2.323\,75$ and $x = 2.323\,85$,

$$f(2.323\,75) = -0.000\,047, \qquad f(2.323\,85) = 0.000\,33.$$

Because one of these values is positive and the other is negative, the Intermediate Value Theorem implies that the solution of equation 1.58b must lie between 2.323 75 and 2.323 85. But every number between 2.323 75 and 2.323 85 rounds to 2.3238. In other words, $x = 2.3238$ is a solution of equation 1.58b correctly rounded to four decimals.

In general, we can say that x is an approximation to a root of the equation $f(x) = 0$, correctly rounded to k decimal places, if x has exactly k digits after the decimal, and

$$f\left(x - \frac{10^{-k}}{2}\right) f\left(x + \frac{10^k}{2}\right) < 0. \tag{1.60}$$

Maximum Possible Error

We are often asked to find an approximation to the solution of an equation such as 1.58b, and be sure that the error is less than some given value ϵ, say $\epsilon = 0.0001$ or $\epsilon = 0.000\,000\,1$. The smaller the value of ϵ, the more accurate must be the approximation. To illustrate, suppose an approximation to the solution of 1.58b is required with error less than $\epsilon = 0.000\,01$. We could verify that $x = 2.323\,762\,277$ has error less than $\epsilon = 0.000\,01$, but there is little point in carrying an approximation with nine decimals when error of 0.000 01 is concerned with the fifth decimal. We suspect that if we round the approximation to five decimals, the result $x = 2.323\,76$ has error less than 0.000 01. To verify this we evaluate

$$f(2.323\,75) = -0.000\,47 \qquad \text{and} \qquad f(2.323\,77) = 0.000\,029.$$

The fact that these values have opposite signs guarantees that the root is between 2.323 75 and 2.323 77. Since the difference between these numbers is 0.000 02 and our approximation 2.323 76 is half-way between them, it follows that the error in 2.323 76 must be less than 0.000 01.

In general, we can say that x is an approximation to a root of the equation $f(x) = 0$, with error less than ϵ, if

$$f(x - \epsilon) f(x + \epsilon) < 0. \tag{1.61}$$

EXAMPLE 1.38

Find an approximation to the smallest root of the equation

$$2x^3e^{-x} + 5x^2 - 1 = 0$$

correctly rounded to six decimal places.

SOLUTION   The graph of $f(x) = 2x^3e^{-x} + 5x^2 - 1$ in Figure 1.116 shows three solutions. Our computer gives $x = -0.766\,051\,059$ as an approximation to the smallest root. To verify that $x = -0.766\,051$ is an approximation, correctly rounded to six decimals, we calculate

$$f(-0.766\,051\,5) = -8.1 \times 10^{-7}, \qquad f(-0.766\,050\,5) = 1.0 \times 10^{-6}.$$

The fact that these values have opposite signs confirms six decimal accuracy of $x = -0.766\,051$.

FIGURE 1.116

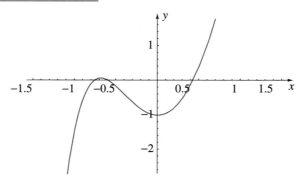

EXAMPLE 1.39

Find an approximation to the largest root of the equation

$$x + 6\sin x = 0$$

with error less than 10^{-8}.

SOLUTION   The graph of $f(x) = x + 6\sin x$ in Figure 1.117 shows five solutions. Our computer yields $x = 5.225\,963\,530$ as an approximation to the largest root. For an approximation with error less than 10^{-8}, we take $x = 5.225\,963\,53$. For verification, we evaluate

$$f(5.225\,963\,52) = -4.1 \times 10^{-8} \qquad \text{and} \qquad f(5.225\,963\,54) = 3.8 \times 10^{-8}.$$

FIGURE 1.117

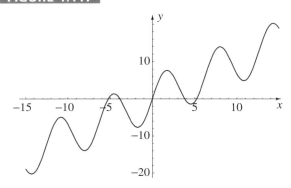

When we have an approximation to the solution of an equation correctly rounded to k decimal places, we can say that we have an approximation with error no greater than $10^{-k}/2$ (compare equations 1.60 and 1.61).

Knowing an approximation with maximum possible error, however, does not guarantee a predictable number of correctly rounded decimal places. Let us illustrate. Suppose that we have used our calculator or computer to approximate the root of an equation $f(x) = 0$ and the result is $3.115\,00$. Suppose further that we know that the error is less than 10^{-5}. Can we give an approximation correctly rounded to two decimal places? No! The fact that the error is less than 10^{-5} allows us to say that the solution satisfies

$$3.115\,00 - 10^{-5} < x < 3.115\,00 + 10^{-5} \qquad \Longrightarrow \qquad 3.114\,99 < x < 3.115\,01.$$

Since left and right sides of the latter inequality round to 3.11 and 3.12, we do not know the approximation correctly rounded to two decimals. However, we do know an approximation correctly rounded to three and four decimals, namely, 3.115 and 3.1150.

As a second example, suppose we know that an approximation to the root of $f(x) = 0$ is 3.435 with error less than 10^{-3}. How many decimal places can we guarantee? We can say that the solution satisfies

$$3.435 - 0.001 < x < 3.435 + 0.001 \qquad \Longrightarrow \qquad 3.434 < x < 3.436.$$

Since the numbers in the right inequality do not both round to the same two decimals, we can guarantee only one correctly rounded decimal, namely, 3.4.

To emphasize this point again, we cannot make generalizations to the effect that a maximum possible error of 10^{-k}, say, guarantees any number of decimal places. In every specific example, we will be able to determine how many decimals are possible, but general statements covering all situations are not possible.

EXAMPLE 1.40

Find points of intersection of the curves

$$y = x^3 - 3x^2 + 2x + 5, \quad y = 6 - 5x^2 - 3x^4.$$

Give coordinates correctly rounded to four decimal places.

SOLUTION Graphs of the curves in Figure 1.118a indicate that whatever points of intersection there are, they are in the interval $-2 \le x \le 2$. Plots on this interval in Figure 1.118b indicate two points of intersection.

FIGURE 1.118a

FIGURE 1.118b

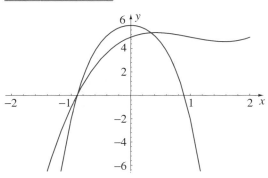

To find x-coordinates of the points of intersection, we solve

$$x^3 - 3x^2 + 2x + 5 = 6 - 5x^2 - 3x^4 \qquad \Longrightarrow \qquad 3x^4 + x^3 + 2x^2 + 2x - 1 = 0.$$

Solutions with four decimals are -0.8924 and 0.3422. To confirm that all four decimals are correct, we calculate the following values of $f(x) = 3x^4 + x^3 + 2x^2 + 2x - 1$,

$$f(-0.892\,45) = 3.1 \times 10^{-4}, \qquad f(-0.892\,35) = -4.6 \times 10^{-4},$$

$$f(0.342\,15) = -4.0 \times 10^{-4}, \qquad f(0.342\,25) = 2.1 \times 10^{-5}.$$

Intersection points on the curves corresponding to $x = -0.8924$ and $x = 0.3422$ can be found by substituting these values into the equations for the curves. If we substitute $x = -0.8924$ into $y = x^3 - 3x^2 + 2x + 5$, we obtain $y = 0.115\,379$, whereas in $y = 6 - 5x^2 - 3x^4$, we get $y = 0.115\,459$. These numbers do not agree to four decimal places. What this points out is that the result of calculations with numbers accurate to four decimals is unlikely to yield numbers accurate to four decimals; accuracy will be lost. How much depends on the nature and the number of calculations? A similar situation arises with $x = 0.3422$; the equation of one curve gives $y = 5.373\,17$ and the other gives $y = 5.373\,36$. They do not agree to four decimal places. What we should have done is carry more decimals in intermediate calculations. For instance, we could carry six decimals with $x = -0.892\,410$ and $x = 0.342\,245$. Verification that $x = -0.8924$ and $x = 0.3422$ are correct to four decimals is the same. But using $x = -0.892\,410$ and $x = 0.342\,245$ leads to corresponding y-values that agree to four decimal places no matter which equation $y = x^3 - 3x^2 + 2x + 5$ or $y = 6 - 5x^2 - 3x^4$ is used, namely $y = 0.1153$ and $y = 5.3732$.

EXAMPLE 1.41

At time $t = 0$, a 5 Ω resistor, a 2 H inductor, and a 0.01 F capacitor are connected with a generator producing an alternating voltage of $10 \sin 5t$ where $t \ge 0$ is time in seconds (Figure 1.119). The current i in the circuit thereafter is $i(t) = f(t) + g(t)$, where

$$f(t) = -e^{-5t/4} \left[\frac{18}{5} \cos\left(\frac{5\sqrt{31}t}{4} \right) + \frac{12\sqrt{31}}{165} \sin\left(\frac{5\sqrt{31}t}{4} \right) \right],$$

FIGURE 1.119

and

$$g(t) = \frac{4}{5}\cos 5t + \frac{2}{5}\sin 5t,$$

are called the **transient** and **steady-state** parts of the current, respectively. Plot graphs of $f(t)$, $g(t)$, and $i(t)$, and explain why the names for $f(t)$ and $g(t)$ are appropriate. Determine the smallest time (correctly rounded to three decimals) at which the magnitude of the current in the circuit is 1 A.

SOLUTION   Plots of $f(t)$ and $g(t)$ are shown in Figures 1.120a and b; their sum is plotted in Figure 1.120c. The exponential factor $e^{-5t/4}$ causes values of $f(t)$ to approach zero within a few seconds. In other words, $f(t)$ is significant only for small t; it dies off quickly, and therefore the adjective "transient" is appropriate. The function $g(t)$ is periodic with amplitude $2/\sqrt{5}$. It remains for all time, and once $f(t)$ becomes insignificant, current $i(t)$ essentially becomes $g(t)$. In other words, the current eventually settles down to $g(t)$, and hence the terminology "steady-state" for $g(t)$ is appropriate.

FIGURE 1.120a

FIGURE 1.120b

FIGURE 1.120c

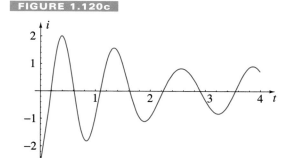

Figure 1.120c makes it clear that there are nine times at which the magnitude of the current is 1 A, four when the current is positive and five when it is negative. The smallest is near $t = 0.15$ s. It satisfies the equation $i(t) = -1$, or, writing the equation in our standard form 1.59,

$$0 = h(t) = -e^{-5t/4}\left[\frac{18}{5}\cos\left(\frac{5\sqrt{31}t}{4}\right) + \frac{12\sqrt{31}}{165}\sin\left(\frac{5\sqrt{31}t}{4}\right)\right]$$
$$+ \frac{4}{5}\cos 5t + \frac{2}{5}\sin 5t + 1.$$

When solving equations of this complexity for one of several roots, calculators and computers usually require the equation and a reasonable approximation to the root, the closer the better. Our computer returned the root $t = 0.146\,206$ when given $t = 0.15$ as initial approximation. To verify that $t = 0.146$ is correctly rounded to three decimals, we calculate that

$$h(0.1455) = -0.012 \quad \text{and} \quad h(0.1465) = 0.0052.$$

EXERCISES 1.7

In Exercises 1–16 use a calculator or computer to find approximations to all roots of the equation accurate to 6 decimal places. Verify the accuracy of each root. Make a plot in order to determine the number of roots.

∗ **1.** $x^2 + 3x + 1 = 0$

∗ **2.** $x^2 - x - 4 = 0$

∗ **3.** $x^3 + x - 3 = 0$

∗ **4.** $x^3 - x^2 + x - 22 = 0$

∗ **5.** $x^3 - 5x^2 - x + 4 = 0$

∗ **6.** $x^5 + x - 1 = 0$

∗ **7.** $x^4 + 3x^2 - 7 = 0$

∗ **8.** $\dfrac{x+1}{x-2} = x^2 + 1$

∗∗ **9.** $x - 10\sin x = 0$

∗∗ **10.** $\sec x = \dfrac{2}{1+x^4}$

∗∗ **11.** $(x+1)^2 = \sin 4x$

∗∗ **12.** $(x+1)^2 = 5\sin 4x$

∗∗ **13.** $x + 4\ln x = 0$

∗∗ **14.** $x\ln x = 6$

∗∗ **15.** $e^x + e^{-x} = 10x$

∗∗ **16.** $x^2 - 4e^{-2x} = 0$

In Exercises 17–24 use a calculator or computer to find approximations to all roots of the equation with error no greater than that specified. Verify the accuracy of each root. Make a plot in order to determine the number of roots.

∗ **17.** $x^3 - 5x - 1 = 0$, $\quad 10^{-3}$

∗ **18.** $x^4 - x^3 + 2x^2 + 6x = 0$, $\quad 10^{-4}$

∗ **19.** $\dfrac{x}{x+1} = x^2 + 2$, $\quad 10^{-5}$

∗ **20.** $(x+1)^2 = x^3 - 4x$, $\quad 10^{-3}$

** **21.** $(x + 1)^2 = 5 \sin 4x$, 10^{-3} ** **22.** $\cos^2 x = x^2 - 1$, 10^{-4}

** **23.** $x + (\ln x)^2 = 0$, 10^{-3} ** **24.** $e^{3x} + e^x = 4$, 10^{-4}

In Exercises 25–28 find all points of intersection for the curves accurate to 4 decimals.

* **25.** $y = x^3$, $y = x + 5$ * **26.** $y = (x + 1)^2$, $y = x^3 - 4x$

* **27.** $y = x^4 - 20$, $y = x^3 - 2x^2$ * **28.** $y = \dfrac{x}{x + 1}$, $y = x^2 + 2$

** **29.** When the beam in the figure below vibrates vertically, there are certain frequencies of vibration, called natural frequencies. They are solutions of the equation

$$\tan x = \frac{e^x - e^{-x}}{e^x + e^{-x}}$$

divided by 20π. Find the two smallest frequencies correct to 4 decimals.

** **30.** A stone of mass 100 gm is thrown vertically upward with speed 20 m/s. Air exerts a resistive force on the stone proportional to its speed, and has magnitude 0.1 N when the speed of the stone is 10 m/s. It can be shown that the height y above the projection point attained by the stone is given by

$$y = -98.1t + 1181 \left(1 - e^{-t/10}\right) \text{ m,}$$

where t is time (measured in seconds with $t = 0$ at the instant of projection).

(a) The time taken for the stone to return to its projection point can be obtained by setting $y = 0$ and solving the equation for t. Do so (correct to 2 decimals).

(b) When air resistance is neglected, the formula for y is

$$y = 20t - 4.905t^2 \text{ m.}$$

What is the elapsed time in this case from the instant the stone is projected until it returns to the projection point?

** **31.** A uniform hydro cable $P = 80$ m long with mass per unit length $\rho = 0.5$ kg/m is hung from two supports at the same level $L = 70$ m apart (figure on page 89). The tension T in the cable at its lowest point must satisfy the equation

$$\frac{\rho g P}{T} = e^{\rho g L/(2T)} - e^{-\rho g L/(2T)},$$

where $g = 9.81$. If we set $z = \rho g/(2T)$, then z must satisfy

$$2Pz = e^{Lz} - e^{-Lz}.$$

Solve this equation for z and hence find T correct to 1 decimal.

$**$ **32.** Planck's law for the energy density E of blackbody radiation at $1000°\,$K states that

$$E = E(\lambda) = \frac{k\lambda^{-5}}{e^{c/\lambda} - 1}$$

where $k > 0$ is a constant and $c = 0.001\,438\,6$. This function is shown in the figure below. The value of λ at which E is a maximum must satisfy the equation

$$(5\lambda - c)e^{c/\lambda} - 5\lambda = 0.$$

Find this value of λ correct to 7 decimals.

$**$ **33.** Let $f(x)$ be a continuous function with domain and range both equal to the interval $[a, b]$. Show that there is at least one value of x in (a, b) for which $f(x) = x$.

$**$ **34.** Use the intermediate value theorem to prove that at any given time there is a pair of points directly opposite each other on the equator of the earth that have exactly the same temperature. Hint: Take the equator to be the circle $x^2 + y^2 = r^2$. Let $f(x)$ be the temperature on the upper semicircle and $g(x)$ be the temperature on the lower semicircle. Consider the function $F(x) = f(x) - g(-x)$.

$***$ **35.** A marathoner runs the 26 odd miles from point A to point B starting at 7:00 a.m. Saturday morning. Starting at 7:00 a.m. Sunday morning he runs the course again, but this time from point B to point A. Prove that there is a point on the course that he passed at exactly the same time on both days.

$**$ **36.**   (a) Use the intermediate value theorem to prove that when the domain of a continuous function is an interval, so also is its range. Hint: Use the idea that a set S of points on the y-axis constitutes an interval if for any two points c and d in S, the points $c < y < d$ are all in S.

   (b) If the domain is an open interval, is the range an open interval?

SUMMARY

Analytic geometry is a combination of geometry and algebra. Algebraic equations are used to describe geometric curves and curves are the geometric representation of equations. The form of an equation dictates the shape of the curve and, conversely, the shape of a curve influences its equation. To illustrate this fact, we discussed straight lines, circles, parabolas, ellipses, and hyperbolas. The most common forms for equations of these curves are as follows:

$$
\text{Straight line}
\begin{cases}
y - y_1 = m(x - x_1) & \text{Point-slope} \\[4pt]
\dfrac{x - x_1}{x_2 - x_1} = \dfrac{y - y_1}{y_2 - y_1} & \text{Two-point} \\[10pt]
y = mx + b & \text{Slope } y\text{-intercept} \\[4pt]
y = m(x - a) & \text{Slope } x\text{-intercept} \\[4pt]
\dfrac{x}{a} + \dfrac{y}{b} = 1 & \text{Two-intercept} \\[8pt]
Ax + By + C = 0 & \text{General}
\end{cases}
$$

$$
\text{Parabola}
\begin{cases}
y = ax^2 + bx + c & \text{Vertical axis of symmetry} \\[4pt]
x = ay^2 + by + c & \text{Horizontal axis of symmetry}
\end{cases}
$$

$$
\text{Circle}
\begin{cases}
x^2 + y^2 + fx + gy + e = 0 \\[4pt]
(x - h)^2 + (y - k)^2 = r^2
\end{cases}
$$

$$
\text{Ellipse}
\begin{cases}
\dfrac{x^2}{a^2} + \dfrac{y^2}{b^2} = 1 \\[10pt]
\dfrac{(x - h)^2}{a^2} + \dfrac{(y - k)^2}{b^2} = 1
\end{cases}
$$

$$
\text{Hyperbola}
\begin{cases}
\dfrac{x^2}{a^2} - \dfrac{y^2}{b^2} = 1 \\[10pt]
\dfrac{y^2}{b^2} - \dfrac{x^2}{a^2} = 1 \\[10pt]
\dfrac{(x - h)^2}{a^2} - \dfrac{(y - k)^2}{b^2} = 1 \\[10pt]
\dfrac{(y - k)^2}{b^2} - \dfrac{(x - h)^2}{a^2} = 1
\end{cases}
$$

Basic to all mathematics is the concept of a function, a rule that assigns to each number x in a domain, a unique number y in the range. A function is simply another way of saying "a quantity y depends on x." The notation $y = f(x)$ for a function immediately suggests that a function can be represented geometrically by a curve—the curve with equation $y = f(x)$—and we call this curve the graph of the function. Polynomials are functions of the form $a_n x^n + \cdots + a_1 x + a_0$, where n is a nonnegative integer and coefficients $a_n, \ldots, a_0$ are constants. Rational functions are quotients of polynomials. Some functions are even, some are odd, most are neither even nor odd, and only $f(x) \equiv 0$ is both even and odd.

We had a glimpse at four of the many types of problems in which calculus proves to be invaluable—displacement and speed, LCR-circuits, optimization, and area. They form the basis for ongoing discussions as calculus unfolds.

Trigonometric functions play a prominent role in many areas of applied mathematics. Particularly important are descriptions of oscillatory systems by the sine and cosine functions. The sine, cosecant, tangent, and cotangent are odd functions, while cosine and secant are even. All are periodic; sine, cosecant, cosine, and secant have period 2π, and tangent and cotangent have period π. Trigonometric functions satisfy many identities. Recognizing when these identities can be used to advantage to simplify expressions or write them in alternative forms is a huge asset.

Exponential and logarithm functions are also important in applications. These functions are not periodic, nor are they even or odd. The exponential function a^x raises a to power x. The logarithm function $\log_a x$ does the reverse. It determines the power that a must be raised to produce x.

In approximating solutions to equations, it is always necessary to indicate the accuracy of the approximation. This can be done by correctly rounding the approximation to a specified number of decimal places, or determining an approximation with maximum error. In both cases, the Intermediate Value Theorem is instrumental in verifying the accuracy.

KEY TERMS

In reviewing this chapter, you should be able to define or discuss the following key terms:

Absolute value	Function
Coordinates	Domain
Quadrant	Range
Length of line segment	Graph
Midpoint of line segment	Rational function
Equation of a curve	Even function
Slope of a line	Odd function
Rise	Translation of a curve
Run	Asymptote
Point-slope formula	Intermediate value theorem
Parallel lines	Accuracy to k decimals
Perpendicular lines	Accuracy with maximum error
Distance from point to line	Trigonometric functions
Parabola	Exponential function
Symmetry about y-axis	Logarithm function
Symmetry about x-axis	Displacement
Circle	Speed
Ellipse	Resistor
Hyperbola	Inductor
Polynomial	Capacitor
Quadratic formula	Area

REVIEW EXERCISES

In Exercises 1 and 2 find the distance between the points and also the midpoint of the line segment joining the points.

∗ **1.** $(-1, 3)$, $(4, 2)$ ∗ **2.** $(2, 1)$, $(-3, -4)$

In Exercises 3–6 find the equation for the line described.

* **3.** Parallel to the line $x - 2y = 4$ and through the point $(2, 3)$

* **4.** Perpendicular to the line joining $(-2, 1)$ to the origin and through the midpoint of the line segment joining $(1, 3)$ and $(-1, 5)$

** **5.** Perpendicular to the line $x = 4y - 11$ and through the point of intersection of this line and $x = \sqrt{y^2 + 9}$

** **6.** Joining the points of intersection of the curves $y = x^2$ and $5x = 6 - y^2$

In Exercises 7–16 find the largest possible domain for the function.

* **7.** $f(x) = \sqrt{x^2 + 5}$

* **8.** $f(x) = \sqrt{x^2 - 5}$

* **9.** $f(x) = \dfrac{1}{x^2 + 3x + 2}$

* **10.** $f(x) = \dfrac{x + 4}{x^3 + 2x^2 + x}$

* **11.** $f(x) = (x^3 - 8)^{1/3}$

* **12.** $f(x) = x^{3/2}$

** **13.** $f(x) = \sqrt{x^2 + 4x - 6}$

** **14.** $f(x) = \dfrac{1}{\sqrt{2x^2 + 4x - 5}}$

** **15.** $f(x) = \sqrt{\dfrac{2x + 1}{x - 3} + 2}$

** **16.** $f(x) = \sqrt{x - \dfrac{1}{x}}$

In Exercises 17–28 identify the curve as a straight line, parabola, circle, ellipse, hyperbola, or none of these.

* **17.** $x + 2y = 4$

* **18.** $x = y^2 - 2y + 3$

* **19.** $y = x^3 + 3$

* **20.** $x^2 + 2y^2 = 4$

* **21.** $y^2 - x^2 = x$

* **22.** $x^2 + y^2 + 5 = 0$

* **23.** $x^2 - 2x + y^2 = 16$

* **24.** $x + y^2 = 3$

* **25.** $x^2 + 2y^2 + y = 2x$

* **26.** $x^2 - x + y^2 + y = 0$

* **27.** $2x^2 + 20x + 38 = 3y^2 + 12y$

* **28.** $2y^2 - x = 3x^2 - y$

In Exercises 29 and 30 find the distance from the point to the line.

* **29.** $(1, -3)$, $\quad y = 2x + 3$

* **30.** $(-2, -5)$, $\quad x = 4 - 3y$

In Exercises 31–52 draw the curve. Then use a calculator or computer to plot the curve as a check.

* **31.** $y = 2x^2 + 3$

* **32.** $x^2 = 4 - y^2$

* **33.** $y = x^3 - 1$

* **34.** $|y| = |x|$

* **35.** $4x^2 + y^2 = 0$

* **36.** $x^2 + 3y^2 = 6$

$*$ **37.** $2y^2 - x^2 = 3$

$*$ **38.** $x^2 - 2x - y^2 + 4y = 1$

$*$ **39.** $y = \sin 3x$

$*$ **40.** $y = \cos(2x + \pi/2)$

$*$ **41.** $y = \cos(2x - \pi/4)$

$*$ **42.** $y = 2\sin(3x + \pi/2)$

$**$ **43.** $x^2 - 4y + 2 = 4x - 2y^2$

$**$ **44.** $y = \sqrt{-x^2 + 4x + 4}$

$**$ **45.** $y = |x| + |x - 1|$

$**$ **46.** $y = \sqrt{|x - 1| - 1}$

$**$ **47.** $x = \tan y$

$**$ **48.** $y = 2\ln(3x + 4)$

$**$ **49.** $x = e^{-y}$

$**$ **50.** $y = \sin|x|$

$**$ **51.** $|y| = |\sin x|$

$**$ **52.** $y = \sqrt{\sin 2x}$

In Exercises 53 and 54 find the angle between the lines.

$**$ **53.** $x + 2y = 4$, $\quad y = 3x - 2$

$**$ **54.** $x = 4y + 2$, $\quad 2x + 3y = 5$

In Exercises 55–58 give an example of a function $y = f(x)$ with the indicated properties.

$**$ **55.** The range of the function consists of one number only.

$**$ **56.** The algebraic formula defining the function cannot be extended beyond $-1 \le x \le 2$.

$**$ **57.** The domain of the function consists of all reals except $x = \pm 1$.

$**$ **58.** The domain of the function is $x \le 0$ and the range is $y \ge 1$.

$**$ **59.** Express $f(x) = \cos 2x - \sin 2x$ in the form $f(x) = A\sin(2x + \phi)$. Use this to draw a graph of the function and find the second smallest positive value of x for which $f(x) = 0$.

$**$ **60.** Is the function $f(x) = 2\sin 2x - 3\cos 3x$ periodic? If so, what is its period?

$**$ **61.** When the mass in the figure below is pulled 5 cm to the right of the position $(x = 0)$ it would occupy were the spring unstretched, and given speed 2 m/s to the right, its position thereafter is given by

$$x(t) = \frac{1}{20}\cos 4t + \frac{1}{2}\sin 4t.$$

Find when the mass passes through $x = 0$ for the first time.

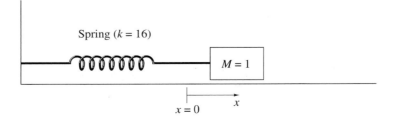

** **62.** A lighthouse is 6 km offshore and a cabin on the straight shoreline is 9 km from the point on the shore nearest the lighthouse (figure below). Show that if a man rows at 3 km/h and walks at 5 km/h, and he beaches the boat at distance x from the near point on the shore, then the total travel time from lighthouse to cabin is

$$t = f(x) = \frac{\sqrt{x^2 + 36}}{3} + \frac{9 - x}{5}, \quad 0 \le x \le 9.$$

Plot a graph of this function.

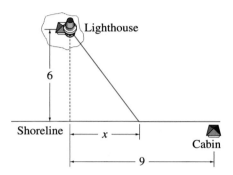

In Exercises 63 and 64 find all solutions of the equation correctly rounded to three decimal places.

* **63.** $x^3 - 2x^2 + 4x - 5 = 0$ ** **64.** $x^2 - 1 = \sin x$

In Exercises 65 and 66 find all solutions of the equation with error less than 10^{-4}.

* **65.** $x^3 + 12x^2 + 4x - 5 = 0$ ** **66.** $x^2 - 1 = 24 \sin x$

** **67.** Prove that the diagonals of a rhombus intersect at right angles. (A rhombus is a parallelogram with all sides of equal length.)

CHAPTER 2

Limits and Continuity

The concept of a "limit" is crucial to calculus, for the two basic operations in calculus are differentiation and integration, each of which is defined in terms of a limit. For this reason we must have a clear understanding of limits from the beginning. In Sections 2.1–2.4 we give an intuitive discussion of limits of functions; in Section 2.5 we show how these ideas can be formalized mathematically.

2.1 Limits

The functions in Figure 2.1 all have value 3 at $x = 1$, but behaviours of the functions close to $x = 1$ are totally different. As x gets closer and closer to 1 in Figure 2.1a, function values get closer and closer to 3. In Figure 2.1b, function values get closer and closer to 3 if x approaches 1 through numbers smaller than 1, but they get closer and closer to 2 if x approaches 1 through numbers larger than 1. Function values approach 2 as x gets closer and closer to 1 in Figure 2.1c whether x approaches 1 through numbers smaller than 1 or larger than 1. In this section we emphasize the distinction between the value of a function at a point, and the values of the function as we approach the point.

FIGURE 2.1a

FIGURE 2.1b

FIGURE 2.1c

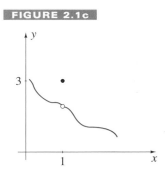

The value of the function $f(x) = x^2 - 4x + 5$ at $x = 2$ is $f(2) = 1$. A completely different consideration is contained in the question, "What number does $f(x) = x^2 - 4x + 5$ get closer and closer to as x gets closer and closer to 2?" Table 2.1 shows that *as x gets closer and closer to 2, $x^2 - 4x + 5$ gets close to 1 and stays close to 1.*

TABLE 2.1

x	x	$f(x) = x^2 - 4x + 5$
1.9	1.01	2.1
1.99	1.000 1	2.01
1.999	1.000 001	2.001
1.9999	1.000 000 01	2.0001

Likewise, the graph of the function (Figure 2.2) clearly shows that $f(x)$ approaches 1 as x approaches 2. This statement is not precise enough for our purposes. For instance, the graph also indicates that as x gets closer and closer to 2, $f(x)$ gets closer and closer to 0. It does not get very close to 0 but, nonetheless, $f(x)$ does get closer and closer to 0 as x gets closer and closer to 2. In fact, we can make this statement for any number less than 1. To distinguish 1 from all numbers less than 1, we say that $x^2 - 4x + 5$ *can be made arbitrarily close to* 1 *by choosing x sufficiently close to* 2.

FIGURE 2.2

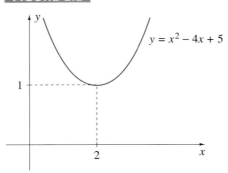

$y = x^2 - 4x + 5$

Numbers less than 1 do not satisfy this statement. For instance, it is not true that $x^2 - 4x + 5$ can be made arbitrarily close to 0. The closest the function gets to 0 is one unit when $x = 2$. In calculus we say that **the limit of** $x^2 - 4x + 5$ **as x approaches** 2 **is** 1 to represent the more lengthy statement "$x^2 - 4x + 5$ can be made arbitrarily close to 1 by choosing x sufficiently close to 2." In addition, we have a notation to represent both statements:

$$\lim_{x \to 2} (x^2 - 4x + 5) = 1. \tag{2.1}$$

This notation is read "the limit of (the function) $x^2 - 4x + 5$ as x approaches 2 is equal to 1," and this stands for the statement "$x^2 - 4x + 5$ can be made arbitrarily close to 1 by choosing x sufficiently close to 2."

We emphasize that limit 2.1 is not concerned with the value of $x^2 - 4x + 5$ at $x = 2$. It is concerned with the number that $x^2 - 4x + 5$ approaches as x approaches 2. These numbers are not always the same.

Generally, we say that a function $f(x)$ has limit L as x approaches a, and write

$$\lim_{x \to a} f(x) = L, \tag{2.2}$$

if $f(x)$ can be made arbitrarily close to L by choosing x sufficiently close to a. Sometimes it is more convenient to write $f(x) \to L$ as $x \to a$ to mean that $f(x)$ approaches L as x approaches a. This is especially so in the middle of a paragraph, as opposed to a displayed equation such as 2.2.

EXAMPLE 2.1

Evaluate $\lim_{x \to 1} (x^2 + 2x + 5)$.

SOLUTION As x gets closer and closer to 1, $x^2 + 2x + 5$ gets closer and closer to 8; therefore, we write

$$\lim_{x \to 1} (x^2 + 2x + 5) = 8.$$

This is corroborated by the graph of the function in Figure 2.3.

FIGURE 2.3

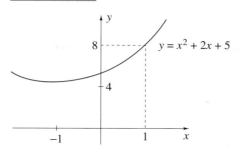

To calculate the limit of a function $f(x)$ as x approaches a, we evaluate $f(x)$ at values of x that get closer and closer to a. For limits of complicated functions such as

$$\lim_{x \to 5} \frac{x^2(3 - x)}{x^3 + x},$$

it would be tedious to evaluate $x^2(3 - x)/(x^3 + x)$ at many values of x approaching 5. The following theorem provides a much easier method.

THEOREM 2.1

If $\lim_{x \to a} f(x) = F$ and $\lim_{x \to a} g(x) = G$, then

(i) $\lim\limits_{x \to a} [f(x) + g(x)] = F + G;$ (2.3a)

(ii) $\lim\limits_{x \to a} [f(x) - g(x)] = F - G;$ (2.3b)

(iii) $\lim\limits_{x \to a} [cf(x)] = cF,$ *when c is a constant;*   (2.3c)

(iv) $\lim\limits_{x \to a} [f(x)g(x)] = FG;$ (2.3d)

(v) $\lim\limits_{x \to a} \dfrac{f(x)}{g(x)} = \dfrac{F}{G},$ *provided* $G \neq 0.$ (2.3e)

What this theorem says is that a limit such as $\lim_{x \to 5} [x^2(3 - x)/(x^3 + x)]$ can be broken down into smaller problems and reassembled later. For instance, since

$$\lim_{x \to 5} x^2 = 25, \qquad \lim_{x \to 5} (3 - x) = -2, \qquad \lim_{x \to 5} x^3 = 125, \qquad \lim_{x \to 5} x = 5,$$

we may write

$$\lim_{x \to 5} \frac{x^2(3 - x)}{x^3 + x} = \frac{25(-2)}{125 + 5} = \frac{-50}{130} = -\frac{5}{13}.$$

Although the results of Theorem 2.1 may seem evident, to prove them mathematically is not a simple task. In fact, because we have not yet given a precise definition for limits, a proof at this time is impossible. When we give definitions for limits in Section 2.5, it will then be possible to verify the theorem (see Exercises 31–35 in Section 2.5).

EXAMPLE 2.2

Evaluate $\displaystyle\lim_{x \to -2} \frac{x+2}{x^2+9}$.

SOLUTION   Since $\lim_{x \to -2} (x+2) = 0$ and $\lim_{x \to -2} (x^2+9) = 13$, part (v) of Theorem 2.1 gives

$$\lim_{x \to -2} \frac{x+2}{x^2+9} = \frac{0}{13} = 0.$$

EXAMPLE 2.3

Evaluate $\displaystyle\lim_{x \to -1} \frac{x^2(1-x^3)}{2x^2+x+1}$.

SOLUTION   Using Theorem 2.1, we can write

$$\lim_{x \to -1} \frac{x^2(1-x^3)}{2x^2+x+1} = \frac{(1)(2)}{2+(-1)+1} = 1.$$

Be sure that you understand how we obtained the expression

$$\frac{(1)(2)}{2+(-1)+1}.$$

In particular, we *did not* set $x = -1$ in $x^2(1-x^3)/(2x^2+x+1)$. Indeed, this is not permitted because to evaluate a limit as x approaches -1, we are not to set $x = -1$; we are to let x get closer and closer to -1. What we did do is take limits of x^2, $1-x^3$, $2x^2$, and x as x approaches -1, and then use Theorem 2.1.

The following example illustrates what can happen if we substitute $x = a$ into $f(x)$ in the evaluation of $\lim_{x \to a} f(x)$.

EXAMPLE 2.4

Evaluate $\displaystyle\lim_{x \to 3} \frac{x^2-9}{x-3}$.

SOLUTION   Because $\lim_{x \to 3} (x-3) = 0$, we cannot use Theorem 2.1. Nor can we set $x = 3$ in $(x^2-9)/(x-3)$ because it is inherent in the limiting procedure that we do not put $x = 3$. Besides, if we did, we would obtain the meaningless expression $0/0$. Figure 2.4a contains a typical graph of the function $(x^2-9)/(x-3)$ on the interval $2 \le x \le 4$, using a calculator or computer. It shows no anomaly in the behaviour of the function at $x = 3$. The graph may, however, be accompanied by a message indicating that the function is undefined at $x = 3$, as indeed it is. But the fact that the function is undefined at $x = 3$ does not concern us here; we are interested in values of the function near $x = 3$, not at $x = 3$, and the graph indicates that as $x \to 3$, values of the function approach 6. To verify this algebraically, we factor x^2-9 into $(x-3)(x+3)$ and divide out a factor of $x-3$ from numerator and denominator:

$$\lim_{x \to 3} \frac{x^2-9}{x-3} = \lim_{x \to 3} \frac{(x-3)(x+3)}{x-3} = \lim_{x \to 3} (x+3) = 6.$$

Dividing by the factor $x - 3$ would not be permissible if $x - 3$ were equal to 0; that is, if x were equal to 3. But once again this cannot happen, because in the limiting operation we let x get closer and closer to 3, but do not set $x = 3$.

FIGURE 2.4a

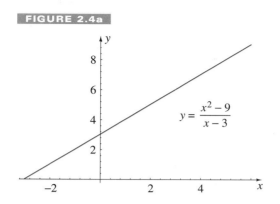

$$y = \frac{x^2 - 9}{x - 3}$$

FIGURE 2.4b

$$y = \frac{x^2 - 9}{x - 3}$$

Note in this example that, although the limit is 6, there is no value of x for which the function $(x^2 - 9)/(x - 3)$ is ever equal to 6. The graph of the function is a straight line with the point at $x = 3$ removed. We have shown this with an open circle in Figure 2.4b.

EXAMPLE 2.5

Evaluate $\displaystyle\lim_{x \to 0} \frac{-2x + 3x^2}{4x - x^2}$.

SOLUTION   The function is undefined at $x = 0$, but

$$\lim_{x \to 0} \frac{-2x + 3x^2}{4x - x^2} = \lim_{x \to 0} \frac{x(-2 + 3x)}{x(4 - x)} = \lim_{x \to 0} \frac{3x - 2}{4 - x} = \frac{-2}{4} = -\frac{1}{2}.$$

Figure 2.5a shows a computer-generated graph of the function on the interval $-2 \le x \le 2$. We have redrawn the graph in Figure 2.5b with a hole at $x = 0$.

FIGURE 2.5a

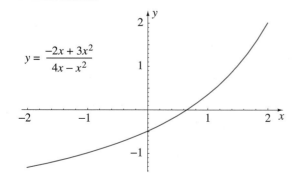

$$y = \frac{-2x + 3x^2}{4x - x^2}$$

FIGURE 2.5b

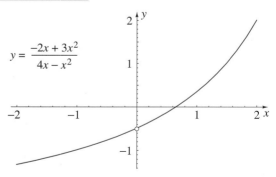

$$y = \frac{-2x + 3x^2}{4x - x^2}$$

EXAMPLE 2.6

Evaluate $\displaystyle\lim_{x\to 0} \frac{\sqrt{1+x}-1}{\sqrt{x}}$.

SOLUTION   Since the limit of the denominator as x approaches 0 is 0, we cannot immediately use Theorem 2.1. The function is undefined for $x \leq 0$; its graph on the interval $0.001 \leq x \leq 1$ in Figure 2.6 suggests, although not conclusively, that the limit is 0. To verify this we rationalize the numerator; that is, rid the numerator of the square root by multiplying numerator and denominator by $\sqrt{1+x}+1$:

$$\lim_{x\to 0} \frac{\sqrt{1+x}-1}{\sqrt{x}} = \lim_{x\to 0}\left(\frac{\sqrt{1+x}-1}{\sqrt{x}}\frac{\sqrt{1+x}+1}{\sqrt{1+x}+1}\right)$$

$$= \lim_{x\to 0}\frac{x}{\sqrt{x}\left(\sqrt{1+x}+1\right)}$$

$$= \lim_{x\to 0}\frac{\sqrt{x}}{\sqrt{1+x}+1}$$

$$= 0.$$

FIGURE 2.6

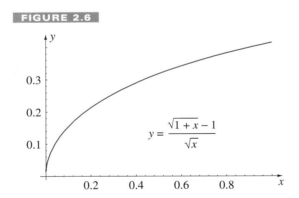

$$y = \frac{\sqrt{1+x}-1}{\sqrt{x}}$$

The following is a second example of this type.

EXAMPLE 2.7

Evaluate $\displaystyle\lim_{x\to 1} \frac{(\sqrt{x+3}-2)(\sqrt{x-1}+3)}{x-1}$.

SOLUTION   Once again we cannot use Theorem 2.1; the denominator has limit 0 as $x \to 1$. The graph of the function for $1.001 \leq x \leq 2$ in Figure 2.7 may suggest a limit, but certainly the

FIGURE 2.7

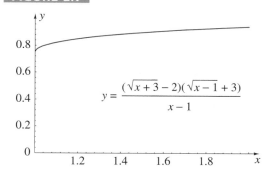

evidence is far from conclusive. Following the lead of the previous example, we rationalize the term $\sqrt{x+3}-2$ in the numerator by multiplying numerator and denominator by $\sqrt{x+3}+2$:

$$\lim_{x \to 1} \frac{(\sqrt{x+3}-2)(\sqrt{x-1}+3)}{x-1} = \lim_{x \to 1} \frac{(\sqrt{x+3}-2)(\sqrt{x-1}+3)(\sqrt{x+3}+2)}{(x-1)(\sqrt{x+3}+2)}$$

$$= \lim_{x \to 1} \frac{(x+3-4)(\sqrt{x-1}+3)}{(x-1)(\sqrt{x+3}+2)}$$

$$= \lim_{x \to 1} \frac{\sqrt{x-1}+3}{\sqrt{x+3}+2}$$

$$= \frac{3}{4}.$$

EXAMPLE 2.8

Evaluate $\quad \lim\limits_{x \to 0} \dfrac{\sin 2x}{\sin x}$.

SOLUTION   Once again we cannot immediately use Theorem 2.1 since the limit of the denominator is zero. But using the double-angle formula $\sin 2x = 2 \sin x \cos x$, we find that

$$\lim_{x \to 0} \frac{\sin 2x}{\sin x} = \lim_{x \to 0} \frac{2 \sin x \cos x}{\sin x} = \lim_{x \to 0} (2 \cos x) = 2.$$

EXAMPLE 2.9

Do the functions

$$f(x) = \sin\left(\frac{1}{x}\right) \qquad \text{and} \qquad g(x) = x \sin\left(\frac{1}{x}\right)$$

have limits as x approaches 0?

SOLUTION   We approach these limits graphically. The computer-generated graph of $\sin(1/x)$ for $-0.01 \leq x \leq 0.01$ in Figure 2.8a is a washout. The graph for $-0.1 \leq x \leq 0.1$ in Figure Figure 2.8b is more instructive.

FIGURE 2.8a

FIGURE 2.8b

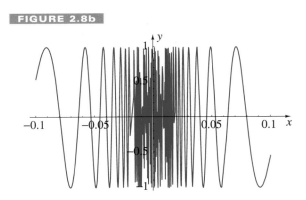

To fully understand this graph, consider dividing the positive x-axis into two parts such as $0 < x \leq 1$ and $1 < x < \infty$. Now the function $\sin x$, with period 2π, has part of a cycle in the interval $0 < x \leq 1$, and an infinite number of oscillations in $1 < x < \infty$ (Figure 2.9).

FIGURE 2.9

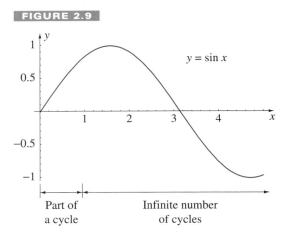

$y = \sin x$

Part of   Infinite number
a cycle   of cycles

The function $\sin(1/x)$ performs two operations; it first inverts a value of x to give $1/x$ and then takes the sine of this number. Given a value of x in the interval $0 < x < 1$, $\sin(1/x)$ first evaluates $1/x$ which is in the interval $1 < 1/x < \infty$, and then takes the sine of $1/x$. Conversely, given an x in $1 < x < \infty$, $\sin(1/x)$ first finds $1/x$, which is in the interval $0 < 1/x < 1$, and then calculates $\sin(1/x)$. The part cycle of $\sin x$ in $0 < x < 1$ is spread out over $1 < x < \infty$ by $\sin(1/x)$. The infinite number of oscillations of $\sin x$ in the interval $1 < x < \infty$ take place in the interval $0 < x < 1$ for $\sin(1/x)$. Notice in particular that $\sin(1/x)$ is equal to 0 when $x = 1/(n\pi)$ for every integer n. Since the function oscillates back and forth between ± 1 more and more rapidly as x approaches 0, it cannot have a limit there.

The function $g(x) = x \sin(1/x)$ has exactly the same number of oscillations as $f(x) = \sin(1/x)$, but the oscillations become smaller and smaller as x approaches 0 (Figure 2.10). In other words, $\lim_{x \to 0} g(x) = 0$.

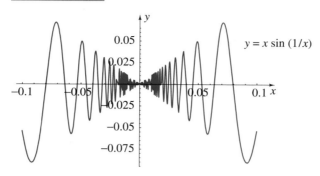

One-Sided Limits

When we write $L = \lim_{x \to a} f(x)$, we mean that $f(x)$ gets arbitrarily close to L as x gets closer and closer to a. But how is x to approach a? Does x approach a through numbers larger than a, or does it approach a through numbers smaller than a? Or does x jump back and forth between numbers larger than a and numbers smaller than a, gradually getting closer and closer to a? We have not previously mentioned "mode" of approach simply because it would have made no difference to our discussion. In each of the preceding examples, all possible modes of approach lead to the same limit. In particular, Table 2.1 and Figure 2.2 illustrate that 1 is the limit of $f(x) = x^2 - 4x + 5$ as x approaches 2 whether x approaches 2 through numbers larger than 2 or through numbers smaller than 2.

Approaching a number a either through numbers larger than a or through numbers smaller than a are two modes of approach that will be very important; therefore, we give them special notations:

$$\lim_{x \to a^-} f(x)$$

indicates that x approaches a through numbers smaller than a (often called a **left-hand limit** since x approaches a along the x-axis from the left of a)

$$\lim_{x \to a^+} f(x)$$

indicates that x approaches a through numbers larger than a (often called a **right-hand limit** since x approaches a along the x-axis from the right of a)

Example 2.6 should, in fact, be designated a right-hand limit,

$$\lim_{x \to 0^+} \frac{\sqrt{1+x} - 1}{\sqrt{x}} = 0,$$

since the presence of $\sqrt{x}$ in the denominator demands that x be positive. Likewise the limit of Example 2.7 should only be right-handed.

A natural question to ask is "What should we conclude if for a function $f(x)$

$$\lim_{x \to a^+} f(x) \neq \lim_{x \to a^-} f(x)?"$$

Our entire discussion has suggested (and indeed it can be proved; see Exercise 20 in Section 2.5) that if a function has a limit as x approaches a, then it has only one such limit; that is, the limit must be the same for every possible method of approach. Consequently, if we arrive at two different results depending on the mode of approach, then we conclude that the function does not have a limit. This situation is illustrated in the following example.

EXAMPLE 2.10

Evaluate, if possible, $\lim\limits_{x \to 0} \dfrac{|x|}{x}$.

SOLUTION   If $x < 0$, then $|x| = -x$, and

$$\lim_{x \to 0^-} \frac{|x|}{x} = \lim_{x \to 0^-} \frac{-x}{x} = -1.$$

FIGURE 2.11

If $x > 0$, then $|x| = x$, and

$$\lim_{x \to 0^+} \frac{|x|}{x} = \lim_{x \to 0^+} \frac{x}{x} = 1.$$

The function has a right-hand limit and a left-hand limit at $x = 0$, but, because they are not the same, $\lim_{x \to 0} (|x|/x)$ does not exist. The graph of $f(x) = |x|/x$ in Figure 2.11 clearly illustrates the situation.

EXAMPLE 2.11

A computer-generated graph of the function $f(x) = (1 - e^{-1/x})/(1 + e^{-1/x})$ is shown in Figure 2.12. What does it say about the right-hand limit, left-hand limit, and limit of $f(x)$ as x approaches 0?

FIGURE 2.12

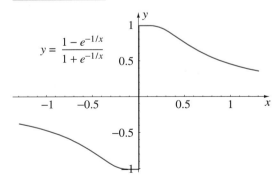

SOLUTION   The graph indicates that the left-hand limit is -1, the right-hand limit is 1, and because these limits are different, the limit as $x \to 0$ does not exist.

EXERCISES 2.1

In Exercises 1–41 find the indicated limit, if it exists.

∗ **1.** $\displaystyle\lim_{x \to 7} \frac{x^2 - 5}{x + 2}$

∗ **2.** $\displaystyle\lim_{x \to -2} \frac{x^3 + 8}{x + 5}$

∗ **3.** $\displaystyle\lim_{x \to -5} \frac{x^2 + 3x + 2}{x^2 + 25}$

∗ **4.** $\displaystyle\lim_{x \to 0} \frac{x^2 + 3x}{3x^2 - 2x}$

∗ **5.** $\displaystyle\lim_{x \to 3^+} \frac{2x - 3}{x^2 - 5}$

∗ **6.** $\displaystyle\lim_{x \to 2^-} \frac{2x - 4}{3x + 2}$

∗ **7.** $\displaystyle\lim_{x \to 0^-} \frac{x^4 + 5x^3}{3x^4 - x^3}$

∗ **8.** $\displaystyle\lim_{x \to 2^+} \frac{x^2 + 2x + 4}{x - 3}$

∗ **9.** $\displaystyle\lim_{x \to 2} \frac{x^2 - 4}{x - 2}$

∗ **10.** $\displaystyle\lim_{x \to 3^+} \frac{x^2 - 9}{x - 3}$

∗ **11.** $\displaystyle\lim_{x \to 5^-} \frac{x^2 - 25}{x - 5}$

∗ **12.** $\displaystyle\lim_{x \to 3} \frac{x^2 - 2x - 3}{3 - x}$

∗ **13.** $\displaystyle\lim_{x \to 2} \frac{x^2 - 4x + 4}{x - 2}$

∗ **14.** $\displaystyle\lim_{x \to 2} \frac{x^3 - 6x^2 + 12x - 8}{x^2 - 4x + 4}$

∗ **15.** $\displaystyle\lim_{x \to 1} \frac{x^3 - 6x^2 + 11x - 6}{x^2 - 3x + 2}$

∗ **16.** $\displaystyle\lim_{x \to 2} \frac{x^3 - 6x^2 + 11x - 6}{x^2 - 3x + 2}$

∗ **17.** $\displaystyle\lim_{x \to 3^+} \frac{x^3 - 6x^2 + 11x - 6}{x^2 - 3x + 2}$

∗ **18.** $\displaystyle\lim_{x \to 3^-} \frac{x^3 - 6x^2 + 11x - 6}{x^2 - 3x + 2}$

∗ **19.** $\displaystyle\lim_{x \to 0} \frac{x^3 - 6x^2 + 11x - 6}{x^2 - 3x + 2}$

∗ **20.** $\displaystyle\lim_{x \to -1} \frac{12x + 5}{x^2 - 2x + 1}$

∗ **21.** $\displaystyle\lim_{x \to 1} \sqrt{\frac{2 - x}{2 + x}}$

∗ **22.** $\displaystyle\lim_{x \to 5} \frac{\sqrt{1 - x^2}}{3x + 2}$

∗∗ **23.** $\displaystyle\lim_{x \to 0} \frac{\sin 4x}{\sin 2x}$

∗∗ **24.** $\displaystyle\lim_{x \to 0^+} \frac{\sin 6x}{\sin 3x}$

∗ **25.** $\displaystyle\lim_{x \to 0} \frac{\tan x}{\sin x}$

∗ **26.** $\displaystyle\lim_{x \to \pi/4} \frac{\sin x}{\tan x}$

∗∗ **27.** $\displaystyle\lim_{x \to 0^+} \frac{\sin 2x}{\tan x}$

∗ **28.** $\displaystyle\lim_{x \to 2} \frac{x - 2}{\sqrt{x} - \sqrt{2}}$

∗∗ **29.** $\displaystyle\lim_{x \to 0} \frac{\sqrt{1 - x} - \sqrt{1 + x}}{x}$

∗∗ **30.** $\displaystyle\lim_{x \to 5^+} \frac{|x^2 - 25|}{x^2 - 25}$

∗∗ **31.** $\displaystyle\lim_{x \to 5^-} \frac{|x^2 - 25|}{x^2 - 25}$

∗∗ **32.** $\displaystyle\lim_{x \to 5} \frac{|x^2 - 25|}{x^2 - 25}$

∗∗ **33.** $\displaystyle\lim_{x \to 0^+} \frac{\sqrt{x + 2} - \sqrt{2}}{\sqrt{x}}$

∗∗ **34.** $\displaystyle\lim_{x \to 0} \frac{1 - \sqrt{x^2 + 1}}{2x^2}$

∗∗ **35.** $\displaystyle\lim_{x \to -2} \frac{\sqrt{-x} + \sqrt{2}}{x + 2}$

∗∗ **36.** $\displaystyle\lim_{x \to 0} \frac{\sqrt{1 + x} - \sqrt{1 - x}}{x}$

∗∗ **37.** $\displaystyle\lim_{x \to -2^+} \frac{\sqrt{x + 3} - \sqrt{-x - 1}}{\sqrt{x + 2}}$

∗∗ **38.** $\displaystyle\lim_{x \to 0} \frac{x}{\sqrt{x + 4} - 2}$

** **39.** $\displaystyle\lim_{x\to 0} \frac{\sqrt{1+x}-\sqrt{1-x}}{\sqrt{2+x}-\sqrt{2-x}}$

** **40.** $\displaystyle\lim_{x\to 0} \frac{\sqrt{x+1}-\sqrt{2x+1}}{\sqrt{3x+4}-\sqrt{2x+4}}$

** **41.** $\displaystyle\lim_{x\to 1} \frac{\sqrt{x+3}-2}{x-1}$

** **42.** Draw a graph of $f(x) = (1/x)\sin x$ on the interval $-\pi \le x \le \pi$. What does it suggest for the limit of the function as $x \to 0$? (This will be confirmed in Section 3.9.)

** **43.** Does the greatest integer function $[\![x]\!]$ of Exercise 50 in Section 1.3 have a limit, a right-hand limit, or a left-hand limit as x approaches integer values?

** **44.** Prove or disprove the following statement: If $f(x) < g(x)$ for all $x \ne a$, then

$$\lim_{x\to a} f(x) \; < \; \lim_{x\to a} g(x).$$

** **45.** If n is a positive integer, evaluate

$$\lim_{h\to 0} \frac{(x+h)^n - x^n}{h}.$$

Hint: Use either the binomial theorem or the result that

$$a^n - b^n = (a-b)(a^{n-1} + a^{n-2}b + \cdots + ab^{n-2} + b^{n-1}).$$

** **46.** Evaluate

$$\lim_{h\to 0} \frac{\sqrt{x+h}-\sqrt{x}}{h}.$$

** **47.** The weight W of an object depends on its distance d from the centre of the earth. If d is less than the radius R of the earth, then W is directly proportional to d; and if d is greater than or equal to R, then W is inversely proportional to d^2. If the weight of the object on the earth's surface is W_0, find a definition for W as a function of d, and draw its graph.

** **48.** At the present time it is impossible for us to calculate algebraically

$$\lim_{x\to 0} \frac{\sin x - x}{x^3}.$$

What can we do?

(a) One suggestion might be to use a calculator or computer to evaluate the function $(\sin x - x)/x^3$ for various values of x that approach 0. Try this with $x = 10^{-n}$, $n = 1, \ldots, 10$, and make any conclusion you feel is justified. How would you feel about the use of the calculator if you knew that the value of the limit were $-1/6$?

(b) Another suggestion might be to plot the function on smaller and smaller intervals around $x = 0$. Do this on the intervals $-0.1 \le x \le 0.1$, $-0.01 \le x \le 0.01$, $-0.001 \le x \le 0.001$, and $-0.0001 \le x \le 0.0001$. What happens?

** **49.** Assume the existence of

$$\lim_{x\to a} \frac{f(x) - g(x)}{x - a}.$$

If $\lim_{x\to a} f(x) = L$, find $\lim_{x\to a} g(x)$.

** **50.** If $f(x)$ is an even function and $\lim_{x\to a} f(x) = L$, find, if possible, $\lim_{x\to -a} f(x)$.

** **51.** If $f(x)$ is an even function and $\lim_{x\to a^+} f(x) = L$, find, if possible, $\lim_{x\to -a^-} f(x)$.

** **52.** If $f(x)$ is an even function and $\lim_{x\to a^+} f(x) = L$, find, if possible, $\lim_{x\to -a^+} f(x)$.

** **53.–55.** Repeat Exercises 50–52 for an odd function $f(x)$.

** **56.** If $\lim_{x\to 0} f(x) = F$, what do you conclude about $\lim_{x\to 0} f(x) \sin(1/x)$? Hint: See Example 2.9.

2.2 Infinite Limits

Functions do not always have limits. Sometimes this is due to the fact that right-hand and left-hand limits are not identical; sometimes it is a result of erratic oscillations. Examples of both of these situations were discussed in Section 2.1 (see Examples 2.9–2.11). Nonexistence of a limit may also be due to excessively large values of the function. For instance, consider the function $f(x) = 1/(x - 2)^2$ which is not defined at $x = 2$. Does it have a limit as x approaches 2? Table 2.2 indicates that as x approaches 2, values of $1/(x - 2)^2$ become very large; in fact, values of the function can be made arbitrarily large by choosing x sufficiently close to 2. Thus, the function does not have a limit as x approaches 2. We express this symbolically in the form

$$\lim_{x\to 2} \frac{1}{(x - 2)^2} = \infty.$$

The symbol ∞ represents what mathematicians call *infinity*. Infinity is not a number; it is simply a symbol that we find convenient to represent various ideas. In the above equation it states that the *limit does not exist*, and indicates that the reason it does not exist is that values of the function become arbitrarily large as x approaches 2.

The graph of $f(x)$ in Figure 2.13 further illustrates this point. We say that the line $x = 2$ is a **vertical asymptote** for the graph.

TABLE 2.2

x	x	$1/(x - 2)^2 \, x$
1.9	10^2	2.1
1.99	10^4	2.01
1.999	10^6	2.001
1.9999	10^8	2.0001
1.99999	10^{10}	2.00001

FIGURE 2.13

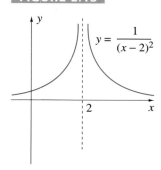

$$y = \frac{1}{(x - 2)^2}$$

EXAMPLE 2.12

FIGURE 2.14

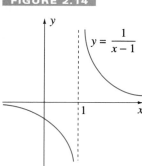

$$y = \frac{1}{x - 1}$$

Evaluate $\displaystyle\lim_{x\to 1} \frac{1}{x - 1}$, if it exists.

SOLUTION   In this example we consider right-hand and left-hand limits as x approaches 1. We find that

$$\lim_{x\to 1^+} \frac{1}{x - 1} = \infty \quad \text{and} \quad \lim_{x\to 1^-} \frac{1}{x - 1} = -\infty,$$

the latter meaning that as x approaches 1 from the left, the function takes on arbitrarily "large" negative values. Either one of these expressions is sufficient to conclude that the function $1/(x - 1)$ does not have a limit as x approaches 1. The function is shown in Figure 2.14; the line $x = 1$ is a vertical asymptote.

EXAMPLE 2.13

Discuss left-hand and right-hand limits of the function

$$f(x) = \frac{x^2 - 9}{x^2 + x - 2}$$

as $x \to 1$ and $x \to -2$.

SOLUTION   The function is undefined at $x = 1$ and $x = -2$ (why?). Its graph is shown in Figure 2.15.

FIGURE 2.15

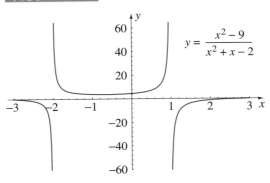

To confirm what we see, it is advantageous to factor numerator and denominator of $f(x)$ as much as possible,

$$f(x) = \frac{(x + 3)(x - 3)}{(x + 2)(x - 1)}.$$

Consider now the (right-hand) limit as $x \to 1^+$. Each of the four factors has a limit as $x \to 1^+$:

$$x + 3 \to 4, \qquad x - 3 \to -2, \qquad x + 2 \to 3, \qquad x - 1 \to 0.$$

Were we to combine them according to Theorem 2.1, we would write

$$\frac{4(-2)}{3(0)}.$$

In spite of the fact that this is not correct—part (v) of the theorem does not allow a 0 in the denominator—this expression does tell us what is happening to $f(x)$ as $x \to 1^+$. It states that as $x \to 1^+$, the numerator of $f(x)$ approaches -8 and the denominator approaches 0. But this means that $f(x)$ must be taking on larger and larger values as $x \to 1^+$. Are these values positive or negative? The numerator is clearly negative, but the sign of the denominator depends on whether the 0 is approached through positive or negative numbers. Recalling that 0 arose from the fact that $x - 1 \to 0$ as $x \to 1^+$, we can be more specific; $x - 1$ must approach 0 through positive numbers since $x > 1$ for $x \to 1^+$. We indicate this by writing $x - 1 \to 0^+$ as $x \to 1^+$. The above fraction is therefore replaced by

$$\frac{4(-2)}{3(0^+)},$$

and it is clearly negative. In other words,

$$\lim_{x \to 1^+} \frac{x^2 - 9}{x^2 + x - 2} = \lim_{x \to 1^+} \frac{(x + 3)(x - 3)}{(x + 2)(x - 1)} = -\infty,$$

as indeed Figure 2.15 indicates. Similarly, as $x \to 1^-$, the fraction

$$\frac{4(-2)}{3(0^-)},$$

which shows limits of the four factors of $f(x)$, indicates that

$$\lim_{x \to 1^-} \frac{(x + 3)(x - 3)}{(x + 2)(x - 1)} = \infty.$$

In the following, the expressions at the right of the page yield limits as $x \to -2^+$ and $x \to -2^-$:

$$\lim_{x \to -2^+} \frac{(x + 3)(x - 3)}{(x + 2)(x - 1)} = \infty, \qquad\qquad \frac{(1)(-5)}{(0^+)(-3)}$$

$$\lim_{x \to -2^-} \frac{(x + 3)(x - 3)}{(x + 2)(x - 1)} = -\infty. \qquad\qquad \frac{(1)(-5)}{(0^-)(-3)}$$

EXERCISES 2.2

In Exercises 1–24 evaluate the limit, if it exists.

* **1.** $\displaystyle\lim_{x \to 2^+} \frac{1}{x - 2}$

* **2.** $\displaystyle\lim_{x \to 2^-} \frac{1}{x - 2}$

* **3.** $\displaystyle\lim_{x \to 2} \frac{1}{x - 2}$

* **4.** $\displaystyle\lim_{x \to 2^+} \frac{1}{(x - 2)^2}$

* **5.** $\displaystyle\lim_{x \to 2^-} \frac{1}{(x - 2)^2}$

* **6.** $\displaystyle\lim_{x \to 2} \frac{1}{(x - 2)^2}$

* **7.** $\displaystyle\lim_{x \to 1} \frac{5x}{(x - 1)^3}$

* **8.** $\displaystyle\lim_{x \to 1/2} \frac{6x^2 + 7x - 5}{2x - 1}$

* **9.** $\displaystyle\lim_{x \to 1} \frac{2x + 3}{x^2 - 2x + 1}$

* **10.** $\displaystyle\lim_{x \to 2} \frac{x - 2}{x^2 - 4x + 4}$

** **11.** $\displaystyle\lim_{x \to \pi/2^+} \tan x$

** **12.** $\displaystyle\lim_{x \to \pi/2^-} \tan x$

** **13.** $\displaystyle\lim_{x \to 0} \csc x$

* **14.** $\displaystyle\lim_{x \to \pi/4} \sec (x - \pi/4)$

** **15.** $\displaystyle\lim_{x \to 3\pi/4} \sec (x - \pi/4)$

* **16.** $\displaystyle\lim_{x \to 0^+} \cot x$

** **17.** $\displaystyle \lim_{x \to 1} \frac{x^2 - 2x + 1}{x^3 - 3x^2 + 3x - 1}$

** **18.** $\displaystyle \lim_{x \to 0} \frac{\sqrt{1 + x} - 1}{x^2}$

** **19.** $\displaystyle \lim_{x \to 0} \frac{2x}{1 - \sqrt{x^2 + 1}}$

** **20.** $\displaystyle \lim_{x \to 4} \frac{|4 - x|}{x^2 - 8x + 16}$

** **21.** $\displaystyle \lim_{x \to 0^+} \ln(4x)$

** **22.** $\displaystyle \lim_{x \to 1} \frac{1}{\ln |x - 1|}$

** **23.** $\displaystyle \lim_{x \to 0} e^{1/x}$

** **24.** $\displaystyle \lim_{x \to 0} e^{1/|x|}$

** **25.** It is not clear whether the limit

$$\lim_{x \to 0^+} x^2 \ln x$$

exists due to the fact that $\lim_{x \to 0^+} x^2 = 0$ and $\lim_{x \to 0^+} \ln x = -\infty$. It depends on which term is more dominant in the product, x^2 or $\ln x$.

    (a) Calculate x^2, $\ln x$, and $x^2 \ln x$ for $x = 10^{-2n}$, $n = 1, \ldots, 5$, and use this information to decide on a value for the limit.

    (b) Plot graphs of $x^2 \ln x$ near $x = 0$ to confirm your calculation in (a).

** **26.** Repeat Exercise 25 for the limit

$$\lim_{x \to 0^+} x^{10} e^{1/x},$$

but pick your own values of x at which to evaluate x^{10} and $e^{1/x}$.

2.3 Limits at Infinity

In many applications we are concerned with the behaviour of a function as its independent variable takes on very large values, positively or negatively. For instance, consider finding, if possible, a number that the function $f(x) = (2x^2 + 3)/(x^2 + 4)$, with domain $x \geq 0$, gets closer and closer to as x becomes larger and larger and larger. The graph of $f(x)$ on the interval $0 \leq x \leq 100$ in Figure 2.16 suggests that function values are approaching 2 for large x. To confirm this algebraically, we divide numerator and denominator of $f(x)$ by x^2,

$$f(x) = \frac{2x^2 + 3}{x^2 + 4} = \frac{2 + \dfrac{3}{x^2}}{1 + \dfrac{4}{x^2}}.$$

For very large x, the terms $3/x^2$ and $4/x^2$ are very close to zero, and therefore $f(x)$ is approximately equal to 2. Indeed, $f(x)$ can be made arbitrarily close to 2 by choosing x sufficiently large. In calculus we express this fact by saying that **the limit of** $(2x^2 + 3)/(x^2 + 4)$ **as** x **approaches infinity is** 2, and we write

$$\lim_{x \to \infty} \frac{2x^2 + 3}{x^2 + 4} = 2.$$

Once again we stress that ∞ is not a number. The notation $x \to \infty$ simply means "as x gets larger and larger and larger." We say that the line $y = 2$ is a **horizontal asymptote**.

FIGURE 2.16

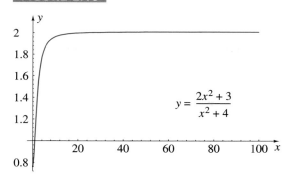

FIGURE 2.16

We can also find limits of functions as x takes on arbitrarily large negative numbers, denoted by $x \to -\infty$.

EXAMPLE 2.14

Evaluate

$$\lim_{x \to -\infty} \frac{5x^4 - 3x + 5}{x^4 - 2x^2 + 5},$$

if it exists.

SOLUTION   Division of numerator and denominator by x^4 leads to

$$\lim_{x \to -\infty} \frac{5 - \dfrac{3}{x^3} + \dfrac{5}{x^4}}{1 - \dfrac{2}{x^2} + \dfrac{5}{x^4}} = 5.$$

If the limit in the above example is taken as $x \to \infty$, the same result is obtained. The graph of this function is shown in Figure 2.17. The line $y = 5$ is a horizontal asymptote.

FIGURE 2.17

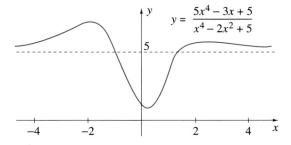

In general we say that a line $y = L$ is a horizontal asymptote for the graph of a function $f(x)$ if either, or both, of the following situations exists:

$$\lim_{x \to -\infty} f(x) = L \qquad \text{or} \qquad \lim_{x \to \infty} f(x) = L.$$

In Example 2.14, both of these conditions are satisfied for $L = 5$.

EXAMPLE 2.15

Evaluate the following limits:

(a) $\displaystyle\lim_{x \to \infty} \frac{2x^3 - 4}{x^3 + x^2 + 2}$ (b) $\displaystyle\lim_{x \to -\infty} \frac{2x^2 - 14}{3x^3 + 5x}$ (c) $\displaystyle\lim_{x \to \infty} \frac{2x^4 - 14}{3x^3 + 5x}$

SOLUTION

(a) To obtain this limit we divide numerator and denominator of the fraction by x^3:

$$\lim_{x \to \infty} \frac{2x^3 - 4}{x^3 + x^2 + 2} = \lim_{x \to \infty} \frac{2 - \dfrac{4}{x^3}}{1 + \dfrac{1}{x} + \dfrac{2}{x^3}} = 2.$$

(b) Once again we divide numerator and denominator by x^3:

$$\lim_{x \to -\infty} \frac{2x^2 - 14}{3x^3 + 5x} = \lim_{x \to -\infty} \frac{\dfrac{2}{x} - \dfrac{14}{x^3}}{3 + \dfrac{5}{x^2}}.$$

Since the numerator approaches 0 and the denominator approaches 3, we conclude that

$$\lim_{x \to -\infty} \frac{2x^2 - 14}{3x^3 + 5x} = 0.$$

We also could have obtained this limit by dividing numerator and denominator by x^2 instead of x^3,

$$\lim_{x \to -\infty} \frac{2x^2 - 14}{3x^3 + 5x} = \lim_{x \to -\infty} \frac{2 - \dfrac{14}{x^2}}{3x + \dfrac{5}{x}}.$$

Now the numerator approaches 2, but since the denominator becomes very large, the fraction once again approaches zero.

(c) Division by x^3 in this case gives

$$\lim_{x \to \infty} \frac{2x^4 - 14}{3x^3 + 5x} = \lim_{x \to \infty} \frac{2x - \dfrac{14}{x^3}}{3 + \dfrac{5}{x^2}}.$$

Since the numerator becomes arbitrarily large as $x \to \infty$ and the denominator approaches 3, it follows that

$$\lim_{x \to \infty} \frac{2x^4 - 14}{3x^3 + 5x} = \infty.$$

EXAMPLE 2.16

Draw a graph of the function

$$f(x) = \frac{1}{x}\sin x, \quad x \geq \pi.$$

Does it have a horizontal asymptote?

SOLUTION   The graph in Figure 2.18 indicates that the positive x-axis is a horizontal asymptote. This is confirmed by the fact that

$$\lim_{x \to \infty} \frac{1}{x}\sin x = 0.$$

We reason this out as follows. The function $\sin x$ does not have a limit as $x \to \infty$, but as $x \to \infty$ its values oscillate back and forth between ± 1. Since these values are multiplied by $1/x$, which is getting smaller and smaller, the product $(1/x)\sin x$ must be getting closer and closer to 0.

FIGURE 2.18

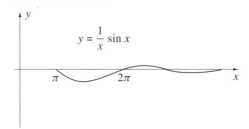

Notice that the graph actually crosses the asymptote an infinite number of times.

Figures 2.16, 2.17, and 2.18 indicate that the graph of a function $y = f(x)$ can approach an asymptote $y = L$ in three ways, from above (Figure 2.17), from below (Figure 2.16), and oscillating about the asymptote gradually getting closer and closer to it (Figure 2.18). A computer-generated graph may or may not always make it clear which situation prevails. This is illustrated in the following example.

EXAMPLE 2.17

Plot a graph of the function $f(x) = (3x - 6)/(x^2 + 5)$. Indicate any horizontal asymptotes and determine how the graph approaches these asymptotes.

SOLUTION   We begin by plotting a graph of the function on the domain $-100 \leq x \leq 100$ (Figure 2.19a). It indicates that $y = 0$ is a horizontal asymptote, but the domain is so large that it is difficult to determine how the graph approaches the asymptote. Figure 2.19b with a smaller domain $-10 \leq x \leq 10$ suggests that the graph approaches $y = 0$ from above when $x \to \infty$ and from below when $x \to -\infty$. We can confirm this in various ways. To verify the situation as $x \to \infty$, we calculate

$$\lim_{x \to \infty} \frac{3x - 6}{x^2 + 5} = \lim_{x \to \infty} \frac{3 - \dfrac{6}{x}}{x + \dfrac{5}{x}} = 0.$$

This confirms that $y = 0$ is indeed a horizontal asymptote as $x \to \infty$. To decide how the asymptote is approached, we note that for large positive x, both $3 - 6/x$ and $x + 5/x$ are positive, and therefore $f(x)$ must approach 0 through positive numbers. We indicate this by writing

$$\lim_{x \to \infty} \frac{3x - 6}{x^2 + 5} = 0^+.$$

We could also have reasoned as follows. The graph of the function crosses the asymptote $y = 0$ when

$$\frac{3x - 6}{x^2 + 5} = 0,$$

and the only solution of this equation is $x = 2$. Combine this with the graph in Figure 2.19b, and we conclude that $f(x) > 0$ for all $x > 2$, and hence the graph must approach $y = 0$ from above as $x \to \infty$.

Similar reasoning shows that the graph approaches $y = 0$ from below as $x \to -\infty$.

FIGURE 2.19a

FIGURE 2.19b

The following example is somewhat more complicated.

EXAMPLE 2.18

Find vertical and horizontal asymptotes for the graph of $f(x) = (x^2 - 16)/(x^2 + x - 6)$. Determine how the graph approaches horizontal asymptotes.

SOLUTION The plot on the interval $-10 \le x \le 10$ in Figure 2.20a suggests vertical asymptotes at $x = 2$ and $x = -3$. These are confirmed with the following limits:

$$\lim_{x \to 2^+} \frac{(x + 4)(x - 4)}{(x + 3)(x - 2)} = -\infty, \qquad \frac{6(-2)}{(5)(0^+)}$$

$$\lim_{x \to 2^-} \frac{(x + 4)(x - 4)}{(x + 3)(x - 2)} = \infty, \qquad \frac{6(-2)}{(5)(0^-)}$$

$$\lim_{x \to -3^+} \frac{(x + 4)(x - 4)}{(x + 3)(x - 2)} = \infty, \qquad \frac{(1)(-7)}{(0^+)(-5)}$$

$$\lim_{x \to -3^-} \frac{(x + 4)(x - 4)}{(x + 3)(x - 2)} = -\infty. \qquad \frac{(1)(-7)}{(0^-)(-5)}$$

The plot in Figure 2.20b suggests that $y = 1$ is a horizontal asymptote as $x \to \infty$. To verify this we calculate

$$\lim_{x \to \infty} \frac{x^2 - 16}{x^2 + x - 6} = \lim_{x \to \infty} \frac{1 - \dfrac{16}{x^2}}{1 + \dfrac{1}{x} - \dfrac{6}{x^2}} = 1.$$

By long division, we can express $f(x)$ in the form

$$f(x) = \frac{x^2 - 16}{x^2 + x - 6} = 1 - \frac{x + 10}{x^2 + x - 6}.$$

Because $(x + 10)/(x^2 + x - 6)$ is positive for large x, it follows that $f(x)$ is less than 1 for large x, and the graph approaches $y = 1$ from below as $x \to \infty$.

The plot in Figure 2.20c indicates that $y = 1$ is also a horizontal asymptote as $x \to -\infty$. Since $(x + 10)/(x^2 + x - 6)$ is negative for "large" negative x, it follows that $f(x)$ is greater than 1 for such x, and the graph approaches $y = 1$ from above.

FIGURE 2.20a

FIGURE 2.20b

FIGURE 2.20c

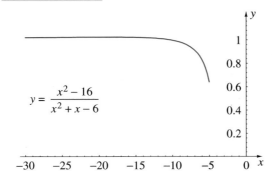

Previous functions in this section have had at most one horizontal asymptote. The following example has two.

EXAMPLE 2.19

Evaluate

$$\lim_{x \to \infty} \frac{\sqrt{2x^2 + 4}}{x + 5},$$

if it exists. What is the limit as $x \to -\infty$?

SOLUTION   When we divide numerator and denominator by x, and take the x inside the square root as x^2,

$$\lim_{x \to \infty} \frac{\sqrt{2x^2 + 4}}{x + 5} = \lim_{x \to \infty} \frac{\frac{1}{x}\sqrt{2x^2 + 4}}{\frac{1}{x}(x + 5)} = \lim_{x \to \infty} \frac{\sqrt{\frac{2x^2 + 4}{x^2}}}{1 + \frac{5}{x}} = \lim_{x \to \infty} \frac{\sqrt{2 + \frac{4}{x^2}}}{1 + \frac{5}{x}} = \sqrt{2}.$$

In evaluating the limit as $x \to -\infty$, we must be extra careful,

$$\lim_{x \to -\infty} \frac{\sqrt{2x^2 + 4}}{x + 5} = \lim_{x \to -\infty} \frac{\frac{\sqrt{2x^2 + 4}}{x}}{1 + \frac{5}{x}}.$$

It is not correct in this case to take x inside the square root as

$$\frac{\sqrt{2x^2 + 4}}{x} = \sqrt{\frac{2x^2 + 4}{x^2}}$$

since, for negative x, the expression on the left is negative and that on the right is positive. In this case, we should replace x by $-\sqrt{x^2}$, and write

$$\frac{\sqrt{2x^2 + 4}}{x} = \frac{\sqrt{2x^2 + 4}}{-\sqrt{x^2}} = -\sqrt{\frac{2x^2 + 4}{x^2}}.$$

Hence,

$$\lim_{x \to -\infty} \frac{\sqrt{2x^2 + 4}}{x + 5} = \lim_{x \to -\infty} \frac{-\sqrt{\frac{2x^2 + 4}{x^2}}}{1 + \frac{5}{x}} = \lim_{x \to -\infty} \frac{-\sqrt{2 + \frac{4}{x^2}}}{1 + \frac{5}{x}} = -\sqrt{2}.$$

The graph of this function (Figure 2.21) confirms these limits; there are two horizontal asymptotes, $y = \sqrt{2}$ and $y = -\sqrt{2}$.

FIGURE 2.21

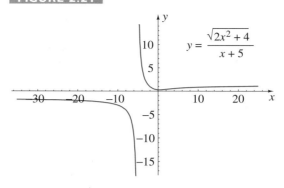

Graphs of functions have horizontal asymptotes if either, or both, of the conditions $\lim_{x \to \pm\infty} f(x) = L$ hold. What this means is that, for large positive or negative values of x, the function is approximately equal to L. The larger the value of x, the better the approximation. Hyperbolas also have asymptotes, but they are not usually horizontal. For example, asymptotes of the hyperbola $x^2 - y^2/9 = 1$ in Figure 2.22 are $y = \pm 3x$. These are often called oblique asymptotes.

FIGURE 2.22

FIGURE 2.23

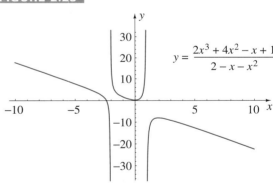

Graphs of functions can also have oblique asymptotes. In Figure 1.37 (page 25), $y = x - 3$ is an oblique asymptote for the graph of $f(x) = (x^3 - 3x^2 + 1)/(x^2 + 1)$. We can see this algebraically if we divide $x^2 + 1$ into $x^3 - 3x^2 + 1$. The result is

$$f(x) = \frac{x^3 - 3x^2 + 1}{x^2 + 1} = x - 3 + \frac{4 - x}{x^2 + 1}.$$

As $x \to \pm\infty$, the term $(4 - x)/(x^2 + 1) \to 0$, and therefore $f(x) \to x - 3$. For large values of x, the function $f(x)$ can be approximated more and more closely by $x - 3$.

Rational functions $P(x)/Q(x)$ always have oblique asymptotes when the degree of polynomial $P(x)$ is one more than the degree of polynomial $Q(x)$. A second example is the function

$f(x) = (2x^3 + 4x^2 - x + 1)/(2 - x - x^2)$ in Figure 2.23. It has an oblique asymptote $y = -2x - 2$ identified from

$$f(x) = \frac{2x^3 + 4x^2 - x + 1}{2 - x - x^2} = -2x - 2 + \frac{x + 5}{2 - x - x^2}.$$

It also has two vertical asymptotes, one at $x = -2$ and the other at $x = 1$.

Functions other than rational functions can also have oblique asymptotes. The graph of $f(x) = x + 2 + 5e^{-x}$ in Figure 2.24 is asymptotic to $y = x + 2$. The straightness of the graph of $f(x) = \sqrt{(2x^4 + 4x^2 - x + 1)/(x^2 + x - 2)}$ for large positive x in Figure 2.25 suggests an asymptote. It is $\sqrt{2}(x - 1/2)$, but to show this algebraically is difficult.

FIGURE 2.24

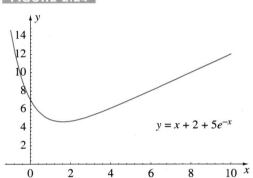

$y = x + 2 + 5e^{-x}$

FIGURE 2.25

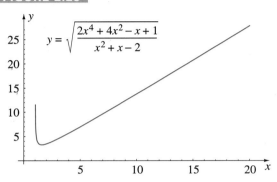

$$y = \sqrt{\frac{2x^4 + 4x^2 - x + 1}{x^2 + x - 2}}$$

EXERCISES 2.3

In Exercises 1–38 evaluate the limit, if it exists.

* **1.** $\displaystyle\lim_{x \to \infty} \frac{x + 1}{2x - 1}$

* **2.** $\displaystyle\lim_{x \to \infty} \frac{1 - x}{3 + 2x}$

* **3.** $\displaystyle\lim_{x \to \infty} \frac{x^2 + 1}{2x^3 + 5}$

* **4.** $\displaystyle\lim_{x \to \infty} \frac{1 - 4x^3}{3 + 2x - x^2}$

* **5.** $\displaystyle\lim_{x \to -\infty} \frac{2 + x - x^2}{3 + 4x^2}$

* **6.** $\displaystyle\lim_{x \to -\infty} \frac{x^3 - 2x^2}{3x^3 + 4x^2}$

* **7.** $\displaystyle\lim_{x \to -\infty} \frac{x^3 - 2x^2 + x + 1}{x^4 + 3x}$

* **8.** $\displaystyle\lim_{x \to -\infty} \frac{x^3 - 2x^2 + x + 1}{x^2 - x + 1}$

** **9.** $\displaystyle\lim_{x \to \infty} \frac{\sqrt{x^2 + 1}}{2x + 1}$

** **10.** $\displaystyle\lim_{x \to \infty} \frac{3x - 1}{\sqrt{5 + 4x^2}}$

** **11.** $\displaystyle\lim_{x \to -\infty} \frac{\sqrt{1 - 2x^2}}{x + 2}$

** **12.** $\displaystyle\lim_{x \to -\infty} \frac{\sqrt{1 - 2x}}{x + 2}$

** **13.** $\displaystyle\lim_{x \to \infty} \sqrt{\frac{2 + x}{x - 2}}$

** **14.** $\displaystyle\lim_{x \to \infty} \frac{\sqrt{3 + x}}{\sqrt{x}}$

** **15.** $\displaystyle\lim_{x \to \infty} (x^2 - x^3)$

** **16.** $\displaystyle\lim_{x \to \infty} \left(x + \frac{1}{x}\right)$

✱✱ **17.** $\displaystyle\lim_{x\to\infty} \frac{x}{\sqrt{x+5}}$

✱✱ **18.** $\displaystyle\lim_{x\to-\infty} \frac{x^2}{\sqrt{3-x}}$

✱✱ **19.** $\displaystyle\lim_{x\to-\infty} \frac{x}{\sqrt[3]{4+x^3}}$

✱✱ **20.** $\displaystyle\lim_{x\to\infty} \frac{3x}{\sqrt[3]{2+4x^3}}$

✱✱ **21.** $\displaystyle\lim_{x\to\infty} \frac{1}{2x}\cos x$

✱✱ **22.** $\displaystyle\lim_{x\to-\infty} \frac{1}{2x}\cos x$

✱✱ **23.** $\displaystyle\lim_{x\to\infty} \frac{\sin 4x}{x^2}$

✱✱ **24.** $\displaystyle\lim_{x\to\infty} \frac{\sin^2 x}{x}$

✱✱ **25.** $\displaystyle\lim_{x\to-\infty} \tan x$

✱✱ **26.** $\displaystyle\lim_{x\to\infty} \frac{1}{x}\tan x$

✱✱ **27.** $\displaystyle\lim_{x\to\infty} \left(\sqrt{x^2+1}-x\right)$

✱✱ **28.** $\displaystyle\lim_{x\to\infty} \left(\sqrt{x^2+4}-x\right)$

✱✱ **29.** $\displaystyle\lim_{x\to\infty} \left(\sqrt{2x^2+1}-x\right)$

✱✱ **30.** $\displaystyle\lim_{x\to-\infty} \left(\sqrt{2x^2+1}-x\right)$

✱✱ **31.** $\displaystyle\lim_{x\to\infty} \frac{\sqrt{3x^2+2}}{x+4}$

✱✱ **32.** $\displaystyle\lim_{x\to\infty} \frac{\sqrt{4x^2+7}}{2x+3}$

✱✱ **33.** $\displaystyle\lim_{x\to-\infty} \frac{\sqrt{3x^2+2}}{x+4}$

✱✱ **34.** $\displaystyle\lim_{x\to-\infty} \frac{\sqrt{4x^2+7}}{2x+3}$

✱✱ **35.** $\displaystyle\lim_{x\to\infty} \left(\sqrt{x^2+4}-\sqrt{x^2-1}\right)$

✱✱ **36.** $\displaystyle\lim_{x\to\infty} \left(\sqrt[3]{1+x}-\sqrt[3]{x}\right)$

✱✱ **37.** $\displaystyle\lim_{x\to\infty} \left(\sqrt{x^2+x}-x\right)$

✱✱ **38.** $\displaystyle\lim_{x\to-\infty} \left(\sqrt{x^2+x}-x\right)$

✱✱ **39.** What is the value of

$$\lim_{x\to-\infty} \frac{\sqrt{ax^2+bx+c}}{dx+e}$$

where $a>0$, and b, c, d, and e are constants?

✱✱✱ **40.** What conditions on the constants a, b, c, d, e, and f will ensure that the

$$\lim_{x\to\infty} \left(\sqrt{ax^2+bx+c}-\sqrt{dx^2+ex+f}\right)$$

exists? What is the value of the limit in this case?

2.4 Continuity

We have noticed that sometimes the limit of a function $f(x)$ as $x\to a$ is the same as the value $f(a)$ of the function at $x=a$. This property is described in the following definition. This is the situation at $x=1$ in Figure 2.1a, but not the case in Figures 2.1b or 2.1c (page 95).

> ### DEFINITION 2.1
> A function $f(x)$ is said to be **continuous** at $x=a$ if
>
> $$\lim_{x\to a} f(x) = f(a). \tag{2.4}$$

There are three conditions imposed by this definition:

1. $f(x)$ must be defined at $x = a$;

2. $\lim_{x \to a} f(x)$ must exist;

3. the numbers in (1) and (2) must be the same.

The graph in Figure 2.26 illustrates a function that is discontinuous (that is, not continuous) at $x = a$, b, c, d, and e. For instance, at $x = a$ conditions (1) and (2) are satisfied but (3) is violated; at $x = b$, (2) is satisfied, but (1) and (3) are not. Figure 2.26 suggests that discontinuities of a function are characterized geometrically by separations in its graph. This is indeed true, and it is often a very informative way to illustrate the nature of a discontinuity.

FIGURE 2.26

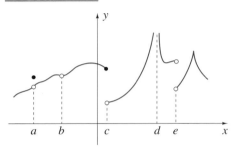

When a function is defined on a closed interval $a \leq x \leq b$, Definition 2.1 must be rephrased in terms of right- and left-hand limits for continuity at $x = a$ and $x = b$. Specifically, we say that $f(x)$ is **continuous from the right** at $x = a$ if $\lim_{x \to a^+} f(x) = f(a)$; and that $f(x)$ is **continuous from the left** at $x = b$ if $\lim_{x \to b^-} f(x) = f(b)$.

A function is said to be **continuous on an interval** if it is continuous at each point of that interval. In the event that the interval is closed, $a \leq x \leq b$, continuity at $x = a$ and $x = b$ is interpreted as continuity from the right and left, respectively. Geometrically, a function is continuous on an interval if a pencil can completely trace its graph without being lifted from the page.

EXAMPLE 2.20

Draw graphs of the following functions, indicating any discontinuities:

(a) $f(x) = \dfrac{x^2 - 16}{x - 4}$ (b) $f(x) = \dfrac{1}{(x - 4)^2}$ (c) $f(x) = \dfrac{|x^2 - 25|}{x^2 - 25}$

SOLUTION

(a) Since $f(x)$ is undefined at $x = 4$, the function is discontinuous there. For $x \neq 4$,

$$f(x) = \frac{(x + 4)(x - 4)}{x - 4} = x + 4.$$

Consequently, the graph of $f(x)$ is a straight line with a hole at $x = 4$ (Figure 2.27a). The computer-generated graph in Figure 2.27b does not display the discontinuity at $x = 4$; depending on the plot interval, the graph may, or may not, be accompanied by an error message about division by 0 at $x = 4$. Note that $\lim_{x \to 4} [(x^2 - 16)/(x - 4)]$ exists and is equal to 8.

FIGURE 2.27a

FIGURE 2.27b

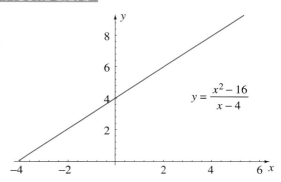

(b) Since $1/(x - 4)^2$ is undefined at $x = 4$, the function is discontinuous there. The limits

$$\lim_{x \to 4^+} \frac{1}{(x - 4)^2} = \infty, \quad \lim_{x \to 4^-} \frac{1}{(x - 4)^2} = \infty, \quad \lim_{x \to \infty} \frac{1}{(x - 4)^2} = 0, \quad \lim_{x \to -\infty} \frac{1}{(x - 4)^2} = 0$$

are displayed in Figure 2.28.

FIGURE 2.28

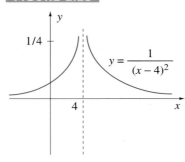

(c) The function $f(x) = |x^2 - 25|/(x^2 - 25)$ is undefined at $x = \pm 5$, and is therefore discontinuous at these values. When $-5 < x < 5$,

$$f(x) = \frac{25 - x^2}{x^2 - 25} = -1,$$

and when $|x| > 5$,

$$f(x) = \frac{x^2 - 25}{x^2 - 25} = 1.$$

The graph is shown in Figure 2.29a; Figure 2.29b is a computer version. Neither of the following limits exists,

$$\lim_{x \to 5} \frac{|x^2 - 25|}{x^2 - 25} \quad \text{or} \quad \lim_{x \to -5} \frac{|x^2 - 25|}{x^2 - 25},$$

although right-hand limits and left-hand limits exist as $x \to 5$ and $x \to -5$.

FIGURE 2.29a

FIGURE 2.29b

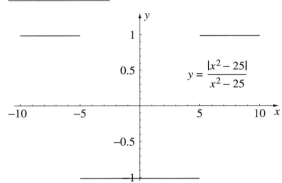

We now mention some general properties of continous functions. According to the following theorem, when continuous functions are added, subtracted, multiplied, and divided, the result is a continuous function.

THEOREM 2.2

If functions $f(x)$ and $g(x)$ are continuous at $x = a$, then so also are the functions $f(x) \pm g(x)$, $f(x)g(x)$, and $f(x)/g(x)$ (provided $g(a) \neq 0$ in the case of division).

This is easily established with Theorem 2.1. For instance, to verify that $f(x) + g(x)$ is continuous, we note that, since $\lim_{x \to a} f(x) = f(a)$ and $\lim_{x \to a} g(x) = g(a)$,

$$\lim_{x \to a} [f(x) + g(x)] = \lim_{x \to a} f(x) + \lim_{x \to a} g(x) \qquad \text{(by Theorem 2.1)}$$

$$= f(a) + g(a) \qquad \text{(by continuity of } f(x) \text{ and } g(x) \text{)}.$$

COROLLARY

If functions $f(x)$ and $g(x)$ are continuous on an interval, then so also are the functions $f(x) \pm g(x)$, $f(x)g(x)$, and $f(x)/g(x)$ (provided $g(x)$ never vanishes in the interval in the case of division).

It is an immediate consequence of this corollary that polynomials are continuous for all real numbers. Rational functions $P(x)/Q(x)$, where P and Q are polynomials, are continuous on intervals in which $Q(x) \neq 0$. The trigonometric functions $\sin x$ and $\cos x$ are continuous for all x; $\sec x$ and $\tan x$ have discontinuities: $x = (2n + 1)\pi/2$, where n is an integer; and $\csc x$ and $\cot x$ have discontinuities: $x = n\pi$ (see Figure 1.91, page 59). The exponential function a^x is continuous for all x, and $\log_a x$ is continuous for $x > 0$.

Besides adding, subtracting, multiplying, and dividing functions, functions can also be "composed," or substituted one into another. For instance, when $f(x) = \sqrt{x + 1/x}$ and $g(x) = x^2 + 3$, the function obtained by replacing each x in $f(x)$ by $g(x)$ is called the **composition** of f and g,

$$f\big(g(x)\big) = \sqrt{x^2 + 3 + \frac{1}{x^2 + 3}}.$$

Composition is often denoted by

$$(f \circ g)(x) = f\big(g(x)\big).$$

Were we required to find the limit of this function as $x \to 1$, say, we would probably write nonchalantly that

$$\lim_{x \to 1} \sqrt{x^2 + 3 + \frac{1}{x^2 + 3}} = \sqrt{4 + \frac{1}{4}} = \frac{\sqrt{17}}{2}.$$

Effectively, we have interchanged the operation of taking square roots and that of taking limits; that is, without thinking we have written

$$\lim_{x \to 1} \sqrt{x^2 + 3 + \frac{1}{x^2 + 3}} = \sqrt{\lim_{x \to 1} \left(x^2 + 3 + \frac{1}{x^2 + 3} \right)} = \sqrt{4 + \frac{1}{4}} = \frac{\sqrt{17}}{2}.$$

This is correct according to the following theorem because the square root function is continuous.

THEOREM 2.3

If $\lim_{x \to a} g(x) = L$, and $f(y)$ is a function that is continuous at $y = L$, then

$$\lim_{x \to a} f\big(g(x)\big) = f\left[\lim_{x \to a} g(x) \right] = f(L). \tag{2.5}$$

An immediate consequence of this result is that the composition of a continuous function with a continuous function yields a continuous function. We state this as a corollary.

COROLLARY

If $g(x)$ is continuous at $x = a$, and $f(y)$ is continuous at $g(a)$, then $f\big(g(x)\big)$ is continuous at $x = a$.

EXAMPLE 2.21

Evaluate

$$\lim_{x \to 2} \sin \left(\frac{x^2 - 2}{2x - 3} \right).$$

SOLUTION   Since the sine function is continuous for all values of its argument, we may take the limit operation inside the function and write

$$\lim_{x \to 2} \sin \left(\frac{x^2 - 2}{2x - 3} \right) = \sin \left[\lim_{x \to 2} \frac{x^2 - 2}{2x - 3} \right] = \sin 2.$$

You may recall that the Intermediate Value Theorem of Section 1.7 requires the function to be continuous. In particular, it states that when $f(x)$ is continuous for $a \leq x \leq b$ and $f(a)f(b) < 0$, there is at least one solution of $f(x) = 0$ between a and b. The function in Figure 2.30a has three solutions. Without continuity, we cannot be sure, in general, whether there are solutions. The function in Figure 2.30b is discontinuous at $x = c$, and there are no solutions of $f(x) = 0$. The function in Figure 2.30c is also discontinuous at $x = c$, but there are two solutions of $f(x) = 0$ between a and b.

FIGURE 2.30a

FIGURE 2.30b

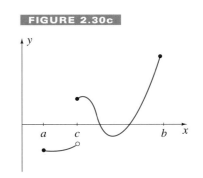

FIGURE 2.30c

Discontinuous functions are encountered in many applications of mathematics. Some are evident, others are hidden. An example of hidden discontinuities is contained in the phenomenon of radioactive disintegration, during which radioactive elements such as uranium-238 emit high-energy particles and in the process become new elements. If we analyze an ore sample that contains, along with various impurities, an amount A_0 of uranium, then disintegrations of uranium nuclei into thorium-234 nuclei gradually reduce the amount of uranium in the sample. Suppose we let $A = f(t)$ represent the amount of uranium in the sample at any given time, and choose time $t = 0$ when $A = A_0$. If we were to carefully measure the amount of uranium in the sample over a very short time interval (perhaps 10^{-10} second), a graph of $A = f(t)$ might appear as shown in Figure 2.31.

Each jump (discontinuity) in this graph signals the disintegration of a uranium nucleus into a thorium nucleus. Were we to complete this graph, and assume that no two disintegrations occur at exactly the same instant, then the total number of discontinuities would be equal to the original number of uranium nuclei in the sample. In practice, we cannot hope to make such infinitesimal measurements. Instead, we take measurements of the amount of uranium in the sample at much larger time intervals, and plot the data as a sequence of points, as in Figure 2.32a. If these points are joined with a smooth curve as in Figure 2.32b, we obtain a reasonable picture of A as a function of time. This macroscopic view (view in the large) of $A = f(t)$ suppresses the true nature of the graph, but it would be entirely adequate if we were interested in knowing, perhaps, when disintegrations would reduce the uranium to half its original amount. Questions such as these will be answered when we continue our discussion of radioactive disintegration in Section 9.2.

FIGURE 2.31

10^{-10} seconds

FIGURE 2.32a

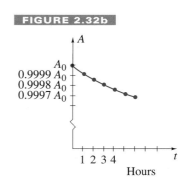

FIGURE 2.32b

A different kind of function arises in industrial economics. An electronics manufacturer sells memory chips at the following prices:

$40 per chip for the first 100 chips;
$35 per chip for each chip after 100 up to 1000;
$30 per chip for each chip after 1000.

If we denote the selling price of the xth chip by $f(x)$, then

$$f(x) = \begin{cases} 40, & 1 \leq x \leq 100 \\ 35, & 101 \leq x \leq 1000 \\ 30, & x \geq 1001. \end{cases}$$

The domain of this function consists of only the positive integers. Its "graph" in Figure 2.33a consists of dots above these integers. In economics, the domain of the function is extended to all positive values of x according to

$$f(x) = \begin{cases} 40, & 0 < x \leq 100 \\ 35, & 100 < x \leq 1000 \\ 30, & x > 1000, \end{cases}$$

with the resulting graph in Figure 2.33b.

FIGURE 2.33a

FIGURE 2.33b

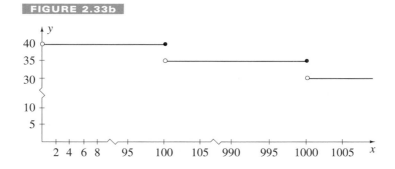

In spite of the fact that $f(x)$ is now defined for all $x > 0$, it still has significance only for x a positive integer. For instance, what sense does it make to say that $f(2.1) = 40$? You cannot buy 2.1 chips. The advantage of extending the domain of $f(x)$ to noninteger values is therefore not one of physical significance. The real reason is that calculus can be applied to the extended function in order to analyze the economics of chip sales. For the present, it is clear that the function $f(x)$ in Figure 2.33b is discontinuous at $x = 100$ and $x = 1000$.

We can derive from $f(x)$ a function that is very useful in economic theory: the average price $r(x)$. It is the average price of a chip when x chips are purchased. Using the fact that average price is total selling price divided by total number of chips, we calculate that

$$r(x) = \begin{cases} 40, & 0 < x \leq 100 \\[2mm] \dfrac{40(100) + 35(x - 100)}{x}, & 100 < x \leq 1000 \\[3mm] \dfrac{40(100) + 35(900) + 30(x - 1000)}{x}, & x > 1000 \end{cases}$$

$$= \begin{cases} 40, & 0 < x \leq 100 \\[2mm] 35 + \dfrac{500}{x}, & 100 < x \leq 1000 \\[3mm] 30 + \dfrac{5500}{x}, & x > 1000. \end{cases}$$

Again, $r(x)$ is meaningful only for x an integer, but as shown in Figure 2.34 we extend its domain to all positive values of x. In addition, note that, unlike $f(x)$, $r(x)$ has no discontinuities for $x > 0$.

FIGURE 2.34

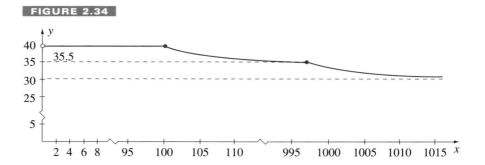

Heaviside Functions and Unit Impulse Functions

One of the simplest, but at the same time most useful, functions in engineering and physics is the Heaviside unit step function; it contains explicit discontinuities. It is defined as follows with graph in Figure 2.35.

$$H(x) = \begin{cases} 0, & x < 0 \\ 1, & x > 0 \end{cases}. \tag{2.6}$$

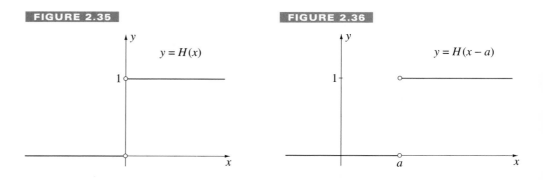

FIGURE 2.35

$y = H(x)$

FIGURE 2.36

$y = H(x - a)$

We have drawn small circles at $x = 0$ to indicate that the function does not have a value there; it is discontinuous at $x = 0$. Closely related to this function is that in Figure 2.36; the jump from value 0 to value 1 takes place at $x = a$ rather than at $x = 0$. Since this simply shifts the graph in Figure 2.35 by a units to the right, it is customary to denote this function by $H(x - a)$. Algebraically we have

$$H(x - a) = \begin{cases} 0, & x < a \\ 1, & x > a \end{cases}. \tag{2.7}$$

For example, consider the Heaviside function $H(x - 4)$. To evaluate this function at $x = 6$ say, we have two choices. Because $6 > 4$, definition 2.7 implies that the value of the function at $x = 6$ is 1. Alternatively, we can substitute $x = 6$ into $H(x - 4)$ to get $H(6 - 4) = H(2)$, and this is equal to 1 by 2.6.

The product of $H(x - a)$ with any other function $f(x)$ results in a function $f(x)H(x - a)$ whose values are equal to those of $f(x)$ for $x > a$, but whose values are 0 for $x < a$. For example, if $f(x) = x^2 + 2$ (Figure 2.37a), a graph of $f(x)H(x + 1) = (x^2 + 2)H(x + 1)$ is shown in Figure 2.37b.

FIGURE 2.37a

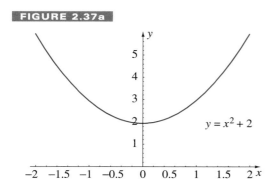

$y = x^2 + 2$

FIGURE 2.37b

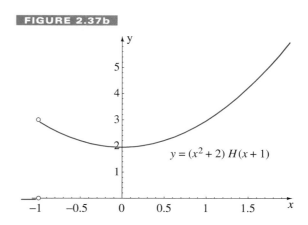

$y = (x^2 + 2)H(x + 1)$

Think of $H(x - a)$ as a switch that turns values of a function $f(x)$ on for $x > a$. For the function $f(x)$ in Figure 2.38a, $f(x)H(x - a)$ is shown in Figure 2.38b. Values are 0 for $x < a$ and those of $f(x)$ for $x > a$.

FIGURE 2.38a

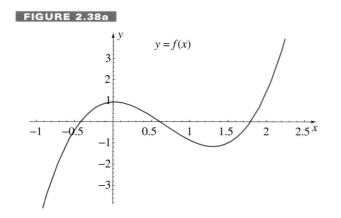

$y = f(x)$

FIGURE 2.38b

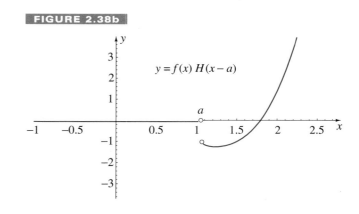

$y = f(x)H(x - a)$

The function $H(x - a) - H(x - b)$ for $b > a$ is shown in Figure 2.39. Algebraically, we have

$$H(x - a) - H(x - b) = \begin{cases} 0, & x < a \\ 1, & a < x < b \\ 0, & x > b \end{cases}. \qquad (2.8)$$

When multiplying $f(x)$ in Figure 2.38a, it yields the function $f(x)[H(x - a) - H(x - b)]$ with graph in Figure 2.40.

FIGURE 2.39

FIGURE 2.40

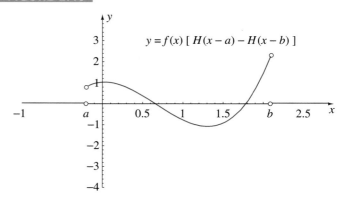

Think of $H(x - a) - H(x - b)$ as an on-off switch. It turns values of other functions on for $x > a$ and off again for $x > b$.

Heaviside functions provide a convenient representation for functions that have different definitions on different intervals. For instance, the function in Figure 2.41 is defined as follows

$$f(x) = \begin{cases} 3x + 3, & x < 0 \\ 4 - x^2, & 0 < x < 2 \\ 2, & x > 2 \end{cases}.$$

FIGURE 2.41

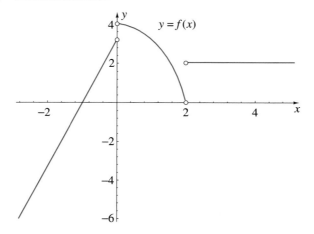

It can be represented algebraically in the form

$$f(x) = (3x + 3)[1 - H(x)] + (4 - x^2)[H(x) - H(x - 2)] + 2H(x - 2)$$

$$= 3x + 3 + (-3x - 3 + 4 - x^2)H(x) + (x^2 - 4 + 2)H(x - 2)$$

$$= 3x + 3 + (1 - 3x - x^2)H(x) + (x^2 - 2)H(x - 2).$$

For a physical example, consider a two-stage rocket wherein the rocket burns fuel during the first stage, jettisons the thrusters, burns fuel in the second stage, achieves orbit, and turns the engines off. Suppose the initial mass of the rocket is M_0 kilograms, of which M_1 kilograms is the first stage. This consists of the mass of the thrusters and fuel. If the thrusters burn fuel at a constant rate of f_1 kilograms per second for 60 seconds, the mass of the rocket during the first stage is

$$M_0 - f_1 t, \qquad 0 < t < 60.$$

After 1 minute, the thrusters are jettisoned, and the remaining engines ignite and burn fuel at a constant rate of f_2 kilograms per second for the next 4 minutes. The mass of the rocket during this time interval is

$$M_0 - M_1 - f_2(t - 60), \qquad 60 < t < 300.$$

Finally, when the rocket achieves orbit and the engines are shut down, the mass of the rocket is

$$M_0 - M_1 - 240 f_2, \qquad t > 300.$$

If $M(t)$ represents the mass of the rocket at time t, it can be represented in terms of Heaviside functions as follows

$$M(t) = (M_0 - f_1 t)[H(t) - H(t - 60)]$$

$$+ [M_0 - M_1 - f_2(t - 60)][H(t - 60) - H(t - 300)]$$

$$+ (M_0 - M_1 - 240 f_2)H(t - 300).$$

By recombining this expression into terms involving $H(t)$, $H(t - 60)$, and $H(t - 300)$, we can also write

$$M(t) = (M_0 - f_1 t)H(t) - [M_1 - 60 f_2 + (f_2 - f_1)t]H(t - 60)$$

$$+ f_2(t - 300)H(t - 300).$$

A graph of this function is shown in Figure 2.42. It has a hole at $t = 300$ because of the fact that $H(t - 300)$ is undefined there. We would simply agree to fill it in.

FIGURE 2.42

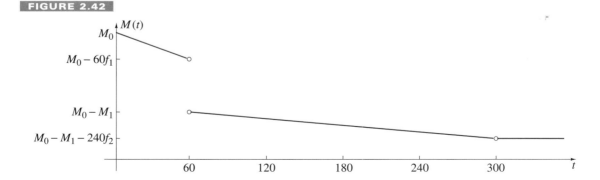

There are many physical examples in which Heaviside functions are very useful. Consider a mass m attached to a spring as shown in Figure 2.43. Motion is initiated at time $t = 0$ by pulling the mass away from its equilibrium position and releasing it. During the subsequent motion, various forces could act on the mass including the spring, friction with the surface on which it slides, air resistance, and others. Suppose that among the others, a force to the left with magnitude 5 N is applied for 3 s beginning at time $t = 1$ s. This force can be represented algebraically as

$$F(t) = -5[H(t-1) - H(t-4)].$$

FIGURE 2.43

The generator in Figure 2.44, were it operational from time $t = 0$, would produce an oscillating voltage $A \sin \omega t$ where A and ω are positive constants. If the generator is indeed turned on at time $t = 0$, by closing the switch, and off again after 10 s by opening the switch, the voltage applied to the circuit can be expressed in the form

$$E(t) = A \sin \omega t \, [H(t) - H(t - 10)].$$

FIGURE 2.44

The beam in Figure 2.45 is made more rigid by attaching a second beam over the middle half. This extra support creates additional loading on the original beam. If the support has mass per unit length m, then its weight per unit length is $-9.81m$, and the extra loading per unit length can be expressed algebraically as

$$F(x) = -9.81m[H(x - L/4) - H(x - 3L/4)].$$

FIGURE 2.45

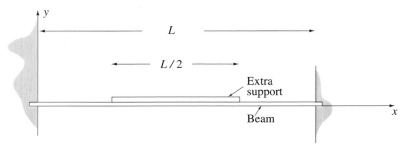

When the area of the rectangle formed by drawing vertical lines at $x = a$ and $x = b$ in Figure 2.39 (page 128) is equal to 1, the height of the horizontal line is $1/(b - a)$ (Figure 2.46a). If b is replaced by $a + \epsilon$ so that ϵ is the width of the nonzero portion of the graph, then the height of the horizontal line must be $1/\epsilon$ in order to create a rectangle of unit area (Figure 2.46b).

FIGURE 2.46a

FIGURE 2.46b

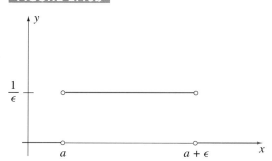

The functions describing these graphs are

$$\frac{1}{b-a}[H(x-a)-H(x-b)], \qquad \frac{1}{\epsilon}[H(x-a)-H(x-a-\epsilon)].$$

Suppose we make ϵ in Figure 2.46b smaller and smaller, so that $a+\epsilon \to a$ and the height of the horizontal line moves upward (in order that the area under the curve will always be 1). A few smaller values of ϵ are shown in Figure 2.47.

FIGURE 2.47

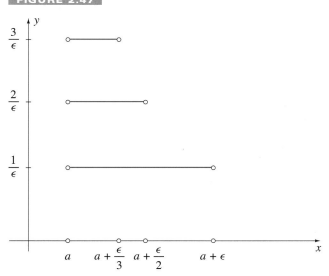

If we take the limit as $\epsilon \to 0$, we obtain what is called the unit impulse function at $x = a$ or the Dirac-delta function. It is denoted by

$$\delta(x-a) = \lim_{\epsilon \to 0} \frac{1}{\epsilon}[H(x-a)-H(x-a-\epsilon)]. \qquad (2.9)$$

Its value is zero for every $x \neq a$, and somehow its value at $x = a$ is "equal to infinity." Clearly this is not a function as we understand functions, since functions cannot have "infinite values." It is called a *generalized function*. Generalized functions have operational properties that make them very useful in handling "point sources" in engineering and physics. For instance, suppose we were to strike the left side of the vibrating mass in Figure 2.43 with a hammer at time $t = t_0$. This would be equivalent to applying a force over a very short time interval. If it were deemed that striking the mass is equivalent not to a 10 N force for 1 s, nor a 20 N force for 1/2 s, nor a 40 N force for 1/4 s, but to the limit of this sequence of forces, we would say that the mass has been struck with an impulse force of 10 N at $t = t_0$, and express the force in the form

$$F(t) = 10\,\delta(t - t_0).$$

If we were to take a sharp object, place it at the centre of the beam in Figure 2.45, and push downward with a force of 200 N, we would represent this force as

$$F(x) = -200\,\delta(x - L/2).$$

It is a point force of 200 N at $x = L/2$.

A unit impulse voltage at time $t = t_0$ for the circuit in Figure 2.44 is $\delta(t - t_0)$. It is equivalent to applying a voltage not equivalent to 1 V for 1 s, nor 2 V for 1/2 s, nor 4 V for 1/4 s, but to a voltage equivalent to the limit of such voltages.

EXERCISES 2.4

In Exercises 1–30 plot a graph of the function indicating any discontinuities.

* **1.** $f(x) = \dfrac{1}{x + 2}$

* **2.** $f(x) = \dfrac{16 - x^2}{x + 4}$

* **3.** $f(x) = |x^2 - 5|$

* **4.** $f(x) = \dfrac{12}{x^2 + 2}$

* **5.** $f(x) = \dfrac{12}{x^2 + 2x}$

* **6.** $f(x) = \dfrac{12}{x^2 + 2x + 2}$

* **7.** $f(x) = \dfrac{3 + 2x - x^2}{x + 1}$

* **8.** $f(x) = \dfrac{x^3 + x^2 - 2x}{x^2 - x}$

* **9.** $f(x) = \dfrac{x^3 - 2x^2 + 5x - 10}{x - 2}$

* **10.** $f(x) = \tan x$

* **11.** $f(x) = \sec 2x$

* **12.** $f(x) = \sin(1/x)$

* **13.** $f(x) = x^2 \sin(1/x)$

* **14.** $f(x) = \dfrac{x + 12}{x^2 - 9}, \quad -5 \le x \le 4$

* **15.** $f(x) = \dfrac{x^2 + 2x}{x^2 - 9}, \quad -3 \le x \le 6$

* **16.** $f(x) = \dfrac{x^3 - 27}{|x - 3|}$

* **17.** $f(x) = \dfrac{x}{x^2 - 1}$

* **18.** $f(x) = \dfrac{3x + 2}{x^2 - x - 2}$

* **19.** $f(x) = \dfrac{x^2 - 3x + 2}{x^2 + 4x - 5}$

* **20.** $f(x) = \dfrac{3x^2 - 6x}{x^2 - 6x - 7}$

* **21.** $f(x) = \dfrac{x^2 + 3x + 2}{x + 4}$

* **22.** $f(x) = \dfrac{x^2 - 2x + 4}{x - 1}$

** **23.** $f(x) = \dfrac{1}{\sqrt{x - 1}}$

** **24.** $f(x) = \dfrac{1}{\sqrt{5 + x}}$

* **25.** $f(x) = \dfrac{1}{x^3 - 4x}$

* **26.** $f(x) = \dfrac{2x^3 - 2}{x^3 + 5x^2}$

* **27.** $f(x) = \dfrac{1}{x^4 + 3x^2}$

* **28.** $f(x) = \dfrac{|3x + 1|}{x + 5}$

** 29. $f(x) = \dfrac{1}{\sqrt{5-x}}$

** 30. $f(x) = \sqrt{\dfrac{x-3}{x+2}}$

In Exercises 31–36 express the function in terms of Heaviside functions. Draw a graph of each function.

** 31. $f(x) = \begin{cases} 1, & x < 0 \\ 2 - x, & 0 < x < 2 \\ 2, & x > 2 \end{cases}$

** 32. $f(x) = \begin{cases} 1 - x^2, & x < 0 \\ x^2, & x > 0 \end{cases}$

** 33. $f(x) = \begin{cases} \sin x, & x < 0 \\ 2 + 2\cos x, & x > 0 \end{cases}$

** 34. $f(x) = \begin{cases} 0, & x < -1 \\ x - 2, & -1 < x < 3 \\ x - 4, & 3 < x < 5 \\ 0, & x > 5 \end{cases}$

** 35. $f(x) = \begin{cases} 0, & x < 0 \\ x, & 0 \le x \le 1 \\ 1 - x, & 1 < x \le 2 \\ 0, & x > 2 \end{cases}$

** 36. $f(x) = \begin{cases} 0, & x < 0 \\ \sin x, & 0 \le x \le \pi \\ 2\sin(x - \pi), & \pi < x \le 2\pi \\ 3\sin(x - 2\pi), & 2\pi < x \le 3\pi \\ 4\sin(x - 3\pi), & 3\pi < x < 4\pi \\ 0, & x \ge 4\pi \end{cases}$

** 37. Is the function

$$f(x) = \begin{cases} x\sin(1/x), & x \ne 0 \\ 0, & x = 0 \end{cases}$$

continuous at $x = 0$?

** 38. If $x = f(t)$ represents the displacement function of a particle moving along the x-axis, can $f(t)$ have discontinuities? Explain with graphs.

** 39. A function $f(x)$ is said to have a *removable discontinuity* at $x = a$ if $\lim_{x \to a} f(x)$ exists, but either $f(a)$ is not defined or $\lim_{x \to a} f(x) \ne f(a)$. Which of the discontinuities in Exercises 1–30 are removable?

** 40. Illustrate graphically that the function $f(x) = x^{-1}\sin x$ has a removable discontinuity at $x = 0$.

** 41. Verify that the condition

$$\lim_{h \to 0} f(a + h) = f(a)$$

is equivalent to 2.4.

** 42. The number of bacteria in a culture is a function of time t, say, $N = f(t)$. Make two graphs of this function, one "in the small" to show birth and death situations, and a second "in the large" to get an overview of the rate of growth of the culture.

* **43.** The *signum function* (or *sign function*), denoted by sgn, is defined by

$$\operatorname{sgn} x = \begin{cases} -1, & x < 0 \\ 0, & x = 0 \\ 1, & x > 0. \end{cases}$$

Draw its graph, indicating any discontinuities.

** **44.** Draw a graph of the function $f(x) = \operatorname{sgn}(x + 1) - \operatorname{sgn}(x - 1)$, indicating any discontinuities. See Exercise 43.

** **45.** Where is the greatest integer function $[\![x]\!]$ of Exercise 50 in Section 1.3 (page 41) discontinuous?

** **46.** (a) Draw a graph of the function $f(x) = [\![10x]\!]/10$. Where is the function discontinuous?

(b) Prove that $f(x)$ truncates positive numbers after the first decimal.

** **47.** What function truncates negative numbers after two decimals?

** **48.** (a) Draw a graph of the function $f(x) = [\![x + 1/2]\!]$. Where is the function discontinuous?

(b) Prove that $f(x)$ rounds positive numbers to the nearest integer.

** **49.** What function rounds positive numbers to:

(a) the nearest tenth;

(b) the nearest hundredth;

(c) 10^{-n} where n is a positive integer?

** **50.** A pump manufacturer has two pricing policies for one of his pumps:

Policy I: $15.00 each for the first 10 pumps,
$12.50 for each pump after 10 up to 100,
$10.00 for each pump after 100;

Policy II: $15.00 each for the first 10 pumps,
10% off the total price on orders of 11 or more,
20% off the total price on orders of 101 or more.

Draw average price functions for each of these policies on the same graph. Which policy is better for a retailer?

** **51.** A retailer of building supplies sells sand by the bag and in bulk. He quotes prices as follows:

$3.00/50 kg bag up to 20 bags,
$50.00/t delivered only in integer tonnes.

If, for example, a customer wants 2.2 t of sand, he receives 2 t in bulk and 4 bags. Find an algebraic expression for the average price function and draw its graph. Hint: The greatest integer function may be useful.

*** **52.** Determine points of continuity, if any, for the function

$$f(x) = \begin{cases} 1, & x \text{ is a rational number} \\ 0, & x \text{ is an irrational number.} \end{cases}$$

*** **53.** Determine points of continuity, if any, for the function

$$f(x) = \begin{cases} x, & x \text{ is a rational number} \\ 0, & x \text{ is an irrational number.} \end{cases}$$

2.5 A Mathematical Definition of Limits

Our work on limits has been intuitive, but some later topics in this book require a precise definition of limits. To obtain such a definition, we begin with our intuitive statement of a limit and make a succession of paraphrases, each of which is one step closer to the definition. We do this because the definition of a limit is at first sight quite overwhelming, and we wish to show that it can be obtained by a fairly straightforward sequence of steps. We hope that this will be a convincing argument that the definition of a limit does indeed describe in mathematical terms the concept of a limit. Our intuitive statement is

- A function $f(x)$ has limit L as x approaches a if $f(x)$ can be made arbitrarily close to L by choosing x sufficiently close to a.

  Next we paraphrase "$f(x)$ can be made arbitrarily close to L."

- A function $f(x)$ has limit L as x approaches a if the difference $|f(x) - L|$ can be made arbitrarily close to zero by choosing x sufficiently close to a.

  Now we take the important step—make "$|f(x) - L|$ can be made arbitrarily close to zero" a mathematical statement.

- A function $f(x)$ has limit L as x approaches a if given any real number $\epsilon > 0$, no matter how small, we can make the difference $|f(x) - L|$ less than ϵ by choosing x sufficiently close to a.

  Penultimately, we paraphrase "by choosing x sufficiently close to a."

- A function $f(x)$ has limit L as x approaches a if, given any $\epsilon > 0$, we can make $|f(x) - L| < \epsilon$ by choosing $|x - a|$ sufficiently close to zero.

  Finally by making "choosing $|x - a|$ sufficiently close to zero" precise, we arrive at the definition of a limit.

> **DEFINITION 2.2**
>
> A function $f(x)$ has limit L as x approaches a if, given any $\epsilon > 0$, we can find a $\delta > 0$ such that
> $$|f(x) - L| < \epsilon$$
> whenever $0 < |x - a| < \delta$.

Notice that by requiring $0 < |x - a|$, this definition states explicitly that, as far as limits are concerned, the value of $f(x)$ at $x = a$ is irrelevant. In taking limits we consider values of x closer and closer to a, but we do not consider the value of $f(x)$ at $x = a$. This definition states in precise terms our intuitive idea of a limit: that $f(x)$ can be made arbitrarily close to L by choosing x sufficiently close to a. A graphical interpretation may help to consolidate the definition. Figure 2.48a indicates a function that has limit L as x approaches a. Let us illustrate what must be done to verify algebraically that $\lim_{x \to a} f(x) = L$. We suppose that we are given a value $\epsilon > 0$, which we should envisage as being very small, but we are never told exactly what it is. We must show that x can be restricted to $0 < |x - a| < \delta$ so that $|f(x) - L| < \epsilon$. The latter inequality is equivalent to $-\epsilon < f(x) - L < \epsilon$ or $L - \epsilon < f(x) < L + \epsilon$, and this describes a horizontal band of width 2ϵ around the line $y = L$ (shaded in Figure 2.48b). What Definition 2.2 requires is that we find an interval of width 2δ around $x = a$, as $|x - a| < \delta$ is equivalent to $a - \delta < x < a + \delta$, such that, whenever x is in this interval, the values of $f(x)$ are all within the shaded horizontal band around $y = L$. Such an interval is shown in Figure 2.48c for the given ϵ. Now Definition 2.2 requires us to verify that the δ-interval can always be found no matter how small ϵ is chosen. This is always possible for the function illustrated in Figure 2.48, and it is clear that the smaller the given value of ϵ the smaller δ will have to be chosen. For instance, for the value of ϵ in Figure 2.48d, δ is smaller than that in Figure 2.48c. In other words, the value of δ depends on the value of ϵ. Herein lies the difficulty in using Definition 2.2. We must determine precisely how δ depends on ϵ in order to ensure that δ can always be found. We illustrate with two examples.

FIGURE 2.48a

FIGURE 2.48b

FIGURE 2.48c

FIGURE 2.48d

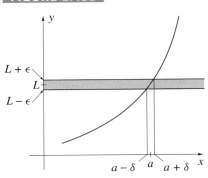

Consider proving first that

$$\lim_{x \to 3} (2x + 4) = 10.$$

Now, it is true that, based on Theorem 2.1, this is quite obvious, but remember, this theorem has not yet been verified. We want to verify the limit assuming that Theorem 2.1 is not at our disposal; verification must be through Definition 2.2. We must show that, given any $\epsilon > 0$, there is a $\delta > 0$ such that

$$|(2x + 4) - 10| < \epsilon$$

whenever $0 < |x - 3| < \delta$; i.e., we must find a δ. Since

$$|2x + 4 - 10| = |2x - 6| = 2|x - 3|,$$

we must find a δ such that

$$2|x - 3| < \epsilon$$

whenever $0 < |x - 3| < \delta$. But this will be true if $\delta = \epsilon/2$; in other words, if we choose x to satisfy $0 < |x - 3| < \delta = \epsilon/2$, then

$$|(2x + 4) - 10| = 2|x - 3| < 2(\epsilon/2) = \epsilon.$$

The verification is now complete. We have shown that we can make $2x + 4$ as close to 10 as we want (within ϵ) by choosing x sufficiently close to 2 (within $\delta = \epsilon/2$).

Verifying that

$$\lim_{x \to 2} (x^2 + 5) = 9$$

is somewhat more complicated, but the idea is the same. We must show that given any $\epsilon > 0$, there is a $\delta > 0$ such that

$$|(x^2 + 5) - 9| < \epsilon \qquad \text{or} \qquad |x^2 - 4| < \epsilon$$

whenever $0 < |x - 2| < \delta$; i.e., we must find a δ. By working backward from the ϵ-inequality we can see how to choose δ. In addition, because the δ-inequality involves x in the combination $x - 2$, we rework the ϵ-inequality into the same combination. Since

$$|x^2 - 4| = |(x - 2)^2 + 4x - 8| = |(x - 2)^2 + 4(x - 2)|,$$

we must choose δ such that

$$|(x - 2)^2 + 4(x - 2)| < \epsilon$$

whenever $0 < |x - 2| < \delta$. Now, all real numbers a and b satisfy the inequality

$$|a + b| \le |a| + |b|. \tag{2.10}$$

With a replaced by $(x - 2)^2$ and b replaced by $4(x - 2)$, we can say that

$$|(x - 2)^2 + 4(x - 2)| \le |x - 2|^2 + 4|x - 2|.$$

As a result, if we can find a $\delta > 0$ such that

$$|x - 2|^2 + 4|x - 2| < \epsilon$$

whenever $0 < |x - 2| < \delta$, then we will have a satisfactory δ. But now we have the key. Suppose we set $z = |x - 2|$ and consider the parabola $Q(z) = z^2 + 4z - \epsilon$ in Figure 2.49.

FIGURE 2.49

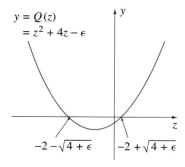

It crosses the z-axis when

$$z^2 + 4z - \epsilon = 0,$$

a quadratic with solutions

$$z = \frac{-4 \pm \sqrt{16 + 4\epsilon}}{2} = -2 \pm \sqrt{4 + \epsilon}.$$

The graph shows that whenever $0 < z < -2 + \sqrt{4 + \epsilon}$, then

$$z^2 + 4z - \epsilon < 0.$$

In other words, if $0 < z < \delta = \sqrt{4 + \epsilon} - 2$, then

$$z^2 + 4z < \epsilon.$$

Since $z = |x - 2|$, we can say that if $0 < |x - 2| < \delta = \sqrt{4 + \epsilon} - 2$, then

$$|x - 2|^2 + 4|x - 2| < \epsilon,$$

and therefore

$$|(x - 2)^2 + 4(x - 2)| < \epsilon.$$

The verification is now complete. We have shown that we can make $x^2 + 5$ as close to 9 as we want (within ϵ) by choosing x sufficiently close to 2 (within $\delta = \sqrt{4 + \epsilon} - 2$).

    These examples have indicated the difference between "evaluation of" and "verification of" a limit. If you have learned only an appreciation for this difference, then our discussion has been worthwhile. If your comprehension goes beyond this and includes a working understanding of the definition of a limit, we could hope for no more. Results in succeeding chapters rely heavily on our intuitive understanding of limits and the ability to calculate limits, but use of Definition 2.2 is kept to a minimum.

EXERCISES 2.5

In Exercises 1–9 use Definition 2.2 to verify the limit.

** **1.** $\lim\limits_{x \to 1} (x + 5) = 6$ ** **2.** $\lim\limits_{x \to 2} (2x - 3) = 1$

** **3.** $\lim\limits_{x \to 0} (x^2 + 3) = 3$ ** **4.** $\lim\limits_{x \to 1} (x^2 + 4) = 5$

** **5.** $\lim\limits_{x \to -2} (3 - x^2) = 1$ ** **6.** $\lim\limits_{x \to 3} (x^2 - 7x) = -12$

** **7.** $\lim\limits_{x \to -1} (x^2 - 3x + 4) = 7$ ** **8.** $\lim\limits_{x \to 1} (x^2 + 3x + 5) = 9$

** **9.** $\lim\limits_{x \to 2} \dfrac{x + 2}{x - 1} = 4$

In Exercises 10–19 give a mathematical definition for each statement.

** **10.** $\lim\limits_{x \to a^+} f(x) = L$ ** **11.** $\lim\limits_{x \to a^-} f(x) = L$

** **12.** $\lim\limits_{x \to \infty} f(x) = L$ ** **13.** $\lim\limits_{x \to -\infty} f(x) = L$

** **14.** $\lim\limits_{x \to a} f(x) = \infty$ ** **15.** $\lim\limits_{x \to a} f(x) = -\infty$

** **16.** $\lim\limits_{x \to \infty} f(x) = \infty$ ** **17.** $\lim\limits_{x \to \infty} f(x) = -\infty$

** **18.** $\lim\limits_{x \to -\infty} f(x) = \infty$ ** **19.** $\lim\limits_{x \to -\infty} f(x) = -\infty$

** **20.** Use Definition 2.2 to prove that a function $f(x)$ cannot have two limits as x approaches a.

In Exercises 21–28 use the appropriate definition from Exercises 10–19 to verify the limit.

** **21.** $\displaystyle\lim_{x \to 1} \frac{1}{(x-1)^2} = \infty$

** **22.** $\displaystyle\lim_{x \to -2} \frac{-1}{(x+2)^2} = -\infty$

** **23.** $\displaystyle\lim_{x \to \infty} (x+5) = \infty$

** **24.** $\displaystyle\lim_{x \to \infty} (5 - x^2) = -\infty$

** **25.** $\displaystyle\lim_{x \to \infty} \frac{x+2}{x-1} = 1$

** **26.** $\displaystyle\lim_{x \to -\infty} \frac{x+2}{x-1} = 1$

** **27.** $\displaystyle\lim_{x \to -\infty} (5 - x) = \infty$

** **28.** $\displaystyle\lim_{x \to -\infty} (3 + x - x^2) = -\infty$

** **29.** Prove that, if $\lim_{x \to a} f(x) = L > 0$, then there exists an open interval I containing a in which $f(x) > 0$ except possibly at $x = a$.

** **30.** Does the function

$$f(x) = \begin{cases} x \sin(1/x) & x \neq 1/(n\pi) \\ 1 & x = 1/(n\pi), \end{cases}$$

where n is an integer, have a limit as $x \to 0$?

In Exercises 31–35 we use Definition 2.2 to prove Theorem 2.1. In each exercise assume that $\lim_{x \to a} f(x) = F$ and that $\lim_{x \to a} g(x) = G$.

** **31.** Show that, given any $\epsilon > 0$, there exist numbers $\delta_1 > 0$ and $\delta_2 > 0$ such that

$$|f(x) - F| < \epsilon/2 \qquad \text{whenever } 0 < |x - a| < \delta_1 \qquad \text{and}$$

$$|g(x) - G| < \epsilon/2 \qquad \text{whenever } 0 < |x - a| < \delta_2.$$

Use these results along with identity 2.10 to prove part (i) of Theorem 2.1.

** **32.** Use a proof similar to that in Exercise 31 to verify part (ii) of Theorem 2.1.

** **33.** Verify part (iii) of Theorem 2.1.

*** **34.** (a) Verify that

$$|f(x)g(x) - FG| \leq |f(x)||g(x) - G| + |G||f(x) - F|.$$

(b) Show that given any $\epsilon > 0$, there exist numbers $\delta_1 > 0$, $\delta_2 > 0$, and $\delta_3 > 0$ such that

$$|f(x)| < |F| + 1 \qquad \text{whenever } 0 < |x - a| < \delta_1,$$

$$|g(x) - G| < \frac{\epsilon}{2(|F| + 1)} \qquad \text{whenever } 0 < |x - a| < \delta_2, \qquad \text{and}$$

$$|f(x) - F| < \frac{\epsilon}{2|G| + 1} \qquad \text{whenever } 0 < |x - a| < \delta_3.$$

(c) Use these results to prove part (iv) of Theorem 2.1.

*** **35.** (a) Verify that when $G \neq 0$,

$$\left| \frac{f(x)}{g(x)} - \frac{F}{G} \right| \leq \frac{|f(x) - F|}{|g(x)|} + \frac{|F||G - g(x)|}{|G||g(x)|}.$$

(b) Show that, given any $\epsilon > 0$, there exist numbers $\delta_1 > 0$, $\delta_2 > 0$, and $\delta_3 > 0$ such that

$$|g(x)| > \frac{|G|}{2} \quad \text{whenever } 0 < |x - a| < \delta_1,$$

$$|f(x) - F| < \frac{\epsilon|G|}{4} \quad \text{whenever } 0 < |x - a| < \delta_2, \quad \text{and}$$

$$|g(x) - G| < \frac{\epsilon|G|^2}{4|F| + 1} \quad \text{whenever } 0 < |x - a| < \delta_3.$$

(c) Now prove part (v) of Theorem 2.1.

SUMMARY

In Section 2.1 we introduced limits of functions. For the most part our discussion was intuitive, beginning with the statement "$\lim_{x \to a} f(x) = L$ if $f(x)$ can be made arbitrarily close to L by choosing x sufficiently close to a." This idea was then extended to include the following:

Right-hand limits:	$\lim_{x \to a^+} f(x)$;
Left-hand limits:	$\lim_{x \to a^-} f(x)$;
Limits at infinity:	$\lim_{x \to \infty} f(x), \quad \lim_{x \to -\infty} f(x)$;
Infinite limits:	$\lim_{x \to a} f(x) = \infty, \quad \lim_{x \to a} f(x) = -\infty.$

Keep in mind that the term "infinite limits" is somewhat of a misnomer since in both situations the limit does not exist.

A function $f(x)$ is continuous at a point $x = a$ if

$$\lim_{x \to a} f(x) = f(a).$$

To be continuous at $x = a$, a function must be defined at $x = a$, have a limit as $x \to a$, and these numbers must be the same. The function is continuous on an interval if it is continuous at each point of that interval. Geometrically, this means that one must be able to completely trace its graph without lifting pencil from page.

We illustrated that discontinuous functions such as the Heaviside and Dirac-delta functions model many physical situations.

In Section 2.5 we developed the mathematical definition of limits.

KEY TERMS

In reviewing this chapter, you should be able to define or discuss the following key terms:

Limit	Oblique asymptote
Right-hand limit	Continuity at a point
Left-hand limit	Continuity on an interval
Infinite limit	Composite function
Vertical asymptote	Heaviside function
Limit at infinity	Dirac-delta function
Horizontal asymptote	

REVIEW EXERCISES

In Exercises 1–20 evaluate the limit, if it exists.

* **1.** $\displaystyle\lim_{x \to 1} \frac{x^2 - 2x}{x + 5}$

* **2.** $\displaystyle\lim_{x \to -1} \frac{x^2 - 1}{x + 1}$

* **3.** $\displaystyle\lim_{x \to -2} \frac{x^2 + 4x + 4}{x + 3}$

* **4.** $\displaystyle\lim_{x \to \infty} \frac{x + 5}{x - 3}$

* **5.** $\displaystyle\lim_{x \to -\infty} \frac{x^2 + 3x + 2}{2x^2 - 5}$

* **6.** $\displaystyle\lim_{x \to -\infty} \frac{5 - x^3}{3 + 4x^3}$

* **7.** $\displaystyle\lim_{x \to \infty} \frac{3x^3 + 2x - 5}{x^2 + 5x}$

* **8.** $\displaystyle\lim_{x \to \infty} \frac{4 - 3x + x^2}{3 + 5x^3}$

* **9.** $\displaystyle\lim_{x \to 2^+} \frac{x^2 - 2x}{x^2 + 2x}$

* **10.** $\displaystyle\lim_{x \to 2^-} \frac{x^2 - 4x + 4}{x - 2}$

* **11.** $\displaystyle\lim_{x \to 0} \frac{x^2 + 2x}{3x - 2x^2}$

* **12.** $\displaystyle\lim_{x \to 1} \frac{x^2 + 5x}{(x - 1)^2}$

* **13.** $\displaystyle\lim_{x \to 1} \frac{\sqrt{x} - 1}{x}$

* **14.** $\displaystyle\lim_{x \to 1} \frac{\sqrt{x} - 1}{x - 1}$

* **15.** $\displaystyle\lim_{x \to 1/2} \frac{(2 - 4x)^3}{x(2x - 1)^2}$

** **16.** $\displaystyle\lim_{x \to \infty} \frac{\cos 5x}{x}$

** **17.** $\displaystyle\lim_{x \to -\infty} x \sin x$

** **18.** $\displaystyle\lim_{x \to -\infty} \frac{\sqrt{3x^2 + 4}}{2x + 5}$

** **19.** $\displaystyle\lim_{x \to \infty} \frac{\sqrt{3x^2 + 4}}{2x + 5}$

** **20.** $\displaystyle\lim_{x \to \infty} \left(\sqrt{2x + 1} - \sqrt{3x - 1} \right)$

In Exercises 21–32 draw or plot a graph of the function, indicating any discontinuities.

* **21.** $f(x) = \dfrac{1}{x - 2}$

* **22.** $f(x) = \dfrac{x}{x - 2}$

* **23.** $f(x) = \dfrac{x^2}{x - 2}$

* **24.** $f(x) = \dfrac{x^2 - 36}{x - 6}$

∗ **25.** $f(x) = \dfrac{x+1}{x-1}$

∗ **26.** $f(x) = \left|\dfrac{x+1}{x-1}\right|$

∗ **27.** $f(x) = \dfrac{|x+1|}{x-1}$

∗ **28.** $f(x) = \dfrac{x+1}{|x-1|}$

∗ **29.** $f(x) = \dfrac{2x}{x^2-3x-4}$

∗ **30.** $f(x) = \dfrac{2x^2}{x^2-3x-4}$

∗ **31.** $f(x) = \dfrac{x^3-3x^2+3x-1}{x-1}$

∗ **32.** $f(x) = \dfrac{x^3-3x^2+3x-1}{x^2-2x+1}$

∗∗ **33.** For what values of x is the function $f(x) = [\![x^2]\!]$ discontinuous? Draw its graph.

Differentiation

Chapters 1 and 2 have prepared the way for calculus. The functions and curves in these chapters yield a wealth of examples for our discussions, and limits from Chapter 2 provide the tool by which calculus is developed. Calculus has two major components, *differentiation* and *integration*. In this chapter we study differentiation. We learn what a *derivative* is algebraically and geometrically, and develop some of its properties; we learn how to differentiate polynomials, and rational, trigonometric, exponential, and logarithm functions; and we see glimpses of the applications that are to follow in Chapter 4.

3.1 The Derivative

Very few quantities in real life remain constant; most are in a state of change. For example, room temperature, the speed of a car, and the angle of elevation of the sun are three commonplace quantities that are constantly changing. A few more technical ones are current in a transmission line, barometric pressure, moisture content of the soil, and stress in vertical members of tall buildings during hurricanes. Rates at which these quantities change are called *derivatives*. To study the concept of a rate of change more thoroughly, we consider two physical situations.

In Section 1.4 we considered a particle moving along the x-axis with displacement function

$$ x = f(t) = t^3 - 27t^2 + 168t + 20, \quad t \geq 0. $$

We showed how the graph of this function gives general information about the motion, but specific detail had to wait for calculus. We are now ready to take a significant step toward the point where we will be able to give a complete analysis of the motion. In Figure 3.1 we have shown the displacement function for the first four seconds $0 \leq t \leq 4$.

FIGURE 3.1

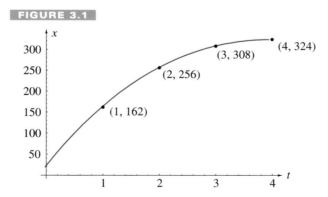

At time $t = 1$ s, the particle is 162 m to the right of the origin. At $t = 4$ s, it is at position $x = 324$ m. During these three seconds, the particle has moved $324 - 162$ or 162 m. When we divide the distance travelled by time taken to travel the distance, we obtain $162/3 = 54$. This is called the **average velocity** of the particle during the time interval. It has units of metres per second, written m/s. During the time interval $1 \leq t \leq 3$, the distance travelled by the particle is 146 m; therefore, its average velocity in this time interval is $146/2$ or 73 m/s. The average velocity during $1 \leq t \leq 2$ is 94 m/s. What we are doing is taking average velocities over shorter and shorter time intervals, all beginning at $t = 1$ s. If we continue the process indefinitely, we are in effect taking the limit of average velocities as the length of the time interval starting at $t = 1$ approaches 0. This will be called the instantaneous velocity of the particle at $t = 1$ s. Let us introduce notation to describe the limiting process, and do so at an arbitrary time t_0 rather than at $t = 1$.

We let Δt represent a small interval of time, often called an **increment** of time. The function value $f(t_0)$ is the position (or displacement) of the particle at time t_0, $f(t_0 + \Delta t)$ is its position at time $t_0 + \Delta t$, and $f(t_0 + \Delta t) - f(t_0)$ is the difference in these displacements. It represents the displacement of the particle at time $t_0 + \Delta t$ relative to its position at time t_0. It may or may not represent the distance travelled by the particle during the time interval Δt. (Can you explain why?) The quotient

$$\frac{f(t_0 + \Delta t) - f(t_0)}{\Delta t}$$

is the average velocity of the particle during the time interval between t_0 and $t_0 + \Delta t$. The limit of this quotient as $\Delta t \to 0$ is called the **instantaneous velocity** of the particle at time t_0; it is denoted by

$$v(t_0) = \lim_{\Delta t \to 0} \frac{f(t_0 + \Delta t) - f(t_0)}{\Delta t}. \tag{3.1}$$

It represents how fast the particle is moving at time t_0; it is the instantaneous rate of change of displacement with respect to time. For example, the instantaneous velocity at $t = 1$ is

$$v(1) = \lim_{\Delta t \to 0} \frac{f(1 + \Delta t) - f(1)}{\Delta t}$$

$$= \lim_{\Delta t \to 0} \frac{(1 + \Delta t)^3 - 27(1 + \Delta t)^2 + 168(1 + \Delta t) + 20 - (1 - 27 + 168 + 20)}{\Delta t}.$$

Simplification of the numerator gives

$$v(1) = \lim_{\Delta t \to 0} \frac{117(\Delta t) - 24(\Delta t)^2 + (\Delta t)^3}{\Delta t},$$

and when we divide numerator and denominator by Δt,

$$v(1) = \lim_{\Delta t \to 0} [117 - 24(\Delta t) + (\Delta t)^2] = 117.$$

The particle is travelling 117 m/s at time $t = 1$ s.

A similar calculation at $t = 5$ s leads to an instantaneous velocity $v(5) = -27$ m/s. The negative sign indicates that the particle is moving to the left. The positive velocity $v(1) = 117$ means that the particle is moving to the right at $t = 1$ s. We shall have much more to say about velocity in Section 3.6 when we relate velocity and speed, and we also introduce acceleration. For now, we simply want you to appreciate that instantaneous velocity is a rate of change; it is the instantaneous rate of change of displacement with respect to time.

For our second situation in which to introduce rates of change, consider an open cylindrical container placed outside a house. The depth D (measured in millimetres) of water in the container during a rainstorm (Figure 3.2) is a function of time t, say $D = f(t)$. A graph of this function might look like that in Figure 3.3. There is no water in the container at time $t = 0$ when the storm begins. As the rain falls, the depth of water increases, and finally when the rain stops at time $t = \bar{t}$, the depth of water remains at a constant level $\overline{D}$ thereafter.

FIGURE 3.2   **FIGURE 3.3**

Intuitively, rain is falling fastest when this curve is steepest, somewhere around the time indicated as $\tilde{t}$. When the curve is flat (just after $t = 0$ and just before $t = \bar{t}$), very little rain is falling. We want to be more specific; we want to be able to say exactly how fast the rain is falling at any given time t_0. We proceed as we did in defining velocity. If we let Δt be a small increment of time, then the difference $f(t_0 + \Delta t) - f(t_0)$ is the number of millimetres of rain that falls during the time interval Δt after t_0. The quotient

$$\frac{f(t_0 + \Delta t) - f(t_0)}{\Delta t}$$

with units of millimetres per minute is called the **average rate** of rainfall during the time interval. The limit

$$\lim_{\Delta t \to 0} \frac{f(t_0 + \Delta t) - f(t_0)}{\Delta t}$$

is called the **instantaneous rate** of rainfall at time t_0. It is the instantaneous rate of change of depth of water in the cylinder with respect to time, at time t_0. If its value were 5, say, then were rain to fall at this rate over an extended period of time, the depth of water in the container would increase by 5 mm every minute.

The rates of change in these two situations are typical of rates of change that occur in a multitude of applications—applications from such diverse fields as engineering, physics, economics, psychology, and medicine, to name a few. We now reformulate them in a mathematical framework that allows us to introduce them in every area of applied mathematics. Suppose a function $f(x)$ is defined for all x, and $x = a$ and $x = a + h$ are two values of x (a is taking the place of t_0 and h is replacing Δt in the previous situations). The quotient

$$\frac{f(a + h) - f(a)}{h}$$

is called the **average rate of change** of $f(x)$ with respect to x in the interval $a \leq x \leq a + h$. The limit of the quotient as h approaches zero,

$$\lim_{h \to 0} \frac{f(a + h) - f(a)}{h}$$

is called the **instantaneous rate of change** of $f(x)$ with respect to x at $x = a$. It is also called the derivative of $f(x)$ at $x = a$.

DEFINITION 3.1

The **derivative** of a function $f(x)$ with respect to x at a point $x = a$, denoted by $f'(a)$, is defined as

$$f'(a) = \lim_{h \to 0} \frac{f(a + h) - f(a)}{h}, \tag{3.2}$$

provided the limit exists.

The operation of taking the derivative of a function is called **differentiation**. We say that we differentiate the function when we find its derivative. It would now be appropriate to use definition 3.1 to calculate derivatives of various functions at various points. Before doing so, however, we feel that it is important to discuss the geometric interpretation of the derivative. It will be prevalent in many applications.

Tangent Lines and the Geometric Interpretation of the Derivative

Algebraically, the derivative of a function is its instantaneous rate of change. Geometrically, derivatives are intimately connected to tangent lines to curves, and most students of calculus have an intuitive idea of what it means for a line to be tangent to a curve. Often it is the idea of "touching." A line is tangent to a curve if it touches the curve at exactly one point. For curves such as circles, ellipses, and parabolas, this notion is adequate, but in general it is unsatisfactory. For instance, in Figure 3.4 we have drawn what would look like the tangent line to the curve $y = x^3$ at the point $P(1, 1)$. But this line intersects the curve again at the point $R(-2, -8)$. The tangent line at P does not touch the curve at precisely one point; it intersects the curve at a second point.

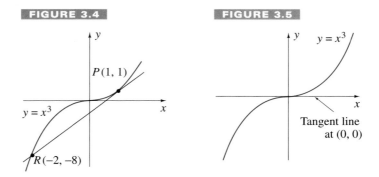

FIGURE 3.4

FIGURE 3.5

You might reply that tangency is a local concept; whether the tangent line at $(1, 1)$ in Figure 3.4 intersects the curve at another point some distance from $(1, 1)$ is irrelevant. True, and perhaps we could remedy the situation by defining the tangent line at $(1, 1)$ as the line that "touches" the curve at $(1, 1)$ and does not cross it there. Unfortunately, this definition does not always work either. For instance, what is the tangent line to $y = x^3$ at $(0, 0)$ (Figure 3.5)? The only reasonable line is the x-axis, but notice that the x-axis crosses the curve at $(0, 0)$. To the right of $(0, 0)$, the curve is above the tangent line, and to the left of $(0, 0)$, it is below the tangent line. This happens quite often as we shall see later.

How then are we to define the tangent line to a curve? The idea that a tangent line touches a curve is good, but it needs to be phrased properly. Consider defining what is meant by the tangent line at the point P on the curve $y = f(x)$ in Figure 3.6. When P is joined to another

FIGURE 3.6

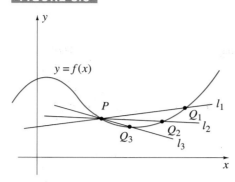

point Q_1 on the curve by a straight line l_1, certainly l_1 is not the tangent line to $y = f(x)$ at P. If we join P to a point Q_2 on $y = f(x)$ closer to P than Q_1, then l_2 is not the tangent line at P either, but it is closer to it than l_1. A point Q_3 even closer to P yields a line l_3 that is even closer to the tangent line than l_2. Repeating this process over and over again leads to a set of lines $l_1, l_2, l_3, \ldots$ which get closer and closer to what we feel is the tangent line to $y = f(x)$ at P. We therefore define the **tangent line** to $y = f(x)$ at P as the limiting position of these lines as points $Q_1, Q_2, Q_3, \ldots$ get arbitrarily close to P.

The line in Figure 3.4 satisfies this definition; it is the limiting position of lines joining P to other points on the curve which approach P. It is irrelevant whether this line intersects the curve again at some distance from P. What is the tangent line to the curve $y = x^3$ at the origin $(0, 0)$? According to the above definition the limiting position of lines joining $(0, 0)$ to other points on the curve is the x-axis; that is, the x-axis is tangent to $y = x^3$ at $(0, 0)$ (Figure 3.5). As mentioned earlier, the tangent line actually crosses from one side of the curve to the other at $(0, 0)$. To the left of $x = 0$, the tangent line is above the curve whereas to the right of $x = 0$, it is below the curve.

Now that we understand what it means for a line to be tangent to a curve, let us make the connection with derivatives. Suppose $P(a, f(a))$ is a point on the curve $y = f(x)$ in Figure 3.7. If $Q(a + h, f(a + h))$ is another point on the curve, then the quotient

$$\frac{f(a + h) - f(a)}{h}$$

in equation 3.2 is the slope of the line joining P and Q. As $h \to 0$, point Q moves along the curve toward P, and the line joining P and Q moves toward the tangent line at P. It follows that limit 3.2, the derivative $f'(a)$, is the slope of the tangent line to the curve $y = f(x)$ at $x = a$ (Figure 3.8).

FIGURE 3.7	**FIGURE 3.8**

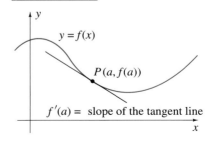

To summarize, algebraically the derivative $f'(a)$ of a function $f(x)$ at $x = a$ is its instantaneous rate of change; geometrically it is the slope of the tangent line to the graph of $f(x)$ at the point $(a, f(a))$.

Let us now use equation 3.2 to calculate derivatives of some simple functions.

EXAMPLE 3.1

Find $f'(1)$ if $f(x) = x^2$.

SOLUTION   According to equation 3.2,

$$f'(1) = \lim_{h \to 0} \frac{f(1+h) - f(1)}{h} = \lim_{h \to 0} \frac{(1+h)^2 - 1}{h}$$

$$= \lim_{h \to 0} \frac{(1 + 2h + h^2) - 1}{h} = \lim_{h \to 0} \frac{2h + h^2}{h} = \lim_{h \to 0} (2 + h) = 2.$$

Algebraically, the instantaneous rate of change of $f(x) = x^2$ when $x = 1$ is equal to 2. Geometrically, the slope of the tangent line to the curve $y = x^2$ at the point $(1, 1)$ is 2 (Figure 3.9).

FIGURE 3.9

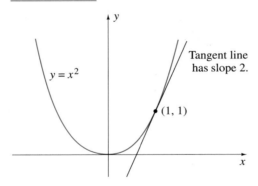

Tangent line has slope 2.

$y = x^2$

$(1, 1)$

EXAMPLE 3.2

Find $f'(2)$ if $f(x) = x^3 - 12x$.

SOLUTION   Using equation 3.2,

$$f'(2) = \lim_{h \to 0} \frac{f(2+h) - f(2)}{h} = \lim_{h \to 0} \frac{[(2+h)^3 - 12(2+h)] - (8 - 24)}{h}$$

$$= \lim_{h \to 0} \frac{6h^2 + h^3}{h} = \lim_{h \to 0} (6h + h^2) = 0.$$

This result is substantiated in Figure 3.10 where we see that the tangent line to $y = x^3 - 12x$ is horizontal (has zero slope) at $x = 2$.

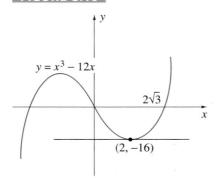

In Examples 3.1 and 3.2 we required the derivative for the function at only one value of a, and therefore set a in equation 3.2 equal to this value. An alternative, which is far more advantageous, especially in an example where the derivative is required at a number of points, is to evaluate $f'(a)$ and then set a to its desired value, or values, later. For instance, in Example 3.1 we calculate that

$$f'(a) = \lim_{h \to 0} \frac{f(a+h) - f(a)}{h} = \lim_{h \to 0} \frac{(a+h)^2 - a^2}{h}$$

$$= \lim_{h \to 0} \frac{2ah + h^2}{h} = \lim_{h \to 0} (2a + h) = 2a.$$

This is the derivative of $f(x) = x^2$ at any value $x = a$. For $a = 1$, we obtain $f'(1) = 2(1) = 2$. But it is also easy to calculate $f'(a)$ at other values of a. For example, $f'(0) = 0$, $f'(-1) = -2$, and $f'(4) = 8$.

We now carry this idea to its logical conclusion. The derivative of $f(x)$ at a is denoted by $f'(a)$. But what is a? It is a specific value of x at which to calculate the derivative, and it can be any value of x. Why not simply drop references to a, and talk about the derivative of $f(x)$ at values of x? Following this suggestion, we denote by $f'(x)$ the derivative of the function $f(x)$ at any value of x. With this notation, equation 3.2 is replaced by

$$f'(x) = \lim_{h \to 0} \frac{f(x+h) - f(x)}{h}, \tag{3.3}$$

for the derivative of $f(x)$ at x. We call $f'(x)$ the **derivative function**.

When a function is represented by the letter y, as in $y = f(x)$, another common notation for the derivative is $\frac{dy}{dx}$. Be careful when using this notation. Do not interpret it as a quotient; it is one symbol representing an accumulation of all the operations in equation 3.3. It is not therefore to be read as "dy divided by dx." Typographically, it is easier to print dy/dx rather than $\frac{dy}{dx}$, and we will take this liberty whenever it is convenient to do so. But remember, dy/dx is not a quotient, it is one symbol representing the limit operation in 3.3.

Sometimes it is more convenient to use parts of each of these notations and write

$$\frac{d}{dx} f(x).$$

In this form we understand that d/dx means to differentiate with respect to x whatever follows it, in this case $f(x)$. Let us use these new notations in calculating two more derivatives.

EXAMPLE 3.3

Find dy/dx if $y = f(x) = (x - 1)/(x + 2)$.

SOLUTION Using 3.3,

$$\frac{dy}{dx} = \lim_{h \to 0} \frac{f(x + h) - f(x)}{h} = \lim_{h \to 0} \frac{1}{h} \left(\frac{x + h - 1}{x + h + 2} - \frac{x - 1}{x + 2} \right).$$

If we bring the terms in parentheses to a common denominator, the result is

$$\frac{dy}{dx} = \lim_{h \to 0} \frac{1}{h} \left[\frac{(x + h - 1)(x + 2) - (x + h + 2)(x - 1)}{(x + h + 2)(x + 2)} \right].$$

When we simplify the numerator, we find

$$\frac{dy}{dx} = \lim_{h \to 0} \frac{1}{h} \left[\frac{3h}{(x + h + 2)(x + 2)} \right]$$

$$= \lim_{h \to 0} \frac{3}{(x + h + 2)(x + 2)}$$

$$= \frac{3}{(x + 2)^2}.$$

EXAMPLE 3.4

Find dv/dt if $v = f(t) = 1/t$.

SOLUTION In terms of variables v and t, equation 3.3 takes the form

$$\frac{dv}{dt} = \lim_{h \to 0} \frac{f(t + h) - f(t)}{h}$$

$$= \lim_{h \to 0} \frac{1}{h} \left(\frac{1}{t + h} - \frac{1}{t} \right)$$

$$= \lim_{h \to 0} \frac{t - (t + h)}{t(t + h)h}$$

$$= \lim_{h \to 0} \frac{-1}{t(t + h)}$$

$$= -\frac{1}{t^2}.$$

In Section 1.1 we proved that the inclination ϕ of a line ℓ is related to its slope m by the equation $\tan\phi = m$ (equation 1.44). When ℓ is the tangent line to a curve $y = f(x)$ at point (x_0, y_0) (Figure 3.11), its slope is $f'(x_0)$. Hence, the inclination ϕ of the tangent line to a curve at a point (x_0, y_0) is given by the equation

$$\tan\phi = f'(x_0). \tag{3.4}$$

FIGURE 3.11

EXAMPLE 3.5

Find the inclination of the tangent line to the curve $y = (x - 1)/(x + 2)$ at the point $(4, 1/2)$.

SOLUTION In Example 3.3, we calculated the slope of the tangent line to this curve at any point as

$$\frac{dy}{dx} = \frac{3}{(x + 2)^2}.$$

According to equation 3.4, the inclination ϕ of the tangent line at $(4, 1/2)$ is given by

$$\tan\phi = \frac{3}{(4 + 2)^2} = \frac{1}{12}.$$

The angle with tangent equal to $1/12$ is $\phi = 0.083$ radians.

In applications, independent and dependent variables of functions $y = f(x)$ represent physical quantities and have units associated with them. Units for the derivative $f'(x)$ are units of y divided by units of x. We see this from equation 3.3 where units of the quotient $[f(x + h) - f(x)]/h$ are clearly units of y divided by units of x, and the limit of this quotient as $h \to 0$ does not alter these units. For example, if x measures length in metres and y measures mass in kilograms, the units of dy/dx are kilograms per metre (kilograms divided by metres).

In Exercises 1–10 use equation 3.3 to find $f'(x)$.

∗ **1.** $f(x) = x + 2$

∗ **2.** $f(x) = 3x^2 + 5$

∗ **3.** $f(x) = 1 + 2x - x^2$

∗ **4.** $f(x) = x^3 + 2x^2$

∗ **5.** $f(x) = x^4 + 4x - 12$

∗ **6.** $f(x) = \dfrac{x + 4}{x - 5}$

∗∗ **7.** $f(x) = \dfrac{x^2 + 2}{x + 3}$

∗∗ **8.** $f(x) = x^2(x + 2)$

∗∗ **9.** $f(x) = \dfrac{3x - 2}{4 - x}$

∗∗ **10.** $f(x) = \dfrac{x^2 - x + 1}{x^2 + x + 1}$

In Exercises 11–14 find the specified rate of change.

∗ **11.** The rate of change of the circumference C of a circle with respect to its radius r

∗ **12.** The rate of change of the area A of a circle with respect to its radius r

∗ **13.** The rate of change of the area A of a sphere with respect to its radius r

∗ **14.** The rate of change of the volume V of a sphere with respect to its radius r

Answer Exercises 15–19 by drawing a graph of the function. Do not calculate the required derivative.

∗ **15.** What is $f'(x)$ if $f(x) = 2x - 4$?

∗ **16.** What is $f'(x)$ if $f(x) = mx + b$, where m and b are constants?

∗ **17.** What is $f'(0)$ if $f(x) = x^2$?

∗ **18.** What is $f'(1)$ if $f(x) = (x - 1)^2$?

∗ **19.** What is $f'(0)$ if $f(x) = x^{1/3}$?

In Exercises 20–23 find the equation of the tangent line to the curve at the given point.

∗ **20.** $y = x^2 + 3$ at $(1, 4)$

∗ **21.** $y = 3 - 2x - x^2$ at $(4, -21)$

∗∗ **22.** $y = 1/x^2$ at $(2, 1/4)$

∗∗ **23.** $y = (x + 1)/(x + 2)$ at $(0, 1/2)$

In Exercises 24–27 find the inclination of the tangent line to the curve at the given point.

∗ **24.** $y = x^2$ at $(1, 1)$

∗ **25.** $y = x^3 - 6x$ at $(2, -4)$

∗∗ **26.** $y = 1/x^2$ at $(2, 1/4)$

∗∗ **27.** $y = 1/(x + 1)$ at $(0, 1)$

In Exercises 28–32 find $f'(x)$.

∗∗ **28.** $f(x) = x^8$

∗∗ **29.** $f(x) = \sqrt{x + 1}$

∗∗ **30.** $f(x) = \dfrac{1}{(x - 2)^4}$

∗∗ **31.** $f(x) = \dfrac{1}{\sqrt{x - 3}}$

∗∗ **32.** $f(x) = x\sqrt{x + 1}$

In Exercises 33 and 34 find the specified rate of change.

∗∗ **33.** The rate of change of the radius r of a circle with respect to its area A

∗∗ **34.** The rate of change of the volume V of a sphere with respect to its area A

∗∗ **35.** Find the angle between the tangent lines to the curves $y = x^2$ and $x = y^2$ at their point of intersection.

∗∗ **36.** Find $f'(x)$ if $f(x) = |x|$.

** **37.** A sphere of radius R has a uniform charge distribution of ρ coulombs per cubic metre. The electrostatic potential V at a distance r from the centre of the sphere is defined by

$$V = f(r) = \begin{cases} \dfrac{\rho}{6\epsilon_0}(3R^2 - r^2), & 0 \leq r \leq R \\[2ex] \dfrac{R^3\rho}{3\epsilon_0 r}, & r > R, \end{cases}$$

where ϵ_0 is a constant. Draw a graph of this function. Does $f(r)$ appear to have a derivative at the surface of the sphere? That is, does $f'(R)$ exist? Prove your conjecture using definition 3.1.

** **38.** Repeat Exercise 37 for the magnitude E of the electrostatic field:

$$E = f(r) = \begin{cases} \dfrac{\rho r}{3\epsilon_0}, & 0 \leq r \leq R \\[2ex] \dfrac{\rho R^3}{3\epsilon_0 r^2}, & r > R. \end{cases}$$

*** **39.** Find $f'(x)$ if $f(x) = x^{1/3}$.

*** **40.** Let $f(x)$ be a function with the property that $f(x + z) = f(x)f(z)$ for all x and z, and such that $f(0) = f'(0) = 1$. Prove that $f'(x) = f(x)$ for all x.

3.2 Rules for Differentiation

Since calculus plays a key role in many branches of applied science, we need to differentiate many types of functions: polynomial, rational, trigonometric, exponential, and logarithm functions, to name a few. To use equation 3.3 each time would be extremely laborious and we might soon doubt the usefulness of differentiation. Fortunately, however, we can develop a number of rules for taking derivatives that eliminate the necessity of using the definition each time. We state each of these formulas as a theorem.

> **THEOREM 3.1**
> If $f(x) = c$, where c is a constant, then $f'(x) = 0$.

PROOF   By equation 3.3,

$$f'(x) = \lim_{h \to 0} \frac{f(x + h) - f(x)}{h} = \lim_{h \to 0} \frac{c - c}{h} = 0. \qquad \blacksquare$$

In short, the derivative of a constant function is zero;

$$\frac{d}{dx}(c) = 0. \tag{3.5}$$

> **THEOREM 3.2**
>
> If $f(x) = x$, then $f'(x) = 1$.

PROOF With equation 3.3,

$$f'(x) = \lim_{h \to 0} \frac{f(x+h) - f(x)}{h} = \lim_{h \to 0} \frac{(x+h) - x}{h} = \lim_{h \to 0} \frac{h}{h} = 1. \qquad\blacksquare$$

In short,

$$\frac{d}{dx}(x) = 1. \tag{3.6}$$

Note that graphs of the functions $f(x) = c$ and $f(x) = x$ in Figure 3.12 confirm the results of equations 3.5 and 3.6. The tangent line to $y = c$ always has slope zero, whereas the tangent line to $y = x$ always has slope equal to one.

FIGURE 3.12

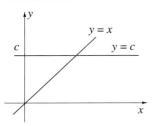

> **THEOREM 3.3**
>
> If $f(x) = x^n$, where n is a positive integer, then $f'(x) = nx^{n-1}$.

PROOF Equation 3.3 gives

$$f'(x) = \lim_{h \to 0} \frac{f(x+h) - f(x)}{h} = \lim_{h \to 0} \frac{(x+h)^n - x^n}{h}.$$

If we expand $(x + h)^n$ by means of the binomial theorem (an alternative proof is given in Exercise 33 on page 160), we have

$$f'(x) = \lim_{h \to 0} \frac{1}{h} \left[x^n + nx^{n-1}h + \frac{n(n-1)}{2}x^{n-2}h^2 + \cdots + nxh^{n-1} + h^n - x^n \right].$$

The first and last terms in the brackets cancel, and dividing h into the remaining terms gives

$$f'(x) = \lim_{h \to 0} \left[nx^{n-1} + \frac{n(n-1)}{2}x^{n-2}h + \cdots + h^{n-1} \right] = nx^{n-1}. \qquad\blacksquare$$

In short,

$$\frac{d}{dx}(x^n) = nx^{n-1}. \tag{3.7}$$

This is called the **power rule** for differentiation. Although we have proved the power rule only for n a positive integer, it is in fact true for every real number n. We will assume that equation 3.7 can be used for any real number n, and will prove this more general result in Section 3.10. For example,

$$\frac{d}{dx}(x^3) = 3x^2 \qquad \text{and} \qquad \frac{d}{dx}(x^{1/3}) = \frac{1}{3}x^{-2/3}.$$

THEOREM 3.4

If $g(x) = cf(x)$, where c is a constant, then

$$g'(x) = cf'(x). \tag{3.8a}$$

PROOF   By equation 3.3 and Theorem 2.1,

$$g'(x) = \lim_{h \to 0} \frac{g(x+h) - g(x)}{h} = \lim_{h \to 0} \frac{cf(x+h) - cf(x)}{h}$$

$$= c \lim_{h \to 0} \frac{f(x+h) - f(x)}{h} = cf'(x). \qquad \blacksquare$$

Thus, for $y = f(x)$, we may write that

$$\frac{d}{dx}(cy) = c\frac{dy}{dx}. \tag{3.8b}$$

THEOREM 3.5

If $p(x) = f(x) + g(x)$, then

$$p'(x) = f'(x) + g'(x). \tag{3.9a}$$

PROOF   Equation 3.3 gives

$$p'(x) = \lim_{h \to 0} \frac{p(x+h) - p(x)}{h}$$

$$= \lim_{h \to 0} \frac{[f(x+h) + g(x+h)] - [f(x) + g(x)]}{h}$$

$$= \lim_{h \to 0} \left[\frac{f(x+h) - f(x)}{h} + \frac{g(x+h) - g(x)}{h} \right]$$

$$= f'(x) + g'(x). \qquad \blacksquare$$

In short, if we set $u = f(x)$ and $v = g(x)$,

$$\frac{d}{dx}(u + v) = \frac{du}{dx} + \frac{dv}{dx}, \tag{3.9b}$$

or, in words, the derivative of a sum is the sum of the derivatives.

We now use these formulas to calculate derivatives in the following examples.

EXAMPLE 3.6

Find dy/dx if

(a) $y = x^4$ (b) $y = 3x^6 - x^{-2}$ (c) $y = \dfrac{x^4 - 6x^2}{3x^3}$

SOLUTION

(a) By equation 3.7,

$$\frac{dy}{dx} = 4x^3.$$

(b) Equation 3.9 allows us to differentiate each term separately; by equations 3.8 and 3.7 it follows that

$$\frac{dy}{dx} = 3(6x^5) - (-2x^{-3}) = 18x^5 + 2x^{-3}.$$

(c) If we write y in the form $y = (x/3) - 2x^{-1}$, we can proceed as in (b):

$$\frac{dy}{dx} = \frac{1}{3}(1) - 2(-x^{-2}) = \frac{1}{3} + \frac{2}{x^2}.$$

EXAMPLE 3.7

If $f(x) = 3x^4 - 2$, evaluate $f'(1)$.

SOLUTION   Since $f'(x) = 12x^3$, it follows that $f'(1) = 12$. Geometrically, 12 is the slope of the tangent line to the curve $y = 3x^4 - 2$ at the point $(1, 1)$ in Figure 3.13. Algebraically, the result implies that at $x = 1$, y changes 12 times as fast as x.

FIGURE 3.13

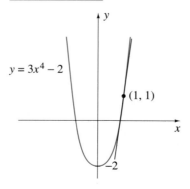

$y = 3x^4 - 2$

$(1, 1)$

EXAMPLE 3.8

Find the equation of the tangent line to the curve $y = f(x) = x^3 + 5x$ at the point $(1, 6)$.

SOLUTION Since $f'(x) = 3x^2 + 5$, the slope of the tangent line to the curve at $(1, 6)$ is

$$f'(1) = 3(1)^2 + 5 = 8.$$

Using point-slope formula 1.4 for a straight line, we obtain for the equation of the tangent line at $(1, 6)$

$$y - 6 = 8(x - 1) \qquad \text{or} \qquad 8x - y = 2.$$

The line through $(1, 6)$ perpendicular to the tangent line in Figure 3.14 is called the **normal line** to the curve at $(1, 6)$. Since two lines are perpendicular only if their slopes are negative reciprocals (see equation 1.6), the normal line at $(1, 6)$ must have slope $-1/8$. The equation of the normal line to $y = x^3 + 5x$ at $(1, 6)$ is therefore

$$y - 6 = -\frac{1}{8}(x - 1), \qquad \text{or} \qquad x + 8y = 49.$$

FIGURE 3.14

EXAMPLE 3.9

Find, accurate to four decimal places, points on the curve $y = x^4 - 4x^3 - x^2 + x$ at which the slope of the tangent line is -1.

SOLUTION We must be careful in trying to estimate where the slope of the graph is -1 in Figure 3.15. Because scales are different on the axes, a line with slope -1 is not inclined at $\pi/4$ radians with respect to the negative x-axis. The line in the figure has slope -1 for reference. The graph suggests three points at which the tangent line is parallel to this line, one just to the left of $x = 0$, one just to the right of $x = 0$, and one near $x = 3$.

FIGURE 3.15

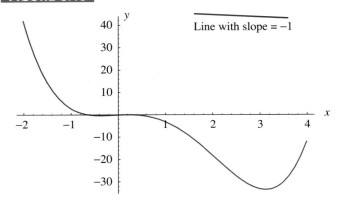

To confirm this, we set the slope of the tangent line to the curve equal to -1:

$$-1 = \frac{dy}{dx} = 4x^3 - 12x^2 - 2x + 1 \qquad \Longrightarrow \qquad 2(2x^3 - 6x^2 - x + 1) = 0.$$

Six-digit solutions of this equation are $-0.459\,261$, $0.350\,168$, and $3.109\,09$. To confirm -0.4593, 0.3502, and 3.1091 as solutions correct to four decimals, we calculate $g(x) = 2x^3 - 6x^2 - x + 1$ at the following values:

$$g(-0.459\,35) \;=\; -5.1 \times 10^{-4}, \quad g(-0.459\,25) \;=\; 6.5 \times 10^{-5},$$

$$g(0.350\,15) \;=\; 8.0 \times 10^{-5}, \qquad g(0.350\,25) \;=\; -3.7 \times 10^{-4},$$

$$g(3.109\,05) \;=\; -8.5 \times 10^{-4}, \qquad g(3.109\,15) \;=\; 1.1 \times 10^{-3}.$$

The Intermediate Value Theorem confirms the four decimal approximations. Using the six decimal approximations, and rounding results to four decimals, corresponding points on the curve are $(-0.4593, -0.2382)$, $(0.3502, 0.0708)$, and $(3.1091, -33.3326)$.

Increment Notation

We have used the letter h to represent a small change in x when calculating derivatives. An alternative notation that is sometimes more suggestive was used in the introductory examples of Section 3.1. A small change in x, called an **increment** in x, is often denoted by Δx. It is pronounced "delta x" since Δ is the capital letter delta in the Greek alphabet. When x is given an increment Δx in the function $y = f(x)$, the corresponding change or increment in y is denoted by Δy. It is equal to

$$\Delta y = f(x + \Delta x) - f(x). \tag{3.10}$$

With this notation, equation 3.3 takes the form

$$f'(x) = \lim_{\Delta x \to 0} \frac{\Delta y}{\Delta x}. \tag{3.11a}$$

The notation dy/dx for the derivative fits very nicely with increment notation,

$$\frac{dy}{dx} = \lim_{\Delta x \to 0} \frac{\Delta y}{\Delta x};$$ (3.11b)

the derivative of y with respect to x is the limit of the change in y divided by the change in x as the change in x approaches zero. We use this notation in the following example.

EXAMPLE 3.10

Use equation 3.11 to calculate the derivative of $y = f(x) = 3x^2 - 2x$ and check your answer using the differentiation rules discussed at the beginning of this section.

SOLUTION Since

$$\Delta y = f(x + \Delta x) - f(x)$$
$$= [3(x + \Delta x)^2 - 2(x + \Delta x)] - [3x^2 - 2x]$$
$$= 3[x^2 + 2x\Delta x + (\Delta x)^2] - 2x - 2\Delta x - 3x^2 + 2x$$
$$= \Delta x(6x - 2 + 3\Delta x),$$

equation 3.11 gives

$$\frac{dy}{dx} = \lim_{\Delta x \to 0} \frac{\Delta y}{\Delta x} = \lim_{\Delta x \to 0} \frac{\Delta x(6x - 2 + 3\Delta x)}{\Delta x} = \lim_{\Delta x \to 0} (6x - 2 + 3\Delta x) = 6x - 2.$$

Power rule 3.7 and rules 3.8 and 3.9 for differentiation of $3x^2 - 2x$ also yield $6x - 2$.

EXERCISES 3.2

In Exercises 1–20 find $f'(x)$.

* **1.** $f(x) = 2x^2 - 3$
* **2.** $f(x) = 3x^3 + 4x + 5$
* **3.** $f(x) = 10x^2 - 3x$
* **4.** $f(x) = 4x^5 - 10x^3 + 3x$
* **5.** $f(x) = 1/x^2$
* **6.** $f(x) = 2/x^3$
* **7.** $f(x) = 5x^4 - 3x^3 + 1/x$
* **8.** $f(x) = -\frac{1}{2x^2} + \frac{3}{x^4}$
* **9.** $f(x) = x^{10} - \frac{1}{x^{10}}$
* **10.** $f(x) = 5x^4 + \frac{1}{4x^5}$
* **11.** $f(x) = 5x^{-4} + \frac{1}{4x^{-5}}$
* **12.** $f(x) = \sqrt{x}$
* **13.** $f(x) = \frac{3}{x^2} + \frac{2}{\sqrt{x}}$
* **14.** $f(x) = \frac{1}{x^{3/2}} + x^{3/2}$
* **15.** $f(x) = 2x^{1/3} - 3x^{2/3}$
* **16.** $f(x) = \pi x^\pi$
* **17.** $f(x) = (x^2 + 2)^2$
* **18.** $f(x) = (4x^6 - x^2)/x^5$
* **19.** $f(x) = x^{5/3} - x^{2/3} + 3$
** **20.** $f(x) = (2x + 5)^3$

In Exercises 21–24 find equations for the tangent and normal lines to the curve at the point indicated. In each case, draw the curve and lines.

* **21.** $y = x^2 - 2x + 5$ at $(2, 5)$
* **22.** $y = \sqrt{x} + 5$ at $(4, 7)$
* **23.** $y = 2x^3 - 3x^2 - 12x$ at $(2, -20)$
** **24.** $x = \sqrt{y + 1}$ at $(3, 8)$

** **25.** Find the points on the curve $y = x^4/4 - 2x^3/3 - 19x^2/2 + 22x$ at which the slope of the tangent line is 2.

** **26.** An electronics firm manufactures a device called a "gadget." The money that it receives from the sale of x gadgets is the total revenue and is denoted by $R(x)$. The total cost $C(x)$ of producing these x gadgets also depends on x. The total profit realized on the sale is

$$P(x) = R(x) - C(x).$$

The derivatives of these functions, $R'(x)$, $C'(x)$, and $P'(x)$ are called, respectively, marginal revenue, marginal cost, and marginal profit. Calculate these marginal functions if

$$R(x) = 75x - \frac{7x^2}{100} \quad \text{and} \quad C(x) = 50x - \frac{x^2}{30} - \frac{x^3}{60\,000},$$

both functions being defined for $1 \le x \le 800$.

** **27.** Draw a graph of the function $x = f(t) = t^3 - 8t^2$. Find the value(s) of t at which the tangent line to this curve is parallel to the line $x = 6t - 3$.

** **28.** At what point(s) on the curve $y = x^3 + x^2 - 22x + 20$ does the tangent line pass through the origin?

** **29.** At what point(s) on the parabola $y = x^2$ does the normal line pass through the point $(2, 5)$? Can you suggest an application of this result?

** **30.** Show that the sum of the x- and y-intercepts of the tangent line to the curve $\sqrt{x} + \sqrt{y} = \sqrt{a}$ is always equal to a.

** **31.** Prove that

$$\frac{d}{dx}(ax + b)^n = an(ax + b)^{n-1}$$

when a and b are constants.

** **32.** Find a formula for

$$\frac{d}{dx}|x|^n,$$

when $n > 1$ is an integer.

** **33.** Give an alternative derivation of power rule 3.7 based on the identity

$$a^n - b^n = (a - b)(a^{n-1} + a^{n-2}b + \cdots + ab^{n-2} + b^{n-1}).$$

*** **34.** Find the two points on the curve $y = x(1 + 2x - x^3)$ that share a common tangent line.

3.3 Differentiability and Continuity

Many functions fail to have a derivative at isolated points. For example, consider the function $f(x) = |x|$ in Figure 3.16. It is clear that for $x > 0$, $f'(x) = 1$, and for $x < 0$, $f'(x) = -1$. At $x = 0$, however, there is a problem. If $f(x)$ is to have a derivative at $x = 0$, it must be given by

$$\lim_{h \to 0} \frac{f(0 + h) - f(0)}{h} = \lim_{h \to 0} \frac{|h|}{h}.$$

But this limit does not exist since

$$\lim_{h \to 0^-} \frac{|h|}{h} = -1 \qquad \text{and} \qquad \lim_{h \to 0^+} \frac{|h|}{h} = 1.$$

Consequently, $f(x) = |x|$ does not have a derivative at $x = 0$.

FIGURE 3.16

FIGURE 3.17

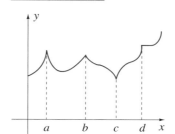

The same conclusion can be drawn at any point at which the graph of a function takes an abrupt change in direction. Such a point is often called a *corner*. As a result, the function in Figure 3.17 does not have a derivative at $x = a, b, c,$ or d. It is also true that a function cannot have a derivative at a point where the function is discontinuous (see, for example, the discontinuities in Figure 2.26). This result is an immediate consequence of the following theorem.

THEOREM 3.6

If a function has a derivative at $x = a$, then the function is continuous at $x = a$.

PROOF   To prove this theorem we show that the existence of $f'(a)$ implies that the $\lim_{x \to a} f(x) = f(a)$, the condition that defines continuity of $f(x)$ at $x = a$ (definition 2.1 in Section 2.4). The derivative $f'(a)$ is defined by

$$f'(a) = \lim_{h \to 0} \frac{f(a + h) - f(a)}{h}.$$

No one will argue with the fact that

$$0 = \lim_{h \to 0} h.$$

But if we multiply left and right sides of these two equations, we obtain

$$0 \cdot f'(a) = \left[\lim_{h \to 0} h \right] \left[\lim_{h \to 0} \frac{f(a + h) - f(a)}{h} \right].$$

Using part (iv) of Theorem 2.1, we may bring the limits on the right together,

$$0 = \lim_{h \to 0} [f(a + h) - f(a)],$$

or

$$\lim_{h \to 0} f(a + h) = f(a).$$

But this is just another way of saying that

$$\lim_{x \to a} f(x) = f(a),$$

(see Exercise 41 in Section 2.4). ∎

As a corollary to this theorem we have the following.

COROLLARY

If $f(x)$ is discontinuous at $x = a$, then $f'(a)$ does not exist.

Students are often heard to say that having a derivative is equivalent to having a tangent line (because derivatives are slopes of tangent lines); that is, a function $f(x)$ has derivative $f'(a)$ at $x = a$ if and only if the graph of $f(x)$ has a tangent line at $x = a$. This is not quite true. For example, consider the function $f(x) = x^{1/3}$ (Figure 3.18). At $(0, 0)$, the tangent line to the graph is the y-axis (use the definition of tangent line in Section 3.1 to convince yourself of this). The derivative of $f(x)$ is $f'(x) = (1/3)x^{-2/3}$; it does not exist at $x = 0$. Thus, we have a tangent line, but no derivative. The reason is that the tangent line is a vertical line, and vertical lines do not have slopes. What can we say then? Two things:

1. If $f'(a)$ exists, then $y = f(x)$ has a tangent line at $\big(a, f(a)\big)$ with slope $f'(a)$.

2. If $f'(a)$ does not exist, then $y = f(x)$ either has a vertical tangent line at $\big(a, f(a)\big)$ or does not have a tangent line when $x = a$.

FIGURE 3.18

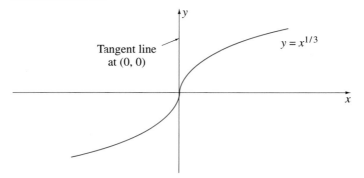

We introduced this section by showing that the derivative of $|x|$ is 1 when $x > 0$; it is -1 when $x < 0$; and it does not exist at $x = 0$. These can be combined into the simple formula

$$\frac{d}{dx}|x| = \frac{|x|}{x}. \tag{3.12}$$

It is straightforward to generalize 3.12 and obtain the derivative of $|f(x)|$ at any point at which $f(x) \neq 0$ and $f'(x)$ exists. When $f(x) > 0$, we may write

$$\frac{d}{dx}|f(x)| = \frac{d}{dx}f(x) = f'(x).$$

On the other hand, when $f(x) < 0$, we have

$$\frac{d}{dx}|f(x)| = \frac{d}{dx}[-f(x)] = -f'(x).$$

Both of these results are contained in the one equation

$$\frac{d}{dx}|f(x)| = \frac{|f(x)|}{f(x)}f'(x). \qquad (3.13)$$

This is the derivative of $|f(x)|$ at any point at which $f(x) \neq 0$. The exceptional case when $f(x) = 0$ is discussed in Exercise 28.

Right-Hand and Left-Hand Derivatives

When a function $f(x)$ is defined on a closed interval $b \leq x \leq c$, equation 3.3 can be used to calculate $f'(x)$ only at points in the open interval $b < x < c$. For instance, it is impossible to evaluate

$$f'(b) = \lim_{h \to 0} \frac{f(b + h) - f(b)}{h}$$

since $f(x)$ is not defined for $x < b$ and therefore $f(b + h)$ is not defined for $h < 0$. When a function $f(x)$ is defined only to the right of a point, we define a right-hand derivative at the point; and when $f(x)$ is defined only to the left of a point, we define its left-hand derivative.

DEFINITION 3.2

The **right-hand derivative** of $f(x)$ with respect to x is defined as

$$f'_+(x) = \lim_{h \to 0^+} \frac{f(x + h) - f(x)}{h}, \qquad (3.14a)$$

provided the limit exists. The **left-hand derivative** of $f(x)$ is

$$f'_-(x) = \lim_{h \to 0^-} \frac{f(x + h) - f(x)}{h}, \qquad (3.14b)$$

if the limit exists.

When a function has a derivative at a point x, its right- and left-hand derivatives both exist at x and are equal to $f'(x)$. It is possible, however, for a function to have both a left-hand and a right-hand derivative at a point but not a derivative. An example of this is the absolute value function $f(x) = |x|$ at $x = 0$ (Figure 3.16, page 161). Its right-hand derivative at $x = 0$ is equal to 1 and its left-hand derivative there is -1.

When a function has a derivative at $x = a$, we say that it is **differentiable** at $x = a$. When it has a derivative at every point in some interval, we say that it is differentiable on that interval. In the event that the interval is closed, $b \leq x \leq c$, we understand that derivatives at $x = b$ and $x = c$ mean right-hand and left-hand derivatives respectively.

EXAMPLE 3.11

What is the derivative of the Heaviside function $H(x - a)$ introduced in Section 2.4?

SOLUTION The tangent line to the graph of the function (Figure 3.19) is horizontal at every point except $x = a$ where the function is undefined. In other words, $H'(x - a) = 0$ except at the discontinuity $x = a$. Right-hand and left-hand derivatives do not exist at $x = a$ either.

FIGURE 3.19

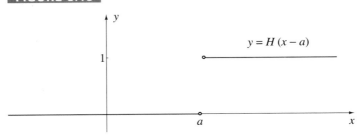

EXAMPLE 3.12

Is the function

$$g(x) = \begin{cases} x \sin (1/x), & x \neq 0 \\ 0, & x = 0 \end{cases}$$

differentiable at $x = 0$?

SOLUTION We encountered the function $x \sin (1/x)$ in Example 2.9 of Section 2.1 and drew its graph (except for the point at $x = 0$) in Figure 2.10 (page 103). According to equation 3.3, the derivative of $g(x)$ at $x = 0$ is

$$g'(0) = \lim_{h \to 0} \frac{g(0 + h) - g(0)}{h},$$

provided the limit exists. When we substitute from the definition of $g(x)$ for $g(h)$ and $g(0)$, this limit takes the form

$$g'(0) = \lim_{h \to 0} \frac{h \sin (1/h) - 0}{h} = \lim_{h \to 0} \sin \left(\frac{1}{h}\right).$$

Since this limit does not exist (see Example 2.9, page 101), $g(x)$ is not differentiable at $x = 0$.

Angle between Intersecting Curves

The angle θ between two curves which intersect at a point (x_0, y_0) (Figure 3.20) is defined as the angle between the tangent lines to the curves at (x_0, y_0). This can be calculated using formula 1.45 once slopes of the tangent lines are known.

FIGURE 3.20

FIGURE 3.21

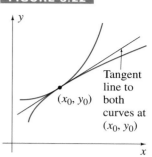

FIGURE 3.22

In the event that $\theta = \pi/2$ (Figure 3.21), the curves are said to be **orthogonal** or perpendicular at (x_0, y_0). Should $\theta = 0$, the curves are said to be tangent at (x_0, y_0) (Figure 3.22).

EXAMPLE 3.13

Find the angle between the curves $y = 5 - x^2$ and $y = 3x/(x+1)$ at their point of intersection.

SOLUTION   Graphs of the curves in Figure 3.23 suggest that they intersect near the point $(2, 2)$. To confirm this we solve

$$5 - x^2 = \frac{3x}{x+1}.$$

The only (real) solution of this equation is $x = 1.757\,28$. The slope of $y = f(x) = 5 - x^2$ at this value of x is[†]

$$f'(1.757\,28) = -2x|_{x=1.757\,28} = -3.514\,56.$$

The slope of $y = g(x) = 3x/(x+1)$ is $g'(x)$, and using equation 3.3 it is straightforward to show that $g'(x) = 3/(x+1)^2$. Thus,

$$g'(1.757\,28) = \frac{3}{(x+1)^2}|_{x=1.757\,28} = 0.394\,602.$$

Using formula 1.45 with $m_1 = -3.514\,56$ and $m_2 = 0.394\,602$, the angle θ between the curves at their point of intersection is given by

$$\tan \theta = \left| \frac{-3.514\,56 - 0.394\,602}{1 + (-3.514\,56)(0.394\,602)} \right| = 10.1050.$$

Hence, $\theta = 1.47$ radians.

[†] When a quantity is suffixed at the subscript level with an expression like $|x = 1.757\,28$, it means that the quantity is to be evaluated at $x = 1.757\,28$.

FIGURE 3.23

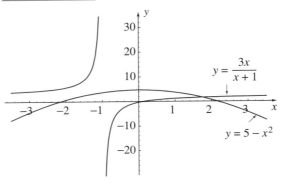

EXERCISES 3.3

In Exercises 1–6 determine whether the function has a right-hand derivative, a left-hand derivative, and a derivative at the given value of x.

* **1.** $f(x) = |x - 5|$ at $x = 5$

* **2.** $f(x) = x^{3/2}$ at $x = 0$

* **3.** $f(x) = |x - 5|^3$ at $x = 5$

* **4.** $f(x) = \operatorname{sgn} x$ at $x = 0$ (see Exercise 43 in Section 2.4, page 134)

* **5.** $f(x) = (x^2 - 1)/(x - 1)$ at $x = 1$

** **6.** $f(x) = [\![x]\!]$ at $x = 1$ (see Exercise 50 in Section 1.3, page 41)

** **7.** Does it make any difference in Example 3.11 if we define $H(a) = 0$?

** **8.** Does it make any difference in Example 3.11 if we define $H(a) = 1$?

** **9.** Does it make any difference in Exercise 4 if $\operatorname{sgn} x$ does not have a value at $x = 0$?

In Exercises 10–13 determine whether the statement is true or false.

* **10.** If a function $f(x)$ has a derivative at $x = a$, then its graph has a tangent line at the point $\big(a, f(a)\big)$.

* **11.** If a function $f(x)$ has a tangent line at a point $\big(a, f(a)\big)$, then it has a derivative at $x = a$.

* **12.** If a function does not have a tangent line at $\big(a, f(a)\big)$, then it does not have a derivative at $x = a$.

* **13.** If a function $f(x)$ does not have a derivative at $x = a$, then it does not have a tangent line at $\big(a, f(a)\big)$.

In Exercises 14–16 show algebraically that $f'(0)$ does not exist. Draw a graph of the function.

** **14.** $f(x) = x^{1/3}$ ** **15.** $f(x) = x^{2/3}$

** **16.** $f(x) = x^{1/4}$

In Exercises 17 and 18 find the angle (or angles) between the curves at their point (or points) of intersection.

** **17.** $y = x^2, \quad x + y = 2$ ** **18.** $y = x^2, \quad y = 1 - x^2$

In Exercises 19 and 20 show that the curves are orthogonal.

* **19.** $x - 2y + 1 = 0, \quad y = 2 - x^2$ * **20.** $y = 3 - x^2, \quad 4y - 7 = x^2$

In Exercises 21 and 22 show that the curves are tangent at the indicated point.

* **21.** $y = x - 2x^2, \quad y = x^3 + 2x$ at $(-1, -3)$

* **22.** $y = x^3, \quad y = x^2 + x - 1$ at $(1, 1)$

** **23.**   (a) The equation $x + 2y = C_1$, where C_1 is a constant, represents a *one-parameter family of curves*. For each value of C_1, the parameter, a different curve in the family is obtained. Draw curves corresponding to $C_1 = -2, -1, 0, 1, 2$.

     (b) Draw a few curves from the one-parameter family $2x - y = C_2$ on the graph in (a).

     (c) Two families of curves are said to be *orthogonal trajectories* if every curve from one family intersects every curve from the other family orthogonally. Are the families in (a) and (b) orthogonal trajectories?

** **24.** For what value of k are the one-parameter families $2x - 3y = C_1$ and $x + ky = C_2$ orthogonal trajectories?

** **25.** Draw a graph of the function $f'(x)$ when $f(x) = [\![x]\!]$ is the greatest integer function (Exercise 50 in Section 1.3).

** **26.** Is the function $f(x) = x|x|$ differentiable at $x = 0$?

** **27.** Find $f'(x)$ if $f(x) = |x| + |x - 1|$. Draw graphs of $f(x)$ and $f'(x)$.

** **28.** If $f(x)$ is a differentiable function, does $|f(x)|$ have a derivative at points where $f(x) = 0$? Hint: Draw some pictures.

** **29.** If $\lim_{x \to \infty} f(x) = L$ (so that $y = L$ is a horizontal asymptote for the graph of the function), is it necessary that $\lim_{x \to \infty} f'(x) = 0$?

** **30.** Is the function

$$f(x) = \begin{cases} x^2 \sin(1/x), & x \neq 0 \\ 0, & x = 0 \end{cases}$$

differentiable at $x = 0$?

*** **31.** For what values of the real number n is the function

$$f(x) = \begin{cases} x^n \sin(1/x), & x \neq 0 \\ 0, & x = 0 \end{cases}$$

differentiable at $x = 0$?

*** **32.** Is the function

$$f(x) = \begin{cases} x^2, & x \text{ a rational number} \\ 0, & x \text{ an irrational number} \end{cases}$$

differentiable at $x = 0$?

*** **33.** Show that the derivative of the function in Exercise 30 is not continuous at $x = 0$.

3.4 Product and Quotient Rules

In this section we add two more formulas to those of Section 3.2 for calculating derivatives. The first is a rule for differentiating a function that is the product of two other functions.

THEOREM 3.7

If $p(x) = f(x)g(x)$, where $f(x)$ and $g(x)$ are differentiable, then

$$p'(x) = f(x)g'(x) + f'(x)g(x). \tag{3.15a}$$

PROOF By equation 3.3,

$$p'(x) = \lim_{h \to 0} \frac{p(x+h) - p(x)}{h}$$

$$= \lim_{h \to 0} \frac{f(x+h)g(x+h) - f(x)g(x)}{h}.$$

To organize this quotient further, we add and subtract the quantity $f(x+h)g(x)$ in the numerator:

$$p'(x) = \lim_{h \to 0} \left\{ \frac{[f(x+h)g(x+h) - f(x+h)g(x)] + [f(x+h)g(x) - f(x)g(x)]}{h} \right\}$$

$$= \lim_{h \to 0} \left[f(x+h) \frac{g(x+h) - g(x)}{h} + g(x) \frac{f(x+h) - f(x)}{h} \right]$$

$$= f(x)g'(x) + g(x)f'(x).$$

In taking the limit of the first term, we have used the fact that $\lim_{h \to 0} f(x+h) = f(x)$, which follows from continuity of $f(x)$ (see Theorem 3.6 and Exercise 41 in Section 2.4). ∎

This result is called the **product rule** for differentiation. If we set $u = f(x)$ and $v = g(x)$, then the product rule may also be expressed in the form

$$\frac{d}{dx}(uv) = u\frac{dv}{dx} + v\frac{du}{dx}. \tag{3.15b}$$

For a change, we use increment notation to prove the **quotient rule**. Use of h in place of Δx works equally well.

THEOREM 3.8

If $p(x) = f(x)/g(x)$, where $f(x)$ and $g(x)$ are differentiable, then

$$p'(x) = \frac{g(x)f'(x) - f(x)g'(x)}{[g(x)]^2}. \tag{3.16a}$$

PROOF Using equation 3.11,

$$p'(x) = \lim_{\Delta x \to 0} \frac{p(x + \Delta x) - p(x)}{\Delta x}$$

$$= \lim_{\Delta x \to 0} \frac{1}{\Delta x} \left[\frac{f(x + \Delta x)}{g(x + \Delta x)} - \frac{f(x)}{g(x)} \right]$$

$$= \lim_{\Delta x \to 0} \frac{f(x + \Delta x)g(x) - g(x + \Delta x)f(x)}{\Delta x \, g(x)g(x + \Delta x)}.$$

To simplify this limit we add and subtract $f(x)g(x)$ in the numerator:

$$p'(x) = \lim_{\Delta x \to 0} \frac{[f(x + \Delta x)g(x) - f(x)g(x)] - [g(x + \Delta x)f(x) - f(x)g(x)]}{\Delta x \, g(x)g(x + \Delta x)}$$

$$= \lim_{\Delta x \to 0} \frac{1}{g(x)g(x + \Delta x)} \left\{ g(x) \left[\frac{f(x + \Delta x) - f(x)}{\Delta x} \right] \right.$$

$$\left. - f(x) \left[\frac{g(x + \Delta x) - g(x)}{\Delta x} \right] \right\}$$

$$= \frac{1}{[g(x)]^2} \left\{ g(x) \lim_{\Delta x \to 0} \frac{f(x + \Delta x) - f(x)}{\Delta x} - f(x) \lim_{\Delta x \to 0} \frac{g(x + \Delta x) - g(x)}{\Delta x} \right\}$$

$$= \frac{g(x)f'(x) - f(x)g'(x)}{[g(x)]^2}. \qquad \blacksquare$$

If we set $u = f(x)$ and $v = g(x)$, then quotient rule 3.16a can be expressed in the form

$$\frac{d}{dx} \left(\frac{u}{v} \right) = \frac{v \dfrac{du}{dx} - u \dfrac{dv}{dx}}{v^2}. \qquad (3.16b)$$

 It makes no difference which term in product rule 3.15 is written first; it does make a difference in the quotient rule. Do not interchange the terms in 3.16.

EXAMPLE 3.14

For the following three functions, find $f'(x)$ in simplified form:

(a) $f(x) = (x^2 + 2)(x^4 + 5x^2 + 1)$ (b) $f(x) = \dfrac{\sqrt{x}}{3x^2 - 2}$ (c) $f(x) = \dfrac{1}{(5x^3 - 2)^2}$

SOLUTION (a) With product rule 3.15a,

$$f'(x) = (x^2 + 2)\frac{d}{dx}(x^4 + 5x^2 + 1) + (x^4 + 5x^2 + 1)\frac{d}{dx}(x^2 + 2)$$

$$= (x^2 + 2)(4x^3 + 10x) + (x^4 + 5x^2 + 1)(2x)$$

$$= 6x^5 + 28x^3 + 22x.$$

(b) With quotient rule 3.16,

$$f'(x) = \frac{(3x^2 - 2)\left(\dfrac{1}{2\sqrt{x}}\right) - \sqrt{x}(6x)}{(3x^2 - 2)^2}$$

$$= \frac{\dfrac{3x^2 - 2 - 12x^2}{2\sqrt{x}}}{(3x^2 - 2)^2}$$

$$= -\frac{9x^2 + 2}{2\sqrt{x}(3x^2 - 2)^2}.$$

(c) If we express $f(x)$ in the form

$$f(x) = \left(\frac{1}{5x^3 - 2}\right)\left(\frac{1}{5x^3 - 2}\right),$$

and use the product rule,

$$f'(x) = \frac{1}{5x^3 - 2}\frac{d}{dx}\left(\frac{1}{5x^3 - 2}\right) + \frac{1}{5x^3 - 2}\frac{d}{dx}\left(\frac{1}{5x^3 - 2}\right)$$

$$= \frac{2}{5x^3 - 2}\frac{d}{dx}\left(\frac{1}{5x^3 - 2}\right).$$

The quotient rule now gives

$$f'(x) = \frac{2}{5x^3 - 2}\left[\frac{(5x^3 - 2)(0) - 1(15x^2)}{(5x^3 - 2)^2}\right]$$

$$= \frac{-30x^2}{(5x^3 - 2)^3}.$$

Differentiation in part (c) of this example was somewhat cumbersome considering the simplicity of the function. Imagine the difficulty had the power been 12 instead of 2. Fortunately, in Section 3.7 we develop a very simple rule for differentiating functions like this.

EXAMPLE 3.15

Plot a graph of the function $y = f(x) = (x - 1)/(x + 2)$ on the interval $-10 \leq x \leq 10$. Calculate dy/dx and show qualitatively that the graph agrees with your calculation.

SOLUTION   A plot is shown in Figure 3.24.

FIGURE 3.24

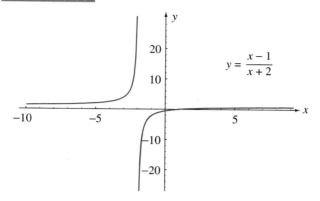

Using quotient rule 3.16, we find

$$\frac{dy}{dx} = \frac{(x+2)(1) - (x-1)(1)}{(x+2)^2} = \frac{3}{(x+2)^2}.$$

The sketch in Figure 3.24 and dy/dx agree that:

(a) the slope of the curve is always positive;

(b) the slope becomes larger and larger as x approaches -2, either from the left or the right, i.e.,

$$\lim_{x \to -2} f'(x) = \infty;$$

(c) the slope approaches zero as x approaches $\pm\infty$, i.e.,

$$\lim_{x \to \pm\infty} f'(x) = 0.$$

Statements (a), (b), and (c) in the above example were arrived at by examining $dy/dx = 3/(x+2)^2$. They could also have been realized by plotting a graph of the derivative function (Figure 3.25). Ordinates of this graph are slopes in Figure 3.24. For example, at $x = 1$, the height of the curve in Figure 3.25 is $1/3$. At $x = 1$ in Figure 3.24, the slope of the tangent line is $1/3$. Notice how clear statements (a), (b), and (c) are from the graph of $f'(x)$ in Figure 3.25. The slope is always positive, $\lim_{x \to -2} f'(x) = \infty$, and $\lim_{x \to \pm\infty} f'(x) = 0$.

FIGURE 3.25

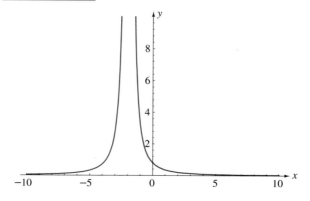

In Chapter 4 when we apply derivatives to geometric and physical problems, we need to know when the derivative exists and when it does not exist, and when it is positive, negative, and zero. A graph of the derivative function is an excellent way to discover and visualize these properties.

EXAMPLE 3.16

Plot a graph of the function $f(x) = (x^2 + 4x - 1)/(x^3 + 2)$ on the interval $-10 \le x \le 10$. Find, to five decimal places, points where the tangent line to the graph is horizontal.

SOLUTION   A graph is shown in Figure 3.26.

FIGURE 3.26

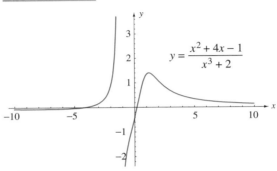

The graph has a vertical asymptote at $x = -2^{1/3}$. Obviously there is a point near $x = 1$ at which the tangent line is horizontal. In addition, because the graph crosses the x-axis near $x = -5$, and $y = 0$ is a horizontal asymptote as $x \to -\infty$, there must be at least one point to the left of $x = -5$ at which the tangent line is horizontal. Whether there is more than one such point is not clear in Figure 3.26. The plot of $f'(x)$ in Figure 3.27a crosses the x-axis near $x = 1$, thus confirming the point in Figure 3.26 where the tangent line is horizontal. It does not make clear, however, the number of points to the left of $x = -5$ at which the graph of $f'(x)$ crosses the x-axis. The graph of $f'(x)$ in Figure 3.27b clearly indicates one, and only one, point to the left of $x = -5$ at which $f'(x) = 0$. To find the two points then where $f'(x) = 0$, we set

$$0 = f'(x) = \frac{(x^3 + 2)(2x + 4) - (x^2 + 4x - 1)(3x^2)}{(x^3 + 2)^2}$$

$$= \frac{-x^4 - 8x^3 + 3x^2 + 4x + 8}{(x^3 + 2)^2}.$$

For a fraction to be equal to 0, its numerator must vanish, and we therefore set

$$-x^4 - 8x^3 + 3x^2 + 4x + 8 = 0.$$

Our computer gives two (real) solutions of this equation, $-8.316\,793$ and $1.238\,656\,0$. The Intermediate Value Theorem of Section 1.7 guarantees that $-8.316\,79$ and $1.238\,66$ are solutions, correct to five decimal places, when we calculate

$$f'(-8.316\,795) = -3.1 \times 10^{-9}, \qquad f'(-8.316\,785) = 1.5 \times 10^{-8},$$

$$f'(1.238\,655) = 2.2 \times 10^{-6}, \qquad f'(1.238\,665) = -1.9 \times 10^{-5}.$$

FIGURE 3.27a

$$f'(x) = \frac{-x^4 - 8x^3 + 3x^2 + 4x + 8}{(x^3 + 2)^2}$$

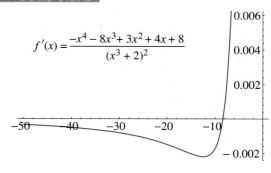

FIGURE 3.27b

$$f'(x) = \frac{-x^4 - 8x^3 + 3x^2 + 4x + 8}{(x^3 + 2)^2}$$

EXAMPLE 3.17

Show that when $f(x)$ is a function, differentiable for all x, and $H(x - a)$ is the Heaviside unit step function of Section 2.4,

$$\frac{d}{dx}[f(x)H(x - a)] = f'(x)H(x - a).$$

SOLUTION   The product rule gives

$$\frac{d}{dx}[f(x)H(x - a)] = f'(x)H(x - a) + f(x)H'(x - a).$$

Because $H'(x - a) = 0$ at all x, except $x = a$ where the derivative does not exist, the required result now follows.

EXERCISES 3.4

In Exercises 1–16 find $f'(x)$ in simplified form.

∗ **1.** $f(x) = (x^2 + 2)(x + 3)$

∗ **2.** $f(x) = (2 - x^2)(x^2 + 4x + 2)$

∗ **3.** $f(x) = \dfrac{x}{3x + 2}$

∗ **4.** $f(x) = \dfrac{x^2}{4x^2 - 5}$

∗ **5.** $f(x) = \dfrac{x^2}{2x - 1}$

∗ **6.** $f(x) = \dfrac{x^3}{4x^2 + 1}$

∗ **7.** $f(x) = \sqrt{x}(x + 1)$

∗ **8.** $f(x) = \dfrac{\sqrt{x}}{3x + 2}$

∗ **9.** $f(x) = \dfrac{2x^2 - 5}{3x + 4}$

∗ **10.** $f(x) = \dfrac{x + 5}{2x^2 - 1}$

$*$ **11.** $f(x) = \dfrac{x(x+1)}{1-3x}$

$*$ **12.** $f(x) = \dfrac{x^2 + 2x + 3}{x^2 - 5x + 1}$

$*$ **13.** $f(x) = \dfrac{x^{1/3}}{1-\sqrt{x}}$

$**$ **14.** $f(x) = \dfrac{\sqrt{x} + 2x}{\sqrt{x} - 4}$

$**$ **15.** $f(x) = \dfrac{1}{x^3 - 3x^2 + 2x + 5}$

$**$ **16.** $f(x) = \dfrac{x^3 + 3x^2 + 3x + 10}{(x+1)^3}$

$*$ **17.** Find equations for the tangent and normal lines to the curve $y = (x+3)/(x-4)$ at the point $(1, -4/3)$. Plot the curve and lines.

$**$ **18.**   (a) Plot a graph of the function $f(x) = x^2/(x^2 + x - 2)$.

   (b) Find $f'(x)$ and show that it agrees with the plot.

$**$ **19.** If the total cost of producing x items of a commodity is given by the equation

$$C(x) = ax\left(\frac{x+b}{x+c}\right),$$

where a, b, and c are constants, show that the marginal cost $C'(x)$ is

$$a\left[1 + \frac{c(b-c)}{(x+c)^2}\right].$$

$**$ **20.** Find a rule for the derivative of the product of three functions $f(x)$, $g(x)$, and $h(x)$.

In Exercises 21 and 22 find the angle (or angles) between the curves at their point (or points) of intersection.

$**$ **21.** $y = x(x-5)$, $y = x\sqrt{4-x}$ $**$ **22.** $y = 2x + 2$, $y = x^2/(x-1)$

$**$ **23.**   (a) A manufacturer's profit from the sale of x kilograms of a commodity per week is given by

$$P(x) = \frac{3x - 200}{x + 400}.$$

Plot a graph of this function.

   (b) The average profit per kilogram when x kilograms are sold is given by

$$p(x) = \frac{P(x)}{x}.$$

Plot a graph of this function.

   (c) If a point (x, P) on the total profit curve is joined to the origin, the slope of this line is the average profit $p(x)$ for that x. Use this idea to find the sales level for highest average profit.

$**$ **24.** Find all points on the curve $y = (5-x)/(6+x)$ at which the tangent line passes through the origin.

$**$ **25.**   (a) Find $f'(x)$ if $f(x) = H(x+1)(x^2+x)$. Draw graphs of $f(x)$ and $f'(x)$.

   (b) Does the derivative exist at $x = -1$ if we define $f(-1) = 0$?

$**$ **26.** Repeat Exercise 25 if $f(x) = H(x+1)(x+1)^2$.

3.5 Higher-Order Derivatives

When $y = f(x)$ is a function of x, its derivative $f'(x)$ is also a function of x. We can therefore take the derivative of the derivative to get what is called the **second derivative** of the function. This can be repeated over and over again. For instance, if $y = f(x) = x^3 + 1/x$, then

$$f'(x) = \frac{dy}{dx} = 3x^2 - \frac{1}{x^2}.$$

We denote the second derivative of y with respect to x by d^2y/dx^2 or $f''(x)$:

$$f''(x) = \frac{d^2y}{dx^2} = \frac{d}{dx}\left(\frac{dy}{dx}\right) = 6x + \frac{2}{x^3}.$$

Similarly,

$$f'''(x) = \frac{d^3y}{dx^3} = 6 - \frac{6}{x^4}.$$

This is called the **third derivative** or the derivative of order three. Clearly we can continue the differentiation process indefinitely to produce derivatives of any positive integer order whatsoever.

EXAMPLE 3.18

How many derivatives does $f(x) = x^{8/3}$ have at $x = 0$?

SOLUTION   Since

$$f'(x) = \frac{8}{3}x^{5/3}, \qquad f''(x) = \left(\frac{8}{3}\right)\left(\frac{5}{3}\right)x^{2/3}, \qquad f'''(x) = \left(\frac{8}{3}\right)\left(\frac{5}{3}\right)\left(\frac{2}{3}\right)x^{-1/3},$$

and $f'''(0)$ is not defined, $f(x)$ has only a first and second derivative at $x = 0$.

EXAMPLE 3.19

Find a formula for the second derivative of a product, $d^2(uv)/dx^2$, if $u = f(x)$ and $v = g(x)$.

SOLUTION   Since

$$\frac{d}{dx}(uv) = v\frac{du}{dx} + u\frac{dv}{dx},$$

then

$$\frac{d^2}{dx^2}(uv) = \frac{d}{dx}\left(v\frac{du}{dx} + u\frac{dv}{dx}\right)$$

$$= v\frac{d^2u}{dx^2} + \frac{dv}{dx}\frac{du}{dx} + u\frac{d^2v}{dx^2} + \frac{du}{dx}\frac{dv}{dx}$$

$$= v\frac{d^2u}{dx^2} + 2\frac{du}{dx}\frac{dv}{dx} + u\frac{d^2v}{dx^2}.$$

In Exercises 1–10 find the indicated derivative.

∗ **1.** $f''(x)$ if $f(x) = x^3 + 5x^4$ ∗ **2.** $f'''(x)$ if $f(x) = x^3 - 3x^2 + 2x + 1$

∗ **3.** $f''(2)$ if $f(x) = (x+1)(x^3 + 3x + 2)$

∗ **4.** $f'''(1)$ if $f(x) = x^4 - 3x^2 + 1/x$

∗ **5.** $f''(x)$ if $f(x) = (x+1)/\sqrt{x}$ ∗ **6.** $f'''(t)$ if $f(t) = t^3 - 1/t^3$

∗ **7.** d^9y/dx^9 if $y = x^{10}$ ∗∗ **8.** $f''(u)$ if $f(u) = \sqrt{u}/(u+1)$

∗∗ **9.** d^6t/dx^6 if $t = x/(2x - 6)$ ∗∗ **10.** $f''(x)$ if $f(x) = x/(\sqrt{x} + 1)$

∗∗ **11.** When a beam of length L and uniform mass m kilograms per metre is supported as shown in the figure below, it bends under its own weight. It can be shown that the deflection $y = f(x)$ from the horizontal must satisfy the equation

$$(EI)\frac{d^2y}{dx^2} = -\frac{mg}{2}(L - x)^2 + \frac{3}{8}mgL(L - x),$$

where E, I, and g are constants. Show that

$$y = f(x) = -\frac{mg}{48EI}(2x^4 - 5Lx^3 + 3L^2x^2)$$

satisfies this equation.

∗∗ **12.** Steady-state temperature T in a region bounded by two concentric spheres of radii a and b (where $a < b$) must satisfy the equation

$$\frac{d^2T}{dr^2} + \frac{2}{r}\frac{dT}{dr} = 0,$$

where r is the radial distance from the common centre of the spheres.

    (a) Show that, for any constants c and d, the function

$$T = f(r) = c + \frac{d}{r}$$

satisfies the equation.

    (b) If temperatures on the two spheres are maintained at constant values T_a and T_b, calculate c and d in terms of a, b, T_a, and T_b.

∗∗ **13.** If heat is generated at the centre of the sphere in Exercise 12 at a constant rate, the equation satisfied by $T(r)$ is

$$\frac{d}{dr}\left(r^2\frac{dT}{dr}\right) = kr^2$$

where k is a constant.

(a) Show that

$$T(r) = \frac{kr^2}{6} + \frac{c}{r} + d$$

satisfies the equation for any constants c and d.

(b) If temperatures on the two spheres are maintained at constant values T_a and T_b, calculate c and d in terms of a, b, T_a, T_b, and k.

** **14.** Find a and b in order that the function $y = f(x) = ax^3 + bx^2$ has its second derivative equal to zero when $x = 1$ and $y = 2$.

** **15.** Find constants a, b, c, and d in order that the function $y = f(x) = ax^3 + bx^2 + cx + d$ has its first derivative equal to 4 when $x = 1$ and $y = 0$, and its second derivative equal to 5 when $x = 2$ and $y = 4$.

** **16.**   (a) Find a formula for $d^3(uv)/dx^3$ if $u = f(x)$ and $v = g(x)$.

(b) On the basis of part (a) and Example 3.19, could you hazard a guess at a formula for $d^4(uv)/dx^4$ (i.e., do you see the pattern emerging)? Exercise 18 asks you to verify the correct formula.

** **17.** Evaluate

$$\frac{d^{2n}}{dx^{2n}}(x^2 - 1)^n$$

for n a positive integer.

*** **18.** If n is a positive integer, and u and v are functions of x, show by mathematical induction that

$$\frac{d^n}{dx^n}(uv) = \sum_{r=0}^{n} \binom{n}{r} \left(\frac{d^r u}{dx^r}\right) \left(\frac{d^{n-r} v}{dx^{n-r}}\right),$$

where $d^0 u/dx^0 = u$, and $\binom{n}{r}$ are the binomial coefficients

$$\binom{n}{r} = \frac{n!}{r!\,(n-r)!}.$$

(If you are not familiar with sigma notation, see Section 6.1 (page 400). Mathematical induction is discussed in Appendix A.)

3.6 Velocity, Speed, and Acceleration

Most applications of differentiation are dealt with in Chapter 4, but velocity, speed, and acceleration are so important in engineering and the physical sciences, it is appropriate to discuss them as soon as possible. In Section 1.4, we used the position function

$$x(t) = t^3 - 27t^2 + 168t + 20, \quad t \geq 0,$$

to define displacement and introduce speed in an intuitive way. In Section 3.1, we defined velocity as the instantaneous rate of change of displacement with respect to time. (Unless we indicate otherwise, velocity means instantaneous velocity, rather than average velocity.) We can now say that velocity is the derivative of displacement,

$$v(t) = \frac{dx}{dt} = 3t^2 - 54t + 168 \text{ m/s}.$$

In terms of the graph of the displacement function (Figure 3.28), velocity is the slope of the tangent line. For instance, $v(3) = 33$ m/s. The particle is moving to the right at 33 m/s; the slope of the tangent line at $(3, 308)$ is 33. (See the triangle on the tangent line and remember that the axes have different scales.) After 5 s, $v(5) = -27$ m/s. The particle is moving to the left at 27 m/s; the slope of the tangent line is -27 at $(5, 310)$. By factoring $v(t)$ in the form

$$v(t) = 3(t - 4)(t - 14),$$

we see that velocity is zero at $t = 4$ and $t = 14$. This is consistent with horizontal tangent lines at $(4, 324)$ and $(14, -176)$.

FIGURE 3.28

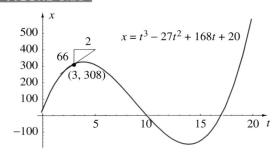

The particle is moving to the right whenever its velocity is positive and to the left when its velocity is negative. Graphically, it moves to the right when the slope of the tangent line is positive ($0 \le t < 4$ and $t > 14$), and moves to the left when the slope of the tangent line is negative ($4 < t < 14$).

Speed is defined to be the magnitude of velocity,

$$\text{speed} = |v(t)|. \tag{3.17}$$

It represents how fast the particle is moving without regard for direction. For instance, at time $t = 0$, velocity and speed are both 168 m/s, but at $t = 5$, velocity is -33 m/s and speed is 33 m/s. Do not confuse velocity and speed; they are not interchangeable. Velocity can be positive or negative depending on direction of motion; speed is never negative.

EXAMPLE 3.20

Figure 3.29 shows the graph of the displacement function $x(t)$ for a particle moving along the x-axis during the time interval $0 \le t \le 10$. Use the graph to answer the following questions:

(a) Is the particle moving to the left or right at $t = 7$ s?

(b) How many times does the particle stop moving?

(c) Is the velocity at $t = 0.5$ s greater or smaller than at $t = 3.5$ s?

(d) Is the speed at $t = 0.5$ s greater or smaller than at $t = 3.5$ s?

(e) Estimate the average velocity of the particle over the interval.

(f) At what times would you say that the velocity is largest and smallest?

(g) At what times would you say that the speed is largest and smallest?

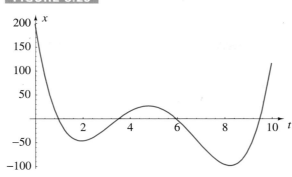

FIGURE 3.29

SOLUTION

(a) Since the slope of the tangent line at $t = 7$ is negative, the velocity is negative, and the particle is moving to the left.

(b) The particle stops moving when its velocity is zero, and this occurs when the tangent line to the graph is horizontal. This happens three times.

(c) At $t = 0.5$ s, the velocity is negative; at $t = 3.5$ s, it is positive. Thus, it is smaller at $t = 0.5$ s.

(d) The tangent line at $t = 0.5$ s is much steeper that it is at $t = 3.5$ s. Since speed measures steepness (slope without regard for sign), speed is greater at 0.5 s.

(e) Average velocity is the difference between initial and final displacements divided by the length of the time interval. If we estimate $x(0) = 200$ and $x(10) = 120$,

$$\text{average velocity} \approx \frac{120 - 200}{10} = -8 \text{ m/s.}$$

(f) Since the slope of the graph appears to be greatest at $t = 10$ s, the velocity must be largest then. The slope and velocity appear to be smallest at $t = 0$ s.

(g) Because speed is the magnitude of velocity, it is never negative. Speed is zero if velocity is zero. Since this happens three times, speed is smallest (and has value 0) at the three times when the tangent line is horizontal. Speed is greatest when steepness of the graph is greatest. This is at either $t = 0$ s or $t = 10$ s. It is difficult to tell from the graph, but we favour a steeper tangent line at $t = 0$ s.

EXAMPLE 3.21

You are now told that the displacement function for the graph in Figure 3.29 is

$$x(t) = t^4 - 20t^3 + \frac{521t^2}{4} - \frac{1243t}{4} + \frac{399}{2}.$$

Answer the following questions:

(a) What are speed and velocity at $t = 1$ s?

(b) Is the velocity equal to zero at $t = 2$ s?

(c) What is the average velocity of the particle for $0 \le t \le 10$?

(d) At what time is speed greatest?

(e) To three decimals, at what times is the speed of the particle equal to 20 m/s?

SOLUTION   The velocity of the particle is

$$v(t) = \frac{dx}{dt} = 4t^3 - 60t^2 + \frac{521t}{2} - \frac{1243}{4} \text{ m/s}.$$

The above questions will be answered algebraically, but it is helpful to visualize responses by plotting the velocity and speed functions. We have done this in Figures 3.30a and b. Figure 3.30b reflects that part of the graph in Figure 3.30a below the t-axis, in the t-axis.

FIGURE 3.30a

$$v = 4t^3 - 60t^2 + \frac{521t}{2} - \frac{1243}{4}$$

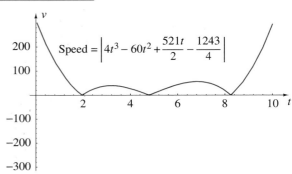

FIGURE 3.30b

$$\text{Speed} = \left| 4t^3 - 60t^2 + \frac{521t}{2} - \frac{1243}{4} \right|$$

(a) At $t = 1$ s, velocity is

$$v(1) = 4(1)^3 - 60(1)^2 + \frac{521}{2} - \frac{1243}{4} = -\frac{425}{4} \text{ m/s}.$$

The speed is 425/4 m/s.

(b) Geometrically, the velocity is zero at $t = 2$ s if the tangent line in Figure 3.29 is horizontal at $t = 2$. It looks close. Equivalently, does the graph in Figure 3.30a cross the t-axis at $t = 2$? Again the decision is not clear. Since $v(2) = 4(2)^3 - 60(2)^2 + (521/2)(2) - 1243/4 = 2.25$ m/s, the velocity is not zero at $t = 2$.

(c) Since displacements at $t = 0$ and $t = 10$ are 399/2 m and 117 m,

$$\text{average velocity} = \frac{117 - 399/2}{10} = -8.25 \text{ m/s}.$$

(d) Figure 3.30b indicates that speed is greatest at $t = 0$ s or $t = 10$ s, favouring $t = 0$ s. Because $v(0) = -1243/4$ m/s and $v(10) = 1177/4$ m/s, it follows that speed is greatest at $t = 0$ s (see also part (g) in Example 3.20).

(e) Solutions can be visualized as times when the height of the graph in Figure 3.30b is 20. Clearly, there are six such times. Alternatively, solutions are times when the ordinate in Figure 3.30a is ±20, the same six times. Solutions of

$$20 = v(t) = 4t^3 - 60t^2 + \frac{521t}{2} - \frac{1243}{4}$$

are 2.306 95, 4.240 43, and 8.452 62. The Intermediate Value Theorem can be used to confirm three-decimal solutions 2.307, 4.240, and 8.453. Three-decimal solutions of $v(t) = -20$ are 1.718, 5.300, and 7.982.

The velocity of a particle moving along the x-axis is the derivative of its displacement function $x(t)$ with respect to t,

$$v(t) = \frac{dx}{dt}. \tag{3.18}$$

The (**instantaneous**) **acceleration** of the particle is defined as the rate of change of velocity with respect to time:

$$a(t) = \frac{dv}{dt} = \frac{d^2x}{dt^2}. \tag{3.19}$$

In actual fact, $x(t)$, $v(t)$, and $a(t)$ are the x-components of the displacement, velocity, and acceleration vectors, respectively. In the absence of a complete discussion of vectors, we omit the terms "vector" and "component," and simply call $x(t)$, $v(t)$, and $a(t)$ displacement, velocity, and acceleration. When distance is measured in metres and time in seconds, velocity is measured in metres per second (m/s). Since acceleration is the time derivative of velocity, its units must be units of velocity divided by units of time, i.e., metres per second per second (m/s/s). Usually we shorten this by saying "metres per second squared," and write m/s^2.

For example, in Figure 3.28 the acceleration of the particle with displacement function $x(t) = t^3 - 27t^2 + 168t + 20$ is

$$a(t) = \frac{d}{dt}(3t^2 - 54t + 168) = 6t - 54 \text{ m/s}^2.$$

The graphical interpretation of acceleration requires the concept of "concavity." This is discussed in Section 4.4.

EXERCISES 3.6

∗ **1.** The figure below shows the graph of the displacement function $x(t)$ of a particle moving along the x-axis during the time interval $0 \le t \le 6$. Answer the following questions.

    (a) Is the particle to the left or right of the origin at times $t = 1$ and $t = 4$?

    (b) Is the particle moving to the right or to the left at times $t = 1/2$ and $t = 3$?

    (c) How many times does the particle change direction?

    (d) Is the velocity greater at $t = 7/2$ or at $t = 9/2$?

    (e) Is the speed greater at $t = 7/2$ or at $t = 9/2$?

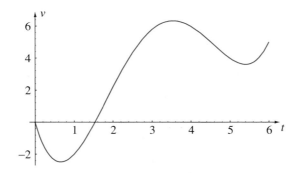

∗ **2.** You are given that the displacement function in Exercise 1 is

$$x(t) = \frac{t^4}{6} - \frac{32t^3}{15} + \frac{25t^2}{3} - \frac{251t}{30}.$$

Confirm each answer in Exercise 1 algebraically.

and we can therefore calculate

$$\frac{dy}{dx} = \frac{(\sqrt{x} + x + 2)\left(\dfrac{1}{2\sqrt{x}}\right) - \sqrt{x}\left(\dfrac{1}{2\sqrt{x}} + 1\right)}{(\sqrt{x} + x + 2)^2}.$$

This can be reduced to

$$\frac{dy}{dx} = \frac{2 - x}{2\sqrt{x}(\sqrt{x} + x + 2)^2}.$$

But notice that if we differentiate the original functions, we obtain

$$\frac{dy}{du} = \frac{(u + 1)(1) - u(1)}{(u + 1)^2} = \frac{1}{(u + 1)^2} \qquad \text{and}$$

$$\frac{du}{dx} = \frac{(x + 2)\left(\dfrac{1}{2\sqrt{x}}\right) - \sqrt{x}(1)}{(x + 2)^2} = \frac{2 - x}{2\sqrt{x}(x + 2)^2}.$$

The product of these derivatives is

$$\frac{dy}{du}\frac{du}{dx} = \frac{2 - x}{2\sqrt{x}(x + 2)^2(u + 1)^2} = \frac{2 - x}{2\sqrt{x}(x + 2)^2\left(\dfrac{\sqrt{x}}{x + 2} + 1\right)^2}$$

$$= \frac{2 - x}{2\sqrt{x}(\sqrt{x} + x + 2)^2}.$$

Consequently, for this example, we can write

$$\frac{dy}{dx} = \frac{dy}{du}\frac{du}{dx}.$$

The derivative of a composite function $y = f\big(g(x)\big)$ can always be calculated by the above formula, according to the following theorem.

THEOREM 3.9   (The Chain Rule)

If $y = f(u)$ and $u = g(x)$ are differentiable functions, then the derivative of the composite function $y = f\big(g(x)\big)$ is

$$\frac{dy}{dx} = \frac{dy}{du}\frac{du}{dx}. \tag{3.20a}$$

Increment notation is particularly useful in proving the chain rule.

PROOF   By equation 3.11, the derivative of the composite function is

$$\frac{dy}{dx} = \lim_{\Delta x \to 0} \frac{\Delta y}{\Delta x} = \lim_{\Delta x \to 0} \frac{f\big(g(x + \Delta x)\big) - f\big(g(x)\big)}{\Delta x}.$$

Now a change Δx in x produces a change $g(x + \Delta x) - g(x)$ in u. If we denote this change by Δu, then it in turn produces the change $f(u + \Delta u) - f(u)$ in y. We may write

$$\frac{dy}{dx} = \lim_{\Delta x \to 0} \frac{f(u + \Delta u) - f(u)}{\Delta x}.$$

Since $f(u)$ is differentiable, its derivative exists and is defined by

$$f'(u) = \lim_{\Delta u \to 0} \frac{f(u + \Delta u) - f(u)}{\Delta u}.$$

An equivalent way to express the fact that this limit is $f'(u)$ is to say that

$$\frac{f(u + \Delta u) - f(u)}{\Delta u} = f'(u) + \epsilon,$$

where ϵ must satisfy the condition $\lim_{\Delta u \to 0} \epsilon = 0$. We may write

$$f(u + \Delta u) - f(u) = [f'(u) + \epsilon]\Delta u,$$

and, if we substitute this into the second expression for dy/dx above, we obtain

$$\frac{dy}{dx} = \lim_{\Delta x \to 0} \frac{[f'(u) + \epsilon]\Delta u}{\Delta x} = \lim_{\Delta x \to 0} \left\{ [f'(u) + \epsilon]\frac{\Delta u}{\Delta x} \right\}.$$

But

$$\lim_{\Delta x \to 0} \frac{\Delta u}{\Delta x} = \lim_{\Delta x \to 0} \frac{g(x + \Delta x) - g(x)}{\Delta x} = \frac{du}{dx}.$$

Furthermore, since $g(x)$ is differentiable, it is continuous (Theorem 3.6), and this implies that $\Delta u \to 0$ as $\Delta x \to 0$. Consequently, $\lim_{\Delta x \to 0} \epsilon = \lim_{\Delta u \to 0} \epsilon = 0$, and these results give

$$\frac{dy}{dx} = f'(u)\frac{du}{dx} = \frac{dy}{du}\frac{du}{dx}. \qquad \blacksquare$$

This result is called the **chain rule** for the derivative of a composite function. It expresses the derivative of a composite function as the product of the derivatives of the functions in the composition. From the point of view of rates of change, the chain rule seems quite reasonable. It states that if a variable is defined in terms of a second variable, which is in turn defined in terms of a third variable, then the rate of change of the first variable with respect to the third is the rate of change of the first with respect to the second multiplied by the rate of change of the second with respect to the third. For example, if car A travels twice as fast as car B, and car B travels three times as fast as car C, then car A travels six times as fast as car C.

It is essential to understand the difference between the derivatives dy/dx and dy/du in rule 3.20a. The second, dy/du, is the derivative of y regarded as a function of u, the given function $y = f(u)$. On the other hand, dy/dx is the derivative of the composite function $f(g(x))$.

The chain rule can also be expressed in terms of the circle notation for composite functions. With $f(g(x))$ denoted by $(f \circ g)(x)$, 3.20a takes the form

$$(f \circ g)'(x) = f'(g(x)) \, g'(x), \qquad (3.20b)$$

or

$$(f \circ g)'(x) = (f' \circ g)(x) \, g'(x). \qquad (3.20c)$$

EXAMPLE 3.22

Find dy/dt at $t = 4$ when $y = x^2 - x$ and $x = \sqrt{t}/(t + 1)$.

SOLUTION By the chain rule

$$\frac{dy}{dt} = \frac{dy}{dx}\frac{dx}{dt} = (2x - 1)\left[\frac{(t + 1)(1/2)t^{-1/2} - \sqrt{t}(1)}{(t + 1)^2}\right].$$

When $t = 4$, we find $x = \sqrt{4}/(4 + 1) = 2/5$. Once again we use the notation $\dfrac{dy}{dt}_{|t=4}$ to represent dy/dt evaluated at $t = 4$,

$$\frac{dy}{dt}_{|t=4} = [2(2/5) - 1]\left[\frac{(4 + 1)(1/2)(4)^{-1/2} - \sqrt{4}}{(4 + 1)^2}\right] = \frac{3}{500}.$$

The Extended Power Rule

When the function $f(u)$ in Theorem 3.9 is a power function, the chain rule gives what we call the **extended power rule**, often considered the most important differentiation formula of calculus.

COROLLARY

When $u = g(x)$ is differentiable,

$$\frac{d}{dx}u^n = nu^{n-1}\frac{du}{dx}. \qquad (3.21)$$

It is essentially power rule 3.7 in Section 3.2 with an extra factor du/dx to account for the fact that what is under the power (u) is not just x, it is a function of x. In the special case that u is equal to x, 3.21 reduces to 3.7. It is important not to confuse these rules. Rule 3.7 can be used only for x^n; if anything other than x is raised to a power, formula 3.21 should be used. The most common error in using 3.21 is to forget the du/dx. With 3.21 it is a simple matter to differentiate the functions in the first paragraph of this section:

$$\frac{d}{dx}(2x^2 - 3)^8 = 8(2x^2 - 3)^7 \frac{d}{dx}(2x^2 - 3) = 8(2x^2 - 3)^7(4x) = 32x(2x^2 - 3)^7$$

and

$$\frac{d}{dx}\left[\frac{1}{(3x^2 + 8)^{12}}\right] = \frac{d}{dx}(3x^2 + 8)^{-12} = -12(3x^2 + 8)^{-13}\frac{d}{dx}(3x^2 + 8)$$

$$= \frac{-12}{(3x^2 + 8)^{13}}(6x) = \frac{-72x}{(3x^2 + 8)^{13}}.$$

Notice that in neither of these examples do you see the letter u, although rule 3.21 is stated in terms of u's. We could have introduced u, in the first example, say, by setting $u = 2x^2 - 3$, and proceeded as follows: With $u = 2x^2 - 3$,

$$\frac{d}{dx}(2x^2 - 3)^8 = \frac{d}{dx}u^8 = \frac{d}{du}(u^8)\frac{du}{dx} = 8u^7(4x) = 32xu^7 = 32x(2x^2 - 3)^7.$$

But this is unnecessary; with an understanding of 3.21, we should proceed directly to the derivatives without defining an intermediate variable. With a little practice, the writing should be shortened even more. For example, calculation of the derivative of $(2x^2 - 3)^8$ should appear as

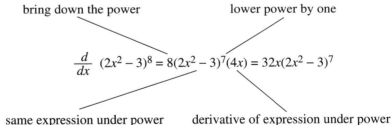

Here is an example to illustrate power rules within power rules within power rules.

EXAMPLE 3.23

Find the derivative of the function $f(x) = \sqrt{1 + \left(2 + \sqrt{1 - 3x}\right)^2}$.

SOLUTION   If we write the function in the form

$$f(x) = \left[1 + \left(2 + \sqrt{1 - 3x}\right)^2\right]^{1/2},$$

then

$$f'(x) = \frac{1}{2}\left[1 + \left(2 + \sqrt{1 - 3x}\right)^2\right]^{-1/2} \frac{d}{dx}\left[1 + \left(2 + \sqrt{1 - 3x}\right)^2\right]$$

$$= \frac{1}{2\sqrt{1 + \left(2 + \sqrt{1 - 3x}\right)^2}} \left[2\left(2 + \sqrt{1 - 3x}\right)\frac{d}{dx}\left(2 + \sqrt{1 - 3x}\right)\right]$$

$$= \frac{2 + \sqrt{1 - 3x}}{\sqrt{1 + \left(2 + \sqrt{1 - 3x}\right)^2}} \left[\frac{1}{2}(1 - 3x)^{-1/2}\frac{d}{dx}(1 - 3x)\right]$$

$$= \frac{2 + \sqrt{1 - 3x}}{2\sqrt{1 - 3x}\sqrt{1 + \left(2 + \sqrt{1 - 3x}\right)^2}}(-3)$$

$$= \frac{-3\left(2 + \sqrt{1 - 3x}\right)}{2\sqrt{1 - 3x}\sqrt{1 + \left(2 + \sqrt{1 - 3x}\right)^2}}.$$

EXERCISES 3.7

In Exercises 1–8 use the chain rule to find dy/dx.

* **1.** $y = t^2 + \dfrac{1}{t}, \quad t = x^2 + 1$

* **2.** $y = \dfrac{u}{u + 1}, \quad u = \sqrt{x} + 1$

* **3.** $y = (u^2 + 1)(u + 1), \quad u = \dfrac{1}{\sqrt{x - 4}}$

* **4.** $y = \dfrac{s}{s^2 - 2}, \quad s = x^2 - 2x + 3$

* **5.** $y = (v^2 + v)(\sqrt{v} + 1), \quad v = \dfrac{x}{x^2 - 1}$

* **6.** $y = \dfrac{t + 3}{t - 4}, \quad t = \dfrac{x - 2}{x + 1}$

* **7.** $y = \dfrac{t^2 + 3}{t - 4}, \quad t = (3x + 2)(x^2 + 4x)$

* **8.** $y = u^2(1 + \sqrt{u}), \quad u = \dfrac{x + 1}{x - x^2}$

In Exercises 9–36 find dy/dx, where $y = f(x)$.

** **9.** $f(x) = x(x^3 + 3)^4$

** **10.** $f(x) = x\sqrt{x + 1}$

** **11.** $f(x) = x^2(2x + 1)^2$

** **12.** $f(x) = \dfrac{x}{\sqrt{2x + 1}}$

** **13.** $f(x) = (x + 2)^2(x^2 + 3)$

** **14.** $f(x) = \dfrac{(2x - 1)^2}{3x + 5}$

** **15.** $f(x) = \dfrac{3x + 5}{(2x - 1)^2}$

** **16.** $f(x) = x^3(2 - 5x^2)^{1/3}$

** **17.** $f(x) = \dfrac{x^3}{(2 - 5x^2)^{1/3}}$

** **18.** $f(x) = (x + 1)^2(3x + 1)^3$

** 19. $f(x) = \dfrac{x^{1/3}}{1 - \sqrt{x}}$

** 20. $f(x) = \dfrac{\sqrt{2 - 3x}}{x^2}$

** 21. $f(x) = \left(\dfrac{x^3 - 1}{2x^3 + 1}\right)^4$

** 22. $f(x) = \sqrt[4]{\dfrac{2 - x}{2 + x}}$

** 23. $f(x) = (x^3 - 2x^2)^3 (x^4 - 2x)^5$

** 24. $f(x) = (x + 5)^4 \sqrt{1 + x^3}$

** 25. $f(x) = \dfrac{x\sqrt{1 - x^2}}{(3 + x)^{1/3}}$

** 26. $f(x) = x(x + 5)^4 \sqrt{1 + x^3}$

** 27. $f(x) = \dfrac{x^2(x^3 + 3)^2}{(x - 2)(x + 5)^2}$

** 28. $f(x) = x\sqrt{1 + x\sqrt{1 + x}}$

** 29. $y = t^2 + \dfrac{1}{t^3}, \quad t = \sqrt{4 - x^2}$

** 30. $y = (2s - s^2)^{1/3}, \quad s = \dfrac{1}{x^2 + 5}$

** 31. $y = \dfrac{v^2}{v^3 - 1}, \quad v = x\sqrt{x^2 - 1}$

** 32. $y = \dfrac{u}{u + 5}, \quad u = \dfrac{\sqrt{x - 1}}{x}$

** 33. $y = u^4(u^3 - 2u)^2, \quad u = \sqrt{x - 2x^2}$

** 34. $y = t + \sqrt{t + \sqrt{t}}, \quad t = \dfrac{x^2 + 1}{x^2 - 1}$

** 35. $y = \left(\dfrac{v^2 + 1}{1 - v^3}\right)^3, \quad v = \dfrac{1}{x^3 + 3x^2 + 2}$

** 36. $y = \dfrac{\sqrt{k}}{1 + k + k^2}, \quad k = x(x^2 + 5)^5$

** 37. If $y = f(u)$, $u = g(s)$, and $s = h(x)$, show that

$$\frac{dy}{dx} = \frac{dy}{du}\frac{du}{ds}\frac{ds}{dx}.$$

** 38. When an electrostatic charge $q = 5 \times 10^{-6}$ coulombs is at a distance r metres from a stationary charge $Q = 3 \times 10^{-6}$ coulombs, the magnitude of the force of repulsion of Q on q is

$$F = \frac{Qq}{4\pi \epsilon_0 r^2} \text{ newtons,}$$

where $\epsilon_0 = 8.85 \times 10^{-12}$. If q is moved directly away from Q at 2 m/s, how fast is F changing when $r = 2$ m?

** 39. When a mass m of 5 kg is r metres from the centre of the earth, the magnitude of the force of attraction of the earth on m is

$$F = \frac{GmM}{r^2} \text{ newtons,}$$

where M is the mass of the earth and $G = 6.67 \times 10^{-11}$. If m is falling at 100 km/h when it is 5 km above the surface of the earth, how fast is F changing? The mean density of the earth is 5.52×10^3 km/m^3, and its mean radius is 6370 km.

** **40.** Show that if an equation can be solved for both y in terms of x ($y = f(x)$) and x in terms of y ($x = g(y)$), then

$$\frac{dy}{dx} = \frac{1}{\dfrac{dx}{dy}}.$$

41. Prove that the derivative of an odd function is an even function, and that the derivative of an even function is an odd function. Recall definition 1.21 in Section 1.3.

** **42.** Determine whether the following very simple proof of the chain rule has a flaw. If Δu denotes the change in $u = g(x)$ resulting from a change Δx in x, and Δy is the change in $y = f(u)$ resulting from Δu, then

$$\frac{dy}{dx} = \lim_{\Delta x \to 0} \frac{\Delta y}{\Delta x} = \lim_{\Delta x \to 0} \frac{\Delta y}{\Delta u}\frac{\Delta u}{\Delta x} = \left(\lim_{\Delta x \to 0} \frac{\Delta y}{\Delta u}\right)\left(\lim_{\Delta x \to 0} \frac{\Delta u}{\Delta x}\right)$$

$$= \left(\lim_{\Delta u \to 0} \frac{\Delta y}{\Delta u}\right)\left(\lim_{\Delta x \to 0} \frac{\Delta u}{\Delta x}\right) = \frac{dy}{du}\frac{du}{dx}.$$

In Exercises 43–46 find d^2y/dx^2.

** **43.** $y = v^2 + v, \quad v = \dfrac{x}{x+1}$

** **44.** $y = (u+1)^3 - \dfrac{1}{u}, \quad u = x + \sqrt{x+1}$

** **45.** $y = \sqrt{t-1}, \quad t = (x+x^2)^2$

** **46.** $y = \dfrac{s}{s+6}, \quad s = \dfrac{\sqrt{x}}{1+\sqrt{x}}$

In Exercises 47–54 assume that $f(u)$ is a differentiable function of u. Find the derivative of the given function with respect to x in as simplified a form as possible. Then set $f(u) = u^3 - 2u$ and simplify further.

** **47.** $f(2x+3)$

** **48.** $[f(3-4x)]^2$

** **49.** $f(1-x^2)$

** **50.** $f(x+1/x)$

** **51.** $f\big(f(x)\big)$

** **52.** $\sqrt{3 - 4[f(1-3x)]^2}$

** **53.** $\dfrac{f(-x)}{3 + 2f(x^2)}$

** **54.** $f\big(x - f(x)\big)$

In Exercises 55–58 show that the families of curves are orthogonal trajectories (see Exercise 23 in Section 3.3). Draw both families of curves.

** **55.** $y = mx, \quad x^2 + y^2 = r^2$

** **56.** $y = ax^2, \quad x^2 + 2y^2 = c^2$

** **57.** $x^2 - y^2 = C_1, \quad xy = C_2$

** **58.** $2x^2 + 3y^2 = C^2, \quad y^2 = ax^3$

** **59.** If $y = f(u)$ and $u = g(x)$, use the chain rule to show that

$$\frac{d^2y}{dx^2} = \frac{d^2u}{dx^2}\frac{dy}{du} + \frac{d^2y}{du^2}\left(\frac{du}{dx}\right)^2.$$

** **60.** Use the result of Exercise 59 to find d^2y/dx^2 at $x = 1$ when

$$y = (u+1)^3, \quad u = 3x - \frac{2}{x^2}.$$

** **61.** Find the rate of change of $y = f(x) = x^9 + x^6$ with respect to x^3.

** **62.** Find the rate of change of $y = f(x) = \sqrt{1 - x^2}$ with respect to $x/(x + 1)$.

** **63.** If $y = f(u)$ and $u = g(x)$, show that

$$\frac{d^3 y}{dx^3} = \frac{d^3 u}{dx^3}\frac{dy}{du} + 3\frac{d^2 y}{du^2}\frac{d^2 u}{dx^2}\frac{du}{dx} + \frac{d^3 y}{du^3}\left(\frac{du}{dx}\right)^3.$$

** **64.** At what point(s) on the hyperbola $x^2 - 16y^2 = 16$ do tangent lines pass through the point $(2, 3)$?

** **65.** Generalize the result of equation 3.13 to find a formula for

$$\frac{d}{dx}|f(x)|^n, \qquad \text{for } n > 1 \text{ an integer.}$$

** **66.** The curve $x^2 y^2 = (x + 1)^2(4 - x^2)$ is called a *conchoid of Nicomedes*. Plot the curve and find points where its tangent line is horizontal.

3.8 Implicit Differentiation

We say that y is defined **explicitly** as a function of x if the dependence of y on x is given in the form

$$y = f(x). \tag{3.22}$$

Examples are $y = x^2$, $y = 3x + \sin x$, and $y = 1/(x + 1)$. In each case, the dependent variable stands alone on the left side of the equation. The differentiation rules in Theorems 3.1 to 3.5, 3.7, and 3.8 are applicable to explicitly defined functions.

An equation in x and y may define y as a function of x even when it is not in form 3.22. Equations for which y has not been separated out are often written in the generic form

$$F(x, y) = 0. \tag{3.23}$$

FIGURE 3.31

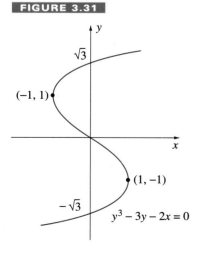

The notation $F(x, y)$ is used to denote an expression that depends on two variables x and y. For example, the volume of a right circular cylinder depends on its radius r and height h; in particular, $V = \pi r^2 h$. In such a case we write that $V = F(r, h) = \pi r^2 h$. Examples of equations of form 3.23 are $y - x^3 = 0$ and $y^3 - 3y - 2x = 0$. Equation 3.23 is said to define y **implicitly** as a function of x in some domain D, if for each x in D there is one, and only one, value of y for which (x, y) satisfies 3.23. The equation $y - x^3 = 0$ defines y implicitly as a function of x for all (real) x. The explicit definition of the function is $y = x^3$. The equation $y^3 - 3y - 2x = 0$ does not define y as a function of x. For each x in the interval $-1 < x < 1$, there are three solutions of the equation for y. For $x < -1$ and $x > 1$, the equation has only one value of y corresponding to each x. This is most easily seen from the plot of the curve in Figure 3.31. We shall have more to say about the equation $y^3 - 3y - 2x = 0$ later in this section.

Each of the following equations defines y as a function of x, and does so implicitly:

$$y + x^2 - 2x + 5 = 0;$$

$$2y + x^2 + y^2 - 25 = 0, \quad y \geq -1;$$

$$x + y^5 + x^2 + y = 0.$$

It is easy to solve the first equation for the explicit definition of the function,

$$y = -x^2 + 2x - 5,$$

and find its derivative,

$$\frac{dy}{dx} = -2x + 2.$$

We can also obtain an explicit definition of the function defined by the second equation, but not so simply. If we write

$$y^2 + 2y + (x^2 - 25) = 0,$$

and use quadratic formula 1.10, we obtain

$$y = \frac{-2 \pm \sqrt{4 - 4(x^2 - 25)}}{2} = -1 \pm \sqrt{26 - x^2}.$$

Since y is required to be greater than or equal to -1, we must choose

$$y = -1 + \sqrt{26 - x^2}.$$

This is the explicit definition of the function, and it is now easy to find the derivative of y with respect to x:

$$\frac{dy}{dx} = \frac{1}{2\sqrt{26 - x^2}}(-2x) = \frac{-x}{\sqrt{26 - x^2}}.$$

The third equation is quite different, for it is impossible to solve this equation for an explicit definition of the function. Does this mean that it is also impossible to obtain dy/dx? The answer is no! To see this, we differentiate both sides of the equation with respect to x, keeping in mind that y is a function of x. Only the term y^5 presents any difficulty, but its derivative can be calculated using extended power rule 3.21,

$$1 + 5y^4\frac{dy}{dx} + 2x + \frac{dy}{dx} = 0.$$

We can now solve this equation for dy/dx by grouping the two terms in dy/dx on one side of the equation, and transposing the remaining two terms:

$$(5y^4 + 1)\frac{dy}{dx} = -1 - 2x.$$

Division by $5y^4 + 1$ gives the required derivative,

$$\frac{dy}{dx} = -\frac{2x + 1}{5y^4 + 1}.$$

This process of differentiating an equation that implicitly defines a function is called **implicit differentiation**. It could also have been used in each of the first two examples. If $y + x^2 - 2x + 5 = 0$ is differentiated with respect to x, we have

$$\frac{dy}{dx} + 2x - 2 = 0,$$

from which, as before,

$$\frac{dy}{dx} = -2x + 2.$$

When $2y + x^2 + y^2 - 25 = 0$ is differentiated with respect to x, we find

$$2\frac{dy}{dx} + 2x + 2y\frac{dy}{dx} = 0,$$

from which,

$$(2 + 2y)\frac{dy}{dx} = -2x,$$

and therefore

$$\frac{dy}{dx} = \frac{-x}{y + 1}.$$

Although this result appears different from the previous expression for dy/dx, when we recall that $y = -1 + \sqrt{26 - x^2}$, we find

$$\frac{dy}{dx} = \frac{-x}{-1 + \sqrt{26 - x^2} + 1} = \frac{-x}{\sqrt{26 - x^2}}.$$

These examples illustrate that if implicit differentiation is used to obtain a derivative, then the result may depend on y as well as x. Naturally, if we require the derivative at a certain value of x, then the y-value used is determined by the original equation defining y implicitly as a function of x.

EXAMPLE 3.24

Assuming that y is defined implicitly as a function of x by the equation

$$x^3y^3 + x^2y + 2x = 12,$$

find dy/dx when $x = 1$.

SOLUTION   When we differentiate both sides of the equation with respect to x, using the product rule on the first two terms, we find

$$3x^2y^3 + 3x^3y^2\frac{dy}{dx} + 2xy + x^2\frac{dy}{dx} + 2 = 0.$$

Thus,

$$(3x^3y^2 + x^2)\frac{dy}{dx} = -(2 + 2xy + 3x^2y^3) \qquad \text{and}$$

$$\frac{dy}{dx} = -\frac{2 + 2xy + 3x^2y^3}{3x^3y^2 + x^2}.$$

When $x = 1$ is substituted into the given equation defining y as a function of x, the result is

$$0 = y^3 + y - 10 = (y - 2)(y^2 + 2y + 5),$$

and the only solution of this equation is $y = 2$. We now substitute $x = 1$ and $y = 2$ into the formula for dy/dx to calculate the derivative at $x = 1$,

$$\frac{dy}{dx}\bigg|_{x=1} = -\frac{2 + 2(1)(2) + 3(1)^2(2)^3}{3(1)^3(2)^2 + (1)^2} = -\frac{30}{13}.$$

Implicit differentiation can also be used to find second and higher order derivatives of functions that are defined implicitly. Calculations can be messy, but principles are the same. For example, when the equation $x^5 + y^3 + y^2 = 1$ defines y implicitly as a function of x, it is straightforward to calculate that

$$\frac{dy}{dx} = \frac{-5x^4}{3y^2 + 2y}.$$

To find the second derivative d^2y/dx^2, we differentiate both sides of this equation with respect to x. We use the quotient rule on the right, and when differentiating the denominator we once again keep in mind that y is a function of x. The result is

$$\frac{d^2y}{dx^2} = \frac{(3y^2 + 2y)(-20x^3) + 5x^4\left(6y\dfrac{dy}{dx} + 2\dfrac{dy}{dx}\right)}{(3y^2 + 2y)^2}.$$

We now replace dy/dx by its expression in terms of x and y,

$$\frac{d^2y}{dx^2} = \frac{-20x^3(3y^2 + 2y) + 5x^4(6y + 2)\left(\dfrac{-5x^4}{3y^2 + 2y}\right)}{(3y^2 + 2y)^2},$$

and bring the two terms in the numerator to a common denominator,

$$\frac{d^2y}{dx^2} = \frac{\dfrac{-20x^3(3y^2 + 2y)^2 - 25x^8(6y + 2)}{(3y^2 + 2y)}}{(3y^2 + 2y)^2}$$

$$= -\frac{20x^3(3y^2 + 2y)^2 + 25x^8(6y + 2)}{(3y^2 + 2y)^3}.$$

It would not be a pleasant task to proceed to the third derivative of y with respect to x.

In the above examples we seem to have adopted the principle that both sides of an equation can be differentiated with respect to the same variable. But can all equations be differentiated? The answer is an emphatic no! For example, if we differentiate both sides of the equation $4x = 2x$, we obtain the ludicrous result that $4 = 2$. Obviously, then, this equation cannot be differentiated with respect to x.

FIGURE 3.32

Possibly the reason differentiation fails in the above example is that the equation contains only one variable, namely, x. Perhaps a more reasonable question might be: "Can every equation containing two variables be differentiated?" To answer this question we consider the situation of a ship heading northeast from a pier. Let us choose east as the positive x-direction and north as the positive y-direction, both originating from the pier (Figure 3.32). We will find the rate of change of the y-coordinate of the ship with respect to its x-coordinate when the ship is 2 km from the pier. Obviously, the answer is 1, since the x- and y-coordinates change at the same rate. But consider the following argument.

When the ship is 2 km from the pier, $x = y = \sqrt{2}$, and therefore

$$x^2 + y^2 = 4.$$

If we differentiate this equation with respect to x, we find that

$$2x + 2y\frac{dy}{dx} = 0, \qquad \text{or} \qquad \frac{dy}{dx} = -\frac{x}{y},$$

and with $x = y = \sqrt{2}$,

$$\frac{dy}{dx} = -1.$$

Differentiation of the equation $x^2 + y^2 = 4$, which contains two variables, has again led to an erroneous result.

These two examples have certainly illustrated that not all equations can be differentiated. What then distinguishes an equation that can be differentiated from one that cannot? The answer is that an equation can be differentiated with respect to a variable only if the equation is valid for a continuous range of values of that variable. The equation $4x = 2x$ cannot be differentiated because it is true only for $x = 0$. The equation $x^2 + y^2 = 4$ in the above example cannot be differentiated because *in that example* it is valid only when the ship is 2 km from the pier; i.e., only when $x = y = \sqrt{2}$. In other situations it might be acceptable to differentiate $x^2 + y^2 = 4$. For example, if at position $(\sqrt{2}, \sqrt{2})$, the ship is following the circular path defined by the equation $x^2 + y^2 = 4$, then it is indeed correct to differentiate $x^2 + y^2 = 4$ to find $dy/dx = -1$. This is an extremely important principle, and we will return to it many times. To emphasize it once again, *we may differentiate an equation with respect to a variable only if the equation is valid for a continuous range of values of that variable.*

Each of the three equations at the beginning of this section was given as defining y as a function of x, and as such defines the function for some domain of values for x. Differentiation of these equations was therefore acceptable according to the above principle.

When we are told that equation 3.23 defines y as a function of x, implicit differentiation leads to the derivative dy/dx. But how can we tell whether an equation $F(x, y) = 0$ defines y implicitly as a function of x? Additionally, given that $F(x, y) = 0$ does define y as a function of x, how do we know that the expression for dy/dx obtained by implicit differentiation is a valid representation of the derivative of the function? After all, functions do not always have derivatives at all points in their domains.

Answers to these questions are intimately related. To see how, suppose an equation $F(x, y) = 0$ in x and y is satisfied by a point (x_0, y_0), and when the equation is differentiated implicitly it leads to a quotient for dy/dx,

$$\frac{dy}{dx} = \frac{P(x, y)}{Q(x, y)}.$$

Mathematicians have shown that if $Q(x_0, y_0) \neq 0$, then the equation $F(x, y) = 0$ defines y implicitly as a function of x for some open interval containing x_0, and the derivative of this function at x_0 is $P(x_0, y_0)/Q(x_0, y_0)$. Proofs are usually given in advanced books on mathematical analysis. When $Q(x_0, y_0) = 0$, two possibilities exist. First, the equation $F(x, y) = 0$ might not define y as a function of x in an interval around x_0. Second, the equation might define a function, but the function does not have a derivative at x_0. To illustrate, consider first the equation $y^3 - 3y - 2x = 0$, which we introduced earlier in this section. The curve defined by this equation is shown in Figure 3.31 (page 191), and it clearly illustrates that the equation does not define y as a function of x. It does define the three functions of x in Figure 3.33. Suppose we differentiate the equation with respect to x,

$$3y^2 \frac{dy}{dx} - 3\frac{dy}{dx} - 2 = 0,$$

and solve for dy/dx,

$$\frac{dy}{dx} = \frac{2}{3(y^2 - 1)}.$$

This derivative is obviously undefined at the points $(-1, 1)$ and $(1, -1)$, and these are precisely the points which separate the original curve into three parts. It is impossible to find a portion of the curve around either of these points which defines y as a function of x. At any other point on the curve, the formula $dy/dx = (2/3)(y^2 - 1)^{-1}$ is a valid representation for the derivative for whichever function of Figure 3.33 contains the point.

FIGURE 3.33

FIGURE 3.34

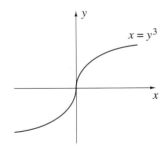

Consider now the equation $x = y^3$, shown graphically in Figure 3.34. Implicit differentiation leads to

$$\frac{dy}{dx} = \frac{1}{3y^2}.$$

Clearly, y is a function of x for all x, but dy/dx is undefined at $(0, 0)$ since the tangent line is vertical at this point.

EXAMPLE 3.25

Plot the curve defined by the equation $y^2 = x^2 - x^4$. Use implicit differentiation to find dy/dx and discuss what happens when the point $(0, 0)$ is substituted into the result.

FIGURE 3.35

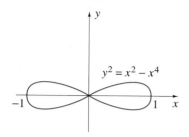

SOLUTION   The curve is shown in Figure 3.35. Implicit differentiation of $y^2 = x^2 - x^4$ gives

$$2y\frac{dy}{dx} = 2x - 4x^3,$$

from which

$$\frac{dy}{dx} = \frac{x(1 - 2x^2)}{y}.$$

This result is undefined at $(0, 0)$, and Figure 3.35 indicates why. The equation does not define y as a function of x around $x = 0$.

EXAMPLE 3.26

Solve Example 1.13 (page 26) using derivatives.

SOLUTION   To the right of the right branch of the hyperbola $x^2 - 4y^2 = 5$ in Figure 3.36 is a dense forest. We are required to find the point at which a person P_2 walking along the line $x = -1$ in the negative y-direction disappears from view of a person P_1 standing at position $(15, 10)$.

FIGURE 3.36

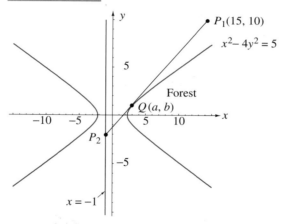

The required point P_2 occurs when line $P_1 P_2$ is tangent to the hyperbola. Let the point of tangency be $Q(a, b)$. Differentiation of $x^2 - 4y^2 = 5$ with respect to x gives

$$2x - 8y\frac{dy}{dx} = 0 \qquad \Longrightarrow \qquad \frac{dy}{dx} = \frac{x}{4y}.$$

The slope of the tangent line at $Q(a, b)$ is therefore $a/(4b)$. Since the slope of $P_1 Q$ is $(b - 10)/(a - 15)$, it follows that

$$\frac{b - 10}{a - 15} = \frac{a}{4b}.$$

Cross multiplication leads to

$$a^2 - 4b^2 = 15a - 40b.$$

Since $Q(a, b)$ is on the hyperbola, it also follows that

$$a^2 - 4b^2 = 5.$$

When we subtract these equations,

$$0 = 15a - 40b - 5 \qquad \Longrightarrow \qquad a = \frac{1}{3}(1 + 8b).$$

Substitution of this into $a^2 - 4b^2 = 5$ gives

$$\frac{1}{9}(1 + 8b)^2 - 4b^2 = 5 \qquad \Longrightarrow \qquad 1 + 16b + 64b^2 - 36b^2 = 45.$$

Thus,

$$0 = 28b^2 + 16b - 44 = 4(7b + 11)(b - 1),$$

and solutions are $b = 1$ and $b = -11/7$. The second solution gives a point on the left branch of the hyperbola which a tangent line passes through $(15, 10)$. The solution $b = 1$ gives the point $(3, 1)$ on the right branch of the hyperbola. The slope of the tangent line at $(3, 1)$ is $3/4$, and therefore its equation is $y - 1 = (3/4)(x - 3)$. It cuts the line $x = -1$ when $y = 1 + (3/4)(-1 - 3) = -2$. Thus, P_2 disappears from view at $(-1, -2)$.

EXERCISES 3.8

In Exercises 1–10 find dy/dx wherever y is defined as a function of x.

∗ **1.** $y^4 + y = 4x^3$ ∗ **2.** $x^4 + y^2 + y^3 = 1$

∗ **3.** $xy + 2x = 4y^2 + 2$ ∗ **4.** $2x^3 - 3xy^4 + 5xy - 10 = 0$

∗ **5.** $x + xy^5 + x^2y^3 = 3$ ∗ **6.** $(x + y)^2 = 2x$

∗ **7.** $x(x - y) - 4y^3 = 2x + 5$ ∗∗ **8.** $\sqrt{x + y} + y^2 = 12x^2 + y$

∗∗ **9.** $\sqrt{1 + xy} - xy = 15$ ∗∗ **10.** $\dfrac{x}{x + y} - \dfrac{y}{x} = 4$

∗ **11.** Find equations for the tangent and normal lines to the curve $xy^2 + y^3 = 2$ at the point $(1, 1)$.

In Exercises 12–15 find d^2y/dx^2 wherever y is defined as a function of x.

∗∗ **12.** $x^2 + y^3 + y = 1$ ∗∗ **13.** $2x^2 - y^3 = 4 - xy$

∗∗ **14.** $y^2 + 2y = 5x$ ∗∗ **15.** $(x + y)^2 = x$

∗∗ **16.** Find dy/dx at $x = 1$ when $x^3y + xy^3 = 2$.

In Exercises 17–20 find dy/dx and d^2y/dx^2.

∗∗ **17.** $(x + y)^2 = x^2 + y^2$ ∗∗ **18.** $x^2y^3 + 2x + 4y = 5$

∗∗ **19.** $xy^2 - 3x^2y = x + 1$ ∗∗ **20.** $x = y\sqrt{1 - y^2}$

** **21.** Find dy/dx when $x = 0$ if $\sqrt{1 - xy} + 3y = 4$.

** **22.** Find dy/dx when $x = 1$ if $x^2 y^3 + xy = 2$.

** **23.** Find dy/dx and $d^2 y/dx^2$ when $x = 2$ if $y^5 + (x - 2)y = 1$.

** **24.** Find dy/dx and $d^2 y/dx^2$ when $y = 1$ if $x^2 + 2xy + 3y^2 = 2$.

** **25.** Find point(s) on the curve $xy^2 + x^2 y = 16$ at which the slope of the tangent line is equal to zero.

** **26.** Find point(s) on the curve $x^2 + y^{2/3} = 2$ at which the second derivative is equal to zero.

** **27.** If a thermal nuclear reactor is built in the shape of a right circular cylinder of radius r and height h, then, according to neutron diffusion theory, r and h must satisfy the equation

$$\left(\frac{2.4048}{r}\right)^2 + \left(\frac{\pi}{h}\right)^2 = k = \text{constant.}$$

Find dr/dh.

** **28.** The elasticity of a function $y = f(x)$ is defined as

$$\frac{Ey}{Ex} = \frac{x}{y}\frac{dy}{dx}.$$

Calculate elasticity for the function defined by each of the following equations:

(a) $f(x) = x\left(\dfrac{x + 1}{x + 2}\right)$ (b) $x = \dfrac{400y + 200}{3 - y}$

** **29.** Show that the elasticity of a function is equal to one if and only if the tangent line to its graph passes through the origin.

** **30.** Find that point $P(a, b)$ on the first-quadrant part of the ellipse $2x^2 + 3y^2 = 14$ at which the tangent line at P is perpendicular to the line joining P and $(2, 5)$.

** **31.** Show that the equation of the tangent line to the hyperbola $b^2 x^2 - a^2 y^2 = a^2 b^2$ at the point (x_0, y_0) is $b^2 x x_0 - a^2 y y_0 = a^2 b^2$.

** **32.** Prove that for any circle $(x - h)^2 + (y - k)^2 = r^2$,

$$\left| \frac{d^2 y/dx^2}{[1 + (dy/dx)^2]^{3/2}} \right| = \frac{1}{r}.$$

** **33.** A solution passes through a conical filter 24 cm deep and 16 cm across the top into a cylindrical vessel of diameter 12 cm. Find an equation relating the depth h of solution in the filter and depth H of solution in the cylinder. What is the rate of change of h with respect to H?

** **34.** If x objects are sold at a price of $r(x)$ per object, the total revenue is $R(x) = xr(x)$. Find the marginal revenue $R'(x)$ if price is defined implicitly by the equation

$$x = 4a^3 - 3ar^2 + r^3,$$

where $a > 0$ is a constant, and $0 < x < 4a^3$.

∗∗ **35.** The general polynomial of degree n is

$$a_0 + a_1 x + a_2 x^2 + \cdots + a_n x^n,$$

where $a_0, a_1, \ldots, a_n$ are constants. Show that two polynomials of degree n can be equal for all x,

$$a_0 + a_1 x + a_2 x^2 + \cdots + a_n x^n = b_0 + b_1 x + b_2 x^2 + \cdots + b_n x^n,$$

if and only if $a_0 = b_0, a_1 = b_1, \ldots, a_n = b_n$.

∗∗ **36.** (a) Find $f'(0)$ if $y = f(x)$ is defined implicitly as a function of x by

$$x\sqrt{1 + 2y} = x^2 - y.$$

(b) Show that by squaring the equation in part (a), we obtain

$$x^2 + 4x^2 y = x^4 + y^2.$$

Differentiate this equation with respect to x to find $f'(0)$. Do you have any difficulties? Explain.

∗∗ **37.–40.** Use implicit differentiation to redo Exercises 55–58 in Section 3.7 (page 190).

∗∗ **41.** (a) Use implicit differentiation to find dy/dx if $y^2 = x^2 - 4x^4$.

(b) Can you calculate dy/dx at $x = 0$ using the result of part (a)? Plot the curve $y^2 = x^2 - 4x^4$ in order to explain this difficulty.

∗∗ **42.** (a) Find dy/dx if $\sqrt{x} + \sqrt{y} = 1$, where $0 \le x \le 1$, defines y implicitly as a function of x.

(b) Draw the curve $\sqrt{|x|} + \sqrt{|y|} = 1$.

∗∗ **43.** What is dy/dx if $x^2 + 4x + y^2 + 6y + 15 = 0$?

∗∗ **44.** Find dy/dx if $y = u/\sqrt{u^2 - 1}$, and $x^2 u^2 + \sqrt{u^2 - 1} = 4$ defines u implicitly as a function of x.

∗∗ **45.** Given that the equations

$$y^4 + y v^3 = 3 \qquad \text{and} \qquad x^2 v + 3x v^2 = 2x^3 y + 1$$

define y as a function of v, and v as a function of x, find dy/dx in terms of x, y, and v.

∗∗ **46.** Show that any function defined implicitly by the equation

$$\frac{x^2}{y^3} - x = C,$$

where C is a constant, satisfies the equation

$$3x^2 \frac{dy}{dx} + y^4 = 2xy.$$

∗∗ **47.** Show that any function defined implicitly by the equation

$$2x^2 - 3y = Cx^2 y^3,$$

where C is a constant, satisfies the equation

$$(xy - x^3)\frac{dy}{dx} + y^2 = 0.$$

** **48.** Verify power rule 3.7 in the case that n is a rational number.

** **49.** Find points on the curve $xy^2 + x^2y = 2$ at which the slope of the tangent line is equal to one.

** **50.** Show that the families of curves $y^2 = x^3/(a - x)$ and $(x^2 + y^2)^2 = b(2x^2 + y^2)$ are orthogonal trajectories.

** **51.** The curve described by the equation $x^3 + y^3 = 3axy$, where $a > 0$ is a constant, is called the *folium of Descartes* (figure below). Find points where the slope of the tangent line is equal to -1.

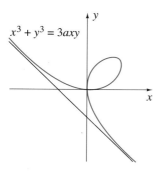

*** **52.** (a) When a point (x, y) moves in the xy-plane so that the product of its distances from the points $(a, 0)$ and $(-a, 0)$ is always equal to a constant which we denote by c^2 $(c > a)$, the curve that it follows is called the *ovals of Cassini*. Verify that the equation for these ovals is $(x^2 + y^2 + a^2)^2 = c^4 + 4a^2x^2$.

(b) Show that when $c < \sqrt{2}a$, there are six points on the ovals at which the tangent line is horizontal, but when $c \geq \sqrt{2}a$, there are only two such points.

*** **53.** Show that the ellipse $b^2x^2 + a^2y^2 = a^2b^2$ and the hyperbola $d^2x^2 - c^2y^2 = c^2d^2$ intersect orthogonally if $a^2 - b^2 = c^2 + d^2$.

3.9 Derivatives of the Trigonometric Functions

The trigonometric functions, their properties, and the identities that they satisfy were discussed in Section 1.5. We emphasize an important convention in calculus. Arguments of trigonometric functions are always measured in radians, never degrees. With this in mind we prove the following theorem.

THEOREM 3.10

$$\lim_{\theta \to 0} \frac{\sin \theta}{\theta} = 1. \tag{3.24}$$

This is stongly suggested by the machine-generated plot of $(\sin \theta)/\theta$ in Figure 3.37. There should be an open circle at the point $(0, 1)$ since the function is undefined at $\theta = 0$.

FIGURE 3.37

FIGURE 3.38

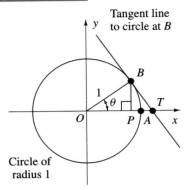

PROOF   If θ is a positive acute angle as shown in Figure 3.38, then

$$\text{area of triangle } BOP \; < \; \text{area of sector } BOA \; < \; \text{area of triangle } OBT.$$

With θ expressed in radians, the area of a sector of a circle is $(r^2\theta)/2$. Consequently,

$$\frac{1}{2}\|BP\|\|OP\| \; < \; \frac{1}{2}(1)^2\theta \; < \; \frac{1}{2}\|OB\|\|BT\|.$$

When we multiply each term in these inequalities by 2 and express lengths of the line segments in terms of θ,

$$(\sin\theta)(\cos\theta) \; < \; \theta \; < \; (1)(\tan\theta).$$

Division by $\sin\theta$ gives

$$\cos\theta \; < \; \frac{\theta}{\sin\theta} \; < \; \frac{1}{\cos\theta},$$

and when each term is inverted, the inequality signs are reversed,

$$\frac{1}{\cos\theta} \; > \; \frac{\sin\theta}{\theta} \; > \; \cos\theta.$$

We now take limits as $\theta \to 0^+$. Since $\cos\theta \to 1$ and $1/\cos\theta \to 1$, it follows that $(\sin\theta)/\theta$ which is between $1/\cos\theta$ and $\cos\theta$ must also approach 1,

$$\lim_{\theta \to 0^+} \frac{\sin\theta}{\theta} = 1.$$

When $\theta < 0$, we set $\phi = -\theta$, in which case

$$\lim_{\theta \to 0^-} \frac{\sin\theta}{\theta} = \lim_{\phi \to 0^+} \frac{\sin(-\phi)}{-\phi} = \lim_{\phi \to 0^+} \frac{\sin\phi}{\phi} = 1.$$

Since left-hand and right-hand limits of $(\sin\theta)/\theta$ are both equal to 1, it follows that

$$\lim_{\theta \to 0} \frac{\sin\theta}{\theta} = 1.$$ ∎

With this result we can now find the derivative of the sine function. Derivatives of the other five trigonometric functions follow easily.

THEOREM 3.11

$$\frac{d}{dx}\sin x = \cos x \qquad (3.25)$$

PROOF If we set $f(x) = \sin x$ in equation 3.3 and use the trigonometric identity

$$\sin A - \sin B = 2\cos\left(\frac{A+B}{2}\right)\sin\left(\frac{A-B}{2}\right),$$

we obtain

$$
\begin{aligned}
f'(x) &= \lim_{h\to 0}\frac{f(x+h)-f(x)}{h} \\[2mm]
&= \lim_{h\to 0}\frac{\sin(x+h)-\sin x}{h} \\[2mm]
&= \lim_{h\to 0}\frac{1}{h}\left[2\cos\left(\frac{2x+h}{2}\right)\sin\left(\frac{h}{2}\right)\right] \\[2mm]
&= \lim_{h\to 0}\left[\cos\left(x+\frac{h}{2}\right)\frac{\sin(h/2)}{h/2}\right] \\[2mm]
&= \lim_{h\to 0}\cos\left(x+\frac{h}{2}\right)\lim_{h/2\to 0}\frac{\sin(h/2)}{h/2} \\[2mm]
&= \cos x.
\end{aligned}
$$

∎

It is a simple application of the chain rule to prove the following corollary.

COROLLARY

If $u = f(x)$ is a differentiable function,

$$\frac{d}{dx}\sin u = \cos u\,\frac{du}{dx}. \qquad (3.26)$$

EXAMPLE 3.27

Find derivatives for the following functions:

(a) $f(x) = \sin 3x$ (b) $f(x) = \sin^3 4x$

SOLUTION

(a) According to formula 3.26,

$$f'(x) = \cos 3x \frac{d}{dx}(3x) = 3\cos 3x.$$

(b) For this function we must use extended power rule 3.21 before equation 3.26:

$$f'(x) = 3\sin^2 4x \frac{d}{dx}\sin 4x = 3\sin^2 4x\cos 4x \frac{d}{dx}(4x) = 12\sin^2 4x\cos 4x.$$

Since the cosine function can be expressed in terms of the sine function, it is straightforward to find its derivative.

THEOREM 3.12

$$\frac{d}{dx}\cos x = -\sin x \qquad (3.27)$$

PROOF Since $\cos x$ can always be expressed in the form $\cos x = \sin(\pi/2 - x)$, it follows that

$$\frac{d}{dx}\cos x = \frac{d}{dx}\sin\left(\frac{\pi}{2} - x\right) = -\cos\left(\frac{\pi}{2} - x\right) = -\sin x. \qquad \blacksquare$$

COROLLARY
If $u = f(x)$ is a differentiable function,

$$\frac{d}{dx}\cos u = -\sin u \frac{du}{dx}. \qquad (3.28)$$

Derivatives of the other trigonometric functions are obtained by expressing them in terms of the sine and cosine functions. For the tangent function, we use the chain rule and the quotient rule to calculate

$$\frac{d}{dx}\tan u = \frac{d}{du}\left(\frac{\sin u}{\cos u}\right)\frac{du}{dx} = \frac{\cos u(\cos u) - \sin u(-\sin u)}{\cos^2 u}\frac{du}{dx} = \frac{1}{\cos^2 u}\frac{du}{dx}.$$

Consequently,

$$\frac{d}{dx}\tan u = \sec^2 u \frac{du}{dx}. \qquad (3.29)$$

A similar calculation gives

$$\frac{d}{dx}\cot u = -\csc^2 u \frac{du}{dx}. \qquad (3.30)$$

For the secant function, we obtain

$$\frac{d}{dx}\sec u = \frac{d}{du}\left(\frac{1}{\cos u}\right)\frac{du}{dx} = (-1)(\cos u)^{-2}(-\sin u)\frac{du}{dx} = \frac{1}{\cos u}\frac{\sin u}{\cos u}\frac{du}{dx}$$

or

$$\frac{d}{dx}\sec u = \sec u \tan u \frac{du}{dx}. \qquad (3.31)$$

Similarly,

$$\frac{d}{dx}\csc u = -\csc u \cot u \frac{du}{dx}. \qquad (3.32)$$

Notice the relationship between derivatives of the *cofunctions*—cosine is the cofunction of sine, cotangent of tangent, and cosecant of secant—and corresponding derivatives for the functions. Each function is replaced by its cofunction and a negative sign is added.

EXAMPLE 3.28

Find dy/dx if y is defined as a function of x in each of the following:

(a) $y = \sin 2x$ (b) $y = 4 \sec (2x^3 + 5)$ (c) $y = \tan^2 4x$

(d) $y = 2 \csc 3x^2 + 5x \sin x$ (e) $x^2 \tan y + y \sin x = 5$

SOLUTION

(a) $\dfrac{dy}{dx} = (\cos 2x)\dfrac{d}{dx}(2x) = 2\cos 2x$

(b) $\dfrac{dy}{dx} = 4\sec (2x^3 + 5)\tan (2x^3 + 5)\dfrac{d}{dx}(2x^3 + 5)$

$\qquad = 24x^2 \sec (2x^3 + 5)\tan (2x^3 + 5)$

(c) $\dfrac{dy}{dx} = 2\tan 4x\dfrac{d}{dx}\tan 4x = 2\tan 4x(4\sec^2 4x) = 8\tan 4x \sec^2 4x$

(d) $\dfrac{dy}{dx} = -2\csc 3x^2 \cot 3x^2\dfrac{d}{dx}(3x^2) + 5\sin x + 5x\cos x$

$\qquad = -12x\csc 3x^2 \cot 3x^2 + 5\sin x + 5x\cos x$

(e) If we differentiate the equation (implicitly) with respect to x, we obtain

$$2x\tan y + x^2 \sec^2 y\frac{dy}{dx} + \frac{dy}{dx}\sin x + y\cos x = 0.$$

When we solve this equation for dy/dx, the result is

$$\frac{dy}{dx} = -\frac{2x\tan y + y\cos x}{x^2 \sec^2 y + \sin x}.$$

EXAMPLE 3.29

Find positive values of x for which the derivative of the function $f(x) = \cos(x + 1/x)$ is equal to zero.

SOLUTION   To get an idea of how many values of x satisfy $f'(x) = 0$ and approximations to them, we plot a graph of $f(x)$. The plot on the interval $0.1 \le x \le 20$ in Figure 3.39a suggests regular behaviour of the function for large values of x, but wild oscillations as $x \to 0$. This is consistent with the fact that for large x, the term $1/x$ becomes less and less significant, and $f(x)$ can be approximated by $\cos x$. On the other hand, when x is close to 0, $f(x)$ should behave much like $\cos(1/x)$ in Figure 2.8b (page 102). This is illustrated in Figure 3.39b where we have plotted the function on the interval $0.01 \le x \le 0.1$.

FIGURE 3.39a

FIGURE 3.39b

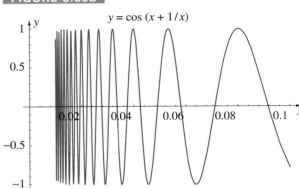

With these ideas in mind, we now solve

$$0 = f'(x) = -\left(1 - \frac{1}{x^2}\right)\sin\left(x + \frac{1}{x}\right).$$

There are two possibilities:

$$1 - \frac{1}{x^2} = 0 \quad \text{or} \quad \sin\left(x + \frac{1}{x}\right) = 0.$$

The first gives $x = \pm 1$, and it is clear that the tangent line at $x = 1$ in Figure 3.39a is indeed horizontal. The second equation above implies that

$$x + \frac{1}{x} = n\pi,$$

where n is an integer. Multiplication by x leads to the quadratic equation

$$x^2 - n\pi x + 1 = 0,$$

with solutions

$$x = \frac{n\pi \pm \sqrt{n^2\pi^2 - 4}}{2}.$$

Clearly, we must choose $n > 0$ (else $n^2\pi^2 - 4 < 0$ when $n = 0$, and $x < 0$ when $n < 0$). When $n = 1$, the solution is $x = (\pi + \sqrt{\pi^2 - 4})/2 \approx 2.8$. This is the first point to the right of $x = 1$ at which the tangent line is horizontal in Figure 3.39a. As n increases, the 4 becomes less and less significant and $x \approx (n\pi \pm n\pi)/2$. When we choose the positive sign, we obtain $x \approx n\pi$, and we can indeed see that points on the graph in Figure 3.38a where the tangent line is horizontal do indeed seem to be multiples of π. When we choose the negative sign, we obtain points to the left of $x = 1$ at which the tangent line is horizontal (Figure 3.39b). For large n, they can be approximated by $1/(n\pi)$, but this is not an easy fact to show.

EXAMPLE 3.30

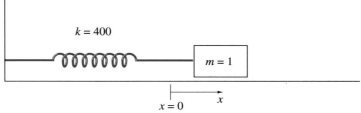

The 1 kg mass in Figure 3.40 is pulled 10 cm to the right of its position when the spring is unstretched (called the equilibrium position), and given velocity 3 m/s to the left. Its displacement from the equilibrium position thereafter is given by

$$x(t) = \frac{1}{10} \cos 20t - \frac{3}{20} \sin 20t \text{ m}, \quad t \geq 0.$$

Find all times when the velocity is zero and all times when the acceleration is zero.

FIGURE 3.40

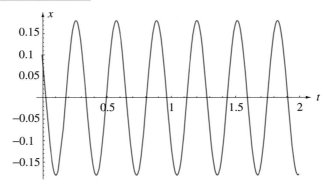

SOLUTION   The displacement function in Figure 3.41 indicates that the mass oscillates back and forth about the equilibrium position. This is called **simple harmonic motion**. It is a direct result of the facts that the mass moves along a frictionless surface and no account has been taken of air resistance.

FIGURE 3.41

To determine times at which velocity and acceleration vanish, it is advantageous to follow the lead of Example 1.26 in Section 1.5 (page 61) and express $x(t)$ in the form

$$\frac{1}{10} \cos 20t - \frac{3}{20} \sin 20t = R \sin(20t + \phi).$$

When we expand the right side with a compound angle formula,

$$\frac{1}{10} \cos 20t - \frac{3}{20} \sin 20t = R(\sin 20t \, \cos \phi + \cos 20t \, \sin \phi).$$

For this to be true for all $t \geq 0$, it must be valid for $t = 0$ and $t = \pi/40$. When we set t equal to these values, two equations in R and ϕ result,

$$\frac{1}{10} = R \, \sin \phi \qquad \text{and} \qquad -\frac{3}{20} = R \, \cos \phi.$$

Squaring and adding these equations eliminates ϕ:

$$\frac{1}{100} + \frac{9}{400} = R^2 \qquad \Longrightarrow \qquad R = \pm\frac{\sqrt{13}}{20}.$$

If we choose to use the positive value, then

$$\sin \phi = \frac{2}{\sqrt{13}} \qquad \text{and} \qquad \cos \phi = -\frac{3}{\sqrt{13}}.$$

When we let $\phi = 2.55$ be the smallest positive solution of these equations, we express $x(t)$ in the form

$$x(t) = \frac{\sqrt{13}}{20} \sin (20t + \phi) \text{ m}.$$

Velocity and acceleration are

$$v(t) = \sqrt{13} \cos (20t + \phi) \text{ m/s} \qquad \text{and} \qquad a(t) = -20\sqrt{13} \sin (20t + \phi) \text{ m/s}^2.$$

The velocity of the mass is zero when the mass is at the points farthest from the equilibrium position. It occurs when

$$0 = \sqrt{13} \cos (20t + \phi) \qquad \Longrightarrow \qquad 20t + \phi = \frac{\pi}{2} + n\pi,$$

where n is an integer. Thus,

$$t = \frac{\pi}{40} + \frac{n\pi}{20} - \frac{\phi}{20},$$

and these are positive for $n \geq 1$. The acceleration vanishes when

$$0 = -20\sqrt{13} \sin (20t + \phi) \qquad \Longrightarrow \qquad 20t + \phi = n\pi,$$

where n is an integer. Thus,

$$t = \frac{n\pi}{20} - \frac{\phi}{20},$$

and this is positive for $n \geq 1$. Notice that acceleration is zero when the mass passes through the equilibrium position.

EXERCISES 3.9

In Exercises 1–30 find dy/dx.

∗ **1.** $y = 2 \sin 3x$

∗ **2.** $y = \cos x - 4 \sin 5x$

∗ **3.** $y = \sin^2 x$

∗ **4.** $y = \tan^{-3} 3x$

∗ **5.** $y = \sec^4 10x$

∗ **6.** $y = \csc (4 - 2x)$

∗ **7.** $y = \sin^2 (3 - 2x^2)$

∗ **8.** $y = x \cot x^2$

∗ **9.** $y = \dfrac{\sin 2x}{\cos 5x}$

∗ **10.** $y = \dfrac{x \sin x}{x + 1}$

∗ **11.** $y = \sqrt{\sin 3x}$

∗ **12.** $y = \left(1 + \tan^3 x\right)^{1/4}$

∗ **13.** $y = \sin^3 x + \cos x$

∗∗ **14.** $y = \sin 2x \cos 2x$

∗∗ **15.** $2 \sin y + 3 \cos x = 1$

∗∗ **16.** $x \cos y - y \cos x = 3$

∗∗ **17.** $4 \sin^2 x - 3 \cos^3 y = 1$

∗∗ **18.** $\tan (x + y) = y$

∗∗ **19.** $x + \sec xy = 5$

∗∗ **20.** $x^3 y + \tan^2 y = 3x$

∗∗ **21.** $x = y^3 \csc^3 y$

∗∗ **22.** $y = \cos (\tan x)$

∗∗ **23.** $y = x^3 - x^2 \cos x + 2x \sin x + 2 \cos x$

∗∗ **24.** $y = \sin^4 x^2 - \cos^4 x^2$

∗∗ **25.** $y = u^3 \sec u, \quad u = x \tan (x + 1)$

∗∗ **26.** $y = \sqrt{3 - \sec v}, \quad v = \tan \sqrt{x}$

∗∗ **27.** $y = \sqrt{t^2 + 1}, \quad t = \sin (\sin x)$

∗∗ **28.** $y = (1 + \sec^3 u)^{1/3}, \quad u = \sqrt{1 + \cos x^2}$

∗∗ **29.** $y = \dfrac{\sin^2 x}{1 + \cos^3 x}$

∗∗ **30.** $y = \dfrac{1 + \tan^3 (3x^2 - 4)}{x^2 \sin x}$

In Exercises 31 and 32 find $d^2 y/dx^2$.

∗∗ **31.** $\sin y = x^2 + y$

∗∗ **32.** $\tan y = x + xy$

In Exercises 33–38 evaluate the limit if it exists.

∗ **33.** $\displaystyle \lim_{x \to 0} \dfrac{\tan x}{x}$

∗∗ **34.** $\displaystyle \lim_{x \to 0} \dfrac{1 - \cos x}{x}$

∗∗ **35.** $\displaystyle \lim_{x \to 0} \dfrac{\sin 2x}{x}$

∗∗ **36.** $\displaystyle \lim_{x \to \infty} \dfrac{\sin (2/x)}{\sin (1/x)}$

∗∗ **37.** $\displaystyle \lim_{x \to 0} \dfrac{(x + 1)^2 \sin x}{3x^3}$

∗∗ **38.** $\displaystyle \lim_{x \to \pi/2} \dfrac{\cos x}{(x - \pi/2)^2}$

∗∗ **39.** Draw graphs of the functions $f(x) = |\sin x|$ and $g(x) = \sin |x|$. Where do these functions fail to be differentiable?

∗∗ **40.** Does $\lim_{x \to 0} g'(x)$ exist for the function $g(x)$ in Example 3.12 of Section 3.3?

** **41.** The angular displacement of the pendulum in the figure below at time t is given by

$$\theta = f(t) = A \cos(\omega t + \phi), \quad t \geq 0,$$

where A, ω, and ϕ are constants. Show that $\theta = f(t)$ satisfies the equation

$$\frac{d^2\theta}{dt^2} + \omega^2\theta = 0.$$

** **42.** When the mass m in the figure below moves vertically on the end of the spring, its displacement y must satisfy the equation

$$m\frac{d^2y}{dt^2} + ky = 0,$$

where $k > 0$ is the constant of elasticity for the spring. Show that the function

$$y = f(t) = A \sin\left(\sqrt{\frac{k}{m}}\, t\right) + B \cos\left(\sqrt{\frac{k}{m}}\, t\right)$$

satisfies this equation for any constants A and B.

$y = 0$ at
equilibrium

** **43.** If the voltage source labelled E in the following diagram is suddenly short-circuited by the switch labelled S, the current i in the circuit thereafter must satisfy the equation

$$L\frac{d^2i}{dt^2} + \frac{1}{C}i = 0,$$

where L and C are the sizes of the inductor and capacitor, respectively, and t is time. Show that

$$i = f(t) = A \cos\left(\frac{t}{\sqrt{LC}}\right) + B \sin\left(\frac{t}{\sqrt{LC}}\right)$$

satisfies this equation for any constants A and B whatsoever.

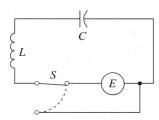

** **44.** Find

$$\frac{d}{dx} |\sin x|.$$

** **45.** If n is a positive integer, find formulas for

$$\frac{d^n}{dx^n}(\sin x) \qquad \text{and} \qquad \frac{d^n}{dx^n}(\cos x).$$

*** **46.** Show that the function

$$f(x) = \begin{cases} x^2 \sin (1/x), & x \neq 0 \\ 0, & x = 0 \end{cases}$$

has a derivative at $x = 0$, but $f'(x)$ is not continuous at $x = 0$.

*** **47.** Any cross-section of the reflector in a car headlight is in the form of a parabola

$$y^2 = 4x, \quad 0 \leq x \leq 2,$$

with the bulb at the point $(1, 0)$ called the *focus* of the parabola. It is a principle of optics that all light rays from the bulb are reflected by the mirror so that the angle between the incident ray and the normal to the mirror is equal to the angle between the reflected ray and the normal. Show that all rays are reflected parallel to the x-axis.

3.10 Derivatives of the Exponential and Logarithm Functions

Exponential functions abound in applied mathematics, and much of this is due to the fact that the derivative of an exponential function always returns the exponential function. We obtain derivatives of exponential and logarithm functions in this section, but many of the applications which rely on these derivatives are delayed until Section 9.2. Exponential and logarithm functions and their properties are discussed in Section 1.6. We assume that the reader is familiar with this material. To differentiate logarithm and exponential functions, we must return to the definition of the derivative (see equation 3.3 in Section 3.1). We can apply the definition to differentiate either $\log_a x$ or a^x, and once one of these derivatives is obtained, implicit differentiation can be used to find the other. We choose to differentiate $\log_a x$.

THEOREM 3.13

The derivative of the logarithm function $\log_a x$ is

$$\frac{d}{dx} \log_a x = \frac{1}{x} \log_a e, \tag{3.33a}$$

where

$$e = \lim_{z \to \infty} \left(1 + \frac{1}{z}\right)^z = \lim_{v \to 0} (1 + v)^{1/v}. \tag{3.33b}$$

PROOF   By equation 3.3,

$$\frac{d}{dx} \log_a x = \lim_{h \to 0} \frac{\log_a(x + h) - \log_a x}{h}$$

$$= \lim_{h \to 0} \frac{1}{h} \left[\log_a \left(\frac{x + h}{x} \right) \right] \qquad \left(\text{since } \log_a b - \log_a c = \log_a (b/c) \right)$$

$$= \lim_{h \to 0} \frac{1}{h} \left[\log_a \left(1 + \frac{h}{x} \right) \right]$$

$$= \frac{1}{x} \lim_{h \to 0} \frac{x}{h} \left(\log_a \left(1 + \frac{h}{x} \right) \right)$$

$$= \frac{1}{x} \lim_{h \to 0} \log_a \left(1 + \frac{h}{x} \right)^{x/h} \qquad \left(\text{since } c \log_a b = \log_a b^c \right).$$

Since the logarithm function is continuous, we may interchange the limit and logarithm operations (see Theorem 2.3 in Section 2.4),

$$\frac{d}{dx} \log_a x = \frac{1}{x} \log_a \left[\lim_{h \to 0} \left(1 + \frac{h}{x} \right)^{x/h} \right].$$

But

$$\lim_{h \to 0} \left(1 + \frac{h}{x} \right)^{x/h} = \lim_{h/x \to 0} \left(1 + \frac{h}{x} \right)^{x/h} = e,$$

and therefore

$$\frac{d}{dx} \log_a x = \frac{1}{x} \log_a e. \qquad \blacksquare$$

If we choose e as the base of logarithms, that is, set $a = e$, derivative 3.33 simplifies to

$$\frac{d}{dx} \log_e x = \frac{1}{x}.$$

Because of the simplicity of this result, logarithms to base e are used more extensively than those to any other base. With the notation $\ln x = \log_e x$ for logarithms to base e, we may state the following corollary.

COROLLARY 1

$$\frac{d}{dx}\ln x = \frac{1}{x} \qquad (3.34)$$

The chain rule gives the next two results.

COROLLARY 2

If $u(x)$ is a differentiable function of x, then

$$\frac{d}{dx}\log_a u = \frac{1}{u}\frac{du}{dx}\log_a e. \qquad (3.35)$$

COROLLARY 3

If $u(x)$ is differentiable, then

$$\frac{d}{dx}\ln u = \frac{1}{u}\frac{du}{dx}. \qquad (3.36)$$

We now use implicit differentiation to find the derivative of the exponential function.

THEOREM 3.14

$$\frac{d}{dx}a^x = a^x \ln a \qquad (3.37)$$

PROOF If we set $y = a^x$, then $x = \log_a y$, and we can differentiate (implicitly) with respect to x using 3.35,

$$1 = \frac{1}{y}\frac{dy}{dx}\log_a e.$$

Thus,

$$\frac{dy}{dx} = \frac{y}{\log_a e} = \frac{a^x}{\log_a e}.$$

When it is necessary to change bases of logarithms, the rule is

$$\log_b x = (\log_a x)(\log_b a),$$

(see equation 1.52 in Section 1.6). If we set $x = b = e$ in this equation, the result is

$$\log_e e = (\log_a e)(\log_e a) \qquad \text{or} \qquad 1 = (\log_a e)(\ln a).$$

Consequently, $1/\log_a e = \ln a$, and our differentiation formula becomes

$$\frac{dy}{dx} = a^x \ln a. \qquad \blacksquare$$

Again it is certainly clear that, should we use exponentials to base e, the differentiation formula 3.37 simplifies.

COROLLARY 1

$$\frac{d}{dx}e^x = e^x \tag{3.38}$$

The exponential function e^x is therefore its own derivative. In fact, for any constant C whatsoever, the function Ce^x differentiates to give itself. In Chapter 5 we shall see that Ce^x is the only function that is its own derivative.

COROLLARY 2

For a differentiable function $u(x)$,

$$\frac{d}{dx}a^u = a^u\frac{du}{dx}\ln a. \tag{3.39}$$

COROLLARY 3

For a differentiable function $u(x)$,

$$\frac{d}{dx}e^u = e^u\frac{du}{dx}. \tag{3.40}$$

EXAMPLE 3.31

Find dy/dx if y is defined as a function of x in each of the following:

(a) $y = 2^{3x}$ (b) $y = \log_{10}(3x^2 + 4)$ (c) $y = x^2e^{-2x}$
(d) $y = (\ln x)/x$ (e) $y = \sqrt{1 + e^{2x}}$

SOLUTION

(a) Using 3.39,

$$\frac{dy}{dx} = 2^{3x}\frac{d}{dx}(3x)\ln 2 = (3\ln 2)2^{3x}.$$

(b) With 3.35,

$$\frac{dy}{dx} = \frac{1}{3x^2 + 4}\frac{d}{dx}(3x^2 + 4)\log_{10}e = \frac{6x}{3x^2 + 4}\log_{10}e.$$

(c) The product rule and 3.40 give

$$\frac{dy}{dx} = 2xe^{-2x} + x^2e^{-2x}(-2) = 2x(1 - x)e^{-2x}.$$

(d) The quotient rule and 3.34 yield

$$\frac{dy}{dx} = \frac{x(1/x) - \ln x}{x^2} = \frac{1 - \ln x}{x^2}.$$

(e) With extended power rule 3.21 and 3.40,

$$\frac{dy}{dx} = \frac{1}{2\sqrt{1 + e^{2x}}} \frac{d}{dx}(1 + e^{2x}) = \frac{2e^{2x}}{2\sqrt{1 + e^{2x}}} = \frac{e^{2x}}{\sqrt{1 + e^{2x}}}.$$

EXAMPLE 3.32

Find values of x for which the first derivative of the function $f(x) = x^2 \ln x$ is equal to zero, and values of x for which the second derivative is equal to zero.

SOLUTION   A graph of the function is shown in Figure 3.42. It suggests that the tangent line is horizontal at a value of x near $1/2$. To find it we set

$$0 = f'(x) = 2x \ln x + x^2 \left(\frac{1}{x}\right) = x(2 \ln x + 1).$$

Since the function is undefined at $x = 0$, we must have

$$2 \ln x + 1 = 0 \qquad \text{or} \qquad \ln x = -\frac{1}{2}.$$

Thus, $x = e^{-1/2} = 1/\sqrt{e}$.

The second derivative is equal to zero when

$$0 = f''(x) = (2 \ln x + 1) + x \left(\frac{2}{x}\right) = 2 \ln x + 3.$$

The only solution of this equation is $x = e^{-3/2}$. The geometric significance of the point on the graph at which $f''(x) = 0$ will be discussed in Section 4.4. Can you see it?

FIGURE 3.42

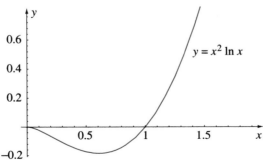

In Section 3.2 we proved power rule 3.7 only in the case where n is a nonnegative integer. It is now easy to prove it for all real n, at least when $x > 0$. To do this we write x^n in the form $x^n = e^{n \ln x}$ (see equation 1.57a). Formulas 3.40 and 3.34 give

$$\frac{d}{dx} x^n = \frac{d}{dx} e^{n \ln x} = e^{n \ln x} \frac{d}{dx} (n \ln x) = x^n \left(\frac{n}{x} \right) = n x^{n-1}.$$

A discussion of the power rule in the case when $x \leq 0$ depends on the value of n (see Exercise 49).

EXAMPLE 3.33

If air resistance proportional to velocity and friction with the surface on which the mass slides are taken into account in the mass-spring system of Example 3.30, the displacement function is of the form

$$x(t) = e^{-t/2} \left[\frac{1}{10} \cos \frac{\sqrt{1599}t}{2} - \frac{1}{8} \sin \frac{\sqrt{1599}t}{2} \right] + \frac{1}{4000},$$

from time $t = 0$ when motion begins until the mass comes to rest for the first time. Find this time.

SOLUTION   The mass comes to rest when its velocity vanishes,

$$0 = -\frac{1}{2} e^{-t/2} \left[\frac{1}{10} \cos \frac{\sqrt{1599}t}{2} - \frac{1}{8} \sin \frac{\sqrt{1599}t}{2} \right]$$
$$+ e^{-t/2} \left[-\frac{\sqrt{1599}}{20} \sin \frac{\sqrt{1599}t}{2} - \frac{\sqrt{1599}}{16} \cos \frac{\sqrt{1599}t}{2} \right].$$

Of the infinity of possible solutions to this equation, only the smallest positive one is acceptable here. To find it we simplify the equation to

$$0 = -\frac{1}{80} e^{-t/2} \left[(4 + 5\sqrt{1599}) \cos \frac{\sqrt{1599}t}{2} + (4\sqrt{1599} - 5) \sin \frac{\sqrt{1599}t}{2} \right].$$

This equation implies that

$$\tan \frac{\sqrt{1599}t}{2} = \frac{4 + 5\sqrt{1599}}{5 - 4\sqrt{1599}}.$$

The smallest positive angle with tangent equal to $(4 + 5\sqrt{1599})/(5 - 4\sqrt{1599})$ is 2.2205 radians. Consequently, the time when the mass stops moving to the left is

$$t = \frac{2}{\sqrt{1599}} (2.2205) = 0.111 \text{ s}.$$

EXAMPLE 3.34

The container in Figure 3.43 represents a chemical reactor in which a chemical is either created or broken down. The chemical enters in the form of a solution at one concentration and leaves the reactor at a different concentration. We assume that flow rates in and out are the same, say q m^3/s, and therefore the volume V of solution in the reactor remains constant. Let us assume that C_i is the concentration in kg/m^3 of the chemical entering the reactor. If $C(t)$ represents the number of kilograms of chemical in the reactor at any given time, then dC/dt is the rate of change of the amount of chemical. This must be equal to the rate at which chemical enters less the rate at which it leaves, plus (or minus) the rate at which it is created (broken down),

$$\frac{dC}{dt} = \left\{ \begin{array}{c} \text{rate at which} \\ \text{chemical enters} \end{array} \right\} - \left\{ \begin{array}{c} \text{rate at which} \\ \text{chemical leaves} \end{array} \right\} + \left\{ \begin{array}{c} \text{rate at which chemical} \\ \text{is created or broken down} \end{array} \right\}.$$

The rate at which chemical enters is qC_i. The rate at which it leaves is $q(C/V)$. It is often assumed that the rate at which chemical is being created or broken down at any given time is proportional to how much chemical is present at that time. Thus, the rate at which it is created or broken down is kC, where k is a constant ($k > 0$ for chemical formation, $k < 0$ for chemical break down). When we fill these expressions into the above equation, we obtain

$$\frac{dC}{dt} = qC_i - \frac{qC}{V} + kC.$$

This is the equation that the amount of chemical $C(t)$ must satisfy at any given time. Show that the function

$$C(t) = \frac{qC_i V}{q - kV} \left[1 - e^{-(q-kV)t/V} \right]$$

satisfies the equation as well as the condition $C(0) = 0$.

FIGURE 3.43

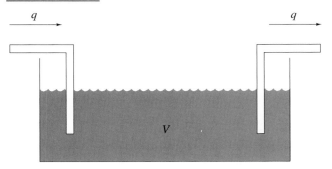

SOLUTION   The derivative of $C(t)$ is

$$\frac{dC}{dt} = \frac{qC_i V}{q - kV} \left[-e^{-(q-kV)t/V} \left(-\frac{q - kV}{V} \right) \right] = qC_i e^{-(q-kV)t/V},$$

and

$$qC_i - \frac{qC}{V} + kC = qC_i + \left(k - \frac{q}{V} \right) \frac{qC_i V}{q - kV} [1 - e^{-(q-kV)t/V}]$$

$$= qC_i - qC_i [1 - e^{-(q-kV)t/V}]$$

$$= qC_i e^{-(q-kV)t/V}.$$

Thus, $C(t)$ satisfies the equation $C' = qC_i - qC/V + kC$ and, furthermore,

$$C(0) = \frac{qC_i V}{q - kV}(1 - e^0) = 0.$$

This example has introduced a type of equation that we encounter many, many times in applications. We call the equation

$$\frac{dC}{dt} = \left(k - \frac{q}{V}\right)C + qC_i$$

a **differential equation**. It is an equation that contains one or more derivatives of a function and the objective is to solve the equation for functions whose derivatives satisfy the equation. For the above equation, we must find functions $C(t)$ whose first derivative is equal to $k - q/V$ times itself plus a constant qC_i. We show how differential equations arise in a multitude of applications and ask you to verify solutions. How these solutions are derived is the subject of Section 5.4 and Chapter 16.

Heat is a form of energy, and many engineering systems are analyzed by setting up what are called *energy balance equations*. We analyze one such here. Figure 3.44 shows a tank into which water enters at temperature $10°C$, the water is heated, and it then leaves at higher temperature. The tank is perfectly insulated so that no heat can escape from its sides, and therefore all heat supplied by the heater raises the temperature of the water. When the tank is full, and we assume that this is always the case, the mass of water is 100 kg. Water enters the tank at a rate of 3/100 kg/s, and leaves at the same rate. The heater adds energy to the water at the rate of 2000 J/s (where J is joules). In reality, temperature of the water in the tank depends on both time and position in the tank. The mixer is added to remove spatial dependence. We assume that the mixer is so efficient that temperature of the water is the same at every point in the tank. Temperature then depends only on time t, denoted by $T(t)$, and this is also the temperature at which water leaves the tank.

Energy balance for this system states that

$$\left\{\begin{array}{c}\text{Rate at which}\\\text{energy enters tank}\\\text{in cool water}\end{array}\right\} + \left\{\begin{array}{c}\text{rate at which}\\\text{energy is added}\\\text{by heater}\end{array}\right\} = \left\{\begin{array}{c}\text{Rate at which}\\\text{energy leaves tank}\\\text{in warmer water}\end{array}\right\} + \left\{\begin{array}{c}\text{Rate at which}\\\text{energy is used to raise}\\\text{temperature of}\\\text{water in tank}\end{array}\right\}.$$

(For simplicity, we have ignored energy associated with the mixer.) The first three rates are known; the fourth can be expressed in terms of dT/dt, the time rate of change of temperature. The second term on the left is 2000. For the remaining terms, we must introduce the *specific heat* $c_p = 4190$ J/kg·C of water. It is the energy required to raise the temperature of 1 kg of water by $1°C$. To raise 5 kg of water $10°$ requires $(5)(10)(4190) = 209\,500$ joules.) Since 3/100 kg of water at temperature $10°C$ enter the tank each second, the rate at which energy enters the tank due to this water is $(3/100)(10)(4190) = 1257$ J/s. (This is the energy required to raise 3/100 kg of water from $0°C$ to $10°C$.) In a similar way, the rate at which energy leaves the tank in the water at temperature T is $(3/100)(T)(4190) = 1257T/10$ J/s. This leaves only the last term on the right side of the energy balance equation. The rate of change of temperature of the

FIGURE 3.44

100 kg of water in the tank is dT/dt; that is, the temperature changes dT/dt degrees each second. It follows that the rate at which energy is used to raise the temperature of this water is $(100)(4190)(dT/dt)$.

When these rates are substituted into the energy balance equation, the result is

$$1257 + 2000 = \frac{1257\,T}{10} + 419\,000\frac{dT}{dt} \qquad \Longrightarrow \qquad \frac{dT}{dt} + \frac{3T}{10\,000} = \frac{3257}{419\,000}.$$

This is another example of a differential equation; find the function $T(t)$ whose derivative satisfies the equation. We show here that the solution is

$$T(t) = \frac{32\,570}{1257} - \frac{20\,000}{1257}e^{-3t/10\,000}$$

in the case where the process begins at time $t = 0$ and the temperature of the water in the tank at this time is $10°$C. (It will be derived in Chapter 16 when we study differential equations in detail, and we shall allow rate at which water enters and leaves the tank, temperature of the water entering the tank, and rate at which energy is added by the heater to be functions of t. See also Exercises 33–35 in this section.) When we substitute this function into the right side of the differential equation and use formula 3.40,

$$\frac{dT}{dt} + \frac{3T}{10\,000} = -\left(\frac{20\,000}{1257}\right)\left(\frac{-3}{10\,000}\right)e^{-3t/10\,000}$$

$$+ \frac{3}{10\,000}\left(\frac{32\,570}{1257} - \frac{20\,000}{1257}e^{-3t/10\,000}\right)$$

$$= \frac{3257}{419\,000}.$$

The graph of $T(t)$ has the shape shown in Figure 3.45. It begins at $T = 10$ and is asymptotic to the line $T = 32\,570/1257$. This is so because

$$\lim_{t\to\infty} T(t) = \frac{32\,570}{1257}.$$

FIGURE 3.45

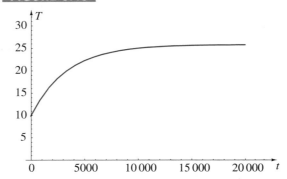

In Exercises 1–30, y is defined as a function of x. Find dy/dx in as simplified a form as possible.

* **1.** $y = 3^{2x}$

* **2.** $y = \ln(3x^2 + 1)$

* **3.** $y = \log_{10}(2x + 1)$

* **4.** $y = e^{1-2x}$

* **5.** $y = xe^{2x}$

* **6.** $y = x \ln x$

* **7.** $y = e^{2 \ln x}$

* **8.** $y = \log_{10}(3 - 4x)$

* **9.** $y = \ln(\sin x)$

* **10.** $y = \ln(3 \cos x)$

* **11.** $y = x \ln(x + 1)$

* **12.** $y = x^2 + x^3 e^{4x}$

* **13.** $y = \dfrac{e^{1-x}}{1 - x}$

* **14.** $y = \sin(e^{2x})$

* **15.** $y = \ln(\ln x)$

* **16.** $y = e^{-2x} \sin 3x$

** **17.** $y = \ln(x^2 e^{4x})$

** **18.** $y = \dfrac{e^x - e^{-x}}{e^x + e^{-x}}$

** **19.** $y = e^{-x} \ln x$

** **20.** $y = x \left[\sin(\ln x) - \cos(\ln x) \right]$

** **21.** $y = e^{\sin u}, \ u = e^{1/x}$

** **22.** $y = \ln(\cos v), \ v = \sin^2 x$

** **23.** $\ln(x + y) = x^2 y$

** **24.** $xe^y + x^2 \ln y + y \sin x = 0$

** **25.** $y = \ln(\sec x + \tan x)$

** **26.** $y = x\sqrt{x^2 + 1} - \ln\left(x + \sqrt{x^2 + 1}\right)$

** **27.** $y = \ln\left(x + 4 + \sqrt{8x + x^2}\right)$

** **28.** $y = x - \dfrac{1}{4} \ln(1 + 5e^{4x})$

** **29.** $e^{xy} = (x + y)^2$

** **30.** $e^{1/x} + e^{1/y} = \dfrac{1}{x} + \dfrac{1}{y}$

** **31.** The figure on page 221 shows a long cylindrical cable. Copper wire runs down the centre of the cable and insulation covers the wire. If r measures radial distance from the centre of the pipe, then steady-state temperature T in the insulation is a function of r that must satisfy the differential equation

$$\frac{d}{dr}\left(r\frac{dT}{dr}\right) = 0.$$

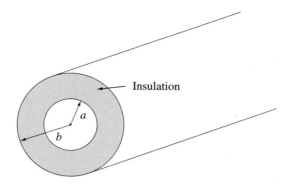

Insulation

(a) Show that the function

$$T(r) = c \ln r + d$$

satisfies the differential equation for any constants c and d.

(b) If temperature on inner and outer edges of the insulation $r = a$ and $r = b$ are constants T_a and T_b, find c and d in terms of a, b, T_a, and T_b.

** **32.** If heat is generated at a constant rate in the copper wire of Exercise 31 (perhaps because of electric current), the differential equation is replaced by

$$\frac{d}{dr}\left(r\frac{dT}{dr}\right) = k,$$

where k is a constant.

(a) Show that the function

$$T(r) = kr + c \ln r + d$$

satisfies the differential equation for any constants c and d.

(b) If temperature on inner and outer edges of the insulation $r = a$ and $r = b$ are constants T_a and T_b, find c and d in terms of a, b, T_a, T_b, and k.

** **33.** (a) What is the differential equation satisfied by temperature of the water in the tank of Figure 3.44 if the rate at which water enters the tank is $100/(t+1)$ kg/s?

(b) Assuming that temperature of the water in the tank is $10°$C at time $t = 0$, show that

$$T(t) = \frac{10t}{t+1} + \frac{t}{419} + 10.$$

Plot a graph of $T(t)$.

** **34.** (a) What is the differential equation satisfied by temperature of the water in the tank of Figure 3.44 if the amount of energy supplied by the heater for the first ten minutes is $q = 20t$ J/s, $0 \le t \le 600$?

(b) Assuming that temperature of the water in the tank is $10°$C at time $t = 0$, show that

$$T(t) = \frac{1}{1257}\left(20t - 7430 + 20\,000e^{-3t/10\,000}\right), \quad 0 \le t \le 600.$$

Plot a graph of $T(t)$.

** **35.** (a) What is the differential equation satisfied by temperature of the water in the tank of Figure 3.44 if the temperature of the incoming water is a function of time, $T_0 = 10e^{-t}$, $t \ge 0$?

(b) Assuming that temperature of the water in the tank is $10°$C at time $t = 0$, show that

$$T(t) = \frac{20\,000}{1257} - \frac{73\,900\,610}{12\,566\,229}e^{-3t/10\,000} - \frac{300}{9997}e^{-t}.$$

Plot a graph of $T(t)$.

** **36.** A spring with constant 5 N/m is attached to a fixed wall on one end and a 1 km mass m on the other (figure below). We choose a coordinate system with x positive to the right and with $x = 0$ at the centre of mass of m when the spring is in the unstretched position. At time $t = 0$, m is pulled 1 m to the left of $x = 0$ and given a speed of 3 m/s to the right. During the subsequent motion, a frictional force equal in newtons to twice the velocity of the mass acts on m. It can be shown that if $x(t)$ is the position of m, then $x(t)$ must satisfy the differential equation

$$\frac{d^2x}{dt^2} + 2\frac{dx}{dt} + 5x = 0$$

and the initial conditions $x(0) = -1$ and $x'(0) = 3$. Verify that $x(t) = e^{-t}(\sin 2t - \cos 2t)$ satisfies the equation and initial conditions.

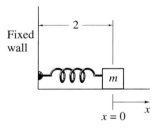

** **37.** The figure on page 223 shows a tank containing 10 000 L of water in which is dissolved 100 kg of salt. Starting at time $t = 0$, a solution containing 1 kg of salt for each 20 L of solution is added to the tank at 4 L per second. Simultaneously, 4 L of solution are removed from the tank each second. It is assumed that the mixer is so efficient that the concentration of salt is the same at all points in the tank, and this is the concentration at which solution leaves the tank. If $S(t)$ represents the number of kilograms of salt in the tank at any given time, then dS/dt is the rate of change of the amount of salt in the tank. It must be equal to the rate at which salt enters less the rate at which salt leaves,

$$\frac{dS}{dt} = \left\{ \begin{array}{c} \text{rate at which} \\ \text{salt enters tank} \end{array} \right\} - \left\{ \begin{array}{c} \text{rate at which} \\ \text{salt leaves tank} \end{array} \right\}.$$

The rate at which salt enters the tank is $(1/20)(4) = 1/5$ kg/s. The rate at which salt leaves is $(S/10\,000)(4) = S/2500$ kg/s. Thus,

$$\frac{dS}{dt} = \frac{1}{5} - \frac{S}{2500}, \qquad S(0) = 100.$$

Show that the solution of this differential equation and initial condition is

$$S(t) = 100(5 - 4e^{-t/2500}) \text{ kg}.$$

Plot its graph. Is it what you would expect?

** **38.** (a) Show that if solution in Exercise 37 is removed at 2 L/s rather than 4 L/s, the differential equation for $S(t)$ is

$$\frac{dS}{dt} = \frac{1}{5} - \frac{2S}{10\,000 + 2t}, \qquad S(0) = 100.$$

(b) Verify that the solution is

$$S(t) = 500 + \frac{t}{10} - \frac{2 \times 10^6}{5000 + t}.$$

Plot a graph of $S(t)$. Does it have an asymptote?

** **39.** (a) Show that if solution in Exercise 37 is removed at 8 L/s rather than 4 L/s, the differential equation for $S(t)$ is

$$\frac{dS}{dt} = \frac{1}{5} - \frac{8S}{10\,000 - 4t}, \qquad S(0) = 100.$$

(b) Verify that the solution is

$$S(t) = 100 + \frac{3t}{25} - \frac{t^2}{15\,625}.$$

For how long is the solution valid? Plot a graph of $S(t)$.

(c) What is the maximum amount of salt in the tank?

** **40.** (a) A beer vat contains 2000 L of beer, 4% of which is alcohol. Beer with 8% alcohol is added to the vat at 2 L/s, and well-stirred mixture is removed at the same rate. Show that if $A(t)$ represents the number of litres of alcohol in the vat, then

$$\frac{dA}{dt} = \frac{4}{25} - \frac{A}{1000}, \qquad A(0) = 80.$$

(b) Show that the solution of this differential equation is

$$A(t) = 160 - 80e^{-t/1000}.$$

Plot a graph of $A(t)$.

(c) When is the beer in the vat 5% alcohol?

** **41.** Two long parallel rectangular loops lying in the same plane have lengths l and L and widths w and W, respectively (figure below). If the loops do not overlap, and the distance between the near sides is s, the mutual inductance between the loops is

$$M = \frac{\mu_0 l}{2\pi} \ln\left[\frac{s + W}{s\left(1 + \dfrac{W}{s + w}\right)}\right]$$

where $\mu_0 > 0$ is a constant. Show that the derivative of M as a function of s is negative for $s > 0$.

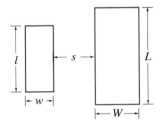

** **42.** Two parallel wires carrying current i (figure below) create a magnetic field. The flux Φ of this field through the loop of dimensions h and w is

$$\Phi = \frac{\mu_0 h i}{2\pi} \ln\left[\frac{R(r + w)}{r(R + w)}\right],$$

where $\mu_0 > 0$ is a constant. Draw a graph of Φ, first as a function of h and then as a function of w.

** **43.** Show that the function

$$y = f(x) = Ax + B\left[\frac{x}{2}\ln\left(\frac{x - 1}{x + 1}\right) + 1\right]$$

satisfies the differential equation

$$(1 - x^2)\frac{d^2y}{dx^2} - 2x\frac{dy}{dx} + 2y = 0$$

for any constants A and B.

** **44.** The equation $x = e^y - e^{-y}$ defines y implicitly as a function of x. Find dy/dx in terms of x in two ways:

  (a) Solve first for an explicit definition for $y = f(x)$.

  (b) Differentiate implicitly with respect to x.

** **45.** The current i in the RC-circuit in the diagram below must satisfy the differential equation

$$R\frac{di}{dt} + \frac{i}{C} = \frac{dE}{dt}.$$

If $E = E_0 \sin \omega t$, where E_0 and ω are constants, show that a solution is

$$i = f(t) = Ae^{-t/(RC)} + \frac{E_0}{Z} \sin (\omega t - \phi),$$

where A is any constant whatsoever, and

$$Z = \sqrt{R^2 + \frac{1}{\omega^2 C^2}}, \qquad \tan \phi = -\frac{1}{\omega C R}.$$

** **46.** The current i in the LR-circuit in the diagram at the top of page 226 must satisfy the differential equation

$$L\frac{di}{dt} + Ri = E.$$

If E is as in Exercise 45, show that a solution is

$$i = f(t) = Ae^{-Rt/L} + \frac{E_0}{Z} \sin (\omega t - \phi),$$

where A is any constant, and

$$Z = \sqrt{R^2 + \omega^2 L^2}, \qquad \tan \phi = \frac{\omega L}{R}.$$

** **47.** The current in the RCL-circuit shown below must satisfy the differential equation

$$L\frac{d^2i}{dt^2} + R\frac{di}{dt} + \frac{i}{C} = \frac{dE}{dt}.$$

If E is as in Exercise 45, show that a solution is

$$i = f(t) = e^{-Rt/(2L)}(A\cos \nu t + B\cos \nu t) + \frac{E_0}{Z}\sin(\omega t - \phi),$$

where A and B are any constants whatsoever, and

$$\nu = \sqrt{\frac{1}{LC} - \frac{R^2}{4L^2}}, \quad Z = \sqrt{R^2 + \left(\omega L - \frac{1}{\omega C}\right)^2}, \quad \tan \phi = \frac{\omega L - 1/(\omega C)}{R}.$$

** **48.**   (a) Show that when a liquid enters the tank in Figure 3.44 (page 219) at $\dot{m}$ kg/s and the mass of liquid in the tank is M kilograms, the temperature of liquid entering the tank is $T_0\,^\circ$C, the specific heat of the liquid is c_p, and the heater adds energy at the rate of q J/s, the differential equation governing temperature $T(t)$ of liquid in the tank is

$$\dot{m}c_pT_0 + q = \dot{m}c_pT + Mc_p\frac{dT}{dt}.$$

(b) Verify that the solution of this differential equation is

$$T(t) = T_0e^{-\dot{m}t/M} + \left(T_0 + \frac{q}{c_p\dot{m}}\right)\left(1 - e^{-\dot{m}t/M}\right)$$

when the temperature of the liquid in the tank is T_0 at time $t = 0$, and $\dot{m}$, T_0, and q are all constants.

** **49.**   (a) The function $f(x) = x^n$ is not defined for $x < 0$ if n is irrational, that is, is not rational. Verify power rule 3.7 for $x < 0$ if n is rational and x^n is defined.

(b) Discuss the derivative of $f(x) = x^n$ at $x = 0$.

3.11 Logarithmic Differentiation

Compare the three functions

$$x^n, \qquad a^x, \qquad x^x.$$

The first, in which the exponent is constant and the base is variable, is called a **power function**. The second, in which the base is constant and the exponent is variable, is called an **exponential function**. We have differentiation rules for power functions and exponential functions, namely

$$\frac{d}{dx}x^n = nx^{n-1} \qquad \text{and} \qquad \frac{d}{dx}a^x = a^x \ln a,$$

respectively. The third function x^x, which we consider only for $x > 0$, is neither a power nor an exponential function; therefore, we cannot use either of the above formulas to find its derivative. Instead, we set $y = f(x) = x^x$, and take natural logarithms,

$$\ln y = x \ln x.$$

Implicit differentiation with respect to x now gives

$$\frac{1}{y}\frac{dy}{dx} = \ln x + \frac{x}{x} = \ln x + 1.$$

Consequently,

$$\frac{dy}{dx} = y(\ln x + 1) \qquad \text{or} \qquad \frac{d}{dx}x^x = x^x(\ln x + 1).$$

This process of taking logarithms and then differentiating is called **logarithmic differentiation**.

EXAMPLE 3.35

Find the derivative of the function $f(x) = x^{\sin x}$ when $x > 0$.

SOLUTION   If we set $y = x^{\sin x}$, then

$$\ln y = \sin x \ln x.$$

Differentiation with respect to x gives

$$\frac{1}{y}\frac{dy}{dx} = \cos x \ln x + \frac{1}{x}\sin x,$$

and this can be solved for dy/dx:

$$\frac{dy}{dx} = x^{\sin x}\left(\cos x \ln x + \frac{1}{x}\sin x\right).$$

Logarithmic differentiation can also be used to differentiate complicated products or quotients, as in the following example.

EXAMPLE 3.36

Find the derivative of the function

$$y = f(x) = \frac{x^3(x^2 + 1)^{2/3}}{\sin^3 x}$$

on the interval $0 < x < \pi$. Extend the result to other values of x.

SOLUTION   When $0 < x < \pi$, we take natural logarithms of both sides of the definition for y:

$$\ln y = 3 \ln x + \frac{2}{3} \ln (x^2 + 1) - 3 \ln (\sin x).$$

Differentiation with respect to x gives

$$\frac{1}{y}\frac{dy}{dx} = \frac{3}{x} + \frac{2}{3}\frac{2x}{x^2 + 1} - \frac{3}{\sin x}\cos x,$$

and therefore

$$\frac{dy}{dx} = y\left[\frac{3}{x} + \frac{4x}{3(x^2 + 1)} - 3\cot x\right] = \frac{x^3(x^2 + 1)^{2/3}}{\sin^3 x}\left[\frac{3}{x} + \frac{4x}{3(x^2 + 1)} - 3\cot x\right].$$

When x is not in the interval $0 < x < \pi$, this derivation may not be valid. For instance, when $x < 0$, it is not acceptable to write $\ln x$, and when x is in the interval $\pi < x < 2\pi$, the term $\ln (\sin x)$ is not defined. These difficulties are easily overcome by first taking absolute values,

$$|y| = \frac{|x|^3(x^2 + 1)^{2/3}}{|\sin x|^3}.$$

Logarithms now give

$$\ln |y| = 3 \ln |x| + \frac{2}{3} \ln (x^2 + 1) - 3 \ln |\sin x|.$$

To differentiate this equation, we use equation 3.13 which states that when $f(x)$ is differentiable, and $f(x) \neq 0$,

$$\frac{d}{dx}|f(x)| = \frac{|f(x)|}{f(x)}f'(x).$$

When this is combined with formula 3.36, we obtain

$$\frac{d}{dx}\ln |f(x)| = \frac{1}{|f(x)|}\frac{d}{dx}|f(x)| = \frac{1}{|f(x)|}\frac{|f(x)|}{f(x)}f'(x) = \frac{f'(x)}{f(x)}. \tag{3.41}$$

Application of this result to the equation in $\ln |y|$ gives

$$\frac{1}{y}\frac{dy}{dx} = \frac{3}{x} + \frac{2}{3}\frac{2x}{x^2 + 1} - 3\frac{\cos x}{\sin x}.$$

This equation is identical to that obtained for dy/dx when $0 < x < \pi$, but its derivation here shows that it is valid even when x is not in the interval $0 < x < \pi$.

EXERCISES 3.11

In Exercises 1–24 use logarithmic differentiation to find $f'(x)$.

** **1.** $f(x) = x^{-x}, \quad x > 0$

** **2.** $f(x) = x^{4\cos x}, \quad x > 0$

** **3.** $f(x) = x^{4x}, \quad x > 0$

** **4.** $f(x) = (\sin x)^x, \quad 0 < x < \pi$

** **5.** $f(x) = \left(1 + \dfrac{1}{x}\right)^x, \quad x > 0$

** **6.** $f(x) = \left(1 + \dfrac{1}{x}\right)^{x^2}, \quad x > 0$

** **7.** $f(x) = \left(\dfrac{1}{x}\right)^{1/x}, \quad x > 0$

** **8.** $f(x) = \left(\dfrac{2}{x}\right)^{3/x}, \quad x > 0$

** **9.** $f(x) = (\sin x)^{\sin x}, \quad 0 < x < \pi$

** **10.** $f(x) = (\ln x)^{\ln x}, \quad x > 1$

** **11.** $f(x) = (x^2 + 3x^4)^3 (x^2 + 5)^4$

** **12.** $f(x) = \dfrac{\sqrt{x}(1 + 2x^2)}{\sqrt{1 + x^2}}$

** **13.** $f(x) = x\sqrt[3]{1 - \sin x}$

** **14.** $f(x) = (x^2 + 3x)^3 (x^2 + 5)^4$

** **15.** $f(x) = x^2 e^{4x}$

** **16.** $f(x) = x^{3/2} e^{-2x}$

** **17.** $f(x) = x^2 \ln x$

** **18.** $f(x) = \dfrac{e^x}{\ln(x - 1)}$

** **19.** $f(x) = (x^3 + 3)^3 (x^2 - 2x)$

** **20.** $f(x) = \dfrac{\sqrt{x}(1 - x^2)}{\sqrt{1 + x^2}}$

** **21.** $f(x) = \dfrac{x^2 - 1}{x\sqrt{1 - 4\tan^2 x}}$

** **22.** $f(x) = x^3(x^2 - 4x)\sqrt{1 + x^3}$

** **23.** $f(x) = \dfrac{\sin^3 3x}{\tan^5 2x}$

** **24.** $f(x) = \dfrac{\sin 2x \sec 5x}{(1 - 2\cot x)^3}$

** **25.** If $u(x)$ is positive for all x, find a formula for the derivative of u^u with respect to x.

** **26.** If a company sells a certain commodity at price r, the market demands

$$x = r^a e^{-b(r+c)},$$

items per week, where $r > a/b$, and a, b, and c are positive constants.

    (a) Show that the demand increases as the price decreases.

    (b) Calculate the elasticity of demand defined by

$$\frac{Er}{Ex} = \frac{x}{r}\frac{dr}{dx}.$$

3.12 Rolle's Theorem and the Mean Value Theorems

Certain results in calculus are immediately seen to be important. For example, the power, product, and quotient rules which eliminate the necessity of using equation 3.3 to calculate derivatives are clearly indispensable. Even the algebraic and geometric interpretations of the derivative itself are recognized as useful. Through various examples and exercises of this chapter, we have hinted at the variety and quantity of applications of the derivative. These will be dealt with at length in Chapter 4. Other results in calculus, especially those of a theoretical nature, are regarded as less important, or even unimportant, often because it is not obvious how they will be used. In this section we consider three very important theorems, without which we would encounter serious difficulty in treating many of the topics in the remainder of this book. The first theorem is needed to prove the second, and the second leads immediately to the third.

THEOREM 3.15   (Rolle's Theorem)

Suppose $f(x)$ is continuous on the closed interval $a \leq x \leq b$ and has a first derivative at each point in the open interval $a < x < b$. If $f(a) = f(b)$, then there exists at least one point c in the open interval at which $f'(c) = 0$.

For the function in Figure 3.46 there are two possible choices for c. Geometrically, Rolle's theorem seems quite evident. Begin with two points, say P and Q in Figure 3.47, which have the same y-coordinate. Now try to join these points by a curve which never has a horizontal tangent line while satisfying the following two conditions:

(a) Do not lift pencil from page—continuity of $f(x)$.

(b) The curve must have a tangent line at all points, and this tangent line must not be vertical—$f'(x)$ exists at all points.

It is impossible; therefore, a point c where $f'(c) = 0$ must exist.

FIGURE 3.46   **FIGURE 3.47**

To verify this theorem directly requires Theorem 4.2 in Section 4.6. Since the latter result is quoted without proof, it seems as reasonable to accept Rolle's theorem on obvious geometric grounds as to base a proof on a theorem which itself is stated without proof. But for those who would like to see a proof based on Theorem 4.2 in Section 4.6, see Exercise 10 in that section.

Rolle's theorem can be used to prove the following result.

THEOREM 3.16   (Cauchy's Generalized Mean Value Theorem)

Suppose functions $f(x)$ and $g(x)$ are continous on the closed interval $a \leq x \leq b$ and differentiable on the open interval $a < x < b$. If $g'(x) \neq 0$ on $a < x < b$, then there exists at least one point c in the open interval for which

$$\frac{f(b) - f(a)}{g(b) - g(a)} = \frac{f'(c)}{g'(c)}. \tag{3.42}$$

PROOF   First note that $g(b) - g(a)$ cannot equal zero. If it did, then $g(a)$ would equal $g(b)$, and Rolle's theorem applied to $g(x)$ on the interval $a \leq x \leq b$ would imply the existence of a point c at which $g'(c) = 0$, contrary to the given assumption. To prove the theorem, we construct a function $h(x)$ to satisfy the conditions of Rolle's theorem. Specifically, we consider

$$h(x) = f(x) - f(a) - \frac{f(b) - f(a)}{g(b) - g(a)}[g(x) - g(a)].$$

Since $f(x)$ and $g(x)$ are continuous for $a \le x \le b$, so too is $h(x)$. In addition,

$$h'(x) = f'(x) - \frac{f(b) - f(a)}{g(b) - g(a)} g'(x);$$

therefore, $h'(x)$ exists for $a < x < b$. Finally, since $h(a) = h(b) = 0$, we may conclude from Rolle's theorem that there exists a number c such that $a < c < b$, and

$$0 = h'(c) = f'(c) - \frac{f(b) - f(a)}{g(b) - g(a)} g'(c);$$

that is,

$$\frac{f'(c)}{g'(c)} = \frac{f(b) - f(a)}{g(b) - g(a)}.$$ ∎

The next theorem states an important special case of this result that occurs when $g(x) = x$.

> **THEOREM 3.17 (The Mean Value Theorem)**
>
> Suppose $f(x)$ is continuous on the closed interval $a \le x \le b$ and has a first derivative at each point of the open interval $a < x < b$. Then there exists at least one point c in the open interval for which
>
> $$f'(c) = \frac{f(b) - f(a)}{b - a}. \tag{3.43}$$

From a geometric point of view, equation 3.43 seems as obvious as Rolle's theorem. Figure 3.48 illustrates that the quotient $[f(b) - f(a)]/(b - a)$ is the slope of the line l joining the points $(a, f(a))$ and $(b, f(b))$ on the graph $y = f(x)$. The mean value theorem states that there is at least one point c between a and b at which the tangent line is parallel to l. In Figure 3.48 there are clearly two such points. Algebraically, 3.43 states that at some point between a and b, the instantaneous rate of change of the function $f(x)$ is equal to its average rate of change over the interval $a \le x \le b$. Similar interpretations of Theorem 3.16 are given in Section 10.6.

FIGURE 3.48

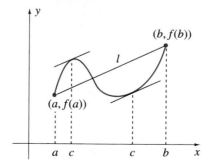

EXAMPLE 3.37

Find all values of c satisfying the mean value theorem for the function $f(x) = x^3 - 4x$ on the interval $-1 \le x \le 3$.

SOLUTION   Since $f(x)$ is differentiable, and therefore continuous, at each point in the interval $-1 \le x \le 3$, we can indeed apply the mean value theorem and claim the existence of at least one number c in $-1 < x < 3$ such that

$$f'(c) = \frac{f(3) - f(-1)}{3 - (-1)};$$

that is,

$$3c^2 - 4 = \frac{15 - 3}{4} = 3.$$

Consequently, $c = \pm\sqrt{7/3}$. Since $-\sqrt{7/3} < -1$, $c = \sqrt{7/3}$ is the only value of c in the interval $-1 < x < 3$ (see Figure 3.49).

FIGURE 3.49

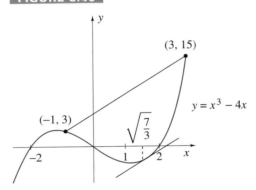

EXAMPLE 3.38

A traffic plane (Figure 3.50) measures the time that it takes a car to travel between points A and B as 15 s, and radios this information to a patrol car. What is the maximum speed at which the police officer can claim the car was travelling between A and B?

SOLUTION   The average speed of the car between A and B is $500/15 = 100/3$ m/s. According to the mean value theorem, the instantaneous speed of the car must also have been $100/3$ m/s at least once. This is the maximum speed attributable to the car between A and B. It may have been travelling much faster at some points, but from the information given, no speed greater than $100/3$ m/s can be claimed by the officer.

FIGURE 3.50

A

B

$\longleftarrow$ 500 m $\longrightarrow$

EXERCISES 3.12

In Exercises 1–14 decide whether the mean value theorem can be applied to the function on the interval. If it cannot, explain why not. If it can, find all values of c in the interval that satisfy equation 3.43.

** **1.** $f(x) = x^2 + 2x,$ $-3 \le x \le 2$ ** **2.** $f(x) = 4 + 3x - 2x^2,$ $1 \le x \le 3$

** **3.** $f(x) = x + 5,$ $2 \le x \le 3$ ** **4.** $f(x) = |x|,$ $-1 \le x \le 1$

** **5.** $f(x) = |x|,$ $0 \le x \le 1$

** **6.** $f(x) = x^3 + 2x^2 - x - 2,$ $-3 \le x \le 2$

** **7.** $f(x) = x^3 + 2x^2 - x - 2,$ $-1 \le x \le 2$

** **8.** $f(x) = (x + 2)/(x - 1),$ $2 \le x \le 4$

** **9.** $f(x) = (x + 1)/(x + 2),$ $-3 \le x \le 2$

** **10.** $f(x) = x^2/(x + 3),$ $-2 \le x \le 3$

** **11.** $f(x) = \sin x,$ $0 \le x \le 2\pi$ ** **12.** $f(x) = \ln(2x + 1),$ $0 \le x \le 2$

** **13.** $f(x) = e^{-x},$ $-1 \le x \le 1$ ** **14.** $f(x) = \sec x,$ $0 \le x \le \pi$

In Exercises 15–18 decide whether Cauchy's generalized mean value theorem can be applied to the functions on the interval. If it cannot, explain why not. If it can, find all values of c in the interval that satisfy equation 3.42.

** **15.** $f(x) = x^2,$ $g(x) = x,$ $1 \le x \le 2$

** **16.** $f(x) = x + 1,$ $g(x) = |x|^{3/2},$ $-1 \le x \le 1$

** **17.** $f(x) = x^2 + 3x - 1,$ $g(x) = x^3 + 5x + 4,$ $0 \le x \le 2$

** **18.** $f(x) = x/(x + 1),$ $g(x) = x/(x - 1),$ $-3 \le x \le -2$

** **19.** Show that if $|f'(x)| \le M$ on $a \le x \le b,$ then

$$|f(b) - f(a)| \le M(b - a).$$

** **20.** Show that the value c that satisfies the mean value theorem for any quadratic function $f(x) = dx^2 + ex + g$ on any interval $a \leq x \leq b$ whatsoever is $c = (a + b)/2$.

** **21.** Use the mean value theorem to show that

$$| \sin a - \sin b| \leq |a - b|$$

for all real a and b. Is the same inequality valid for the cosine function?

** **22.** Let $f(x)$ and $g(x)$ be two functions that are differentiable at each point of the interval $a \leq x \leq b$. Prove that if $f(a) = g(a)$ and $f(b) = g(b)$, then there exists c in the open interval $a < x < b$ for which $f'(c) = g'(c)$.

** **23.** Verify that, for a cubic polynomial $f(x) = dx^3 + ex^2 + gx + h$ defined on any interval $a \leq x \leq b$, the values of c that satisfy equation 3.43 are equidistant from $x = -e/(3d)$.

SUMMARY

In this chapter we defined the derivative of a function $y = f(x)$ as

$$\frac{dy}{dx} = f'(x) = \lim_{h \to 0} \frac{f(x + h) - f(x)}{h}.$$

Algebraically, it is the instantaneous rate of change of y with respect to x; geometrically, it is the slope of the tangent line to the graph of $f(x)$. To eliminate the necessity of using this definition over and over again, we derived the sum, product, quotient, and power rules:

$$\frac{d}{dx}(u + v) = \frac{du}{dx} + \frac{dv}{dx},$$

$$\frac{d}{dx}(uv) = u\frac{dv}{dx} + v\frac{du}{dx},$$

$$\frac{d}{dx}\left(\frac{u}{v}\right) = \frac{v\dfrac{du}{dx} - u\dfrac{dv}{dx}}{v^2},$$

$$\frac{d}{dx}(u^n) = nu^{n-1}\frac{du}{dx}.$$

These four simple rules are fundamental to all calculus.

When a function $y = f(x)$ is defined implicitly by some equation $F(x, y) = 0$, we use implicit differentiation to find its derivative. We differentiate each term in the equation with respect to x, and then solve the resulting equation for dy/dx. We pointed out that care must be taken in differentiating equations. An equation can be differentiated with respect to a variable only if it is valid for a continuous range of values of that variable.

The chain rule defines the derivative of a composite function $y = f(g(x))$ as the product of the derivatives of $y = f(u)$ and $u = g(x)$:

$$\frac{dy}{dx} = \frac{dy}{du}\frac{du}{dx}.$$

These rules and techniques form the basis for the rest of differential calculus. When they are combined with the derivatives of the trigonometric functions, the exponential and logarithm functions, and the functions of Chapter 8, we will be well prepared to handle those applications of calculus that involve differentiation. Derivative formulas for the trigonometric, exponential, and logarithm functions are listed below:

$$\frac{d}{dx}\sin u = \cos u \, \frac{du}{dx}, \qquad\qquad \frac{d}{dx}\cos u = -\sin u \, \frac{du}{dx},$$

$$\frac{d}{dx}\tan u = \sec^2 u \, \frac{du}{dx}, \qquad\qquad \frac{d}{dx}\cot u = -\csc^2 u \, \frac{du}{dx},$$

$$\frac{d}{dx}\sec u = \sec u \tan u \, \frac{du}{dx}, \qquad \frac{d}{dx}\csc u = -\csc u \cot u \, \frac{du}{dx},$$

$$\frac{d}{dx}\log_a u = \frac{1}{u}\frac{du}{dx}\log_a e, \qquad \frac{d}{dx}\ln u = \frac{1}{u}\frac{du}{dx},$$

$$\frac{d}{dx}a^u = a^u \frac{du}{dx}\ln a, \qquad\qquad \frac{d}{dx}e^u = e^u \frac{du}{dx}.$$

Velocity and acceleration have finally been given formal definitions. Velocity is the derivative of displacement, and acceleration is the derivative of velocity, or the second derivative of displacement,

$$v(t) = \frac{dx}{dt}, \qquad a(t) = \frac{dv}{dt} = \frac{d^2x}{dt^2}.$$

We completed the chapter by using Rolle's theorem to prove two mean value theorems. When $f(x)$ and $g(x)$ are continuous for $a \leq x \leq b$ and differentiable for $a < x < b$, Cauchy's generalized mean value theorem guarantees the existence of at least one point c between a and b such that

$$\frac{f(b) - f(a)}{g(b) - g(a)} = \frac{f'(c)}{g'(c)},$$

provided also that $g'(x) \neq 0$ for $a < x < b$. When $g(x) = x$, we obtain as a special case, the mean value theorem

$$f'(c) = \frac{f(b) - f(a)}{b - a}.$$

KEY TERMS

In reviewing this chapter, you should be able to define or discuss the following key terms:

Average rate of change	Angle between curves
Instantaneous rate of change	Orthogonal
Derivative	Chain rule
Tangent line	Extended power rule
Derivative function	Explicit definition of a function
Power rule	Implicit definition of a function
Normal line	Implicit differentiation
Increment	Logarithmic differentiation
Differentiable	Rolle's theorem
Right-hand derivative	Generalized mean value theorem
Left-hand derivative	Mean value theorem
Product rule	Velocity
Quotient rule	Acceleration

REVIEW EXERCISES

In Exercises 1–48 assume that y is defined as a function of x and find dy/dx in as simplified a form as possible.

* **1.** $y = x^3 + \dfrac{1}{x^2}$

* **2.** $y = 3x^2 + 2x + \dfrac{1}{x}$

* **3.** $y = 2x - \dfrac{1}{3x^2} + \dfrac{1}{\sqrt{x}}$

* **4.** $y = x^{1/3} - \dfrac{2}{3}x^{5/3}$

* **5.** $y = x(x^2 + 5)^4$

* **6.** $y = (x^2 + 2)^2(x^3 - 3)^3$

* **7.** $y = \dfrac{3x^2}{x^3 - 5}$

* **8.** $y = \dfrac{3x - 2}{x + 5}$

* **9.** $y = \dfrac{x^2 + 2x + 2}{x^2 + 2x - 1}$

* **10.** $y = \dfrac{4x}{x^2 + 5x - 2}$

* **11.** $xy + 3y^3 = x + 1$

* **12.** $\dfrac{x}{y} + \dfrac{y}{x} = x$

* **13.** $x^2 y^2 - 3y \sin x = 14$

* **14.** $x^2 y + y\sqrt{1 + x} = 3$

* **15.** $y = \tan^3(3x + 2)$

* **16.** $y = \sec^2(1 - 4x)$

* **17.** $y = \dfrac{\sin 2x}{\cos 3x}$

* **18.** $y = \sec(\tan 2x)$

* **19.** $y = x^2 \cos x^2$

* **20.** $y = \sin^2 x \cos^2 x$

* **21.** $y = u^2 - 2u, \quad u = (1 + 2x)^{5/3}$

* **22.** $y = t + \cos 2t, \quad t = x - \cos 2x$

* **23.** $y = \sqrt{1 - t^3}, \quad t = \sqrt{1 + x^2}$

* **24.** $y = v \cos^2 v, \quad v = \sqrt{1 - x^2}$

* **25.** $y = \sqrt{1 + \sqrt{1 + x}}$

* **26.** $x = e^{2y}$

* **27.** $\dfrac{2xy}{3x + 4} = x^2 + 2$

* **28.** $y = (x^2 + 1) \ln(x^2 - 1)$

* **29.** $x \sin y + 2xy = 4$

* **30.** $5 \cos(x - y) = 1$

** **31.** $x = \sqrt{1 + x \cot y^2}$

** **32.** $x = \sqrt{\dfrac{4 + y}{4 - y}}$

** **33.** $y = \sqrt{\dfrac{4 + x^2}{4 - x^2}}$

** **34.** $y = \dfrac{x^2 \sqrt{1 - x}}{x + 5}$

** **35.** $y = \sqrt{7 - \sqrt{7 - \sqrt{x}}}$

** **36.** $\dfrac{x}{x + y} = \dfrac{y}{x - y}$

** **37.** $y = \sqrt{\dfrac{4 + t}{4 - t}}, \quad t = \tan x$

** **38.** $x = \dfrac{y^2 - 2y}{y^3 + 4y + 6}$

** **39.** $y = \dfrac{x^3 - 6x^2 + 12x - 8}{x^2 - 4x + 4}$

** **40.** $y = \dfrac{x - 2}{\sqrt{x} - \sqrt{2}}$

** **41.** $y = x^{2x}$

** **42.** $y = (\cos x)^x, \, 0 < x < \pi/2$

** **43.** $y = \dfrac{e^x}{e^x + 1}$

** **44.** $y = \log_{10} (\log_{10} x)$

** **45.** $y = e^x \ln x$

** **46.** $x = e^y + e^{-y}$

** **47.** $xy e^{xy} = 1$

** **48.** $x^2 y + \ln (x + y) = x + 2$

In Exercises 49–52 find equations for the tangent and normal lines to the curve at the point indicated.

* **49.** $y = x^3 + 3x - 2$ at $(1, 2)$

* **50.** $y = \dfrac{1}{x + 5}$ at $(0, 1/5)$

* **51.** $y = \cos 2x$ at $(\pi/2, -1)$

* **52.** $y = \dfrac{x^2 + 3x}{2x - 5}$ at $(1, -4/3)$

In Exercises 53–55 find $d^2 y/dx^2$ assuming that y is defined implicitly as a function of x.

** **53.** $x^2 - y^2 + 2(x - y) = y^3$

** **54.** $(x - y)^2 = 3xy$

** **55.** $\sin (x + y) = x$

** **56.** Draw a graph of $f(x) = \sin x^2$. Is it periodic?

** **57.** Find all points on the curve $y = x^3 + x^2$ at which the tangent line passes through the origin.

** **58.** Find that point on the curve $x = y^2 - 4$ at which the normal line passes through the point $(-6, 7)$. What application could be made of this result?

** **59.**   (a) How many functions with domain $-1 \le x \le 1$ are defined implicitly by the equation $x^2 + y^2 = 1$?

   (b) How many continuous functions with domain $-1 \le x \le 1$ are defined implicitly by the equation?

** **60.** What is the rate of change of the area A of an equilateral triangle with respect to its side length ℓ?

** **61.** In a heated house, the temperature varies as the thermostat continually engages and disengages the furnace. Suppose that at the thermostat the temperature T in degrees Celsius over a four-hour time interval $0 \le t \le 4$ is given by

$$T = f(t) = 20 + 3 \sin (4\pi t - \pi/2).$$

   (a) Sketch a graph of $f(t)$.

   (b) How many times is the furnace on during the four-hour period?

   (c) What is the maximum time rate of change of temperature?

** **62.** The curve defined by the equation $(x^2 + y^2)^2 = x^2 - y^2$ and shown below is called a *lemniscate*. Find the four points at which the tangent line is horizontal.

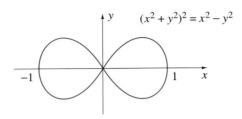

** **63.** Find all points c in the interval $3 \leq x \leq 6$ that satisfy equation 3.43 when $f(x) = x^3 + 3x - 2$.

** **64.** Find all points c in the interval $-1 \leq x \leq 1$ that satisfy equation 3.42 when $f(x) = 3x^2 - 2x + 4$ and $g(x) = x^3 + 2x$.

Applications of Differentiation

Our discussions in Chapter 3 hinted at some of the applications of the derivative; in this chapter we deal with them in detail. In Section 4.1 we show that derivatives provide one of the most powerful methods for solving equations. Sections 4.2–4.6 are devoted to the topic of optimization, its theory and applications. Sections 4.6–4.9, with their wealth of applied problems, show the power of calculus in applied mathematics. In Section 4.6 we illustrate the simplicity that calculus brings to solving applied maxima-minima problems; in Section 4.7 we develop a deeper understanding for the already familiar notions of velocity and acceleration; in Section 4.8 we use the interpretation of the derivative as a rate of change to investigate interdependences of related quantities in a wide variety of applications; and in Section 4.9 we apply derivatives to LCR-circuits. The calculation of many otherwise intractable limits becomes relatively straightforward with L'Hôpital's rule in Section 4.10. In Section 4.11 we discuss differentials, quantities essential to the topic of integration which begins in Chapter 5. We learn to approximate functions by means of polynomials in Section 4.12. To close the chapter, in Section 4.13 we offer suggestions on how to organize and solve complicated multistage problems.

4.1 Newton's Iterative Procedure for Solving Equations

In almost every area of applied mathematics, it is necessary to solve equations. When the problem is to solve one equation in one unknown, the equation can be expressed in the form

$$f(x) = 0, \tag{4.1}$$

where $f(x)$ is usually a differentiable function of x. It might be a polynomial, a trigonometric function, an exponential or a logarithm function, or a complicated combination of these. The equation may have one solution or many solutions; these solutions are also called roots of the equation or zeros of the function. Few equations can be solved by formula. Even when $f(x)$ is a polynomial, the only simple formula is quadratic formula 1.10 which solves the equation for second degree polynomials. Functions other than polynomials rarely have formulas for their zeros.

Since we can seldom find exact solutions of 4.1, we consider approximating the solutions. Sophisticated calculators have routines for approximating solutions to equations; computer software packages have one or more commands for doing this. Unfortunately, with calculators and computers, you really have no idea what is going on inside them. You supply an equation and out comes a number. In this section we develop one of the most powerful techniques for approximating solutions to equations. The method is calculus-based and it, or a modification of it, is used by many calculators and computers to solve equations. If your machine uses this technique you will now understand what it is doing.

FIGURE 4.1

To introduce the technique, consider the problem of finding the depth of water that half fills (by volume) the hemispherical tank in Figure 4.1. It can be shown (see Exercise 15 in Section 7.4) that when the water is x metres deep, the volume of water in the tank is given by the formula

$$\frac{\pi}{3}(15x^2 - x^3).$$

Since the volume of one-half the tank is one-quarter that of a sphere of radius 5, it follows that the tank is half full when x satisfies the equation

$$\frac{1}{4}\left[\frac{4}{3}\pi(5)^3\right] = \frac{\pi}{3}(15x^2 - x^3),$$

and this equation simplifies to

$$f(x) = x^3 - 15x^2 + 125 = 0.$$

How do we solve this equation when we know intuitively that the solution we want is somewhere around $x = 3$? A sketch of $f(x)$ between $x = 3$ and $x = 3.5$ is shown in Figure 4.2. The curvature has been exaggerated in order to more clearly depict the following geometric construction. If we set $x_1 = 3$, then x_1 is an approximation to the solution of the equation— not a good approximation, but an approximation nonetheless. Suppose we draw the tangent line to $y = f(x)$ at $(x_1, f(x_1))$. If x_2 is the point of intersection of this tangent line with the x-axis, it is clear that x_2 is a better approximation to the solution of the equation than x_1. If we draw the tangent line to $y = f(x)$ at $(x_2, f(x_2))$, its intersection point x_3 with the x-axis is an even better approximation. Continuation of this process leads to a succession of numbers $x_1, x_2, x_3, \ldots$, each of which is closer to the solution of the equation $f(x) = 0$ than the preceding numbers. This procedure for finding better and better approximations to the solution of an equation is called Newton's iterative procedure (or the Newton-Raphson iterative procedure). We say that the numbers $x_1, x_2, x_3, \ldots$ converge to the root of the equation.

FIGURE 4.2

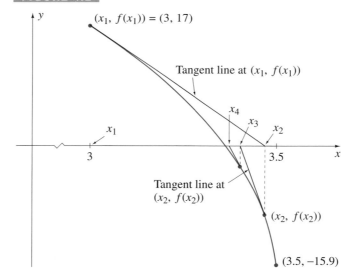

What we need now is an algebraic formula by which to calculate the numbers $x_2, x_3, x_4, \ldots$. The equation of the tangent line to $y = f(x)$ at $(x_1, f(x_1))$ is

$$y - f(x_1) = f'(x_1)(x - x_1).$$

To find the point of intersection of this tangent line with the x-axis, we set $y = 0$,

$$-f(x_1) = f'(x_1)(x - x_1),$$

and solve for x,

$$x = x_1 - \frac{f(x_1)}{f'(x_1)}.$$

But the point of intersection of the tangent line at $(x_1, f(x_1))$ with the x-axis is the second approximation x_2; that is,

$$x_2 = x_1 - \frac{f(x_1)}{f'(x_1)}.$$

To find x_3 we repeat this procedure with x_2 replacing x_1; the result is

$$x_3 = x_2 - \frac{f(x_2)}{f'(x_2)}.$$

As we repeat this process over and over again, the following formula for the $(n + 1)^{\text{th}}$ approximation x_{n+1} in terms of the nth approximation x_n emerges:

$$x_{n+1} = x_n - \frac{f(x_n)}{f'(x_n)}. \tag{4.2}$$

This formula defines each approximation in Newton's iterative procedure in terms of its predecessor.

Let us use this procedure to find the solution of $f(x) = x^3 - 15x^2 + 125 = 0$ in the above example. The derivative of $f(x)$ is $f'(x) = 3x^2 - 30x$, and formula 4.2 becomes

$$x_{n+1} = x_n - \frac{x_n^3 - 15x_n^2 + 125}{3x_n^2 - 30x_n}.$$

Calculation of the first five approximations beginning with $x_1 = 3$ gives

$$x_2 = x_1 - \frac{x_1^3 - 15x_1^2 + 125}{3x_1^2 - 30x_1} = 3 - \frac{3^3 - 15(3)^2 + 125}{3(3)^2 - 30(3)} = 3.269\,84;$$

$$x_3 = x_2 - \frac{x_2^3 - 15x_2^2 + 125}{3x_2^2 - 30x_2} = 3.263\,52;$$

$$x_4 = x_3 - \frac{x_3^3 - 15x_3^2 + 125}{3x_3^2 - 30x_3} = 3.263\,518\,2;$$

$$x_5 = x_4 - \frac{x_4^3 - 15x_4^2 + 125}{3x_4^2 - 30x_4} = 3.263\,518\,2.$$

Newton's iterative procedure has therefore produced $3.263\,518$ as an approximate solution to the equation $x^3 - 15x^2 + 125 = 0$. In spite of the fact that we have written six decimals in this final answer, our analysis in no way guarantees this degree of accuracy; we have simply judged on the basis of $x_4 = x_5$ that $3.263\,518$ might be accurate to six decimals. The Intermediate Value Theorem confirms this when we calculate

$$f(3.263\,517\,5) = 4.8 \times 10^{-5} \quad \text{and} \quad f(3.263\,518\,5) = -1.8 \times 10^{-5}.$$

EXAMPLE 4.1

Use Newton's iterative procedure to find the only positive root of the equation

$$3x^4 + 15x^3 - 125x - 1500 = 0$$

accurate to 5 decimals. (This equation will be encountered in Exercise 44 of Section 4.6.)

SOLUTION   The plot of $f(x) = 3x^4 + 15x^3 - 125x - 1500$ in Figure 4.3 shows that the solution is between $x = 4$ and $x = 5$. The approximations predicted by Newton's method are defined by

$$x_{n+1} = x_n - \frac{f(x_n)}{f'(x_n)} = x_n - \frac{3x_n^4 + 15x_n^3 - 125x_n - 1500}{12x_n^3 + 45x_n^2 - 125}.$$

Suppose we choose $x_1 = 4$ as the initial approximation; this is closer to the root than $x = 5$. We find that

$$x_2 = 4.199\,56, \qquad x_3 = 4.187\,268, \qquad x_4 = 4.187\,218\,7, \qquad x_5 = 4.187\,218\,7.$$

Since

$$f(4.187\,215) = -5.7 \times 10^{-3} \qquad \text{and} \qquad f(4.187\,225) = 9.7 \times 10^{-3},$$

it follows that the root is $x = 4.187\,22$, accurate to five decimals.

FIGURE 4.3

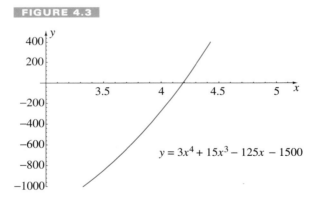

$$y = 3x^4 + 15x^3 - 125x - 1500$$

In the following example we again use Newton's iterative procedure to approximate the solution of an equation but specify the required accuracy as maximum error.

EXAMPLE 4.2

Use Newton's method to find the smallest root of the cubic equation $x^3 - 3x + 1 = 0$ with error less than 10^{-5}.

SOLUTION   The graph in Figure 4.4 indicates that the smallest root is just to the right of $x = -2$. We take $x_1 = -2$ and use Newton's iterative procedure to define approximations:

$$x_{n+1} = x_n - \frac{f(x_n)}{f'(x_n)} = x_n - \frac{x_n^3 - 3x_n + 1}{3x_n^2 - 3}.$$

The next four approximations are

$$x_2 = -1.888\,89, \qquad x_3 = -1.879\,45, \qquad x_4 = -1.879\,385, \qquad x_5 = -1.879\,385\,2.$$

For an approximation with error less than 10^{-5}, we suggest $-1.879\,39$. The Intermediate Value Theorem confirms this when we calculate

$$f(-1.879\,40) = -1.1 \times 10^{-4} \qquad \text{and} \qquad f(-1.879\,38) = 4.0 \times 10^{-5}.$$

FIGURE 4.4

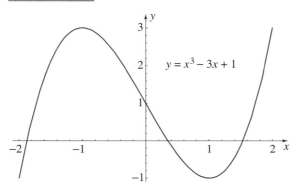

$y = x^3 - 3x + 1$

The value of a technique for approximating solutions of an equation depends on two factors, applicability and rate of convergence. Newton's method scores well in both categories. The method can be applied to any equation 4.1, provided $f(x)$ is differentiable in an interval containing the root. It converges to a solution $x = a$, provided $f'(a) \neq 0$ and x_1 is chosen sufficiently close to a (see Exercise 33 in Section 11.2). The condition $f'(a) \neq 0$ is included so that the denominator in equation 4.2 does not approach 0 as $x_n \to a$. However, $f(x_n)$ usually approaches 0 faster than $f'(x_n)$ so that even when $f'(a) = 0$, Newton's method may be successful. It is impossible to indicate how close x_1 must be to a in order to guarantee convergence to a. In some examples x_1 can be any number whatsoever, but we are also aware of examples where $|x_1 - a|$ must be less than 0.01 (see Example 4.3, page 244). Provided x_1 is sufficiently close to the root, convergence of the approximations to the root is rapid. This makes Newton's method valuable from the point of view of the second criterion, rate of convergence.

When x_1 is not sufficiently close to a, Newton's iterative procedure may not converge, or may converge to a solution other than expected. For instance, if we attempt to approximate the largest root in Example 4.2, and inadvertently choose $x_1 = 1$, we cannot find the second approximation: algebraically because $3x_1^2 - 3 = 0$, and geometrically because at $x_1 = 1$, the tangent line is horizontal, and does not intersect the x-axis. The function $f(x)$ in Figure 4.5 has zeros near $x = 1/2$ and $x = 1$. If we attempt to find the smaller zero using an initial

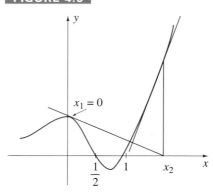

approximation $x_1 = 0$, we find the second approximation x_2 to be larger than 1. Further iterations then converge to the zero near $x = 1$, not the zero near $x = 1/2$. We conclude, therefore, that the initial approximation in Newton's iterative procedure is most important. A poor choice for x_1 may lead to numbers that either converge to the wrong root or do not converge at all.

EXAMPLE 4.3

FIGURE 4.6

A uniform hydro cable $P = 80$ m long with mass per unit length $\rho = 0.5$ kg/m is hung from two supports at the same level $L = 70$ m apart (Figure 4.6). The tension T in the cable at its lowest point must satisfy the equation

$$\frac{\rho g P}{T} = e^{\rho g L/(2T)} - e^{-\rho g L/(2T)},$$

where $g = 9.81$. If we set $z = \rho g/(2T)$, then z must satisfy

$$2Pz = e^{Lz} - e^{-Lz}.$$

Solve this equation for z and hence find T correct to one decimal.

SOLUTION   When we substitute $P = 80$ and $L = 70$, the equation for z becomes

$$160z = e^{70z} - e^{-70z}.$$

The exponential function e^{70z} grows very rapidly for $z > 0$, and e^{-70z} approaches zero very quickly. It follows that the solution of the equation must be quite small. If we plot $f(z) = e^{70z} - e^{-70z} - 160z$ on the interval $0 \le z \le 0.02$ (Figure 4.7), we capture the solution. With $z_1 = 0.013$, Newton's iterative procedure defines further approximations by

$$z_1 = 0.013, \qquad z_{n+1} = z_n - \frac{e^{70z_n} - e^{-70z_n} - 160z_n}{70e^{70z_n} + 70e^{-70z_n} - 160}.$$

To get an idea of how accurate z should be in order to give T correct to one decimal, we note that for $z = 0.013$, tension is $T \approx 189$. Thus, we should determine z to around five or six figures. Iteration of Newton's procedure gives

$$z_2 = 0.012\,957\,3, \qquad z_3 = 0.012\,957\,0, \qquad z_4 = 0.012\,957\,0.$$

With z approximated by $0.012\,957\,0$, tension is $T = 189.3$ N. We could verify that this is accurate to one decimal by defining $g(T) = e^{\rho g L/(2T)} - e^{-\rho g L/(2T)} - \rho g P/T$ and evaluating $g(189.25)$ and $g(189.35)$. The first is positive and the second is negative.

FIGURE 4.7

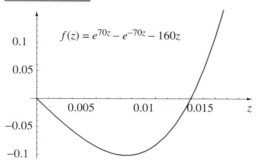

$$f(z) = e^{70z} - e^{-70z} - 160z$$

EXERCISES 4.1

In Exercises 1–16 use Newton's iterative procedure to find approximations to all roots of the equation accurate to 6 decimal places. In each case, make a plot in order to obtain an initial approximation to each root.

* **1.** $x^2 + 3x + 1 = 0$ 　　　　　　* **2.** $x^2 - x - 4 = 0$

* **3.** $x^3 + x - 3 = 0$ 　　　　　　* **4.** $x^3 - x^2 + x - 22 = 0$

* **5.** $x^3 - 5x^2 - x + 4 = 0$ 　　　* **6.** $x^5 + x - 1 = 0$

* **7.** $x^4 + 3x^2 - 7 = 0$ 　　　　　* **8.** $\dfrac{x+1}{x-2} = x^2 + 1$

* **9.** $x - 10\sin x = 0$ 　　　　　　* **10.** $\sec x = \dfrac{2}{1 + x^4}$

* **11.** $(x+1)^2 = \sin 4x$ 　　　　　* **12.** $(x+1)^2 = 5\sin 4x$

* **13.** $x + 4\ln x = 0$ 　　　　　　* **14.** $x\ln x = 6$

* **15.** $e^x + e^{-x} = 10x$ 　　　　　* **16.** $x^2 - 4e^{-2x} = 0$

In Exercises 17–24 use Newton's iterative procedure to find approximations to all roots of the equation with error no greater than that specified.

* **17.** $x^3 - 5x - 1 = 0, \quad 10^{-3}$ 　　* **18.** $x^4 - x^3 + 2x^2 + 6x = 0, \quad 10^{-4}$

* **19.** $\dfrac{x}{x+1} = x^2 + 2, \quad 10^{-5}$ 　　* **20.** $(x+1)^2 = x^3 - 4x, \quad 10^{-3}$

* **21.** $(x+1)^2 = 5\sin 4x, \quad 10^{-3}$ 　* **22.** $\cos^2 x = x^2 - 1, \quad 10^{-4}$

* **23.** $x + (\ln x)^2 = 0, \quad 10^{-3}$ 　　* **24.** $e^{3x} + e^x = 4, \quad 10^{-4}$

In Exercises 25–28 find all points of intersection for the curves accurate to four decimals.

** **25.** $y = x^3, \quad y = x + 5$ 　　　** **26.** $y = (x+1)^2, \quad y = x^3 - 4x$

** **27.** $y = x^4 - 20, \quad y = x^3 - 2x^2$ 　** **28.** $y = \dfrac{x}{x+1}, \quad y = x^2 + 2$

** **29.** When the beam in the figure below vibrates vertically, there are certain frequencies of vibration called natural frequencies. They are solutions of the equation

$$\tan x = \frac{e^x - e^{-x}}{e^x + e^{-x}}$$

divided by 20π. Find the two smallest frequencies correct to four decimals.

** **30.** A stone of mass 100 gm is thrown vertically upward with speed 20 m/s. Air exerts a resistive force on the stone proportional to its speed, and has magnitude 0.1 N when the speed of the stone is 10 m/s. It can be shown that the height y above the projection point attained by the stone is given by

$$y = -98.1t + 1181\left(1 - e^{-t/10}\right) \text{ m,}$$

where t is time (measured in seconds with $t = 0$ at the instant of projection).

(a) The time taken for the stone to return to its projection point can be obtained by setting $y = 0$ and solving the equation for t. Do so (correct to 2 decimals).

(b) When air resistance is neglected, the formula for y is

$$y = 20t - 4.905t^2 \text{ m.}$$

What is the elapsed time in this case from the instant the stone is projected until it returns to the projection point?

** **31.** Planck's law for the energy density E of blackbody radiation at $1000°\text{K}$ states that

$$E = E(\lambda) = \frac{k\lambda^{-5}}{e^{c/\lambda} - 1}$$

where $k > 0$ is a constant and $c = 0.001\,438\,6$. This function is shown in the figure below. The value of λ at which E is a maximum must satisfy the equation

$$(5\lambda - c)e^{c/\lambda} - 5\lambda = 0.$$

Find this value of λ correct to seven decimals.

4.2 Increasing and Decreasing Functions

Many mathematical concepts have their origins in intuitive ideas. In this section we analyze the intuitive idea of one quantity "getting larger" and another "getting smaller." To describe what it means for a quantity to be increasing (getting larger) or decreasing (getting smaller), we first suppose that the quantity is represented by some function $f(x)$ of a variable x. The mathematical definition for $f(x)$ to be increasing or decreasing is as follows.

> **DEFINITION 4.1**
>
> A function $f(x)$ is said to be **increasing** on an interval I if for all $x_1 > x_2$ in I
>
> $$f(x_1) > f(x_2). \tag{4.3a}$$
>
> A function $f(x)$ is said to be **decreasing** on I if for all $x_1 > x_2$ in I
>
> $$f(x_1) < f(x_2). \tag{4.3b}$$

The continuous function in Figure 4.8 is increasing on the intervals

$$a \leq x \leq b, \qquad c \leq x \leq d, \qquad e \leq x \leq f,$$

and decreasing on the intervals

$$b \leq x \leq c, \qquad d \leq x \leq e.$$

FIGURE 4.8

FIGURE 4.9

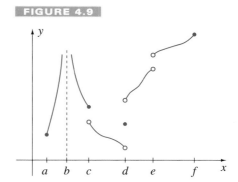

When a function has points of discontinuity, such as in Figure 4.9, the situation is somewhat more complicated. This function is increasing on the intervals

$$a \leq x < b, \qquad d \leq x < e, \qquad e < x \leq f,$$

and decreasing for

$$b < x < d.$$

Pay special attention to whether end points of each interval are included.

Figures 4.8 and 4.9 indicate that the sign of $f'(x)$ determines whether a function is increasing or decreasing on an interval. The following test is easily verifiable.

FIGURE 4.10

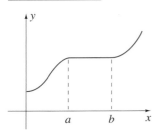

Increasing and Decreasing Test

(i) A function $f(x)$ is increasing on an interval I if on I

$$f'(x) \geq 0 \tag{4.4a}$$

and is equal to zero at only a finite number of points.

(ii) A function $f(x)$ is decreasing on an interval I if on I

$$f'(x) \leq 0 \tag{4.4b}$$

and is equal to zero at only a finite number of points.

FIGURE 4.11

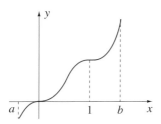

Figures 4.10 and 4.11 illustrate why we permit $f'(x)$ to vanish at only a finite number of points and do not, therefore, allow it to vanish on an interval. In Figure 4.10, $f'(x)$ is equal to zero on the interval $a \leq x \leq b$, and certainly $f(x)$ is not increasing on any interval that contains these points. The function $f(x)$ in Figure 4.11 has $f'(0) = 0 = f'(1)$, and yet $f(x)$ is increasing on the interval $a \leq x \leq b$.

Conditions 4.4 are sufficient to guarantee that a function is increasing or decreasing on an interval; they are not necessary. For example, the function $f(x) = x - \sin x$ in Figure 4.12 has derivative equal to 0 at the infinity of values $x = 2n\pi$ where n is an integer, yet the function is increasing on the interval $-\infty < x < \infty$. What we are saying is that tests more general than 4.4 can be formulated, but we feel that the extra complexity is not worth the gain. For a proof of 4.4, do Exercise 47.

FIGURE 4.12

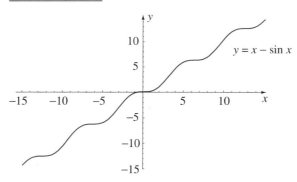

Intervals on which a function is increasing and decreasing are separated by points where the derivative is equal to zero or does not exist. Keep this in mind in the following example.

EXAMPLE 4.4

Find intervals on which the following functions are increasing and decreasing:

(a) $f(x) = 2x^3 + 3x^2 - 5x + 4$ (b) $f(x) = \dfrac{x^2 - 9}{x^2 + x - 2}$

SOLUTION

(a) The graph of the function in Figure 4.13a suggests that the two points where $f'(x) = 0$, one near $x = -2$, and the other near $x = 1$, separate the x-axis into three intervals on which the function is either increasing or decreasing. To find these points we solve

$$0 = f'(x) = 6x^2 + 6x - 5 \quad\Longrightarrow\quad x = \frac{-6 \pm \sqrt{36 + 120}}{12} = \frac{-3 \pm \sqrt{39}}{6}.$$

The graph of $f'(x) = 6x^2 + 6x - 5$ in Figure 4.13b is a parabola. We can see that $f'(x) \geq 0$ for $x \leq (-3 - \sqrt{39})/6 \approx -1.5$ and $x \geq (-3 + \sqrt{39})/6 \approx 0.5$, and $f'(x) \leq 0$ for $(-3 - \sqrt{39})/6 \leq x \leq (-3 + \sqrt{39})/6$. Hence, $f(x)$ is increasing on the intervals $x \leq (-3 - \sqrt{39})/6$ and $x \geq (-3 + \sqrt{39})/6$, and decreasing on $(-3 - \sqrt{39})/6 \leq x \leq (-3 + \sqrt{39})/6$.

FIGURE 4.13a

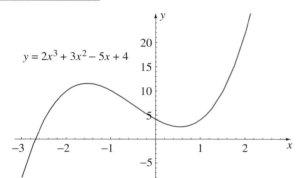

$y = 2x^3 + 3x^2 - 5x + 4$

FIGURE 4.13b

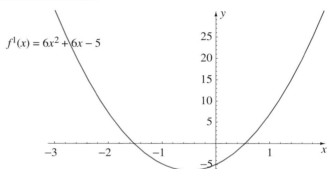

$f'(x) = 6x^2 + 6x - 5$

(b) The graph of $f(x)$ in 4.14a indicates that discontinuities at $x = -2$ and $x = 1$ and the point between $x = -2$ and $x = 0$ where $f'(x) = 0$ separate intervals on which the function is increasing and decreasing. Whether there are other points far out on the x-axis where $f'(x) = 0$ is not clear. To find all points where $f'(x) = 0$, we set

$$0 = f'(x) = \frac{(x^2 + x - 2)(2x) - (x^2 - 9)(2x + 1)}{(x^2 + x - 2)^2} = \frac{x^2 + 14x + 9}{(x^2 + x - 2)^2}.$$

This implies that $x^2 + 14x + 9 = 0$, and there are two points where $f'(x) = 0$, namely $x = -7 \pm 2\sqrt{10}$. Figure 4.14b is an exaggerated version of the graph of $f(x)$ to the left of $x = -2$; it is not a computer plot. It illustrates that $f'(x) = 0$ at $x = -7 - 2\sqrt{10}$ and that the graph is asymptotic to the line $y = 1$ (from above). We can now say that the function is increasing on the intervals

$$-\infty < x \leq -7 - 2\sqrt{10}, \qquad -7 + 2\sqrt{10} \leq x < 1, \qquad 1 < x < \infty,$$

and decreasing for

$$-7 - 2\sqrt{10} \leq x < -2, \qquad -2 < x \leq -7 + 2\sqrt{10}.$$

FIGURE 4.14a

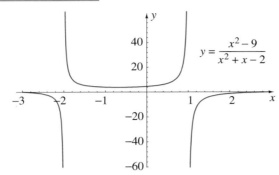

$$y = \frac{x^2 - 9}{x^2 + x - 2}$$

FIGURE 4.14b

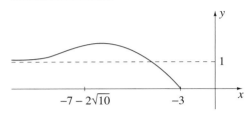

EXERCISES 4.2

In Exercises 1–26 determine intervals on which the function is increasing and decreasing.

* **1.** $f(x) = 2x - 3$

* **2.** $f(x) = 4 - 5x$

* **3.** $f(x) = x^2 - 3x + 4$

* **4.** $f(x) = -2x^2 + 5x$

* **5.** $f(x) = 3x^2 + 6x - 2$

* **6.** $f(x) = 5 + 2x - 4x^2$

* **7.** $f(x) = 2x^3 - 18x^2 + 48x + 1$

* **8.** $f(x) = x^3 + 6x^2 + 12x + 5$

* **9.** $f(x) = 4x^3 - 18x^2 + 1$

* **10.** $f(x) = 4 - 18x - 9x^2 - 2x^3$

* **11.** $f(x) = 3x^4 + 4x^3 - 24x + 2$

* **12.** $f(x) = 3x^4 - 4x^3 + 24x^2 - 48x$

* **13.** $f(x) = x^4 - 4x^3 - 8x^2 + 48x + 24$ * **14.** $f(x) = x^5 - 5x + 2$

** **15.** $f(x) = x + \dfrac{1}{x}$

** **16.** $f(x) = x^2 + \dfrac{1}{x^2}$

** **17.** $f(x) = \dfrac{x}{2 - x}$

** **18.** $f(x) = \dfrac{x^2 + 4}{x^2 - 1}$

** **19.** $f(x) = \dfrac{x^3}{x + 1}$

** **20.** $f(x) = |x^2 - 1| + 1$

** **21.** $f(x) = xe^{-x}$

** **22.** $f(x) = x^2 e^{-x}$

** **23.** $f(x) = \ln(x^2 + 5)$

** **24.** $f(x) = x \ln x$

** **25.** $f(x) = \dfrac{x^2 - 9}{x - 3}$

** **26.** $f(x) = \dfrac{x^3 + 2x^2 - x - 2}{2 - x - x^2}$

** **27.** If the price of a certain car is set at r, then the market demands x cars per year, where

$$x = 4a^3 - 3ar^2 + r^3,$$

where $a > 0$ is a constant, and $0 < r < 2a$. Show that the price function $r = f(x)$ defined implicitly by this equation is a decreasing function.

Figure (a) below contains the graph of a function $f(x)$, and figure (b) the graph of $f'(x)$. Note that $f'(x) \geq 0$ when $f(x)$ is increasing and $f'(x) \leq 0$ when $f(x)$ is decreasing. In addition, the corner in $f(x)$ at $x = 2$ is reflected in the discontinuity in $f'(x)$. In Exercises 28–35 draw similar graphs for the function and its derivative.

(a)

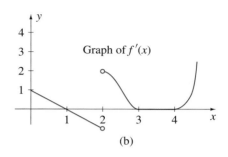

(b)

∗∗ 28. $f(x) = x^2 + 2x$

∗∗ 29. $f(x) = x^4 - x^2$

∗∗ 30. $f(x) = \dfrac{x - 1}{x + 2}$

∗∗ 31. $f(x) = \dfrac{x - 4}{x^2}$

∗∗ 32. $f(x) = \dfrac{|x|}{x}$

∗∗ 33. $f(x) = |x^2 - 4|$

∗∗ 34. $f(x) = \dfrac{|x^2 - 4|}{x - 2}$

∗∗ 35. $f(x) = \sin 3x$

In Exercises 36–39 find (accurate to four decimals) intervals on which the function is increasing and decreasing.

∗∗ 36. $f(x) = x^4 + 2x^2 - 6x + 5$

∗∗ 37. $f(x) = 3x^4 - 20x^3 - 24x^2 + 48x$

∗∗ 38. $f(x) = x^2 \sin x, \quad -\pi \leq x \leq \pi$

∗∗ 39. $f(x) = \tan x - x(x + 2), \quad -\pi/2 < x < \pi/2$

∗∗ 40. Show that $\sin x < x$ for all $x > 0$. Hint: Calculate $f(0)$ and $f'(x)$ for $f(x) = x - \sin x$.

∗∗ 41. Show that $\cos x > 1 - x^2/2$ for all $x > 0$. Hint: See the technique of Exercise 40.

∗∗ 42. Use the result of Exercise 40 to prove that for $x > 0$

$$\sin x > x - \frac{x^3}{6}.$$

∗∗ 43. Use the result of Exercise 41 to prove that for $x > 0$

$$\cos x < 1 - \frac{x^2}{2} + \frac{x^4}{24}.$$

∗∗ 44. Use the technique of Exercise 40 to verify that for $x > 0$

$$\frac{1}{\sqrt{1 + 3x}} > 1 - \frac{3x}{2}.$$

** **45.** If $f(x)$ and $g(x)$ are differentiable and increasing on an interval I, is $f(x)g(x)$ increasing on I?

** **46.** If positive functions $f(x)$ and $g(x)$ are differentiable and increasing on an interval I, is $f(x)g(x)$ increasing on I?

** **47.** Verify test 4.4.

** **48.** A number x_0 is called a fixed point of a function $f(x)$ if $f(x_0) = x_0$. How many fixed points can a function have if $f'(x) < 1$ for all x?

4.3 Relative Maxima and Minima

One of the most important applications of calculus is in the field of optimization, the study of maxima and minima. In this section we begin discussions of this topic, which continue through to Section 4.6. Fundamental to discussions on optimization are critical points.

> **DEFINITION 4.2**
>
> A **critical point** of a function is a point in the domain of the function at which the first derivative either is equal to zero or does not exist.

Specifically, $x = c$ is a critical point for $f(x)$ if $f'(c) = 0$ or $f'(c)$ does not exist but, in the latter case, $f(c)$ must exist. With the interpretation of the derivative as the slope of a tangent line, we can state that a critical point of a function is a point at which the graph of the function has a horizontal tangent line, a vertical tangent line, or no tangent line at all. For example, the eight points a through h on the x-axis in Figure 4.15 are all critical. At a, b, c, and d, the tangent line is horizontal, at e and f, the tangent line is vertical, and at g and h, there is no tangent line.

FIGURE 4.15

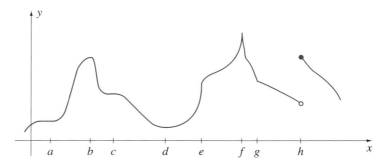

When the domain of a function $f(x)$ is a closed interval $a \leq x \leq b$, the end point $x = a$ is said to be critical if the right-hand derivative of $f(x)$ at $x = a$ is equal to zero or does not exist. Likewise, $x = b$ is critical if the left-hand derivative vanishes or does not exist there. This is consistent with the definition of differentiability in Section 3.3. The function in Figure 4.16 has domain $1 \leq x \leq 3$. Both end points are critical—$x = 1$ because $f'_+(1)$ does not exist, and $x = 3$ because $f'_-(3) = 0$.

The function in Figure 4.15 is discontinuous at $x = h$, and we know that a function cannot have a derivative at a point of discontinuity (Theorem 3.6). This does not mean that every point of discontinuity of a function is critical. The function must be defined at a point for that point to be critical. Thus, points of discontinuity are critical only if the function is defined at the point. In the remainder of this section we consider only critical points at which a function is continuous and which are not end points of its domain of definition.

FIGURE 4.16

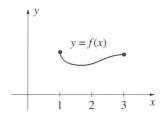

EXAMPLE 4.5

Find critical points for the following functions:

(a) $f(x) = x^3 - 7x^2 + 11x + 6$ (b) $f(x) = \dfrac{x^2}{x^3 - 1}$ (c) $f(x) = x \ln x$ (d) $f(x) = |x|$

SOLUTION

(a) For critical points we first solve

$$0 = f'(x) = 3x^2 - 14x + 11 = (3x - 11)(x - 1),$$

and obtain $x = 11/3$ and $x = 1$. These are the only critical points since there are no points where $f'(x)$ does not exist.

(b) For critical points we calculate

$$f'(x) = \frac{(x^3 - 1)(2x) - x^2(3x^2)}{(x^3 - 1)^2} = \frac{-x(x^3 + 2)}{(x^3 - 1)^2}.$$

Clearly $f'(x) = 0$ when $x = 0$, and when $x^3 + 2 = 0$ which implies that $x = -2^{1/3}$. The derivative does not exist when $x^3 - 1 = 0$, i.e., when $x = 1$. But $f(x)$ is not defined at $x = 1$ either, and therefore $x = 1$ is not a critical point. There are only two critical points: $x = 0$ and $x = -2^{1/3}$.

(c) For critical points we first solve

$$0 = f'(x) = \ln x + \frac{x}{x} = \ln x + 1.$$

The only solution of this equation is $x = 1/e$. Since $f'(x)$ exists for all x in the domain of $f(x)$, namely, $x > 0$, the function has no other critical points.

(d) The only critical point of $f(x) = |x|$ is $x = 0$. Its derivative is equal to 1 when $x > 0$, -1 when $x < 0$, but does not exist at $x = 0$.

At the critical points b and f in Figure 4.17, the graph of the function has "high" points. They are described in the following definition.

FIGURE 4.17

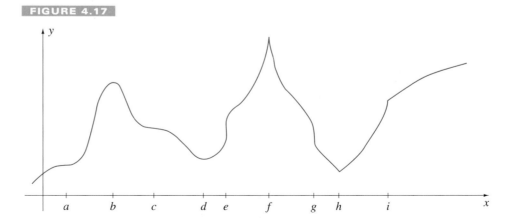

> **DEFINITION 4.3**
>
> A function $f(x)$ is said to have a **relative maximum** $f(x_0)$ at $x = x_0$ if there exists an open interval I containing x_0 such that for all x in I
>
> $$f(x) \leq f(x_0). \tag{4.5}$$

Since such intervals can be drawn around $x = b$ and $x = f$, relative or local maxima occur at these points. At a relative or local maximum, the graph of the function is highest relative to nearby points.

Critical points d and h in Figure 4.17, where the graph has "low" points, are described in a similar definition.

> **DEFINITION 4.4**
>
> A function $f(x)$ is said to have a **relative minimum** $f(x_0)$ at $x = x_0$ if there exists an open interval I containing x_0 such that for all x in I
>
> $$f(x) \geq f(x_0). \tag{4.6}$$

Relative (or local) minima therefore occur at $x = d$ and $x = h$ in Figure 4.17.

The critical points $x = a$ and $x = c$, where $f'(x) = 0$, and $x = e$ and $x = g$, where $f'(x)$ does not exist, will be discussed in Section 4.4. At $x = i$, the graph takes an abrupt change in direction. The function has a left-hand derivative and a right-hand derivative, but $f'(i)$ does not exist. In Section 3.3 we called the corresponding point on the curve a corner. Corners can sometimes be relative extrema ($x = h$ yields a corner and a relative minimum).

Relative maxima and minima represent high and low points on the graph of a function relative to points near them. It is not coincidence in Figure 4.17 that the two relative maxima and the two relative minima occur at critical points. According to the following theorem, this is always the case.

> **THEOREM 4.1**
>
> Relative maxima and relative minima of a function must occur at critical points of the function.

PROOF To verify this we prove that at any point at which the derivative $f'(x)$ of a function $f(x)$ exists and is not zero, it is impossible for $f(x)$ to have a relative extremum, that is, a relative maximum or a relative minimum. Suppose that at some point $x = a$, the derivative $f'(a)$ exists and is positive (Figure 4.18),

$$0 < f'(a) = \lim_{h \to 0} \frac{f(a + h) - f(a)}{h}.$$

According to the proof developed in Exercise 29 in Section 2.5 (page 139), there exists an open interval $I : b < x < c$ around $x = a$ in which

$$\frac{f(a + h) - f(a)}{h} > 0.$$

This implies that when $h > 0$ (and $a + h$ is in I), $f(a + h) - f(a)$ must also be positive, and therefore $f(a + h) > f(a)$. But when $h < 0$ (and $a + h$ is in I), $f(a + h) - f(a)$ must be negative, and therefore $f(a + h) < f(a)$. There is an interval $b < x < a$ in which $f(x) < f(a)$, and an interval $a < x < c$ in which $f(x) > f(a)$. Thus, $x = a$ cannot yield a relative extremum.

A similar proof holds when $f'(a) < 0$.

FIGURE 4.18

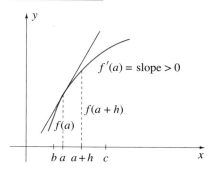

Although every relative extremum of a function must occur at a critical point, not all critical points give relative extrema. For continuous functions, there is a simple test to determine whether a critical point gives a relative maximum or a relative minimum. To understand this test, consider the critical points in Figure 4.17 as you read the statements.

First-Derivative Test for Relative Extrema of Continuous Functions

(i) If $f'(x)$ (slope of a graph) changes from a positive quantity to a negative quantity as x increases through a critical point, then that critical point yields a relative maximum for $f(x)$.

(ii) If $f'(x)$ changes from a negative quantity to a positive quantity as x increases through a critical point, then that critical point yields a relative minimum for $f(x)$.

If $f'(x)$ does not change sign as x increases through a critical point, we can say nothing. There might be a relative maximum, there might be a relative minimum, or there might be neither of these (see Exercises 62–64).

EXAMPLE 4.6

Find relative maxima and relative minima for the following functions:

(a) $f(x) = 2x^3 - 9x^2 - 23x + 6$ (b) $f(x) = 3.01x - \sin(3x + 1)$, $0 < x < 5$

(c) $f(x) = x^{5/3} - x^{2/3}$

SOLUTION

(a) The graph of the function in Figure 4.19a indicates a relative maximum near $x = -1$ and a relative minimum near $x = 4$. To locate them precisely, we find critical points by solving

$$0 = f'(x) = 6x^2 - 18x - 23 \quad \Longrightarrow \quad x = \frac{18 \pm \sqrt{324 + 24 \cdot 23}}{12} = \frac{9 \pm \sqrt{219}}{6}.$$

These are the only critical points as there are no points at which $f'(x)$ is undefined. Figure 4.19a indicates that a relative maximum occurs at $x = (9 - \sqrt{219})/6$ and a relative minimum at $x = (9 + \sqrt{219})/6$. This can be confirmed algebraically or with the graph of $f'(x)$ in Figure 4.19b. It shows that as x increases through $(9 - \sqrt{219})/6$, $f'(x)$ changes from positive to negative; therefore, $x = (9 - \sqrt{219})/6$ yields a relative maximum of $f((9 - \sqrt{219})/6) \approx 18.02$. Since $f'(x)$ changes from negative to positive as x increases through $(9 + \sqrt{219})/6$, this value of x yields a relative minimum of $f((9 + \sqrt{219})/6) \approx -102.0$.

FIGURE 4.19a

FIGURE 4.19b

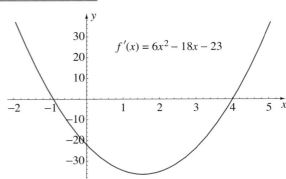

(b) The graph in Figure 4.20 does not make it clear whether the function has relative extrema; the tangent line may become horizontal at or near $x = 2$ and $x = 4$, but we cannot be sure. Critical points of the function are given by

$$0 = f'(x) = 3.01 - 3\cos(3x + 1) \qquad \Longrightarrow \qquad \cos(3x + 1) = \frac{3.01}{3}.$$

Since $3.01/3 > 1$, this equation has no solutions. Hence, $f(x)$ has no critical points and there can be no relative maxima or minima.

FIGURE 4.20

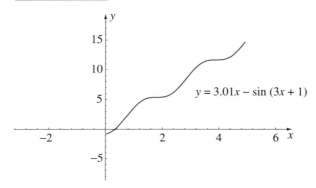

(c) The graph of the function in Figure 4.21 makes it clear that a relative minimum occurs just to the left of $x = 1/2$. It also suggests that $f(0) = 0$ is a relative maximum. Confirmation is provided by critical points, found by first solving

$$0 = f'(x) = \frac{5}{3}x^{2/3} - \frac{2}{3}x^{-1/3} = \frac{5x - 2}{3x^{1/3}}.$$

Clearly, $x = 2/5$ is a critical point. The relative minimum at this point is $f(2/5) \approx -0.33$. In addition, because $f'(0)$ does not exist, but $f(0) = 0$, it follows that $x = 0$ is also a critical point. There are two reasons, each sufficient by itself to conclude that $f(0) = 0$ is a relative maximum. First, $f(x)$ is continuous at $x = 0$, and $f'(x)$ changes from a positive quantity to a negative quantity as x increases through 0. Secondly, since $f(x) = x^{2/3}(x - 1)$ is negative for all $x < 1$, except at $x = 0$ where $f(0) = 0$, there must be a relative maximum at $x = 0$. It is also interesting to note that

$$\lim_{x \to 0^-} f'(x) = \infty \qquad \text{and} \qquad \lim_{x \to 0^+} f'(x) = -\infty.$$

This means that the graph has a sharp point at $(0, 0)$.

FIGURE 4.21

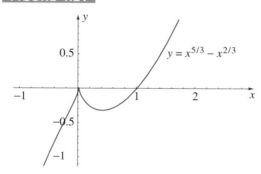

$y = x^{5/3} - x^{2/3}$

EXAMPLE 4.7

The Beattie-Bridgeman equation of state for an ideal gas relates pressure P and volume V according to

$$P = \frac{RT}{V^2}\left(1 - \frac{CT}{V^3}\right)\left[V + B\left(1 - \frac{b}{V}\right)\right] - \frac{A}{V^2}\left(1 - \frac{a}{V}\right),$$

where $R = 0.082\,06$ is the universal gas constant, T is temperature of the gas, and a, A, b, B, and C are constants. For air, $a = 0.019\,31$, $A = 1.3012$, $b = -0.001\,101$, $B = 0.046\,11$, and $C = 4.34 \times 10^4$. For $T = 300$, plot a graph of P as a function of V on the interval $300 \leq V \leq 400$. Verify that, to one decimal accuracy, $V = 373.5$ gives a relative maximum.

SOLUTION   The plot of the function in Figure 4.22 shows a relative maximum near $V = 370$. To verify that 373.5 is the critical point yielding this maximum we calculate

$$P'(V) = RT\left(-\frac{2}{V^3} + \frac{5CT}{V^6}\right)\left[V + B\left(1 - \frac{b}{V}\right)\right]$$
$$+ RT\left(\frac{1}{V^2} - \frac{CT}{V^5}\right)\left[1 + \frac{Bb}{V^2}\right] - A\left(-\frac{2}{V^3} + \frac{3a}{V^4}\right).$$

Since $P'(373.45) = 2.29 \times 10^{-8}$ and $P'(373.55) = -1.19 \times 10^{-7}$, it follows (by the Intermediate Value Theorem) that, to one decimal place, the critical point is 373.5.

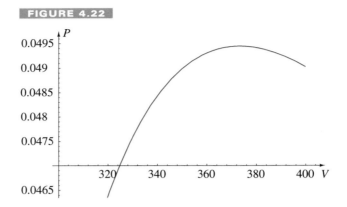

FIGURE 4.22

In Exercises 1–40 plot a graph of the function. Find all critical points of the function and confirm algebraically with the first derivative test which critical points give relative maxima and relative minima.

* **1.** $f(x) = x^2 - 2x + 6$

* **2.** $f(x) = 2x^3 + 15x^2 + 24x + 1$

* **3.** $f(x) = x^4 - 4x^3/3 + 2x^2 - 24x$

* **4.** $f(x) = (x - 1)^5$

* **5.** $f(x) = \dfrac{x + 1}{x^2 + 8}$

* **6.** $f(x) = \dfrac{x^2 + 1}{x - 1}$

* **7.** $f(x) = \dfrac{x}{\sqrt{1 - x}}$

* **8.** $f(x) = x^3 - 6x^2 + 12x + 9$

* **9.** $f(x) = x^3/3 - x^2/2 - 2x$

* **10.** $f(x) = x^{1/3}$

** **11.** $f(x) = \sin^2 x$

** **12.** $f(x) = \dfrac{x^3}{x^4 + 1}$

** **13.** $f(x) = 3x^4 - 16x^3 + 18x^2 + 2$

** **14.** $f(x) = x + \dfrac{1}{x}$

** **15.** $f(x) = 2x^3 - 15x^2 + 6x + 4$

** **16.** $f(x) = |x| + x$

** **17.** $f(x) = (x - 1)^{2/3}$

** **18.** $f(x) = \dfrac{(x - 1)^3}{(x + 1)^4}$

** **19.** $f(x) = (x + 2)^3(x - 4)^3$

** **20.** $f(x) = x + 2\sin x$

** **21.** $f(x) = x^{1/5} + x$

** **22.** $f(x) = x^2 + \dfrac{25x^2}{(x - 2)^2}$

** **23.** $f(x) = \left(\dfrac{x + 8}{x}\right)\sqrt{x^2 + 100}$

** **24.** $f(x) = \dfrac{1 + x + x^2 + x^3}{1 + x^3}$

** **25.** $f(x) = x^{5/4} - x^{1/4}$
 ** **26.** $f(x) = \dfrac{x^2}{x^2 - 4}$

** **27.** $f(x) = \dfrac{x^2}{x^3 - 1}$
 ** **28.** $f(x) = \dfrac{(2x - 1)(x - 8)}{(x - 1)(x - 4)}$

** **29.** $f(x) = \sin^2 x \cos x, \quad 0 \le x \le 2\pi$

** **30.** $f(x) = \sin x + \cos x$

** **31.** $f(x) = 2 \csc x - \cot x, \quad 0 < x < \pi/2$

** **32.** $f(x) = \csc x + 8 \sec x, \quad 0 < x < \pi/2$

** **33.** $f(x) = \dfrac{\tan x}{x}$
 ** **34.** $f(x) = x + \sin^2 x, \quad 0 < x < 2\pi$

** **35.** $f(x) = e^{1/x}$
 ** **36.** $f(x) = x \ln x$

** **37.** $f(x) = x^2 \ln x$
 ** **38.** $f(x) = xe^{-2x}$

** **39.** $f(x) = xe^{-x^2}$
 ** **40.** $f(x) = x^2 e^{3x}$

** **41.** Does the function $f(x) = x^2$ defined only for $0 \le x \le 1$ have a relative minimum of $f(0) = 0$ and a relative maximum of $f(1) = 1$?

** **42.** If $f(x) = x^2$ is defined only for $0 \le x \le 1$, are $x = 0$ and $x = 1$ critical points?

In Exercises 43–50 determine whether the statement is true or false.

** **43.** Points of discontinuity of a function are critical if, and only if, the function is defined at the point of discontinuity.

** **44.** When a function is defined only on the interval $a \le x \le b$, the ends $x = a$ and $x = b$ must yield relative maxima or minima for the function.

** **45.** A function is discontinuous at a point if, and only if, it has no derivative at the point.

** **46.** A function can have a relative maximum and a relative minimum at the same point.

** **47.** If the derivative of a function changes sign when passing through a point, the point must yield a relative extremum for the function.

** **48.** If a function has two relative maxima, it must have a relative minimum between them.

** **49.** It is possible for every point in the domain of a function to be critical.

** **50.** On an interval of finite length, a nonconstant function can have only a finite number of critical points at which its derivative is equal to zero.

** **51.** Verify that the function $f(x) = x^3 + \cos x$ has two critical points, one of which is $x = 0$. Find the other critical point correct to three decimal places.

** **52.** (a) Plot a graph of the function $f(x) = e^{-x} \sin x, \ x \ge 0$.

   (b) For what values of x does $f(x)$ have relative extrema?

In Exercises 53–56 find all critical points of the function correct to four decimals. Determine whether each critical point yields a relative maximum or a relative minimum.

** **53.** $f(x) = x^4 + 6x^2 + 4x + 1$
 ** **54.** $f(x) = x^4 - 10x^2 - 4x + 5$

** **55.** $f(x) = x^3 - 2 \cos x$
 ** **56.** $f(x) = \dfrac{x^2 - 4}{(x^2 - 5x + 4)^2}$

✱✱ **57.** The distance from the origin to any point (x, y) on the curve $C : y = f(x)$ is given by $D = \sqrt{x^2 + y^2}$. Show that if $P(x_0, y_0)$ is a point on C for which D has a relative extremum, then the line OP is perpendicular to the tangent line to C at P. Assume that $f(x)$ is differentiable, and that the curve does not pass through the origin.

✱✱ **58.** An experiment is performed n times, resulting in n measurements: $x_1, x_2, \ldots, x_n$ of the variable x. To obtain the best estimate of the value of x, we choose that value $\overline{x}$ which makes the sum of the squares of the differences between the x_i and $\overline{x}$ as small as possible; i.e., we choose $\overline{x}$ to minimize

$$S(x) = (x - x_1)^2 + (x - x_2)^2 + \cdots + (x - x_n)^2.$$

Show that $\overline{x}$ is the ordinary arithmetic mean of the x_i:

$$\overline{x} = \frac{1}{n}(x_1 + x_2 + \cdots + x_n).$$

✱✱ **59.** A 200 g mass is attached to a spring with constant 20 N/m in the figure below. If the mass is pulled 10 cm below the position where it would hang motionless ($y = 0$), and given upward velocity of 2 m/s, then its position as a function of time t (in seconds) is given by

$$y(t) = -\frac{1}{10} \cos 10t + \frac{1}{5} \sin 10t \text{ m}.$$

(a) Find the maximum distance from $y = 0$ attained by the mass by calculating critical points.

(b) Determine the distance in (a) by expressing $y(t)$ in the form suggested by Example 1.26 in Section 1.5 (page 61).

$y = 0$ at equilibrium

200 g

✱✱ **60.** The equation $(x^2 + y^2 + x)^2 = x^2 + y^2$ describes a *cardioid* (figure below). Find the maximum y-coordinate for points on the curve.

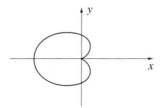

✱✱ **61.** The equation $(x^2 + y^2)^2 = x^2 y$ describes a *bifolium* (figure at the top of page 261). Find the points on the curve farthest from the origin.

** **62.** (a) Verify that $x = 0$ is a critical point for the function

$$f(x) = \begin{cases} x \sin(1/x), & x \neq 0, \\ 0 & x = 0, \end{cases}$$

because $f'(0)$ does not exist.
(b) Show that $f'(x)$ does not change sign as x increases through $x = 0$.
(c) Does $x = 0$ yield a relative maximum or minimum for $f(x)$?

** **63.** Repeat Exercise 62 for the function

$$f(x) = \begin{cases} |x \sin(1/x)|, & x \neq 0, \\ 0 & x = 0, \end{cases}$$

** **64.** Repeat Exercise 62 for the function

$$f(x) = \begin{cases} -|x \sin(1/x)|, & x \neq 0, \\ 0 & x = 0, \end{cases}$$

*** **65.** Show that the function $f(x) = x^3 + px + q$ has three distinct zeros if and only if $4p^3 + 27q^2 < 0$.

4.4 Concavity and Points of Inflection

In the first three sections of this chapter we concentrated on the first derivative of a function. In this section we turn our attention to the second derivative.

Consider the function in Figure 4.23. If we draw tangent lines to the graph at the five points $c_1, c_2, c_3, c_4,$ and c_5, it is clear that the slope is greater at c_5 than it is at c_4, greater at c_4 than at c_3, and so on. In fact, given any two points x_1 and x_2 in the interval $a < x < b$, where $x_2 > x_1$, the slope at x_2 is greater than at x_1,

$$f'(x_2) > f'(x_1).$$

What we are saying is that the function $f'(x)$ is increasing on the interval $a < x < b$.

FIGURE 4.23

FIGURE 4.24

FIGURE 4.25

The first derivative of the function $f(x)$ in Figure 4.24 is decreasing on the interval $b < x < c$. In Figure 4.25, we have pieced together the functions in Figures 4.23 and 4.24 to form a function which has $f'(x)$ increasing on $a < x \leq b$ and decreasing on $b \leq x < c$. Since the left-hand derivative $f'_-(b)$ in Figure 4.23 and the right-hand derivative $f'_+(b)$ in Figure 4.24 are the same, it follows that $f'(b)$ exists in Figure 4.25, and that is why b is included in the intervals in which $f'(x)$ is increasing and decreasing. We give names to intervals on which the first derivative of a function is increasing and decreasing, and points which separate such intervals in the following definition.

DEFINITION 4.5

The graph of a function $f(x)$ is said to be **concave upward** on an interval I if $f'(x)$ is increasing on I, and **concave downward** on I if $f'(x)$ is decreasing on I. Points on a graph that separate intervals of opposite concavity are called **points of inflection**.

The graph in Figure 4.26 is concave upward on the intervals

$$b \leq x \leq c, \qquad d \leq x < e, \qquad e < x \leq f,$$

and concave downward for

$$a \leq x \leq b, \qquad c \leq x \leq d, \qquad f \leq x < g, \qquad g < x \leq h.$$

The points $x = e$ and $x = g$ are not included in these intervals because $f'(x)$ is not defined at these points. Points on the curve corresponding to $x = b, c, d$, and f are points of inflection; those at $x = e$ and g are not. They do not separate intervals of opposite concavity.

FIGURE 4.26

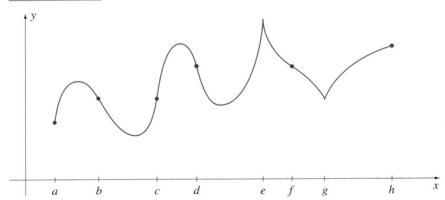

The tangent line has a peculiar property at a point of inflection. To see it, draw the tangent line at the four points of inflection in Figure 4.26. The tangent line actually crosses from one side of the curve to the other (something many readers think a tangent line should not do).

The critical points $x = a$ and $x = c$ in Figure 4.17 (page 253), where $f'(x) = 0$, and the critical points $x = e$ and $x = g$ where $f'(x)$ does not exist, do not yield relative extrema for $f(x)$. Critical points of these types are illustrated again in Figures 4.27 and 4.28. The tangent line at $x = a$ is horizontal in Figure 4.27 ($f'(a) = 0$). Since the graph is concave downward to the left of $x = a$ and concave upward to the right, the point $(a, f(a))$ is a point of inflection. We call it a **horizontal point of inflection**. The point $(a, f(a))$ in Figure 4.28 is also a point of inflection. As the tangent line at this point is vertical, we call it a **vertical point of inflection**.

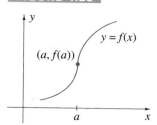

In Section 4.2 we stated that a function is increasing on an interval if its first derivative is greater than or equal to zero on that interval, and equal to zero at only a finite number of points; it is decreasing if its first derivative is less than or equal to zero. Furthermore, points that separate intervals on which the function is increasing and decreasing are points at which the first derivative is either equal to zero (and changes sign) or does not exist. We can use these ideas to locate analogously points of inflection and intervals on which the graph of a function is concave upward and concave downward.

Test for Concavity

(i) The graph of a function $f(x)$ is concave upward on an interval I if on I

$$f''(x) \geq 0 \qquad (4.7a)$$

and is equal to zero at only a finite number of points.

(ii) The graph of a function $f(x)$ is concave downward on an interval I if on I

$$f''(x) \leq 0 \qquad (4.7b)$$

and is equal to zero at only a finite number of points.

(iii) Points of inflection occur where

$$f''(x) = 0 \qquad (4.7c)$$

and $f''(x)$ changes sign, or perhaps also where $f''(x)$ does not exist.

Consequently, just as the first derivative of a function is used to determine its relative extrema and intervals on which it is increasing and decreasing, the second derivative is used to find points of inflection and intervals on which the graph of the function is concave upward and concave downward. This parallelism between the first and second derivative is reiterated in the following table.

First Derivative	**Second Derivative**
1. A relative extremum occurs at a point where $f'(x) = 0$ and $f'(x)$ changes sign as x increases through the point.	1. A point of inflection occurs at a point where $f''(x) = 0$ and $f''(x)$ changes sign as x increases through the point.
2. A relative extremum may also occur at a point where $f'(x)$ does not exist.	2. A point of inflection may also occur at a point where $f''(x)$ does not exist.
3. $f(x)$ is increasing on an interval if $f'(x) \geq 0$; $f(x)$ is decreasing on an interval if $f'(x) \leq 0$. $f'(x)$ may be equal to zero at only a finite number of points in either case.	3. $f(x)$ is concave upward on an interval if $f''(x) \geq 0$; $f(x)$ is concave downward on an interval if $f''(x) \leq 0$. $f''(x)$ may be equal to zero at only a finite number of points in either case.

These discussions lead to what is often called the *second-derivative test* for determining whether a critical point at which $f'(x) = 0$ yields a relative maximum or a relative minimum.

Second-Derivative Test for Relative Extrema

    (i) If $f'(x_0) = 0$ and $f''(x_0) > 0$, then $x = x_0$ yields a relative minimum for $f(x)$.

    (ii) If $f'(x_0) = 0$ and $f''(x_0) < 0$, then $x = x_0$ yields a relative maximum for $f(x)$.

    (iii) If $f'(x_0) = 0$ and $f''(x_0) = 0$, then no conclusion can be made.

In (i), $f'(x_0) = 0$ implies that $x = x_0$ is a critical point with a horizontal tangent line; $f''(x_0) > 0$ implies that around $x = x_0$ the curve is concave upward, and $x = x_0$ therefore yields a relative minimum. Similarly, $f''(x_0) < 0$ at a critical point in (ii) implies that the curve is concave downward and therefore has a relative maximum.

To show that the test fails in case (iii), consider three functions $f(x) = x^4$, $f(x) = -x^4$, and $f(x) = x^3$ in Figures 4.29, 4.30, and 4.31 respectively. For each, $f'(0) = f''(0) = 0$; yet the first has a relative minimum, the second a relative maximum, and the third a horizontal point of inflection at $x = 0$.

FIGURE 4.29　　　　**FIGURE 4.30**　　　　**FIGURE 4.31**

$y = x^4$　　　　$y = -x^4$　　　　$y = x^3$

The second-derivative test very quickly determines the nature of a critical point x_0 when $f'(x_0) = 0$ and $f''(x_0) \neq 0$. For instance, the second derivative of the function $f(x) = 2x^3 - 9x^2 - 23x + 6$ in Example 4.6(a) of Section 4.3 (page 255) is $f''(x) = 12x - 18$. Since $f''\left((9 - \sqrt{219})/6\right) < 0$, the critical point $x = (9 - \sqrt{219})/6$ yields a relative maximum, and because $f''\left((9 + \sqrt{219})/6\right) > 0$, $x = (9 + \sqrt{219})/6$ gives a relative minimum.

A more refined test to determine the nature of a critical point x_0 when $f'(x_0) = f''(x_0) = 0$ is discussed in Exercise 35.

EXAMPLE 4.8 ▼

For each of the following functions, find all points of inflection and intervals on which the graph of the function is concave upward and concave downward.

    (a) $\displaystyle f(x) = \frac{2}{3}x^3 - 6x^2 + 16x + 1$ (b) $\displaystyle f(x) = x^{6/5} + x^{1/5}$

SOLUTION

    (a) The graph of the function in Figure 4.32 indicates that there is one point of inflection. To find it, we solve

$$0 = f''(x) = \frac{d}{dx}(2x^2 - 12x + 16) = 4x - 12,$$

and obtain $x = 3$. Since $f''(x)$ changes sign as x passes through 3, $x = 3$ gives the point of inflection $(3, 13)$. The graph $y = f(x)$ is concave downward for $x \leq 3$ (since $f''(x) \leq 0$) and concave upward for $x \geq 3$ ($f''(x) \geq 0$). Since $f''(x)$ is defined for all x, there can be no other points of inflection.

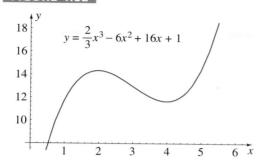

$$y = \frac{2}{3}x^3 - 6x^2 + 16x + 1$$

(b) The plot of the function in Figure 4.33a suggests a point of inflection between $x = 0$ and $x = 1$; the curve seems to be concave downward for small positive x and concave upward for larger x. The graph on the smaller interval in Figure 4.33b does not make this any clearer, but it suggests that $(0, 0)$ is a point of inflection. To confirm these suspicions, we first solve

$$0 = f''(x) = \frac{d}{dx}\left(\frac{6}{5}x^{1/5} + \frac{1}{5}x^{-4/5}\right) = \frac{6}{25}x^{-4/5} - \frac{4}{25}x^{-9/5} = \frac{6x - 4}{25x^{9/5}},$$

and find $x = 2/3$. Since $f''(x)$ changes sign as x passes through $2/3$, $x = 2/3$ gives a point of inflection with $f(2/3) \approx 1.5$.

We also note that $f'(x)$ and $f''(x)$ do not exist at $x = 0$, although $f(0)$ does ($= 0$). Since $f''(x)$ changes sign as x passes through 0, $(0, 0)$ must also be a point of inflection. The fact that

$$\lim_{x \to 0} f'(x) = \lim_{x \to 0}\left(\frac{6}{5}x^{1/5} + \frac{1}{5}x^{-4/5}\right) = \lim_{x \to 0}\frac{6x + 1}{5x^{4/5}} = \infty,$$

indicates that $(0, 0)$ is a vertical point of inflection. With the points of inflection now in place, we can say that the graph is concave upward for $x < 0$ and $x \geq 2/3$, and concave downward for $0 < x \leq 2/3$.

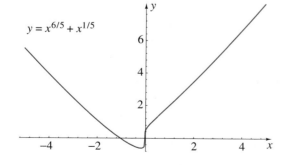

$$y = x^{6/5} + x^{1/5}$$

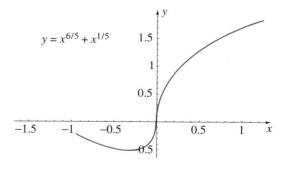

$$y = x^{6/5} + x^{1/5}$$

In Section 4.6 we will see the importance of critical points in the solution of applied maxima-minima problems. In some problems the significance of inflection points is equally important. As an example, consider how an infectious disease spreads through a population of susceptible individuals. To simplify the problem, we assume that all individuals are equally susceptible and that the disease is introduced to the population by one infective. Contact between individuals spreads the disease slowly at first, then more rapidly as more individuals become infected and, finally, when most of the population has become infected, the spread tapers off. To describe this process in more mathematical terms, we denote by N the total number of individuals in the population, and by $x(t)$ the number of infectives at any given time t. Since the disease is spread through contacts between infectives and noninfectives, the rate at which individuals become infected depends on both the number of infectives x and the number of noninfectives $N - x$ at any given time. In practice, the rate of infection, represented by dx/dt, is often assumed proportional to the product of x and $N - x$:

$$\frac{dx}{dt} = kx(N - x),$$

where k is a constant. Given that this equation describes the rate at which the disease spreads, it is possible to determine the number of infectives $x(t)$ present at any time. We cannot solve this problem completely at this point, but we can obtain considerable information about its graph.

First, we note that dx/dt is always nonnegative so that the graph of $x(t)$ is increasing. Second, since points of inflection on the graph are defined by a vanishing second derivative, we set

$$0 = x''(t) = \frac{d}{dt}[kNx - kx^2] = kN\frac{dx}{dt} - 2kx\frac{dx}{dt} = k(N - 2x)\frac{dx}{dt}.$$

But $k \neq 0$, and $dx/dt = 0$ only when the population is either totally infected or not at all infected; therefore, a point of inflection occurs only when $N - 2x = 0$ or $x = N/2$, that is, when the population is half infected. Since $x''(t) > 0$ when $x < N/2$ and $x''(t) < 0$ when $x > N/2$, the graph of $x(t)$ is concave upward for $x < N/2$ and concave downward for $x > N/2$. These are shown in Figure 4.34; we have chosen a time convention of $t = 0$ when the first infective is introduced. In other words, we have completely determined the shape of the graph of $x(t)$. What remains is the insertion of a time scale and this we do in Section 9.7.

FIGURE 4.34

At this point the significance of the point of inflection at $x = N/2$ is clear. Since the slope of the curve (dx/dt) is increasing for $x < N/2$ and decreasing for $x > N/2$, it follows that when $x = N/2$, the rate of spread of the disease is at its peak.

EXERCISES 4.4

In Exercises 1–14 determine where the graph of the function is concave upward, is concave downward, and has points of inflection.

* **1.** $f(x) = x^3 - 3x^2 - 3x + 5$

* **2.** $f(x) = 3x^4 + 4x^3 - 24x + 2$

** **3.** $f(x) = x^2 + \dfrac{1}{x^2}$

** **4.** $f(x) = \dfrac{x^2 + 4}{x^2 - 1}$

** **5.** $f(x) = x + \cos x, \quad |x| < 2\pi$

** **6.** $f(x) = x^2 - \sin x$

** **7.** $f(x) = x^2 - 2\sin x$

** **8.** $f(x) = x^2 - 4\sin x, \quad |x| < 2\pi$

** **9.** $f(x) = x \ln x$

** **10.** $f(x) = x^2 \ln x$

** **11.** $f(x) = e^{1/x}$

** **12.** $f(x) = xe^{-2x}$

** **13.** $f(x) = x^2 e^{3x}$

** **14.** $f(x) = x^2 - e^{-x}$

In Exercises 15–22 use the second-derivative test to determine whether critical points where $f'(x) = 0$ yield relative maxima or relative minima.

* **15.** $f(x) = x^3 - 3x^2 - 3x + 5$

* **16.** $f(x) = x + \dfrac{1}{x}$

* **17.** $f(x) = 3x^4 - 16x^3 + 18x^2 + 2$

* **18.** $f(x) = x^{5/4} - x^{1/4}$

** **19.** $f(x) = x \ln x$

** **20.** $f(x) = x^2 \ln x$

** **21.** $f(x) = xe^{2x}$

** **22.** $f(x) = x^2 e^{-2x}$

** **23.** Is the graph of a function concave upward, concave downward, both, or neither on an interval I if on I its second derivative is always equal to zero?

** **24.** Prove that the curve $y = 2\cos x$ passes through all points of inflection of the curve $y = x \sin x$.

** **25.** Show that every cubic polynomial has exactly one point of inflection on its graph.

Draw the graph of a function $f(x)$ that is defined everywhere on the interval $I : a \le x \le b$, and that possesses the properties in Exercises 26–31. In each case assume that $a < x_0 < b$.

** **26.** $f(x)$ is increasing on I and discontinuous at x_0.

** **27.** $f''(x) \ge 0$ on I and $f(x)$ is not concave upward on I.

** **28.** $f(x)$ is decreasing on I, $f(x)$ is continuous at x_0, but $f'(x)$ is discontinuous at x_0.

** **29.** $f(x)$ is increasing on $a \le x \le x_0$, increasing on $x_0 < x \le b$, but not increasing on I.

** **30.** $f(x)$ is concave downward on $a \le x < x_0$, concave downward on $x_0 < x \le b$, but not concave downward on I.

** **31.** $f(x)$ is increasing on $a \le x \le x_0$, concave downward on $a \le x < x_0$, decreasing on $x_0 \le x \le b$, and concave upward on $x_0 < x \le b$.

** **32.** If $f(x)$ and $g(x)$ are twice differentiable and concave upward on an interval I, is $f(x)g(x)$ concave upward on I?

** **33.** If the functions in Exercise 32 are also increasing on I, is $f(x)g(x)$ concave upward on I?

** **34.** In thermodynamics, van der Waals equation for an ideal gas relates pressure P and specific volume V by

$$P = \frac{RT}{V - b} - \frac{a}{V^2},$$

where T is temperature of the gas, R is the universal gas constant, and a and b are constants that are to be determined. They are determined by the condition that there are values P_c, T_c, and V_c at which the graph of P as a function of V has a horizontal point of inflection. Show that this condition implies that

$$a = \frac{27 R^2 T_c^2}{64 P_c}, \qquad b = \frac{RT_c}{8 P_c}, \qquad V_c = \frac{3 RT_c}{8 P_c}.$$

** **35.** The second-derivative test fails to classify a critical point x_0 of a function $f(x)$ as yielding a relative maximum, a relative minimum, or a horizontal point of inflection if $f'(x_0) = f''(x_0) = 0$. The following test can be used in such cases. Suppose $f(x)$ has derivatives of all orders in an open interval around x_0, and the first n derivatives all vanish at x_0, but the $(n + 1)^{\text{th}}$ derivative at x_0 is not zero. If we denote the nth derivative of $f(x)$ at x_0 by $f^{(n)}(x_0)$, these conditions are

$$0 = f'(x_0) = f''(x_0) = \cdots = f^{(n)}(x_0), \qquad f^{(n+1)}(x_0) \neq 0.$$

Then:

    (i) if n is even, $f(x)$ has a horizontal point of inflection at x_0;
    (ii) if n is odd and $f^{(n+1)}(x_0) > 0$, $f(x)$ has a relative minimum at x_0;
    (iii) if n is odd and $f^{(n+1)}(x_0) < 0$, $f(x)$ has a relative maximum at x_0.

A proof of this result requires the use of material from Chapter 11 and is therefore delayed until that time (see Exercise 17 in Section 11.3). Note that the second-derivative test is the special case when $n = 1$. Use this test to determine whether critical points of the following functions yield relative maxima, relative minima, or horizontal points of inflection.

    (a) $f(x) = (x^2 - 1)^3$ (b) $f(x) = x^2 \sqrt{1 - x}$

** **36.** Show that the points of inflection of $f(x) = (k - x)/(x^2 + k^2)$, where k is a constant, all lie on a straight line.

** **37.** Prove that if a cubic polynomial has both a relative maximum and a relative minimum, then the point of inflection between these extrema is the midpoint of the line segment joining them.

*** **38.** (a) In Exercise 30 of Section 3.3 (page 167), the derivative of the function

$$f(x) = \begin{cases} x^2 \sin\left(\dfrac{1}{x}\right), & x \neq 0 \\ 0, & x = 0 \end{cases}$$

was shown to be zero at $x = 0$, that is, $f'(0) = 0$. It follows that the graph of the function has a horizontal tangent line at $(0, 0)$. Plot the graph and the tangent line.

(b) Show that $f''(0)$ does not exist. Hint: See Exercise 33 in Section 3.3 (page 167).

(c) Is the point $(0, 0)$ a relative maximum, a relative minimum, a horizontal point of inflection, or none of these?

*** **39.** Show analytically that the equation

$$2 \sin \theta = \theta (1 + \cos \theta)$$

has no solution in the interval $0 < \theta < \pi$.

4.5 Curve Sketching

We have made extensive use of graphing calculators and/or computers to graph functions and curves, but we have pointed out, and illustrated on several occasions, that electronic output can sometimes be inconclusive and even misleading (see for instance Example 1.16). Electronic devices can never replace sound mathematical analysis. Whenever there is a question about electronic output, we use calculus to find the answer.

In this chapter we have defined critical points, relative extrema, increasing and decreasing functions, concavity, and points of inflection, and each of these gives information about the graph of a function. In the first four examples of this section we compare what might be called precomputer and postcomputer approaches to graphing. In the precomputer approach we use the concepts in the first sentence of this paragraph to draw a graph of the function, assuming therefore no access to graphing calculators or computers. In the postcomputer approach, we use the computer to plot a graph, and then calculus to answer any questions posed by the plot.

EXAMPLE 4.9

▼

(a) Use calculus to draw a graph of the function $f(x) = (x - 4)/x^2$.

(b) Plot a graph of the function, and then use calculus to pinpoint significant information concerning the graph.

SOLUTION

(a) With an x-intercept equal to 4, and the limits

$$\lim_{x \to 0^-} f(x) = -\infty, \qquad \lim_{x \to 0^+} f(x) = -\infty, \qquad \lim_{x \to -\infty} f(x) = 0^-, \qquad \lim_{x \to \infty} f(x) = 0^+,$$

we begin our sketch as shown in Figure 4.35a. This information would lead us to suspect that the graph should be completed as shown in Figure 4.35b. To verify this we find critical points for $f(x)$:

$$0 = f'(x) = \frac{x^2(1) - (x - 4)(2x)}{x^4} = \frac{x(8 - x)}{x^4} = \frac{8 - x}{x^3}.$$

Clearly, $x = 8$ is the only critical point of the function, and the first derivative test indicates that it gives a relative maximum. Our sketch is as shown in Figure 4.35c.

Figure 4.35c makes it clear that there is a point of inflection to the right of $x = 8$, which we could pinpoint with $f''(x)$:

$$f''(x) = \frac{x^3(-1) - (8 - x)(3x^2)}{x^6} = \frac{2(x - 12)}{x^4}.$$

Since $f''(12) = 0$ and $f''(x)$ changes sign as x passes through 12, $(12, 1/18)$ is the point of inflection. Figure 4.35d contains the final graph.

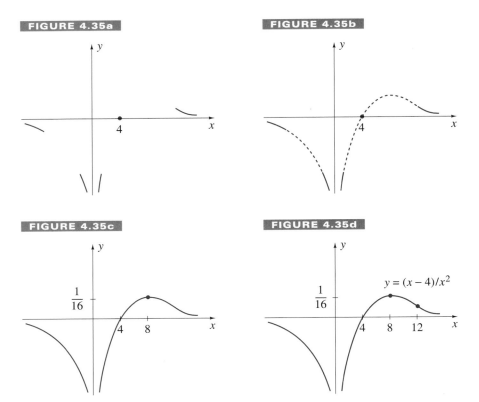

(b) A plot of the function on the interval $-10 \le x \le 10$ in Figure 4.36a does not show a lot. It certainly indicates that the y-axis is a vertical asymptote, and this is confirmed by the fact that $\lim_{x \to 0} f(x) = -\infty$. The plot also suggests that the x-axis is a horizontal asymptote, confirmed by

$$\lim_{x \to -\infty} f(x) = 0^-, \qquad \lim_{x \to \infty} f(x) = 0^+.$$

The fact that the graph approaches $y = 0$ from above as $x \to \infty$ means that the graph must cross the x-axis at least once. It does so exactly once since $f(x) = 0$ only when $x = 4$. To get a better visualization of the graph for $x > 0$, we now plot it on the interval $1 \le x \le 10$, at the same time restricting the y-values to $-1 \le y \le 1$ (Figure 4.36b). There must be a relative maximum to the right of $x = 4$ (perhaps more than one). To find it (or them), we solve

$$0 = f'(x) = \frac{x^2(1) - (x - 4)(2x)}{x^4} = \frac{8 - x}{x^3}.$$

The only solution is $x = 8$. This yields a relative maximum. With the graph concave downward at $x = 8$, and concave upward for large x (recall that $y = 0$ is a horizontal asymptote), there is a point of inflection to the right of $x = 8$. To find it we solve

$$0 = f''(x) = \frac{x^3(-1) - (8 - x)(3x^2)}{x^6} = \frac{2(x - 12)}{x^4}.$$

Since $f''(12) = 0$ and $f''(x)$ changes sign as x passes through 12, there is a point of inflection at $(12, 1/18)$. By restricting y-values even more and plotting on the interval $-20 \le x \le 20$, we obtain the plot in Figure 4.36c; it shows all of the above information.

FIGURE 4.36a

FIGURE 4.36b

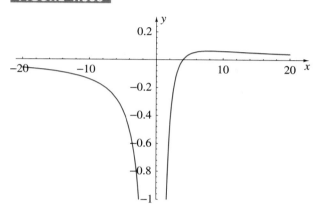

FIGURE 4.36c

EXAMPLE 4.10

Repeat Example 4.9 for the function $f(x) = x^{5/3} - x^{2/3}$.

SOLUTION

(a) We first note that the three points $(0, 0)$, $(1, 0)$, and $(-1, -2)$ are on the graph. Next we add the facts that

$$\lim_{x \to -\infty} f(x) = -\infty \qquad \text{and} \qquad \lim_{x \to \infty} f(x) = \infty,$$

as shown in Figure 4.37a. We cannot be sure that the concavity is as indicated, but this will be verified shortly. To find critical points of $f(x)$, we first solve

$$0 = f'(x) = \frac{5}{3}x^{2/3} - \frac{2}{3}x^{-1/3} = \frac{5x - 2}{3x^{1/3}}.$$

The only solution is $x = 2/5$, and because $f'(x)$ changes from negative to positive as x increases through $2/5$, we have a relative minimum of $f(2/5) = (2/5)^{5/3} - (2/5)^{2/3} = -0.33$. Since $f'(x)$ does not exist at $x = 0$, this is also a critical point, and we calculate that

$$\lim_{x \to 0^-} f'(x) = \infty \qquad \text{and} \qquad \lim_{x \to 0^+} f'(x) = -\infty.$$

This information, along with the relative minimum, is shown in Figure 4.37b. We now join these parts smoothly to produce the sketch in Figure 4.37c. To verify that the concavity is as indicated we calculate

$$f''(x) = \left(\frac{5}{3}\right)\left(\frac{2}{3}\right)x^{-1/3} - \left(\frac{2}{3}\right)\left(-\frac{1}{3}\right)x^{-4/3} = \frac{2(5x + 1)}{9x^{4/3}}.$$

Clearly there is only one point of inflection, namely when $x = -1/5$. The sign of $f''(x)$ on each side of this point indicates that the concavity is indeed as shown in Figure 4.37c, and the final sketch is in Figure 4.37d.

FIGURE 4.37a

FIGURE 4.37b

FIGURE 4.37c

FIGURE 4.37d

(b) The software package in our computer cannot plot a graph of the function $f(x) = x^{5/3} - x^{2/3}$ for negative values of x. (It interprets $x^{5/3}$ and $x^{2/3}$ for negative x as complex numbers.) Your calculator or computer may do the same. To rectify this, define $f(x) = -(-x)^{5/3} - (-x)^{2/3}$ for $x < 0$. A plot then looks like that in Figure 4.38 on the interval $-1 \le x \le 2$.

A relative minimum is indicated just to the left of $x = 1/2$. To locate it, we solve

$$0 = f'(x) = \frac{5}{3}x^{2/3} - \frac{2}{3}x^{-1/3} = \frac{5x - 2}{3x^{1/3}}.$$

FIGURE 4.38

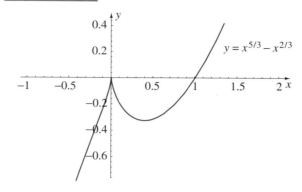

$$y = x^{5/3} - x^{2/3}$$

The only solution is $x = 2/5$, and because $f'(x)$ changes from negative to positive as x increases through $2/5$, we do indeed have a relative minimum of $f(2/5) = (2/5)^{5/3} - (2/5)^{2/3} = -0.33$. The derivative $f'(x)$ does not exist at $x = 0$, but $f(0) = 0$. We calculate that

$$\lim_{x \to 0^-} f'(x) = \lim_{x \to 0^-} \frac{5x - 2}{3x^{1/3}} = \infty \qquad \text{and} \qquad \lim_{x \to 0^+} f'(x) = -\infty.$$

This means that there is a very sharp point at $(0, 0)$ where the function has a relative maximum. Its tangent line is vertical at this point. It is difficult to get a sense of the concavity of the graph for $x < 0$, it appears so straight. To assess this we consider

$$f''(x) = \left(\frac{5}{3}\right)\left(\frac{2}{3}\right)x^{-1/3} - \left(\frac{2}{3}\right)\left(-\frac{1}{3}\right)x^{-4/3} = \frac{2(5x + 1)}{9x^{4/3}}.$$

Since $f''(x)$ changes sign as x passes through $-1/5$, we have a point of inflection $(-1/5, -0.418)$. The graph is concave downward for $x \le -1/5$ and concave upward for $-1/5 \le x < 0$. It is also concave upward for $0 < x < \infty$, so that $(0, 0)$ is not a point of inflection.

EXAMPLE 4.11

Repeat Example 4.9 for the function $f(x) = x + \sin x$.

SOLUTION

(a) The sine function creates oscillations around $y = x$. They might look like Figure 4.39b, c, or d. To decide which, we find critical points,

$$0 = f'(x) = 1 + \cos x.$$

The solutions of this equation are $x = (2n + 1)\pi$, where n is an integer. Since $f'(x)$ does not change sign as x passes through these points, they cannot yield relative maxima or minima. They must give horizontal points of inflection. The correct graph is Figure 4.39c.

FIGURE 4.39a

FIGURE 4.39b

FIGURE 4.39c

FIGURE 4.39d

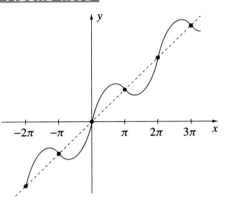

(b) A plot of $f(x) = x + \sin x$ is shown in Figure 4.40. It is not clear whether it has relative extrema. Critical points will provide the answer. We solve $0 = f'(x) = 1 + \cos x$ for $x = (2n+1)\pi$, where n is an integer. Because $f'(x)$ does not change sign as x passes through these points, they cannot yield relative maxima or minima. They must give horizontal points of inflection.

FIGURE 4.40

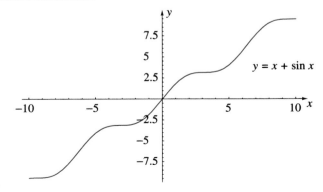

EXAMPLE 4.12

Repeat Example 4.9 for the function $f(x) = (x^2 + 1)/(x - 1)$.

SOLUTION

(a) The function is discontinuous at $x = 1$. We calculate that

$$\lim_{x \to 1^+} \frac{x^2 + 1}{x - 1} = \infty \qquad \text{and} \qquad \lim_{x \to 1^-} \frac{x^2 + 1}{x - 1} = -\infty.$$

If we divide $x^2 + 1$ by $x - 1$, we write $f(x)$ in the form

$$f(x) = x + 1 + \frac{2}{x - 1}.$$

This shows that $y = x + 1$ is an oblique asymptote for the graph (Figure 4.41a). Clearly, we should find critical points by solving

$$0 = f'(x) = 1 - \frac{2}{(x - 1)^2} = \frac{(x - 1)^2 - 2}{(x - 1)^2}.$$

This implies that $(x - 1)^2 = 2$, from which $x = 1 \pm \sqrt{2}$. Because $f'(x)$ changes from a positive quantity to a negative quantity as x increases through $1 - \sqrt{2}$, there is a relative maximum at $x = 1 - \sqrt{2}$ of $f(1 - \sqrt{2}) = 2 - 2\sqrt{2}$. Similarly, a relative minimum of $f(1 + \sqrt{2}) = 2 + 2\sqrt{2}$ occurs at $x = 1 + \sqrt{2}$. The final graph is shown in Figure 4.41b.

FIGURE 4.41a

FIGURE 4.41b

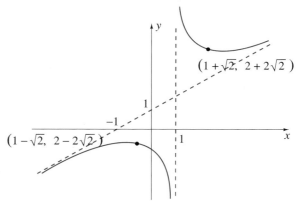

(b) Figure 4.42 is a plot of the function. It shows a vertical asymptote at $x = 1$, confirmed by right- and left-hand limits as $x \to 1$. The relative maximum to the left of $x = 0$ and the relative minimum to the right of $x = 2$ can be located with critical points as in part (a). The straightness of the graph for large $|x|$ suggests an oblique asymptote which is identified by expressing $f(x)$ in the form $f(x) = x + 1 + 2/(x - 1)$.

FIGURE 4.42

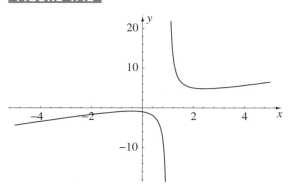

Electronic devices cannot plot curves that contain unspecified parameters. For example, we can use a graphing calculator or computer to plot $y = e^{-ax^2}$ for any given value of a, but we cannot plot the curve without specifying a value for a. What is appropriate is to conjecture the shape of $y = e^{-ax^2}$ from plots with various values of a, and verify the conjecture with calculus. Let us illustrate with some examples.

EXAMPLE 4.13

The logistic model for the size N of a population (a specific animal in a specific area, say) is

$$\frac{dN}{dt} = kN\left(1 - \frac{N}{C}\right).$$

Constant $k > 0$ is the *relative growth rate* of the population (birth rate less death rate due to natural causes), and $C > 0$ is the *carrying capacity* of the environment. It is the limiting population that the environment will sustain because of many factors such as food and water supply. The solution of this equation can be expressed in the form

$$N(t) = \frac{C}{1 + \left(\dfrac{C - N_0}{N_0}\right)e^{-kt}}$$

where N_0 is the size of the population at $t = 0$. Draw a graph of this function assuming $N_0 < C$ and $0 < k < 1$.

SOLUTION Let us get an idea of shape by plotting the function with values $C = 100\,000$, $N_0 = 1000$, and $k = 1/2$. After experimenting with various domains, we arrive at Figure 4.43 as the most informative. The graph has no relative extrema, one point of inflection, and is asymptotic to the line $N = 100\,000 = C$. We now determine whether these features remain true for all values of C, N_0, and k.

FIGURE 4.43

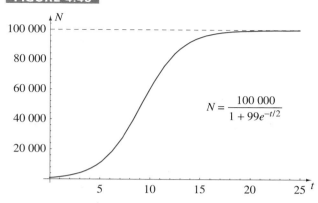

Critical points of $N(t)$ are given by

$$0 = N'(t) = \frac{-C}{\left[1 + \left(\dfrac{C - N_0}{N_0}\right) e^{-kt}\right]^2} \left[-k \left(\frac{C - N_0}{N_0}\right) e^{-kt}\right].$$

There are no solutions, and $N'(t)$ always exists, so that $N(t)$ has no relative maxima or minima. For points of inflection, we consider

$$0 = N''(t) = kC \left(\frac{C - N_0}{N_0}\right) \left\{ \frac{-ke^{-kt}}{\left[1 + \left(\dfrac{C - N_0}{N_0}\right) e^{-kt}\right]^2} \right.$$

$$\left. - \frac{2e^{-kt}}{\left[1 + \left(\dfrac{C - N_0}{N_0}\right) e^{-kt}\right]^3} \left[-k \left(\frac{C - N_0}{N_0}\right) e^{-kt}\right] \right\}$$

$$= \frac{-k^2 C \left(\dfrac{C - N_0}{N_0}\right) e^{-kt}}{\left[1 + \left(\dfrac{C - N_0}{N_0}\right) e^{-kt}\right]^3} \left[1 + \left(\frac{C - N_0}{N_0}\right) e^{-kt} - 2 \left(\frac{C - N_0}{N_0}\right) e^{-kt}\right]$$

$$= \frac{-k^2 C \left(\dfrac{C - N_0}{N_0}\right) e^{-kt}}{\left[1 + \left(\dfrac{C - N_0}{N_0}\right) e^{-kt}\right]^3} \left[1 - \left(\frac{C - N_0}{N_0}\right) e^{-kt}\right].$$

For this to be true, we must have

$$1 - \left(\frac{C - N_0}{N_0}\right) e^{-kt} = 0 \qquad \Longrightarrow \qquad t = -\frac{1}{k} \ln \left(\frac{N_0}{C - N_0}\right).$$

For this value of t to be positive, $N_0/(C - N_0)$ must be less than 1, and this requires $N_0 < C/2$. In other words, a point of inflection occurs if the beginning population is less than half the carrying capacity. When this is the case, the point of inflection is at $t = -(1/k)\ln[N_0/(C - N_0)]$, and N for this value of t is

$$N\left[-\frac{1}{k}\ln\left(\frac{N_0}{C - N_0}\right)\right] = \frac{C}{1 + 1} = \frac{C}{2}.$$

It is clear that

$$\lim_{t\to\infty} N(t) = \lim_{t\to\infty} \frac{C}{1 + \left(\dfrac{C - N_0}{N_0}\right)e^{-kt}} = C^-,$$

so that the graph is asymptotic to $N = C$ (from below). Thus, graphs of $N(t)$ in the two cases $N_0 < \frac{C}{2}$ and $N_0 > \frac{C}{2}$ are shown in Figures 4.44.

FIGURE 4.44a

FIGURE 4.44b

EXAMPLE 4.14

When a drug is injected into the blood at time $t = 0$, it is sometimes assumed in biomedical engineering that the concentration of the drug in a nearby organ is given by a function of the form

$$f(t) = k(e^{-at} - e^{-bt})$$

where k, a, and b are positive constants with $b > a$. Draw a graph of this function indicating any relative extrema and points of inflection.

SOLUTION   For critical points we solve

$$0 = f'(t) = k(-ae^{-at} + be^{-bt}).$$

Thus,

$$ae^{-at} = be^{-bt} \qquad \Longrightarrow \qquad e^{(b-a)t} = \frac{b}{a},$$

the solution of which is $t = (b - a)^{-1} \ln (b/a)$. To test this for maximum or minimum, and find points of inflection, we use

$$f''(t) = k(a^2 e^{-at} - b^2 e^{-bt}).$$

It is not obvious whether

$$f''[(b - a)^{-1} \ln (b/a)] = k \left[a^2 e^{-\frac{a}{b-a} \ln (b/a)} - b^2 e^{-\frac{b}{b-a} \ln (b/a)} \right]$$

is positive or negative. If we note, however, that at the critical point, $ae^{-at} = be^{-bt}$, and substitute this into $f''(t)$, we can see that

$$f''[(b - a)^{-1} \ln (b/a)] = k[a^2 e^{-at} - b(ae^{-at})] = kae^{-at}(a - b).$$

This is clearly negative (since $b > a$), and therefore $f(t)$ has a relative maximum at $t = (b - a)^{-1} \ln (b/a)$. For points of inflection we set

$$0 = a^2 e^{-at} - b^2 e^{-bt} \quad \Longrightarrow \quad t = \frac{2}{b - a} \ln \left(\frac{b}{a} \right).$$

To confirm that this value of t gives a point of inflection we should check that $f''(t)$ changes sign as x passes through this t. This is not obvious. We can, however, conclude that a point of inflection must occur here if we examine the information in Figure 4.45a. The graph begins at the origin and achieves a relative maximum at $t = (b - a)^{-1} \ln (b/a)$ where the graph is concave downward. It is also asymptotic to the positive t-axis since $\lim_{t \to \infty} f(t) = 0^+$. Hence, there must be a change in concavity to the right of the maximum, and this can occur only at $t = 2(b - a)^{-1} \ln (b/a)$. The final graph is shown in Figure 4.45b.

FIGURE 4.45a

FIGURE 4.45b

EXERCISES 4.5

In Exercises 1–10 find all relative maxima and minima for the function and points of inflection on its graph. Use this information to draw a graph of the function.

* **1.** $f(x) = \dfrac{x^3}{3} - \dfrac{x^2}{2} - 2x$ 　　　　* **2.** $f(x) = x^3 - 6x^2 + 12x + 9$

* **3.** $f(x) = 3x^4 - 16x^3 + 18x^2 + 2$ 　　* **4.** $f(x) = 2x^3 - 15x^2 + 6x + 4$

** **5.** $f(x) = x - 3x^{1/3}$ 　　　　　　** **6.** $f(x) = \dfrac{x}{x^2 + 4}$

** **7.** $f(x) = (x - 2)^3(x + 2)$ 　　　　** **8.** $f(x) = x^{2/3}(8 - x)$

** **9.** $f(x) = \dfrac{(x + 2)^2}{x^3}$ 　　　　　** **10.** $f(x) = 2x^{3/2} - 9x + 12x^{1/2}$

In Exercises 11–20 plot a graph of the function. Use derivatives to discuss any significant features of the plot.

** **11.** $f(x) = \dfrac{x^3 + 16}{x}$ 　　　　　** **12.** $f(x) = \dfrac{x^2 + x + 1}{x}$

** **13.** $f(x) = x + \dfrac{1}{x}$ 　　　　　　** **14.** $f(x) = \dfrac{x^3}{x^2 - 4}$

** **15.** $f(x) = \dfrac{2x^2}{x^2 - 8x + 12}$ 　　　** **16.** $f(x) = \dfrac{x^2 + 1}{x^2 - 1}$

** **17.** $f(x) = \dfrac{(x - 1)^3}{(x + 1)^4}$ 　　　　** **18.** $f(x) = (x - 1)^{2/3}$

** **19.** $f(x) = (x + 2)^3(x - 4)^3$ 　　　** **20.** $f(x) = x + 2\sin x$

** **21.** An emf device producing constant voltage V and with constant internal resistance r maintains current i through a circuit with resistance R in the figure below, where

$$V = i(r + R).$$

The power (work per unit time) necessary to maintain this current in r and R is given by

$$P = i^2(r + R) = i^2 r + i^2 R.$$

(a) If we define $P_R = i^2 R$ and $P_r = i^2 r$ as the power dissipated in R and r respectively, draw P_R and P_r as functions of R.

(b) Draw a graph of $P(R) = P_R(R) + P_r(R)$.

** **22.** The rate of photosynthesis P in a leaf depends on the intensity of light I on the leaf according to

$$P = \dfrac{MI}{I + K} - R,$$

where $M > R$ and K are all positive constants. Draw a graph of this function.

In Exercises 23–34 plot a graph of the function. Identify all relative extrema.

** **23.** $f(x) = x/(x^2 + 3)$

** **24.** $f(x) = x^4 + 10x^3 + 6x^2 - 64x + 5$

** **25.** $f(x) = x^4 - 2x^3 + 2x$

** **26.** $f(x) = \dfrac{x^2 - 8}{x - 5}$

** **27.** $f(x) = \dfrac{x^2}{x^2 - 4}$

** **28.** $f(x) = x^{5/4} - x^{1/4}$

** **29.** $f(x) = |x^2 - 9| + 2$

** **30.** $f(x) = \dfrac{(2x - 1)(x - 8)}{(x - 1)(x - 4)}$

** **31.** $f(x) = \sin^2 x \cos x$, $\quad 0 \le x \le 2\pi$ ** **32.** $f(x) = \sin x + \cos x$

** **33.** $f(x) = \dfrac{x}{\sqrt{1 - x}}$

** **34.** $f(x) = \left(\dfrac{x + 8}{x}\right)\sqrt{x^2 + 100}$

** **35.** (a) A company produces x kilograms of a commodity per day at a total cost of

$$C(x) = \frac{x^2}{300}\left(\frac{x + 100}{x + 300}\right) + 60, \quad 1 \le x \le 200.$$

Plot a graph of the function. Show that it is always concave upward.

(b) The average production cost per kilogram when x kilograms are produced is given by $c(x) = C(x)/x$. Plot a graph of this function.

(c) Show that the output at which the average cost is least satisfies the equation

$$(x + 300)^2(x^2 - 18\,000) - 60\,000x^2 = 0.$$

Do this in two ways:

(i) by finding critical points for $c(x)$;

(ii) by noting that the average cost is the slope of the line joining a point $(x, C(x))$ to the origin; hence minimum average cost occurs when the tangent line to the $C(x)$ graph passes through the origin.

** **36.** The radial probability density function for the ground state of the hydrogen atom is

$$P(r) = \left(\frac{4r^2}{a^2}\right)e^{-2r/a}, \quad r \ge 0,$$

where $a > 0$ is a constant. Draw a graph of the function identifying its relative maximum and points of inflection.

** **37.** The radial probability density function for the second state of the hydrogen atom is

$$P(r) = \left(\frac{r^2}{8a^3}\right)\left(2 - \frac{r}{a}\right)^2 e^{-r/a}, \quad r \ge 0,$$

where $a > 0$ is a constant. Draw a graph of the function identifying its relative extrema.

** **38.** In quantum mechanics, the probability that an energy level E will be occupied is

$$P(E) = \frac{1}{e^{(E - E_f)/(kT)} + 1},$$

where E_f is the Fermi energy, k is the Boltzmann constant, and T is temperature. Draw a graph of the function identifying relative extrema and points of inflection.

THEOREM 4.3

If a function has absolute extrema on an interval, then they occur either at critical points or at the ends of the interval.

A plot of a function on an interval normally makes it clear where absolute extrema occur, provided of course that they exist. It is then a matter of evaluating the function at the appropriate critical point(s) or end point(s). When the function has discontinuities or the interval is not of finite length, careful consideration of the function at discontinuities and limits as $x \to \pm\infty$ may be required.

EXAMPLE 4.15

Find the absolute maximum and minimum of the function

$$f(x) = \frac{x^3 - 2x^2 + x + 20}{x^2 + 5}$$

on the interval $0 \leq x \leq 6$, if they exist.

SOLUTION   Since $f(x)$ is continuous on the closed interval $0 \leq x \leq 6$, the function must have absolute extrema. The plot in Figure 4.50 makes it clear that the absolute minimum is at the critical point to the left of $x = 3$. The absolute maximum appears to be at $x = 6$, but there is a relative maximum near $x = 0$ that should be investigated. For critical points we solve

$$0 = f'(x) = \frac{(x^2 + 5)(3x^2 - 4x + 1) - (x^3 - 2x^2 + x + 20)(2x)}{(x^2 + 5)^2}$$

$$= \frac{x^4 + 14x^2 - 60x + 5}{(x^2 + 5)^2}.$$

To four decimal places, solutions of this equation are $x = 0.0850$ and $x = 2.7188$. The absolute minimum of the function is $f(2.7188) = 2.262$, and because $f(0.0850) = 4.008$ and $f(6) = 4.146$, the absolute maximum is 4.146.

FIGURE 4.50

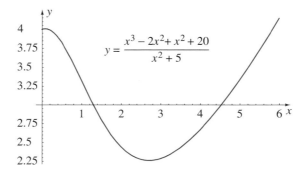

$$y = \frac{x^3 - 2x^2 + x^2 + 20}{x^2 + 5}$$

EXAMPLE 4.16

Find the absolute maximum and minimum for the function

$$f(x) = \frac{2x^2 + 3x}{x^2 + 4}$$

on the intervals (a) $5 \leq x < \infty$ and (b) $1 \leq x < \infty$, if they exist.

SOLUTION

(a) The graph in Figure 4.51a appears to be asymptotic to the line $y = 2$ and to have a relative maximum at or near $x = 6$. The asymptote is confirmed by

$$\lim_{x \to \infty} \frac{2x^2 + 3x}{x^2 + 4} = \lim_{x \to \infty} \frac{2 + \dfrac{3}{x}}{1 + \dfrac{4}{x^2}} = 2.$$

For critical points, we solve

$$0 = f'(x) = \frac{(x^2 + 4)(4x + 3) - (2x^2 + 3x)(2x)}{(x^2 + 4)^2} = \frac{-(x - 6)(3x + 2)}{(x^2 + 4)^2}.$$

The only critical point in $5 \leq x < \infty$ is indeed $x = 6$. Consequently, the absolute maximum of the function is $f(6) = 9/4$, and the function does not have an absolute minimum.

FIGURE 4.51a

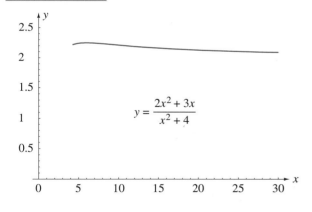

$$y = \frac{2x^2 + 3x}{x^2 + 4}$$

FIGURE 4.51b

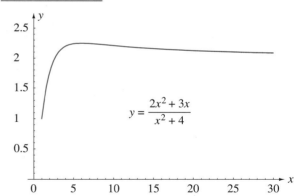

$$y = \frac{2x^2 + 3x}{x^2 + 4}$$

(b) The graph on the interval $1 \leq x < \infty$ in Figure 4.51b indicates that the absolute maximum is still $f(6) = 9/4$, and the function now has an absolute minimum of $f(1) = 1$.

EXAMPLE 4.17

Find absolute extrema of the function

$$f(x) = \frac{x^2 - 9}{x^2 + x - 2}$$

on the intervals (a) $1 < x < \infty$ and (b) $1 < a \leq x \leq b < \infty$, if they exist.

SOLUTION   A graph of the function in Figure 4.52 suggests that $y = 1$ is a horizontal asymptote and $x = 1$ is a vertical asymptote. This is confirmed by the limits

$$\lim_{x \to 1^+} \frac{x^2 - 9}{(x + 2)(x - 1)} = \infty \quad \text{and} \quad \lim_{x \to \infty} \frac{x^2 - 9}{x^2 + x - 2} = 1.$$

We now calculate

$$f'(x) = \frac{(x^2 + x - 2)(2x) - (x^2 - 9)(2x + 1)}{(x^2 + x - 2)^2} = \frac{x^2 + 14x + 9}{(x^2 + x - 2)^2}.$$

This is positive for $1 < x < \infty$, and therefore $f(x)$ is increasing on the interval $1 < x < \infty$. It follows that $f(x)$ does not have absolute extrema on this interval. For $1 < a \le x \le b < \infty$, the absolute minimum is $f(a)$ and the absolute maximum is $f(b)$.

FIGURE 4.52

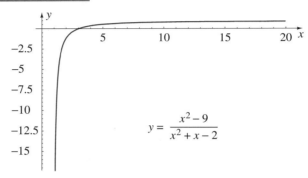

$$y = \frac{x^2 - 9}{x^2 + x - 2}$$

We now consider applied maxima-minima problems.

EXAMPLE 4.18

A rectangular field is to be fenced on three sides with 1000 m of fencing (the fourth side being a straight river's edge). Find the dimensions of the field in order that the area be as large as possible.

FIGURE 4.53

River's edge

SOLUTION   Since the area of the field is to be maximized, we first define a function representing this area. The area of a field of width w and length l (Figure 4.53) is

$$A = lw.$$

This represents the area of a field with arbitrary length l and arbitrary width w. Our problem states that only 1000 m of fencing are available for the three sides; therefore, l and w must satisfy the equation

$$2w + l = 1000.$$

With this equation we can express A completely in terms of w (or l):

$$A(w) = w(1000 - 2w) = 1000w - 2w^2.$$

To maximize the area of the field, we must therefore maximize the function $A(w)$. But what are the values of w under consideration? Clearly, w cannot be negative, and in order to satisfy the restriction $2w + l = 1000$, w cannot exceed 500. The physical problem has now been modelled mathematically. Find the absolute maximum of the (continuous) function $A(w)$ on the (closed) interval $0 \leq w \leq 500$.

The graph of this function in Figure 4.54 clearly indicates that area is maximized by the critical point at or near $x = 250$. To find the critical point, we solve

$$0 = A'(w) = 1000 - 4w;$$

the only solution is $w = 250$. The largest possible area is obtained when the width of the field is 250 m and its length is 500 m.

FIGURE 4.54

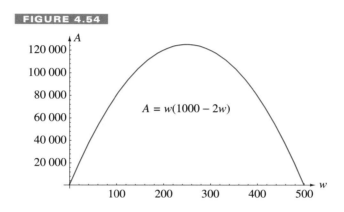

Pop cans that must hold 300 mL are made in the shape of right-circular cylinders. Find dimensions of the can that minimize its surface area.

SOLUTION   The surface area of the can (Figure 4.55a) consists of a circular top, an identical bottom, and a rectangular piece formed into the side of the can. The total area of these three pieces is

$$A = 2\pi r^2 + 2\pi rh,$$

where we measure r and h in centimetres (see Figure 4.55b). Since each can holds 300 mL of pop, it follows that

$$\pi r^2 h = 300 \qquad \text{or} \qquad h = \frac{300}{\pi r^2}.$$

This equation can be used to express A completely in terms of r,

$$A(r) = 2\pi r^2 + 2\pi r \left(\frac{300}{\pi r^2} \right).$$

For what values of r is $A(r)$ defined? The radius of the can must be positive and therefore $r > 0$. How large can r be? In effect, it can be as large as desired. The height of the cylinder can always be chosen sufficiently small when r is large to satisfy the volume condition $\pi r^2 h = 300$.

EXAMPLE 4.19

FIGURE 4.55a

FIGURE 4.55b

The can may not be esthetically pleasing for large r and small h, but no mathematical difficulties occur in this situation. To solve the problem we must therefore maximize

$$A(r) = 2\pi r^2 + \frac{600}{r}, \quad r > 0.$$

The graph of the function in Figure 4.56, together with the facts that

$$\lim_{r \to 0^+} A(r) = \infty \quad \text{and} \quad \lim_{r \to \infty} A(r) = \infty,$$

shows that area is minimized at the critical point near $r = 3.5$. To find this critical point, we solve

$$0 = A'(r) = 4\pi r - \frac{600}{r^2} \quad \text{or} \quad 4\pi r^3 - 600 = 0.$$

The only solution is $r = (150/\pi)^{1/3}$. Thus, minimum surface area occurs when the radius of the can is $(150/\pi)^{1/3}$ cm and its height, obtained from the fact that $h = 300/(\pi r^2)$, is $2(150/\pi)^{1/3}$ cm.

FIGURE 4.56

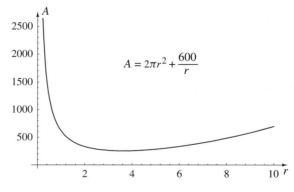

$$A = 2\pi r^2 + \frac{600}{r}$$

On the basis of these two examples, we suggest the following steps in solving applied maxima-minima problems:

1. Sketch a diagram illustrating the situation.

2. Identify the quantity that is to be maximized or minimized, choose a letter to represent it, and find an expression for this quantity.

3. If necessary, use information in the problem to rewrite the expression in 2. as a function of only one variable.

4. Determine the domain of the function in 3.

5. Maximize or minimize the function in 3. on the domain in 4.

6. Interpret the maximum or minimum values in terms of the original problem.

Step 2. is crucial. Do not consider subsidiary information in the problem until the quantity to be maximized or minimized is clearly identified, labelled, and an expression found for it. Only then should other information be considered. For instance, in Example 4.18, the restriction $2w + l = 1000$ for the length of fencing available was not introduced until the expression $A = lw$ had been identified. Likewise, in Example 4.19 volume condition $\pi r^2 h = 300$ was introduced after surface area $A = 2\pi r^2 + 2\pi rh$.

We now take a look at three additional examples.

EXAMPLE 4.20

A lighthouse is 6 km off shore and a cabin on the straight shoreline is 9 km from the point on the shore nearest the lighthouse. If a man rows at a rate of 3 km/h, and walks at a rate of 5 km/h, where should he beach his boat in order to get from the lighthouse to the cabin as quickly as possible? Repeat the problem in the case that the cabin is 4 km along the shoreline.

SOLUTION   Figure 4.57 illustrates the path followed by the man when he beaches the boat x kilometres from the point on land closest to the lighthouse. His travel time t for this complete journey is the sum of his time t_1 in the water and his time t_2 on land. Since velocities are constant, each of these times may be calculated by dividing distance by velocity; that is, $t_1 = \sqrt{x^2 + 36}/3$ and $t_2 = (9 - x)/5$. To minimize t for some value of x between 0 and 9, we minimize the function

$$t(x) = t_1 + t_2 = \frac{\sqrt{x^2 + 36}}{3} + \frac{9 - x}{5}, \quad 0 \le x \le 9.$$

The graph in Figure 4.58 indicates that $t(x)$ has one critical point and the function has its absolute minimum at this critical point. To find it, we solve

$$0 = t'(x) = \frac{x}{3\sqrt{x^2 + 36}} - \frac{1}{5} \quad \text{or} \quad 5x = 3\sqrt{x^2 + 36}.$$

Thus,

$$25x^2 = 9(x^2 + 36),$$

from which we accept only the positive solution $x = 9/2$. The boat should therefore be beached 4.5 km from the cabin.

FIGURE 4.58

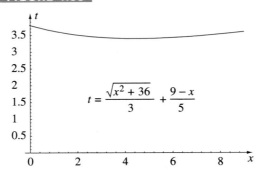

When the cabin is 4 km along the shoreline,

$$t(x) = \frac{\sqrt{x^2 + 36}}{3} + \frac{4 - x}{5}, \quad 0 \le x \le 4.$$

The graph in Figure 4.59 indicates that $f(x)$ does not have any critical points in the interval $0 \le x \le 4$. This is confirmed by the fact that critical points are again given by the equation

$$\frac{x}{3\sqrt{x^2 + 36}} - \frac{1}{5} = 0,$$

but the solution $x = 9/2$ must be rejected since it does not fall in the interval $0 \le x \le 4$. Travel time is therefore minimized by heading directly to shore or directly to the cabin, and Figure 4.59 suggests the cabin. Verification is provided by

$$t(0) = \frac{\sqrt{36}}{3} + \frac{4}{5} = \frac{14}{5}, \qquad t(4) = \frac{\sqrt{16 + 36}}{3} = \frac{2\sqrt{13}}{3} < \frac{14}{5}.$$

FIGURE 4.59

EXAMPLE 4.21

A rectangular beam is to be cut from a circular log of radius 25 cm. In structural engineering, it is shown that the strength of a beam is proportional to the product of its width and the square of its depth. As a result, a beam should be deeper than it is wide. (For instance, place a standard $2'' \times 4'' \times 8'$ piece of lumber between two supports 2 m apart, and sit on it. It will probably support you if you sit on the narrow edge, but not if you sit on the wide edge. In other words, the 2 × 4 is much stronger when its depth is more than its width.) Find the dimensions of the strongest beam that can be cut from the log.

SOLUTION For a typical beam with width w and depth d as shown in Figure 4.60a, its strength is given by

$$S = kwd^2,$$

where $k > 0$ is a constant. If we join the centre of the log to one of the corners of the beam (Figure 4.60b), then from the right-angled triangle

$$\left(\frac{w}{2}\right)^2 + \left(\frac{d}{2}\right)^2 = 25^2 \qquad \Longrightarrow \qquad d^2 = 2500 - w^2.$$

Thus,

$$S = kw(2500 - w^2).$$

FIGURE 4.60a

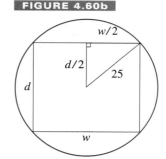

FIGURE 4.60b

This function must be maximized on the interval $0 \le w \le 50$. The graph in Figure 4.61 clearly shows that the maximum occurs at the one and only critical point of the function. It can be found by solving

$$0 = \frac{dS}{dw} = k(2500 - 3w^2) \qquad \Longrightarrow \qquad w = \frac{50}{\sqrt{3}}.$$

Thus, for maximum strength, the beam should be $50/\sqrt{3}$ cm wide and $\sqrt{2500 - 2500/3} = 50\sqrt{2/3}$ cm deep.

FIGURE 4.61

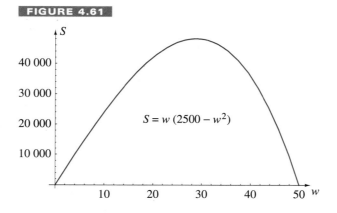

EXAMPLE 4.22

A wall of a building is to be braced by a beam that must pass over a parallel wall 5 m high and 2 m from the building (Figure 4.62). Find the length of the shortest beam that can be used if dimensions of the beam and wall are ignored.

FIGURE 4.62

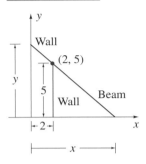

SOLUTION   If we introduce coordinates as shown in Figure 4.62, we see that the shortest beam is the shortest of all line segments that stretch from points on the positive x-axis to points on the positive y-axis, and pass through the point $(2, 5)$. The length of the line segment that has intercepts x and y is given by

$$D = \sqrt{x^2 + y^2}.$$

To express D in terms of only one variable, we must find an equation relating x and y. Since slopes of the line segments joining $(0, y)$ to $(x, 0)$ and $(2, 5)$ to $(x, 0)$ must be identical, we have

$$\frac{y - 0}{0 - x} = \frac{0 - 5}{x - 2} \qquad \text{or} \qquad y = \frac{5x}{x - 2}.$$

Thus,

$$D = \sqrt{x^2 + \frac{25x^2}{(x - 2)^2}}.$$

Since D is a minimum when D^2 is a minimum, we consider minimizing

$$D^2 = f(x) = x^2 + \frac{25x^2}{(x - 2)^2}, \quad x > 2.$$

A graph of the function is shown in Figure 4.63. As expected, D^2 becomes very large as $x \to 2^+$ and $x \to \infty$. It is minimized at the critical point near $x = 5$. To find this critical point, we solve

$$0 = f'(x) = 2x + \frac{50x}{(x - 2)^2} - \frac{50x^2}{(x - 2)^3}$$

$$= \frac{2x[(x - 2)^3 + 25(x - 2) - 25x]}{(x - 2)^3}$$

$$= \frac{2x}{(x - 2)^3}[(x - 2)^3 - 50].$$

FIGURE 4.63

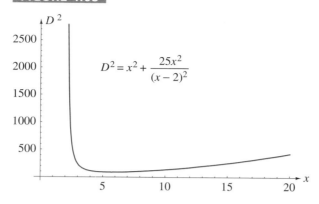

$$D^2 = x^2 + \frac{25x^2}{(x-2)^2}$$

Since x cannot be equal to zero, we set

$$(x-2)^3 = 50,$$

and this implies that $x = 2 + 50^{1/3}$. The shortest beam has length

$$\sqrt{f\left(2 + 50^{1/3}\right)} = \left(2 + 50^{1/3}\right)\sqrt{1 + \frac{25}{50^{2/3}}} = 9.6 \text{ m}.$$

We can also solve this problem by introducing the angle θ in Figure 4.64. The length D of the beam can be expressed in terms of θ by

$$D = x \sec \theta.$$

Since $x - 2 = 5 \cot \theta$, it follows that

$$D = g(\theta) = \sec \theta (2 + 5 \cot \theta), \quad 0 < \theta < \pi/2.$$

FIGURE 4.64

FIGURE 4.65

$$D = \sec \theta (2 + 5 \cot \theta)$$

The graph in Figure 4.65 clearly shows $g(\theta) \to \infty$ as $\theta \to 0^+$ and $\theta \to \pi/2^-$. The minimum is at the critical point near $\theta = 1$, and we can find this point by solving

$$0 = g'(\theta) = \sec \theta \tan \theta (2 + 5 \cot \theta) + \sec \theta (-5 \csc^2 \theta).$$

When this equation is expressed in the form

$$\frac{\cos\theta}{\sin^2\theta}(2\tan^3\theta - 5) = 0,$$

we see that the only critical point is given by

$$\tan\theta = \left(\frac{5}{2}\right)^{1/3}.$$

The angle between 0 and $\pi/2$ satisfying this equation is 0.9358 radians, and the shortest beam has length $g(0.9358) = 9.6$ m.

EXAMPLE 4.23

Repeat Example 4.22 but take into account the fact that the beam is 10 cm on each side. The beam is to sit flush against the building and on the floor (Figure 4.66). The top of the wall can be cut at any angle to accommodate flush contact with the beam.

FIGURE 4.66

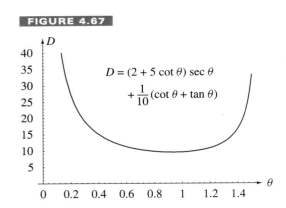

FIGURE 4.67

$$D = (2 + 5\cot\theta)\sec\theta + \frac{1}{10}(\cot\theta + \tan\theta)$$

SOLUTION   The length of beam from A to B is

$$D = \|EF\| + \|AC\| + \|BG\| = x\sec\theta + \frac{1}{10}\cot\theta + \frac{1}{10}\tan\theta.$$

Since $x - 2 = 5\cot\theta$,

$$D = g(\theta) = (2 + 5\cot\theta)\sec\theta + \frac{1}{10}(\cot\theta + \tan\theta), \quad 0 < \theta < \frac{\pi}{2}.$$

The graph of this function in Figure 4.67 once again shows that the absolute minimum occurs at the critical point near $\theta = 1$. To find it we solve

$$0 = g'(\theta) = \sec\theta\tan\theta(2 + 5\cot\theta) + \sec\theta(-5\csc^2\theta) + \frac{1}{10}(-\csc^2\theta + \sec^2\theta)$$

$$= \frac{\sin\theta}{\cos^2\theta}\left(2 + \frac{5\cos\theta}{\sin\theta}\right) - \frac{5}{\sin^2\theta\cos\theta} + \frac{1}{10}\left(\frac{1}{\cos^2\theta} - \frac{1}{\sin^2\theta}\right).$$

Multiplication by $10 \sin^2 \theta \, \cos^2 \theta$ gives

$$
\begin{aligned}
0 &= 10 \sin^3 \theta \left(2 + \frac{5 \cos \theta}{\sin \theta} \right) - 50 \cos \theta + \sin^2 \theta - \cos^2 \theta \\
&= 20 \sin^3 \theta + 50 \cos \theta \, \sin^2 \theta - 50 \cos \theta + \sin^2 \theta - (1 - \sin^2 \theta) \\
&= 20 \sin^3 \theta + 50 \cos \theta (1 - \cos^2 \theta) - 50 \cos \theta + 2 \sin^2 \theta - 1 \\
&= 20 \sin^3 \theta - 50 \cos^3 \theta + 2 \sin^2 \theta - 1.
\end{aligned}
$$

To solve this equation we use Newton's iterative procedure. Approximations are defined by

$$
\theta_{n+1} = \theta_n - \frac{20 \sin^3 \theta_n - 50 \cos^3 \theta_n + 2 \sin^2 \theta_n - 1}{60 \sin^2 \theta_n \, \cos \theta_n + 150 \cos^2 \theta_n \, \sin \theta_n + 4 \sin \theta_n \, \cos \theta_n}.
$$

If we use the result $\theta_1 = 0.9358$ from Example 4.22 as the first approximation, the next two approximations are $\theta_2 = 0.9314$ and $\theta_3 = 0.9314$. Consequently, the shortest beam is $g(0.9314) = 9.8$ m.

EXERCISES 4.6

In Exercises 1–6 find absolute extrema, if they exist, for the function on the interval. Do so without plotting a graph of the function.

* **1.** $f(x) = x^3 - x^2 - 5x + 4, \quad -2 \leq x \leq 3$

* **2.** $f(x) = \dfrac{x - 4}{x + 1}, \quad 0 \leq x \leq 10$

** **3.** $f(x) = x + \dfrac{1}{x}, \quad \dfrac{1}{2} \leq x \leq 5$ ** **4.** $f(x) = x - 2 \sin x, \quad 0 \leq x \leq 4\pi$

** **5.** $f(x) = x\sqrt{x + 1}, \quad -1 \leq x \leq 1$ ** **6.** $f(x) = \dfrac{12}{x^2 + 2x + 2}, \quad x < 0$

In Exercises 7–9 find absolute extrema, if they exist, for the function on the interval. Use a plot of the graph of the function as an aid.

* **7.** $f(x) = \dfrac{x + 1}{x - 1}, \quad x > 1$ * **8.** $f(x) = \dfrac{x}{x^2 + 3}, \quad x > 0$

* **9.** $f(x) = \dfrac{(2x - 1)(x - 8)}{(x - 1)(x - 4)}, \quad x \leq -2$

** **10.** Use Theorem 4.2 (page 283) to prove Theorem 3.15 in Section 3.12 (page 230).

** **11.** One end of a uniform beam of length L is built into a wall, and the other end is simply supported (figure below). If the beam has constant mass per unit length m, its deflection y from the horizontal at a distance x from the built-in end is given by

$$(48EI)y = -mg(2x^4 - 5Lx^3 + 3L^2x^2),$$

where E and I are constants depending on the material and cross-section of the beam, and $g > 0$ is a constant. How far from the built-in end does maximum deflection occur?

** **12.** An agronomist wishes to fence eight rectangular plots for experimentation as shown in the figure below. If each plot must contain 9000 m^2, find the minimum amount of fencing that can be used.

** **13.** A closed box is to have length equal to three times its width and total surface area of 30 m^2. Find the dimensions that produce maximum volume.

** **14.** A square box with open top is to have a volume of 6000 L. Find the dimensions of the box that minimize the amount of material used.

** **15.** A cherry orchard has 255 trees, each of which produces on the average 25 baskets of cherries. For each additional tree planted in the same area, the yield per tree decreases by one-twelfth of a basket. How many more trees will produce a maximum crop?

** **16.** A high school football field is being designed to accommodate a 400 m track. The track runs along the length of the field and has semicircles beyond the end zones with diameters equal to the width of the field. If the end zones must be 10 m deep, find the dimensions of the field that maximize playing area (not including the end zones).

** **17.** When a manufacturing company sells x objects per month, it sets the price r of each object at

$$r(x) = 100 - \frac{x^2}{10\,000}.$$

The total cost C of producing the x objects per month is

$$C(x) = \frac{x^2}{10} + 2x + 20.$$

Find the number of objects the company should sell per month in order to realize maximum profits.

** **18.** The base of an isosceles triangle, which is not one of the equal sides, has length b, and its altitude has length a. Find the area of the largest rectangle that can be placed inside the triangle if one of the sides of the rectangle must lie on the base of the triangle.

** **19.** A manufacturer builds cylindrical metal cans that hold $1000 \, \text{cm}^3$. There is no waste involved in cutting material for the curved surface of the can. However, each circular end piece is cut from a square piece of metal, leaving four waste pieces. Find the height and radius of the can that uses the least amount of metal, including all waste materials.

** **20.** Sides AB and AC of an isosceles triangle have equal length. The base BC has length $2a$, as does the altitude AD from A to BC. Find the height of a point P on AD at which the sum of the distances AP, BP, and CP is a minimum.

** **21.** Two poles are driven into the ground 3 m apart. One pole protrudes 2 m above the ground and the other pole 1 m above the ground. A single piece of rope is attached to the top of one pole, passed through a loop on the ground, pulled taut, and attached to the top of the other pole. Where should the loop be placed in order that the rope be as short as possible?

** **22.** In designing pages for a book, a publisher decides that the rectangular printed region on each page must have area $150 \, \text{cm}^2$. If the page must have 2.5 cm margins on each side and 3.75 cm margins at top and bottom, find the dimensions of the page of smallest possible area.

** **23.** Find the points on the hyperbola $y^2 - x^2 = 9$ closest to $(4, 0)$.

** **24.** Find the point on the parabola $y = x^2$ closest to $(-2, 5)$.

** **25.** A light source is to be placed directly above the centre of a circular area of radius r (figure below). The illumination at any point on the edge of the circle is directly proportional to the cosine of the angle θ and inversely proportional to the square of the distance d from the source. Find the height h above the circle at which illumination on the edge of the table is maximized.

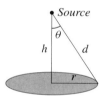

** **26.** Among all line segments that stretch from points on the positive x-axis to points on the positive y-axis and pass through the point $(2, 5)$, find that one which makes with the positive x- and y-axes the triangle with least possible area.

** **27.** A long piece of metal one metre wide is to be bent in two places to form a spillway so that its cross section is an isosceles trapezoid (figure below). Find the angle θ at which the bends should be formed in order to obtain maximum possible flow along the spillway if the lengths of AB, BC, and CD are all one-third metre.

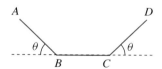

** **28.** A printer contracts to print 200 000 copies of a membership card. It costs \$10 per hour to run the press, and the press produces 1000 impressions per hour. The printer is at liberty to choose the number of set types per impression to a limit of 40, each set type costing \$2. If x set types are chosen, then each impression yields x cards. How many set types should be used?

** **29.** At noon a ship S_1 is 20 km north of ship S_2. If S_1 sails south at 6 km/h and S_2 sails east at 8 km/h, find when the two ships are closest together.

∗∗ **48.** A packing company wishes to form the 1 m by 2 m piece of cardboard in figure (a) below into a box as shown in figure (b). Cuts are to be made along solid lines and folds along dotted lines, and two sides are to be taped together as shown. If the outer flaps on top and bottom must meet in the centre but the inner flaps need not, find the dimensions of the box holding the most volume. How far apart will the inner flaps be?

(a)

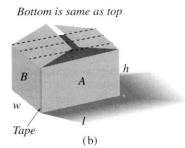

Bottom is same as top

(b)

∗∗ **49.** The cost of fuel per hour for running a ship varies directly as the cube of the speed, and is $\$B$ per hour when the speed is b kilometres per hour. There are also fixed costs of $\$A$ per hour. Find the most economical speed at which to make a trip.

∗∗ **50.** A paper drinking cup in the form of a right-circular cone can be made from a circular piece of paper by removing a sector and joining edges OA and OB, as shown in the figure below. If the radius of the circle is R, what choice of θ yields a cup of maximum volume?

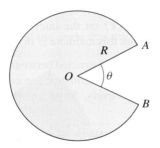

∗∗ **51.** A truck driver estimates that the cost per kilometre of operating his rig is $40 + (50 - v)^2/10$ cents, where v is his average speed in kilometres per hour. In order to cover this cost he charges $\$20$/h for rental of his truck. For his driving he charges $\$10$/h. For a 1000 km nonstop haul, what average speed would you recommend to the driver?

∗∗ **52.** Verify formula 1.7 (page 8) by minimizing the distance function from a point to a line.

∗∗ **53.** A window is in the form of a rectangle surmounted by a semicircle with diameter equal to the width of the rectangle. The rectangle is of clear glass costing $\$a$ per unit area, while the semicircle is of coloured glass costing $\$b$ per unit area. The coloured glass transmits only a fraction p $(0 < p < 1)$ as much light per unit area as the clear glass. In addition, the curved portion of the window is surmounted by a special frame at a cost of $\$c$ per unit length. If the total cost of the window must not exceed $\$A$, find the dimensions of the window that admit the most light.

∗∗ **54.** Find the point on the ellipse $4x^2 + 9y^2 = 36$ closest to $(4, 13\sqrt{5}/6)$.

** **55.** The frame for a kite is to be made from six pieces of wood as shown in the figure below. The four outside pieces have predetermined lengths a and b ($b > a$). Yet to be cut are the two diagonal pieces. How long should they be in order to make the area of the kite as large as possible?

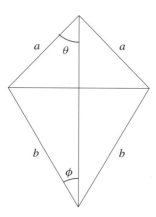

** **56.** Suppose that the line $Ax + By + C = 0$ intersects the parabola $y = ax^2 + bx + c$ at points P and Q (figure below). Find the point R on the parabola between P and Q that maximizes the area of triangle PQR.

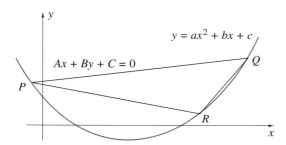

*** **57.** There are n red stakes securely driven into the ground in a straight line. A blue stake is to be added to the line, and each red stake is to be joined to the blue one by a string. Where should the blue stake be placed in order that the total length of all strings will be as small as possible?

*** **58.** When blood flows through a vein or artery, it encounters resistance because of friction with the walls of the blood vessel and the viscosity of the blood itself. *Poiseuille's law* for laminar blood flow states that, for a circular vessel, resistance R to blood flow is proportional to the length L of the blood vessel and inversely proportional to the fourth power of its radius r:

$$R = k\frac{L}{r^4},$$

where k is a constant. The figure at the top of page 302 shows a blood vessel of radius r_1 from A to B and a branching vessel from D to C of radius $r_2 < r_1$. The resistance encountered by the blood in flowing from A to C is given by

$$R = k\frac{L_1}{r_1^4} + k\frac{L_2}{r_2^4},$$

where L_1 and L_2 are the lengths of AD and DC respectively.

(a) If B is assumed to the right of E, show that R can be expressed in terms of θ as

$$R = f(\theta) = \frac{k}{r_1^4}(X - Y \cot \theta) + \frac{k}{r_2^4}Y \csc \theta,$$

where X and Y are the lengths of AE and CE respectively.

(b) Show that $f(\theta)$ has only one critical point $\overline{\theta}$ in the range $0 < \theta < \pi/2$, and $\overline{\theta}$ is defined by

$$\cos \overline{\theta} = \frac{r_2^4}{r_1^4}.$$

(c) Verify that $\overline{\theta}$ yields a relative minimum for $f(\theta)$.

(d) Show that

$$f(\overline{\theta}) = \frac{kX}{r_1^4} + \frac{kY}{r_2^4}\sqrt{1 - \left(\frac{r_2}{r_1}\right)^8} < f(\pi/2).$$

(e) Does $\overline{\theta}$ provide an absolute minimum for $f(\theta)$?

∗∗∗ **59.** If the fencing in Example 4.18 (page 286) is to form the arc of a circle, what is the maximum possible area?

∗∗∗ **60.** The upper left corner of a piece of paper a units wide and b units long $(b > a)$ is folded to the right edge as shown in the figure below. Calculate length x in order that the length y of the fold be a minimum.

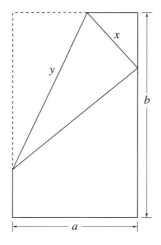

*** **61.** The line $y = mx + c$ intersects the ellipse $b^2x^2 + a^2y^2 = a^2b^2$ in two points P and Q. Find the point R on the ellipse in order that triangle PQR have maximum possible area.

*** **62.** A sheet of metal a m wide and L m long is to be bent into a trough as shown in the figure below. End pieces are also to be attached. If edge AB must be the arc of a circle, determine the radius of the circle in order that the trough hold the biggest possible volume.

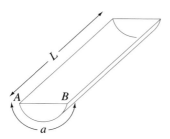

4.7 Velocity and Acceleration

Kinematics is the study of motion—relationships among the displacement, velocity, and acceleration of a body, particularly as they pertain to forces acting on the body. In this section we bring together for a final discussion everything we have learned about displacement, velocity, and acceleration, but do so mainly in a one-dimensional setting, and only from a differentiation point of view; given the position of a particle moving along a straight line, find its velocity and acceleration and use these quantities to describe the motion of the particle. In Section 5.2 we reverse these operations; beginning with the acceleration, we find velocity and position. This process is essential to physics and engineering, where the acceleration of a particle is determined by the forces acting on it.

When the position of a particle moving along the x-axis is known as a function of time t, say, $x(t)$, its **instantaneous velocity** is the derivative of $x(t)$ with respect to t,

$$v(t) = \frac{dx}{dt}. \tag{4.9}$$

The **instantaneous acceleration** of the particle is defined as the rate of change of velocity with respect to time:

$$a(t) = \frac{dv}{dt} = \frac{d^2x}{dt^2}. \tag{4.10}$$

In Section 3.6 we discussed the motion of a particle moving along the x-axis with position function

$$x(t) = t^3 - 27t^2 + 168t + 20, \quad t \geq 0,$$

where x is measured in metres and t in seconds. We now add a final touch to the discussion. The best way to describe the motion of the particle is with a graph of the displacement function (Figure 4.68). With calculus we can show how geometric properties of the graph reflect important features about the velocity and acceleration of the particle. The velocity and acceleration of the particle are

$$v(t) = \frac{dx}{dt} = 3t^2 - 54t + 168 = 3(t - 4)(t - 14) \text{ m/s},$$

$$a(t) = \frac{d^2x}{dt^2} = 6t - 54 = 6(t - 9) \text{ m/s}^2.$$

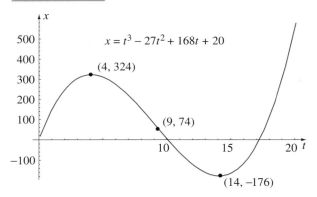

FIGURE 4.68

$x = t^3 - 27t^2 + 168t + 20$

(4, 324)

(9, 74)

(14, −176)

Ignoring the physical interpretations of dx/dt and d^2x/dt^2 as velocity and acceleration for the moment, and concentrating only on the fact that they are the first and second derivatives of the function $x(t)$, we immediately find that $x(t)$ has a relative maximum of $x(4) = 324$ and a relative minimum of $x(14) = -176$ at the critical points $t = 4$ and $t = 14$. The graph has a point of inflection at $(9, 74)$.

Let us now discuss what the graph tells us about the motion of the particle. Ordinates represent horizontal distances of the particle from the origin $x = 0$. When an ordinate is positive, the particle is that distance to the right of the origin; when an ordinate is negative, the particle is that distance to the left of the origin. For instance, at time $t = 0$, we calculate $x = 20$ and therefore the particle begins 20 m to the right of $x = 0$. At time $t = 4$, it is 324 m to the right of $x = 0$, and at $t = 14$, it is 176 m to the left of the origin.

The slope of the graph represents velocity of the particle. When slope is positive, namely, in the intervals $0 < t < 4$ and $t > 14$, the particle is moving to the right along the x-axis; for $4 < t < 14$, velocity is negative, indicating that the particle is moving to the left. At times $t = 4$ and $t = 14$, the particle is instantaneously at rest.

The concavity of the graph reflects the sign of the acceleration. For $0 < t < 9$, the graph is concave downward, its slope is decreasing. Physically this means that it has a negative acceleration; that is, its velocity is decreasing. For $t > 9$, the graph is concave upward, its slope is increasing. Physically, acceleration is positive; that is, velocity is increasing. At the point of inflection $(9, 74)$, the acceleration changes sign. Notice that the acceleration is not zero at $t = 4$ and $t = 14$. In spite of the fact that the velocity is zero at these times, the acceleration does not vanish. You might ask yourself what feature of the graph would reflect coincident zeros for velocity and acceleration (see Exercise 16).

Speed is the magnitude of velocity. It represents how fast the particle is moving without regard for direction. For instance, at time $t = 0$, the velocity and speed are both 168 m/s, whereas at time $t = 10$, velocity is -72 m/s and speed is 72 m/s. Geometrically, speed is represented by the slope of the graph without regard for sign.

With the above ideas in mind, let us detail the history of the particle's motion. At time $t = 0$, it begins 20 m to the right of the origin, moving to the right with velocity of 168 m/s. Since the acceleration is negative (to the left), the particle is slowing down (both velocity and speed are decreasing), until at time $t = 4$ s, it comes to an instantaneous stop 324 m to the right of the origin. Because the acceleration continues to be negative, the particle moves to the left, its velocity decreasing, but its speed increasing. At time $t = 9$ s, when the particle is 74 m to the right of the origin, the acceleration changes sign. At this instant, the velocity has attained a (relative) minimum value, but speed is a (relative) maximum. With acceleration to the right for $t > 9$ s, the particle continues to move left, but slows down, its velocity increasing until at time $t = 14$ s, when it once again comes to a stop 176 m to the left of the origin. For time $t > 14$ s, it moves to the right, picking up speed, and passes through the origin just after $t = 17$ s.

Further analysis of the interdependences of displacement, velocity, and acceleration is contained in the following example.

EXAMPLE 4.24

The position of a particle moving along the x-axis is given by the function

$$x(t) = 3t^4 - 32t^3 + 114t^2 - 144t + 40, \quad 0 \le t \le 5,$$

where x is measured in metres and t in seconds. Answer the following questions concerning its motion:

(a) What are its velocity and speed at $t = 1/2$ s?

(b) When is its acceleration increasing?

(c) Is the velocity increasing or decreasing at $t = 2$ s?

(d) Is the particle speeding up or slowing down at $t = 2$ s?

(e) What is the maximum velocity of the particle in the time interval $0 \le t \le 2$?

(f) What is the maximum distance the particle ever attains from the origin?

SOLUTION   We use the graph of the displacement function in Figure 4.69 to suggest answers when possible, and calculus to confirm them, when necessary.

FIGURE 4.69

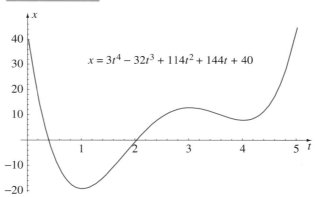

$x = 3t^4 - 32t^3 + 114t^2 + 144t + 40$

The velocity and acceleration are

$$v(t) = 12t^3 - 96t^2 + 228t - 144 = 12(t - 1)(t - 3)(t - 4) \text{ m/s},$$

$$a(t) = 12(3t^2 - 16t + 19) \text{ m/s}^2.$$

(a) The slope at $t = 1/2$ is negative, indicating a negative velocity. It is $v(1/2) = -105/2$ m/s. The speed is $105/2$ m/s.

(b) The graph does not tell us when acceleration is increasing. Algebraically, acceleration is increasing when its derivative is nonnegative. Since $da/dt = 12(6t - 16) = 24(3t - 8)$, acceleration is increasing for $8/3 \le t \le 5$.

(c) The graph appears to be concave downward at $t = 2$, implying that velocity is decreasing. Confirmation is provided by the fact that $a(2) = -12$ m/s^2.

(d) The graph seems to be becoming less steep around $t = 2$, indicating that the particle is slowing down. Since $v(2) = 24$ m/s, the particle is moving to the right. But according to part (c), its acceleration is to the left at this time. With velocity and acceleration in opposite directions, the particle is slowing down.

(e) Maximum velocity on $0 \le t \le 2$ occurs at the point of inflection to the left of $t = 2$. To find it, we set

$$0 = a(t) = 12(3t^2 - 16t + 19).$$

Of the two solutions $(8 \pm \sqrt{7})/3$, only $(8 - \sqrt{7})/3$ is in the interval $0 \le t \le 2$. Maximum velocity is therefore $v\big((8 - \sqrt{7})/3\big) = 25.35$ m/s.

(f) Maximum distance from the origin is represented by the point on the graph farthest from the t-axis. This is at $t = 5$ s, for which $x(5) = 45$ m.

EXAMPLE 4.25

The mass M in Figure 4.70 is pulled x_0 metres to the right of its position where the spring is unstretched, and then released. If at time $t = t_0$, during the subsequent oscillations, it is struck with a force of F newtons to the right, its position $x(t)$ must satisfy

$$M\frac{d^2x}{dt^2} + kx = F\delta(t - t_0), \quad x(0) = x_0, \quad x'(0) = 0,$$

where $k > 0$ is the spring constant, and $\delta(t - t_0)$ is the Dirac-delta function of Section 2.4. This assumes that air resistance due to motion is negligible as is friction with the surface on which oscillations take place.

(a) Show that

$$x(t) = x_0 \cos\sqrt{\frac{k}{M}}t + \frac{F}{\sqrt{kM}} \sin\sqrt{\frac{k}{M}}(t - t_0) H(t - t_0),$$

where $H(t - t_0)$ is the Heaviside unit step function, satisfies these conditions at every $t \ne t_0$.

(b) Show that the velocity of the mass changes by F/M m/s as a result of being struck at time t_0.

FIGURE 4.70

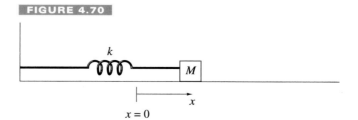

$x = 0$

SOLUTION

(a) Since the derivative of $H(t - t_0)$ equals 0 for any $t \neq t_0$, we find that for $t \neq t_0$,

$$x'(t) = -\sqrt{\frac{k}{M}} x_0 \sin \sqrt{\frac{k}{M}} t + \frac{F}{M} \cos \sqrt{\frac{k}{M}} (t - t_0) H(t - t_0),$$

and

$$x''(t) = -\frac{k x_0}{M} \cos \sqrt{\frac{k}{M}} t - \frac{F \sqrt{k}}{M^{3/2}} \sin \sqrt{\frac{k}{M}} (t - t_0) H(t - t_0).$$

Thus, for $t \neq t_0$,

$$M \frac{d^2 x}{dt^2} + kx = M \left[-\frac{k x_0}{M} \cos \sqrt{\frac{k}{M}} t - \frac{F \sqrt{k}}{M^{3/2}} \sin \sqrt{\frac{k}{M}} (t - t_0) H(t - t_0) \right]$$

$$+ k \left[x_0 \cos \sqrt{\frac{k}{M}} t + \frac{F}{\sqrt{kM}} \sin \sqrt{\frac{k}{M}} (t - t_0) H(t - t_0) \right]$$

$$= 0;$$

that is, $x(t)$ satisfies the differential equation $M x'' + kx = F \delta(t - t_0)$ for $t \neq t_0$. In addition, $x(0) = x_0$, and

$$x'(0) = \left[-\sqrt{\frac{k}{M}} x_0 \sin \sqrt{\frac{k}{M}} t + \frac{F}{M} \cos \sqrt{\frac{k}{M}} (t - t_0) H(t - t_0) \right]_{t=0} = 0.$$

(b) The change in velocity at $t = t_0$ is

$$\lim_{t \to t_0^+} x'(t) - \lim_{t \to t_0^-} x'(t) = \lim_{t \to t_0^+} \left[-\sqrt{\frac{k}{M}} x_0 \sin \sqrt{\frac{k}{M}} t + \frac{F}{M} \cos \sqrt{\frac{k}{M}} (t - t_0) \right]$$

$$- \lim_{t \to t_0^-} \left[-\sqrt{\frac{k}{M}} x_0 \sin \sqrt{\frac{k}{M}} t \right]$$

$$= -\sqrt{\frac{k}{M}} x_0 \sin \sqrt{\frac{k}{M}} t_0 + \frac{F}{M} + \sqrt{\frac{k}{M}} x_0 \sin \sqrt{\frac{k}{M}} t_0$$

$$= \frac{F}{M}.$$

EXAMPLE 4.26

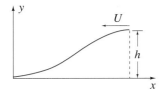

FIGURE 4.71

A landing approach is to be shaped generally as shown in Figure 4.71. The following conditions are imposed on the approach pattern:

(a) Altitude must be h metres when descent commences.

(b) Smooth touchdown must occur at $x = 0$.

(c) Constant horizontal speed U m/s must be maintained throughout.

(d) At no time must vertical acceleration in absolute value exceed some fixed positive constant k.

(e) The glide path must be a cubic polynomial.

Find when descent should commence.

SOLUTION   If we let $y = f(x) = ax^3 + bx^2 + cx + d$ denote the cubic polynomial for the landing curve, then d must be zero for the curve to pass through $(0, 0)$. For a smooth touchdown, we also demand that

$$0 = f'(0) = (3ax^2 + 2bx + c)_{|x=0} = c.$$

Suppose we denote horizontal distance from $x = 0$ to the point at which descent commences by L. Then descent begins at the point (L, h), and

$$h = f(L) = aL^3 + bL^2 \quad \text{and} \quad 0 = f'(L) = 3aL^2 + 2bL.$$

These equations can be solved for $a = -2h/L^3$ and $b = 3h/L^2$, so that the glide path has equation

$$y = f(x) = -\frac{2hx^3}{L^3} + \frac{3hx^2}{L^2}.$$

Since a constant horizontal speed U must be maintained, we can say that $dx/dt = -U$ (the negative sign is necessary because $dx/dt < 0$). It follows that

$$\frac{dy}{dt} = \frac{dy}{dx}\frac{dx}{dt} = \left(-\frac{6hx^2}{L^3} + \frac{6hx}{L^2}\right)(-U).$$

Vertical acceleration is therefore

$$\frac{d^2y}{dt^2} = \frac{d}{dt}\left(\frac{dy}{dt}\right) = \frac{d}{dx}\left(\frac{dy}{dt}\right)\frac{dx}{dt} = \left(\frac{-12hx}{L^3} + \frac{6h}{L^2}\right)(-U)^2 = \frac{6hU^2}{L^2}\left(1 - \frac{2x}{L}\right).$$

Figure 4.72 shows that acceleration is positive for $0 < x < L/2$ and negative for $L/2 < x < L$. In addition, in magnitude, it is largest at $x = 0$ and $x = L$, where

$$\left|\frac{d^2y}{dt^2}\right| = \frac{6hU^2}{L^2}.$$

Consequently, we must have

$$\frac{6hU^2}{L^2} \leq k \quad \implies \quad L \geq \sqrt{\frac{6hU^2}{k}};$$

that is, the plane must begin descent when it is at least $\sqrt{6hU^2/k}$ metres from touchdown.

FIGURE 4.72

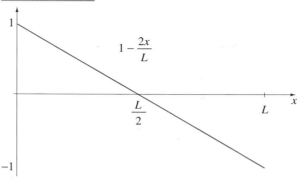

EXERCISES 4.7

In Exercises 1–10 find the velocity and acceleration of an object that moves along the x-axis with the given position function. In each exercise, discuss the motion, including in your discussion a graph of the function $x(t)$. Assume that x is measured in metres and t in seconds.

∗∗ **1.** $x(t) = 2t + 5, \quad t \geq 5$ ∗∗ **2.** $x(t) = t^2 - 7t + 6, \quad t \geq 0$

∗∗ **3.** $x(t) = t^2 + 5t + 10, \quad t \geq 1$ ∗∗ **4.** $x(t) = -2t^3 + 2t^2 + 16t - 1, \quad t \geq 0$

∗∗ **5.** $x(t) = t^3 - 9t^2 + 15t + 3, \quad t \geq 2$ ∗∗ **6.** $x(t) = 3\cos 4t, \quad t \geq 0$

∗∗ **7.** $x(t) = 1/t, \quad t \geq 1$ ∗∗ **8.** $x(t) = t + 4/t, \quad t \geq 1$

∗∗ **9.** $x(t) = (t - 4)/t^2, \quad t \geq 2$ ∗∗ **10.** $x(t) = (t - 1)^2 \sqrt{t}, \quad t \geq 1$

∗∗ **11.** An object moving along the x-axis has position given by

$$x(t) = t^3 - 9t^2 + 24t + 1, \quad 0 \leq t \leq 6,$$

where x is measured in metres and $t \geq 0$ is time in seconds. Determine:

(a) whether speed is increasing or decreasing at $t = 1$ s;

(b) maximum and minimum velocity;

(c) maximum and minimum speed;

(d) maximum and minimum acceleration;

(e) when acceleration is increasing.

∗∗ **12.** Repeat Exercise 11 if

$$x(t) = 2 - 15t + 9t^2 - t^3, \quad 0 \leq t \leq 6.$$

∗∗ **13.** A particle moves along the x-axis in such a way that its position as a function of time t is given by

$$x(t) = t \sin t, \quad 0 \leq t \leq 9.$$

(a) Plot a graph of $x(t)$.

(b) Use Newton's iterative procedure to find when velocity is zero.

(c) Use Newton's iterative procedure to find when acceleration is equal to 1.

✳✳ **14.** An object moving along the x-axis has position function given by

$$x(t) = 3t^4 - 16t^3 + 18t^2 + 2, \quad 0 \le t \le 4,$$

where x is measured in metres and t in seconds. Determine:

(a) on what intervals the velocity is increasing and decreasing;

(b) on what intervals the speed is increasing and decreasing;

(c) when the velocity is a maximum and a minimum;

(d) when the speed is a maximum and a minimum;

(e) the maximum distance from the origin achieved by the particle;

(f) the maximum distance from the point $x = 5$ achieved by the particle.

✳✳ **15.** Repeat Exercise 14 for the position function

$$x(t) = \frac{14t^3}{45} - \frac{101t^2}{45} + \frac{132t}{45} + 2, \quad 0 \le t \le 6.$$

✳✳ **16.** What feature on the displacement graph would indicate a time when the velocity and acceleration are simultaneously zero?

✳✳ **17.** Are critical points for the velocity function the same as those for the speed function? Explain, using graphs.

✳✳ **18.** What is the maximum speed of the particle in Example 4.24 over time interval $0 \le t \le 2$?

✳✳ **19.** In many velocity and acceleration problems it is more convenient to express acceleration in terms of a derivative with respect to position, as opposed to a derivative with respect to time. Show that acceleration can be written in the form

$$a = v\frac{dv}{dx}.$$

In Exercises 20–23 assume that a particle moves along the x-axis in such a way that its position, velocity, and acceleration are continuous functions on the interval $a \le t \le b$. Discuss the validity of each statement.

✳✳ **20.** When position has a relative maximum so also does the absolute value of the distance from the origin to the particle.

✳✳ **21.** When position has a relative minimum so also does the absolute value of the distance from the origin to the particle.

✳✳ **22.** When velocity has a relative minimum so also does speed.

✳✳ **23.** When velocity has a relative maximum so also does speed.

4.8 Related Rates

Many interesting and practical problems involving rates of change are commonly referred to as *related-rate problems*. In these problems, two or more quantities are related to each other and rates at which some of them change are known. It is required to find rates at which the others change. Related-rate problems deal almost exclusively with rates of change of quantities with respect to time. To solve these problems we first consider three examples. These will suggest the general procedures by which all related rate problems can be analyzed. We shall then discuss two somewhat more complicated problems.

EXAMPLE 4.27

A man 2 m tall walks directly away from a streetlight that is 8 m high at the rate of $3/2$ m/s. How fast is the length of his shadow changing?

SOLUTION   When x denotes the distance between the man and the light post (Figure 4.73), the fact that he walks directly away from the light at $3/2$ m/s means that x is changing at a rate of $3/2$ m/s, that is, $dx/dt = 3/2$ m/s. If s represents the length of the man's shadow, then we are searching for ds/dt.

FIGURE 4.73

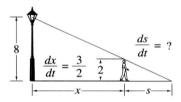

Similar triangles in Figure 4.73 enable us to relate s and x,

$$\frac{x + s}{s} = \frac{8}{2},$$

and this equation can be solved for s in terms of x:

$$s = \frac{x}{3}.$$

Now, s and x are each functions of time t,

$$s = f(t) \quad \text{and} \quad x = g(t),$$

although we have not calculated the exact form of these functions. Indeed, the essence of the related-rate problem is to find ds/dt without ever knowing $f(t)$ explicitly. To do this we note that, since the equation $s = x/3$ is valid at any time t when the man is walking away from the light, we may differentiate with respect to t to obtain

$$\frac{ds}{dt} = \frac{1}{3}\frac{dx}{dt}.$$

This equation relates the known rate $dx/dt = 3/2$ with the unknown rate ds/dt. It follows that

$$\frac{ds}{dt} = \left(\frac{1}{3}\right)\left(\frac{3}{2}\right) = \frac{1}{2},$$

and the man's shadow is therefore getting longer at the rate of $1/2$ m/s.

Knowing dx/dt in this example, we have calculated ds/dt, and have done so without finding s explicitly as a function of time t. This is the essence of a related rate problem. Since dx/dt is a constant value, it is quite easy to find s as a function of t, and hence ds/dt. Indeed, if we choose time $t = 0$ when the man starts to walk away from the streetlight, then his distance from the light at any given time is $x = 3t/2$ metres. Combine this with the fact that $s = x/3$ and we may write

$$s = \frac{1}{3}\left(\frac{3t}{2}\right) = \frac{t}{2} \text{ m.}$$

With this explicit formula for s, it is clear that $ds/dt = 1/2$. What is important to realize is that the solution in this paragraph is possible only because the man walks at a constant rate. Were his speed not constant, it might be impossible to find s explicitly in terms of t. The next example illustrates this point in that the given rate is known only at one instant in time.

| **EXAMPLE 4.28** |

A ladder leaning against a house (Figure 4.74) is prevented from moving by a young child. Suddenly, something distracts the child and he releases the ladder. The ladder begins slipping down the wall of the house, picking up speed as it falls. If the top end of the ladder is moving at 1 m/s when the lower end is 15 m from the house, how fast is the foot of the ladder moving away from the house at this instant?

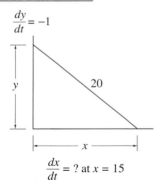

$$\frac{dy}{dt} = -1$$

$$\frac{dx}{dt} = ? \text{ at } x = 15$$

SOLUTION   Figure 4.74 indicates that when y denotes the height of the top of the ladder above the ground, then $dy/dt = -1$ m/s when $x = 15$ m (the negative sign because y is decreasing). We emphasize here that $dy/dt = -1$ *only* when $x = 15$. What is required is dx/dt when $x = 15$ m. Because the triangle in the figure is right-angled, we may write

$$x^2 + y^2 = 20^2,$$

and this equation is valid at any time during which the ladder is slipping. If we differentiate with respect to time t, using extended power rule 3.21,

$$2x\frac{dx}{dt} + 2y\frac{dy}{dt} = 0.$$

When $x = 15$, we calculate that $y = \sqrt{400 - 225} = 5\sqrt{7}$, and therefore at this instant

$$15\frac{dx}{dt} + 5\sqrt{7}(-1) = 0.$$

This yields $dx/dt = \sqrt{7}/3$, and we can say that when the foot of the ladder is 15 m from the wall it is moving away from the wall at $\sqrt{7}/3$ m/s.

EXAMPLE 4.29

FIGURE 4.75

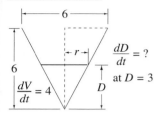

A tank in the form of a right circular cone with altitude 6 m and base radius 3 m (Figure 4.75) is being filled with water at a rate of 4000 L/min. How fast is the surface of the water rising when the depth is 3 m?

SOLUTION   Figure 4.75, which illustrates a cross-section of the tank, indicates that when the depth of water in the tank is D, the volume V of water is

$$V = \frac{1}{3}\pi r^2 D.$$

Of the three variables V, r, and D in this equation, we are concerned only with V and D, since dV/dt is given and dD/dt is what we want. This suggests that we eliminate r using similar triangles. Since $r/D = 3/6$, we have $r = D/2$; therefore,

$$V = \frac{1}{3}\pi \left(\frac{D}{2}\right)^2 D = \frac{1}{12}\pi D^3.$$

Because this result is valid for all time t during the filling process, we can differentiate with respect to t, once again using extended power rule 3.21:

$$\frac{dV}{dt} = \frac{1}{4}\pi D^2 \frac{dD}{dt}.$$

Since $dV/dt = 4$ m^3/min (converted from the rate of 4000 L/min since the litre is not an acceptable unit of measure for volume), we find that when $D = 3$,

$$4 = \frac{1}{4}\pi (3)^2 \frac{dD}{dt},$$

from which $dD/dt = 16/(9\pi)$. The surface is therefore rising at a rate of $16/(9\pi)$ m/min.

These examples illustrate the following general procedure for solving related-rate problems:

1. Sketch a diagram illustrating all given information, especially given rates of change and desired rates of change. Do not draw the diagram at the instant in question; draw it slightly before or slightly after.

2. Find an equation valid for all time (in some interval about the instant in question) that involves only variables whose rates of change are given or required.

3. Differentiate the equation in 2. and solve for the required rate.

Steps 1. and 3. are usually quite straightforward; step 2., on the other hand, may tax your ingenuity. To find the equation in the appropriate variables, it may be necessary to introduce and substitute for additional variables. Finding these substitutions requires you to analyze the problem very closely.

 Be careful not to substitute numerical data that represent the instant at which the derivative is required before differentiation has taken place. Numerical data must be substituted after differentiation. For instance, in Example 4.29, radius r of the surface of the water when $D = 3$ is $3/2$. If we substitute this into $V = \pi r^2 D/3$, we obtain a function $V = \pi (3/2)^2 D/3 = 3\pi D/4$ which is valid only when $D = 3$. It cannot therefore be differentiated; only equations that are valid for a range of values of t can be differentiated with respect to t.

We now apply this procedure to two further examples. The first is an extension of Example 4.27.

| EXAMPLE 4.30

A man 2 m tall walks along the edge of a straight road 10 m wide. On the other edge of the road stands a streetlight 8 m high. If the man walks at 3/2 m/s, how fast is his shadow lengthening when he is 10 m from the point directly opposite the light?

SOLUTION   First we draw Figure 4.76 wherein the man's speed is represented as the time rate of change of his distance x from the point on his side of the road directly opposite the light. What is required is the rate of change ds/dt of the length of his shadow when $x = 10$.

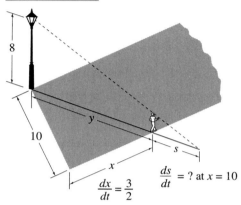

FIGURE 4.76

To find an equation relating x and s, we first use similar (vertical) triangles to write

$$\frac{y+s}{s} = \frac{8}{2} = 4,$$

from which $y = 3s$, an equation that relates s to y, rather than s to x. However, since

$$y^2 = x^2 + 100,$$

we substitute to obtain

$$9s^2 = x^2 + 100.$$

The derivative of this equation with respect to time t gives

$$18s\frac{ds}{dt} = 2x\frac{dx}{dt}.$$

When $x = 10$, we obtain $s = \sqrt{100 + 100}/3 = 10\sqrt{2}/3$, and, at this instant,

$$18\left(\frac{10\sqrt{2}}{3}\right)\frac{ds}{dt} = 2(10)\left(\frac{3}{2}\right).$$

Thus,

$$\frac{ds}{dt} = \frac{30}{60\sqrt{2}} = \frac{\sqrt{2}}{4},$$

and the man's shadow is therefore lengthening at the rate of $\sqrt{2}/4$ m/s.

It is worthwhile noting in this example that were we to solve for ds/dt before substituting $x = 10$ and $s = 10\sqrt{2}/3$, then we would have

$$\frac{ds}{dt} = \frac{x}{9s}\frac{dx}{dt}.$$

Since dx/dt is always equal to 3/2, and $s = \sqrt{x^2 + 100}/3$, we can write

$$\frac{ds}{dt} = \frac{x}{3\sqrt{x^2 + 100}}\left(\frac{3}{2}\right) = \frac{x}{2\sqrt{x^2 + 100}},$$

a general formula for ds/dt. The limit of this rate as x becomes very large is

$$\lim_{x\to\infty}\frac{ds}{dt} = \lim_{x\to\infty}\frac{x}{2\sqrt{x^2 + 100}} = \frac{1}{2}.$$

But for very large x, the man essentially walks directly away from the light, and this answer, as we might expect, is identical to that in Example 4.27.

| **EXAMPLE 4.31**

One end of a rope is tied to a box. The other end is passed over a pulley 5 m above the floor and tied at a level 1 m above the floor to the back of a truck. If the rope is taut and the truck moves at 1/2 m/s, how fast is the box rising when the truck is 3 m from the plumb line through the pulley?

FIGURE 4.77

SOLUTION   In Figure 4.77, we have represented the speed of the truck as the rate of change of length x, $dx/dt = 1/2$ m/s. What is required is dy/dt when $x = 3$.

To find an equation relating x and y, we first use the fact that the length z of rope between pulley and truck is the hypotenuse of a right-angled triangle with sides of lengths x and 4,

$$z^2 = x^2 + 16.$$

For an equation relating y and z, we note that the length of the rope, call it L, remains constant, and is equal to the sum of z and $5 - y$,

$$L = z + (5 - y).$$

These two equations can be combined into

$$(L - 5 + y)^2 = x^2 + 16,$$

and differentiation with respect to time t now gives

$$2(L - 5 + y)\frac{dy}{dt} = 2x\frac{dx}{dt}.$$

When $x = 3$, we may write that

$$(L - 5 + y)^2 = 9 + 16 = 25.$$

We could solve this equation for y (in terms of L), but it is really not y that is needed to obtain dy/dt from the previous equation. It is $L - 5 + y$, and this is clearly equal to 5. Thus, when $x = 3$, we have

$$2(5)\frac{dy}{dt} = 2(3)\left(\frac{1}{2}\right);$$

that is, $dy/dt = 3/10$, and the box is rising at a rate of 3/10 m/s.

The reader should compare this example with the problem in Exercise 2. They may appear similar, but are really quite different.

EXERCISES 4.8

** **1.** A convertible is travelling along a straight highway at 100 km/h. A child in the car accidentally releases a helium-filled balloon, which then rises vertically at 10 m/s. How fast are the child and balloon separating 4 s after the balloon is released?

** **2.** A rope passes over a pulley and one end is attached to a cart as shown in the figure below. If the rope is pulled vertically downward at 2 m/s, how fast is the cart moving when $s = 6$ m?

** **3.** A light is on the ground 20 m from a building. A man 2 m tall walks from the light directly toward the building at 3 m/s. How fast is the length of his shadow on the building changing when he is 8 m from the building?

** **4.** A funnel in the shape of a right circular cone is 15 cm across the top and 30 cm deep. A liquid is flowing in at the rate of 80 mL/s and flowing out at 15 mL/s. At what rate is the surface of the liquid rising when the liquid fills the funnel to a depth of 20 cm?

** **5.** A water tank is in the form of a right circular cylinder of diameter 3 m and height 3 m on top of a right circular cone of diameter 3 m and height 1 m. If water is being drawn from the bottom at the rate of 1 L/min, how fast is the water level falling when

   (a) it is 1 m from the top of the tank?

   (b) it is 3.5 m from the top of the tank?

** **6.** A point P moves along the curve $y = x^2 + x + 4$ where x and y are measured in metres. Its x-coordinate decreases at 2 m/s. If the perpendicular from P to the x-axis intersects this axis at point Q, how fast is the area of the triangle with vertices P, Q, and the origin changing when the x-coordinate of P is 2 m?

** **7.** Water is being pumped into a swimming pool which is 10 m wide, 20 m long, 1 m deep at the shallow end, and 3 m deep at the deep end. If the water level is rising at 1 cm/min when the depth is 1 m at the deep end, at what rate is water being pumped into the pool?

** **8.** Boyle's law for a perfect gas states that the pressure exerted by the gas on its containing vessel is inversely proportional to the volume occupied by the gas. If, when the volume is 10 L and the pressure is 50 N/m^2, the volume is increasing at 1/2 L/s, find the rate of change of the pressure of the gas.

** **9.** A woman driving 100 km/h along a straight highway notes that the shadow of a cloud is keeping pace with her. What can she conclude about the speed of the cloud?

** **10.** A fisherman is trolling at a rate of 2 m/s with his lure 100 m behind the boat and on the surface. Suddenly a fish strikes and dives vertically at a rate of 3 m/s. If the fisherman permits the line to run freely and it always remains straight, how fast is the line being played out when the reel is 50 m from its position at the time of the strike?

** **11.** Air expands adiabatically in accordance with the law $PV^{7/5} =$ constant. If, at a given time, the volume V is 100 L and the pressure P is 40 N/cm², at what rate is the pressure changing when the volume is decreasing at 1 L/s?

** **12.** Sand is poured into a right circular cylinder of radius 1/2 m along its axis (figure below). Once sand completely covers the bottom, a right circular cone is formed on the top of the sand.

(a) If 0.02 m³ of sand enters the container every minute, how fast is the top of the sand pile rising?

(b) How fast is the sand rising along the side of the cylinder?

Sand

1

** **13.** A balloon has the shape of a right circular cylinder of radius r and length l with a hemisphere at each end of radius r. The balloon is being filled at a rate of 10 mL/s in such a way that l increases twice as fast as r. Find the rate of change of r when $r = 8$ cm and $l = 20$ cm.

** **14.** An oval racetrack has a straight stretch 100 m long and two semicircles, each with radius 50 m (figure below). Car 1, on the infield, moves along the x-axis from O to B. It accelerates from rest at O, attains a speed of 10 m/s at C, and maintains this speed along CB. Car 2 travels along the quarter oval $ADEB$. It is at D when Car 1 is at C. Between D and B, Car 2 maintains the same rate of change of its x-coordinate as does Car 1.

(a) Find a formula for the rate of change of the y-coordinate of Car 2 between D and B.

(b) How fast is the y-coordinate of Car 2 changing when it is at point E?

(c) If the cars collide at B, which car suffers the most damage?

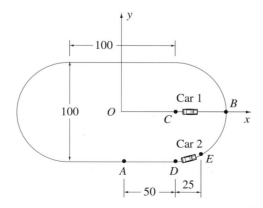

** **15.** A ship is 1 km north of a pier and is travelling N30°E at 3 km/h. A second ship is 3/4 km east of the pier and is travelling east at 7 km/h. How fast are the ships separating?

** **16.** The circle in the figure below represents a long-playing record which is rotating clockwise at 100/3 revolutions per minute. A bug is walking away from the centre of the record directly toward point P on the rim of the record at 1 cm/s. When the bug is at position R, 10 cm from O, angle θ is $\pi/4$ radians. Find the rate at which the distance from the bug to the fixed point Q is changing when the bug is at R.

** **17.** Eight skaters form a "whip." Show that the seventh person on the whip travels twice as fast as the fourth person.

** **18.** A particle moves counterclockwise around a circle of radius 5 cm centred at the origin making 4 revolutions each second. How fast is the particle moving away from the point with coordinates $(5, 6)$ when it is at position $(-3, 4)$?

** **19.** Two people (A and B in the figure below) walk along opposite sides of a road 10 m wide. A walks to the right at 1 m/s, and B walks to the left at 2 m/s. A third person C walks along a sidewalk 5 m from the road in such a way that B is always on the line joining A and C. Find the speed of C.

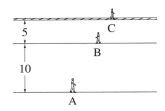

** **20.** Let $P(x, y)$ be a point on the first quadrant portion of the hyperbola $x^2 - y^2 = 1$. Let R be the foot of the perpendicular from P to the x-axis, and $Q(x^*, 0)$ be the x-intercept of the normal line to the hyperbola at P.

    (a) Show that $x^* = 2x$.

    (b) If P moves along the hyperbola so that its x-coordinate is decreasing at 3 units per unit time, how fast is the area of triangle QPR changing when $x = 4$?

** **21.** A solution passes from a conical filter 24 cm deep and 16 cm across the top into a cylindrical container of diameter 12 cm. When the depth of solution in the filter is 12 cm, its level is falling at the rate of 1 cm/min. How fast is the level of solution rising in the cylinder at this instant?

** **22.** A light is at the top of a pole 25 m high and a ball is dropped at the same height from a point 10 m from the light. How fast is the shadow of the ball moving along the ground 1 s later? The distance fallen by the ball t seconds after it has been dropped is $d = 4.905t^2$ metres.

** **23.** A point moves along the parabola $y = x^2 - 3x$ (x and y measured in metres) in such a way that its x-coordinate changes at the rate of 2 m/s. How fast is its distance from the point $(1, 2)$ changing when it is at $(4, 4)$?

** **24.** Repeat Exercise 23 given that the parabola is replaced by the curve $(x + y)^2 = 16x$.

** **25.** The volume of wood in the trunk of a tree is sometimes calculated by considering it as a frustrum of a right circular cone (figure on page 321).

    (a) Verify that the volume of the trunk is

$$V = \frac{1}{3}\pi h(R^2 + rR + r^2).$$

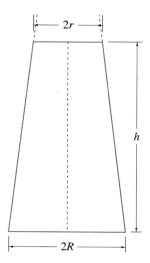

(b) Suppose that at the present time the radii of the top and bottom are $r = 10$ and $R = 50$ cm, and the height is $h = 30$ m. If the tree continues to grow so that ratios r/R and r/h always remain the same as they are now, and R increases at a rate of $1/2$ cm/year, how fast will the volume be changing in 2 years?

✱✱ **26.** Sand is poured into a right circular cone of radius 2 m and height 3 m along its axis (figure below). The sand forms two cones of equal height h, one inverted on top of the other.

(a) If 0.02 m^3 of sand enters the container every minute, how fast is the top of the pile rising when it is just level with the top of the container?

(b) How fast is the sand rising along the side of the container at this instant?

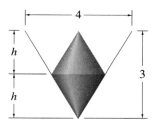

✱✱ **27.** The figure below shows a rod OB rotating counterclockwise in the xy-plane about O at a rate of 2 revolutions per second. Attached to OB is a rod AB, and A is confined to sliding horizontally along the x-axis. We are given that OB has length l metres, and AB has length L metres $(L > 2l)$.

(a) Find the velocity of A as a function of x and θ.

(b) Find the acceleration of A.

✱✱ **28.** If to the mechanism in Exercise 27 we add a rod AC (figure on page 320), where C is confined to sliding vertically, find the velocity of C in terms of x, θ, and y.

** **29.** In the figure below the boy's feet make 1 revolution per second around a sprocket of radius R metres. The chain travels around a sprocket of radius r metres on the back wheel, which itself has radius $\overline{R}$ metres. If a stone imbedded in the tire becomes dislodged, how fast is it travelling when it leaves the tire?

** **30.** (a) If θ is the angle formed by the minute and hour hands of a clock, what is the time rate of change of θ (in radians per minute)?

(b) If the lengths of the hands on the clock are 10 cm and 7.5 cm, find the rate at which their tips approach each other at 3:00 (i.e., find dz/dt in the figure below).

(c) Repeat (b) but replace time 3:00 with 8:05.

** **31.** If the sides of a triangle have lengths a, b, and c, its area is given by

$$A = \sqrt{s(s-a)(s-b)(s-c)},$$

where $s = (a+b+c)/2$ is one-half its perimeter. If the length of each side increases at a rate of 1 cm/min, how fast is A changing when $a = 3$ cm, $b = 4$ cm, and $c = 5$ cm?

4.9 *LCR*-Circuits

In Section 1.4 we introduced, in an intuitive way, how resistors, capacitors, and inductors relate to charge and current in electric circuits. With derivatives now in place, we can be explicit and precise.

Since current is the rate of flow of charge, and rates are derivatives, we can say that current is the derivative of charge with respect to time,

$$i = \frac{dQ}{dt}. \tag{4.11}$$

When there is a capacitor in the circuit, it is often convenient to think of $Q(t)$ as the charge on the capacitor, in which case i represents the rate of change of this charge. When there is no capacitor, or even when there is, we can think of charge flowing past any specific point in the circuit. How much flows past this point per unit time is represented by i.

Voltage V_C across the terminals of a capacitor is charge divided by capacitance (see equation 1.24),

$$V_c = \frac{Q}{C}. \tag{4.12}$$

Voltage V_R across the terminals of a resistor is current multiplied by resistance (see equation 1.25),

$$V_R = iR. \tag{4.13}$$

Voltage V_L across the terminals of an inductor is inductance multiplied by the rate of change of current (see equation 1.27). Since di/dt represents rate of change of current,

$$V_L = L\frac{di}{dt}. \tag{4.14a}$$

We can also state that

$$V_L = L\frac{d^2Q}{dt^2}. \tag{4.14b}$$

When all three elements are placed in a circuit with an emf device producing voltage V, Kirchhoff's loop rule (Section 1.4) implies that

$$L\frac{di}{dt} + Ri + \frac{Q}{C} = V. \tag{4.15a}$$

If i is replace by dQ/dt, we have

$$L\frac{d^2Q}{dt^2} + R\frac{dQ}{dt} + \frac{Q}{C} = V, \tag{4.15b}$$

and if this equation is differentiated with respect to t, we also have

$$L\frac{d^2i}{dt^2} + R\frac{di}{dt} + \frac{i}{C} = \frac{dV}{dt}. \tag{4.15c}$$

EXAMPLE 4.32

At time $t = 0$, a 6 Ω resistor, a 1 H inductor, and a 0.04 F capacitor are connected with a generator producing a voltage of $10 \sin 5t$, where $t \geq 0$ is time in seconds, by closing the switch (Figure 4.78).

(a) What equation must current in the circuit satisfy for $t > 0$?

(b) Show that

$$i(t) = \frac{5}{3} \sin 5t - \frac{1}{3} e^{-3t} (5 \cos 4t + 10 \sin 4t)$$

satisfies the equation in (a).

(c) Plot a graph of $i(t)$ and explain the significance of each term.

FIGURE 4.78

SOLUTION

(a) With the particular values for L, C, R, and V in Figure 4.78, equation 4.15c becomes

$$\frac{d^2i}{dt^2} + 6\frac{di}{dt} + 25i = 50 \cos 5t.$$

(b) Since

$$\frac{di}{dt} = \frac{25}{3} \cos 5t + e^{-3t} (5 \cos 4t + 10 \sin 4t)$$

$$- \frac{1}{3} e^{-3t} (-20 \sin 4t + 40 \cos 4t)$$

$$= \frac{25}{3} \cos 5t + \frac{1}{3} e^{-3t} (-25 \cos 4t + 50 \sin 4t),$$

and

$$\frac{d^2i}{dt^2} = -\frac{125}{3} \sin 5t - e^{-3t} (-25 \cos 4t + 50 \sin 4t)$$

$$+ \frac{1}{3} e^{-3t} (100 \sin 4t + 200 \cos 4t)$$

$$= -\frac{125}{3} \sin 5t + \frac{1}{3} e^{-3t} (275 \cos 4t - 50 \sin 4t),$$

we find that

$$\frac{d^2 i}{dt^2} + 6\frac{di}{dt} + 25i = -\frac{125}{3}\sin 5t + \frac{1}{3}e^{-3t}(275\cos 4t - 50\sin 4t)$$

$$+ 50\cos 5t + 2e^{-3t}(-25\cos 4t + 50\sin 4t)$$

$$+ \frac{125}{3}\sin 5t - \frac{25}{3}e^{-3t}(5\cos 4t + 10\sin 4t)$$

$$= 50\cos 5t.$$

Thus, $i(t)$ does indeed satisfy $i'' + 6i' + 25i = 50\cos 5t$.

(c) The plot of $i(t)$ in Figure 4.79 is composed of two functions, $(5/3)\sin 5t$ and $-(1/3)e^{-3t}(5\cos 4t + 10\sin 4t)$. Just after the switch is closed, both parts contribute significantly to $i(t)$. Within a few seconds, however, the exponential factor e^{-3t} causes the second term of $i(t)$ to become negligible. This is called the transient part of the current; it persists for a very short time interval. The term $(5/3)\sin 5t$ remains for all time, and once the transient part of the current becomes insignificant, $i(t)$ is essentially $(5/3)\sin 5t$. This is called the steady-state part of the current.

FIGURE 4.79

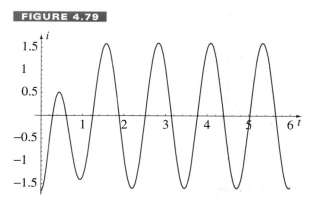

EXAMPLE 4.33

When an inductor and capacitor are connected to an emf device (Figure 4.80), the charge on the capacitor must satisfy equation 4.15b with $R = 0$, i.e.,

$$L\frac{d^2 Q}{dt^2} + \frac{Q}{C} = V.$$

(a) Show that if the emf device is a battery producing constant voltage V, beginning at time $t = 0$, and the capacitor has no initial charge, then the function

$$Q(t) = CV\left[1 - \cos\left(\frac{t}{\sqrt{LC}}\right)\right]$$

satisfies this equation, and the conditions $Q(0) = 0$ and $i(0) = 0$.

(b) Draw a graph of $Q(t)$ and interpret it in terms of charge on the capacitor and current in the circuit.

FIGURE 4.80

SOLUTION

(a) Since $\dfrac{dQ}{dt} = CV\left[\dfrac{1}{\sqrt{LC}}\sin\left(\dfrac{t}{\sqrt{LC}}\right)\right]$, it follows that

$$L\dfrac{d^2Q}{dt^2} + \dfrac{Q}{C} = L\left[\dfrac{CV}{LC}\cos\left(\dfrac{t}{\sqrt{LC}}\right)\right] + \dfrac{CV}{C}\left[1 - \cos\left(\dfrac{t}{\sqrt{LC}}\right)\right] = V.$$

Clearly, $Q(0) = 0$ and $i(0) = Q'(0) = 0$ also.

(b) To graph $Q(t)$, we first draw $-\cos\left(t/\sqrt{LC}\right)$ in Figure 4.81a. Shifting this curve upward 1 unit and changing the scale on the vertical axis gives the graph of $Q(t)$ in Figure 4.81b.

FIGURE 4.81a

FIGURE 4.81b

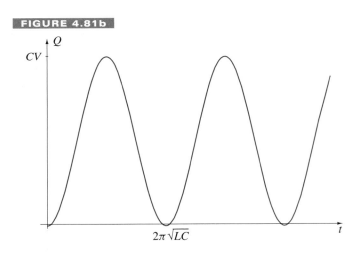

When the circuit is closed at $t = 0$, there is no charge on the capacitor or current in the circuit. The battery immediately begins charging the capacitor, but because there is little charge on the capacitor, the voltage across its terminals is small. The remainder of the voltage (making up V which is constant for all time) is across the inductor. This is consistent with the fact that concavity is relatively large here and voltage across the inductor is proportional to the second derivative of Q, and the second derivative is positive. As time approaches $\pi\sqrt{LC}/2$, charge on the capacitor approaches CV, what would normally be its capacity if the inductor were not a part of the circuit. At time $t = \pi\sqrt{LC}/2$, voltage across the capacitor is V, and that across the inductor vanishes. There is a point of inflection at which $d^2Q/dt^2 = 0$. Current is now at a maximum in the circuit and the capacitor continues to accumulate charge. The voltage across the capacitor exceeds V, and therefore that across the inductor is negative. This agrees with the fact that the curve is concave downward and therefore $d^2Q/dt^2 < 0$. At $t = \pi\sqrt{LC}$, charge has reached a maximum value of $2CV$, voltage across the capacitor is $2V$, and that across the inductor is $-V$. Charge now begins to flow in the reverse direction and current (slope) is negative. The capacitor discharges completely at $t = 2\pi \sqrt{LC}$, and the cycle repeats.

EXAMPLE 4.34

The emf device in the RC-circuit of Figure 4.82 produces a constant voltage of V volts. If the switch is closed at time $t = 0$, and then opened again at $t = t_0$, charge on the capacitor must satisfy

$$R\frac{dQ}{dt} + \frac{Q}{C} = V[1 - H(t - t_0)], \quad t > 0,$$

where $H(t - t_0)$ is the Heaviside unit step function of Section 2.4.

(a) Show that

$$Q(t) = CV[1 - e^{-t/(RC)}] - CV[1 - e^{-(t-t_0)/(RC)}]H(t - t_0)$$

satisfies this equation for all $t \neq t_0$.

(b) Draw a graph of $Q(t)$. What is the initial charge on the capacitor?

(c) Are charge on the capacitor and current in the circuit continuous?

FIGURE 4.82

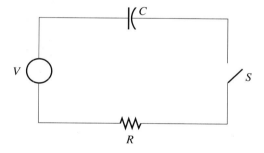

SOLUTION

(a) Since the derivative of the Heaviside function is zero for every $t \neq t_0$,

$$\frac{dQ}{dt} = CV\left(\frac{1}{RC}\right)e^{-t/(RC)} - CV\left(\frac{1}{RC}\right)e^{-(t-t_0)/(RC)}H(t - t_0)$$

$$= \frac{V}{R}\left[e^{-t/(RC)} - e^{-(t-t_0)/(RC)}H(t - t_0)\right].$$

Hence,

$$R\frac{dQ}{dt} + \frac{Q}{C} = V\left[e^{-t/(RC)} - e^{-(t-t_0)/(RC)}H(t - t_0)\right] + V\left[1 - e^{-t/(RC)}\right]$$

$$- V\left[1 - e^{-(t-t_0)/(RC)}\right]H(t - t_0)$$

$$= V[1 - H(t - t_0)].$$

(b) To graph $Q(t)$, we write it in the form

$$Q(t) = CV \begin{cases} 1 - e^{-t/(RC)}, & 0 \le t < t_0 \\ e^{-(t-t_0)/(RC)} - e^{-t/(RC)}, & t > t_0 \end{cases}$$

$$= CV \begin{cases} 1 - e^{-t/(RC)}, & 0 \le t < t_0 \\ -\left[1 - e^{t_0/(RC)}\right]e^{-t/(RC)}, & t > t_0 \end{cases}.$$

For $0 \le t < t_0$, we first draw $e^{-t/(RC)}$ as in Figure 4.83a, turn it upside down, shift it vertically one unit, and change the scale on the vertical axis (Figure 4.83b). As $t \to t_0^-$, the graph approaches $CV[1 - e^{-t_0/(RC)}]$.

For $t > t_0$, the graph declines exponentially. As $t \to t_0^+$, it approaches $CV[1 - e^{-t_0/(RC)}]$, and it is asymptotic to the t-axis (Figure 4.83c). Combining Figures 4.83b and c gives the final graph in Figure 4.83d. The capacitor has no initial charge.

(c) If we define the value of the charge on the capacitor at t_0 to be $CV[1 - e^{-t_0/(RC)}]$, it is continuous. Since current in the circuit is the slope of the curve, it is undefined at t_0 and therefore current is discontinuous at time t_0. It suddenly reverses direction when the emf device is disconnected.

FIGURE 4.83a

FIGURE 4.83b

FIGURE 4.83c

FIGURE 4.83d

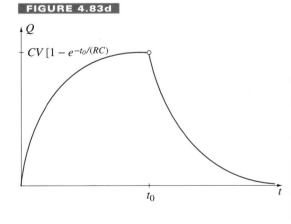

✳✳ **1.** If the switch in the *RC*-circuit below is closed at time $t = 0$, equation 4.15b for charge on the capacitor becomes

$$R\frac{dQ}{dt} + \frac{Q}{C} = V, \quad t > 0.$$

(a) Verify that when V is constant, the function

$$Q(t) = De^{-t/(RC)} + CV$$

satisfies the equation for any constant D.

(b) If Q_0 is the charge on the capacitor when the switch is closed, show that

$$Q(t) = CV[1 - e^{-t/(RC)}] + Q_0 e^{-t/(RC)}.$$

(c) Draw a graph of the function in (b) when $Q_0 = 0$.

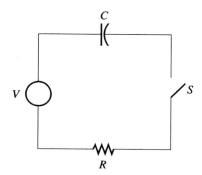

✳✳ **2.** If the switch in the *LR*-circuit below is closed at time $t = 0$, equation 4.15a for current in the circuit becomes

$$L\frac{di}{dt} + Ri = V, \quad t > 0.$$

(a) Verify that when V is constant, the function

$$i(t) = De^{-Rt/L} + V/R$$

satisfies the equation for any constant D.

(b) Using the fact that $i(0) = 0$, determine D, and draw a graph of $i(t)$.

** **3.** If the switch in the LC-circuit below is closed at time $t = 0$, equation 4.15b for charge on the capacitor becomes

$$L\frac{d^2Q}{dt^2} + \frac{Q}{C} = V, \quad t > 0.$$

(a) Verify that when $V = A \sin \omega t$, where A and ω are constants, the function

$$Q(t) = D \cos \frac{t}{\sqrt{LC}} + E \sin \frac{t}{\sqrt{LC}} - \frac{A/\omega}{\omega L - \frac{1}{\omega C}} \sin \omega t$$

satisfies the equation for any constants D and E.

(b) If Q_0 is the charge on the capacitor when the switch is closed, show that

$$Q(t) = Q_0 \cos \frac{t}{\sqrt{LC}} - \frac{A\sqrt{LC}}{\omega L - \frac{1}{\omega C}} \sin \frac{t}{\sqrt{LC}} - \frac{A/\omega}{\omega L - \frac{1}{\omega C}} \sin \omega t.$$

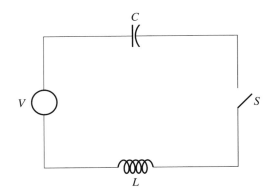

** **4.** (a) If $V = A \sin \omega t$, where A and ω are constants, in Exercise 1, and there is no charge on the capacitor at time $t = 0$, verify that

$$Q(t) = \frac{CA}{1 + \omega^2 R^2 C^2} \sin \omega t + \frac{\omega R C^2 A}{1 + \omega^2 R^2 C^2}[e^{-t/(RC)} - \cos \omega t]$$

satisfies $RQ' + Q/C = V$.

(b) Show that $Q(t)$ can be expressed in the form

$$Q(t) = \frac{\omega R C^2 A}{1 + \omega^2 R^2 C^2}e^{-t/(RC)} + \frac{A/\omega}{Z} \cos (\omega t - \phi)$$

where

$$Z = \sqrt{R^2 + \frac{1}{\omega^2 C^2}} \quad \text{and} \quad \tan \phi = -\frac{1}{\omega C R}.$$

** **5.** (a) If $V = A \sin \omega t$, where A and ω are constants, in Exercise 2, verify that

$$i(t) = \frac{\omega L A}{R^2 + \omega^2 L^2}(e^{-Rt/L} - \cos \omega t) + \frac{RA}{R^2 + \omega^2 L^2} \sin \omega t$$

satisfies $Li' + Ri = V$.

(b) Show that $i(t)$ can be expressed in the form

$$i(t) = \frac{\omega L A}{R^2 + \omega^2 L^2} e^{-Rt/L} + \frac{A}{Z} \sin(\omega t - \phi)$$

where

$$Z = \sqrt{R^2 + \omega^2 L^2} \quad \text{and} \quad \tan \phi = \frac{\omega L}{R}.$$

** **6.** The current in the RCL-circuit below must satisfy equation 4.15c. If $V = A \sin \omega t$, where A and ω are constants, show that a solution is

$$i(t) = e^{-Rt/(2L)}(D \cos vt + E \sin vt) + \frac{A}{Z} \sin(\omega t - \phi)$$

where D and E are any constants whatsoever, and

$$v = \sqrt{\frac{1}{LC} - \frac{R^2}{4L^2}}, \qquad Z = \sqrt{R^2 + \left(\omega L - \frac{1}{\omega C}\right)^2}, \qquad \tan \phi = \frac{\omega L - \frac{1}{\omega C}}{R}.$$

** **7.** An inductor L, a resistor R, and a capacitor C are connected with a generator, producing an oscillatory voltage $V = V_0 \cos \omega t$, for $t \geq 0$ (figure on page 330). If L, C, R, V_0, and ω are all constants, steady-state current $i(t)$ in the circuit is given by

$$i(t) = \frac{V_0}{R^2 + \left(\omega L - \frac{1}{\omega C}\right)^2} \left[R \cos \omega t + \left(\omega L - \frac{1}{\omega C}\right) \sin \omega t \right].$$

Find the value of ω that makes the amplitude of the current a maximum. Do this in two ways:

(a) Express $i(t)$ in the form

$$i(t) = \frac{V_0}{Z} \cos(\omega t - \phi)$$

in which case amplitude V_0/Z must be maximized.

(b) Find critical points for $i(t)$ as given.

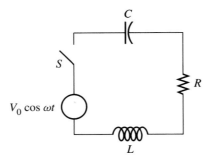

4.10 Indeterminate Forms and L'Hôpital's Rule

Derivatives are instantaneous rates of change, defined as limits of average rates of change. As a result, it was necessary to discuss limits in Chapter 2 prior to the introduction of derivatives in Section 3.1. Now that we have derivatives, it may seem quite surprising that they can be used to evaluate many limits.

The Indeterminate Form $0/0$

If we let x approach zero in numerator and denominator of the limit

$$\lim_{x \to 0^+} \frac{\sqrt{1+x} - 1}{\sqrt{x}}, \qquad (4.16)$$

we find that

$$\lim_{x \to 0^+} \left(\sqrt{1+x} - 1\right) = 0 \quad \text{and} \quad \lim_{x \to 0^+} \sqrt{x} = 0.$$

We say that limit 4.16 is of the **indeterminate form** $0/0$. Likewise, the limit

$$\lim_{x \to 2} \frac{x^3 - x^2 - 8x + 12}{x^2 - 4x + 4} \qquad (4.17)$$

is of the indeterminate form $0/0$ since both numerator and denominator approach zero as x approaches 2.

In Section 2.1 we evaluated limit 4.16 by rationalizing the numerator (see Example 2.6, page 100). Factoring numerator and denominator in limit 4.17 gives

$$\lim_{x \to 2} \frac{x^3 - x^2 - 8x + 12}{x^2 - 4x + 4} = \lim_{x \to 2} \frac{(x - 2)^2(x + 3)}{(x - 2)^2} = \lim_{x \to 2} (x + 3) = 5.$$

These two examples illustrate the "trickery" to which we resorted in Chapter 2 in order to evaluate limits. With Cauchy's generalized mean value theorem from Section 3.12, however, we can prove a result called L'Hôpital's rule, which makes evaluation of many limits of the indeterminate form $0/0$ quite simple.

THEOREM 4.4   (L'Hôpital's Rule)

Let $f(x)$ and $g(x)$ be functions that are differentiable on an open interval I except possibly at the point $x = a$ in I. Suppose further that $g'(x) \neq 0$ in I, except possibly at a. If $\lim_{x \to a} f(x) = 0$ and $\lim_{x \to a} g(x) = 0$, and if

$$\lim_{x \to a} \frac{f'(x)}{g'(x)} = L,$$

then

$$\lim_{x \to a} \frac{f(x)}{g(x)} = L.$$

PROOF   We define two functions $F(x)$ and $G(x)$ that are identical to $f(x)$ and $g(x)$ on I but have value zero at $x = a$:

$$F(x) = \begin{cases} f(x) & x \neq a \\ 0 & x = a, \end{cases} \qquad G(x) = \begin{cases} g(x) & x \neq a \\ 0 & x = a. \end{cases}$$

Since $F'(x) = f'(x)$ and $G'(x) = g'(x)$ for x in I except possibly at $x = a$, $F(x)$ and $G(x)$ are therefore differentiable on I except possibly at $x = a$. In addition, differentiability of a function implies continuity of the function (Theorem 3.6), so that $F(x)$ and $G(x)$ must certainly be continuous on I except possibly at $x = a$. But

$$\lim_{x \to a} F(x) = \lim_{x \to a} f(x) = 0 = F(a),$$

and the same is true for $G(x)$; hence, $F(x)$ and $G(x)$ are continuous for all x in I. Consequently, $F(x)$ and $G(x)$ are identical to $f(x)$ and $g(x)$ in every respect, except that they have been assigned a value at $x = a$ to guarantee their continuity there. This extra condition permits us to apply Cauchy's generalized mean value theorem to $F(x)$ and $G(x)$ on the interval between a and x, so long as x is in I. There exists a number c between a and x such that

$$\frac{F(x) - F(a)}{G(x) - G(a)} = \frac{F'(c)}{G'(c)},$$

or, since $F(a) = G(a) = 0$,

$$\frac{F(x)}{G(x)} = \frac{F'(c)}{G'(c)}.$$

Since x and c are points in I, we can also write that $F(x) = f(x)$, $F'(c) = f'(c)$, $G(x) = g(x)$, $G'(c) = g'(c)$, and therefore

$$\frac{f(x)}{g(x)} = \frac{F(x)}{G(x)} = \frac{F'(c)}{G'(c)} = \frac{f'(c)}{g'(c)}.$$

If we now let x approach a, then c must also approach a since it is always between a and x. Consequently,

$$\lim_{x \to a} \frac{f(x)}{g(x)} = \lim_{c \to a} \frac{f'(c)}{g'(c)} = \lim_{x \to a} \frac{f'(x)}{g'(x)},$$

and if

$$L = \lim_{x \to a} \frac{f'(x)}{g'(x)},$$

it follows that

$$\lim_{x \to a} \frac{f(x)}{g(x)} = L. \qquad \blacksquare$$

This theorem is also valid if L is replaced by ∞ or $-\infty$. The only difference in the proof is to make the same change in the last sentence.

Theorem 4.4 is also valid if $x \to a$ is replaced by either a right-hand limit, $x \to a^+$, or a left-hand limit, $x \to a^-$. The only difference in these cases is that interval I is replaced by open intervals $a < x < b$ and $b < x < a$ respectively, and the proofs are almost identical. In addition, the following theorem indicates that $x \to a$ can be replaced by $x \to \infty$ (or $x \to -\infty$).

THEOREM 4.5 (L'Hôpital's Rule)

Let $f(x)$ and $g(x)$ be functions that are differentiable for some interval $x > b > 0$, and suppose that $g'(x) \neq 0$ on this interval. If $\lim_{x \to \infty} f(x) = 0$ and $\lim_{x \to \infty} g(x) = 0$, and if

$$\lim_{x \to \infty} \frac{f'(x)}{g'(x)} = L \quad (\text{or} \pm\infty),$$

then

$$\lim_{x \to \infty} \frac{f(x)}{g(x)} = L \quad (\text{or} \pm\infty).$$

In other words, L'Hôpital's rule applies to any type of limit that yields the indeterminate form $0/0$ (be it $x \to a$, $x \to a^+$, $x \to a^-$, $x \to \infty$, or $x \to -\infty$). A common error when using L'Hôpital's rule is to differentiate $f(x)/g(x)$ with the quotient rule and then take the limit of the resulting derivative. L'Hôpital's rule calls for the limit of $f'(x)/g'(x)$; $f(x)$ and $g(x)$ are differentiated separately.

If we use L'Hôpital's rule on limit 4.16, we find

$$\lim_{x \to 0^+} \frac{\sqrt{1+x} - 1}{\sqrt{x}} = \lim_{x \to 0^+} \frac{\dfrac{1}{2\sqrt{1+x}}}{\dfrac{1}{2\sqrt{x}}} = \lim_{x \to 0^+} \frac{\sqrt{x}}{\sqrt{1+x}} = 0.$$

For limit 4.17, we have

$$\lim_{x \to 2} \frac{x^3 - x^2 - 8x + 12}{x^2 - 4x + 4} = \lim_{x \to 2} \frac{3x^2 - 2x - 8}{2x - 4},$$

which is still a limit of the indeterminate form $0/0$. Note that this is a conditional equation; that is, it says that the limit on the left is equal to the limit on the right, provided the limit on the right exists. If we apply L'Hôpital's rule a second time, to the limit on the right, we obtain

$$\lim_{x \to 2} \frac{x^3 - x^2 - 8x + 12}{x^2 - 4x + 4} = \lim_{x \to 2} \frac{6x - 2}{2} = 5.$$

EXAMPLE 4.35

Evaluate the following limits:

$$\text{(a)} \quad \lim_{x \to -3} \frac{3 + x}{\sqrt{3} - \sqrt{-x}} \qquad \text{(b)} \quad \lim_{x \to 0} \frac{\tan x}{x}$$

$$\text{(c)} \quad \lim_{x \to 4} \frac{x - 4}{x^2 - 8x + 16} \qquad \text{(d)} \quad \lim_{x \to 4} \frac{x^3 - 4x^2 + 9x - 36}{x^2 + 5}$$

SOLUTION

(a) Since we have the indeterminate form $0/0$, we use L'Hôpital's rule to write

$$\lim_{x \to -3} \frac{3 + x}{\sqrt{3} - \sqrt{-x}} = \lim_{x \to -3} \frac{1}{\dfrac{1}{2\sqrt{-x}}} = \lim_{x \to -3} 2\sqrt{-x} = 2\sqrt{3}.$$

(b) If we use L'Hôpital's rule, we have

$$\lim_{x \to 0} \frac{\tan x}{x} = \lim_{x \to 0} \frac{\sec^2 x}{1} = 1.$$

(c) By L'Hôpital's rule,

$$\lim_{x \to 4} \frac{x - 4}{x^2 - 8x + 16} = \lim_{x \to 4} \frac{1}{2x - 8}.$$

Since

$$\lim_{x \to 4^+} \frac{1}{2x - 8} = \infty \qquad \text{and} \qquad \lim_{x \to 4^-} \frac{1}{2x - 8} = -\infty,$$

we conclude that

$$\lim_{x \to 4^+} \frac{x - 4}{x^2 - 8x + 16} = \infty \qquad \text{and} \qquad \lim_{x \to 4^-} \frac{x - 4}{x^2 - 8x + 16} = -\infty.$$

(d) This limit is not of the indeterminate form $0/0$ since $\lim_{x \to 4} (x^2 + 5) = 21$; thus we cannot use L'Hôpital's rule. Since $\lim_{x \to 4} (x^3 - 4x^2 + 9x - 36) = 0$,

$$\lim_{x \to 4} \frac{x^3 - 4x^2 + 9x - 36}{x^2 + 5} = 0.$$

Had we used L'Hôpital's rule in part (d) of this example, we would have obtained an incorrect answer:

$$\lim_{x \to 4} \frac{x^3 - 4x^2 + 9x - 36}{x^2 + 5} = \lim_{x \to 4} \frac{3x^2 - 8x + 9}{2x} = \frac{25}{8}.$$

In other words, L'Hôpital's rule is not to be used indiscriminately; *it must be used only on the indeterminate forms for which it is designed.*

The Indeterminate Form ∞/∞

The limit

$$\lim_{x \to \infty} \frac{1 + \sqrt{x - 1}}{2x + 5}$$

is said to be of the **indeterminate form** ∞/∞ since numerator and denominator become increasingly large as $x \to \infty$. Theorems 4.4 and 4.5 for L'Hôpital's rule can be adapted to this indeterminate form also; hence we calculate that

$$\lim_{x \to \infty} \frac{1 + \sqrt{x - 1}}{2x + 5} = \lim_{x \to \infty} \frac{\dfrac{1}{2\sqrt{x - 1}}}{2} = \lim_{x \to \infty} \frac{1}{4\sqrt{x - 1}} = 0.$$

EXAMPLE 4.36

Evaluate the following limits, if they exist:

$$\text{(a)} \quad \lim_{x \to \infty} \frac{x^2}{e^x} \qquad \text{(b)} \quad \lim_{x \to -\infty} \frac{\sqrt{2x^2 + 3x + 2}}{1 - x}$$

SOLUTION

(a) Since this limit exhibits the indeterminate form ∞/∞, we use L'Hôpital's rule to write

$$\lim_{x \to \infty} \frac{x^2}{e^x} = \lim_{x \to \infty} \frac{2x}{e^x}.$$

Since this limit is still of the form ∞/∞, we use L'Hôpital's rule again,

$$\lim_{x \to \infty} \frac{x^2}{e^x} = \lim_{x \to \infty} \frac{2}{e^x} = 0.$$

(b) By L'Hôpital's rule,

$$\lim_{x \to -\infty} \frac{\sqrt{2x^2 + 3x + 2}}{1 - x} = \lim_{x \to -\infty} \frac{\dfrac{4x + 3}{2\sqrt{2x^2 + 3x + 2}}}{-1} = \lim_{x \to -\infty} \frac{-(4x + 3)}{2\sqrt{2x^2 + 3x + 2}}.$$

This limit is also of the indeterminate form ∞/∞. Further applications of L'Hôpital's rule do not lead to a simpler form for the limit. Thus, L'Hôpital's rule does not prove advantageous on this limit. It is better to divide numerator and denominator by x,

$$\lim_{x \to -\infty} \frac{\sqrt{2x^2 + 3x + 2}}{1 - x} = \lim_{x \to -\infty} \frac{\dfrac{\sqrt{2x^2 + 3x + 2}}{x}}{\dfrac{1}{x} - 1} = \lim_{x \to -\infty} \frac{-\sqrt{2 + \dfrac{3}{x} + \dfrac{2}{x^2}}}{\dfrac{1}{x} - 1}$$

$$= \sqrt{2}.$$

EXAMPLE 4.37

Show that L'Hôpital's rule cannot be used to evaluate

$$\lim_{x \to \infty} \frac{x - \cos x}{x}.$$

What is the value of the limit?

SOLUTION   The limit is of the indeterminate form ∞/∞. If we apply L'Hôpital's rule we obtain

$$\lim_{x \to \infty} \frac{x - \cos x}{x} = \lim_{x \to \infty} \frac{1 + \sin x}{1}.$$

But this limit does not exist, and therefore L'Hôpital's rule has failed. But we do not need the rule since division of numerator and denominator by x gives

$$\lim_{x \to \infty} \frac{x - \cos x}{x} = \lim_{x \to \infty} \left(1 - \frac{\cos x}{x}\right) = 1.$$

The Indeterminate Form $0 \cdot \infty$

The limits

$$\lim_{x \to \infty} x e^{-2x} \qquad \text{and} \qquad \lim_{x \to 0^+} x^2 \ln x$$

are said to be of the **indeterminate form** $0 \cdot \infty$. L'Hôpital's rule can again be used if we first rearrange the limits into one of the forms $0/0$ or ∞/∞:

$$\lim_{x \to \infty} x e^{-2x} = \lim_{x \to \infty} \frac{x}{e^{2x}} = \lim_{x \to \infty} \frac{1}{2e^{2x}} = 0;$$

$$\lim_{x \to 0^+} x^2 \ln x = \lim_{x \to 0^+} \frac{\ln x}{\dfrac{1}{x^2}} = \lim_{x \to 0^+} \frac{\dfrac{1}{x}}{\dfrac{-2}{x^3}} = \lim_{x \to 0^+} \left(-\frac{x^2}{2}\right) = 0.$$

Note that, had we converted the second limit into the $0/0$ form, we would have had

$$\lim_{x \to 0^+} x^2 \ln x = \lim_{x \to 0^+} \frac{x^2}{\dfrac{1}{\ln x}} = \lim_{x \to 0^+} \frac{2x}{\dfrac{-1}{x (\ln x)^2}} = \lim_{x \to 0^+} \left(-2x^2\right) (\ln x)^2.$$

Although this is correct, the limit on the right is more difficult to evaluate than the original. In other words, we must be judicious in converting a limit from the $0 \cdot \infty$ indeterminate form to either $0/0$ or ∞/∞.

EXAMPLE 4.38

Evaluate the following limits if they exist:

(a) $\displaystyle \lim_{x \to \pi/2} (x - \pi/2) \sec x$ (b) $\displaystyle \lim_{x \to 0^+} x e^{1/x}$

SOLUTION

(a) $\displaystyle\lim_{x\to\pi/2}(x-\pi/2)\sec x = \lim_{x\to\pi/2}\frac{x-\pi/2}{\cos x} = \lim_{x\to\pi/2}\frac{1}{-\sin x} = -1$

(b) $\displaystyle\lim_{x\to0^+}xe^{1/x} = \lim_{x\to0^+}\frac{e^{1/x}}{1/x} = \lim_{x\to0^+}\frac{e^{1/x}(-1/x^2)}{-1/x^2} = \lim_{x\to0^+}e^{1/x} = \infty.$

The Indeterminate Forms 0^0, 1^∞, ∞^0, and $\infty - \infty$

Various other indeterminate forms arise in the evaluation of limits, and many of these can be reduced to the $0/0$ and ∞/∞ forms by introducing logarithms. In particular, the limits

$$\lim_{x\to0^+} x^x, \quad \lim_{x\to\infty}\left(1+\frac{1}{x}\right)^{x^2}, \quad \lim_{x\to\pi/2^-}(\sec x)^{\cos x}, \quad\text{and}\quad \lim_{x\to\pi/2}(\sec x - \tan x) \qquad (4.18)$$

are said to display the indeterminate forms 0^0, 1^∞, ∞^0, and $\infty - \infty$, respectively. To evaluate $\lim_{x\to0^+} x^x$, we set

$$L = \lim_{x\to0^+} x^x$$

and take natural logarithms of both sides,

$$\ln L = \ln\left(\lim_{x\to0^+} x^x\right).$$

As the logarithm function is continuous, we may interchange the limit and logarithm operations (see Theorem 2.3):

$$\ln L = \lim_{x\to0^+}(\ln x^x) = \lim_{x\to0^+} x\ln x = \lim_{x\to0^+}\frac{\ln x}{\dfrac{1}{x}}.$$

We are now in a position to use L'Hôpital's rule:

$$\ln L = \lim_{x\to0^+}\frac{\dfrac{1}{x}}{\dfrac{-1}{x^2}} = \lim_{x\to0^+}(-x) = 0.$$

Exponentiation of both sides of $\ln L = 0$ now gives $L = e^0 = 1$; that is,

$$\lim_{x\to0^+} x^x = 1.$$

For the second limit in 4.18 we again set

$$L = \lim_{x\to\infty}\left(1+\frac{1}{x}\right)^{x^2}$$

and take natural logarithms:

$$\ln L = \ln \left[\lim_{x \to \infty} \left(1 + \frac{1}{x} \right)^{x^2} \right] = \lim_{x \to \infty} \left[\ln \left(1 + \frac{1}{x} \right)^{x^2} \right]$$

$$= \lim_{x \to \infty} \left[x^2 \ln \left(1 + \frac{1}{x} \right) \right] = \lim_{x \to \infty} \left[\frac{\ln \left(\dfrac{x+1}{x} \right)}{\dfrac{1}{x^2}} \right].$$

By L'Hôpital's rule, we have

$$\ln L = \lim_{x \to \infty} \left[\frac{\dfrac{x}{x+1} \left(\dfrac{-1}{x^2} \right)}{\dfrac{-2}{x^3}} \right] = \lim_{x \to \infty} \frac{x^2}{2(x+1)} = \infty.$$

Consequently,

$$L = \lim_{x \to \infty} \left(1 + \frac{1}{x} \right)^{x^2} = \infty.$$

In the third limit of 4.18 we set

$$L = \lim_{x \to \pi/2^-} (\sec x)^{\cos x},$$

in which case

$$\ln L = \ln \left[\lim_{x \to \pi/2^-} (\sec x)^{\cos x} \right] = \lim_{x \to \pi/2^-} \left[\cos x \ln (\sec x) \right]$$

$$= \lim_{x \to \pi/2^-} \left[\frac{\ln (\sec x)}{\sec x} \right] = \lim_{x \to \pi/2^-} \left[\frac{\dfrac{1}{\sec x} \sec x \tan x}{\sec x \tan x} \right]$$

$$= \lim_{x \to \pi/2^-} \cos x = 0.$$

Thus,

$$L = \lim_{x \to \pi/2^-} (\sec x)^{\cos x} = e^0 = 1.$$

Finally, the last limit in 4.18 is evaluated by rewriting it in the 0/0 form,

$$\lim_{x \to \pi/2} (\sec x - \tan x) = \lim_{x \to \pi/2} \left(\frac{1 - \sin x}{\cos x} \right) = \lim_{x \to \pi/2} \frac{-\cos x}{-\sin x} = 0.$$

EXAMPLE 4.39

Plot a graph of the function $f(x) = x^2 \ln x$. Find limits of $f(x)$ and $f'(x)$ as $x \to 0^+$. Where is the point of inflection on the graph?

SOLUTION The plot in Figure 4.84 suggests that $f(x)$ and $f'(x)$ approach 0 as $x \to 0^+$. To confirm this we use L'Hôpital's rule to calculate

$$\lim_{x \to 0^+} f(x) = \lim_{x \to 0^+} x^2 \ln x = \lim_{x \to 0^+} \frac{\ln x}{\dfrac{1}{x^2}} = \lim_{x \to 0^+} \frac{\dfrac{1}{x}}{-\dfrac{2}{x^3}} = \lim_{x \to 0^+} \left(\frac{-x^2}{2} \right) = 0^-,$$

and

$$\lim_{x \to 0^+} f'(x) = \lim_{x \to 0^+} \left(2x \ln x + \frac{x^2}{x} \right) = 2 \lim_{x \to 0^+} \frac{\ln x}{\dfrac{1}{x}}$$

$$= 2 \lim_{x \to 0^+} \frac{\dfrac{1}{x}}{-\dfrac{1}{x^2}} = 2 \lim_{x \to 0^+} (-x) = 0^-.$$

For the point of inflection between $x = 0$ and $x = 1/2$, we solve

$$0 = f''(x) = 2 \ln x + \frac{2x}{x} + 1 = 2 \ln x + 3.$$

The only solution is $x = e^{-3/2}$. Since $f''(x)$ changes sign as x passes through $e^{-3/2}$, there is a point of inflection at $(e^{-3/2}, -3e^{-3}/2)$.

FIGURE 4.84

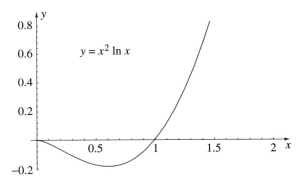

$y = x^2 \ln x$

EXERCISES 4.10

In Exercises 1–42 evaluate the limit if it exists.

* **1.** $\displaystyle \lim_{x \to 0} \frac{x^2 + 3x}{x^3 + 5x^2}$

* **2.** $\displaystyle \lim_{x \to 3} \frac{x^2 - 9}{x - 3}$

* **3.** $\displaystyle \lim_{x \to -\infty} \frac{x^3 + 3x - 2}{x^2 + 5x + 1}$

* **4.** $\displaystyle \lim_{x \to \infty} \frac{2x^2 + 3x}{5x^3 + 4}$

* **5.** $\displaystyle\lim_{x\to 5}\frac{x^2-10x+25}{x^3-125}$

* **6.** $\displaystyle\lim_{x\to 1}\frac{1}{(x-1)^2}$

* **7.** $\displaystyle\lim_{x\to\infty}\frac{\sqrt{x^2+1}}{2x+5}$

* **8.** $\displaystyle\lim_{x\to-\infty}\frac{\sin x}{x}$

* **9.** $\displaystyle\lim_{x\to\infty}\frac{\sin(2/x)}{\sin(1/x)}$

* **10.** $\displaystyle\lim_{x\to\pi/2}\frac{\cos x}{(x-\pi/2)^2}$

* **11.** $\displaystyle\lim_{x\to 1^+}\frac{(1-1/x)^3}{\sqrt{x-1}}$

* **12.** $\displaystyle\lim_{x\to\infty}\frac{\sin(1/x)}{1/x^2}$

* **13.** $\displaystyle\lim_{x\to 9^-}\frac{\sqrt{x}-3}{\sqrt{9-x}}$

* **14.** $\displaystyle\lim_{x\to 0}\frac{\sqrt{5+x}-\sqrt{5-x}}{x}$

* **15.** $\displaystyle\lim_{x\to 0}\frac{x-\sin x}{x^3}$

* **16.** $\displaystyle\lim_{x\to a}\frac{x^n-a^n}{x-a}$

* **17.** $\displaystyle\lim_{x\to 0}\frac{(1-\cos x)^2}{3x^2}$

* **18.** $\displaystyle\lim_{x\to 0}\frac{\tan x}{x}$

* **19.** $\displaystyle\lim_{x\to 0}\frac{\sin 3x}{\tan 2x}$

* **20.** $\displaystyle\lim_{x\to 1}\frac{\left(1-\sqrt{2-x}\right)^{3/2}}{x-1}$

** **21.** $\displaystyle\lim_{x\to 0}\frac{\sqrt{x+1}-\sqrt{2x+1}}{\sqrt{3x+4}-\sqrt{2x+4}}$

** **22.** $\displaystyle\lim_{x\to 0}\frac{(1-\cos x)^2}{3x^4}$

** **23.** $\displaystyle\lim_{x\to\infty}x\sin\left(\frac{1}{x}\right)$

** **24.** $\displaystyle\lim_{x\to 2}\frac{(x-2)^{10}}{\left(\sqrt{x}-\sqrt{2}\right)^{10}}$

** **25.** $\displaystyle\lim_{x\to 0}\left(\frac{4}{x^2}-\frac{2}{1-\cos x}\right)$

** **26.** $\displaystyle\lim_{x\to\infty}xe^x$

** **27.** $\displaystyle\lim_{x\to\infty}x^2e^{-4x}$

** **28.** $\displaystyle\lim_{x\to-\infty}x\sin\left(\frac{4}{x}\right)$

** **29.** $\displaystyle\lim_{x\to 0}x\cot x$

** **30.** $\displaystyle\lim_{x\to 0}\csc x\,(1-\cos x)$

** **31.** $\displaystyle\lim_{x\to 0^+}(\sin x)^x$

** **32.** $\displaystyle\lim_{x\to 0^+}x^{\sin x}$

** **33.** $\displaystyle\lim_{x\to\infty}\left(\frac{x+5}{x+3}\right)^x$

** **34.** $\displaystyle\lim_{x\to 0}(1+x)^{\cot x}$

** **35.** $\displaystyle\lim_{x\to\infty}x^{1/x}$

** **36.** $\displaystyle\lim_{x\to 0^+}|\ln x|^{\sin x}$

** **37.** $\displaystyle\lim_{x\to 0^+}xe^{1/x}$

** **38.** $\displaystyle\lim_{x\to 0}(\tan x-\csc x)$

** **39.** $\displaystyle\lim_{x\to 0}(\csc x-\cot x)$

** **40.** $\displaystyle\lim_{x\to 1}\left(\frac{x}{\ln x}-\frac{1}{x\ln x}\right)$

** **41.** $\displaystyle\lim_{x\to 1}\left(\frac{x}{x-1}-\frac{1}{\ln x}\right)$

** **42.** $\displaystyle\lim_{x\to 0}\left(\frac{1}{x^2}-\frac{1}{\sin^2 x}\right)$

In Exercises 43–54 draw a graph of the function using the tools of this chapter. When you have finished, check with a plot.

** **43.** $f(x) = xe^{-2x}$

** **44.** $f(x) = x^2 e^{3x}$

** **45.** $f(x) = xe^{-x^2}$

** **46.** $f(x) = e^{1/x}$

** **47.** $f(x) = \dfrac{\ln x}{x}$

** **48.** $f(x) = x^2 \ln x$

** **49.** $f(x) = xe^{1/x}$

** **50.** $f(x) = \dfrac{x^2}{\ln x}$

** **51.** $f(x) = x^x, \quad x > 0$

** **52.** $f(x) = x^{10} e^{-x}$

** **53.** $f(x) = e^{-x} \ln x, \quad x > 0$

** **54.** $f(x) = 2\csc x - \cot x, \quad 0 < x < \pi/2$

** **55.** Evaluate

$$\lim_{x \to \infty} \left(\frac{x + a}{x + b} \right)^{cx}$$

for any constants a, b, and c.

** **56.** The indeterminate forms 0^0, 1^∞, and ∞^0 are often evaluated by introducing logarithms. Show that the limit

$$\lim_{x \to \infty} (x - \ln x)$$

can be evaluated by introducing exponentials.

** **57.** When an electrostatic field E is applied to a gaseous or liquid polar dielectric, a net dipole moment P per unit volume is set up, where

$$P(E) = \frac{e^E + e^{-E}}{e^E - e^{-E}} - \frac{1}{E}.$$

Show that $\lim_{E \to 0^+} P(E) = 0$.

** **58.** Planck's law for the energy density ψ of blackbody radiation states that

$$\psi = \psi(\lambda) = \frac{k\lambda^{-5}}{e^{c/\lambda} - 1},$$

where k and c are positive constants and λ is the wavelength of the radiation.

(a) Show that

$$\lim_{\lambda \to 0^+} \psi(\lambda) = 0 \quad \text{and} \quad \lim_{\lambda \to \infty} \psi(\lambda) = 0.$$

(b) Show that $\psi(\lambda)$ has one critical point which must satisfy the equation

$$(5\lambda - c)e^{c/\lambda} = 5\lambda.$$

Find the critical point accurate to seven decimals when $c = 0.001\,438\,6$.

(c) Plot a graph of the function $\psi(\lambda)$, when $c = 0.001\,438\,6$ and $k = 1$.

** **59.** The following limit arises in the calculation of the electric field intensity for a half-wave antenna:

$$\lim_{\theta \to 0} f(\theta),$$

where

$$f(\theta) = \sin\theta \left\{ \frac{\sin\left[\pi/2(\cos\theta - 1)\right]}{\cos\theta - 1} + \frac{\sin\left[\pi/2(\cos\theta + 1)\right]}{\cos\theta + 1} \right\}.$$

Evaluate this limit by first showing that $f(\theta)$ can be written in the form

$$f(\theta) = \frac{2\cos\left(\frac{\pi}{2}\cos\theta\right)}{\sin\theta},$$

and then using L'Hôpital's rule.

** **60.** Find all values of a, b, and c for which

$$\lim_{x \to 0} \frac{e^{ax} - bx - \cos(x + cx^2)}{2x^3 + 5x^2} = 5.$$

** **61.** The maximum flow rate of gas through a nozzle is governed by the function

$$f(x) = x\left(\frac{2}{x+1}\right)^{(x+1)/(x-1)}.$$

(a) Plot a graph of $f(x)$ on the interval $0 \le x \le 20$. Did you get any error messages? Should you have? Show that the function is discontinuous at $x = 1$, but $\lim_{x \to 1} f(x) = 1/e$.

(b) Plot the function on the interval $0 \le x \le 200$. Does the graph appear to have a horizontal asymptote? Show that $\lim_{x \to \infty} f(x) = 2$.

*** **62.** (a) Draw a graph of the function

$$f(x) = \begin{cases} e^{-1/x^2} & x \ne 0 \\ 0 & x = 0. \end{cases}$$

(b) Show that for every positive integer n,

$$\lim_{x \to 0} \frac{e^{-1/x^2}}{x^n} = 0.$$

(c) Prove by mathematical induction that $f^{(n)}(0) = 0$, where $f^{(n)}(0)$ is the nth derivative of $f(x)$ evaluated at $x = 0$.

4.11 Differentials

In Section 3.1 we pointed out that the notation dy/dx for the derivative of a function $y = f(x)$ should not be considered a quotient. In this section, we change this. It is not that we lied to you in Chapter 3; it's just that we were not prepared to tell you the whole story. Besides, nothing in Chapters 3 and 4 would have been any easier had we regarded dy/dx as "dy divided by dx." Beginning in Chapter 5, however, it is essential that we be able to do this, and therefore in this section we define "differentials" dx and dy so that dy/dx can be regarded as a quotient.

When we use the notation

$$\frac{dy}{dx} = \lim_{\Delta x \to 0} \frac{f(x + \Delta x) - f(x)}{\Delta x}$$

for the derivative of a function $y = f(x)$, we call Δx an increment in x. It represents a change in the value of the independent variable from some value x to another value $x + \Delta x$. This change can be positive or negative depending on whether we want $x + \Delta x$ to be larger or smaller than x. When the independent variable changes from x to $x + \Delta x$, the dependent variable changes by an amount Δy where

$$\Delta y = f(x + \Delta x) - f(x). \tag{4.19}$$

In other words, Δy is the change in y resulting from the change Δx in x. For the function in Figure 4.85, Δy is positive when Δx is positive, and Δy is negative when Δx is negative.

FIGURE 4.85a

FIGURE 4.85b

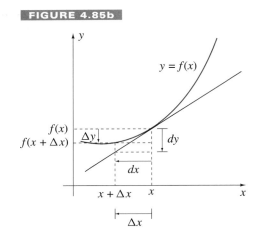

For example, when $y = x^2 - 2x$, the change Δy in y when x is changed from 3 to 3.2 is

$$\Delta y = [(3.2)^2 - 2(3.2)] - [3^2 - 2(3)] = 0.84.$$

The function increases by 0.84 when x increases from 3 to 3.2.

For purposes of integration, a topic which begins in Chapter 5 and continues in every chapter thereafter, an alternative notation for an increment in the independent variable x is more suggestive.

DEFINITION 4.7

An increment Δx in the independent variable x is denoted by

$$dx = \Delta x, \tag{4.20}$$

and when written as dx, it is called the **differential** of x.

The differential dx is synonymous with the increment Δx; it represents a change in x (in most applications a very small change). The differential of the dependent variable is *not* synonymous with Δy.

> **DEFINITION 4.8**
>
> The **differential** of $y = f(x)$, corresponding to the differential dx in x, is denoted by dy and is defined by
>
> $$dy = f'(x)\,dx. \qquad (4.21)$$

The difference between Δy and dy is most easily seen in Figure 4.85. We know that Δy is the exact change in the function $y = f(x)$ when x is changed by an amount Δx or dx. It is the difference in the height of the curve at x and $x + \Delta x = x + dx$. Now the slope of the tangent line to the graph at the point (x, y) is $f'(x)$. Definition 4.8 indicates that dy can be interpreted as the difference in the height of this tangent line at x and at $x + dx$. In other words, dy is the change in y corresponding to the change dx in x if we follow the tangent line to $y = f(x)$ at (x, y) rather than the curve itself.

Figure 4.85 also suggests that when dx is very small (close to zero), dy is approximately equal to Δy; that is,

$$dy \approx \Delta y \qquad \text{when } dx \approx 0.$$

This is illustrated in the following numerical example.

EXAMPLE 4.40

Find Δy and dy for the function $y = f(x) = \sqrt{x^2 + 1}$ when $x = 2$ and $dx = 0.1$.

SOLUTION   According to equation 4.19, the change in y as x increases from 2 to 2.1 is

$$\Delta y = f(2.1) - f(2) = \sqrt{(2.1)^2 + 1} - \sqrt{2^2 + 1} = 0.089\,87.$$

Since $f'(x) = x/\sqrt{x^2 + 1}$, the differential of y for $x = 2$ and $dx = 0.1$ is

$$dy = f'(2)(0.1) = \frac{2}{\sqrt{2^2 + 1}}(0.1) = 0.089\,44.$$

The difference between dy and Δy is, therefore, $0.000\,43$, a difference of 43 parts in 8987.

Before the invention of electronic calculators, differentials were used to approximate a function near points at which it was easily evaluated. For example, imagine trying to evaluate the function $f(x) = x^{1/3}$ at $x = 126$ without a calculator. We could use differentials to approximate $126^{1/3}$ as follows:

$$126^{1/3} = f(126) = f(125) + \Delta y \approx f(125) + dy$$

$$= 5 + f'(5)(1) = 5 + \frac{1}{3(125)^{2/3}} = 5 + \frac{1}{75} = \frac{376}{75}.$$

With the advent of the electronic calculator, problems of this type are archaic. On the other hand, differentials are indispensable when we examine changes in a function without specifying values for the independent variable. Very prominent in this context are "relative" and "percentage" changes.

When a quantity y undergoes a change Δy, then its *relative change* is defined as

$$\frac{\Delta y}{y}, \tag{4.22}$$

and its *percentage change* is given by

$$100\frac{\Delta y}{y}. \tag{4.23}$$

Sometimes relative and percentage changes are more important than actual changes. To illustrate this, consider the function $V = 4\pi r^3/3$ which represents the volume of a sphere. If the radius of the sphere is increased from 0.10 m to 0.11 m, then the change in the volume of the sphere is

$$\Delta V = \frac{4}{3}\pi(0.11)^3 - \frac{4}{3}\pi(0.10)^3 = 4.4\pi \times 10^{-4} \text{ m}^3.$$

This is not a very large quantity, but in relation to the original size of the sphere, we have a relative change of

$$\frac{\Delta V}{V} = \frac{4.4\pi \times 10^{-4}}{4\pi(0.10)^3/3} = 0.33,$$

and a percentage change of

$$100\frac{\Delta V}{V} = 33\%.$$

Suppose the same increase of 0.01 m is applied to a sphere with radius 100 m. The change in the volume is

$$\Delta V = \frac{4}{3}\pi(100.01)^3 - \frac{4}{3}\pi(100)^3 = 4.0\pi \times 10^2 \text{ m}^3.$$

This is quite a large change in volume, but the relative change is

$$\frac{\Delta V}{V} = \frac{4.0\pi \times 10^2}{4\pi(100)^3/3} = 3.0 \times 10^{-4},$$

and the percentage change is

$$100\frac{\Delta V}{V} = 0.03\%.$$

Although the change 400π in V when $r = 100$ is much larger than the change 0.00044π when $r = 0.1$, the relative and percentage changes are much smaller when $r = 100$. We see, then, that in certain cases it may be relative and percentage changes that are significant rather than actual changes.

In the above example, when $r = 100$ m, the change $dr = 0.01$ is certainly small compared to r. Therefore, we should be able to use the differential $dV = V'(r)\,dr = 4\pi r^2\,dr$ to approximate ΔV. With $r = 100$ and $dr = 0.01$, we have

$$dV = 4\pi(100)^2(0.01) = 4.0\pi \times 10^2 \text{ m}^3.$$

(To two significant figures, dV is equal to ΔV.) Thus, relative change in V is approximately equal to dV/V and percentage change $100dV/V$. So, for small changes in an independent variable, the differential of the dependent variable may be used in place of its increment in the calculation of relative and percentage changes; that is, equations 4.22 and 4.23 can be replaced by

$$\frac{dy}{y} \quad \text{and} \quad 100\frac{dy}{y}.$$

We do this in the following example.

EXAMPLE 4.41

When a pendulum swings, the frequency (number of cycles per second) of its oscillations is given by

$$f = h(l) = 2\pi\sqrt{\frac{g}{l}},$$

where l is the length of the pendulum and $g > 0$ is the acceleration due to gravity. If the length of the pendulum is increased by $\frac{1}{4}\%$, calculate the approximate percentage change in f.

SOLUTION   The approximate change in f is given by

$$df = h'(l)\,dl = 2\pi\sqrt{g}\left(-\frac{1}{2}\right)l^{-3/2}\,dl = -\pi\sqrt{g}\frac{dl}{l^{3/2}};$$

hence, the approximate percentage change in f is

$$100\frac{df}{f} = 100\left(\frac{-\pi\sqrt{g}\,dl}{l^{3/2}}\right)\left(\frac{l^{1/2}}{2\pi\sqrt{g}}\right) = -\frac{1}{2}\left(100\frac{dl}{l}\right).$$

But because l increases by $\frac{1}{4}\%$, it follows that $100(dl/l) = 1/4$, and

$$100\frac{df}{f} = -\frac{1}{8}.$$

Therefore, the frequency changes by $-\frac{1}{8}\%$, the negative sign indicating that because l increases, f decreases.

The differential dy cannot always be used as an approximation for the actual change Δy in a function $y = f(x)$. Sometimes it cannot be used even when dx is very close to zero. For example, if $f(x) = 2x^3 + 9x^2 - 24x + 6$, then

FIGURE 4.86

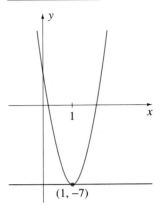

$$dy = f'(x)\,dx = (6x^2 + 18x - 24)\,dx.$$

If x is changed from 1 to 1.01, then the approximate change in y as predicted by the differential is

$$dy = (6 + 18 - 24)(0.01) = 0.$$

In fact, for any dx whatsoever, we find that $dy = 0$. Geometrically speaking, we can see why. Since $f'(1) = 0$, $x = 1$ is a critical point of $f(x)$ (Figure 4.86); therefore, dy, which is the tangent line approximation to Δy, will always be zero. We cannot use differentials to approximate function changes at critical points.

The differential for the function $y = f(x) = x^{100}$ is

$$dy = 100x^{99}\,dx.$$

If x is changed by 1% from $x = 1$ to $x = 1.01$, then $dx = 0.01$ and

$$dy = 100(1)^{99}(0.01) = 1.0.$$

The actual change in y is

$$\Delta y = (1.01)^{100} - 1^{100} = 1.7.$$

We would hardly regard this dy as a very good approximation for Δy even though the change in x is only one per cent.

This latter example begs the question."How small, in general, must dx be in order that dy be a reasonable approximation for Δy?" This is not a simple question to answer. We will discuss approximations in more detail in Section 4.12 and Chapter 11, and then be able to answer the more important question, "How good an approximation to Δy is dy?" After all, it is not much use to say that $dy \approx \Delta y$ if we cannot say to how many decimal places the approximation is accurate. Suffice it to say now that use of the differential dy to approximate Δy is to be regarded with some reservation. This is not to say that differentials are useless. We will see in Chapters 5–7 that differentials are indispensable to the topic of integration.

We make one last comment before leaving this section. If equation 4.21 is divided by differential dx, then

$$\frac{dy}{dx} = f'(x).$$

Now, the left side of this equation is the differential of y divided by the differential of x. The quotient of differentials dy and dx is equal to the derivative $f'(x)$. The entity dy/dx can henceforth be regarded either as "the derivative of y with respect to x," or "dy divided by dx," whichever is appropriate for the discussion at hand.

EXERCISES 4.11

In Exercises 1–10 find dy in terms of x and dx.

* **1.** $y = x^2 + 3x - 2$

* **2.** $y = \dfrac{x+1}{x-1}$

* **3.** $y = \sqrt{x^2 - 2x}$

* **4.** $y = \sin(x^2 + 2) - \cos x$

* **5.** $y = x^{1/3} - x^{5/3}$

* **6.** $y = x^3\sqrt{3 - 4x^2}$

* **7.** $y = x^2 \sin x$

* **8.** $y = \dfrac{x^3 - 3x^2 + 3x + 5}{x^2 - 2x + 1}$

* **9.** $y = \sqrt{1 + \sqrt{1 - x}}$

* **10.** $y = \dfrac{x^2(x - 2)}{x^3 + 5x}$

** **11.** The momentum M and kinetic energy K of a mass m moving with speed v are given by

$$M = mv \quad \text{and} \quad K = \frac{1}{2}mv^2.$$

If v is changed by 1%, what are approximate percentage changes in M and K?

** **12.** The magnitude of the gravitational force of attraction between two point masses m and M is given by

$$F = \frac{GmM}{r^2},$$

where $G > 0$ is a constant, and r is the distance between the masses. If r changes by 2%, by how much does F change approximately?

** **13.** Under adiabatic expansion, a gas obeys the law

$$PV^{7/5} = \text{a constant},$$

where P is pressure and V is volume. If the pressure is increased by 2%, find the approximate percentage change in the volume.

** **14.** The magnitude of the gravitational force of attraction between two point masses m and M is defined in Exercise 12. If the earth is considered a perfect sphere (radius 6.37×10^6 m), then this law predicts a gravitational attraction of $9.81m$ N on a mass m on its surface. Use differentials to determine the height above the surface of the earth at which the gravitational attraction decreases to $9.80m$ N.

4.12 Taylor Polynomials

FIGURE 4.87

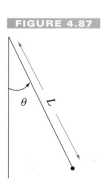

The material in this section is optional to the extent that it will not be used until Chapter 11. If not studied now, it will be necessary to do so before covering Section 11.3.

Figure 4.87 shows a pendulum consisting of a mass on the end of a string of length L. If the mass is pulled slightly to the side and released, it swings back and forth for some time to come. The position of the mass can be described by the angle θ that the string makes with the vertical.

By analyzing the forces acting on the mass, it can be shown that when air resistance is neglected, θ, as a function of time t, must satisfy the equation

$$\frac{d^2\theta}{dt^2} + \frac{g}{L}\sin\theta = 0, \tag{4.24}$$

where $g = 9.81$ is the acceleration due to gravity. Thus, $\theta(t)$ is a function whose second derivative is $-g/L$ times the sine of itself. This is an extremely difficult equation to solve; in fact, no combination of the simple functions with which we are familiar satisfies 4.24. What is sometimes done in examples like this is replace $\sin\theta$ with a simpler function $f(\theta)$ that simultaneously approximates $\sin\theta$ and for which the equation

$$\frac{d^2\theta}{dt^2} + \frac{g}{L}f(\theta) = 0 \tag{4.25}$$

is solvable. Although the solution of 4.25 only approximates the solution of 4.24, if $f(\theta)$ approximates $\sin\theta$ very closely, the solution of 4.25 may be sufficiently close to the solution of 4.24 to give real insight into the motion of the pendulum. In particular, as we shall show shortly, when θ is very small, $\sin\theta$ can be approximated by θ itself. In other words, for small oscillations of the pendulum, 4.24 can be replaced by

$$\frac{d^2\theta}{dt^2} + \frac{g}{L}\theta = 0. \tag{4.26}$$

It is straightforward to show that

$$\theta(t) = A\cos\sqrt{\frac{g}{L}}t + B\sin\sqrt{\frac{g}{L}}t$$

is a solution of 4.26 for any constants A and B whatsoever. If motion is initiated at time $t = 0$ by pulling the mass an amount θ_0 to the right and then releasing it, then $\theta(t)$ must satisfy the additional conditions

$$\theta(0) = \theta_0, \qquad \theta'(0) = 0.$$

These imply that $A = \theta_0$ and $B = 0$, and hence

$$\theta(t) = \theta_0\cos\sqrt{\frac{g}{L}}t.$$

This function yields simple harmonic motion for the pendulum, as we expect.

In this section we show how to approximate complicated functions by polynomials. The interval on which the approximation is required and the accuracy of the approximation dictate the degree of the polynomial. To begin with, we recall the Mean Value Theorem. It states that when $f(x)$ is continuous on the closed interval between c and some given value of x, and $f'(x)$ exists on the open interval between c and x, there exists a number between c and x, call it z_1, such that

$$\frac{f(x) - f(c)}{x - c} = f'(z_1),$$

or

$$f(x) = f(c) + f'(z_1)(x - c). \tag{4.27}$$

This theorem was a corollary to Cauchy's generalized mean value theorem in Section 3.12. We give a direct proof of it here, because a similar but more complicated argument leads to extensions of the Mean Value Theorem.

Consider the following function of y for fixed c and x:

$$F(y) = f(x) - f(y) - \left(\frac{x - y}{x - c}\right)[f(x) - f(c)].$$

Because f is continuous on the closed interval between c and x, $F(y)$ is continuous for y in this interval. Furthermore,

$$F'(y) = -f'(y) + \frac{1}{x - c}[f(x) - f(c)].$$

Since $f'(x)$ exists on the open interval between c and x, so also does $F'(y)$. Finally, $F(0) = 0 = F(x)$. Rolle's Theorem (Theorem 3.15 in Section 3.12) implies that there exists a number z_1 between c and x at which

$$0 = F'(z_1) = -f'(z_1) + \frac{1}{x - c}[f(x) - f(c)],$$

and this gives 4.27.

Suppose now that $f(x)$ and $f'(x)$ are continuous on the closed interval between c and x, and $f''(x)$ exists on the open interval between c and x. Consider the following function of y,

$$F(y) = f(x) - f(y) - f'(y)(x - y) - \left(\frac{x - y}{x - c}\right)^2[f(x) - f(c) - f'(c)(x - c)].$$

Because $f(x)$ and $f'(x)$ are continuous on the closed interval between c and x, $F(y)$ is continuous for y in this interval. Furthermore,

$$F'(y) = -f'(y) - f''(y)(x - y) + f'(y) + \frac{2(x - y)}{(x - c)^2}[f(x) - f(c) - f'(c)(x - c)].$$

Since $f'(x)$ is continuous on the closed interval between c and x, and $f''(x)$ exists on the open interval between c and x, it follows that $F'(y)$ exists on the open interval. Finally, $F(c) = 0 = F(x)$, and therefore Rolle's Theorem implies the existence of a number z_2 between c and x at which

$$0 = F'(z_2)$$

$$= -f'(z_2) - f''(z_2)(x - z_2) + f'(z_2) + \frac{2(x - z_2)}{(x - c)^2}[f(x) - f(c) - f'(c)(x - c)].$$

This can be rearranged into the form

$$f(x) = f(c) + f'(c)(x - c) + \frac{f''(z_2)}{2}(x - c)^2. \qquad (4.28)$$

By assuming that $f(x)$, $f'(x)$, and $f''(x)$ are continuous on the closed interval between c and x, and $f'''(x)$ exists on the open interval, it can be shown in a similar way that there exists a number z_3 between c and x such that

$$f(x) = f(c) + f'(c)(x - c) + \frac{f''(c)}{2}(x - c)^2 + \frac{f'''(z_3)}{3!}(x - c)^3. \qquad (4.29)$$

These results can be extended indefinitely if we assume that $f(x)$ has derivatives of all orders on the closed interval between c and x. For example, there exists z_4 such that

$$f(x) = f(c) + f'(c)(x - c) + \frac{f''(c)}{2}(x - c)^2 + \frac{f'''(c)}{3!}(x - c)^3$$
$$+ \frac{f''''(z_4)}{4!}(x - c)^4. \tag{4.30}$$

The last terms in equations 4.27–4.30 are called **Taylor remainders**, and when we denote them by $R_0, R_1, R_2, \ldots$, we have

$$f(x) = f(c) + R_0, \tag{4.31a}$$

$$f(x) = f(c) + f'(c)(x - c) + R_1, \tag{4.31b}$$

$$f(x) = f(c) + f'(c)(x - c) + \frac{f''(c)}{2!}(x - c)^2 + R_2, \tag{4.31c}$$

$$f(x) = f(c) + f'(c)(x - c) + \frac{f''(c)}{2!}(x - c)^2 + \frac{f'''(c)}{3!}(x-c)^3 + R_3, \tag{4.31d}$$

and, in general,

$$f(x) = f(c) + f'(c)(x - c) + \frac{f''(c)}{2!}(x - c)^2 + \cdots$$
$$+ \frac{f^{(n-1)}(c)}{(n - 1)!}(x - c)^{n-1} + R_{n-1}, \tag{4.32a}$$

where

$$R_{n-1} = \frac{f^{(n)}(z_n)}{n!}(x - c)^n, \tag{4.32b}$$

and the notation $f^{(n)}(z_n)$ stands for the nth derivative of $f(x)$ evaluated at z_n.

What happens for most functions is that the remainders get smaller and smaller as n increases, sometimes for all x, sometimes only for x in some interval around c. What this means is that if we drop the remainders from 4.31 and write

$$f(x) \approx f(c), \tag{4.33a}$$

$$f(x) \approx f(c) + f'(c)(x - c), \tag{4.33b}$$

$$f(x) \approx f(c) + f'(c)(x - c) + \frac{f''(c)}{2!}(x - c)^2, \tag{4.33c}$$

$$f(x) \approx f(c) + f'(c)(x - c) + \frac{f''(c)}{2!}(x - c)^2 + \frac{f'''(c)}{3!}(x - c)^3, \tag{4.33d}$$

etc., we have polynomial approximations for $f(x)$ that get better and better as more and more terms are included. They are called the **Taylor polynomials** of $f(x)$ about c. They provide approximate values for $f(x)$ in terms of values of $f(x)$ and its derivatives at c. We illustrate with some examples.

EXAMPLE 4.42

Find the first six Taylor polynomials for $\sin x$ about $x = 0$. Plot the polynomials and $\sin x$ to show how the approximations improve as more terms are included.

SOLUTION   The value of $f(x) = \sin x$ and its first five derivatives at $x = 0$ are

$$f(0) = 0, \; f'(0) = \cos 0 = 1, \; f''(0) = -\sin 0 = 0, \; f'''(0) = -\cos 0 = -1,$$

$$f^{(4)}(0) = \sin 0 = 0, \; f^{(5)}(0) = \cos 0 = 1.$$

If we denote the polynomials in 4.33 by $P_0(x), \; P_1(x), \ldots,$ the first six are

$$P_0(x) = f(0) = 0,$$
$$P_1(x) = f(0) + f'(0)(x - 0) = x,$$
$$P_2(x) = f(0) + f'(0)(x - 0) + \frac{f''(0)}{2!}(x - 0)^2 = x,$$
$$P_3(x) = f(0) + f'(0)(x - 0) + \frac{f''(0)}{2!}(x - 0)^2 + \frac{f'''(0)}{3!}(x - 0)^3 = x - \frac{x^3}{3!},$$
$$P_4(x) = x - \frac{x^3}{3!},$$
$$P_5(x) = x - \frac{x^3}{3!} + \frac{x^5}{5!}.$$

Because odd-ordered derivatives of $\sin x$ vanish, even-numbered polynomials are the same as their odd predecessors. We have shown polynomials $P_1(x)$, $P_3(x)$, and $P_5(x)$ along with $\sin x$ for positive values of x in Figure 4.88. We have included $P_7(x)$ and $P_9(x)$ to further emphasize the following facts. The higher the degree of the polynomial, the more closely it approximates $\sin x$ for small values of x, and the larger the interval on which the approximation is reasonable.

FIGURE 4.88

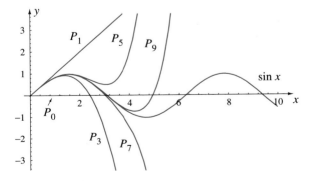

This is also illustrated in the following table.

TABLE 4.1

x	0.1	0.5	1	2
$\sin x$	0.099 83	0.479 43	0.841 47	0.91
$P_1(x)$	0.1	0.5	1	2
$P_3(x)$	0.099 83	0.479 17	0.833 33	0.666 67
$P_5(x)$	0.099 83	0.479 43	0.841 67	0.933 33
$P_7(x)$	0.099 83	0.479 43	0.841 47	0.907 94
$P_9(x)$	0.099 83	0.479 43	0.841 47	0.909 35

EXAMPLE 4.43

Find the first five Taylor polynomials of e^x about $x = 0$. Plot the polynomials and e^x.

SOLUTION   Since all derivatives of e^x at $x = 0$ are equal to 1, the first five polynomials are

$$P_0(x) = 1,$$
$$P_1(x) = 1 + x,$$
$$P_2(x) = 1 + x + \frac{x^2}{2!},$$
$$P_3(x) = 1 + x + \frac{x^2}{2!} + \frac{x^3}{3!},$$
$$P_4(x) = 1 + x + \frac{x^2}{2!} + \frac{x^3}{3!} + \frac{x^4}{4!}.$$

They are illustrated in Figure 4.89. When $x > 0$, the polynomials are always less than e^x and approach e^x as more terms are included. For $x < 0$, the polynomials are alternately higher and lower than e^x, gradually getting closer and closer.

FIGURE 4.89

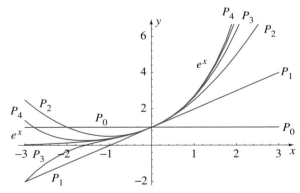

EXAMPLE 4.44

Find the first five Taylor polynomials for $\ln x$ about $x = 1$. Plot the polynomials and $\ln x$.

SOLUTION   Since

$$\ln x_{|x=1} = 0, \quad \frac{d}{dx}\ln x_{|x=1} = \frac{1}{x}\bigg|_{x=1} = 1, \quad \frac{d^2}{dx^2}\ln x_{|x=1} = -\frac{1}{x^2}\bigg|_{x=1} = -1,$$

$$\frac{d^3}{dx^3}\ln x_{|x=1} = \frac{2}{x^3}\bigg|_{x=1} = 2, \quad \frac{d^4}{dx^4}\ln x_{|x=1} = \frac{-3!}{x^3}\bigg|_{x=1} = -3!,$$

the first five polynomials are

$$P_0(x) = 0,$$

$$P_1(x) = x - 1,$$

$$P_2(x) = (x - 1) - \frac{1}{2}(x - 1)^2,$$

$$P_3(x) = (x - 1) - \frac{1}{2}(x - 1)^2 + \frac{1}{3}(x - 1)^3,$$

$$P_4(x) = (x - 1) - \frac{1}{2}(x - 1)^2 + \frac{1}{3}(x - 1)^3 - \frac{1}{4}(x - 1)^4.$$

They are plotted in Figure 4.90. For $0 < x < 1$, the polynomials are greater than $\ln x$ and approach $\ln x$ as more terms are included. For $1 < x \le 2$, the polynomials are alternately higher and lower than $\ln x$, gradually getting closer and closer.

FIGURE 4.90

FIGURE 4.91

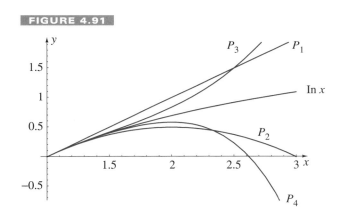

The plot of the Taylor polynomials on the interval $1 \le x \le 3$ in Figure 4.91 indicates that they do not approach $\ln x$ for $x > 2$.

To make this clearer, below is a list of the first 30 Taylor polynomials evaluated at $x = 2.1$. At first these numbers seem to be getting closer together, but then they separate. We have shown them in Figure 4.92.

0.0	1.1	0.495	0.93867	0.57264	0.89474
0.59948	0.87787	0.60992	0.87192	0.61254	0.87192
0.61038	0.87594	0.60569	0.88317	0.59599	0.89331
0.58443	0.90632	0.56994	0.92233	0.55232	0.94164
0.53123	0.96462	0.50623	0.99179	0.47675	1.02375

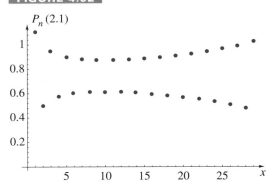

FIGURE 4.92

In Chapter 11 we show that when Taylor polynomials of a function $f(x)$ about a point $x = c$ approximate the function more and more closely in some interval, then c is always the midpoint of the interval. In other words, the interval always takes the form $c - R < x < c + R$. The end points $x = c \pm R$ are sometimes included also. In addition, the interval never extends to include a point at which $f(x)$ or any of its derivatives fails to exist. For instance, in Example 4.44 approximations are valid in the interval $0 < x \leq 2$. The interval extends one unit on either side of the point of expansion $x = 1$, and $x = 0$ is the first point at which $\ln x$ fails to be defined.

EXERCISES 4.12

In Exercises 1–12 find the first five Taylor polynomials for the function $f(x)$ about the point c. Plot the polynomials and function on the same graph.

∗ **1.** $f(x) = e^{2x}$, $c = 0$ ∗ **2.** $f(x) = \cos 3x$, $c = 0$

∗ **3.** $f(x) = 1/(1 - x)$, $c = 0$ ∗ **4.** $f(x) = \sin x$, $c = \pi/2$

∗ **5.** $f(x) = x/(1 + 2x)$, $c = 0$ ∗ **6.** $f(x) = 1/(2 - x)$, $c = 1$

∗ **7.** $f(x) = 1/(1 + 3x)^2$, $c = 0$ ∗ **8.** $f(x) = xe^{-x}$, $c = 0$

∗ **9.** $f(x) = \ln x$, $c = 2$ ∗ **10.** $f(x) = \sqrt{1 + 3x}$, $c = 0$

∗ **11.** $f(x) = \tan^{-1} x$, $c = 0$ ∗ **12.** $f(x) = 1/(4 + x)^{1/3}$, $c = 2$

∗∗ **13.** If the fifth Taylor polynomial in Exercise 1 is used to approximate e^{2x} on the interval $0 \leq x \leq 0.1$, what is the maximum error in doing so?

∗∗ **14.** If the third Taylor polynomial in Exercise 6 is used to approximate $1/(2 - x)$ on the interval $0.9 \leq x \leq 1.1$, what is the maximum error in doing so?

4.13 Problem Solving

Differentiation is a powerful tool for solving problems in many fields—physics, chemistry, engineering, psychology, medicine, and economics, to name a few. Many of these applications involve related rate and applied maxima-minima problems, and are examples of *word* problems. In this section we give some suggestions on how to handle word problems, which may be helpful in other mathematical problems as well. As we follow these suggestions through the examples, we illustrate the logical thought processes that lead from problem to solution.

Three suggestions important to many mathematical problems, and in particular to those that are calculus-based, are:

Suggestion 1:   Make a diagram whenever possible to illustrate all given information.
Suggestion 2:   Identify the nature of the problem. Is it a related-rate problem, a maxima-minima problem, or some other kind of problem? There are always key words in the statement of the problem to help make this decision. Find them.

This suggestion is extremely important because further steps in the solution are intimately connected to the nature of the problem. However, most problems will also follow the next suggestion.

Suggestion 3:   Choose letters to represent key variables and find appropriate formulas or equations involving these variables.

Problem 1   The slope and y-intercept of the line $y = mx + b$ are such that the line intersects the parabola $y = x^2$ in two points P and Q (Figure 4.93a). A point R on the parabola between P and Q is joined to P and Q to create a triangle PQR. Where should R be chosen in order that the area of the triangle be as large as possible?

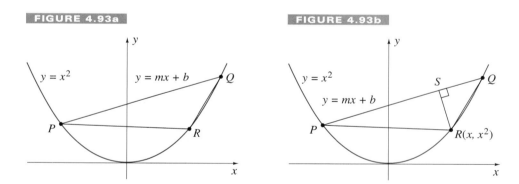

FIGURE 4.93a   **FIGURE 4.93b**

Certainly we can follow Suggestion 1 and begin with Figure 4.93a. To identify the nature of the problem (Suggestion 2), we key on the last four words "as large as possible"; they identify that we have a maximum problem. Find the position of R in order to maximize the area of triangle PQR. According to Suggestion 3 we should choose a letter, say A, to represent the area of triangle PQR, and now find a formula for A. What should A be expressed in terms of? Since we want to find the position of R that maximizes A, A should be expressed in terms of R. How do we do this? By giving coordinates to R. If its x-coordinate is denoted by x, then its y-coordinate is x^2 (since it is on the parabola). Notice that we have not assigned coordinates to P and Q; we will do this only if it becomes necessary. We now express A in terms of x since x identifies the position of R. Because the area of a triangle is one-half the product of its base and height, we drop a perpendicular from R to PQ to intersect PQ at S (Figure 4.93b). Then

$$A = \frac{1}{2} \|PQ\| \|RS\|.$$

A key point to realize here is that P and Q are fixed and, therefore, so also is length $\|PQ\|$ of the base of the triangle. It is not necessary to find a formula for $\|PQ\|$; simply regard it as a constant. We must find $\|RS\|$ in terms of x. With the equation of the line expressed in the form $mx - y + b = 0$, formula 1.7 gives the distance from $R(x, x^2)$ to S

$$\|RS\| = \frac{|mx - y + b|}{\sqrt{m^2 + 1}} = \frac{|mx - x^2 + b|}{\sqrt{m^2 + 1}}.$$

Thus,

$$A = \frac{1}{2}\|PQ\|\frac{|mx - x^2 + b|}{\sqrt{m^2 + 1}}.$$

Are we ready to differentiate? No. This is an absolute maximum problem—among all points on the parabola between P and Q, find the one that maximizes A. It follows that x must be restricted to the interval $x_P \leq x \leq x_Q$ where x_P and x_Q are x-coordinates of P and Q. Our mathematical problem, therefore, is to find the absolute maximum of the function

$$A(x) = \frac{\|PQ\|}{2\sqrt{m^2 + 1}}|mx - x^2 + b|, \quad x_P \leq x \leq x_Q.$$

To find critical points of $A(x)$, we first set $A'(x) = 0$, using differentiation formula 3.13,

$$0 = A'(x) = \frac{\|PQ\|}{2\sqrt{m^2 + 1}}\frac{|mx - x^2 + b|}{mx - x^2 + b}(m - 2x).$$

The only solution of this equation is $x = m/2$. We should now evaluate A at x_P, at the critical point $m/2$, and at x_Q. Since the triangle degenerates to a straight line when R is at either P or Q (that is, when $x = x_P$ or $x = x_Q$), it follows that $x = m/2$ must yield a maximum for A. Thus, R should be chosen as the point $(m/2, m^2/4)$.

Problem 2 A hemispherical tank of radius 3 m has a light on its upper edge as in Figure 4.94a. A stone falls vertically along the axis of symmetry of the tank, and when it is 1 m from the bottom of the tank it is falling at 2 m/s. How fast is its shadow moving along the surface of the tank at this instant?

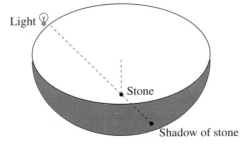

FIGURE 4.94a

Light

Stone

Shadow of stone

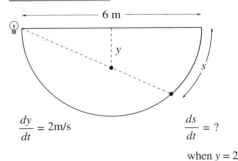

FIGURE 4.94b

6 m

y

s

$\dfrac{dy}{dt} = 2\text{m/s}$

$\dfrac{ds}{dt} = ?$

when $y = 2$

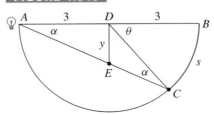

FIGURE 4.94c

A 3 D 3 B

α θ

y

E α

s

C

Although Figure 4.94a would seem to fullfil Suggestion 1, there is really no need to draw a 3-dimensional diagram. The shadow follows a circular path down the side of the tank which we have shown in Figure 4.94b.

The nature of the problem is identified in the last sentence "How fast is the shadow moving?" It is a related-rate problem; we want the rate at which the shadow is moving, given the rate at which the stone is falling.

To follow Suggestion 3, we introduce the distance s along the circle as the distance that the shadow has travelled since the stone passed the diameter of the circle. Its rate of change with respect to time t, namely ds/dt, is what is required. To express the rate at which the stone is falling as a derivative, we let y be the distance from the diameter of the circle to the stone. Then $dy/dt = 2$ m/s. These facts have been added to Figure 4.94b.

For a related-rate problem, we must find an equation relating y and s. First, arc length along a circle is the product of the radius of the circle and the angle θ that it subtends at the centre of the circle. Therefore, we name some key points in Figure 4.94c, join C and D, and write

$$s = 3\theta.$$

We wanted an equation relating s and y, not s and θ. We therefore must relate y and θ. Since triangle ADC is isosceles, angles DAC and DCA are equal—called α in Figure 4.94c, and $\theta = 2\alpha$. Hence

$$s = 3(2\alpha) = 6\alpha.$$

From triangle ADE, we may write $y = 3 \tan \alpha$. Combine this with $s = 6\alpha$, and we finally relate s and y:

$$y = 3 \tan \left(\frac{s}{6} \right).$$

Differentiation with respect to t gives

$$\frac{dy}{dt} = 3 \sec^2 \left(\frac{s}{6} \right) \left(\frac{1}{6} \frac{ds}{dt} \right).$$

When the stone is 1 m from the bottom of the tank, $y = 2$ m, and therefore

$$2 = 3 \tan \left(\frac{s}{6} \right).$$

It follows that at this instant

$$\sec^2 \left(\frac{s}{6} \right) = 1 + \tan^2 \left(\frac{s}{6} \right) = 1 + \left(\frac{2}{3} \right)^2 = \frac{13}{9}.$$

Substitution of this into the equation that relates dy/dt and ds/dt, along with $dy/dt = 2$, yields

$$2 = 3 \left(\frac{13}{9} \right) \left(\frac{1}{6} \frac{ds}{dt} \right).$$

Hence, $ds/dt = 36/13$, and the shadow is moving at $36/13$ m/s.

Problem 3   Prove that when $0 < a < b < \pi/2$,

$$\frac{\tan b}{\tan a} > \frac{b}{a}. \tag{4.34}$$

This is not a word problem; it is included to once again illustrate the logic by which we proceed from problem to solution. We do not begin with $0 < a < b < \pi/2$; the important requirement is to verify the tangent inequality. As there is no obvious connection between the two sides of the inequality, perhaps it would be beneficial to rewrite it in the form

$$\frac{1}{b} \tan b > \frac{1}{a} \tan a. \tag{4.35}$$

This looks promising; both sides now have exactly the same form. Effectively, we have multiplied 4.34 by $b^{-1} \tan a$ to obtain 4.35. Inequality 4.35 remains in the same direction only if $b^{-1} \tan a$ is positive, and this is indeed the case because we are considering only values of a and b between 0 and $\pi/2$.

We have noted that both sides of 4.35 have the same form. In fact, they can be obtained by setting $x = b$ and $x = a$ in the function

$$f(x) = \frac{1}{x} \tan x.$$

We can now paraphrase the problem: Show that for any two points a and b in the interval $0 < x < \pi/2$, with $b > a$,

$$f(b) > f(a).$$

In other words, we must show that $f(x)$ is increasing on the interval $I : 0 < x < \pi/2$. But according to 4.4, this is established if $f'(x) \geq 0$ on I. Using the product rule we calculate

$$f'(x) = -\frac{1}{x^2} \tan x + \frac{1}{x} \sec^2 x = \frac{1}{x^2}(x \sec^2 x - \tan x)$$

$$= \frac{1}{x^2}\left(\frac{x}{\cos^2 x} - \frac{\sin x}{\cos x}\right) = \frac{1}{x^2 \cos^2 x}(x - \sin x \cos x).$$

Since $x^2 \cos^2 x$ is positive on I, the sign of $f'(x)$ is determined by that of $x - \sin x \cos x$. Therefore, we must show that on I,

$$0 \leq x - \sin x \cos x.$$

FIGURE 4.95

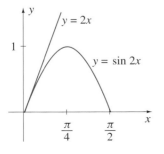

This inequality can be simplified if it is multiplied by 2,

$$0 \leq 2x - 2 \sin x \cos x = 2x - \sin 2x \quad \text{or} \quad 2x \geq \sin 2x. \tag{4.36}$$

Inequality 4.34 is verified once 4.36 is established. Graphs of the functions $2x$ and $\sin 2x$ on I are shown in Figure 4.95. Important to this picture are the facts that the line $y = 2x$ and the curve $y = \sin 2x$ both have slope 2 at $x = 0$, but the slope of $y = \sin 2x$ is less than 2 for $x > 0$. Clearly, then, $2x \geq \sin 2x$, and our proof is complete.

We have solved Problem 3 backwards. We have replaced inequality 4.34 with 4.35, which is tantamount to showing that $f(x) = x^{-1}\tan x$ is increasing on $I : 0 < x < \pi/2$. But this is established once 4.36 is verified, and Figure 4.95 verifies 4.36.

We could now rewrite the solution to this problem in a very elegant way. We could begin by saying:

"Consider the function $f(x) = x^{-1}\tan x$ on the interval $0 < x < \pi/2$." We could show that $f(x)$ is increasing and 4.35 and 4.34 would immediately follow. This is exactly how the solution would appear in most calculus books and/or solutions manuals. The author would appear brilliant; out of nowhere he would produce the function $f(x) = x^{-1}\tan x$. He actually solved the problem backwards, as above, and then rewrote it.

Problems 1, 2, and 3 have been chosen because they are involved, multistage problems to emphasize the importance of Suggestions 1, 2, and 3, and to illustrate how to organize our thinking into a logical sequence. We hope that these suggestions prove useful as you solve more and more involved problems in calculus and other applied mathematics courses.

SUMMARY

In this chapter we discussed a number of applications of differentiation, the first of which was Newton's iterative procedure for approximating the roots of equations. It is perhaps the most popular of all approximation methods, because of its speed, simplicity, and accuracy.

In Section 4.3 we defined a critical point of a function as a point in its domain where its first derivative either vanishes or does not exist. Geometrically, this corresponds to a point where the graph of the function has a horizontal tangent line, a vertical tangent line, or no tangent line at all. The first derivative test indicates if critical points yield relative maxima or relative minima. A function is increasing (or decreasing) on an interval if its graph slopes upward to the right (respectively left), and this is characterized by a nonnegative (respectively nonpositive) derivative. It is concave upward (or downward) if its slope is increasing (respectively decreasing), and consequently if its second derivative is nonnegative (respectively nonpositive). Points that separate intervals of opposite concavity are called points of inflection.

In Section 4.6 we illustrated that many applied extrema problems require absolute extrema rather than relative extrema. Absolute extrema of a continuous function on a closed interval must occur at either critical points or the ends of the interval. This fact implies that to find the absolute extrema of a continuous function $f(x)$ on a closed interval $a \le x \le b$, we evaluate $f(x)$ at its critical points between a and b and at a and b. The largest and smallest of these values are the absolute extrema of $f(x)$ on $a \le x \le b$. Plotting $f(x)$ can also prove valuable.

When an object moves along a straight line, its velocity and acceleration are the first and second derivatives respectively of its displacement with respect to time. In other words, if we observe straight-line motion of an object, and record its position as a function of time, then we can calculate the velocity and acceleration of that object at any instant.

Changes in a number of interrelated quantities usually produce changes in the others—sometimes small, sometimes large. How the rates of change of these variables relate to each other was the subject of Section 4.8. Related-rate problems made us acutely aware of the importance of differentiating an equation with respect to a variable only if the equation is valid for a continuous range of values of that variable.

Cauchy's generalized mean value theorem enabled us to develop L'Hôpital's rule in Section 4.10 for evaluation of various indeterminate forms such as $0/0$, ∞/∞, $0 \cdot \infty$, 0^0, 1^∞, ∞^0, and $\infty - \infty$.

Complicated functions can sometimes be approximated by their Taylor polynomials; the higher the degree of the polynomial, the better the approximation.

KEY TERMS

In reviewing this chapter, you should be able to define or discuss the following key terms:

Newton's iterative procedure
Increasing function
Decreasing function
Critical point
Relative maximum
Relative minimum
First derivative test
Concave upward
Concave downward
Point of inflection
Second derivative test
Horizontal point of inflection
Vertical point of inflection

Absolute maximum
Absolute minimum
Instantaneous velocity
Instantaneous acceleration
Instantaneous speed
Related rates
Indeterminate form
L'Hôpital's rule
Differential
Relative change
Percentage change
Taylor polynomial

REVIEW EXERCISES

** **1.** (a) Prove that the area of the isosceles triangle in the figure below is

$$A = \frac{l^2 \sin \theta}{2}.$$

(b) If the angle θ is increasing at $1/2$ radian per minute, but l remains constant, how fast is the area of the triangle changing? What does your answer predict when $\theta = 0$, $\pi/2$, and π?

(c) When does A change most rapidly and most slowly; that is, when is $|dA/dt|$ largest and smallest?

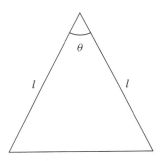

** **2.** Plot a graph of each of the following functions, indicating all relative maxima and minima and points of inflection.

(a) $f(x) = 4x^3 + x^2 - 2x + 1$ (b) $f(x) = \dfrac{x^2 - 2x + 4}{x^2 - 2x + 1}$

** **3.** Of all pairs of positive numbers that add to some given constant $c > 0$, find that pair which has the largest product.

** **4.** Solve Exercise 3 for the smallest product.

** **5.** Of all pairs of positive numbers that multiply to some given constant $c > 0$, find that pair which has the smallest sum.

** **6.** Solve Exercise 5 for the largest sum.

** **7.** Two sides of the triangle in the figure below maintain constant lengths of 3 cm and 4 cm but the length l of the third side decreases at the rate of 1 cm/min. How fast is angle θ changing when l is 4 cm?

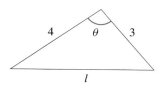

In Exercises 8–19, evaluate the limit if it exists.

* **8.** $\displaystyle \lim_{x \to 0} \frac{3x^2 + 2x^3}{3x^3 - 2x^2}$

* **9.** $\displaystyle \lim_{x \to \infty} \frac{\sin 3x}{2x}$

* **10.** $\displaystyle \lim_{x \to 4} \frac{x^2 - 16}{x - 4}$

* **11.** $\displaystyle \lim_{x \to 0} \frac{\sin 3x}{2x}$

* **12.** $\displaystyle \lim_{x \to -\infty} \frac{\sin x^2}{2x}$

** **13.** $\displaystyle \lim_{x \to 2^+} \frac{\sqrt{x} - 2}{\sqrt{x} - \sqrt{2}}$

** **14.** $\displaystyle \lim_{x \to \infty} x^2 e^{-3x}$

** **15.** $\displaystyle \lim_{x \to 0^+} x^{2x}$

** **16.** $\displaystyle \lim_{x \to 0^+} x^4 \ln x$

** **17.** $\displaystyle \lim_{x \to 0} \frac{\sin 2x}{\tan 3x}$

** **18.** $\displaystyle \lim_{x \to \infty} \left(\frac{x + 1}{x - 1} \right)^x$

** **19.** $\displaystyle \lim_{x \to -\infty} x e^x$

** **20.** Use Newton's method to find all critical points for the following functions accurate to six decimals.

(a) $f(x) = x^4 + 3x^2 - 2x + 5$ (b) $f(x) = \dfrac{x^3 + 1}{3x^3 + 5x + 1}$

** **21.** An object moves along the x-axis with its position defined as a function of time t by

$$x = x(t) = t^4 - \frac{44}{3}t^3 + 62t^2 - 84t, \quad t \ge 0.$$

Plot a graph of this function, indicating times when the velocity and acceleration of the object are equal to zero.

** **22.** If the graph in the figure below represents the position $x(t)$ of an object moving along the x-axis, what could physically cause the corner at time t_0?

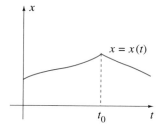

∗∗ **23.** If an object moves along the x-axis with constant acceleration, is it possible for a graph of its position function $x(t)$ to have a point of inflection?

∗∗ **24.** An open box is formed from a square piece of cardboard (l units long on each side) by cutting out a square at each corner and folding up the sides. What is the maximum possible volume for the box?

∗∗ **25.** (a) If the average speed of a car for a trip is 80 km/h, must it at some time have had an instantaneous speed of 80 km/h? Explain.

  (b) Must the car at some instant have had a speed of 83 km/h?

∗∗ **26.** Plot graphs of the following functions, indicating all relative maxima and minima and points of inflection:

$$\text{(a)} \quad f(x) = \frac{x^3}{x^2 - 1} \qquad \text{(b)} \quad f(x) = x^2 + \sin^2 x$$

∗∗ **27.** If at some instant of time sides a and b of the triangle in the figure below form a right angle, and if these sides are increasing at equal rates, does it remain a right-angled triangle?

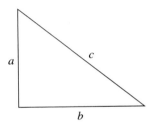

∗∗ **28.** A football team presently sells tickets at prices of $8, $9, and $10 per seat, depending on the position of the seat. At these prices it averages sales of 10 000 at $10, 20 000 at $9, and 30 000 at $8. The team wishes to raise the price of each ticket by the same amount, but feels that for every dollar the price is raised, 10% fewer tickets of each type will be sold. What price increase per ticket will maximize revenue?

∗∗ **29.** Repeat Exercise 28 given that the team takes into account the fact that profit from concession sales for each person at the game is fifty cents.

∗∗ **30.** If a particle moves away from the origin along the positive x-axis with a constant speed of 10 m/s, how fast is its distance from the curve $y = x^2$ changing when it is at $x = 3$ m?

∗∗ **31.** Each evening a cow in a pasture returns to its barn at point B (figure below). But it always does so by first walking to the river for a drink. If the cow walks at 2 km/h and stops to drink for 2 minutes, what is the minimum time it takes for the cow to get from the pasture to the barn? What is the minimum time if the cow walks twice as fast?

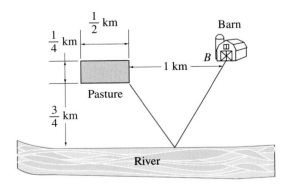

** **32.** A farmer has 100 hectares to plant in corn and potatoes. Undamaged corn yields $\$p$ per hectare and potatoes $\$q$ per hectare. For each crop, the loss due to disease and pests per unit hectare is directly proportional to the area planted. If the farmer plants x hectares of corn, the loss due to disease and pests is equal to ax per hectare. The total loss of corn is therefore ax^2 hectares. Similarly, the total loss of potatoes is by^2, if the area planted in potatoes is y hectares. Find the areas that should be planted in corn and potatoes in order to minimize monetary loss. Substitute sample values for a, b, p, and q to see whether your results look reasonable.

CHAPTER 5

The Indefinite Integral or Antiderivative

In Chapter 3 we introduced the derivative and ways to differentiate functions defined explicitly and implicitly. We then discussed various applications of calculus, including velocity and acceleration, related rates, maxima and minima, Newton's iterative procedure, and L'Hôpital's rule. In this chapter, we reverse the differentiation process. Instead of giving you a function and asking for its derivative, we give you the derivative and ask you to find the function. This process of "backwards differentiation" or "antidifferentiation" has such diverse applications that antidifferentiation is as important to calculus as differentiation. In many problems we find ourselves differentiating at one stage and antidifferentiating at another.

Antidifferentiation is a much more difficult process than differentiation. All but one of the rules for differentiation were developed in Chapter 3; we have a few more functions to differentiate in Chapter 8, and one more rule in Section 10.1. In contrast to this, the list of formulas and techniques for finding antiderivatives is endless. In this chapter we introduce the three simplest, but most important, techniques; in Chapter 9 we discuss many others.

5.1 The Reverse Operation of Differentiation

In our discussions on velocity and acceleration in Section 4.7, we showed that when an object moves along the x-axis with its position described by the function $x(t)$, its velocity and acceleration are the first and second derivatives of $x(t)$ with respect to time t:

$$v(t) = \frac{dx}{dt} \quad \text{and} \quad a(t) = \frac{d^2x}{dt^2}.$$

For example, if $x(t) = t^3 + 3t^2$, then

$$v(t) = 3t^2 + 6t \quad \text{and} \quad a(t) = 6t + 6.$$

When engineers and physicists study the motions of objects, a more common type of problem is to determine the position of an object; that is, the position is not given. What they might know, however, is the acceleration of the object (perhaps through Newton's second law, which states that acceleration is proportional to the resultant force on the object). So the question we must now ask is: If we know the acceleration $a(t)$ of an object as a function of time, can we obtain its velocity and position by reversing the differentiations? In the above example, if we know that

$$a(t) = \frac{dv}{dt} = 6t + 6,$$

can we find the function $v(t)$ that differentiates to give $6t + 6$? We know that to arrive at the terms $6t$ and 6 after differentiation, $v(t)$ might have contained the terms $3t^2$ and $6t$. In other words, one possible velocity function that differentiates to give $a(t) = 6t + 6$ is $v(t) = 3t^2 + 6t$.

It is not, however, the only one; $v(t) = 3t^2 + 6t + 10$ and $v(t) = 3t^2 + 6t - 22$ also have derivative $6t + 6$. In fact, for any constant C whatsoever, the derivative of

$$v(t) = 3t^2 + 6t + C$$

is $a(t) = 6t + 6$. In Theorem 5.1 we shall show that this velocity function represents all functions that have $6t + 6$ as their derivative. We shall also demonstrate how to evaluate the constant C.

For the present, let us set $C = 0$ so that

$$v(t) = \frac{dx}{dt} = 3t^2 + 6t.$$

We now ask what position function $x(t)$ differentiates to give $3t^2 + 6t$. One possibility is $x(t) = t^3 + 3t^2$, and for the same reason as above, so is

$$x(t) = t^3 + 3t^2 + D$$

for any constant D whatsoever.

Thus, by reversing the differentiation operation in this example we have proceeded from the acceleration $a(t) = 6t + 6$ to possible velocity functions $v(t)$, and then to possible position functions $x(t)$. This process of antidifferentiation has applications far beyond velocity and acceleration problems, and we shall see many of them as we progress through this book. We begin our formal study of antidifferentiation with the following definition.

DEFINITION 5.1

A function $F(x)$ is called an **antiderivative** or **indefinite integral** of $f(x)$ on an interval I if on I

$$F'(x) = f(x). \tag{5.1}$$

For example, since

$$\frac{d}{dx}x^4 = 4x^3,$$

we say that x^4 is an antiderivative or an indefinite integral of $4x^3$ for all x. But for any constant C, the function $x^4 + C$ is also an antiderivative of $4x^3$. The following theorem indicates that these are the only antiderivatives of $4x^3$.

THEOREM 5.1

If $F(x)$ is an indefinite integral of $f(x)$ on an interval I, then every indefinite integral of $f(x)$ on I is of the form

$$F(x) + C, \qquad \text{where } C \text{ is a constant.}$$

PROOF   Suppose that $F(x)$ and $G(x)$ are two antiderivatives of $f(x)$ on I. If we define a function $D(x) = G(x) - F(x)$, then on I

$$D'(x) = G'(x) - F'(x) = f(x) - f(x) = 0.$$

If x_1 and x_2 are any two points in the interval I, then certainly $D'(x) = 0$ on the interval $x_1 \leq x \leq x_2$. But differentiability of $D(x)$ on $x_1 \leq x \leq x_2$ implies continuity of $D(x)$ thereon also (Theorem 3.6). We may therefore apply the mean value theorem (Theorem 3.17) to $D(x)$ on the interval $x_1 \leq x \leq x_2$, and conclude that there exists a number c between x_1 and x_2 such that

$$D'(c) = \frac{D(x_2) - D(x_1)}{x_2 - x_1}.$$

But because $D'(c) = 0$, it follows that $D(x_1) = D(x_2)$. Since x_1 and x_2 are any two points in I, we conclude that $D(x)$ must have the same value at every point in I; that is, on I,

$$D(x) = G(x) - F(x) = C,$$

where C is a constant. Consequently,

$$G(x) = F(x) + C,$$

and our proof is complete.   ∎

Because of this theorem, if we find one antiderivative $F(x)$ of $f(x)$ by any means whatsoever, then we have found every antiderivative of $f(x)$, since every antiderivative can be written as $F(x)$ plus a constant C. As every indefinite integral of $f(x)$ is of the form $F(x) + C$, we call $F(x) + C$ *the* indefinite integral of $f(x)$. The operation of taking the indefinite integral is denoted by

$$\int f(x)\, dx = F(x) + C, \tag{5.2}$$

where the differential dx indicates that antidifferentiation is with respect to x. We call $f(x)$ the **integrand** of the indefinite integral. For example, we write

$$\int x^2\, dx = \frac{x^3}{3} + C \qquad \text{and} \qquad \int \frac{1}{x^3}\, dx = -\frac{1}{2x^2} + C.$$

The following theorem is fundamental to the calculation of antiderivatives. Its proof is a straightforward exercise in differentiation and the use of Definition 5.1.

THEOREM 5.2

If $f(x)$ and $g(x)$ have antiderivatives on an interval I, then on I:

$$(i) \qquad \int [f(x) + g(x)]\, dx = \int f(x)\, dx + \int g(x)\, dx; \tag{5.3a}$$

$$(ii) \qquad \int k f(x)\, dx = k \int f(x)\, dx, \qquad k \text{ a constant.} \tag{5.3b}$$

For example, to evaluate the antiderivative of $2x^3 - 4x$, we write

$$\int (2x^3 - 4x)\, dx = \int 2x^3\, dx + \int -4x\, dx \qquad \text{(by part (i) of Theorem 5.2)}$$

$$= 2\int x^3\, dx - 4\int x\, dx \qquad \text{(by part (ii) of 5.2)}$$

$$= 2\left(\frac{x^4}{4}\right) - 4\left(\frac{x^2}{2}\right) + C$$

$$= \frac{x^4}{2} - 2x^2 + C.$$

Every differentiation formula developed in Chapter 3 can be expressed as an integration formula. In fact, equations 5.3 are integral counterparts of equations 3.9 and 3.8. Some of the differentiation formulas for trigonometric, exponential, and logarithm functions are restated below as integration formulas.

$$\int \sin x\, dx = -\cos x + C, \tag{5.4a}$$

$$\int \cos x\, dx = \sin x + C, \tag{5.4b}$$

$$\int \sec^2 x\, dx = \tan x + C, \tag{5.4c}$$

$$\int \sec x \tan x\, dx = \sec x + C, \tag{5.4d}$$

$$\int \csc^2 x\, dx = -\cot x + C, \tag{5.4e}$$

$$\int \csc x \cot x\, dx = -\csc x + C, \tag{5.4f}$$

$$\int e^x\, dx = e^x + C, \tag{5.4g}$$

$$\int a^x\, dx = a^x \log_a e + C, \tag{5.4h}$$

$$\int \frac{1}{x}\, dx = \ln |x| + C, \tag{5.4i}$$

The last equation follows immediately from 3.41 with $f(x) = x$.

Perhaps the most important integration formula is the counterpart of power rule 3.7. Since

$$\frac{d}{dx} x^{n+1} = (n + 1)x^n,$$

it follows that

$$\int (n + 1)x^n\, dx = x^{n+1} + \overline{C}.$$

With property 5.3b, the $n + 1$ can be removed from the integral and taken to the other side of the equation,

$$\int x^n \, dx = \frac{1}{n+1} x^{n+1} + C, \qquad n \neq -1, \tag{5.5}$$

where $C = \overline{C}/(n+1)$. As indicated, this result is valid provided $n \neq -1$, but formula 5.4i takes care of this exceptional case.

EXAMPLE 5.1

Evaluate

$$\int \left(2x + \frac{1}{x^2} \right) dx.$$

SOLUTION   Using Theorem 5.2 and formula 5.5, we find

$$\int \left(2x + \frac{1}{x^2} \right) dx = 2 \int x \, dx + \int \frac{1}{x^2} \, dx = 2 \left(\frac{x^2}{2} \right) - \frac{1}{x} + C = x^2 - \frac{1}{x} + C.$$

Where is $x^2 - 1/x + C$, the indefinite integral of $2x + 1/x^2$? It is valid on the intervals $x < 0$ and $x > 0$, but not for all x because the function is not defined at $x = 0$. To make this clear, we should write

$$\int \left(2x + \frac{1}{x^2} \right) dx = \begin{cases} x^2 - \dfrac{1}{x} + C_1, & x < 0 \\[2mm] x^2 - \dfrac{1}{x} + C_2, & x > 0 \end{cases}$$

where the constants C_1 and C_2 need not be the same. For brevity we often write

$$\int \left(2x + \frac{1}{x^2} \right) dx = x^2 - \frac{1}{x} + C,$$

thereby suppressing the complete description of the indefinite integral. When the context demands that we distinguish various intervals on which the indefinite integral is defined, we shall be careful to give the extended version.

EXAMPLE 5.2

Find a curve that passes through the point $(1, 5)$ and whose tangent line at each point (x, y) has slope $5x^4 - 3x^2 + 2$.

SOLUTION   If $y = f(x)$ is the equation of the curve, then

$$\frac{dy}{dx} = 5x^4 - 3x^2 + 2.$$

If we take antiderivatives of both sides of this equation with respect to x, we obtain

$$\int \frac{dy}{dx} \, dx = \int (5x^4 - 3x^2 + 2) \, dx \qquad \text{or}$$

$$y = x^5 - x^3 + 2x + C.$$

Since $(1, 5)$ is a point on the curve, its coordinates must satisfy the equation of the curve:

$$5 = 1^5 - 1^3 + 2(1) + C.$$

Thus $C = 3$, and the required curve is

$$y = x^5 - x^3 + 2x + 3.$$

In taking the antiderivative of each side of the equation

$$\frac{dy}{dx} = 5x^4 - 3x^2 + 2$$

in the above example, we added an arbitrary constant C to the right side. You might question why we did not add a constant to the left side. Had we done so, the result would have been

$$y + D = x^5 - x^3 + 2x + E.$$

If we then wrote

$$y = x^5 - x^3 + 2x + (E - D)$$

and defined $C = E - D$, we would have obtained exactly the same result. Hence, nothing is gained by adding an arbitrary constant to both sides; a constant on one side is sufficient.

The problem in Example 5.2 was geometric: find the equation of a curve satisfying certain properties. We quickly recast it as the problem of finding the function $y = f(x)$ that satisfies the equation

$$\frac{dy}{dx} = 5x^4 - 3x^2 + 2,$$

subject to the additional condition that $f(1) = 5$. Once again this is a differential equation. We discuss differential equations briefly in Section 5.4, and study them in detail in Chapter 16. When we solve a differential equation with no subsidiary conditions, say,

$$\frac{dy}{dx} = 4x^2 + 7x,$$

we do not get a function, but rather a one-parameter family of functions. For this differential equation, we obtain

$$y = \frac{4}{3}x^3 + \frac{7}{2}x^2 + C,$$

FIGURE 5.1

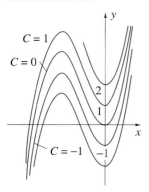

a one-parameter family of cubic polynomials, C being the parameter. Geometrically, we have the one-parameter family of curves in Figure 5.1. Parameter C represents a vertical shift of one curve relative to another. Note that if a vertical line is drawn at any position x to intersect these curves, then at the points of intersection, every curve has exactly the same slope, namely, $4x^2 + 7x$. For example, the slope of each cubic at $x = 0$ is zero. If an extra condition is added to the differential equation, such as to demand that y be equal to 4 when $x = 2$, then

$$4 = \frac{4}{3}(2)^3 + \frac{7}{2}(2)^2 + C,$$

or $C = -62/3$. This condition singles out one particular function from the family, namely, $y = 4x^3/3 + 7x^2/2 - 62/3$. Geometrically, it determines that curve in the family which passes through the point $(2, 4)$.

In this chapter, we discuss three basic ways to find antiderivatives. First, some antiderivatives are obvious, and the better you are at differentiation, the more likely this will be. For example, you should have no trouble recognizing that

$$\int 7x^5 \, dx = \frac{7x^6}{6} + C \quad \text{and} \quad \int \frac{1}{x^3} \, dx = \frac{-1}{2x^2} + C.$$

Our second method results from the answer to the following question: How do we check that a function $F(x)$ is an antiderivative of $f(x)$? We differentiate $F(x)$, of course. This simple fact suggests an approach to slightly more complex problems, say,

$$\int (2x + 3)^5 \, dx.$$

We might reason that in order to have $2x + 3$ raised to power 5 after differentiation, we had $2x + 3$ to power 6 before differentiation; that is, a logical "guess" for the antiderivative is $(2x + 3)^6$. Differentiation of this function gives

$$\frac{d}{dx}(2x + 3)^6 = 6(2x + 3)^5(2) = 12(2x + 3)^5,$$

and we see that $(2x + 3)^6$ is not the correct antiderivative. It has produced $(2x + 3)^5$, as required, but it has also given an undesirable factor of 12. We therefore "fix up" our original guess by multiplying it by $1/12$; that is, the correct antiderivative is

$$\int (2x + 3)^5 \, dx = \frac{1}{12}(2x + 3)^6 + C.$$

This is what we call "guess and fix up": we guess the antiderivative closely enough to fix it up. Our guess must be correct within a *multiplicative constant*. We cannot be out by x's. Let us illustrate this with two very similar problems,

$$\int \frac{1}{(5x + 2)^5} \, dx \quad \text{and} \quad \int \frac{1}{(5x^2 + 2)^5} \, dx.$$

For the first problem we might guess at $(5x + 2)^{-4}$. Differentiation gives

$$\frac{d}{dx}\left[\frac{1}{(5x + 2)^4}\right] = \frac{-4}{(5x + 2)^5}(5) = \frac{-20}{(5x + 2)^5}.$$

Since we are out by a factor of -20, we fix up the original guess,

$$\int \frac{1}{(5x + 2)^5} \, dx = \frac{-1}{20(5x + 2)^4} + C.$$

It might seem just as logical to guess at $(5x^2 + 2)^{-4}$ for the second problem, but differentiation yields

$$\frac{d}{dx}\left[\frac{1}{(5x^2 + 2)^4}\right] = \frac{-4}{(5x^2 + 2)^5}(10x) = \frac{-40x}{(5x^2 + 2)^5}.$$

This time the discrepancy is $-40x$, which involves x. We must not try to fix up the original guess; it must be abandoned totally. This indefinite integral is quite difficult, and we will have to wait for the more powerful techniques of Chapter 9. To emphasize once again, do not try to fix up x's, only constants.

EXAMPLE 5.3

Evaluate the following indefinite integrals:

$$\text{(a)} \quad \int \cos 3x \, dx \qquad \text{(b)} \quad \int e^{-2x} \, dx \qquad \text{(c)} \quad \int \frac{x}{3x^2 - 4} \, dx$$

SOLUTION

(a) To obtain $\cos 3x$ after differentiation, we guess $\sin 3x$ as the antiderivative. Since

$$\frac{d}{dx} \sin 3x = 3 \cos 3x,$$

it is necessary to fix up a 3,

$$\int \cos 3x \, dx = \frac{1}{3} \sin 3x + C.$$

(b) With an initial guess of e^{-2x} for the antiderivative based on the fact that the derivative of an exponential function always returns the same exponential, we calculate

$$\frac{d}{dx} e^{-2x} = -2e^{-2x}.$$

Consequently, we fix up a -2,

$$\int e^{-2x} \, dx = -\frac{1}{2} e^{-2x} + C.$$

(c) Since

$$\frac{d}{dx} \ln (3x^2 - 4) = \frac{1}{3x^2 - 4}(6x),$$

the required antiderivative is

$$\int \frac{x}{3x^2 - 4} \, dx = \frac{1}{6} \ln |3x^2 - 4| + C.$$

We have inserted absolute values as suggested by formula 5.4i.

The third technique for finding antiderivatives is discussed in Section 5.3.

EXAMPLE 5.4

Find the antiderivative of the Heaviside unit step function $H(x - a)$ introduced in Section 2.4.

SOLUTION Because the function has two values (Figure 5.2), we subdivide the integration into two parts. For $x < a$, $H(x - a) = 0$, and the antiderivative is a constant. For $x > a$, $H(x - a) = 1$, and the antiderivative is x plus a constant. Thus,

$$\int H(x - a)\, dx = \begin{cases} C, & x < a \\ x + D, & x > a \end{cases}.$$

Constants C and D need not be the same.

FIGURE 5.2

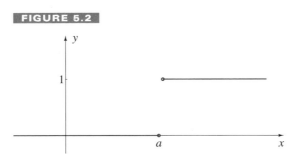

We use these ideas to discuss deflections of beams. When a beam that might otherwise be horizontal is subjected to loads, it does not remain horizontal. By analyzing internal forces and moments, it can be shown that the shape $y(x)$ of a uniform beam with constant cross section (Figure 5.3) is governed by the equation

$$\frac{d^4 y}{dx^4} = \frac{F(x)}{EI} \tag{5.6}$$

where E is a constant called Young's modulus of elasticity (depending on the material of the beam), and I is also a constant (the moment of inertia of the cross section of the beam). Quantity $F(x)$ is the load placed on the beam; it is the vertical force per unit length in the x-direction, placed at position x, including the weight of the beam itself. For example, if a beam has mass 100 kg and length 10 m (Figure 5.4), then the load due to its weight is a constant $F(x) = -9.81(100/10) = -98.1$ N/m at every point of the beam.

FIGURE 5.3

Beam

FIGURE 5.4

Mass = 100 kg

10 m

FIGURE 5.5

Mass = 50 kg

5   10

Mass = 100 kg

Suppose a block with mass 50 kg, uniform in cross-section, and length 5 m is placed on the left half of the beam in Figure 5.4 (see Figure 5.5). It adds an additional load of $9.81(10) = 98.1$ N/m over the interval $0 < x < 5$. The total load can be represented in terms of Heaviside unit step functions as

$$F(x) = -98.1 - 98.1[H(x) - H(x - 5)].$$

(Recall that the function $H(x) - H(x - 5)$ turns the function (-98.1 in this case) in front of it on at $x = 0$ and off at $x = 5$.) Because the beam extends from $x = 0$ to $x = 10$, and $H(x) = 1$ thereon, we can write that

$$F(x) = -196.2 + 98.1\, H(x - 5), \qquad 0 < x < 10.$$

Accompanying equation 5.6 will be four *boundary conditions* defining the type of support (if any) at each end of the beam. Three types of supports are common. We discuss them at the left end of the beam, but they also occur at the right end.

1. Simple Support

The end of a beam is **simply-supported** when it cannot move vertically, but it is free to rotate. Visualize that a horizontal pin perpendicular to the xy-plane passes through a hole in the end of the beam at $x = 0$ (Figure 5.6). The pin is fixed, but the end of the beam can rotate on the pin. In this case, $y(x)$ must satisfy the **boundary conditions**

$$y(0) = y''(0) = 0. \tag{5.7a}$$

2. Built-in End

If the end $x = 0$ of the beam is permanently fixed in a horizontal position (Figure 5.7), $y(x)$ satisfies

$$y(0) = y'(0) = 0. \tag{5.7b}$$

3. Free "Support"

If the end $x = 0$ of the beam is not supported (Figure 5.8), $y(x)$ satisfies

$$y''(0) = y'''(0) = 0. \tag{5.7c}$$

FIGURE 5.6

FIGURE 5.7

FIGURE 5.8

When two boundary conditions at each end of a beam accompany differential equation 5.6, we have what is called a **boundary-value problem**. For example, if the end $x = 0$ of the beam in Figure 5.5 is horizontally built-in, and the right end is free, just like a diving board (Figure 5.9), the boundary-value problem for deflections of the beam is

FIGURE 5.9

$$\frac{d^4 y}{dx^4} = \frac{1}{EI}[-196.2 + 98.1\, H(x - 5)],$$

$$0 < x < 10, \quad y(0) = y'(0) = 0, \quad y''(10) = y'''(10) = 0.$$

Integration of the differential equation four times on the intervals $0 < x < 5$ and $5 < x < 10$ gives

$$y(x) = \frac{1}{EI} \begin{cases} -8.175x^4 + Ax^3 + Bx^2 + Cx + D, & 0 < x < 5, \\ -4.0875x^4 + Px^3 + Qx^2 + Rx + S, & 5 < x < 10. \end{cases}$$

To evaluate the eight constants, we impose the four boundary conditions, and also demand that $y(x)$ and its first three derivatives be continuous at $x = 5$. This means that left- and right-hand limits of $EIy(x)$, $EIy'(x)$, $EIy''(x)$, and $EIy'''(x)$ be the same at $x = 5$.

$$EIy(0) = 0 \quad \Longrightarrow \quad D = 0,$$

$$EIy'(0) = 0 \quad \Longrightarrow \quad C = 0$$

$$EIy''(10) = 0 \quad \Longrightarrow \quad -49.05(10)^2 + 6P(10) + 2Q = 0,$$

$$EIy'''(10) = 0 \quad \Longrightarrow \quad -98.1(10) + 6P = 0.$$

$$\lim_{x \to 5^-} EIy(x) = \lim_{x \to 5^+} EIy(x) \quad \Longrightarrow$$

$$-8.175(5)^4 + 125A + 25B + 5C + D = -4.0875(5)^4 + 125P + 25Q + 5R + S,$$

$$\lim_{x \to 5^-} EIy'(x) = \lim_{x \to 5^+} EIy'(x) \quad \Longrightarrow$$

$$-32.7(5)^3 + 75A + 10B + C = -16.35(5)^3 + 75P + 10Q + R,$$

$$\lim_{x \to 5^-} EIy''(x) = \lim_{x \to 5^+} EIy''(x) \quad \Longrightarrow$$

$$-98.1(5)^2 + 30A + 2B = -49.05(5)^2 + 30P + 2Q,$$

$$\lim_{x \to 5^-} EIy'''(x) = \lim_{x \to 5^+} EIy'''(x) \quad \Longrightarrow$$

$$-196.2(5) + 6A = -98.1(5) + 6P.$$

Solutions of these equations are

$$A = 245.25, \quad B = -3065.625, \quad C = 0, \quad D = 0,$$

$$P = 163.5, \quad Q = -2452.5, \quad R = -2043.75, \quad S = 2554.6875.$$

The function describing deflections of the beam is, therefore,

$$y(x) =$$

$$\frac{1}{EI} \begin{cases} -8.175x^4 + 245.25x^3 - 3065.625x^2, & 0 < x < 5, \\ -4.0875x^4 + 163.5x^3 - 2452.5x^2 - 2043.75x + 2554.6875, & 5 < x < 10. \end{cases}$$

A graph of this function for $EI = 10^6$ is shown in Figure 5.10.

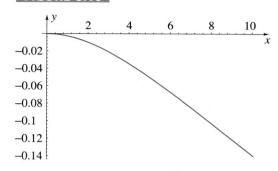

FIGURE 5.10

Be consoled by the fact that there are easier ways to solve this problem. One such is to use what are called *Laplace transforms*. You will learn them in advanced calculus courses.

In Exercises 1–20 evaluate the indefinite integral.

∗ **1.** $\displaystyle\int (x^3 - 2x)\,dx$ ∗ **2.** $\displaystyle\int (x^4 + 3x^2 + 5x)\,dx$

∗ **3.** $\displaystyle\int (2x^3 - 3x^2 + 6x + 6)\,dx$ ∗ **4.** $\displaystyle\int \sin x\,dx$

∗ **5.** $\displaystyle\int 3\cos x\,dx$ ∗ **6.** $\displaystyle\int \sqrt{x}\,dx$

∗ **7.** $\displaystyle\int \left(x^{10} - \frac{1}{x^3}\right)dx$ ∗ **8.** $\displaystyle\int \left(\frac{1}{x^2} - \frac{2}{x^4}\right)dx$

∗ **9.** $\displaystyle\int \left(x^{3/2} - x^{2/7}\right)dx$ ∗ **10.** $\displaystyle\int \left(\frac{1}{x^2} + \frac{1}{2\sqrt{x}}\right)dx$

∗ **11.** $\displaystyle\int \left(\frac{4}{x^{3/2}} + 2x^{1/3}\right)dx$ ∗ **12.** $\displaystyle\int \left(-\frac{1}{2x^2} + 3x^3\right)dx$

∗ **13.** $\displaystyle\int \frac{1}{x^\pi}\,dx$ ∗ **14.** $\displaystyle\int \left(2\sqrt{x} + 3x^{3/2} - 5x^{5/2}\right)dx$

∗ **15.** $\displaystyle\int x^2(x^2 - 3)\,dx$ ∗ **16.** $\displaystyle\int \sqrt{x}(x + 1)\,dx$

∗ **17.** $\displaystyle\int \left(\frac{x - 2}{x^3}\right)dx$ ∗ **18.** $\displaystyle\int x^2(1 + x^2)^2\,dx$

∗ **19.** $\displaystyle\int (x^2 + 1)^3\,dx$ ∗ **20.** $\displaystyle\int \frac{(x - 1)^2}{\sqrt{x}}\,dx$

In Exercises 21–24 find the curve $y = f(x)$ that passes through the given point and whose slope at each point (x, y) is defined by the indicated derivative.

∗ **21.** $dy/dx = x^2 - 3x + 2$, $(2, 1)$ ∗ **22.** $dy/dx = 2x^3 + 4x$, $(0, 5)$

∗ **23.** $dy/dx = -2x^4 + 3x^2 + 6$, $(1, 0)$ ∗ **24.** $dy/dx = 2 - 4x + 8x^7$, $(1, 1)$

✳✳ **25.** Find the equation of the curve that has a second derivative equal to $6x^2$ and passes through the points $(0, 2)$ and $(-1, 3)$.

✳✳ **26.** Find a function $f(x)$ that has a relative maximum $f(2) = 3$ and has a second derivative equal to $-5x$.

✳✳ **27.** Is it possible to find a function $f(x)$ that has a relative minimum $f(2) = 3$ and has a second derivative equal to $-5x$?

In Exercises 28–59 use "guess and fix up" to evaluate the antiderivative.

✳ **28.** $\displaystyle\int \sqrt{x + 2}\, dx$

✳ **29.** $\displaystyle\int (x + 5)^{3/2}\, dx$

✳ **30.** $\displaystyle\int \sqrt{2 - x}\, dx$

✳ **31.** $\displaystyle\int \frac{1}{\sqrt{4x + 3}}\, dx$

✳ **32.** $\displaystyle\int (2x - 3)^{3/2}\, dx$

✳ **33.** $\displaystyle\int (3x + 1)^5\, dx$

✳ **34.** $\displaystyle\int (1 - 2x)^7\, dx$

✳ **35.** $\displaystyle\int \frac{1}{(x + 4)^2}\, dx$

✳ **36.** $\displaystyle\int \frac{1}{(1 + 3x)^6}\, dx$

✳ **37.** $\displaystyle\int x(x^2 + 1)^3\, dx$

✳ **38.** $\displaystyle\int x^2(2 + 3x^3)^7\, dx$

✳ **39.** $\displaystyle\int \frac{x}{(2 + x^2)^2}\, dx$

✳ **40.** $\displaystyle\int \cos 2x\, dx$

✳✳ **41.** $\displaystyle\int \cos^2 x \sin x\, dx$

✳✳ **42.** $\displaystyle\int 3 \sin 2x \cos 2x\, dx$

✳✳ **43.** $\displaystyle\int \sec 12x \tan 12x\, dx$

✳✳ **44.** $\displaystyle\int \csc^2 4x\, dx$

✳✳ **45.** $\displaystyle\int e^{4x}\, dx$

✳✳ **46.** $\displaystyle\int x e^{-x^2}\, dx$

✳✳ **47.** $\displaystyle\int \frac{e^{3/x}}{x^2}\, dx$

✳✳ **48.** $\displaystyle\int e^{4x - 3}\, dx$

✳✳ **49.** $\displaystyle\int \frac{1}{3x + 2}\, dx$

✳✳ **50.** $\displaystyle\int \frac{2}{7 - 5x}\, dx$

✳✳ **51.** $\displaystyle\int \frac{x}{1 - x^2}\, dx$

✳✳ **52.** $\displaystyle\int \frac{3x^2}{1 - 4x^3}\, dx$

✳✳ **53.** $\displaystyle\int 2^x\, dx$

✳✳ **54.** $\displaystyle\int 3^{2x}\, dx$

✳✳ **55.** $\displaystyle\int \frac{e^x}{e^x + 1}\, dx$

✳✳ **56.** $\displaystyle\int \sin x(1 + \cos x)^4\, dx$

✳✳ **57.** $\displaystyle\int \frac{\cos x}{\sin^3 x}\, dx$

✳✳ **58.** $\displaystyle\int e^{2x}(1 + e^{2x})^3\, dx$

✳✳ **59.** $\displaystyle\int \frac{\sec^2 x}{\tan^2 x}\, dx$

In Exercises 60–65 find a one-parameter family of functions satisfying the differential equation.

** **60.** $\dfrac{dy}{dx} = x^3 - \dfrac{1}{x^2}$

** **61.** $\dfrac{dy}{dx} = \sqrt{3 - 4x}$

** **62.** $\dfrac{dy}{dx} = \dfrac{1}{(3x + 5)^{3/2}}$

** **63.** $\dfrac{dy}{dx} = x^2(2x^3 + 4)^4$

** **64.** $\dfrac{dy}{dx} = \dfrac{x^3}{(2 + 3x^4)^2}$

** **65.** $\dfrac{dy}{dx} = \sin x(1 + \cos^2 x)$

** **66.** Find a function $y = f(x)$ that satisfies the differential equation

$$\frac{dy}{dx} = \frac{1}{x^2}$$

and passes through the two points $(1, 1)$ and $(-1, -2)$.

** **67.** (a) Find the indefinite integral of the signum function of Exercise 43 in Section 2.4 (page 134).

(b) Prove that sgn x cannot have an antiderivative on any interval containing $x = 0$.

** **68.** Find deflections of a uniform beam with mass m and length L when both ends are simply supported.

** **69.** Repeat Exercise 68 if both ends are fixed horizontally.

** **70.** Repeat Exercise 68 if the left end $x = 0$ is fixed horizontally and the right end is free.

** **71.** Repeat Exercise 68 if the left end is fixed horizontally and the right end is simply supported.

** **72.** Find deflections of the beam in Figure 5.5 if the 50 kg block is on the right half of the beam. Is the deflection at the right end larger or smaller than in Figure 5.5?

** **73.** Find deflections of the beam in Figure 5.5 if the 50 kg block is centred on the beam. How does the deflection at the right end compare to that in Figure 5.5 and Exercise 72?

5.2 Integrating Velocity and Acceleration

In Section 4.7 we discussed relationships among position, velocity, and acceleration from the viewpoint of derivatives. We now consider these relationships through antiderivatives, an approach providing a far more practical viewpoint when it comes to applications.

For motion along the x-axis, velocity is the derivative of position with respect to time: $v(t) = dx/dt$. We can say, therefore, that position is the antiderivative of velocity:

$$x(t) = \int v(t)\,dt. \tag{5.8}$$

Similarly, as acceleration is the derivative of velocity,

$$v(t) = \int a(t)\,dt. \tag{5.9}$$

Thus, given the acceleration of an object moving along a straight line, we can antidifferentiate to find its velocity and antidifferentiate again for its position. Since each antidifferentiation introduces an arbitrary constant, additional information must be specified in order to evaluate these constants.

EXAMPLE 5.5

The car in Figure 5.11 accelerates from rest when the light turns green. Initially, the acceleration is 10 m/s^2, but it decreases linearly to zero after 10 s. Find the velocity and position of the car during this time interval.

FIGURE 5.11

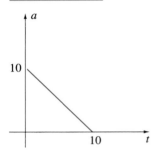

SOLUTION   Let us set up the coordinate system in Figure 5.11 and choose time $t = 0$ when the car pulls away from the light. Given this time convention, the acceleration of the car as a function of time t (see Figure 5.12) is

$$a(t) = 10 - t, \quad 0 \le t \le 10.$$

If $v(t)$ denotes the velocity of the car during the time interval, then

$$\frac{dv}{dt} = 10 - t,$$

and antidifferentiation gives

$$v(t) = 10t - \frac{t^2}{2} + C.$$

FIGURE 5.12

By our time convention, the velocity of the car is zero at time $t = 0$; that is, $v(0) = 0$, and from this we obtain

$$0 = 10(0) - \frac{(0)^2}{2} + C.$$

Consequently, $C = 0$, and

$$v(t) = 10t - \frac{t^2}{2}.$$

This is the velocity of the car during the time interval $0 \le t \le 10$, measured in metres per second.

Since the position of the car with respect to its original position at time $t = 0$ is denoted by x,

$$v = \frac{dx}{dt} = 10t - \frac{t^2}{2}.$$

Thus,

$$x(t) = 5t^2 - \frac{t^3}{6} + D.$$

Since $x(0) = 0$, the constant D must also be zero, and the position of the car indicated by its distance in metres from the stoplight, for $0 \le t \le 10$, is

$$x(t) = 5t^2 - \frac{t^3}{6}.$$

EXAMPLE 5.6

A stone is thrown vertically upward over the edge of a cliff at 25 m/s. When does it hit the base of the cliff if the cliff is 100 m high?

FIGURE 5.13

SOLUTION Let us measure y as positive upward, taking $y = 100$ and $t = 0$ at the point and instant of projection (Figure 5.13). A law of physics states that when an object near the earth's surface is acted on by gravity alone, it experiences an acceleration whose magnitude is 9.81 m/s². If a denotes the acceleration of the stone and v its velocity, then

$$a = \frac{dv}{dt} = -9.81$$

(a is negative since it is in the negative y-direction). To obtain v we take antiderivatives with respect to t:

$$v(t) = -9.81t + C.$$

By our time convention, $v(0) = 25$ so that

$$25 = -9.81(0) + C.$$

Thus, $C = 25$, and

$$v(t) = -9.81t + 25.$$

We now have the velocity of the stone at any given instant. To find the position of the stone we set $v = dy/dt$,

$$\frac{dy}{dt} = -9.81t + 25,$$

and take antiderivatives once again:

$$y(t) = -4.905t^2 + 25t + D.$$

Since $y = 100$ when $t = 0$,

$$100 = -4.905(0)^2 + 25(0) + D.$$

Hence, $D = 100$, and

$$y(t) = -4.905t^2 + 25t + 100.$$

We have found the equation that tells us exactly where the stone is at any given time. To determine when the stone strikes the base of the cliff, we set $y(t) = 0$; that is,

$$0 = -4.905t^2 + 25t + 100,$$

a quadratic equation with solutions

$$t = \frac{-25 \pm \sqrt{25^2 - 4(-4.905)(100)}}{-9.81} = \frac{25 \pm \sqrt{2587}}{9.81} = 7.7 \quad \text{or} \quad -2.6.$$

Since the negative root must be rejected, we find that the stone strikes the base of the cliff after 7.7 s.

EXAMPLE 5.7

Two stones are thrown vertically upward one second apart over the edge of the cliff in Example 5.6. The first is thrown at 25 m/s, the second at 20 m/s. Determine if and when they ever pass each other.

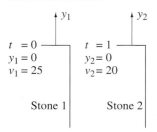

SOLUTION   Let us choose a different coordinate system to that of Example 5.6 by taking y as positive upward with $y = 0$ and $t = 0$ at the point and instant of projection of the first stone (Figure 5.14). The acceleration of stone 1 is

$$a_1 = -9.81 = \frac{dv_1}{dt},$$

from which

$$v_1(t) = -9.81t + C.$$

Since $v_1(0) = 25$, it follows that $25 = C$, and

$$v_1(t) = -9.81t + 25, \quad t \geq 0.$$

Thus,

$$\frac{dy_1}{dt} = -9.81t + 25,$$

from which we have

$$y_1(t) = -4.905t^2 + 25t + D.$$

Since $y_1(0) = 0$, we find that $0 = D$, and

$$y_1(t) = -4.905t^2 + 25t, \quad t \geq 0.$$

The acceleration of stone 2 is also -9.81; hence we have

$$a_2 = -9.81 = \frac{dv_2}{dt},$$

from which

$$v_2(t) = -9.81t + E.$$

Because $v_2(1) = 20$, we must have $20 = -9.81(1) + E$, or $E = 29.81$.

Thus,

$$v_2(t) = -9.81t + 29.81, \quad t \geq 1.$$

Consequently,

$$\frac{dy_2}{dt} = -9.81t + 29.81,$$

from which

$$y_2(t) = -4.905t^2 + 29.81t + F.$$

Since $y_2(1) = 0$, it follows that $0 = -4.905(1)^2 + 29.81(1) + F$, or $F = -24.905$.

Thus,

$$y_2(t) = -4.905t^2 + 29.81t - 24.905, \quad t \geq 1.$$

The stones will pass each other if y_1 and y_2 are ever equal for the same time t; that is, if

$$-4.905t^2 + 25t = -4.905t^2 + 29.81t - 24.905.$$

The solution of this equation is

$$t = \frac{24.905}{4.81} = 5.2.$$

Because the first stone does not strike the base of the cliff for 7.7 s (Example 5.6), it follows that the stones do indeed pass each other 5.2 s after the first stone is projected.

An important point to note in the above examples is that the coordinate system and time convention were specified immediately; that is, we decided, and did so at the start of the problem, where to place the origin of our coordinate system, which direction to choose as positive, and when to choose $t = 0$. Only then were we able to specify the correct sign for acceleration. Furthermore, we determined constants from antidifferentiations using initial conditions expressed in terms of our coordinate system and time convention. Throughout the solutions, we were careful to refer everything to our choice of coordinates and time. Remember, then, to specify clearly the coordinate system and time convention at the beginning of a problem.

We should also note that in each of the examples we antidifferentiated with respect to time, *only those equations that were valid for a range of values of time.* It is a common error to antidifferentiate equations that are valid only at one instant. For instance, students would reason in Example 5.6 that the initial velocity is 25, $v = dy/dt = 25$, and hence $y(t) = 25t + C$. This is incorrect, because the equation $dy/dt = 25$ is valid only at time $t = 0$, and cannot be antidifferentiated.

EXERCISES 5.2

In Exercises 1–8 we have defined acceleration $a(t)$ of an object moving along the x-axis during some time interval and specified the initial conditions $x(0)$ and $v(0)$. Find the velocity $v(t)$ and position $x(t)$ of the object as functions of time.

* **1.** $a(t) = t + 2$, $0 \le t \le 3$; $v(0) = 0, x(0) = 0$

* **2.** $a(t) = 6 - 2t$, $0 \le t \le 3$; $v(0) = 5, x(0) = 0$

* **3.** $a(t) = 6 - 2t$, $0 \le t \le 4$; $v(0) = 5, x(0) = 0$

* **4.** $a(t) = 120t - 12t^2$, $0 \le t \le 10$; $v(0) = 0, x(0) = 4$

* **5.** $a(t) = t^2 + 1$, $0 \le t \le 5$; $v(0) = -1, x(0) = 1$

* **6.** $a(t) = t^2 + 5t + 4$, $0 \le t \le 15$; $v(0) = -2, x(0) = -3$

* **7.** $a(t) = \cos t$, $t \ge 0$; $v(0) = 0, x(0) = 0$

* **8.** $a(t) = 3 \sin t$, $t \ge 0$; $v(0) = 1, x(0) = 4$

** **9.** The velocity of an object moving along the x-axis is given in metres per second by

$$v(t) = 3t^2 - 9t + 6$$

where t is time in seconds. If the object starts from position $x = 1$ m at time $t = 0$, answer each of the following questions:

(a) What is the acceleration of the object at $t = 5$ s?

(b) What is the position of the object at $t = 2$ s?

(c) Is the object speeding up or slowing down at $t = 5/4$ s?

(d) What is the closest the object ever comes to the origin?

5.3 Change of Variable in the Indefinite Integral

In Section 5.1 we suggested two methods for evaluating indefinite integrals—"recognition" and "guess and fix up." In this section we show how a "change of variable" can often replace a complex integration problem with a simpler one.

Consider the indefinite integral

$$\int x\sqrt{2x + 1}\, dx.$$

What is annoying about this integrand is the sum of two terms $2x + 1$ under a square root. This can be changed by setting $u = 2x + 1$. As a result, $\sqrt{2x + 1} = \sqrt{u}$, and the x in front of the square root is equal to $(u - 1)/2$. Now the differential dx is used to indicate antidifferentiation in the problem with respect to x. But surely there must be another reason why we have chosen the differential to denote this. If we regard $x\sqrt{2x + 1}\, dx$ as a product of $x\sqrt{2x + 1}$ and dx, then perhaps we should obtain an expression for dx in terms of du. To do this we note that since $u = 2x + 1$, then

$$\frac{du}{dx} = 2.$$

As derivatives can be regarded as quotients of differentials, we can rewrite this equation in the form

$$dx = \frac{du}{2}.$$

If we make all these substitutions into the indefinite integral

$$\int x\sqrt{2x + 1}\, dx,$$

the result is an integration problem in the variable u, which is easy to evaluate:

$$\int \left(\frac{u - 1}{2}\right)\sqrt{u}\,\frac{du}{2} = \frac{1}{4}\int (u^{3/2} - u^{1/2})\, du = \frac{1}{4}\left(\frac{2}{5}u^{5/2} - \frac{2}{3}u^{3/2}\right) + C.$$

If u is now replaced by $2x + 1$, the result is

$$\frac{1}{10}(2x + 1)^{5/2} - \frac{1}{6}(2x + 1)^{3/2} + C.$$

Differentiation of this function quickly indicates that its derivative is indeed $x\sqrt{2x + 1}$ and, therefore,

$$\int x\sqrt{2x + 1}\, dx = \frac{1}{10}(2x + 1)^{5/2} - \frac{1}{6}(2x + 1)^{3/2} + C.$$

In this example, the substitution $u = 2x + 1$ replaces a complex integration in x with a simple one in u. Once the problem in u is solved, replacement of u's with x's gives the solution to the original indefinite integral. This method is generally applicable and is justified in the following theorem.

THEOREM 5.3

Suppose the change of variable $u = h(x) \Leftrightarrow x = g(u)$ replaces the indefinite integral

$$\int f(x)\,dx \quad \text{with} \quad \int f\big(g(u)\big)\,g'(u)\,du,$$

where $g(u)$ is differentiable on some interval, and $f(x)$ is continuous on the range of $g(u)$. If $F(u)$ is an antiderivative of $f\big(g(u)\big)\,g'(u)$, then

$$\int f(x)\,dx = F\big(h(x)\big) + C.$$

EXAMPLE 5.8

Evaluate the following indefinite integrals:

(a) $\displaystyle \int \sqrt[5]{2x+4}\,dx$ (b) $\displaystyle \int \frac{x}{\sqrt{x+1}}\,dx$ (c) $\displaystyle \int \sin^3 x \cos^2 x\,dx$

SOLUTION

(a) We could use "guess and fix up" in this case. With an initial guess of $(2x+4)^{6/5}$, we calculate

$$\frac{d}{dx}(2x+4)^{6/5} = \frac{6}{5}(2x+4)^{1/5}(2).$$

Thus,

$$\int (2x+4)^{1/5}\,dx = \frac{5}{12}(2x+4)^{6/5} + C.$$

Alternatively, if we set $u = 2x+4$, then $du = 2dx$, and

$$\int (2x+4)^{1/5}\,dx = \int u^{1/5}\frac{du}{2} = \frac{1}{2}\left\{\frac{5}{6}u^{6/5}\right\} + C = \frac{5}{12}(2x+4)^{6/5} + C.$$

(b) If we set $u = x+1$, then $du = dx$, and

$$\int \frac{x}{\sqrt{x+1}}\,dx = \int \frac{(u-1)}{\sqrt{u}}\,du = \int (u^{1/2} - u^{-1/2})\,du = \frac{2}{3}u^{3/2} - 2u^{1/2} + C$$

$$= \frac{2}{3}(x+1)^{3/2} - 2(x+1)^{1/2} + C.$$

A different substitution is also possible. If we set $u = \sqrt{x+1}$, then

$$du = \frac{1}{2\sqrt{x+1}}\,dx \quad \text{and}$$

$$\int \frac{x}{\sqrt{x+1}}\,dx = \int (u^2 - 1)(2du) = 2\left(\frac{u^3}{3} - u\right) + C$$

$$= \frac{2}{3}(x+1)^{3/2} - 2(x+1)^{1/2} + C.$$

(c) If we set $u = \cos x$, then $du = -\sin x\, dx$ and

$$\int \sin^3 x \cos^2 x\, dx = \int \sin^2 x \cos^2 x \sin x\, dx$$

$$= \int (1 - \cos^2 x) \cos^2 x \sin x\, dx$$

$$= \int (1 - u^2) u^2 (-du) = \int (u^4 - u^2)\, du$$

$$= \frac{u^5}{5} - \frac{u^3}{3} + C = \frac{1}{5} \cos^5 x - \frac{1}{3} \cos^3 x + C.$$

EXERCISES 5.3

In Exercises 1–30 evaluate the indefinite integral.

* **1.** $\displaystyle\int (5x + 14)^9\, dx$

* **2.** $\displaystyle\int \sqrt{1 - 2x}\, dx$

* **3.** $\displaystyle\int \frac{1}{(3y - 12)^{1/4}}\, dy$

* **4.** $\displaystyle\int \frac{5}{(5 - 42x)^{1/4}}\, dx$

* **5.** $\displaystyle\int x^2 (3x^3 + 10)^4\, dx$

* **6.** $\displaystyle\int \frac{x}{(x^2 + 4)^2}\, dx$

* **7.** $\displaystyle\int \sin^4 x \cos x\, dx$

* **8.** $\displaystyle\int \frac{x^2}{(x - 2)^4}\, dx$

* **9.** $\displaystyle\int z\sqrt{1 - 3z}\, dz$

* **10.** $\displaystyle\int \frac{x}{\sqrt{2x + 3}}\, dx$

* **11.** $\displaystyle\int \frac{1 + \sqrt{x}}{\sqrt{x}}\, dx$

* **12.** $\displaystyle\int s^3 \sqrt{s^2 + 5}\, ds$

* **13.** $\displaystyle\int \sin^2 x \cos^3 x\, dx$

* **14.** $\displaystyle\int \sqrt{1 - \cos x} \sin x\, dx$

* **15.** $\displaystyle\int \frac{x^3}{(3 - x^2)^3}\, dx$

* **16.** $\displaystyle\int y^2 \sqrt{y - 4}\, dy$

* **17.** $\displaystyle\int \frac{(1 + \sqrt{u})^{1/2}}{\sqrt{u}}\, du$

* **18.** $\displaystyle\int x^8 (3x^3 - 5)^6\, dx$

* **19.** $\displaystyle\int \frac{1 + z^{1/4}}{\sqrt{z}}\, dz$

* **20.** $\displaystyle\int \frac{x + 1}{(x^2 + 2x + 2)^{1/3}}\, dx$

* **21.** $\displaystyle\int \frac{(x - 1)(x + 2)}{\sqrt{x}}\, dx$

** **22.** $\displaystyle\int \frac{\cos^3 x}{(3 - 4\sin x)^4}\, dx$

** **23.** $\displaystyle\int \sqrt{1 + \sin 4t} \cos^3 4t\, dt$

** **24.** $\displaystyle\int \sqrt{1 + \sqrt{x}}\, dx$

** **25.** $\displaystyle\int \tan^2 x \sec^2 x\, dx$ ** **26.** $\displaystyle\int \tan x \sec^2 x\, dx$

** **27.** $\displaystyle\int \frac{e^{2x}}{e^{2x}+1}\, dx$ ** **28.** $\displaystyle\int \frac{\ln x}{x}\, dx$

** **29.** $\displaystyle\int \frac{1}{x \ln x}\, dx$ ** **30.** $\displaystyle\int \frac{x}{(x^2+1)[\ln (x^2+1)]^2}\, dx$

** **31.** Consider the expression

$$\int \sin^3 x \cos^3 x\, dx$$

(a) Evaluate it by making the substitution $u = \sin x$;

(b) Evaluate it by making the substitution $u = \cos x$.

(c) Verify that these answers are the same.

** **32.** Evaluate

$$\int \sqrt{\frac{x^2}{1+x}}\, dx.$$

In Exercises 33–36 use the suggested change of variable to evaluate the indefinite integral.

** **33.** $\displaystyle\int \frac{\sqrt{4x - x^2}}{x^3}\, dx;$ set $u = 2/x$

** **34.** $\displaystyle\int \frac{\sqrt{x - x^2}}{x^4}\, dx;$ set $u = 1/x$

** **35.** $\displaystyle\int \frac{x}{(5 - 4x - x^2)^{3/2}}\, dx;$ set $u^2 = (5+x)/(1-x)$

** **36.** $\displaystyle\int \frac{1}{3(1 - x^2) - (5 + 4x)\sqrt{1 - x^2}}\, dx;$ set $u^2 = (1-x)/(1+x)$

** **37.** Show that the substitution $u - x = \sqrt{x^2 + x + 4}$ replaces the integral

$$\int \sqrt{x^2 + x + 4}\, dx$$

with the integral of a rational function of u.

** **38.** Show that the substitution $(x + 1)u = \sqrt{4 + 3x - x^2}$ replaces the integral

$$\int \frac{1}{\sqrt{4 + 3x - x^2}}\, dx$$

with the integral of a rational function of u.

5.4 An Introduction to Differential Equations

A differential equation contains derivatives of some unknown function; the equation must be solved for this function. In this section we use calculus to derive mathematical models to describe four applied problems. These models turn out to be differential equations that are reasonably easy to solve.

Before doing so, however, let us express extended power rule 3.21 in an integral form that is particularly useful for differential equations. Equation 3.21 states that when $u(x)$ is a differentiable function of x,

$$\frac{d}{dx}u^n = nu^{n-1}\frac{du}{dx}.$$

It follows, therefore, that the antiderivative of $nu^{n-1}du/dx$ is u^n,

$$\int nu^{n-1}\frac{du}{dx}\,dx = u^n + C.$$

Property 5.3b permits us to take the constant outside the integral, and the equation can then be divided by this quantity:

$$\int u^{n-1}\frac{du}{dx}\,dx = \frac{1}{n}u^n + C, \quad n \neq 0.$$

Now n is an arbitrary number in this formula, and therefore there is no loss in generality if we replace each n with $n + 1$. The result is

$$\int u^n\frac{du}{dx}\,dx = \frac{1}{n+1}u^{n+1} + C, \quad n \neq -1. \tag{5.10a}$$

The $n = -1$ case is the integral counterpart of formula 3.41,

$$\int \frac{1}{u}\frac{du}{dx}\,dx = \ln|u| + C. \tag{5.10b}$$

EXAMPLE 5.9

Find all curves that satisfy the condition that at every point on the curve the normal line passes through the origin.

SOLUTION   If $y = f(x)$ is the equation of a curve, then the slope of the tangent line at a point (x, y) on this curve is dy/dx, and the slope of the normal line is the negative reciprocal thereof. As derivatives can now be regarded as quotients of differentials, the negative reciprocal of dy/dx is $-dx/dy$. But when the normal line passes through the origin, its slope must also be given by the quotient $(y - 0)/(x - 0)$ (Figure 5.15). It follows that the required curve $y = f(x)$ must satisfy

$$\frac{y}{x} = -\frac{dx}{dy} \quad \text{or} \quad \frac{dy}{dx} = -\frac{x}{y}.$$

To solve this differential equation, we write

$$y\frac{dy}{dx} = -x$$

and antidifferentiate both sides with respect to x:

$$\int \left(y \frac{dy}{dx} \right) dx = \int -x \, dx.$$

The right integral is obviously $-x^2/2$, and, by formula 5.10a, the antiderivative on the left is $y^2/2$. When a constant of integration is added, we have

$$\frac{y^2}{2} = -\frac{x^2}{2} + C.$$

If we set $D^2 = 2C$, then

$$x^2 + y^2 = D^2;$$

thus, the only curves satisfying the given condition are circles centred at the origin.

FIGURE 5.15

EXAMPLE 5.10

Figure 5.16 shows a liquid container with a hole in its bottom. It is interesting, and perhaps surprising, to find that the speed at which liquid exits through the hole is independent of the shape of the container; it depends only on the depth of liquid. We show this here, and then discover a technique for finding the depth of liquid in the container as a function of time.

FIGURE 5.16

If we consider a thin surface layer of the liquid, then during a small interval of time the depth of liquid in the container drops by an amount equal to the thickness of the layer. Simultaneously, a volume of liquid equal to that in the layer exits through the hole and, in so doing, causes the gravitational potential energy of the layer, due to its elevated position relative to the hole, to be converted into kinetic energy of the liquid passing through the hole. Suppose we let v be the speed at which the liquid leaves the container when the depth of liquid is y. The potential energy of the layer relative to the bottom of the container is

$$\text{(mass of layer)}(g)(y),$$

where $g = 9.81$. The kinetic energy of an equal amount of liquid as it leaves the container is

$$\frac{1}{2}\text{(mass of layer)}v^2.$$

When we equate these energies, and cancel the mass of the layer, we obtain

$$\frac{1}{2}v^2 = gy.$$

In other words, the speed at which liquid exits through the hole is

$$v = \sqrt{2gy}. \tag{5.11}$$

This result is known as Torricelli's law. It results from an idealized situation in which all potential energy is converted into kinetic energy. Experience suggests that exit speed depends on other factors as well—the size of the hole, for one. Water leaves more slowly through a small hole than through a large one. It is often assumed that v is somewhat less than $\sqrt{2gy}$, and 5.11 is replaced by

$$v = c\sqrt{2gy}, \tag{5.12}$$

where $0 < c < 1$ is a constant called the *discharge coefficient*. Let us call 5.12 the modified Torricelli law. In the remainder of this example, use 5.12 to find a differential equation satisfied by depth of liquid in the container. Solve for depth when the container is a right-circular cylinder of radius r and height h with vertical axis.

SOLUTION   Suppose the area of the surface of the liquid is a function of depth y denoted by $A(y)$, and $V(t)$ is the volume of liquid in the container at any time t. Since depth of liquid changes at rate dy/dt, the rate at which the volume of liquid in the container changes is

$$\frac{dV}{dt} = A(y)\frac{dy}{dt}.$$

Both dV/dt and dy/dt are negative since V and y are decreasing. On the other hand, the rate at which liquid exits through the hole is the product av of the area a of the hole and exit speed v. It follows then that

$$A(y)\frac{dy}{dt} = -av = -ac\sqrt{2gy}.$$

Once the shape of the container is specified, then $A(y)$ is known, and this equation becomes a differential equation for $y(t)$.

If the container is a right circular cylinder with radius r, then $A(y) = \pi r^2$, in which case

$$\pi r^2\frac{dy}{dt} = -ac\sqrt{2gy} \qquad \Longrightarrow \qquad \frac{1}{\sqrt{y}}\frac{dy}{dt} = -\frac{\sqrt{2g}\,ac}{\pi r^2}.$$

Integration of both sides with respect to time t gives

$$2\sqrt{y} = -\frac{\sqrt{2g}\,ac}{\pi r^2}t + D,$$

where D is the constant of integration. If the cylinder is originally full and liquid exits starting at time $t = 0$, then $2\sqrt{h} = D$, and

$$2\sqrt{y} = -\frac{\sqrt{2g}ac}{\pi r^2}t + 2\sqrt{h} \quad \Longrightarrow \quad y = \left(\sqrt{h} - \frac{\sqrt{g}act}{\sqrt{2}\pi r^2}\right)^2.$$

It is now a simple matter to determine how long the cylinder takes to empty. Setting $y(t) = 0$ and solving for t gives

$$t = \frac{\pi r^2}{ac}\sqrt{\frac{2h}{g}}.$$

This was a particularly simple example in that the cross-sectional area $A(y)$ of the container was constant. Containers with variable cross-sections are discussed in the exercises.

EXAMPLE 5.11

When a mothball is exposed to the air, it slowly evaporates. If the mothball remains spherical at all times, and its radius after one year is only half of its original radius, when will it completely disappear?

SOLUTION   The size of the mothball can be measured in various ways—by its volume, its surface area, or its radius. Since we were given information about its radius, let us denote by $r(t)$ the radius of the mothball as a function of time t, and choose time $t = 0$ when the mothball is originally exposed to the air. If we let R represent the original radius of the mothball, then we can say that $r(0) = R$ and $r(1) = R/2$. This is the only information about $r(t)$ that we were given, and it certainly is not enough to solve the problem.

We need a statement on the instantaneous rate of evaporation of the mothball. Without such a statement, we must make our own reasonable assumption about how fast the naphthalene evaporates from the surface of the mothball. Since evaporation takes place only at the surface, let us assume that the amount of naphthalene evaporating per unit time is proportional to the surface area of the mothball at that time. If V and A represent volume and area of the mothball at any time, then this assumption becomes

$$\frac{dV}{dt} = kA,$$

where k is a constant. We have four variables in the problem: t, r, A, and V. By substituting $V = 4\pi r^3/3$ and $A = 4\pi r^2$, we can eliminate V and A:

$$\frac{d}{dt}\left(\frac{4}{3}\pi r^3\right) = 4\pi r^2 k,$$

and this equation reduces to

$$\frac{dr}{dt} = k.$$

Now we have a differential equation for $r(t)$, with solution

$$r(t) = kt + C.$$

Using the conditions $r(0) = R$ and $r(1) = R/2$, we find $C = R$ and $k = -R/2$. Consequently,

$$r(t) = R - \frac{Rt}{2} = R\left(1 - \frac{t}{2}\right).$$

The mothball completely disappears when $r = 0$, and this occurs when $t = 2$ years.

EXAMPLE 5.12

A tank originally contains 1000 L of water, in which 5 kg of salt has been dissolved.

(a) If a brine mixture containing 2 kg of salt for each 100 L of solution is poured into the tank at 10 mL/s, find the amount of salt in the tank as a function of time.

(b) If at the same time as brine is being added the mixture in the tank is being drawn off at 5 mL/s, find the amount of salt in the tank as a function of time. Assume that the mixture is stirred constantly.

SOLUTION

(a) Suppose we let $S(t)$ represent the number of grams of salt in the tank at time t. If we choose time $t = 0$ at the instant the brine mixture begins entering the original solution, then $S(0) = 5000$. Since 10 mL of mixture enter the tank each second, and each millilitre contains 0.02 gm of salt, it follows that 0.2 gm of salt enter the tank each second. Consequently, after t seconds, $0.2t$ gm of salt have been added to the tank, and the total amount of salt in the solution is, in grams,

$$S(t) = 5000 + 0.2t.$$

(b) Once again we let $S(t)$ represent the number of grams of salt in the tank at time t. Its derivative dS/dt, the rate of change of $S(t)$, is the difference between how fast salt is being added to the tank in the brine, and how fast salt is being removed as the mixture is removed. As in part (a), salt is being added at the constant rate of 0.2 gm/s. The rate at which salt leaves the tank, on the other hand, is not constant; it depends on the concentration of salt in the tank. Since the tank originally contained 10^6 mL of solution, and a net 5 mL enters each second, the amount of solution in the tank at time t is $10^6 + 5t$ mL. It follows that the concentration of salt in the solution at time t is, in grams per millilitre,

$$\frac{S}{10^6 + 5t},$$

where $S = S(t)$ is the amount of salt in the tank at that time. As solution is being drawn off at the rate of 5 mL/s, the rate at which salt leaves the tank is, in grams per second,

$$\frac{S}{10^6 + 5t}(5).$$

The net rate at which salt enters the tank is therefore

$$0.2 - \frac{5S}{10^6 + 5t}.$$

Since dS/dt also represents the rate of change of the amount of salt in the tank, we set

$$\frac{dS}{dt} = \frac{1}{5} - \frac{5S}{10^6 + 5t}.$$

To find $S(t)$ we must solve this differential equation subject to the condition that $S(0) = 5000$. We can clear the term $10^6 + 5t$ from the denominator if we multiply the equation by this factor:

$$(10^6 + 5t)\frac{dS}{dt} + 5S = \frac{1}{5}(10^6 + 5t).$$

But note that the left-hand side of this differential equation is the derivative of the product of S and $10^6 + 5t$; that is,

$$\frac{d}{dt}[S(10^6 + 5t)] = (10^6 + 5t)\frac{dS}{dt} + 5S.$$

Consequently, we may express the differential equation in the form

$$\frac{d}{dt}[S(10^6 + 5t)] = \frac{1}{5}(10^6 + 5t),$$

and take antiderivatives with respect to t on both sides:

$$S(10^6 + 5t) = \frac{1}{5}\int (10^6 + 5t)\,dt = \frac{1}{50}(10^6 + 5t)^2 + C.$$

Since $S(0) = 5000$, we find

$$5000(10^6) = \frac{1}{50}(10^6)^2 + C,$$

and hence, $C = -15 \times 10^9$. Thus, $S(t)$ is defined implicitly by

$$S(10^6 + 5t) = \frac{1}{50}(10^6 + 5t)^2 - 15 \times 10^9,$$

or explicitly by

$$S(t) = \frac{1}{50}(10^6 + 5t) - \frac{15 \times 10^9}{10^6 + 5t}.$$

A plot of this function is shown in Figure 5.17. It is asymptotic to the line

$$S = (10^6 + 5t)/50.$$

FIGURE 5.17

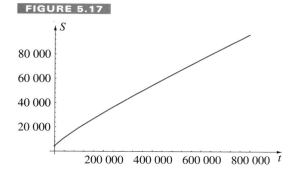

EXERCISES 5.4

✳✳ **1.** A boy lives 6 km from school. He decides to walk to school at a speed that is always proportional to the square of his distance from the school. If he is halfway to school after one hour, find his distance from school at any time. How long does it take him to reach school?

✳✳ **2.** A diving board of length L and uniform mass per unit length m is fixed horizontally (figure below). If a diver of mass M stands on the end of the board, the deflection $y = f(x)$ of the board must satisfy the differential equation

$$(EI)\frac{d^2y}{dx^2} = -Mg(L - x) - \frac{mg}{2}(L - x)^2,$$

where E and I are constants depending on the shape and material of the board, and $g > 0$ is the acceleration due to gravity.

(a) Find $y = f(x)$.

(b) What is the maximum deflection of the board from the horizontal?

✳✳ **3.** The water trough in the figure below is 4 m long. Its cross-section is an isosceles triangle with a half-metre base and a half-metre altitude. Water leaks out through a hole of area 1 cm^2 in the bottom with speed in metres per second given by $v = \sqrt{gD/2}$, where D (in metres) is the depth of water in the trough, and $g > 0$ is the acceleration due to gravity. Find how long a full trough takes to empty.

✳✳ **4.** The y-intercept of the normal line to a curve at any point is always equal to 2. If the curve passes through the point $(3, 4)$, find its equation.

✳✳ **5.** A tank has 100 L of solution containing 4 kg of sugar. A mixture with 10 gm of sugar per litre of solution is added at 200 mL/min. At the same time, 100 mL of well-stirred mixture is removed each minute. Find the amount of sugar in the tank as a function of time.

✳✳ **6.** A tank in the form of an inverted right circular cone of height H and radius R is filled with water. Water escapes through a hole of cross-sectional area a at the vertex. Use the modified Torricelli law 5.12 to find a formula for the time the tank takes to empty.

✷✷ **7.** A spring of negligible mass and elasticity constant $k > 0$ is attached to a wall at one end and a mass M at the other (figure below). The mass is free to slide horizontally along a frictionless surface. If $x = 0$ is taken as the position of M when the spring is unstretched and M is set into motion, the differential equation describing the position $x(t)$ of M is

$$\frac{d^2x}{dt^2} = -\frac{k}{M}x.$$

If motion is initiated by imparting a speed v_0 in the positive x-direction to M at position $x = 0$, find the velocity of M as a function of position. Note that

$$\frac{d^2x}{dt^2} = \frac{dv}{dt} = \frac{dv\,dx}{dx\,dt} = v\frac{dv}{dx}.$$

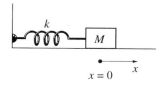

✷✷ **8.** When a mass m falls under the influence of gravity alone, it experiences an acceleration d^2r/dt^2 described by

$$m\frac{d^2r}{dt^2} = -\frac{GmM}{r^2},$$

where M is the mass of the earth, G is a positive constant, and r is the distance from m to the centre of the earth (figure below). If m is dropped from a height h above the surface of the earth, find its velocity when it strikes the earth. What is the maximum attainable speed of m?

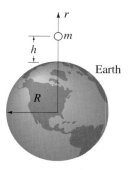

✷✷ **9.** A container in the form of an inverted right circular cone of radius 4 cm and height 10 cm is full of water. Water evaporates from the surface at a rate proportional to the area of the surface. If the water level drops 1 cm in the first 5 days, how long will it take for the water to evaporate completely?

✷✷ **10.** Allometry is the quantitative study of the size of one part of a biological object in relation to the size of another part. For example, a researcher might be interested in the rate at which a child's eyes grow in relation to the rate of growth of his or her head. It is clear that the head grows more rapidly than the eyes, but just how much more rapidly? If $x(t)$ and $y(t)$ represent volumes of the eye and head respectively, at any given time t, then $(1/x)dx/dt$ and $(1/y)dy/dt$ are called *relative growth rates*. Numerous experimental results have suggested that these relative growth rates are proportional

$$\frac{1}{y}\frac{dy}{dt} = k\frac{1}{x}\frac{dx}{dt}.$$

This is called the *allometric law*. If we eliminate t from this equation, we obtain a differential equation for y as a function of x:

$$\frac{dy}{dx} = k\frac{y}{x}.$$

Solve this differential equation. (Hint: Multiply by x^k.)

** **11.** In psychology, it is sometimes assumed that if R represents the reaction of a subject to an amount S of stimulus, then relative increases of R and S are proportional. In other words, if S is changed at rate dS/dt, and the resulting rate of change of R is dR/dt, then

$$\frac{1}{R}\frac{dR}{dt} = k\frac{1}{S}\frac{dS}{dt}.$$

Although this is often called the Brentano-Stevens law, it is essentially the allometric law of Exercise 10 in a different setting. Find R as a function of S.

** **12.** Three boys of the same height carry a board of length L and uniform mass per unit length m horizontally as shown in the figure below. The weight of the board causes it to bend, and its shape is the same on either side of the middle boy. Between the first two boys, for $0 \le x \le L/2$, the displacement of the board from the horizontal $y = f(x)$ must satisfy the differential equation

$$(EI)\frac{d^2y}{dx^2} = Ax - \frac{mg}{2}x^2,$$

where E and I are constants depending on the cross-section and material of the board, A is the weight supported by the boy at $x = 0$, and $g > 0$ is the acceleration due to gravity.

(a) Solve this equation along with the conditions $f(0) = f(L/2) = 0$ to find $f(x)$ as a function of x, E, I, m, g, and A.

(b) Find a condition that permits determination of A.

** **13.** The steady-state temperature T in a region bounded by two concentric spheres of radii 1 m and 2 m must satisfy the differential equation

$$\frac{d^2T}{dr^2} + \frac{2}{r}\frac{dT}{dr} = 0,$$

where r is the radial distance from the common centre of the spheres. If temperatures on the inner and outer spheres are maintained at $10°C$ and $20°C$ respectively, find the temperature distribution $T(r)$ between the spheres.

** **14.** A room with volume 100 m^3 initially contains 0.1% carbon dioxide. Beginning at time $t = 0$, fresher air containing 0.05% carbon dioxide flows into the room at 5 m^3/min. The well-mixed air in the room flows out at the same rate. Find the amount of carbon dioxide in the room as a function of time. What is the limit of the function as $t \to \infty$?

** **15.** Find all curves such that, if the normal is drawn at any point (x, y) on the curve, then that part of the normal between (x, y) and the x-axis is bisected by the y-axis.

** **16.** It is sometimes assumed that the density ρ of the atmosphere is related to height h above sea level according to the differential equation

$$\frac{d\rho}{dh} = -\frac{\rho^{2-\delta}}{k\delta},$$

where $\delta > 1$ and $k > 0$ are constants.

    (a) Show that if ρ_0 is density at sea level, then

$$\rho^{\delta-1} = -\frac{h}{k}\left(\frac{\delta - 1}{\delta}\right) + \rho_0^{\delta-1}.$$

    (b) If air pressure P and density are related by the equation $P = k\rho^\delta$, prove that

$$P^{1-1/\delta} = P_0^{1-1/\delta} - [(1 - 1/\delta)\rho_0 P_0^{-1/\delta}]h,$$

where P_0 is air pressure at sea level.

    (c) Show that the effective height of the atmosphere is

$$\frac{\delta P_0}{(\delta - 1)\rho_0}.$$

SUMMARY

The indefinite integral or antiderivative of a function $f(x)$ is a family of functions, each of which has $f(x)$ as its first derivative. If we can find one antiderivative of $f(x)$, then every other antiderivative in the family differs from this one by at most an additive constant. According to Theorem 5.2, the antiderivative of a sum of two functions is the sum of their antiderivatives, and multiplicative constants may be by-passed when finding antiderivatives

$$\int [f(x) + g(x)]\,dx = \int f(x)\,dx + \int g(x)\,dx; \qquad \int cf(x)\,dx = c\int f(x)\,dx.$$

The most important integration formula is for powers of x,

$$\int x^n\,dx = \begin{cases} \dfrac{1}{n+1}x^{n+1} + C, & n \neq -1 \\[2mm] \ln|x| + C, & n = -1. \end{cases}$$

Other integration formulas that arise from differentiations of trigonometric and exponential functions are listed in equations 5.4.

    In this chapter we have studied three ways to evaluate indefinite integrals. (Others will follow in Chapter 9.) First, due to our expertise in differentiation, some antiderivatives are immediately recognizable. Second, sometimes an antiderivative can be guessed to within a multiplicative constant, and then fixed up. Third, a change of variable can often replace a complex integration problem with a simpler one.

The Definite Integral

There are two aspects to calculus: differentiation and integration. We dealt with differentiation and its applications in Chapters 3 and 4. Integration in Chapter 5 was synonymous with antidifferentiation. In this chapter we investigate a new type of integral called the "definite integral." Before doing that, we introduce sigma notation, a compact notation for sums of terms, particularly useful for definite integrals. In Section 6.2 we discuss four problems that motivate the concept of the definite integral. In subsequent sections we develop the definition for the integral and discuss various ways to evaluate it. Section 6.6 presents an application of the definite integral—finding the average value of a function—and Chapter 7 develops a multitude of physical applications.

The definite integral is, by definition, very different from the indefinite integral, yet the two are intimately related through the Fundamental Theorem of Integral Calculus (Sections 6.4 and 6.5).

6.1 Sigma Notation

One of the most important notations in calculus, *sigma notation*, is used to represent a sum of terms, all of which are similar in form. For example, the six terms in the sum

$$\frac{1}{1 + 2^2} + \frac{2}{1 + 3^2} + \frac{3}{1 + 4^2} + \frac{4}{1 + 5^2} + \frac{5}{1 + 6^2} + \frac{6}{1 + 7^2}$$

are all formed in the same way: each is an integer divided by 1 plus the square of the next integer. If k represents an integer, we can say that every term has the form $k/[1 + (k + 1)^2]$; the first term is obtained by setting $k = 1$, the second by setting $k = 2$, and so on. As $k/[1 + (k + 1)^2]$ represents each and every term in the sum, we can describe the sum in words by saying, "Assign k in $k/[1 + (k + 1)^2]$ the integer values between 1 and 6, inclusively, and add the resulting numbers together." The notation used to represent this statement is

$$\sum_{k=1}^{6} \frac{k}{1 + (k + 1)^2}.$$

The Σ symbol is the Greek letter sigma which, in this case, means "sum." Summed are expressions of the form $k/[1 + (k + 1)^2]$, and the "$k = 1$" and "6" indicate that every integer from 1 to 6 is substituted into $k/[1 + (k + 1)^2]$. We call $k/[1 + (k + 1)^2]$ the **general term** of the sum, since it represents each and every term therein. The letter k is called the **index of summation** or **variable of summation**, and 1 and 6 are called the **limits of summation**. Any letter may be used to represent the index of summation; most commonly used are i, j, k, l, m, and n. Here is another example:

$$\sum_{n=5}^{15} n^3 = 5^3 + 6^3 + 7^3 + 8^3 + 9^3 + 10^3 + 11^3 + 12^3 + 13^3 + 14^3 + 15^3.$$

In summing a large number of terms, it is quite cumbersome to write them all down. One way around this difficulty is to write the first few terms to indicate the pattern by which the terms are formed, three dots, and the last term. For example, to indicate the sum of the cubes of the positive integers less than or equal to 100, we write

$$1^3 + 2^3 + 3^3 + 4^3 + 5^3 + \cdots + 100^3,$$

where the dots indicate that all numbers between 5^3 and 100^3 are to be filled in according to the same pattern suggested. Obviously, it would be preferable to express the sum in sigma notation

$$\sum_{k=1}^{100} k^3,$$

which is compact and leaves no doubt as to the pattern by which terms are formed.

EXAMPLE 6.1

Write each of the following sums in sigma notation:

(a) $\dfrac{1}{2 \cdot 3} + \dfrac{4}{3 \cdot 4} + \dfrac{9}{4 \cdot 5} + \dfrac{16}{5 \cdot 6} + \cdots + \dfrac{169}{14 \cdot 15}$;

(b) $\dfrac{16}{\sqrt{2}} + \dfrac{32}{\sqrt{3}} + \dfrac{64}{\sqrt{4}} + \dfrac{128}{\sqrt{5}} + \cdots + \dfrac{4096}{\sqrt{10}}$.

SOLUTION

(a) We examine each term carefully to determine the pattern. If we use i as the index of summation, the general term is $i^2/[(i + 1)(i + 2)]$. Consequently,

$$\frac{1}{2 \cdot 3} + \frac{4}{3 \cdot 4} + \frac{9}{4 \cdot 5} + \frac{16}{5 \cdot 6} + \cdots + \frac{169}{14 \cdot 15} = \sum_{i=1}^{13} \frac{i^2}{(i + 1)(i + 2)}.$$

(b) If n is chosen as the index of summation, then

$$\frac{16}{\sqrt{2}} + \frac{32}{\sqrt{3}} + \frac{64}{\sqrt{4}} + \frac{128}{\sqrt{5}} + \cdots + \frac{4096}{\sqrt{10}} = \sum_{n=4}^{12} \frac{2^n}{\sqrt{n - 2}}.$$

The representations for the sums in Example 6.1 in terms of sigma notation are not unique. In fact, there is an infinite number of representations for each sum. Consider, for example, the sum represented by

$$\sum_{i=1}^{9} \frac{2^{i+3}}{\sqrt{i + 1}}.$$

If we write out some of the terms in the summation, we find

$$\sum_{i=1}^{9} \frac{2^{i+3}}{\sqrt{i + 1}} = \frac{16}{\sqrt{2}} + \frac{32}{\sqrt{3}} + \frac{64}{\sqrt{4}} + \frac{128}{\sqrt{5}} + \cdots + \frac{4096}{\sqrt{10}},$$

the same sum as that in Example 6.1(b). This sum can also be represented by

$$\sum_{j=-2}^{6} \frac{2^{j+6}}{\sqrt{j+4}} \quad \text{and} \quad \sum_{m=15}^{23} \frac{2^{m-11}}{\sqrt{m-13}}.$$

We can transform any one of these representations into any other by making a change of variable of summation. For example, if in the summation of Example 6.1(b) we set $i = n - 3$, then $n = i + 3$, and

$$\frac{2^n}{\sqrt{n-2}} = \frac{2^{i+3}}{\sqrt{i+1}}.$$

For the limits, we find that $i = 1$ when $n = 4$, and $i = 9$ when $n = 12$. It follows that

$$\sum_{n=4}^{12} \frac{2^n}{\sqrt{n-2}} = \sum_{i=1}^{9} \frac{2^{i+3}}{\sqrt{i+1}}.$$

Similarly, the changes $j = n - 6$ and $m = n + 11$ transform

$$\sum_{n=4}^{12} \frac{2^n}{\sqrt{n-2}}$$

into

$$\sum_{j=-2}^{6} \frac{2^{j+6}}{\sqrt{j+4}} \quad \text{and} \quad \sum_{m=15}^{23} \frac{2^{m-11}}{\sqrt{m-13}}.$$

EXAMPLE 6.2

Change each of the following summations into representations that are initiated with the integer 1:

$$\text{(a)} \quad \sum_{i=4}^{26} \frac{i^{2/3}}{i^2 + i + 1} \qquad \text{(b)} \quad \sum_{j=-3}^{102} \frac{j^2 + 2j + 5}{\sin(j + 5)}$$

SOLUTION

(a) To initiate the summation at 1, we want $n = 1$ when $i = 4$, so we set $n = i - 3$. Then $i = n + 3$, and by substitution we have

$$\sum_{i=4}^{26} \frac{i^{2/3}}{i^2 + i + 1} = \sum_{n=1}^{23} \frac{(n+3)^{2/3}}{(n+3)^2 + (n+3) + 1} = \sum_{n=1}^{23} \frac{(n+3)^{2/3}}{n^2 + 7n + 13}.$$

(b) In this case, we set $n = j + 4$. Then, $j = n - 4$, and

$$\sum_{j=-3}^{102} \frac{j^2 + 2j + 5}{\sin(j + 5)} = \sum_{n=1}^{106} \frac{(n-4)^2 + 2(n-4) + 5}{\sin(n + 1)} = \sum_{n=1}^{106} \frac{n^2 - 6n + 13}{\sin(n + 1)}.$$

If we examine the results of Example 6.2 and the summations immediately preceding this example, we soon come to realize that there is a very simple way to change variables. To illustrate, consider once again

$$\sum_{n=4}^{12} \frac{2^n}{\sqrt{n-2}}.$$

Should we wish to initiate the summation with 1, rather than 4, we lower both limits by 3. To compensate, we replace each n in the general term by $n+3$; the result is

$$\sum_{n=1}^{9} \frac{2^{n+3}}{\sqrt{(n+3)-2}} = \sum_{n=1}^{9} \frac{2^{n+3}}{\sqrt{n+1}}.$$

Likewise, for simplicity in the summation

$$\sum_{n=1}^{10} \frac{(n+4)^2}{e^{n+4}},$$

it would be advisable to lower each n in the general term by 4. This can be done provided we raise each limit by 4,

$$\sum_{n=1}^{10} \frac{(n+4)^2}{e^{n+4}} = \sum_{n=5}^{14} \frac{n^2}{e^n}.$$

Every summation represented in sigma notation is of the form

$$\sum_{i=m}^{n} f(i), \tag{6.1}$$

where m and n are integers $(n > m)$, and $f(i)$ is some function of the index of summation i. In Example 6.2(a), $f(i) = i^{2/3}/(i^2 + i + 1)$, $m = 4$, and $n = 26$; in Example 6.1(a), $f(i) = i^2/[(i+1)(i+2)]$, $m = 1$, and $n = 13$. The following properties of sigma notation are easily proved by writing out each summation.

THEOREM 6.1

If $f(i)$ and $g(i)$ are functions of i, and m and n are positive integers such that $n > m$, then

$$\sum_{i=m}^{n} [f(i) + g(i)] = \sum_{i=m}^{n} f(i) + \sum_{i=m}^{n} g(i); \tag{6.2a}$$

$$\sum_{i=m}^{n} cf(i) = c\sum_{i=m}^{n} f(i), \tag{6.2b}$$

if c is a constant independent of i.

Compare this theorem with Theorem 5.2 in Section 5.1 (page 366); notice the similarities between properties of summations in sigma notation and those of indefinite integrals.

We emphasize that sigma notation is simply a concise symbolism used to represent a sum of terms; it does not evaluate the sum. In the following discussion, we develop formulas for sums that prove useful in future work.

Summation $\sum_{i=1}^{n} i$ represents the sum of the first n integers:

$$\sum_{i=1}^{n} i = 1 + 2 + 3 + 4 + \cdots + (n - 1) + n.$$

If we write the terms on the right in reverse order, we have

$$\sum_{i=1}^{n} i = n + (n - 1) + (n - 2) + \cdots + 4 + 3 + 2 + 1.$$

Addition of these two equations gives us

$$2 \sum_{i=1}^{n} i = (n + 1) + (n + 1) + (n + 1) + \cdots + (n + 1) + (n + 1) = n(n + 1).$$

Consequently,

$$\sum_{i=1}^{n} i = \frac{n(n + 1)}{2}. \tag{6.3}$$

This result can be used to develop formulas for the sums of the squares, cubes, etc., of the positive integers. To find the sum of the squares of the first n integers, we note that by expansion and simplification

$$i^3 - (i - 1)^3 = 3i^2 - 3i + 1$$

for any integer i whatsoever. It follows that

$$\sum_{i=1}^{n} [i^3 - (i - 1)^3] = \sum_{i=1}^{n} (3i^2 - 3i + 1).$$

But if we write the left side in full, we find

$$\sum_{i=1}^{n} [i^3 - (i - 1)^3] = [1^3 - 0^3] + [2^3 - 1^3] + [3^3 - 2^3] + \cdots [n^3 - (n - 1)^3].$$

Most of these terms cancel one another, leaving only n^3; that is,

$$\sum_{i=1}^{n} [i^3 - (i - 1)^3] = n^3.$$

Thus,

$$n^3 = \sum_{i=1}^{n} (3i^2 - 3i + 1)$$

$$= 3 \sum_{i=1}^{n} i^2 - 3 \sum_{i=1}^{n} i + \sum_{i=1}^{n} 1 \qquad \text{(using Theorem 6.1)}$$

$$= 3 \sum_{i=1}^{n} i^2 - 3\frac{n(n+1)}{2} + n \qquad \text{(using formula 6.3)}.$$

We can solve this equation for $\sum_{i=1}^{n} i^2$:

$$\sum_{i=1}^{n} i^2 = \frac{1}{3}\left[n^3 + \frac{3n(n+1)}{2} - n \right] = \frac{n(n+1)(2n+1)}{6}. \qquad (6.4)$$

A similar procedure beginning with the identity $i^4 - (i-1)^4 = 4i^3 - 6i^2 + 4i - 1$ yields

$$\sum_{i=1}^{n} i^3 = \frac{n^2(n+1)^2}{4}. \qquad (6.5)$$

These results can also be established independently of one another by mathematical induction.

EXERCISES 6.1

In Exercises 1–10 express the sum in sigma notation. Initiate the summation with the integer 1.

∗ **1.** $2 \cdot 3 + 3 \cdot 4 + 4 \cdot 5 + 5 \cdot 6 + \cdots 99 \cdot 100$

∗ **2.** $\dfrac{1}{2} + \dfrac{2}{4} + \dfrac{3}{8} + \dfrac{4}{16} + \dfrac{5}{32} + \cdots + \dfrac{10}{1024}$

∗ **3.** $\dfrac{16}{14+15} + \dfrac{17}{15+16} + \dfrac{18}{16+17} + \cdots + \dfrac{199}{197+198}$

∗ **4.** $1 + \sqrt{2} + \sqrt{3} + 2 + \sqrt{5} + \sqrt{6} + \sqrt{7} + \sqrt{8} + 3 + \cdots + 121$

∗ **5.** $1 + \dfrac{1}{2} + \dfrac{1}{2 \cdot 3} + \dfrac{1}{2 \cdot 3 \cdot 4} + \cdots + \dfrac{1}{2 \cdot 3 \cdot 4 \cdot 5 \cdot \cdots \cdot 16}$

∗ **6.** $-2 + 3 - 4 + 5 - 6 + 7 - 8 + \cdots - 1020$

∗ **7.** $\dfrac{2 \cdot 3}{1 \cdot 4} + \dfrac{6 \cdot 7}{5 \cdot 8} + \dfrac{10 \cdot 11}{9 \cdot 12} + \dfrac{14 \cdot 15}{13 \cdot 16} + \cdots \dfrac{414 \cdot 415}{413 \cdot 416}$

∗ **8.** $\dfrac{\tan 1}{2} + \dfrac{\tan 2}{1 + 2^2} + \dfrac{\tan 3}{1 + 3^2} + \dfrac{\tan 4}{1 + 4^2} + \cdots + \dfrac{\tan 225}{1 + 225^2}$

∗ **9.** $4^3 + 5^2 + 6 + 1 + \dfrac{1}{8} + \dfrac{1}{9^2} + \cdots + \dfrac{1}{25^{18}}$

∗ **10.** $0.9 + 0.99 + 0.999 + \cdots + 0.999\,999\,999$

In Exercises 11–15 verify by a change of variable of summation that the two summations are identical.

* **11.** $\displaystyle\sum_{n=1}^{24} \frac{n^2}{2n+1}$ $\qquad\displaystyle\sum_{i=4}^{27} \frac{i^2-6i+9}{2i-5}$

* **12.** $\displaystyle\sum_{k=2}^{101} \frac{3k-k^2}{\sqrt{k+5}}$ $\qquad\displaystyle\sum_{m=0}^{99} \frac{2-m-m^2}{\sqrt{7+m}}$

* **13.** $\displaystyle\sum_{n=5}^{20} (-1)^n \frac{2^n}{n^2+1}$ $\qquad\displaystyle\sum_{j=1}^{16} 16(-1)^j \frac{2^j}{j^2+8j+17}$

* **14.** $\displaystyle\sum_{i=0}^{37} \frac{3^{3i}}{i!}$ $\qquad\displaystyle\sum_{m=2}^{39} \frac{3^{3m}}{729(m-2)!}$

* **15.** $\displaystyle\sum_{r=15}^{225} \frac{1}{r^2-10r}$ $\qquad\displaystyle\sum_{n=10}^{220} \frac{1}{n^2-25}$

In Exercises 16–21 use Theorem 6.1 and formulas 6.3–6.5 to evaluate the sum.

* **16.** $\displaystyle\sum_{n=1}^{12} (3n+2)$ $\qquad\qquad$ * **17.** $\displaystyle\sum_{j=1}^{21} (2j^2+3j)$

* **18.** $\displaystyle\sum_{m=1}^{n} (4m-2)^2$ $\qquad\qquad$ * **19.** $\displaystyle\sum_{k=2}^{29} (k^3-3k^2)$

* **20.** $\displaystyle\sum_{n=1}^{25} (n+5)(n-4)$ $\qquad\qquad$ * **21.** $\displaystyle\sum_{i=1}^{n} i(i-3)^2$

** **22.** Verify Theorem 6.1.

** **23.** Find a formula for

$$\sum_{k=1}^{n} \frac{1}{k(k+1)}.$$

Hint: $\dfrac{1}{k(k+1)} = \dfrac{1}{k} - \dfrac{1}{k+1}.$

** **24.** If $f(x)$ is a function of x, defined for all x, simplify the sum

$$\sum_{i=1}^{n} [f(i) - f(i-1)].$$

** **25.** Prove formula 6.5.

** **26.** Is

$$\sum_{i=1}^{n} [f(i)g(i)] \quad \text{equal to} \quad \left[\sum_{i=1}^{n} f(i)\right]\left[\sum_{i=1}^{n} g(i)\right]?$$

** **27.** A finite geometric series is a sum of terms of the form

$$a + ar + ar^2 + ar^3 + \cdots + ar^{n-1}.$$

There is a first term a, and every term thereafter is obtained by multiplying the preceding term by r (called the *common ratio*).

(a) If S_n represents this sum, express S_n in sigma notation.

(b) Prove that

$$S_n = \frac{a(1 - r^n)}{1 - r}.$$

Use the formula in Exercise 27 to sum the finite geometric series in Exercises 28–31.

** **28.** $\dfrac{1}{8} + \dfrac{1}{16} + \dfrac{1}{32} + \dfrac{1}{64} + \cdots + \dfrac{1}{1\,048\,576}$

** **29.** $1 - \dfrac{1}{3} + \dfrac{1}{9} - \dfrac{1}{27} + \dfrac{1}{81} - \cdots - \dfrac{1}{19\,683}$

** **30.** $40(0.99) + 40(0.99)^2 + 40(0.99)^3 + \cdots 40(0.99)^{15}$

** **31.** $\sqrt{0.99} + 0.99 + (0.99)^{3/2} + (0.99)^2 + \cdots + (0.99)^{10}$

** **32.** Prove that

$$\left| \sum_{i=1}^{n} f(i) \right| \leq \sum_{i=1}^{n} |f(i)|.$$

*** **33.** Express the following summation in sigma notation:

$$1 + \frac{1}{2} - \frac{1}{4} - \frac{1}{8} + \frac{1}{16} + \frac{1}{32} - \frac{1}{64} - \frac{1}{128} + \cdots + \frac{1}{4096}.$$

6.2 The Need for the Definite Integral

In this section we consider four inherently different problems: one on area, one on volume, one on blood flow, and one on work; but we shall see that a common method of solution exists for all four of them. This common theme leads to the definition of the definite integral in Section 6.3.

Problem 1

We first consider the problem of finding the area A in Figure 6.1a. At present, we have formulas for areas of very few geometric shapes—rectangles, triangles, polygons, and circles— and the shape in Figure 6.1a is not one of them. We can, however, find an approximation to the area by constructing rectangles in the following way. Between a and b pick $n - 1$ points x_1, $x_2, x_3, \ldots, x_{n-1}$ on the x-axis such that

$$a = x_0 < x_1 < x_2 < \cdots < x_{n-1} < x_n = b.$$

Draw vertical lines through each of these $n - 1$ points to intersect the curve $y = f(x)$ and form rectangles, as shown in Figure 6.1b.

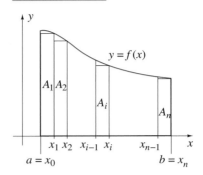

FIGURE 6.1a

FIGURE 6.1b

If A_i denotes the area of the ith such rectangle ($i = 1, \ldots, n$), then

$$A_i = \text{(height of rectangle)(width of rectangle)}$$

$$= f(x_i)(x_i - x_{i-1}).$$

The sum of these n rectangular areas is an approximation to the required area; that is, A is approximately equal to

$$\sum_{i=1}^{n} A_i = \sum_{i=1}^{n} f(x_i)(x_i - x_{i-1}).$$

If we let the number of these rectangles get larger and larger (and at the same time require each to have smaller and smaller width that eventually approaches zero), the approximation appears to get better and better. In fact, we expect that

$$A = \lim_{n \to \infty} \sum_{i=1}^{n} A_i = \lim_{n \to \infty} \sum_{i=1}^{n} f(x_i)(x_i - x_{i-1}). \tag{6.6}$$

Problem 2

If the area in Figure 6.1a is rotated around the x-axis, it traces out a volume V. This is certainly not a standard shape for which we have a volume formula, and we therefore consider finding an approximation to V. We take the rectangles in Figure 6.1b (which approximate the area) and rotate them about the x-axis (Figure 6.2). Each A_i traces out a disc of volume V_i ($i = 1, \ldots, n$), where

$$V_i = \text{(surface area of disc)(thickness of disc)}$$

$$= \pi(\text{radius of disc})^2(\text{thickness of disc})$$

$$= \pi[f(x_i)]^2(x_i - x_{i-1}).$$

An approximation to the required volume is then

$$\sum_{i=1}^{n} V_i = \sum_{i=1}^{n} \pi[f(x_i)]^2(x_i - x_{i-1}).$$

FIGURE 6.2

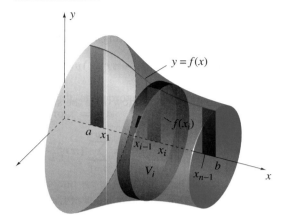

Again we feel intuitively that as the number of discs becomes larger and larger (and the width of each approaches zero), the approximation becomes better and better, and

$$V = \lim_{n \to \infty} \sum_{i=1}^{n} V_i = \lim_{n \to \infty} \sum_{i=1}^{n} \pi [f(x_i)]^2 (x_i - x_{i-1}). \tag{6.7}$$

Problem 3

When blood flows through a vein or artery, it encounters resistance due to friction with the walls of the blood vessel and due to the viscosity of the blood itself. As a result, the velocity of the blood is not constant across a cross-section of the vessel; blood flows more quickly near the centre of the vessel than near its walls. It has been shown that for laminar blood flow in a vessel of circular cross-section (Figure 6.3), the velocity of blood is given by

$$v = v(r) = c(R^2 - r^2), \quad 0 \le r \le R,$$

where $c > 0$ is a constant, R is the radius of the blood vessel, and r is radial distance measured from the centre of the vessel. We wish to find the rate of blood flow through the vessel, that is, the volume of blood flowing through the cross-section per unit time.

FIGURE 6.3

Cross section of
blood vessel

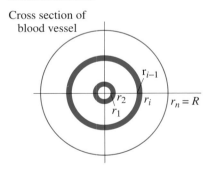

If v were constant over the cross-section, then flow per unit time would be the product of v and the cross-sectional area. Unfortunately, this is not the case, but we can still use the idea that flow is velocity multiplied by area. We divide the cross-section into rings with radii

$$0 = r_0 < r_1 < r_2 < \cdots < r_{n-1} < r_n = R.$$

** **13.** Use the formula

$$\sum_{i=1}^{n} \cos i\theta = \frac{\cos \frac{(n+1)\theta}{2} \sin \frac{n\theta}{2}}{\sin \frac{\theta}{2}}$$

to evaluate

$$\int_{0}^{\pi/2} \cos x \, dx.$$

*** **14.** In this exercise we evaluate

$$\int_{a}^{b} x^k \, dx$$

for $b > a > 0$ and any $k \neq -1$.

(a) Let $h = (b/a)^{1/n}$ and subdivide the interval $a \leq x \leq b$ into n subintervals by the points

$$x_0 = a, \quad x_1 = ah, \quad x_2 = ah^2, \ldots, x_i = ah^i, \ldots, x_n = ah^n = b.$$

Show that with the choice of $x_i^* = x_i$ in the ith subinterval $x_{i-1} \leq x \leq x_i$, equation 6.10 gives

$$\int_{a}^{b} x^k \, dx = a^{k+1} \lim_{n \to \infty} \left[\left(\frac{h-1}{h} \right) \sum_{i=1}^{n} (h^{k+1})^i \right].$$

(b) Use the result of Exercise 27 in Section 6.1 (page 406) to write the summation in closed form, and hence show that

$$\int_{a}^{b} x^k \, dx = (b^{k+1} - a^{k+1}) \lim_{n \to \infty} \frac{\left(\frac{b}{a} \right)^{k/n} \left[\left(\frac{b}{a} \right)^{1/n} - 1 \right]}{\left(\frac{b}{a} \right)^{(k+1)/n} - 1}.$$

(c) Use L'Hôpital's rule to evaluate this limit, and hence obtain

$$\int_{a}^{b} x^k \, dx = \frac{b^{k+1} - a^{k+1}}{k+1}.$$

6.4 The Fundamental Theorem of Integral Calculus

In Section 6.3 we demonstrated how to evaluate definite integrals using definition 6.10. Integrands x^2 and $5x - 2$ in Examples 6.3 and 6.4 are very simple polynomials, as are the integrands in questions 1–9 of Exercise set 6.3, but in spite of this, calculations were frequently laborious, and invariably required summation formulas from Section 6.1. We promised a very simple technique that would replace these calculations, and this is the substance of the *fundamental theorem of integral calculus*.

THEOREM 6.3   (The Fundamental Theorem of Integral Calculus)

If $f(x)$ is continuous on the interval $a \leq x \leq b$, and $F(x)$ is an antiderivative of $f(x)$ thereon, then

$$\int_a^b f(x)\,dx = F(b) - F(a). \tag{6.11}$$

PROOF   Since $f(x)$ is continuous on $a \leq x \leq b$, the definite integral of $f(x)$ from $x = a$ to $x = b$ exists and is defined by equation 6.10, where we are at liberty to choose the Δx_i and x_i^* in any way whatsoever. For any choice of Δx_i, a convenient choice for the x_i^* can be found by applying the mean value theorem n times to $F(x)$, once on each subinterval $x_{i-1} \leq x \leq x_i$. This is possible since $F'(x) = f(x)$ is continuous for $a \leq x \leq b$. The mean value theorem states that for each subinterval, there exists at least one point c_i between x_{i-1} and x_i such that

$$\frac{F(x_i) - F(x_{i-1})}{x_i - x_{i-1}} = F'(c_i), \qquad i = 1, \ldots, n.$$

But $F'(c_i) = f(c_i)$, so that

$$F(x_i) - F(x_{i-1}) = f(c_i)\Delta x_i, \qquad i = 1, \ldots, n.$$

If we now choose $x_i^* = c_i$, then

$$f(x_i^*)\Delta x_i = F(x_i) - F(x_{i-1}),$$

and equation 6.10 gives

$$\int_a^b f(x)\,dx = \lim_{\|\Delta x_i\| \to 0} \sum_{i=1}^n f(x_i^*)\Delta x_i = \lim_{\|\Delta x_i\| \to 0} \sum_{i=1}^n [F(x_i) - F(x_{i-1})].$$

When we write out all terms in the summation, we find that many cancellations take place,

$$\int_a^b f(x)\,dx = \lim_{\|\Delta x_i\| \to 0} \{[F(x_1) - F(x_0)] + [F(x_2) - F(x_1)] + \cdots +$$

$$[F(x_n) - F(x_{n-1})]\}$$

$$= \lim_{\|\Delta x_i\| \to 0} \{F(x_n) - F(x_0)\}$$

$$= \lim_{\|\Delta x_i\| \to 0} \{F(b) - F(a)\}$$

$$= F(b) - F(a). \qquad \blacksquare$$

If we introduce the notation

$$\{F(x)\}_a^b$$

to represent the difference $F(b) - F(a)$, then Theorem 6.3 can be expressed in the form

$$\int_a^b f(x)\,dx = \left\{ \int f(x)\,dx \right\}_a^b. \tag{6.12}$$

This is a fantastic result. No longer is it necessary to consider limits of summations in order to evaluate definite integrals. We simply find an antiderivative of the integrand, substitute $x = b$ and $x = a$, and subtract. For instance, to evaluate the definite integral in Example 6.3, we easily write

$$\int_0^1 x^2 \, dx = \left\{\frac{x^3}{3}\right\}_0^1 = \frac{1}{3} - 0 = \frac{1}{3}.$$

Note that had we used *the* indefinite integral $x^3/3 + C$ for x^2, we would have had

$$\int_0^1 x^2 \, dx = \left\{\frac{x^3}{3} + C\right\}_0^1 = \left\{\frac{1}{3} + C\right\} - C = \frac{1}{3}.$$

Because the arbitrary constant always vanishes in the evaluation of definite integrals, we omit the constant in this context.

Similarly, for Example 6.4, we obtain

$$\int_{-1}^1 (5x - 2) \, dx = \left\{\frac{5x^2}{2} - 2x\right\}_{-1}^1 = \left\{\frac{5}{2} - 2\right\} - \left\{\frac{5}{2} + 2\right\} = -4.$$

EXAMPLE 6.5 ▼

Evaluate the following definite integrals:

$$\text{(a)} \quad \int_1^2 (3x^2 - x + 4) \, dx \qquad \text{(b)} \quad \int_{-2}^4 \sqrt{x + 4} \, dx$$

SOLUTION

(a) $\quad \displaystyle\int_1^2 (3x^2 - x + 4) \, dx = \left\{x^3 - \frac{x^2}{2} + 4x\right\}_1^2$

$$= \{8 - 2 + 8\} - \left\{1 - \frac{1}{2} + 4\right\} = \frac{19}{2}$$

(b) $\quad \displaystyle\int_{-2}^4 \sqrt{x + 4} \, dx = \left\{\frac{2}{3}(x + 4)^{3/2}\right\}_{-2}^4 = \frac{2}{3}(8)^{3/2} - \frac{2}{3}(2)^{3/2} = \frac{28\sqrt{2}}{3}$

EXAMPLE 6.6 ▼

Evaluate

$$\int_1^3 \left(\frac{1}{x^2} + 3x^3\right) dx.$$

SOLUTION Since $-1/x + 3x^4/4$ is an antiderivative for $1/x^2 + 3x^3$,

$$\int_1^3 \left(\frac{1}{x^2} + 3x^3\right) dx = \left\{-\frac{1}{x} + \frac{3x^4}{4}\right\}_1^3 = \left\{-\frac{1}{3} + \frac{243}{4}\right\} - \left\{-1 + \frac{3}{4}\right\} = \frac{182}{3}.$$

EXAMPLE 6.7

Can the following definite integral be evaluated with Theorem 6.3,

$$\int_{-1}^{1} \frac{1}{x^2}\, dx\ ?$$

SOLUTION   No! Theorem 6.3 requires the integrand $1/x^2$ to be continuous on the interval $-1 \le x \le 1$, and this is not the case. The function is discontinuous at $x = 0$.

There is a difficulty with Theorem 6.3. It is subtle, but important. The theorem states that, to evaluate definite integrals of continuous functions, we use antiderivatives. But how do we know that continuous functions have antiderivatives? We don't yet. This fact will be established in Section 6.5 when we verify an alternative version of the fundamental theorem. You might ask why this alternative version is not proved first. Would it not be more logical to first establish existence of antiderivatives, and then use this fact to prove Theorem 6.3? From a logic point of view, the answer is yes. However, from a practical point of view, Theorem 6.3 is infinitely more useful than its alternative version, and we therefore want to give it every possible emphasis. To prove the alternative version first would detract from the importance and simplicity of Theorem 6.3.

Before we move on to the alternative version of the fundamental theorem, we present the following theorems which describe properties of the definite integral.

THEOREM 6.4

If $f(x)$ is continuous on $a \le x \le b$, then:

(i) $$\int_{b}^{a} f(x)\, dx = -\int_{a}^{b} f(x)\, dx;$$ (6.13a)

(ii) $$\int_{a}^{b} f(x)\, dx = \int_{a}^{c} f(x)\, dx + \int_{c}^{b} f(x)\, dx.$$ (6.13b)

THEOREM 6.5

If $f(x)$ and $g(x)$ are continuous on $a \le x \le b$, then:

(i) $$\int_{a}^{b} [f(x) + g(x)]\, dx = \int_{a}^{b} f(x)\, dx + \int_{a}^{b} g(x)\, dx;$$ (6.14a)

(ii) $$\int_{a}^{b} kf(x)\, dx = k \int_{a}^{b} f(x)\, dx,$$ (6.14b)

when k is a constant.

Properties 6.14 are analogous to 5.3 for indefinite integrals. Theorems 6.4 and 6.5 can be proved using either Theorem 6.3 or equation 6.10.

We can also establish the following property.

> ### THEOREM 6.6
>
> When $f(x)$ is continuous on $a \leq x \leq b$ and $m \leq f(x) \leq M$ on this interval,
>
> $$m(b - a) \leq \int_a^b f(x)\, dx \leq M(b - a). \tag{6.15}$$

PROOF By equation 6.10, we can write

$$
\begin{aligned}
\int_a^b f(x)\, dx &= \lim_{\|\Delta x_i\| \to 0} \sum_{i=1}^n f(x_i^*) \Delta x_i \\
&\leq \lim_{\|\Delta x_i\| \to 0} \sum_{i=1}^n M \Delta x_i \\
&= M \lim_{\|\Delta x_i\| \to 0} \sum_{i=1}^n \Delta x_i \\
&= M \lim_{\|\Delta x_i\| \to 0} [(x_1 - a) + (x_2 - x_1) + (x_3 - x_2) + \cdots + (b - x_{n-1})] \\
&= M \lim_{\|\Delta x_i\| \to 0} (b - a) \\
&= M(b - a).
\end{aligned}
$$

A similar proof establishes the inequality involving m. ∎

EXAMPLE 6.8

Use Theorem 6.6 to find a maximum possible value for

$$\int_1^4 \frac{\sin x^2}{1 + x^2}\, dx.$$

SOLUTION Clearly $\sin x^2 \leq 1$ for all x, and on the interval $1 \leq x \leq 4$, the largest value of $1/(1 + x^2)$ is $1/2$. Consequently, $(\sin x^2)/(1 + x^2) \leq 1/2$ for $1 \leq x \leq 4$ and, by Theorem 6.6,

$$\int_1^4 \frac{\sin x^2}{1 + x^2}\, dx \leq \frac{1}{2}(4 - 1) = \frac{3}{2}.$$

Suppose that $v(t) = 3t^2 - 6t - 105$ represents the velocity (in metres per second) of a particle moving along the x-axis beginning at time $t = 0$. We can easily calculate the definite integral of $v(t)$ between any two times, say $t = 0$ and $t = 12$. By doing so, we get our first glimpse of definite integrals at work in applied problems.

$$\int_0^{12} v(t)\, dt = \int_0^{12} (3t^2 - 6t - 105)\, dt = \left\{ t^3 - 3t^2 - 105t \right\}_0^{12} = 36.$$

Realizing that integration is a limit-summation, and what is being added are products of velocities $v(t)$ multiplied by small time increments dt, we interpret 36 as the displacement of the particle at time $t = 12$ s relative to its displacement at $t = 0$ s. Although we do not have enough information to determine where the particle is at any given time, we can say that at $t = 12$ s, it is 36 m to the right of where it is at $t = 0$ s. In general, when $v(t)$ is the velocity of a particle moving along the x-axis,

$$\int_a^b v(t)\, dt \tag{6.16}$$

is the displacement of the particle at time $t = b$ relative to its displacement at time $t = a$. If the definite integral is positive, then at time $t = b$ the particle is to the right of its position at $t = a$; and if the integral is negative, then at $t = b$, the particle is to the left of its position at $t == a$.

Speed is the magnitude of velocity. If we integrate speed between the same limits we get a different result,

$$\int_0^{12} |v(t)|\, dt = \int_0^{12} |3t^2 - 6t - 105|\, dt.$$

Since $3t^2 - 6t - 105 < 0$ for $0 \le t < 7$, and $3t^2 - 6t - 105 > 0$ for $7 < t \le 12$, we divide the integration into two parts,

$$\int_0^{12} |v(t)|\, dt = \int_0^7 -(3t^2 - 6t - 105)\, dt + \int_7^{12} (3t^2 - 6t - 105)\, dt$$

$$= \left\{ -t^3 + 3t^2 + 105t \right\}_0^7 + \left\{ t^3 - 3t^2 - 105t \right\}_7^{12}$$

$$= 1114.$$

Since this integral adds products of speed $|v(t)|$ and time increments dt, which we interpret as distance travelled, 1114 m must be the distance travelled by the particle between $t = 0$ s and $t = 12$ s. In general,

$$\int_a^b |v(t)|\, dt \tag{6.17}$$

is the distance travelled between times $t = a$ and $t = b$.

These ideas are reinforced by graphs of velocity and speed in Figure 6.8. Because velocity is negative for $0 < t < 7$, the particle is moving to the left at these times. Products $v(t)\, dt$ are negative during this time interval, and therefore contribute negatively to the definite integral of $v(t)$. The particle stops at $t = 7$ s, and then moves to the right from $t = 7$ s to $t = 12$ s. Products $v(t)\, dt$ then contribute positively to the definite integral with the ultimate result being 36 m. At time $t = 12$ s, the particle is 36 m to the right of its position at $t = 0$ s. On the other hand, speed is always positive, and therefore products $|v(t)|\, dt$ always contribute positively to the integral of speed.

FIGURE 6.8a

FIGURE 6.8a

FIGURE 6.8b

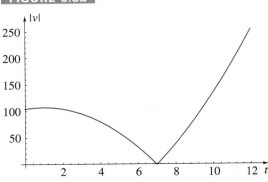

EXERCISES 6.4

In Exercises 1–34 evaluate the definite integral.

∗ **1.** $\displaystyle\int_3^4 (x^3 + 3)\, dx$

∗ **2.** $\displaystyle\int_1^3 (x^2 - 2x + 3)\, dx$

∗ **3.** $\displaystyle\int_{-1}^1 (4x^3 + 2x)\, dx$

∗ **4.** $\displaystyle\int_{-3}^{-1} \frac{1}{x^2}\, dx$

∗ **5.** $\displaystyle\int_4^2 \left(x^2 + \frac{3}{x^3}\right) dx$

∗ **6.** $\displaystyle\int_0^{\pi/2} \sin x\, dx$

∗ **7.** $\displaystyle\int_{-1}^1 (x^2 - 1 - x^4)\, dx$

∗ **8.** $\displaystyle\int_{-1}^{-2} \left(\frac{1}{x^2} - 2x\right) dx$

∗ **9.** $\displaystyle\int_1^2 (x^4 + 3x^2 + 2)\, dx$

∗ **10.** $\displaystyle\int_0^1 x(x^2 + 1)\, dx$

∗ **11.** $\displaystyle\int_0^1 x^2(x^2 + 1)^2\, dx$

∗ **12.** $\displaystyle\int_0^{2\pi} \cos 2x\, dx$

∗ **13.** $\displaystyle\int_1^3 \frac{x^2 + 3}{x^2}\, dx$

∗ **14.** $\displaystyle\int_0^1 (x^{2.2} - x^\pi)\, dx$

∗ **15.** $\displaystyle\int_{-1}^1 x^2(x^3 - x)\, dx$

∗ **16.** $\displaystyle\int_3^4 \frac{(x^2 - 1)^2}{x^2}\, dx$

∗ **17.** $\displaystyle\int_1^2 \left(\sqrt{x} - \frac{1}{\sqrt{x}}\right) dx$

∗ **18.** $\displaystyle\int_{-2}^3 (x - 1)^3\, dx$

∗ **19.** $\displaystyle\int_2^4 \frac{(x^2 - 1)(x^2 + 1)}{x^2}\, dx$

∗ **20.** $\displaystyle\int_0^{\pi/4} 3 \cos x\, dx$

∗ **21.** $\displaystyle\int_0^{\pi/4} \sec^2 x\, dx$

∗ **22.** $\displaystyle\int_{\pi/2}^{\pi} \sin x \cos x\, dx$

∗ **23.** $\displaystyle\int_{\pi/6}^{\pi/3} \csc^2 3x\, dx$

∗ **24.** $\displaystyle\int_{-\pi/4}^{\pi/4} \sec x \tan x\, dx$

* **25.** $\displaystyle\int_0^2 2^x \, dx$

* **26.** $\displaystyle\int_{-1}^2 e^x \, dx$

* **27.** $\displaystyle\int_0^1 e^{3x} \, dx$

* **28.** $\displaystyle\int_{-3}^{-2} \frac{1}{x} \, dx$

* **29.** $\displaystyle\int_1^3 \frac{(x+1)^2}{x} \, dx$

* **30.** $\displaystyle\int_0^1 3^{4x} \, dx$

* **31.** $\displaystyle\int_0^5 |x| \, dx$

* **32.** $\displaystyle\int_0^4 x|x+1| \, dx$

** **33.** $\displaystyle\int_{-5}^5 |x| \, dx$

** **34.** $\displaystyle\int_{-2}^1 x|x+1| \, dx$

In Exercises 35–40, $v(t)$ represents the velocity of a particle moving along the x-axis. Calculate the definite integral of $v(t)$ and $|v(t)|$ between the two times shown. Interpret each number and plot (or draw) a graph of the velocity function to corroborate results.

** **35.** $v(t) = 3t^2 - 6t - 105$, $t = 0$ to $t = 9$

** **36.** $v(t) = 3t^2 - 6t - 105$, $t = 0$ to $t = 5$

** **37.** $v(t) = -t^2 + 3t - 2$, $t = 0$ to $t = 2$

** **38.** $v(t) = -t^2 + 3t - 2$, $t = 1$ to $t = 2$

** **39.** $v(t) = t^3 - 3t^2 + 2t$, $t = 0$ to $t = 2$

** **40.** $v(t) = t^3 - 3t^2 + 2t$, $t = 0$ to $t = 3$

In Exercises 41–46 use Thoerem 6.6 to find maximum and minimum values for the integral.

** **41.** $\displaystyle\int_0^{\pi/4} \frac{\sin x}{1+x^2} \, dx$

** **42.** $\displaystyle\int_0^{\pi/2} \frac{\sin x}{1+x} \, dx$

** **43.** $\displaystyle\int_0^{\pi} \frac{\sin x}{2+x^2} \, dx$

** **44.** $\displaystyle\int_{\pi/4}^{\pi/2} \frac{\sin 2x}{10+x^2} \, dx$

** **45.** $\displaystyle\int_0^1 (1+4x^4)\cos(x^2) \, dx$

** **46.** $\displaystyle\int_1^3 \sqrt{4+x^3} \, dx$

6.5 An Alternative Version of the Fundamental Theorem

The fundamental theorem of integral calculus in Section 6.4 allows us to use antiderivatives to evaluate definite integrals. To complete the picture, we now show that, in some sense, definite integrals can be regarded as antiderivatives.

When $f(t)$ is continuous for $a \le t \le b$, its definite integral

$$\int_a^b f(t) \, dt$$

is a number. If b is changed but a is kept fixed, the value of the definite integral changes; for each value of b, there is a new value for the definite integral. In other words, the value of the definite integral is a function of its upper limit. Suppose we replace b by x, and denote the resulting function by $F(x)$,

$$F(x) = \int_a^x f(t) \, dt.$$

We now show that the derivative of $F(x)$ is $f(x)$; that is, $F(x)$ is an antiderivative of $f(x)$.

THEOREM 6.7 (Alternative Version of the Fundamental Theorem)

When $f(x)$ is continuous for $a \leq x \leq b$, the function

$$F(x) = \int_a^x f(t)\,dt \qquad (6.18)$$

is differentiable for $a \leq x \leq b$, and $F'(x) = f(x)$.

PROOF If x is any point in the open interval $a < x < b$, then h can always be chosen sufficiently small that $x + h$ is also in the interval $a < x < b$. By definition 3.3, the derivative of $F(x)$ at this x is defined as

$$
\begin{aligned}
F'(x) &= \lim_{h \to 0} \frac{F(x+h) - F(x)}{h} \\
&= \lim_{h \to 0} \frac{1}{h}\left[\int_a^{x+h} f(t)\,dt - \int_a^x f(t)\,dt\right] \\
&= \lim_{h \to 0} \frac{1}{h}\left[\int_a^{x+h} f(t)\,dt + \int_x^a f(t)\,dt\right] \qquad \text{(by property 6.13a)} \\
&= \lim_{h \to 0} \frac{1}{h}\int_x^{x+h} f(t)\,dt \qquad \text{(by property 6.13b)}.
\end{aligned}
$$

According to property 6.15,

$$mh \leq \int_x^{x+h} f(t)\,dt \leq Mh$$

where m and M are the minimum and maximum of $f(x)$ on the interval between x and $x + h$. Division by h gives

$$m \leq \frac{1}{h}\int_x^{x+h} f(t)\,dt \leq M.$$

Consider what happens as we let $h \to 0$. The limit of the middle term is $F'(x)$. Furthermore, the numbers m and M must approach one another, and in the limit must both be equal to $f(x)$; they are minimum and maximum values of $f(x)$ on the interval between x and $x + h$, and h is approaching zero. We conclude therefore that

$$F'(x) = \lim_{h \to 0} \int_x^{x+h} f(t)\,dt = f(x).$$

This argument can also be used to establish that $F(x)$ has a right-hand derivative $f(a)$ at $x = a$, and a left-hand derivative $f(b)$ at $x = b$. ∎

Recall that Theorem 6.3 assumed the existence of antiderivatives of continuous functions. Theorem 6.7 establishes this fact, and states that definite integral 6.18 with a variable upper limit is one such antiderivative. In practice, we do not use 6.18 to find antiderivatives, but the theorem does establish existence of antiderivatives of continuous functions.

Symbolically, we may write the result of Theorem 6.7 in the form

$$\frac{d}{dx} \int_a^x f(t)\, dt = f(x). \tag{6.19}$$

The lower limit in this result need not be a; it can also be b:

$$\frac{d}{dx} \int_b^x f(t)\, dt = f(x).$$

The lower limit can also be any number c between a and b:

$$\frac{d}{dx} \int_c^x f(t)\, dt = f(x).$$

EXAMPLE 6.9

Evaluate $\dfrac{d}{dx} \displaystyle\int_0^x \sqrt{1 - t^2}\, dt$.

SOLUTION   According to 6.18,

$$\frac{d}{dx} \int_0^x \sqrt{1 - t^2}\, dt = \sqrt{1 - x^2}.$$

This is valid for $-1 \le x \le 1$.

EXAMPLE 6.10

Evaluate $\dfrac{d}{dx} \displaystyle\int_1^{2x^2} \frac{\sin t}{1 + t^2}\, dt$.

SOLUTION   For this problem we set $u = 2x^2$, and invoke the chain rule:

$$\frac{d}{dx} \int_1^{2x^2} \frac{\sin t}{1 + t^2}\, dt = \frac{d}{dx} \int_1^u \frac{\sin t}{1 + t^2}\, dt = \left[\frac{d}{du} \int_1^u \frac{\sin t}{1 + t^2}\, dt \right] \frac{du}{dx}$$

$$= \frac{\sin u}{1 + u^2}(4x) = \frac{4x \sin(2x^2)}{1 + 4x^4}.$$

EXAMPLE 6.11

Evaluate $\dfrac{d}{dx} \displaystyle\int_x^5 (1 + t^3)^{2/3}\, dt$.

If we set $x = 1$, and note that 6.20 implies that $\ln 1 = 0$, we obtain $\ln x_1 = 0 + C$. Consequently,

$$\ln(x_1 x) = \ln x + \ln x_1,$$

and for $x = x_2$ we obtain identity 1.56a,

$$\ln(x_1 x_2) = \ln x_1 + \ln x_2. \tag{6.24}$$

Differentiation formula 6.22 implies that for $x > 0$,

$$\frac{d}{dx} \ln\left(\frac{1}{x}\right) = \frac{1}{1/x}\left(-\frac{1}{x^2}\right) = -\frac{1}{x}.$$

When this is added to 6.21,

$$\frac{d}{dx}\left[\ln x + \ln\left(\frac{1}{x}\right)\right] = \frac{1}{x} - \frac{1}{x} = 0.$$

Because this is valid for $x > 0$, Theorem 5.1 implies that

$$\ln x + \ln\left(\frac{1}{x}\right) = C.$$

Substitution of $x = 1$ gives $C = 0$, and therefore

$$\ln\left(\frac{1}{x}\right) = -\ln x. \tag{6.25}$$

We can now verify property 1.56b,

$$\ln\left(\frac{x_1}{x_2}\right) = \ln\left[x_1\left(\frac{1}{x_2}\right)\right]$$

$$= \ln x_1 + \ln\left(\frac{1}{x_2}\right) \qquad \text{(by 6.24)}$$

$$= \ln x_1 - \ln x_2 \qquad \text{(by 6.25)}.$$

For fixed $x_2 > 0$, equation 6.22 gives

$$\frac{d}{dx} \ln(x^{x_2}) = \frac{1}{x^{x_2}} x_2 x^{x_2-1} = \frac{x_2}{x} = x_2 \frac{d}{dx} \ln x = \frac{d}{dx}(x_2 \ln x).$$

Since this is valid for all $x > 0$, Theorem 5.1 once again gives

$$\ln(x^{x_2}) = x_2 \ln x + C.$$

If we set $x = 1$, we obtain $C = 0$, and for $x = x_1$,

$$\ln(x_1^{x_2}) = x_2 \ln x_1. \tag{6.26}$$

With the logarithm function defined by 6.20, there is a natural way to introduce the number e. We define e as the number whose natural logarithm is 1,

$$\ln e = 1 \quad \Rightarrow \quad \int_1^e \frac{1}{t}\,dt = 1. \tag{6.27}$$

The resulting exponential function e^x would satisfy properties 1.54 since all exponential functions satisfy properties of this type. Furthermore, it follows from 6.26 and 6.27 that

$$\ln(e^x) = x \ln e = x. \tag{6.28}$$

This is property 1.57b. We can also establish 1.57a by setting $y = e^{\ln x}$. When we take logarithms of both sides, and use 6.27 and 6.28,

$$\ln y = \ln\left(e^{\ln x}\right) = \ln x(\ln e) = \ln x. \tag{6.29}$$

The graph in Figure 6.9 indicates that for any given y-value on the curve, there is only one value of x that gives that y. In other words, if $\ln y = \ln x$ as in 6.29, it follows that $y = x$. Since $y = e^{\ln x}$, we have $x = e^{\ln x}$.

EXERCISES 6.5

In Exercises 1–20 differentiate the definite integral with respect to x.

∗ **1.** $\displaystyle\int_0^x (3t^2 + t)\,dt$

∗ **2.** $\displaystyle\int_1^x \frac{1}{\sqrt{t^2 + 1}}\,dt$

∗ **3.** $\displaystyle\int_x^2 \sin(t^2)\,dt$

∗ **4.** $\displaystyle\int_x^{-1} t^3 \cos t\,dt$

∗ **5.** $\displaystyle\int_0^{3x} (2t - t^4)^2\,dt$

∗ **6.** $\displaystyle\int_1^{2x} \sqrt{t + 1}\,dt$

∗ **7.** $\displaystyle\int_4^{3x^2} \sin(3t + 4)\,dt$

∗ **8.** $\displaystyle\int_{-2}^{5x+4} \sqrt{t^3 + 1}\,dt$

∗ **9.** $\displaystyle\int_x^{2x} (3\sqrt{t} - 2t)\,dt$

∗ **10.** $\displaystyle\int_{4x}^{4x+4} \left(t^3 - \frac{1}{\sqrt{t}}\right)\,dt$

∗ **11.** $\displaystyle\int_{-2x}^x \tan(3t + 1)\,dt$

∗ **12.** $\displaystyle\int_{-x^2}^{-2x^2} \sec(1 - t)\,dt$

∗ **13.** $\displaystyle\int_0^{\sin x} \cos(t^2)\,dt$

∗ **14.** $\displaystyle\int_{\cos x}^{\sin x} \frac{1}{\sqrt{t + 1}}\,dt$

∗ **15.** $\displaystyle\int_0^{2\sqrt{x}} \sqrt{t}\,dt$

∗ **16.** $\displaystyle\int_{\sqrt{x}}^{2\sqrt{x}} \sqrt{t}\,dt$

∗ **17.** $\displaystyle\int_1^{x^2} t^2 e^{4t}\,dt$

∗ **18.** $\displaystyle\int_x^2 \ln(t^2 + 1)\,dt$

∗ **19.** $\displaystyle\int_x^{2x} t \ln t\,dt$

∗ **20.** $\displaystyle\int_{-2x}^{3x} e^{-4t^2}\,dt$

** **21.** Verify that when $a(x)$ and $b(x)$ are differentiable functions of x,

$$\frac{d}{dx}\int_{a(x)}^{b(x)} f(t)\, dt = f[b(x)]\frac{db}{dx} - f[a(x)]\frac{da}{dx}.$$

Is equation 6.18 a special case of this result?

* **22.–28.**   Use the result of Exercise 21 to redo Exercises 8, 10, 12, 14, 16, 18, and 20.

6.6 Average Values

The average value of two numbers c and d is defined as $(c+d)/2$. The average value of a set of n numbers $y_1, y_2, \ldots, y_n$ is $(y_1 + y_2 + \cdots + y_n)/n$. We would like to extend this idea to define the average value of a function $f(x)$ over an interval $a \leq x \leq b$. By beginning with some simple functions we can see how to do this.

The function $f(x)$ in Figure 6.10a is equal to 1 for the first third of the interval shown, 2 for the second third, and 3 for the last third. We all probably feel that its average value over the interval $0 \leq x \leq 6$ is 2. The function in Figure 6.10b takes on the same function values, namely, 1, 2, and 3, but not on the same subintervals. The fact that it has value 2 for $2 \leq x \leq 5$ and value 3 for $5 \leq x \leq 6$ suggests that its average value should be somewhat less than the average value of 2 for the function in Figure 6.10a. The average value of the function in Figure 6.10c should be even less than that in Figure 6.10b. These three functions indicate that two factors are important when considering average values of functions: values that the function takes on, and lengths of the intervals on which they take these values. Perhaps what should be done to calculate average values for these functions is to add together the products obtained by multiplying each of the function values by the length of the interval in which it has this value, and then divide this sum by the length of the overall interval.

FIGURE 6.10a   FIGURE 6.10b   FIGURE 6.10c

For the function in Figure 6.10a this yields an average value of

$$\frac{1}{6-0}[1(2-0) + 2(4-2) + 3(6-4)] = 2;$$

for the function in Figure 6.10b,

$$\frac{1}{6-0}[1(2-0) + 2(5-2) + 3(6-5)] = \frac{11}{6};$$

and for the function in Figure 6.10c,

$$\frac{1}{6-0}[1(4-0) + 2(5-4) + 3(6-5)] = \frac{3}{2}.$$

This procedure is applicable only to a function whose domain can be subdivided into a finite number of subintervals inside each of which the function has a constant value. Such a function is said to be *piecewise constant*. What shall we do for functions which are not piecewise constant? By rephrasing the above procedure, it will become obvious. The same three average values are obtained if we adopt the following approach (illustrated for the function $f(x)$ in Figure 6.10c). Divide the interval $0 \leq x \leq 6$ into three subintervals $0 < x \leq 4$, $4 < x \leq 5$, and $5 < x \leq 6$. Pick a point in each subinterval; call the points x_1^*, x_2^*, and x_3^*. Evaluate $f(x)$ at each point and multiply by the length of the subinterval in which the point is found. Add these results, and divide by the length of the interval to obtain the average value,

$$\frac{1}{6}[f(x_1^*)(4 - 0) + f(x_2^*)(5 - 4) + f(x_3^*)(6 - 5)] = \frac{1}{6}[1(4) + 2(1) + 3(1)] = \frac{3}{2}.$$

But this is the procedure used to define the definite integral of a function; it lacks the limit because the function is piecewise constant. In other words, for a function $f(x)$ which has a value at every point in the interval $a \leq x \leq b$ (but is not necessarily piecewise constant), we define its average value as

$$\text{Average value} = \frac{1}{b - a} \int_a^b f(x)\,dx = \frac{1}{b - a} \lim_{\|\Delta x_i\| \to 0} \sum_{i=1}^n f(x_i^*)\Delta x_i. \qquad (6.30)$$

EXAMPLE 6.13

What is the average value of the function $f(x) = x^2$ on the interval $0 \leq x \leq 2$?

SOLUTION   By equation 6.30,

$$\text{Average value} = \frac{1}{2} \int_0^2 x^2\,dx = \frac{1}{2}\left\{\frac{x^3}{3}\right\}_0^2 = \frac{1}{2}\left(\frac{8}{3}\right) = \frac{4}{3}.$$

EXAMPLE 6.14

Find the average value of $f(x) = \sin x$ on the intervals: (a) $0 \leq x \leq \pi/2$ (b) $0 \leq x \leq \pi$
(c) $0 \leq x \leq 2\pi$.

SOLUTION   We calculate average values on these intervals as:

FIGURE 6.11

(a) $\dfrac{1}{\pi/2} \displaystyle\int_0^{\pi/2} \sin x\,dx = \dfrac{2}{\pi}\{-\cos x\}_0^{\pi/2} = \dfrac{2}{\pi}[0 + 1] = \dfrac{2}{\pi};$

(b) $\dfrac{1}{\pi} \displaystyle\int_0^{\pi} \sin x\,dx = \dfrac{1}{\pi}\{-\cos x\}_0^{\pi} = \dfrac{1}{\pi}[1 + 1] = \dfrac{2}{\pi};$

(c) $\dfrac{1}{2\pi} \displaystyle\int_0^{2\pi} \sin x\,dx = \dfrac{1}{2\pi}\{-\cos x\}_0^{2\pi} = \dfrac{1}{2\pi}[-1 + 1] = 0.$

The graph of $\sin x$ in Figure 6.11 also suggests that the average values in (a) and (b) should be the same, and that the average value in (c) should be zero.

EXAMPLE 6.15

What is the average value of a linear function?

SOLUTION   The average value of the linear function $f(x) = mx + c$ over an interval $a \leq x \leq b$ is

$$
\begin{aligned}
\frac{1}{b-a} \int_a^b (mx+c)\,dx &= \frac{1}{b-a}\left\{\frac{mx^2}{2}+cx\right\}_a^b \\
&= \frac{1}{b-a}\left[\left(\frac{mb^2}{2}+cb\right)-\left(\frac{ma^2}{2}+ca\right)\right] \\
&= \frac{1}{b-a}\left[\frac{m}{2}(b^2-a^2)+c(b-a)\right] \\
&= \frac{m}{2}(b+a)+c \\
&= m\left(\frac{b+a}{2}\right)+c.
\end{aligned}
$$

This is the value of the function midway between $x = a$ and $x = b$.

Theorem 6.7 can be used to establish the next theorem.

THEOREM 6.8   (Mean Value Theorem for Definite Integrals)

If $f(x)$ is continuous for $a \leq x \leq b$, then there exists at least one number c between a and b such that

$$
\int_a^b f(x)\,dx = (b-a)f(c). \tag{6.31}
$$

PROOF   By Theorem 6.7, the function

$$
F(x) = \int_a^x f(t)\,dt
$$

is continuous for $a \leq x \leq b$, and has derivative $f(x)$ for $a < x < b$. Mean Value Theorem 3.17 applied to $F(x)$ guarantees at least one number c between a and b such that

$$
F(b) - F(a) = (b-a)F'(c).
$$

By substitution,

$$
\int_a^b f(t)\,dt - \int_a^a f(t)\,dt = (b-a)f(c).
$$

Since the second integral vanishes, we have

$$
\int_a^b f(x)\,dx = (b-a)f(c).
$$

By writing equation 6.31 in the form

$$f(c) = \frac{1}{b-a} \int_a^b f(x)\, dx,$$

Theorem 6.8 states that the function must take on its average value at least once in the interval.

EXAMPLE 6.16

Find all values of c satisfying Theorem 6.8 for the function $f(x) = x^2$ on the interval $0 \le x \le 1$.

SOLUTION   Substituting $a = 0$ and $b = 1$ in equation 6.31 gives

$$(1-0)c^2 = \int_0^1 x^2\, dx = \left\{ \frac{x^3}{3} \right\}_0^1 = \frac{1}{3} - 0 = \frac{1}{3}.$$

Of the two solutions $c = \pm 1/\sqrt{3}$ for this equation, only the positive one is between 0 and 1. Thus, only $c = 1/\sqrt{3}$ satisfies Theorem 6.8 for the function $f(x) = x^2$ on the interval $0 \le x \le 1$.

EXAMPLE 6.17

Find all values of c satisfying Theorem 6.8 for the function $f(x) = \sin x$ on the interval $\pi/4 \le x \le 3\pi/4$.

SOLUTION   Substituting $a = \pi/4$ and $b = 3\pi/4$ in equation 6.31 gives

$$\left(\frac{3\pi}{4} - \frac{\pi}{4} \right) \sin c = \int_{\pi/4}^{3\pi/4} \sin x\, dx = \{ -\cos x \}_{\pi/4}^{3\pi/4} = \frac{1}{\sqrt{2}} + \frac{1}{\sqrt{2}} = \sqrt{2}.$$

Thus,

$$\sin c = \frac{2\sqrt{2}}{\pi}.$$

There are two angles between $\pi/4$ and $3\pi/4$ with a sine equal to $2\sqrt{2}/\pi$, namely $c = 1.12$ and $c = 2.02$.

EXERCISES 6.6

In Exercises 1–20 find the average value of the function over the interval.

* **1.** $f(x) = x^2 - 2x, \quad 0 \le x \le 2$ * **2.** $f(x) = x^3 - x, \quad -1 \le x \le 1$

* **3.** $f(x) = x^3 - x, \quad 0 \le x \le 1$ * **4.** $f(x) = x^4, \quad 1 \le x \le 2$

* **5.** $f(x) = \sqrt{x+1}, \quad 0 \le x \le 1$ * **6.** $f(x) = \sqrt{x+1}, \quad -1 \le x \le 1$

* **7.** $f(x) = x^4 - 1, \quad 0 \le x \le 1$ * **8.** $f(x) = x^4 - 1, \quad 0 \le x \le 2$

* **9.** $f(x) = \cos x, \quad -\pi/2 \le x \le \pi/2$ * **10.** $f(x) = \cos x, \quad 0 \le x \le \pi/2$

* **11.** $f(x) = |x|, \quad -2 \le x \le 2$ * **12.** $f(x) = |x|, \quad 0 \le x \le 2$

** **13.** $f(x) = |x^2 - 4|, \quad 0 \le x \le 3$ ** **14.** $f(x) = |x^2 - 4|, \quad -3 \le x \le 3$

** **15.** $f(x) = \operatorname{sgn} x, \quad -1 \le x \le 1$ (see exercise 43 in Section 2.4, page 134)

** **16.** $f(x) = \operatorname{sgn} x, \quad -1 \le x \le 3$

** **17.** $f(x) = H(x - 1), \quad 0 \le x \le 2$ (see Section 2.4)

** **18.** $f(x) = H(x - 4), \quad 0 \le x \le 2$ (see Section 2.4)

** **19.** $f(x) = [\![x]\!], \quad 0 \le x \le 3$ (see exercise 50 in Section 1.3, page 41)

** **20.** $f(x) = [\![x]\!], \quad 0 \le x \le 3.5$ (see exercise 50 in Section 1.3)

** **21.** If a particle moving along the x-axis is at position x_1 at time t_1 and at position x_2 at time t_2, its average velocity over this time interval is

$$\frac{x_2 - x_1}{t_2 - t_1}.$$

Verify that this is the same as the average of the velocity function over the time interval $t_1 \le t \le t_2$.

** **22.** The velocity v of blood flowing through a circular vein or artery of radius R at a distance r from the centre of the blood vessel is

$$v(r) = c(R^2 - r^2).$$

(See Problem 3 in Section 6.2, page 409.) What is the average value of $v(r)$ with respect to r?

In Exercises 23–30 find all values of c satisfying equation 6.31 for the function $f(x)$ on the specified interval.

* **23.** $f(x) = 2x - x^2, \quad 0 \le x \le 2$ * **24.** $f(x) = x^3 - 8x, \quad -2 \le x \le 2$

* **25.** $f(x) = \cos x, \quad 0 \le x \le \pi/2$ * **26.** $f(x) = \cos x, \quad 0 \le x \le \pi$

* **27.** $f(x) = \sqrt{x+1}, \quad 1 \le x \le 3$ ** **28.** $f(x) = x^2(x+1), \quad 0 \le x \le 1$

** **29.** $f(x) = x\sqrt{x^2 + 1}, \quad 0 \le x \le 2$ ** **30.** $f(x) = 1/x^2 + 1/x^3, \quad 1 \le x \le 2$

6.7 Change of Variable in the Definite Integral

The fundamental theorem of integral calculus indicates that to evaluate the definite integral

$$\int_{-4}^{-1} \frac{x}{\sqrt{x+5}} \, dx$$

we should first find an indefinite integral for $x/\sqrt{x+5}$. To do this we set $u = x+5$, in which case $du = dx$, and

$$\int \frac{x}{\sqrt{x+5}} \, dx = \int \frac{u-5}{\sqrt{u}} \, du = \int (u^{1/2} - 5u^{-1/2}) \, du$$

$$= \frac{2}{3} u^{3/2} - 10u^{1/2} + C = \frac{2}{3}(x+5)^{3/2} - 10(x+5)^{1/2} + C.$$

Consequently,

$$\int_{-4}^{-1} \frac{x}{\sqrt{x+5}} \, dx = \left\{ \frac{2}{3}(x+5)^{3/2} - 10(x+5)^{1/2} \right\}_{-4}^{-1}$$

$$= \left[\frac{2}{3}(4)^{3/2} - 10(4)^{1/2} \right] - \left[\frac{2}{3}(1)^{3/2} - 10(1)^{1/2} \right]$$

$$= -\frac{16}{3}.$$

An alternative approach, that usually turns out to be less work, is to make the change of variable $u = x + 5$ directly in the definite integral. In this case we again replace

$$\frac{x}{\sqrt{x+5}} \, dx \qquad \text{by} \qquad \frac{u-5}{\sqrt{u}} \, du.$$

In addition, we replace the limits $x = -4$ and $x = -1$ by those values of u that correspond to these values of x, namely, $u = 1$ and $u = 4$ respectively. We then obtain

$$\int_{-4}^{-1} \frac{x}{\sqrt{x+5}} \, dx = \int_{1}^{4} \frac{u-5}{\sqrt{u}} \, du = \int_{1}^{4} (u^{1/2} - 5u^{-1/2}) \, du$$

$$= \left\{ \frac{2}{3} u^{3/2} - 10u^{1/2} \right\}_{1}^{4}$$

$$= \left[\frac{2}{3}(4)^{3/2} - 10(4)^{1/2} \right] - \left[\frac{2}{3}(1)^{3/2} - 10(1)^{1/2} \right]$$

$$= -\frac{16}{3}.$$

That this method is generally acceptable is proved in the following theorem.

THEOREM 6.9

Suppose $f(x)$ is continuous on $a \leq x \leq b$, and we set $x = g(u)$, where $a = g(\alpha)$ and $b = g(\beta)$. Then

$$\int_a^b f(x)\, dx = \int_\alpha^\beta f\big(g(u)\big)g'(u)\, du, \qquad (6.32)$$

if $g'(u)$ is continuous on $\alpha \leq u \leq \beta$, and if $g(u)$ is between a and b when u is between α and β.

PROOF If we define

$$F(x) = \int f(x)\, dx \qquad \text{and} \qquad G(u) = \int f\big(g(u)\big)g'(u)\, du,$$

then by Theorem 6.3,

$$\int_a^b f(x)\, dx = F(b) - F(a) \qquad \text{and} \qquad \int_\alpha^\beta f\big(g(u)\big)g'(u)\, du = G(\beta) - G(\alpha).$$

So we must show that

$$F(b) - F(a) = G(\beta) - G(\alpha).$$

Now the chain rule applied to $y = F(x)$, $x = g(u)$ yields

$$\frac{dy}{du} = \frac{dy}{dx}\frac{dx}{du} = f(x)g'(u);$$

that is,

$$\frac{d}{du}F\big(g(u)\big) = f\big(g(u)\big)g'(u).$$

Since this is an identity in u, we may integrate with respect to u from $u = \alpha$ to $u = \beta$:

$$\int_\alpha^\beta \frac{d}{du}F\big(g(u)\big)\, du = \int_\alpha^\beta f\big(g(u)\big)g'(u)\, du.$$

Thus,

$$\big\{F\big(g(u)\big)\big\}_\alpha^\beta = \{G(u)\}_\alpha^\beta \qquad \text{or}$$

$$G(\beta) - G(\alpha) = F\big(g(\beta)\big) - F\big(g(\alpha)\big) = F(b) - F(a). \qquad \blacksquare$$

EXAMPLE 6.18

Evaluate

$$\int_2^4 \frac{\sqrt{x}}{(\sqrt{x} - 1)^4}\, dx.$$

SOLUTION   If we set $u = \sqrt{x} - 1$, then $du = [1/(2\sqrt{x})]\, dx$, and

$$\int_2^4 \frac{\sqrt{x}}{(\sqrt{x} - 1)^4}\, dx = \int_{\sqrt{2}-1}^1 \frac{u+1}{u^4} 2(u+1)\, du$$

$$= 2\int_{\sqrt{2}-1}^1 \left(\frac{1}{u^2} + \frac{2}{u^3} + \frac{1}{u^4} \right) du$$

$$= 2\left\{ -\frac{1}{u} - \frac{1}{u^2} - \frac{1}{3u^3} \right\}_{\sqrt{2}-1}^1$$

$$= -2\left[1 + 1 + \frac{1}{3} \right] + 2\left[\frac{1}{\sqrt{2}-1} + \frac{1}{(\sqrt{2}-1)^2} + \frac{1}{3(\sqrt{2}-1)^3} \right]$$

$$= 21.2.$$

What follows is a discussion leading to the well-known formula for the area of a circle. Although you should be able to follow the calculations, it is highly likely that you will not understand the source of some of the ideas. Do not be alarmed; in time, we will deal with all aspects of the derivation in detail. For now, simply take the example as an illustration of the power of integral calculus, and an indication of things to come.

In Section 9.5 we show that the area of the circle in Figure 6.12 is given by the definite integral

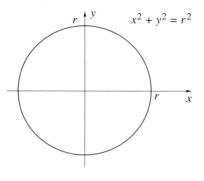

FIGURE 6.12

r, y, $x^2 + y^2 = r^2$, r, x

$$A = 4\int_0^r \sqrt{r^2 - x^2}\, dx.$$

The integral itself actually gives the first quadrant area; the 4 provides a quadrupling factor for the other three quadrants. It is not possible to guess an antiderivative for $\sqrt{r^2 - x^2}$. In Section 9.5 we learn that an appropriate change of variable is to set $x = r \sin \theta$. The integrand becomes

$$\sqrt{r^2 - x^2} = \sqrt{r^2 - r^2 \sin^2 \theta} = r\sqrt{1 - \sin^2 \theta} = r\sqrt{\cos^2 \theta} = r|\cos \theta|,$$

absolute values being necessary to ensure positivity for $\sqrt{\cos^2 \theta}$. Since $dx = r \cos \theta\, d\theta$, we have

$$A = 4\int_{\theta_1}^{\theta_2} r|\cos \theta|\, r \cos \theta\, d\theta,$$

where θ_1 and θ_2 are values of θ corresponding to $x = 0$ and $x = r$. When $x = 0$, the equation $x = r \sin \theta$ requires $0 = r \sin \theta$. We choose $\theta_1 = 0$ as the solution of this equation. (There are others, and in Section 9.5 we shall see why this choice is made.) Similarly, when $x = r$, we choose the solution $\theta_2 = \pi/2$ of $r = r \sin \theta$. Then,

$$A = 4\int_0^{\pi/2} r|\cos \theta|\, r \cos \theta\, d\theta.$$

For θ in the interval $0 \leq \theta \leq \pi/2$, $\cos \theta$ is positive, and absolute values may be dropped,

$$A = 4r^2 \int_0^{\pi/2} \cos^2 \theta \, d\theta.$$

To find an antiderivative for $\cos^2 \theta$, we solve double-angle formula 1.39c for $\cos^2 \theta$,

$$\cos 2\theta = 2\cos^2 \theta - 1 \qquad \Longrightarrow \qquad \cos^2 \theta = \frac{1 + \cos 2\theta}{2}.$$

Thus,

$$A = 4r^2 \int_0^{\pi/2} \left(\frac{1 + \cos 2\theta}{2} \right) d\theta = 2r^2 \int_0^{\pi/2} (1 + \cos 2\theta) \, d\theta$$

$$= 2r^2 \left\{ \theta + \frac{1}{2} \sin 2\theta \right\}_0^{\pi/2} = \pi r^2,$$

the formula for the area of a circle. Be reminded that you are not expected to solve other problems like this yet. In time, yes, but not now. What we hope is that the example gives you a glimpse of the power of integral calculus and the importance of the tools that we have developed in Chapters 5 and 6.

EXERCISES 6.7

In Exercises 1–22 evaluate the definite integral.

∗ 1. $\displaystyle\int_1^2 x(3x^2 - 2)^4 \, dx$

∗ 2. $\displaystyle\int_0^1 z\sqrt{1 - z} \, dz$

∗ 3. $\displaystyle\int_{-1}^0 \frac{x}{\sqrt{x + 3}} \, dx$

∗ 4. $\displaystyle\int_{\pi/4}^{\pi/3} \cos^5 x \sin x \, dx$

∗ 5. $\displaystyle\int_1^3 x^3\sqrt{9 - x^2} \, dx$

∗ 6. $\displaystyle\int_{-5}^6 \frac{x}{\sqrt{x^2 - 12}} \, dx$

∗ 7. $\displaystyle\int_4^5 y^2\sqrt{y - 4} \, dy$

∗ 8. $\displaystyle\int_{1/2}^1 \sqrt{\frac{x^2}{1 + x}} \, dx$

∗ 9. $\displaystyle\int_1^4 \frac{\sqrt{1 + \sqrt{u}}}{\sqrt{u}} \, du$

∗ 10. $\displaystyle\int_{-2}^1 \frac{x + 1}{(x^2 + 2x + 2)^{1/3}} \, dx$

∗ 11. $\displaystyle\int_3^4 \frac{x^2}{(x - 2)^4} \, dx$

∗ 12. $\displaystyle\int_0^{\pi/6} \sqrt{2 + 3\sin x} \cos x \, dx$

∗ 13. $\displaystyle\int_{\pi/4}^{\pi/2} \frac{\sin^3 x}{(1 + \cos x)^4} \, dx$

∗ 14. $\displaystyle\int_1^4 \frac{(x + 1)(x - 1)}{\sqrt{x}} \, dx$

** **15.** $\int_{4}^{9} \sqrt{1 + \sqrt{x}}\, dx$

** **16.** $\int_{-1/2}^{1} \sqrt{\dfrac{x^2}{1+x}}\, dx$

** **17.** $\int_{-1}^{1} \dfrac{|x|}{(x+2)^3}\, dx$

** **18.** $\int_{-1}^{1} \left| \dfrac{x}{(x+2)^3} \right|\, dx$

** **19.** $\int_{0}^{1} x^2 e^{x^3}\, dx$

** **20.** $\int_{1}^{2} \dfrac{(\ln x)^2}{x}\, dx$

** **21.** $\int_{2}^{4} \dfrac{1}{x \ln x}\, dx$

** **22.** $\int_{-\pi/4}^{\pi/4} \dfrac{\sec^2 x}{\sqrt{4 + 3 \tan x}}\, dx$

** **23.** Show that if $f(x)$ is an odd function, then

$$\int_{-a}^{a} f(x)\, dx = 0;$$

and that if $f(x)$ is an even function,

$$\int_{-a}^{a} f(x)\, dx = 2 \int_{0}^{a} f(x)\, dx.$$

In Exercises 24–26 use the suggested substitution to evaluate the definite integral.

** **24.** $\int_{1}^{3} \dfrac{1}{x^{3/2}\sqrt{4-x}}\, dx, \quad u = \dfrac{1}{\sqrt{x}}$

** **25.** $\int_{-6}^{-1} \dfrac{\sqrt{x^2 - 6x}}{x^4}\, dx, \quad u = \dfrac{1}{x}$

** **26.** $\int_{-4}^{0} \dfrac{x}{(5 - 4x - x^2)^{3/2}}\, dx, \quad u^2 = \dfrac{1-x}{5+x}$

*** **27.** Show that

$$\int_{0}^{\pi} \dfrac{\cos^2\left[(\pi/2)\cos\theta\right]}{\sin\theta}\, d\theta = \dfrac{\pi}{2} \int_{0}^{2\pi} \dfrac{1 - \cos\phi}{\phi(2\pi - \phi)}\, d\phi.$$

This integral is used in calculating radiated power from a half-wave antenna.

SUMMARY

The definite integral of a continuous function $f(x)$ from $x = a$ to $x = b$ is a number, one that depends on the function $f(x)$ and the limits a and b. We have defined the definite integral by subdividing the interval $a \leq x \leq b$ into n parts by $n + 1$ points $a = x_0 < x_1 < \cdots < x_{n-1} < x_n = b$, and choosing a point x_i^* in each subinterval $x_{i-1} \leq x \leq x_i$. The definite integral is then the limit of the summation

$$\int_{a}^{b} f(x)\, dx = \lim_{\|\Delta x_i\| \to 0} \sum_{i=1}^{n} f(x_i^*)\Delta x_i,$$

where $\Delta x_i = x_i - x_{i-1}$. Since all calculations in this limit take place on the x-axis, we regard the definite integral as an integration along the x-axis from $x = a$ to $x = b$.

The fundamental theorem of integral calculus allows us to calculate definite integrals using antiderivatives,

$$\int_{a}^{b} f(x)\, dx = \left\{ \int f(x)\, dx \right\}_{a}^{b}.$$

This presupposes that a continuous function has an antiderivative, as verified in Section 6.5. It was shown that when the definite integral is given a variable upper limit x, the resulting function

$$F(x) = \int_a^x f(t)\,dt$$

is an antiderivative of $f(x)$, that is, $F'(x) = f(x)$.

The average value of a function $f(x)$ over an interval $a \le x \le b$ is defined as

$$\frac{1}{b-a}\int_a^b f(x)\,dx.$$

The mean value theorem for definite integrals guarantees the existence of at least one number c between a and b, at which $f(x)$ takes on its average value,

$$f(c) = \frac{1}{b-a}\int_a^b f(x)\,dx.$$

Evaluation of a definite integral with a complex integrand can sometimes be simplified with an appropriate change of variable.

KEY TERMS

In reviewing this chapter, you should be able to define or discuss the following key terms:

Sigma notation   Average value of a function
Definite integral   Mean value theorem for integrals
Norm of a partition   Change of variable
Fundamental theorem

REVIEW EXERCISES

In Exercises 1–20 evaluate the definite integral.

* **1.** $\displaystyle\int_0^3 (x^2 + 3x - 2)\,dx$ * **2.** $\displaystyle\int_{-1}^1 (x^2 - x^4)\,dx$

* **3.** $\displaystyle\int_{-1}^1 (x^3 - 3x)\,dx$ * **4.** $\displaystyle\int_0^2 (x^2 - 2x)\,dx$

* **5.** $\displaystyle\int_1^2 (x + 1)^2\,dx$ * **6.** $\displaystyle\int_{-3}^{-2} \frac{1}{x^2}\,dx$

* **7.** $\displaystyle\int_4^9 \left(\frac{1}{\sqrt{x}} - \sqrt{x}\right) dx$

* **8.** $\displaystyle\int_0^\pi \cos x\, dx$

* **9.** $\displaystyle\int_{-1}^1 x(x+1)^2\, dx$

* **10.** $\displaystyle\int_1^2 x^2(x^2+3)\, dx$

* **11.** $\displaystyle\int_0^3 \sqrt{x+1}\, dx$

* **12.** $\displaystyle\int_1^5 x\sqrt{x^2-1}\, dx$

* **13.** $\displaystyle\int_1^4 \left(\frac{\sqrt{x}+1}{\sqrt{x}}\right) dx$

* **14.** $\displaystyle\int_{-1}^0 x\sqrt{x+1}\, dx$

* **15.** $\displaystyle\int_1^2 \frac{x^2+1}{(x+1)^4}\, dx$

* **16.** $\displaystyle\int_{-4}^{-2} x^2\sqrt{2-x}\, dx$

* **17.** $\displaystyle\int_0^{\pi/4} \frac{\cos x}{(1+\sin x)^2}\, dx$

* **18.** $\displaystyle\int_2^3 x(1+2x^2)^4\, dx$

** **19.** $\displaystyle\int_1^8 \frac{(1+x^{1/3})^2}{x^{2/3}}\, dx$

** **20.** $\displaystyle\int_{-4}^4 |x+2|\, dx$

** **21.** Use equation 6.10 to evaluate the following definite integrals:

(a) $\displaystyle\int_0^2 (x-5)\, dx$ (b) $\displaystyle\int_0^3 (x^2+3)\, dx$

** **22.** Prove that each of the following answers is incorrect, but do so without evaluating the definite integral. Think about what the definite integral represents.

(a) $\displaystyle\int_0^4 (x^2+3x)\, dx = -3$ (b) $\displaystyle\int_{-3}^{-2} \frac{1}{x}\, dx = 5$

In Exercises 23–26 find the average value of the function on the interval.

* **23.** $f(x) = \sqrt{x+4}, \quad 0 \le x \le 1$

* **24.** $f(x) = 1/x^2 - x, \quad -2 \le x \le -1$

** **25.** $f(x) = x\sqrt{x+1}, \quad 0 \le x \le 1$

** **26.** $f(x) = \cos^3 x \sin^2 x, \quad 0 \le x \le \pi/2$

In Exercises 27–32 differentiate the integral with respect to x.

* **27.** $\displaystyle\int_1^x t\sqrt{t^3+1}\, dt$

* **28.** $\displaystyle\int_x^{-3} t^2(t+1)^3\, dt$

* **29.** $\displaystyle\int_1^{x^2} \sqrt{t^2+1}\, dt$

* **30.** $\displaystyle\int_{2x}^4 t\cos t\, dt$

* **31.** $\displaystyle\int_{2x+3}^{1-x} \frac{1}{t^2+1}\, dt$

* **32.** $\displaystyle\int_{-x^2}^{x^2} \sin^2 t\, dt$

In Exercises 33 and 34, $v(t)$ represents the velocity of a particle moving along the x-axis. Calculate the definite integral of $v(t)$ and $|v(t)|$ between the two times shown. Interpret each number physically and draw a graph of $v(t)$ to corroborate results.

** **33.** $v(t) = t^3 - 6t^2 + 11t - 6$, $t = 0$ to $t = 1$

** **34.** $v(t) = t^3 - 6t^2 + 11t - 6$, $t = 1$ to $t = 3$

In Exercises 35–42 evaluate the definite integral.

** **35.** $\displaystyle\int_0^1 \frac{x^3}{(x^2 + 1)^{3/2}}\, dx$

** **36.** $\displaystyle\int_0^{\pi/6} \frac{\cos^3 x}{\sqrt{1 + \sin x}}\, dx$

** **37.** $\displaystyle\int_{-1}^1 x^3\sqrt{1 - x^2}\, dx$

** **38.** $\displaystyle\int_{-1}^2 \left|\frac{x}{\sqrt{3 + x}}\right|\, dx$

** **39.** $\displaystyle\int_{-1}^2 x^2(4 - x^3)^5\, dx$

** **40.** $\displaystyle\int_1^5 \frac{6x^2 + 8x + 2}{\sqrt{x^3 + 2x^2 + x}}\, dx$

** **41.** $\displaystyle\int_1^2 \frac{x - 25}{\sqrt{x} - 5}\, dx$

** **42.** $\displaystyle\int_0^1 \frac{1}{\sqrt{2 + x} + \sqrt{x}}\, dx$

Applications of the Definite Integral

In Chapter 6 we defined the definite integral of a function $f(x)$ as the limit of a summation

$$\int_a^b f(x)\,dx = \lim_{\|\Delta x_i\|\to 0} \sum_{i=1}^n f(x_i^*)\Delta x_i.$$

The limit-summation on the right defines the value of the definite integral on the left. In this chapter we think of this equation in the reverse direction in order to evaluate geometric and physical quantities. Each quantity is expressed as a limit-summation of the form in this equation. The limit-summation can immediately be interpreted as a definite integral. The definite integral can then be evaluated by means of the corresponding indefinite integral.

7.1 Area

We have formulas for the areas of very few shapes—squares, rectangles, triangles, and polygons of any shape, since they can be divided into rectangles and triangles. Consider finding the area in Figure 7.1a. Each of us has intuitive ideas about this area. In this section we take these intuitive ideas of what the area *ought to be*, and make a precise mathematical definition of what the area *is*.

FIGURE 7.1a

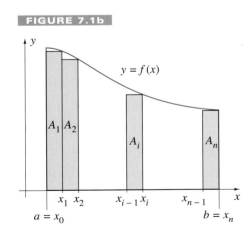

FIGURE 7.1b

Recall from Problem 1 of Section 6.2 that the area in Figure 7.1a can be approximated by rectangles. Specifically, we partition the interval $a \le x \le b$ into n parts by points

$$a = x_0 < x_1 < x_2 < \cdots < x_{n-1} < x_n = b,$$

and construct rectangles as shown in Figure 7.1b. The area A_i of the ith rectangle is

$$A_i = f(x_i)(x_i - x_{i-1}) = f(x_i)\Delta x_i,$$

and an approximation to the required area A is therefore

$$\sum_{i=1}^{n} A_i = \sum_{i=1}^{n} f(x_i)\Delta x_i.$$

If we let the number of rectangles get larger and larger, and at the same time require each rectangle to have smaller and smaller width that eventually approaches zero, we feel that better and better approximations can be obtained. In fact, if we take the limit as the norm $\|\Delta x_i\|$ of the partition approaches zero, we should get A. Since we have no formal definition for the area of odd-shaped figures, we take this opportunity to make our own. We define the area A in Figure 7.1a as

$$A = \lim_{\|\Delta x_i\| \to 0} \sum_{i=1}^{n} f(x_i)\Delta x_i. \tag{7.1}$$

Area A then has been defined as the limit of a sum of rectangular areas.

The right side of equation 7.1 is strikingly similar to the definition of the definite integral of $f(x)$ from $x = a$ to $x = b$ (equation 6.10),

$$\int_a^b f(x)\,dx = \lim_{\|\Delta x_i\| \to 0} \sum_{i=1}^{n} f(x_i^*)\Delta x_i. \tag{6.10}$$

The only difference is the absence of *'s in 7.1. But when we recall that in 6.10, x_i^* may be chosen as any point in the subinterval $x_{i-1} \le x \le x_i$, we see that by choosing $x_i^* = x_i$,

$$\int_a^b f(x)\,dx = \lim_{\|\Delta x_i\| \to 0} \sum_{i=1}^{n} f(x_i)\Delta x_i.$$

It follows that area A of Figure 7.1a may be calculated by means of the definite integral

$$A = \int_a^b f(x)\,dx. \tag{7.2}$$

It is important to realize that equation 7.2 does not imply that a definite integral should always be thought of as an area; on the contrary, we will find that definite integrals can represent many quantities. What we have said is that the area in Figure 7.1a is defined by limit 7.1. But this limit may also be interpreted as the definite integral of $f(x)$ with respect to x from $x = a$ to $x = b$; hence, the area may be calculated by definite integral 7.2. Since definite integrals can be evaluated using antiderivatives, it seems that we have a very simple way to find areas.

It is simple to extend this result to the problem of finding the area in Figure 7.2. With our interpretation of equation 7.2, we can state that the area under the curve $y = f(x)$, above the x-axis and between the vertical lines $x = a$ and $x = b$, is given by

$$A_2 = \int_a^b f(x)\,dx.$$

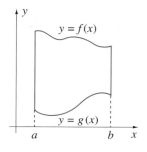

FIGURE 7.2

Since the area under the curve $y = g(x)$ is given similarly by

$$A_1 = \int_a^b g(x)\,dx,$$

it follows that the required area is

$$A = A_2 - A_1 = \int_a^b f(x)\,dx - \int_a^b g(x)\,dx = \int_a^b [f(x) - g(x)]\,dx. \qquad (7.3)$$

We now have two formulas for finding areas: equation 7.2 for the area under a curve, above the x-axis, and between two vertical lines; and equation 7.3 for the area between two curves and two vertical lines. Note that 7.2 is a special case of 7.3. At this point we could solve a number of area problems using these two results, but to do so would strongly suggest that integration should be approached from a "formula" point of view; and if there is any point of view that we wish to adopt, it is completely the opposite. By the end of this chapter we hope to have developed a sufficiently clear understanding of the limit-summation process that use of integration in situations other than those discussed here will be straightforward.

To illustrate how we arrive at the correct definite integral for area problems without memorizing either 7.2 or 7.3, consider again finding the area in Figure 7.1a. We draw a rectangle of width dx at position x as shown in Figure 7.3. The area of this rectangle is

$$f(x)\,dx.$$

We visualize this rectangle as a representative for a large number of such rectangles between a and b. To find the required area we add together all such rectangular areas and take the limit as their widths approach zero. But this is precisely the concept of the definite integral, so that we write for the limit-summation process

$$A = \int_a^b f(x)\,dx.$$

Limits a and b identify x-positions of first and last rectangles, respectively.

FIGURE 7.3

FIGURE 7.4

Similarly, for the area in Figure 7.4 we draw a rectangle of width dx and length $f(x) - g(x)$ and, therefore, of area

$$[f(x) - g(x)]\,dx.$$

To add areas of all such rectangles between a and b and, at the same time, to take the limit as their widths approach zero, we once again use the definite integral,

$$A = \int_a^b [f(x) - g(x)]\, dx.$$

For area problems then, we start with the area of a representative rectangle and proceed to the required area by summation with the definite integral. We express this symbolically as follows:

$$(7.4)$$

| **EXAMPLE 7.1**

Find the area enclosed by the curves $y = x^2$ and $y = x^3$.

FIGURE 7.5

SOLUTION   Area of the representative rectangle in Figure 7.5 is

$$(x^2 - x^3)\, dx;$$

hence,

$$A = \int_0^1 (x^2 - x^3)\, dx = \left\{ \frac{x^3}{3} - \frac{x^4}{4} \right\}_0^1 = \frac{1}{12}.$$

In the next two examples, expressions for areas of representative rectangles vary within required areas. In such cases, we set up different integrals corresponding to different representative rectangles.

| **EXAMPLE 7.2**

Find the area bounded by the x-axis and the curve $y = x^3 - x$.

FIGURE 7.6

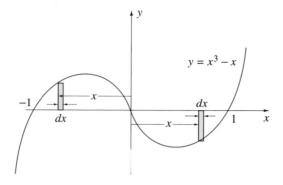

SOLUTION Areas of rectangles between $x = -1$ and $x = 0$ (Figure 7.6) are

$$[(x^3 - x) - 0]\,dx,$$

whereas between $x = 0$ and $x = 1$ areas are

$$[0 - (x^3 - x)]\,dx.$$

Consequently,

$$A = \int_{-1}^{0} (x^3 - x)\,dx + \int_{0}^{1} (-x^3 + x)\,dx$$

$$= \left\{ \frac{x^4}{4} - \frac{x^2}{2} \right\}_{-1}^{0} + \left\{ -\frac{x^4}{4} + \frac{x^2}{2} \right\}_{0}^{1}$$

$$= -\left(\frac{1}{4} - \frac{1}{2} \right) + \left(-\frac{1}{4} + \frac{1}{2} \right)$$

$$= \frac{1}{2}.$$

We could have saved ourselves some calculations in this example by noting that because of the symmetry of the diagram ($x^3 - x$ is an odd function), the two areas are identical. Hence, we could find the left (or right) area and double it,

$$A = 2 \int_{-1}^{0} (x^3 - x)\,dx = 2\left\{ \frac{x^4}{4} - \frac{x^2}{2} \right\}_{-1}^{0} = -2\left(\frac{1}{4} - \frac{1}{2} \right) = \frac{1}{2}.$$

EXAMPLE 7.3

Find the area of the triangle with edges $y = x$, $y = -x/2$, and $y = 5x - 44$.

FIGURE 7.7

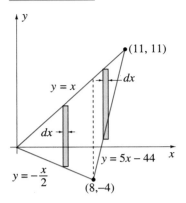

SOLUTION   Areas of representative rectangles to the left and right of $x = 8$ (Figure 7.7) are, respectively,

$$[x - (-x/2)]\, dx \qquad \text{and} \qquad [x - (5x - 44)]\, dx;$$

therefore

$$A = \int_0^8 [x - (-x/2)]\, dx + \int_8^{11} [x - (5x - 44)]\, dx$$

$$= \frac{3}{2} \int_0^8 x\, dx + \int_8^{11} (44 - 4x)\, dx$$

$$= \frac{3}{2} \left\{ \frac{x^2}{2} \right\}_0^8 + \left\{ 44x - 2x^2 \right\}_8^{11}$$

$$= 48 + (484 - 242) - (352 - 128)$$

$$= 66.$$

We see from the above examples that the length of a representative rectangle in equation 7.4 as "upper y minus lower y" is valid whether the rectangle is in the first quadrant (Figure 7.5), the second and fourth quadrants (Figure 7.6), or partially in the first and partially in the fourth (Figure 7.7). In fact it is valid for rectangles in all quadrants. Remember this; we use it in many applications.

There is nothing special about vertical rectangles. Sometimes it is more convenient to subdivide an area into horizontal rectangles. For example, to find the area in Figure 7.8, we draw a representative rectangle at position y of width dy. Its length is $f(y) - g(y)$, and therefore its area is

$$[f(y) - g(y)]\, dy.$$

Adding over all rectangles gives

$$A = \int_a^b [f(y) - g(y)]\, dy. \tag{7.5}$$

Corresponding to 7.4, we could write that for horizontal rectangles

$$A = \int_{\substack{y\text{-position of} \\ \text{first rectangle}}}^{\substack{y\text{-position of} \\ \text{last rectangle}}} \overbrace{\underbrace{\left[\begin{matrix} x\text{-coordinate} \\ \text{of right end} \\ \text{of rectangle} \end{matrix} - \begin{matrix} x\text{-coordinate} \\ \text{of left end} \\ \text{of rectangle} \end{matrix} \right]}_{\text{length of horizontal rectangle}} \underbrace{dy}_{\substack{\text{width of} \\ \text{horizontal} \\ \text{rectangle}}}}^{\text{area of horizontal rectangle}} \quad . \tag{7.6}$$

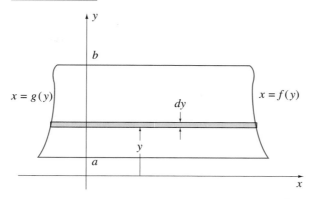

EXAMPLE 7.4 ▼

Find the area bounded by the curves $y = \sqrt{x + 14}$, $x = \sqrt{y}$, and $y = 0$.

SOLUTION   Subdivision of the area (Figure 7.9) into vertical rectangles results in two integrations: one to the left and the other to the right of the y-axis. On the other hand, throughout the required area, the area of a horizontal rectangle is

$$[\sqrt{y} - (y^2 - 14)]\, dy,$$

and therefore

$$A = \int_0^4 \left(\sqrt{y} - y^2 + 14\right) dy = \left\{ \frac{2}{3}y^{3/2} - \frac{1}{3}y^3 + 14y \right\}_0^4 = \frac{16}{3} - \frac{64}{3} + 56 = 40.$$

▲

In choosing between horizontal and vertical rectangles, consider two objectives:

1. to minimize the number of integrations; and
2. to obtain simple definite integrals.

For instance, by choosing one type of rectangle we may obtain only one definite integral, but it may be very difficult to evaluate. If the other type of rectangle leads to two simple definite integrals, then it would be wise to choose the two simple integrals.

EXAMPLE 7.5 ▼

Find the area enclosed by the curves

$$y = \frac{x}{\sqrt{x^2 - 16}}, \qquad y = \frac{x^2}{15}, \qquad y = -x^2, \qquad x = 6, \qquad x \geq 0.$$

SOLUTION   Examination of Figure 7.10 indicates that horizontal rectangles necessitate three integrals. In addition, we would have to solve equations $y = x^2/15$ and $y = x/\sqrt{x^2 - 16}$ for x in terms of y. We therefore opt for vertical rectangles.

$$A = \int_0^5 \left[\frac{x^2}{15} - (-x^2) \right] dx + \int_5^6 \left[\frac{x}{\sqrt{x^2 - 16}} - (-x^2) \right] dx$$

$$= \frac{16}{15} \int_0^5 x^2 \, dx + \int_5^6 \left[\frac{x}{\sqrt{x^2 - 16}} + x^2 \right] dx$$

$$= \frac{16}{15} \left\{ \frac{x^3}{3} \right\}_0^5 + \left\{ \sqrt{x^2 - 16} + \frac{x^3}{3} \right\}_5^6$$

$$= \frac{16}{15} \left(\frac{125}{3} \right) + \left(\sqrt{20} + 72 \right) - \left(3 + \frac{125}{3} \right)$$

$$\approx 76.2.$$

FIGURE 7.10

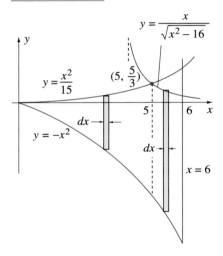

The mean value theorem for definite integrals (Theorem 6.8) states that there is a number c between a and b such that

$$\int_a^b f(x) \, dx = (b - a) f(c).$$

It has a very simple interpretation in terms of area when $f(x) \geq 0$ for $a \leq x \leq b$. Since $f(x) \geq 0$ for $a \leq x \leq b$, the definite integral may be interpreted as the area under the curve $y = f(x)$, above the x-axis, and between vertical lines at $x = a$ and $x = b$ (Figure 7.11). The right side is the area of the rectangle of width $b - a$ and height $f(c)$ shaded in Figure 7.11. The mean value theorem guarantees at least one point c between a and b for which the rectangular area is equal to the area under the curve. For the curve in Figure 7.11, there is exactly one such point c; for the curve in Figure 7.12, there are three choices for c.

FIGURE 7.11

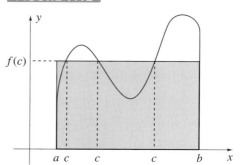

FIGURE 7.12

EXERCISES 7.1

In Exercises 1–16 find the area bounded by the curves.

* **1.** $y^2 = 4x$, $x^2 = 4y$ * **2.** $y = x^3 + 8$, $y = 4x + 8$

* **3.** $yx^2 = 4$, $y = 5 - x^2$ * **4.** $x = y(y - 2)$, $x + y = 12$

* **5.** $x = 4y - 4y^2$, $y = x - 3$, $y = 1$, $y = 0$

* **6.** $y = e^{3x}$, $x = 1$, $x = 2$, $y = -x$

* **7.** $12y = 7 - x^2$, $y = 1/(2x)$ * **8.** $y = \sqrt{x + 4}$, $y = (x + 4)^2/8$

* **9.** $y = (x - 1)^5$, $x = 0$, $y = 0$ * **10.** $y = \sin x$ $(0 \le x \le \pi)$, $y = 0$

* **11.** $y = x^5 - x$, $y = 0$

* **12.** $x + y = 1$, $x + y = 5$, $y = 2x + 1$, $y = 2x + 6$

* **13.** $x = \sec^2 y$, $y = 0$, $y = \pi/4$, $x = 0$

* **14.** $xy = e$, $y = x^2$, $y = 2$ (smaller area)

* **15.** $y = e^2(2 - x^2)$, $y = e^{2x}$, $y = e^{-2x}$, $y \ge 1$

* **16.** $x = |y| + 1$, $x + (y - 1)^2 = 4$, $y = 0$ (above the x-axis)

** **17.**   (a) Find the area A bounded by the curves

$$y = 16 - x^2, \quad y = 0, \quad x = 1, \quad x = 3.$$

   In the remainder of this problem we approximate A by rectangles and show that the accuracy of the approximation increases as the number of rectangles increases. To do this you will need a large graph of the function $f(x) = 16 - x^2$ on the interval $0 \le x \le 3$.

   (b) On the graph draw two rectangles of equal width (one unit) and with heights determined in the same way as in Figure 7.1b (page 443). If A_2 denotes the sum of the areas of these two rectangles, what is A_2? What is the error in the approximation of A by A_2? Illustrate this error on the graph.

   (c) Repeat part (b) with four rectangles all of width one-half, denoting the sum of the areas of the four rectangles by A_4. Show on the graph the extra precision of A_4 over A_2.

   (d) Repeat part (b) with eight rectangles all of equal width, denoting the sum of the areas of the eight rectangles by A_8. Show on the graph the extra precision of A_8 over A_4.

This discussion is continued in Exercise 46.

** **18.** Repeat Exercise 17 with the area bounded by the curves

$$y = x^3 + 1, \quad x = 1, \quad x = 3, \quad y = 0.$$

In Exercises 19–22 set up (but do not evaluate) definite integral(s) for the area bounded by the curves.

* **19.** $x = 1/\sqrt{4 - y^2}, \quad 4x = -y^2, \quad y = -1, \quad y = 1$

* **20.** $x^2 + y^2 = 4, \quad x^2 + y^2 = 4x$ (interior to both)

* **21.** $x^2 + y^2 = 4, \quad x^2 + y^2 = 6x$ (interior to both)

* **22.** $x^2 + y^2 = 16, \quad x = y^2$ (smaller area)

** **23.** Find the area bounded by $y^2 = 4ax$ and $x^2 = 4ay$, where $a > 0$ is a constant.

In Exercises 24–35 find the area bounded by the curves.

** **24.** $y = x/\sqrt{x + 3}, \quad x = 1, \quad x = 6, \quad y = -x^2$

** **25.** $x = y^2 + 2, \quad x = -(y - 4)^2, \quad y = -x + 4, \quad y = 0$

** **26.** $y = x^3 - x, \quad x + y + 1 = 0, \quad x = \sqrt{y + 1}$

** **27.** $y = \left| \dfrac{x}{(x - 2)^3} \right|, \quad y = 0, \quad x = -1, \quad x = 1$

** **28.** $x = 2ye^{-y^2}, \quad y = x$ ** **29.** $y = \sin^3 x, \quad y = 1/8, \quad 0 \le x \le 2\pi$

** **30.** $y = \ln x^2, \quad y = 1 - x^2, \quad y = 1$ ** **31.** $|x|^{1/2} + |y|^{1/2} = 1$

** **32.** $y^2 = x^2(4 - x^2)$ ** **33.** $y^2 = x^4(9 + x)$

** **34.** $y^2 = x^2(x^2 - 4), \quad x = 5$ ** **35.** $(2x - y)^2 = x^3, \quad x = 4$

In Exercises 36–43, it is necessary to use a calculator or computer to find points of intersection of the curves. Find the area bounded by the curves (to 3 decimal places).

** **36.** $y = x^3 + 3x^2 + 2x + 1, \quad x = 0, \quad y = 0$

** **37.** $y = x^3 - 4x, \quad y = 2 - x - x^2$

** **38.** $y = x^4 - 5x^2 + 5, \quad y = 0$ ** **39.** $y = e^x, \quad y = 2 - x^2$

** **40.** $y = \cos x, \quad 4y = x + 2$ ** **41.** $y = \dfrac{2}{x + 2}, \quad y = x^3 + 3x - 1, \quad x = 0$

** **42.** $y = x^3 - 3x^2 + 4x - 2, \quad x = 4 - y^2$

** **43.** $x = y^3 - y^2 - 2y, \quad x = \sqrt{2y + 1}$

** **44.** For what values of m do the curves

$$y = \frac{x}{3x^2 + 1} \quad \text{and} \quad y = mx$$

bound an area? Find the area.

** **45.** Find a point (a, b) on the curve $y = x/\sqrt{x^2 + 1}$ such that the area bounded by this curve, the x-axis, and the line $x = a$ is equal to twice that area bounded by the curve, the y-axis, and the line $y = b$.

** **46.** If 2^n $(n = 1, 2, \ldots)$ rectangles (all of equal width) are drawn in Exercise 17 to approximate A, and A_{2^n} denotes the sum of the areas of these rectangles, show that

$$A_{2^n} = A - \left[\frac{1}{2^{n-2}} + \frac{1}{6} \left(\frac{3}{2^{n-3}} + \frac{1}{2^{2n-3}} \right) \right],$$

and hence that

$$\lim_{n \to \infty} A_{2^n} = A.$$

** **47.** Show that curves

$$y = \frac{x^3}{x^4 + 16} \quad \text{and} \quad 204y = 13x^2 - 1$$

bound three areas. Find the largest area.

** **48.** Let P be a point on the cubic curve $y = f(x) = ax^3$. Let the tangent line at P intersect $y = f(x)$ again at Q, and let A be the area of the region bounded by $y = f(x)$ and the line PQ. Let B be the area of the region defined in the same way by starting with Q instead of P. Show that B is 16 times as large as A.

** **49.** Prove that the result in Exercise 48 is valid for any cubic $y = f(x) = ax^3 + bx^2 + cx + d$.

*** **50.** A circular pasture has radius R. A cow is tied to a stake at the edge of the pasture. What length of rope permits the cow to graze on half the pasture?

7.2 Volumes of Solids of Revolution

In Section 6.2 we discussed the idea of rotating flat areas around coplanar lines to produce volumes of solids of revolution. To find the volume generated when the area in Figure 7.1a is revolved about the x-axis, we again approximate the area by n rectangles as in Figure 7.1b. If each of these rectangles is rotated around the x-axis, then n discs are formed. Since the radius of the ith disc is $f(x_i)$ (Figure 7.13), its volume is given by

$$V_i = \pi [f(x_i)]^2 (x_i - x_{i-1}) = \pi [f(x_i)]^2 \Delta x_i.$$

FIGURE 7.13

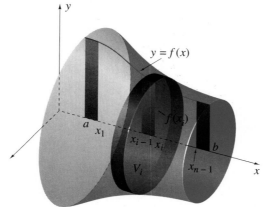

An approximation to the required volume V is therefore

$$\sum_{i=1}^{n} V_i = \sum_{i=1}^{n} \pi [f(x_i)]^2 \Delta x_i.$$

If we let the number of rectangles get larger and larger, and at the same time require their widths to get smaller and smaller, it seems reasonable that we will obtain better and better approximations. Furthermore, if we take the limit as the norm of the partition approaches zero, we should get what we think is V. Since we have no formal definition for such volumes, we make our own. We define

$$V = \lim_{\|\Delta x_i\| \to 0} \sum_{i=1}^{n} \pi [f(x_i)]^2 \Delta x_i. \tag{7.7}$$

But we can interpret the right side of this definition as the definite integral of $\pi [f(x)]^2$ with respect to x from $x = a$ to $x = b$; that is,

$$\int_{a}^{b} \pi [f(x)]^2 dx = \lim_{\|\Delta x_i\| \to 0} \sum_{i=1}^{n} \pi [f(x_i)]^2 \Delta x_i.$$

Consequently, we can calculate the volume of the solid of revolution generated by rotating the area in Figure 7.1a around the x-axis, by means of the definite integral

$$V = \int_{a}^{b} \pi [f(x)]^2 \, dx. \tag{7.8}$$

The volume of the solid of revolution has been defined by limit 7.7, but for evaluation of this limit we use definite integral 7.8.

To avoid memorizing 7.8 as a formula, we use the technique introduced in Section 7.1. For the volume obtained by rotating the area in Figure 7.1a about the x-axis, we construct a rectangle of width dx at position x, as shown in either Figure 7.3 or Figure 7.14. When this rectangle is rotated around the x-axis, the volume of the disc generated is

$$\pi [f(x)]^2 \, dx,$$

where dx is the thickness of the disc and $\pi [f(x)]^2$ is the area of its flat surface. This disc is pictured as representing a large number of such discs between a and b. We find the required volume by adding volumes of all such discs and taking the limit as their widths approach zero. But this is the concept of the definite integral, so that we write for the limit-summation process

$$V = \int_{a}^{b} \pi [f(x)]^2 \, dx,$$

where limits $x = a$ and $x = b$ identify x-positions of first and last discs respectively. This is called the **disc method** for finding the volume of a solid of revolution.

A slightly more general problem is that of finding the volume of the solid of revolution generated by rotating an area bounded by two curves and two vertical lines as shown in Figure 7.15 around the x-axis. If a representative rectangle of width dx at position x is rotated around the x-axis, the volume formed is a **washer**. Since outer and inner radii of this washer are $f(x)$ and $g(x)$ respectively, its volume is

$$\{\pi [f(x)]^2 - \pi [g(x)]^2\} \, dx.$$

FIGURE 7.14

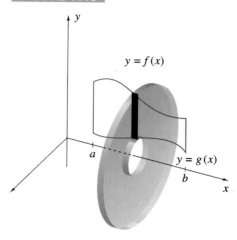

FIGURE 7.15

To add volumes of all such washers between a and b, and at the same time take limits as their widths approach zero, we use the definite integral

$$V = \int_a^b \{\pi[f(x)]^2 - \pi[g(x)]^2\} \, dx. \tag{7.9}$$

EXAMPLE 7.6

Prove that the volume of a sphere of radius r is $4\pi r^3/3$.

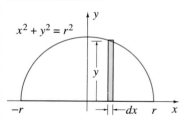

FIGURE 7.16

SOLUTION   A sphere of radius r is formed when the semicircle $x^2 + y^2 \leq r^2$ ($y \geq 0$) is rotated around the x-axis. Volume of the representative disc generated by rotating the rectangle in Figure 7.16 around the x-axis is

$$\pi y^2 \, dx = \pi(r^2 - x^2) \, dx.$$

Since the volume formed by the left quarter circle is the same as for the right quarter circle, we calculate the volume generated by the right quarter and double the result:

$$V = 2\int_0^r \pi(r^2 - x^2) \, dx = 2\pi \left\{ r^2x - \frac{x^3}{3} \right\}_0^r = \frac{4}{3}\pi r^3.$$

EXAMPLE 7.7

Prove that the volume of a right-circular cone of base radius r and height h is $\pi r^2 h/3$.

FIGURE 7.17

SOLUTION The cone can be generated by rotating the triangle in Figure 7.17 around the x-axis. Volume of the representative disc formed by rotating the rectangle shown is

$$\pi y^2 \, dx = \pi \left(\frac{rx}{h}\right)^2 dx;$$

hence,

$$V = \int_0^h \frac{\pi r^2 x^2}{h^2} \, dx = \frac{\pi r^2}{h^2} \left\{\frac{x^3}{3}\right\}_0^h = \frac{1}{3}\pi r^2 h.$$

EXAMPLE 7.8

Find the volume of the solid of revolution obtained by rotating the area bounded by the curves $y = 1 - x^2$ and $y = 4 - 4x^2$ around (a) the x-axis and (b) the line $y = -1$.

FIGURE 7.18

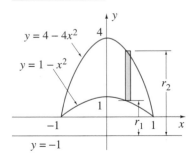

SOLUTION

(a) When the rectangle in Figure 7.18 is rotated around the x-axis, volume of the washer formed is

$$(\pi y_2^2 - \pi y_1^2) \, dx = \pi[(4 - 4x^2)^2 - (1 - x^2)^2] \, dx = 15\pi(1 - 2x^2 + x^4) \, dx.$$

Because of the symmetry of the area, we rotate only the right half and double the result:

$$V = 2\int_0^1 15\pi(1 - 2x^2 + x^4) \, dx = 30\pi\left\{x - \frac{2x^3}{3} + \frac{x^5}{5}\right\}_0^1 = 16\pi.$$

FIGURE 7.19

(b) When the rectangle in Figure 7.19 is rotated around the line $y = -1$, inner and outer radii of the washer are r_1 and r_2 respectively. Now r_1 is a length in the y-direction, and in Section 7.1 we learned that to calculate lengths in the y-direction we take upper y minus lower y. Hence, $r_1 = (1 - x^2) - (-1) = 2 - x^2$. Similarly, $r_2 = (4 - 4x^2) - (-1) = 5 - 4x^2$. Volume of the washer is therefore

$$(\pi r_2^2 - \pi r_1^2) \, dx = [\pi(5 - 4x^2)^2 - \pi(2 - x^2)^2] \, dx = 3\pi(7 - 12x^2 + 5x^4) \, dx.$$

Consequently,

$$V = 2\int_0^1 3\pi(7 - 12x^2 + 5x^4) \, dx = 6\pi\left\{7x - 4x^3 + x^5\right\}_0^1 = 24\pi.$$

We can also rotate horizontal rectangles around vertical lines to produce discs and washers as illustrated in the next two examples.

EXAMPLE 7.9

Find the volume of the solid of revolution when the area enclosed by the curves $y = \ln x$, $y = 0$, $y = 1$, and $x = 0$ is revolved around the y-axis.

SOLUTION   When the rectangle in Figure 7.20 is rotated around the y-axis, volume of the disc formed is

$$\pi x^2 \, dy = \pi (e^y)^2 \, dy = \pi e^{2y} \, dy.$$

The required volume is therefore

$$V = \int_0^1 \pi e^{2y} \, dy = \pi \left\{ \frac{1}{2} e^{2y} \right\}_0^1 = \frac{\pi}{2}(e^2 - 1).$$

FIGURE 7.20

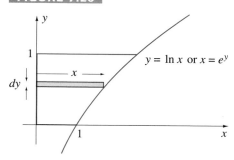

EXAMPLE 7.10

Find the volume of the solid of revolution obtained by rotating the area enclosed by the curves $y = x^2 - 1$ and $y = 0$ around the line $x = 5$.

FIGURE 7.21

SOLUTION   When the horizontal rectangle in Figure 7.21 is rotated around the line $x = 5$, volume of the washer formed is

$$(\pi r_2^2 - \pi r_1^2) \, dy = \pi [(5 - x_2)^2 - (5 - x_1)^2] \, dy.$$

Since x_1 and x_2 are x-coordinates of points on the curve $y = x^2 - 1$, we solve this equation for $x = \pm\sqrt{y+1}$. Since $x_1 > 0$ and $x_2 < 0$, we set $x_1 = \sqrt{y+1}$ and $x_2 = -\sqrt{y+1}$. Thus, the volume of the washer can be expressed as

$$\pi \left[\left(5 + \sqrt{y+1}\right)^2 - \left(5 - \sqrt{y+1}\right)^2 \right] dy = 20\pi \sqrt{y+1} \, dy.$$

The required volume is therefore

$$V = \int_{-1}^0 20\pi \sqrt{y+1} \, dy = 20\pi \left\{ \frac{2}{3}(y+1)^{3/2} \right\}_{-1}^0 = \frac{40\pi}{3}.$$

It follows that

$$V = \int_a^b 2\pi y[f(y) - g(y)] \, dy.$$

The method described is called the **cylindrical-shell method** for finding the volume of a solid of revolution.

EXAMPLE 7.11

Find the volume of the solid of revolution when the area enclosed by $x = 2y - y^2$ and the y-axis is revolved around the x-axis.

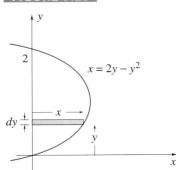

SOLUTION   Volume of the cylindrical shell formed by rotating the rectangle in Figure 7.25 around the x-axis is approximately

$$(2\pi y)(x)dy = 2\pi y(2y - y^2)dy = 2\pi(2y^2 - y^3)dy.$$

Hence,

$$V = \int_0^2 2\pi(2y^2 - y^3) \, dy = 2\pi \left\{ \frac{2y^3}{3} - \frac{y^4}{4} \right\}_0^2 = 2\pi \left(\frac{16}{3} - 4 \right) = \frac{8\pi}{3}.$$

EXAMPLE 7.12

Use cylindrical shells to calculate the volume in Example 7.10.

SOLUTION   When we rotate the rectangle in Figure 7.26 around the line $x = 5$, the volume of the cylindrical shell is approximately

$$2\pi(5 - x)(0 - y)dx = 2\pi(5 - x)(1 - x^2)dx = 2\pi(x^3 - 5x^2 - x + 5)dx.$$

Total volume is therefore

$$V = \int_{-1}^{1} 2\pi(x^3 - 5x^2 - x + 5)\,dx = 2\pi\left\{\frac{x^4}{4} - \frac{5x^3}{3} - \frac{x^2}{2} + 5x\right\}_{-1}^{1}$$

$$= 2\pi\left(\frac{1}{4} - \frac{5}{3} - \frac{1}{2} + 5\right) - 2\pi\left(\frac{1}{4} + \frac{5}{3} - \frac{1}{2} - 5\right) = \frac{40\pi}{3}.$$

This is the result obtained by the washer method in Example 7.10.

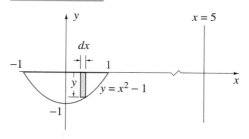

Even though the area in this example is symmetric about the y-axis, the volume generated by the right half is less than that generated by the left. For this reason we cannot integrate from $x = 0$ to $x = 1$ and double the result. If rotation were performed about the x-axis, we could indeed integrate over either half of the interval and double. Finally, for rotation about the y-axis, we would integrate over only one half of the interval and neglect the other half in order to eliminate duplications.

EXAMPLE 7.13

Find the volume of the solid of revolution obtained by rotating the area enclosed by the curves $y = x^3$, $y = -x$, and $y = 1$ around the line $y = -2$.

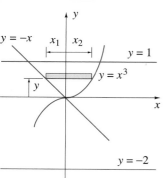

SOLUTION   When we rotate the rectangle in Figure 7.27, volume of the representative cylindrical shell is approximately

$$2\pi(y + 2)(x_2 - x_1)dy = 2\pi(y + 2)(y^{1/3} + y)dy.$$

Total volume is therefore

$$V = \int_{0}^{1} 2\pi(y + 2)(y^{1/3} + y)\,dy = 2\pi\int_{0}^{1}(y^{4/3} + 2y^{1/3} + y^2 + 2y)\,dy$$

$$= 2\pi\left\{\frac{3}{7}y^{7/3} + \frac{3}{2}y^{4/3} + \frac{y^3}{3} + y^2\right\}_{0}^{1} = 2\pi\left(\frac{3}{7} + \frac{3}{2} + \frac{1}{3} + 1\right) = \frac{137\pi}{21}.$$

Equations 7.9 and 7.11 provide two methods for calculating volumes of solids of revolution: washers and shells. Where applicable both methods give the same results. We illustrate this for the area bounded by the curves in Figure 7.28. As indicated in the figure, each curve defines y as a function of x, and x as a function of y. When this area is rotated around the x-axis, cylindrical shells can be used to calculate the volume of the solid of revolution

$$V = \int_c^d 2\pi y[q(y) - p(y)]\,dy = 2\pi \int_c^d yq(y)\,dy - 2\pi \int_c^d yp(y)\,dy. \qquad (7.12)$$

FIGURE 7.28

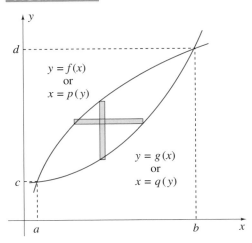

Suppose we make the change of variable $x = q(y)$ in the first integral. When this equation is solved for y in terms of x, the result is $y = g(x)$, and therefore $dy = g'(x)\,dx$. Because $x = a$ when $y = c$ and $x = b$ when $y = d$, we obtain

$$\int_c^d yq(y)\,dy = \int_a^b g(x)xg'(x)\,dx.$$

Now, the product rule for differentiation gives

$$\frac{d}{dx}\left\{x[g(x)]^2\right\} = 2xg(x)g'(x) + [g(x)]^2,$$

and therefore

$$g(x)xg'(x) = \frac{1}{2}\frac{d}{dx}\{x[g(x)]^2\} - \frac{1}{2}[g(x)]^2.$$

So

$$\int_c^d yq(y)\,dy = \int_a^b \left\{\frac{1}{2}\frac{d}{dx}\left\{x[g(x)]^2\right\} - \frac{1}{2}[g(x)]^2\right\} dx$$

$$= \left\{\frac{1}{2}x[g(x)]^2\right\}_a^b - \frac{1}{2}\int_a^b [g(x)]^2\,dx$$

$$= \frac{1}{2}b[g(b)]^2 - \frac{1}{2}a[g(a)]^2 - \frac{1}{2}\int_a^b [g(x)]^2\,dx$$

$$= \frac{1}{2}bd^2 - \frac{1}{2}ac^2 - \frac{1}{2}\int_a^b [g(x)]^2\,dx.$$

Similarly, the change of variable $x = p(y)$ on the second integral in 7.12 leads to

$$\int_c^d yp(y)\, dy = \frac{1}{2}bd^2 - \frac{1}{2}ac^2 - \frac{1}{2}\int_a^b [f(x)]^2\, dx.$$

Substitution of these into 7.12 gives

$$V = 2\pi \left\{ \frac{1}{2}bd^2 - \frac{1}{2}ac^2 - \frac{1}{2}\int_a^b [g(x)]^2\, dx \right\}$$

$$- 2\pi \left\{ \frac{1}{2}bd^2 - \frac{1}{2}ac^2 - \frac{1}{2}\int_a^b [f(x)]^2\, dx \right\}$$

$$= -\pi \int_a^b [g(x)]^2\, dx + \pi \int_a^b [f(x)]^2\, dx$$

$$= \int_a^b \left\{ \pi[f(x)]^2 - \pi[g(x)]^2 \right\}\, dx.$$

But this is precisely the integral obtained when the washer method is used to find the volume.

EXERCISES 7.2

In Exercises 1–12 use the disc or washer method to find the volume of the solid of revolution obtained by rotating the area bounded by the curves about the line.

* **1.** $x^2 + y^2 = 36$, about $y = 0$

* **2.** $y^2 = 5 - x$, $x = 0$, about $x = 0$

* **3.** $y = x^2 + 4$, $y = 2x^2$, about $y = 0$

* **4.** $x - y^2 = 16$, $x = 20$, about $x = 0$

* **5.** $x - 1 = y^2$, $x = 5$, about $x = 1$

* **6.** $x + y + 1 = 0$, $2y = x - 2$, $y = 0$, about $y = 0$

* **7.** $y = 4x^2 - 4x$, $y = x^3$, about $y = -2$

* **8.** $x = 2y - y^2 - 2$, $x = -5$, about $x = 0$

* **9.** $y^2 = 5 - x$, $x = 0$, about $x = 6$

* **10.** $y = x^2 - 2x$, $y = 2x - x^2$, about $y = 2$

* **11.** $y = \csc x$, $y = 0$, $x = \pi/4$, $x = 3\pi/4$, about $y = 0$

* **12.** $y = \ln(x + 1)$, $y = 1$, $x = 0$, about $x = 0$

In Exercises 13–24 use the cylindrical shell method to find the volume of the solid of revolution obtained by rotating the area bounded by the curves about the line.

$*$ **13.** $y = 1 - x^3$, $x = 0$, $y = 0$,   about $x = 0$

$*$ **14.** $y = -\sqrt{4 - x}$, $x = 0$, $y = 0$,   about $y = 0$

$*$ **15.** $y = (x - 1)^2$, $y = 1$,   about $x = 0$

$*$ **16.** $x + y = 4$, $y = 2\sqrt{x - 1}$, $y = 0$,   about $y = 0$

$*$ **17.** $y = 3x - x^2$, $y = x^2 - 3x$,   about $x = 4$

$*$ **18.** $y = 2 - |x|$, $y = 0$,   about $y = -1$

$*$ **19.** $x = y^3$, $x = 2 - y^2$, $y = 0$,   about $y = 1$

$*$ **20.** $y = x^2$, $y = -x^2$, $x = -1$,   about $x = -1$

$*$ **21.** $y = -\sqrt{9 - x}$, $x = 0$, $y = 0$,   about $y = 0$

$*$ **22.** $y = x$, $xy = 9$, $x + y = 10$, $(x \geq y)$,   about $x = 0$

$*$ **23.** $y = 0$, $(x + 1)y = \sin x$, $0 \leq x \leq 2\pi$,   about $x = -1$

$*$ **24.** $y = 10 - x^2$, $x^2 y = 9$,   about $y = 0$

In Exercises 25–32 use the most appropriate method to find the volume of the solid of revolution obtained by rotating the area bounded by the curves about the line.

$**$ **25.** $(x^2 + 1)^2 y = 4$, $y = 1$,   about $x = 0$

$**$ **26.** $y = (x - 1)^2 - 4$, $5y = 12x$, $x = 0$, $(x \geq 0)$,   about $x = 0$

$**$ **27.** $x^2 - y^2 = 5$, $9y = x^2 + 9$, $9y + x^2 + 9 = 0$, $(-3 \leq x \leq 3)$,   about $x = 0$

$**$ **28.** $y = |x^2 - 1|$, $x = -2$, $x = 2$, $y = -1$,   about $y = -1$

$**$ **29.** $y = x^2 - 2$, $y = 0$,   about $y = -1$

$**$ **30.** $x = \sqrt{4 + 12y^2}$, $x - 20y = 24$, $y = 0$,   about $y = 0$

$**$ **31.** $y = (x + 1)^{1/4}$, $y = -(x + 1)^2$, $x = 0$,   about $x = 0$

$**$ **32.** $y = x^4 - 3$, $y = 0$,   about $y = -1$

In Exercises 33–36 a calculator or computer is needed to find the points of intersection of the curves. Find the volume of the solid of revolution when the area bounded by the curves is rotated about the line, correct to three decimal places.

$**$ **33.** $y = x^3 - x$, $y = \sqrt{x}$,   about $x = 0$

$**$ **34.** $y = e^{-2x}$, $y = 4 - x^2$,   about $y = -1$

$**$ **35.** $y = \dfrac{1}{\sqrt{x - 1}}$, $y = 16 - x^2$,   about $y = -1$

$**$ **36.** $y = \sqrt{4 - x}$, $y = x^3 + 1$, $y = 0$,   about $y = 0$

$**$ **37.** A tapered rod of length L has circular cross-sections. If the radii of its ends are a and b, what is the volume of the rod?

$**$ **38.** During one revolution an airplane propeller displaces an amount of air that can be calculated as a volume of a solid of revolution. If the area yielding the volume is that bounded by the curves $64x = y(y - 4)$ and $64x = y(4 - y)$, and is rotated about the x-axis, calculate the volume of air displaced.

$**$ **39.** If a sphere of radius r is sliced a distance h from its centre, show that the volume of the smaller piece is

$$V = \frac{\pi}{3}(r - h)^2(2r + h).$$

** **40.** An embankment is to be built around the circular wading pool in figure (a) below. Figure (b) shows a cross-section of the embankment.

(a) Find a cubic polynomial $y = ax^3 + bx^2 + cx + d$ to fit the three points $(0, 0)$, $(4, 2)$, and $(6, 0)$.

(b) Determine the amount of fill required to build the embankment.

(a)   (b)

*** **41.** Find the volume of the donut obtained by rotating the circle in the figure below about the y-axis.

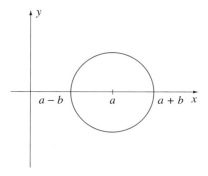

*** **42.** A cylindrical hole is bored through the centre of a sphere, the length of the hole being L. Show that no matter what the radius of the sphere, the volume of the sphere that remains is always the same and equal to the volume of a sphere of diameter L.

*** **43.** Water half fills a cylindrical pail of radius a and height L. When the pail is rotated about its axis of symmetry with angular speed ω (figure below), the surface of the water assumes a parabolic shape, the cross-section of which is given by

$$y = H + \frac{\omega^2 x^2}{2g},$$

where $g > 0$ is the acceleration due to gravity and H is a constant. Find the speed ω in terms of L, a, and g, at which water spills over the top.

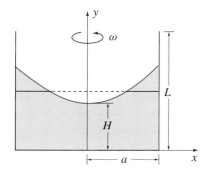

*** **44.** A right circular cone of height H and base radius r has its vertex at the centre of a sphere of radius R ($R < H$). Find that part of the volume of the sphere inside the cone.

*** **45.** Devise a way to calculate the volume of the solid of revolution when the area in the first quadrant bounded by

$$y = 1 - x^2, \quad x = 0, \quad y = 0$$

is rotated about the line $y = x + 1$. Hint: Distance formula 1.7.

7.3 Lengths of Curves

Formula 1.1 defines the length of the straight-line segment joining two points (x_1, y_1) and (x_2, y_2). The formula $s = r\theta$ gives the length of the arc of a circle of radius r subtended by an angle θ at the centre of the circle. In this section we derive a result that will theoretically enable us to find the length of any curve.

Consider finding the length of the curve $C : y = f(x)$ in Figure 7.29a joining points $A(a, f(a))$ and $B(b, f(b))$. To find its length L we begin by approximating C with a series of straight-line segments. Specifically, we choose $n - 1$ consecutive points on C between A and B,

$$A = P_0, P_1, P_2, \ldots, P_{n-1}, P_n = B,$$

and join each P_{i-1} to P_i $(i = 1, \ldots, n)$ by means of a straight-line segment, as in Figure 7.29b. If coordinates of P_i are denoted by (x_i, y_i), then length of the line segment joining P_{i-1} and P_i is

$$\|P_{i-1}P_i\| = \sqrt{(x_i - x_{i-1})^2 + (y_i - y_{i-1})^2}.$$

For a large number of these line segments, it is reasonable to approximate L by the sum of the lengths of the segments:

$$\sum_{i=1}^{n} \|P_{i-1}P_i\| = \sum_{i=1}^{n} \sqrt{(x_i - x_{i-1})^2 + (y_i - y_{i-1})^2}.$$

In fact, as we increase n and at the same time decrease the length of each segment, we expect the approximation to become more and more accurate. We therefore define

$$L = \lim_{\|\Delta x_i\| \to 0} \sum_{i=1}^{n} \|P_{i-1}P_i\| = \lim_{\|\Delta x_i\| \to 0} \sum_{i=1}^{n} \sqrt{(x_i - x_{i-1})^2 + (y_i - y_{i-1})^2}. \quad (7.13a)$$

FIGURE 7.29a

FIGURE 7.29b

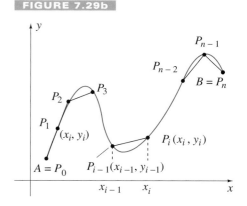

If we set $\Delta x_i = x_i - x_{i-1}$, and note that $y_{i-1} = f(x_{i-1})$ and $y_i = f(x_i)$, we have

$$L = \lim_{\|\Delta x_i\| \to 0} \sum_{i=1}^{n} \sqrt{(\Delta x_i)^2 + [f(x_i) - f(x_{i-1})]^2}$$

$$= \lim_{\|\Delta x_i\| \to 0} \sum_{i=1}^{n} \sqrt{1 + \left[\frac{f(x_i) - f(x_{i-1})}{\Delta x_i}\right]^2} \, \Delta x_i. \tag{7.13b}$$

In order that 7.13b be a useful definition from the point of view of calculation, we must find a convenient way to evaluate the limit-summation. To do this we assume that $f'(x)$ exists at every point in the interval $a \le x \le b$, and apply the mean value theorem to $f(x)$ on each subinterval $x_{i-1} \le x \le x_i$ (Figure 7.30). The theorem guarantees the existence of at least one point x_i^* between x_{i-1} and x_i such that

$$f'(x_i^*) = \frac{f(x_i) - f(x_{i-1})}{x_i - x_{i-1}} = \frac{f(x_i) - f(x_{i-1})}{\Delta x_i}.$$

Consequently, using these points x_i^*, we can express the length of C in the form

$$L = \lim_{\|\Delta x_i\| \to 0} \sum_{i=1}^{n} \sqrt{1 + [f'(x_i^*)]^2} \, \Delta x_i. \tag{7.14}$$

But the right side of this equation is the definition of the definite integral of the function $\sqrt{1 + [f'(x)]^2}$ with respect to x from $x = a$ to $x = b$:

$$\int_a^b \sqrt{1 + [f'(x)]^2} \, dx = \lim_{\|\Delta x_i\| \to 0} \sum_{i=1}^{n} \sqrt{1 + [f'(x_i^*)]^2} \, \Delta x_i.$$

Thus we can calculate the length of C by the definite integral

$$L = \int_a^b \sqrt{1 + \left(\frac{dy}{dx}\right)^2} \, dx. \tag{7.15}$$

FIGURE 7.30

In this derivation we assumed that $f'(x)$ was defined at each point in the interval $a \leq x \leq b$, necessary for the mean value theorem to apply. But in order to guarantee existence of the definite integral in 7.15, Theorem 6.2 requires continuity of the integrand $\sqrt{1 + (dy/dx)^2}$. Consequently, to ensure that the length of a curve can be calculated by means of 7.15, we assume that $f(x)$ has a continuous first derivative on the interval $a \leq x \leq b$.

EXAMPLE 7.14

Find the length of the curve $y = x^{3/2}$ from $(1, 1)$ to $(2, 2\sqrt{2})$ (Figure 7.31).

SOLUTION   Since $dy/dx = (3/2)x^{1/2}$, equation 7.15 gives

$$
L = \int_1^2 \sqrt{1 + \left(\frac{3}{2}x^{1/2}\right)^2}\, dx = \int_1^2 \sqrt{1 + \frac{9}{4}x}\, dx
$$

$$
= \left\{\frac{8}{27}\left(1 + \frac{9}{4}x\right)^{3/2}\right\}_1^2 = \frac{8}{27}\left[\left(\frac{11}{2}\right)^{3/2} - \left(\frac{13}{4}\right)^{3/2}\right] = \frac{1}{27}(22^{3/2} - 13^{3/2}).
$$

FIGURE 7.31

The result in 7.15 is clearly useful when the equation of the curve is expressed in the form $y = f(x)$. If the curve is defined in the form $x = g(y)$ (Figure 7.32), a similar analysis gives

$$
L = \int_c^d \sqrt{1 + \left(\frac{dx}{dy}\right)^2}\, dy. \tag{7.16}
$$

FIGURE 7.32

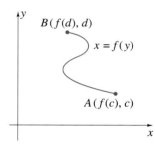

For example, we may write the equation of the curve in Example 7.14 in the form $x = y^{2/3}$, in which case $dx/dy = (2/3)y^{-1/3}$, and

$$
L = \int_1^{2\sqrt{2}} \sqrt{1 + \left(\frac{2}{3}y^{-1/3}\right)^2}\, dy = \int_1^{2\sqrt{2}} \sqrt{1 + \frac{4}{9}y^{-2/3}}\, dy
$$

$$
= \int_1^{2\sqrt{2}} \sqrt{\frac{4 + 9y^{2/3}}{9y^{2/3}}}\, dy = \int_1^{2\sqrt{2}} \frac{\sqrt{4 + 9y^{2/3}}}{3y^{1/3}}\, dy.
$$

If we set $u = 4 + 9y^{2/3}$, then $du = 6y^{-1/3}dy$, and

$$
L = \frac{1}{3}\int_{13}^{22} \sqrt{u}\, \frac{du}{6} = \frac{1}{18}\left\{\frac{2}{3}u^{3/2}\right\}_{13}^{22} = \frac{1}{27}(22^{3/2} - 13^{3/2}).
$$

Integrals 7.15 and 7.16 can both be interpreted geometrically. This interpretation will be useful when we study parametric equations of curves in Section 10.1, and line integrals in Chapter 15. Figure 7.33 shows the curve C of Figure 7.29a. At a point (x, y) on C, we draw the tangent line to the curve. If dx is a short length along the x-axis at position x, length of that part of the tangent line between vertical lines at x and $x + dx$ is given by

$$
\sqrt{(dx)^2 + (dy)^2}.
$$

In the derivation of 7.20 we assumed continuity of $F(x)$, and this guarantees existence of the definite integral.

EXAMPLE 7.16

Find the work necessary to expand a spring from a stretch of 5 cm to a stretch of 15 cm if a force of 200 N stretches it 10 cm.

SOLUTION   Let the spring be stretched in the positive x-direction, and let $x = 0$ correspond to the free end of the spring in the unstretched position in the top half of Figure 7.37. In the bottom half of this figure we have shown the spring stretched to an intermediate position.

FIGURE 7.37

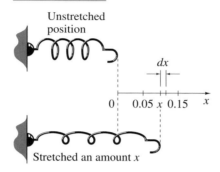

According to Hooke's law, when the spring is stretched an amount x, the restoring force in the spring is proportional to x:

$$F_s = -kx,$$

where $k > 0$ is a constant. The negative sign indicates that the force is in the negative x-direction. Since $F_s = -200$ N when $x = 0.10$ m,

$$-200 = -k(0.10),$$

and it follows that $k = 2000$ N/m. The force required to counteract the restoring force of the spring when it is stretched an amount x is therefore

$$F(x) = 2000x.$$

Work done by this force in stretching the spring from position x a further amount dx is approximately

$$2000x\, dx,$$

and hence total work to increase the stretch from 5 cm to 15 cm is

$$W = \int_{0.05}^{0.15} 2000x\, dx = 2000 \left\{ \frac{x^2}{2} \right\}_{0.05}^{0.15} = 20.$$

Total work done is therefore 20 N·m (or joules).

EXAMPLE 7.17

A cable hangs vertically from the top of a building so that a length of 100 m, having a mass of 300 kg, is hanging from the edge of the roof. What work is required to lift the entire cable to the top of the building?

FIGURE 7.38

SOLUTION   When the cable has been lifted to an intermediate point where its lower end is y metres above its original position (Figure 7.38), the length of cable still hanging is $(100 - y)$ m. Since each metre of cable has mass 3 kg, the mass of $(100 - y)$ m of cable is $3(100 - y)$ kg. It follows that the force that must be exerted to overcome gravity on this much cable is $9.81(3)(100 - y)$ N. The work that this force does in raising the end of the cable an additional amount dy is approximately

$$9.81(3)(100 - y)\,dy,$$

and total work to raise the entire cable is therefore

$$W = \int_0^{100} 9.81(3)(100 - y)\,dy = 29.43\left\{100y - \frac{y^2}{2}\right\}_0^{100} = 147\,150 \text{ J.}$$

EXAMPLE 7.18

A tank in the form of an inverted right circular cone of depth 10 m and radius 4 m is full of water. Find the work required to pump the water to a level 1 m above the top of the tank.

SOLUTION   A cross-section of the tank is shown in Figure 7.39. Suppose we approximate water in the tank with circular discs formed by rotating the rectangle shown around the y-axis. The force of gravity on the representative disc at position y is its volume $\pi x^2\,dy$, multiplied by the density of water (1000 kg/m^3), multiplied by the acceleration due to gravity (-9.81),

$$-9.81(1000)\pi x^2\,dy.$$

It is negative because gravity acts in the negative y-direction. The work done by an equal and opposite force in lifting the disc to a level 1 m above the top of the tank (a distance $11 - y$) is approximately

$$(11 - y)9.81(1000)\pi x^2\,dy.$$

The total work to empty the tank is therefore

$$W = \int_0^{10} (11 - y)9810\pi x^2\,dy.$$

To express x in terms of y, we note that x and y are coordinates of points on the straight line through the origin and the point $(4, 10)$. Since the equation of this line is $y = 5x/2$, we obtain

$$W = 9810\pi \int_0^{10} (11 - y)\left(\frac{2y}{5}\right)^2 dy = 1569.6\pi \int_0^{10} (11y^2 - y^3)\,dy$$

$$= 1569.6\pi \left\{\frac{11y^3}{3} - \frac{y^4}{4}\right\}_0^{10} = 5.75 \times 10^6 \text{ J.}$$

FIGURE 7.39

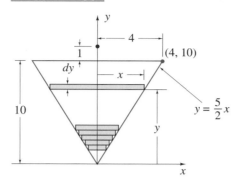

Two observations are noteworthy:

1. In Examples 7.16 and 7.17, the differential represents distance moved and the integrand represents force, and we applied equation 7.20 directly. In Example 7.18, the differential is part of the force, and the distance moved is part of the integrand. This is why we stated earlier that it is not advisable to use equation 7.20 as a formula. The integrand is not always force and the differential is not always distance moved.

FIGURE 7.40

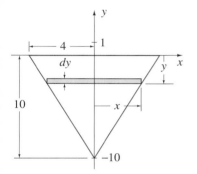

2. In each of these examples we set up the coordinate system; it was not given. We can use any coordinate system whatsoever; but once we have chosen our coordinates, we must refer everything to that system. For instance, were we to use the coordinate system in Figure 7.40 for Example 7.18, the definite integral would be

$$W = \int_{-10}^{0} (1 - y)9810\pi \left[\frac{2}{5}(10 + y) \right]^2 dy.$$

Evaluation of this definite integral would again lead to the result $W = 5.75 \times 10^6$ J.

EXERCISES 7.4

* **1.** A spring requires a 10 N force to stretch it 3 cm. Find the work to increase the stretch of the spring from 5 cm to 7 cm.

* **2.** Find the work to increase the stretch of the spring in Exercise 1 from 7 cm to 9 cm.

* **3.** A cage of mass M kilograms is to be lifted from the bottom of a mine shaft h metres deep. If the mass of the cable used to hoist the cage is m kilograms per metre, find the work done.

* **4.** A uniform cable of length 50 m and mass 100 kg hangs vertically from the top of a building 100 m high. How much work is required to get 10 m of the cable on top of the building?

* **5.** A 2 m chain of mass 20 kg lies on the floor. If friction between floor and chain is ignored, how much work is required to lift one end of the chain 2 m straight up?

* **6.** How much work is required to lift the end of the chain in Exercise 5 only 1 m?

* **7.** How much work is required to lift the end of the chain in Exercise 5 a distance of 4 m?

* **8.** A 5 m chain of mass 15 kg hangs vertically. It is required to lift the lower end of the chain 5 m so that it is level with the upper end. Calculate the work done using each of the coordinate systems in the figure below.

* **9.** A tank filled with water has the form of a paraboloid of revolution with vertical axis (figure below). If the depth of the tank is 12 m and the diameter of the top is 8 m, find the work in pumping the water to the top of the tank.

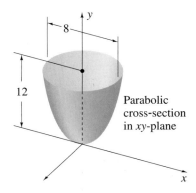

Parabolic cross-section in xy-plane

* **10.** In Exercise 9, find the work to empty the tank through an outlet 2 m above the top of the tank.

** **11.** The ends of a trough are isosceles triangles with width 2 m and depth 1 m. The trough is 5 m long and it is full of water. How much work is required to lift all of the water to the top of the trough?

** **12.** How much work is required to lift the water in the trough of Exercise 11 to a height of 2 m above the top of the trough?

** **13.** A rectangular swimming pool full of water is 25 m long and 10 m wide. The depth is 3 m for the first 10 m of length, decreasing linearly to 1 m at the shallow end. How much work is required to lower the level of the water in the pool by 1/2 m?

** **14.** How much work is required to empty the pool of Exercise 13 over its edge?

** **15.** The force of repulsion between two point charges of like sign, one of size q and the other of size Q, has magnitude

$$F = \frac{qQ}{4\pi\epsilon_0 r^2},$$

where ϵ_0 is a constant and r is the distance between the charges. If Q is placed at the origin, and q is moved along the x-axis from $x = 2$ to $x = 5$, find the work done by the electrostatic force.

** **16.** Two positive charges q_1 and q_2 are placed at positions $x = 5$ and $x = -2$ on the x-axis. A third positive charge q_3 is moved along the x-axis from $x = 1$ to $x = -1$. Find the work done by the electrostatic forces of q_1 and q_2 on q_3 (see Exercise 15).

** **17.** If the chain in Exercise 5 is stretched out straight on the floor, and if the coefficient of friction between floor and chain is 0.01, what work is required to raise one end of the chain 2 m? The force of friction on that part of the chain on the floor is 0.01 times the weight of the chain on the floor.

** **18.** Two similar springs, each 1 m long in an unstretched position, have spring constant k N/m. The springs are joined together at P (figure below), and their free ends are fastened to two posts 4 m apart. What work is done in moving the midpoint P a distance b m to the right?

** **19.** If the force of a crossbow (in newtons) is proportional to the draw (in metres), and it is 200 N at a full draw of 50 cm, what work is required to fully draw the crossbow?

** **20.** A bucket of water with mass 100 kg is on the ground attached to one end of a cable with mass per unit length 5 kg/m. The other end of the cable is attached to a windlass 100 m above the bucket. If the bucket is raised at a constant speed, water runs out through a hole in the bottom at a constant rate to the extent that the bucket has mass 80 kg when it reaches the top. To further complicate matters, a pigeon of mass 2 kg lands on the bucket when it is 50 m above the ground. He immediately begins taking a bath, splashing water over the side of the bucket at the rate of 1 kg/m. Find the work done by the windlass in raising the bucket 100 m.

** **21.** A hemispherical tank with diametral plane at the top has radius 5 m.

    (a) Show that if the depth of oil in the tank is h metres when the tank is one-half full by volume, then h must satisfy the equation

$$h^3 - 15h^2 + 125 = 0.$$

    Find h to four decimals.

    (b) Find the work to empty the half-full tank if the oil has density 750 kg/m^3, and the outlet is at the top of the tank.

** **22.** Newton's universal law of gravitation states that the force of attraction between two point masses m and M has magnitude

$$F = \frac{GmM}{r^2},$$

where r is the distance between the masses and $G = 6.67 \times 10^{-11}$ N·m^2/kg^2 is a constant.

    (a) If M represents the mass of the earth and we regard it as a point mass concentrated at its centre, show that Newton's universal law of gravitation at the earth's surface reduces to $F = mg$, where $g = 9.82$ m/s^2. Assume for the calculation of M that the earth is a sphere with radius 6370 km and mean density 5.52×10^3 kg/m^3.

    (b) Use the original law $F = GmM/r^2$ with the earth regarded as a point mass to calculate the work required to lift a mass of 10 kg from the earth's surface to a height of 10 km.

    (c) Calculate the work in (b) using the constant gravitational force $F = mg$ in (a). Is there a significant difference?

** **23.** A right-circular cylinder with horizontal axis has radius r metres and length h metres. If it is full of oil with density ρ kilograms per cubic metre, how much work is required to empty the tank through an outlet at its top?

** **24.** A gas is confined in a cylinder, closed on one end with a piston on the other. If the temperature is held constant, then $PV = C$, where P is the pressure of the gas in the cylinder, V is its volume, and C is a constant. The force needed to move the piston is $F = PA$, where A is the area of the face of the piston. If the radius of the cylinder is 3 units, find the work done in moving the piston from a point 10 units to a point 5 units from the closed end.

** **25.** A gas is confined in a cylinder, closed on one end with a piston on the other. If the gas expands adiabatically, then the pressure P and volume V of the gas obey the law

$$PV^{7/5} = C,$$

where C is a constant. The force on the piston face due to the pressure of the gas is given by $F = PA$, where A is the area of the face of the piston. Show that if the piston is moved so as to reduce the volume occupied by the gas from V_0 to $V_0/2$, then the work done by the piston is

$$\frac{5}{2}\left(2^{2/5} - 1\right)CV_0^{-2/5}.$$

*** **26.** A diving bell of mass 10 000 kg is attached to a chain of mass 5 kg/m, and the bell sits on the bottom of the ocean 100 m below the surface (figure below). How much work is required to lift the bell to deck level 5 m above the surface? Take into account the fact that when the bell and chain are below the surface, they weigh less than when above. The apparent loss in weight is equal to the weight of water displaced by the bell and chain. Assume that the bell is a perfect cube, 2 m on each side; therefore, when completely submerged, it displaces 8 m^3 of water. Assume also that the chain displaces 1 L of water per metre of length.

7.5 Energy

Many engineering systems are analyzed on the basis of energy, and there are many kinds of energy—potential, kinetic, thermal, and electrical, to name a few. Often one kind of energy is transformed into another, or others, as a system evolves. We consider a number of examples of this here.

In lifting a mass vertically, work is done against gravity. We say that the mass gains gravitational potential energy as a result of the lift, the amount being equal to the work done. For instance, when a mass m is lifted an amount $h > 0$ against gravity in Figure 7.41a, the

FIGURE 7.41a   FIGURE 7.41b

force is constant (provided h is relatively small), and the work done is $W = mgh$, where $g = 9.81$. We say that the mass has gained gravitational potential energy as a result of the lift,

$$\text{GPE} = mgh. \qquad (7.21)$$

We also say that the difference in gravitational potential energies between initial and final positions P_1 and P_2 is mgh. In Figure 7.41b, we have shown the same situation, but initial position P_1 is at height h_0 above some other reference point denoted by P_0. At position P_1, the mass has GPE mgh_0 relative to P_0; at position P_2, it has GPE $mg(h + h_0)$ relative to P_0. The difference in these GPEs is still $mg(h + h_0) - mgh_0 = mgh$.

We have mentioned gains in GPE, differences in GPE, and GPE at one point relative to another, and this is what is important. Often, however, it is convenient to talk about GPE at a point, rather than the difference in GPE between points. This can be accomplished by choosing some specific point as having zero GPE and saying that the GPE at any other point is the difference in GPE between the two points. For instance, if GPE is set equal to zero at P_0 in Figure 7.41b, then GPE is mgh_0 at P_1 and $mg(h + h_0)$ at P_2. The GPE at any point below P_0 is negative.

If m is allowed to drop from position P_2, it loses GPE as it falls. It picks up speed, and energy is associated with the speed. This is called **kinetic energy**; it is defined by

$$\text{KE} = \frac{1}{2}mv^2, \qquad (7.22)$$

where v is the speed of m. Provided air drag on m is ignored, whatever GPE m loses, it gains in KE. To put it another way, the sum of its GPE and KE is always the same:

$$\text{GPE} + \text{KE} = C = mgh + \frac{1}{2}mv^2. \qquad (7.23)$$

This is called **conservation of energy**. Energy simply changes from gravitational potential to kinetic during the fall. We apply conservation of energy in the following example.

EXAMPLE 7.19   ◂▸

A stone is thrown vertically upward with speed 20 m/s at the edge of a 50 m cliff. How fast is it travelling when it strikes the base of the cliff?

SOLUTION   Let us take the base of the cliff (Figure 7.42) as our place of zero gravitational potential energy. When the stone is initially thrown upward, its GPE is $mg(50)$ and its KE is $m(20)^2/2$. If we substitute these into 7.23,

$$C = 50mg + 200m,$$

FIGURE 7.42

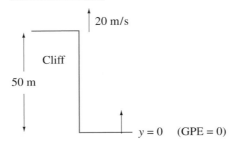

Substitution of this back into 7.23 and replacing h by y gives

$$mgy + \frac{1}{2}mv^2 = 50mg + 200m \qquad \Longrightarrow \qquad 2gy + v^2 = 100g + 400.$$

This equation relates speed and height. At the bottom of the cliff $y = 0$, in which case

$$v^2 = 100g + 400 \qquad \Longrightarrow \qquad v = \sqrt{100g + 400} = 37.2.$$

The stone therefore strikes the bottom of the cliff at 37.2 m/s.

We can also solve the problem using the technique from Section 5.2. Since acceleration of the stone is $a = dv/dt = -9.81$,

$$v = -9.81t + C.$$

(Here, v is velocity rather than speed.) If we choose $t = 0$ when the stone is thrown upward, then $v(0) = 20$, and this implies that $C = 20$. Hence,

$$\frac{dy}{dt} = v = -9.81t + 20.$$

A second integration gives

$$y = -4.905t^2 + 20t + D.$$

Since $y(0) = 50$, it follows that $D = 50$, and

$$y = -4.905t^2 + 20t + 50.$$

The stone strikes the base of the cliff when

$$0 = y(t) = -4.905t^2 + 20t + 50 \qquad \Longrightarrow \qquad t = \frac{-20 \pm \sqrt{400 - 4(-4.905)(50)}}{-9.81}.$$

The positive solution is $t = 5.827$ s, and at this time

$$v = -9.81(5.827) + 20 = -37.2 \text{ m/s}.$$

The energy solution related velocity v to position y; the solution using integration relates velocity to time t. Which is preferable depends on the problem.

But V is the product of the length of the column, d, and the (unit) cross-sectional area; that is, $V = (1)d = d$, and hence

$$W = 9.81\rho d. \qquad (7.26)$$

A comparison of equations 7.25 and 7.26 suggests that pressure at any point is precisely the weight of a column of fluid of unit cross-sectional area above that point.

FIGURE 7.45

Surface of fluid

d

FIGURE 7.46

y

Surface of fluid

$x = f(y)$

x

b

dy

y

x_1

$x = g(y)$

a

x_2

We now consider the problem of determining total force on one side of a flat plate (Figure 7.46), immersed vertically in a fluid of density ρ. If the area of the plate is subdivided into horizontal rectangles of width dy, then pressure at each point of this rectangle is approximately $P = -9.81\rho y$. This is an approximation because slight variations in pressure do occur over vertical displacements within the rectangle. It follows that force on the representative rectangle is approximately equal to its area multiplied by $-9.81\rho y$,

$$-9.81\rho y(x_2 - x_1)\,dy = -9.81\rho y[f(y) - g(y)]\,dy.$$

Total force on the plate is found by adding forces on all such rectangles and taking the limit as their widths approach zero. We obtain the required force, therefore, as

$$F = \int_a^b -9.81\rho y[f(y) - g(y)]\,dy. \qquad (7.27)$$

Once again we do not suggest that equation 7.27 be memorized because it is associated with the choice of coordinates in Figure 7.46. For a different coordinate system, the definite integral would be correspondingly different (see Example 7.21). What is important is the procedure: subdivide the area of the plate into horizontal rectangles, find the force on a representative rectangle, and finally add over all rectangles with a definite integral.

Note that vertical rectangles cannot be used without further discussion since it is not evident how to calculate the force on such a rectangle. Consideration of vertical rectangles is given in Exercise 8.

EXAMPLE 7.21

The vertical face of a dam is parabolic with breadth 100 m and height 50 m. Find the total force due to fluid pressure on the face.

SOLUTION   If we set up the coordinate system in Figure 7.47, we see that the edge of the dam has an equation of the form $y = kx^2$. Since $(50, 50)$ is a point on this curve, it follows that $k = 1/50$, and $y = x^2/50$. Area of the representative rectangle is $2x\,dy$, and it is at depth $50 - y$ below the surface of the water. Since density of water is 1000 kg/m^3, force on the representative rectangle is approximately equal to

$$(9.81)(1000)(50 - y)2x\,dy = 19\,620(50 - y)\sqrt{50y}\,dy.$$

Total force on the dam must therefore be

$$F = \int_0^{50} 19\,620(50 - y)5\sqrt{2}y^{1/2}\,dy = 98\,100\sqrt{2}\int_0^{50}(50y^{1/2} - y^{3/2})\,dy$$

$$= 98\,100\sqrt{2}\left\{\frac{100y^{3/2}}{3} - \frac{2y^{5/2}}{5}\right\}_0^{50} = 6.54 \times 10^8\ \text{N}.$$

FIGURE 7.47

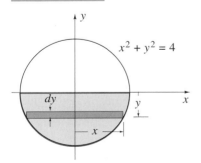

FIGURE 7.48

EXAMPLE 7.22

A tank in the form of a right circular cylinder of radius 2 m and length 10 m lies on its side. If it is half filled with oil of density ρ kilograms per cubic metre, find the force on each end of the tank.

SOLUTION   Force on the representative rectangle in Figure 7.48 is approximately equal to the pressure $9.81\rho(-y)$ multiplied by area $2x\,dy$ of the rectangle,

$$9.81\rho(-y)2x\,dy = -19.62\rho y\sqrt{4 - y^2}\,dy.$$

Total force on each end of the tank is therefore

$$F = \int_{-2}^{0} -19.62\rho y\sqrt{4 - y^2}\,dy = 19.62\rho\left\{\frac{1}{3}(4 - y^2)^{3/2}\right\}_{-2}^{0} = 52.32\rho\ \text{N}.$$

EXERCISES 7.6

* **1.** A tropical fish tank has length 1 m, width 0.5 m, and depth 0.5 m. Find the force due to water pressure on each of the sides and bottom when the tank is full.

* **2.** The vertical surface of a dam exposed to the water of a lake has the shape shown below. Find the force of the water on the face of the dam.

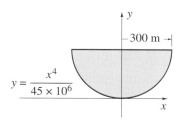

** **3.** The vertical end of a water trough is an isosceles triangle with width 2 m and depth 1 m. Find the force of the water on each end when the trough is half full by volume.

** **4.** A square plate, 2 m on each side, has one diagonal vertical. If it is one-half submerged in water, what is the force due to water pressure on each side of the plate?

** **5.** A cylindrical oil tank of radius r and height h has its axis vertical. If the density of the oil is ρ, find the force on the bottom of the tank when it is full.

** **6.** A rectangular swimming pool full of water is 25 m long and 10 m wide. The depth is 3 m for the first 10 m at the deep end, decreasing linearly to 1 m at the shallow end. Find the force due to the weight of the water on each of the sides and ends of the pool.

** **7.** The vertical face of a dam across a river has the shape of a parabola 36 m across the top and 9 m deep at the centre. What is the force that the river exerts on the dam if the water is 0.5 m from the top?

** **8.** Show that the force due to fluid pressure on the vertical rectangle in the figure below is

$$F = \frac{9.81\rho}{2}h(y_1^2 - y_2^2),$$

where ρ is the density of the fluid.

** **9.** A flat plate in the shape of a trapezoid is submerged vertically in a fluid with density ρ. The plate has two parallel vertical sides of lengths 6 and 8 and a third side of length 5 that is perpendicular to the parallel sides and at a depth of 1 below the surface of the fluid (figure on page 488). Find the force due to fluid pressure on each side of the plate, using both horizontal and vertical rectangles (see Exercise 8 for vertical rectangles).

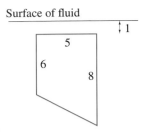

Surface of fluid

5

6

8

1

** **10.** The base of a triangular plate, of length a, lies in the surface of a fluid of density ρ. The third vertex of the triangle is at depth b below the surface (figure below). Show that the force due to fluid pressure on each side of the plate is $9.81\rho ab^2/6$, no matter what the shape of the triangle.

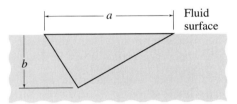

** **11.** Set up (but do not evaluate) a definite integral(s) to find the force due to water pressure on each side of the flat vertical plate in the figure below.

** **12.** In Exercise 6, find the force due to water pressure on each part of the bottom of the pool.

** **13.** The bow of a landing barge (figure below) consists of a rectangular flat plate A metres wide and B metres long. When the barge is stationary, this plate makes an angle of $\pi/6$ radians with the surface of the water. Find the maximum force of the water on the bow.

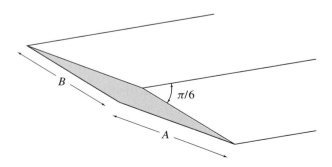

** **14.** A water tank is to be built in the form of a rectangular box with all six sides welded along their joins (figure below). Sides $ABCD$ and $EFGH$ are to be square, and all sides except $ABCD$ are supported from the outside. If the maximum force that $ABCD$ can withstand is 20 000 N, what is the largest cross-section that can be built, assuming that at some stage the tank will be full?

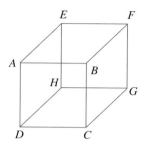

** **15.** A cylindrical oil tank of radius r and length h has its axis horizontal. If the density of the oil is ρ, find the force on each end of the tank when it is half full.

** **16.** A right circular cylinder of radius r and height h is immersed in a fluid of density ρ with its axis vertical. Show that the buoyant force on the cylinder due to the pressure of the fluid is equal to the weight of the fluid displaced. This is known as *Archimedes' principle*, and is valid for an object of any shape.

** **17.** The lower half of a cubical tank 2 m on each side is occupied by water, and the upper half by oil (density 0.90 gm/cm^3).

    (a) What is the force on each side of the tank due to the pressure of the water and the oil?

    (b) If the oil and water are stirred to create a uniform mixture, does the force on each side change? If not explain why not. If so, by how much does it increase or decrease?

** **18.** Archimedes' principle states that the buoyant force on an object when immersed or partially immersed in a fluid is equal to the weight of the fluid displaced by the object.

    (a) Show that if an object floats partially submerged in water, the percentage of the volume of the object above water is

$$100\frac{\rho_w - \rho_o}{\rho_w},$$

    where ρ_o and ρ_w are the densities of the object and water respectively.

    (b) If the densities of ice and water are 915 and 1000 kg/m^3 respectively, show that only 8.5% of the volume of an iceberg is above water.

** **19.** If a full tube of mercury is inverted in a large container of mercury, the level of mercury in the tube will fall, but it will stabilize at a point higher than that in the container (figure on page 490). This is due to the fact that air pressure acts on the surface of the mercury in the container but not on the surface of the mercury in the tube. The extra column of mercury, of height h, creates a force at A that counteracts the atmospheric pressure transmitted through the mercury in the container to the tube so that the total pressure at A is equal to the total pressure at B.

    (a) Show that if the density of mercury is 13.6 gm/mL, then the atmospheric pressure at the surface of the mercury in the container is $1.33 \times 10^5 h$ N/m^2, provided h is measured in metres.

    (b) If h is measured as 761 mm, find the atmospheric pressure.

** **20.** (a) A block of wood (density 0.40 gm/cm^3) is cubical (0.25 m on each side). If it floats in water, how deep is its lowest point below the surface? Refer to Archimedes' principle in Exercise 18.

(b) Repeat part (a) given that the block is a sphere of radius 0.25 m.

** **21.** A tank is to be built in the form of a right circular cylinder with horizontal axis. The ends are to be joined to the cylindrical side by a continuous weld. One end of the tank and the cylindrical side are supported from the outside. The remaining unsupported end can withstand a total force of 40 000 N less 1000 N for each metre of weld on that end. What is the maximum radius for the tank if it is to hold a fluid with density 1.019×10^3 kg/m^3?

** **22.** Find the ratio L/R such that the forces due to fluid pressure on the rectangular and semi-circular parts of the plate in the figure below are equal.

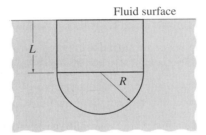

7.7 Centres of Mass and Centroids

Everyone is acquainted with the action of a teeter-totter or seesaw. Two children of unequal masses can pass many hours rocking, provided the child with greater mass sits closer to the fulcrum. In this section we discuss the mathematics of the seesaw. This requires a definition of "moments of masses," and moments lead to the idea of the "centre of mass" of distributions of masses, lumped or continuous.

To discuss the mathematics of a seesaw, we consider in Figure 7.49 a uniform seesaw of length $2L$ balanced at its centre, with a child of mass m at one end. If a second child of equal mass is placed at the other end, the ideal seesaw situation is created. If, however, the mass of the second child is $M > m$, then this child must be placed somewhat closer to the fulcrum. To find the exact position, we must determine what might be called *rocking power* of a mass. A little experimentation shows that when $M = 2m$, M must be placed halfway between the end and the fulcrum; when $M = 3m$, M must be placed a distance $L/3$ from the fulcrum; and in general, when $M = am$, M must be placed L/a from the fulcrum. Now the rocking power of the child of mass m is constant, and for each mass $M = am$ we have found an equal and opposite rocking power if M is placed at L/a.

FIGURE 7.49

Clearly, then, rocking power depends on both mass and distance from the fulcrum. A little thought shows that the mathematical quantity that remains constant for the various masses $M = am$ is the product of M and distance to the fulcrum; in each case, this product is $(am)(L/a) = mL$, the same product as for the child of mass m. It would appear, then, that rocking power should be defined as the product of mass and distance. We do this in the following definition, and at the same time give rocking power a new name.

DEFINITION 7.1

The **first moment** of a point mass m about a point P is the product md where d is the directed distance from P to m.

If directed distances to the right of point P in Figure 7.50 are chosen as positive and distances to the left are negative, then d_1 is positive and d_2 is negative. Mass m_1 has a positive first moment $m_1 d_1$ about P, and m_2 has a negative first moment $m_2 d_2$.

FIGURE 7.50

In Figure 7.51, we have placed five children of masses m_1, m_2, m_3, m_4, and m_5 on the same seesaw. A sixth child of mass m_6 is to be placed somewhere on the seesaw so that all six children form the ideal seesaw.

FIGURE 7.51

To find the appropriate position for the sixth child, we let x be directed distance from the fulcrum to the point where this child should be placed. Total first moment of all six children about the fulcrum, choosing distances to the right as positive and to the left as negative, is

$$m_1(-L) + m_2(-3L/4) + m_3(-L/2) + m_4(L/4) + m_5(L) + m_6(x).$$

We regard this as the resultant first moment of all six children attempting to turn the seesaw—clockwise if the moment is positive, counterclockwise if the moment is negative. Balance occurs if this resultant first moment is zero:

$$0 = -m_1 L - \frac{3}{4}m_2 L - \frac{1}{2}m_3 L + \frac{1}{4}m_4 L + m_5 L + m_6 x.$$

We may solve this equation for the position of m_6:

$$x = \frac{L}{4m_6}(4m_1 + 3m_2 + 2m_3 - m_4 - 4m_5).$$

We now turn this problem around and place the six children at distances from the left end as shown in Figure 7.52. If the mass of the seesaw itself is neglected, where should the fulcrum be placed in order to create the ideal seesaw? To solve this problem, we let the distance that the fulcrum should be placed from the left end be represented by $\overline{x}$. In order for balance to occur, the total first moment of all six children about the fulcrum must vanish; hence,

$$0 = m_1(x_1 - \overline{x}) + m_2(x_2 - \overline{x}) + m_3(x_3 - \overline{x}) + m_4(x_4 - \overline{x}) + m_5(x_5 - \overline{x}) + m_6(x_6 - \overline{x}).$$

The solution of this equation is

$$\overline{x} = \frac{m_1 x_1 + m_2 x_2 + m_3 x_3 + m_4 x_4 + m_5 x_5 + m_6 x_6}{m_1 + m_2 + m_3 + m_4 + m_5 + m_6} = \frac{1}{M}\sum_{i=1}^{6} m_i x_i, \qquad (7.28)$$

where $M = \sum_{i=1}^{6} m_i$ is the total mass of all six children. This point $\overline{x}$ at which the fulcrum creates a balancing position is called the *centre of mass* for the six children. It is a point where masses to the right are balanced by masses to the left.

FIGURE 7.52

FIGURE 7.53

In the remainder of this section we extend the idea of a centre of mass of point masses along a line (the seesaw) to the centre of mass of a distribution of point masses in a plane, and then to the centre of mass of a continuous distribution of mass. Our first step is to define the first moment of a point mass in a plane about a line in the plane (Figure 7.53).

DEFINITION 7.2

The first moment of a mass m about a line l is md where d is the directed distance from l to m.

Once again directed distances are used in calculating first moments, and therefore distances on one side of the line must be chosen as positive and distances on the other side as negative. For vertical and horizontal lines, there is a natural convention for doing this. Distances to the right of a vertical line are chosen as positive, and distances to the left are negative. Distance upward from a horizontal line are positive, and distances downward are negative. In particular, when a mass m is located at position (x, y) in the xy-plane, its first moments about the x- and y-axes are my and mx, respectively.

First moments of a system of n point masses $m_1, m_2, \ldots, m_n$ located at points (x_1, y_1), $(x_2, y_2), \ldots, (x_n, y_n)$ respectively (Figure 7.54a) about the x- and y-axes are defined as the sums of the first moments of the individual masses about these lines:

$$\text{First moment of system about } x\text{-axis} = \sum_{i=1}^{n} m_i y_i, \qquad (7.29a)$$

$$\text{First moment of system about } y\text{-axis} = \sum_{i=1}^{n} m_i x_i. \qquad (7.29b)$$

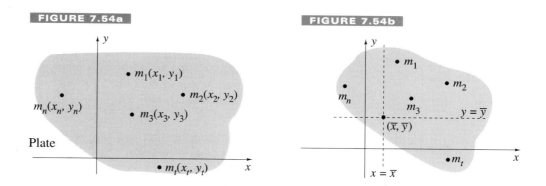

FIGURE 7.54a

Plate

FIGURE 7.54b

What is the physical meaning of these first moments? Do they, for instance, play the same role that first moments did for the seesaw? To see this we imagine that each point mass is embedded in a thin plastic plate in the xy-plane. The plate itself is massless and extends to include all n masses m_i. Picture now that the plate is horizontal, and a sharp edge is placed along the y-axis. Does the plate rotate about this edge, or does it balance? It is clear that the plate balances if the first moment of the system about the y-axis is equal to zero, and rotates otherwise, the direction depending on whether the first moment is positive or negative. Similarly, the plate balances on a sharp edge placed along the x-axis if the first moment of the system about the x-axis vanishes. In general, the plate balances along any straight edge if the first moment of the system about that edge vanishes.

We defined the centre of mass for a distribution of children on the seesaw as the point at which to place the fulcrum in order to obtain balance. Analogously, we define the *centre of mass* $(\overline{x}, \overline{y})$ of the distribution of point masses in Figure 7.54a as the position to place a sharp point in order to obtain balance. Remember that the plastic itself is massless and only the point masses m_i can create moments. Balance occurs at a point $(\overline{x}, \overline{y})$ if balance occurs about every straight line through $(\overline{x}, \overline{y})$. In particular, balance must occur about the lines $x = \overline{x}$ and $y = \overline{y}$ parallel to the y- and x-axes as in Figure 7.54b. Since balance occurs about $x = \overline{x}$ if the total first moment of the system about this line vanishes, we obtain the condition that

$$0 = m_1(x_1 - \overline{x}) + m_2(x_2 - \overline{x}) + \cdots + m_n(x_n - \overline{x}),$$

which can be solved for $\overline{x}$,

$$\overline{x} = \frac{1}{M} \sum_{i=1}^{n} m_i x_i, \tag{7.30}$$

where $M = \sum_{i=1}^{n} m_i$ is the total mass of the system. Note that this equation is identical to 7.28. Similarly, for balance about $y = \overline{y}$, we find that $\overline{y}$ must be

$$\overline{y} = \frac{1}{M} \sum_{i=1}^{n} m_i y_i. \tag{7.31}$$

We have obtained a unique point $(\overline{x}, \overline{y})$ based on conditions of balance about the lines $x = \overline{x}$ and $y = \overline{y}$. Does this necessarily imply that balance occurs about every straight line through $(\overline{x}, \overline{y})$? The answer is yes (see Exercise 34).

Every planar point mass distribution has a centre of mass $(\overline{x}, \overline{y})$ defined by equations 7.30 and 7.31. Our derivation has shown that the first moment of the system about any line through $(\overline{x}, \overline{y})$ must be equal to zero. This point is significant in another way. If a particle of mass M (the total mass of the system) is located at the centre of mass $(\overline{x}, \overline{y})$, its first moment about the y-axis is $M\overline{x}$. But from equation 7.30, we have

$$M\overline{x} = \sum_{i=1}^{n} m_i x_i, \tag{7.32}$$

and we conclude that the first moment of this fictitious particle M about the y-axis is exactly the same as the first moment of the system about the y-axis. Similarly, the first moment of M about the x-axis is $M\overline{y}$, and from 7.31,

$$M\overline{y} = \sum_{i=1}^{n} m_i y_i. \tag{7.33}$$

Thus, the centre of mass of a system of point masses m_i is a point at which a single particle of mass $M = \sum_{i=1}^{n} m_i$ has the same first moments about the x- and y-axes as the system. It can be shown further (Exercise 34) that the first moment of M about any line is the same as the first moment of the system about that line.

In summary, we defined the centre of mass of a system of point masses as a balance point. We found as a result that the centre of mass is a point at which a single particle of mass equal to the total mass of the system has the same first moment about any line as the system itself. This argument is reversible. Were we to define the centre of mass as a point to place the mass of the system for equivalent first moments, it would be a balance point. In other words, we have two equivalent definitions of the centre of mass of a system of point masses—a balance point or an equivalent point for first moments.

We now make the transition from a discrete system of particles to a continuous distribution of mass in the form of a thin plate of constant mass per unit area ρ (Figure 7.55). In order to find the mass of the plate, we proceed in exactly the same way that we did for areas. We divide the plate into vertical rectangles, the mass in a representative rectangle of width dx at position x being

$$\rho[f(x) - g(x)]\,dx.$$

FIGURE 7.55

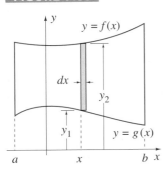

To find the total mass of the plate, we add over all such rectangles of ever-diminishing widths:

$$M = \int_a^b \rho[f(x) - g(x)]\,dx. \tag{7.34}$$

Based on our discussion for systems of point masses, we define the centre of mass of a continuous distribution of mass as that point $(\overline{x}, \overline{y})$ where a particle of mass M has the same first moments about the x- and y-axes as the distribution. In algebraic terms, we note that $M\overline{x}$ is the first moment about the y-axis of a particle of mass M at $(\overline{x}, \overline{y})$. To this we must equate the first moment of the original distribution about the y-axis. Now each point in the representative rectangle in Figure 7.55 is approximately the same distance x from the y-axis—approximately, because the rectangle does have finite, though very small, width. The first moment, then, of this rectangle about the y-axis is approximately

$$x\rho[f(x) - g(x)]\,dx.$$

We find the first moment of the entire plate about the y-axis by adding first moments of all such rectangles and taking the limit as their widths approach zero. But once again this is the process defined by the definite integral, and we obtain, therefore, for the first moment of the plate about the y-axis

$$\int_a^b x\rho[f(x) - g(x)]\,dx.$$

Consequently,

$$M\overline{x} = \int_a^b x\rho[f(x) - g(x)]\,dx, \tag{7.35}$$

and this equation can be solved for $\overline{x}$ once M and the integral on the right have been evaluated. Equation 7.35 represents for continuous distributions what equation 7.32 does for discrete distributions.

To find $\overline{y}$ we must equate the product $M\overline{y}$ to the first moment of the plate about the x-axis. If we consider the representative rectangle in Figure 7.55 we see that not all points therein are the same distance from the x-axis. To circumvent this problem, we consider all of the mass of the rectangle to be concentrated at its centre of mass. Since the centre of mass is the midpoint of the rectangle—a point distant $[f(x) + g(x)]/2$ from the x-axis—it follows that the first moment of this rectangle about the x-axis is

$$\frac{1}{2}[f(x) + g(x)]\rho[f(x) - g(x)]\,dx.$$

The total first moment of the plate about the x-axis is the definite integral of this expression, and we set

$$M\overline{y} = \int_a^b \frac{1}{2}[f(x) + g(x)]\rho[f(x) - g(x)]\,dx. \tag{7.36}$$

This equation is used to evaluate $\overline{y}$.

Equations 7.35 and 7.36 can be memorized as formulas for $\overline{x}$ and $\overline{y}$, but it is easier to perform the above operations mentally and arrive at these equations. Besides, for various shapes of plates, we might use horizontal rectangles or combinations of horizontal and vertical rectangles, and in such cases 7.35 and 7.36 would have to be modified.

On the basis of definitions 7.35 and 7.36 for $(\overline{x}, \overline{y})$, we can show that the first moment of M at $(\overline{x}, \overline{y})$ about any line is the same as the first moment of the plate about that same line. This implies that the plate balances along any line through $(\overline{x}, \overline{y})$ and therefore at $(\overline{x}, \overline{y})$.

EXAMPLE 7.23

Find the centre of mass of a thin plate of constant mass per unit area ρ if its edges are defined by the curves

$$y = 2 - x^2, \qquad y = 0, \qquad x = 0, \qquad x \geq 0.$$

FIGURE 7.56

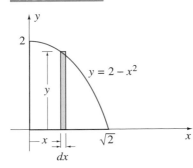

SOLUTION   Using vertical rectangles (Figure 7.56), we find

$$M = \int_0^{\sqrt{2}} \rho y \, dx = \rho \int_0^{\sqrt{2}} (2 - x^2) \, dx = \rho \left\{ 2x - \frac{x^3}{3} \right\}_0^{\sqrt{2}} = \frac{4\sqrt{2}\rho}{3}.$$

If $(\overline{x}, \overline{y})$ is the centre of mass of the plate, then $M\overline{x}$ is the first moment of the single particle of mass M about the y-axis. This must be equated to the first moment of the plate about the y-axis. Since $x\rho(2 - x^2) \, dx$ is approximately the first moment of the rectangle in Figure 7.56 about the y-axis, the following integral gives the first moment of the plate about the y-axis,

$$\int_0^{\sqrt{2}} x\rho(2 - x^2) \, dx = \rho \int_0^{\sqrt{2}} (2x - x^3) \, dx = \rho \left\{ x^2 - \frac{x^4}{4} \right\}_0^{\sqrt{2}} = \rho.$$

Hence we set $M\overline{x} = \rho$, and solve this equation for $\overline{x}$:

$$\overline{x} = \frac{\rho}{M} = \rho \frac{3}{4\sqrt{2}\rho} = \frac{3}{4\sqrt{2}}.$$

To find $\overline{y}$, we calculate the first moment of the plate about the x-axis. Since the centre of mass of the rectangle in Figure 7.56 is $y/2$ units above the x-axis, it follows that the first moment of this rectangle about the x-axis is $(y/2)\rho y \, dx$. When we integrate this to find the first moment of the plate about the x-axis, and equate it to $M\overline{y}$, the result is

$$M\overline{y} = \int_0^{\sqrt{2}} \frac{y}{2} \rho y \, dx = \frac{\rho}{2} \int_0^{\sqrt{2}} (2 - x^2)^2 \, dx = \frac{\rho}{2} \int_0^{\sqrt{2}} (4 - 4x^2 + x^4) \, dx$$

$$= \frac{\rho}{2} \left\{ 4x - \frac{4x^3}{3} + \frac{x^5}{5} \right\}_0^{\sqrt{2}} = \frac{16\sqrt{2}\rho}{15}.$$

Thus,

$$\overline{y} = \frac{16\sqrt{2}\rho}{15} \frac{3}{4\sqrt{2}\rho} = \frac{4}{5}.$$

In Exercises 23–26 use the technique suggested in Exercise 22 to find the centroid of the area.

** **23.**

** **24.**

** **25.**

** **26.**

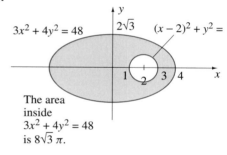

The area inside $3x^2 + 4y^2 = 48$ is $8\sqrt{3}\,\pi$.

In Exercises 27–30 find coordinates of the centroid of the area accurate to three decimal places.

** **27.** $y = x^3 + 3x^2 + 2x + 1$, $x = 0$, $y = 0$

** **28.** $y = x^4 - 5x^2 + 5$, $y = 0$ (above the x-axis)

** **29.** $x = y^3 - y^2 - 2y$, $x = \sqrt{2y + 1}$

** **30.** $y = x^3 - x$, $y = \sqrt{x}$

** **31.** Prove the following *theorem of Pappus*: If a plane area is revolved about a coplanar axis not crossing the area, the volume generated is equal to the product of the area and the circumference of the circle described by the centroid of the area.

** **32.** Use the result of Exercise 31 to find the volume of the donut in Exercise 41 of Section 7.2 (page 465).

** **33.** Use the result of Exercise 31 to find the volume in Exercise 45 of Section 7.2 (page 466).

** **34.** (a) Show that the first moment of the system of point masses in Figure 7.54b (page 493) about any line is the same as the first moment of a point mass $M = \sum_{i=1}^{n} m_i$ at $(\overline{x}, \overline{y})$ about that line. Hint: Use formula 1.7 for the distance from a point to a line.

  (b) Does it follow that the first moment of the system about any line through $(\overline{x}, \overline{y})$ is zero?

** **35.** A thin flat plate of area A is immersed vertically in a fluid of density ρ. Show that the total force due to fluid pressure on each side of the plate is equal to the product of 9.81ρ, A, and the depth of the centroid of the plate below the surface of the fluid.

*** **36.** Show that centroid of a triangle with vertices (x_1, y_1), (x_2, y_2), and (x_3, y_3) is

$$\overline{x} = \frac{x_1 + x_2 + x_3}{3}, \qquad \overline{y} = \frac{y_1 + y_2 + y_3}{3}.$$

7.8 Moments of Inertia

Newton's second law $F = ma$ is fundamental to the study of translational motion of bodies. For rotational motion of bodies, its counterpart is $\tau = I\alpha$ where τ is torque, α is angular acceleration, and I is the *moment of inertia* of the body. The kinetic energy of a body of mass m moving with velocity v is $mv^2/2$. The kinetic energy of a body rotating with angular velocity ω is $I\omega^2/2$. Thus, for rotational motion, there is a quantity called the *moment of inertia* of a body which is analogous to mass in translational equations. In this section we define and calculate moments of inertia.

To define moments of inertia of bodies, we begin with the moment of inertia of a point mass.

> **DEFINITION 7.3**
>
> The **moment of inertia** or **second moment** of a point mass m about a line l (Figure 7.53, page 492) is the product md^2 where d is the directed distance from l to m.

In particular, if m is at position (x, y) in the xy-plane, its moments of inertia about the x- and y-axes are my^2 and mx^2 respectively. For a system of n particles of masses $m_1, m_2, \ldots, m_n$ located at points (x_1, y_1), (x_2, y_2), $\ldots$, (x_n, y_n) as in Figure 7.54a (page 493), moments of inertia of the system about the x- and y-axes are sums of the moments of inertia of the particles about the x- and y-axes:

$$\text{Moment of inertia about } x\text{-axis} = \sum_{i=1}^{n} m_i y_i^2, \tag{7.39a}$$

$$\text{Moment of inertia about } y\text{-axis} = \sum_{i=1}^{n} m_i x_i^2. \tag{7.39b}$$

The transition from the discrete case to a continuous distribution in the form of a thin plate with constant mass per unit area ρ is not always so simple as for first moments. First consider the moment of inertia of the plate in Figure 7.55 in Section 7.7 (page 494) about the y-axis. The mass of the representative rectangle is

$$\rho[f(x) - g(x)]\,dx,$$

and each point of the rectangle is approximately the same distance x from the y-axis. The moment of inertia, then, of this rectangle about the y-axis is approximately

$$x^2\rho[f(x) - g(x)]\,dx.$$

The moment of inertia of the plate about the y-axis is found by adding moments of inertia of all such rectangles and taking the limit as their widths approach zero. But again this process defines a definite integral, and therefore the moment of inertia of the plate in Figure 7.55 about the y-axis is

$$I = \int_a^b x^2 \rho[f(x) - g(x)] \, dx. \tag{7.40}$$

EXAMPLE 7.26

Find the moment of inertia about the y-axis of a thin plate with constant mass per unit area ρ if its edges are defined by the curves

$$y = x^3, \qquad y = \sqrt{2 - x}, \qquad x = 0.$$

SOLUTION   Since the moment of inertia of the vertical rectangle in Figure 7.59 is

$$x^2 \rho(y_2 - y_1) \, dx = \rho x^2 \left(\sqrt{2 - x} - x^3\right) dx,$$

the moment of inertia of the plate is

$$I = \int_0^1 \rho x^2 \left(\sqrt{2 - x} - x^3\right) dx = \rho \int_0^1 x^2 \sqrt{2 - x} \, dx - \rho \int_0^1 x^5 \, dx.$$

In the first integral we set $u = 2 - x$, in which case $du = -dx$, and

$$I = \rho \int_2^1 (2 - u)^2 u^{1/2}(-du) - \rho \left\{\frac{x^6}{6}\right\}_0^1 = \rho \int_1^2 (4u^{1/2} - 4u^{3/2} + u^{5/2}) \, du - \frac{\rho}{6}$$

$$= \rho \left\{\frac{8u^{3/2}}{3} - \frac{8u^{5/2}}{5} + \frac{2u^{7/2}}{7}\right\}_1^2 - \frac{\rho}{6} = \frac{256\sqrt{2} - 319}{210}\rho.$$

FIGURE 7.59

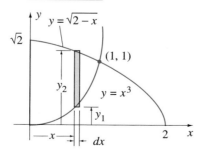

EXAMPLE 7.27

Find the moment of inertia about the line $y = -1$ of a thin plate of constant mass per unit area ρ if its edges are defined by the curves

$$x = y^2, \qquad x = 2y.$$

SOLUTION   Since the directed distance from the line $y = -1$ to all points in the horizontal rectangle in Figure 7.60 is approximately $y + 1$, the moment of inertia of the rectangle about the line $y = -1$ is approximately $(y + 1)^2 \rho(x_2 - x_1)\, dy$. It follows that the moment of inertia of the plate is

$$I = \int_0^2 (y + 1)^2 \rho(x_2 - x_1)\, dy = \rho \int_0^2 (y + 1)^2 (2y - y^2)\, dy$$

$$= \rho \int_0^2 (-y^4 + 3y^2 + 2y)\, dy = \rho \left\{ -\frac{y^5}{5} + y^3 + y^2 \right\}_0^2 = \frac{28\rho}{5}.$$

FIGURE 7.60

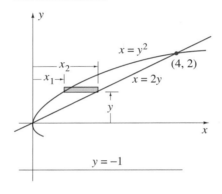

In Examples 7.26 and 7.27, and in the discussion leading to equation 7.40, we chose rectangles that had lengths parallel to the line about which we required the moment of inertia. This is not just coincidence; the use of perpendicular rectangles is more complicated. Consider, for instance, finding the moment of inertia about the x-axis of the plate in Example 7.26. To use the vertical rectangles in Figure 7.59, we first require the moment of inertia of such a rectangle about the x-axis. The mass of the rectangle, as in Example 7.26, must be multiplied by the square of the distance from the x-axis to the rectangle. Unfortunately, different points in the rectangle are at different distances. One suggestion might be to concentrate all of the mass of the rectangle at its centre of mass and use the distance from the x-axis to the centre of mass. This is *incorrect*. The centre of mass is a point at which mass can be concentrated if we are discussing first moments. We are discussing second moments. (In Exercise 12 we show that the moment of inertia of the rectangle in Figure 7.61 cannot be obtained by concentrating its mass at its centre of mass.) What are we to do then? To use this type of rectangle, we must first develop a formula for its moment of inertia. To obtain this formula, let us consider the moment of inertia about the x-axis of the rectangle of width h and length $y_2 - y_1$ in Figure 7.61. If we subdivide this rectangle into smaller rectangles of width dy, the moment of inertia of the tiny rectangle about the x-axis is approximately

$$y^2 \rho h\, dy.$$

FIGURE 7.61

The moment of inertia of the long, vertical rectangle can be obtained by adding over all the tiny rectangles as their widths dy approach zero,

$$\int_{y_1}^{y_2} y^2 \rho h\, dy = \rho h \int_{y_1}^{y_2} y^2\, dy = \frac{\rho h}{3}(y_2^3 - y_1^3). \tag{7.41}$$

We can use this formula to state that the moment of inertia about the x-axis of the vertical rectangle in Figure 7.59 is

$$\frac{\rho}{3}(y_2^3 - y_1^3)\, dx = \frac{\rho}{3}[(2-x)^{3/2} - x^9]\, dx.$$

The moment of inertia of the plate about the x-axis is therefore

$$I = \int_0^1 \frac{\rho}{3}[(2-x)^{3/2} - x^9]\, dx = \frac{\rho}{3}\left\{ -\frac{2}{5}(2-x)^{5/2} - \frac{x^{10}}{10} \right\}_0^1 = \frac{(16\sqrt{2} - 5)\rho}{30}.$$

The alternative procedure for this problem is to use horizontal rectangles which are parallel to the x-axis and obtain two definite integrals:

$$I = \int_0^1 y^2 \rho y^{1/3}\, dy + \int_1^{\sqrt{2}} y^2 \rho (2 - y^2)\, dy.$$

In summary, we have two methods for determining moments of inertia of thin plates:

1. Choose rectangles parallel to the line about which the moment of inertia is required, in which case only the basic idea of mass times distance squared is needed.

2. Choose rectangles perpendicular to the line about which the moment of inertia is required, in which case formula 7.41, or a similar formula, is needed.

As for finding centres of mass, we could, in the special case of uniform mass distributions, drop all references to mass and talk about second moments of area about a line.

** **17.** The polar moment of inertia of a point mass m at (x, y) is defined as the product of m and the square of its distance from the origin

$$J_0 = m(x^2 + y^2).$$

For the thin plate (with constant mass per unit area) in the figure below, let I_x and I_y be its moments of inertia about the x- and y-axes. Show that

$$J_0 = I_x + I_y.$$

In Exercises 18–20 the curves define a plate with mass per unit area equal to 2. Find its moment of inertia about the line accurate to three decimals.

** **18.** $y = 1 - x^2$, $x = y^2$ about $y = 0$

** **19.** $y = x^3 - x$, $y = \sqrt{x}$ about $x = 0$

** **20.** $y = x^3 - x$, $y = \sqrt{x}$ about $y = 0$

*** **21.** Find the second moment of area of a rectangle about its diagonal.

7.9 Additional Applications

Volumes by Slicing

If we can represent the area of parallel cross-sections of a volume as a function of one variable, we can use a definite integral to calculate the volume. In particular, when we use the disc or washer method to determine the volume of a solid of revolution, parallel cross-sections are circles. In the following example, parallel cross-sections are squares.

EXAMPLE 7.29

A uniformly tapered rod of length 2 m has square cross-sections. If the areas of its ends are 4 and 16 cm^2 as in Figure 7.63a, what is the volume of the rod?

SOLUTION   If we define an x-coordinate perpendicular to the square cross-sections as in Figure 7.63b, then the side lengths of the cross-sections at $x = 0$ and $x = 2$ are 0.02 m and 0.04 m respectively. Since the rod is uniformly tapered, side length of the cross-section at x is

$$0.02 + x \left(\frac{0.04 - 0.02}{2} \right) = 0.02 + 0.01x = \frac{2 + x}{100}.$$

The area of the cross-section at x is therefore

$$A(x) = 10^{-4}(2 + x)^2.$$

If we construct at x a "slab" of cross-sectional area $A(x)$ and width dx, the volume of the slab is

$$A(x)\,dx = 10^{-4}(2+x)^2\,dx.$$

To obtain the volume of the rod, we add volumes of all such slabs between $x = 0$ and $x = 2$, and take the limit as their widths approach zero:

$$V = \int_0^2 A(x)\,dx = \int_0^2 10^{-4}(2+x)^2\,dx$$

$$= 10^{-4}\left\{\frac{(2+x)^3}{3}\right\}_0^2 = \frac{10^{-4}}{3}(64-8) = \frac{56}{3} \times 10^{-4}\,\text{m}^3.$$

FIGURE 7.63a

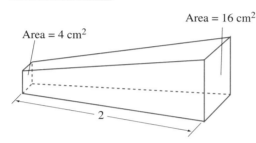
Area = 4 cm²   Area = 16 cm²
2

FIGURE 7.63b

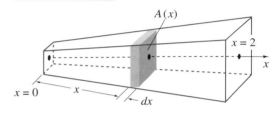
$A(x)$
$x = 2$
x
$x = 0$ x
dx

Area of a Surface of Revolution

If a curve in the xy-plane is rotated about the x- or y-axis (or a line parallel to the x- or y-axis) in order to produce a surface, we can calculate the area of this surface using lengths along curves discussed in Section 7.3.

EXAMPLE 7.30

If that part of the parabola $y = x^2$ between $x = 0$ and $x = 1$ is rotated around the y-axis, find the area of the surface of revolution traced out by the curve (Figure 7.64).

FIGURE 7.64

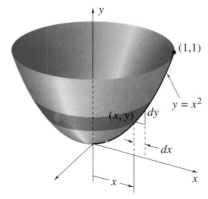
y
$(1,1)$
$y = x^2$
(x, y) dy
dx
x
x

SOLUTION   We approximate length along the parabola corresponding to a change dx in x by the tangential straight-line length:

$$\sqrt{(dx)^2 + (dy)^2} = \sqrt{1 + \left(\frac{dy}{dx}\right)^2}\, dx = \sqrt{1 + (2x)^2}\, dx = \sqrt{1 + 4x^2}\, dx.$$

If this straight-line segment is rotated around the y-axis, each point follows a circular path of radius approximately equal to x; therefore, the area traced out by the line segment is approximately equal to

$$(2\pi x)\left(\sqrt{1 + 4x^2}\, dx\right).$$

We find total surface area by adding all such areas, and taking the limit as widths dx approach zero:

$$A = \int_0^1 2\pi x\sqrt{1 + 4x^2}\, dx = 2\pi \left\{\frac{(1 + 4x^2)^{3/2}}{12}\right\}_0^1 = \frac{(5\sqrt{5} - 1)\pi}{6}.$$

EXAMPLE 7.31

Find area of the surface of revolution traced out by rotating that part of the curve $y = x^3$ between $x = 1$ and $x = 2$ about the x-axis.

SOLUTION   We approximate length along the cubic corresponding to a change dx in x by the tangential straight-line length:

$$\sqrt{(dx)^2 + (dy)^2} = \sqrt{1 + \left(\frac{dy}{dx}\right)^2}\, dx = \sqrt{1 + (3x^2)^2}\, dx = \sqrt{1 + 9x^4}\, dx.$$

If this straight-line segment is rotated about the x-axis (Figure 7.65), each point follows a circular path of radius approximately equal to y; therefore, the area traced out by the line segment is approximately equal to

$$(2\pi y)\left(\sqrt{1 + 9x^4}\, dx\right).$$

By adding over all such areas,

$$A = \int_1^2 2\pi y\sqrt{1 + 9x^4}\, dx = 2\pi \int_1^2 x^3\sqrt{1 + 9x^4}\, dx$$

$$= 2\pi \left\{\frac{(1 + 9x^4)^{3/2}}{54}\right\}_1^2 = \frac{(145^{3/2} - 10^{3/2})\pi}{27}.$$

FIGURE 7.65

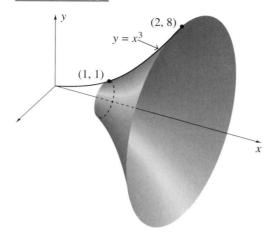

Rates of Flow

In Problem 3 of Section 6.2 (page 409) we considered laminar blood flow in a circular vessel. Specifically, velocity of blood through the cross-section in Figure 7.66a is a function of radial distance r from the centre of the vessel:

$$v(r) = c(R^2 - r^2), \quad 0 \le r \le R,$$

where $c > 0$ is a constant and R is the radius of the vessel. If we construct, at radius r, a thin ring of width dr as in Figure 7.66b, then area of this ring is approximately $(2\pi r)\,dr$. Since v does not vary greatly over this ring, the amount of blood flowing through the ring per unit time is approximately $v(r)$ multiplied by the area of the ring:

$$v(r)(2\pi r\,dr).$$

We can find total flow through the blood vessel by adding flows through all rings and taking the limit as widths dr of the rings approach zero:

$$F = \int_0^R v(r)2\pi r\,dr = 2\pi \int_0^R rc(R^2 - r^2)\,dr = 2\pi c \left\{ \frac{R^2 r^2}{2} - \frac{r^4}{4} \right\}_0^R = \frac{\pi c R^4}{2}.$$

FIGURE 7.66a

FIGURE 7.66b

Elongation of Rods

Hooke's law states that forces exerted by springs are proportional to stretch and compression. In particular, suppose x measures distance to the right in Figure 7.67 and $x = 0$ corresponds to the right end of the spring when the spring is unstretched and uncompressed. Hooke's law then states that a force F applied to the spring as shown causes stretch

$$ x = \frac{F}{k}, $$

where k is the spring constant.

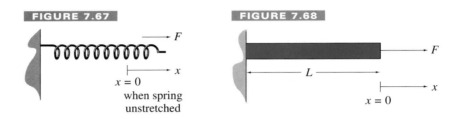

FIGURE 7.67

$x = 0$
when spring
unstretched

FIGURE 7.68

Figure 7.68 shows a rod of natural length L. When a force F is applied to the right end of the rod, the rod acts like a spring in that the rod stretches. It stretches only minutely even when F is large; but it does stretch. It is shown in the area of "strength of materials" that the amount of stretch is related to F by the equation

$$ x = \frac{FL}{AE}, \tag{7.42} $$

where A is the (constant) cross-sectional area of the rod, and E is Young's modulus of elasticity, a constant that depends on the material of the rod. In effect, AE/L plays the role of spring constant k.

Now suppose the rod is turned vertically and hung from its top end (Figure 7.69). Because of its weight, the rod stretches and, using 7.42, we can calculate how much. If we consider a small length dy at position y, the force on each cross section in this element is approximately the same, and equal to the weight of that part of the rod below it,

$$ \rho g (L - y) A $$

where ρ is the density of the material in the rod. According to 7.42, the element dy stretches by

$$ \frac{\rho g (L - y) A \, dy}{AE} = \frac{\rho g (L - y) \, dy}{E}. $$

Total stretch in the rod is therefore

$$ \int_0^L \frac{\rho g (L - y)}{E} dy = \frac{\rho g}{E} \left\{ Ly - \frac{y^2}{2} \right\}_0^L = \frac{\rho g L^2}{2E}. $$

This may be somewhat surprising in that stretch does not depend on the cross-sectional area of the rod. This is explained by the fact that the weight of rod below any cross-section is

FIGURE 7.69

$y = 0$

y

L

dy

proportional to cross-sectional area A, but stretch in 7.42 is inversely proportional to A. The two factors compensate. To get an idea of the magnitude of this stretch, suppose that $L = 2$ m, $\rho = 8000$ kg/m^3, and $E = 2 \times 10^{11}$ N/m^2. Then

$$\frac{\rho g L^2}{2E} = \frac{8000(9.81)(2)^2}{2(2 \times 10^{11})} = 7.85 \times 10^{-7} \text{ m.}$$

Suppose now that cross-sectional area A of the rod is not constant; the rod is tapered, say, with circular cross-sections (Figure 7.70). Radius of the large end is r and the rod tapers to a point at $y = L$. The force on cross-sections of the element of width dy at position y is again the weight of rod below it, namely,

$$\frac{1}{3}\pi x^2 (L - y)\rho g,$$

where x is the radius of the rod at position y. Since the equation of the side of the rod is

$$y = \frac{-L}{r}(x - r) \qquad \Longrightarrow \qquad x = r\left(1 - \frac{y}{L}\right),$$

the weight below y is

$$\frac{1}{3}\pi r^2 \left(1 - \frac{y}{L}\right)^2 (L - y)\rho g = \frac{\pi r^2 \rho g}{3L^2}(L - y)^3.$$

The element dy therefore stretches

$$\frac{\pi r^2 \rho g}{3L^2}(L - y)^3 \frac{dy}{\pi x^2 E} = \frac{r^2 \rho g (L - y)^3 dy}{3L^2 E r^2 (1 - y/L)^2} = \frac{\rho g (L - y)}{3E} dy.$$

This is one-third the stretch for the nontapered rod, and hence total stretch of the tapered rod is $\rho g L^2 / (6E)$.

FIGURE 7.70

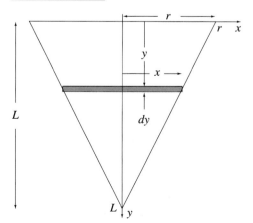

Repairs, Depreciation, and Replacement

In most businesses, depreciation and eventual replacement of equipment must be considered. Take, for example, a consulting engineer who must decide when to replace his car. When he buys a new car, he knows that its value depreciates quickly at first and less rapidly after a few years. On the other hand, the longer he keeps the car, the more costly repairs become. The problem then is to decide on the optimum time to replace the car. Let us consider a situation in which the rate of depreciation of the car is $f(t)$, and the rate of accumulation of maintenance costs is $g(t)$. We have shown a possible $f(t)$ in Figure 7.71, where the ordinate at time t represents the number of dollars per year the car depreciates at that time. If we draw a rectangle of width dt at time t, then the area of this rectangle, $f(t)\,dt$, represents the amount the car depreciates during the time interval dt. Consequently, the total depreciation of the car from the time of purchase ($t = 0$) to time T is

$$\int_0^T f(t)\,dt.$$

In a similar way, the total maintenance cost from time $t = 0$ to $t = T$ (Figure 7.72) is

$$\int_0^T g(t)\,dt.$$

The cost associated with owning and maintaining the car over a period of T years is

$$\int_0^T f(t)\,dt + \int_0^T g(t)\,dt = \int_0^T [f(t) + g(t)]\,dt,$$

and the average yearly cost is

$$C(T) = \frac{1}{T} \int_0^T [f(t) + g(t)]\,dt.$$

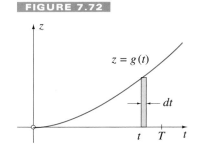

The best time to replace the car is that value T which minimizes average yearly cost. Critical points of $C(T)$ are defined by

$$0 = C'(T) = -\frac{1}{T^2} \int_0^T [f(t) + g(t)]\,dt + \frac{1}{T}\frac{d}{dT} \int_0^T [f(t) + g(t)]\,dt.$$

Equation 6.18 in Section 6.5 can be used to evaluate the derivative of the integral; the result is

$$0 = -\frac{1}{T^2} \int_0^T [f(t) + g(t)]\,dt + \frac{1}{T}[f(T) + g(T)]$$

or
$$\frac{1}{T}\int_0^T [f(t) + g(t)]\, dt = f(T) + g(T).$$

Since the left side of this equation is $C(T)$, we can say that critical points of $C(T)$ occur when

$$C(T) = f(T) + g(T); \tag{7.43}$$

that is, when average yearly costs are equal to the rate at which costs are changing. At one of these points, if there is more than one, $C(T)$ will take on its minimum value.

EXAMPLE 7.32

Find the optimum time to replace a car if rates of depreciation and accumulation of maintenance costs are, respectively,

$$f(t) = -\frac{V}{50}(t - 10) \qquad \text{and} \qquad g(t) = \frac{V}{400}t^2,$$

where V is the value of the new car. Assume that the car must be replaced by the time it has depreciated to no value.

SOLUTION   Since total depreciation of the car in T years is given by

$$\int_0^T f(t)\, dt = \int_0^T -\frac{V}{50}(t - 10)\, dt = -\frac{V}{50}\left\{\frac{t^2}{2} - 10t\right\}_0^T = -\frac{V}{100}(T^2 - 20T),$$

it follows that the car completely depreciates when

$$V = -\frac{V}{100}(T^2 - 20T).$$

The only solution of this equation is $T = 10$ years. Do not make the mistake of assuming that complete depreciation occurs when $f(t) = 0$; this is not always true. Complete depreciation occurs when the area under $f(t)$ is equal to the initial value of the car.

The average yearly cost associated with the car if it is kept for T years is

$$C(T) = \frac{1}{T}\int_0^T [f(t) + g(t)]\, dt$$

$$= \frac{1}{T}\int_0^T \left(-\frac{Vt}{50} + \frac{V}{5} + \frac{Vt^2}{400}\right) dt$$

$$= \frac{1}{T}\left(-\frac{VT^2}{100} + \frac{VT}{5} + \frac{VT^3}{1200}\right)$$

$$= -\frac{VT}{100} + \frac{V}{5} + \frac{VT^2}{1200}, \qquad 0 < T \le 10.$$

For critical points of $C(T)$, we solve

$$0 = C'(T) = -\frac{V}{100} + \frac{VT}{600},$$

the solution of which is $T = 6$. We could also have obtained $T = 6$ by solving equation 7.43. Since

$$\lim_{T \to 0^+} C(T) = \frac{V}{5}, \qquad C(6) = \frac{17V}{100}, \qquad C(10) = \frac{11V}{60},$$

it follows that the car should be replaced in 6 years.

EXERCISES 7.9

* **1.** Verify that the surface area of a sphere of radius r is $4\pi r^2$.

* **2.** Find the area of the curved surface of a right circular cone of radius r and height h.

* **3.** Calculate the rate of flow of blood through a circular vessel of radius R if the velocity profile is (a) $v = f(r) = cR\sqrt{R^2 - r^2}$ and (b) $v = f(r) = (c/R^2)(R^2 - r^2)^2$.

* **4.** Find the volume of the pyramid in the figure below.

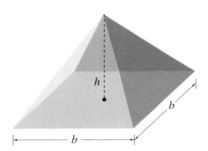

* **5.** The amount of water consumed by a community varies throughout the day, peaking, naturally, around meal hours. During the six-hour period between 12:00 noon ($t = 0$) and 6:00 P.M. ($t = 6$), we find that the number of cubic metres of water consumed per hour at time t is given by the function

$$f(t) = 5000 + 21.65t^2 - 249.7t^3 + 97.52t^4 - 9.680t^5.$$

Find the total consumption during this 6-hour period.

** **6.** The number of bees per unit area at a distance x from a hive is given by

$$\rho(x) = \frac{600\,000}{31\pi R^5}(R^3 + 2R^2x - Rx^2 - 2x^3), \quad 0 \le x \le R,$$

where R is the maximum distance travelled by the bees.

    (a) What is the number of bees in the colony?

    (b) How many bees are within a distance $R/2$ of the hive?

** **7.** If the radius of the blood vessel in Figure 7.66 (page 511) is reduced to $R/2$ because of arteriosclerosis, the velocity profile is

$$v(r) = c(R^2 - 4r^2), \quad 0 \leq r \leq R/2.$$

What percentage of the normal flow $(\pi c R^4/2)$ gets through the hardened vessel?

** **8.** A tree trunk of diameter 50 cm (figure below) has a wedge cut from it by two planes. The lower plane is perpendicular to the axis of the trunk, and together the planes make an angle $\pi/3$ radians, meeting along a diameter of the circular cross section of the trunk. Find the volume of the wedge.

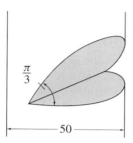

** **9.** Find the area of the surface of revolution formed by rotating about the x-axis that part of the curve $24xy = x^4 + 48$ between $x = 1$ and $x = 2$.

** **10.** Find the area of the surface of revolution generated by rotating the curve $8y^2 = x^2(1 - x^2)$ about the x-axis.

** **11.** Find the area of the surface of revolution generated by rotating the loop of the curve $9y^2 = x(3 - x)^2$ about (a) the y-axis and (b) the x-axis.

** **12.** The base of a solid is the circle $x^2 + y^2 = r^2$, and every plane section perpendicular to the x-axis is an isosceles triangle (figure below). Find the volume of the solid.

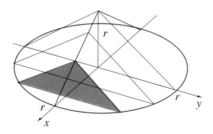

** **13.** The end of the rod at $x = 0$ in the figure below is rigidly fixed. If a force with magnitude F is applied to the right end, how long is the rod?

(natural length)

** **14.** In the figure below, a mass M is placed on a vertical rod. If the length of the compressed rod is L, what is its length if M is removed and the rod is turned horizontally? Assume that M is so large that the weight of the rod can be neglected in comparison.

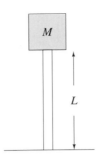

** **15.** What is the answer to Exercise 14 if the weight of the rod is taken into account?

** **16.** Suppose the rod in Exercise 13 is turned vertically so that its top end is fixed, and the force F pulls vertically downward on the lower end. How long is the rod if the weight of the rod is also taken into account?

** **17.** What happens when you attempt to find the stretch of the rod in Figure 7.70 if the pointed end is at $y = 0$ and the larger end is at $y = L$?

** **18.** Suppose a mass M is attached to the lower end of the tapered rod in Figure 7.70 (page 513). It is so large that the mass of the rod is negligible by comparison. What happens when you attempt to find the stretch in the rod?

** **19.** A tapered rod of length L has square cross-sections. The squares on the ends have dimensions a and b ($b > a$). The rod is placed in a vertical position with larger end below smaller end, and smaller end fixed in position (figure below). What is the length of the rod in this position?

** **20.** What is the length of the rod in Exercise 19 if the rod is turned upside down?

** **21.** Repeat Exercise 20 if a mass M is distributed over the bottom of the rod.

** **22.** Repeat Exercise 19 if a mass M is distributed over the bottom of the rod.

** **23.** Electrons are fired from an electron gun at a target (figure on page 519). The probability that an electron strikes the target in a ring of unit area at a distance x from the centre of the target is given by

$$p = f(x) = \frac{5}{3\pi R^5}(R^3 - x^3), \quad 0 \le x \le R.$$

What percentage of a cascade of electrons hits within a distance r from the centre of the target?

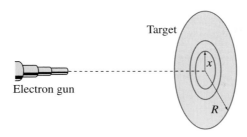

In Exercises 24–26, $f(t)$ and $g(t)$ represent rate of depreciation and rate of accumulation of maintenance costs for a piece of machinery. Draw graphs of these functions, and show that the total area under $y = f(t)$ is equal to V (the original value of the machine). Determine the optimum time at which to replace the machine. Draw a graph of $C(T)$ (the average yearly expense) to illustrate your answer.

** **24.** $f(t) = \begin{cases} \dfrac{V}{10} - \dfrac{Vt}{200}, & 0 < t \le 20 \\ 0, & t > 20 \end{cases}$ $\qquad g(t) = \dfrac{Vt^2}{4000}$

** **25.** $f(t) = \begin{cases} -\dfrac{V}{50}(t - 10), & 0 < t \le 10 \\ 0, & t > 10 \end{cases}$ $\qquad g(t) = \dfrac{Vt}{400}$

** **26.** $f(t) = \begin{cases} \dfrac{V}{10} - \dfrac{6Vt^2}{15\,625}, & 0 < t \le 12.5 \\ 0, & t > 12.5 \end{cases}$ $\qquad g(t) = \dfrac{Vt^2}{3000}$

** **27.** Two right circular cylinders, each of radius r, have axes that intersect at right angles. Find the volume common to the two cylinders.

** **28.** What is the volume of air trapped in the attic of a house if the roof has shape shown in the figure below? The peak of the roof is 2 m above the base.

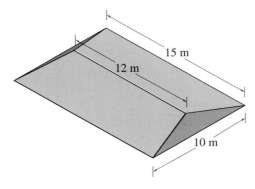

** **29.** Suppose that at time $t = 0$ the number of people in a certain population is N_0, and the probability that each individual lives to age t is $p(t)$. For time $t > 0$, the birth rate is $r(t)$ individuals per unit time. Show that at time T the number of people in the population is

$$N(T) = N_0 p(T) + \int_0^T p(T - t) r(t) \, dt.$$

*** **30.** The depth of water in the container in the figure below is originally 10 cm. Water evaporates from the surface at a rate proportional to the area of the surface. If the water level drops 1 cm in 5 days, how long does it take for the water to evaporate completely? Note that Exercise 9 in Section 5.4 is this same problem given a container that is a right circular cone. Now we ask you to repeat the problem with no knowledge of the shape of the container.

*** **31.** An appliance retailer must be concerned with her inventory costs. For example, given that she sells N refrigerators per year, she must decide whether to order the year's supply at one time, in which case a large storage area would be necessary, or to make periodic orders throughout the year, perhaps of $N/12$ at the first of each month. In the latter case she would incur costs due to paper work, delivery charges, etc. Let us suppose that the retailer decides to order in equal lot sizes, x, at equally spaced intervals, N/x times per year.

(a) If each order has fixed costs of F dollars, plus f dollars for each refrigerator, what are the total yearly ordering costs?

(b) Between successive deliveries, the retailer's stock dwindles from x to 0. If she assumes that the number of refrigerators in stock decreases linearly in time, and the yearly stocking cost per refrigerator is p dollars, what are the yearly stocking costs?

(c) If the retailer's total yearly inventory costs are her ordering costs in part (a) plus her stocking costs in part (b), what value of x minimizes inventory costs?

7.10 Improper Integrals

According to the fundamental theorem of integral calculus (Theorem 6.3), definite integrals are evaluated with indefinite integrals,

$$\int_a^b f(x)\, dx = \left\{ \int f(x)\, dx \right\}_a^b,$$

provided $f(x)$ is continuous on $a \le x \le b$. In this section we investigate what to do when $f(x)$ is not continuous on $a \le x \le b$, or when either a or b is infinite.

Consider whether a reasonable meaning can be given to the integrals

$$\int_1^\infty \frac{1}{x^2}\, dx \qquad \text{and} \qquad \int_1^\infty \frac{1}{\sqrt{x}}\, dx.$$

Both integrals are *improper* in the sense that their upper limits are not finite, and we therefore call them **improper integrals**. If $b > 1$, there is no difficulty with evaluation and interpretation of

$$\int_1^b \frac{1}{x^2}\, dx \qquad \text{and} \qquad \int_1^b \frac{1}{\sqrt{x}}\, dx.$$

Clearly,

$$\int_1^b \frac{1}{x^2}\, dx = \left\{ -\frac{1}{x} \right\}_1^b = 1 - \frac{1}{b} \quad \text{and}$$

$$\int_1^b \frac{1}{\sqrt{x}}\, dx = \left\{ 2\sqrt{x} \right\}_1^b = 2\sqrt{b} - 2.$$

We can interpret the first integral as the area under the curve $y = 1/x^2$, above the x-axis, and between the vertical lines $x = 1$ and $x = b$ (Figure 7.73). The second integral has exactly the same interpretation but uses the curve $y = 1/\sqrt{x}$ in place of $y = 1/x^2$ (Figure 7.74).

FIGURE 7.73

FIGURE 7.74

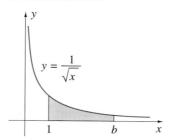

In Table 7.1 we have listed values of these definite integrals corresponding to various values of b. It is clear that the two integrals display completely different characteristics. As b is made very large, the integral of $1/x^2$ is always less than 1, but gets closer to 1 as b increases. In other words,

$$\lim_{b \to \infty} \int_1^b \frac{1}{x^2}\, dx = 1.$$

Geometrically, the area in Figure 7.73 is always less than 1 but approaches 1 as b approaches infinity.

Contrast this with the integral of $1/\sqrt{x}$, which becomes indefinitely large as b increases:

$$\lim_{b \to \infty} \int_1^b \frac{1}{\sqrt{x}}\, dx = \infty;$$

that is, the area in Figure 7.74 can be made as large as desired by choosing b sufficiently large.

TABLE 7.1

b	$\int_1^b \frac{1}{x^2}dx$	$\int_1^b \frac{1}{\sqrt{x}}dx$
100	0.99	18
10 000	0.999 9	198
1 000 000	0.999 999	1 998
100 000 000	0.999 999 99	19 998

On the basis of these calculations, we would like to say that the improper integral of $1/x^2$ has value 1, whereas the improper integral of $1/\sqrt{x}$ has no value. In other words, if these improper integrals are to have values, they should be defined by the limits

$$\int_1^\infty \frac{1}{x^2}\, dx = \lim_{b \to \infty} \int_1^b \frac{1}{x^2}\, dx \quad \text{and} \quad \int_1^\infty \frac{1}{\sqrt{x}}\, dx = \lim_{b \to \infty} \int_1^b \frac{1}{\sqrt{x}}\, dx.$$

In the second case the limit does not exist, and we interpret this to mean that the improper integral does not exist or has no value.

In general, then, an improper integral with an infinite upper limit is defined in terms of limits as follows.

DEFINITION 7.4

If $f(x)$ is continuous for $x \geq a$, we define

$$\int_a^\infty f(x)\, dx = \lim_{b \to \infty} \int_a^b f(x)\, dx, \tag{7.44}$$

provided the limit exists.

If the limit exists, we say that the improper integral is equal to the limit, or that the improper integral *converges* to the limit. If the limit does not exist, we say that the improper integral has no value, or that it *diverges*.

EXAMPLE 7.33

Determine whether the following improper integrals converge or diverge:

(a) $\displaystyle\int_3^\infty \frac{1}{(x-2)^3}\, dx$ (b) $\displaystyle\int_0^\infty \frac{x^2}{\sqrt{1+x^3}}\, dx$ (c) $\displaystyle\int_1^\infty \frac{2x^2+6}{3x^2+5}\, dx$

SOLUTION

(a) Using equation 7.44,

$$\int_3^\infty \frac{1}{(x-2)^3}\, dx = \lim_{b \to \infty} \int_3^b \frac{1}{(x-2)^3}\, dx = \lim_{b \to \infty} \left\{ \frac{-1}{2(x-2)^2} \right\}_3^b$$

$$= \lim_{b \to \infty} \left[\frac{1}{2} - \frac{1}{2(b-2)^2} \right] = \frac{1}{2},$$

and the improper integral therefore converges to $1/2$.

(b) Once again 7.44 gives

$$\int_0^\infty \frac{x^2}{\sqrt{1+x^3}}\, dx = \lim_{b \to \infty} \int_0^b \frac{x^2}{\sqrt{1+x^3}}\, dx = \lim_{b \to \infty} \left\{ \frac{2}{3}\sqrt{1+x^3} \right\}_0^b$$

$$= \lim_{b \to \infty} \left[\frac{2}{3}\sqrt{1+b^3} - \frac{2}{3} \right] = \infty,$$

and the improper integral therefore diverges.

FIGURE 7.75

(c) Though we cannot at present find an antiderivative for $f(x) = (2x^2+6)/(3x^2+5)$, we can solve this problem by interpreting the improper integral as a limit summation. The graph of $f(x)$ in Figure 7.75 indicates that the curve is asymptotic to the line $y = 2/3$. Clearly, then, the improper integral

$$\int_1^\infty \frac{2x^2 + 6}{3x^2 + 5}\, dx$$

can have no value since the area under the curve is larger than the area of a rectangle of width 2/3 and infinite length.

We can define improper integrals with infinite lower limits in exactly the same way as for improper integrals with infinite upper limits. Specifically, if $f(x)$ is continuous for $x \le b$, we define

$$\int_{-\infty}^b f(x)\, dx = \lim_{a \to -\infty} \int_a^b f(x)\, dx, \tag{7.45}$$

provided the limit exists. For example,

$$\int_{-\infty}^5 \frac{1}{(x - 6)^5}\, dx = \lim_{a \to -\infty} \int_a^5 \frac{1}{(x - 6)^5}\, dx = \lim_{a \to -\infty} \left\{ \frac{-1}{4(x - 6)^4} \right\}_a^5$$

$$= \lim_{a \to -\infty} \left[-\frac{1}{4} + \frac{1}{4(a - 6)^4} \right] = -\frac{1}{4}.$$

When $f(x)$ is continuous for all x, we define

$$\int_{-\infty}^\infty f(x)\, dx = \lim_{a \to -\infty} \int_a^c f(x)\, dx + \lim_{b \to \infty} \int_c^b f(x)\, dx, \tag{7.46}$$

provided both limits exist. The number c is arbitrary; existence of the limits is independent of c (see Exercise 33). For example, if we choose $c = 0$, then

$$\int_{-\infty}^\infty \frac{x}{(1 + x^2)^2}\, dx = \lim_{a \to -\infty} \int_a^0 \frac{x}{(1 + x^2)^2}\, dx + \lim_{b \to \infty} \int_0^b \frac{x}{(1 + x^2)^2}\, dx$$

$$= \lim_{a \to -\infty} \left\{ \frac{-1}{2(1 + x^2)} \right\}_a^0 + \lim_{b \to \infty} \left\{ \frac{-1}{2(1 + x^2)} \right\}_0^b$$

$$= \lim_{a \to -\infty} \left[\frac{1}{2(1 + a^2)} - \frac{1}{2} \right] + \lim_{b \to \infty} \left[\frac{1}{2} - \frac{1}{2(1 + b^2)} \right]$$

$$= -\frac{1}{2} + \frac{1}{2} = 0.$$

EXAMPLE 7.34

(a) Is it possible to paint the area bounded by the curves $y = 1/x$, $x = 1$, and $y = 0$?

(b) Rotate the area around the x-axis to form a funnel. Does it have finite volume?

FIGURE 7.76

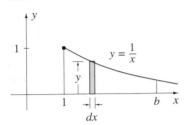

SOLUTION

(a) Painting is possible if the area of the region bounded by the curves is finite (Figure 7.76). Since

$$\int_1^\infty \frac{1}{x}\, dx = \lim_{b \to \infty} \int_1^b \frac{1}{x}\, dx = \lim_{b \to \infty} \{\ln |x|\}_1^b = \lim_{b \to \infty} \ln b = \infty,$$

the area is not finite, and cannot therefore be painted.

(b) We use discs to determine whether the funnel has finite volume:

$$\int_1^\infty \pi y^2\, dx = \lim_{b \to \infty} \int_1^b \frac{\pi}{x^2}\, dx = \lim_{b \to \infty} \left\{\frac{-\pi}{x}\right\}_1^b = \lim_{b \to \infty} \left(\pi - \frac{\pi}{b}\right) = \pi.$$

The volume is finite. Consider this now. The funnel can be filled with paint, but the flat cross-sectional area inside the funnel cannot be painted. How can this be?

EXAMPLE 7.35

Find the escape velocity of a projectile from the earth's surface.

FIGURE 7.77

Earth

SOLUTION   Suppose the projectile is fired from the earth's surface with speed v in the x-direction (Figure 7.77). If the projectile is to escape the earth's gravitational pull, its initial kinetic energy must be greater than or equal to the work done against gravity as the projectile travels from the earth's surface ($x = R$) to a point where it is free of gravity ($x = \infty$). When the projectile is a distance x from the centre of the earth, the force of attraction on it is

$$F(x) = -\frac{GMm}{x^2},$$

where $G > 0$ is a constant, m is the mass of the projectile, and M is the mass of the earth. This is known as *Newton's universal law of gravitation.* The work done against gravity in travelling from $x = R$ to $x = \infty$ is

$$W = \int_R^\infty \frac{GmM}{x^2}\, dx = \lim_{b \to \infty} \int_R^b \frac{GmM}{x^2}\, dx$$

$$= \lim_{b \to \infty} \left\{-\frac{GmM}{x}\right\}_R^b = \lim_{b \to \infty} \left[\frac{GmM}{R} - \frac{GmM}{b}\right] = \frac{GmM}{R}.$$

Since the initial kinetic energy of the projectile is $mv^2/2$, it escapes the gravitational pull of the earth if

$$\frac{1}{2}mv^2 \geq \frac{GmM}{R};$$

that is, if

$$v \geq \sqrt{\frac{2GM}{R}}.$$

Hence $\sqrt{2GM/R}$ is the escape velocity (notice it is independent of the mass of the projectile). If we take the mean radius of the earth as 6370 km, its mean density as $\rho = 5.52 \times 10^3$ kg/m^3, and $G = 6.67 \times 10^{-11}$, we obtain $v = 11.2$ km/s. (Compare this with a 308 Winchester that fires a 150-grain bullet with a muzzle velocity of only 0.81 km/s.)

A second type of improper integral occurs when the integrand is discontinuous at a point or points in the interval of integration. For example, integrands of the improper integrals

$$\int_1^5 \frac{1}{\sqrt{x-1}}\, dx \quad \text{and} \quad \int_1^5 \frac{1}{(x-1)^4}\, dx$$

have infinite discontinuities at $x = 1$. If c is a number between 1 and 5, it is easy to calculate

$$\int_c^5 \frac{1}{\sqrt{x-1}}\, dx = \left\{ 2\sqrt{x-1} \right\}_c^5 = 4 - 2\sqrt{c-1} \quad \text{and}$$

$$\int_c^5 \frac{1}{(x-1)^4}\, dx = \left\{ \frac{-1}{3(x-1)^3} \right\}_c^5 = \frac{1}{3(c-1)^3} - \frac{1}{192},$$

and one possible interpretation of these definite integrals is the areas in Figures 7.78 and 7.79.

| FIGURE 7.78 | FIGURE 7.79 |

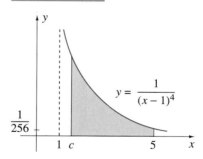

If we let c approach 1 from the right in the first integral, we find that

$$\lim_{c \to 1^+} \int_c^5 \frac{1}{\sqrt{x-1}}\, dx = \lim_{c \to 1^+} \left(4 - 2\sqrt{c-1} \right) = 4;$$

and in the second integral,

$$\lim_{c \to 1^+} \int_c^5 \frac{1}{(x-1)^4}\, dx = \lim_{c \to 1^+} \left[\frac{1}{3(c-1)^3} - \frac{1}{192} \right] = \infty.$$

It would seem reasonable then to define the area under the curve $y = 1/\sqrt{x-1}$, above the x-axis, and between the vertical lines $x = 1$ and $x = 5$, as 4 square units. On the other hand, the area bounded by $y = 1/(x-1)^4$, $y = 0$, $x = 5$, and $x = c$ becomes increasingly large as $x = c$ moves closer to $x = 1$, and we cannot therefore assign a value to this area when $c = 1$. In other words, if we define

$$\int_1^5 \frac{1}{\sqrt{x-1}}\,dx = \lim_{c \to 1^+} \int_c^5 \frac{1}{\sqrt{x-1}}\,dx \quad \text{and}$$

$$\int_1^5 \frac{1}{(x-1)^4}\,dx = \lim_{c \to 1^+} \int_c^5 \frac{1}{(x-1)^4}\,dx,$$

then the first improper integral has a value of 4, whereas the second has no value.

These improper integrals are examples of the general situation described in the following definition and illustrated in Figure 7.80.

FIGURE 7.80a

FIGURE 7.80b

FIGURE 7.80c

DEFINITION 7.5

If $f(x)$ is continuous at every point in the interval $a \le x \le b$ except at $x = a$, then

$$\int_a^b f(x)\,dx = \lim_{c \to a^+} \int_c^b f(x)\,dx, \tag{7.47}$$

provided the limit exists (Figure 7.80a). If $f(x)$ is continuous at every point in the interval $a \le x \le b$ except $x = b$, then

$$\int_a^b f(x)\,dx = \lim_{c \to b^-} \int_a^c f(x)\,dx, \tag{7.48}$$

provided the limit exists (Figure 7.80b). If $f(x)$ is continuous at every point in the interval $a \le x \le b$ except $x = d$ where $a < d < b$, then

$$\int_a^b f(x)\,dx = \lim_{c \to d^-} \int_a^c f(x)\,dx + \lim_{c \to d^+} \int_c^b f(x)\,dx, \tag{7.49}$$

provided both limits exist (Figure 7.80c).

EXAMPLE 7.36

Determine whether the following improper integrals converge or diverge:

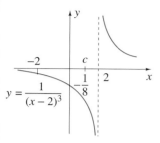

FIGURE 7.81

$$\text{(a)} \quad \int_{-2}^{2} \frac{1}{(x-2)^3}\,dx \qquad \text{(b)} \quad \int_{-1}^{1} \frac{1}{x^4}\,dx \qquad \text{(c)} \quad \int_{1}^{5} \frac{x}{\sqrt{x-1}}\,dx$$

SOLUTION

(a) The graph of the integrand (Figure 7.81) shows a discontinuity at $x = 2$. Hence,

$$\int_{-2}^{2} \frac{1}{(x-2)^3}\,dx = \lim_{c \to 2^-} \int_{-2}^{c} \frac{1}{(x-2)^3}\,dx = \lim_{c \to 2^-} \left\{ \frac{-1}{2(x-2)^2} \right\}_{-2}^{c}$$

$$= \lim_{c \to 2^-} \left[\frac{-1}{2(c-2)^2} + \frac{1}{32} \right] = -\infty.$$

The improper integral therefore diverges.

(b) Since the discontinuity is interior to the interval of integration (Figure 7.82), we use equation 7.49:

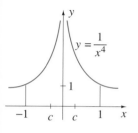

FIGURE 7.82

$$\int_{-1}^{1} \frac{1}{x^4}\,dx = \lim_{c \to 0^-} \int_{-1}^{c} \frac{1}{x^4}\,dx + \lim_{c \to 0^+} \int_{c}^{1} \frac{1}{x^4}\,dx$$

$$= \lim_{c \to 0^-} \left\{ \frac{-1}{3x^3} \right\}_{-1}^{c} + \lim_{c \to 0^+} \left\{ \frac{-1}{3x^3} \right\}_{c}^{1}$$

$$= \lim_{c \to 0^-} \left[-\frac{1}{3} - \frac{1}{3c^3} \right] + \lim_{c \to 0^+} \left[\frac{1}{3c^3} - \frac{1}{3} \right].$$

Since neither of these limits exists, the improper integral diverges.

(c) Since the integrand is discontinuous only at $x = 1$ (Figure 7.83),

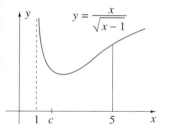

FIGURE 7.83

$$\int_{1}^{5} \frac{x}{\sqrt{x-1}}\,dx = \lim_{c \to 1^+} \int_{c}^{5} \frac{x}{\sqrt{x-1}}\,dx.$$

If we set $u = x - 1$, then $du = dx$, and

$$\int \frac{x}{\sqrt{x-1}}\,dx = \int \frac{u+1}{\sqrt{u}}\,du = \int (u^{1/2} + u^{-1/2})\,du$$

$$= \frac{2}{3}u^{3/2} + 2u^{1/2} + C = \frac{2}{3}(x-1)^{3/2} + 2\sqrt{x-1} + C.$$

Thus,

$$\int_{1}^{5} \frac{x}{\sqrt{x-1}}\,dx = \lim_{c \to 1^+} \left\{ \frac{2}{3}(x-1)^{3/2} + 2\sqrt{x-1} \right\}_{c}^{5}$$

$$= \lim_{c \to 1^+} \left[\frac{28}{3} - \frac{2}{3}(c-1)^{3/2} - 2\sqrt{c-1} \right] = \frac{28}{3},$$

and the improper integral converges.

Various types of improper integrals may occur in the same problem. For example, Figure 7.84 shows that the integral of $f(x) = 1/(x^2 - 1)$ from $x = -3$ to $x = \infty$ involves the use of five limits:

$$\int_{-3}^{\infty} \frac{1}{x^2 - 1}\, dx = \lim_{c \to -1^-} \int_{-3}^{c} \frac{1}{x^2 - 1}\, dx + \lim_{c \to -1^+} \int_{c}^{0} \frac{1}{x^2 - 1}\, dx$$

$$+ \lim_{c \to 1^-} \int_{0}^{c} \frac{1}{x^2 - 1}\, dx + \lim_{c \to 1^+} \int_{c}^{10} \frac{1}{x^2 - 1}\, dx$$

$$+ \lim_{b \to \infty} \int_{10}^{b} \frac{1}{x^2 - 1}\, dx.$$

If just one of these limits fails to exist, then the improper integral diverges.

FIGURE 7.84

FIGURE 7.85

If an integrand has a finite discontinuity, such as the function $f(x)$ in Figure 7.85, then the improper integral of $f(x)$ from $x = a$ to $x = b$ is defined in terms of two limits,

$$\int_{a}^{b} f(x)\, dx = \lim_{c \to d^-} \int_{a}^{c} f(x)\, dx + \lim_{c \to d^+} \int_{c}^{b} f(x)\, dx,$$

but there is no question in this case that the improper integral converges.

EXERCISES 7.10

In Exercises 1–18 determine whether the improper integral converges or diverges. Find the value for each convergent integral.

∗ **1.** $\displaystyle\int_{3}^{\infty} \frac{1}{(x + 4)^2}\, dx$

∗ **2.** $\displaystyle\int_{3}^{\infty} \frac{1}{(x + 4)^{1/3}}\, dx$

∗ **3.** $\displaystyle\int_{-\infty}^{-4} \frac{x}{\sqrt{x^2 - 2}}\, dx$

∗ **4.** $\displaystyle\int_{-\infty}^{-4} \frac{x}{(x^2 - 2)^4}\, dx$

∗ **5.** $\displaystyle\int_{-\infty}^{\infty} \frac{10^{10} x^3}{(x^4 + 5)^2}\, dx$

∗ **6.** $\displaystyle\int_{-\infty}^{\infty} \frac{x^3}{(x^4 + 5)^{1/4}}\, dx$

* **7.** $\displaystyle\int_0^1 \frac{1}{(1-x)^{5/3}}\,dx$

* **8.** $\displaystyle\int_0^1 \frac{1}{\sqrt{1-x}}\,dx$

* **9.** $\displaystyle\int_1^\infty x\sqrt{x^2-1}\,dx$

* **10.** $\displaystyle\int_2^5 \frac{x}{\sqrt{x^2-4}}\,dx$

* **11.** $\displaystyle\int_{-1}^1 \frac{x}{(1-x^2)^2}\,dx$

* **12.** $\displaystyle\int_{-\infty}^\infty \frac{1}{x^2}\,dx$

* **13.** $\displaystyle\int_0^\infty \frac{1}{\sqrt{x}}\,dx$

* **14.** $\displaystyle\int_{-\infty}^{\pi/2} \frac{x}{(x^2-4)^2}\,dx$

* **15.** $\displaystyle\int_4^\infty \cos x\,dx$

* **16.** $\displaystyle\int_{-\infty}^\infty \sin x\,dx$

** **17.** $\displaystyle\int_0^\infty \frac{x}{\sqrt{x+3}}\,dx$

** **18.** $\displaystyle\int_2^3 \frac{x^3}{\sqrt{x^2-4}}\,dx$

** **19.** If $f(x)$ is a continuous odd function, is it necessarily true that

$$\int_{-\infty}^\infty f(x)\,dx = 0?$$

** **20.** (a) Is it possible to assign a number to represent the area bounded by the curves

$$y = 1 - x^{-1/4}, \qquad y = 1, \qquad x = 1\,?$$

   (b) If the region in (a) is rotated about the line $y = 1$, is it possible to assign a number to represent its volume?

** **21.** Repeat Exercise 20 if $y = 1 - x^{-1/4}$ is replaced by $y = 1 - x^{-2/3}$.

** **22.** Repeat Exercise 20 if $y = 1 - x^{-1/4}$ is replaced by $y = 1 - x^{-3}$.

** **23.** A function $f(x)$ qualifies as a *probability density function* (pdf) on the interval $x \geq 0$ if it satisfies two conditions: $f(x) \geq 0$ for $x \geq 0$, and

$$\int_0^\infty f(x)\,dx = 1.$$

   (a) Show that each of the following functions qualifies as a pdf:

$$f(x) = \frac{6x}{(1+3x^2)^2}; \qquad f(x) = \frac{2x}{(1+x)^3}$$

   (b) If a variable x has pdf $f(x)$ defined for $x \geq 0$, then the probability that x lies in an interval I is the definite integral of $f(x)$ over I. In particular, the probability that x is greater than or equal to a is

$$P(x \geq a) = \int_a^\infty f(x)\,dx.$$

   Calculate $P(x \geq 3)$ for each pdf in part (a).

** **24.** Verify that

$$\int_1^\infty \frac{1}{x^p}\,dx$$

converges if $p > 1$ and diverges if $p \leq 1$.

The key, then, to the use of the definite integral is to divide the region into smaller elements, calculate the required quantity for a representative element, and then use a definite integral to add over all elements.

An important point to keep in mind is that for lengths in the y-direction, we always take "upper y minus lower y," and for lengths in the x-direction, we use "larger x minus smaller x." This way many sign errors can be avoided when calculating *lengths*. For instance, note that this rule is used to find the length of a rectangle (in areas, volumes, fluid pressure), the radius of a circle (in volumes, surface area), depth (fluid pressure), and some distances (work).

An improper integral is a definite integral with an infinite limit and/or a point of discontinuity in the interval of integration. The value of an improper integral is defined by first calculating it over a finite interval in which the integrand is continuous, and then letting the limit on the integral approach either infinity or the point of discontinuity.

KEY TERMS

In reviewing this chapter, you should be able to define or discuss the following key terms:

Area	First moment of mass
Representative rectangle	First moment of area
Volume of solid of revolution	Centre of mass
Disc method	Centroid
Washer method	Moment of inertia
Cylindrical shell method	Second moment
Length of a curve	Volume by slicing
Work	Area of surface of revolution
Fluid pressure	Improper integral
Fluid force	

REVIEW EXERCISES

In Exercises 1–5, calculate for the region bounded by the curves: (a) its area; (b) the volumes of the solids of revolution obtained by rotating the region about the x- and y-axes; (c) its centroid; and (d) its second moments of area about the x- and y-axes.

∗ **1.** $y = 9 - x^2$, $y = 0$

∗ **2.** $y = x^3$, $y = 2 - x$, $x = 0$

∗ **3.** $x = y^2 - 1$, $x = 4 - 4y^2$

∗ **4.** $2y = x$, $y + 1 = x$, $y = 0$

∗∗ **5.** $2x + y = 2$, $y - 2x = 2$, $x = y + 1$, $y + x + 1 = 0$

In Exercises 6–10 find the volume of the solid of revolution obtained by rotating the area bounded by the curves about the line.

∗ **6.** $y = |x| + 2$, $y = 3$,   about $y = 1$

∗ **7.** $x = y^4$, $x = 4$,   about $x = 4$

∗ **8.** $y = x^3 - x$, $y = 0$,   about $y = 0$

∗ **9.** $yx^2 = 1$, $x = 1$, $x = 2$, $y = 0$ about $y = -1$

∗∗ **10.** $y = x + 2$, $y = \sqrt{4 - 2x}$, $y = 0$,   about $y = 1$

In Exercises 11–15 find the first and second moments of area of the region bounded by the curves about the line.

* **11.** $y = 4x^2$, $y = 2 + 2x^2$,   about $y = -1$

* **12.** $2x = \sqrt{y}$, $\sqrt{2}x = \sqrt{y - 2}$, $x = 0$,   about $x = 1$

* **13.** $3y = 2x + 6$, $2y + x = 4$, $y = 0$,   about $y = -2$

* **14.** $3y = 2x + 6$, $2y + x = 4$, $y = 0$,   about $y = 2$

** **15.** $y = x^2 - 2x$ $(y \leq 0)$, $y = 2x$, $y = 4 - 2x$,   about $x = 1$

* **16.** A water tank in the form of a right circular cylinder with radius 1 m and height 3 m has its axis vertical. If it is half full, how much work is required to empty the tank through an outlet at the top of the tank?

** **17.** A car runs off a bridge into a river and submerges. If the tops of its front and rear side windows are 2 m below the surface of the water (figure below), what is the force due to water pressure on each window?

* **18.** Find the surface area of the volume of the solid of revolution obtained by rotating the area bounded by the curves

$$2x + y = 2, \qquad y - 2x = 2, \qquad y = 0$$

about the x-axis.

** **19.** A tank in the form of an inverted right circular cone of depth 10 m and radius 4 m is full of water. How much work is required to lower the water level in the tank by 2 m by pumping water out through an outlet 1 m above the top of the tank?

** **20.** Show that if it takes W units of work to stretch a spring from equilibrium to a certain stretch, then it requires $3W$ more units of work to double that stretch.

In Exercises 21–28 determine whether the improper integral converges or diverges. Evaluate each convergent integral.

** **21.** $\displaystyle \int_1^\infty \frac{1}{x^{3/2}}\, dx$

** **22.** $\displaystyle \int_0^3 \frac{1}{\sqrt{3 - x}}\, dx$

** **23.** $\displaystyle \int_{-1}^0 \frac{1}{(x + 1)^2}\, dx$

** **24.** $\displaystyle \int_{-2}^2 \frac{x}{\sqrt{4 - x^2}}\, dx$

** **25.** $\displaystyle\int_{-\infty}^{\infty} \frac{1}{(x+3)^3}\, dx$ ** **26.** $\displaystyle\int_{-\infty}^{-3} \frac{1}{\sqrt{-x}}\, dx$

** **27.** $\displaystyle\int_{-6}^{\infty} x\sqrt{x^2+4}\, dx$ ** **28.** $\displaystyle\int_{-\infty}^{\infty} \frac{x}{(x^2-1)^2}\, dx$

** **29.** A uniformly tapered rod of length 1 m has circular cross-sections, the radii of its ends being 1 cm and 2 cm. Find the volume of the rod by:

   (a) using the fact that every cross-section is circular;

   (b) considering the rod as a volume of a solid of revolution.

** **30.** In Exercise 16, how much work is required to empty the tank if its axis is horizontal and the outlet is at the top of the tank?

Further Transcendental Functions and Their Derivatives

In Chapter 3 we differentiated functions that involve a finite number of additions, subtractions, multiplications, divisions, and roots. They are examples of what are called *algebraic functions*. A function $y = f(x)$ is said to be **algebraic** if, for all x in its domain, it satisfies an equation of the form

$$P_0(x)y^n + P_1(x)y^{n-1} + \cdots + P_{n-1}(x)y + P_n(x) = 0, \qquad (8.1)$$

where $P_0(x), \ldots, P_n(x)$ are polynomials in x, and n is a positive integer. A polynomial $P(x)$ is itself algebraic since it satisfies $y - P(x) = 0$; that is, it satisfies 8.1 with $n = 1$, $P_0(x) = 1$, and $P_1(x) = -P(x)$. Rational functions $P(x)/Q(x)$ are also algebraic ($n = 1$, $P_0(x) = Q(x)$, and $P_1(x) = -P(x)$). The function $f(x) = x^{1/3}$ is algebraic since it satisfies $y^3 - x = 0$. The equation $y^3 + y = x$ defines y implicitly as a function of x. We cannot find an explicit form for the function, but according to 8.1, the function so defined is algebraic.

A function that is not algebraic is called **transcendental**. The trigonometric, exponential, and logarithm functions which we differentiated in Chapter 3 are transcendental. In this chapter we consider two other classes of transcendental functions: inverse trigonometric and hyperbolic functions.

8.1 Inverse Functions

When we speak of a function $y = g(x)$, it is inherent that there is a unique y associated with each x; that is, given a value of x in the domain of $g(x)$, the function associates one—and only one—value of y in the range. However, it may happen, and quite often does, that a value of y in the range of the function may be associated with more than one value of x. For example, each $y > 0$ in the range of the function $y = g(x) = x^2$ (Figure 8.1) is associated with two values of x, namely, $x = \pm\sqrt{y}$.

FIGURE 8.1

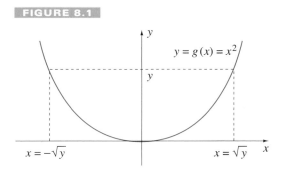

Some functions have the property that each value of y in the range arises from only one x in the domain. For instance, given any value y in the range of the function $y = f(x) = x^3$ in Figure 8.2, there is a unique x such that $y = x^3$, namely, $x = y^{1/3}$. Such a function is said to be "one-to-one." Formally, we say that a function $f(x)$ is **one-to-one** if, for any two values x_1 and x_2 in the domain of $f(x)$, it follows that $f(x_1) \neq f(x_2)$. To repeat, a one-to-one function $f(x)$ has the property that given any y in its range, there is one—and only one—x in its domain for which $y = f(x)$. We can therefore define a function which maps values in the range of $f(x)$ onto values in the domain, a function that maps y onto x if x is mapped by $f(x)$ onto y. We call this function the inverse function of $f(x)$. It reverses the action of $f(x)$. For the function $f(x) = x^3$ in Figure 8.2, the inverse function is the function that takes cube roots of real numbers.

FIGURE 8.2

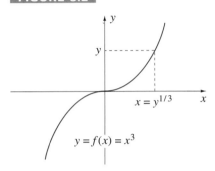

$x = y^{1/3}$

$y = f(x) = x^3$

Unless there is a good reason to do otherwise, it is our custom to denote the independent variable of a function by the letter x and the dependent variable by y. For the inverse function of a function $y = f(x)$, there is a natural tendency to denote the independent variable by y and the dependent variable by x since the inverse function maps the range of $f(x)$ onto its domain. Not for obvious reasons, it is usually, but not always, better to maintain our usual practice of using x as independent variable and y as dependent variable even for inverse functions. When a function $f(x)$ has an inverse function, we adopt the notation $f^{-1}(x)$ to represent the inverse function. For example, when $f(x) = x^3$, the inverse function is $f^{-1}(x) = x^{1/3}$. Be careful with this notation. Do not interpret the "-1" as a power and write $f^{-1}(x)$ as $1/f(x)$. This is not correct. The notation f^{-1} represents a function, just as tan represents the tangent function and $\sqrt{}$ represents the positive square root function.

The inverse function $f^{-1}(x)$ of a function $f(x)$ "undoes" what $f(x)$ "does"; it reverses the effect of $f(x)$. For example, the function $f(x) = x^2$, $x \geq 0$ in Figure 8.3 is one-to-one; it squares nonnegative numbers. Its inverse is $f^{-1}(x) = \sqrt{x}$, the positive square root function (Figure 8.4). Squaring an x and then taking the positive square root of the result returns the original x.

FIGURE 8.3

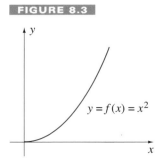

$y = f(x) = x^2$

FIGURE 8.4

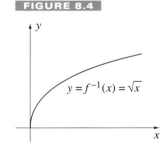

$y = f^{-1}(x) = \sqrt{x}$

FIGURE 8.5   FIGURE 8.6

Likewise, the inverse of $g(x) = x^2$, $x \leq 0$ (Figure 8.5) is the negative square root function $g^{-1}(x) = -\sqrt{x}$ (Figure 8.6).

The fact that $f^{-1}(x)$ undoes $f(x)$ can be stated algebraically as follows. For each x in the domain of $f(x)$,

$$f^{-1}\big(f(x)\big) = x. \tag{8.2}$$

This is the defining relation for an inverse function. We understand that the domain of $f^{-1}(x)$ is the same as the range of $f(x)$ so that $f^{-1}(x)$ operates on all outputs of $f(x)$.

EXAMPLE 8.1

What is the inverse function of $f(x) = (x + 1)/(x - 2)$?

SOLUTION   To find the algebraic definition of $f^{-1}(x)$, we solve $y = (x + 1)/(x - 2)$ for x in terms of y. First we cross-multiply,

$$x + 1 = y(x - 2);$$

then group terms in x,

$$x(y - 1) = 2y + 1;$$

and finally divide by $y - 1$,

$$x = \frac{2y + 1}{y - 1}.$$

What have we accomplished by solving for x in terms of y? If we take an x in the domain of $f(x)$, then $f(x)$ produces $y = (x + 1)/(x - 2)$. Now this same x and y satisfy the equation $x = (2y+1)/(y-1)$ because this equation is simply a rearrangement of $y = (x+1)/(x-2)$. Consequently, if we substitute y into the right side of the equation,

$$x = \frac{2y + 1}{y - 1},$$

we obtain the original x. This equation must therefore define the inverse function of $f(x)$. In other words, if we denote the independent variable by x, the inverse function is

$$f^{-1}(x) = \frac{2x + 1}{x - 1}.$$

Example 8.1 has illustrated that to find the inverse of a function $y = f(x)$, the equation should be solved for x in terms of y, and then variables should be renamed. If the equation does not have a unique solution for x in terms of y, then $f(x)$ does not have an inverse function. Such is the case for the function $g(x) = x^2$ in Figure 8.1. Solving $y = g(x) = x^2$ for x gives two solutions $x = \pm\sqrt{y}$.

So far our discussion of inverse functions has been algebraic. The geometry of inverse functions is most revealing. You may have noticed a relationship between the graphs of $f(x)$ and $f^{-1}(x)$ in Figures 8.3 and 8.4, and of $g(x)$ and $g^{-1}(x)$ in Figures 8.5 and 8.6. The inverse function is the mirror image of the function in the line $y = x$; that is, graphs of inverse pairs are symmetric about the line $y = x$. These two examples are not mere coincidence, but as we now show, graphs of inverse functions are always mirror images of each other in the line $y = x$.

Suppose $f^{-1}(x)$ is the inverse of $f(x)$, and that (x, y) is any point on the graph of $y = f(x)$. Since $f^{-1}(x)$ is the inverse of $f(x)$, the result of substituting y into $f^{-1}(x)$ is x; that is, the point (y, x) is on the graph of $f^{-1}(x)$. But the point (y, x) is the mirror image of (x, y) in the line $y = x$. When this is combined with the fact that the domain of $f^{-1}(x)$ is the range of $f(x)$, our conjecture is verified.

This suggests a very simple way to determine graphically the inverse of a function $f(x)$: take its mirror image in the line $y = x$. Note, too, that if the mirror image does not represent a function, then no inverse function for $f(x)$ exists. For example, the mirror image of $y = x^2$, $-\infty < x < \infty$, in the line $y = x$ is shown in Figure 8.7 and does not represent a function. We conclude as before that this function does not have an inverse.

If $y = f^{-1}(x)$ is the mirror image of $y = f(x)$ in the line $y = x$, then $y = f(x)$ is the mirror image of $y = f^{-1}(x)$. This means that $f(x)$ is the inverse of $f^{-1}(x)$, and that $f(x)$ undoes what $f^{-1}(x)$ does:

$$f\left(f^{-1}(x)\right) = x. \tag{8.3}$$

Geometrically, a function $f(x)$ has an inverse if the reflection of its graph $y = f(x)$ in the line $y = x$ represents a function. But we know that a curve represents a function if every vertical line that intersects it does so at exactly one point. Furthermore, a vertical line intersects the reflected curve at exactly one point only if its horizontal reflection intersects $y = f(x)$ at exactly one point. These two facts enable us to state that a function $f(x)$ has an inverse function if—and only if—every horizontal line that intersects it does so at exactly one point. This is a geometric interpretation of a function being one-to-one. See, for example, the functions in Figures 8.3 and 8.5. Horizontal lines that intersect the curves do so at exactly one point. The function in Figure 8.1, which has no inverse, is intersected in two points by every horizontal line $y = c > 0$. When a function is increasing on an interval I (Figure 8.8), it passes the horizontal line test (and is one-to-one). Likewise, a function which is decreasing on I (Figure 8.9) is one-to-one. A function which is either increasing on an interval I or decreasing on I is said to be **monotonic** on I. What we have shown is the following result.

FIGURE 8.7

Mirror image of
$y = x^2$ in $y = x$

FIGURE 8.8

FIGURE 8.9

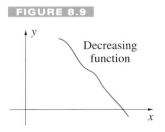

> **THEOREM 8.1**
>
> A function that is monotonic on an interval has an inverse function on that interval.

It is important to realize that monotonicity is a sufficient condition for existence of an inverse function; that is, if a function is monotonic, then it has an inverse. It is not, however, a necessary condition. The function in Figure 8.10a is not monotonic on the interval $-1 \leq x \leq 1$, but it does have the inverse function in Figure 8.10b.

FIGURE 8.10a

FIGURE 8.10b

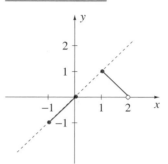

EXAMPLE 8.2

Does the function $f(x) = x^2 - 2x$ have an inverse? Does it have an inverse on the interval $x \geq 1$, and on the interval $x \leq 1$?

SOLUTION   The graph of $f(x)$ in Figure 8.11 indicates that $f(x)$ does not have an inverse. When restricted to the interval $x \geq 1$, the function is one-to-one, and does have an inverse. To find it, we solve $y = x^2 - 2x$ for x in terms of y. The quadratic formula applied to

$$x^2 - 2x - y = 0$$

gives

$$x = \frac{2 \pm \sqrt{4 + 4y}}{2} = 1 \pm \sqrt{y + 1}.$$

Since x must be greater than or equal to one, we must choose $x = 1 + \sqrt{y + 1}$. Hence, the inverse of $f(x) = x^2 - 2x$, $x \geq 1$ is

$$f^{-1}(x) = 1 + \sqrt{x + 1}.$$

Similarly, when restricted to the interval $x \leq 1$, $f(x)$ has inverse

$$f^{-1}(x) = 1 - \sqrt{x + 1}.$$

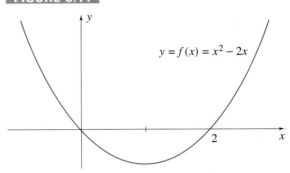

FIGURE 8.11

$y = f(x) = x^2 - 2x$

This example illustrates that when a function is not monotonic on an interval I, the interval can usually be subdivided into subintervals on which the function is monotonic, and on each such subinterval the function has an inverse.

EXERCISES 8.1

In Exercises 1–14 determine graphically whether the function has an inverse. Find each inverse function.

∗ **1.** $f(x) = 2x + 3$

∗ **2.** $f(x) = \sqrt{x + 1}$

∗ **3.** $f(x) = x^2 + x$

∗ **4.** $f(x) = \dfrac{x + 5}{2x + 4}$

∗ **5.** $f(x) = 1/x$

∗ **6.** $f(x) = 3x^3 + 2$

∗ **7.** $f(x) = \sqrt{4 - x^2}, \quad 0 \le x \le 2$

∗ **8.** $f(x) = 2x + |x|$

∗ **9.** $f(x) = x + |x|$

∗ **10.** $f(x) = \sqrt{1 - x^2}$

∗ **11.** $f(x) = x^4 + 2x^2 + 2, \quad x \le 0$

∗ **12.** $f(x) = x^2 - 2x + 4, \quad x \ge 1$

∗ **13.** $f(x) = \left(\dfrac{x + 2}{x - 2}\right)^3$

∗ **14.** $f(x) = \dfrac{x}{3 + x^2}$

In Exercises 15–20 show that the function does not have an inverse. Subdivide its domain of definition into subintervals on which the function has an inverse, and find the inverse function on each subinterval.

∗∗ **15.** $f(x) = x^4$

∗∗ **16.** $f(x) = 1/x^4$

∗∗ **17.** $f(x) = x^2 + 2x + 3$

∗∗ **18.** $f(x) = x^4 + 4x^2 + 2$

∗∗ **19.** $f(x) = \dfrac{x^2}{x^2 + 4}$

∗∗ **20.** $f(x) = \dfrac{x^4}{x^2 + 4}$

** **21.** If a manufacturing firm sells x objects of a certain commodity per week, it sells them at a price of r per object, and r depends on x, $r = f(x)$. In economic theory this function is usually considered decreasing, as shown in the left figure below; hence it has an inverse function $x = f^{-1}(r)$. In this function x depends on r, indicating that if the price of the object is set at r, then the market will demand x of them per week. This function is therefore called the *demand function* (right figure below). Find demand functions if

(a) $r = \dfrac{a}{x+b} + c$, a, b, and c positive constants; and

(b) $r = \dfrac{a}{x^2+b} + c$, a, b, and c positive constants.

Draw the demand function $x = f^{-1}(r)$ and given function $r = f(x)$ in each case.

 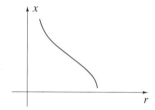

** **22.** Show that the demand function

$$x = f(r) = 4a^3 - 3ar^2 + r^3, \quad 0 < r < 2a$$

has an inverse function $r = f^{-1}(x)$. What is the domain of $f^{-1}(x)$? Draw its graph.

** **23.** Find the inverse function for

$$f(x) = \frac{x^2}{(1+x)^2}, \quad -1 < x \le 0.$$

8.2 The Inverse Trigonometric Functions

The function $f(x) = \sin x$, defined for all real x, does not have an inverse. This is evident from the fact that the function is not one-to-one, or that it fails the horizontal line test. By restricting the domain of $\sin x$, however, the function can be made one-to-one, and this can be done in many ways. In particular, that part of $\sin x$ on the interval $-\pi/2 \le x \le \pi/2$ is one-to-one, and therefore has an inverse function. This function, denoted by

$$y = \operatorname{Sin}^{-1} x, \tag{8.4}$$

and called the **inverse sine function**, is shown in Figure 8.12. The range of the function is

$$-\frac{\pi}{2} \le \operatorname{Sin}^{-1} x \le \frac{\pi}{2}. \tag{8.5}$$

The values that fall in this range are called **principal values** of the inverse sine function. They have resulted from our restriction of the domain of $\sin x$ to this same interval.

FIGURE 8.12

FIGURE 8.13

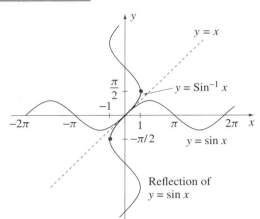

An equivalent way to derive the inverse sine function is as follows. The reflection of the graph of $y = \sin x$ in the line $y = x$ does not represent a function, but by restricting the range of values of the reflected curve, we can produce a single-valued function (Figure 8.13). Once again this can be done in many ways, and when we do so by restricting the y-values to the interval $[-\pi/2, \pi/2]$, the resulting function is called the inverse sine function $\text{Sin}^{-1} x$. Note very carefully that $\text{Sin}^{-1} x$ is not the inverse function of $f(x) = \sin x$, because the sine function has no inverse. It is, however, the inverse of the sine function restricted to the domain $-\pi/2 \leq x \leq \pi/2$. It is perhaps then a misnomer to call the function $\text{Sin}^{-1} x$ the inverse sine function, but this has become the accepted terminology.

We now know what the inverse sine function looks like graphically, but what does it mean to say that $y = \text{Sin}^{-1} x$? Certainly given any value of x, we can push a few buttons on an electronic calculator and find $\text{Sin}^{-1} x$ for that x. To use inverse trigonometric functions in practice, we must have a feeling for what they do. To obtain this insight, we note that if (x, y) is a point on the curve $y = \text{Sin}^{-1} x$, then (y, x) is a point on the sine curve; that is,

$$y = \text{Sin}^{-1} x \quad \text{only if} \quad x = \sin y. \tag{8.6}$$

In the latter equation y is an angle and x is the sine of that angle. Thus, when we see $y = \text{Sin}^{-1} x$, we should not read "y equals inverse sine x," but rather "y is an angle whose sine is x." For instance, if $x = 1/2$, then $y = \text{Sin}^{-1}(1/2)$ is read "y is an angle whose sine is $1/2$." Clearly the angle whose sine is $1/2$ is $\pi/6$, and we write $y = \text{Sin}^{-1}(1/2) = \pi/6$. Remember that we must choose the principal value $\pi/6$; there are many angles that have a sine equal to $1/2$, but the inverse sine function demands that we choose that angle in the range $-\pi/2 \leq y \leq \pi/2$, and state it with a number in this interval. For example, $2\pi + \pi/6$ may represent the same angle with the positive x-axis as $\pi/6$, but the number $2\pi + \pi/6$ does not lie between $-\pi/2$ and $\pi/2$. In summary, the function $\sin x$ regards x as an angle and assigns to x the sine of the angle; the inverse sine function $\text{Sin}^{-1} x$ regards x as the sine of an angle and assigns to x the angle with that sine.

Another notation that is commonly used for the inverse sine function is $\arcsin x$. One reason the notation $\arcsin x$ is preferable to $\text{Sin}^{-1} x$ is the possible misinterpretation of $\text{Sin}^{-1} x$. Sometimes students regard the "-1" as a power and write $\text{Sin}^{-1} x$ as $1/\sin x$. This is *not* correct, a fact that you were warned about in Section 8.1. Sin^{-1} is the name of a function, just as sin is the name of the sine function, and $\sqrt{\ }$ is the notation for the positive square root function. The capital S in Sin^{-1} should also warn you that this is the inverse sine function.

EXAMPLE 8.3

Simplify each of the following expressions:

(a) $\mathrm{Sin}^{-1}\,(-\sqrt{3}/2)$ (b) $\mathrm{Sin}^{-1}\,(1)$ (c) $\mathrm{Sin}^{-1}\,(3)$

(d) $\mathrm{Sin}^{-1}\,(3/5) + \mathrm{Sin}^{-1}\,(4/5)$ (e) $\sin\left[\mathrm{Sin}^{-1}\,(\sqrt{3}/2)\right]$

SOLUTION

(a) $\mathrm{Sin}^{-1}\,(-\sqrt{3}/2)$ asks for the angle whose sine is equal to $-\sqrt{3}/2$. Clearly,

$$\mathrm{Sin}^{-1}\,(-\sqrt{3}/2) = -\pi/3.$$

(b) $\mathrm{Sin}^{-1}\,(1) = \pi/2$.

(c) $\mathrm{Sin}^{-1}\,(3)$ is not defined since the domain of $\mathrm{Sin}^{-1}\,x$ is $-1 \le x \le 1$.

(d)   If $\phi = \mathrm{Sin}^{-1}\,(3/5)$, then ϕ is illustrated in the triangle in Figure 8.14. Since the third side must have length 4, it follows that $\mathrm{Sin}^{-1}\,(4/5) = \pi/2 - \phi$, and

$$\mathrm{Sin}^{-1}\left(\frac{3}{5}\right) + \mathrm{Sin}^{-1}\left(\frac{4}{5}\right) = \phi + \left(\frac{\pi}{2} - \phi\right) = \frac{\pi}{2}.$$

(e) $\sin\left[\mathrm{Sin}^{-1}\,(\sqrt{3}/2)\right] = \sin\,(\pi/3) = \sqrt{3}/2$.

FIGURE 8.14

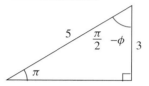

EXAMPLE 8.4

Find the values of x for which the following are valid:

(a) $\sin\left(\mathrm{Sin}^{-1}\,x\right) = x$ (b) $\mathrm{Sin}^{-1}\,(\sin x) = x$

SOLUTION   These two equations express the fact that the sine function and the inverse sine function are inverses, provided we are careful about domains:

(a) The equation $\sin\left(\mathrm{Sin}^{-1}\,x\right) = x$ is valid for $-1 \le x \le 1$. Given an x in this interval, $\mathrm{Sin}^{-1}\,x$ finds that angle (in the principal value range) which has x as its sine. Then $\sin\left(\mathrm{Sin}^{-1}\,x\right)$ takes the sign of this angle. Naturally it returns the original number x.

(b) The composite function $\mathrm{Sin}^{-1}\,(\sin x)$ is defined for all x, but only on the domain $-\pi/2 \le x \le \pi/2$ is it equal to x.

Our analysis of the inverse sine function is now complete. We could repeat a similar discussion five times, once for each of the inverse functions associated with the other five trigonometric functions. However, we shall give an abbreviated version for the inverse cosine function and tabulate results for the remaining four functions.

The reflection of the graph of the function $f(x) = \cos x$ is shown in Figure 8.15, and it does not represent a function. If we restrict the y-values on the reflected curve to

$$0 \leq y \leq \pi, \tag{8.7}$$

then we do obtain a function called the inverse cosine function, denoted by

$$y = \cos^{-1} x. \tag{8.8}$$

The values in equation 8.7 are called the principal values of the inverse cosine function. Note again that $\cos^{-1} x$ is not the inverse function of $\cos x$, but of $f(x) = \cos x,\ 0 \leq x \leq \pi$.

FIGURE 8.15

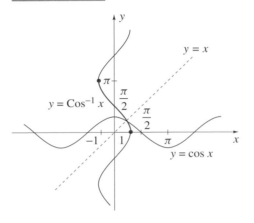

When we write $y = \cos^{-1} x$, we do not read this as "y equals inverse cosine x," but rather as "y is an angle whose cosine is x," for if $y = \cos^{-1} x$, then $x = \cos y$.

The remaining four inverse trigonometric functions, along with those of inverse sine and cosine, are shown in Figure 8.16. Black and blue curves represent reflections in the line $y = x$ of the trigonometric functions $\tan x$, $\cot x$, $\csc x$, and $\sec x$. Blue curves represent principal values of the inverse trigonometric functions $\tan^{-1} x$, $\cot^{-1} x$, $\csc^{-1} x$, and $\sec^{-1} x$. Principal values of the six inverse trigonometric functions are listed in Table 8.1.

TABLE 8.1

Inverse Trigonometric Function	Principal Values
$\sin^{-1} x$	$-\dfrac{\pi}{2} \leq y \leq \dfrac{\pi}{2}$
$\tan^{-1} x$	$-\dfrac{\pi}{2} < y < \dfrac{\pi}{2}$
$\cos^{-1} x$	$0 \leq y \leq \pi$
$\cot^{-1} x$	$0 < y < \pi$
$\csc^{-1} x$	$-\pi < y \leq -\dfrac{\pi}{2},\ 0 < y \leq \dfrac{\pi}{2}$
$\sec^{-1} x$	$-\pi \leq y < -\dfrac{\pi}{2},\ 0 \leq y < \dfrac{\pi}{2}$

FIGURE 8.16a

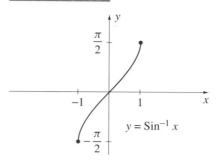

$y = \text{Sin}^{-1} x$

FIGURE 8.16b

$y = \text{Cos}^{-1} x$

FIGURE 8.16c

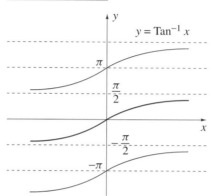

$y = \text{Tan}^{-1} x$

FIGURE 8.16d

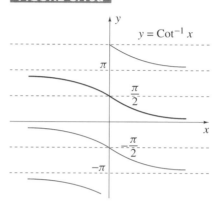

$y = \text{Cot}^{-1} x$

FIGURE 8.16e

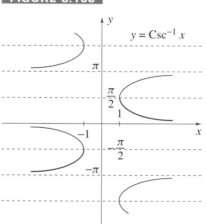

$y = \text{Csc}^{-1} x$

FIGURE 8.16f

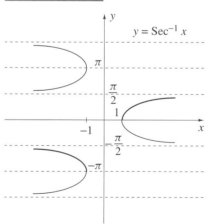

$y = \text{Sec}^{-1} x$

It is reasonable to ask why the principal values of $\text{Sec}^{-1} x$ were not chosen as $0 \le y < \pi/2$, $\pi/2 < y \le \pi$. Had they been chosen in this way, they would have been very similar to those of $\text{Cos}^{-1} x$ and $\text{Cot}^{-1} x$. Likewise, why are the principal values of $\text{Csc}^{-1} x$ not $-\pi/2 \le y < 0$, $0 < y \le \pi/2$? The answer is that they could have been selected in this way, and some authors do indeed make this choice; it is simply a matter of preference. Each choice does, however, create corresponding changes in later work. Specifically, in the exercises of Section 8.3, if principal values of $\text{Sec}^{-1} x$ and $\text{Csc}^{-1} x$ are selected in this alternative way, then derivatives of these functions are modified correspondingly.

In the remainder of this section we solve problems that make use of inverse trigonometric functions.

EXAMPLE 8.5

Find all solutions of the equation

$$\sin^2 x + 3\sin x = 2.$$

SOLUTION   The quadratic formula applied to the equation

$$(\sin x)^2 + 3(\sin x) - 2 = 0$$

yields

$$\sin x = \frac{-3 \pm \sqrt{9+8}}{2} = \frac{-3 \pm \sqrt{17}}{2}.$$

Since $\sin x$ can take on only values in the interval $-1 \le \sin x \le 1$, the possibility that $\sin x = (-3 - \sqrt{17})/2$ must be rejected, leaving

$$\sin x = \frac{-3 + \sqrt{17}}{2}.$$

One solution of this equation is

$$x = \arcsin\left(\frac{\sqrt{17}-3}{2}\right) = 0.60 \text{ radians.}$$

This is not the only solution, however; there are many angles with a sine equal to $(\sqrt{17}-3)/2$. Since $\sin x$ is 2π-periodic, the angles $2n\pi + 0.60$, for n any integer, all have sine equal to $(\sqrt{17}-3)/2$. Because $\sin(\pi - x) = \sin x$, it follows that $\sin(\pi - 0.60) = (\sqrt{17}-3)/2$, and therefore $\pi - 0.60$ is another solution. When multiples of 2π are added to this angle, $2n\pi + (\pi - 0.60)$ are also solutions. Thus, the complete set of solutions is

$$2n\pi + 0.60, \quad 2n\pi + (\pi - 0.60),$$

where n is an integer. Figure 8.17 suggests that this set of numbers can be represented more compactly as an initial rotation of $\pi/2$, plus or minus $\pi/2 - 0.60 = 0.97$, and possible multiples of 2π; that is,

$$\frac{\pi}{2} \pm 0.97 + 2n\pi = \left(\frac{4n+1}{2}\right)\pi \pm 0.97.$$

FIGURE 8.17

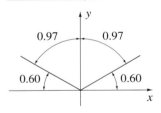

Often in problems like this, $x = \arcsin[(\sqrt{17}-3)/2] = 0.60$ is the only solution given. We can see the reasoning behind this conclusion. There is only one principal value of the inverse sine function, and therefore one solution to the equation. But the equation $\sin x = (\sqrt{17}-3)/2$ says nothing about the inverse sine function. We have introduced it simply as a convenience. When we obtain the solution 0.60 we must ask whether there are other solutions to the original equation. For the above example, there are other solutions; in some problems, the principal value is the only acceptable solution. Remember, then, if we introduce an inverse trigonometric function into a problem along with its corresponding principal values, we must ask whether there are other possibilities besides the principal values. This point is emphasized again in the following example.

EXAMPLE 8.6 ▼

Find all solutions of the equation

$$\tan (\cos x) = \frac{1}{\sqrt{3}}.$$

SOLUTION   If we set $y = \cos x$, then $\tan y = 1/\sqrt{3}$, and one solution of this equation is $y = \text{Tan}^{-1} (1/\sqrt{3}) = \pi/6$. But there are many other solutions for y, namely,

$$y = \frac{\pi}{6} + n\pi,$$

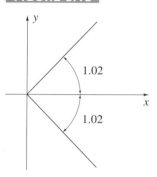

where n is an integer. But $y = \cos x$, and $\cos x$ must take on values in $-1 \le \cos x \le 1$. There is only one possibility for n, namely, $n = 0$; hence,

$$\cos x = \frac{\pi}{6}.$$

From the solution $x = \text{Cos}^{-1} (\pi/6) = 1.02$ and Figure 8.18, we obtain

$$x = \pm 1.02 + 2n\pi,$$

where n is an integer, as the complete set of solutions.

▲

EXAMPLE 8.7 ▼

Find constants $R > 0$ and $0 < \phi < \pi$ so that the function $f(x) = 3 \cos \omega x - 4 \sin \omega x$, where $\omega > 0$ is a constant, can be expressed in the form $R \sin (\omega x + \phi)$ for all real x.

SOLUTION   If we expand $R \sin (\omega x + \phi)$ by means of compound angle formula 1.34a, and equate it to $f(x)$, we have

$$R[\sin \omega x \cos \phi + \cos \omega x \sin \phi] = 3 \cos \omega x - 4 \sin \omega x.$$

This equation will be valid for all x if we can find values of R and ϕ so that

$$R \cos \phi = -4 \qquad \text{and} \qquad R \sin \phi = 3.$$

When we square and add these equations, the result is

$$R^2 \cos^2 \phi + R^2 \sin^2 \phi = R^2 = (-4)^2 + (3)^2 = 25.$$

Consequently, $R = 5$, and

$$\cos \phi = -\frac{4}{5} \qquad \text{and} \qquad \sin \phi = \frac{3}{5}.$$

The only angle in the range $0 < \phi < \pi$ satisfying these equations is $\phi = \arccos(-4/5) = 2.50$ radians. Notice that $\arcsin(3/5)$ does not give this angle. Thus, $f(x)$ can be expressed in the form

$$f(x) = 5 \sin (\omega x + 2.50).$$

▲

EXAMPLE 8.8

The triangle in Figure 8.19a represents cross sections of a prism (the length of the prism being perpendicular to the page). A ray of light (in the plane of the page) is incident on the prism at angle i relative to the normal to the prism. The light is refracted at the two faces of the prism and leaves an angle of deviation ψ relative to the incident direction. Snell's law relates angles of incident and refracted light at each of the faces. It states that

$$n \sin \alpha = \sin i \quad \text{and} \quad n \sin \beta = \sin \phi,$$

where n is the index of refraction of the material of the prism. Find ψ as an explicit function of i.

FIGURE 8.19a

FIGURE 8.19b

SOLUTION   If we introduce angles δ, ϵ, and γ in Figure 8.19b, then

$$\gamma + \delta + \epsilon = \pi, \quad \alpha + \delta = \pi/2, \quad \beta + \epsilon = \pi/2.$$

From these,

$$\gamma = \pi - (\delta + \epsilon) = \pi - (\pi/2 - \alpha + \pi/2 - \beta) = \alpha + \beta,$$

and

$$\psi = (i - \alpha) + (\phi - \beta) = i + \phi - (\alpha + \beta) = i + \phi - \gamma.$$

Angle γ is known. It remains to find ϕ explicitly in terms of i. First,

$$\sin \phi = n \sin \beta = n \sin (\gamma - \alpha) = n(\sin \gamma \, \cos \alpha - \cos \gamma \sin \alpha).$$

Now, $\sin \alpha = (1/n) \sin i$, and because α is an acute angle,

$$\cos \alpha = \sqrt{1 - \sin^2 \alpha} = \sqrt{1 - \frac{\sin^2 i}{n^2}}.$$

Thus,

$$\sin \phi = n \left(\sin \gamma \sqrt{1 - \frac{\sin^2 i}{n^2}} - \cos \gamma \frac{\sin i}{n} \right) = \sin \gamma \sqrt{n^2 - \sin^2 i} - \cos \gamma \, \sin i.$$

Since ϕ is between 0 and $\pi/2$, we may write that

$$\psi(i) = i - \gamma + \phi = i - \gamma + \text{Sin}^{-1}\left(\sin\gamma\sqrt{n^2 - \sin^2 i} - \cos\gamma \sin i\right).$$

In Exercise 40 of Section 8.3 we show how this function can be used to determine the index of refraction of the prism.

EXERCISES 8.2

In Exercises 1–16 evaluate the expression (if it has a value).

* **1.** $\text{Tan}^{-1}(-1/3)$

* **2.** $\text{Sin}^{-1}(1/4)$

* **3.** $\text{Sec}^{-1}(\sqrt{3})$

* **4.** $\text{Csc}^{-1}(-2/\sqrt{3})$

* **5.** $\text{Cot}^{-1}(1)$

* **6.** $\text{Cos}^{-1}(3/2)$

* **7.** $\text{Sin}^{-1}(\pi/2)$

* **8.** $\text{Tan}^{-1}(-1)$

* **9.** $\sin\left(\text{Tan}^{-1}\sqrt{3}\right)$

* **10.** $\tan\left(\text{Sin}^{-1}3\right)$

* **11.** $\text{Sin}^{-1}[\tan(1/6)]$

* **12.** $\text{Tan}^{-1}[\sin(1/6)]$

* **13.** $\sec\left[\text{Cos}^{-1}(1/2)\right]$

* **14.** $\text{Sin}^{-1}[\sin(3\pi/4)]$

* **15.** $\sin\left[\text{Sin}^{-1}(1/\sqrt{2})\right]$

* **16.** $\text{Sin}^{-1}\left[\cos\left(\text{Sec}^{-1}(-\sqrt{2})\right)\right]$

In Exercises 17–22 find all solutions of the equation.

* **17.** $\sin x = 1/3$

* **18.** $\tan x = -1.2$

* **19.** $4\sin^2 x - 2\cos^2 x = 1$

* **20.** $4\sin^2 x + 2\cos^2 x = 1$

* **21.** $\cos^2 x - 3\cos x + 1 = 0$

* **22.** $\sin^2 x - 3\sin x - 5 = 0$

In Exercises 23–26 draw a graph of the function.

* **23.** $f(x) = 2 + \left(\text{Csc}^{-1}x\right)^2$

* **24.** $f(x) = \sqrt{\text{Tan}^{-1}x} + \sqrt{\text{Sec}^{-1}x}$

* **25.** $f(x) = \text{Sin}^{-1}(x - 3)$

* **26.** $f(x) = \text{Sin}^{-1}x + \text{Csc}^{-1}x$

** **27.** If ϕ is the angle formed by AB and AO in the figure below, find ϕ as a function of θ.

In Exercises 28–33 find all solutions of the equation.

** **28.** $\sin x \tan^2 x - 3 + \tan^2 x - 3 \sin x = 0$

** **29.** $\sin x + \cos x = 1$

** **30.** $\sec (\sin x) = -\sqrt{2}$ ** **31.** $\cos \left(\operatorname{Sin}^{-1} x\right) = 1/2$

** **32.** $\sec (\tan x) = -\sqrt{2}$ ** **33.** $\operatorname{Cos}^{-1} [\tan (x^2 + 4)] = 2\pi - 5$

** **34.** Draw graphs of the following functions:

$$\text{(a)} \quad f(x) = \sin \left(\operatorname{Sin}^{-1} x\right) \qquad \text{(b)} \quad f(x) = \operatorname{Sin}^{-1} (\sin x)$$

** **35.** Draw graphs of the following functions:

$$\text{(a)} \quad f(x) = \cos \left(\operatorname{Cos}^{-1} x\right) \qquad \text{(b)} \quad f(x) = \operatorname{Cos}^{-1} (\cos x)$$

** **36.** Express the function $f(x) = 4 \sin 2x + \cos 2x$ in the form $R \sin (2x + \phi)$ where $R > 0$ and $0 < \phi < \pi$.

** **37.** Express the function $f(x) = -2 \sin 3x + 4 \cos 3x$ in the form $R \cos (3x + \phi)$ where $R > 0$ and $0 < \phi < \pi$.

** **38.** Repeat Exercise 36 for $f(x) = -2 \sin 2x + 4 \cos 2x$.

** **39.** Repeat Exercise 37 for $f(x) = -4 \sin 3x + 5 \cos 3x$.

** **40.** A pendulum consists of a mass m suspended from a string of length L (figure below). At time $t = 0$, the mass is pulled through a small angle θ_0—to the right when $\theta_0 > 0$, and to the left when $\theta_0 < 0$—and given an initial speed v_0 to the right. Its subsequent angular displacements are given by

$$\theta = \theta(t) = \theta_0 \cos \omega t + \frac{v_0}{\omega L} \sin \omega t, \quad t \geq 0,$$

where $\omega = \sqrt{9.81/L}$, provided any resistance due to the air is neglected. Show that $\theta(t)$ can be expressed in the form

$$\theta(t) = \sqrt{\theta_0^2 + \frac{v_0^2}{\omega^2 L^2}} \sin (\omega t + \phi),$$

where $\phi = \operatorname{Tan}^{-1} (\omega L \theta_0 / v_0)$.

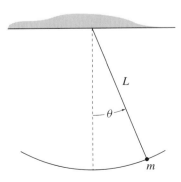

** **41.** What changes, if any, must be made in Exercise 40 if the initial speed v_0 is to the left rather than the right?

** **42.** A mass m is suspended from a spring with constant k (figure below). If, at time $t = 0$, the mass is given an initial displacement y_0 and an upward speed v_0, its subsequent displacements are given by

$$y = y(t) = y_0 \cos \omega t + \frac{v_0}{\omega} \sin \omega t, \quad t \geq 0,$$

where $\omega = \sqrt{k/m}$, provided any resistance due to the air is neglected. Show that $y(t)$ can be expressed in the form

$$y(t) = \sqrt{y_0^2 + \frac{v_0^2}{\omega^2}} \sin (\omega t + \phi),$$

where $\phi = \text{Tan}^{-1} (\omega y_0 / v_0)$.

$y = 0$ at equilibrium

** **43.** What changes, if any, must be made in Exercise 42 if the initial speed v_0 is downward rather than upward?

** **44.** An inductance L, resistance R, and capacitance C are connected in series with a generator producing an oscillatory voltage $E = E_0 \cos \omega t$, $t \geq 0$ (figure below). If $L, C, R, E_0 > 0$, and $\omega > 0$ are all constants, the steady-state current I in the circuit is

$$I = \frac{E_0}{\sqrt{R^2 + \left(\omega L - \frac{1}{\omega C} \right)^2}} \left[R \cos \omega t + \left(\omega L - \frac{1}{\omega C} \right) \sin \omega t \right].$$

Express I in the form

$$I = A \cos (\omega t - \phi),$$

where $A > 0$ and $-\pi/2 \leq \phi \leq \pi/2$.

** **45.** Show that

$$\text{Csc}^{-1} x = \begin{cases} \text{Sin}^{-1} \left(\dfrac{1}{x} \right), & x \geq 1 \\[2ex] -\pi - \text{Sin}^{-1} \left(\dfrac{1}{x} \right), & x \leq -1. \end{cases}$$

** **46.** Show that

$$\operatorname{Sin}^{-1} x = \begin{cases} -\operatorname{Cos}^{-1} \sqrt{1 - x^2}, & -1 \leq x < 0 \\ \operatorname{Cos}^{-1} \sqrt{1 - x^2}, & 0 \leq x \leq 1. \end{cases}$$

** **47.** Prove that

$$\operatorname{Sec}^{-1} x = \begin{cases} \operatorname{Cos}^{-1} \left(\dfrac{1}{x} \right), & x \geq 1 \\ -\operatorname{Cos}^{-1} \left(\dfrac{1}{x} \right), & x \leq -1. \end{cases}$$

** **48.** Prove that

$$\operatorname{Cot}^{-1} x = \begin{cases} \operatorname{Tan}^{-1} \left(\dfrac{1}{x} \right), & x > 0 \\ \pi + \operatorname{Tan}^{-1} \left(\dfrac{1}{x} \right), & x < 0. \end{cases}$$

** **49.** Verify that

$$\operatorname{Sec}^{-1} x = \begin{cases} \operatorname{Tan}^{-1} \sqrt{x^2 - 1}, & x \geq 1 \\ -\pi + \operatorname{Tan}^{-1} \sqrt{x^2 - 1}, & x \leq -1. \end{cases}$$

** **50.** Verify that

$$\operatorname{Csc}^{-1} x = \begin{cases} \operatorname{Cot}^{-1} \sqrt{x^2 - 1}, & x \geq 1 \\ -\pi + \operatorname{Cot}^{-1} \sqrt{x^2 - 1}, & x \leq -1. \end{cases}$$

** **51.** Verify that if $0 \leq x < 1$, then

$$2 \operatorname{Tan}^{-1} \sqrt{\frac{1 + x}{1 - x}} = \pi - \operatorname{Cos}^{-1} x.$$

** **52.** Find intervals on which the function

$$f(x) = x + \sec \left(\frac{x}{2} - 1 \right), \quad 0 \leq x \leq 5,$$

is increasing and decreasing.

8.3 Derivatives of the Inverse Trigonometric Functions

Derivatives of inverse trigonometric functions are most easily obtained with implicit differentiation. We begin with the inverse sine function.

THEOREM 8.2

$$\frac{d}{dx} \operatorname{Sin}^{-1} x = \frac{1}{\sqrt{1 - x^2}} \tag{8.9}$$

PROOF If we set $y = \text{Sin}^{-1} x$, then $x = \sin y$, and we can differentiate implicitly with respect to x:

$$1 = \cos y \frac{dy}{dx}.$$

Thus,

$$\frac{dy}{dx} = \frac{1}{\cos y}.$$

To express $\cos y$ in terms of x we can proceed in two ways:

(i) From the trigonometric identity $\sin^2 y + \cos^2 y = 1$, we obtain

$$\cos y = \pm\sqrt{1 - \sin^2 y} = \pm\sqrt{1 - x^2}.$$

We know $y = \text{Sin}^{-1} x$ and the principal values of $\text{Sin}^{-1} x$ are $-\pi/2 \leq y \leq \pi/2$. Therefore y is an angle in either the first or fourth quadrant. It follows that $\cos y \geq 0$, and we must choose

$$\cos y = \sqrt{1 - x^2}.$$

FIGURE 8.20

(ii) We use the triangle in Figure 8.20 to replace the trigonometric identity. The triangle is designed to fit the equation $x = \sin y$. The third side is then $\sqrt{1 - x^2}$. It follows that

$$\cos y = \sqrt{1 - x^2}.$$

To ensure $\cos y$ is indeed positive, we resort again to the fact that $y = \text{Sin}^{-1} x$ and principal values are $-\pi/2 \leq y \leq \pi/2$. Finally, then, we have

$$\frac{dy}{dx} = \frac{1}{\sqrt{1 - x^2}}. \qquad \blacksquare$$

It is a straightforward application of the chain rule to obtain the following corollary.

COROLLARY

If $u(x)$ is a differentiable function, then

$$\frac{d}{dx}\text{Sin}^{-1} u = \frac{1}{\sqrt{1 - u^2}} \frac{du}{dx}. \qquad (8.10)$$

Next we obtain the derivative of the inverse cosine function.

THEOREM 8.3

$$\frac{d}{dx}\text{Cos}^{-1} x = \frac{-1}{\sqrt{1 - x^2}} \qquad (8.11)$$

PROOF   If we set $y = \text{Cos}^{-1} x$, then $x = \cos y$, and we can differentiate with respect to x,

$$1 = -\sin y \frac{dy}{dx},$$

and solve for

$$\frac{dy}{dx} = \frac{-1}{\sin y}.$$

The triangle in Figure 8.21, obtained from $x = \cos y$, yields $\sin y = \sqrt{1 - x^2}$. Since the principal values of $y = \text{Cos}^{-1} x$ are $0 \leq y \leq \pi$, it follows that $\sin y$ is indeed nonnegative, and therefore

FIGURE 8.21

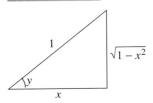

$$\frac{dy}{dx} = \frac{-1}{\sqrt{1 - x^2}}. \qquad \blacksquare$$

COROLLARY
If $u(x)$ is differentiable, then

$$\frac{d}{dx} \text{Cos}^{-1} u = \frac{-1}{\sqrt{1 - u^2}} \frac{du}{dx}. \qquad (8.12)$$

Derivatives for the other inverse trigonometric functions can be derived in a similar way. We list them below and include derivatives of $\text{Sin}^{-1} u$ and $\text{Cos}^{-1} u$ for completeness:

$$\frac{d}{dx} \text{Sin}^{-1} u = \frac{1}{\sqrt{1 - u^2}} \frac{du}{dx}; \qquad (8.13a)$$

$$\frac{d}{dx} \text{Cos}^{-1} u = \frac{-1}{\sqrt{1 - u^2}} \frac{du}{dx}; \qquad (8.13b)$$

$$\frac{d}{dx} \text{Tan}^{-1} u = \frac{1}{1 + u^2} \frac{du}{dx}; \qquad (8.13c)$$

$$\frac{d}{dx} \text{Cot}^{-1} u = \frac{-1}{1 + u^2} \frac{du}{dx}; \qquad (8.13d)$$

$$\frac{d}{dx} \text{Sec}^{-1} u = \frac{1}{u\sqrt{u^2 - 1}} \frac{du}{dx}; \qquad (8.13e)$$

$$\frac{d}{dx} \text{Csc}^{-1} u = \frac{-1}{u\sqrt{u^2 - 1}} \frac{du}{dx}. \qquad (8.13f)$$

Note that the derivative of an inverse cofunction is the negative of the derivative of the corresponding inverse function; that is, derivatives of $\text{Cos}^{-1} u$, $\text{Cot}^{-1} u$, and $\text{Csc}^{-1} u$ are the negatives of the derivatives of $\text{Sin}^{-1} u$, $\text{Tan}^{-1} u$, and $\text{Sec}^{-1} u$.

| EXAMPLE 8.9

Find dy/dx if y is defined as a function of x in each of the following:

(a) $y = \text{Sec}^{-1}(x^2)$ (b) $y = 2x \, \text{Cot}^{-1}(3x)$ (c) $y = \left[\text{Sin}^{-1}(x^3)\right]^2$

(d) $y = \dfrac{\text{Sin}^{-1} x}{\text{Cos}^{-1} x}$ (e) $\text{Cos}^{-1}(xy) + x^2 y + 5 = 0$

SOLUTION

(a) $\dfrac{dy}{dx} = \dfrac{1}{x^2\sqrt{(x^2)^2 - 1}}\dfrac{d}{dx}(x^2) = \dfrac{2x}{x^2\sqrt{x^4 - 1}} = \dfrac{2}{x\sqrt{x^4 - 1}}$

(b) $\dfrac{dy}{dx} = 2\,\mathrm{Cot}^{-1}(3x) + 2x\left(\dfrac{-1}{1 + 9x^2}\right)(3) = 2\mathrm{Cot}^{-1}(3x) - \dfrac{6x}{1 + 9x^2}$

(c) $\dfrac{dy}{dx} = 2\,\mathrm{Sin}^{-1}(x^3)\dfrac{d}{dx}\mathrm{Sin}^{-1}(x^3) = 2\mathrm{Sin}^{-1}(x^3)\dfrac{1}{\sqrt{1 - x^6}}(3x^2) = \dfrac{6x^2\mathrm{Sin}^{-1}(x^3)}{\sqrt{1 - x^6}}$

(d) $\dfrac{dy}{dx} = \dfrac{\mathrm{Cos}^{-1}x\left(\dfrac{1}{\sqrt{1 - x^2}}\right) - \mathrm{Sin}^{-1}x\left(\dfrac{-1}{\sqrt{1 - x^2}}\right)}{\left(\mathrm{Cos}^{-1}x\right)^2} = \dfrac{\mathrm{Cos}^{-1}x + \mathrm{Sin}^{-1}x}{\sqrt{1 - x^2}\left(\mathrm{Cos}^{-1}x\right)^2}$

(e) Differentiation with respect to x gives

$$\dfrac{-1}{\sqrt{1 - x^2y^2}}\left(y + x\dfrac{dy}{dx}\right) + 2xy + x^2\dfrac{dy}{dx} = 0.$$

Thus,

$$\dfrac{dy}{dx}\left(\dfrac{-x}{\sqrt{1 - x^2y^2}} + x^2\right) = \dfrac{y}{\sqrt{1 - x^2y^2}} - 2xy,$$

from which

$$\dfrac{dy}{dx} = \dfrac{y - 2xy\sqrt{1 - x^2y^2}}{\sqrt{1 - x^2y^2}}\dfrac{\sqrt{1 - x^2y^2}}{-x + x^2\sqrt{1 - x^2y^2}} = \dfrac{y - 2xy\sqrt{1 - x^2y^2}}{-x + x^2\sqrt{1 - x^2y^2}}.$$

EXAMPLE 8.10

Find the derivative of the function $f(x) = \mathrm{Sin}^{-1}x + \mathrm{Cos}^{-1}x$. What is your conclusion?

SOLUTION The derivative of the function is

$$f'(x) = \dfrac{1}{\sqrt{1 - x^2}} + \dfrac{-1}{\sqrt{1 - x^2}} = 0.$$

Since this result is valid for the entire domain $-1 \le x \le 1$ of $f(x)$, we can antidifferentiate both sides of the equation,

$$f(x) = C = \text{a constant.}$$

Since $f(0) = \mathrm{Sin}^{-1}(0) + \mathrm{Cos}^{-1}(0) = 0 + \pi/2 = \pi/2$, it follows that $C = \pi/2$. Consequently,

$$\mathrm{Sin}^{-1}x + \mathrm{Cos}^{-1}x = \dfrac{\pi}{2}.$$

EXAMPLE 8.11

A road information sign specifying distance to the next city has markings 1 m high. Bottoms of markings are 5 m above road level (Figure 8.22). If the average motorist's eye level is 1.5 m above ground level, at what distance from the sign do the letters appear tallest?

SOLUTION   The letters appear tallest when the angle θ that they subtend at the eye (Figure 8.22) is greatest. Since

$$\theta = \beta - \alpha = \text{Tan}^{-1}\left(\frac{9/2}{x}\right) - \text{Tan}^{-1}\left(\frac{7/2}{x}\right), \quad 0 < x < \infty,$$

critical points of $\theta(x)$ are given by

$$0 = \frac{d\theta}{dx} = \frac{1}{1 + \left(\frac{9}{2x}\right)^2}\left(-\frac{9}{2x^2}\right) - \frac{1}{1 + \left(\frac{7}{2x}\right)^2}\left(-\frac{7}{2x^2}\right)$$

$$= \frac{-18}{4x^2 + 81} + \frac{14}{4x^2 + 49}.$$

Equivalently,

$$18(4x^2 + 49) = 14(4x^2 + 81),$$

which has only one positive solution, $x = 3\sqrt{7}/2$. Since $\theta(x)$ approaches zero as x approaches zero and as x becomes very large, it follows that θ is maximized when the motorist is $3\sqrt{7}/2$ m from the sign.

FIGURE 8.22

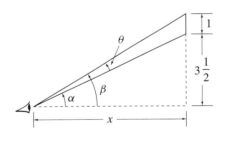

EXAMPLE 8.12

Figure 8.23 contains an optimization application in seismic prospecting. A source at A sends a signal into a layer of soil of some thickness D. That part of the signal that travels in the surface of the soil does so with speed v_1, and v_1 can be calculated by dividing s by the time t_1 that it takes the signal to reach the receiver at B, $v_1 = s/t_1$.

FIGURE 8.23

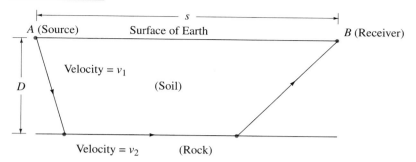

It is known that there is a horizontal layer of rock below the soil, and it is depth D of this rock that is to be determined. A signal emitted at A travels through the soil, and strikes the rock interface. Some of the signal is reflected back into the soil, some penetrates the rock, and some travels basically in the surface of the rock at unknown speed $v_2 > v_1$. Each part of the surface of the rock acts as an emitter sending the signal back into the soil, some of which reaches the receiver B. This happens in many ways and the receiver records an accumulation of many signals. But there is a shortest time for signals to do this, and by calculating and recording this time we can find D. The quickest possible path is shown in Figure 8.24 for some unknown angle θ. Show that the minimum time is

$$\frac{2D}{v_1}\sqrt{1 - \left(\frac{v_1}{v_2}\right)^2} + \frac{s}{v_2}.$$

FIGURE 8.24

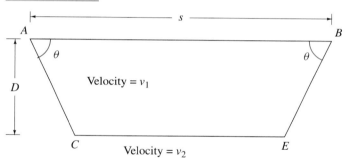

SOLUTION   The time for the signal to travel from A to B in Figure 8.24 is

$$t = \frac{2D}{v_1}\csc\theta + \frac{s - 2D\cot\theta}{v_2}.$$

The smallest possible value for θ occurs when C and E become coincident in which case $\theta = \text{Tan}^{-1}(2D/s)$. Thus, the domain of function $t(\theta)$ is $\text{Tan}^{-1}(2D/s) \leq \theta \leq \pi/2$. Critical points of $t(\theta)$ are given by

$$0 = \frac{dt}{d\theta} = -\frac{2D}{v_1}\csc\theta\,\cot\theta + \frac{2D}{v_2}\csc^2\theta$$

$$= \frac{-2D\cos\theta}{v_1\sin^2\theta} + \frac{2D}{v_2\sin^2\theta}$$

$$= \frac{-2D}{v_1 v_2 \sin^2\theta}(v_2\cos\theta - v_1).$$

∗∗ **33.** (a) Show that if the principal values of $f(x) = \text{Sec}^{-1} x$ are chosen as $0 \le y \le \pi$, $y \ne \pi/2$, then the derivative of the function is

$$\frac{d}{dx} \text{Sec}^{-1} x = \frac{1}{|x|\sqrt{x^2 - 1}}.$$

(b) What is the derivative of $f(x) = \text{Csc}^{-1} x$ if its principal values are chosen as $-\pi/2 \le y \le \pi/2$, $y \ne 0$?

∗∗ **34.** Find the angle between the curves $y = \text{Sin}^{-1} x$ and $y = \text{Cos}^{-1} x$ at their point of intersection.

∗∗ **35.** Verify the results in equations 8.13c–f.

∗∗ **36.** When a force F is applied to the object of mass m in the figure below, three other forces act on m: the force of gravity mg ($g > 0$) directly downward, a reactional force of the supporting surface, and a horizontal, frictional force opposing F. The least force F that will overcome friction and produce motion is given by

$$F = \frac{\mu mg}{\cos \theta + \mu \sin \theta},$$

where μ is a constant called the *coefficient of static friction*. Find the angle θ for which F is minimal.

∗∗ **37.** Find d^2y/dx^2 if y is defined implicitly as a function of x by

$$\ln (x^2 + y^2) = 2 \text{Tan}^{-1}\left(\frac{y}{x}\right).$$

∗∗∗ **38.** A runner moves counterclockwise around the track in the figure below at a rate of 4 m/s. A camera at the centre of the track is placed on a swivel so that it can follow the runner. Find the rate at which the camera turns when (a) the runner is at A and (b) the runner is at B.

∗∗ **39.** A rope with a ring at one end is passed through two fixed rings at the same level (figure follows). The end of the rope without the ring is then passed through the ring at the other end, and a mass m is attached to it. If the rope moves so as to maximize the distance from m to the line through the fixed rings, find angle θ.

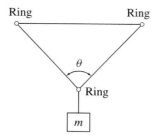

*** **40.** Show that the angle of deviation ψ in Example 8.8 is a minimum when $i = \phi$. Prove that if ψ_m is the minimum angle of deviation, then

$$n = \frac{\sin [(\psi_m + \gamma)/2]}{\sin (\gamma/2)}.$$

This equation can be used to determine n experimentally. Angle of incidence i is varied until ψ_m is achieved and measured. With γ also known, n can be calculated.

8.4 Hyperbolic Functions and Their Derivatives

Certain combinations of the exponential function occur so often in physical applications that they are given special names. Specifically, half the difference of e^x and e^{-x} is defined as the hyperbolic sine function and half their sum is the hyperbolic cosine function. These functions are denoted as follows:

$$\sinh x = \frac{e^x - e^{-x}}{2} \qquad \text{and} \qquad \cosh x = \frac{e^x + e^{-x}}{2}. \qquad (8.14)$$

The names of these functions and their notations bear a striking resemblance to those for the trigonometric functions, and there are reasons for this. First, the hyperbolic functions $\sinh x$ and $\cosh x$ are related to the curve $x^2 - y^2 = 1$ (see Figure 8.29, page 567), called the *unit hyperbola*, in much the same way as the trigonometric functions $\sin x$ and $\cos x$ are related to the unit circle $x^2 + y^2 = 1$. We will point out two of these similarities in Examples 8.16 and 8.17. Second, for each identity satisfied by the trigonometric functions, there is a corresponding identity satisfied by the hyperbolic functions—not the same identity, but one very similar. For example, using equations 8.14, we have

$$(\cosh x)^2 - (\sinh x)^2 = \left(\frac{e^x + e^{-x}}{2}\right)^2 - \left(\frac{e^x - e^{-x}}{2}\right)^2$$

$$= \frac{1}{4} \left[\left(e^{2x} + 2 + e^{-2x}\right) - \left(e^{2x} - 2 + e^{-2x}\right) \right]$$

$$= 1.$$

Thus the hyperbolic sine and cosine functions satisfy the identity

$$\cosh^2 x - \sinh^2 x = 1, \qquad (8.15)$$

which is reminiscent of the identity $\cos^2 x + \sin^2 x = 1$ for the trigonometric functions.

Just as four other trigonometric functions are defined in terms of $\sin x$ and $\cos x$, four corresponding hyperbolic functions are defined as follows:

$$\tanh x = \frac{\sinh x}{\cosh x} = \frac{e^x - e^{-x}}{e^x + e^{-x}}, \qquad \coth x = \frac{\cosh x}{\sinh x} = \frac{e^x + e^{-x}}{e^x - e^{-x}},$$

(8.16)

$$\operatorname{sech} x = \frac{1}{\cosh x} = \frac{2}{e^x + e^{-x}}, \qquad \operatorname{csch} x = \frac{1}{\sinh x} = \frac{2}{e^x - e^{-x}}.$$

These definitions and 8.15 immediately imply that

$$1 - \tanh^2 x = \operatorname{sech}^2 x,$$

(8.17a)

$$\coth^2 x - 1 = \operatorname{csch}^2 x,$$

(8.17b)

analogous to $1 + \tan^2 x = \sec^2 x$ and $1 + \cot^2 x = \csc^2 x$ respectively.

Graphs of the six hyperbolic functions are shown in Figure 8.26. The functions $\cosh x$ and $\operatorname{sech} x$ are even, the other four are odd.

FIGURE 8.26a

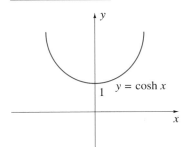

$y = \cosh x$

FIGURE 8.26b

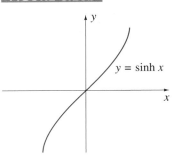

$y = \sinh x$

FIGURE 8.26c

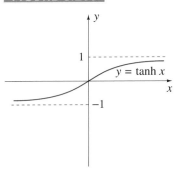

$y = \tanh x$

FIGURE 8.26d

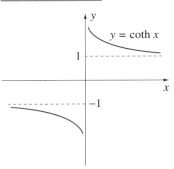

$y = \coth x$

FIGURE 8.26e

$y = \operatorname{sech} x$

FIGURE 8.26f

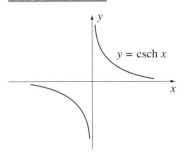

$y = \operatorname{csch} x$

Most trigonometric identities can be derived from the compound angle formulas for $\sin(A \pm B)$ and $\cos(A \pm B)$. It is easy to verify similar formulas for the hyperbolic functions:

$$\sinh(A \pm B) = \sinh A \cosh B \pm \cosh A \sinh B, \tag{8.18a}$$

$$\cosh(A \pm B) = \cosh A \cosh B \pm \sinh A \sinh B. \tag{8.18b}$$

For example, equations 8.14 give

$$\cosh A \cosh B - \sinh A \sinh B = \left(\frac{e^A + e^{-A}}{2}\right)\left(\frac{e^B + e^{-B}}{2}\right)$$

$$- \left(\frac{e^A - e^{-A}}{2}\right)\left(\frac{e^B - e^{-B}}{2}\right)$$

$$= \frac{1}{4}\left[\left(e^{A+B} + e^{A-B} + e^{B-A} + e^{-A-B}\right)\right.$$

$$\left. - \left(e^{A+B} - e^{A-B} - e^{B-A} + e^{-A-B}\right)\right]$$

$$= \frac{1}{2}\left[e^{A-B} + e^{-(A-B)}\right]$$

$$= \cosh(A - B).$$

With these formulas, we can derive hyperbolic identities analogous to trigonometric identities 1.38–1.41:

$$\tanh(A \pm B) = \frac{\tanh A \pm \tanh B}{1 \pm \tanh A \tanh B}, \tag{8.18c}$$

$$\sinh 2A = 2 \sinh A \cosh A, \tag{8.18d}$$

$$\cosh 2A = \cosh^2 A + \sinh^2 A \tag{8.18e}$$

$$= 2 \cosh^2 A - 1 \tag{8.18f}$$

$$= 1 + 2 \sinh^2 A, \tag{8.18g}$$

$$\tanh 2A = \frac{2 \tanh A}{1 + \tanh^2 A}, \tag{8.18h}$$

$$\sinh A \sinh B = \frac{1}{2} \cosh(A + B) - \frac{1}{2} \cosh(A - B), \tag{8.18i}$$

$$\sinh A \cosh B = \frac{1}{2} \sinh(A + B) + \frac{1}{2} \sinh(A - B), \tag{8.18j}$$

$$\cosh A \cosh B = \frac{1}{2} \cosh(A + B) + \frac{1}{2} \cosh(A - B), \tag{8.18k}$$

$$\sinh A + \sinh B = 2 \sinh\left(\frac{A + B}{2}\right) \cosh\left(\frac{A - B}{2}\right), \tag{8.18l}$$

$$\sinh A - \sinh B = 2 \cosh\left(\frac{A + B}{2}\right) \sinh\left(\frac{A - B}{2}\right), \tag{8.18m}$$

$$\cosh A + \cosh B = 2 \cosh\left(\frac{A + B}{2}\right) \cosh\left(\frac{A - B}{2}\right), \tag{8.18n}$$

$$\cosh A - \cosh B = 2 \sinh\left(\frac{A + B}{2}\right) \sinh\left(\frac{A - B}{2}\right). \tag{8.18o}$$

Since the hyperbolic sine and cosine are defined in terms of the exponential function, for which we know the derivative, and the remaining hyperbolic functions are defined in terms of the hyperbolic sine and cosine, it follows that calculation of the derivatives of the hyperbolic functions should be straightforward. Indeed, if $u(x)$ is a differentiable function of x, then

$$\frac{d}{dx}\sinh u = \cosh u \frac{du}{dx}, \tag{8.19a}$$

$$\frac{d}{dx}\cosh u = \sinh u \frac{du}{dx}, \tag{8.19b}$$

$$\frac{d}{dx}\tanh u = \operatorname{sech}^2 u \frac{du}{dx}, \tag{8.19c}$$

$$\frac{d}{dx}\coth u = -\operatorname{csch}^2 u \frac{du}{dx}, \tag{8.19d}$$

$$\frac{d}{dx}\operatorname{sech} u = -\operatorname{sech} u \tanh u \frac{du}{dx}, \tag{8.19e}$$

$$\frac{d}{dx}\operatorname{csch} u = -\operatorname{csch} u \coth u \frac{du}{dx}. \tag{8.19f}$$

EXAMPLE 8.13

Find dy/dx if y is defined as a function of x by:

(a) $y = \operatorname{sech}(3x^2)$ (b) $y = \tanh(1 - 4x)$
(c) $y = \cos 2x \sinh 2x$ (d) $y = \cosh\left(\operatorname{Tan}^{-1} x^2\right)$

SOLUTION

(a) $\dfrac{dy}{dx} = -\operatorname{sech}(3x^2)\tanh(3x^2)\dfrac{d}{dx}(3x^2) = -6x\operatorname{sech}(3x^2)\tanh(3x^2)$

(b) $\dfrac{dy}{dx} = \operatorname{sech}^2(1 - 4x)\dfrac{d}{dx}(1 - 4x) = -4\operatorname{sech}^2(1 - 4x)$

(c) $\dfrac{dy}{dx} = \cos 2x(2\cosh 2x) + (-2\sin 2x)\sinh 2x$

$\phantom{\dfrac{dy}{dx}} = 2(\cos 2x \cosh 2x - \sin 2x \sinh 2x)$

(d) $\dfrac{dy}{dx} = \sinh\left(\operatorname{Tan}^{-1} x^2\right)\dfrac{d}{dx}\operatorname{Tan}^{-1} x^2 = \sinh\left(\operatorname{Tan}^{-1} x^2\right)\dfrac{2x}{1 + x^4}$

EXAMPLE 8.14

When a uniform cable is hung between two fixed supports (Figure 8.27), the shape of the curve $y = f(x)$ must satisfy the differential equation

$$\frac{d^2 y}{dx^2} = \frac{\rho g}{H}\sqrt{1 + \left(\frac{dy}{dx}\right)^2},$$

FIGURE 8.27

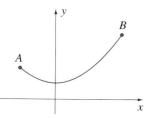

where ρ is the mass per unit length of the cable, $g > 0$ is the acceleration due to gravity, and $H > 0$ is the tension in the cable at its lowest point. Show that a solution of the equation is

$$y = f(x) = \frac{H}{\rho g} \cosh\left(\frac{\rho g x}{H}\right) + C,$$

where C is a constant.

SOLUTION   In Exercise 41 of Section 9.5 we derive this solution. For now we simply wish to verify that the hyperbolic cosine is indeed a solution. The first derivative of the function is

$$\frac{dy}{dx} = \sinh\left(\frac{\rho g x}{H}\right),$$

and therefore its second derivative is

$$\frac{d^2 y}{dx^2} = \frac{\rho g}{H} \cosh\left(\frac{\rho g x}{H}\right).$$

On the other hand,

$$\frac{\rho g}{H}\sqrt{1 + \left(\frac{dy}{dx}\right)^2} = \frac{\rho g}{H}\sqrt{1 + \sinh^2\left(\frac{\rho g x}{H}\right)} = \frac{\rho g}{H}\sqrt{\cosh^2\left(\frac{\rho g x}{H}\right)} = \frac{\rho g}{H}\cosh\left(\frac{\rho g x}{H}\right).$$

Thus, the function $y = \left[H/(\rho g)\right]\cosh\left(\rho g x / H\right) + C$ does indeed satisfy the given differential equation.

Example 8.14 shows that the many telephone and hydro wires crisscrossing the country hang in the form of hyperbolic cosines. Engineers often call this curve a *catenary*.

EXAMPLE 8.15

In studying wave guides, the electrical engineer often encounters the differential equation

$$\frac{d^2 y}{dx^2} - ky = 0,$$

where $k > 0$ is a constant. Show that $y = f(x) = A \cosh \sqrt{k}x + B \sinh \sqrt{k}x$ is a solution for any constants A and B.

EXERCISES 8.4

In Exercises 1–10 evaluate the expression (if it has a value).

∗ **1.** $3 \cosh 1$ ∗ **2.** $\sinh (\pi/2)$

∗ **3.** $\tanh \sqrt{1 - \sin 3}$ ∗ **4.** $\mathrm{Sin}^{-1} (\mathrm{sech}\, 10)$

∗ **5.** $\mathrm{Cos}^{-1} (2 \,\mathrm{csch}\, 1)$ ∗ **6.** $\coth (\sinh 5)$

∗ **7.** $\sqrt{\ln |\sinh (-3)|}$ ∗ **8.** $\mathrm{sech} [\sec (\pi/3)]$

∗ **9.** $e^{-2 \cosh e}$ ∗ **10.** $\sinh [\mathrm{Cot}^{-1} (-3\pi/10)]$

In Exercises 11–20 y is defined as a function of x. Find dy/dx in as simplified a form as possible.

∗ **11.** $y = \mathrm{csch}\, (2x + 3)$ ∗ **12.** $y = x \sinh (x/2)$

∗ **13.** $y = \sqrt{1 - \mathrm{sech}\, x}$ ∗ **14.** $y = \tanh (\ln x)$

∗ **15.** $\cosh (x + y) = 2x$ ∗ **16.** $y + \coth x = \sqrt{1 + y}$

∗ **17.** $y = u \cosh u, \ u = e^x + e^{-x}$ ∗ **18.** $y = \tan (\cosh t), \ t = \cos (\tanh x)$

∗ **19.** $y = \mathrm{Tan}^{-1} (\sinh x)$ ∗ **20.** $y = \ln \sqrt{\tanh 2x}$

∗ **21.** Verify the results in identities 8.18c–o (page 563).

∗ **22.** Verify the differentiation formulas 8.19 (page 564).

∗ **23.** Show that $y = \sinh x$ and $y = \tanh x$ have points of inflection at $x = 0$.

∗∗ **24.** To analyze vertical vibrations of the beam in the figure below, we must solve the differential equation

$$\frac{d^4 y}{dx^4} - k^4 y = 0,$$

where $k > 0$ is a constant.

(a) Show that a solution is

$$y = f(x) = A \cos kx + B \sin kx + C \cosh kx + D \sinh kx$$

for any constants A, B, C, and D.

(b) If the left end $(x = 0)$ is fastened horizontally and the right end $(x = L)$ is pinned, then $f(x)$ must satisfy the conditions

$$f(0) = f'(0) = f(L) = f''(L) = 0.$$

Show that these restrictions imply that $C = -A$, $D = -B$, and A and B must satisfy the equations

$$A(\cos kL - \cosh kL) + B(\sin kL - \sinh kL) = 0,$$

$$A(\cos kL + \cosh kL) + B(\sin kL + \sinh kL) = 0.$$

(c) Eliminate A and B between these equations to show that k must satisfy the condition

$$\tan kL = \tanh kL.$$

** **25.** Show that $y = \operatorname{sech} x$ has points of inflection for $x = \pm \ln(\sqrt{2}+1)$.

** **26.** A uniform hydro cable P metres long with mass ρ kilograms per metre of length is hung from two supports at the same level L metres apart (figure below). If the supports are b metres higher than the lowest point in the cable and H is the tension in the cable at the lowest point, it can be shown that

$$\rho g P = 2H \sinh \frac{\rho g L}{2H}.$$

(a) If $L = 70$ m, $\rho = 0.5$ kg/m, and $P = 80$ m, find H. Hint: Set $z = \rho g/(2H)$, and use Newton's iterative procedure with $z_1 = 0.01$ to solve $Pz = \sinh(Lz)$ for z.

(b) Use the result in (a) to find b. See Example 8.14 for the equation of the curve describing the cable.

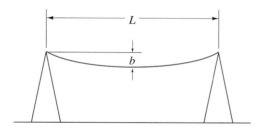

** **27.** Each hyperbolic function has associated with it an inverse hyperbolic function.

(a) Draw the inverse functions $\operatorname{Sinh}^{-1} x$, $\operatorname{Tanh}^{-1} x$, $\operatorname{Coth}^{-1} x$, and $\operatorname{Csch}^{-1} x$ for $\sinh x$, $\tanh x$, $\coth x$, and $\operatorname{csch} x$.

(b) Why do $\cosh x$ and $\operatorname{sech} x$ not have inverse functions? It is customary to associate functions $\operatorname{Cosh}^{-1} x$ and $\operatorname{Sech}^{-1} x$ with $\cosh x$ and $\operatorname{sech} x$ by restricting their domains to nonnegative numbers. Draw graphs of $\operatorname{Cosh}^{-1} x$ and $\operatorname{Sech}^{-1} x$.

(c) Obtain the following derivatives of the inverse hyperbolic functions:

$$\frac{d}{dx} \operatorname{Sinh}^{-1} x = \frac{1}{\sqrt{x^2+1}};$$

$$\frac{d}{dx} \operatorname{Cosh}^{-1} x = \frac{1}{\sqrt{x^2-1}};$$

$$\frac{d}{dx} \operatorname{Tanh}^{-1} x = \frac{1}{1-x^2}, \quad |x| < 1;$$

$$\frac{d}{dx} \operatorname{Coth}^{-1} x = \frac{1}{1-x^2}, \quad |x| > 1;$$

$$\frac{d}{dx} \operatorname{Sech}^{-1} x = \frac{-1}{x\sqrt{1-x^2}};$$

$$\frac{d}{dx} \operatorname{Csch}^{-1} x = \frac{-1}{|x|\sqrt{1+x^2}}.$$

(d) Show that

$$\text{Sinh}^{-1} x = \ln\left(x + \sqrt{x^2 + 1}\right);$$

$$\text{Cosh}^{-1} x = \ln\left(x + \sqrt{x^2 - 1}\right);$$

$$\text{Tanh}^{-1} x = \frac{1}{2}\ln\left(\frac{1+x}{1-x}\right), \quad |x| < 1.$$

✳✳✳ **28.** When an object of mass m falls from rest at time $t = 0$ under the influence of gravity and an air resistance proportional to the square of velocity, its velocity v as a function of time is defined by the equation

$$\frac{1}{2}\sqrt{\frac{\beta}{mg}}\ln\left(\frac{\sqrt{mg/\beta} - v}{\sqrt{mg/\beta} + v}\right) = -\frac{\beta t}{m},$$

where $\beta > 0$ is a constant and $g > 0$ is the acceleration due to gravity.

(a) Show that when this equation is solved for v in terms of t, the result is

$$v(t) = \sqrt{\frac{mg}{\beta}}\tanh\left(\sqrt{\frac{\beta g}{m}}\,t\right).$$

(b) Determine the limit of v for large t, called the *limiting velocity*.

SUMMARY

With derivatives of trigonometric, exponential, and logarithm functions in Chapter 3, and derivatives of inverse trigonometric and hyperbolic functions in this chapter, we have completed the list of differentiation formulas.

Inverse trigonometric functions reverse the roles of trigonometric functions. A trigonometric function such as the sine function associates a value called $\sin x$ with a real number (angle) x. The corresponding inverse function, $\text{Sin}^{-1} x$, regards x as the sine of an angle, and yields the angle.

Hyperbolic functions are special combinations of exponential functions that arise sufficiently often in applications to warrant special consideration. They satisfy identities and have derivatives very similar to the trigonometric functions.

The transcendental functions that we have discussed fall into three pairs: trigonometric and inverse trigonometric, exponential and logarithm (the inverse of the exponential), and hyperbolic and inverse hyperbolic (Exercise 27 in Section 8.4). The relationship between a function and its inverse is very important. For instance, evaluation of the indeterminate forms 0^0, 1^∞, and ∞^0 in Section 4.10 was based on the fact that taking logarithms and exponentials are inverse operations. By taking logarithms, we transformed each indeterminate form into a limit to which we could apply L'Hôpital's rule. We then used exponentials to recapture the original limit. In Section 9.2, the inverse character of exponential and logarithm functions is indispensable to the solution of many differential equations. The integration technique of Section 9.5 is based on the relation of an inverse trigonometric function to the corresponding trigonometric function.

KEY TERMS

In reviewing this chapter, you should be able to define or discuss the following key terms:

Algebraic function
Transcendental function
One-to-one function
Inverse function
Horizontal line test

Monotonic function
Inverse trigonometric function
Principal values
Hyperbolic function

REVIEW EXERCISES

In Exercises 1–10 find all solutions of the equation.

* **1.** $\cos^2 x + 5 \cos x - 1 = 0$

* **2.** $4 \sin 2x = 1$

* **3.** $\csc^2 (x + 1) = 3$

* **4.** $\text{Tan}^{-1} (3x + 2) = 5 - 2\pi$

* **5.** $\cos 2x = \sin x$

* **6.** $\ln (\sin x) + \ln (1 + \sin x) = \ln 3 - \ln 2$

* **7.** $3 \, \text{Sin}^{-1} \left(e^{x+2}\right) = 2$

* **8.** $3 \sin \left(e^{x+2}\right) = 2$

* **9.** $\tan (x \cosh 2) = 1/4$

* **10.** $\sinh x = 4$

In Exercises 11–20 y is defined as a function of x. Find dy/dx in as simplified a form as possible.

* **11.** $y = \text{Sin}^{-1} (2 - 3x)$

* **12.** $y = 3 \sinh (x^2)$

* **13.** $y = \dfrac{\text{Cos}^{-1} x}{\text{Sin}^{-1} x}$

* **14.** $y = \text{Tan}^{-1}\left(\dfrac{1}{x} + x\right)$

* **15.** $y = e^{\cosh x}$

* **16.** $\sinh y = \sin x$

* **17.** $\text{Sec}^{-1} (x + y) = xy$

* **18.** $y = x \, \text{Csc}^{-1}\left(\dfrac{1}{x^2}\right)$

* **19.** $y = e^{2x} \cosh 2x$

* **20.** $\ln \left[\text{Tan}^{-1} (x + y)\right] = 1$

** **21.** Draw graphs of the following functions:

(a) $f(x) = \tan \left(\text{Tan}^{-1} x\right)$ (b) $f(x) = \text{Tan}^{-1} (\tan x)$

Techniques of Integration

In Chapter 5, we discussed three techniques for evaluating antiderivatives. First, since every differentiation formula can be restated as an antidifferentiation formula, it follows that the more competent we are at differentiation, the more likely we are to recognize an antiderivative. Our second technique was called "guess and fix up." It is applicable to sufficiently simple integrands that we can immediately guess the antiderivative to within a multiplicative constant. It is then a matter of adjusting the constant. The third technique was to change the variable of integration. By a suitable transformation, replace a complicated antiderivative with a simpler one.

In this chapter we develop additional techniques applicable to more complex integration problems. At the same time we take the opportunity to review the applications of definite integrals in Chapter 7.

9.1 Integration Formulas and Substitutions

Every differentiation formula can be restated as an antidifferentiation formula. A partial list of frequently encountered formulas is as follows:

$$\int u^n \, du = \frac{u^{n+1}}{n+1} + C, \quad n \neq -1; \tag{9.1a}$$

$$\int \frac{1}{u} \, du = \ln|u| + C; \tag{9.1b}$$

$$\int e^u \, du = e^u + C; \tag{9.1c}$$

$$\int a^u \, du = a^u \log_a e + C; \tag{9.1d}$$

$$\int \cos u \, du = \sin u + C; \tag{9.1e}$$

$$\int \sin u \, du = -\cos u + C; \tag{9.1f}$$

$$\int \sec^2 u \, du = \tan u + C; \tag{9.1g}$$

$$\int \csc^2 u \, du = -\cot u + C; \tag{9.1h}$$

$$\int \sec u \tan u \, du = \sec u + C; \tag{9.1i}$$

$$\int \csc u \cot u \, du = -\csc u + C; \tag{9.1j}$$

$$\int \tan u \, du \; = \; \ln|\sec u| + C \; = \; -\ln|\cos u| + C; \tag{9.1k}$$

$$\int \cot u \, du \; = \; -\ln|\csc u| + C \; = \; \ln|\sin u| + C; \tag{9.1l}$$

$$\int \sec u \, du \; = \; \ln|\sec u + \tan u| + C; \tag{9.1m}$$

$$\int \csc u \, du \; = \; \ln|\csc u - \cot u| + C. \tag{9.1n}$$

Formulas 9.1a–j are restatements of differentiation formulas 5.4 and 5.5 in Chapter 5. Formulas 9.1k and 9.1m are verified in Example 9.1 below. Formulas 9.1l and 9.1n are similar.

EXAMPLE 9.1

Verify formulas 9.1k and 9.1m.

SOLUTION   To verify 9.1k we express $\tan u$ in terms of $\sin u$ and $\cos u$:

$$\int \tan u \, du \; = \; \int \frac{\sin u}{\cos u} \, du \; = \; -\ln|\cos u| + C \; = \; \ln|\sec u| + C.$$

If integration to $\ln|\cos u|$ is not obvious, try the substitution $v = \cos u$.

To obtain the antiderivative of $\sec u$ requires a subtle trick. We first multiply numerator and denominator by $\sec u + \tan u$,

$$\int \sec u \, du \; = \; \int \sec u \frac{\sec u + \tan u}{\sec u + \tan u} \, du \; = \; \int \frac{\sec^2 u + \sec u \tan u}{\sec u + \tan u} \, du.$$

If we now set $v = \sec u + \tan u$, then $dv = (\sec u \tan u + \sec^2 u) \, du$, and

$$\int \sec u \, du \; = \; \int \frac{1}{v} \, dv \; = \; \ln|v| + C \; = \; \ln|\sec u + \tan u| + C.$$

We could list many other integration formulas, but one of the purposes of this chapter is to develop integration techniques that eliminate the need for excessive memorization of formulas. Our emphasis is on development of antiderivatives rather than memorization of formulas, and we prefer to keep the list of formulas short.

In Chapter 5 we demonstrated how a change of variable could sometimes replace a complex integration problem with a simpler one. When the integrand contains $x^{1/n}$, where n is a positive integer, it is often useful to set $u = x^{1/n}$, or, equivalently, $x = u^n$. We illustrate in the following example.

EXAMPLE 9.2

Evaluate the following indefinite integrals:

$$\text{(a)} \quad \int \frac{2 - \sqrt{x}}{2 + \sqrt{x}} \, dx \qquad \text{(b)} \quad \int \frac{x^{1/3}}{1 + x^{2/3}} \, dx \qquad \text{(c)} \quad \int \frac{\sqrt{x}}{1 + x^{1/3}} \, dx$$

SOLUTION

(a) If we set $u = \sqrt{x}$ or $x = u^2$, then $dx = 2u\,du$, and

$$\int \frac{2 - \sqrt{x}}{2 + \sqrt{x}}\,dx = \int \frac{2 - u}{2 + u}(2u\,du) = 2\int \frac{2u - u^2}{2 + u}\,du.$$

Long division of $-u^2 + 2u$ by $u + 2$ immediately gives

$$\int \frac{2 - \sqrt{x}}{2 + \sqrt{x}}\,dx = 2\int\left(-u + 4 - \frac{8}{u + 2}\right)du$$

$$= 2\left(-\frac{u^2}{2} + 4u - 8\ln|u + 2|\right) + C$$

$$= -x + 8\sqrt{x} - 16\ln\left(2 + \sqrt{x}\right) + C.$$

(b) If we set $u = x^{1/3}$ or $x = u^3$, then $dx = 3u^2\,du$, and

$$\int \frac{x^{1/3}}{1 + x^{2/3}}\,dx = \int \frac{u}{1 + u^2}(3u^2\,du) = 3\int \frac{u^3}{1 + u^2}\,du.$$

Division of u^3 by $u^2 + 1$ leads to

$$\int \frac{x^{1/3}}{1 + x^{2/3}}\,dx = 3\int\left(u - \frac{u}{u^2 + 1}\right)du$$

$$= 3\left(\frac{u^2}{2} - \frac{1}{2}\ln|u^2 + 1|\right) + C$$

$$= \frac{3}{2}x^{2/3} - \frac{3}{2}\ln\left(x^{2/3} + 1\right) + C.$$

(c) The purpose of the substitution $u = x^{1/n}$ is to rid the integrand of fractional powers. In the presence of both $x^{1/2}$ and $x^{1/3}$, we set $u = x^{1/6}$ or $x = u^6$. Then $dx = 6u^5\,du$, and

$$\int \frac{\sqrt{x}}{1 + x^{1/3}}\,dx = \int \frac{u^3}{1 + u^2}(6u^5\,du) \quad \text{(and by long division)}$$

$$= 6\int\left(u^6 - u^4 + u^2 - 1 + \frac{1}{1 + u^2}\right)du$$

$$= 6\left(\frac{u^7}{7} - \frac{u^5}{5} + \frac{u^3}{3} - u + \mathrm{Tan}^{-1}\,u\right) + C$$

$$= \frac{6}{7}x^{7/6} - \frac{6}{5}x^{5/6} + 2x^{1/2} - 6x^{1/6} + 6\,\mathrm{Tan}^{-1}\left(x^{1/6}\right) + C.$$

EXERCISES 9.1

In Exercises 1–22 evaluate the indefinite integral.

* **1.** $\displaystyle\int \frac{x^2}{5 - 3x^3}\, dx$

* **2.** $\displaystyle\int x e^{-2x^2}\, dx$

* **3.** $\displaystyle\int \frac{x}{(x^2 + 2)^{1/3}}\, dx$

* **4.** $\displaystyle\int \frac{e^x}{1 + e^x}\, dx$

* **5.** $\displaystyle\int \frac{4t + 8}{t^2 + 4t + 5}\, dt$

* **6.** $\displaystyle\int x^2 \sqrt{1 - 3x^3}\, dx$

* **7.** $\displaystyle\int (x + 1)(x^2 + 2x)^{1/3}\, dx$

* **8.** $\displaystyle\int \frac{x^2}{(1 + x^3)^3}\, dx$

* **9.** $\displaystyle\int \frac{\sqrt{x}}{1 - \sqrt{x}}\, dx$

* **10.** $\displaystyle\int \frac{1 - \sqrt{x}}{\sqrt{x}}\, dx$

* **11.** $\displaystyle\int \frac{x + 2}{x + 1}\, dx$

* **12.** $\displaystyle\int \frac{x^2 + 2}{x^2 + 1}\, dx$

* **13.** $\displaystyle\int \frac{\sin \theta}{\cos \theta - 1}\, d\theta$

* **14.** $\displaystyle\int \frac{x + 3}{\sqrt{2x + 4}}\, dx$

* **15.** $\displaystyle\int \frac{e^x}{1 + e^{2x}}\, dx$

* **16.** $\displaystyle\int \sin^3 2x \cos 2x\, dx$

* **17.** $\displaystyle\int x^5 (2x^2 - 5)^4\, dx$

* **18.** $\displaystyle\int \frac{x^3}{(x + 5)^2}\, dx$

* **19.** $\displaystyle\int z^2 \sqrt{3 - z}\, dz$

* **20.** $\displaystyle\int \tan 3x\, dx$

* **21.** $\displaystyle\int \frac{(x - 3)^{2/3}}{(x - 3)^{2/3} + 1}\, dx$

* **22.** $\displaystyle\int \frac{\sqrt{x}}{1 + x^{1/4}}\, dx$

** **23.** Find the length of the curve $y = \ln(\cos x)$ from $(0, 0)$ to $(\pi/4, -(\ln 2)/2)$.

** **24.** Find the area bounded by the curves

$$y = \frac{x^2 + 1}{x + 1}, \qquad x + 3y = 7.$$

** **25.** (a) Find the centroid of the area bounded by the curves

$$y = (3 + x)^{3/2}, \qquad x = 1, \qquad y = 0.$$

  (b) What is its second moment of area about the line $x = 1$?

** **26.** If $f(x)$ is a continuous even function, prove that

$$\int_{-a}^{a} f(x)\, dx = 2 \int_{0}^{a} f(x)\, dx;$$

and that if $f(x)$ is a continuous odd function,

$$\int_{-a}^{a} f(x)\, dx = 0.$$

** **27.**   (a) Show that the function $f(x) = \lambda e^{-\lambda x}$, where $\lambda > 0$ is a constant, qualifies as a probability density function on the interval $x \geq 0$. See Exercise 23 in Section 7.10 for the definition of a pdf.

(b) If the probability that $x \geq 3$ is 0.5, what is λ?

** **28.** A paratrooper and his parachute fall from rest with a combined mass of 100 kg. At any instant during descent, the parachute has an air resistance force acting on it that in newtons is equal to one-half its speed (in metres per second). Assuming that the paratrooper falls vertically downward and that the parachute is already open when the jump takes place, the differential equation describing his motion is

$$200\frac{dv}{dt} = 1962 - v,$$

where v is velocity and t is time.

(a) By writing this equation in the form

$$\frac{1}{1962 - v}\frac{dv}{dt} = \frac{1}{200},$$

and antidifferentiating with respect to t, find v as a function of t.

(b) If x measures vertical distance fallen, then $v = dx/dt$. Use this to find the position of the paratrooper as a function of time.

** **29.** Evaluate

$$\int_0^{4\pi} \sqrt{1 + \cos x}\, dx.$$

Hint: $\cos x = 2\cos^2(x/2) - 1$.

** **30.** Evaluate

$$\int \frac{1}{x(3 + 2x^n)}\, dx$$

for any $n > 0$. Hint: Find A and B in order that

$$\frac{1}{x(3 + 2x^n)} = \frac{A}{x} + \frac{Bx^{n-1}}{3 + 2x^n}.$$

*** **31.** Use the substitution $u - x = \sqrt{x^2 + 3x + 4}$ to evaluate

$$\int \frac{1}{(x^2 + 3x + 4)^{3/2}}\, dx.$$

*** **32.** Show that the substitution $u - x = \sqrt{x^2 + bx + c}$ replaces the integral

$$\int \sqrt{x^2 + bx + c}\, dx$$

with the integral of a rational function of u.

*** **33.** Show that when the quadratic $c + bx - x^2$ factors as $c + bx - x^2 = (p + x)(q - x)$, then either of the substitutions $(p + x)u = \sqrt{c + bx - x^2}$ or $(q - x)u = \sqrt{c + bx - x^2}$ replaces the integral

$$\int \frac{1}{\sqrt{c + bc - x^2}}\, dx$$

with the integral of a rational function of u.

*** **34.** If a, b, and n are constants, evaluate

$$\int x(a + bx)^n \, dx.$$

*** **35.** The function

$$f(x) = \frac{1}{\sqrt{2\pi}\sigma} e^{-(x-\mu)^2/(2\sigma^2)}, \quad -\infty < x < \infty,$$

where μ and σ are constants, is called the *normal probability density function*. Consequently,

$$\int_{-\infty}^{\infty} f(x) \, dx = 1.$$

If the expected value of x is defined as

$$\int_{-\infty}^{\infty} x f(x) \, dx,$$

show that μ is the expected value.

*** **36.** Evaluate, if possible,

$$\int_{-\infty}^{\infty} e^{-|a-x|} e^{-|x|} \, dx,$$

where a is a constant.

9.2 Exponential Growth and Decay

Exponential functions play a prominent role in many areas of applied mathematics. This is often due to the fact that the derivative of the exponential function e^{kt} is ke^{kt}; that is, the derivative of e^{kt} is proportional to itself. In this section we exploit this property in a wide variety of applications. When $k > 0$, a quantity represented by e^{kt} is said to experience exponential growth; and when $k < 0$, it experiences exponential decay. We consider an example of each.

Suppose that at some initial time t, which we take as $t = 0$, the number of bacteria in a culture is 2500. Given a proper environment, the culture will grow; by studying this growth over a limited time span, we may be able to predict the number $N(t)$ of bacteria present at any given time. If we carefully observe the culture over a small time interval, perhaps 40 seconds, starting at time $t = 0$, we could draw a graph of $N = N(t)$ as shown in Figure 9.1. Each jump (discontinuity) in this graph indicates the birth of a new bacterium. Analyses at this level are called *microscopic*. We are interested in a more *macroscopic* analysis, with which we might be able to predict when the number of bacteria increases by, say, 25%, or doubles. In order to do so, we take measurements of the number of bacteria at much larger time intervals than those in Figure 9.1. The result is a sequence of points such as that in Table 9.1 which we have plotted in Figure 9.2a.

TABLE 9.1

t (minutes)	0	5	10	15	20	25	30	35	40	45	50
N (number)	2500	2650	2810	2980	3150	3340	3540	3750	3980	4210	4470

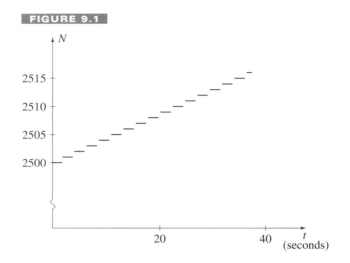

FIGURE 9.1

FIGURE 9.2a   FIGURE 9.2b

If we join these points with a smooth curve as in Figure 9.2b, we obtain a reasonable overview of N as a function of time. Figure 9.2b is, however, only a graphical representation of physical measurements. It is useful to fit a mathematical formula to this graph; that is, to find an algebraic expression for $N(t)$. To do this often requires a great deal of experimentation with different types of curves. For this particular problem, the key is in the following measurements. At any time t we measure, for a very small interval Δt, the corresponding change $N(t + \Delta t) - N(t)$ in N (Figure 9.3).

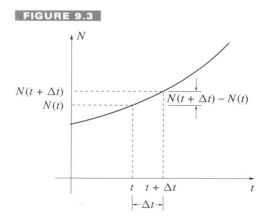

FIGURE 9.3

If we take a number of such measurements, we find that $N(t + \Delta t) - N(t)$ is always proportional to the product of N and Δt; that is,

$$N(t + \Delta t) - N(t) = kN \, \Delta t, \tag{9.2}$$

where $k > 0$ is some constant. In Table 9.2 we have used $\Delta t = 5$ to calculate $N(t+\Delta t)-N(t)$ and $N\Delta t = 5N$. In each case, $N(t + 5) - N(t)$ is approximately 0.012 times $5N$.

TABLE 9.2

t (minutes)	0	5	10	15	20	25	30	35	40	45
$N(t + 5) - N(t)$	150	160	170	170	190	200	210	230	230	250
$5N$	12500	13250	14050	14900	15750	16700	17700	18750	19900	21050
$\frac{N(t+5)-N(t)}{5N}$	0.012	0.012	0.012	0.011	0.012	0.012	0.012	0.012	0.012	0.012

Experiments have shown that this is not a peculiarity of just one type of bacterium, but is characteristic of the growth pattern for many bacteria, the value of k varying from bacterium to bacterium. If we divide both sides of equation 9.2 by Δt,

$$\frac{N(t + \Delta t) - N(t)}{\Delta t} = kN,$$

and take limits as Δt approaches zero, we obtain

$$\frac{dN}{dt} = kN. \tag{9.3}$$

Experimental measurements have therefore led to mathematical law 9.3 for the rate of growth of bacteria in a culture. The time rate of change of the number of bacteria is proportional to the number of bacteria present at that time. The more bacteria are present, the faster their number increases.

To find $N(t)$, we must solve this differential equation subject to the condition: $N(0) = N_0$. If we rewrite equation 9.3 in the form

$$k = \frac{1}{N} \frac{dN}{dt}$$

and take antiderivatives with respect to t, formula 9.1b gives

$$kt + C = \ln |N| = \ln N.$$

We may dispense with the absolute values since N is always positive. Exponentiation of both sides to base e yields

$$e^{kt+C} = e^{\ln N} \qquad \text{or} \qquad N = e^C e^{kt} = De^{kt},$$

where we have set $D = e^C$. The initial condition $N(0) = N_0$ implies that $D = N_0$, and hence,

$$N = N_0 e^{kt}.$$

On the basis of experimental evidence, we have shown that the number of bacteria in a culture increases exponentially. To obtain the constant k, we must specify one additional piece of information. For example, if measurements indicate that the number of bacteria increases by 10% in 50 minutes, then $N(50) = (11/10)N_0$; that is,

$$\frac{11}{10}N_0 = N_0 e^{50k}.$$

When we divide by N_0 on each side and take natural logarithms, we have

$$\ln\left(\frac{11}{10}\right) = 50k, \qquad \text{or} \qquad k = \frac{1}{50}\ln\left(\frac{11}{10}\right) = 0.0019.$$

Finally, the number of bacteria in this particular culture is given by

$$N(t) = N_0 e^{(t/50)\ln(1.1)} = N_0 e^{0.0019t}.$$

An important result of this example is that *all* solutions of the differential equation

$$\frac{dN}{dt} = kN \tag{9.4a}$$

take the form

$$N = Ce^{kt}, \tag{9.4b}$$

where C is a constant. This is simply another statement of the property of the exponential function e^{kt} that its derivative is k times itself. But our derivation shows that the *only* functions that satisfy this property are constants multiplied by e^{kt}; no other function has a derivative which is proportional to itself.

As an example of a decay problem, we continue the discussion of radioactive disintegration begun in Section 2.4. Consider an ore sample that contains, along with various impurities, an amount A_0 of radioactive material, say, uranium. Disintegrations gradually reduce this amount of uranium. To describe the rate at which these disintegrations occur, we let $A(t)$ represent the amount of uranium in the sample at any given time, and choose time $t = 0$ when $A = A_0$. If we were to take measurements and perform an analysis similar to that for bacteria in a culture (see also Section 2.4), we would obtain the following *law of radioactive disintegration*: The time rate of change of the amount of radioactive material is proportional at any instant to the amount of radioactive material present. Algebraically, therefore,

$$\frac{dA}{dt} = kA, \tag{9.5a}$$

where k is a constant. Since A is decreasing, dA/dt and hence k must be negative. Except for the sign of k, this is differential equation 9.4a, and its solution is therefore given by 9.4b,

$$A = Ce^{kt}.$$

The initial condition $A(0) = A_0$ requires $C = A_0$, and hence,

$$A = A_0 e^{kt}. \tag{9.5b}$$

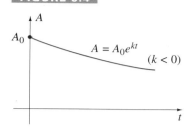

FIGURE 9.4

The amount of radioactive material therefore decreases exponentially, as shown in Figure 9.4. To find k we need to know A at one additional time. For example, if we know that one ten-millionth of one percent of the original amount of uranium decays in 6.5 years, then we have

$$0.999\,999\,999 A_0 = A_0 e^{13k/2}.$$

If we solve this equation for k, we obtain

$$k = \frac{2}{13} \ln(0.999\,999\,999) = -1.54 \times 10^{-10},$$

and, therefore,

$$A = A_0 e^{-1.54 \times 10^{-10} t}.$$

The law of radioactive disintegration has an important application in the dating of once-living plants and animals. All living tissue contains two isotopes of carbon: C^{14} (carbon–14), which is radioactive, and C^{12} (carbon–12), which is stable. In living tissue, the ratio of the amount of C^{14} to that of C^{12} is $1/10\,000$ for all fragments of the tissue. When the tissue dies, however, the ratio changes due to the fact that no more carbon is produced, and the original C^{14} present decays radioactively into an element other than C^{12}. Thus, as the dead tissue ages, the ratio of C^{14} to C^{12} decreases, and by measuring this ratio, it is possible to predict how long ago the tissue was alive. Suppose, for example, the present ratio of C^{14} to C^{12} in a specimen is $1/100\,000$, that is, one-tenth that for a living tissue. Then 90% of the original amount of C^{14} in the specimen has disintegrated.

If we let $A(t)$ be the amount of C^{14} in the specimen at time t, taking $A = A_0$ to be the amount present in the living specimen at its death ($t = 0$), then $A = A_0 e^{kt}$. To determine k, we use the fact that the half-life of C^{14} is approximately 5550 years. (The *half-life* of a radioactive element is the time required for one-half an original sample of the material to disintegrate.) For carbon–14, this means that A is equal to $A_0/2$ when $t = 5550$,

$$\frac{A_0}{2} = A_0 e^{5550k}.$$

When we divide by A_0 and take natural logarithms,

$$k = -\frac{1}{5550} \ln 2 = -0.000\,125.$$

Consequently, the amount of C^{14} in the specimen of dead tissue at any time t is given by

$$A = A_0 e^{-0.000\,125t}.$$

If T is the present time, when the amount of C^{14} is known to be 10% of its original amount, then at this time

$$0.1 A_0 = A_0 e^{-0.000\,125T}.$$

The solution of this equation is

$$T = \frac{\ln 10}{0.000\,125} = 18\,400,$$

and we conclude that the tissue died about 18 400 years ago.

When a hot (or cold) object is placed in an environment that has a different temperature, the object cools down (or heats up). For example, when a hot cup of coffee is placed on a table it cools down due to colder room temperature. Suppose, for example, that the cup of coffee is initially at temperature $95°$ C and the room stays at constant temperature $20°$ C. Newton's law of cooling states that the rate of change of the temperature of the coffee is proportional to the difference between temperature of the coffee and that of the room. If $T(t)$ denotes temperature of the coffee, then

$$\frac{dT}{dt} = k(T - 20), \tag{9.6}$$

where $k < 0$ is a constant. It is negative because $T - 20 > 0$ and $dT/dt < 0$. According to 9.6, the coffee cools quickly at first because $T - 20$ is large, but more and more slowly as T approaches 20. We can solve this differential equation for $T(t)$ by writing

$$\frac{1}{T - 20}\frac{dT}{dt} = k,$$

and integrating with respect to t,

$$\int \frac{1}{T - 20}\frac{dT}{dt}dt = \int k\,dt.$$

The result is

$$\ln|T - 20| = kt + C.$$

We can drop absolute values since temperature of the coffee is never less than $20°$ C. Exponentiating both sides of this equation gives

$$T - 20 = e^{kt+C} = e^C e^{kt} \quad\Longrightarrow\quad T = 20 + De^{kt},$$

where $D = e^C$. If we choose $t = 0$ when the temperature of the coffee is $95°$ C, then $95 = 20 + D \Longrightarrow D = 75$. Thus, temperature of the coffee is

$$T(t) = 20 + 75e^{kt}.$$

To find k we need to know temperature of the coffee at one other time. For example, if we knew that temperature dropped to $50°$ C in 5 min, then

$$50 = 20 + 75e^{5k} \quad\Longrightarrow\quad k = \frac{1}{5}\ln(2/5) = -0.183.$$

Coffee temperature is

$$T(t) = 20 + 75e^{-0.183t}.$$

We can also express $T(t)$ in the form

$$T(t) = 20 + 75e^{(t/5)\ln(2/5)} = 20 + 75e^{\ln[(2/5)^{t/5}]} = 20 + 75\left(\frac{2}{5}\right)^{t/5}.$$

The plot of these functions in Figure 9.5 indicates that coffee temperature never reaches $20°$ C; the graph is asymptotic to the line $T = 20$. However, temperature of the coffee is within $5°$ of $20°$ C in about 15 min, and within $1°$ in about 24 min.

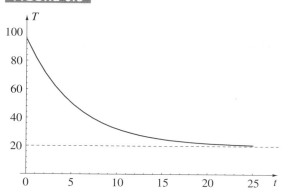

FIGURE 9.5

Admittedly, Newton's law of cooling is an approximation to what really happens physically. It assumes that there is no temperature variation within the cup of coffee and that room temperature remains constant. Idealistic as this is, it is a good starting position for further discussions.

Other examples of quantities that experience exponential growth and decay are discussed in the exercises.

EXERCISES 9.2

∗ **1.** Bacteria in a culture increase at a rate proportional to the number present. If the original number increases by 25% in 2 h, when will it double?

∗ **2.** If the number of bacteria in a culture doubles in 3 h, when will it triple?

∗ **3.** If one-half of a sample of radioactive substance decays in 15 d, how long does it take for 90% of the sample to decay?

∗ **4.** If 10% of a sample of radioactive material decays in 3 s, what is its half-life?

∗ **5.** After 4 half-lives of a radioactive substance, what percentage of the original amount remains?

∗ **6.** In psychology, it is sometimes assumed that if R represents the reaction of a subject to an amount S of stimulus, then dR/dS is inversely proportional to S; that is,

$$\frac{dR}{dS} = \frac{k}{S},$$

where $k > 0$ is a constant. This is called the *Weber-Fechner law*. Find R as a function of S if S_0 is the smallest stimulus detectable by the subject.

∗ **7.** If the Weber-Fechner law in Exercise 6 is replaced by the Brentano-Stevens law,

$$\frac{dR}{dS} = k\frac{R}{S},$$

find $R(S)$. Can the detection threshold S_0 of Exercise 6 be used to determine the constant of integration in this case?

∗ **8.** Suppose the amount of a drug injected into the body decreases at a rate proportional to the amount still present. If a dose decreases by 5% in the first hour, when will it decrease to one-half its original amount?

* **9.** In the arctic, the snowy owl preys on the lemming; the numbers of owls and lemmings in any given area fluctuate in time. If the number of lemmings increases, so does the number of owls, but with some delay. Eventually, an overabundance of owls results, and a scarcity of food (lemmings) to support such a population causes a decrease in the number of owls. The lemming population consequently begins to rise and the cycle repeats. One model to describe the relationship between L, the number of lemmings, and s, the number of snowy owls, states that $s(L)$ must satisfy the differential equation

$$\frac{ds}{dL} = \frac{s(AL - B)}{L(C - Ds)},$$

where A, B, C, and D are positive constants. Solve this differential equation for an implicit definition of $s(L)$.

** **10.** A sugar cube 1 cm on each side is dropped into a cup of coffee. If the sugar dissolves in such a way that the cube always remains a cube, compare the times for the cube to completely dissolve under the following conditions:

(a) Dissolving occurs at a rate proportional to the surface area of the remaining cube; and

(b) Dissolving occurs at a rate proportional to the amount of sugar remaining.

** **11.** Solve Exercise 10 if the sugar is not in the form of a cube, but rather in free form from the sugar bowl. Assume that the sugar consists of n spherical particles each of radius r_0 cm.

** **12.** An analysis of a sample of fossil remains shows that it contains only 1.51% of the original C^{14} in the living creature. When did the creature die?

** **13.** If a fossilized creature died 100 000 years ago, what percentage of the original C^{14} remains?

** **14.** Glucose is administered intravenously to the bloodstream at a constant rate of R units per unit time. As the glucose is added, it is converted by the body into other substances at a rate proportional to the amount of glucose in the blood at that time. Show that the amount of glucose in the blood as a function of time t is given by

$$C(t) = \frac{R}{k}(1 - e^{-kt}) + C_0 e^{-kt},$$

where k is a constant and C_0 is the amount at time $t = 0$ when the intravenous feeding is initiated. Sketch a graph of this function for $C_0 < R/k$ and $C_0 > R/k$.

** **15.** Prove that if a quantity decreases at a rate proportional to its present amount, and if its percentage decrease in some interval of time is $i\%$, then its percentage decrease in any interval of time of the same length is $i\%$.

** **16.** Water at temperature $90°C$ is placed in a room at constant temperature $20°C$. Newton's law of cooling states that the time rate of change of the temperature T of the water is proportional to the difference between T and the temperature of the environment:

$$\frac{dT}{dt} = k(T - 20),$$

where k is a constant. If the water cools to $60°C$ in 40 min, find T as a function of t.

The exponential growth of bacteria as defined by the equation $N = N_0 e^{kt}$ is unrealistic for large t in that there is no maximum for $N(t)$. Draw the growth curve in Exercises 17–20 and show that each has a maximum.

** **17.** The *logistic* growth curve

$$N(t) = \frac{N_0(1 + b)}{1 + be^{-kt}}, \quad t \geq 0,$$

where b, k, and N_0 are positive constants.

** **18.** The *Gompertz* growth curve

$$N(t) = Ae^{-be^{-kt}}, \quad t \geq 0,$$

where A, b, and k are positive constants.

** **19.** The *monomolecular* growth curve

$$N(t) = \frac{N_0}{1 - b}(1 - be^{-kt}), \quad t \geq 0,$$

where $b < 1$, k, and N_0 are positive constants.

** **20.** The *Von Bertalanffy* growth curve

$$N(t) = N_0(1 - be^{-kt})^3, \quad t \geq 0,$$

where $b < 1$, k, and N_0 are positive constants.

** **21.** The differential equation

$$\frac{dN}{dt} = kN(t)\cos t, \quad t \geq 0,$$

where $k > 0$ is a constant, describes a population that undergoes periodic growths and declines. If $N(0) = N_0$, find $N(t)$ and draw its graph.

** **22.** Draw a graph of the sigmoid survival curve

$$s(x) = 1 - \left(1 - e^{-kx}\right)^n, \quad x \geq 0,$$

where $k > 0$ is a constant and n is a positive integer.

** **23.** When a deep-sea diver inhales air, his body tissues absorb extra amounts of nitrogen. Suppose the diver enters the water at time $t = 0$, drops very quickly to depth d, and remains at this depth for a very long time. The amount N of nitrogen in his body tissues increases as he remains at this depth until a maximum amount $\overline{N}$ is reached. The time rate of change of N is proportional to the difference $\overline{N} - N$. Show that if N_0 is the amount of nitrogen in his body tissues when he enters the water, then

$$N = N_0 e^{-kt} + \overline{N}(1 - e^{-kt}),$$

where $k > 0$ is a constant.

** **24.** A dead body is found after the final calculus examination. The coroner arrives on the scene at 9:00 A.M. and immediately records the temperature of the body as $30.1°$C. One hour later she notes that the body's temperature has dropped to $29.2°$C, and that, during the hour, room temperature has remained at $20°$C. Assuming that the body's temperature obeys Newton's law of cooling (Exercise 16), determine when death occurred. (Normal body temperature is $37°$C.)

** **25.** (a) In Section 7.6 we proved that the rate of change of pressure P with respect to vertical distance y within a fluid with density ρ is $dP/dy = -\rho g$, where $g = 9.81$. Since air may be considered a fluid, air pressure above sea level ($y = 0$) must satisfy the equation

$$\frac{dP}{dy} = -\rho g,$$

where $\rho = \rho(y)$ is air density at altitude y. If air is assumed to be a perfect gas, the pressure of a given volume V of gas must satisfy the equation

$$P = \rho R T,$$

where $T = T(y)$ is the temperature of the air and $R > 0$ is a constant. To a reasonable approximation, temperature varies linearly with respect to y:

$$T = T_0 - \lambda y$$

where T_0 is sea level temperature and λ is a constant (about $6.10 \times 10^{-4}°$C per 100 m). Show that $P(y)$ must satisfy the differential equation

$$\frac{dP}{dy} = -\frac{Pg}{R(T_0 - \lambda y)},$$

and solve the equation for

$$P(y) = P_0 \left(1 - \frac{\lambda y}{T_0}\right)^{g/(\lambda R)},$$

where P_0 is sea level pressure.

(b) Find the altitude at which air pressure is one one-millionth that of sea level pressure, if $T_0 = 293°$ K, $g = 9.81$ m/s^2, and $R = 287$ m·N/kg·$°$ K.

** **26.** When light passes through a material, such as water or glass, its intensity decreases with depth of penetration. The Bouguer-Lambert law for absorption of light states that the amount of light absorbed by a small thickness of the material is proportional to the product of this thickness and the intensity of light incident on it.

(a) Show that if I is the intensity of light, then

$$I = I(x) = I_0 e^{-kx},$$

where $k > 0$ is a constant, I_0 is the intensity at the surface of the material, and x is the penetration distance from the surface of the material (figure below).

(b) Give an alternative statement of the Bouguer-Lambert law in terms of rates of change.

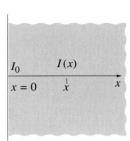

Surface
of
material
I_0 $I(x)$

$x = 0$ x x

*** **27.** (a) To perform a one-hour operation on a dog, a veterinarian anesthetizes the dog with sodium pentobarbital. During the operation the dog's body breaks down the drug at a rate proportional to the amount present, and only half the original dose remains after 5 h. If the dog has mass 20 kg, and 0.02 gm of sodium pentobarbital per kilogram of body mass are required to maintain surgical anesthesia, what original dose is required?

(b) If, after the hour, the operation is unfinished and the dog begins to stir, a further dose of anesthetic must be administered. What extra dose will guarantee surgical anesthesia for an additional 20 min?

9.3 Integration by Parts

One of the most powerful techniques for finding antiderivatives is that called **integration by parts**. It results from the product rule for differentiation:

$$\frac{d}{dx}(uv) = u\frac{dv}{dx} + v\frac{du}{dx}.$$

If we take antiderivatives of both sides of this equation with respect to x, we obtain

$$\int \frac{d}{dx}(uv)\,dx = \int u\frac{dv}{dx}\,dx + \int v\frac{du}{dx}\,dx.$$

Since the antiderivative on the left is uv, this equation may be rewritten in the form

$$\int u\,dv = uv - \int v\,du. \tag{9.7}$$

Equation 9.7 is called the **integration-by-parts formula**. It says that, given an unknown integral (the left side of 9.7), if we arbitrarily divide the function after the integral sign into two parts— one called u and the other called dv—then we can set the unknown integral equal to the right side of 9.7. What this does is replace the integral

$$\int u\,dv \qquad \text{with} \qquad \int v\,du.$$

The problem is to choose u and dv so that the new integral is simpler than the original. For many examples, it is advantageous to let dv be the *most complicated part of the integrand (plus differential) that we can integrate mentally*. The example

$$\int xe^x\,dx$$

will clarify these ideas. To use integration by parts on this problem we must define u and dv in such a way that $u\,dv = xe^x\,dx$. There are four possibilities:

$$
\begin{aligned}
u &= 1, & dv &= xe^x\,dx; \\
u &= x, & dv &= e^x\,dx; \\
u &= e^x, & dv &= x\,dx; \\
u &= xe^x, & dv &= dx.
\end{aligned}
$$

Our rule suggests the choice

$$u = x, \qquad dv = e^x \, dx,$$

in which case

$$du = dx, \qquad v = e^x.$$

Formula 9.7 then gives

$$\int xe^x \, dx = xe^x - \int e^x \, dx.$$

We have therefore replaced integration of xe^x with integration of e^x, a definite simplification. The final solution is therefore

$$\int xe^x \, dx = xe^x - e^x + C.$$

This example illustrates that integration by parts replaces an integral with two others: one mental integration (from dv to v) and a second integration that is hopefully simpler than the original. We should also note that in the evaluation of the mental integration we did not include a constant of integration. For example, in the integration from $dv = e^x \, dx$ to $v = e^x$, we did not write $v = e^x + D$. Had we done so, the solution would have proceeded as follows:

$$\int xe^x \, dx = x(e^x + D) - \int (e^x + D) \, dx$$

$$= x(e^x + D) - e^x - Dx + C$$

$$= xe^x - e^x + C.$$

Constant D has therefore disappeared from the eventual solution. This always occurs and, consequently, we do not include an arbitrary constant in the mental integration for v.

EXAMPLE 9.3

Evaluate the following integrals:

$$\text{(a)} \int \ln x \, dx \qquad \text{(b)} \int x^2 \cos x \, dx \qquad \text{(c)} \int e^x \sin x \, dx$$

SOLUTION

(a) Integration by parts leaves no choice for u and dv in this integral; they must be

$$u = \ln x, \qquad dv = dx,$$

in which case

$$du = \frac{1}{x} \, dx, \qquad v = x.$$

Then we have

$$\int \ln x \, dx = x \ln x - \int x \frac{1}{x} \, dx = x \ln x - x + C.$$

(b) If we set

$$u = x^2, \qquad dv = \cos x \, dx,$$

then

$$du = 2x \, dx, \qquad v = \sin x, \qquad \text{and}$$

$$\int x^2 \cos x \, dx = x^2 \sin x - \int 2x \sin x \, dx.$$

Integration by parts has therefore reduced the power on x from 2 to 1. To eliminate the x in front of the trigonometric function completely, we perform integration by parts once again, this time with

$$u = x, \qquad dv = \sin x \, dx.$$

Then,

$$du = dx, \qquad v = -\cos x, \qquad \text{and}$$

$$\int x^2 \cos x \, dx = x^2 \sin x - 2 \left(-x \cos x - \int -\cos x \, dx \right)$$

$$= x^2 \sin x + 2x \cos x - 2 \sin x + C.$$

(c) If we set

$$u = e^x, \qquad dv = \sin x \, dx,$$

then

$$du = e^x \, dx, \qquad v = -\cos x, \qquad \text{and}$$

$$\int e^x \sin x \, dx = -e^x \cos x - \int -e^x \cos x \, dx.$$

Integration by parts appears to have led nowhere; the integration of $e^x \cos x$ is essentially as difficult as that of $e^x \sin x$. If, however, we persevere and integrate by parts once again with

$$u = e^x, \qquad dv = \cos x \, dx,$$

then

$$du = e^x \, dx, \qquad v = \sin x, \qquad \text{and}$$

$$\int e^x \sin x \, dx = -e^x \cos x + e^x \sin x - \int e^x \sin x \, dx.$$

This equation can now be solved for the unknown integral. By transposing the last term on the right to the left, we find that

$$2 \int e^x \sin x \, dx = e^x (\sin x - \cos x),$$

and therefore,

$$\int e^x \sin x \, dx = \frac{1}{2} e^x (\sin x - \cos x) + C,$$

to which we have added the arbitrary constant C.

The technique in part (c) of this example is often called **integration by reproduction**. Two applications of integration by parts enabled us to reproduce the unknown integral and then solve the equation for it.

EXAMPLE 9.4 ▼

Evaluate

$$\int \frac{x^3}{\sqrt{9+x^2}}\, dx.$$

SOLUTION METHODS

(i) If we set $u = 9 + x^2$, then $du = 2x\,dx$, and

$$\int \frac{x^3}{\sqrt{9+x^2}}\, dx = \int \frac{x^2}{\sqrt{9+x^2}}(x\,dx) = \int \frac{u-9}{\sqrt{u}}\frac{du}{2} = \frac{1}{2}\int (u^{1/2} - 9u^{-1/2})\, du$$

$$= \frac{1}{2}\left(\frac{2}{3}u^{3/2} - 18u^{1/2}\right) + C = \frac{1}{3}(9+x^2)^{3/2} - 9\sqrt{9+x^2} + C.$$

(ii) If we set $u = \sqrt{9+x^2}$, then $du = \left(x/\sqrt{9+x^2}\right)dx$, and

$$\int \frac{x^3}{\sqrt{9+x^2}}\, dx = \int x^2\left(\frac{x}{\sqrt{9+x^2}}\right)dx = \int (u^2 - 9)\, du = \frac{u^3}{3} - 9u + C$$

$$= \frac{1}{3}(9+x^2)^{3/2} - 9\sqrt{9+x^2} + C.$$

(iii) If we set

$$u = x^2, \qquad dv = \frac{x}{\sqrt{9+x^2}}\, dx,$$

then

$$du = 2x\,dx, \qquad v = \sqrt{9+x^2}, \qquad \text{and}$$

$$\int \frac{x^3}{\sqrt{9+x^2}}\, dx = x^2\sqrt{9+x^2} - \int 2x\sqrt{9+x^2}\, dx$$

$$= x^2\sqrt{9+x^2} - 2\left[\frac{1}{3}(9+x^2)^{3/2}\right] + C$$

$$= x^2\sqrt{9+x^2} - \frac{2}{3}(9+x^2)^{3/2} + C.$$

▲

This example illustrates that more than one technique may be effective on an indefinite integral, and the answer may appear different depending on the technique used. According to Theorem 5.1, these solutions can only differ by an additive constant. To illustrate this in the above example, note that the solution obtained by method (iii) can be written as

$$\sqrt{9+x^2}\left[x^2 - \frac{2}{3}(9+x^2)\right] + C = \frac{1}{3}\sqrt{9+x^2}\left[-18 + x^2\right] + C$$

$$= \frac{1}{3}\sqrt{9+x^2}\left[-27 + (9+x^2)\right] + C$$

$$= -9\sqrt{9+x^2} + \frac{1}{3}(9+x^2)^{3/2} + C,$$

which is the solution obtained by the other two methods. It is also worth noting that the three methods work very well provided the power on x in the numerator of the integrand is odd. When it is even, they fail miserably. In such cases, trigonometric substitutions (Section 9.5) come to the rescue (see Example 9.12).

Integration-by-parts formula 9.7 applies to indefinite integrals. For definite integrals, it is replaced by

$$\int_a^b u \, dv = \left\{uv\right\}_a^b - \int_a^b v \, du. \tag{9.8}$$

EXAMPLE 9.5

Evaluate

$$\int_0^4 \frac{x}{\sqrt{x+8}} \, dx.$$

SOLUTION   If we set

$$u = x, \qquad dv = \frac{1}{\sqrt{x+8}} \, dx,$$

then

$$du = dx, \qquad v = 2\sqrt{x+8}, \qquad \text{and}$$

$$\int_0^4 \frac{x}{\sqrt{x+8}} \, dx = \left\{2x\sqrt{x+8}\right\}_0^4 - \int_0^4 2\sqrt{x+8} \, dx$$

$$= 8\sqrt{12} - 2\left\{\frac{2}{3}(x+8)^{3/2}\right\}_0^4$$

$$= \frac{16}{3}\left(4\sqrt{2} - 3\sqrt{3}\right).$$

EXERCISES 9.3

In Exercises 1–18 evaluate the indefinite integral.

* **1.** $\displaystyle\int x \sin x \, dx$

* **2.** $\displaystyle\int x^2 e^{2x} \, dx$

* **3.** $\displaystyle\int x^4 \ln x \, dx$

* **4.** $\displaystyle\int \sqrt{x} \ln (2x) \, dx$

* **5.** $\displaystyle\int z \sec^2 (z/3) \, dz$

* **6.** $\displaystyle\int x\sqrt{3 - x} \, dx$

* **7.** $\displaystyle\int \operatorname{Sin}^{-1} x \, dx$

* **8.** $\displaystyle\int x^2 \sqrt{x + 5} \, dx$

* **9.** $\displaystyle\int \frac{x}{\sqrt{2 + x}} \, dx$

* **10.** $\displaystyle\int \frac{x^2}{\sqrt{2 + x}} \, dx$

* **11.** $\displaystyle\int \frac{x}{\sqrt{2 + x^2}} \, dx$

* **12.** $\displaystyle\int (x - 1)^2 \ln x \, dx$

* **13.** $\displaystyle\int e^x \cos x \, dx$

* **14.** $\displaystyle\int \operatorname{Tan}^{-1} x \, dx$

* **15.** $\displaystyle\int \cos (\ln x) \, dx$

* **16.** $\displaystyle\int e^{2x} \cos 3x \, dx$

* **17.** $\displaystyle\int \frac{x^3}{\sqrt{5 + 3x^2}} \, dx$

* **18.** $\displaystyle\int \ln (x^2 + 4) \, dx$

** **19.** Consider the evaluation of $\int x^5 e^x \, dx$. Were we to use integration by parts, the answer would eventually be of the form

$$\int x^5 e^x \, dx = Ax^5 e^x + Bx^4 e^x + Cx^3 e^x + Dx^2 e^x + Exe^x + Fe^x + G.$$

Differentiate this equation, and thereby obtain A, B, C, D, E, and F (with no integration).

** **20.** The face of a dam is shown in the figure below, where the curve has equation

$$y = e^{k|x|} - 1, \qquad k = \frac{1}{100} \ln 201.$$

Find the total force on the dam due to water pressure when the water level on the dam is 100 m.

** **21.** If we set $u = x^{-1}$ and $dv = dx$, then $du = -x^{-2}\, dx$ and $v = x$, and

$$\int \frac{1}{x}\, dx = \left(\frac{1}{x}\right) x - \int x \left(-\frac{1}{x^2}\, dx\right) = 1 + \int \frac{1}{x}\, dx.$$

Subtraction of $\int x^{-1}\, dx$ from each side of this equation now gives $0 = 1$. What is wrong with the argument?

Fourier series is a topic of mathematics fundamental to all branches of engineering. We are constantly evaluating integrals of the form

$$\int_{-L}^{L} f(x)\, \sin \frac{n\pi x}{L}\, dx \qquad \text{and} \qquad \int_{-L}^{L} f(x)\, \cos \frac{n\pi x}{L}\, dx$$

where $L > 0$ is a fixed constant and n is a positive integer. In Exercises 22–25, evaluate these integrals for the given function $f(x)$.

** **22.** $f(x) = x$ ** **23.** $f(x) = x^2$

** **24.** $f(x) = 1 - 2x$ ** **25.** $f(x) = 2x^2 - 3x$

** **26.** The *gamma function* $\Gamma(n)$ for $n > 0$ is defined by the improper integral

$$\Gamma(n) = \int_{0}^{\infty} x^{n-1} e^{-x}\, dx.$$

Show that when n is a positive integer, $\Gamma(n) = (n - 1)!$.

The *Laplace transform* of a function $f(t)$ is the function $F(s)$ defined by the improper integral

$$F(s) = \int_{0}^{\infty} e^{-st} f(t)\, dt.$$

In Exercises 27–30 find the Laplace transform of the function.

** **27.** $f(t) = e^{3t}$ ** **28.** $f(t) = t^2$

** **29.** $f(t) = \sin t$ ** **30.** $f(t) = te^{-t}$

** **31.** Show that the function

$$f(x) = \frac{x^{\alpha-1} e^{-x/\beta}}{\Gamma(\alpha)\beta^{\alpha}}$$

for positive constants α and β qualifies as a pdf on the interval $x \geq 0$. See Exercise 23 in Section 7.10 for the definition of a pdf and Exercise 26 above for the definition of $\Gamma(\alpha)$.

In the study of *signals and communications* in electrical engineering, the Fourier transform of a function $f(t)$ is

$$F(\omega) = \int_{-\infty}^{\infty} f(t)e^{-i\omega t} dt$$

where i is the complex number whose square is -1; that is, $i^2 = -1$. In Exercises 32–35 find the Fourier transform of the function $f(t)$.

** **32.** $f(t) = \begin{cases} 1, & |t| < L/2 \\ 0, & |t| > L/2 \end{cases}$ $L > 0$ a constant

** **33.** $f(t) = \begin{cases} 0, & |t| > T \\ 1 + t/T, & -T \le t \le 0 \\ 1 - t/T, & 0 \le t \le T \end{cases}$ $T > 0$ a constant

** **34.** $f(t) = e^{-a|t|}, \quad a > 0$ a constant

** **35.** $f(t) = \begin{cases} 0, & t < a \\ 1, & a < t < b \\ 0, & t > b \end{cases}$ a and b constants

In Exercises 36–38 evaluate the indefinite integral.

** **36.** $\displaystyle\int \mathrm{Tan}^{-1}\sqrt{x}\, dx$ ** **37.** $\displaystyle\int x^2 \cos^2 x\, dx$

** **38.** $\displaystyle\int xe^x \sin x\, dx$

*** **39.** Use the substitution $x = \sin^2\theta$ and integration by parts to find a formula for

$$\int_0^1 x^n(1-x)^m dx.$$

*** **40.** Verify that for $n \ge 1$ an integer,

$$\int x^n \cos x\, dx = \sin x \sum_{r=0}^{[n/2]} \frac{(-1)^r n!}{(n-2r)!} x^{n-2r}$$

$$+ \cos x \sum_{r=0}^{[(n-1)/2]} \frac{(-1)^r n!}{(n-2r-1)!} x^{n-2r-1} + C,$$

where the square brackets indicate the greatest integer function of Exercise 50 in Section 1.3.

9.4 Trigonometric Integrals

Differentiation formulas for trigonometric functions suggest three pairings of these functions: sine and cosine, tangent and secant, and cotangent and cosecant. The derivative of either function in a given pair involves at most functions of that pair. Note that integration formulas 9.1e–n also substantiate this pairing. Each antiderivative gives functions in the same pairing. Because of this, a first step in every integral involving trigonometric functions is to rewrite the integrand so that each term involves only one pair. For example, to evaluate

$$\int \frac{\tan x + \sec x \cot^2 x}{\sin^3 x}\, dx,$$

we would first rewrite it in the form

$$\int \left(\frac{\csc^3 x}{\cot x} + \frac{\cos x}{\sin^5 x} \right) dx.$$

The first term contains cosecants and cotangents, and the second has sines and cosines. We now consider integrals involving each pair.

Integrals Involving the Sine and Cosine

We frequently encounter integrals of the form

$$\int \cos^n x \sin^m x \, dx,$$

where m and n may or may not be integers. The key to evaluation of this type of integral is recognition of the fact that when either $n = 1$ or $m = 1$ the integrations become straightforward. To illustrate this, consider differentiation of $\sin^{n+1} x$ by power rule 3.21,

$$\frac{d}{dx} \sin^{n+1} x = (n + 1) \sin^n x \cos x.$$

This means that differentiation of a power of $\sin x$ leads to a power of $\sin x$ one lower than the original, multiplied by $\cos x$. In order to reverse the procedure and antidifferentiate a power of $\sin x$, we need a factor of $\cos x$. It then becomes a matter of raising the power on $\sin x$ by one and dividing by the new power; that is, we can say

$$\int \sin^n x \cos x \, dx = \frac{1}{n + 1} \sin^{n+1} x + C, \quad n \neq -1. \tag{9.9a}$$

Similarly,

$$\int \cos^n x \sin x \, dx = -\frac{1}{n + 1} \cos^{n+1} x + C, \quad n \neq -1. \tag{9.9b}$$

Formulas 9.1l and 9.1k contain the $n = -1$ cases for these integrals. Of course, a substitution can always be made to evaluate integrals 9.9a and b. For example, if in 9.9a we set $u = \sin x$, then $du = \cos x \, dx$, and

$$\int \sin^n x \cos x \, dx = \int u^n \, du = \frac{u^{n+1}}{n + 1} + C = \frac{\sin^{n+1} x}{n + 1} + C.$$

Also useful are the formulas

$$\int \sin nx \, dx = -\frac{1}{n} \cos nx + C, \qquad \text{and} \tag{9.9c}$$

$$\int \cos nx \, dx = \frac{1}{n} \sin nx + C. \tag{9.9d}$$

Trigonometric identities are used to write integrands in the forms contained in equations 9.9. Particularly helpful are

$$\sin^2 x + \cos^2 x = 1, \tag{9.10a}$$

$$\sin 2x = 2 \sin x \cos x, \tag{9.10b}$$

$$\cos 2x = 2 \cos^2 x - 1, \tag{9.10c}$$

$$\cos 2x = 1 - 2 \sin^2 x. \tag{9.10d}$$

EXAMPLE 9.6   ▼

Evaluate the following integrals:

(a) $\displaystyle\int \sin^3 x \, dx$ (b) $\displaystyle\int \sin^3 x \cos^5 x \, dx$ (c) $\displaystyle\int \sin 2x \cos^6 2x \, dx$

(d) $\displaystyle\int \sqrt{\sin x} \cos^3 x \, dx$ (e) $\displaystyle\int \frac{\cos^3 3x}{\sin^2 3x} \, dx$ (f) $\displaystyle\int \sin^2 x \, dx$

(g) $\displaystyle\int \cos^4 x \, dx$ (h) $\displaystyle\int \sin^2 x \cos^2 x \, dx$

SOLUTION

(a) With identity 9.10a, we may write

$$\int \sin^3 x \, dx = \int \sin x (1 - \cos^2 x) \, dx = \int (\sin x - \cos^2 x \sin x) \, dx,$$

and the second term is in form 9.9b. Thus,

$$\int \sin^3 x \, dx = -\cos x + \frac{1}{3} \cos^3 x + C.$$

(b) Once again we use 9.10a to rewrite the integrand as two easily integrated terms:

$$\int \sin^3 x \cos^5 x \, dx = \int \sin x (1 - \cos^2 x) \cos^5 x \, dx$$

$$= \int (\cos^5 x \sin x - \cos^7 x \sin x) \, dx$$

$$= -\frac{1}{6} \cos^6 x + \frac{1}{8} \cos^8 x + C.$$

(c) This integrand is already in a convenient form for integration. Guess and fix up leads to

$$\int \sin 2x \cos^6 2x \, dx = -\frac{1}{14} \cos^7 2x + C.$$

(d) Using 9.10a, we obtain

$$\int \sqrt{\sin x}\, \cos^3 x \, dx = \int \sin^{1/2} x \cos x \, (1 - \sin^2 x) \, dx$$

$$= \int (\sin^{1/2} x \cos x - \sin^{5/2} x \cos x) \, dx$$

$$= \frac{2}{3} \sin^{3/2} x - \frac{2}{7} \sin^{7/2} x + C.$$

(e) Again we use 9.10a,

$$\int \frac{\cos^3 3x}{\sin^2 3x} \, dx = \int \frac{\cos 3x \,(1 - \sin^2 3x)}{\sin^2 3x} \, dx = \int \left(\frac{\cos 3x}{\sin^2 3x} - \cos 3x \right) dx$$

$$= \frac{-1}{3 \sin 3x} - \frac{1}{3} \sin 3x + C.$$

In these five examples, at least one of the powers on the sine and cosine functions was odd, and in each case we used identity 9.10a. In the next three examples, where all powers are even, identities 9.10b,c,d are useful.

(f) Double-angle formula 9.10d can be rearranged as $\sin^2 x = (1 - \cos 2x)/2$, and therefore

$$\int \sin^2 x \, dx = \frac{1}{2} \int (1 - \cos 2x) \, dx.$$

Formula 9.9d can now be used on the second term

$$\int \sin^2 x \, dx = \frac{1}{2} \left(x - \frac{1}{2} \sin 2x \right) + C.$$

(g) With double-angle formula 9.10c rewritten in the form $\cos^2 x = (1 + \cos 2x)/2$, we obtain

$$\int \cos^4 x \, dx = \frac{1}{4} \int (1 + \cos 2x)^2 \, dx = \frac{1}{4} \int (1 + 2 \cos 2x + \cos^2 2x) \, dx.$$

When x is replaced by $2x$ in 9.10c, the result is $\cos 4x = 2 \cos^2 2x - 1$. This can be rearranged as $\cos^2 2x = (1 + \cos 4x)/2$, and therefore

$$\int \cos^4 x \, dx = \frac{1}{4} \int \left(1 + 2 \cos 2x + \frac{1 + \cos 4x}{2} \right) dx.$$

We now use 9.9d on the two cosine terms,

$$\int \cos^4 x \, dx = \frac{1}{4} \left(\frac{3x}{2} + \sin 2x + \frac{1}{8} \sin 4x \right) + C.$$

(h) In this integral we use identities 9.10b and d,

$$\int \sin^2 x \cos^2 x \, dx = \int (\sin x \cos x)^2 \, dx = \frac{1}{4} \int \sin^2 2x \, dx$$

$$= \frac{1}{4} \int \left(\frac{1 - \cos 4x}{2} \right) dx = \frac{1}{8} \left(x - \frac{1}{4} \sin 4x \right) + C.$$

EXAMPLE 9.7

Evaluate

$$\int \sin 5x \cos 2x \, dx.$$

SOLUTION   Unlike the previous example in which all trigonometric functions had the same argument, this integrand is the product of trigonometric functions with different arguments. Our first step is to rewrite the integrand so that we do not have such a product. Note that a sum is obviously integrable, but not a product. The product formula

$$\sin A \cos B = \frac{1}{2} [\sin (A + B) + \sin (A - B)]$$

can be used to advantage here. We can rewrite the integrand in the form

$$\int \sin 5x \cos 2x \, dx = \int \frac{1}{2} (\sin 7x + \sin 3x) \, dx,$$

and now integration is easily handled by 9.9c,

$$\int \sin 5x \cos 2x \, dx = \frac{1}{2} \left(-\frac{1}{7} \cos 7x - \frac{1}{3} \cos 3x \right) + C = -\frac{1}{14} \cos 7x - \frac{1}{6} \cos 3x + C.$$

Two integration by parts (and reproduction) can also be used to evaluate this integral, but the above method is much simpler.

Integrals Involving Tangent and Secant

For integrals involving tangent and secant, we have two alternatives: rewrite the integrand in terms of sines and cosines, or express the integrand in terms of easily integrated combinations of tangent and secant. It is usually more fruitful to investigate the second possibility before expressing the integrand completely in terms of sines and cosines. In this regard we note that combinations of tangent and secant that can be integrated immediately are

$$\int \tan^n x \sec^2 x \, dx = \frac{\tan^{n+1} x}{n + 1} + C, \quad n \neq -1; \tag{9.11a}$$

$$\int \sec^n x \tan x \, dx = \frac{\sec^n x}{n} + C, \quad n \neq 0. \tag{9.11b}$$

Once again these results are suggested by differentiations of powers of $\tan x$ and $\sec x$. Alternatively, these integrals can be evaluated with substitutions: $u = \tan x$ for 9.11a and $u = \sec x$ for 9.11b.

To rearrange integrands into these integrable combinations, we use the identity

$$1 + \tan^2 x = \sec^2 x. \tag{9.12}$$

EXAMPLE 9.8

Evaluate the following integrals:

(a) $\displaystyle\int \tan^2 x \, dx$ (b) $\displaystyle\int \tan^3 x \, dx$ (c) $\displaystyle\int \sec^4 x \, dx$

(d) $\displaystyle\int \tan^4 3x \sec^2 3x \, dx$ (e) $\displaystyle\int \tan^3 x \sec^3 x \, dx$ (f) $\displaystyle\int \frac{\sec^4 x}{\tan^2 x} \, dx$

SOLUTION

(a) $$\int \tan^2 x \, dx = \int (\sec^2 x - 1) \, dx = \tan x - x + C$$

(b) $$\int \tan^3 x \, dx = \int \tan x \, (\sec^2 x - 1) \, dx$$

$$= \int (\sec^2 x \tan x - \tan x) \, dx$$

$$= \frac{1}{2} \sec^2 x + \ln |\cos x| + C, \text{ or } \frac{1}{2} \tan^2 x + \ln |\cos x| + C$$

(c) $$\int \sec^4 x \, dx = \int \sec^2 x \, (1 + \tan^2 x) \, dx$$

$$= \int (\sec^2 x + \tan^2 x \sec^2 x) \, dx$$

$$= \tan x + \frac{1}{3} \tan^3 x + C$$

(d) $$\int \tan^4 3x \sec^2 3x \, dx = \frac{1}{15} \tan^5 3x + C$$

(e) $$\int \tan^3 x \sec^3 x \, dx = \int \tan x \, (\sec^2 x - 1) \sec^3 x \, dx$$

$$= \int (\sec^5 x \tan x - \sec^3 x \tan x) \, dx$$

$$= \frac{1}{5} \sec^5 x - \frac{1}{3} \sec^3 x + C$$

(f)
$$\int \frac{\sec^4 x}{\tan^2 x} \, dx = \int \frac{\sec^2 x \, (1 + \tan^2 x)}{\tan^2 x} \, dx$$

$$= \int \left(\frac{\sec^2 x}{\tan^2 x} + \sec^2 x \right) dx$$

$$= -\frac{1}{\tan x} + \tan x + C$$

EXAMPLE 9.9

Evaluate

$$\int \sec^3 x \, dx.$$

SOLUTION   We use integration by parts and reproduction. By setting

$$u = \sec x, \qquad dv = \sec^2 dx,$$

then

$$du = \sec x \tan x \, dx, \qquad v = \tan x,$$

and formula 9.7 gives

$$\int \sec^3 x \, dx = \sec x \tan x - \int \sec x \tan^2 x \, dx.$$

Since $\tan^2 x = \sec^2 x - 1$, we can write

$$\int \sec^3 x \, dx = \sec x \tan x - \int \sec x \, (\sec^2 x - 1) \, dx$$

$$= \sec x \tan x - \int \sec^3 x \, dx + \ln | \sec x + \tan x |.$$

We now solve for

$$\int \sec^3 x \, dx = \frac{1}{2} (\sec x \tan x + \ln | \sec x + \tan x |) + C.$$

Integrands in the following example cannot be transformed by means of identity 9.12 into forms 9.11, and we therefore express them in terms of sines and cosines.

EXAMPLE 9.10

Evaluate the following integrals:

$$\text{(a)} \int \frac{\tan^2 2x}{\sec^3 2x}\, dx \qquad \text{(b)} \int \frac{1}{\sec x \tan^2 x}\, dx$$

SOLUTION

(a)
$$\int \frac{\tan^2 2x}{\sec^3 2x}\, dx = \int \sin^2 2x \cos 2x\, dx = \frac{1}{6}\sin^3 2x + C$$

(b)
$$\int \frac{1}{\sec x \tan^2 x}\, dx = \int \frac{\cos^3 x}{\sin^2 x}\, dx = \int \frac{\cos x\,(1 - \sin^2 x)}{\sin^2 x}\, dx$$
$$= \int \left(\frac{\cos x}{\sin^2 x} - \cos x \right) dx = -\frac{1}{\sin x} - \sin x + C$$

Integrals Involving Cotangent and Cosecant

Since derivatives of the cotangent and cosecant pair

$$\frac{d}{dx}\cot x = -\csc^2 x, \qquad \frac{d}{dx}\csc x = -\csc x \cot x$$

are analogous to those of the tangent and secant pair

$$\frac{d}{dx}\tan x = \sec^2 x, \qquad \frac{d}{dx}\sec x = \sec x \tan x,$$

a discussion parallel to that concerning the tangent and secant could be made here. Aside from the fact that cotangent replaces tangent and cosecant replaces secant, there are also sign changes. Examples can be found in the exercises.

EXERCISES 9.4

In Exercises 1–20 evaluate the indefinite integral.

* **1.** $\displaystyle\int \cos^3 x \sin x \, dx$

* **2.** $\displaystyle\int \frac{\cos x}{\sin^3 x} \, dx$

* **3.** $\displaystyle\int \tan^5 x \sec^2 x \, dx$

* **4.** $\displaystyle\int \csc^3 x \cot x \, dx$

* **5.** $\displaystyle\int \cos^3 (x + 2) \, dx$

* **6.** $\displaystyle\int \sqrt{\tan x} \sec^4 x \, dx$

* **7.** $\displaystyle\int \frac{1}{\sin^4 t} \, dt$

* **8.** $\displaystyle\int \sec^6 3x \tan 3x \, dx$

* **9.** $\displaystyle\int \cos^2 x \, dx$

* **10.** $\displaystyle\int \frac{\tan^3 x \sec^2 x}{\sin^2 x} \, dx$

* **11.** $\displaystyle\int \sin^3 y \cos^2 y \, dy$

* **12.** $\displaystyle\int \frac{\csc^2 \theta}{\cot^2 \theta} \, d\theta$

* **13.** $\displaystyle\int \frac{\sin \theta}{1 + \cos \theta} \, d\theta$

* **14.** $\displaystyle\int \frac{\sec^2 x}{\sqrt{1 + \tan x}} \, dx$

* **15.** $\displaystyle\int \cos \theta \sin 2\theta \, d\theta$

* **16.** $\displaystyle\int \frac{3 + 4 \csc^2 x}{\cot^2 x} \, dx$

* **17.** $\displaystyle\int \sin^5 x \cos^5 x \, dx$

* **18.** $\displaystyle\int \sin^4 x \, dx$

* **19.** $\displaystyle\int \frac{\tan^3 x}{\sec^4 x} \, dx$

* **20.** $\displaystyle\int \frac{\csc^4 x}{\cot^3 x} \, dx$

* **21.** Find the area between the x-axis and $y = \sin x$ for $0 \le x \le \pi$.

** **22.** Find the volume of the solid of revolution that is obtained by rotating the area bounded by $y = \tan x$, $y = 0$, and $x = \pi/4$ about the line $y = -1$.

** **23.** Evaluate

$$\int_0^\pi \sqrt{1 - \sin^2 x} \, dx.$$

In Exercises 24–31 evaluate the indefinite integral.

** **24.** $\displaystyle\int \cot^4 z \, dz$

** **25.** $\displaystyle\int \frac{\cos^3 \theta}{3 + \sin \theta} \, d\theta$

** **26.** $\displaystyle\int \frac{\cos^4 \theta}{1 + \sin \theta} \, d\theta$

** **27.** $\displaystyle\int \sin^4 x \cos^2 x \, dx$

** **28.** $\displaystyle\int \cos 6x \cos 2x \, dx$

** **29.** $\displaystyle\int \cos^2 2x \sin 3x \, dx$

** **30.** $\displaystyle\int \frac{1}{\sin x \cos^2 x} \, dx$

** **31.** $\displaystyle\int \sec^5 x \, dx$

** **32.** The alternating current in a power line is given by

$$I = f(t) = A \cos \omega t + B \sin \omega t,$$

where A, B, and ω are constants, and t is time. The root-mean-square (rms) current I_{rms} is defined by

$$(I_{rms})^2 = \frac{1}{T} \int_0^T I^2 \, dt,$$

where the interval $0 \le t \le T$ represents any number of complete oscillations of the current. Show that I_{rms} is $1/\sqrt{2}$ times the amplitude of the current.

The dc-value of a periodic waveform $f(t)$ with period p in the theory of signals is defined as

$$F_{dc} = \frac{1}{p} \int_{-p/2}^{p/2} f(t) \, dt.$$

In Exercises 33–36 find the dc-value for the waveform.

** **33.** $f(t)$ of Exercise 32

** **34.** $f(t) = A + B \cos \omega t$, A, B, ω constants

** **35.** $f(t) = \sin^2 \omega t$, ω constant

** **36.** $f(t) = t$, $0 < t < 1$, $f(t+1) = f(t)$

*** **37.** Evaluate

$$\int (1 + \sin \theta + \sin^2 \theta + \sin^3 \theta + \cdots) \, d\theta.$$

*** **38.** If n is a positive even integer, show that

$$\int \sec^n x \, dx = \sum_{r=0}^{n/2-1} \binom{n/2 - 1}{r} \frac{\tan^{2r+1} x}{2r + 1} + C.$$

*** **39.** A set of functions $f_1(x)$, $f_2(x)$, $\ldots$ is said to be orthonormal over an interval $a \le x \le b$ if

$$\int_a^b f_n(x) f_m(x) \, dx = \begin{cases} 0 & m \ne n \\ 1 & m = n. \end{cases}$$

Show that the set of functions

$$\frac{1}{\sqrt{2\pi}}, \quad \frac{1}{\sqrt{\pi}} \sin x, \quad \frac{1}{\sqrt{\pi}} \cos x, \quad \frac{1}{\sqrt{\pi}} \sin 2x, \quad \frac{1}{\sqrt{\pi}} \cos 2x, \quad \frac{1}{\sqrt{\pi}} \sin 3x, \quad \ldots$$

is orthonormal over the interval $0 \le x \le 2\pi$. Orthonormality of these functions is the basis for *Fourier series*, a topic of fundamental importance in all branches of engineering.

9.5 Trigonometric Substitutions

Physical and geometric problems arising from circles, ellipses and hyperbolas often give rise to integrals involving the square roots $\sqrt{a^2 - b^2x^2}$, $\sqrt{a^2 + b^2x^2}$, and $\sqrt{b^2x^2 - a^2}$. *Trigonometric substitutions* replace them with integrals involving trigonometric functions.

Consider first two integrals that involve square roots of the form $\sqrt{a^2 - b^2 x^2}$:

$$\text{(i)} \quad \int \frac{1}{\sqrt{1 - x^2}} \, dx \quad \text{and} \quad \text{(ii)} \quad \int \frac{1}{\sqrt{4 - 9x^2}} \, dx.$$

Integral (i) is evident from differentiation of the inverse trigonometric functions in Section 8.3,

$$\int \frac{1}{\sqrt{1 - x^2}} \, dx = \text{Sin}^{-1} x + C.$$

We now show that a substitution can be made on this integral which is applicable to much more difficult problems involving the square root $\sqrt{a^2 - b^2 x^2}$, including (ii). If instead of the square root $\sqrt{1 - x^2}$ in (i) we had the square root $\sqrt{1 - \sin^2 \theta}$, we could immediately simplify the latter root to $\sqrt{\cos^2 \theta} = |\cos \theta|$. This prompts us to make the substitution $x = \sin \theta$, from which we have $dx = \cos \theta \, d\theta$. Integral (i) then becomes

$$\int \frac{1}{\sqrt{1 - x^2}} \, dx = \int \frac{1}{|\cos \theta|} \cos \theta \, d\theta.$$

To eliminate the absolute values, we need to know whether $\cos \theta$ is positive or negative. But this means that we must know the possible values of θ. There is a problem. The equation $x = \sin \theta$ does not really define θ; for any given x, we do not have a unique θ, but an infinite number of possibilities. We must therefore restrict the values of θ, and we do this by specifying that $\theta = \text{Sin}^{-1} x$. In other words, although we have used the equation $x = \sin \theta$ to change variables, and will continue to do so, it is really the equation $\theta = \text{Sin}^{-1} x$ that properly defines the change. Since θ so defined must lie in the interval $-\pi/2 \leq \theta \leq \pi/2$ (the principal values of the inverse sine function), it follows that $\cos \theta \geq 0$, and absolute values may be dropped. Consequently,

$$\int \frac{1}{\sqrt{1 - x^2}} \, dx = \int \frac{\cos \theta}{|\cos \theta|} \, d\theta = \int d\theta = \theta + C = \text{Sin}^{-1} x + C.$$

For integral (ii), the substitution $x = \sin \theta$ is of little use since the square root $\sqrt{4 - 9 \sin^2 \theta}$ does not simplify. If, however, the square root were $\sqrt{4 - 4 \sin^2 \theta}$, it would immediately reduce to $\sqrt{4 - 4 \sin^2 \theta} = \sqrt{4 \cos^2 \theta} = 2|\cos \theta|$. We can obtain this result if the substitution is modified to $x = (2/3) \sin \theta$ (or, more properly, $\theta = \text{Sin}^{-1} (3x/2)$). Since $dx = (2/3) \cos \theta \, d\theta$, we have

$$\int \frac{1}{\sqrt{4 - 9x^2}} \, dx = \int \frac{(2/3) \cos \theta}{\sqrt{4 - 4 \sin^2 \theta}} \, d\theta = \frac{2}{3} \int \frac{\cos \theta}{2|\cos \theta|} \, d\theta.$$

Once again absolute values may be dropped because $-\pi/2 \leq \theta \leq \pi/2$:

$$\int \frac{1}{\sqrt{4 - 9x^2}} dx = \frac{1}{3} \int d\theta = \frac{\theta}{3} + C = \frac{1}{3} \text{Sin}^{-1} \left(\frac{3x}{2} \right) + C.$$

For integrals containing square roots of the form $\sqrt{a^2 + b^2 x^2}$, consider the pair of integrals

$$(iii) \quad \int \frac{1}{1 + x^2} \, dx \qquad (iv) \quad \int \frac{1}{(3 + 5x^2)^{3/2}} \, dx.$$

Integral (iii) obviously has answer $\text{Tan}^{-1} x + C$. Alternatively, if we had $1 + \tan^2 \theta$ instead of $1 + x^2$, the two terms in the denominator would simplify to the one term $\sec^2 \theta$. We therefore substitute $x = \tan \theta$ or, more properly, $\theta = \text{Tan}^{-1} x$, from which we have $dx = \sec^2 \theta \, d\theta$. With this substitution, we obtain

$$\int \frac{1}{1 + x^2} \, dx = \int \frac{\sec^2 \theta}{\sec^2 \theta} \, d\theta = \int d\theta = \theta + C = \text{Tan}^{-1} x + C.$$

The substitution $x = \tan \theta$ in (iv) yields $(3 + 5x^2)^{3/2} = (3 + 5\tan^2 \theta)^{3/2}$. For simplification, we need a 3 in front of the $\tan^2 \theta$ and not a 5. This can be accomplished by modification of our substitution to

$$x = \sqrt{\frac{3}{5}} \tan \theta,$$

from which

$$dx = \sqrt{\frac{3}{5}} \sec^2 \theta \, d\theta.$$

Then,

$$\int \frac{1}{(3 + 5x^2)^{3/2}} \, dx = \int \frac{1}{(3 + 3\tan^2 \theta)^{3/2}} \sqrt{\frac{3}{5}} \sec^2 \theta \, d\theta = \frac{1}{3\sqrt{5}} \int \frac{\sec^2 \theta}{(\sec^2 \theta)^{3/2}} \, d\theta$$

$$= \frac{1}{3\sqrt{5}} \int \frac{1}{\sec \theta} \, d\theta = \frac{1}{3\sqrt{5}} \int \cos \theta \, d\theta = \frac{1}{3\sqrt{5}} \sin \theta + C.$$

FIGURE 9.6

To express $\sin \theta$ in terms of x, we draw the triangle in Figure 9.6 to fit the change of variable $\sqrt{5}x/\sqrt{3} = \tan \theta$. Since the hypotenuse of the triangle is $\sqrt{3 + 5x^2}$, we obtain

$$\int \frac{1}{(3 + 5x^2)^{3/2}} \, dx = \frac{1}{3\sqrt{5}} \frac{\sqrt{5}x}{\sqrt{3 + 5x^2}} + C = \frac{x}{3\sqrt{3 + 5x^2}} + C.$$

For integrals containing square roots of the form $\sqrt{b^2 x^2 - a^2}$, consider

$$\int \frac{1}{\sqrt{x^2 - 4}} \, dx.$$

If we had $\sqrt{4\sec^2 \theta - 4}$ instead of $\sqrt{x^2 - 4}$, an immediate simplification would occur. We therefore substitute $x = 2\sec \theta$, from which $dx = 2\sec \theta \tan \theta \, d\theta$. Once again the real substitution is $\theta = \text{Sec}^{-1}(x/2)$, in which case θ is an angle in either the first or third quadrant. With this change,

$$\sqrt{x^2 - 4} = \sqrt{4\sec^2 \theta - 4} = \sqrt{4\tan^2 \theta} = 2|\tan \theta| = 2\tan \theta$$

and

$$\int \frac{1}{\sqrt{x^2 - 4}} \, dx = \int \frac{2\sec \theta \tan \theta}{2\tan \theta} \, d\theta = \int \sec \theta \, d\theta = \ln|\sec \theta + \tan \theta| + C.$$

FIGURE 9.7

To express $\tan\theta$ in terms of x, we draw the triangle in Figure 9.7 to fit the change of variable $x/2 = \sec\theta$. Since the third side is $\sqrt{x^2 - 4}$, we have

$$\int \frac{1}{\sqrt{x^2 - 4}}\, dx = \ln\left|\frac{x}{2} + \frac{\sqrt{x^2 - 4}}{2}\right| + C = \ln\left|x + \sqrt{x^2 - 4}\right| + D,$$

where $D = C - \ln 2$.

We have illustrated by examples that trigonometric substitutions can be useful in the evaluation of integrals involving the square roots $\sqrt{a^2 - b^2 x^2}$, $\sqrt{a^2 + b^2 x^2}$, and $\sqrt{b^2 x^2 - a^2}$. Essentially, the method replaces terms under the square root by a perfect square and thereby rids the integrand of the square root. To obtain the trigonometric substitution appropriate to each square root, we suggest the following procedure. We have been working with the trigonometric identities:

$$1 - \sin^2\theta = \cos^2\theta,$$

$$1 + \tan^2\theta = \sec^2\theta,$$

$$\sec^2\theta - 1 = \tan^2\theta.$$

To determine the trigonometric substitution appropriate to, say $\sqrt{a^2 - b^2 x^2}$, we mentally set $a = b = 1$ and obtain $\sqrt{1 - x^2}$. We note that $1 - x^2$ resembles $1 - \sin^2\theta$, and immediately set $x = \sin\theta$. To simplify the original square root $\sqrt{a^2 - b^2 x^2}$, we then modify this substitution to

$$x = \frac{a}{b}\sin\theta \qquad \left[\text{or } \theta = \text{Sin}^{-1}\left(\frac{bx}{a}\right)\right].$$

Similarly, for the square root $\sqrt{a^2 + b^2 x^2}$, we initially set $x = \tan\theta$, having mentally set $a = b = 1$, and noted that $1 + x^2$ resembles $1 + \tan^2\theta$. We then modify the substitution to

$$x = \frac{a}{b}\tan\theta \qquad \left[\text{or } \theta = \text{Tan}^{-1}\left(\frac{bx}{a}\right)\right].$$

Finally, for the square root $\sqrt{b^2 x^2 - a^2}$, we initially set $x = \sec\theta$ and then modify to

$$x = \frac{a}{b}\sec\theta \qquad \left[\text{or } \theta = \text{Sec}^{-1}\left(\frac{bx}{a}\right)\right].$$

The square roots $\sqrt{a^2 - b^2 x^2}$, $\sqrt{a^2 + b^2 x^2}$, and $\sqrt{b^2 x^2 - a^2}$ are replaced respectively by

$$\begin{array}{ccc}
\sqrt{a^2 - a^2\sin^2\theta} & \sqrt{a^2 + a^2\tan^2\theta} & \sqrt{a^2\sec^2\theta - a^2} \\
= \sqrt{a^2\cos^2\theta} & = \sqrt{a^2\sec^2\theta} & \sqrt{a^2\tan^2\theta} \\
= a|\cos\theta| & = a|\sec\theta| & = a|\tan\theta| \\
= a\cos\theta; & = a\sec\theta; & = a\tan\theta.
\end{array}$$

In each case it is our choice of principal values for the inverse trigonometric functions that enables us to neglect absolute values.

We now consider more complex examples.

EXAMPLE 9.11

Evaluate the following integrals:

$$\text{(a)} \int \frac{1}{x^2(9x^2+2)}\,dx \qquad \text{(b)} \int x^2\sqrt{3-x^2}\,dx$$

SOLUTION

(a) If we set $x = \left(\sqrt{2}/3\right)\tan\theta$, then $dx = \left(\sqrt{2}/3\right)\sec^2\theta\,d\theta$, and

$$\int \frac{1}{x^2(9x^2+2)}\,dx = \int \frac{1}{\left(\dfrac{2}{9}\tan^2\theta\right)(2\tan^2\theta + 2)}\left(\frac{\sqrt{2}}{3}\right)\sec^2\theta\,d\theta$$

$$= \frac{3}{2\sqrt{2}}\int \frac{\sec^2\theta}{\tan^2\theta\ \sec^2\theta}\,d\theta$$

$$= \frac{3}{2\sqrt{2}}\int \cot^2\theta\,d\theta = \frac{3}{2\sqrt{2}}\int (\csc^2\theta - 1)\,d\theta$$

$$= \frac{3}{2\sqrt{2}}(-\cot\theta - \theta) + C$$

$$= \frac{-3}{2\sqrt{2}}\left[\frac{\sqrt{2}}{3x} + \text{Tan}^{-1}\!\left(\frac{3x}{\sqrt{2}}\right)\right] + C \qquad\text{(Figure 9.8)}$$

$$= \frac{-1}{2x} - \frac{3}{2\sqrt{2}}\text{Tan}^{-1}\!\left(\frac{3x}{\sqrt{2}}\right) + C.$$

FIGURE 9.8

$\sqrt{9x^2+2}$

$3x$

θ

$\sqrt{2}$

(b) If we set $x = \sqrt{3}\sin\theta$, then $dx = \sqrt{3}\cos\theta\,d\theta$, and

$$\int x^2\sqrt{3-x^2}\,dx = \int (3\sin^2\theta)\sqrt{3 - 3\sin^2\theta}\,(\sqrt{3}\cos\theta\,d\theta)$$

$$= 9\int \sin^2\theta\sqrt{1 - \sin^2\theta}\,\cos\theta\,d\theta$$

$$= 9\int \sin^2\theta\,\cos^2\theta\,d\theta$$

$$= \frac{9}{4}\int \sin^2 2\theta\,d\theta \qquad\text{(by 9.10b)}$$

$$= \frac{9}{4}\int \left(\frac{1 - \cos 4\theta}{2}\right)d\theta \qquad\text{(by 9.10d)}$$

$$= \frac{9}{8}\left(\theta - \frac{1}{4}\sin 4\theta\right) + C \qquad\text{(by 9.9d)}$$

$$= \frac{9}{8}\text{Sin}^{-1}\!\left(\frac{x}{\sqrt{3}}\right) - \frac{9}{32}(2\sin 2\theta\cos 2\theta) + C \qquad\text{(by 9.10b)}$$

$$= \frac{9}{8}\text{Sin}^{-1}\!\left(\frac{x}{\sqrt{3}}\right) - \frac{9}{16}(2\sin\theta\cos\theta)(1 - 2\sin^2\theta) + C$$

$$\text{(by 9.10b,d)}$$

FIGURE 9.9

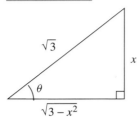

$$= \frac{9}{8}\text{Sin}^{-1}\left(\frac{x}{\sqrt{3}}\right) - \frac{9}{8}\left(\frac{x}{\sqrt{3}}\right)\frac{\sqrt{3-x^2}}{\sqrt{3}}\left(1 - \frac{2x^2}{3}\right) + C \quad \text{(Figure 9.9)}$$

$$= \frac{9}{8}\text{Sin}^{-1}\left(\frac{x}{\sqrt{3}}\right) - \frac{x}{8}\sqrt{3-x^2}(3-2x^2) + C.$$

EXAMPLE 9.12

Use a trigonometric substitution to evaluate the integral in Example 9.4.

SOLUTION   If we set $x = 3\tan\theta$, then $dx = 3\sec^2\theta\, d\theta$, and

$$\int \frac{x^3}{\sqrt{9+x^2}}\, dx = \int \frac{27\tan^3\theta}{\sqrt{9+9\tan^2\theta}}3\sec^2\theta\, d\theta$$

$$= 27\int \frac{\tan^3\theta\ \sec^2\theta}{\sec\theta}\, d\theta$$

$$= 27\int \tan\theta\,(\sec^2\theta - 1)\sec\theta\, d\theta \qquad \text{(by 9.12)}$$

$$= 27\int (\sec^3\theta\tan\theta - \sec\theta\tan\theta)\, d\theta$$

$$= 27\left(\frac{1}{3}\sec^3\theta - \sec\theta\right) + C \qquad \text{(by 9.11b)}$$

FIGURE 9.10

$$= 9\left(\frac{\sqrt{9+x^2}}{3}\right)^3 - 27\frac{\sqrt{9+x^2}}{3} + C \quad \text{(Figure 9.10)}$$

$$= \frac{1}{3}(9+x^2)^{3/2} - 9\sqrt{9+x^2} + C.$$

EXAMPLE 9.13

In Section 6.7 we led you through an integration that verified the formula πr^2 for the area of a circle. You now have all the tools to do it on your own. Do so.

SOLUTION   If we consider the circle $x^2 + y^2 = r^2$ in Figure 9.11, we see that its area is four times that in the first quadrant:

$$A = 4 \int_0^r \sqrt{r^2 - x^2}\, dx.$$

If we set $x = r \sin \theta$, then $dx = r \cos \theta\, d\theta$, and

$$A = 4 \int_0^{\pi/2} \sqrt{r^2 - r^2 \sin^2 \theta}\, r \cos \theta\, d\theta$$

$$= 4r^2 \int_0^{\pi/2} \cos^2 \theta\, d\theta$$

$$= 4r^2 \int_0^{\pi/2} \left(\frac{1 + \cos 2\theta}{2} \right) d\theta \qquad \text{(by 9.10c)}$$

$$= 2r^2 \left\{ \theta + \frac{1}{2} \sin 2\theta \right\}_0^{\pi/2} \qquad \text{(by 9.9d)}$$

$$= 2r^2 (\pi/2) = \pi r^2.$$

FIGURE 9.11

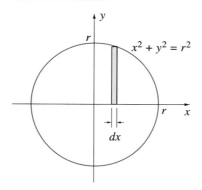

In Exercises 1–20 evaluate the indefinite integral.

∗ **1.** $\displaystyle \int \frac{1}{x\sqrt{2x^2 - 4}}\, dx$

∗ **2.** $\displaystyle \int \frac{1}{\sqrt{9 - 5x^2}}\, dx$

∗ **3.** $\displaystyle \int \frac{1}{10 + x^2}\, dx$

∗ **4.** $\displaystyle \int \frac{1}{x^2 \sqrt{4 - x^2}}\, dx$

∗ **5.** $\displaystyle \int \sqrt{7 - x^2}\, dx$

∗ **6.** $\displaystyle \int x\sqrt{5x^2 + 3}\, dx$

* **7.** $\displaystyle\int x^3\sqrt{4+x^2}\,dx$

* **8.** $\displaystyle\int \frac{1}{1-x^2}\,dx$

* **9.** $\displaystyle\int \frac{1}{\sqrt{x^2-5}}\,dx$

* **10.** $\displaystyle\int \frac{x+5}{10x^2+2}\,dx$

* **11.** $\displaystyle\int \frac{1}{x\sqrt{x^2+3}}\,dx$

* **12.** $\displaystyle\int \frac{\sqrt{4-x^2}}{x}\,dx$

* **13.** $\displaystyle\int \frac{x^2}{(2-9x^2)^{3/2}}\,dx$

* **14.** $\displaystyle\int \frac{\sqrt{x^2-16}}{x^2}\,dx$

* **15.** $\displaystyle\int \frac{1}{x^2\sqrt{2x^2+7}}\,dx$

* **16.** $\displaystyle\int \frac{1}{x^3\sqrt{x^2-4}}\,dx$

* **17.** $\displaystyle\int \frac{\sqrt{9-z^2}}{z^4}\,dz$

* **18.** $\displaystyle\int \frac{y^3}{\sqrt{y^2+4}}\,dy$

* **19.** $\displaystyle\int \frac{1}{(4x^2-9)^{3/2}}\,dx$

* **20.** $\displaystyle\int \frac{x^2+2}{x^3+x}\,dx$

** **21.** Show that if $a>0$ then

$$\int \frac{1}{a^2-x^2}\,dx = \frac{1}{2a}\ln\left|\frac{a+x}{a-x}\right| + C.$$

** **22.** Verify that the area inside the ellipse $b^2x^2 + a^2y^2 = a^2b^2$ is πab.

** **23.** Find the area common to the circles $x^2 + y^2 = 4$ and $x^2 + y^2 = 4x$.

** **24.** A boy initially at O (figure below) walks along the edge of a pier (the y-axis) towing his sailboat by a string of length L.

    (a) If the boat starts at Q and the string always remains straight, show that the equation of the curved path $y = f(x)$ followed by the boat must satisfy the differential equation

$$\frac{dy}{dx} = -\frac{\sqrt{L^2-x^2}}{x}.$$

    (b) Find $y = f(x)$.

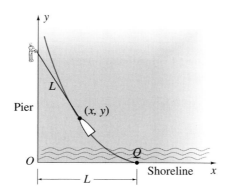

** **25.** A dog at position $(L, 0)$ in the xy-plane spots a rabbit at position $(0, 0)$ running in the positive y-direction. If the dog runs at the same speed as the rabbit and always moves directly toward the rabbit, find the equation of the path followed by the dog.

** **26.** The parabola $x = y^2$ divides the circle $x^2 + y^2 = 4$ into two parts. Find the second moment of area of the smaller part about the x-axis.

** **27.** Find the centroid of the area in Exercise 26.

** **28.** Find the horizontal line that divides the ellipse $b^2x^2 + a^2y^2 = a^2b^2$ into two parts so that the area of the lower part is twice the area of the upper part.

** **29.** If a thin circular plate has radius 2 units and constant mass per unit area ρ, find its moment of inertia about any tangent line to its edge.

** **30.** A cylindrical oil can with horizontal axis has radius r and length h. If the density of the oil is ρ, find the force on each end of the can when it is full.

In Exercises 31–37 evaluate the indefinite integral.

** **31.** $\displaystyle\int \frac{2x^4 - x^2}{2x^2 + 1} \, dx$

** **32.** $\displaystyle\int (7 - x^2)^{3/2} \, dx$

** **33.** $\displaystyle\int \frac{1}{x - x^3} \, dx$

** **34.** $\displaystyle\int \frac{1}{x^3(4x^2 - 1)^{3/2}} \, dx$

** **35.** $\displaystyle\int \sqrt{x^2 - 4} \, dx$

** **36.** $\displaystyle\int \sqrt{1 + 3x^2} \, dx$

** **37.** $\displaystyle\int \frac{x^2}{\sqrt{x^2 - 5}} \, dx$

** **38.** Find the length of the curve $8y^2 = x^2(1 - x^2)$.

** **39.** Find the length of the parabola $y = x^2$ from $(0, 0)$ to $(1, 1)$.

** **40.** At what distance from the centre of a circle should a line be drawn in order that the second moment of area of the circle about that line will be equal to twice the second moment of the circle about a line through its centre?

** **41.** When a flexible cable of constant mass per unit length ρ hangs between two fixed points A and B (figure on page 612), the shape $y = f(x)$ of the cable must satisfy the differential equation

$$\frac{d^2y}{dx^2} = \frac{\rho g}{H}\sqrt{1 + \left(\frac{dy}{dx}\right)^2},$$

where $g > 0$ and $H > 0$ are constants. If we set $k = \rho g/H$ and $p = dy/dx$, then

$$\frac{dp}{dx} = k\sqrt{1 + p^2} \qquad \text{or} \qquad \frac{1}{\sqrt{1 + p^2}}\frac{dp}{dx} = k.$$

It follows that

$$\int \frac{1}{\sqrt{1 + p^2}} \, dp = kx + C.$$

(a) Evaluate the integral shown to find $p = dy/dx$.

(b) Integrate once more to find the shape of the cable (see also Example 8.14 in Section 8.4).

*** **42.** Find the area inside the loop of the strophoid $y^2(a + x) = x^2(a - x)$.

*** **43.**   (a) By rationalizing the denominator, show that

$$\frac{1}{x + 1 + \sqrt{x^2 + 1}} = \frac{x + 1 - \sqrt{x^2 + 1}}{2x}.$$

(b) Use (a) to prove that

$$\int_0^1 \frac{1}{x + 1 + \sqrt{x^2 + 1}} \, dx = 1 - \frac{\sqrt{2}}{2} + \frac{1}{2} \ln \left(\frac{1 + \sqrt{2}}{2} \right).$$

*** **44.** Show that

$$\int_{-1}^1 \sqrt{\frac{1 + x}{1 - x}} \, dx = \pi.$$

9.6 Completing the Square and Trigonometric Substitutions

Trigonometric substitutions reduce integrals containing square roots of the form $\sqrt{a^2 - b^2 x^2}$, $\sqrt{a^2 + b^2 x^2}$, and $\sqrt{b^2 x^2 - a^2}$ to trigonometric integrals. For integrals containing square roots of the form $\sqrt{ax^2 + bx + c}$ we can again reduce the integral to a trigonometric integral by a trigonometric substitution, if we first complete the square.

Consider the integral

$$\int \frac{1}{(x^2 + 2x + 5)^{3/2}} \, dx.$$

If we complete the square in the denominator, we obtain

$$\int \frac{1}{(x^2 + 2x + 5)^{3/2}} \, dx = \int \frac{1}{[(x + 1)^2 + 4]^{3/2}} \, dx.$$

Had the denominator been $(x^2 + 4)^{3/2}$, we would have set $x = 2 \tan \theta$. It is natural, then, for the denominator $[(x + 1)^2 + 4]^{3/2}$ to set

$$x + 1 = 2 \tan \theta,$$

in which case $dx = 2\sec^2\theta\,d\theta$, and

$$\int \frac{1}{(x^2 + 2x + 5)^{3/2}}\,dx = \int \frac{1}{(4\tan^2\theta + 4)^{3/2}} 2\sec^2\theta\,d\theta = \int \frac{2\sec^2\theta}{8\sec^3\theta}\,d\theta$$

$$= \frac{1}{4}\int \cos\theta\,d\theta = \frac{1}{4}\sin\theta + C$$

$$= \frac{x + 1}{4\sqrt{x^2 + 2x + 5}} + C. \qquad\qquad \text{(Figure 9.12)}$$

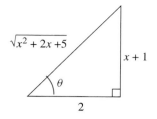

FIGURE 9.12

Consider another example,

$$\int \sqrt{4 + x - x^2}\,dx.$$

Completion of the square leads to

$$\int \sqrt{4 + x - x^2}\,dx = \int \sqrt{-(x - 1/2)^2 + 17/4}\,dx.$$

If we now set

$$x - \frac{1}{2} = \frac{\sqrt{17}}{2}\sin\theta,$$

then

$$dx = \frac{\sqrt{17}}{2}\cos\theta\,d\theta, \qquad \text{and}$$

$$\int \sqrt{4 + x - x^2}\,dx = \int \sqrt{-\frac{17}{4}\sin^2\theta + \frac{17}{4}}\frac{\sqrt{17}}{2}\cos\theta\,d\theta = \int \frac{\sqrt{17}}{2}\cos\theta\frac{\sqrt{17}}{2}\cos\theta\,d\theta$$

$$= \frac{17}{4}\int \cos^2\theta\,d\theta = \frac{17}{4}\int \left(\frac{1 + \cos 2\theta}{2}\right)d\theta$$

FIGURE 9.13

$$= \frac{17}{8}\left(\theta + \frac{1}{2}\sin 2\theta\right) + C = \frac{17}{8}(\theta + \sin\theta\cos\theta) + C$$

$$= \frac{17}{8}\operatorname{Sin}^{-1}\left(\frac{2x - 1}{\sqrt{17}}\right) + \frac{17}{8}\left[\frac{2}{\sqrt{17}}\left(x - \frac{1}{2}\right)\right]\frac{\sqrt{4 + x - x^2}}{\sqrt{17}/2} + C$$

$$\text{(Figure 9.13)}$$

$$= \frac{17}{8}\operatorname{Sin}^{-1}\left(\frac{2x - 1}{\sqrt{17}}\right) + \frac{1}{4}(2x - 1)\sqrt{4 + x - x^2} + C.$$

As a final illustrative example, we consider

$$\int \frac{x}{\sqrt{2x^2 + 3x - 6}}\,dx = \frac{1}{\sqrt{2}}\int \frac{x}{\sqrt{x^2 + \dfrac{3x}{2} - 3}}\,dx = \frac{1}{\sqrt{2}}\int \frac{x}{\sqrt{\left(x + \dfrac{3}{4}\right)^2 - \dfrac{57}{16}}}\,dx.$$

If we set

$$x + \frac{3}{4} = \frac{\sqrt{57}}{4} \sec \theta,$$

then

$$dx = \frac{\sqrt{57}}{4} \sec \theta \tan \theta \, d\theta, \qquad \text{and}$$

$$\int \frac{x}{\sqrt{2x^2 + 3x - 6}} \, dx = \frac{1}{\sqrt{2}} \int \frac{(\sqrt{57}/4) \sec \theta - 3/4}{\sqrt{(57/16) \sec^2 \theta - 57/16}} \frac{\sqrt{57}}{4} \sec \theta \tan \theta \, d\theta$$

$$= \frac{1}{\sqrt{2}} \int \frac{(\sqrt{57}/4) \sec \theta - 3/4}{(\sqrt{57}/4) \tan \theta} \frac{\sqrt{57}}{4} \sec \theta \tan \theta \, d\theta$$

$$= \frac{1}{\sqrt{2}} \int \left(\frac{\sqrt{57}}{4} \sec^2 \theta - \frac{3}{4} \sec \theta \right) d\theta$$

$$= \frac{\sqrt{57}}{4\sqrt{2}} \tan \theta - \frac{3}{4\sqrt{2}} \ln |\sec \theta + \tan \theta| + C$$

$$= \frac{\sqrt{57}}{4\sqrt{2}} \left[\frac{\sqrt{x^2 + (3/2)x - 3}}{\sqrt{57}/4} \right]$$

$$- \frac{3}{4\sqrt{2}} \ln \left| \frac{x + 3/4}{\sqrt{57}/4} + \frac{\sqrt{x^2 + (3/2)x - 3}}{\sqrt{57}/4} \right| + C$$

(Figure 9.14)

$$= \frac{\sqrt{2x^2 + 3x - 6}}{2}$$

$$- \frac{3}{4\sqrt{2}} \ln \left| 4x + 3 + 2\sqrt{2}\sqrt{2x^2 + 3x - 6} \right| + D,$$

FIGURE 9.14

where

$$D = C + \frac{3}{8\sqrt{2}} \ln 57.$$

We have shown that completing the square is the natural generalization of the technique of trigonometric substitutions. It reduces integrals containing square roots of the form $\sqrt{ax^2 + bx + c}$ to trigonometric integrals.

EXAMPLE 9.14

Find the volume of a donut.

SOLUTION   A donut is generated (as the volume of the solid of revolution) when the circle

$$(x - a)^2 + y^2 = b^2, \qquad (a > b),$$

is rotated around the y-axis (Figure 9.15). Cylindrical shells for the vertical rectangles shown yield the volume generated by the upper semicircle, which can be doubled to give

$$V = 2 \int_{a-b}^{a+b} 2\pi x y \, dx.$$

If we solve the equation of the circle for $y = \pm\sqrt{b^2 - (x-a)^2}$, then

$$V = 2 \int_{a-b}^{a+b} 2\pi x \sqrt{b^2 - (x-a)^2} \, dx.$$

We now set $x - a = b \sin\theta$, in which case $dx = b\cos\theta \, d\theta$, and

$$V = 4\pi \int_{-\pi/2}^{\pi/2} (a + b\sin\theta) b\cos\theta \, b\cos\theta \, d\theta$$

$$= 4\pi b^2 \int_{-\pi/2}^{\pi/2} (a\cos^2\theta + b\cos^2\theta \sin\theta) \, d\theta$$

$$= 4\pi b^2 \int_{-\pi/2}^{\pi/2} \left[a\left(\frac{1 + \cos 2\theta}{2}\right) + b\cos^2\theta \sin\theta \right] d\theta$$

$$= 4\pi b^2 \left\{ \frac{a}{2}\left(\theta + \frac{1}{2}\sin 2\theta\right) - \frac{b}{3}\cos^3\theta \right\}_{-\pi/2}^{\pi/2}$$

$$= 4\pi b^2 \left[\frac{a}{2}\left(\frac{\pi}{2} + \frac{\pi}{2}\right) \right] = 2\pi^2 ab^2.$$

FIGURE 9.15

$(x-a)^2 + y^2 = b^2$

x

$a-b$ a $a+b$ x

y

FIGURE 9.16

x_2

x_1

y

a x

$(x-a)^2 + y^2 = b^2$

With horizontal rectangles (Figure 9.16), the washer method yields

$$V = 2 \int_0^b (\pi x_2^2 - \pi x_1^2) \, dy.$$

We can solve the equation of the circle for $x = a \pm \sqrt{b^2 - y^2}$; hence,

$$x_1 = a - \sqrt{b^2 - y^2}, \qquad x_2 = a + \sqrt{b^2 - y^2}.$$

Thus

$$V = 2\pi \int_0^b \left[\left(a + \sqrt{b^2 - y^2}\right)^2 - \left(a - \sqrt{b^2 - y^2}\right)^2 \right] dy,$$

and this simplifies to

$$V = 2\pi \int_0^b 4a\sqrt{b^2 - y^2} \, dy.$$

If we set $y = b \sin \theta$, then $dy = b \cos \theta \, d\theta$, and

$$V = 8\pi a \int_0^{\pi/2} b \cos \theta \, b \cos \theta \, d\theta$$

$$= 8\pi ab^2 \int_0^{\pi/2} \left(\frac{1 + \cos 2\theta}{2} \right) d\theta$$

$$= 4\pi ab^2 \left\{ \theta + \frac{1}{2} \sin 2\theta \right\}_0^{\pi/2}$$

$$= 4\pi ab^2 (\pi/2) = 2\pi^2 ab^2.$$

EXERCISES 9.6

In Exercises 1–10 evaluate the indefinite integral.

∗ **1.** $\displaystyle \int \frac{x}{\sqrt{27 + 6x - x^2}} \, dx$

∗ **2.** $\displaystyle \int \frac{1}{\sqrt{x^2 + 2x + 2}} \, dx$

∗ **3.** $\displaystyle \int \frac{1}{(y^2 + 4y)^{3/2}} \, dy$

∗ **4.** $\displaystyle \int \frac{1}{3x - x^2 - 4} \, dx$

∗ **5.** $\displaystyle \int \frac{\sqrt{x^2 + 2x - 3}}{x + 1} \, dx$

∗ **6.** $\displaystyle \int \frac{x}{(4x - x^2)^{3/2}} \, dx$

∗ **7.** $\displaystyle \int \sqrt{-y^2 + 6y} \, dy$

∗ **8.** $\displaystyle \int \frac{2x - 3}{x^2 + 6x + 13} \, dx$

∗ **9.** $\displaystyle \int \frac{5 - 4x}{\sqrt{12x - 4x^2 - 8}} \, dx$

∗∗ **10.** $\displaystyle \int \frac{1}{x\sqrt{6 + 4\ln x + (\ln x)^2}} \, dx$

∗∗ **11.** Use the substitution $z = 1/x$ to evaluate

$$\int \frac{1}{x\sqrt{x^2 + 6x + 3}} \, dx.$$

∗∗ **12.** One of the gates in a dam is circular with radius 1 m. If the gate is closed and the surface of the water is 3 m above the top of the gate, find the force due to water pressure on the gate.

∗∗ **13.** Evaluate

$$\int \frac{1}{3x - x^2} \, dx$$

(a) by completing the square, and setting $x - 3/2 = (3/2) \sin \theta$;

(b) by multiplying numerator and denominator by -1, completing the square, and setting $x - 3/2 = (3/2) \sec \theta$.

(c) Explain the difference in the two answers.

In Exercises 14–16 evaluate the integral.

∗∗ **14.** $\displaystyle \int \sqrt{x^2 - 2x - 3} \, dx$

∗∗∗ **15.** $\displaystyle \int \frac{1}{x\sqrt{2x - x^2}} \, dx$

∗∗∗ **16.** $\displaystyle \int \frac{1}{(2x + 5)\sqrt{2x - 3 + 8x - 12}} \, dx$

9.7 Partial Fractions

Partial fractions is a method that we apply to rational functions, integrals of the form

$$\int \frac{N(x)}{D(x)} \, dx, \tag{9.13}$$

where $N(x)$ and $D(x)$ are polynomials in x, and the degree of $N(x)$ is less than the degree of $D(x)$. When the degree of $N(x)$ is greater than or equal to that of $D(x)$, we divide $D(x)$ into $N(x)$. For example, in the integral

$$\int \frac{x^4 + 4x^3 + 2x + 4}{x^3 + 1} \, dx,$$

the numerator has degree 4 and the denominator degree 3. By long division, we obtain

$$\int \frac{x^4 + 4x^3 + 2x + 4}{x^3 + 1} \, dx = \int \left(x + 4 + \frac{x}{x^3 + 1} \right) dx.$$

The first two terms on the right, namely, $x + 4$, can be integrated immediately, and partial fractions can be applied to the remaining term $x/(x^3 + 1)$.

As a general rule then: If the degree of $N(x)$ is greater than or equal to the degree of $D(x)$, divide $D(x)$ into $N(x)$ to produce a quotient polynomial $Q(x)$ and a remainder polynomial $R(x)$ of degree less than that of $D(x)$; that is,

$$\frac{N(x)}{D(x)} = Q(x) + \frac{R(x)}{D(x)},$$

where deg R < deg D. Integral 9.13 can then be written in the form

$$\int \frac{N(x)}{D(x)} \, dx = \int Q(x) \, dx + \int \frac{R(x)}{D(x)} \, dx.$$

The first integral on the right is trivial. This leaves integration of the rational function

$$\int \frac{R(x)}{D(x)} \, dx, \tag{9.14}$$

where deg R < deg D.

To use partial fractions on integral 9.14 we must factor the denominator $D(x)$. It is an established fact of algebra that every real polynomial, and specifically $D(x)$, can be factored into real linear factors $ax + b$ and irreducible quadratic factors $ax^2 + bx + c$. A quadratic factor is irreducible if $b^2 - 4ac < 0$. In the complete factorization of $D(x)$, these factors may or may not be repeated; that is, the factorization of $D(x)$ contains terms of the form

$$(ax + b)^n \qquad \text{and} \qquad (ax^2 + bx + c)^n,$$

where $n \geq 1$ is an integer. When $n = 1$ the factor is nonrepeated, and when $n > 1$, it is repeated. For example, in the factorization

$$D(x) = (x - 1)(2x + 1)^3(3x^2 + 4x + 5)(x^2 + 1)^2,$$

$x - 1$ and $3x^2 + 4x + 5$ are nonrepeated, and $2x + 1$ and $x^2 + 1$ are repeated.

Having factored $D(x)$, we can separate the rational function in 9.14 into fractional components. We call this the **partial fraction decomposition** of the integrand. We illustrate with the rational function $(x^2 + 2)/[(x - 1)(2x + 1)^3(3x^2 + 4x + 5)(x^2 + 1)^2]$, and then state general rules for all decompositions. The partial fraction decomposition of this particular rational function is

$$\frac{x^2 + 2}{(x - 1)(2x + 1)^3(3x^2 + 4x + 5)(x^2 + 1)^2}$$

$$= \frac{A}{x - 1} + \frac{B}{2x + 1} + \frac{C}{(2x + 1)^2} + \frac{D}{(2x + 1)^3}$$

$$+ \frac{Ex + F}{3x^2 + 4x + 5} + \frac{Gx + H}{x^2 + 1} + \frac{Ix + J}{(x^2 + 1)^2},$$

where A, B, ..., J are constants. The first term corresponds to the nonrepeated linear factor $x - 1$, the next three correspond to the repeated linear factor $2x + 1$, the fifth term corresponds to the nonrepeated quadratic factor $3x^2 + 4x + 5$, and the last two terms result from the repeated quadratic $x^2 + 1$.

Let us now state general rules for the partial fraction decomposition of rational functions $R(x)/D(x)$. There are three rules:

1. For each repeated or nonrepeated linear factor $(ax + b)^n$ in $D(x)$, include the following terms in the decomposition:

$$\frac{A_1}{ax + b} + \frac{A_2}{(ax + b)^2} + \cdots + \frac{A_n}{(ax + b)^n}. \tag{9.15a}$$

The number of terms corresponds to the power n on $(ax + b)^n$.

2. For each repeated or nonrepeated irreducible quadratic factor $(ax^2 + bx + c)^n$ in $D(x)$, include the following terms in the decomposition

$$\frac{B_1x + C_1}{ax^2 + bx + c} + \frac{B_2x + C_2}{(ax^2 + bx + c)^2} + \cdots + \frac{B_nx + C_n}{(ax^2 + bx + c)^n}. \tag{9.15b}$$

Again the number of terms corresponds to the power n on $(ax^2 + bx + c)^n$.

3. The complete decomposition is the sum of all terms in 9.15a and all terms in 9.15b.

The following examples will clarify these rules.

EXAMPLE 9.15

What form do partial fraction decompositions for the following rational functions take?

(a) $\dfrac{x^2 + 2x + 3}{3x^3 - x^2 - 3x + 1}$

(b) $\dfrac{x^2 + 3x - 1}{x^4 + x^3 + x^2 + x}$

(c) $\dfrac{x^3 + 3x^2 - x}{x^5 + x^4 + 2x^3 + 2x^2 + x + 1}$

(d) $\dfrac{3x^5 - 1}{(3x^2 + 5)(x^2 + 2x + 3)(2x - 1)^2}$

SOLUTION

(a) Since $3x^3 - x^2 - 3x + 1 = (3x - 1)(x - 1)(x + 1)$, we have nonrepeated linear factors. Rules 1. and 3. give the partial fraction decomposition

$$\frac{x^2 + 2x + 3}{3x^3 - x^2 - 3x + 1} = \frac{A}{3x - 1} + \frac{B}{x - 1} + \frac{C}{x + 1}.$$

(b) The factorization $x^4 + x^3 + x^2 + x = x(x+1)(x^2+1)$ has two nonrepeated linear factors, and a nonrepeated quadratic factor. The three rules lead to

$$\frac{x^2 + 3x - 1}{x^4 + x^3 + x^2 + x} = \frac{A}{x} + \frac{B}{x+1} + \frac{Cx + D}{x^2 + 1}.$$

(c) Since $x^5 + x^4 + 2x^3 + 2x^2 + x + 1 = (x+1)(x^2+1)^2$, the partial fraction decomposition is

$$\frac{x^3 + 3x^2 - x}{x^5 + x^4 + 2x^3 + 2x^2 + x + 1} = \frac{A}{x+1} + \frac{Cx + D}{x^2 + 1} + \frac{Ex + F}{(x^2 + 1)^2}.$$

(d) Since $x^2 + 2x + 3$ can be factored no further, the decomposition is

$$\frac{3x^5 - 1}{(3x^2 + 5)(x^2 + 2x + 3)(2x - 1)^2} = \frac{Ax + B}{3x^2 + 5} + \frac{Cx + D}{x^2 + 2x + 3} + \frac{E}{2x - 1} + \frac{F}{(2x - 1)^2}.$$

To illustrate how to calculate coefficients in partial fraction decompositions, we use part (b) of Example 9.15. We bring the right side of the decomposition to a common denominator,

$$\frac{x^2 + 3x - 1}{x^4 + x^3 + x^2 + x} = \frac{A}{x} + \frac{B}{x+1} + \frac{Cx + D}{x^2 + 1}$$
$$= \frac{A(x+1)(x^2+1) + Bx(x^2+1) + x(x+1)(Cx+D)}{x^4 + x^3 + x^2 + x},$$

and equate numerators

$$x^2 + 3x - 1 = A(x+1)(x^2+1) + Bx(x^2+1) + x(x+1)(Cx+D).$$

There are two methods for finding the constants $A, B, C,$ and D.

Method 1   First, we gather together terms in the various powers of x on the right side of the equation,

$$x^2 + 3x - 1 = (A + B + C)x^3 + (A + C + D)x^2 + (A + B + D)x + A.$$

Now, Exercise 35 in Section 3.8 states that two polynomials of the same degree can be equal for all values of x if and only if coefficients of corresponding powers of x are identical. Since we have equal cubic polynomials in the above equation, we equate coefficients of $x^3, x^2, x,$ and x^0 (meaning terms with no x's):

$$x^3: \quad 0 = A + B + C,$$
$$x^2: \quad 1 = A + C + D,$$
$$x: \quad 3 = A + B + D,$$
$$x^0: \quad -1 = A.$$

The solution of these four linear equations in four unknowns is $A = -1$, $B = 3/2$, $C = -1/2$, and $D = 5/2$. Hence, the partial fraction decomposition of $(x^2 + 3x - 1)/(x^4 + x^3 + x^2 + x)$ is

$$\frac{x^2 + 3x - 1}{x^4 + x^3 + x^2 + x} = -\frac{1}{x} + \frac{3/2}{x + 1} + \frac{-x/2 + 5/2}{x^2 + 1}.$$

Method 2 In this method we substitute convenient values of x into the equation

$$x^2 + 3x - 1 = A(x + 1)(x^2 + 1) + Bx(x^2 + 1) + x(x + 1)(Cx + D).$$

Clearly $x = 0$ is most convenient, since it yields the value of A,

$$-1 = A(1)(1),$$

that is, $A = -1$. Convenient also is $x = -1$,

$$(-1)^2 + 3(-1) - 1 = B(-1)(2);$$

it gives $B = 3/2$. The values $x = 1$ and $x = 2$ yield the equations

$$1 + 3(1) - 1 = A(2)(2) + B(1)(2) + 1(2)(C + D),$$

$$4 + 3(2) - 1 = A(3)(5) + B(2)(5) + 2(3)(2C + D).$$

When $A = -1$ and $B = 3/2$ are substituted into these, the resulting equations are

$$C + D = 2,$$

$$4C + 2D = 3.$$

The solution of these is $C = -1/2$ and $D = 5/2$.

It is also possible to use a combination of Methods 1 and 2 for finding coefficients in a partial fraction decomposition; that is, substitute some values of x and equate some coefficients. The total number must be equal to the number of unknown coefficients in the decomposition.

Once we have completed the partial fraction decomposition of a rational function, we can integrate the function by finding antiderivatives of the component fractions. For example, with the partial fraction decomposition of $(x^2 + 3x - 1)/(x^4 + x^3 + x^2 + x)$, it is very simple to find its antiderivative,

$$\int \frac{x^2 + 3x - 1}{x^4 + x^3 + x^2 + x}\, dx = \int \left(-\frac{1}{x} + \frac{3/2}{x + 1} + \frac{-x/2 + 5/2}{x^2 + 1} \right) dx$$

$$= -\int \frac{1}{x}\, dx + \frac{3}{2} \int \frac{1}{x + 1}\, dx$$

$$- \frac{1}{2} \int \frac{x}{x^2 + 1}\, dx + \frac{5}{2} \int \frac{1}{x^2 + 1}\, dx$$

$$= -\ln |x| + \frac{3}{2} \ln |x + 1| - \frac{1}{4} \ln (x^2 + 1) + \frac{5}{2} \text{Tan}^{-1} x + C.$$

Other integrations by partial fraction decompositions are illustrated in the following example.

EXAMPLE 9.16

Evaluate the following indefinite integrals:

(a) $\displaystyle\int \frac{x}{x^4 + 6x^2 + 5}\, dx$

(b) $\displaystyle\int \frac{x^2 + 1}{2x^3 - 5x^2 + 4x - 1}\, dx$

(c) $\displaystyle\int \frac{1}{x^3 + 8}\, dx$

(d) $\displaystyle\int \frac{x^2}{x^3 + 8}\, dx$

SOLUTION

(a) Since $x^4 + 6x^2 + 5 = (x^2 + 1)(x^2 + 5)$, the partial fraction decomposition of $x/(x^4 + 6x^2 + 5)$ has the form

$$\frac{x}{x^4 + 6x^2 + 5} = \frac{Ax + B}{x^2 + 1} + \frac{Cx + D}{x^2 + 5}.$$

When we bring the right side to a common denominator and equate numerators, the result is

$$x = (Ax + B)(x^2 + 5) + (Cx + D)(x^2 + 1).$$

We now multiply out the right side and equate coefficients of like powers of x,

$$x^3 : \quad 0 = A + C,$$
$$x^2 : \quad 0 = B + D,$$
$$x : \quad 1 = 5A + C,$$
$$x^0 : \quad 0 = 5B + D.$$

The solution of these equations is $A = 1/4$, $B = 0$, $C = -1/4$, and $D = 0$, and therefore

$$\int \frac{x}{x^4 + 6x^2 + 5}\, dx = \int \left(\frac{x/4}{x^2 + 1} + \frac{-x/4}{x^2 + 5} \right) dx$$

$$= \frac{1}{8} \ln (x^2 + 1) - \frac{1}{8} \ln (x^2 + 5) + C.$$

(b) Since $2x^3 - 5x^2 + 4x - 1 = (2x - 1)(x - 1)^2$, the partial fraction decomposition takes the form

$$\frac{x^2 + 1}{2x^3 - 5x^2 + 4x - 1} = \frac{A}{2x - 1} + \frac{B}{x - 1} + \frac{C}{(x - 1)^2}$$

$$= \frac{A(x - 1)^2 + B(x - 1)(2x - 1) + C(2x - 1)}{(2x - 1)(x - 1)^2}.$$

When we equate numerators,

$$x^2 + 1 = A(x - 1)^2 + B(x - 1)(2x - 1) + C(2x - 1).$$

We now set $x = 0$, $x = 1$, and $x = 1/2$:

$$x = 0 : \quad 1 \quad = A + B - C,$$

$$x = 1 : \quad 2 \quad = C(1),$$

$$x = 1/2 : \quad 5/4 = A(1/4).$$

These give $A = 5$, $B = -2$, and $C = 2$, and therefore

$$\int \frac{x^2 + 1}{2x^3 - 5x^2 + 4x - 1}\, dx = \int \left(\frac{5}{2x - 1} - \frac{2}{x - 1} + \frac{2}{(x - 1)^2} \right) dx$$

$$= \frac{5}{2} \ln |2x - 1| - 2 \ln |x - 1| - \frac{2}{x - 1} + C.$$

(c) Since $x^3 + 8 = (x + 2)(x^2 - 2x + 4)$, we set

$$\frac{1}{x^3 + 8} = \frac{1}{(x + 2)(x^2 - 2x + 4)} = \frac{A}{x + 2} + \frac{Bx + C}{x^2 - 2x + 4}$$

$$= \frac{A(x^2 - 2x + 4) + (Bx + C)(x + 2)}{(x + 2)(x^2 - 2x + 4)},$$

and now equate numerators,

$$1 = A(x^2 - 2x + 4) + (Bx + C)(x + 2).$$

We set $x = -2$ and equate coefficients of x^2 and 1:

$$x = -2 : \quad 1 = 12A,$$

$$x^2 : \quad 0 = A + B,$$

$$1 : \quad 1 = 4A + 2C.$$

These give $A = 1/12$, $B = -1/12$, and $C = 1/3$, and therefore

$$\int \frac{1}{x^3 + 8}\, dx = \int \left(\frac{1/12}{x + 2} + \frac{-x/12 + 1/3}{x^2 - 2x + 4} \right) dx$$

$$= \frac{1}{12} \ln |x + 2| + \frac{1}{12} \int \frac{4 - x}{(x - 1)^2 + 3}\, dx.$$

For the integral on the right, we set $x - 1 = \sqrt{3} \tan \theta$, in which case $dx = \sqrt{3} \sec^2 \theta\, d\theta$, and

$$\int \frac{1}{x^3 + 8}\, dx = \frac{1}{12} \ln |x + 2| + \frac{1}{12} \int \frac{4 - (1 + \sqrt{3} \tan \theta)}{3 \sec^2 \theta} \sqrt{3} \sec^2 \theta\, d\theta$$

$$= \frac{1}{12} \ln |x + 2| + \frac{1}{12\sqrt{3}} \int \left(3 - \sqrt{3} \tan \theta \right) d\theta$$

$$= \frac{1}{12} \ln |x + 2| + \frac{1}{12\sqrt{3}} \left(3\theta + \sqrt{3} \ln |\cos \theta| \right) + C$$

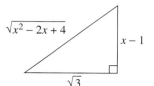

FIGURE 9.17

$$= \frac{1}{12} \ln |x + 2| + \frac{\sqrt{3}}{12} \text{Tan}^{-1}\left(\frac{x - 1}{\sqrt{3}}\right)$$

$$+ \frac{1}{12} \ln \left| \frac{\sqrt{3}}{\sqrt{x^2 - 2x + 4}} \right| + C \qquad \text{(Figure 9.17)}$$

$$= \frac{1}{12} \ln |x + 2| + \frac{\sqrt{3}}{12} \text{Tan}^{-1}\left(\frac{x - 1}{\sqrt{3}}\right) - \frac{1}{24} \ln (x^2 - 2x + 4) + D,$$

where $D = C + (1/24) \ln 3$.

(d) Do not be misled into partial fractions in this example; the rational function is immediately integrable,

$$\int \frac{x^2}{x^3 + 8} dx = \frac{1}{3} \ln |x^3 + 8| + C.$$

In Section 4.4 we discussed the rate at which an infectious disease might spread through a population of susceptible individuals. If N represents the number of individuals in the population and $x(t)$ the number infected at any given time t, it is often assumed that the rate of infection is proportional to the product of the numbers of infected and not infected:

$$\frac{dx}{dt} = kx(N - x),$$

where k is a constant. Thus, $x(t)$ must satisfy this differential equation subject to an initial condition such as, perhaps, $x(0) = 1$, where one infected individual is introduced into the population at time $t = 0$. To solve this differential equation, we write

$$\frac{1}{x(N - x)} \frac{dx}{dt} = k,$$

and determine the partial fraction decomposition for the left side, namely,

$$\left(\frac{1/N}{x} + \frac{1/N}{N - x}\right) \frac{dx}{dt} = k.$$

Integration of both sides of this equation with respect to time gives

$$\frac{1}{N} (\ln |x| - \ln |N - x|) = kt + C.$$

Since x and $N - x$ are both positive, we may drop the absolute values and write

$$\ln \left(\frac{x}{N - x}\right) = N(kt + C).$$

To solve this equation for an explicit definition of $x(t)$, we first exponentiate both sides of the equation,

$$\frac{x}{N-x} = De^{Ft},$$

where we have substituted $D = e^{NC}$ and $F = kN$. Multiplication by $N - x$ gives

$$x = De^{Ft}(N-x) \quad \text{or} \quad x\left(1 + De^{Ft}\right) = NDe^{Ft}.$$

Thus,

$$x(t) = \frac{NDe^{Ft}}{1 + De^{Ft}} = \frac{ND}{D + e^{-Ft}}.$$

Since $x(0) = 1$,

$$1 = \frac{ND}{1 + D},$$

which implies that $D = 1/(N-1)$, and

$$x(t) = \frac{\dfrac{N}{N-1}}{\dfrac{1}{N-1} + e^{-Ft}} = \frac{N}{1 + (N-1)e^{-Ft}}.$$

The shape of this curve is illustrated in Figure 9.18a (see also the discussion in Section 4.4). The time scale can be affixed to the diagram once one further piece of information is given. For instance, if it is known that one-half the population will be infected in 20 days, then $x(20) = N/2$, and the time scale is as shown in Figure 9.18b. This added information determines F, since

$$\frac{N}{2} = \frac{N}{1 + (N-1)e^{-20F}},$$

and this equation gives

$$F = \frac{1}{20}\ln(N-1).$$

FIGURE 9.18a

FIGURE 9.18b

EXERCISES 9.7

In Exercises 1–16 evaluate the indefinite integral.

* **1.** $\displaystyle\int \frac{x+2}{x^2-2x+1}\,dx$

* **2.** $\displaystyle\int \frac{1}{y^3+3y^2+3y+1}\,dy$

* **3.** $\displaystyle\int \frac{1}{z^3+z}\,dz$

* **4.** $\displaystyle\int \frac{x^2+2x-4}{x^2-2x-8}\,dx$

* **5.** $\displaystyle\int \frac{x}{(x-4)^2}\,dx$

* **6.** $\displaystyle\int \frac{y+1}{y^3+y^2-6y}\,dy$

* **7.** $\displaystyle\int \frac{3x+5}{x^3-x^2-x+1}\,dx$

* **8.** $\displaystyle\int \frac{x^3}{(x^2+2)^2}\,dx$

* **9.** $\displaystyle\int \frac{1}{x^2-3}\,dx$

* **10.** $\displaystyle\int \frac{y^2}{y^2+3y+2}\,dy$

* **11.** $\displaystyle\int \frac{z^2+3z-2}{z^3+5z}\,dz$

* **12.** $\displaystyle\int \frac{y^2+6y+4}{y^4+5y^2+4}\,dy$

* **13.** $\displaystyle\int \frac{x}{x^4+7x^2+6}\,dx$

* **14.** $\displaystyle\int \frac{x^2+3}{x^4+x^2-2}\,dx$

* **15.** $\displaystyle\int \frac{3t+4}{t^4-3t^3+3t^2-t}\,dt$

* **16.** $\displaystyle\int \frac{x^3+6}{x^4+2x^3-3x^2-4x+4}\,dx$

** **17.** Find the length of the curve $y = \ln(1-x^2)$ from $x = 0$ to $x = 1/2$.

** **18.** Natural factors such as space and available food supply inhibit the growth of any population. Let N represent the population of a species and M the maximum size of a stable population for which the death rate is just equal to the birth rate. If the time rate of change of N is assumed proportional to both N and $M - N$, then N must satisfy the differential equation

$$\frac{dN}{dt} = kN(M-N).$$

If $N(0) = N_0 < M$, solve this equation for $N(t)$, and draw its graph.

** **19.** Find the centroid of the area bounded by the curves

$$(x+2)^2 y = 4 - x, \qquad x = 0, \qquad y = 0, \qquad (x, y \ge 0).$$

** **20.** Two substances A and B react to form a third substance C in such a way that 1 gm of A reacts with 1 gm of B to produce 2 gm of C. The rate at which C is formed is proportional to the product of the amounts of A and B present in the mixture.

(a) Show that if 10 gm of A and 15 gm of B are brought together at time $t = 0$, the amount $x(t)$ of C present in the mixture must satisfy the differential equation

$$\frac{dx}{dt} = k(20-x)(30-x), \qquad x(0) = 0,$$

where k is a constant.

(b) Solve this equation for $x(t)$.

In Exercises 21–26 evaluate the indefinite integral.

** **21.** $\displaystyle\int \frac{x^3+x+2}{x^5+2x^3+x}\,dx$

** **22.** $\displaystyle\int \frac{1}{x^5+x^4+2x^3+2x^2+x+1}\,dx$

** **23.** $\displaystyle\int \frac{1}{(x^2+5)(x^2+2x+3)}\,dx$ ** **24.** $\displaystyle\int \frac{1}{1+x^3}\,dx$

** **25.** $\displaystyle\int \frac{\sin x}{\cos x(1+\cos^2 x)}\,dx$ ** **26.** $\displaystyle\int \frac{x^4+8x^3-x^2+2x+1}{x^5+x^4+x^2+x}\,dx$

** **27.** If an integrand is a rational function of $\sin x$ and $\cos x$, it can be reduced to a rational function of t by the substitution

$$t = \tan\left(\frac{x}{2}\right).$$

Show that with this substitution

$$dx = \frac{2}{1+t^2}\,dt, \qquad \sin x = \frac{2t}{1+t^2}, \qquad \cos x = \frac{1-t^2}{1+t^2}.$$

In Exercises 28–31 use the substitution of Exercise 27 to evaluate the integral.

** **28.** $\displaystyle\int \sec x\,dx$ ** **29.** $\displaystyle\int \frac{1}{3+5\sin x}\,dx$

** **30.** $\displaystyle\int \frac{1}{1-2\cos x}\,dx$ *** **31.** $\displaystyle\int \frac{1}{\sin x + \cos x}\,dx$

** **32.** Show that the answer to Exercise 28 can be expressed in the usual form $\ln|\sec x + \tan x| + C$.

*** **33.**   (a) Use the change of variable in Exercise 27 to show that

$$\int \frac{1}{5-4\cos x}\,dx = \frac{2}{3}\operatorname{Tan}^{-1}\left(3\tan\frac{x}{2}\right) + C.$$

(b) What happens when the antiderivative in (a) is used to evaluate

$$\int_0^{2\pi} \frac{1}{5-4\cos x}\,dx?$$

(c) Show that

$$\frac{x}{3} + \frac{2}{3}\operatorname{Tan}^{-1}\left(\frac{\sin x}{2-\cos x}\right)$$

is also an antiderivative of $(5-4\cos x)^{-1}$. Use it to evaluate the definite integral in (b).

*** **34.** Evaluate

$$\int \frac{x^2+x+3}{x^4+x^3+2x^2+11x-5}\,dx.$$

*** **35.** Evaluate

$$\int \frac{2x^3+8x^2-3x+5}{x^4+3x^3+x^2+2x-12}\,dx.$$

9.8 Using Tables and Reduction Formulas

For antiderivatives more difficult than we have considered in this chapter, it may be necessary to consult a set of integration tables; and even for integrals which are amenable to our techniques, it may be more expeditious to use tables. There are many sets of integration tables. For example, there is *Standard Mathematical Tables* published by the Chemical Rubber Publishing Company, and *Handbook of Mathematical, Scientific, and Engineering Formulas, Tables, Functions, Graphs, Transforms* published by the Research and Education Association.

To illustrate the use of tables, consider the integral

$$\int \frac{1}{x^2(3 - 4x)^2} \, dx.$$

Certainly we could evaluate this integral by finding the partial fraction decomposition of the integrand. Alternatively, in tables we find the integration formula

$$\int \frac{1}{x^2(a + bx)^2} \, dx = -\frac{a + 2bx}{a^2 x(a + bx)} + \frac{2b}{a^3} \ln\left(\frac{a + bx}{x}\right).$$

Hence, if we set $a = 3$ and $b = -4$ in this formula, we obtain

$$\int \frac{1}{x^2(3 - 4x)^2} \, dx = -\frac{3 - 8x}{9x(3 - 4x)} - \frac{8}{27} \ln\left(\frac{3 - 4x}{x}\right) + C.$$

Tables do not normally include a constant of integration. Remember to add it. There is an additional difficulty with tables; they do not always place absolute value signs around arguments of logarithm functions. The above answer should read

$$\int \frac{1}{x^2(3 - 4x)^2} \, dx = \frac{8x - 3}{9x(3 - 4x)} - \frac{8}{27} \ln\left|\frac{3 - 4x}{x}\right| + C$$

in order to accommodate definite integrals wherein $(3 - 4x)/x$ is negative.

EXAMPLE 9.17

Use tables to evaluate the integral

$$\int \frac{1}{x^2 + 2x + 3} \, dx.$$

SOLUTION   Tables indicate that for an integral of the form

$$\int \frac{1}{X} \, dx$$

where $X = ax^2 + bx + c$ and $q = 4ac - b^2$, there are three possibilities:

$$\int \frac{1}{X} \, dx = \frac{2}{\sqrt{q}} \, \mathrm{Tan}^{-1}\left(\frac{2ax + b}{\sqrt{q}}\right), \tag{9.16a}$$

$$\int \frac{1}{X} \, dx = \frac{-2}{\sqrt{-q}} \, \mathrm{Tanh}^{-1}\left(\frac{2ax + b}{\sqrt{-q}}\right), \tag{9.16b}$$

$$\int \frac{1}{X} \, dx = \frac{1}{\sqrt{-q}} \, \ln\left(\frac{2ax + b - \sqrt{-q}}{2ax + b + \sqrt{-q}}\right). \tag{9.16c}$$

No explanation is given as to when each should be used; it is up to the reader to make this decision. Certainly, we would use 9.16a when $q > 0$, and 9.16b and c when $q < 0$ in order to keep $\sqrt{q}$ and $\sqrt{-q}$ real. In our example, $q = 4(1)(3) - (2)^2 = 8$, and we therefore use 9.16a,

$$\int \frac{1}{x^2 + 2x + 3} \, dx = \frac{2}{\sqrt{8}} \, \text{Tan}^{-1}\left(\frac{2(1)x + 2}{\sqrt{8}}\right) + C = \frac{1}{\sqrt{2}} \, \text{Tan}^{-1}\left(\frac{x + 1}{\sqrt{2}}\right) + C.$$

EXAMPLE 9.18

Use tables to evaluate the integral

$$\int \frac{1}{x\sqrt{3 - 2x - 2x^2}} \, dx.$$

SOLUTION   It should be noted that none of our integration techniques seems particularly suited to this integral. Tables list the formulas

$$\int \frac{1}{x\sqrt{X}} \, dx = -\frac{1}{\sqrt{c}} \ln\left(\frac{\sqrt{X} + \sqrt{c}}{x} + \frac{b}{2\sqrt{c}}\right), \qquad \text{and} \qquad (9.17a)$$

$$\int \frac{1}{x\sqrt{X}} \, dx = \frac{1}{\sqrt{-c}} \text{Sin}^{-1}\left(\frac{bx + 2c}{x\sqrt{b^2 - 4ac}}\right), \qquad (9.17b)$$

for $X = ax^2 + bx + c$. Since $c = 3$ in our example, we use 9.17a,

$$\int \frac{1}{x\sqrt{3 - 2x - 2x^2}} \, dx = -\frac{1}{\sqrt{3}} \ln\left(\frac{\sqrt{3 - 2x - 2x^2} + \sqrt{3}}{x} + \frac{-2}{2\sqrt{3}}\right) + C.$$

To check that 9.17a is the correct formula, and that we have used it correctly, we calculate the derivative of this function:

$$\frac{d}{dx}\left[\frac{-1}{\sqrt{3}} \ln\left(\frac{\sqrt{3 - 2x - 2x^2} + \sqrt{3}}{x} - \frac{1}{\sqrt{3}}\right)\right]$$

$$= \frac{-1/\sqrt{3}}{\dfrac{\sqrt{3 - 2x - 2x^2} + \sqrt{3}}{x} - \dfrac{1}{\sqrt{3}}} \times$$

$$\left[\frac{\dfrac{x(-2 - 4x)}{2\sqrt{3 - 2x - 2x^2}} - (\sqrt{3 - 2x - 2x^2} + \sqrt{3})(1)}{x^2}\right]$$

$$= \frac{-1}{\sqrt{3}} \frac{\sqrt{3}x}{\sqrt{3}\sqrt{3 - 2x - 2x^2} + 3 - x} \times$$

$$\left[\frac{-x(1 + 2x) - (3 - 2x - 2x^2) - \sqrt{3}\sqrt{3 - 2x - 2x^2}}{x^2\sqrt{3 - 2x - 2x^2}} \right]$$

$$= \frac{-x}{\sqrt{3}\sqrt{3 - 2x - 2x^2} + 3 - x} \left[\frac{-3 + x - \sqrt{3}\sqrt{3 - 2x - 2x^2}}{x^2\sqrt{3 - 2x - 2x^2}} \right]$$

$$= \frac{1}{x\sqrt{3 - 2x - 2x^2}}.$$

Once again, until we can be sure that the argument of the logarithm function is positive, it would be wise to insert absolute value signs:

$$\int \frac{1}{x\sqrt{3 - 2x - 2x^2}}\, dx = -\frac{1}{\sqrt{3}} \ln \left| \frac{\sqrt{3 - 2x - 2x^2} + \sqrt{3}}{x} - \frac{1}{\sqrt{3}} \right| + C.$$

Integration tables contain entries called **reduction formulas**. To introduce these, consider the integral

$$\int x^3 e^{2x}\, dx.$$

We could use integration by parts three times to eliminate the x^3 before e^{2x}. A better procedure is to use integration by parts on

$$\int x^n e^{2x}\, dx$$

where n represents a positive integer. With

$$u = x^n, \qquad dv = e^{2x}\, dx, \qquad du = nx^{n-1}\, dx, \qquad v = \frac{1}{2}e^{2x},$$

integration by parts formula 9.7 gives

$$\int x^n e^{2x}\, dx = \frac{1}{2}x^n e^{2x} - \int \frac{n}{2}x^{n-1}e^{2x}\, dx$$

$$= \frac{1}{2}x^n e^{2x} - \frac{n}{2}\int x^{n-1}e^{2x}\, dx. \tag{9.18}$$

We now use this formula with $n = 3$ on the given integral,

$$\int x^3 e^{2x}\, dx = \frac{1}{2}x^3 e^{2x} - \frac{3}{2}\int x^2 e^{2x}\, dx.$$

But formula 9.18 can be used again with $n = 2$ to simplify the integral on the right,

$$\int x^3 e^{2x} \, dx = \frac{1}{2} x^3 e^{2x} - \frac{3}{2} \left(\frac{1}{2} x^2 e^{2x} - \frac{2}{2} \int x e^{2x} \, dx \right)$$

$$= \frac{1}{2} x^3 e^{2x} - \frac{3}{4} x^2 e^{2x} + \frac{3}{2} \int x e^{2x} \, dx.$$

One more application of 9.18 with $n = 1$ rids the integrand of the pesky factor in front of the exponential,

$$\int x^3 e^{2x} \, dx = \frac{1}{2} x^3 e^{2x} - \frac{3}{4} x^2 e^{2x} + \frac{3}{2} \left(\frac{1}{2} x e^{2x} - \frac{1}{2} \int e^{2x} \, dx \right)$$

$$= \frac{1}{2} x^3 e^{2x} + \frac{3}{4} x^2 e^{2x} - \frac{3}{4} x e^{2x} - \frac{3}{8} e^{2x} + C.$$

Equation 9.18 is an example of a reduction formula. By using it on the antiderivative of $x^n e^{2x}$, we reduce the power on x from n to $n - 1$. Successive reductions eventually eliminate this term.

Many reduction formulas can be found in tables. They have the same purpose as 9.18: to reduce and, perhaps, eliminate powers of certain factors in an integrand.

EXAMPLE 9.19

Use a reduction formula to evaluate

$$\int \cos^6 x \, dx.$$

SOLUTION   The following reduction formula can be found in tables,

$$\int \cos^n x \, dx = \frac{1}{n} \cos^{n-1} x \sin x + \frac{n-1}{n} \int \cos^{n-2} x \, dx. \tag{9.19}$$

It reduces the power on $\cos x$ by two each time it is used. We begin with $n = 6$,

$$\int \cos^6 x \, dx = \frac{1}{6} \cos^5 x \sin x + \frac{5}{6} \int \cos^4 x \, dx.$$

We now set $n = 4$ in 9.19,

$$\int \cos^6 x \, dx = \frac{1}{6} \cos^5 x \sin x + \frac{5}{6} \left(\frac{1}{4} \cos^3 x \sin x + \frac{3}{4} \int \cos^2 x \, dx \right).$$

Finally, with $n = 2$ we obtain

$$\int \cos^6 x \, dx = \frac{1}{6} \cos^5 x \sin x + \frac{5}{24} \cos^3 x \sin x + \frac{5}{8} \left(\frac{1}{2} \cos x \sin x + \frac{1}{2} \int dx \right)$$

$$= \frac{1}{6} \cos^5 x \sin x + \frac{5}{24} \cos^3 x \sin x + \frac{5}{16} \cos x \sin x + \frac{5x}{16} + C.$$

Sometimes it is necessary to use more than one entry from tables on an integral, as the following example illustrates.

EXAMPLE 9.20

Evaluate

$$\int \frac{\sqrt{6x - x^2}}{x} \, dx.$$

SOLUTION   We use the reduction formula

$$\int \frac{\sqrt{2ax - x^2}}{x^n} \, dx = \frac{(2ax - x^2)^{3/2}}{(3 - 2n)ax^n} + \frac{n - 3}{(2n - 3)a} \int \frac{\sqrt{2ax - x^2}}{x^{n-1}} \, dx.$$

If we set $n = 1$ and $a = 3$ in this identity,

$$\int \frac{\sqrt{6x - x^2}}{x} \, dx = \frac{(6x - x^2)^{3/2}}{(3 - 2)(3)x} + \frac{1 - 3}{(2 - 3)(3)} \int \sqrt{6x - x^2} \, dx$$

$$= \frac{(6x - x^2)^{3/2}}{3x} + \frac{2}{3} \int \sqrt{6x - x^2} \, dx.$$

We can now use the formula

$$\int \sqrt{2ax - x^2} \, dx = \frac{1}{2}\left[(x - a)\sqrt{2ax - x^2} + a^2 \text{Sin}^{-1}\left(\frac{x - a}{|a|}\right)\right]$$

to finish the integration:

$$\int \frac{\sqrt{6x - x^2}}{x} \, dx = \frac{(6x - x^2)^{3/2}}{3x} + \frac{2}{3}\frac{1}{2}\left[(x - 3)\sqrt{6x - x^2} + 9\text{Sin}^{-1}\left(\frac{x - 3}{3}\right)\right] + C$$

$$= \frac{(6x - x^2)^{3/2}}{3x} + \frac{1}{3}(x - 3)\sqrt{6x - x^2} + 3\text{Sin}^{-1}\left(\frac{x - 3}{3}\right) + C$$

$$= \sqrt{6x - x^2} + 3\text{Sin}^{-1}\left(\frac{x - 3}{3}\right) + C.$$

Tables are an excellent companion to partial fractions. Partial fractions decompose complicated rational functions into simpler ones which can often be found in tables. This is illustrated in the following example.

EXAMPLE 9.21

Evaluate

$$\int \frac{3x^4 - x^3 + 18x^2 - 5x + 17}{x^5 - x^4 + 6x^3 - 6x^2 + 9x - 9} \, dx.$$

SOLUTION   Since $x^5 - x^4 + 6x^3 - 6x^2 + 9x - 9 = (x - 1)(x^2 + 3)^2$, the partial fraction decomposition of the integrand is

$$\frac{3x^4 - x^3 + 18x^2 - 5x + 17}{(x - 1)(x^2 + 3)^2} = \frac{A}{x - 1} + \frac{Bx + C}{x^2 + 3} + \frac{Dx + E}{(x^2 + 3)^2}.$$

We find $A = 2$, $B = 1$, $C = 0$, $D = 3$, and $E = 1$, so that

$$\int \frac{3x^4 - x^3 + 18x^2 - 5x + 17}{x^5 - x^4 + 6x^3 - 6x^2 + 9x - 9}\, dx = \int \left(\frac{2}{x - 1} + \frac{x}{x^2 + 3} + \frac{3x + 1}{(x^2 + 3)^2} \right) dx$$

$$= 2\ln|x - 1| + \frac{1}{2}\ln(x^2 + 3) - \frac{3}{2(x^2 + 3)}$$

$$+ \int \frac{1}{(x^2 + 3)^2}\, dx.$$

Using the reduction formula

$$\int \frac{1}{(a^2 + x^2)^n}\, dx = \frac{x}{2a^2(n - 1)(a^2 + x^2)^{n-1}}$$

$$+ \frac{2n - 3}{2a^2(n - 1)} \int \frac{1}{(a^2 + x^2)^{n-1}}\, dx, \quad n \geq 2,$$

(with n set equal to 2), we obtain

$$\int \frac{1}{(x^2 + 3)^2}\, dx = \frac{x}{2(3)(1)(3 + x^2)} + \frac{1}{2(3)(1)} \int \frac{1}{3 + x^2}\, dx.$$

With the integration formula

$$\int \frac{1}{a^2 + x^2}\, dx = \frac{1}{a}\text{Tan}^{-1}\!\left(\frac{x}{a}\right),$$

we can finish the problem,

$$\int \frac{3x^4 - x^3 + 18x^2 - 5x + 17}{x^5 - x^4 + 6x^3 - 6x^2 + 9x - 9}\, dx = 2\ln|x - 1| + \frac{1}{2}\ln(x^2 + 3) - \frac{3}{2(x^2 + 3)}$$

$$+ \frac{x}{6(x^2 + 3)} + \frac{1}{6}\left[\frac{1}{\sqrt{3}}\text{Tan}^{-1}\!\left(\frac{x}{\sqrt{3}}\right) \right] + C$$

$$= 2\ln|x - 1| + \frac{1}{2}\ln(x^2 + 3) + \frac{x - 9}{6(x^2 + 3)}$$

$$+ \frac{1}{6\sqrt{3}}\text{Tan}^{-1}\!\left(\frac{x}{\sqrt{3}}\right) + C.$$

EXERCISES 9.8

In Exercises 1–20 use tables to evaluate the indefinite integral.

* **1.** $\displaystyle\int \frac{1}{x^2+9}\,dx$

* **2.** $\displaystyle\int \frac{1}{x^2-16}\,dx$

* **3.** $\displaystyle\int \frac{1}{4-3x^2}\,dx$

* **4.** $\displaystyle\int \frac{1}{x\sqrt{x^2+5}}\,dx$

* **5.** $\displaystyle\int \frac{x}{\sqrt{3x-2}}\,dx$

* **6.** $\displaystyle\int \frac{\sqrt{2x+3}}{x}\,dx$

* **7.** $\displaystyle\int \frac{1}{\sqrt{x^2+4x}}\,dx$

* **8.** $\displaystyle\int \frac{1}{\sqrt{4x-9x^2}}\,dx$

* **9.** $\displaystyle\int \sqrt{3+2x+4x^2}\,dx$

* **10.** $\displaystyle\int \sqrt{3+2x-4x^2}\,dx$

* **11.** $\displaystyle\int \frac{1}{3+\sin x}\,dx$

* **12.** $\displaystyle\int \frac{1}{1+3\sin x}\,dx$

* **13.** $\displaystyle\int x\,\mathrm{Sin}^{-1}x\,dx$

* **14.** $\displaystyle\int \frac{1}{x^2}\,\mathrm{Sin}^{-1}x\,dx$

* **15.** $\displaystyle\int e^{4x}\cos 2x\,dx$

* **16.** $\displaystyle\int e^{-2x}\sin 3x\,dx$

* **17.** $\displaystyle\int x^7\ln x\,dx$

* **18.** $\displaystyle\int (\ln x)^2\,dx$

* **19.** $\displaystyle\int \frac{1}{e^x+2e^{-2x}}\,dx$

* **20.** $\displaystyle\int \frac{1}{2+x^3}\,dx$

In Exercises 21–30 use a reduction formula to aid in the evaluation of the integral.

** **21.** $\displaystyle\int x^4 e^{3x}\,dx$

** **22.** $\displaystyle\int x^3\cos 2x\,dx$

** **23.** $\displaystyle\int \frac{x}{\sqrt{9+x+x^2}}\,dx$

** **24.** $\displaystyle\int \frac{x}{\sqrt{9+x-x^2}}\,dx$

** **25.** $\displaystyle\int (4+x^2)^{3/2}\,dx$

** **26.** $\displaystyle\int \frac{x^2}{\sqrt{2x-x^2}}\,dx$

** **27.** $\displaystyle\int \sin^6 x\,dx$

** **28.** $\displaystyle\int \frac{1}{\cos^6 x}\,dx$

** **29.** $\displaystyle\int \cot^5 x\,dx$

** **30.** $\displaystyle\int x^3\,\mathrm{Sin}^{-1}x\,dx$

** **31.** Evaluate

$$\int \sec^9 x\,dx$$

by first developing the reduction formula

$$\int \sec^n x\,dx = \frac{\tan x\ \sec^{n-2}x}{n-1} + \frac{n-2}{n-1}\int \sec^{n-2}x\,dx.$$

*** **32.** (a) Use a trigonometric substitution to evaluate

$$\int \frac{5x + 2}{(3 + x^2)^4} \, dx.$$

(b) Verify the reduction formula in 9.21and use it to evaluate the integral in (a).

*** **33.** If n is a positive integer, define

$$I_n = \int_{-1}^{1} (1 - x^2)^n \, dx.$$

(a) Show that

$$(2n + 1)I_n = 2n I_{n-1}, \quad n \geq 2.$$

(b) Prove that

$$I_n = \frac{2^{2n+1}(n!)^2}{(2n + 1)!}, \quad n \geq 1.$$

(c) Evaluate

$$\int_0^{\pi/2} \cos^{2n+1} x \, dx.$$

*** **34.** Verify that

$$\int_0^{\pi/2} \sin^n x \, dx = \int_0^{\pi/2} \cos^n x \, dx = \begin{cases} \frac{2 \cdot 4 \cdot 6 \cdots (n-1)}{1 \cdot 3 \cdot 5 \cdots (n)}, & n \text{ an odd integer} \\ \frac{1 \cdot 3 \cdot 5 \cdots (n-1)}{2 \cdot 4 \cdot 6 \cdots (n)} \frac{\pi}{2}, & n \text{ an even integer} \end{cases}$$

9.9 Numerical Integration

To evaluate the definite integral of a continuous function $f(x)$ with respect to x from $x = a$ to $x = b$, we have used the fundamental theorem of integral calculus: Find an indefinite integral for $f(x)$, substitute the limits $x = b$ and $x = a$, and subtract. The evaluation procedure depends on our ability to produce an indefinite integral for $f(x)$. When a function $f(x)$ is complicated, it may be difficult, or even impossible to find its antiderivative. In such a case, it may be necessary to approximate the definite integral of $f(x)$ on some interval $a \leq x \leq b$, rather than evaluate it analytically. We discuss three methods for doing this, the rectangular, trapezoidal, and Simpson's rules. Each method divides the interval $a \leq x \leq b$ into subintervals and approximates $f(x)$ with an easily integrated function on each subinterval.

If $f(x) \geq 0$ on $a \leq x \leq b$ (Figure 9.19), the definite integral of $f(x)$ with respect to x can be interpreted as the area bounded by $y = f(x)$, $y = 0$, $x = a$, and $x = b$. We have carefully pointed out that it is not always wise to think of a definite integral as area, but for our discussion here it is convenient to do so. Our problem is to approximate the area in Figure 9.19 when it is difficult or impossible to find an antiderivative for $f(x)$.

FIGURE 9.19

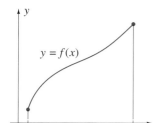

Rectangular Rule

The first method is to return to definition 7.1 for area in terms of approximating rectangles. We subdivide the interval $a \leq x \leq b$ into n subintervals by points $a = x_0 < x_1 < \cdots < x_{n-1} < x_n = b$. For simplicity, we choose n equal subdivisions, in which case

$$x_i = a + i \left(\frac{b - a}{n} \right),$$

and denote the width of each subinterval by $h = (b - a)/n$. Area under $y = f(x)$, above $y = 0$, and between $x = a$ and $x = b$, which is given by the definite integral of $f(x)$ from a to b, is approximated by the rectangles in Figure 9.20,

$$\int_a^b f(x)\, dx \approx \sum_{i=1}^n f(x_i)\, h = h \sum_{i=1}^n f(x_i). \tag{9.20}$$

This is the **rectangular rule** for approximating the definite integral; we have replaced the area under $y = f(x)$ with n rectangles. Another way of looking at it is to say that we have replaced the original function $f(x)$ by a function that is constant on each subinterval (Figure 9.21), but the constant value varies from subinterval to subinterval. Such a function is called a *step function*. The definite integral of this function from $x = a$ to $x = b$ is the right side of 9.20. It approximates the definite integral of $f(x)$.

Graphically, it is reasonable to expect that an increase in the number of subdivisions results in an increase in the accuracy of the approximation.

FIGURE 9.20

FIGURE 9.21

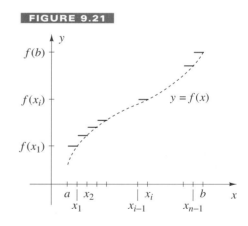

EXAMPLE 9.22

Use a subdivision of the interval $1 \leq x \leq 3$ into 5, 10, and 20 equal parts to approximate the definite integral

$$\int_1^3 \sin x\, dx$$

with the rectangular rule.

SOLUTION   With 5 equal parts, $h = 2/5$ and $x_i = 1 + 2i/5$. Consequently,

$$\int_1^3 \sin x\, dx \approx \frac{2}{5} \sum_{i=1}^{5} \sin\left(1 + \frac{2i}{5}\right)$$

$$= \frac{2}{5}\left(\sin\frac{7}{5} + \sin\frac{9}{5} + \sin\frac{11}{5} + \sin\frac{13}{5} + \sin 3\right)$$

$$= 1.370.$$

With 10 equal parts, $h = 2/10 = 1/5$ and $x_i = 1 + i/5$ so that

$$\int_1^3 \sin x\, dx \approx \frac{1}{5} \sum_{i=1}^{10} \sin\left(1 + \frac{i}{5}\right)$$

$$= \frac{1}{5}\left(\sin\frac{6}{5} + \sin\frac{7}{5} + \sin\frac{8}{5} + \sin\frac{9}{5} + \sin 2 + \sin\frac{11}{5}\right.$$

$$\left. + \sin\frac{12}{5} + \sin\frac{13}{5} + \sin\frac{14}{5} + \sin 3\right)$$

$$= 1.455.$$

With 20 equal parts we have

$$\int_1^3 \sin x\, dx \approx \frac{1}{10} \sum_{i=1}^{20} \sin\left(1 + \frac{i}{10}\right)$$

$$= \frac{1}{10}\left(\sin\frac{11}{10} + \sin\frac{6}{5} + \sin\frac{13}{10} + \cdots + \sin\frac{29}{10} + \sin 3\right)$$

$$= 1.494.$$

When we compare these results with the correct answer

$$\int_1^3 \sin x\, dx = \left\{-\cos x\right\}_1^3 = -\cos 3 + \cos 1 = 1.530\,294\,8,$$

we see that as n increases from 5 to 10 to 20, the approximation improves, but it must be increased even further to give a reasonable approximation to this integral.

Trapezoidal Rule

Regarding area, it is clear in Figure 9.20 that, were we to join successive points $(x_i, f(x_i))$ on the curve $y = f(x)$ with straight line segments as in Figure 9.22a, the area under this broken straight line would be a much better approximation to the area under $y = f(x)$ than that provided by the rectangular rule. Effectively, we now approximate the area by n trapezoids.

FIGURE 9.22a

FIGURE 9.22b

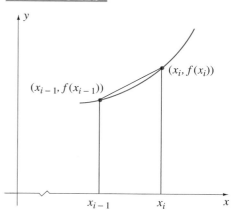

Since the area of a trapezoid is its width multiplied by the average of its parallel lengths, it follows that the area of the ith trapezoid in Figure 9.22b is given by

$$h \left[\frac{f(x_i) + f(x_{i-1})}{2} \right],$$

where again $h = x_i - x_{i-1}$. As a result, the area under $y = f(x)$ can be approximated by the sum

$$h \left[\frac{f(x_1) + f(x_0)}{2} + \frac{f(x_2) + f(x_1)}{2} + \cdots + \frac{f(x_n) + f(x_{n-1})}{2} \right]$$

$$= \frac{h}{2} [f(a) + 2f(x_1) + 2f(x_2) + \cdots + 2f(x_i) + \cdots + 2f(x_{n-1}) + f(b)]$$

$$= \frac{h}{2} \left[f(a) + 2 \sum_{i=1}^{n-1} f(x_i) + f(b) \right].$$

We write therefore that

$$\int_a^b f(x)\,dx \approx \frac{h}{2} \left[f(a) + 2 \sum_{i=1}^{n-1} f(x_i) + f(b) \right], \qquad (9.21\text{a})$$

where $x_i = a + ih = a + i(b - a)/n$, and call this the **trapezoidal rule** for approximating a definite integral. Note that if 9.21a is written in the form

$$\int_a^b f(x)\,dx \approx h \left[\frac{f(a) - f(b)}{2} \right] + h \sum_{i=1}^{n} f(x_i), \qquad (9.21\text{b})$$

the summation on the right, except for the first two terms, is the rectangular rule. In other words, the extra numerical calculation involved in using the trapezoidal rule rather than the rectangular rule is minimal, but it would appear that the accuracy is significantly increased. For this reason, the trapezoidal rule supplants the rectangular rule in most applications.

EXAMPLE 9.23

Use the trapezoidal rule to approximate the definite integral in Example 9.22.

SOLUTION   With 5 equal partitions,

$$\int_1^3 \sin x \, dx \approx \frac{(2/5)}{2}\left[\sin 1 + 2\sum_{i=1}^{4} \sin\left(1 + \frac{2i}{5}\right) + \sin 3\right] = 1.5098.$$

With 10 equal parts,

$$\int_1^3 \sin x \, dx \approx \frac{(1/5)}{2}\left[\sin 1 + 2\sum_{i=1}^{9} \sin\left(1 + \frac{i}{5}\right) + \sin 3\right] = 1.5252.$$

Finally, with $n = 20$, we have

$$\int_1^3 \sin x \, dx \approx \frac{(1/10)}{2}\left[\sin 1 + 2\sum_{i=1}^{19} \sin\left(1 + \frac{i}{10}\right) + \sin 3\right] = 1.5290.$$

As expected, these approximations are significantly better than corresponding results using the rectangular rule.

Simpson's Rule

The rectangular rule replaces a function $f(x)$ with a step function; the trapezoidal rule replaces $f(x)$ with a succession of linear functions—geometrically, a broken straight line. So far as ease of integration is concerned, the next simplest function is a quadratic function. Consider, then, replacing the curve $y = f(x)$ by a succession of parabolas on the subintervals $x_{i-1} \le x \le x_i$. Now the equation of a parabola with a vertical axis of symmetry is of the form $y = ax^2 + bx + c$ with three constants a, b, and c to be determined. If this parabola is to approximate $y = f(x)$ on $x_{i-1} \le x \le x_i$, we should have the parabola pass through the end points $(x_{i-1}, f(x_{i-1}))$ and $(x_i, f(x_i))$. But this imposes only two conditions on a, b, and c, not three. To take advantage of this flexibility, we demand that the parabola also pass through the point $(x_{i+1}, f(x_{i+1}))$ (Figure 9.23).

FIGURE 9.23

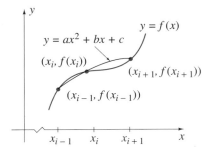

In other words, instead of replacing $y = f(x)$ with n parabolas, one on each subinterval $x_{i-1} \le x \le x_i$, we replace it with $n/2$ parabolas, one on each pair of subintervals

$x_{i-1} \le x \le x_{i+1}$. Note that this requires n to be an even integer. These three conditions imply that a, b, and c must satisfy the equations

$$ax_{i-1}^2 + bx_{i-1} + c = f(x_{i-1}), \tag{9.22a}$$

$$ax_i^2 + bx_i + c = f(x_i), \tag{9.22b}$$

$$ax_{i+1}^2 + bx_{i+1} + c = f(x_{i+1}). \tag{9.22c}$$

These equations determine the values for a, b, and c, but it will not be necessary to actually solve them. Suppose for the moment that we have solved equations 9.22 for a, b, and c, and we continue our main discussion. With the parabola $y = ax^2 + bx + c$ replacing the curve $y = f(x)$ on the interval $x_{i-1} \le x \le x_{i+1}$, we approximate the area under $y = f(x)$ with that under the parabola, namely,

$$\int_{x_{i-1}}^{x_{i+1}} (ax^2 + bx + c)\, dx = \frac{a}{3}(x_{i+1}^3 - x_{i-1}^3) + \frac{b}{2}(x_{i+1}^2 - x_{i-1}^2) + c(x_{i+1} - x_{i-1})$$

$$= \frac{x_{i+1} - x_{i-1}}{6}[2a(x_{i+1}^2 + x_{i+1}x_{i-1} + x_{i-1}^2)$$

$$+ 3b(x_{i+1} + x_{i-1}) + 6c].$$

Now, if we use $x_{i+1} - x_{i-1} = 2h$, $x_{i+1} = x_i + h$, and $x_{i-1} = x_i - h$ to write everything in terms of h and x_i, we obtain

$$\int_{x_{i-1}}^{x_{i+1}} (ax^2 + bx + c)\, dx = \frac{2h}{6}\big\{2a\big[(x_i + h)^2 + (x_i + h)(x_i - h) + (x_i - h)^2\big]$$

$$+ 3b\big[(x_i + h) + (x_i - h)\big] + 6c\big\}$$

$$= \frac{h}{3}\big\{a(6x_i^2 + 2h^2) + 6bx_i + 6c\big\}.$$

But, if we add equation 9.22a, equation 9.22c, and four times equation 9.22b, we find that

$$f(x_{i-1}) + 4f(x_i) + f(x_{i+1}) = a(x_{i-1}^2 + 4x_i^2 + x_{i+1}^2) + b(x_{i-1} + 4x_i + x_{i+1}) + 6c$$

$$= a\big[(x_i - h)^2 + 4x_i^2 + (x_i + h)^2\big]$$

$$+ b\big[(x_i - h) + 4x_i + (x_i + h)\big] + 6c$$

$$= a(6x_i^2 + 2h^2) + 6bx_i + 6c.$$

Thus, the area under the parabola may be written in the form

$$\int_{x_{i-1}}^{x_{i+1}} (ax^2 + bx + c)\, dx = \frac{h}{3}\big[f(x_{i-1}) + 4f(x_i) + f(x_{i+1})\big],$$

and the right side is free of a, b, and c. The same expression would have resulted had we solved equations 9.22 for a, b, and c and then evaluated the integral of $ax^2 + bx + c$ from x_{i-1} to x_{i+1}. The above derivation, however, is much simpler.

When we add all such integrals over the $n/2$ subintervals $x_{i-1} \leq x \leq x_{i+1}$ between $x = a$ and $x = b$, we obtain

$$\frac{h}{3}\left[f(x_0) + 4f(x_1) + f(x_2)\right] + \frac{h}{3}\left[f(x_2) + 4f(x_3) + f(x_4)\right] + \cdots$$

$$+ \frac{h}{3}\left[f(x_{n-2}) + 4f(x_{n-1}) + f(x_n)\right]$$

$$= \frac{h}{3}\left[f(x_0) + 4f(x_1) + 2f(x_2) + 4f(x_3) + \cdots\right.$$

$$\left. + 2f(x_{n-2}) + 4f(x_{n-1}) + f(x_n)\right].$$

In other words, the definite integral of $f(x)$ from $x = a$ to $x = b$ can be approximated by

$$\int_a^b f(x)\,dx \approx \frac{h}{3}\left[f(a) + 4f(x_1) + 2f(x_2) + 4f(x_3) + \cdots\right.$$

$$\left. + 2f(x_{n-2}) + 4f(x_{n-1}) + f(b)\right], \tag{9.23}$$

where $x_i = a + ih = a + i(b-a)/n$. This result is called **Simpson's rule** for approximating a definite integral. Although the formula does not explicitly display it (except by counting terms), *do not forget that n must be an even integer.*

EXAMPLE 9.24

Approximate the definite integral in Example 9.22 using Simpson's rule with $n = 10$ and $n = 20$.

SOLUTION   With $n = 10$,

$$\int_1^3 \sin x\,dx \approx \frac{1/5}{3}\left[\sin 1 + 4\sin\frac{6}{5} + 2\sin\frac{7}{5} + 4\sin\frac{8}{5} + \cdots\right.$$

$$\left. + 2\sin\frac{13}{5} + 4\sin\frac{14}{5} + \sin 3\right]$$

$$= 1.530\,308\,5.$$

With $n = 20$,

$$\int_1^3 \sin x\,dx \approx \frac{1/10}{3}\left[\sin 1 + 4\sin\frac{11}{10} + 2\sin\frac{6}{5} + 4\sin\frac{13}{10} + \cdots\right.$$

$$\left. + 2\sin\frac{14}{5} + 4\sin\frac{29}{10} + \sin 3\right]$$

$$= 1.530\,295\,7.$$

Table 9.3 lists the approximations in Examples 9.22–9.24. The correct answer for the integral is $1.530\,294\,8$ (to 7 decimals). It is clear that each method gives a better approximation as the value of n increases, and that Simpson's rule is by far the most accurate.

TABLE 9.3

	Rectangular rule	Trapezoidal rule	Simpson's rule
$n = 5$	1.370	1.5098	
$n = 10$	1.455	1.5252	1.530 308 5
$n = 20$	1.494	1.5290	1.530 295 7

In practice, we use the rectangular, trapezoidal, and Simpson's rules to approximate definite integrals that cannot be handled analytically, and we will not therefore have the correct answer with which to compare the approximation. We would still like to make some statement about the accuracy of the approximation, however, since what good is the approximation otherwise? The following two theorems give error estimates for the trapezoidal rule and for Simpson's rule.

THEOREM 9.1

If $f''(x)$ exists on $a \leq x \leq b$, and T_n is the error in approximating the definite integral of $f(x)$ from $x = a$ to $x = b$ using the trapezoidal rule with n equal subdivisions,

$$T_n = \int_a^b f(x)\,dx - \frac{h}{2}\left[f(a) + 2\sum_{i=1}^{n-1} f(x_i) + f(b) \right],$$

then

$$|T_n| \leq \frac{M(b-a)^3}{12n^2}, \tag{9.24}$$

where M is the maximum value of $|f''(x)|$ on $a \leq x \leq b$.

THEOREM 9.2

If $f''''(x)$ exists on $a \leq x \leq b$, and S_n is the error in approximating the definite integral of $f(x)$ from $x = a$ to $x = b$ using Simpson's rule with n equal subdivisions,

$$S_n = \int_a^b f(x)\,dx - \frac{h}{3}\left[f(a) + 4\sum_{i=1}^{n/2} f(x_{2i-1}) + 2\sum_{i=1}^{n/2-1} f(x_{2i}) + f(b) \right],$$

then

$$|S_n| \leq \frac{M(b-a)^5}{180n^4}, \tag{9.25}$$

where M is the maximum value of $|f''''(x)|$ on $a \leq x \leq b$.

Proofs of these theorems can be found in books on numerical analysis. Note that because of the n^4 factor in the denominator of 9.25, the accuracy of Simpson's rule increases much more rapidly than does that of the trapezoidal rule.

For the function $f(x) = \sin x$ in Examples 9.22–9.24,

$$f''(x) = -\sin x \quad \text{and} \quad f''''(x) = \sin x.$$

Consequently, in both cases we can state that $M = 1$, and therefore

$$|T_n| \leq \frac{(3-1)^3}{12n^2} = \frac{2}{3n^2} \quad \text{and} \quad |S_n| \leq \frac{(3-1)^5}{180n^4} = \frac{8}{45n^4}.$$

For $n = 10$ and $n = 20$, we find

$$|T_{10}| \leq \frac{2}{3(10)^2} < 0.0067 \quad \text{and} \quad |S_{10}| \leq \frac{8}{45(10)^4} < 0.000\,018;$$

$$|T_{20}| \leq \frac{2}{3(20)^2} < 0.0017 \quad \text{and} \quad |S_{20}| \leq \frac{8}{45(20)^4} < 0.000\,012.$$

Differences between the correct value for the integral and the approximations listed in Table 9.3 corroborate these predictions.

EXAMPLE 9.25

What is the maximum possible error in using the trapezoidal rule with 100 equal subdivisions to approximate

$$\int_1^3 \frac{1}{\sqrt{1+x^3}}\, dx?$$

SOLUTION   According to formula 9.24, if T_{100} is the maximum possible error, then

$$|T_{100}| \leq \frac{M(3-1)^3}{12(100)^2} = \frac{2M}{3 \times 10^4},$$

where M is the maximum of (the absolute value of) the second derivative of the integrand $1/\sqrt{1+x^3}$ on $1 \leq x \leq 3$. Now,

$$\frac{d^2}{dx^2} \frac{1}{\sqrt{1+x^3}} = \frac{d}{dx}\left[\frac{-3x^2}{2(1+x^3)^{3/2}}\right] = \frac{3x(5x^3-4)}{4(1+x^3)^{5/2}}.$$

Instead of maximizing the absolute value of this function on the interval $1 \leq x \leq 3$, which would require another derivative, we note that the maximum value of the numerator is obtained for $x = 3$, and the minimum value of the denominator occurs at $x = 1$. It follows therefore that the second derivative cannot possibly be larger than

$$\frac{3(3)[5(3)^3 - 4]}{4(1+1)^{5/2}} = \frac{1179}{16\sqrt{2}}.$$

Thus, M must be less than or equal to $1179/(16\sqrt{2})$, and we can state that

$$|T_{100}| \leq \frac{2}{3 \times 10^4}\left(\frac{1179}{16\sqrt{2}}\right) \leq 0.0035;$$

that is, the error in using the trapezoidal rule with 100 equal subdivisions to approximate the definite integral cannot be any larger than 0.0035.

EXAMPLE 9.26

How many equal subdivisions of the interval $0 \leq x \leq 2$ guarantee an error of less than 10^{-5} in the approximation of the definite integral

$$\int_0^2 e^{-x^2} \, dx$$

using Simpson's rule?

SOLUTION   According to formula 9.25, the error in using Simpson's rule with n equal subdivisions to approximate this definite integral is

$$|S_n| \leq \frac{M(2-0)^5}{180n^4},$$

where M is the maximum of the (absolute value of the) fourth derivative of e^{-x^2} on the interval $0 \leq x \leq 2$. It is a short calculation to find

$$\frac{d^4}{dx^4} \left(e^{-x^2} \right) = 4(3 - 12x^2 + 4x^4)e^{-x^2}.$$

Instead of maximizing the absolute value of this function, we note that e^{-x^2} has a maximum value of 1 (when $x = 0$). Furthermore, because $|3 - 12x^2 + 4x^4| \leq 3 + 12x^2 + 4x^4$, which has a maximum at $x = 2$, it follows that

$$M \leq 4[3 + 12(2)^2 + 4(2)^4](1) = 460.$$

Consequently,

$$|S_n| \leq \frac{460(2)^5}{180n^4} = \frac{3680}{45n^4}.$$

The error is less than 10^{-5} if n is chosen sufficiently large that

$$\frac{3680}{45n^4} < 10^{-5};$$

that is, if

$$n > \left(\frac{3680}{45 \times 10^{-5}} \right)^{1/4} = 53.5.$$

Since n must be an even integer, the required accuracy is guaranteed if n is chosen greater than or equal to 54.

Numerical techniques are indispensable in situations where we know that y is a function of x, say $y = f(x)$, but we do not know the exact form of the function $f(x)$. What might be available is a set of experimental data listing values of y corresponding to various values of x (Table 9.4). We have plotted these points in Figure 9.24.

TABLE 9.4

x	1.0	1.2	1.4	1.6	1.8	2.0	2.2	2.4	2.6	2.8	3.0
y	0.84	1.12	1.40	1.60	1.75	1.82	1.79	1.62	1.34	0.94	0.42

FIGURE 9.24

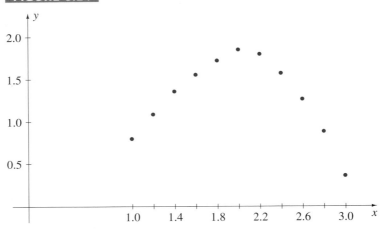

The trapezoidal rule and Simpson's rule can be used to approximate the definite integral of $f(x)$ from $x = 1$ to $x = 3$ even though we do not have a formula for $f(x)$. Indeed, formulas 9.21 and 9.23 do not use the form of the function $f(x)$ to be integrated, only its values at the subdivision points x_i. We have these points in Table 9.4 or Figure 9.24. Since there are 11 points and, therefore, 10 subdivisions of the interval $1 \le x \le 3$, the trapezoidal rule gives

$$\int_1^3 f(x)\,dx \approx \frac{1/5}{2}\Big[f(1.0) + 2f(1.2) + 2f(1.4) + \cdots$$

$$+2f(2.6) + 2f(2.8) + f(3.0)\Big]$$

$$= \frac{1}{10}\Big[0.84 + 2(1.12) + 2(1.40) + \cdots + 2(1.34) + 2(0.94) + 0.42\Big]$$

$$= 2.80.$$

Simpson's rule gives

$$\int_1^3 f(x)\,dx \approx \frac{1/5}{3}\Big[f(1.0) + 4f(1.2) + 2f(1.4) + \cdots$$

$$+2f(2.6) + 4f(2.8) + f(3.0)\Big]$$

$$= \frac{1}{15}\Big[0.84 + 4(1.12) + 2(1.40) + \cdots + 2(1.34) + 4(0.94) + 0.42\Big]$$

$$= 2.81.$$

What we lose in this type of application is the ability to predict a maximum possible error in the approximation since we cannot estimate M in formulas 9.24 and 9.25.

The error bounds in formulas 9.24 and 9.25 are somewhat idealistic in the sense that they are error predictions based on the use of exact numbers. For instance, they predict that by increasing n indefinitely, any degree of accuracy is attainable. Theoretically this is true, but practically it is not. No matter how we choose to evaluate the summations in 9.21 and 9.23, be it by hand, by an electronic hand calculator, or by a high-speed computer, each calculation is rounded off to a certain number of decimals. The final sum takes into account many, many of these "approximate numbers," and must therefore be inherently inaccurate. We call this *round-off error* and it is very difficult to predict how extensive it is. It depends on both the number and nature of the operations involved in 9.21 and 9.23. In the approximation of a definite integral by the trapezoidal rule or Simpson's rule, there are two sources of error. Formulas 9.24 and 9.25 predict errors due to the methods themselves; round-off errors may also be appreciable for large n.

We should emphasize once again that, although we have used area as a convenient vehicle by which to explain the approximation of definite integrals by the rectangular, trapezoidal, and Simpson's rules, it is not necessary for $f(x)$ to be nonnegative. All three methods can be used to approximate the definite integral of a function $f(x)$, be it positive, negative, or sometimes positive and sometimes negative on the interval of integration. The only condition that we have imposed is that $f(x)$ be continuous. In view of our discussion of improper integrals in Section 7.10, even this is not always necessary.

EXERCISES 9.9

In Exercises 1–10 use the trapezoidal rule and Simpson's rule with 10 equal subdivisions to approximate the definite integral. In each case, evaluate the integral analytically to get an idea of the accuracy of the approximation.

* **1.** $\displaystyle\int_1^2 \frac{1}{x}\, dx$

* **2.** $\displaystyle\int_2^3 \frac{1}{\sqrt{x+2}}\, dx$

* **3.** $\displaystyle\int_0^1 \tan x\, dx$

* **4.** $\displaystyle\int_0^{1/2} e^x\, dx$

* **5.** $\displaystyle\int_{-1}^1 \sqrt{x+1}\, dx$

* **6.** $\displaystyle\int_{-3}^{-2} \frac{1}{x^3}\, dx$

* **7.** $\displaystyle\int_{1/2}^1 \cos x\, dx$

* **8.** $\displaystyle\int_0^1 \frac{1}{3+x^2}\, dx$

* **9.** $\displaystyle\int_1^3 \frac{1}{x^2+x}\, dx$

* **10.** $\displaystyle\int_0^{1/2} xe^{x^2}\, dx$

In Exercises 11–14 use the trapezoidal rule and Simpson's rule with 10 equal subdivisions to approximate the definite integral.

* **11.** $\displaystyle\int_0^2 \frac{1}{1+x^3}\, dx$

* **12.** $\displaystyle\int_0^1 e^{x^2}\, dx$

* **13.** $\displaystyle\int_1^2 \sqrt{1+x^4}\, dx$

* **14.** $\displaystyle\int_{-1}^0 \sin(x^2)\, dx$

* **15.** Show graphically that if $y = f(x)$ is concave downward on the interval $a \le x \le b$, then the trapezoidal rule underestimates the definite integral of $f(x)$ on this interval.

* **16.** What happens to the errors in 9.24 and 9.25 when the number of partitions is doubled?

∗ **17.** The definite integral

$$\int_a^b e^{-x^2}\, dx$$

is very important in mathematical statistics. Use Simpson's rule with $n = 16$ to evaluate the integral for $a = 0$ and $b = 1$.

∗∗ **18.** Use Simpson's rule with 10 equal intervals to approximate the definite integral for the length of the parabola $y = x^2$ between $x = 0$ and $x = 1$. Compare the answer to that of Exercise 39 in Section 9.5.

∗∗ **19.** Use the trapezoidal rule and Simpson's rule with 10 equal subdivisions to approximate the definite integral for the length of the curve $y = \sin x$ from $x = 0$ to $x = \pi/2$.

∗∗ **20.** The numerical techniques of this section can also be used to approximate many improper integrals. Consider

$$\int_0^4 \frac{e^x}{\sqrt{x}}\, dx.$$

    (a) Why can the trapezoidal rule and Simpson's rule not be used directly to approximate this integral?

    (b) Show that the change of variable $u = \sqrt{x}$ replaces this improper integral with an integral that is not improper and use Simpson's rule with 20 equal subdivisions to approximate its value.

    (c) Could you use the rectangular rule on the improper integral?

∗∗ **21.** (a) Show that the definite integral

$$L = \frac{4}{3}\int_0^3 \sqrt{\frac{81 - 5x^2}{9 - x^2}}\, dx$$

represents the length L of the ellipse $4x^2 + 9y^2 = 36$.

    (b) Show that when we set $x = \sin\theta$, the θ-integral is no longer improper.

    (c) Use the trapezoidal rule and Simpson's rule with 8 equal subdivisions to approximate the θ-integral in (b).

∗∗ **22.** To approximate the improper integral

$$\int_1^\infty \frac{1}{1 + x^4}\, dx$$

set $x = 1/t$, and use the trapezoidal rule and Simpson's rule with 10 equal subdivisions on the resulting integral.

∗∗ **23.** Use the technique of Exercise 22 to approximate

$$\int_1^\infty \frac{x^2}{x^4 + x^2 + 1}\, dx.$$

∗∗ **24.** Use both the trapezoidal rule and Simpson's rule to approximate the definite integral from $x = -1$ to $x = 4$ for the function tabulated in the following table.

x	-1.0	-0.5	0	0.5	1.0	1.5	2.0	2.5	3.0	3.5	4.0
y	2.287	0.395	0	0.145	0.310	0.334	1.819	0.123	0.021	-0.037	-0.055

** **25.** When the truss in the figure below is subjected to a force F at its centre, point A deflects an amount y from its equilibrium position. Forces (in kilonewtons) required to produce deflections from 0 to 5 cm at intervals of 0.5 cm are listed in the accompanying table. Plot these points on a set of axes with y horizontal and F vertical. Use Simpson's rule to approximate the work done by F in deflecting A by a full 5 cm.

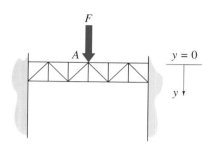

y	0	0.5	1.0	1.5	2.0	2.5	3.0	3.5	4.0	4.5	5.0
F	0	1.45	2.90	4.40	5.90	7.43	9.05	10.7	13.2	15.3	18.0

** **26.** Show that when $f(x)$ is a cubic polynomial, evaluation of the definite integral of $f(x)$ from $x = a$ to $x = b$ by Simpson's rule always gives the *exact* answer. Illustrate with an example.

In Exercises 27–30 how many equal subdivisions of the interval of integration guarantee an error of less than 10^{-4} in the approximation of the definite integral using (a) the trapezoidal rule and (b) Simpson's rule?

** **27.** $\displaystyle\int_{1}^{4} \frac{1}{x}\, dx$

** **28.** $\displaystyle\int_{0}^{\pi/4} \cos x\, dx$

** **29.** $\displaystyle\int_{0}^{1/3} e^{2x}\, dx$

** **30.** $\displaystyle\int_{4}^{5} \frac{1}{\sqrt{x+2}}\, dx$

SUMMARY

Antidifferentiation is a far more complicated process than differentiation; it is not possible to state a set of rules and formulas that will suffice for most functions. Certainly there are integration formulas that we must learn, but most of those given in this chapter have been differentiation formulas listed in Chapters 3 and 8 written in terms of integrals rather than derivatives. We have stressed the importance of knowing these obvious integration formulas but, beyond this, it becomes an organizational problem—organizing a difficult integral into a form that utilizes these simple formulas. The three most important techniques for doing this are substitutions, integration by parts, and partial fractions.

There is a variety of substitutions for the evaluation of indefinite integrals, many suggested by the form of the integrand. For example, a term in $x^{1/n}$ (n an integer) suggests the substitution $u = x^{1/n}$. Trigonometric substitutions are most important. They eliminate square roots of the form $\sqrt{a^2 \pm b^2x^2}$ and $\sqrt{b^2x^2 - a^2}$, and the general root $\sqrt{ax^2 + bx + c}$.

Integration by parts is a powerful integration technique. It is used to evaluate antiderivatives of transcendental functions that are multiplied by powers of x, it leads to the method of integration by reproduction, and it is used to develop reduction formulas.

The method of partial fractions decomposes complicated rational functions into simple fractions that are either immediately integrable or amenable to other methods such as trigonometric substitutions.

Tables of integrals provide formulas for simple antiderivatives as well as very complex ones. They contain reduction formulas and are an excellent companion to partial fractions.

Even with the techniques that we have studied and tables, there are many functions that either cannot be antidifferentiated at all or can be antidifferentiated only with extreme difficulty. To approximate definite integrals of such functions, numerical techniques such as the rectangular rule, the trapezoidal rule, and Simpson's rule are essential.

KEY TERMS

In reviewing this chapter, you should be able to discuss the following key terms:

Exponential growth
Exponential decay
Integration by parts
Integration by reproduction
Trigonometric substitution

Partial fraction decomposition
Rectangular rule
Trapezoidal rule
Simpson's rule

REVIEW EXERCISES

In Exercises 1–50 evaluate the indefinite integral.

* **1.** $\displaystyle\int \sqrt{2-x}\,dx$

* **2.** $\displaystyle\int \frac{1}{(x+3)^2}\,dx$

* **3.** $\displaystyle\int \frac{x^2+3}{x}\,dx$

* **4.** $\displaystyle\int \frac{x^2+3}{x+1}\,dx$

* **5.** $\displaystyle\int \frac{x^2+3}{x^2+1}\,dx$

* **6.** $\displaystyle\int \frac{x}{\sqrt{x+3}}\,dx$

* **7.** $\displaystyle\int \sin^2 x \cos^3 x\,dx$

* **8.** $\displaystyle\int x \sin x\,dx$

* **9.** $\displaystyle\int \tan^2 (2x)\,dx$

* **10.** $\displaystyle\int \frac{x}{x^2+2x-3}\,dx$

* **11.** $\displaystyle\int \frac{1}{\sqrt{4-3x^2}}\,dx$

* **12.** $\displaystyle\int \frac{2-\sqrt{x}}{\sqrt{x}+5}\,dx$

* **13.** $\displaystyle\int \frac{x}{3x^2+4}\,dx$

* **14.** $\displaystyle\int \frac{e^x}{\sqrt{1-e^{2x}}}\,dx$

* **15.** $\displaystyle\int x^2 \ln x\,dx$

* **16.** $\displaystyle\int \frac{x}{(x^2+1)^2}\,dx$

* **17.** $\displaystyle\int \frac{x^2}{(x^2+1)^2}\,dx$

* **18.** $\displaystyle\int \frac{x^3}{(x^2+1)^2}\,dx$

* **19.** $\displaystyle\int \frac{x+1}{x^3-4x}\,dx$

* **20.** $\displaystyle\int \left(\frac{x+1}{x-1}\right)^2 dx$

* **21.** $\displaystyle\int \frac{x^2}{(1+3x^3)^4}\,dx$

* **22.** $\displaystyle\int \mathrm{Cos}^{-1}x\,dx$

* **23.** $\displaystyle\int \sin x \cos 2x\,dx$

* **24.** $\displaystyle\int \sin x \cos 5x\,dx$

* **25.** $\displaystyle\int e^{3x}\cos 2x\,dx$

* **26.** $\displaystyle\int \frac{1}{\sqrt{x^2+4x-5}}\,dx$

* **27.** $\displaystyle\int \frac{1}{x^2+4x-5}\,dx$

* **28.** $\displaystyle\int x^3\sqrt{4-x^2}\,dx$

* **29.** $\displaystyle\int \frac{\cos 2x}{1-\sin 2x}\,dx$

* **30.** $\displaystyle\int \frac{6x}{4-x^2}\,dx$

* **31.** $\displaystyle\int \frac{1}{x\sqrt{\ln x}}\,dx$

* **32.** $\displaystyle\int \frac{1}{x^2+4x-4}\,dx$

* **33.** $\displaystyle\int \frac{\sin x}{1+\cos^2 x}\,dx$

* **34.** $\displaystyle\int \frac{1}{x^4+x^3}\,dx$

* **35.** $\displaystyle\int x\sec^2(3x)\,dx$

* **36.** $\displaystyle\int \frac{1}{\sqrt{16-3x+x^2}}\,dx$

* **37.** $\displaystyle\int \frac{\sqrt{x^2-4}}{x^2}\,dx$

* **38.** $\displaystyle\int x^2\mathrm{Tan}^{-1}x\,dx$

* **39.** $\displaystyle\int \frac{x^2}{x^3+3x^2+3x+1}\,dx$

* **40.** $\displaystyle\int \frac{\ln x}{x}\,dx$

* **41.** $\displaystyle\int \frac{x^2}{1+4x^6}\,dx$

* **42.** $\displaystyle\int \frac{1}{x(9+x^2)^2}\,dx$

* **43.** $\displaystyle\int \frac{x^2+2}{x^3+5x^2+4x}\,dx$

* **44.** $\displaystyle\int \frac{x^2+2}{x^3+4x^2+4x}\,dx$

* **45.** $\displaystyle\int \frac{x^2+2}{x^3+x^2+4x}\,dx$

* **46.** $\displaystyle\int \frac{3x^2+2x+4}{x^3+x^2+4x}\,dx$

* **47.** $\displaystyle\int x\mathrm{Sin}^{-1}x\,dx$

* **48.** $\displaystyle\int \sqrt{\cot x}\,\csc^4 x\,dx$

* **49.** $\displaystyle\int \ln(\sqrt{x}+1)\,dx$

* **50.** $\displaystyle\int \frac{1}{(4x-x^2)^{3/2}}\,dx$

In Exercises 51–54 use the trapezoidal rule and Simpson's rule with ten equal partitions to approximate the definite integral.

* **51.** $\displaystyle\int_1^2 \frac{\sin x}{x}\,dx$

* **52.** $\displaystyle\int_0^1 \sqrt{\sin x}\,dx$

* **53.** $\displaystyle\int_2^4 \frac{1}{\ln x}\,dx$

* **54.** $\displaystyle\int_{-1}^3 \frac{1}{1+e^x}\,dx$

* **55.** If 10% of a sample of radioactive material decays in 10 days, what is its half-life?

In Exercises 56–71 evaluate the indefinite integral.

** 56. $\displaystyle\int \frac{1}{x^{1/3} - \sqrt{x}}\, dx$

** 57. $\displaystyle\int \ln\,(1 + x^2)\, dx$

** 58. $\displaystyle\int \frac{x}{x^4 + 16}\, dx$

** 59. $\displaystyle\int \csc^3 x\, dx$

** 60. $\displaystyle\int \frac{1}{(3x - x^2)^{3/2}}\, dx$

** 61. $\displaystyle\int \frac{1}{x^3\sqrt{x^2 - 9}}\, dx$

** 62. $\displaystyle\int \sin\sqrt{x}\, dx$

** 63. $\displaystyle\int \sin\,(\ln x)\, dx$

** 64. $\displaystyle\int x \cos x \sin 3x\, dx$

** 65. $\displaystyle\int \frac{x^4 + 3x^2 + 1}{x(x^2 + 1)^2}\, dx$

** 66. $\displaystyle\int \frac{1}{1 + \cos 2x}\, dx$

** 67. $\displaystyle\int \frac{x^4 + 3x^2 - 2x + 5}{x^2 - 3x + 7}\, dx$

** 68. $\displaystyle\int \sin^2 x \cos 3x\, dx$

** 69. $\displaystyle\int \frac{1}{x^3(4 - x^2)^{3/2}}\, dx$

** 70. $\displaystyle\int \sqrt{1 - x^2}\,\operatorname{Sin}^{-1} x\, dx$

** 71. $\displaystyle\int \frac{1}{x + \sqrt{x^2 + 4}}\, dx$

** 72. (a) Use the substitution $z^2 = (1 + x)/(1 - x)$ to show that

$$\int \sqrt{\frac{1 + x}{1 - x}}\, dx = 2\operatorname{Tan}^{-1}\sqrt{\frac{1 + x}{1 - x}} - \sqrt{1 - x^2} + C.$$

(b) Evaluate the integral in (a) by multiplying numerator and denominator of the integrand by $\sqrt{1 + x}$. Verify that this answer is the same as that in (a).

Parametric Equations and Polar Coordinates

In this chapter we introduce parametric representations for curves, and polar coordinates. Parametric representations of curves offer an alternative to the explicit and implicit forms used in Chapters 1–9. Polar coordinates provide an alternative way to identify points in a plane. They are more efficient than Cartesian coordinates in many applications. In particular, we shall see how they provide a unified approach to conic sections.

10.1 Parametric Equations

A curve is defined explicitly by equations of the form $y = f(x)$ or $x = g(y)$, and implicitly by an equation $F(x, y) = 0$. A third method is described in the following definition.

> **DEFINITION 10.1**
>
> A curve is said to be defined **parametrically** if it is given in the form
>
> $$x = x(t), \qquad y = y(t), \qquad \alpha \leq t \leq \beta. \qquad (10.1)$$

Variable t is called a *parameter*; it is a connecting link between x and y. Each value of t in the interval $\alpha \leq t \leq \beta$ is substituted into equations 10.1, and the pair $(x, y) = \big(x(t), y(t)\big)$ represents a point on the curve. When the interval for t is unspecified, we assume that it consists of all values for which both $x(t)$ and $y(t)$ are defined.

Graphing calculators and computers are adept at plotting curves defined parametrically. Given the functions $x(t)$ and $y(t)$, and the interval $\alpha \leq t \leq \beta$, they plot sufficiently many points $\big(x(t), y(t)\big)$ and join them with straight lines to give an excellent rendering of the curve.

EXAMPLE 10.1

Plot the curve

$$x = 2 - t, \qquad y = t + 5, \qquad -6 \leq t \leq 5.$$

What does it look like? Verify this analytically.

SOLUTION   The plot in Figure 10.1 appears to be a straight line joining the points $(8, -1)$ and $(-3, 10)$. This is easily verified by solving $x = 2 - t$ for $t = 2 - x$ and substituting into $y = t + 5$,

$$y = (2 - x) + 5 = -x + 7.$$

651

FIGURE 10.1

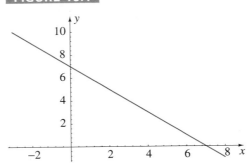

When $x(t)$ and $y(t)$ in 10.1 are linear functions, the curve is always a straight line or line segment. If α and β are finite, they define ends of the line segment; if they are not finite, the whole line results. Line segments can also be defined by nonlinear functions (see Exercise 31).

EXAMPLE 10.2

Plot the curve

$$x = t^2 - 2t + 4, \qquad y = 3 - 2t, \qquad -2 \le t \le 2.$$

What does it look like? Verify this analytically.

SOLUTION   The plot in Figure 10.2 appears to be a parabola joining the points $(12, 7)$ and $(4, -1)$. To verify this, we solve the second equation for $t = (3 - y)/2$, and substitute into the first,

$$x = \left(\frac{3 - y}{2}\right)^2 - 2\left(\frac{3 - y}{2}\right) + 4 = \frac{y^2}{4} - \frac{y}{2} + \frac{13}{4}.$$

FIGURE 10.2

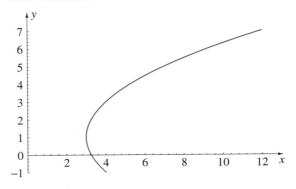

EXAMPLE 10.3

Discuss the curve defined parametrically by

$$x = 2\cos\theta, \quad y = 2\sin\theta, \quad 0 \leq \theta < 2\pi.$$

SOLUTION If the given equations are squared and added, the result is

$$x^2 + y^2 = 4\cos^2\theta + 4\sin^2\theta = 4,$$

an implicit definition of the curve. The given equations therefore define the circle parametrically. Seldom is it possible to give a geometric interpretation for parameter t in equations 10.1. Parameter θ for a circle is an exception; it can be interpreted as the angle in Figure 10.3. Values $0 \leq \theta < 2\pi$ describe the complete circle in a counterclockwise direction beginning at $(2, 0)$. Additional values of θ duplicate existing points. For example, using $0 \leq \theta < 4\pi$ traces the circle twice.

FIGURE 10.3

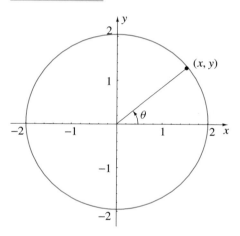

EXAMPLE 10.4

Find an implicit definition for the curve

$$x = t - t^3, \quad y = t + t^3,$$

and plot the curve.

SOLUTION By adding and subtracting the given equations, we have

$$x + y = 2t \quad \text{and} \quad y - x = 2t^3.$$

It follows from the first of these that $t = (x + y)/2$, and when this is substituted into the second, we obtain an implicit definition of the curve

$$y - x = \frac{(x + y)^3}{4}.$$

Unfortunately, as is always the case when an electronic device is used to plot parametric equations, Figure 10.4a does not indicate which values of t give which points on the curve. We can remedy this by relating graphs of $x(t)$ and $y(t)$ in Figures 10.4b–c to the curve in Figure 10.4a. For example, beginning with $t = 0$, Figures 10.4b and c give $x = 0$ and $y = 0$, and therefore the point $(0, 0)$ in Figure 10.4a. As t increases from $t = 0$ to $t = 1$, values of x increase from 0 to a maximum value $2/(3\sqrt{3})$ at $t = 1/\sqrt{3}$, and then decrease to 0. Simultaneously, values of y increase steadily from 0 to 2. This gives the first quadrant part of the curve in Figure 10.4a. As t increases beyond 1, values of x decrease through negative numbers while y continues to increase. This is reflected in that part of the graph in Figure 10.4a in the second quadrant. Because $x = t - t^3$ and $y = t + t^3$ are odd functions, replacing t by $-t$ reverses the signs of x and y. This means that corresponding to each point (x, y) on the graph for which $t > 0$, we must have the point $(-x, -y)$ corresponding to $-t$. This gives third and fourth quadrant parts of the curve in Figure 10.4a.

FIGURE 10.4a

FIGURE 10.4b

FIGURE 10.4c

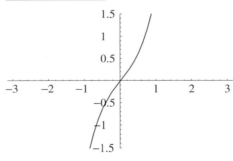

It is worthwhile noting that, were calculators and/or computers not programmed to do parametric plots, we would draw parametric curves precisely as outlined in this example. Piece the curve together from separate graphs of $x(t)$ and $y(t)$.

EXAMPLE 10.5

Draw the curve defined parametrically by

$$x = a\cos t, \qquad y = b\sin 2t,$$

where $a > 0$ and $b > 0$ are constants.

SOLUTION Because $a \cos t$ and $b \sin 2t$ are periodic with periods 2π and π, respectively, it follows that all values of x and y are obtained for $0 \le t \le 2\pi$. We begin by plotting the curve using specific values for a and b, say $a = 2$ and $b = 1$. The result is shown in Figure 10.5a. The required curve can be obtained by changing scales on the x- and y-axes (Figure 10.5b).

FIGURE 10.5a

FIGURE 10.5b

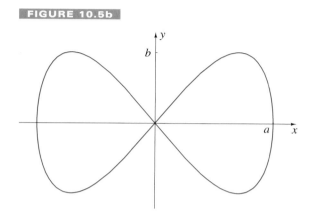

To confirm this, we draw the curve using the technique of Example 10.4. We draw graphs of $x = a \cos t$ and $y = b \sin 2t$ (Figures 10.6a and b). Value $t = 0$ gives the point $(a, 0)$ in Figure 10.6c. As t increases from $t = 0$ to $t = \pi/2$, values of x decrease from a to 0 and y increases from 0 to b, and then decreases from b to 0. This gives the first quadrant part of the curve in Figure 10.6c. As t increases from $\pi/2$ to π, values of x decrease from 0 to $-a$ and y decreases from 0 to $-b$, and then increases from $-b$ to 0. This adds the third quadrant part of the curve in Figure 10.6d. Continuation leads to the full curve in Figure 10.5b.

FIGURE 10.6a

FIGURE 10.6b

FIGURE 10.6c

FIGURE 10.6d

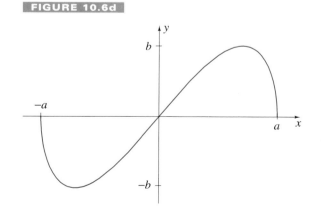

Parametric equations are frequently used in problems concerning the motion of objects. For example, suppose a stone is thrown horizontally over the edge of a cliff 100 m above a river (Figure 10.7). If the initial speed of the stone is 30 m/s, the path followed by the stone is described parametrically by

$$x = 30t, \qquad y = 100 - 4.905t^2, \qquad 0 \le t \le \sqrt{\frac{100}{4.905}},$$

where t is time in seconds and x and y are in metres. For these equations, t has been chosen equal to zero at the instant of projection, and $t = \sqrt{100/4.905}$ is the time at which the stone strikes the river.

FIGURE 10.7

100

y

x

When parameter t is eliminated from this parametric representation of the trajectory, the resulting equation is

$$y = 100 - 4.905 \left(\frac{x}{30}\right)^2 = 100 - 0.005\,45x^2.$$

It clearly indicates that the stone follows a parabolic path. However, if we were to discuss the motion of the stone as regards, say, velocity and acceleration, we would find it necessary to return to the parametric representation which describes the motion in horizontal and vertical directions.

We say that equations 10.1 define y **parametrically as a function of** x if the curve so defined represents a function, that is, if every vertical line that intersects the curve does so exactly once. The parametric equations in Example 10.1 therefore define a function parametrically, but in Examples 10.2–10.5 they do not. However, each of the curves in these latter examples can be divided into subcurves that do represent functions. For instance, in Example 10.4, each of the subcurves

$$
\begin{array}{llr}
x = t - t^3, & y = t + t^3, & t \le -1/\sqrt{3}; \\
x = t - t^3, & y = t + t^3, & -1/\sqrt{3} \le t \le 1/\sqrt{3}; \\
x = t - t^3, & y = t + t^3, & t \ge 1/\sqrt{3};
\end{array}
$$

defines a function, and does so parametrically (Figure 10.8); together all three curves make up the original curve.

FIGURE 10.8a

FIGURE 10.8b

FIGURE 10.8c

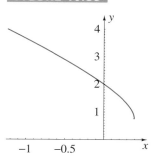

When equations 10.1 do define y parametrically as a function of x, it is straightforward to find the derivative of the function. If we denote this function by $y = f(x)$, then the chain rule applied to $y = f(x)$, $x = x(t)$ gives

$$\frac{dy}{dt} = \frac{dy}{dx}\frac{dx}{dt}.$$

Provided then that $dx/dt \neq 0$, we may solve for

$$\frac{dy}{dx} = \frac{\dfrac{dy}{dt}}{\dfrac{dx}{dt}}. \tag{10.2}$$

This is called the **parametric rule** for differentiation of a function defined parametrically by equations 10.1; it defines the derivative of y with respect to x in terms of derivatives of the given functions $x(t)$ and $y(t)$ with respect to t.

For example, the curve in Figure 10.9 is defined parametrically by

$$x(t) = \frac{1}{t^2 + 1}, \qquad y(t) = \frac{2}{t(t^2 + 1)}.$$

FIGURE 10.9

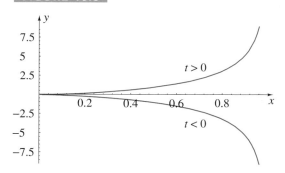

It can be subdivided into two parts, one corresponding to values of $t > 0$ and the other to values of $t < 0$, and each defines y as a function of x. The derivative of either function is

$$\frac{dy}{dx} = \frac{\dfrac{dy}{dt}}{\dfrac{dx}{dt}} = \frac{\dfrac{-2(3t^2 + 1)}{(t^3 + t)^2}}{\dfrac{-2t}{(t^2 + 1)^2}} = \frac{3t^2 + 1}{t^3}.$$

Neither function has a relative maximum or minimum, and this is consistent with the fact that dy/dx never vanishes. The derivative is undefined for $t = 0$, as is y. We can also calculate d^2y/dx^2 in spite of the fact that dy/dx is in terms of t rather than x. Once again it is the chain rule that comes to the rescue:

$$\frac{d^2y}{dx^2} = \frac{d}{dx}\left(\frac{dy}{dx}\right) = \frac{d}{dt}\left(\frac{dy}{dx}\right)\frac{dt}{dx} = \frac{\dfrac{d}{dt}\left(\dfrac{dy}{dx}\right)}{\dfrac{dx}{dt}}$$

$$= \frac{\dfrac{t^3(6t) - (3t^2 + 1)(3t^2)}{t^6}}{\dfrac{-2t}{(t^2 + 1)^2}} = \frac{-3t^4 - 3t^2}{t^6} \cdot \frac{(t^2 + 1)^2}{-2t} = \frac{3(t^2 + 1)^3}{2t^5}.$$

This derivative is positive for $t > 0$ and negative for $t < 0$, agreeing with the fact that the curve in Figure 10.9 is concave upward when $t > 0$ and concave downward when $t < 0$.

EXAMPLE 10.6

Plot the strophoid

$$x = \frac{1 - t^2}{1 + t^2}, \qquad y = \frac{t(1 - t^2)}{1 + t^2}$$

and find points at which its tangent line is horizontal.

SOLUTION   The curve is shown in Figure 10.10.

FIGURE 10.10

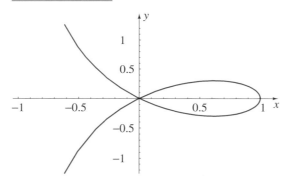

Points at which the tangent line is horizontal can be found by solving

$$0 = \frac{dy}{dx} = \frac{\dfrac{dy}{dt}}{\dfrac{dx}{dt}} = \frac{\dfrac{(1 + t^2)(1 - 3t^2) - (t - t^3)(2t)}{(1 + t^2)^2}}{\dfrac{(1 + t^2)(-2t) - (1 - t^2)(2t)}{(1 + t^2)^2}} = \frac{1 - 4t^2 - t^4}{-4t}.$$

But this implies that

$$t^4 + 4t^2 - 1 = 0,$$

a quadratic equation in t^2,

$$t^2 = \frac{-4 \pm \sqrt{16 + 4}}{2} = -2 \pm \sqrt{5}.$$

Since t^2 must be nonnegative, it follows that $t^2 = \sqrt{5} - 2$, and therefore

$$t = \pm\sqrt{\sqrt{5} - 2}.$$

For these values of t, $x = 0.62$ and $y = \pm 0.30$. Points with a horizontal tangent line are therefore $(0.62, \pm 0.30)$.

EXAMPLE 10.7

The tire of a car rolling along the x-axis without slipping picks up a stone at the origin. The path followed by the stone is called a *cycloid*. Show that parametric equations for the cycloid in terms of the angle θ through which the tire has rotated since picking up the stone are

$$x = R(\theta - \sin\theta), \qquad y = R(1 - \cos\theta),$$

where R is the radius of the tire. Plot the cycloid for $0 \leq \theta \leq 4\pi$. In what direction is the stone travelling when it meets the road?

SOLUTION   The x-coordinate of P in Figure 10.11 is equal to $\|OB\|$ minus $\|PC\|$. Since length $\|PB\|$ along the circle is equal to $\|OB\|$ and $\|PC\| = \|PA\| \sin\theta = R \sin\theta$, it follows that

$$x = R\theta - R\sin\theta = R(\theta - \sin\theta).$$

Furthermore,

$$y = \|AB\| - \|AC\| = R - R\cos\theta = R(1 - \cos\theta).$$

FIGURE 10.11

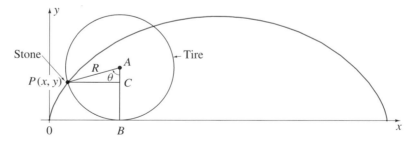

The plot in Figure 10.12 for $0 \leq \theta \leq 4\pi$ shows the path of the stone for two revolutions of the tire. (It was plotted for $R = 1$ and scales were then changed from 1 to R.)

FIGURE 10.12

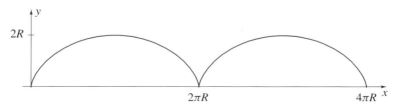

The slope of the cycloid is given by

$$\frac{dy}{dx} = \frac{dy/d\theta}{dx/d\theta} = \frac{R(\sin\theta)}{R(1-\cos\theta)} = \frac{\sin\theta}{1-\cos\theta}.$$

It is undefined when $\theta = 2n\pi$, values of θ at which the stone meets the road. Using L'Hôpital's rule, we calculate

$$\lim_{\theta\to 2n\pi^-} \frac{\sin\theta}{1-\cos\theta} = \lim_{\theta\to 2n\pi^-} \frac{\cos\theta}{\sin\theta} = -\infty, \quad \text{and} \quad \lim_{\theta\to 2n\pi^+} \frac{\sin\theta}{1-\cos\theta} = \infty.$$

These show that the stone is moving vertically downward as it meets the road, and then vertically upward as it leaves the road.

Integrals 7.17 in Section 7.3 define length of a curve. We express them in a slightly different form,

$$L = \int_A^B \sqrt{(dx)^2 + (dy)^2},$$

FIGURE 10.13

where A and B are the initial and final points on the curve (Figure 10.13). When the curve is defined parametrically by 10.1, this integral becomes

$$L = \int_\alpha^\beta \sqrt{\left(\frac{dx}{dt}\right)^2 + \left(\frac{dy}{dt}\right)^2}\, dt. \tag{10.3}$$

Verification of this formula is left to Exercise 43.

EXAMPLE 10.8

Find the length of the curve

$$x = e^t \cos t, \qquad y = e^t \sin t, \qquad 0 \le t \le \pi/2.$$

SOLUTION   According to 10.3,

$$L = \int_0^{\pi/2} \sqrt{(e^t \cos t - e^t \sin t)^2 + (e^t \sin t + e^t \cos t)^2}\, dt$$

$$= \int_0^{\pi/2} \sqrt{e^{2t}(2\cos^2 t + 2\sin^2 t)}\, dt$$

$$= \int_0^{\pi/2} \sqrt{2}\, e^t\, dt = \sqrt{2}\left\{e^t\right\}_0^{\pi/2} = \sqrt{2}(e^{\pi/2} - 1).$$

EXERCISES 10.1

In Exercises 1–12 first draw the curve and then plot it.

* **1.** $x = 2 + t, \ y = 3t - 1$

* **2.** $x = t^2 + 3t + 4, \ y = 1 - t$

* **3.** $x = 1 + 2\cos t, \ y = 2 + 2\sin t, \quad 0 \le t \le 2\pi$

* **4.** $x = -2 + 4\cos t, \ y = 3 + 4\sin t, \quad 0 \le t \le \pi$

* **5.** $x = 1 + \cos t, \ y = -1 - \sin t, \quad 0 \le t \le \pi/4$

* **6.** $x = t + 1/t, \ y = t - 1/t$

* **7.** $x = 2\cos t, \ y = 4\sin t, \quad 0 \le t < 2\pi$

* **8.** $x = 1 + 3\cos t, \ y = -2 + 2\sin t, \quad 0 \le t \le \pi$

* **9.** $x = -1 + \sin t, \ y = -1 - 3\cos t, \ 0 \le t \le \pi/2$

* **10.** $x = t - t^2, \ y = t + t^2$

* **11.** $x = t^2 + 1, \ y = t^3 + 3$

* **12.** $x = 2\cot\theta, \ y = 2\sin^2\theta, \quad -\pi/2 \le \theta \le \pi/2, \ \theta \ne 0$

In Exercises 13–20 assume that y is defined parametrically as a function of x, and find dy/dx.

* **13.** $x = t^3 + 3t - 2, \ y = t^2 - 1$ * **14.** $x = \dfrac{u}{u - 1}, \ y = \dfrac{u^2}{u^2 - 1}$

* **15.** $x = (v^3 + 2)\sqrt{v - 1}, \ y = 2v^3 + 3$ * **16.** $x = \sqrt{\dfrac{2 + t}{2 - t}}, \ y = \sqrt{\dfrac{2 - t}{2 + t}}$

* **17.** $x = s^{3/2} - s^{2/3}, \ y = s^2 + 2s$ * **18.** $x = (2t + 3)^4, \ y = \dfrac{t}{t + 6}$

* **19.** $x = \left(\dfrac{1 + u}{1 - u}\right)^{1/3}, \ y = \left(\dfrac{1 - u}{1 + u}\right)^4$ * **20.** $x = \sqrt{-t^2 + 3t + 5}, \ y = \dfrac{1}{t^2 + 2t - 5}$

* **21.** Find equations for the tangent and normal lines to the curve

$$x = t + \frac{1}{t}, \quad y = t - \frac{1}{t}$$

at the point corresponding to $t = 4$.

* **22.** Find point(s) on the curve

$$x = \frac{t^3}{3} - 3t, \quad y = \frac{3t^2}{2} + t$$

where the slope of the tangent line to the curve is equal to 1.

In Exercises 23–26 assume that y is defined parametrically as a function of x, and find dy/dx and d^2y/dx^2.

** **23.** $x = t^2 + \dfrac{1}{t}, \ y = t^2 - \dfrac{1}{t}$ ** **24.** $x = \sqrt{t - 1}, \ y = \sqrt{t + 1}$

** **25.** $x = 2u + 5, \ y = 7 - 14u$ ** **26.** $x = v^2 + 2v + 3, \ y = 2v - 4$

through any number of complete revolutions, bringing it back to its original position, and then rotate a further $\pi/6$ radians. Alternatively, we could rotate in a clockwise direction any number of complete revolutions, and then a further $-11\pi/6$ radians. In other words, polar coordinate θ for Q could be any of the values $\pi/6 + 2n\pi$, where n is an integer. Possible values of θ for point R in Figure 10.15 are $3\pi/4 + 2n\pi$. For point Q, $r = 2$ and for R, $r = 1$. Polar coordinates r and θ for a point are written in the form (r, θ) so that possible polar coordinates for Q and R are $(2, \pi/6 + 2n\pi)$ and $(1, 3\pi/4 + 2n\pi)$.

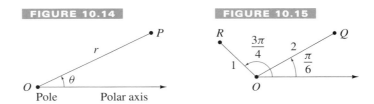

FIGURE 10.14   FIGURE 10.15

This situation is not like that for Cartesian coordinates where each point has only one set of coordinates (x, y), and every ordered pair of real numbers specifies one point. With polar coordinates, every ordered pair of real numbers (r, θ), where r must be nonnegative, represents one and only one point; but every point has an infinity of possible representations. We should point out that in some applications this is not a desirable situation. For instance, in the branch of mathematics called *tensor analysis*, it is necessary that polar coordinates assign exactly one pair of coordinates to each point. This can be accomplished in any region that does not contain the pole by demanding, for instance, that $-\pi < \theta \le \pi$. When we use polar coordinates to find areas in Section 10.4, we must also be particular about our choice of θ. For now, however, no advantage is gained by imposing restrictions on θ, and we therefore accept the fact that if (r, θ) are polar coordinates of a point, so are $(r, \theta + 2n\pi)$ for any integer n. Note also that polar coordinates for the pole are $(0, \theta)$ for any θ whatsoever.

If we introduce into a plane both a system of Cartesian coordinates (x, y) and a system of polar coordinates (r, θ), then relations exist between the two. Suppose the pole of polar coordinates and the origin of Cartesian coordinates are chosen as the same point, and that the polar axis is chosen as the positive x-axis (Figure 10.16). In this case, Cartesian and polar coordinates of any point P are related by the equations

$$x = r\cos\theta, \quad y = r\sin\theta. \tag{10.4}$$

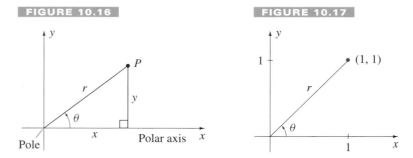

FIGURE 10.16   FIGURE 10.17

These equations define Cartesian coordinates of a point in terms of its polar coordinates; that is, given its polar coordinates (r, θ), we can calculate its Cartesian coordinates (x, y) by means of 10.4. For example, if polar coordinates of a point are $(3, 2)$, then its Cartesian coordinates are

$$x = 3\cos 2 = -1.25, \quad y = 3\sin 2 = 2.73.$$

Equations 10.4 implicitly define polar coordinates of a point in terms of its Cartesian coordinates. For instance, if Cartesian coordinates of a point are $(1, 1)$ (Figure 10.17), its polar coordinates must satisfy

$$1 = r\cos\theta, \quad 1 = r\sin\theta.$$

If we square and add these equations, we have

$$1 + 1 = r^2\cos^2\theta + r^2\sin^2\theta = r^2,$$

and therefore $r = \sqrt{2}$. It follows that $\cos\theta = \sin\theta = 1/\sqrt{2}$, from which we determine $\theta = \pi/4 + 2n\pi$. Thus, polar coordinates of the point are $(\sqrt{2}, \pi/4 + 2n\pi)$.

Equations 10.4 define r and θ implicitly in terms of x and y, but obviously it would be preferable to have explicit definitions. There is no problem expressing r explicitly in terms of x and y,

$$r = \sqrt{x^2 + y^2}, \tag{10.5}$$

but the case for θ is not so simple. If we substitute expression 10.5 for r into equations 10.4, we obtain

$$\cos\theta = \frac{x}{\sqrt{x^2 + y^2}}, \quad \sin\theta = \frac{y}{\sqrt{x^2 + y^2}}. \tag{10.6}$$

Except for the pole, these two equations determine all possible values of θ for given x and y. What we would like to do is obtain one equation, if possible, that defines θ. If we divide the second of these equations by the first, we have

$$\tan\theta = \frac{y}{x}, \tag{10.7}$$

and this equation suggests that we set

$$\theta = \text{Tan}^{-1}\left(\frac{y}{x}\right). \tag{10.8}$$

$P(-1, 1)$

Unfortunately, neither equation 10.7 nor 10.8 is satisfactory. For instance, given the point P with Cartesian coordinates $(-1, 1)$ (Figure 10.18), equation 10.7 yields $\tan\theta = -1$, the solutions of which are $\theta = -\pi/4 + n\pi$. These are angles in the second and fourth quadrants, so only half of them are acceptable polar angles for P. Equation 10.8 gives $\theta = -\pi/4$, which is not a possible polar angle for P.

We suggest all angles satisfying 10.7 be found, and then a diagram be used to determine those angles that are acceptable values for θ. We illustrate in the following example.

EXAMPLE 10.9

Find all polar coordinates for points with the following Cartesian coordinates:

(a) $(1, 2)$ (b) $(-2, 3)$ (c) $(3, -1)$ (d) $(-2, -4)$

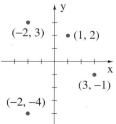

SOLUTION

(a) $r = \sqrt{1^2 + 2^2} = \sqrt{5}$. Angles which satisfy $\tan\theta = 2/1$ are $1.11 + n\pi$. Since the point is in the first quadrant (Figure 10.19), acceptable values for θ are $1.11 + 2n\pi$. Polar coordinates of the point are therefore $(\sqrt{5}, 1.11 + 2n\pi)$.

(b) $r = \sqrt{(-2)^2 + 3^2} = \sqrt{13}$. Angles satisfying $\tan\theta = -3/2$ are $-0.98 + n\pi$. Since the point is in the second quadrant, possible values for θ are $(\pi - 0.98) + 2n\pi = 2.16 + 2n\pi$. Polar coordinates are therefore $(\sqrt{13}, 2.16 + 2n\pi)$.

(c) $r = \sqrt{3^2 + (-1)^2} = \sqrt{10}$. Since values of θ satisfying $\tan \theta = -1/3$ are $-0.32 + n\pi$, and the point is in the fourth quadrant, it follows that polar coordinates are $(\sqrt{10}, -0.32 + 2n\pi)$.

(d) $r = \sqrt{(-2)^2 + (-4)^2} = 2\sqrt{5}$. Since angles satisfying $\tan \theta = 2$ are $1.11 + n\pi$, and the point is in the third quadrant, polar coordinates are $(2\sqrt{5}, 4.25 + 2n\pi)$.

The results in equations 10.4–10.8 are valid only when the pole and origin coincide, and the polar axis and positive x-axis are identical. For a different arrangement, these relations must be changed accordingly. For example, if the pole is at the point with Cartesian coordinates (h, k) and the polar axis is as shown in Figure 10.20, equations 10.4 are replaced by

$$x = h + r \cos(\theta + \alpha), \quad y = k + r \sin(\theta + \alpha). \tag{10.9}$$

You will verify these equations in Exercise 13.

FIGURE 10.20

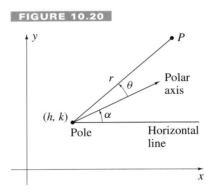

The usual choice of pole and polar axis is that in Figure 10.16, and unless otherwise stipulated, we assume this to be the case.

EXERCISES 10.2

In Exercises 1–8 plot the point having the given set of Cartesian coordinates, and find all possible polar coordinates.

* **1.** $(1, -1)$ * **2.** $(-1, \sqrt{3})$

* **3.** $(4, 3)$ * **4.** $(-2\sqrt{3}, 2)$

* **5.** $(2, 6)$ * **6.** $(-1, -4)$

* **7.** $(7, -5)$ * **8.** $(-5, 2)$

In Exercises 9–12 plot the point having the given set of polar coordinates, and find its Cartesian coordinates.

* **9.** $(2, \pi/4)$ * **10.** $(6, -\pi/6)$

* **11.** $(7, 1)$ * **12.** $(3, -2.4)$

* **13.** Verify that polar and Cartesian coordinates as shown in Figure 10.20 are related by equations 10.9.

10.3 Curves in Polar Coordinates

A curve is defined explicitly in Cartesian coordinates by equations $y = f(x)$ or $x = g(y)$, and implicitly by $F(x, y) = 0$. A point is on a curve if and only if its Cartesian coordinates (x, y) satisfy the equation of the curve.

Analogously, a curve is defined explicitly in polar coordinates if its equation is expressed in either of the forms

$$r = f(\theta) \qquad \text{or} \qquad \theta = g(r), \tag{10.10}$$

and is defined implicitly when its equation is given in the form

$$F(r, \theta) = 0. \tag{10.11}$$

A point is on a curve if at least one of its sets of polar coordinates (r, θ) satisfies the equation of the curve. All sets of polar coordinates for a point need not satisfy the equation. For example, the origin or pole has polar coordinates $(0, \theta)$ for any θ whatsoever. But only those coordinates of the form $\big(0, (2n+1)\pi\big)$ satisfy the equation $r = 1 + \cos\theta$. Likewise, the polar coordinates $(1, \pi)$ satisfy the equation $r = \sin(\theta/2)$, but the coordinates $(1, 3\pi)$ of the same point do not satisfy this equation.

Often we are required to transform the equation of a curve from Cartesian coordinates to polar coordinates, and vice versa. To transform from Cartesian to polar is straightforward: replace each x with $r\cos\theta$ and each y with $r\sin\theta$. For example, the equation $x^2 + y^2 = 9$ describes a circle centred at the origin with radius 3. In polar coordinates, its equation is

$$9 = (r\cos\theta)^2 + (r\sin\theta)^2 = r^2\cos^2\theta + r^2\sin^2\theta = r^2.$$

Consequently, $r = 3$ is the equation of this circle in polar coordinates, a much simpler equation that $x^2 + y^2 = 9$.

EXAMPLE 10.10

Find equations in polar coordinates for the following curves:

(a) $2x + 3y = 3$ (b) $x^2 - 2x + y^2 = 0$ (c) $x^2 + y^2 = \sqrt{x^2 + y^2} - 4x$

SOLUTION

(a) For $2x + 3y = 3$, we obtain

$$3 = 2r\cos\theta + 3r\sin\theta, \qquad \text{or} \qquad r = \frac{3}{2\cos\theta + 3\sin\theta}.$$

(b) For the equation $x^2 - 2x + y^2 = 0$, we have

$$0 = -2x + (x^2 + y^2) = -2r\cos\theta + r^2 = r(r - 2\cos\theta).$$

Thus,

$$r = 0 \qquad \text{or} \qquad r = 2\cos\theta.$$

Since $r = 0$ defines the pole, and this point also satisfies $r = 2\cos\theta$ (for $\theta = \pi/2$), it follows that we need only write $r = 2\cos\theta$.

(c) For the curve with equation $x^2 + y^2 = \sqrt{x^2 + y^2} - 4x$, we obtain

$$r^2 = r - 4r\cos\theta = r(1 - 4\cos\theta).$$

Thus,

$$0 = r^2 - r(1 - 4\cos\theta) = r(r - 1 + 4\cos\theta),$$

from which we have

$$r = 0 \qquad \text{or} \qquad r = 1 - 4\cos\theta.$$

Again the pole satisfies the second of these equations, and therefore the equation of the curve in polar coordinates is $r = 1 - 4\cos\theta$.

To transform equations of curves from polar to Cartesian coordinates can sometimes be more difficult, principally because we have no substitution for θ. If, however, the equation involves $\cos\theta$ and/or $\sin\theta$, we use equations 10.6.

EXAMPLE 10.11

Find equations in Cartesian coordinates for the following curves:

$$\text{(a)} \quad r = 1 + \cos\theta \qquad \text{(b)} \quad r^2\cos 2\theta = 1 \qquad \text{(c)} \quad r^2 = 9\sin 2\theta$$

SOLUTION

(a) We use equations 10.5 and 10.6 to write

$$\sqrt{x^2 + y^2} = 1 + \frac{x}{\sqrt{x^2 + y^2}},$$

and multiplication by $\sqrt{x^2 + y^2}$ gives

$$x^2 + y^2 = x + \sqrt{x^2 + y^2}.$$

(b) For $r^2\cos 2\theta = 1$, we use double-angle formula 1.39c to write the equation in terms of $\cos\theta$ rather than $\cos 2\theta$, and then use equations 10.5 and 10.6:

$$1 = r^2(2\cos^2\theta - 1) = (x^2 + y^2)\left(2\frac{x^2}{x^2 + y^2} - 1\right) = (x^2 + y^2)\frac{x^2 - y^2}{x^2 + y^2} = x^2 - y^2.$$

(c) This time we use double-angle formula 1.39a on $\sin 2\theta$:

$$x^2 + y^2 = 18\sin\theta\cos\theta = 18\frac{y}{\sqrt{x^2 + y^2}}\frac{x}{\sqrt{x^2 + y^2}} \qquad \text{or}$$

$$(x^2 + y^2)^2 = 18xy.$$

Examples 10.10 and 10.11 illustrate that equations for some curves are simpler when expressed in polar coordinates. These polar representations can prove very efficient in producing graphs, whether we are plotting with an electronic device or drawing by hand. We illustrate with the curve $r = 1 + \cos\theta$ of Example 10.11. To plot this curve by graphing calculator or computer, we supply the function $r = 1 + \cos\theta$ and the range of values of θ. Because $1 + \cos\theta$ is 2π-periodic, the range $-\pi \le \theta \le \pi$ suffices. Other values of θ create duplications. The result is shown in Figure 10.21; it is called a *cardioid*.

FIGURE 10.21

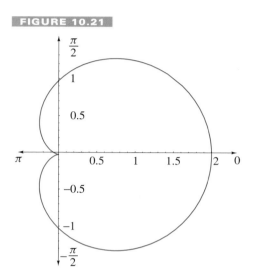

To draw the cardioid by hand we first create a Cartesian coordinate system consisting of a horizontal θ-axis and a vertical r-axis. On this set of axes we graph the *function $r = f(\theta) = 1 + \cos\theta$* (Figure 10.22). This is *not* the required curve; it is a graph of the function $f(\theta)$, illustrating values of r for various values of θ. It represents an "infinite table of values" for r as a function of θ.

FIGURE 10.22

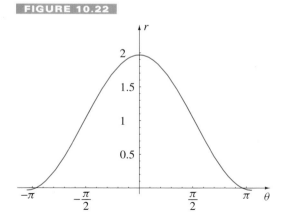

To draw the cardioid $r = 1 + \cos\theta$, we now read pairs of polar coordinates (r, θ) from Figure 10.22, interpreting r as radial distance and θ as rotation. Suppose we begin with the two points $(2, 0)$ and $(1, \pi/2)$. Figure 10.22 indicates that as θ increases from 0 to $\pi/2$, values of r decrease from 2 to 1. This means that as we rotate from the $\theta = 0$ line to the $\theta = \pi/2$

line in the first quadrant, radial distances from the origin become smaller. This is shown in Figure 10.23a. As rotation is increased from $\pi/2$ to π, Figure 10.22 shows that radial distances continue to decrease, eventually reaching 0 at an angle of π radians as in Figure 10.23b. Notice that the line $\theta = \pi$ is tangent to the curve at the pole, reflecting the fact that the pole is attained for an angle of π radians. Consideration of the graph in Figure 10.22 to the left of $\theta = 0$ leads to that part of the cardioid below the $\theta = 0$ and $\theta = \pi$ lines in Figure 10.23c. The symmetry of Figure 10.22 about the r-axis is reflected in the symmetry of Figure 10.23c about the $\theta = 0$ and $\theta = \pi$ lines. Since the function $r = f(\theta) = 1 + \cos\theta$ is 2π-periodic, only values of θ in the interval $-\pi < \theta \le \pi$ need be considered. Values outside this interval retrace previous points.

FIGURE 10.23a

FIGURE 10.23b

FIGURE 10.23c

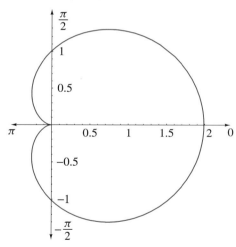

EXAMPLE 10.12

Plot the curve $r^2 = 9\sin 2\theta$.

SOLUTION The explicit definition of the curve is $r = 3\sqrt{\sin 2\theta}$. For $\sin 2\theta$ to be nonnegative, θ must be restricted to the intervals $-\pi \le \theta \le -\pi/2$ and $0 \le \theta \le \pi/2$. When these

FIGURE 10.24

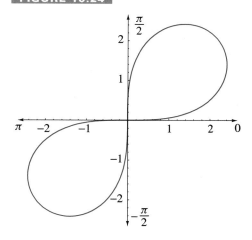

are submitted to whatever electronic device you use to make polar plots, the result is shown in Figure 10.24.

Lacking a device that does polar plots we can draw the curve by first drawing a graph of the function $\sin 2\theta$ in Figure 10.25a. A graph of the function $r = 3\sqrt{\sin 2\theta}$ then follows (Figure 10.25b). Reading pairs of polar coordinates from this graph gives the curve in Figure 10.24. It is called a *lemniscate*.

FIGURE 10.25a

FIGURE 10.25b

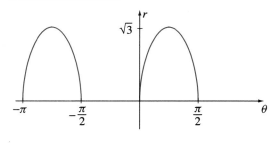

EXAMPLE 10.13

Draw the curve $r = |a - b\cos\theta|$ where $b > a > 0$ are constants.

SOLUTION   We begin by drawing the function $-b\cos\theta$ in Figure 10.26a. A shift of a units vertically gives the graph in Figure 10.26b. Absolute values lead to Figure 10.26c. Interpreting r and θ as distance and rotation leads to the curve in Figure 10.26d. Angles at which $r = 0$ are $\theta = \pm\text{Cos}^{-1}(a/b)$.

FIGURE 10.26a

FIGURE 10.26b

FIGURE 10.26c

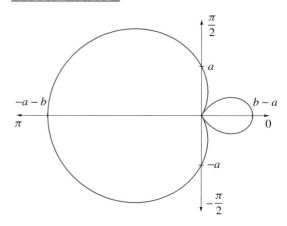

FIGURE 10.26d

To find points of intersection of two curves whose equations are given in Cartesian coordinates, we solve the equations simultaneously for all (real) solutions. Each solution represents a distinct point of intersection. For curves whose equations are given in polar coordinates, the situation is somewhat more complex because we have multiple names for points. To find points of intersection, we again solve the equations simultaneously for all solutions. Each solution represents a point of intersection; but as points have many sets of polar coordinates, some of these solutions may represent the same point. In addition, it may also happen that one set of polar coordinates for a point of intersection satisfies one equation, whereas a different set satisfies the other equation. Particularly troublesome in this respect is the pole, which has so many sets of polar coordinates. The best way to handle these difficulties is to graph the curves.

EXAMPLE 10.14

Find points of intersection for the curves $r = \sin\theta$ and $r = 1 - \sin\theta$.

SOLUTION   If we set $\sin\theta = 1 - \sin\theta$, then $\sin\theta = 1/2$. All solutions of this equation are defined by

$$\theta = \begin{cases} \dfrac{\pi}{6} + 2n\pi \\[2mm] \dfrac{5\pi}{6} + 2n\pi \end{cases}$$

where n is an integer. Graphs of the curves in Figure 10.27 indicate that these values of θ give the points of intersection $(1/2, \pi/6)$ and $(1/2, 5\pi/6)$. The figure also indicates that the origin is a point of intersection of the curves. We did not obtain this point by solving $r = \sin\theta$ and $r = 1 - \sin\theta$ because different values of θ yield $r = 0$ in $r = \sin\theta$ and $r = 1 - \sin\theta$. To obtain $r = 0$ from $r = \sin\theta$, θ must be one of the values $n\pi$, whereas to obtain $r = 0$ from $r = 1 - \sin\theta$, θ must be one of the values $\pi/2 + 2n\pi$. Thus, both curves pass through the pole, but the pole cannot be obtained by simultaneously solving the equations of the curves.

FIGURE 10.27

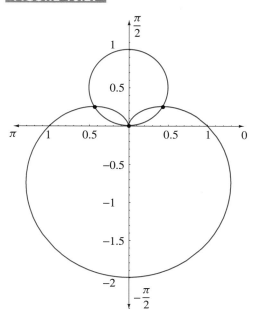

Slopes of Curves in Polar Coordinates

When a curve has polar equation $r = f(\theta), \alpha \leq \theta \leq \beta$, substitution into equations 10.4 gives

$$x = f(\theta)\cos\theta, \quad y = f(\theta)\sin\theta, \quad \alpha \leq \theta \leq \beta. \tag{10.12}$$

These are parametric equations for the curve, where the parameter is the polar angle θ. Equation 10.2 gives

$$\frac{dy}{dx} = \frac{\dfrac{dy}{d\theta}}{\dfrac{dx}{d\theta}} = \frac{f'(\theta)\sin\theta + f(\theta)\cos\theta}{f'(\theta)\cos\theta - f(\theta)\sin\theta}. \tag{10.13}$$

This formula defines the slope of the tangent line to a curve which has polar equation $r = f(\theta)$.

EXAMPLE 10.15

Find points on the cardioid $r = 1 + \sin\theta$ at which the tangent line is horizontal.

SOLUTION   The graph of the cardioid in Figure 10.28 indicates three points at which the tangent line is horizontal. To find them we use equation 10.13 to write

$$0 = \frac{dy}{dx} = \frac{(\cos\theta)\sin\theta + (1 + \sin\theta)\cos\theta}{(\cos\theta)\cos\theta - (1 + \sin\theta)\sin\theta}.$$

Since the numerator must vanish, we set

$$0 = \cos\theta\sin\theta + \cos\theta + \sin\theta\cos\theta = \cos\theta(1 + 2\sin\theta).$$

From $\cos\theta = 0$, we choose $\theta = \pi/2$ and from $1 + 2\sin\theta = 0$, we take $\theta = -\pi/6$ and $-5\pi/6$. Thus, points at which the cardioid has a horizontal tangent line have Cartesian coordinates $(0, 2)$ and $(\pm\sqrt{3}/4, -1/4)$.

FIGURE 10.28

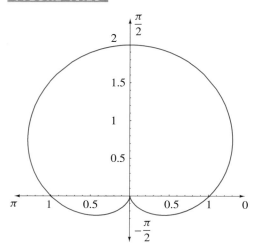

Lengths of Curves in Polar Coordinates

The length of a curve defined parametrically by $x = x(t)$ and $y = y(t)$ is given by formula 10.3:

$$L = \int_\alpha^\beta \sqrt{\left(\frac{dx}{dt}\right)^2 + \left(\frac{dy}{dt}\right)^2}\, dt.$$

If we substitute from equations 10.12 into this formula with t replaced by θ, we obtain

$$L = \int_\alpha^\beta \sqrt{[f'(\theta)\cos\theta - f(\theta)\sin\theta]^2 + [f'(\theta)\sin\theta + f(\theta)\cos\theta]^2}\, d\theta$$

$$= \int_\alpha^\beta \sqrt{[f'(\theta)]^2 + [f(\theta)]^2}\, d\theta.$$

Thus, we may write for the length of a curve $r = f(\theta)$, $\alpha \le \theta \le \beta$,

$$L = \int_{\alpha}^{\beta} \sqrt{r^2 + \left(\frac{dr}{d\theta}\right)^2}\, d\theta. \tag{10.14}$$

EXAMPLE 10.16

Find the length of the cardioid $r = 1 - \cos\theta$.

SOLUTION   According to equation 10.14 (see Figure 10.29),

$$
\begin{aligned}
L &= \int_0^{2\pi} \sqrt{(1 - \cos\theta)^2 + (\sin\theta)^2}\, d\theta \\
&= \int_0^{2\pi} \sqrt{2}\sqrt{1 - \cos\theta}\, d\theta \\
&= \sqrt{2} \int_0^{2\pi} \sqrt{1 - \left[1 - 2\sin^2\left(\frac{\theta}{2}\right)\right]}\, d\theta \\
&= 2 \int_0^{2\pi} \sin\left(\frac{\theta}{2}\right)\, d\theta \\
&= 2\left\{-2\cos\left(\frac{\theta}{2}\right)\right\}_0^{2\pi} \\
&= 8.
\end{aligned}
$$

FIGURE 10.29

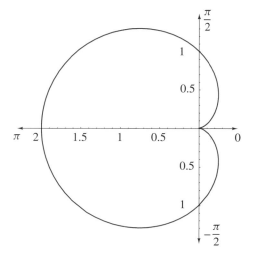

EXERCISES 10.3

In Exercises 1–10 find an equation for the curve in polar coordinates. Draw, and then plot, the curve.

* **1.** $x + 2y = 5$

* **2.** $y = -x$

* **3.** $x^2 + y^2 = 3$

* **4.** $x^2 - 2x + y^2 - 2y + 1 = 0$

* **5.** $y = 4x^2$

* **6.** $x^2 + 2y^2 = 3$

* **7.** $x^2 + y^2 = x$

* **8.** $x^2 + y^2 = \sqrt{x^2 + y^2} - x$

* **9.** $(x^2 + y^2)^2 = x$

* **10.** $y = 1/x^2$

In Exercises 11–20 find the equation of the curve in Cartesian coordinates. Draw, and then plot, the curve.

* **11.** $r = 5$

* **12.** $\theta = 1$

* **13.** $r = 3 \sin \theta$

* **14.** $r^2 = 4 \sin 2\theta$

* **15.** $r = 3 + 3 \sin \theta$

* **16.** $r = 2 \sin 2\theta$

* **17.** $r^2 = -4 \cos 2\theta$

* **18.** $r = 3 - 4 \cos \theta$

* **19.** $r = 5 \csc \theta$

* **20.** $r = \cot^2 \theta \csc \theta$

* **21.** Plot and draw the curves:

(a) $r = 2 + 2 \sin \theta$ (b) $r = 2 + 4 \sin \theta$ (c) $r = 4 + 2 \sin \theta$

In Exercises 22–25 find all points of intersection for the curves.

* **22.** $r = 2$, $\quad r^2 = 8 \cos 2\theta$

* **23.** $r = \cos \theta$, $\quad r = 1 + \cos \theta$

* **24.** $r = 1 + \cos \theta$, $\quad r = 2 - 2 \cos \theta$

* **25.** $r = 1$, $\quad r = 2 \cos 2\theta$

In Exercises 26–29 find the slope of the curve at the given value of θ.

* **26.** $r = 9 \cos 2\theta$ at $\theta = \pi/6$

* **27.** $r^2 = 9 \sin 2\theta$ at $\theta = -5\pi/6$

* **28.** $r = 3 - 5 \cos \theta$ at $\theta = 3\pi/4$

* **29.** $r = 2 \cos (\theta/2)$ at $\theta = \pi/2$

* **30.** Find the slope of the tangent line to the curve $r = 3/(1 - \sin \theta)$ at the point with polar coordinates $(6, \pi/6)$ in two ways:

(a) by using 10.12;

(b) by finding the equation of the curve in Cartesian coordinates, and calculating dy/dx.

** **31.** Show that if $f(\theta)$ is an even function, then the curve $r = f(\theta)$ is symmetric about the lines $\theta = 0$ and $\theta = \pi$ (or x-axis). Illustrate with two examples.

In Exercises 32–39 draw, and then plot, the curve.

** **32.** $r = \sin 3\theta$

** **33.** $r = \cos 2\theta$

** **34.** $r = \sin 4\theta$

** **35.** $r^2 = \theta$

** **36.** $r = e^\theta$

** **37.** $r = 2 \sin (\theta/2)$

** **38.** $r = -2 \cos (\theta/2)$

** **39.** $r = 1 + \cos (\theta + \pi/6)$

** **40.** Find the length of the cardioid $r = a(1 + \sin \theta)$. (a is a constant.)

** **41.** (a) The electrostatic charge distribution consisting of a charge $q > 0$ at the point with polar coordinates $(s, 0)$ and a charge $-q$ at (s, π) is called a *dipole*. When s is very small, the lines of force for the dipole are defined by the equation $r = A \sin^2 \theta$, where each value of the constant $A > 0$ defines a particular line of force. Plot or draw lines of force for $A = 1, 2$, and 3.

(b) The equipotential lines for the dipole are defined by $r^2 = B \cos \theta$, where B is a constant. Plot or draw equipotential lines for $B = \pm 1, \pm 2$, and ± 3.

** **42.** Draw, and then plot, the *bifolium* $(x^2 + y^2)^2 = x^2 y$.

** **43.** Curves with equations of the form $r = a(1 \pm \cos \theta)$ or $r = a(1 \pm \sin \theta)$ ($a > 0$ a constant) are called *cardioids*.

(a) Draw all such curves.

(b) Find equations for the cardioids in Cartesian coordinates.

** **44.** Curves with equations of the form $r^2 = a^2 \cos 2\theta$ or $r^2 = a^2 \sin 2\theta$ ($a > 0$ a constant) are called *lemniscates*.

(a) Draw all such curves.

(b) Find equations for the lemniscates in Cartesian coordinates.

** **45.** (a) Draw the curves $r = b \pm a \cos \theta$ and $r = b \pm a \sin \theta$ where a and b are positive constants in the three cases $a < b$, $a = b$, and $a > b$.

(b) Find equations for the curves in Cartesian coordinates.

(c) Compare these curves with the cardioids of Exercise 43 when $a = b$.

** **46.** A curve with equation of the form $r = a \sin n\theta$ or $r = a \cos n\theta$ where $a > 0$ is a constant and $n > 0$ is an integer is called a *rose*. Show that the rose has n petals.

** **47.** Show that the roses $r = |a \sin n\theta|$ and $r = |a \cos n\theta|$, where $a > 0$ is a constant and $n > 0$ is an integer, have $2n$ petals.

*** **48.** (a) A patrol boat at point A in the figure below spots a submarine at point B at a time that we call $t = 0$. The submarine, unaware of the patrol boat, follows a straight line path at constant speed v along some angle ϕ relative to BA (unknown to the patrol boat). The patrol boat heads directly toward point B at speed $V > v$ for $k/(v + V)$ units of time, arriving at point C. Show that the submarine and patrol boat are equidistant from B at $t = k/(v + V)$.

(b) We set up a system of polar coordinates with B as pole and BA as polar axis. Let the distance $\|BC\|$ be denoted by r_0. Suppose that the patrol boat now follows the logarithmic spiral $r = r_0 e^{\theta/\alpha}$, still at speed V, where $\alpha = \sqrt{V^2/v^2 - 1}$. Show that the patrol boat must intercept the submarine.

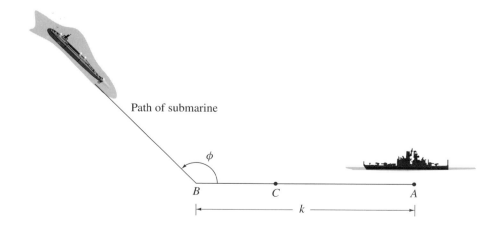

Path of submarine

10.4 Areas in Polar Coordinates

FIGURE 10.30

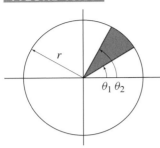

In Section 7.1 we used definite integrals to find areas bounded by curves whose equations are conveniently expressed in Cartesian coordinates. In this section we indicate how to find areas bounded by curves whose equations are expressed in polar coordinates. We require the formula

$$\frac{1}{2}r^2(\theta_2 - \theta_1) \qquad\qquad (10.15)$$

for the area of the shaded sector of the circle in Figure 10.30. This formula results from the fact that the area of the sector is the fractional part $(\theta_2 - \theta_1)/(2\pi)$ of the area πr^2 of the circle.

Consider finding the area in Figure 10.31a bounded by the radial lines $\theta = \alpha$ and $\theta = \beta$ and the curve $r = f(\theta)$. We divide the area into subareas by means of $n + 1$ radial lines $\theta = \theta_i$ where

$$\alpha = \theta_0 < \theta_1 < \theta_2 < \cdots < \theta_{n-1} < \theta_n = \beta.$$

On that part of the curve $r = f(\theta)$ between $\theta = \theta_{i-1}$ and $\theta = \theta_i$, we pick any point with polar coordinates $\left(f(\theta_i^*), \theta_i^*\right)$ as in Figure 10.31b. If between the lines $\theta = \theta_{i-1}$ and $\theta = \theta_i$ we draw the arc of a circle with centre at the pole and radius $f(\theta_i^*)$, a sector is formed with area

$$\Delta A_i = \frac{1}{2}[f(\theta_i^*)]^2 \Delta\theta_i,$$

where $\Delta\theta_i = \theta_i - \theta_{i-1}$. Since this sector approximates that part of the required area between the radial lines $\theta = \theta_{i-1}$ and $\theta = \theta_i$, we can say that an approximation to the required area is

$$\sum_{i=1}^{n} \Delta A_i = \sum_{i=1}^{n} \frac{1}{2}[f(\theta_i^*)]^2 \Delta\theta_i.$$

FIGURE 10.31a

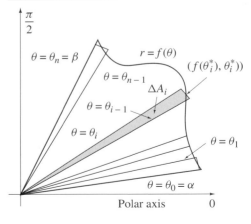

FIGURE 10.31b

By increasing the number of sectors indefinitely, and at the same time requiring each $\Delta\theta_i$ to approach zero, we obtain a better and better approximation, and in the limit

$$\text{area} = \lim_{\|\Delta\theta_i\|\to 0} \sum_{i=1}^{n} \frac{1}{2}[f(\theta_i^*)]^2 \Delta\theta_i.$$

But this limit is the definition of the definite integral of the function $(1/2)[f(\theta)]^2$ with respect to θ from $\theta = \alpha$ to $\theta = \beta$, and we therefore write

$$\text{area} = \int_{\alpha}^{\beta} \frac{1}{2}[f(\theta)]^2 \, d\theta. \tag{10.16}$$

In order to arrive at this integral in any given problem, without memorizing it, we use the procedure adopted for definite integrals in Cartesian coordinates discussed in Chapter 7. We draw at angle θ a representative sector of angular width $d\theta$ and radius r (Figure 10.32). The area of this sector is

$$\frac{1}{2}r^2 \, d\theta = \frac{1}{2}[f(\theta)]^2 \, d\theta.$$

If areas of all such sectors from angle α to angle β are added together, and the limit is taken as their widths approach zero, the required area is obtained. But this is the process defined by the definite integral, and we therefore write equation 10.16 for the area. Definite integral 10.16 exists when $f(\theta)$ is continuous for $\alpha \le \theta \le \beta$.

FIGURE 10.32

EXAMPLE 10.17

Find the area inside the cardioid $r = 1 + \sin\theta$.

SOLUTION   The area of the representative sector in Figure 10.33 is

$$\frac{1}{2}r^2 \, d\theta = \frac{1}{2}(1 + \sin\theta)^2 \, d\theta,$$

FIGURE 10.33

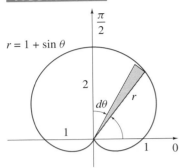

and we must add over all sectors interior to the cardioid. Since areas on either side of the $\theta = \pi/2$ line are identical, we calculate the area to the right and double the result. To find the area to the right of the line $\theta = \pi/2$, we must identify angular positions of the first and last sectors. The first sector is at the pole, and the equation of the cardioid indicates that $r = 0$ when $\sin\theta = -1$, that is, when $\theta = -\pi/2 + 2n\pi$. But which of these values of θ shall we choose? Similarly, the last sector occurs when $r = 2$, in which case $\sin\theta = 1$, and θ could be any of the values $\pi/2 + 2n\pi$. Again, which shall we choose? If we choose $\alpha = -\pi/2$ and $\beta = \pi/2$, then all values of θ in the interval $-\pi/2 \le \theta \le \pi/2$ yield points on the right half of the cardioid with no duplications. Consequently,

$$
\text{area} = 2\int_{-\pi/2}^{\pi/2} \frac{1}{2}(1+\sin\theta)^2\,d\theta = \int_{-\pi/2}^{\pi/2}(1+2\sin\theta+\sin^2\theta)\,d\theta
$$

$$
= \int_{-\pi/2}^{\pi/2}\left(1+2\sin\theta+\frac{1-\cos 2\theta}{2}\right)d\theta
$$

$$
= \left\{\frac{3\theta}{2}-2\cos\theta-\frac{\sin 2\theta}{4}\right\}_{-\pi/2}^{\pi/2}
$$

$$
= \frac{3\pi}{2}.
$$

EXAMPLE 10.18

Find the area common to the circles $x^2 + y^2 = 4$ and $x^2 + y^2 = 4x$.

FIGURE 10.34

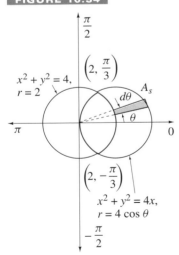

SOLUTION   Equations for the circles in polar coordinates are $r = 2$ and $r = 4\cos\theta$, and they intersect in the points with polar coordinates $(2, \pm\pi/3)$ (Figure 10.34). If A_s is the area above the x-axis, outside $x^2 + y^2 = 4$ and inside $x^2 + y^2 = 4x$, then the area common to the circles is the area of either circle less twice A_s:

$$
\text{area} = \pi(2)^2 - 2A_s.
$$

The area of the representative element is the difference in the areas of two sectors:

$$
\frac{1}{2}(4\cos\theta)^2\,d\theta - \frac{1}{2}(2)^2\,d\theta = 2(4\cos^2\theta - 1)\,d\theta.
$$

Since all sectors in A_s can be identified by values of θ between 0 and $\pi/3$, the required area is

$$
\text{area} = 4\pi - 2\int_0^{\pi/3} 2(4\cos^2\theta - 1)\,d\theta = 4\pi - 4\int_0^{\pi/3}[2(1+\cos 2\theta) - 1]\,d\theta
$$

$$
= 4\pi - 4\{\theta + \sin 2\theta\}_0^{\pi/3} = 4\pi - 4\left(\frac{\pi}{3} + \frac{\sqrt{3}}{2}\right) = \frac{8\pi}{3} - 2\sqrt{3}.
$$

EXERCISES 10.4

In Exercises 1–10 find the area enclosed by the curve.

* **1.** $r = 3 \sin \theta$

* **2.** $r = -6 \cos \theta$

* **3.** $r = 2 \sin 2\theta$

* **4.** $r^2 = 2 \sin 2\theta$

* **5.** $r^2 = -\cos \theta$

* **6.** $r = 2 - 2 \cos \theta$

* **7.** $r = 4 - 4 \cos \theta$

* **8.** $r = 4 - 2 \cos \theta$

* **9.** $r = \sin 3\theta$

* **10.** $r = 2(\cos \theta + \sin \theta)$

In Exercises 11–21 find the indicated area.

* **11.** Outside $r = 3$ but inside $r = 6 \sin \theta$

* **12.** Inside both $r = 1$ and $r = 1 - \sin \theta$

* **13.** Inside $r = 2 \sin 2\theta$ but outside $r = 1$

* **14.** Inside both $r = 2 + 2 \cos \theta$ and $r = 2 - 2 \cos \theta$

* **15.** Inside both $r = \sin \theta$ and $r = \cos \theta$

* **16.** Inside both $r = \cos \theta$ and $r = 1 - \cos \theta$

* **17.** Inside $r = 1 - 4 \cos \theta$

* **18.** Inside $r = 4 + 3 \sin \theta$ but outside $r = 2$

* **19.** Inside $r = |1 - 4 \cos \theta|$

* **20.** Inside the bifolium $r = \sin \theta \cos^2 \theta$

* **21.** Bounded by $\theta = 0$ and $r = \theta$, $0 \le \theta \le \pi$

*** **22.** (a) Show that in polar coordinates the strophoid

$$y^2 = x^2 \frac{a - x}{a + x},$$

where $a > 0$ is a constant, takes the form $r = a \cos 2\theta \sec \theta$.

(b) Draw or plot the curve and find the area inside its loop.

10.5 Definitions of Conic Sections

In Section 1.2 we used conic sections to illustrate the algebraic-geometric interplay of plane analytic geometry. They can be visualized as curves of intersection of a plane with a pair of right circular cones (Figure 10.35). Certainly this suggests why the conic sections are so-named, but because of the three-dimensional nature of the cone, an analysis of conic sections from this point of view is not yet possible. In this section we use plane analytic geometry to develop definitions for parabolas, ellipses, and hyperbolas.

FIGURE 10.35

(a) Circle: Plane perpend-icular to axis of cones

(b) Ellipse: Plane cuts completely across one cone but not perpendicular to axis

(c) Parabola: Plane cuts only one cone but not completely across

(d) Hyperbola: Plane cuts both cones

(e) Pair of straight lines: Plane passes through vertex and cuts through both cones

(f) One straight line: Plane passes through vertex and touches both cones

The Parabola

DEFINITION 10.2

A **parabola** is the curve traced out by a point that moves in a plane so that its distances from a fixed point called the *focus* and a fixed line called the *directrix* are always the same.

Suppose the focus of a parabola is the point (p, q) and the directrix is a line $y = r$ parallel to the x-axis as in Figure 10.36a. If $P(x, y)$ is any point on the parabola, then the fact that its distance from (p, q) must be equal to its distance from $y = r$ is expressed as

$$\sqrt{(x - p)^2 + (y - q)^2} = |y - r|. \tag{10.17}$$

With the absolute values, this equation includes the case of a directrix above the focus as in Figure 10.36b.

FIGURE 10.36a

FIGURE 10.36b

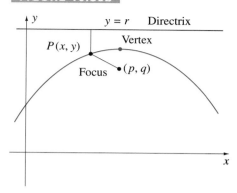

If we square both sides of the equation and rearrange terms, we obtain

$$(x - p)^2 = (y - r)^2 - (y - q)^2 = 2y(q - r) + r^2 - q^2.$$

We can solve this equation for y in terms of x; the result is

$$y = \frac{1}{2(q - r)}[(x - p)^2 + (q^2 - r^2)]. \qquad (10.18)$$

We could rewrite this equation in our accustomed form $y = ax^2 + bx + c$ for a parabola, but the present form is more informative. Firstly, the line $x = p$ through the focus and perpendicular to the directrix is the line of symmetry for the parabola. Secondly, the parabola opens upward if $q > r$, in which case the focus is above the directrix, and opens downward if $q < r$. Finally, the vertex of the parabola is found by setting $x = p$, in which case $y = (q + r)/2$, halfway between the focus and directrix.

A similar analysis shows that when the directrix is parallel to the y-axis (Figure 10.37) the equation of the parabola is of the form

$$x = \frac{1}{2(p - r)}[(y - q)^2 + (p^2 - r^2)]. \qquad (10.19)$$

FIGURE 10.37

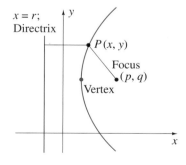

Given the focus and directrix (parallel to a coordinate axis) of a parabola, we can easily find its equation: use formulas 10.18 or 10.19, or follow the algebraic steps leading from 10.17 to 10.18. Conversely, given the equation of a parabola in the form $y = ax^2 + bx + c$ or $x = ay^2 + by + c$, we can identify its focus and directrix (see Exercises 52 and 53).

EXAMPLE 10.19

Find the equation of the parabola that has focus $(2, 4)$ and directrix $x = 6$.

SOLUTION   If (x, y) is any point on the parabola (Figure 10.38), the fact that its distance from $(2, 4)$ is equal to its distance from $x = 6$ is expressed as

$$\sqrt{(x - 2)^2 + (y - 4)^2} = 6 - x.$$

If we square both sides and simplify, the result is $x = (16 + 8y - y^2)/8$.

FIGURE 10.38

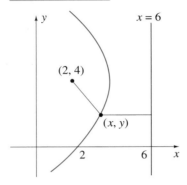

The Ellipse

DEFINITION 10.3

An **ellipse** is the curve traced out by a point that moves in a plane so that the sum of its distances from two fixed points called *foci* remains constant.

The equation of an ellipse is simplest when the foci lie on either the x-axis or the y-axis and are equidistant from the origin. Suppose the foci are $(c, 0)$ and $(-c, 0)$ (Figure 10.39), and the sum of the distances from these foci to a point on the ellipse is $2a$, where $a > c \geq 0$.

FIGURE 10.39

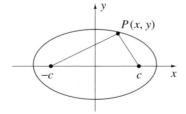

If $P(x, y)$ is any point on the ellipse, Definition 10.3 implies that

$$\sqrt{(x + c)^2 + y^2} + \sqrt{(x - c)^2 + y^2} = 2a. \tag{10.20}$$

If we transpose the second term to the right side and square both sides, we obtain

$$(x + c)^2 + y^2 = 4a^2 - 4a\sqrt{(x - c)^2 + y^2} + (x - c)^2 + y^2,$$

and this equation simplifies to

$$a^2 - cx = a\sqrt{(x - c)^2 + y^2}.$$

Squaring once again leads to

$$a^4 - 2a^2cx + c^2x^2 = a^2(x^2 - 2cx + c^2 + y^2) \qquad \text{or}$$

$$x^2(a^2 - c^2) + a^2y^2 = a^4 - a^2c^2.$$

Division by $a^2(a^2 - c^2)$ gives

$$\frac{x^2}{a^2} + \frac{y^2}{a^2 - c^2} = 1. \tag{10.21}$$

It is customary to denote y-intercepts of an ellipse by $\pm b$ ($b > 0$) (Figure 10.40a), in which case $b^2 = a^2 - c^2$, and the equation of the ellipse becomes

$$\frac{x^2}{a^2} + \frac{y^2}{b^2} = 1, \qquad \text{or} \qquad b^2x^2 + a^2y^2 = a^2b^2. \tag{10.22}$$

FIGURE 10.40a

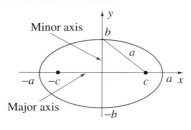

Minor axis

Major axis

FIGURE 10.40b

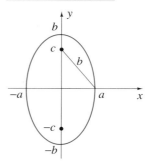

The line segment across the ellipse and through the foci is called the **major axis** of the ellipse; it has length $2a$ (see Figure 10.40a). The midpoint of the major axis is called the **centre** of the ellipse. The line segment across the ellipse, through its centre, and perpendicular to the major axis is called the **minor axis**; it has length $2b$. Note that the line segment joining either end of the minor axis to a focus (Figure 10.40a) has length a, and the triangle formed specifies the relationship among a, b, and c, namely, $a^2 = b^2 + c^2$.

A similar analysis shows that when the foci of the ellipse are on the y-axis, equidistant from the origin, the equation of the ellipse is again in form 10.22. In this case, $2b$ is the length of the major axis, $2a$ is the length of the minor axis, and $b^2 = a^2 + c^2$ (Figure 10.40b).

What we should remember is that an equation of form 10.22 always specifies an ellipse. Foci are on the longer axis and can be located using $c^2 = |a^2 - b^2|$. The length of the major axis represents the sum of the distances from any point on the ellipse to the foci.

EXAMPLE 10.20

Draw the ellipse $16x^2 + 9y^2 = 144$, indicating its foci.

SOLUTION   If we write the ellipse in the form $x^2/9 + y^2/16 = 1$, its x- and y-intercepts are ± 3 and ± 4. A sketch of the ellipse is therefore as shown in Figure 10.41. The foci must lie on the y-axis at distances $\pm c = \pm\sqrt{4^2 - 3^2} = \pm\sqrt{7}$ from the origin.

FIGURE 10.41

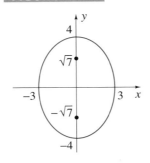

If $a = b$ in equation 10.22, then $x^2 + y^2 = a^2$, and this is the equation for a circle with radius a and centre at the origin. But if $a = b$, the distance from the origin to each focus of the ellipse must be $c = 0$. In other words, a circle may be regarded as a degenerate ellipse whose foci are at one and the same point. It is also true that when c is very small compared to half the length of the major axis (Figure 10.42), the ellipse is shaped very much like a circle. On the other hand, when these lengths are almost equal (Figure 10.43), the ellipse is long and narrow.

FIGURE 10.42

FIGURE 10.43

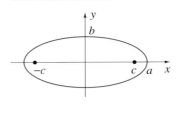

When the centre of an ellipse is at point (h, k) and the foci lie on either the line $x = h$ or $y = k$ (Figure 10.44), the equation for the ellipse is somewhat more complex than 10.22. If $2a$ and $2b$ are again the lengths of the axes of the ellipse, a calculation similar to that leading from 10.20 to 10.22 gives (see Exercise 55)

$$\frac{(x - h)^2}{a^2} + \frac{(y - k)^2}{b^2} = 1. \tag{10.23}$$

Alternatively, the curves in Figure 10.44 are those in Figures 10.39 and 10.40b translated h units in the x-direction and k units in the y-direction. According to Section 1.3, equations for the translated ellipses can be obtained by replacing x and y in $x^2/a^2 + y^2/b^2 = 1$ by $x - h$ and $y - k$ respectively.

FIGURE 10.44a

FIGURE 10.44b

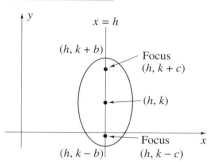

EXAMPLE 10.21

Draw the ellipse $16x^2 + 25y^2 - 160x + 50y = 1175$.

SOLUTION   If we complete the squares on the x- and y-terms, we obtain

$$16(x - 5)^2 + 25(y + 1)^2 = 1600 \qquad \text{or}$$

$$\frac{(x - 5)^2}{100} + \frac{(y + 1)^2}{64} = 1.$$

The centre of the ellipse is $(5, -1)$, and lengths of its major and minor axes are 20 and 16 respectively (Figure 10.45).

FIGURE 10.45

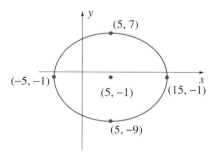

The Hyperbola

DEFINITION 10.4

A **hyperbola** is the path traced out by a point that moves in a plane so that the difference between its distances from two fixed points called *foci* remains constant.

Like the ellipse, the simplest hyperbolas have foci on either the x- or y-axis, equidistant from the origin. Suppose the foci are $(\pm c, 0)$ (Figure 10.46) and the difference in the distances from $P(x, y)$ to these foci is $2a$. Then Definition 10.4 implies that

$$\left| \sqrt{(x + c)^2 + y^2} - \sqrt{(x - c)^2 + y^2} \right| = 2a. \tag{10.24}$$

This equation can be simplified by a calculation similar to that leading to 10.22; the result is

$$\frac{x^2}{a^2} - \frac{y^2}{b^2} = 1, \tag{10.25}$$

where $b^2 = c^2 - a^2$. The hyperbola has x-intercepts equal to $\pm a$.

When the foci are on the y-axis (Figure 10.47), the equation of the hyperbola becomes

$$\frac{y^2}{b^2} - \frac{x^2}{a^2} = 1, \tag{10.26}$$

where $2b$ is the constant difference in the distances from a point (x, y) to the foci, and $a^2 = c^2 - b^2$. This hyperbola intersects the y-axis at $\pm b$.

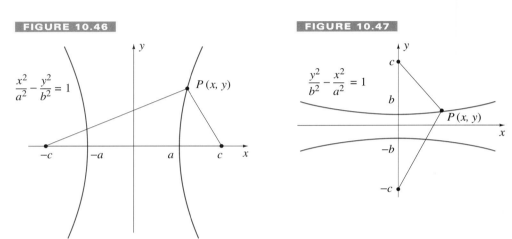

FIGURE 10.46

FIGURE 10.47

That part of the line segment joining the foci of a hyperbola which is between the two branches of the curve is called the **transverse axis** of the hyperbola; it has length $2a$ in Figure 10.48a and $2b$ in Figure 10.48b. The midpoint of the transverse axis is called the **centre** of the hyperbola. The line segment perpendicular to the transverse axis, through its centre, and of length $2b$ in Figure 10.48a and $2a$ in Figure 10.48b, is called the **conjugate axis**. Asymptotes of both hyperbolas are the lines $y = \pm bx/a$.

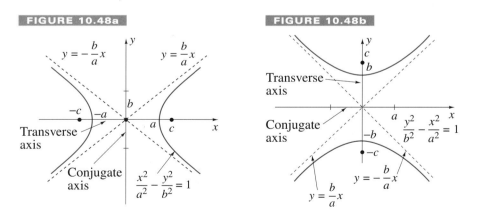

FIGURE 10.48a

FIGURE 10.48b

What we should remember is that an equation of form 10.25 or 10.26 specifies a hyperbola. The foci lie on the extension of the transverse axis and can be located using $c^2 = a^2 + b^2$. The length of the transverse axis represents the difference between the distances from any point on the hyperbola to the foci.

EXAMPLE 10.22

Draw the hyperbola $16x^2 - 9y^2 = 144$, indicating its foci.

SOLUTION   If we express the hyperbola in the form $x^2/9 - y^2/16 = 1$, its x-intercepts are ± 3. With asymptotes $y = \pm 4x/3$, we obtain Figure 10.49. The foci lie on the x-axis at distances $\pm c = \pm\sqrt{4^2 + 3^2} = \pm 5$ from the origin.

FIGURE 10.49

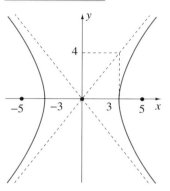

When the centre of a hyperbola is at point (h, k) and its foci are on the lines $x = h$ or $y = k$, equations 10.25 and 10.26 are modified in exactly the same way as equation 10.22 was modified for an ellipse. We replace each x by $x - h$ and each y by $y - k$ (see also Exercise 56). Consequently, equations for the hyperbolas in Figures 10.50 are

$$\frac{(x - h)^2}{a^2} - \frac{(y - k)^2}{b^2} = 1 \quad \text{and} \quad \frac{(y - k)^2}{b^2} - \frac{(x - h)^2}{a^2} = 1. \quad (10.27)$$

FIGURE 10.50a

FIGURE 10.50b

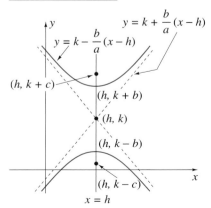

EXAMPLE 10.23

Draw the hyperbola $x^2 - y^2 + 4x + 10y = 5$.

SOLUTION   If we complete squares on x- and y-terms, we find

$$(x + 2)^2 - (y - 5)^2 = -16 \qquad \text{or}$$

$$\frac{(y - 5)^2}{16} - \frac{(x + 2)^2}{16} = 1.$$

The centre of the hyperbola is $(-2, 5)$, and the length of its transverse axis (along $x = -2$) is 8. If we solve the equation for y, we obtain

$$y = 5 \pm \sqrt{(x + 2)^2 + 16};$$

asymptotes of the hyperbola are then $y = 5 \pm (x + 2)$. The hyperbola can now be sketched as in Figure 10.51. Its foci are at the points $(-2, 5 + 4\sqrt{2})$ and $(-2, 5 - 4\sqrt{2})$.

FIGURE 10.51

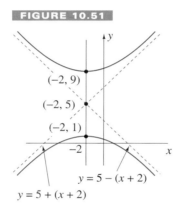

If P is a point on a conic section, the *focal radii* at P are the lines joining P to the foci (Figure 10.52). As a result, a parabola has one focal radius at each point, and an ellipse and hyperbola have two each. One of the properties of conics that makes them so useful is the fact that the normal line to the conic at any point bisects the angle between the focal radii. For the parabola, the normal bisects the angle between the focal radius and the line through P parallel to the axis of symmetry of the parabola. We will verify these facts in Exercises 59 and 60. To obtain one physical significance of these results, suppose each conic in Figure 10.52 is rotated about the x-axis to form a surface of revolution, which we regard as a mirror. It is a law of optics that, when a ray of light strikes a reflecting surface, the angle between incident light and the normal to the surface is always equal to the angle between reflected light and the normal. Consequently, if a beam of light travels in the negative x-direction and strikes the parabolic mirror in Figure 10.52a, all light is reflected toward the focus. Conversely, if F is a source of light, all light striking the mirror is reflected parallel to the x-axis. If either focus of the ellipse in Figure 10.52b is a light source, all light striking the elliptic mirror is reflected toward the other focus. Similarly, if either focus of the hyperbola in Figure 10.52c is a source, all light striking the mirror is reflected in a direction that would make it seem to originate at the other focus. Conversely, if light that is directed at one focus first strikes the mirror, it is reflected toward the other focus. This is precisely why we have parabolic reflectors in automobile headlights and searchlights, parabolic and hyperbolic reflectors in telescopes, and elliptic ceilings in whispering rooms.

FIGURE 10.52a FIGURE 10.52b FIGURE 10.52c

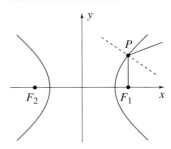

Thus far, we have defined parabolas, ellipses, and hyperbolas in terms of distances; for the parabola we use a focus and a directrix, and for the ellipse and hyperbola two foci. In Section 10.6 we show that ellipses and hyperbolas can also be defined in terms of a focus and directrix.

EXERCISES 10.5

In Exercises 1–14 identify the equation as representing a straight line, circle, parabola, ellipse, hyperbola, or none of these.

∗ **1.** $2x + 3y = y^2$

∗ **2.** $x^2 + y^2 - 3x + 2y = 25$

∗ **3.** $2x - y = 3$

∗ **4.** $x^2 + y^3 = 3x + 2$

∗ **5.** $5x^2 = 11 - 2y^2$

∗ **6.** $2x^2 - 3y^2 + 5 = 0$

∗ **7.** $y^2 - x + 3y = 14 - x^2$

∗ **8.** $x^2 + 2x = 3y + 4$

∗ **9.** $y^2 + x^2 - 2x + 6y + 15 = 0$

∗ **10.** $x^2 + 2y^2 + 24 = 0$

∗ **11.** $5 + y^2 = 3x^2$

∗ **12.** $x^2 + 2y^2 = 24$

∗ **13.** $y^3 = 3x + 4$

∗ **14.** $3 - x = 4y$

In Exercises 15–36 draw the curve. Identify foci for each ellipse and hyperbola.

∗ **15.** $y = 2x^2 - 1$

∗ **16.** $\dfrac{x^2}{25} + \dfrac{y^2}{36} = 1$

∗ **17.** $x^2 - \dfrac{y^2}{16} = 1$

∗ **18.** $3x = 4y^2 - 1$

∗ **19.** $\dfrac{y^2}{4} - \dfrac{x^2}{25} = 1$

∗ **20.** $7x^2 + 3y^2 = 16$

∗ **21.** $x + y^2 = 1$

∗ **22.** $2y^2 + x = 3y + 5$

∗ **23.** $9x^2 + 289y^2 = 2601$

∗ **24.** $y^2 = 10(2 - x^2)$

∗ **25.** $3x^2 - 4y^2 = 25$

∗ **26.** $y^2 - x^2 = 5$

∗ **27.** $y = -x^2 + 6x - 9$

∗ **28.** $2x^2 - 3y^2 = 5$

∗ **29.** $3x^2 + 6y^2 = 21$

∗ **30.** $x^2 + 16y^2 = 2$

∗ **31.** $y^2 - 3x^2 = 1$

∗ **32.** $x = -(4 + y)^2$

∗ **33.** $x^2 + 2x + 4y^2 - 16y + 13 = 0$

∗ **34.** $x^2 - 6x - 4y^2 - 24y = 11$

∗ **35.** $9x^2 + y^2 - 18x - 6y = 0$

∗ **36.** $9x^2 - 16y^2 - 18x - 64y = 91$

* **37.** Find the equation of a hyperbola that passes through the point $(1, 2)$ and has asymptotes $y = \pm 4x$.

* **38.** Find the equation of an ellipse through the points $(-2, 4)$ and $(3, 1)$.

** **39.** Find the width of the elliptic arch in the figure below.

** **40.** Find the height of the parabolic arch in the figure below.

** **41.** Explain how an ellipse can be drawn with a piece of string, two tacks, and a pencil.

** **42.** Find the equation of the ellipse traced out by a point that moves so that the sum of its distances from $(\pm 4, 0)$ is always equal to 10:

  (a) by using 10.22 with suitable values for a and b;

  (b) by establishing and simplifying an equation similar to 10.20.

** **43.** Find the equation of the hyperbola traced out by a point that moves so that the difference between its distances from $(0, \pm 3)$ is always equal to 1:

  (a) by using equation 10.26 with suitable values for a and b;

  (b) by establishing and simplifying an equation similar to 10.24.

** **44.** Show that the equation of every straight line, every circle, and every conic section discussed in this section can be obtained by appropriate choices of constants A, C, D, E, and F in the equation

$$Ax^2 + Cy^2 + Dx + Ey + F = 0.$$

** **45.** Show that the equation of the tangent line to the ellipse $b^2x^2 + a^2y^2 = a^2b^2$ at a point (x_0, y_0) is $b^2xx_0 + a^2yy_0 = a^2b^2$.

** **46.** Show that the equation of the tangent line to the hyperbola $b^2x^2 - a^2y^2 = a^2b^2$ at a point (x_0, y_0) is $b^2xx_0 - a^2yy_0 = a^2b^2$.

** **47.** Find the point P on that part of the ellipse $2x^2 + 3y^2 = 14$ in the first quadrant where the tangent line at P is perpendicular to the line joining P and $(2, 5)$.

** **48.** Find the area inside the ellipse $b^2x^2 + a^2y^2 = a^2b^2$.

** **49.** Among all rectangles that can be inscribed inside the ellipse $b^2x^2 + a^2y^2 = a^2b^2$ and have sides parallel to the axes, find the one with largest possible area.

** **50.** A *prolate spheroid* is the solid of revolution obtained by rotating an ellipse about its major axis. An *oblate spheroid* is obtained by rotating the ellipse about its minor axis. Find volumes for the prolate and oblate spheroids generated by the ellipse $b^2x^2 + a^2y^2 = a^2b^2$ if $a > b$.

** **51.** A sharp noise originating at one focus F_1 of an ellipse is reflected by the ellipse toward the other focus F_2. Explain why all reflected noise arrives at F_2 at exactly the same time.

** **52.** Use equation 10.18 to show that when a parabola is written in the form $y = ax^2 + bx + c$, the following formulas identify its focus (p, q) and directrix $y = r$:

$$p = -\frac{b}{2a}, \qquad q = \frac{1}{4a}(1 + 4ac - b^2), \qquad r = \frac{1}{4a}(-1 + 4ac - b^2).$$

** **53.** What are formulas for the focus (p, q) and directrix $x = r$ for a parabola of the type $x = ay^2 + by + c$?

** **54.** Use the formulas in Exercises 52 and 53 to identify the focus and directrix for any parabolas in Exercises 15–36.

** **55.** Show that when the centre of an ellipse is at point (h, k) and its foci are on the line $x = h$ or $y = k$, Definition 10.3 leads to equation 10.23.

** **56.** Show that when the centre of a hyperbola is at point (h, k) and its foci are on the line $x = h$ or $y = k$, Definition 10.4 leads to equation 10.27.

** **57.** Prove that the normal line to the parabola $x = ay^2$ in the figure below bisects the angle between the focal radius FP and the line PQ parallel to the x-axis. Hint: Draw the tangent line PR at P and show that $\|PF\| = \|RF\|$.

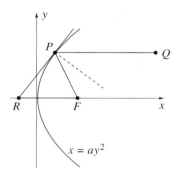

*** **58.** When a beam of light travelling in the negative x-direction (figure below) strikes a parabolic mirror with cross-section represented by $x = ay^2 + c$, all light rays are reflected to the focus F of the mirror. Show that all photons that pass simultaneously through $x = d$ arrive at F at the same time.

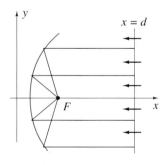

*** **59.** Prove that the normal line to an ellipse or hyperbola bisects the angle between the focal radii.

*** **60.** Prove that the normal line to the parabola $y = ax^2 + bx + c$ at any point P bisects the angle between the focal radius and the line through P parallel to the y-axis.

*** **61.** A line segment through the focus of a parabola with ends on the parabola is called a *focal chord*. It is an established result that tangents to a parabola at the ends of a focal chord are perpendicular to each other and intersect on the directrix. Prove this for the parabola $y = ax^2$.

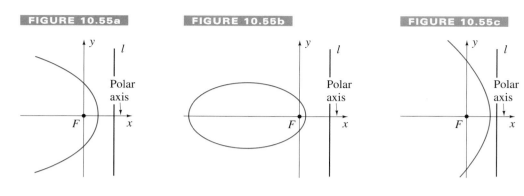

The conic sections in Figures 10.56 and 10.57 have equations of the form, respectively,

$$r = \frac{\epsilon d}{1 - \epsilon \sin \theta} \quad \text{and} \quad r = \frac{\epsilon d}{1 + \epsilon \sin \theta}. \qquad (10.34\text{b})$$

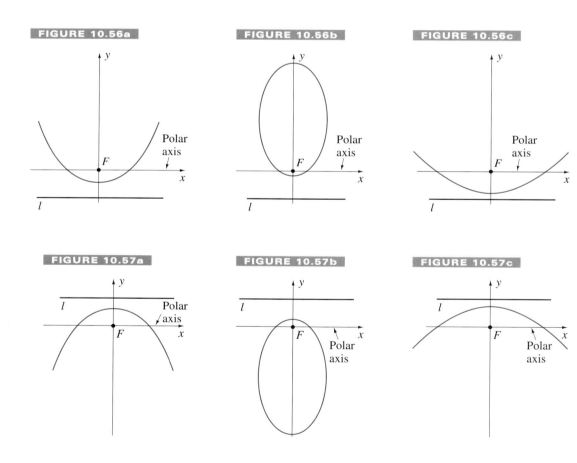

In Figures 10.54–10.57, one focus of the conic is chosen as the pole. In other words, the simplicity of equations 10.29 and 10.34 to describe conic sections is a direct consequence of the fact that the pole is at a focus, and the directrix is either parallel or perpendicular to the polar axis.

EXAMPLE 10.24

Draw the curve $r = 15/(3 + 2\cos\theta)$.

SOLUTION   If we write the equation in form 10.34a,

$$r = \frac{5}{1 + \left(\frac{2}{3}\right)\cos\theta},$$

the eccentricity $\epsilon = 2/3$ indicates that the curve is an ellipse. Both foci lie on the x-axis, and one is at the origin. The ends of the major axis occur when $\theta = 0$ and $\theta = \pi$, and for these values $r = 3$ and $r = 15$ (Figure 10.58). It now follows that the centre of the ellipse is at $x = -6$, and its other focus is at $x = -12$. If $b > 0$ denotes half the length of the minor axis (it is also the maximum y-value on the ellipse, occurring when $x = -6$), then $b^2 = a^2 - c^2 = 9^2 - 6^2 = 45$. Consequently, $b = 3\sqrt{5}$, and the ellipse is as shown in Figure 10.58. This information now permits us to write the equation of the ellipse in Cartesian coordinates:

$$\frac{(x+6)^2}{81} + \frac{y^2}{45} = 1.$$

FIGURE 10.58

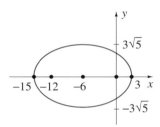

An alternative approach in the above Example 10.24 would be to substitute $r = \sqrt{x^2 + y^2}$ and $\cos\theta = x/\sqrt{x^2+y^2}$ into the polar equation for the conic, simplify it to $(x+6)^2/81+y^2/45 = 1$, and then draw the ellipse from this equation. We illustrate this method in the following example.

EXAMPLE 10.25

Draw the curve $r = 2/(3 - 4\sin\theta)$.

SOLUTION   If we set $r = \sqrt{x^2 + y^2}$ and $\sin\theta = y/\sqrt{x^2 + y^2}$, then

$$\sqrt{x^2 + y^2} = \frac{2}{3 - \dfrac{4y}{\sqrt{x^2 + y^2}}} = \frac{2\sqrt{x^2 + y^2}}{3\sqrt{x^2 + y^2} - 4y}.$$

Division by $\sqrt{x^2 + y^2}$ leads to

$$3\sqrt{x^2 + y^2} - 4y = 2 \qquad \text{or}$$

$$3\sqrt{x^2 + y^2} = 4y + 2.$$

If we now square both sides, we obtain

$$9(x^2 + y^2) = 4 + 16y + 16y^2 \qquad \text{or}$$

$$9x^2 - 7y^2 - 16y = 4.$$

If we complete the square on the y-terms, we obtain

$$9x^2 - 7\left(y + \frac{8}{7}\right)^2 = 4 - \frac{64}{7} = -\frac{36}{7}.$$

Division by $-36/7$ yields the equation

$$\frac{\left(y + \dfrac{8}{7}\right)^2}{\dfrac{36}{49}} - \frac{x^2}{\dfrac{4}{7}} = 1.$$

This equation describes a hyperbola with its centre at $(0, -8/7)$ and its y-intercepts equal to $-8/7 \pm 6/7 = -2/7, -2$. Its asymptotes are

$$y = -\frac{8}{7} \pm \frac{6}{7}\sqrt{\frac{7x^2}{4}} = -\frac{8}{7} \pm \frac{3x}{\sqrt{7}}.$$

The hyperbola is shown in Figure 10.59, but only the top half is described by $r = 2/(3 - 4\sin\theta)$. The equation $3\sqrt{x^2 + y^2} = 4y + 2$ does not permit $y \le -2$.

FIGURE 10.59

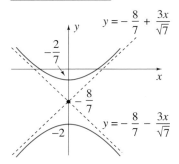

EXAMPLE 10.26

Find a polar representation for the ellipse

$$\frac{(x-1)^2}{4} + \frac{y^2}{9} = 1.$$

FIGURE 10.60

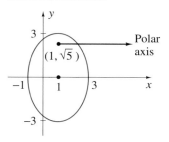

SOLUTION   To find a polar representation for the ellipse (Figure 10.60), we could use the usual polar coordinates defined by $x = r \cos \theta$ and $y = r \sin \theta$, but the resulting equation would not be simple. Try it. We know that a simple polar representation must result if the pole is chosen at a focus of the ellipse and polar axis either parallel or perpendicular to the directrix. For this ellipse, foci are on the line $x = 1$ at distances $\pm c = \pm \sqrt{9 - 4} = \pm \sqrt{5}$ from the x-axis. Let us choose the pole at position $(1, \sqrt{5})$ and the polar axis parallel to the x-axis, and therefore parallel to the directrix. According to equation 10.9, polar and Cartesian coordinates are related by $x = 1 + r \cos \theta$ and $y = \sqrt{5} + r \sin \theta$. If we substitute these into the equation for the ellipse, we obtain

$$\frac{r^2 \cos^2 \theta}{4} + \frac{(\sqrt{5} + r \sin \theta)^2}{9} = 1,$$

or

$$9r^2 \cos^2 \theta + 4(5 + 2\sqrt{5}r \sin \theta + r^2 \sin^2 \theta) = 36.$$

This equation can be expressed as a quadratic equation in r:

$$r^2(9 \cos^2 \theta + 4 \sin^2 \theta) + r(8\sqrt{5} \sin \theta) - 16 = 0.$$

Solutions for r are

$$r = \frac{-8\sqrt{5} \sin \theta \pm \sqrt{320 \sin^2 \theta + 64(9 \cos^2 \theta + 4 \sin^2 \theta)}}{2(9 \cos^2 \theta + 4 \sin^2 \theta)}$$

$$= \frac{-8\sqrt{5} \sin \theta \pm \sqrt{576(\cos^2 \theta + \sin^2 \theta)}}{2(9 \cos^2 \theta + 4 \sin^2 \theta)}$$

$$= \frac{4(\pm 3 - \sqrt{5} \sin \theta)}{9 \cos^2 \theta + 4 \sin^2 \theta}.$$

Since r must be nonnegative, we must choose $+3$, and not -3, and therefore

$$r = \frac{4(3 - \sqrt{5} \sin \theta)}{9(1 - \sin^2 \theta) + 4 \sin^2 \theta}$$

$$= \frac{4(3 - \sqrt{5} \sin \theta)}{9 - 5 \sin^2 \theta}$$

$$= \frac{4(3 - \sqrt{5} \sin \theta)}{(3 - \sqrt{5} \sin \theta)(3 + \sqrt{5} \sin \theta)}$$

$$= \frac{4}{3 + \sqrt{5} \sin \theta}.$$

EXERCISES 10.6

In Exercises 1–10 draw the conic section.

∗ **1.** $r = \dfrac{3}{1 + \cos \theta}$

∗ **2.** $r = \dfrac{16}{3 + 5 \cos \theta}$

∗ **3.** $r = \dfrac{4}{3 - 3 \sin \theta}$

∗ **4.** $r = \dfrac{16}{5 + 3 \cos \theta}$

∗ **5.** $r = \dfrac{4}{3 - 4 \sin \theta}$

∗ **6.** $r = \dfrac{4}{4 - 3 \sin \theta}$

∗ **7.** $r = \dfrac{1}{2 - 2 \cos \theta}$

∗ **8.** $r = \dfrac{1}{2 + \sin \theta}$

∗ **9.** $r = \dfrac{\sec \theta}{3 + 6 \sec \theta}$

∗ **10.** $r = \dfrac{4 \csc \theta}{7 \csc \theta - 2}$

In Exercises 11–16 find the Cartesian equation for the curve.

∗ **11.** $r = \dfrac{3}{1 - \sin \theta}$

∗ **12.** $r = \dfrac{1}{3 + \cos \theta}$

∗ **13.** $r = \dfrac{1}{1 + 2 \cos \theta}$

∗ **14.** $r = \dfrac{2}{1 - 3 \cos \theta}$

∗ **15.** $r = \dfrac{4}{6 - 3 \sin \theta}$

∗ **16.** $r = \dfrac{4}{5 + 5 \cos \theta}$

In Exercises 17–20 find a polar equation in one of the four forms of 10.29 or 10.34 for the conic.

∗∗ **17.** $x^2 - 16y^2 = 16$

∗∗ **18.** $4x^2 + 9y^2 = 36$

∗∗ **19.** $(x - 1)^2 + \dfrac{(y + 1)^2}{4} = 1$

∗∗ **20.** $x^2 - 9(y - 2)^2 = 9$

∗∗ **21.**   (a) Show that the eccentricity ϵ of an ellipse or hyperbola is always equal to the distance from its centre to either focus, divided by half the length of the major or transverse axis.

      (b) Discuss the eccentricities of the ellipses in Figures 10.42 and 10.43 (page 686).

∗∗ **22.** A circle has been described as a degenerate ellipse in the sense that its foci coincide. What happens to the eccentricity of an ellipse if the distance between its foci approaches zero? Hint: See Exercise 21.

10.7 Translation and Rotation of Axes

In Section 1.3 we discussed horizontal and vertical translations of curves and how they affect equations of the curves. For example, when the ellipse $4x^2 + 9y^2 = 36$ in Figure 10.61a is shifted one unit to the left and four units upward, the result is the ellipse in Figure 10.61b with equation $4(x + 1)^2 + 9(y - 4)^2 = 36$.

FIGURE 10.61a

FIGURE 10.61b

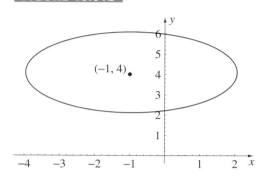

Another way of viewing this process does prove beneficial in this section. The equation $x^2 - y^2 + 4x + 10y = 5$ defines the hyperbola in Figure 10.62a. This is easily seen by completing squares on the x- and y-terms, $(y - 5)^2 - (x + 2)^2 = 16$. The equation is complicated because the hyperbola is not symmetric with respect to the x- and y-axes. If we choose a new coordinate system with origin $(-2, 5)$, and X- and Y-axes parallel to the x- and y-axes (Figure 10.62b), the equation of the hyperbola in this coordinate system is much simpler, $Y^2 - X^2 = 16$.

FIGURE 10.62a

FIGURE 10.62b

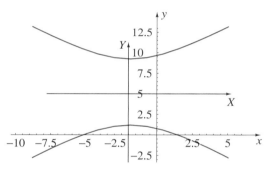

Transforming equations of curves between Cartesian coordinate systems whose axes are translated relative to one another is easily formulated in a general way. If (x, y) and (X, Y) are coordinates of point P in Figure 10.63, then

$$x = X + h, \qquad y = Y + k. \qquad (10.35)$$

FIGURE 10.63

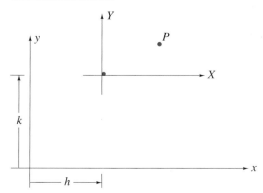

Given a curve with equation $F(x, y) = 0$ in the xy-coordinate system, its equation in the XY-system is $F(X + h, Y + k) = 0$. Conversely, a curve with equation $G(X, Y) = 0$ in the XY-system has equation $G(x - h, y - k) = 0$ in the xy-system.

EXAMPLE 10.27

Find a coordinate system in which the equation of the parabola $x = 2y^2 - 4y - 1$ takes the form $X = 2Y^2$.

SOLUTION   The vertex of the parabola $x = 2y^2 - 4y - 1 = 2(y - 1)^2 - 3$ (Figure 10.64a) is $(-3, 1)$. If we choose X- and Y-axes through this point (Figure 10.64b), then $x = X - 3$ and $y = Y + 1$. When we substitute these into $x = 2(y - 1)^2 - 3$,

$$X - 3 = 2(Y)^2 - 3 \qquad \Rightarrow X = 2Y^2.$$

FIGURE 10.64a

FIGURE 10.64b

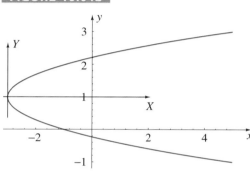

Sometimes the equation of a curve can be simplified by rotating the axes, as opposed to translating them. For example, the curve in Figure 10.65 has equation $x^2 + y^2 - 2xy - \sqrt{2}x - \sqrt{2}y + 2 = 0$; it appears to be a parabola. To be sure, we need to choose coordinate axes so that one axis is parallel to what appears to be the axis of symmetry of the curve. We discuss this in general before turning to this example.

FIGURE 10.65

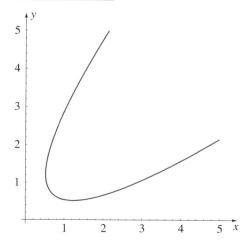

Suppose X- and Y-axes are obtained by rotating x- and y-axes by angle θ radians (Figure 10.66). If (x, y) and (X, Y) are coordinates of a point P relative to the two sets of axes, then

$$
\begin{aligned}
x &= r\cos(\theta + \phi) & y &= r\sin(\theta + \phi) \\
&= r(\cos\theta\cos\phi - \sin\theta\sin\phi) & &= r(\sin\theta\cos\phi + \cos\theta\sin\phi) \\
&= (r\cos\phi)\cos\theta - (r\sin\phi)\sin\theta & &= (r\cos\phi)\sin\theta + (r\sin\phi)\cos\theta \\
&= X\cos\theta - Y\sin\theta & &= X\sin\theta + Y\cos\theta.
\end{aligned}
$$

These equations

$$
x = X\cos\theta - Y\sin\theta, \qquad y = X\sin\theta + Y\cos\theta \tag{10.36}
$$

define (x, y)-coordinates in terms of (X, Y)-coordinates. By inverting them (or by replacing θ by $-\theta$), we get formulas for X and Y in terms of x and y,

$$
X = x\cos\theta + y\sin\theta, \qquad Y = -x\sin\theta + y\cos\theta. \tag{10.37}
$$

FIGURE 10.66

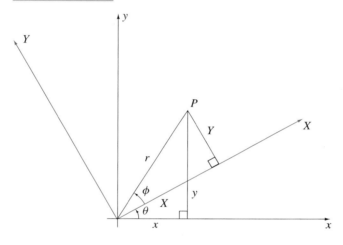

EXAMPLE 10.28

Show that the curve in Figure 10.65 is the parabola $x = y^2 + 1$ rotated $\pi/4$ radians counterclockwise around the origin.

SOLUTION   To show this we find the equation of the curve with respect to the axes in Figure 10.67. According to 10.36,

$$
x = \frac{X}{\sqrt{2}} - \frac{Y}{\sqrt{2}}, \qquad y = \frac{X}{\sqrt{2}} + \frac{Y}{\sqrt{2}},
$$

and, when these are substituted into $x^2 + y^2 - \sqrt{2}x - \sqrt{2}y - 2xy + 2 = 0$,

$$
\left(\frac{X}{\sqrt{2}} - \frac{Y}{\sqrt{2}}\right)^2 + \left(\frac{X}{\sqrt{2}} + \frac{Y}{\sqrt{2}}\right)^2 - \sqrt{2}\left(\frac{X}{\sqrt{2}} - \frac{Y}{\sqrt{2}}\right) - \sqrt{2}\left(\frac{X}{\sqrt{2}} + \frac{Y}{\sqrt{2}}\right)
$$

$$
- 2\left(\frac{X}{\sqrt{2}} - \frac{Y}{\sqrt{2}}\right)\left(\frac{X}{\sqrt{2}} + \frac{Y}{\sqrt{2}}\right) + 2 = 0.
$$

This equation simplifies to $X = Y^2 + 1$, as required.

FIGURE 10.67

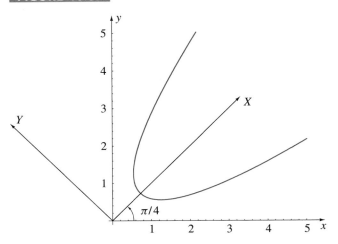

EXAMPLE 10.29

Find the equation of the hyperbola $x^2 - y^2 = 1$ when it is rotated by $\pi/6$ radians counter-clockwise around the origin.

SOLUTION   Rotating the curve $\pi/6$ radians is equivalent to rotating axes by $-\pi/6$ radians (Figure 10.68a), in which case

$$x = \frac{\sqrt{3}X}{2} + \frac{Y}{2}, \qquad y = -\frac{X}{2} + \frac{\sqrt{3}Y}{2}.$$

Substitution of these into $x^2 - y^2 = 1$ gives

$$\left(\frac{\sqrt{3}X}{2} + \frac{Y}{2}\right)^2 - \left(-\frac{X}{2} + \frac{\sqrt{3}Y}{2}\right)^2 = 1,$$

which simplifies to $X^2 + 2\sqrt{3}XY - Y^2 = 2$. This is the equation of the hyperbola with respect to the XY-coordinates. It follows that the equation of the hyperbola in Figure 10.68b is $x^2 + 2\sqrt{3}xy - y^2 = 2$.

Equivalently, we could write the equation of the hyperbola in Figure 10.68c with respect to the XY-axes as $X^2 - Y^2 = 1$, and then use 10.37 with $\theta = \pi/6$,

$$X = \frac{\sqrt{3}x}{2} + \frac{y}{2}, \qquad Y = -\frac{x}{2} + \frac{\sqrt{3}y}{2}.$$

Substitution of these into $X^2 - Y^2 = 1$ once again leads to $x^2 + 2\sqrt{3}xy - y^2 = 2$.

FIGURE 10.68a

FIGURE 10.68a

FIGURE 10.68b

FIGURE 10.68c

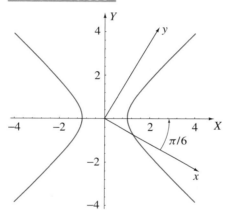

When an XY-coordinate system is translated h units in the x-direction and k units in the y-direction, and then rotated counterclockwise by θ radians (Figure 10.69), relations among x, y, X, and Y are a combination of 10.35 and 10.36,

$$x = (X + h)\cos\theta - (Y + k)\sin\theta, \qquad y = (X + h)\sin\theta + (Y + k)\cos\theta. \quad (10.38)$$

FIGURE 10.69

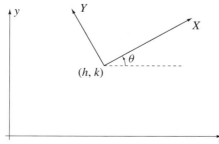

For example, to find the equation of the ellipse $4x^2 + 9y^2 = 36$ (Figure 10.70a) when it is translated 2 units to the right, 1 unit down, and rotated counterclockwise $\pi/3$ radians, we draw Figure 10.70b.

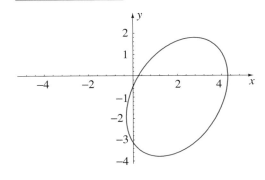

Coordinates are related by

$$x = \frac{1}{2}(X - 2) + \frac{\sqrt{3}}{2}(Y + 1), \qquad y = -\frac{\sqrt{3}}{2}(X - 2) + \frac{1}{2}(Y + 1).$$

Substitution of these into $4x^2 + 9y^2 = 36$ gives

$$4\left[\frac{1}{2}(X - 2) + \frac{\sqrt{3}}{2}(Y + 1)\right]^2 + 9\left[-\frac{\sqrt{3}}{2}(X - 2) + \frac{1}{2}(Y + 1)\right]^2 = 36,$$

and this simplifies to

$$31(X - 2)^2 - 10\sqrt{3}(X - 2)(Y + 1) + 21(Y + 1)^2 = 144.$$

The equation of the ellipse in Figure 10.70c is therefore

$$31(x - 2)^2 - 10\sqrt{3}(x - 2)(y + 1) + 21(y + 1)^2 = 144.$$

We have considered conic sections whose equations can be expressed in the form

$$Ax^2 + Cy^2 + Dx + Ey + F = 0, \tag{10.39}$$

(see Exercise 44 in Section 10.5). Parabolas, ellipses, and hyperbolas defined by 10.39 have axes of symmetry parallel to the x- and y-axes. The most general second-degree polynomial equation in x and y is

$$Ax^2 + Bxy + Cy^2 + Dx + Ey + F = 0. \tag{10.40}$$

This equation also describes conic sections; the xy-term represents a rotation of the conic so that axes of symmetry are no longer parallel to the coordinate axes. To demonstrate this, we rotate axes to find a coordinate system in which the xy-term vanishes. What should remain is an equation of form 10.39.

The equation of curve 10.40 in an XY-coordinate system with axes rotated by angle θ relative to the x- and y-axes can be obtained by substituting 10.36 into 10.40,

$$A(X \cos\theta - Y \sin\theta)^2 + B(X \cos\theta - Y \sin\theta)(X \sin\theta + Y \cos\theta)$$

$$+ C(X \sin\theta + Y \cos\theta)^2 + D(X \cos\theta - Y \sin\theta)$$

$$+ E(X \sin\theta + Y \cos\theta) + F = 0.$$

This can be rearranged into the form

$$A'X^2 + B'XY + C'Y^2 + D'X + E'Y + F' = 0, \tag{10.41a}$$

where

$$A' = A \cos^2\theta + B \sin\theta \cos\theta + C \sin^2\theta,$$

$$B' = B(\cos^2\theta - \sin^2\theta) + 2(C - A) \sin\theta \cos\theta,$$

$$C' = A \sin^2\theta - B \sin\theta \cos\theta + C \cos^2\theta, \tag{10.41b}$$

$$D' = D \cos\theta + E \sin\theta,$$

$$E' = -D \sin\theta + E \cos\theta,$$

$$F' = F.$$

To eliminate the XY-term, θ must be chosen so that

$$0 = B' = B(\cos^2\theta - \sin^2\theta) + 2(C - A) \sin\theta \cos\theta$$

$$= B \cos 2\theta + (C - A) \sin 2\theta,$$

or

$$\cot 2\theta = \frac{A - C}{B}. \tag{10.42}$$

Once θ is chosen to satisfy 10.42, equation 10.41a reduces to form 10.39. To facilitate the change, in practice, we need values for $\sin\theta$ and $\cos\theta$ (for 10.41b or 10.36). We do this by restricting values of θ. In general, we can restrict θ to $-\pi < \theta < \pi$. For purposes of eliminating B', we can do more. Figures 10.71a–b show rotations by angles in the intervals $0 < \theta < \pi/2$ and $-\pi < \phi < -\pi/2$ where $\phi = \theta - \pi$. They produce exactly the same B' since $B' = B(\cos^2\theta - \sin^2\theta) + 2(C - A) \sin\theta \cos\theta$. This is also true for the rotations in 10.71c–d where $\phi = \theta + \pi$. What this means is that θ can be restricted to the interval $-\pi/2 < \theta < \pi/2$.

FIGURE 10.71a

FIGURE 10.71a

FIGURE 10.71b

FIGURE 10.71c

FIGURE 10.71d

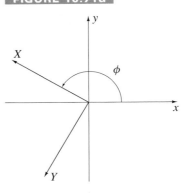

Now consider the rotations as shown in Figures 10.72a–b where $\phi = \theta - \pi/2$. Since $\cos(\theta - \pi/2) = \sin\theta$ and $\sin(\theta - \pi/2) = -\cos\theta$, it follows that these rotations produce B'-terms with opposite signs. That is, θ can be restricted to the interval $0 < \theta < \pi/2$ for purposes of eliminating B'.

FIGURE 10.72a

FIGURE 10.72b

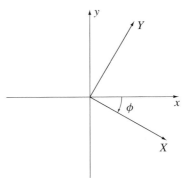

When $0 < \theta < \pi/2$, double-angle formulas 1.39d and 1.39c can be solved for

$$\sin\theta = \sqrt{\frac{1 - \cos 2\theta}{2}}, \qquad \cos\theta = \sqrt{\frac{1 + \cos 2\theta}{2}}, \tag{10.43}$$

and these equations define exact values for $\sin\theta$ and $\cos\theta$ in terms of $\cos 2\theta$. Equation 10.42 determines $\cot 2\theta$, and therefore $\cos 2\theta$, in terms of A, B, and C. Since $\cos 2\theta$ and $\cot 2\theta$ have the same sign $(0 < 2\theta < \pi)$, the triangle in Figure 10.73 leads to the formula

$$\cos 2\theta = \text{sgn}\left(\frac{A-C}{B}\right)\frac{|A-C|}{\sqrt{B^2+(A-C)^2}}, \tag{10.44a}$$

where $\text{sgn}(x)$ is the signum function of Exercise 43 in Section 2.4,

$$\text{sgn}(x) = \begin{cases} -1, & x < 0 \\ 0 & x = 0 \\ 1 & x > 0 \end{cases}.$$

Because 2θ is in the principal value range of the inverse cosine function, we can write explicitly that

$$\theta = \frac{1}{2}\text{Cos}^{-1}\left[\text{sgn}\left(\frac{A-C}{B}\right)\frac{|A-C|}{\sqrt{B^2+(A-C)^2}}\right]. \tag{10.44b}$$

FIGURE 10.73

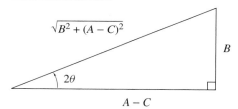

Let us use formulas 10.43 and 10.44 to eliminate the xy-term from some second-degree polynomial equations.

EXAMPLE 10.30

Rotate axes to eliminate the xy-term from the equation $7x^2 - 6\sqrt{3}xy + 13y^2 = 16$. Plot the curve.

SOLUTION The angle of rotation θ that eliminates the xy-term is defined by equation 10.44b,

$$\theta = \frac{1}{2}\text{Cos}^{-1}\left[\text{sgn}\left(\frac{7-13}{-6\sqrt{3}}\right)\frac{|7-13|}{\sqrt{108+(7-13)^2}}\right] = \frac{1}{2}\text{Cos}^{-1}\left(\frac{1}{2}\right) = \frac{\pi}{6}.$$

Axes should therefore be rotated counterclockwise $\pi/6$ radians. To find the equation of the curve with xy-term eliminated, we can use equations 10.41 or return to transformation 10.36. We illustrate both. With $A = 7$, $B = -6\sqrt{3}$, $C = 13$, and $F = -16$ in 10.41b,

$$A' = 7\left(\sqrt{3}/2\right)^2 - 6\sqrt{3}(1/2)\left(\sqrt{3}/2\right) + 13(1/2)^2 = 4,$$

$$B' = -6\sqrt{3}(3/4 - 1/4) + 2(13 - 7)(1/2)\left(\sqrt{3}/2\right) = 0,$$

$$C' = 7(1/2)^2 + 6\sqrt{3}(1/2)\left(\sqrt{3}/2\right) + 13\left(\sqrt{3}/2\right)^2 = 16,$$

$$D' = 0,$$

$$E' = 0,$$

$$F' = -16.$$

The equation of the curve with respect to rotated coordinates X and Y is

$$4X^2 + 16Y^2 - 16 = 0 \qquad \Rightarrow \qquad X^2 + 4Y^2 = 4.$$

Alternatively, if we substitute from 10.36, namely,

$$x = \frac{\sqrt{3}X}{2} - \frac{Y}{2}, \qquad y = \frac{X}{2} + \frac{\sqrt{3}Y}{2},$$

into $7x^2 - 6\sqrt{3}xy + 13y^2 = 16$,

$$7\left(\frac{\sqrt{3}X}{2} - \frac{Y}{2}\right)^2 - 6\sqrt{3}\left(\frac{\sqrt{3}X}{2} - \frac{Y}{2}\right)\left(\frac{X}{2} + \frac{\sqrt{3}Y}{2}\right) + 13\left(\frac{X}{2} + \frac{\sqrt{3}Y}{2}\right)^2 = 16,$$

and this simplifies to $X^2 + 4Y^2 = 4$. Figure 10.74 contains a plot of the ellipse.

FIGURE 10.74

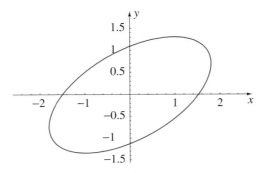

EXAMPLE 10.31

Rotate axes to eliminate the xy-term from the equation $4x^2 - 4xy + y^2 = 5x$. Plot the curve.

SOLUTION   The angle of rotation θ that eliminates the xy-term is defined by 10.44a,

$$\cos 2\theta = \operatorname{sgn}\left(\frac{4-1}{-4}\right)\frac{|4-1|}{\sqrt{16+(4-1)^2}} = -\frac{3}{5}.$$

Axes should therefore be rotated through angle θ where, according to 10.43,

$$\sin\theta = \sqrt{\frac{1-(-3/5)}{2}} = \frac{2}{\sqrt{5}}, \qquad \cos\theta = \sqrt{\frac{1+(-3/5)}{2}} = \frac{1}{\sqrt{5}}.$$

Transformation 10.36 becomes

$$x = \frac{X}{\sqrt{5}} - \frac{2Y}{\sqrt{5}}, \qquad y = \frac{2X}{\sqrt{5}} + \frac{Y}{\sqrt{5}},$$

and when these are substituted into $4x^2 - 4xy + y^2 = 5x$,

$$4\left(\frac{X}{\sqrt{5}} - \frac{2Y}{\sqrt{5}}\right)^2 - 4\left(\frac{X}{\sqrt{5}} - \frac{2Y}{\sqrt{5}}\right)\left(\frac{2X}{\sqrt{5}} + \frac{Y}{\sqrt{5}}\right) + \left(\frac{2X}{\sqrt{5}} + \frac{Y}{\sqrt{5}}\right)^2 = 5\left(\frac{X}{\sqrt{5}} - \frac{2Y}{\sqrt{5}}\right).$$

This equation simplifies to $X = \sqrt{5}Y^2 + 2Y$, the parabola in Figure 10.75.

FIGURE 10.75

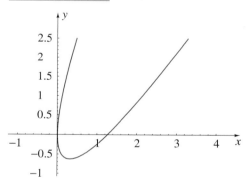

Classification of Conics

The xy-term in the general second-degree polynomial equation

$$Ax^2 + Bxy + Cy^2 + Dx + Ey + F = 0 \tag{10.45}$$

can be eliminated by rotating axes. The resulting equation then takes the form

$$A'X^2 + C'Y^2 + D'X + E'Y + F' = 0. \tag{10.46}$$

It is a straightforward calculation using 10.41b to show that in general

$$B'^2 - 4A'C' = B^2 - 4AC. \tag{10.47}$$

We say that $B^2 - 4AC$ is invariant under a rotation of axes. It is also invariant under translations. This fact gives us a quick way to determine whether equation 10.45 represents a parabola, an ellipse, or a hyperbola. According to Exercise 44 in Section 10.5, except for special cases, equation 10.46 is a parabola if $A'C' = 0$, an ellipse if $A'C' < 0$, and a hyperbola if $A'C' > 0$. It follows therefore that equation 10.45 should describe a parabola when $B^2 - 4AC = 0$, an ellipse when $B^2 - 4AC < 0$, and a hyperbola when $B^2 - 4AC > 0$. This is often true, but there also exist degenerate cases when 10.45 describes a pair of straight lines, a point, or nothing at all. Let us state results as follows: Except for degenerate cases, $Ax^2 + Bxy + Cy^2 + Dx + Ey + F = 0$ describes

$$\text{an ellipse when } B^2 - 4AC < 0,$$

$$\text{a parabola when } B^2 - 4AC = 0, \tag{10.48}$$

$$\text{a hyperbola when } B^2 - 4AC > 0.$$

Because of the possibility of degeneracy, we recommend that 10.48 not be used to determine whether 10.45 describes a parabola, an ellipse, or a hyperbola. It is preferable to transform 10.45 by means of a rotation of axes into form 10.46. We illustrate with some examples.

EXAMPLE 10.32

Determine what is described by each of the following equations:

(a) $4x^2 - 4xy + y^2 = 5x$ (b) $4x^2 - 4xy + y^2 = 5y - 3$

(c) $4x^2 - 4xy + y^2 = 5$ (d) $4x^2 - 4xy + y^2 = -2$

SOLUTION In all four cases, $B^2 - 4AC = 16 - 16 = 0$, suggesting parabolas. This is indeed the case for $4x^2 - 4xy + y^2 = 5x$ in part (a) (see Example 10.31). With the transformation

$$x = \frac{X}{\sqrt{5}} - \frac{2Y}{\sqrt{5}}, \qquad y = \frac{2X}{\sqrt{5}} + \frac{Y}{\sqrt{5}}$$

that rotates axes through the angle θ defined by $\cos 2\theta = -3/5$, we replace the quadratic terms $4x^2 - 4xy + y^2$ by

$$4\left(\frac{X}{\sqrt{5}} - \frac{2Y}{\sqrt{5}}\right)^2 - 4\left(\frac{X}{\sqrt{5}} - \frac{2Y}{\sqrt{5}}\right)\left(\frac{2X}{\sqrt{5}} + \frac{Y}{\sqrt{5}}\right) + \left(\frac{2X}{\sqrt{5}} + \frac{Y}{\sqrt{5}}\right)^2 = 5Y^2.$$

(b) With the above result, $4x^2 - 4xy + y^2 = 5y - 3$ is replaced by

$$5Y^2 = 5\left(\frac{2X}{\sqrt{5}} + \frac{Y}{\sqrt{5}}\right) - 3 \quad \Rightarrow \quad X = \frac{1}{2\sqrt{5}}\left(5Y^2 - \sqrt{5}Y + 3\right).$$

This is the parabola in Figure 10.76a.

(c) The equation $4x^2 - 4xy + y^2 = 5$ is transformed to $5Y^2 = 5$, or $Y = \pm 1$, the pair of straight lines in Figure 10.76b. We could also have arrived at this conclusion by writing the equation $4x^2 - 4xy + y^2 = 5$ in the form

$$(2x - y)^2 = 5 \quad \Rightarrow \quad 2x - y = \pm\sqrt{5}.$$

These are the lines with equations $Y = \pm 1$.

(d) Equation $4x^2 - 4xy + y^2 = -2$ is replaced by $5Y^2 = -2$, an impossibility. No points satisfy $4x^2 - 4xy + y^2 = -2$. This is also obvious from the fact that $(2x - y)^2 = -2$ is an impossibility.

FIGURE 10.76a

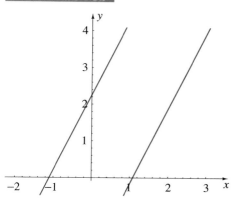

FIGURE 10.76b

EXAMPLE 10.33

Determine what is described by the equations:

$$\text{(a)}\quad x^2 - 3xy + y^2 + x + y = 2 \qquad \text{(b)}\quad x^2 - 3xy + y^2 + x + y = 1$$

SOLUTION   To eliminate the xy-term, we rotate axes by the angle θ defined by

$$\cos 2\theta = \operatorname{sgn}\left(\frac{1 - 1}{-3}\right)\frac{|1 - 1|}{\sqrt{9 + (1 - 1)^2}} = 0.$$

Thus, $2\theta = \pi/2$ or $\theta = \pi/4$. With the rotation

$$x = \frac{X}{\sqrt{2}} - \frac{Y}{\sqrt{2}}, \qquad y = \frac{X}{\sqrt{2}} + \frac{Y}{\sqrt{2}},$$

$x^2 - 3xy + y^2$ is transformed to

$$\left(\frac{X}{\sqrt{2}} - \frac{Y}{\sqrt{2}}\right)^2 - 3\left(\frac{X}{\sqrt{2}} - \frac{Y}{\sqrt{2}}\right)\left(\frac{X}{\sqrt{2}} + \frac{Y}{\sqrt{2}}\right) + \left(\frac{X}{\sqrt{2}} + \frac{Y}{\sqrt{2}}\right)^2 = \frac{1}{2}\left(4Y^2 - X^2\right).$$

(a) Equation $x^2 - 3xy + y^2 + x + y = 2$ is transformed to

$$\frac{1}{2}\left(4Y^2 - X^2\right) + \left(\frac{X}{\sqrt{2}} - \frac{Y}{\sqrt{2}}\right) + \left(\frac{X}{\sqrt{2}} + \frac{Y}{\sqrt{2}}\right) = 2 \quad \Rightarrow \quad 4Y^2 - X^2 + 2\sqrt{2}X = 4.$$

When the square is completed on the X-term, the hyperbola

$$2Y^2 - \frac{\left(X - \sqrt{2}\right)^2}{2} = 1$$

in Figure 10.77a results.

(b) The equation $x^2 - 3xy + y^2 + x + y = 1$ reduces to $2Y^2 - \left(X - \sqrt{2}\right)^2 = 0$. This describes the two straight lines $Y = \pm\left(X - \sqrt{2}\right)/\sqrt{2}$ in Figure 10.77b.

FIGURE 10.77a

FIGURE 10.77b

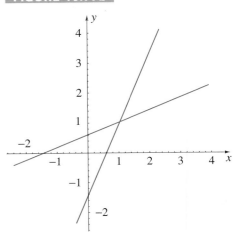

EXERCISES 10.7

In Exercises 1–18 eliminate the xy-term from the equation by a suitable rotation of axes. Identify, and plot or draw, whatever is described by the equation.

* **1.** $x^2 - 6xy + 9y^2 + x = 4$ * **2.** $x^2 - 6xy + 9y^2 = 4$

* **3.** $x^2 - 6xy + 9y^2 = 0$ * **4.** $x^2 - 6xy + 9y^2 = -3$

* **5.** $x^2 + 3xy + 5y^2 = 3$ * **6.** $x^2 + 3xy + 5y^2 + x = 4$

* **7.** $x^2 + 3xy + 5y^2 = 0$ * **8.** $x^2 + 3xy + 5y^2 = -10$

* **9.** $x^2 + 4xy + y^2 = 1$ * **10.** $x^2 + 4xy + y^2 = x - y - 1$

* **11.** $x^2 + 4xy + y^2 = x - y + 1/2$ * **12.** $x^2 + 4xy + y^2 = 0$

* **13.** $9x^2 + 24xy + 16y^2 + 90x - 130y = 0$

* **14.** $x^2 + 6xy + 9y^2 = 16$

* **15.** $2x^2 - 2\sqrt{2}xy + 3y^2 = 4$ * **16.** $x^2 - 10xy + y^2 + x + y + 1 = 0$

* **17.** $2x^2 - 4xy + 5y^2 + x - y = 4$ * **18.** $9x^2 + 24xy + 16y^2 + 80x - 60y = 0$

* **19.** Draw the curve $xy - 2y - 4x = 0$ by (a) solving for y in terms of x and (b) rotating axes to eliminate the xy-term.

* **20.** By factoring $x^2 - 6xy + 9y^2$, show that the equation in Exercise 14 describes two straight lines.

** **21.** Show that $2x^2 - 5xy + y^2 + y - 3x = 0$ describes two straight lines by (a) factoring and (b) rotating axes.

** **22.** Draw the curve $y^2 = 4 + \sqrt{3}xy$ by (a) solving for y in terms of x and (b) rotating axes to eliminate the xy-term.

SUMMARY

In Chapter 1 we defined curves explicitly and implicitly. In this chapter we added a third description—the parametric definition

$$x = x(t), \quad y = y(t), \quad \alpha \le t \le \beta.$$

When such a curve defines a function, the derivative of the function can be calculated using the parametric rule:

$$\frac{dy}{dx} = \frac{\frac{dy}{dt}}{\frac{dx}{dt}};$$

second and higher-order derivatives can be calculated using the chain rule. The length of such a curve is defined by the definite integral

$$\int_\alpha^\beta \sqrt{\left(\frac{dx}{dt}\right)^2 + \left(\frac{dy}{dt}\right)^2}\, dt.$$

Polar coordinates provide an alternative way to identify the positions of points in a plane. They use distance from a point, called the pole, and rotation of a half line, called the polar axis. Many curves that have complex equations in Cartesian coordinates can be represented very simply using polar coordinates. Particularly simple are multileaved roses, cardioids, lemniscates, and some circles. Polar coordinates also provide a unified approach to parabolas, ellipses, and hyperbolas. Each curve can be described as the path traced out by a point that moves so that the ratio ϵ of its distances from a fixed point and a fixed line remains constant. The curve is an ellipse, a parabola, or a hyperbola depending on whether $0 < \epsilon < 1$, $\epsilon = 1$, or $\epsilon > 1$ respectively. Ellipses and hyperbolas can also be defined as curves traced out by a point that moves so that the sum and difference of its distances from two fixed points remains constant.

When a curve $r = f(\theta)$ in polar coordinates encloses a region R, the definite integral

$$\int_\alpha^\beta \frac{1}{2}[f(\theta)]^2\, d\theta$$

with appropriate choices of α and β can be used to find the area of R.

Parametric equations for a polar curve $r = f(\theta)$, $\alpha \le \theta \le \beta$ are

$$x(\theta) = f(\theta)\cos\theta, \quad y(\theta) = f(\theta)\sin\theta, \quad \alpha \le \theta \le \beta.$$

Its slope is given by

$$\frac{dy}{dx} = \frac{f'(\theta)\sin\theta + f(\theta)\cos\theta}{f'(\theta)\cos\theta - f(\theta)\sin\theta},$$

and its length can be calculated with the definite integral

$$\int_\alpha^\beta \sqrt{r^2 + \left(\frac{dr}{d\theta}\right)^2}\, d\theta.$$

Except for degenerate cases, the general second-degree polynomial equation in x and y,

$$Ax^2 + Bxy + Cy^2 + Dx + Ey + F = 0$$

represents an ellipse, parabola, or hyperbola. The presence of the xy-term (Bxy) indicates that the conic is rotated so that axes of symmetry are not parallel to the coordinate axes. The curve is an ellipse, parabola, or hyperbola as the quantity $B^2 - 4AC$ is negative, zero, or positive.

KEY TERMS

In reviewing this chapter, you should be able to define or discuss the following key terms:

Parametric equations   Focus
Parametric rule   Directrix
Length of parametric curve   Ellipse
Pole   Major axis
Polar axis   Minor axis
Polar coordinates   Hyperbola
Slope of polar curve   Transverse axis
Length of polar curve   Conjugate axis
Area in polar coordinates   Eccentricity
Parabola   Rotation of axes

REVIEW EXERCISES

In Exercises 1–30 draw, and then plot, the curve.

* **1.** $r = \cos\theta$

* **2.** $r = -\sin\theta$

* **3.** $r = \dfrac{3}{1 + 2\sin\theta}$

* **4.** $r = \dfrac{3}{2 + \sin\theta}$

* **5.** $r = \dfrac{1}{1 + \sin\theta}$

* **6.** $r = \dfrac{3}{2 - 2\cos\theta}$

* **7.** $r + 1 = 2\sin\theta$

* **8.** $r^2 = 4\cos\theta$

* **9.** $x = 2\cos t, \ \ y = 3\sin t, \ \ 0 \le t < 2\pi$

* **10.** $x = 4 + t, \ \ y = 5 - 3t^2$

* **11.** $x = \sin^2 t, \ \ y = \cos^2 t, \ \ 0 \le t \le 2\pi$

* **12.** $x = 1 + 4\sin t, \ \ y = -2 + 4\cos t, \ \ 0 \le t \le \pi$

* **13.** $(x^2 + y^2)^3 = x$

* **14.** $x^2 + y^2 = 2\sqrt{x^2 + y^2}$

* **15.** $r^2 = 4\cos^2\theta$

* **16.** $r = 3\cos 2\theta$

* **17.** $x + y = \sqrt{x^2 + y^2}$

* **18.** $x^2 + y^2 = x + y$

* **19.** $r = 4\cos 3\theta$

* **20.** $r = \sin^2\theta - \cos^2\theta$

* **21.** $x^2 + y^2 - 3x + 2y = 1$

* **22.** $x^2 - 3y^2 + 4 = 0$

* **23.** $2x^2 + 3y^2 - 6y = 0$

* **24.** $y^2 + x + 2y = 3$

* **25.** $r(2\cos\theta - \sin\theta) = 3$

* **26.** $r = \cos(\theta/2)$

* **27.** $r = \sin^2\theta$

* **28.** $x^2 = y^2 + y$

* **29.** $x = e^{-t}, \ \ y = \ln t, \ \ t > 0$

* **30.** $x = \sin\theta, \ \ y = \sin 2\theta$

In Exercises 31–35 find the indicated area.

$*$ **31.** Inside $r = 2 + 2\cos\theta$

$*$ **32.** Inside $r = 4$ but outside $r = 4\sin 2\theta$

$*$ **33.** Common to $r = 2$ and $r^2 = 9\cos 2\theta$

$*$ **34.** Inside $r = \sin^2\theta$

$*$ **35.** Common to $r = 1 + \sin\theta$ and $r = 2 - 2\sin\theta$

In Exercises 36 and 37 assume that y is defined as a function of x, and find dy/dx and d^2y/dx^2.

$**$ **36.** $x = t^3 + 2t, \quad y = 3t - t^3$ $**$ **37.** $x = 2\sin u, \quad y = 3\cos u$

In Exercises 38–40 find the length of the curve.

$**$ **38.** $r = 2 + 2\cos\theta$ $**$ **39.** $x = t^2, \quad y = t^3, \quad 0 \le t \le 1$

$**$ **40.** $x = e^t\cos t, \quad y = e^t\sin t, \quad 0 \le t \le \pi/2$

$**$ **41.** Find the equation of the tangent line to the curve $r = 2 - 2\sin\theta$ at the point with polar coordinates $(1, \pi/6)$.

In Exercises 42–45 determine whether the curve is an ellipse, parabola, hyperbola, or none of these. Find X- and Y-axes so that the equation of the the curve in X and Y does not have a product term XY.

$*$ **42.** $xy + x - 2y + 3 = 0$ $*$ **43.** $x^2 - \sqrt{3}xy + 2y^2 = 1$

$*$ **44.** $9x^2 + 6xy + y^2 + y = 2$ $*$ **45.** $4x^2 + 3xy + 4y^2 + 1 = 0$

Infinite Sequences and Series

Sequences and series play an important role in many areas of applied mathematics. *Sequences of numbers* were first encountered in Section 4.1, although we did not use the term *sequences* at the time. Newton's iterative procedure produces a set of numbers x_1, x_2, x_3,... each of which approximates a root of an equation $f(x) = 0$. The first number is chosen as some initial approximation to the solution of the equation, and subsequent numbers defined by the formula

$$x_{n+1} = x_n - \frac{f(x_n)}{f'(x_n)}$$

are better and better approximations. This ordered set of numbers is called a **sequence**. Each number in the set corresponds to a positive integer, and each is calculated according to a stated formula.

A series of numbers is the sum of the numbers in a sequence. If the numbers are x_1, x_2, x_3,..., the corresponding **series** is denoted symbolically by

$$\sum_{n=1}^{\infty} x_n = x_1 + x_2 + x_3 + \cdots,$$

where the three dots indicate that the addition is never-ending. It is very well to write an expression like this, but it does not have meaning. No matter how fast we add, or how fast a calculator adds, or even how fast a supercomputer adds, an infinity of numbers can never be added together in a finite amount of time. We shall give meaning to such expressions, and show that they are really the only sensible way to define many of the more common transcendental functions such as trigonometric, exponential, and hyperbolic.

The first two sections of this chapter and Section 11.8 are devoted to sequences and the remaining ten to series. This is not to say that series are more important than sequences; they are not. Discussions on series invariably become discussions on sequences associated with series. We have found that difficulties with this chapter can usually be traced back to a failure to distinguish between the two concepts. Special attention to the material in Sections 11.1, 11.2, and 11.8 will be rewarded; a cursory treatment leads to confusion in other sections.

11.1 Infinite Sequences of Numbers

Sequences of numbers are defined as follows.

> **DEFINITION 11.1**
>
> An **infinite sequence of numbers** is a function f whose domain is the set of positive integers.

For example, when $f(n) = 1/n$, the following numbers are associated with the positive integers,

$$1, \quad \frac{1}{2}, \quad \frac{1}{3}, \quad \frac{1}{4}, \quad \ldots .$$

The word *infinite* simply indicates that an infinity of numbers is defined by the sequence, as there is an infinity of positive integers, but it indicates nothing about the nature of the numbers. Often we write the numbers $f(n)$ in a line separated by commas,

$$f(1), \ f(2), \ \ldots, \ f(n), \ldots \tag{11.1a}$$

and refer to this array as the sequence rather than the rule by which it is formed. Since this notation is somewhat cumbersome, we adopt a notation similar to that used for the sequence defined by Newton's iterative procedure. We set $c_1 = f(1)$, $c_2 = f(2)$, $\ldots$, $c_n = f(n)$, $\ldots$, and write for 11.1a

$$c_1, c_2, c_3, \ \ldots, \ c_n, \ \ldots . \tag{11.1b}$$

The first number c_1 is called the first **term** of the sequence, c_2 the second term, and for general n, c_n is called the nth term (or general term) of the sequence. For the above example, we have

$$c_1 = 1, \quad c_2 = \frac{1}{2}, \quad c_3 = \frac{1}{3}, \quad \text{etc.}$$

In some applications, it is more convenient to define a sequence as a function whose domain is the set of integers larger than or equal to some fixed integer N, and N can be positive, negative, or zero. Later in this chapter, we find it convenient to initiate the assignment with $N = 0$. For now we prefer to use Definition 11.1 where $N = 1$, in which case we have the natural situation where the first term of the sequence corresponds to $n = 1$, the second term to $n = 2$, etc.

EXAMPLE 11.1

The general terms of four sequences are:

$$\text{(a)} \ \frac{1}{2^{n-1}} \qquad \text{(b)} \ \frac{n}{n+1} \qquad \text{(c)} \ (-1)^n |n - 3| \qquad \text{(d)} \ (-1)^{n+1}$$

Write out the first six terms of each sequence.

SOLUTION   The first six terms of these sequences are:

$$\text{(a)} \quad 1, \ \frac{1}{2}, \ \frac{1}{4}, \ \frac{1}{8}, \ \frac{1}{16}, \ \frac{1}{32}. \qquad\qquad \text{(b)} \quad \frac{1}{2}, \ \frac{2}{3}, \ \frac{3}{4}, \ \frac{4}{5}, \ \frac{5}{6}, \ \frac{6}{7}.$$

$$\text{(c)} \quad -2, \ 1, \ 0, \ 1, \ -2, \ 3. \qquad\qquad \text{(d)} \quad 1, \ -1, \ 1, \ -1, \ 1, \ -1.$$

The sequences in Example 11.1 are said to be defined **explicitly**; we have an explicit formula for the nth term of the sequence in terms of n. This allows easy determination of any term in the sequence. For instance, to find the one-hundredth term, we simply replace n by 100 and perform the resulting arithmetic. Contrast this with the sequence in the following example.

EXAMPLE 11.2

The first term of a sequence is $c_1 = 1$ and every other term is to be obtained from the formula

$$c_{n+1} = 5 + \sqrt{2 + c_n}, \quad n \geq 1.$$

Calculate c_2, c_3, c_4, and c_5.

SOLUTION   To obtain c_2 we set $n = 1$ in the formula:

$$c_{1+1} = c_2 = 5 + \sqrt{2 + c_1} = 5 + \sqrt{2 + 1} = 5 + \sqrt{3} = 6.732.$$

To find c_3, we set $n = 2$:

$$c_3 = 5 + \sqrt{2 + c_2} = 5 + \sqrt{2 + \left(5 + \sqrt{3}\right)} = 5 + \sqrt{7 + \sqrt{3}} = 7.955.$$

Similarly,

$$c_4 = 5 + \sqrt{2 + c_3} = 5 + \sqrt{7 + \sqrt{7 + \sqrt{3}}} = 8.155, \qquad \text{and}$$

$$c_5 = 5 + \sqrt{2 + c_4} = 5 + \sqrt{7 + \sqrt{7 + \sqrt{7 + \sqrt{3}}}} = 8.187.$$

When the terms of a sequence are defined by a formula such as the one in Example 11.2, the sequence is said to be defined **recursively**. The terms for a sequence obtained from Newton's iterative procedure are so defined. To find the 100th term of a recursively-defined sequence, we must know the 99th; to find the 99th, we must know the 98th; to find the 98th, we need the 97th; and so on down the line. In other words, to find a term in the sequence, we must first find every term that precedes it. Obviously, it is much more convenient to have an explicit definition for c_n in terms of n, but this is not always possible. It can be very difficult to find an explicit formula for the nth term of a sequence which is defined recursively.

Sometimes it is impossible to give an algebraic formula for the terms of a sequence. This is illustrated in the following example.

EXAMPLE 11.3

The nth term of a sequence is the nth prime integer (greater than 1) when all such primes are listed in ascending order. List its first ten terms.

SOLUTION   The first ten terms of the sequence are

$$2, \ 3, \ 5, \ 7, \ 11, \ 13, \ 17, \ 19, \ 23, \ 29.$$

No one has yet discovered a formula for the prime integers; therefore, it is presently impossible to express terms of the sequence explicitly or recursively. In spite of this, the sequence is well-defined; and with enough perseverence, and perhaps a high-powered computer, we could find any number of terms in this sequence.

When the general term of a sequence is known explicitly, any term in the sequence is obtained by substituting the appropriate value of n. In other words, the general term specifies every term in the sequence. We therefore use the general term to abbreviate the notation for a sequence by writing the general term in braces. Specifically, for the sequence in Example 11.1a, we write

$$\left\{ \frac{1}{2^{n-1}} \right\}_1^\infty = 1, \ \frac{1}{2}, \ \frac{1}{4}, \ \frac{1}{8}, \ \ldots, \ \frac{1}{2^{n-1}}, \ \ldots,$$

where 1 and ∞ indicate that the first term corresponds to the integer $n = 1$, and that there is an infinite number of terms in the sequence. In general, we write

$$\{c_n\}_1^\infty = c_1, \ c_2, \ c_3, \ \ldots, \ c_n, \ \ldots. \tag{11.2}$$

If, as is the case in this section, the first term of a sequence corresponds to the integer $n = 1$, we abbreviate the notation further and simply write $\{c_n\}$ in place of $\{c_n\}_1^\infty$.

Since a sequence $\{c_n\}$ is a function whose domain is the set of positive integers, we can represent $\{c_n\}$ graphically. The sequences of Example 11.1 are shown in Figure 11.1.

FIGURE 11.1a

FIGURE 11.1b

FIGURE 11.1c

FIGURE 11.1d

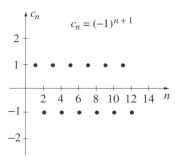

In most applications of sequences we are interested in a number called the **limit** of the sequence. Intuitively, a number L is called the limit of a sequence $\{c_n\}$ if, as we go farther and farther out in the sequence, the terms get arbitrarily close to L, and stay close to L. If such a number L exists, we write

$$L = \lim_{n \to \infty} c_n, \tag{11.3}$$

and say that the sequence $\{c_n\}$ **converges** to L. If no such number exists, we say that the sequence does not have a limit, or that the sequence **diverges**.

For the sequences of Example 11.1, it is evident that:

(a) $\lim\limits_{n\to\infty} \dfrac{1}{2^{n-1}} = 0;$

(b) $\lim\limits_{n\to\infty} \dfrac{n}{n+1} = 1;$

(c) $\lim\limits_{n\to\infty} (-1)^n |n-3|$ does not exist;

(d) $\lim\limits_{n\to\infty} (-1)^{n+1}$ does not exist;

and in Example 11.3,

$$\lim\limits_{n\to\infty} c_n \text{ does not exist,}$$

(see Exercise 31). Note how the points on the graphs in Figures 11.1a and b cluster around the limits 0 and 1 as n gets larger and larger. No such clustering occurs in the remaining two figures.

It is usually, but not always, easy to determine whether an explicitly defined sequence has a limit, and what that limit is. It is like finding the limit of a function $f(x)$ as $x \to \infty$. For example, if we divide numerator and denominator of the sequence $\{(n^2 + n - 3)/(2n^2 + n + 4)\}$ by n^2,

$$\lim\limits_{n\to\infty} \frac{n^2 + n - 3}{2n^2 + n + 4} = \lim\limits_{n\to\infty} \frac{1 + \dfrac{1}{n} - \dfrac{3}{n^2}}{2 + \dfrac{1}{n} + \dfrac{4}{n^2}} = \frac{1}{2}.$$

Seldom is it obvious algebraically whether a recursively-defined sequence has a limit. For instance, it is not at all clear whether the recursive sequence of Example 11.2 has a limit. In spite of the fact that differences between successive terms are approaching zero, and terms of the sequence are therefore getting closer together, the sequence might not have a limit. For instance, the sequence $\{\sqrt{n}\}$ does not have a limit, yet it is easy to show that differences between successive terms get smaller and smaller as n increases. (See Exercise 30 for another example.) There are ways to verify that a sequence has a limit, and some of these will be discussed in Section 11.8.

When a sequence arises in applications, it may be perfectly clear that the sequence is convergent. Such is often the case in the field of numerical analysis, where recursive sequences commonly arise in the form of iterative procedures. We have already encountered Newton's iterative procedure as one example. When this method is applied to the equation

$$f(x) = x^3 - 3x + 1 = 0$$

with initial approximation $x_1 = 0.7$ to find the root between 0 and 1, the sequence obtained is

$$x_1 = 0.7, \qquad x_{n+1} = x_n - \frac{f(x_n)}{f'(x_n)} = x_n - \frac{x_n^3 - 3x_n + 1}{3x_n^2 - 3}, \qquad n \geq 1.$$

The first three terms of this sequence are illustrated in Figure 11.2, and the tangent line construction by which they are obtained makes it clear that the sequence must converge to the solution of the equation between 0 and 1. In other words, it is not necessary for us to verify convergence of the sequence algebraically; it is obvious geometrically.

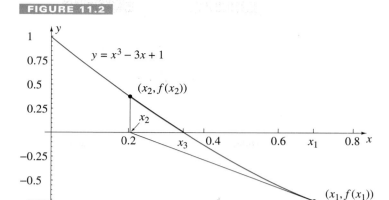

FIGURE 11.2

To find the solution of the equation we evaluate terms of the sequence algebraically until they repeat:

$$x_1 = 0.7, \qquad x_2 = 0.205, \qquad x_3 = 0.342,$$
$$x_4 = 0.347\,285, \qquad x_5 = 0.347\,296\,355, \qquad x_6 = 0.347\,296\,355.$$

Obviously, $x_6 = 0.347\,296\,355$ must be close to the root of $x^3 - 3x + 1 = 0$ between 0 and 1. How close can be verified with the Intermediate Value Theorem from Section 1.7. For instance, to verify that $0.347\,296$ is an approximation to the solution correctly rounded to six decimal places, we calculate

$$f(0.347\,295\,5) = 2.3 \times 10^{-6} \qquad \text{and} \qquad f(0.347\,296\,5) = -3.8 \times 10^{-7}.$$

The fact that function values are opposite in sign verifies that $0.347\,296$ is correct to six decimal places.

Because Figure 11.2 illustrated that the sequence defined by Newton's iterative procedure must converge to a root of $f(x) = x^3 - 3x + 1 = 0$, it was unnecessary to verify existence of a limit of the sequence algebraically. Are we proposing that a graph of the function $f(x)$ should always be drawn when using Newton's method to solve equations? Not really, although we are of the philosophy that pictures should be drawn whenever possible. Numerical analysts have proved that under very mild restrictions, the sequence defined by Newton's method always converges to a root of the equation provided the initial approximation is sufficiently close to that root. See Exercise 63 for further discussion of this point. In practice, we approximate the solution of an equation $f(x) = 0$ using Newton's method exactly as we did in Section 4.1, and as illustrated above. We set up the appropriate sequence, choose an initial approximation, and iterate to find further approximations. We do not verify, algebraically or geometrically, that the sequence has a limit. If terms get closer and closer together, we suspect that the sequence is convergent, that is, that terms are getting closer and closer to the limit. We do not know this for sure, nor do we try to verify it. The real problem is to solve the equation $f(x) = 0$. We have what we believe is an approximation to the solution, and we verify this directly with the Intermediate Value Theorem.

Another way to solve for the same root of $x^3 - 3x + 1 = 0$ is to rewrite the equation in the form

$$x = \frac{1}{3}(x^3 + 1),$$

and use this to define the following recursive sequence

$$x_1 = \frac{1}{3}, \qquad x_{n+1} = \frac{1}{3}(x_n^3 + 1), \qquad n \geq 1.$$

The initial term was chosen somewhat arbitrarily. In practice, it should be as close to the root as possible. The first eight terms of the sequence are:

$x_1 = 1/3,$	$x_5 = 0.347\,293\,53,$
$x_2 = 0.345\,679,$	$x_6 = 0.347\,296\,01,$
$x_3 = 0.347\,102\,19,$	$x_7 = 0.347\,296\,31,$
$x_4 = 0.347\,272\,95,$	$x_8 = 0.347\,296\,35.$

The fact that x_7 and x_8 agree to six decimal places leads us to believe that the sequence converges and that its limit is approximately equal to $0.347\,296$. We cannot be positive of this, nor do we attempt to verify it. What we want is an approximation to the root of $x^3 - 3x + 1 = 0$ between 0 and 1. We can verify directly that $x = 0.347\,296$ is an approximation accurate to six decimal places, as above.

This method of finding the root of an equation is often called the **method of successive approximations**, or **fixed point iteration**. What we do is rearrange the equation $f(x) = 0$ into the form $x = g(x)$, and define a recursive sequence

$$x_1 = A, \qquad x_{n+1} = g(x_n), \qquad n \geq 1,$$

where A is some initial approximation to the root (the closer the better). We iterate, hoping that terms of the sequence get closer together. If they do, we terminate iterations when we suspect that we have an approximation to the solution of the equation with the required accuracy, and verify this directly with the Intermediate Value Theorem. If terms do not seem to converge, we may try a different rearrangement of the equation into the form $x = g(x)$, or another method.

We can visualize the method of successive approximations geometrically. In terms of the above example, we are finding where the graph of the function $f(x) = x^3 - 3x + 1$ crosses the x-axis (Figure 11.3). There are three points, and we have been concentrating on the intercept between $x = 0$ and $x = 1$. When we rewrite the equation in the form $x = (x^3 + 1)/3$, an alternative interpretation is possible. We are searching for x-coordinates of points of intersection of the curves $y = x$ and $y = (x^3 + 1)/3$ (Figure 11.4).

FIGURE 11.3

FIGURE 11.4

FIGURE 11.5

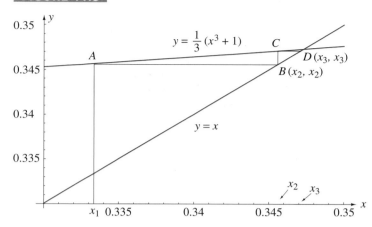

To show how the method of successive approximations works, we have expanded that part of Figure 11.4 between $x = 0.33$ and $x = 0.35$ (Figure 11.5).

To find x_2 we substitute $x_1 = 1/3$ into $(x^3 + 1)/3$. It is the y-coordinate of the point A on the curve $y = (x^3 + 1)/3$. If we proceed horizontally from A to line $y = x$, then point B has coordinates (x_2, x_2). To find x_3 algebraically, we evaluate $(x^3 + 1)/3$ at x_2. Geometrically, x_3 is the y-coordinate of point C. If we move horizontally to the line $y = x$, coordinates of D are (x_3, x_3). Continuation leads to the sequence converging to the required root.

We illustrate the method again in the following example.

EXAMPLE 11.4

Use successive approximations to approximate the solution of $f(x) = x^3 + 25x - 50 = 0$ between $x = 1$ and $x = 2$ with error less than 10^{-6}. Illustrate the sequence graphically.

SOLUTION   The equation can be written in the form $x = (50 - x^3)/25$ leading to the sequence

$$x_1 = 1, \qquad x_{n+1} = \frac{50 - x_n^3}{25}, \qquad n \geq 1.$$

The next fifteen terms are

$$\begin{aligned}
&x_2 = 1.96 &&x_3 = 1.70 &&x_4 = 1.80 \\
&x_5 = 1.77 &&x_6 = 1.778\,2 &&x_7 = 1.775\,1 \\
&x_8 = 1.776\,3 &&x_9 = 1.775\,81 &&x_{10} = 1.776\,00 \\
&x_{11} = 1.775\,93 &&x_{12} = 1.775\,954 &&x_{13} = 1.775\,945 \\
&x_{14} = 1.775\,948\,2 &&x_{15} = 1.775\,946\,9 &&x_{16} = 1.775\,947\,4.
\end{aligned}$$

The terms certainly appear to be getting closer together (agreeing to six decimals at this stage), therefore suggesting that the sequence has a limit. We do not verify this. We have what we feel is an approximation to the required solution of $x^3 + 25x - 50 = 0$ with sufficient accuracy.

FIGURE 11.6

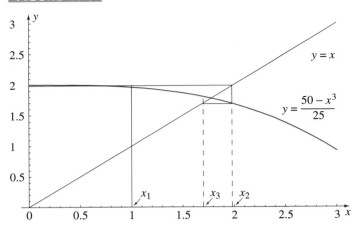

To verify that 1.775 947 is a solution with error less than 10^{-6}, we use the Intermediate Value Theorem with the calculations

$$f(1.775\,946) = -4.4 \times 10^{-5}, \qquad f(1.775\,948) = 2.5 \times 10^{5}.$$

The terms of the sequence are shown in Figure 11.6.

EXERCISES 11.1

In Exercises 1–20 determine whether the sequence is convergent or divergent. Find limits for convergent sequences.

* **1.** $\left\{\dfrac{1}{n}\right\}$ * **2.** $\{3^{n} + 1\}$

* **3.** $\{3\}$ * **4.** $\left\{\left(\dfrac{3}{4}\right)^{n+1}\right\}$

* **5.** $\left\{\left(\dfrac{4}{3}\right)^{n+1}\right\}$ * **6.** $\left\{\left(-\dfrac{15}{16}\right)^{n+5}\right\}$

* **7.** $\left\{\sin\left(\dfrac{n\pi}{2}\right)\right\}$ * **8.** $\left\{\dfrac{n}{n^{2} + n + 2}\right\}$

* **9.** $\left\{\dfrac{(-1)^{n}}{n}\right\}$ * **10.** $\{\operatorname{Tan}^{-1} n\}$

* **11.** $\left\{\dfrac{\ln n}{n^{2} + 1}\right\}$ * **12.** $\left\{(-1)^{n}\sqrt{n^{2} + 1}\right\}$

$* \ 13. \ c_1 = 2, \quad c_{n+1} = \dfrac{2}{c_n - 1}, \quad n \geq 1$ $* \ 14. \ c_1 = 4, \quad c_{n+1} = -\dfrac{c_n}{n^2}, \quad n \geq 1$

$* \ 15. \ \left\{ \dfrac{n+1}{2n+3} \right\}$ $* \ 16. \ \left\{ \dfrac{2n+3}{n^2 - 5} \right\}$

$* \ 17. \ \left\{ \dfrac{n^2 + 5n - 4}{n^2 + 2n - 2} \right\}$ $* \ 18. \ \left\{ ne^{-n} \right\}$

$* \ 19. \ \left\{ \dfrac{\ln n}{n} \right\}$ $* \ 20. \ \left\{ \dfrac{n}{n+1} \operatorname{Tan}^{-1} n \right\}$

In Exercises 21–25 find an explicit formula for the general term of the sequence. In each case assume that the remaining terms follow the pattern suggested by the given terms.

$* \ 21. \ \dfrac{1}{2}, \ \dfrac{3}{4}, \ \dfrac{7}{8}, \ \dfrac{15}{16}, \ \dfrac{31}{32}, \ldots$ $* \ 22. \ 4, \ \dfrac{7}{4}, \ \dfrac{10}{9}, \ \dfrac{13}{16}, \ \dfrac{16}{25}, \ \dfrac{19}{36}, \ldots$

$* \ 23. \ \dfrac{\ln 2}{\sqrt{2}}, \ -\dfrac{\ln 3}{\sqrt{3}}, \ \dfrac{\ln 4}{\sqrt{4}}, \ \dfrac{-\ln 5}{\sqrt{5}}, \ldots$ $** \ 24. \ 1, \ 0, \ 1, \ 0, \ 1, \ 0, \ \ldots$

$** \ 25. \ 1, \ 1, \ -1, \ -1, \ 1, \ 1, \ -1, \ -1, \ \ldots$

In Exercises 26–29 show how L'Hôpital's rule can be used to evaluate the limit for the sequence.

$* \ 26. \ \left\{ \dfrac{\ln n}{\sqrt{n}} \right\}$ $* \ 27. \ \left\{ \dfrac{n^3 + 1}{e^n} \right\}$

$** \ 28. \ \left\{ n \sin\left(\dfrac{4}{n} \right) \right\}$ $** \ 29. \ \left\{ \left(\dfrac{n+5}{n+3} \right)^n \right\}$

$** \ 30.$ If a sequence converges, then differences between successive terms in the sequence must approach zero. The converse is not always true. The following example is an illustration: Show that differences between successive terms of the sequence $\{\ln n\}$ approach zero, but the sequence itself diverges.

$** \ 31.$ Justify the statement that the sequence of Example 11.3 does not have a limit.

In Exercises 32–43 use Newton's iterative procedure with the given initial approximation x_1 to define a sequence of approximations to a solution of the equation. Determine graphically whether the sequence has a limit. Approximate any limit that exists to seven decimals.

$* \ 32. \ x_1 = 1, \quad x^2 + 3x + 1 = 0$ $* \ 33. \ x_1 = -1, \quad x^2 + 3x + 1 = 0$

$* \ 34. \ x_1 = -1.5, \quad x^2 + 3x + 1 = 0$ $* \ 35. \ x_1 = -3, \quad x^2 + 3x + 1 = 0$

$* \ 36. \ x_1 = 4, \quad x^3 - x^2 + x - 22 = 0$ $* \ 37. \ x_1 = 2, \quad x^3 - x^2 + x - 22 = 0$

$* \ 38. \ x_1 = 2, \quad x^5 - 3x + 1 = 0$ $* \ 39. \ x_1 = 1, \quad x^5 - 3x + 1 = 0$

$* \ 40. \ x_1 = 0, \quad x^5 - 3x + 1 = 0$ $* \ 41. \ x_1 = 4/5, \quad x^5 - 3x + 1 = 0$

$* \ 42. \ x_1 = 0.85, \quad x^5 - 3x + 1 = 0$ $* \ 43. \ x_1 = -2, \quad x^5 - 3x + 1 = 0$

In Exercises 44–48 illustrate that the method of successive approximations with the suggested rearrangement of the equation $f(x) = 0$, along with the initial approximation x_1, leads to a sequence which converges to a root of the equation. Find the root accurate to 4 decimals.

* **44.** $f(x) = x^2 - 2x - 1$; $x_1 = 2$;   use $x = 2 + 1/x$

* **45.** $f(x) = x^3 + 6x + 3$; $x_1 = -1$;   use $x = -(1/6)(x^3 + 3)$

* **46.** $f(x) = x^4 - 120x + 20$; $x_1 = 0$;   use $x = (x^4 + 20)/120$

* **47.** $f(x) = x^3 - 2x^2 - 3x + 1$; $x_1 = 3$;   use $x = (2x^2 + 3x - 1)/x^2$

* **48.** $f(x) = 8x^3 - x^2 - 1$; $x_1 = 0$;   use $x = (1/2)(1 + x^2)^{1/3}$

In Exercises 49–54 find a rearrangement of the equation which leads, through the method of successive approximations, to a 4-decimal approximation to the root of the equation.

** **49.** $x^3 - 6x^2 + 11x - 7 = 0$ between $x = 3$ and $x = 4$

** **50.** $x^4 - 3x^2 - 3x + 1 = 0$ between $x = 0$ and $x = 1$

** **51.** $x^4 + 4x^3 - 50x^2 + 100x - 50 = 0$ between $x = 0$ and $x = 1$

** **52.** $\sin^2 x = 1 - x^2$ between $x = 0$ and $x = 1$

** **53.** $\sec x = 2/(1 + x^4)$ between $x = 0$ and $x = 1$

** **54.** $e^x + e^{-x} - 10x = 0$ between $x = 0$ and $x = 1$

** **55.** (a) Use Newton's iterative procedure with $x_1 = 1$ to approximate the root, between 0 and 1, of $x^4 - 15x + 2 = 0$.

  (b) Use the method of successive approximations with $x_1 = 1$ and $x_{n+1} = (x_n^4 + 2)/15$ to approximate the root in (a).

  (c) Use Newton's method to approximate the root between 2 and 3.

  (d) What happens if the sequence in (b) is used to approximate the root between 2 and 3 with $x_1 = 2$ and $x_1 = 3$?

** **56.** A superball is dropped from the top of a building 20 m high. Each time it strikes the ground, it rebounds to 99% of the height from which it fell.

  (a) If d_n denotes the distance travelled by the ball between the nth and $(n+1)^{\text{th}}$ bounces, find a formula for d_n.

  (b) If t_n denotes the time between the nth and $(n + 1)^{\text{th}}$ bounces, find a formula for t_n.

** **57.** A dog sits at a farmhouse patiently watching for his master to return from the fields. When the farmer is 1 km from the farmhouse, the dog immediately takes off for the farmer. When he reaches the farmer, he turns and runs back to the farmhouse, whereupon he again turns and runs to the farmer. The dog continues this frantic action until the farmer reaches the farmhouse. If the dog runs twice as fast as the farmer, find the distance d_n run by the dog from the point when he reaches the farmer for the nth time to the point when he reaches the farmer for the $(n + 1)^{\text{th}}$ time. Ignore any accelerations of the dog in the turns.

** **58.** The equilateral triangle in the left figure at the top of page 729 has perimeter P. If each side of the triangle is divided into three equal parts, an equilateral triangle is drawn on the middle segment of each side, and the figure transformed into the middle figure, what is the perimeter P_1 of this figure? If each side of this figure is now subdivided into three equal portions and equilateral triangles are similarly constructed to result in the right figure, what is the perimeter P_2 of this figure? If this subdivision process is continued indefinitely, what is the perimeter P_n after the nth subdivision? What is $\lim_{n \to \infty} P_n$?

** **59.** A stone of mass 100 gm is thrown vertically upward with speed 20 m/s. Air exerts a resistive force on the stone proportional to its speed, and has magnitude $1/10$ N when the speed of the stone is 10 m/s. It can be shown that the height y (in metres) above the projection point attained by the stone is given by

$$y = -98.1t + 1181\left(1 - e^{-t/10}\right),$$

where t is time (measured in seconds with $t = 0$ at the instant of projection).

(a) The time taken for the stone to return to its projection point can be obtained by setting $y = 0$ and solving the equation for t. Do so (correct to 2 decimals).

(b) Find the time for the stone to return if air resistance is ignored.

** **60.** When the beam in the figure below vibrates vertically, there are certain frequencies of vibration, called *natural frequencies of the system*. They are solutions of the equation

$$\tan x = \frac{e^x - e^{-x}}{e^x + e^{-x}}$$

divided by 20π. Find the two smallest natural frequencies.

** **61.** If A_n is the area of the figure with perimeter P_n in Exercise 58, find a formula for A_n.

** **62.** (a) Show that, if α is the only root of the equation $x = g(x)$ and the left figure below is a graph of $g(x)$, then the right figure exhibits geometrically the sequence of approximations of α determined by the method of successive approximations.

FIGURE 11.9

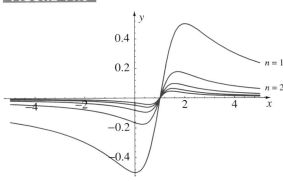

The sequence in the next example is indispensable to future discussions.

EXAMPLE 11.6

Show that the limit of the sequence of functions $\{|x|^n/n!\}$ is zero for all x; that is, verify that

$$\lim_{n \to \infty} \frac{|x|^n}{n!} = 0, \quad -\infty < x < \infty.$$

SOLUTION   Suppose that x is any fixed value, and let $m = [\![|x|]\!]$, where $[\![x]\!]$ is the greatest integer function of Exercise 50 in Section 1.3. It is the largest integer that does not exceed x. (If $x = 3.4$, then $[\![|3.4|]\!] = 3$; if $x = -22.6$, then $[\![|-22.6|]\!] = [\![22.6]\!] = 22$; and if $x = -4$, then $[\![|-4|]\!] = [\![4]\!] = 4$.) Since $|x|^n$ is $|x|$ multiplied by itself n times, and $n!$ has n multiplications, we may write that

$$\frac{|x|^n}{n!} = \left(\frac{|x|}{1}\right)\left(\frac{|x|}{2}\right)\left(\frac{|x|}{3}\right) \cdots \left(\frac{|x|}{n}\right).$$

If n is chosen greater than $m = [\![|x|]\!]$, then

$$\frac{|x|^n}{n!} = \left(\frac{|x|}{1}\right)\left(\frac{|x|}{2}\right)\left(\frac{|x|}{3}\right) \cdots \left(\frac{|x|}{m}\right)\left(\frac{|x|}{m+1}\right) \cdots \left(\frac{|x|}{n}\right).$$

Suppose we let

$$M = \left(\frac{|x|}{1}\right)\left(\frac{|x|}{2}\right)\left(\frac{|x|}{3}\right) \cdots \left(\frac{|x|}{m}\right),$$

a fixed constant. Then,

$$\frac{|x|^n}{n!} = M\left(\frac{|x|}{m+1}\right)\left(\frac{|x|}{m+2}\right) \cdots \left(\frac{|x|}{n}\right)$$

$$< M\left(\frac{|x|}{m+1}\right)\left(\frac{|x|}{m+1}\right) \cdots \left(\frac{|x|}{m+1}\right),$$

where we have replaced $m+2, m+3, \ldots, n$ in the denominators by the smaller integer $m+1$. Since there are $n - m$ bracketed terms on the right, all equal to each other, we have

$$\frac{|x|^n}{n!} < M \left(\frac{|x|}{m+1} \right)^{n-m} = M \left(\frac{|x|}{m+1} \right)^{-m} \left(\frac{|x|}{m+1} \right)^{n}.$$

Now, $M/[|x|/(m+1)]^{-m}$ is a fixed number as far as the limit on n is concerned. Furthermore, $|x|/(m+1)$ is also constant and is less than 1. Consequently,

$$\lim_{n \to \infty} \frac{|x|^n}{n!} \leq \lim_{n \to \infty} M \left(\frac{|x|}{m+1} \right)^{-m} \left(\frac{|x|}{m+1} \right)^{n}$$

$$= M \left(\frac{|x|}{m+1} \right)^{-m} \lim_{n \to \infty} \left(\frac{|x|}{m+1} \right)^{n}$$

$$= M \left(\frac{|x|}{m+1} \right)^{-m} (0)$$

$$= 0.$$

Since $|x|^n/n! \geq 0$, it follows that $\lim_{n \to \infty} |x|^n/n! = 0$.

◀◀◀◀

EXERCISES 11.2

In Exercises 1–14, $f_n(x)$ is the nth term in a sequence of functions $\{f_n(x)\}$. Plot graphs of the first five functions in the sequence. Determine whether the sequence has a limit.

* **1.** $f_n(x) = \dfrac{nx}{1 + n^2 x^2}, 0 \leq x \leq 1$

* **2.** $f_n(x) = \dfrac{n^2 x}{1 + n^3 x^2}, 0 \leq x \leq 1$

* **3.** $f_n(x) = \dfrac{nx^2}{1 + nx}, 0 \leq x \leq 1$

* **4.** $f_n(x) = \dfrac{1}{x} + \dfrac{1}{n} \sin \left(\dfrac{1}{nx} \right), 1 \leq x \leq 2$

* **5.** $f_n(x) = nx^n(1 - x), 0 \leq x \leq 1$

* **6.** $f_n(x) = nx^n(1 - x), 0 \leq x \leq 2$

* **7.** $f_n(x) = n^2 x^n (2 - x), 0 \leq x \leq 2$

* **8.** $f_n(x) = n^2 x^n (1 - x^2), -1 \leq x \leq 1$

* **9.** $f_n(x) = \dfrac{2 + nx^2}{1 + nx}, 0 \leq x \leq 2$

** **10.** $f_n(x) = (\sin x)^{1/n}, 0 < x < \pi$

** **11.** $f_n(x) = (\sin x)^{1/n}, 0 \leq x \leq \pi$

** **12.** $f_n(x) = \left(\dfrac{\sin x}{x} \right)^{1/n}, 0 < x < \pi$

** **13.** $f_n(x) = \left(\dfrac{\sin x}{x} \right)^{1/n}, 0 < x \leq \pi$

** **14.** $f_n(x) = n^2 x e^{-nx}, 0 \leq x < \infty$

** **15.** For what values of x does the sequence of functions $\{(1 - x^n)/(1 - x)\}$ have a limit?

11.3 Taylor Polynomials, Remainders, and Series

Section 4.12 introduced Taylor polynomials as polynomial approximations to functions. If you have not studied this section previously, it is appropriate to do so now. The results are summarized in the following theorem.

THEOREM 11.1 (Taylor's Remainder Formula)

If $f(x)$ and its first $n - 1$ derivatives are continuous on the closed interval between c and x, and if $f(x)$ has an nth derivative on the open interval between c and x, then there exists a point z_{n-1} between c and x such that

$$f(x) = f(c) + f'(c)(x - c) + \frac{f''(c)}{2!}(x - c)^2 + \frac{f'''(c)}{3!}(x - c)^3 + \cdots$$

$$+ \frac{f^{(n-1)}(c)}{(n - 1)!}(x - c)^{n-1} + \frac{f^{(n)}(z_{n-1})}{n!}(x - c)^n. \qquad (11.4)$$

The notation $f^{(n)}(z_{n-1})$ represents the nth derivative of $f(x)$ evaluated at $x = z_{n-1}$.

In this section we use this result to introduce Taylor series for a function. We assume that $f(x)$ has derivatives of all orders on the closed interval between c and some value of x. Taylor's remainder formula can then be written down for all values of n. For $n = 1, 2, 3$, and 4, we obtain

$$f(x) = f(c) + f'(z_0)(x - c), \qquad (11.5a)$$

$$f(x) = f(c) + f'(c)(x - c) + \frac{f''(z_1)}{2!}(x - c)^2, \qquad (11.5b)$$

$$f(x) = f(c) + f'(c)(x - c) + \frac{f''(c)}{2!}(x - c)^2 + \frac{f'''(z_2)}{3!}(x - c)^3, \qquad (11.5c)$$

$$f(x) = f(c) + f'(c)(x - c) + \frac{f''(c)}{2!}(x - c)^2 + \frac{f'''(c)}{3!}(x - c)^3$$

$$+ \frac{f''''(z_3)}{4!}(x - c)^4. \qquad (11.5d)$$

The last terms in these equations are called **Taylor remainders**, and when they are denoted by R_0, R_1, R_2, and R_3 respectively,

$$f(x) = f(c) + R_0, \qquad (11.6a)$$

$$f(x) = f(c) + f'(c)(x - c) + R_1, \qquad (11.6b)$$

$$f(x) = f(c) + f'(c)(x - c) + \frac{f''(c)}{2!}(x - c)^2 + R_2, \qquad (11.6c)$$

$$f(x) = f(c) + f'(c)(x - c) + \frac{f''(c)}{2!}(x - c)^2 + \frac{f'''(c)}{3!}(x - c)^3 + R_3, \qquad (11.6d)$$

and, in general,

$$f(x) = f(c) + f'(c)(x - c) + \frac{f''(c)}{2!}(x - c)^2 + \cdots$$

$$+ \frac{f^{(n)}(c)}{n!}(x - c)^n + R_n. \qquad (11.6e)$$

We call R_n the remainder for the simple reason that, if we drop the remainders from equations 11.6, then what remains is a sequence of polynomial approximations to $f(x)$, the Taylor polynomials of Section 4.12:

$$f(x) \approx P_0(x) = f(c), \tag{11.7a}$$

$$f(x) \approx P_1(x) = f(c) + f'(c)(x - c), \tag{11.7b}$$

$$f(x) \approx P_2(x) = f(c) + f'(c)(x - c) + \frac{f''(c)}{2!}(x - c)^2, \tag{11.7c}$$

$$f(x) \approx P_3(x) = f(c) + f'(c)(x - c) + \frac{f''(c)}{2!}(x - c)^2$$
$$+ \frac{f'''(c)}{3!}(x - c)^3. \tag{11.7d}$$

The nth remainder is

$$R_n = R_n(c, x) = \frac{f^{(n+1)}(z_n)}{(n + 1)!}(x - c)^{n+1}, \tag{11.8}$$

where z_n (which, as the notation suggests, varies with n) is always between c and x. We have written $R_n(c, x)$ to emphasize the fact that remainders depend on the point of expansion c, the value of x being considered, and the choice of n. The sequence of remainders $\{R_n(c, x)\}$ is a sequence of functions. Suppose the sequence of remainders has limit zero on some interval,

$$\lim_{n \to \infty} R_n(c, x) = 0. \tag{11.9}$$

What this means is that the sequence of polynomials in 11.7 approximates $f(x)$ more and more closely as n gets larger and larger. It is tedious to write groups of equations like 11.7 and say that, the more and more terms that are added, the better and better is the approximation. We invent a notation to stand for this. We write

$$f(x) = f(c) + f'(c)(x - c) + \frac{f''(c)}{2!}(x - c)^2 + \frac{f'''(c)}{3!}(x - c)^3 + \cdots . \tag{11.10}$$

The $\cdots$ indicate that additional terms of the same form are to be added indefinitely. This is called the **Taylor series** of $f(x)$ about the point c. Be absolutely clear on what we mean by 11.10, because it cannot be taken literally. There is an infinity of terms on the right side of the equation, and substitution of a value for x (and c) gives an infinity of numbers to add. No matter how fast you add, your calculator adds, or a supercomputer adds, an infinity of numbers cannot be added in a finite amount of time. Equation 11.10 does not therefore mean that adding the terms on the right gives $f(x)$. What it means is that

$$\lim_{n \to \infty} \left| f(x) - \sum_{k=0}^{n} \frac{f^{(k)}(c)}{k!}(x - c)^k \right| = 0.$$

In words, this translates to any of the following three equivalent statements:

1. The more and more terms that are added on the right, the closer and closer the sum gets to $f(x)$.

2. By adding sufficiently many terms, we can make the sum on the right as close to $f(x)$ as desired.

3. If the Taylor series is truncated at any number of terms, the resulting Taylor polynomial approximates $f(x)$; the larger n, the better is the approximation.

To use the terminology from Section 11.2, we say that the Taylor polynomials $P_n(x)$ converge to $f(x)$, meaning again that they get closer and closer to $f(x)$, the larger the value of n. We also say that the Taylor series converges to $f(x)$.

When $c = 0$, the Taylor series becomes

$$f(x) = f(0) + f'(0)x + \frac{f''(0)}{2!}x^2 + \frac{f'''(0)}{3!}x^3 + \cdots . \tag{11.11}$$

We call this the Maclaurin series for $f(x)$. It is a polynomial in powers of x rather than $x - c$.

To obtain the Taylor (or Maclaurin) series for a function $f(x)$ is quite simple; evaluate all derivatives of $f(x)$ at $x = c$ (or $x = 0$), and use formula 11.10 (or 11.11). To show that the series converges to $f(x)$ is a different matter. In Section 4.12 we calculated Taylor polynomials for $\sin x$, e^x, and $\ln x$, and saw graphically how the polynomials approximate the functions more and more closely as more and more terms are included. Sometimes Taylor polynomials converge to the function for all x, sometimes they converge only on some interval around the point of expansion $(x = c)$. This may not always be obvious graphically, and we should therefore consider finding this interval algebraically. It is called the **interval of convergence** for the Taylor series. It consists of all values of x for which Taylor remainders approach zero. We therefore turn our attention to examination of

$$R_n(c, x) = \frac{f^{(n+1)}(z_n)}{(n+1)!}(x - c)^{n+1}.$$

We must determine values of x for which the limit of the sequence $\{R_n(c, x)\}$ is equal to zero. The difficulty is that $R_n(c, x)$ is defined in terms of z_n, an unknown value except that it must be between c and x. This problem is usually circumvented by finding a maximum value for $|R_n(c, x)|$, and showing that this maximum value approaches zero. We illustrate in the next three examples.

EXAMPLE 11.7

Find the Maclaurin series for $\sin x$ and show that it converges to $\sin x$ for all x.

SOLUTION   If $f(x) = \sin x$, then:

$$f(0) = \sin x_{|x=0} = 0;$$

$$f'(0) = \cos x_{|x=0} = 1;$$

$$f''(0) = -\sin x_{|x=0} = 0;$$

$$f'''(0) = -\cos x_{|x=0} = -1;$$

$$f''''(0) = \sin x_{|x=0} = 0; \quad \text{etc.}$$

Taylor's remainder formula for $\sin x$ and $c = 0$ yields

$$\sin x = x - \frac{x^3}{3!} + \frac{x^5}{5!} - \frac{x^7}{7!} + \cdots + \frac{d^n}{dx^n}(\sin x)_{|x=0}\frac{x^n}{n!} + R_n(0, x),$$

where

$$R_n(0, x) = \frac{d^{n+1}}{dx^{n+1}}(\sin x)_{|x=z_n}\frac{x^{n+1}}{(n+1)!}.$$

But the $(n + 1)^{\text{th}}$ derivative of $\sin x$ is $\pm \sin x$ or $\pm \cos x$, so that

$$\left|\frac{d^{n+1}}{dx^{n+1}}(\sin x)_{|x=z_n}\right| \leq 1.$$

Hence,

$$|R_n(0, x)| \leq \frac{|x|^{n+1}}{(n+1)!}.$$

But according to Example 11.6, $\lim_{n\to\infty} |x|^n/n! = 0$ for any x whatsoever, and therefore

$$\lim_{n\to\infty} R_n(0, x) = 0.$$

The Maclaurin series for $\sin x$ therefore converges to $\sin x$ for all x, and we may write

$$\sin x = x - \frac{x^3}{3!} + \frac{x^5}{5!} - \frac{x^7}{7!} + \cdots$$

$$= \sum_{n=0}^{\infty} \frac{(-1)^n}{(2n+1)!} x^{2n+1}, \qquad -\infty < x < \infty.$$

In Figure 11.10 we show some of the Taylor polynomials (about $x = 0$) for $\sin x$. As more and more terms of the series are included in the polynomials, two things happen. The polynomials approximate $\sin x$ more and more accurately for small x, and the interval on which the polynomials approximate $\sin x$ with any degree of accuracy gets larger and larger.

FIGURE 11.10

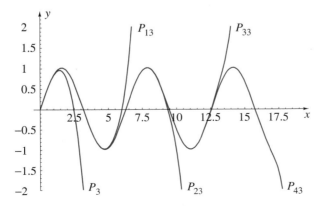

In a similar way it can be shown that the Maclaurin series for $\cos x$ is

$$\cos x = \sum_{n=0}^{\infty} \frac{(-1)^n}{(2n)!} x^{2n} = 1 - \frac{x^2}{2!} + \frac{x^4}{4!} - \frac{x^6}{6!} + \cdots, \quad -\infty < x < \infty.$$

Sometimes it is necessary to consider points on either side of the point of expansion separately. This is illustrated in the next example.

EXAMPLE 11.8

Find the Maclaurin series for e^x and show that it converges to e^x for all x.

SOLUTION   Since

$$\frac{d^n}{dx^n}(e^x)_{|x=0} = e^x_{|x=0} = 1,$$

Taylor's remainder formula for e^x and $c = 0$ gives

$$e^x = 1 + x + \frac{x^2}{2!} + \frac{x^3}{3!} + \cdots + \frac{x^n}{n!} + R_n(0, x),$$

where

$$R_n(0, x) = \frac{d^{n+1}}{dx^{n+1}}(e^x)_{|x=z_n} \frac{x^{n+1}}{(n + 1)!} = e^{z_n} \frac{x^{n+1}}{(n + 1)!}.$$

Now, if $x < 0$, then $x < z_n < 0$, and

$$|R_n(0, x)| < e^0 \frac{|x|^{n+1}}{(n + 1)!},$$

which approaches zero as n becomes infinite (see Example 11.6). If $x > 0$, then $0 < z_n < x$, and

$$|R_n(0, x)| < e^x \frac{|x|^{n+1}}{(n + 1)!},$$

which again has limit zero as n approaches infinity. Thus, for *any* x, $\lim_{n \to \infty} R_n(0, x) = 0$, and the Maclaurin series for e^x converges to e^x:

$$e^x = 1 + x + \frac{x^2}{2!} + \frac{x^3}{3!} + \cdots = \sum_{n=0}^{\infty} \frac{1}{n!} x^n, \quad -\infty < x < \infty.$$

FIGURE 11.11

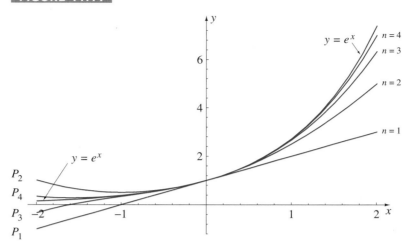

Separate discussions for $x < 0$ and $x > 0$ are reflected in how the Taylor polynomials converge to e^x in Figure 11.11. For $x > 0$, polynomials are always less than e^x and increase toward e^x. For $x < 0$, polynomials are alternately larger and smaller than e^x, gradually approaching e^x.

The following example is more difficult. Remainders do not approach zero for all x so that the Taylor series does not converge to the function for all x. The problem would have been even more difficult had we not suggested values of x to consider.

EXAMPLE 11.9

Find the Taylor series for $\ln x$ about the point 1 and show that, for $1/2 \le x \le 2$, it converges to $\ln x$.

SOLUTION   If $f(x) = \ln x$, then:

$$f(1) = \ln x_{|x=1} = 0;$$

$$f'(1) = \frac{1}{x}\bigg|_{x=1} = 1;$$

$$f''(1) = \frac{-1}{x^2}\bigg|_{x=1} = -1;$$

$$f'''(1) = \frac{2}{x^3}\bigg|_{x=1} = 2;$$

$$f''''(1) = \frac{-3!}{x^4}\bigg|_{x=1} = -3!; \quad \text{etc.}$$

Taylor's remainder formula for $\ln x$ and $c = 1$ yields

$$\ln x = (x - 1) - \frac{1}{2!}(x - 1)^2 + \frac{2!}{3!}(x - 1)^3 + \cdots$$

$$+ \frac{(-1)^{n+1}(n - 1)!}{n!}(x - 1)^n + R_n(1, x),$$

where

$$R_n(1, x) = \frac{d^{n+1}}{dx^{n+1}}(\ln x)_{|x=z_n} \frac{(x - 1)^{n+1}}{(n + 1)!} = \frac{(-1)^n n!}{(z_n)^{n+1}} \frac{(x - 1)^{n+1}}{(n + 1)!}$$

$$= \frac{(-1)^n}{n + 1}\left(\frac{x - 1}{z_n}\right)^{n+1},$$

and z_n is between 1 and x.

If $1 < x \le 2$, then the largest value of $x - 1$ is 1. Furthermore, z_n must be larger than 1. It follows that

$$|R_n(1, x)| < \frac{1}{n + 1}\left(\frac{1}{1}\right)^{n+1} = \frac{1}{n + 1}$$

and therefore

$$\lim_{n \to \infty} R_n(1, x) = 0.$$

If $1/2 \le x < 1$, then $-1/2 \le x - 1 < 0$. Combine this with $x < z_n < 1$, and we can state that $-1 < (x - 1)/z_n < 0$. Then,

$$|R_n(1, x)| < \frac{1}{n + 1} \quad \text{and} \quad \lim_{n \to \infty} R_n(1, x) = 0.$$

Thus, for $1/2 \le x \le 2$, the sequence of remainders $\{R_n(1, x)\}$ approaches zero, and the Taylor series converges to $\ln x$ for those values of x:

$$\ln x = (x - 1) - \frac{1}{2}(x - 1)^2 + \frac{1}{3}(x - 1)^3 + \cdots$$

$$= \sum_{n=1}^{\infty} \frac{(-1)^{n+1}}{n}(x - 1)^n, \quad \frac{1}{2} \le x \le 2.$$

This series actually converges to $\ln x$ on the larger interval $0 < x \le 2$. We will consider this in Example 11.18. We can see it in Figure 11.12. For $1 < x \le 2$, polynomials are alternately above and below $\ln x$. For $0 < x < 1$, they approach $\ln x$ from above.

FIGURE 11.12

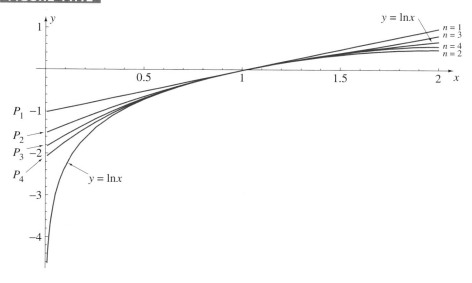

The above examples were chosen for their simplicity. It is simple to find n^{th} derivatives of $\sin x$, e^x, and $\ln x$; that remainders approach zero follows immediately. You may disagree that the examples were simple, but this only serves to emphasize the following point. If these are the simplest possible examples, and they are not simple, imagine the difficulty in finding and showing that Taylor series of other functions converge to the function. The problem is so great that we do our utmost to avoid calculating derivatives and considering remainders. Based on discussions in Section 11.4, we will find alternatives in Section 11.5.

EXERCISES 11.3

In Exercises 1–5 use Taylor's remainder formula to show that the Taylor series for the function $f(x)$ about the point indicated converges to $f(x)$ for all x. In each case, plot the first six Taylor polynomials to show how they approximate $f(x)$ more closely as the degree of the polynomial increases.

** **1.** $f(x) = \cos x$ about $x = 0$ ** **2.** $f(x) = e^{5x}$ about $x = 0$

** **3.** $f(x) = \sin(10x)$ about $x = 0$ ** **4.** $f(x) = \sin x$ about $x = \pi/4$

** **5.** $f(x) = e^{2x}$ about $x = 1$

In Exercises 6–15 find the Taylor series for the function $f(x)$ about the point c. Plot sufficiently many polynomials and the function to determine the interval on which the series converges to the function.

One of the easiest, and perhaps the most important power series is

$$\sum_{n=0}^{\infty} x^n = 1 + x + x^2 + x^3 + \cdots . \tag{11.15}$$

It is so important in applications that it is given a special name; it is called a geometric series. A **geometric series** is a series in which every term after the first is obtained by multiplying the previous term by the same amount, called the **common ratio**. For geometric series 11.15, the common ratio is x; each term is x times the previous term. To show that the series converges we must show that the sequence of polynomials

$$P_n(x) = 1 + x + x^2 + \cdots + x^n$$

has a limit. If we multiply $P_n(x)$ by x, then

$$x \, P_n(x) = x + x^2 + x^3 + \cdots + x^{n+1}.$$

When these are subtracted,

$$P_n(x) - x \, P_n(x) = 1 - x^{n+1},$$

from which

$$P_n(x) = \frac{1 - x^{n+1}}{1 - x}. \tag{11.16}$$

This is a simplified form for $P_n(x)$. Since

$$\lim_{n \to \infty} x^{n+1} = \begin{cases} 0, & -1 < x < 1 \\ 1, & x = 1 \\ \text{does not exist,} & \text{otherwise} \end{cases}$$

it follows that when the sequence $\{P_n(x)\}$ is restricted to the interval $|x| < 1$, it has a limit,

$$\lim_{n \to \infty} P_n(x) = \frac{1}{1 - x}, \qquad |x| < 1.$$

We can therefore write that

$$\sum_{n=0}^{\infty} x^n = 1 + x + x^2 + \cdots = \frac{1}{1 - x}, \qquad |x| < 1. \tag{11.17}$$

The geometric series has sum $1/(1 - x)$ on the interval (of convergence) $|x| < 1$. It does not have a sum for $|x| \geq 1$.

Figure 11.13 shows some of the polynomials $P_n(x)$ and $1/(1 - x)$. They illustrate how polynomials approximate $1/(1 - x)$ more closely as n increases. Polynomials are defined for all x and $1/(1 - x)$ is defined for all $x \neq 1$, but curves $y = P_n(x)$ approach $y = 1/(1 - x)$ only for $|x| < 1$. For $x > 0$, polynomials approach $1/(1 - x)$ from below; for $x < 0$, they oscillate about $1/(1 - x)$, but gradually approach $1/(1 - x)$.

The series

$$\sum_{n=0}^{\infty} \frac{(-1)^n 2^n}{(2n)!} x^{2n+2} = x^2 - \frac{2}{2!} x^4 + \frac{2^2}{4!} x^6 - \frac{2^3}{6!} x^8 + \cdots$$

can be expressed in the form

$$\sum_{n=0}^{\infty} \frac{(-1)^n 2^n}{(2n)!} x^{2n+2} = x^2 \left[1 - \frac{(\sqrt{2}x)^2}{2!} + \frac{(\sqrt{2}x)^4}{4!} - \frac{(\sqrt{2}x)^6}{6!} + \cdots \right].$$

It is x^2 times the Maclaurin series for $\cos x$ with x replaced by $\sqrt{2}x$. In other words,

$$\sum_{n=0}^{\infty} \frac{(-1)^n 2^n}{(2n)!} x^{2n+2} = x^2 \cos \sqrt{2}x,$$

valid for $-\infty < \sqrt{2}x < \infty$, or $-\infty < x < \infty$.

Other methods for finding sums of power series are discussed in Section 11.6.

For many power series there is no known function to which the power series converges. In such cases we write

$$f(x) = \sum_{n=0}^{\infty} a_n x^n, \tag{11.20}$$

and say that the power series defines the value of $f(x)$ at each x in the interval of convergence. For instance, a very important series in engineering and physics is

$$\sum_{n=0}^{\infty} \frac{(-1)^n}{2^{2n} (n!)^2} x^{2n},$$

which converges for all x. It arises so often in applications that it is given a special name, the *Bessel function of the first kind of order zero*, and is denoted by $J_0(x)$. In other words, we write

$$J_0(x) = \sum_{n=0}^{\infty} \frac{(-1)^n}{2^{2n} (n!)^2} x^{2n},$$

and say that the power series defines the value of $J_0(x)$ for each x. We get an idea of what $J_0(x)$ looks like by plotting the polynomial approximations obtained by adding more and more terms of the series. Figure 11.14 shows the first five polynomials for $x \geq 0$,

$$P_1(x) = 1,$$

$$P_2(x) = 1 - \frac{x^2}{2^2},$$

$$P_3(x) = 1 - \frac{x^2}{2^2} + \frac{x^4}{2^4 (2!)^2},$$

$$P_4(x) = 1 - \frac{x^2}{2^2} + \frac{x^4}{2^4 (2!)^2} - \frac{x^6}{2^6 (3!)^2},$$

$$P_5(x) = 1 - \frac{x^2}{2^2} + \frac{x^4}{2^4 (2!)^2} - \frac{x^6}{2^6 (3!)^2} + \frac{x^8}{2^8 (4!)^2}.$$

FIGURE 11.14

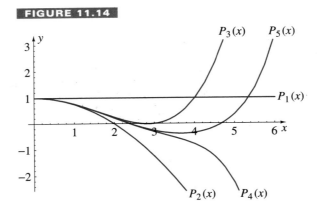

They are even functions so their graphs would be symmetric about the y-axis. As the number of terms is increased, the polynomials approximate $J_0(x)$ in Figure 11.15 more and more closely.

FIGURE 11.15

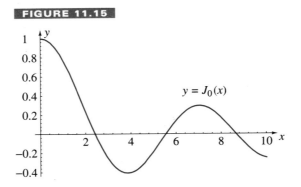

We have called 11.12 a power series in x. It is also said to be a *power series about* 0 (meaning the point 0 on the x-axis), where we note that for any interval of convergence whatsoever, 0 is always at its centre. This suggests that power series about other points might be considered, and this is indeed the case. The general power series about a point c on the x-axis is

$$\sum_{n=0}^{\infty} a_n (x - c)^n = a_0 + a_1(x - c) + a_2(x - c)^2 + \cdots \qquad (11.21)$$

and is said to be a power series in $x - c$.

A power series in $x - c$ has an interval of convergence and a radius of convergence analogous to a power series in x. In particular, every power series in $x - c$ has a radius of convergence R such that, if $R = 0$, the power series converges only for $x = c$; if $R = \infty$, the power series converges for all x; and if $0 < R < \infty$, the series converges for $|x - c| < R$, diverges for $|x - c| > R$, and may or may not converge for $x = c \pm R$. The radius of convergence is again given by equations 11.18, provided the limits exist or are equal to infinity. For power series in $x - c$, then, the point c is the centre of the interval of convergence.

EXAMPLE 11.14

Find the open interval of convergence of the power series

$$2^2(x + 2) + 2^3(x + 2)^2 + 2^4(x + 2)^3 + \cdots + 2^{n+1}(x + 2)^n + \cdots .$$

SOLUTION   By writing the power series in sigma notation,

$$\sum_{n=1}^{\infty} 2^{n+1}(x + 2)^n,$$

we can use equation 11.18a (or 11.18b) to calculate its radius of convergence,

$$R = \lim_{n \to \infty} \left| \frac{a_n}{a_{n+1}} \right| = \lim_{n \to \infty} \frac{2^{n+1}}{2^{n+2}} = \frac{1}{2}.$$

Since this is a power series about -2, the open interval of convergence is $-5/2 < x < -3/2$. We can do better. This is a geometric series with common ratio $2(x+2)$. According to equation 11.19a, its sum is

$$\sum_{n=1}^{\infty} 2^{n+1}(x + 2)^n = 2^2(x + 2)[1 + 2(x + 2) + 2^2(x + 2)^2 + \cdots]$$

$$= \frac{4(x + 2)}{1 - 2(x + 2)} = \frac{-4(x + 2)}{2x + 3},$$

valid for $-1 < 2(x + 2) < 1 \implies -5/2 < x < -3/2$. This is now known to be the interval of convergence, not just the open interval of convergence.

EXERCISES 11.4

In Exercises 1–25 find the open interval of convergence for the power series.

* **1.** $\displaystyle\sum_{n=1}^{\infty} \frac{1}{n} x^n$

* **2.** $\displaystyle\sum_{n=1}^{\infty} n^2 x^n$

* **3.** $\displaystyle\sum_{n=0}^{\infty} \frac{1}{(n + 1)^3} x^n$

* **4.** $\displaystyle\sum_{n=0}^{\infty} n^2 3^n x^n$

* **5.** $\displaystyle\sum_{n=0}^{\infty} \frac{1}{2^n} (x - 1)^n$

* **6.** $\displaystyle\sum_{n=4}^{\infty} (-1)^n n^3 (x + 3)^n$

$*$ **7.** $\displaystyle\sum_{n=1}^{\infty} \frac{1}{\sqrt{n}}(x+2)^n$

$*$ **8.** $\displaystyle\sum_{n=2}^{\infty} 2^n \left(\frac{n-1}{n+2}\right)^2 (x-4)^n$

$**$ **9.** $\displaystyle\sum_{n=1}^{\infty} \frac{1}{n^2} x^{2n}$

$**$ **10.** $\displaystyle\sum_{n=0}^{\infty} (-1)^n x^{3n}$

$**$ **11.** $\displaystyle\sum_{n=0}^{\infty} \frac{n-1}{n+1}(2x)^n$

$**$ **12.** $\displaystyle\sum_{n=0}^{\infty} \frac{1}{\sqrt{n+1}} x^{3n+1}$

$**$ **13.** $\displaystyle\sum_{n=0}^{\infty} \frac{(-1)^n}{3^n} x^{2n+1}$

$**$ **14.** $\displaystyle\sum_{n=1}^{\infty} \frac{(-e)^n}{n^2} x^n$

$**$ **15.** $\dfrac{x}{9} + \dfrac{4}{3^4}x^2 + \dfrac{9}{3^6}x^3 + \cdots + \dfrac{n^2}{3^{2n}}x^n + \cdots$

$**$ **16.** $x + 2^2 x^2 + 3^3 x^3 + \cdots + n^n x^n + \cdots$

$**$ **17.** $\dfrac{1}{36}(x+10)^6 + \dfrac{1}{49}(x+10)^7 + \dfrac{1}{64}(x+10)^8 + \cdots + \dfrac{1}{n^2}(x+10)^n + \cdots$

$**$ **18.** $3x + 8(3x)^2 + 27(3x)^3 + \cdots + n^3(3x)^n + \cdots$

$**$ **19.** $\dfrac{3}{4}x^2 + x^4 + \dfrac{27}{16}x^6 + \cdots + \dfrac{3^n}{(n+1)^2}x^{2n} + \cdots$

$**$ **20.** $1 + \dfrac{1}{5}x^3 + \dfrac{1}{25}x^6 + \cdots + \dfrac{1}{5^n}x^{3n} + \cdots$

$**$ **21.** $\displaystyle\sum_{n=2}^{\infty} \frac{1}{\ln n} x^n$

$**$ **22.** $\displaystyle\sum_{n=2}^{\infty} \frac{1}{n^2 \ln n} x^n$

$**$ **23.** $\displaystyle\sum_{n=1}^{\infty} \frac{(n!)^3}{(3n)!} x^n$

$**$ **24.** $\displaystyle\sum_{n=1}^{\infty} \frac{2 \cdot 4 \cdot 6 \cdot \cdots \cdot (2n)}{3 \cdot 5 \cdot 7 \cdot \cdots \cdot (2n+1)} x^n$

$**$ **25.** $\displaystyle\sum_{n=1}^{\infty} \frac{[1 \cdot 3 \cdot 5 \cdot \cdots \cdot (2n+1)]^2}{2^{2n}(2n)!} x^n$

In Exercises 26–37 find the sum of the power series.

$*$ **26.** $\displaystyle\sum_{n=0}^{\infty} \frac{1}{4^n} x^{3n}$

$*$ **27.** $\displaystyle\sum_{n=1}^{\infty} (-e)^n x^n$

$*$ **28.** $\displaystyle\sum_{n=1}^{\infty} \frac{1}{3^{2n}}(x-1)^n$

$*$ **29.** $\displaystyle\sum_{n=2}^{\infty} (x+5)^{2n}$

$**$ **30.** $\displaystyle\sum_{n=0}^{\infty} \frac{(-1)^n}{(2n)!} x^{4n}$

$**$ **31.** $\displaystyle\sum_{n=0}^{\infty} \frac{5^n}{n!} x^n$

$**$ **32.** $\displaystyle\sum_{n=0}^{\infty} \frac{(-1)^n}{3^{2n+1}(2n+1)!} x^{2n+2}$

$**$ **33.** $\displaystyle\sum_{n=0}^{\infty} \frac{(-3)^n}{n!}(x+1)^n$

$**$ **34.** $\displaystyle\sum_{n=1}^{\infty} \frac{(-1)^n}{n!} x^n$

$**$ **35.** $\displaystyle\sum_{n=0}^{\infty} \frac{(-1)^{n+1}}{(2n+1)!}(x+1)^{2n+3}$

$**$ **36.** $\displaystyle\sum_{n=0}^{\infty} \frac{2^n}{n!}(x-1/2)^n$

$**$ **37.** $\displaystyle\sum_{n=0}^{\infty} \frac{(-1)^n}{2^{2n}(2n)!} x^{4n+4}$

** **38.** If m is a nonnegative integer, the Bessel function of order m of the first kind is defined by the power series

$$J_m(x) = \sum_{n=0}^{\infty} \frac{(-1)^n}{2^{2n+m} n! (n+m)!} x^{2n+m}.$$

(a) Write out the first five terms of $J_0(x)$, $J_1(x)$, and $J_m(x)$.

(b) Find the interval of convergence for each $J_m(x)$.

39. The hypergeometric series is

$$1 + \frac{\alpha\beta}{\gamma} x + \frac{\alpha(\alpha+1)\beta(\beta+1)}{2!\,\gamma(\gamma+1)} x^2 + \frac{\alpha(\alpha+1)(\alpha+2)\beta(\beta+1)(\beta+2)}{3!\,\gamma(\gamma+1)(\gamma+2)} x^3 + \cdots,$$

where α, β, and γ are all constants.

(a) Write this series in sigma notation.

(b) What is the radius of convergence of the hypergeometric series if γ is not zero or a negative integer?

11.5 Taylor Series Expansions of Functions

Let us summarize what we have seen in Sections 11.3 and 11.4. Given a function $f(x)$ and a point c, the Taylor series of $f(x)$ about c is

$$f(x) = \sum_{n=0}^{\infty} \frac{f^{(n)}(c)}{n!} (x-c)^n,$$

and the series converges to $f(x)$ at all values of x for which Taylor remainders approach zero. Given a power series $\sum_{n=0}^{\infty} a_n (x-c)^n$, there is an interval of convergence inside of which the series has a sum. If this sum is $f(x)$, we write

$$f(x) = \sum_{n=0}^{\infty} a_n (x-c)^n, \quad x \text{ in the interval of convergence.}$$

Clearly, Taylor series are power series. A power series would be a Taylor series if a_n were equal to $f^{(n)}(c)/n!$. If this were the case, then power series and Taylor series would be one and the same. The following theorem allows us to prove this.

THEOREM 11.3

If $f(x) = \sum_{n=0}^{\infty} a_n (x-c)^n$, and the radius of convergence R is greater than zero, then each of the following series has radius of convergence R:

$$f'(x) = \sum_{n=0}^{\infty} n a_n (x-c)^{n-1}, \tag{11.22a}$$

$$\int f(x)\,dx = \sum_{n=0}^{\infty} \frac{a_n}{n+1} (x-c)^{n+1} + C. \tag{11.22b}$$

Because of the difficulty in proving this theorem, and in order to preserve the continuity of our discussion, we omit a proof. Note that the theorem is stated in terms of radii of convergence rather than intervals of convergence. This is due to the fact that in differentiating a power series we may lose the end points of the original interval of convergence, and in integrating we may pick them up.

The next theorem implies that power series and Taylor series are one and the same, that every power series is a Taylor series, the Taylor series of its sum.

THEOREM 11.4

If $f(x)$ is the sum of the power series $\sum_{n=0}^{\infty} a_n(x - c)^n$ with $R > 0$, then the series is the Taylor series of $f(x)$.

PROOF   When we set $x = c$ in

$$f(x) = \sum_{n=0}^{\infty} a_n(x - c)^n = a_0 + a_1(x - c) + a_2(x - c)^2 + \cdots$$

we obtain $f(c) = a_0$. If we differentiate the power series according to Theorem 11.3, we obtain

$$f'(x) = a_1 + 2a_2(x - c) + 3a_3(x - c)^2 + \cdots .$$

When we substitute $x = c$, the result is

$$f'(c) = a_1.$$

If we differentiate the power series for $f'(x)$, we obtain

$$f''(x) = 2a_2 + 3 \cdot 2a_3(x - c) + 4 \cdot 3a_4(x - c)^2 + \cdots ,$$

and substitute $x = c$,

$$f''(c) = 2a_2 \quad \text{or} \quad a_2 = \frac{f''(c)}{2!}.$$

Continued differentiation and substitution leads to the result that, for all n,

$$a_n = \frac{f^{(n)}(c)}{n!}.$$

The power series $f(x) = \sum_{n=0}^{\infty} a_n(x - c)^n$ is therefore the Taylor series of $f(x)$. ∎

The following corollary is an immediate consequence of this theorem.

COROLLARY

If two power series $\sum_{n=0}^{\infty} a_n(x - c)^n$ and $\sum_{n=0}^{\infty} b_n(x - c)^n$ with positive radii of convergence have identical sums,

$$\sum_{n=0}^{\infty} a_n(x - c)^n = \sum_{n=0}^{\infty} b_n(x - c)^n,$$

then $a_n = b_n$ for all n.

Theorem 11.4 shows that Sections 11.3 and 11.4 were dealing with the same problem but coming at it from different directions. In Section 11.3, $f(x)$ and $x = c$ were given, and we developed the Taylor series for $f(x)$ about $x = c$. Theorem 11.4 shows that this is the only power series for $f(x)$ about $x = c$. In Section 11.4, a power series $\sum_{n=0}^{\infty} a_n (x - c)^n$ was given, and we determined its interval of convergence and sum $f(x)$. Theorem 11.4 indicates that the power series is actually the Taylor series of $f(x)$ about $x = c$. This simplifies the problem of finding the Taylor series for a function $f(x)$ about a point $x = c$ immeasurably. Instead of finding $f^{(n)}(c)$ and showing that Taylor remainders approach zero, we can proceed as follows. If, by any method whatsoever, we can find a power series $\sum_{n=0}^{\infty} a_n (x - c)^n$ that has sum $f(x)$, then it must be the Taylor series of $f(x)$ about $x = c$. In the remainder of this section we show how easy it is to do this. In essence, we take series with known sums, such as 11.19, and construct other series from them.

EXAMPLE 11.15

Find (a) the Maclaurin series for $1/(4 + 5x)$; (b) the Taylor series about 5 for $1/(13 - 2x)$.

SOLUTION

(a) We write

$$\frac{1}{4 + 5x} = \frac{1}{4\left(1 + \dfrac{5x}{4}\right)}$$

and interpret the right side as one-quarter of the sum of a geometric series with first term equal to 1 and common ratio $-5x/4$. Equation 11.19a then gives

$$\frac{1}{4 + 5x} = \frac{1}{4}\left[1 + \left(-\frac{5x}{4}\right) + \left(-\frac{5x}{4}\right)^2 + \cdots\right], \qquad \left|-\frac{5x}{4}\right| < 1,$$

$$= \sum_{n=0}^{\infty} \frac{(-1)^n 5^n}{4^{n+1}} x^n, \qquad |x| < \frac{4}{5}.$$

(b) By a similar procedure, we have

$$\frac{1}{13 - 2x} = \frac{1}{3 - 2(x - 5)}$$

$$= \frac{1}{3\left[1 - \dfrac{2}{3}(x - 5)\right]}$$

$$= \frac{1}{3}\left[1 + \frac{2}{3}(x - 5) + \left(\frac{2}{3}\right)^2 (x - 5)^2 + \cdots\right], \qquad \left|\frac{2}{3}(x - 5)\right| < 1,$$

$$= \sum_{n=0}^{\infty} \frac{2^n}{3^{n+1}} (x - 5)^n, \qquad |x - 5| < \frac{3}{2}.$$

In both examples, properties of geometric series gave not only the required series, but also their intervals of convergence. To appreciate the simplicity of these solutions, we suggest using Taylor remainders in an attempt to obtain the series with the same intervals of convergence. You will quickly abort.

Addition and Subtraction of Power Series

According to the following theorem, convergent power series can be added and subtracted in their common interval of convergence.

THEOREM 11.5

If $f(x) = \sum\limits_{n=0}^{\infty} a_n(x-c)^n$ and $g(x) = \sum\limits_{n=0}^{\infty} b_n(x-c)^n$ have positive radii of convergence, then

$$f(x) \pm g(x) = \sum_{n=0}^{\infty} (a_n \pm b_n)(x-c)^n, \qquad (11.23)$$

valid for every x that is common to the intervals of convergence of the two series.

We use this result in the following example.

EXAMPLE 11.16

Find the Maclaurin series for $f(x) = 5x/(x^2 - 3x - 4)$.

SOLUTION   We decompose $f(x)$ into its partial fractions,

$$f(x) = \frac{5x}{x^2 - 3x - 4} = \frac{4}{x-4} + \frac{1}{x+1},$$

and expand each of these terms in a Maclaurin series,

$$\frac{4}{x-4} = \frac{-1}{1 - \dfrac{x}{4}} = -\left(1 + \frac{x}{4} + \frac{x^2}{4^2} + \cdots\right), \qquad |x| < 4, \qquad \text{and}$$

$$\frac{1}{1+x} = 1 - x + x^2 - x^3 + \cdots, \qquad |x| < 1.$$

Addition of these series within their common interval of convergence gives the Maclaurin series for $f(x)$:

$$\frac{5x}{x^2 - 3x - 4} = \left(-1 - \frac{x}{4} - \frac{x^2}{4^2} - \frac{x^3}{4^3} - \cdots\right) + \left(1 - x + x^2 - x^3 + \cdots\right)$$

$$= \left(-1 - \frac{1}{4}\right)x + \left(1 - \frac{1}{4^2}\right)x^2 + \left(-1 - \frac{1}{4^3}\right)x^3 + \cdots$$

$$= \sum_{n=1}^{\infty} \left[(-1)^n - \frac{1}{4^n}\right]x^n, \qquad |x| < 1.$$

Differentiation and Integration of Power Series

Perhaps the most useful technique for generating Taylor series is to differentiate or integrate known expansions according to Theorem 11.3.

EXAMPLE 11.17

Find Maclaurin series for the following functions:

$$(a) \ \cos x \qquad (b) \ \frac{1}{(1-x)^3}$$

SOLUTION

(a) We derived the Maclaurin series for $\sin x$ in Example 11.7:

$$\sin x = x - \frac{x^3}{3!} + \frac{x^5}{5!} - \frac{x^7}{7!} - \cdots .$$

We then stated that the Maclaurin series for $\cos x$ could be derived in a similar way. Term-by-term differentiation of the sine series is faster,

$$\cos x = \frac{d}{dx}(\sin x) = \frac{d}{dx}\left(x - \frac{x^3}{3!} + \frac{x^5}{5!} - \frac{x^7}{7!} - \cdots \right)$$

$$= 1 - \frac{x^2}{2!} + \frac{x^4}{4!} - \frac{x^6}{6!} + \cdots$$

$$= \sum_{n=0}^{\infty} \frac{(-1)^n}{(2n)!} x^{2n}, \qquad -\infty < x < \infty.$$

(b) Term-by-term differentiation of the series

$$\frac{1}{1-x} = 1 + x + x^2 + x^3 + \cdots + x^n + \cdots , \qquad |x| < 1$$

gives

$$\frac{1}{(1-x)^2} = 1 + 2x + 3x^2 + \cdots + nx^{n-1} + \cdots , \qquad |x| < 1.$$

Another differentiation yields

$$\frac{2}{(1-x)^3} = 2 + 3 \cdot 2x + 4 \cdot 3x^2 + \cdots + n(n-1)x^{n-2} + \cdots , \qquad |x| < 1.$$

Division by 2 now gives

$$\frac{1}{(1-x)^3} = 1 + 3x + 6x^2 + \cdots + \frac{n(n-1)}{2}x^{n-2} + \cdots$$

$$= \sum_{n=2}^{\infty} \frac{n(n-1)}{2} x^{n-2}$$

$$= \frac{1}{2} \sum_{n=0}^{\infty} (n+2)(n+1)x^n, \qquad |x| < 1.$$

EXAMPLE 11.18

Find the Taylor series about 1 for $\ln x$.

SOLUTION Noting that $\ln x$ is an antiderivative of $1/x$, we first expand $1/x$ in a Taylor series about 1:

$$\frac{1}{x} = \frac{1}{(x-1)+1} = 1 - (x-1) + (x-1)^2 - (x-1)^3 + \cdots, \qquad |x-1| < 1.$$

If we integrate this series term-by-term, we have

$$\ln|x| = \left[x - \frac{1}{2}(x-1)^2 + \frac{1}{3}(x-1)^3 - \frac{1}{4}(x-1)^4 + \cdots\right] + C.$$

Substitution of $x = 1$ implies that $0 = 1 + C$; that is, $C = -1$, and hence,

$$\ln|x| = (x-1) - \frac{1}{2}(x-1)^2 + \frac{1}{3}(x-1)^3 - \cdots$$

$$= \sum_{n=1}^{\infty} \frac{(-1)^{n+1}}{n}(x-1)^n.$$

According to Theorem 11.3, the radius of convergence of this series is also $R = 1$; that is, the open interval of convergence is $0 < x < 2$.

Comparison of the solutions in Examples 11.9 and 11.18 indicates once again the advantage of avoiding the use of Taylor's remainder formula.

Multiplication of Power Series

In Example 11.16 we added the Maclaurin series for $4/(x-4)$ and $1/(1+x)$ to obtain the Maclaurin series for $5x/(x^2 - 3x - 4)$. An alternative procedure might be to multiply the two series since

$$\frac{5x}{x^2 - 3x - 4} = 5x\left(\frac{1}{x-4}\right)\left(\frac{1}{x+1}\right)$$

$$= \frac{5x}{-4}\left(1 + \frac{x}{4} + \frac{x^2}{4^2} + \frac{x^3}{4^3} + \cdots\right)\left(1 - x + x^2 - x^3 + \cdots\right).$$

The rules of algebra demand that we multiply every term of the first series by every term of the second. If we do this and group all products with like powers of x, we obtain

$$\frac{5x}{x^2 - 3x - 4} = \frac{5x}{-4}\left[1 + \left(-1 + \frac{1}{4}\right)x + \left(1 - \frac{1}{4} + \frac{1}{4^2}\right)x^2 \right.$$

$$\left. + \left(-1 + \frac{1}{4} - \frac{1}{4^2} + \frac{1}{4^3}\right)x^3 + \cdots\right].$$

It is clear that the coefficient of x^n is a finite geometric series to which we can apply formula 11.16:

$$(-1)^n \left[1 - \frac{1}{4} + \frac{1}{4^2} - \cdots + \frac{(-1)^n}{4^n} \right] = (-1)^n \left[\frac{1 - \left(-\frac{1}{4}\right)^{n+1}}{1 + \frac{1}{4}} \right]$$

$$= (-1)^n \frac{4}{5} \left[1 - \left(-\frac{1}{4}\right)^{n+1} \right].$$

Consequently,

$$\frac{5x}{x^2 - 3x - 4} = \frac{5x}{-4} \sum_{n=0}^{\infty} (-1)^n \frac{4}{5} \left[1 - \left(-\frac{1}{4}\right)^{n+1} \right] x^n$$

$$= \sum_{n=0}^{\infty} (-1)^{n+1} \left[1 - \frac{(-1)^{n+1}}{4^{n+1}} \right] x^{n+1}$$

$$= \sum_{n=0}^{\infty} \left[(-1)^{n+1} - \frac{1}{4^{n+1}} \right] x^{n+1}$$

$$= \sum_{n=1}^{\infty} \left[(-1)^n - \frac{1}{4^n} \right] x^n.$$

For this example, then, multiplication as well as addition of power series leads to the Maclaurin series. Clearly, addition of power series is much simpler for this example, but we have at least demonstrated that power series can be multiplied together. That this is generally possible is stated in the following theorem.

THEOREM 11.6

If $f(x) = \sum_{n=0}^{\infty} a_n(x - c)^n$ and $g(x) = \sum_{n=0}^{\infty} b_n(x - c)^n$ have positive radii of convergence R_1 and R_2, respectively, then

$$f(x)g(x) = \sum_{n=0}^{\infty} d_n(x - c)^n, \tag{11.24a}$$

where

$$d_n = \sum_{i=0}^{n} a_i b_{n-i} = a_0 b_n + a_1 b_{n-1} + \cdots + a_{n-1} b_1 + a_n b_0, \tag{11.24b}$$

and the radius of convergence is the smaller of R_1 and R_2.

EXAMPLE 11.19

Find the Maclaurin series for $f(x) = [1/(x-1)]\ln(1-x)$.

SOLUTION   If we integrate the Maclaurin series

$$\frac{1}{1-x} = 1 + x + x^2 + x^3 + \cdots, \qquad |x| < 1,$$

we find

$$-\ln|1-x| = \left(x + \frac{x^2}{2} + \frac{x^3}{3} + \frac{x^4}{4} + \cdots\right) + C.$$

By setting $x = 0$, we obtain $C = 0$, and

$$\ln(1-x) = -x - \frac{x^2}{2} - \frac{x^3}{3} - \frac{x^4}{4} - \cdots.$$

We have dropped absolute value signs since the radius of convergence of the series is 1. We now form the Maclaurin series for $f(x)$:

$$\frac{1}{x-1}\ln(1-x) = \frac{-1}{1-x}\ln(1-x)$$

$$= \left(1 + x + x^2 + x^3 + \cdots\right)\left(x + \frac{x^2}{2} + \frac{x^3}{3} + \cdots\right)$$

$$= x + \left(1 + \frac{1}{2}\right)x^2 + \left(1 + \frac{1}{2} + \frac{1}{3}\right)x^3 + \cdots$$

$$= \sum_{n=1}^{\infty}\left(1 + \frac{1}{2} + \frac{1}{3} + \cdots + \frac{1}{n}\right)x^n.$$

Since both of the multiplied series have radius of convergence 1, so also does the Maclaurin series for $(x-1)^{-1}\ln(1-x)$. In other words, its open interval of convergence is $-1 < x < 1$.

EXAMPLE 11.20

Find the first three nonzero terms in the Maclaurin series for $\tan x$.

SOLUTION   If $\tan x = \sum a_n x^n$, then by setting $\tan x = \sin x / \cos x$, we have

$$\sin x = \cos x \sum_{n=0}^{\infty} a_n x^n.$$

We now substitute Maclaurin series for $\sin x$ and $\cos x$:

$$x - \frac{x^3}{3!} + \frac{x^5}{5!} - \frac{x^7}{7!} + \cdots = \left(1 - \frac{x^2}{2!} + \frac{x^4}{4!} - \frac{x^6}{6!} + \cdots\right)\left(a_0 + a_1 x + a_2 x^2 + \cdots\right).$$

According to the corollary to Theorem 11.4, two power series can be identical only if corresponding coefficients are equal. We therefore multiply the right side and equate coefficients of like powers of x:

$$x^0 : \quad 0 = a_0;$$

$$x : \quad 1 = a_1;$$

$$x^2 : \quad 0 = a_2 - \frac{a_0}{2!}, \quad \text{which implies } a_2 = 0;$$

$$x^3 : \quad -\frac{1}{3!} = a_3 - \frac{a_1}{2!}, \quad \text{which implies } a_3 = \frac{1}{2!} - \frac{1}{3!} = \frac{1}{3};$$

$$x^4 : \quad 0 = a_4 - \frac{a_2}{2!} + \frac{a_0}{4!}, \quad \text{from which } a_4 = 0;$$

$$x^5 : \quad \frac{1}{5!} = a_5 - \frac{a_3}{2!} + \frac{a_1}{4!}, \quad \text{from which } a_5 = \frac{1}{5!} + \frac{1}{6} - \frac{1}{4!} = \frac{2}{15}.$$

The first three nonzero terms in the Maclaurin series for $\tan x$ are therefore

$$\tan x = x + \frac{1}{3}x^3 + \frac{2}{15}x^5 + \cdots .$$

Binomial Expansion

One of the most widely used power series is the binomial expansion. We are well acquainted with the binomial theorem, which predicts the product $(a + b)^m$ for any positive integer m:

$$(a + b)^m = \sum_{n=0}^{m} \binom{m}{n} a^n b^{m-n}. \tag{11.25}$$

With the usual definition of the binomial coefficients,

$$\binom{m}{n} = \frac{m!}{(m-n)!\, n!} = \frac{m(m-1)(m-2)\cdots(m-n+1)}{n!},$$

the binomial theorem becomes

$$(a + b)^m = a^m + ma^{m-1}b + \frac{m(m-1)}{2!}a^{m-2}b^2 + \cdots + mab^{m-1} + b^m. \tag{11.26}$$

Even when m is not a positive integer, this form for the binomial theorem remains almost intact. To show this, we consider the power series

$$1 + \sum_{n=1}^{\infty} \frac{m(m-1)(m-2)\cdots(m-n+1)}{n!} x^n$$

for any real number m except a nonnegative integer. The radius of convergence of this power series is

$$R = \lim_{n \to \infty} \left| \frac{m(m-1)(m-2)\cdots(m-n+1)}{n!} \frac{(n+1)!}{m(m-1)(m-2)\cdots(m-n)} \right|$$

$$= \lim_{n \to \infty} \left| \frac{n+1}{m-n} \right| = 1.$$

The open interval of convergence is therefore $|x| < 1$. Whether the series converges at the end points $x = \pm 1$ depends on the value of m. For the time being, we will work on the interval $|x| < 1$, and at the end of the discussion, we will state the complete result. Let us denote the sum of the series by

$$f(x) = 1 + \sum_{n=1}^{\infty} \frac{m(m-1)(m-2)\cdots(m-n+1)}{n!} x^n, \qquad |x| < 1.$$

If we differentiate this series term-by-term according to Theorem 11.3,

$$f'(x) = \sum_{n=1}^{\infty} \frac{m(m-1)\cdots(m-n+1)}{(n-1)!} x^{n-1}, \qquad |x| < 1,$$

and then multiply both sides by x, we have

$$xf'(x) = \sum_{n=1}^{\infty} \frac{m(m-1)\cdots(m-n+1)}{(n-1)!} x^n, \qquad |x| < 1.$$

If we add these results, we obtain

$$f'(x) + xf'(x) = \sum_{n=1}^{\infty} \frac{m(m-1)\cdots(m-n+1)}{(n-1)!} x^{n-1}$$

$$+ \sum_{n=1}^{\infty} \frac{m(m-1)\cdots(m-n+1)}{(n-1)!} x^n.$$

We now change the variable of summation in the first sum:

$$(1+x)f'(x) = \sum_{n=0}^{\infty} \frac{m(m-1)\cdots(m-n)}{n!} x^n + \sum_{n=1}^{\infty} \frac{m(m-1)\cdots(m-n+1)}{(n-1)!} x^n.$$

When these summations are added over their common range, beginning at $n = 1$, and the $n = 0$ term in the first summation is written out separately, the result is

$$(1+x)f'(x) = m + \sum_{n=1}^{\infty} \frac{m(m-1)\cdots(m-n+1)}{(n-1)!} \left(\frac{m-n}{n} + 1 \right) x^n$$

$$= m \left[1 + \sum_{n=1}^{\infty} \frac{m(m-1)\cdots(m-n+1)}{n!} x^n \right]$$

$$= mf(x).$$

Consequently, the function $f(x)$ must satisfy the differential equation

$$\frac{f'(x)}{f(x)} = \frac{m}{1 + x}.$$

Integration immediately gives

$$\ln |f(x)| = m \ln |1 + x| + C, \qquad \text{or}$$

$$f(x) = D(1 + x)^m.$$

To evaluate D, we note that, from the original definition of $f(x)$ as the sum of the power series, $f(0) = 1$, and this implies that $D = 1$. Thus,

$$f(x) = (1 + x)^m,$$

and we may write finally that

$$(1 + x)^m = 1 + \sum_{n=1}^{\infty} \frac{m(m - 1)(m - 2) \cdots (m - n + 1)}{n!} x^n \qquad (11.27\text{a})$$

$$= 1 + mx + \frac{m(m - 1)}{2!} x^2 + \frac{m(m - 1)(m - 2)}{3!} x^3 + \cdots \quad (11.27\text{b})$$

valid for $|x| < 1$. This is called the binomial expansion of $(1 + x)^m$; it is the Maclaurin series for $(1 + x)^m$. We have verified the binomial expansion for m any real number except a nonnegative integer, but in the case of a nonnegative integer, the series terminates after $m + 1$ terms and is therefore valid for these values of m also. We mentioned earlier that the binomial expansion may also converge at the end points $x = \pm 1$ depending on the value of m. The complete result states that 11.27 is valid for

$$-\infty < x < \infty \quad \text{if } m \text{ is a nonnegative integer,}$$

$$-1 < x < 1 \quad \text{if } m \leq -1,$$

$$-1 < x \leq 1 \quad \text{if } -1 < m < 0,$$

$$-1 \leq x \leq 1 \quad \text{if } m > 0 \text{ but not an integer.}$$

It is not difficult to generalize this result to expand $(a + b)^m$ for real m. If $|b| < |a|$, we write

$$(a + b)^m = a^m \left(1 + \frac{b}{a}\right)^m$$

and now expand the bracketed term by means of 11.27:

$$(a + b)^m = a^m \left[1 + m\left(\frac{b}{a}\right) + \frac{m(m - 1)}{2!}\left(\frac{b}{a}\right)^2 + \cdots\right] \qquad |b| < |a|,$$

$$= a^m + ma^{m-1}b + \frac{m(m - 1)}{2!} a^{m-2}b^2 + \cdots \qquad |b| < |a|, \quad (11.28)$$

which, as we predicted, is equation 11.26 except that the series does not terminate.

EXAMPLE 11.21

Use the binomial expansion to find the Maclaurin series for $1/(1-x)^3$.

SOLUTION By 11.27b, we have

$$\frac{1}{(1-x)^3} = (1-x)^{-3}$$

$$= 1 + (-3)(-x) + \frac{(-3)(-4)}{2!}(-x)^2 + \frac{(-3)(-4)(-5)}{3!}(-x)^3 + \cdots$$

$$= 1 + 3x + \frac{3 \cdot 4}{2}x^2 + \frac{4 \cdot 5}{2}x^3 + \frac{5 \cdot 6}{2}x^4 + \cdots$$

$$= \frac{1}{2}\sum_{n=0}^{\infty}(n+1)(n+2)x^n, \qquad |x| < 1.$$

This result was also obtained in Example 11.17 by differentiation of the Maclaurin series for $1/(1-x)$.

EXAMPLE 11.22

Find the Maclaurin series for $\operatorname{Sin}^{-1} x$.

SOLUTION By the binomial expansion, we have

$$\frac{1}{\sqrt{1-x^2}} = 1 + \left(-\frac{1}{2}\right)(-x^2) + \frac{\left(-\frac{1}{2}\right)\left(-\frac{3}{2}\right)}{2!}(-x^2)^2 + \frac{\left(-\frac{1}{2}\right)\left(-\frac{3}{2}\right)\left(-\frac{5}{2}\right)}{3!}(-x^2)^3 + \cdots$$

$$= 1 + \frac{1}{2}x^2 + \frac{3}{2^2 2!}x^4 + \frac{3 \cdot 5}{2^3 3!}x^6 + \frac{3 \cdot 5 \cdot 7}{2^4 4!}x^8 + \cdots, \qquad |x| < 1.$$

Integration of this series gives

$$\operatorname{Sin}^{-1} x = \left(x + \frac{1}{2 \cdot 3}x^3 + \frac{3}{2^2 2! \, 5}x^5 + \frac{3 \cdot 5}{2^3 3! \, 7}x^7 + \frac{3 \cdot 5 \cdot 7}{2^4 4! \, 9}x^9 + \cdots\right) + C.$$

Evaluation of both sides of this equation at $x = 0$ gives $C = 0$. According to Theorem 11.3, the radius of convergence of this series must be 1, and we can write

$$\operatorname{Sin}^{-1} x = x + \sum_{n=1}^{\infty} \frac{1 \cdot 3 \cdot 5 \cdot \cdots \cdot (2n-1)}{2^n n! \, (2n+1)} x^{2n+1}$$

$$= x + \sum_{n=1}^{\infty} \frac{1 \cdot 2 \cdot 3 \cdot 4 \cdot 5 \cdot \cdots \cdot (2n-2)(2n-1)(2n)}{2 \cdot 4 \cdot \cdots \cdot (2n)2^n n! \, (2n+1)} x^{2n+1}$$

$$= x + \sum_{n=1}^{\infty} \frac{(2n)!}{(2n+1)2^{2n}(n!)^2} x^{2n+1}$$

$$= \sum_{n=0}^{\infty} \frac{(2n)!}{(2n+1)2^{2n}(n!)^2} x^{2n+1}, \qquad |x| < 1.$$

EXAMPLE 11.23

When measuring the velocity v of gas flow using a Pitot tube, the following equation for pressure P of the gas is encountered:

$$P(v) = P_0 \left[1 - \left(\frac{k-1}{2} \right) \left(\frac{v}{c} \right)^2 \right]^{k/(k-1)}$$

where P_0, c, and k are constants. The first two terms in the binomial expansion of P represent the situation for an incompressible gas. The third term in the expansion is regarded as the error in using the tube to determine velocity for compressible flow. Find the first three terms in the binomial expansion.

SOLUTION   Using equation 11.27,

$$P(v) = P_0 \left[1 - \left(\frac{k}{k-1} \right) \left(\frac{k-1}{2} \right) \left(\frac{v}{c} \right)^2 \right.$$

$$\left. + \frac{1}{2} \left(\frac{k-1}{2} \right) \left(\frac{k}{k-1} - 1 \right) \left(\frac{k-1}{2} \right)^2 \left(\frac{v}{c} \right)^4 - \cdots \right]$$

$$= P_0 \left(1 - \frac{k}{2c^2} v^2 + \frac{k}{8c^4} v^4 - \cdots \right).$$

EXERCISES 11.5

In Exercises 1–30 find the Maclaurin or Taylor series of the function about the indicated point.

* **1.** $f(x) = \dfrac{1}{3x+2}$ about $x = 0$ * **2.** $f(x) = \dfrac{1}{4+x^2}$ about $x = 0$

* **3.** $f(x) = \cos(x^2)$ about $x = 0$ * **4.** $f(x) = e^{5x}$ about $x = 0$

* **5.** $f(x) = e^x$ about $x = 3$ * **6.** $f(x) = e^{1-2x}$ about $x = 0$

* **7.** $f(x) = e^{1-2x}$ about $x = -1$ * **8.** $f(x) = \cosh x$ about $x = 0$

* **9.** $f(x) = \sinh x$ about $x = 0$

* **10.** $f(x) = x^4 + 3x^2 - 2x + 1$ about $x = 0$

* **11.** $f(x) = x^4 + 3x^2 - 2x + 1$ about $x = -2$

** **12.** $f(x) = \dfrac{1}{x+3}$ about $x = 2$

** **13.** $f(x) = \dfrac{x}{2x+5}$ about $x = 1$ ** **14.** $f(x) = \dfrac{x^2}{3-4x}$ about $x = 2$

** **15.** $f(x) = \dfrac{1}{\sqrt{1+x}}$ about $x = 0$ ** **16.** $f(x) = \ln(1+2x)$ about $x = 0$

** **17.** $f(x) = (1+3x)^{3/2}$ about $x = 0$

** **18.** $f(x) = \ln x$ about $x = 2$

** **19.** $f(x) = \ln(x+3)$ about $x = -1$

** **20.** $f(x) = 1/x$ about $x = 4$

** **21.** $f(x) = \dfrac{1}{(x+2)^3}$ about $x = 0$ ** **22.** $f(x) = \dfrac{1}{(2-x)^2}$ about $x = 3$

** **23.** $f(x) = \dfrac{1}{(x+3)^2}$ about $x = 1$ ** **24.** $f(x) = \dfrac{1}{x^2+8x+15}$ about $x = 0$

** **25.** $f(x) = \text{Tan}^{-1} x$ about $x = 0$ ** **26.** $f(x) = \sqrt{x+3}$ about $x = 0$

** **27.** $f(x) = \sqrt{x+3}$ about $x = 2$ ** **28.** $f(x) = (1-2x)^{1/3}$ about $x = 1$

** **29.** $f(x) = \dfrac{x^2}{(1+x^2)^2}$ about $x = 0$ ** **30.** $f(x) = x(1-x)^{1/3}$ about $x = 0$

In Exercises 31–33 find the first four nonzero terms in the Maclaurin series for the function.

** **31.** $f(x) = \tan 2x$ ** **32.** $f(x) = \sec x$

** **33.** $f(x) = e^x \sin x$

** **34.** Find the Maclaurin series for $\cos^2 x$.

In Exercises 35–38 find the Maclaurin series for the function.

** **35.** $f(x) = \dfrac{1}{x^6 - 3x^3 - 4}$ ** **36.** $f(x) = \text{Sin}^{-1}(x^2)$

** **37.** $f(x) = \dfrac{2x^2 + 4}{x^2 + 4x + 3}$ ** **38.** $f(x) = \ln\left[\dfrac{1 + x/\sqrt{2}}{1 - x/\sqrt{2}}\right]$

** **39.** Prove the corollary to Theorem 11.4.

** **40.** Prove that if a power series with positive radius of convergence has a sum of zero, $\sum_{n=0}^{\infty} a_n(x-c)^n = 0$, then $a_n = 0$ for all n.

** **41.** If, during a working day, one person drinks from a fountain every 30 seconds (on the average), then the probability that exactly n people drink in a time interval of length t seconds is given by the *Poisson distribution*:

$$P_n(t) = \frac{1}{n!}\left(\frac{t}{30}\right)^n e^{-t/30}.$$

Calculate $\sum_{n=0}^{\infty} P_n(t)$ and interpret the result.

** **42.** A certain experiment is to be performed until it is successful. The probability that it will be successful in any given attempt is p $(0 < p < 1)$, and therefore the probability that it will fail is $q = 1 - p$. The expected number of times that the experiment must be performed in order to be successful can be shown to be represented by the infinite series

$$\sum_{n=1}^{\infty} npq^{n-1} = \sum_{n=1}^{\infty} np(1-p)^{n-1}.$$

(a) What is the sum of this series?

(b) If p is the probability that a single die will come up 6, is the answer in (a) what you would expect?

** **43.** Find the Maclaurin series for the *error function* erf(x) defined by

$$\text{erf}(x) = \frac{2}{\sqrt{\pi}} \int_0^x e^{-t^2}\, dt.$$

** **44.** Find Maclaurin series for the *Fresnel integrals* $C(x)$ and $S(x)$ defined by

$$C(x) = \int_0^x \cos(\pi t^2/2)\, dt, \qquad S(x) = \int_0^x \sin(\pi t^2/2)\, dt.$$

In Exercises 45 and 46 use Maclaurin series to find a formula for the nth derivative of the function at $x = 0$.

** **45.** $f(x) = \dfrac{x}{(4 + 3x)^2}$

** **46.** $f(x) = xe^{-2x}$

In Exercises 47 and 48 use Taylor series to find a formula for the nth derivative of the function at $x = 2$.

** **47.** $f(x) = \dfrac{1}{3 + x}$

** **48.** $f(x) = xe^{-x}$

** **49.** Show that even-order derivatives of $x^2 \sin 2x$ at $x = 0$ are equal to zero.

** **50.** Show that odd-order derivatives of e^{-x^2} at $x = 0$ vanish.

Show that Bessel functions of the first kind (defined in Exercise 38 of Section 11.4) satisfy the properties in Exercises 51 and 52.

** **51.** $2m J_m(x) - x J_{m-1}(x) = x J_{m+1}(x)$

** **52.** $J_{m-1}(x) - J_{m+1}(x) = 2J'_m(x)$

** **53.** If the function $(1 - 2\mu x + x^2)^{-1/2}$ is expanded in a Maclaurin series in x,

$$\frac{1}{\sqrt{1 - 2\mu x + x^2}} = \sum_{n=0}^{\infty} P_n(\mu)x^n,$$

the coefficients $P_n(\mu)$ are called the *Legendre polynomials*. Find $P_0(\mu)$, $P_1(\mu)$, $P_2(\mu)$, and $P_3(\mu)$.

** **54.** (a) If we define $f(x) = x/(e^x - 1)$ at $x = 0$ as $f(0) = 1$, it turns out that $f(x)$ has a Maclaurin series expansion with positive radius of convergence. When this expansion is expressed in the form

$$\frac{x}{e^x - 1} = 1 + B_1 x + \frac{B_2}{2!}x^2 + \frac{B_3}{3!}x^3 + \cdots,$$

the coefficients $B_1, B_2, B_3, \ldots$ are called the *Bernoulli numbers*. Write this equation in the form

$$x = (e^x - 1)\left(1 + B_1 x + \frac{B_2}{2!}x^2 + \cdots\right),$$

and substitute the Maclaurin series for e^x to find the first five Bernoulli numbers.

(b) Show that the odd Bernoulli numbers all vanish for $n \geq 3$.

** **55.** Show that

$$e^{x(t-1/t)/2} = \sum_{n=0}^{\infty} J_n(x)t^n.$$

For a definition of $J_n(x)$, see Exercises 38 in Section 11.4.

11.6 Sums of Power Series

Theorem 11.3 provides an important technique for finding sums of power series. If a series with unknown sum can be reduced to a series with known sum by differentiations or integrations, then the unknown sum can be obtained when these operations are reversed.

EXAMPLE 11.24

Find the sum of the series $\displaystyle\sum_{n=0}^{\infty} (n+1)x^n$.

SOLUTION   The radius of convergence of the series is

$$R = \lim_{n\to\infty} \left| \frac{n+1}{n+2} \right| = 1.$$

If we denote the sum of the series by $S(x)$,

$$S(x) = \sum_{n=0}^{\infty} (n+1)x^n,$$

and using Theorem 11.3 to integrate the series term-by-term,

$$\int S(x)\,dx = \sum_{n=0}^{\infty} x^{n+1} + C.$$

But the series on the right is a geometric series with sum $x/(1-x)$, provided $|x| < 1$, and we may write therefore

$$\int S(x)\,dx = \frac{x}{1-x} + C, \qquad |x| < 1.$$

If we now differentiate this equation, we obtain

$$S(x) = \frac{(1-x)(1) - x(-1)}{(1-x)^2} = \frac{1}{(1-x)^2},$$

and therefore

$$\sum_{n=0}^{\infty} (n+1)x^n = \frac{1}{(1-x)^2}.$$

The open interval of convergence is $-1 < x < 1$. If we set $x = 1$, the series reduces to $\sum_{n=0}^{\infty} (n+1) = 1 + 2 + 3 + \cdots$, which clearly does not have a sum. At $x = -1$, we have $1 - 2 + 3 - 4 + \cdots$ which diverges also. The interval of convergence is therefore $-1 < x < 1$.

EXAMPLE 11.25

Find the sum of the series $\displaystyle\sum_{n=1}^{\infty} x^n/n$.

SOLUTION The radius of convergence of the series is

$$R = \lim_{n\to\infty} \left| \frac{\dfrac{1}{n}}{\dfrac{1}{n+1}} \right| = 1.$$

If we denote the sum of the series by

$$S(x) = \sum_{n=1}^{\infty} \frac{1}{n} x^n,$$

and differentiate the series term-by-term,

$$S'(x) = \sum_{n=1}^{\infty} x^{n-1}.$$

This is a geometric series with sum $1/(1-x)$, so that

$$S'(x) = \frac{1}{1-x}.$$

Antidifferentiation now gives

$$S(x) = -\ln(1-x) + C.$$

Since $S(0) = 0$, it follows that

$$0 = -\ln(1) + C.$$

Hence, $C = 0$, and $S(x) = -\ln(1-x)$. We have shown therefore that

$$\sum_{n=1}^{\infty} \frac{1}{n} x^n = -\ln(1-x).$$

The open interval of convergence is $-1 < x < 1$.

In Exercises 1–10 find the sum of the power series.

** **1.** $\displaystyle\sum_{n=1}^{\infty} nx^{n-1}$

** **2.** $\displaystyle\sum_{n=2}^{\infty} n(n-1)x^{n-2}$

** **3.** $\displaystyle\sum_{n=1}^{\infty} (n+1)x^{n-1}$

** **4.** $\displaystyle\sum_{n=1}^{\infty} n^2 x^{n-1}$

** **5.** $\displaystyle\sum_{n=1}^{\infty} (n^2+2n)x^n$

** **6.** $\displaystyle\sum_{n=0}^{\infty} \frac{1}{n+1}x^n$

** **7.** $\displaystyle\sum_{n=0}^{\infty} \frac{(-1)^n}{2n+1}x^{2n+1}$

** **8.** $\displaystyle\sum_{n=1}^{\infty} \frac{(-1)^n}{n}x^{2n}$

** **9.** $\displaystyle\sum_{n=2}^{\infty} n3^n x^{2n}$

** **10.** $\displaystyle\sum_{n=0}^{\infty} \left(\frac{n+1}{n+2}\right)x^n$

11.7 Applications of Taylor Series and Taylor's Remainder Formula

When a function is approximated by its nth-degree Taylor polynomial

$$f(x) \approx f(c) + f'(c)(x-c) + \frac{f''(c)}{2!}(x-c)^2 + \cdots + \frac{f^{(n)}(c)}{n!}(x-c)^n,$$

the error in doing so is the Taylor remainder

$$R_n(c,x) = \frac{f^{(n+1)}(z_n)}{(n+1)!}(x-c)^{n+1},$$

where z_n is between c and x. The smaller R_n is, the better the approximation. Because z_n is unknown, we cannot find $R_n(c,x)$. What we do is replace $f^{(n+1)}(z_n)$ by some larger value, thereby obtaining a maximum value for the error. For instance, consider using the first three terms of the Maclaurin series for e^x to approximate e^x on the interval $0 \le x \le 1/2$. Taylor's remainder formula, with $c = 0$, states that

$$e^x = 1 + x + \frac{x^2}{2} + R_2(0,x)$$

where

$$R_2(0,x) = \frac{d^3}{dx^3}(e^x)_{|x=z}\frac{x^3}{3!} = \frac{e^z x^3}{6},$$

and z is between 0 and x. Although z is unknown—except that it is between 0 and x—we can say that, because only the values $0 \le x \le \frac{1}{2}$ are under consideration, z must be less than $\frac{1}{2}$. It follows that

$$R_2(0,x) < \frac{e^{1/2}x^3}{6} \le \frac{\sqrt{e}(1/2)^3}{6} < 0.035.$$

Thus the quadratic function $1 + x + x^2/2$ approximates e^x on the interval $0 \leq x \leq 1/2$ with error no greater than 0.035.

In the following example, we determine the number of terms of a Maclaurin series required to guarantee a certain accuracy.

EXAMPLE 11.26

How many terms in the Maclaurin series for $\ln(1+x)$ guarantee a truncation error of less than 10^{-6} for any x in the interval $0 \leq x \leq 1/2$?

SOLUTION   The nth derivative of $\ln(1+x)$ is

$$\frac{d^n}{dx^n} \ln(1+x) = \frac{(-1)^{n+1}(n-1)!}{(x+1)^n}, \qquad n \geq 1,$$

and therefore Taylor's remainder formula with $c = 0$ states that

$$\ln(1+x) = x - \frac{x^2}{2} + \frac{x^3}{3} - \frac{x^4}{4} + \cdots + \frac{(-1)^{n+1}}{n}x^n + R_n(0,x),$$

where

$$R_n(0,x) = \frac{d^{n+1}}{dx^{n+1}} \ln(1+x)_{|x=z_n} \frac{x^{n+1}}{(n+1)!} = \frac{(-1)^n n!}{(z_n+1)^{n+1}} \frac{x^{n+1}}{(n+1)!}$$

$$= \frac{(-1)^n}{(n+1)(z_n+1)^{n+1}}x^{n+1}.$$

Since z_n is between 0 and x and $0 \leq x \leq 1/2$, we can state that x must be less than or equal to $1/2$, and z_n must be greater than 0. Hence,

$$|R_n(0,x)| < \frac{1}{(n+1)(1)^{n+1}}(1/2)^{n+1} = \frac{1}{(n+1)2^{n+1}},$$

This is less than 10^{-6} if

$$\frac{1}{(n+1)\,2^{n+1}} < 10^{-6} \qquad \text{or} \qquad (n+1)2^{n+1} > 10^6.$$

A calculator quickly indicates that the smallest value of n for which this is true is $n = 15$. Consequently, if $\ln(1+x)$ is approximated by the 15th degree polynomial

$$\ln(1+x) \approx x - \frac{x^2}{2} + \frac{x^3}{3} - \frac{x^4}{4} + \cdots + \frac{x^{15}}{15}$$

on the interval $0 \leq x \leq 1/2$, the truncation error is less than 10^{-6}.

Limits

We have customarily used L'Hôpital's rule to evaluate limits of the indeterminate form $0/0$. Maclaurin and Taylor series can sometimes be used to advantage. Consider

$$\lim_{x \to 0} \frac{x - \sin x}{x^3}.$$

Three applications of L'Hôpital's rule give a limit of $1/6$. Alternatively, if we substitute the Maclaurin series for $\sin x$,

$$\lim_{x \to 0} \frac{x - \sin x}{x^3} = \lim_{x \to 0} \frac{1}{x^3} \left[x - \left(x - \frac{x^3}{3!} + \frac{x^5}{5!} - \cdots \right) \right]$$

$$= \lim_{x \to 0} \left[\frac{1}{6} - \frac{x^2}{5!} + \cdots \right]$$

$$= \frac{1}{6}.$$

Here is another example.

EXAMPLE 11.27

Evaluate

$$\lim_{\lambda \to 0^+} \frac{\lambda^{-5}}{e^{c/\lambda} - 1}$$

where $c > 0$ is a constant (see also Exercise 58 in Section 4.10).

SOLUTION   We begin by making the change of variable $v = 1/\lambda$ in the limit,

$$\lim_{\lambda \to 0^+} \frac{\lambda^{-5}}{e^{c/\lambda} - 1} = \lim_{v \to \infty} \frac{v^5}{e^{cv} - 1}.$$

We now expand e^{cv} into its Maclaurin series,

$$\lim_{\lambda \to 0^+} \frac{\lambda^{-5}}{e^{c/\lambda} - 1} = \lim_{v \to \infty} \frac{v^5}{\left(1 + cv + \dfrac{c^2 v^2}{2!} + \cdots \right) - 1}$$

$$= \lim_{v \to \infty} \frac{v^5}{cv + \dfrac{c^2 v^2}{2!} + \cdots}.$$

If we now divide numerator and denominator by v^5,

$$\lim_{\lambda \to 0^+} \frac{\lambda^{-5}}{e^{c/\lambda} - 1} = \lim_{v \to \infty} \frac{1}{\dfrac{c}{v^4} + \dfrac{c^2}{2v^3} + \dfrac{c^3}{3! \, v^2} + \dfrac{c^4}{4! \, v} + \dfrac{c^5}{5!} + \dfrac{c^6 v}{6!} + \cdots} = 0.$$

Evaluation of Definite Integrals

In Section 9.9 we developed three numerical techniques for approximating definite integrals of functions $f(x)$ that have no obvious antiderivatives: the rectangular rule, the trapezoidal rule, and Simpson's rule. Each method divides the interval of integration into a number of subintervals and approximates $f(x)$ by a more elementary function on each subinterval. The rectangular rule replaces $f(x)$ by a step function, the trapezoidal rule uses a succession of linear functions, and Simpson's rule uses quadratic functions.

Another possibility is to replace $f(x)$ by a truncated power series (a polynomial) over the entire interval of integration. For instance, consider the definite integral

$$\int_0^{1/2} \frac{\sin x}{x}\, dx,$$

where $(\sin x)/x$ is defined as 1 at $x = 0$. The integral defines the area in Figure 11.16.

FIGURE 11.16

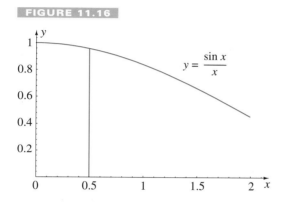

Taylor's remainder formula for $\sin x$ gives

$$\sin x = x - \frac{x^3}{3!} + \frac{x^5}{5!} - \cdots + \frac{d^n}{dx^n}(\sin x)_{|x=0}\frac{x^n}{n!} + R_n(0, x)$$

where

$$R_n(0, x) = \frac{d^{n+1}(\sin x)}{dx^{n+1}}\bigg|_{x=z_n} \frac{x^{n+1}}{(n + 1)!}$$

and z_n is between 0 and x. Therefore

$$\frac{\sin x}{x} = 1 - \frac{x^2}{3!} + \frac{x^4}{5!} - \cdots + \frac{1}{x}R_n(0, x),$$

where the term before $x^{-1}R_n(0, x)$ is $\dfrac{d^n}{dx^n}(\sin x)_{|x=0}\dfrac{x^{n-1}}{n!}$. When we take definite integrals,

$$\int_0^{1/2} \frac{\sin x}{x}dx = \int_0^{1/2}\left[1 - \frac{x^2}{3!} + \frac{x^4}{5!} - \cdots + \frac{1}{x}R_n(0, x)\right]dx$$

$$= \left\{x - \frac{x^3}{3\cdot 3!} + \frac{x^5}{5\cdot 5!} - \cdots\right\}_0^{1/2} + \int_0^{1/2}\frac{1}{x}R_n(0, x)\, dx,$$

$$= \frac{1}{2} - \frac{(1/2)^3}{3\cdot 3!} + \frac{(1/2)^5}{5\cdot 5!} - \cdots + \int_0^{1/2}\frac{1}{x}R_n(0, x)\, dx.$$

Now,

$$\left| \int_0^{1/2} \frac{1}{x} R_n(0, x) \, dx \right| \le \int_0^{1/2} \left| \frac{1}{x} \frac{d^{n+1}(\sin x)}{dx^{n+1}} \Big|_{x=z_n} \frac{x^{n+1}}{(n+1)!} \right| dx.$$

Since $\left| \dfrac{d^{n+1}(\sin x)}{dx^{n+1}} \Big|_{x=z_n} \right| \le 1$, it follows that

$$\left| \int_0^{1/2} \frac{1}{x} R_n(0, x) \, dx \right| \le \int_0^{1/2} \frac{x^n}{(n+1)!} \, dx$$

$$= \left\{ \frac{x^{n+1}}{(n+1)(n+1)!} \right\}_0^{1/2} = \frac{(1/2)^{n+1}}{(n+1)(n+1)!}.$$

Thus, if we write

$$\int_0^{1/2} \frac{\sin x}{x} \, dx \approx \frac{1}{2} - \frac{(1/2)^3}{3 \cdot 3!} + \frac{(1/2)^5}{5 \cdot 5!} = 0.493\,107\,639,$$

then the maximum error is

$$\frac{(1/2)^6}{6 \cdot 6!} \quad \text{or} \quad \frac{(1/2)^7}{7 \cdot 7!},$$

depending on whether we regard $x - x^3/3! + x^5/5!$ as a fifth or sixth degree approximation for $\sin x$. Since $(1/2)^7/(7 \cdot 7!) < 0.000\,000\,222$ gives a smaller error, we can say that

$$0.493\,107\,639 - 0.000\,000\,222 < \int_0^{1/2} \frac{\sin x}{x} \, dx < 0.493\,107\,693 + 0.000\,000\,222;$$

that is,

$$0.493\,107\,417 < \int_0^{1/2} \frac{\sin x}{x} \, dx < 0.493\,107\,861.$$

Consequently, using only three terms of the Maclaurin series for $(\sin x)/x$, we can say that to five decimal places

$$\int_0^{1/2} \frac{\sin x}{x} \, dx = 0.493\,11.$$

An easier analysis is given in Example 11.49 of Section 11.13. It uses *alternating series* instead of Taylor's remainder formula.

EXAMPLE 11.28

A very important function in engineering and physics is the error function $\text{erf}(x)$ defined by

$$\text{erf}(x) = \frac{2}{\sqrt{\pi}} \int_0^x e^{-t^2} \, dt.$$

Calculate $\text{erf}(1)$ correct to three decimal places.

SOLUTION   The value of erf(1) is defined by

$$\text{erf}(1) = \frac{2}{\sqrt{\pi}} \int_0^1 e^{-t^2} dt.$$

We do not relish the idea of using Taylor's remainder formula on e^{-t^2} because of the complexity of finding the nth derivative of e^{-t^2}. Instead, we transform the integral by setting $w = t^2$, in which case

$$\text{erf}(1) = \frac{2}{\sqrt{\pi}} \int_0^1 e^{-w} \frac{dw}{2\sqrt{w}} = \frac{1}{\sqrt{\pi}} \int_0^1 \frac{e^{-w}}{\sqrt{w}} dw.$$

Taylor's remainder formula applied to e^{-w} gives

$$e^{-w} = 1 - w + \frac{w^2}{2!} - \frac{w^3}{3!} + \cdots + \frac{(-1)^n w^n}{n!} + R_n(0, w)$$

where

$$R_n(0, w) = \frac{d^{n+1}}{dw^{n+1}}(e^{-w})|_{w=w_n} \frac{w^{n+1}}{(n+1)!} = \frac{(-1)^{n+1} e^{-w_n} w^{n+1}}{(n+1)!}.$$

Consequently,

$$\text{erf}(1) = \frac{1}{\sqrt{\pi}} \int_0^1 \frac{1}{\sqrt{w}} \left[1 - w + \frac{w^2}{2!} - \frac{w^3}{3!} + \cdots + \frac{(-1)^n w^n}{n!} + R_n(0, w) \right] dw$$

$$= \frac{1}{\sqrt{\pi}} \int_0^1 \left[\frac{1}{\sqrt{w}} - \sqrt{w} + \frac{w^{3/2}}{2!} - \frac{w^{5/2}}{3!} + \cdots \right.$$

$$\left. + \frac{(-1)^n w^{n-1/2}}{n!} + \frac{1}{\sqrt{w}} R_n(0, w) \right] dw$$

$$= \frac{1}{\sqrt{\pi}} \left\{ 2\sqrt{w} - \frac{2w^{3/2}}{3} + \frac{2w^{5/2}}{5 \cdot 2!} - \frac{2w^{7/2}}{7 \cdot 3!} + \cdots + \frac{2(-1)^n w^{n+1/2}}{(2n+1)n!} \right\}_0^1$$

$$+ \frac{1}{\sqrt{\pi}} \int_0^1 \frac{R_n(0, w)}{\sqrt{w}} dw$$

$$= \frac{2}{\sqrt{\pi}} \left[1 - \frac{1}{3} + \frac{1}{5 \cdot 2!} - \frac{1}{7 \cdot 3!} + \cdots + \frac{(-1)^n}{(2n+1)n!} \right]$$

$$+ \frac{1}{\sqrt{\pi}} \int_0^1 \frac{R_n(0, w)}{\sqrt{w}} dw.$$

Now,

$$\left| \int_0^1 \frac{R_n(0, w)}{\sqrt{w}} dw \right| \leq \int_0^1 \frac{1}{\sqrt{w}} \left| \frac{(-1)^{n+1} e^{-w_n} w^{n+1}}{(n+1)!} \right| dw$$

$$= \frac{1}{(n+1)!} \int_0^1 e^{-w_n} w^{n+1/2} dw.$$

Since $0 < w_n < w < 1$, we can say $e^{-w_n} \leq 1$. Thus,

$$\left| \int_0^1 \frac{R_n(0, w)}{\sqrt{w}} dw \right| \leq \frac{1}{(n+1)!} \left\{ \frac{2w^{n+3/2}}{2n+3} \right\}_0^1 = \frac{2}{(2n+3)(n+1)!}.$$

Consequently, when the series for erf(1) is truncated as

$$\text{erf}(1) = \frac{1}{\sqrt{\pi}} \left[1 - \frac{1}{3} + \frac{1}{5 \cdot 2!} - \frac{1}{7 \cdot 3!} + \cdots + \frac{(-1)^n}{(2n+1)n!} \right],$$

the maximum error is $2/[\sqrt{\pi}(2n+3)(n+1)!]$. When $n = 6$, the maximum error is $2/[15\sqrt{\pi}(7!)] < 0.000\,015$. The error in using

$$\text{erf}(1) \approx \frac{2}{\sqrt{\pi}} \left(1 - \frac{1}{3} + \frac{1}{5 \cdot 2!} - \frac{1}{7 \cdot 3!} + \frac{1}{9 \cdot 4!} - \frac{1}{11 \cdot 5!} + \frac{1}{13 \cdot 6!} \right) = 0.842\,714$$

is therefore less than $0.000\,015$; that is,

$$0.842\,714 - 0.000\,015 < \text{erf}(1) < 0.842\,714 + 0.000\,015,$$

or

$$0.842\,699 < \text{erf}(1) < 0.842\,729.$$

Thus, to three decimal places erf(1) = 0.843.

Alternating series can also be used in this example (see Example 11.50).

Differential Equations

Many differential equations arising in physics and engineering have solutions that can be expressed only in terms of infinite series. One such equation is Bessel's differential equation of order zero for a function $y = f(x)$:

$$xy'' + y' + xy = 0.$$

Before considering this somewhat difficult differential equation, we introduce the ideas through an easier example.

| **EXAMPLE 11.29**

Determine whether the differential equation

$$\frac{dy}{dx} - 2y = x$$

has a solution which can be expressed as a power series $y = \sum_{n=0}^{\infty} a_n x^n$ with positive radius of convergence.

SOLUTION   If

$$y = f(x) = \sum_{n=0}^{\infty} a_n x^n = a_0 + a_1 x + a_2 x^2 + \cdots$$

is to be a solution of the differential equation, we may substitute the power series into the differential equation,

$$\left(a_1 + 2a_2 x + 3a_3 x^2 + 4a_4 x^3 + \cdots\right) - 2\left(a_0 + a_1 x + a_2 x^2 + \cdots\right) = x.$$

We now gather together like terms in the various powers of x,

$$0 = (a_1 - 2a_0) + (2a_2 - 2a_1 - 1)x + (3a_3 - 2a_2)x^2 + (4a_4 - 2a_3)x^3 + \cdots.$$

Since the power series on the right has sum zero, its coefficients must all vanish (see Exercise 40 in Section 11.5), and therefore we must set

$$a_1 - 2a_0 = 0,$$

$$2a_2 - 2a_1 - 1 = 0,$$

$$3a_3 - 2a_2 = 0,$$

$$4a_4 - 2a_3 = 0,$$

and so on. These equations imply that

$$a_1 = 2a_0;$$

$$a_2 = \frac{1}{2}(1 + 2a_1) = \frac{1}{2}(1 + 4a_0);$$

$$a_3 = \frac{2}{3}a_2 = \frac{2}{3!}(1 + 4a_0);$$

$$a_4 = \frac{2}{4}a_3 = \frac{2^2}{4!}(1 + 4a_0).$$

The pattern emerging is

$$a_n = \frac{2^{n-2}}{n!}(1 + 4a_0), \qquad n \geq 2.$$

Thus,

$$f(x) = a_0 + 2a_0 x + \frac{1}{2}(1 + 4a_0)x^2 + \cdots + \frac{2^{n-2}}{n!}(1 + 4a_0)x^n + \cdots$$

$$= a_0 + 2a_0 x + \frac{1}{4}(1 + 4a_0)\left(\frac{2^2}{2!}x^2 + \frac{2^3}{3!}x^3 + \cdots + \frac{2^n}{n!}x^n + \cdots\right).$$

We can find the sum of the series in parentheses by noting that the Maclaurin series for e^{2x} is

$$e^{2x} = 1 + (2x) + \frac{(2x)^2}{2!} + \frac{(2x)^3}{3!} + \cdots.$$

Therefore the solution of the differential equation is

$$y = f(x) = a_0 + 2a_0 x + \frac{1}{4}(1 + 4a_0)\left[e^{2x} - 1 - 2x\right]$$

$$= -\frac{1}{4} - \frac{x}{2} + \frac{1}{4}(1 + 4a_0)e^{2x}$$

$$= Ce^{2x} - \frac{1}{4} - \frac{x}{2}.$$

Using power series to solve the differential equation in Example 11.29 is certainly not the most expedient method. A far simpler method will be discussed in Section 16.3. But the example clearly illustrated the procedure by which power series are used to solve differential equations. We now apply the procedure to Bessel's differential equation of order zero.

EXAMPLE 11.30

Find a power series solution $y = \sum_{n=0}^{\infty} a_n x^n$, with positive radius of convergence, for Bessel's differential equation of order zero,

$$xy'' + y' + xy = 0.$$

SOLUTION   In this example we abandon the $\cdots$ notation of Example 11.29, and maintain sigma notation throughout. When we substitute $y = \sum_{n=0}^{\infty} a_n x^n$ into the differential equation, we obtain

$$0 = x\sum_{n=2}^{\infty} n(n-1)a_n x^{n-2} + \sum_{n=1}^{\infty} na_n x^{n-1} + x\sum_{n=0}^{\infty} a_n x^n$$

$$= \sum_{n=2}^{\infty} n(n-1)a_n x^{n-1} + \sum_{n=1}^{\infty} na_n x^{n-1} + \sum_{n=0}^{\infty} a_n x^{n+1}.$$

In order to bring these three summations together as one, and combine terms in like powers of x, we lower the index of summation in the last term by 2:

$$0 = \sum_{n=2}^{\infty} n(n-1)a_n x^{n-1} + \sum_{n=1}^{\infty} na_n x^{n-1} + \sum_{n=2}^{\infty} a_{n-2} x^{n-1}.$$

We now combine the three summations over their common interval, beginning at $n = 2$, and write separately the $n = 1$ term in the second summation:

$$0 = a_1 + \sum_{n=2}^{\infty} [n(n-1)a_n + na_n + a_{n-2}]x^{n-1}.$$

But the only way a power series can be equal to zero is for all of its coefficients to be equal to zero; that is,

$$a_1 = 0; \qquad n(n-1)a_n + na_n + a_{n-2} = 0, \quad n \geq 2.$$

Thus,

$$a_n = -\frac{a_{n-2}}{n^2}, \qquad n \geq 2,$$

a recursive relation defining the unknown coefficient a_n of x^n in terms of the coefficient a_{n-2} of x^{n-2}. Since $a_1 = 0$, it follows that

$$0 = a_1 = a_3 = a_5 = \cdots .$$

For $n = 2$,

$$a_2 = -\frac{a_0}{2^2}.$$

For $n = 4$,

$$a_4 = -\frac{a_2}{4^2} = \frac{a_0}{2^2 4^2} = \frac{a_0}{2^4 (2!)^2}.$$

For $n = 6$,

$$a_6 = -\frac{a_4}{6^2} = \frac{-a_0}{2^4 (2!)^2 6^2} = -\frac{a_0}{2^6 (3!)^2}.$$

The solution is therefore

$$y = a_0 - \frac{a_0}{2^2} x^2 + \frac{a_0}{2^4 (2!)^2} x^4 - \frac{a_0}{2^6 (3!)^2} x^6 + \cdots$$

$$= a_0 \sum_{n=0}^{\infty} \frac{(-1)^n}{2^{2n} (n!)^2} x^{2n}.$$

The function defined by the infinite series

$$J_0(x) = \sum_{n=0}^{\infty} \frac{(-1)^n}{2^{2n} (n!)^2} x^{2n}, \qquad -\infty < x < \infty$$

is called the zero-order Bessel function of the first kind.

EXERCISES 11.7

In Exercises 1–10 find a maximum possible error in using the given terms of the Taylor series to approximate the function on the interval specified.

** **1.** $e^x \approx 1 + x + \dfrac{x^2}{2} + \dfrac{x^3}{6}$ for $0 \le x \le 0.01$

** **2.** $e^x \approx 1 + x + \dfrac{x^2}{2} + \dfrac{x^3}{6}$ for $0 \le x < 0.01$

** **3.** $e^x \approx 1 + x + \dfrac{x^2}{2} + \dfrac{x^3}{6}$ for $-0.01 \le x \le 0$

** **4.** $e^x \approx 1 + x + \dfrac{x^2}{2} + \dfrac{x^3}{6}$ for $|x| \le 0.01$

** **5.** $\sin x \approx x - \dfrac{x^3}{3!}$ for $0 \le x \le 1$

** **6.** $\cos x \approx 1 - \dfrac{x^2}{2!} + \dfrac{x^4}{4!}$ for $|x| \le 0.1$

** **7.** $\ln(1 - x) \approx -x - \dfrac{x^2}{2} - \dfrac{x^3}{3}$ for $0 \le x \le 0.01$

** **8.** $\dfrac{1}{(1 - x)^3} \approx 1 + 3x + 6x^2 + 10x^3$ for $|x| < 0.2$

** **9.** $\sin 3x \approx 3x - \dfrac{9x^3}{2} + \dfrac{81x^5}{40}$ for $|x| < \pi/100$

** **10.** $\ln x \approx (x - 1) - \dfrac{1}{2}(x - 1)^2 + \dfrac{1}{3}(x - 1)^3 - \dfrac{1}{4}(x - 1)^4$ for $|x - 1| \le 1/2$

In Exercises 11–14 evaluate the integral correct to three decimals.

** **11.** $\displaystyle\int_0^1 \dfrac{\sin x}{x}\,dx$ ** **12.** $\displaystyle\int_0^{1/2} \cos(x^2)\,dx$

** **13.** $\displaystyle\int_{-1}^1 x^{11} \sin x\,dx$ ** **14.** $\displaystyle\int_0^{0.3} e^{-x^2}\,dx$

In Exercises 15–20 use series to evaluate the limit.

** **15.** $\displaystyle\lim_{x \to 0} \dfrac{\tan x}{x}$ ** **16.** $\displaystyle\lim_{x \to 0} \dfrac{1 - \cos x}{x^2}$

** **17.** $\displaystyle\lim_{x \to 0} \dfrac{(1 - \cos x)^2}{3x^4}$ ** **18.** $\displaystyle\lim_{x \to 0} \dfrac{\sqrt{1 + x} - 1}{x}$

** **19.** $\displaystyle\lim_{x \to \infty} x \sin\left(\dfrac{1}{x}\right)$ ** **20.** $\displaystyle\lim_{x \to 0} \left(\dfrac{e^x + e^{-x}}{e^x - e^{-x}} - \dfrac{1}{x}\right)$

In Exercises 21–24 determine where the Maclaurin series for the function may be truncated in order to guarantee the accuracy indicated.

** **21.** $\sin(x/3)$ on $|x| \leq 4$ with error less than 10^{-3}

** **22.** $1/\sqrt{1+x^3}$ on $0 < x < 1/2$ with error less than 10^{-4}

** **23.** $\ln(1-x)$ on $|x| < 1/3$ with error less than 10^{-2}

** **24.** $\cos^2 x$ on $|x| < 0.1$ with error less than 10^{-3}

In Exercises 25–30 find a series solution in powers of x for the differential equation.

** **25.** $y' + 3y = 4$ ** **26.** $y'' + y' = 0$

** **27.** $xy' - 4y = 3x$ ** **28.** $4xy'' + 2y' + y = 0$

** **29.** $y'' + y = 0$ ** **30.** $xy'' + y = 0$

** **31.** Find the natural logarithm of 0.999 999 999 9 accurate to 15 decimal places.

** **32.** In special relativity theory, the kinetic energy K of an object moving with speed v is defined by

$$K = c^2(m - m_0),$$

where c is a constant (the speed of light), m_0 is the rest mass of the object, and m is its mass when moving with speed v. The masses m and m_0 are related by

$$m = \frac{m_0}{\sqrt{1 - v^2/c^2}}.$$

Use the binomial expansion to show that

$$K = \frac{1}{2}m_0 v^2 + m_0 c^2 \left(\frac{3}{8}\frac{v^4}{c^4} + \frac{5}{16}\frac{v^6}{c^6} + \cdots \right),$$

and hence, to a first approximation, kinetic energy is defined by the classical expression $m_0 v^2/2$.

** **33.** The ellipse $b^2 x^2 + a^2 y^2 = a^2 b^2$ can be represented parametrically by

$$x = a\cos t, \quad y = b\sin t, \quad 0 \leq t < 2\pi.$$

(a) Show that the length of the circumference of the ellipse is defined by the definite integral

$$L = 4b \int_0^{\pi/2} \sqrt{1 - k^2 \sin^2 t}\, dt, \quad k^2 = 1 - \frac{a^2}{b^2}.$$

(b) Use the binomial expansion to show that

$$L = 2\pi b \left(1 - \frac{k^2}{4} - \frac{3k^4}{64} - \cdots \right)$$

so that to a first approximation, L is the circumference of a circle of radius b.

** **34.** Planck's law for the energy density Ψ of blackbody radiation of wavelength λ states that

$$\Psi(\lambda) = \frac{8\pi ch\lambda^{-5}}{e^{ch/(\lambda kT)} - 1},$$

where $h > 0$ is Planck's constant, c is the (constant) speed of light, and T is temperature, also assumed constant. Show that, for long wavelengths, Planck's law reduces to the Rayleigh-Jeans law:

$$\Psi(\lambda) = \frac{8\pi kT}{\lambda^4}.$$

** **35.** In the figure below charges of $q > 0$ and $-q$ coulombs are a distance d apart. The configuration is called an electric dipole.

(a) The electric field at point P due to these charges is

$$E = \frac{q}{4\pi\epsilon_0(x - d/2)^2} - \frac{q}{4\pi\epsilon_0(x + d/2)^2}$$

where ϵ_0 is a constant. Verify that E can be expressed in the form

$$E = \frac{q}{4\pi\epsilon_0 x^2}\left[\left(1 - \frac{d}{2x}\right)^{-2} - \left(1 + \frac{d}{2x}\right)^{-2}\right].$$

(b) Use the binomial expansion to show that when d is very much less than x, E can be approximated by

$$E = \frac{qd}{2\pi\epsilon_0 x^3}.$$

** **36.** The potential at point P in the diagram below due to the electric dipole is

$$V = \frac{1}{4\pi\epsilon_0}\left(\frac{q}{r_1} - \frac{q}{r_2}\right)$$

where ϵ_0 is a constant.

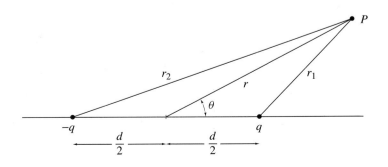

(a) Use the cosine law to express V in the form

$$V = \frac{1}{4\pi\epsilon_0} \left(\frac{1}{\sqrt{r^2 + d^2/4 - rd\cos\theta}} - \frac{1}{\sqrt{r^2 + d^2/4 + rd\cos\theta}} \right).$$

(b) Use the binomial expansion to show that when d is very much less than r, V can be approximated by

$$V = \frac{qd\cos\theta}{4\pi\epsilon_0 r^2}.$$

11.8 Convergence of Sequences of Numbers

In Section 11.1 we introduced the basic ideas of convergence for sequences of numbers. Examples were so simple that we had no difficulty determining whether sequences converged or diverged. As this is not always the case, we discuss two important ways to show that a sequence converges. Our concern is particularly with recursively-defined sequences, since it is usually straightforward to determine whether an explicit sequence is convergent or divergent. For instance, consider the sequence

$$c_1 = 2, \qquad c_{n+1} = 5 + \sqrt{2 + c_n}, \qquad n \geq 1,$$

of Example 11.2 in Section 11.1. Its next four terms are

$$c_2 = 6.732, \qquad c_3 = 7.955, \qquad c_4 = 8.155, \qquad c_5 = 8.187.$$

We suspect that the sequence has a limit. Why? Because terms are rapidly getting closer and closer to each other, or, to put it another way, differences between terms are rapidly approaching zero. But this does not guarantee convergence. Differences between terms of the sequences $\{\sqrt{n}\}$ and $\{\ln n\}$ get smaller and smaller as n increases, but neither sequence has a limit. The following two definitions lead to Theorem 11.7, which can be very useful in verifying convergence of sequences like this.

DEFINITION 11.2

A sequence $\{c_n\}$ is said to be

$\quad$ (i) **increasing** if $\quad c_{n+1} > c_n \quad$ for all $n \geq 1$; $\qquad$ (11.29a)

$\quad$ (ii) **nondecreasing** if $\quad c_{n+1} \geq c_n \quad$ for all $n \geq 1$; $\qquad$ (11.29b)

$\quad$ (iii) **decreasing** if $\quad c_{n+1} < c_n \quad$ for all $n \geq 1$; $\qquad$ (11.29c)

$\quad$ (iv) **nonincreasing** if $\quad c_{n+1} \leq c_n \quad$ for all $n \geq 1$. $\qquad$ (11.29d)

If a sequence satisfies any one of these four properties, it is said to be **monotonic**.

> **DEFINITION 11.3**
>
> A sequence $\{c_n\}$ is said to have an **upper bound** U (be bounded above by U) if
>
> $$c_n \leq U \qquad\qquad (11.30\text{a})$$
>
> for all $n \geq 1$. It has a **lower bound** V (is bounded below by V) if
>
> $$c_n \geq V \qquad\qquad (11.30\text{b})$$
>
> for all $n \geq 1$.

Note that if U is an upper bound for a sequence, then any number greater than U is also an upper bound. If V is a lower bound, so too is any number smaller than V.

Let us illustrate these definitions with some simple explicit sequences before stating our first convergence theorem and applying it to the above recursive sequence. The sequence

$$\left\{ \frac{1}{2^{n-1}} \right\} = 1, \ \frac{1}{2}, \ \frac{1}{4}, \ \frac{1}{8}, \ \cdots$$

is decreasing, has an upper bound $U = 1$, and a lower bound $V = 0$. The sequence

$$\left\{ \frac{n}{n+1} \right\} = \frac{1}{2}, \ \frac{2}{3}, \ \frac{3}{4}, \ \frac{4}{5}, \ \cdots$$

is increasing, has an upper bound $U = 5$, and a lower bound $V = -2$. The sequence

$$\{(-1)^{n+1}\} = 1, \ -1, \ 1, \ -1, \ \cdots$$

is not monotonic, has an upper bound $U = 1$, and a lower bound $V = -3$.

The above recursive sequence $c_1 = 1$, $c_{n+1} = 5 + \sqrt{2 + c_n}$ appears to be increasing, $V = 0$ is obviously a lower bound, and $U = 10$ appears to be an upper bound. The reason why we would verify these conjectures is contained in the following theorem.

> **THEOREM 11.7**
>
> A bounded, monotonic sequence has a limit.

To expand on this statement somewhat, consider a sequence $\{c_n\}$ whose terms are illustrated graphically in Figure 11.17a. Suppose that the sequence is increasing and therefore monotonic, and that U is an upper bound for the sequence. We have shown the upper bound as a horizontal line in the figure; c_1 is a lower bound. Our intuition suggests that because the terms in the sequence always increase, and they never exceed U, the sequence must have a limit. Theorem 11.7 confirms this. The theorem does not suggest the value of the limit, but obviously it must be less than or equal to U.

Similarly, when a sequence is decreasing or nonincreasing and has a lower bound V (Figure 11.17b), it must approach a limit that is greater than or equal to V.

FIGURE 11.17a

FIGURE 11.17b

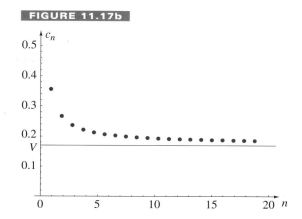

Another way of stating Theorem 11.7 is as follows.

COROLLARY
A monotonic sequence has a limit if and only if it is bounded.

We are now prepared to give a complete and typical discussion for the recursive sequence $c_1 = 1$, $c_{n+1} = 5 + \sqrt{2 + c_n}$.

EXAMPLE 11.31

Verify that the sequence

$$c_1 = 1, \qquad c_{n+1} = 5 + \sqrt{2 + c_n}, \qquad n \geq 1,$$

has a limit, and find it.

SOLUTION   The first five terms of the sequence are

$$c_1 = 1, \quad c_2 = 6.732, \quad c_3 = 7.955, \quad c_4 = 8.155, \quad c_5 = 8.187.$$

They suggest that the sequence is increasing; that is, $c_{n+1} > c_n$. To prove this we use mathematical induction (see Appendix A). Certainly the inequality is valid for $n = 1$ since $c_2 > c_1$. Suppose k is an integer for which $c_{k+1} > c_k$. Then

$$2 + c_{k+1} > 2 + c_k,$$

from which

$$\sqrt{2 + c_{k+1}} > \sqrt{2 + c_k}.$$

It follows that

$$5 + \sqrt{2 + c_{k+1}} > 5 + \sqrt{2 + c_k}.$$

The left side is c_{k+2} and the right side is c_{k+1}. Therefore we have proved that $c_{k+2} > c_{k+1}$. Hence, by mathematical induction $c_{n+1} > c_n$ for all $n \geq 1$. Since the sequence is increasing, its first term $c_1 = 1$ must be a lower bound. Certainly any upper bound, if one exists, must be at least 8.187 (c_5). We can take any number greater than 8.187 and use mathematical induction to test whether it is indeed an upper bound. It appears that $U = 10$ might be a reasonable guess for an upper bound for this sequence, and we verify this by induction as follows. Clearly $c_1 < 10$. We suppose that k is some integer for which $c_k < 10$. Then

$$c_{k+1} = 5 + \sqrt{2 + c_k} < 5 + \sqrt{2 + 10} = 5 + \sqrt{12} < 10.$$

By mathematical induction, $c_n < 10$ for $n \geq 1$.

Since the sequence is monotonic and bounded, Theorem 11.7 guarantees that it has a limit, call it L. To evaluate L, we take limits on each side of the equation defining the sequence recursively:

$$\lim_{n \to \infty} c_{n+1} = \lim_{n \to \infty} \left(5 + \sqrt{2 + c_n} \right).$$

It is important to note that this cannot be done until the conditions of Theorem 11.7 have been checked. Since terms c_n of the sequence approach L as $n \to \infty$, it follows that $5 + \sqrt{2 + c_n}$ approaches $5 + \sqrt{2 + L}$. Furthermore, as $n \to \infty$, c_{n+1} must also approach L. Do not make the mistake of saying that c_{n+1} approaches $L + 1$ as $n \to \infty$. Think about what $\lim_{n \to \infty} c_{n+1}$ means. We conclude therefore that

$$L = 5 + \sqrt{2 + L}.$$

If we transpose the 5 and square both sides of the equation, we obtain the quadratic equation

$$L^2 - 11L + 23 = 0,$$

with solutions

$$L = \frac{11 \pm \sqrt{29}}{2}.$$

Only the positive square root satisfies the original equation $L = 5 + \sqrt{2 + L}$ defining L, so that $L = (11 + \sqrt{29})/2$. The other root, $(11 - \sqrt{29})/2 \approx 2.8$ can also be eliminated on the grounds that all terms beyond the first are greater than 6.

Sequences do not have to be monotonic to be convergent. Convergence can occur for other reasons. The following example illustrates a second common way for sequences to converge. Again we have chosen a recursive sequence as illustration because for an explicit sequence the limit is usually obvious. Consider the recursive sequence

$$c_1 = 2, \qquad c_{n+1} = 2 + \frac{1}{c_n}, \qquad n \geq 1.$$

The first six terms of the sequence are

$$c_1 = 2, \quad c_2 = 2.5, \quad c_3 = 2.4, \quad c_4 = 2.417, \quad c_5 = 2.413\,8, \quad c_6 = 2.414\,29.$$

The terms seem to be clustering around a number close to 2.414, one larger, one smaller, one larger, etc. (Figure 11.18). Unfortunately, lack of monotony precludes the possibility of using Theorem 11.7 to discuss convergence. In Theorem 11.8, we discuss properties that imply convergence for sequences of this type, but it is helpful to first illustrate these properties with a specific example such as the one above. What are the properties that lead us to believe that this

FIGURE 11.18

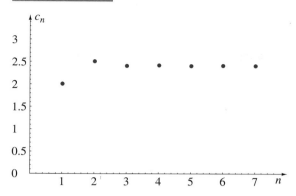

sequence converges? First, the terms are not monotonic; they are "up-down-up-down-up-down."
How do we say this mathematically? Differences between successive terms in the sequence are

$$c_2 - c_1 = 2.5 - 2 = 0.5,$$

$$c_3 - c_2 = 2.4 - 2.5 = -0.1,$$

$$c_4 - c_3 = 2.417 - 2.4 = 0.017,$$

$$c_5 - c_4 = 2.413\,8 - 2.417 = -0.003\,2,$$

$$c_6 - c_5 = 2.414\,29 - 2.413\,8 = 0.000\,49.$$

The fact that these differences are alternately positive and negative implies that the terms in the
sequence $\{c_n\}$ are "up-down-up-down-up-down." This is not enough to guarantee convergence,
however. For example, the sequence $1, -1, 1, -1, 1, -1, \ldots$ is "up-down-up-down-up-
down," but it does not converge. The added feature of the above sequence is that absolute values
of the differences

$$|c_2 - c_1| = 0.5, \quad |c_3 - c_2| = 0.1, \quad |c_4 - c_3| = 0.017,$$

$$|c_5 - c_4| = 0.003\,2, \quad |c_6 - c_5| = 0.000\,49$$

seem to form a decreasing sequence with limit zero. We have said "seem" because we have
not proved that these properties hold for all differences, only the first five. We shall provide
a general verification after stating the next theorem. However, it is the "up-down-up-down-
up-down" nature of the sequence together with the fact that absolute values of the differences
decrease and approach zero that lead us to believe that the sequence $\{c_n\}$ has a limit. This is
formalized in the following theorem.

THEOREM 11.8

Suppose a sequence $\{c_n\}$ has the following properties:

1. differences $c_{n+1} - c_n$ alternate in sign;
2. absolute values of these differences $|c_{n+1} - c_n|$ are decreasing;
3. absolute values of the differences $|c_{n+1} - c_n|$ approach 0.

Then the sequence $\{c_n\}$ converges, and its limit lies between any two successive terms in
the sequence.

A sequence that satisfies condition 1. of this theorem is said to be an **oscillating sequence**. Its terms are "up-down-up-down-up-down." When terms also satisfy conditions 2. and 3., they oscillate about a limit, one below, one above, one below, one above, etc., but gradually they get closer and closer to the limit.

We now verify that the above sequence $c_1 = 2$, $c_{n+1} = 2 + 1/c_n$ is oscillating and convergent. The difference between the $(n + 1)^{\text{th}}$ and n^{th} terms is

$$c_{n+1} - c_n = \left(2 + \frac{1}{c_n}\right) - \left(2 + \frac{1}{c_{n-1}}\right) = \frac{-(c_n - c_{n-1})}{c_n c_{n-1}}.$$

Since all terms in the sequence are clearly positive, the denominator of this expression is positive. It follows that the difference $c_{n+1} - c_n$ must have the opposite sign to the previous difference $c_n - c_{n-1}$; that is, the differences $c_{n+1} - c_n$ alternate in sign. Furthermore, since 2 is a lower bound for all terms in the sequence, it follows that for $n \geq 2$,

$$|c_{n+1} - c_n| = \frac{|c_n - c_{n-1}|}{c_n c_{n-1}} < \frac{|c_n - c_{n-1}|}{(2)(2)} = \frac{|c_n - c_{n-1}|}{4}.$$

But if each difference is less than one-quarter the previous difference, the differences must be decreasing and have limit zero. By Theorem 11.8 this sequence has a limit L which we obtain by taking limits in the recursive definition,

$$\lim_{n \to \infty} c_{n+1} = \lim_{n \to \infty} \left(2 + \frac{1}{c_n}\right) \quad \Longrightarrow \quad L = 2 + \frac{1}{L}.$$

Of the two solutions $1 \pm \sqrt{2}$ of this equation, only $L = 1 + \sqrt{2}$ lies between c_1 and c_2.

Notice that we did not use mathematical induction to verify that the above sequence is oscillating and convergent. This is characteristic of oscillating sequences; mathematical induction is not required to verify the three properties of Theorem 11.8. It may be necessary to use induction to prove ancillary results (such as a lower bound for the sequence), but induction is not needed to verify the properties of Theorem 11.8.

An oscillating sequence results when the method of successive approximations is applied to the equation $f(x) = x^3 + 25x - 50 = 0$ for the root between $x = 1$ and $x = 2$. With

$$x_1 = 1, \qquad x_{n+1} = \frac{50 - x_n^3}{25}, \qquad n \geq 1,$$

the next fifteen terms are

$$
\begin{array}{lll}
x_2 = 1.96 & x_3 = 1.70 & x_4 = 1.80 \\
x_5 = 1.77 & x_6 = 1.778\,2 & x_7 = 1.775\,1 \\
x_8 = 1.776\,3 & x_9 = 1.775\,81 & x_{10} = 1.776\,00 \\
x_{11} = 1.775\,93 & x_{12} = 1.775\,954 & x_{13} = 1.775\,945 \\
x_{14} = 1.775\,948\,2 & x_{15} = 1.775\,946\,9 & x_{16} = 1.775\,947\,4.
\end{array}
$$

The terms are indeed oscillating but appear to be approaching a limit. It is not necessary to verify the conditions of Theorem 11.8. We have what we feel is a potential candidate for the solution of the equation $x^3 + 25x - 50 = 0$ between $x = 1$ and $x = 2$. To verify that 1.775 947 is the solution correctly rounded to six decimal places, we use the Intermediate Value Theorem with the calculations

$$f(1.775\,946\,5) = -2.7 \times 10^{-5}, \qquad f(1.775\,947\,5) = 7.3 \times 10^{6}.$$

Now that we know what it means for a sequence to have a limit, and how to find limits, we can be more precise. To give a mathematical definition for the limit of a sequence, we start with our intuitive description and make a succession of paraphrases, each of which is one step closer to a precise definition:

A sequence $\{c_n\}$ has limit L if terms get arbitrarily close to L, and stay close to L, as n gets larger and larger.

A sequence $\{c_n\}$ has limit L if terms can be made arbitrarily close to L by choosing n sufficiently large.

A sequence $\{c_n\}$ has limit L if differences $|c_n - L|$ can be made arbitrarily close to 0 by choosing n sufficiently large.

A sequence $\{c_n\}$ has limit L if given any real number $\epsilon > 0$, no matter how small, we can make differences $|c_n - L|$ less than ϵ by choosing n sufficiently large.

Finally, we arrive at the following definition.

DEFINITION 11.4

A sequence $\{c_n\}$ has **limit** L if, for any given $\epsilon > 0$, there exists an integer N such that for all $n > N$

$$|c_n - L| < \epsilon.$$

This definition puts our intuitive idea of a limit in precise terms. For those who have studied Section 2.5, note the similarity between Definition 11.4 and Definition 2.2. For a better understanding of Definition 11.4, it is helpful to consider its geometric interpretation. The inequality $|c_n - L| < \epsilon$, when written in the form $L - \epsilon < c_n < L + \epsilon$ is interpreted as a horizontal band of width 2ϵ around L (Figure 11.19a). Definition 11.4 requires that no matter how small ϵ, we can find a stage, denoted by N, beyond which all terms in the sequence are contained in the horizontal band. For the sequence and ϵ in Figure 11.19a, N must be chosen as shown. For the same sequence, but a smaller ϵ, N must be chosen correspondingly larger (Figure 11.19b). Proofs of some results in the rest of this chapter require a working knowledge of this definition. As an example, we use it to verify that a sequence cannot have two limits, a fact that we have implicitly assumed throughout our discussions.

FIGURE 11.19a

FIGURE 11.19b

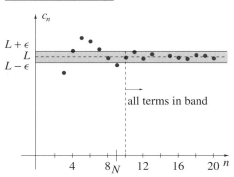

THEOREM 11.9

A sequence can have at most one limit.

PROOF We prove this by showing that a sequence cannot have two distinct limits. Suppose to the contrary that a sequence $\{c_n\}$ has two distinct limits L_1 and L_2, where $L_2 > L_1$, and let $L_2 - L_1 = \delta$. If we set $\epsilon = \delta/3$, then according to Definition 11.4, there exists an integer N_1 such that for all $n > N_1$

$$|c_n - L_1| < \epsilon = \delta/3;$$

that is, for $n > N_1$, all terms in the sequence are within a distance $\delta/3$ of L_1.

But since L_2 is also supposed to be a limit, there exists an N_2 such that for $n > N_2$, all terms in the sequence are within a distance $\epsilon = \delta/3$ of L_2:

$$|c_n - L_2| < \epsilon = \delta/3.$$

But this is impossible (Figure 11.20), if L_1 and L_2 are a distance δ apart. This contradiction therefore implies that L_1 and L_2 are the same; that is, $\{c_n\}$ cannot have two limits. ∎

FIGURE 11.20

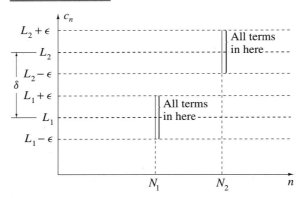

The following theorem, which states some of the properties of convergent sequences, can also be proved using Definition 11.4. Only the first two parts, however, are straightforward (see Exercises 58, 65, and 66).

THEOREM 11.10

If sequences $\{c_n\}$ and $\{d_n\}$ have limits C and D, then:
  (i) $\{kc_n\}$ has limit kC if k is a constant;
  (ii) $\{c_n \pm d_n\}$ has limit $C \pm D$;
  (iii) $\{c_n d_n\}$ has limit CD;
  (iv) $\{c_n/d_n\}$ has limit C/D provided that $D \neq 0$, and none of the $d_n = 0$.

EXERCISES 11.8

In Exercises 1–24 determine whether the statement is true or false. Verify any statement that is true, and give a counter example to any statement that is false.

∗ **1.** A sequence can be increasing and nondecreasing.

∗ **2.** A monotonic sequence must be bounded.

∗ **3.** An increasing sequence must have a lower bound.

∗ **4.** A decreasing sequence must have a lower bound.

∗ **5.** An increasing sequence with a lower bound must have a limit.

∗ **6.** An increasing sequence with an upper bound must have a limit.

∗ **7.** If a sequence diverges, then either $\lim_{n\to\infty} c_n = \infty$, or $\lim_{n\to\infty} c_n = -\infty$.

∗ **8.** A sequence cannot be both increasing and decreasing.

∗ **9.** A sequence can be both nonincreasing and nondecreasing.

∗ **10.** If a sequence is monotonic and has a limit, it must be bounded.

∗ **11.** If a sequence is bounded and has a limit, it must be monotonic.

∗ **12.** If a sequence is bounded, but not monotonic, it cannot have a limit.

∗ **13.** If a sequence has a limit, it must be bounded.

∗ **14.** If a sequence is not monotonic, it cannot have a limit.

∗ **15.** If all terms of a sequence $\{c_n\}$ are less than U, and $L = \lim_{n\to\infty} c_n$ exists, then L must be less than U.

∗ **16.** A sequence $\{c_n\}$ has a limit if and only if $\{c_n^2\}$ has a limit.

∗ **17.** An oscillating sequence must converge.

∗ **18.** If terms of a sequence are alternately positive and negative, the sequence is oscillating.

∗∗ **19.** A sequence $\{c_n\}$ of positive numbers converges if $c_{n+1} < c_n/2$.

∗∗ **20.** A sequence $\{c_n\}$ of numbers converges if $c_{n+1} < c_n/2$.

∗∗ **21.** If an increasing sequence has a limit L, then L must be equal to the smallest upper bound for the sequence.

∗∗ **22.** If an infinite number of terms of a sequence all have the same value a, then either a is the limit of the sequence or the sequence has no limit.

∗∗ **23.** If absolute values of differences between terms of an oscillating sequence decrease, the sequence converges.

∗∗ **24.** If terms of an oscillating sequence approach zero, then terms must be alternately positive and negative.

In Exercises 25–41 discuss, with proofs, whether the sequence is monotonic, and whether it has an upper bound, a lower bound, and a limit.

** **25.** $c_1 = 1$, $c_{n+1} = \dfrac{1}{10}\left(c_n^3 + 12\right)$, $n \geq 1$

** **26.** $c_1 = 0$, $c_{n+1} = \dfrac{1}{12}\left(c_n^4 + 5\right)$, $n \geq 1$

** **27.** $c_1 = 3$, $c_{n+1} = \sqrt{5 + c_n}$, $n \geq 1$

** **28.** $c_1 = 1$, $c_{n+1} = \sqrt{5 + c_n}$, $n \geq 1$

** **29.** $c_1 = 5$, $c_{n+1} = 1 + \sqrt{6 + c_n}$, $n \geq 1$

** **30.** $c_1 = 3$, $c_{n+1} = 1 + \sqrt{6 + c_n}$, $n \geq 1$

** **31.** $c_1 = 1$, $c_{n+1} = 4 - \sqrt{5 - c_n}$, $n \geq 1$

** **32.** $c_1 = 4$, $c_{n+1} = 4 - \sqrt{5 - c_n}$, $n \geq 1$

** **33.** $c_1 = 2$, $c_{n+1} = \dfrac{1}{3 - c_n}$, $n \geq 1$

** **34.** $c_1 = 1$, $c_{n+1} = \dfrac{1}{4 - 2c_n}$, $n \geq 1$

** **35.** $c_1 = 1$, $c_{n+1} = \dfrac{7}{16 - 8c_n^2}$, $n \geq 1$

** **36.** $c_1 = 0$, $c_{n+1} = \dfrac{7}{16 - 8c_n^2}$, $n \geq 1$

** **37.** $c_1 = 1$, $c_{n+1} = \dfrac{4c_n}{4 + c_n}$, $n \geq 1$

** **38.** $c_1 = 4$, $c_{n+1} = \dfrac{3c_n}{2 + c_n}$, $n \geq 1$

** **39.** $c_1 = 2$, $c_{n+1} = \dfrac{2c_n^2}{3 + c_n}$, $n \geq 1$

** **40.** $c_1 = \dfrac{3}{2}$, $c_{n+1} = \dfrac{c_n + 2}{4 - c_n}$, $n \geq 1$

** **41.** $c_1 = 0$, $c_{n+1} = \dfrac{3 - c_n}{5 - 2c_n}$, $n \geq 1$

** **42.** Show that the sequence

$$c_1 = 1, \quad c_{n+1} = \frac{1}{4 - c_n - c_n^2}, \quad n \geq 1$$

is monotonic and bounded. Find an approximation to its limit accurate to 5 decimals.

** **43.** (a) Prove that, if a sequence converges, then differences between successive terms in the sequence must approach zero.

  (b) The following example illustrates that the converse is not true: Show that differences between successive terms of the sequence $\{\ln n\}$ approach zero, but the sequence itself diverges.

In Exercises 44–47 show that the sequence is convergent and find its limit.

** **44.** $c_1 = 2, \quad c_{n+1} = \dfrac{8 - c_n}{5}, \quad n \geq 1$

** **45.** $c_1 = 20, \quad c_{n+1} = 12 + \dfrac{3}{c_n}, \quad n \geq 1$

** **46.** $c_1 = 1, \quad c_{n+1} = \dfrac{1}{1 + c_n}, \quad n \geq 1$

** **47.** $c_1 = 10, \quad c_{n+1} = \dfrac{1}{3 + 2c_n}, \quad n \geq 1$

** **48.** Show that the sequence

$$c_1 = 1, \qquad c_{n+1} = \frac{3}{1 + c_n}, \qquad n \geq 1,$$

is convergent and find its limit. Hint: You may find it helpful to prove that $1 \leq c_n \leq 2$ for $n \geq 1$. Do this by mathematical induction.

** **49.** Repeat Exercise 48 if $c_1 = 0$. In this case you might want to show that $1 \leq c_n \leq 2$ for $n \geq 4$.

** **50.** Show that the sequence

$$c_1 = 1, \qquad c_{n+1} = \frac{4}{2 + 5c_n}, \qquad n \geq 1,$$

is convergent and find its limit. Hint: You may find it helpful to prove that $1/2 \leq c_n \leq 1$ for $n \geq 1$. Do this by mathematical induction.

** **51.** Show that the sequence

$$c_1 = 1, \qquad c_{n+1} = \sqrt{26 - c_n}, \qquad n \geq 1,$$

is convergent and find its limit. Hint: Show that $4 \leq c_n \leq 5$ for $n \geq 2$, and then consider $(c_{n+1})^2 - (c_n)^2$.

** **52.** Show that the sequence

$$c_1 = 4, \qquad c_{n+1} = \sqrt{20 - 3c_n}, \qquad n \geq 1,$$

is convergent and find its limit.

** **53.** Show that the sequence

$$c_1 = 2, \qquad c_{n+1} = \frac{c_n}{4c_n - 1}, \qquad n \geq 1,$$

does not converge.

** **54.** Show that the sequence in Exercise 53 diverges for any first term except $c_1 = 0$.

** **55.** Show that the sequence

$$c_1 = 1, \qquad c_{n+1} = \frac{5c_n}{2c_n - 1}, \qquad n \geq 1,$$

is convergent and find its limit. Hint: Express c_{n+1} in the form $c_{n+1} = 5/(2 - 1/c_n)$ and show that $2 \leq c_n \leq 4$ for $n \geq 3$.

** **56.** Show that the sequence

$$c_1 = 3, \qquad c_{n+1} = \sqrt[3]{10 - c_n}, \qquad n \geq 1,$$

is convergent and find its limit.

** **57.** Show that the sequence

$$c_1 = a, \qquad c_{n+1} = \frac{c_n}{bc_n - 1}, \qquad n \geq 1,$$

where a and b are nonzero constants diverges.

** **58.** Prove parts (i) and (ii) of Theorem 11.10.

** **59.** Find bounds for the sequence

$$c_1 = 1, \quad c_{n+1} = \frac{1 + c_n}{1 + 2c_n}, \quad n \geq 1.$$

** **60.** Determine whether the following sequence is monotonic, has bounds, and has a limit,

$$c_1 = -30, \quad c_2 = -20, \quad c_{n+1} = 5 + \frac{c_n}{2} + \frac{c_{n-1}}{3}, \quad n \geq 2.$$

** **61.** A sequence $\{c_n\}$ has only positive terms. Prove that:
    (a) if $\lim_{n \to \infty} c_n = L < 1$, there exists an integer N such that for all $n \geq N$

$$c_n < \frac{L + 1}{2}.$$

    (b) if $\lim_{n \to \infty} c_n = L > 1$, there exists an integer N such that for all $n \geq N$,

$$c_n > \frac{L + 1}{2}.$$

These results are used in Theorem 11.16, Section 11.11.

** **62.** Prove that if $\lim_{n \to \infty} c_n = L$, there exists an integer N such that for all $n \geq N$,

$$c_n < L + 1.$$

This result is used in Theorem 11.15, Section 11.10.

** **63.** The *Fibonacci sequence* found in many areas of applied mathematics is defined by

$$c_1 = 1, \quad c_2 = 1, \quad c_{n+1} = c_n + c_{n-1}, \quad n \geq 2.$$

    (a) Evaluate the first ten terms of this sequence.
    (b) Is the sequence monotonic, is it bounded, and does it have a limit?
    (c) Prove that
$$c_n^2 - c_{n-1}c_{n+1} = (-1)^{n+1}, \quad n \geq 2.$$
    (d) Verify that an explicit formula for c_n is

$$c_n = \frac{1}{\sqrt{5}}\left[\left(\frac{1 + \sqrt{5}}{2}\right)^n - \left(\frac{1 - \sqrt{5}}{2}\right)^n\right].$$

(e) Define a sequence $\{b_n\}$ as the ratio of terms in the Fibonacci sequence

$$b_n = \frac{c_{n+1}}{c_n}.$$

Is this sequence monotonic? Does it have a limit?

∗∗ **64.** Prove that the sequence

$$c_1 = d, \quad c_{n+1} = \sqrt{a + 2bc_n}, \quad n \geq 1$$

(a, b, and d all positive constants) is increasing if and only if $d < b + \sqrt{a + b^2}$. What happens when $d = b + \sqrt{a + b^2}$?

∗∗ **65.** In this exercise we outline a proof of part (iii) of Theorem 11.10.

(a) Verify that

$$|c_n d_n - CD| \leq |c_n||d_n - D| + |D||c_n - C|.$$

(b) Show that, given any $\epsilon > 0$, there exist positive integers N_1, N_2, and N_3 such that

$$|c_n| < |C| + 1 \qquad \text{whenever } n > N_1,$$

$$|d_n - D| < \frac{\epsilon}{2(|C| + 1)} \qquad \text{whenever } n > N_2,$$

$$|c_n - C| < \frac{\epsilon}{2|D| + 1} \qquad \text{whenever } n > N_3.$$

(c) Use these results to prove part (iii) of Theorem 11.10.

∗∗ **66.** In this exercise we outline a proof for part (iv) of Theorem 11.10.

(a) Verify that when $d_n \neq 0$ and $D \neq 0$,

$$\left| \frac{c_n}{d_n} - \frac{C}{D} \right| \leq \frac{|c_n - C|}{|d_n|} + \frac{|C||d_n - D|}{|D||d_n|}.$$

(b) Show that, given any $\epsilon > 0$, there exist positive integers N_1, N_2, and N_3 such that

$$|d_n| > \frac{|D|}{2} \qquad \text{whenever } n > N_1,$$

$$|c_n - C| < \frac{\epsilon|D|}{4} \qquad \text{whenever } n > N_2,$$

$$|d_n - D| < \frac{\epsilon|D|^2}{4|C| + 1} \qquad \text{whenever } n > N_3.$$

(c) Now prove part (iv) of Theorem 11.10.

∗∗ **67.** Find an explicit formula for the recursive sequence

$$c_1 = 1, \quad c_2 = 2, \quad c_{n+1} = \frac{c_n + c_{n-1}}{2}, \quad n \geq 2.$$

11.9 Infinite Series of Numbers

Infinite series of numbers arise in two ways. When $\{c_n\}$ is an infinite sequence of numbers, the expression

$$\sum_{n=1}^{\infty} c_n = c_1 + c_2 + c_3 + \cdots + c_n + \cdots \qquad (11.31)$$

is called an **infinite series of numbers**, or simply a series. From the sequences of Example 11.1 we may form the following series of numbers.

EXAMPLE 11.32

Write out the first six terms of the following series:

(a) $\displaystyle\sum_{n=1}^{\infty} \frac{1}{2^{n-1}}$ (b) $\displaystyle\sum_{n=1}^{\infty} \frac{n}{n+1}$ (c) $\displaystyle\sum_{n=1}^{\infty} (-1)^n |n-3|$ (d) $\displaystyle\sum_{n=1}^{\infty} (-1)^{n+1}$

SOLUTION   The first six terms of these series are shown below:

(a) $\displaystyle\sum_{n=1}^{\infty} \frac{1}{2^{n-1}} = 1 + \frac{1}{2} + \frac{1}{4} + \frac{1}{8} + \frac{1}{16} + \frac{1}{32} + \cdots$

(b) $\displaystyle\sum_{n=1}^{\infty} \frac{n}{n+1} = \frac{1}{2} + \frac{2}{3} + \frac{3}{4} + \frac{4}{5} + \frac{5}{6} + \frac{6}{7} + \cdots$

(c) $\displaystyle\sum_{n=1}^{\infty} (-1)^n |n-3| = -2 + 1 - 0 + 1 - 2 + 3 - \cdots$

(d) $\displaystyle\sum_{n=1}^{\infty} (-1)^{n+1} = 1 - 1 + 1 - 1 + 1 - 1 + \cdots$

Infinite series of numbers also arise when specific values of x are substituted into power or Taylor series. For example, if we substitute $x = 1$ into the power series $\sum_{n=1}^{\infty} x^n/(n2^n)$, we obtain

$$\sum_{n=1}^{\infty} \frac{1}{n\,2^n} = \frac{1}{2} + \frac{1}{2 \cdot 2^2} + \frac{1}{3 \cdot 2^3} + \cdots + \frac{1}{n\,2^n} + \cdots \,;$$

if we set $x = -2$, we obtain

$$\sum_{n=1}^{\infty} \frac{(-1)^n}{n\,2^n} = -1 + \frac{1}{2} - \frac{1}{3} + \frac{1}{4} - \cdots + \frac{(-1)^n}{n} + \cdots .$$

As mentioned earlier in this chapter, it is not possible to add an infinity of numbers in a finite amount of time, and therefore expression 11.31 is as yet meaningless. To attach a meaning we take the same approach as we did for Taylor series and power series. We illustrate with a simple example, and then give a formal definition. Consider the infinite series of Example 11.32(a),

$$\sum_{n=1}^{\infty} \frac{1}{2^{n-1}} = 1 + \frac{1}{2} + \frac{1}{4} + \frac{1}{8} + \cdots .$$

If we start adding terms, a pattern soon emerges. Indeed, if we denote by S_n the sum of the first n terms of this series, we find that

$$S_1 = 1,$$

$$S_2 = 1 + \frac{1}{2} = \frac{3}{2}, \qquad \text{(sum of first two terms)}$$

$$S_3 = 1 + \frac{1}{2} + \frac{1}{4} = \frac{7}{4}, \qquad \text{(sum of first three terms)}$$

$$S_4 = 1 + \frac{1}{2} + \frac{1}{4} + \frac{1}{8} = \frac{15}{8}, \qquad \text{(sum of first four terms)}$$

$$S_5 = 1 + \frac{1}{2} + \frac{1}{4} + \frac{1}{8} + \frac{1}{16} = \frac{31}{16}, \qquad \text{(sum of first five terms)}$$

and so on. We see that the sum of the first n terms of the series is given by the formula

$$S_n = \frac{2^n - 1}{2^{n-1}} = 2 - \frac{1}{2^{n-1}}.$$

As we add more and more terms of this series together, the sum S_n gets closer and closer to 2. It is always less than 2, but S_n can be made arbitrarily close to 2 by choosing n sufficiently large. If this series is to have a sum, the only reasonable sum is 2. In practice, this is precisely what we do; we define the sum of the series $\sum_{n=1}^{\infty} 1/2^{n-1}$ to be 2.

Let us now take this idea and define sums for all infinite series. We begin by defining a sequence $\{S_n\}$ as follows:

$$S_1 = c_1,$$
$$S_2 = c_1 + c_2,$$
$$S_3 = c_1 + c_2 + c_3,$$
$$\vdots \qquad \vdots$$
$$S_n = c_1 + c_2 + c_3 + \cdots + c_n,$$
$$\vdots \qquad \vdots$$

It is called the **sequence of partial sums** of the series $\sum_{n=1}^{\infty} c_n$. The nth term of the sequence $\{S_n\}$ represents the sum of the first n terms of the series. If this sequence has a limit, say, S, then the more terms of the series that we add together, the closer the sum gets to S. It seems reasonable, then, to call S the sum of the series. We therefore make the following definition.

DEFINITION 11.5

Let $S_n = \sum_{k=1}^{n} c_k$ be the nth partial sum of a series $\sum_{n=1}^{\infty} c_n$. If the sequence of partial sums $\{S_n\}$ has limit S,

$$S = \lim_{n \to \infty} S_n,$$

we call S the sum of the series and write

$$\sum_{n=1}^{\infty} c_n = S;$$

if $\{S_n\}$ does not have a limit, we say that the series does not have a sum.

If a series has sum S, we say that the series **converges** to S, which means that its sequence of partial sums converges to S. If a series does not have a sum, we say that the series **diverges**, which means that its sequence of partial sums diverges. Partial sums are to infinite series of numbers what Taylor polynomials are to Taylor series. Before proceeding with examples we need to make one comment about terminology and one about notation. In the next few sections we will be concentrating on "series of numbers" as opposed to "power series" or "series of functions." We shall usually write "series" instead of "series of numbers"; the context will always make it clear when we are discussing series of numbers. Whenever the first term of a series corresponds to $n = 1$, we shall drop the limits $n = 1$ and ∞ on the sigma notation when the notation appears in text. For instance, we shall write $\sum c_n$ in place of $\sum_{n=1}^{\infty} c_n$. Limits will always be retained when sigma notation appears in a displayed equation, or when the lower limit is not $n = 1$.

According to this definition, the series $\sum 1/2^{n-1}$ of Example 11.32(a) has sum 2. Since every term of the series $\sum n/(n+1)$ in Example 11.32(b) is greater than or equal to $1/2$, it follows that the sum of the first n terms is $S_n \geq n(1/2) = n/2$. As the sequence of partial sums is therefore unbounded, it cannot possibly have a limit (see the corollary to Theorem 11.7). The series therefore does not have a sum; it diverges. Examination of the first few partial sums of the series $\sum (-1)^n |n - 3|$ in Example 11.32(c) leads to the result that for $n > 1$,

$$S_n = \begin{cases} \dfrac{n-4}{2}, & \text{if } n \text{ is even} \\[2mm] \dfrac{1-n}{2}, & \text{if } n \text{ is odd.} \end{cases}$$

Since the sequence $\{S_n\}$ does not have a limit, the series diverges. The partial sums of the series $\sum (-1)^{n+1}$ are

$$S_1 = 1, \quad S_2 = 0, \quad S_3 = 1, \quad S_4 = 0, \quad \cdots.$$

Since this sequence does not have a limit, the series does not have a sum.

Perhaps the most important series of numbers, and certainly the series that occurs most frequently in applications is the **geometric series**. We discussed it in the context of power series in Section 11.4, but it is worthwhile repeating the ideas here, unencumbered by other considerations. Geometric series are of the form

$$\sum_{n=1}^{\infty} ar^{n-1} = a + ar + ar^2 + ar^3 + \cdots. \tag{11.32}$$

Each term is obtained by multiplying the previous term by the same constant r, the **common ratio**. If $\{S_n\}$ is the sequence of partial sums for this series, then

$$S_n = a + ar + ar^2 + \cdots + ar^{n-1}.$$

If we multiply this equation by r,

$$r S_n = ar + ar^2 + ar^3 + \cdots + ar^n.$$

When we subtract these equations, the result is

$$S_n - r S_n = a - ar^n.$$

Hence, for $r \neq 1$, we obtain

$$S_n = \frac{a(1 - r^n)}{1 - r}.$$

Furthermore, when $r = 1$,

$$S_n = na,$$

and therefore the sum of the first n terms of a geometric series is

$$S_n = \begin{cases} \dfrac{a(1 - r^n)}{1 - r}, & r \neq 1 \\ na, & r = 1. \end{cases} \tag{11.33a}$$

To determine whether a geometric series has a sum, we consider the limit of this sequence. Certainly, $\lim_{n \to \infty} na$ does not exist, unless trivially $a = 0$. In addition, $\lim_{n \to \infty} r^n = 0$ when $|r| < 1$, and does not exist when $|r| > 1$. Nor does it exist when $r = -1$. Thus, we may state that

$$\lim_{n \to \infty} S_n = \begin{cases} \dfrac{a}{1 - r}, & |r| < 1 \\ \text{does not exist}, & |r| \geq 1. \end{cases}$$

The geometric series therefore has sum

$$\sum_{n=1}^{\infty} ar^{n-1} = \frac{a}{1 - r} \tag{11.33b}$$

if $|r| < 1$, but otherwise diverges.

The series in parts (a) and (d) of Example 11.32 are geometric. The first has common ratio $1/2$ and therefore converges to $1/(1 - 1/2) = 2$; the second has common ratio -1 and therefore diverges.

We encounter the **harmonic series**

$$\sum_{n=1}^{\infty} \frac{1}{n} = 1 + \frac{1}{2} + \frac{1}{3} + \frac{1}{4} + \cdots \tag{11.34}$$

quite often in our work. It does not have a sum, and we can show this by considering the following partial sums of the series:

$$S_1 = 1,$$

$$S_2 = 1 + \frac{1}{2} = \frac{3}{2},$$

$$S_4 = S_2 + \frac{1}{3} + \frac{1}{4} > \frac{3}{2} + \frac{1}{4} + \frac{1}{4} = \frac{4}{2},$$

$$S_8 = S_4 + \frac{1}{5} + \frac{1}{6} + \frac{1}{7} + \frac{1}{8} > \frac{4}{2} + \frac{1}{8} + \frac{1}{8} + \frac{1}{8} + \frac{1}{8} = \frac{5}{2},$$

$$S_{16} = S_8 + \frac{1}{9} + \frac{1}{10} + \cdots + \frac{1}{16} > \frac{5}{2} + \frac{1}{16} + \frac{1}{16} + \cdots + \frac{1}{16} = \frac{6}{2}.$$

This procedure can be continued indefinitely, and shows that the sequence of partial sums is unbounded. The harmonic series therefore diverges (see once again the corollary to Theorem 11.7).

In each of the above examples, we used Definition 11.5 for the sum of a series to determine whether the series converges or diverges; that is, we formed the sequence of partial sums $\{S_n\}$ in order to consider its limit. For most examples, it is either too difficult or impossible to evaluate S_n in a simple form, and in such cases consideration of the limit of the sequence $\{S_n\}$ is impractical. Consequently, we must develop alternative ways to decide on the convergence of a series. We do this in Sections 11.10 to 11.12.

We now discuss some fairly simple but important results on convergence of series. If a series $\sum c_n$ has a finite number of its terms altered in any fashion whatsoever, the new series converges if and only if the original series converges. The new series may converge to a different sum, but it converges if the original series converges. For example, if we double the first three terms of the geometric series in Example 11.32(a), but do not change the remaining terms, the new series is

$$2 + 1 + \frac{1}{2} + \frac{1}{8} + \frac{1}{16} + \cdots .$$

It is not geometric. But its nth partial sum, call it S_n, is very closely related to that of the geometric series, call it T_n. In fact for $n \geq 4$, we can say that $S_n = T_n + 7/4$ (7/4 is the total change in the first three terms). Since $\lim_{n \to \infty} T_n = 2$, it follows that $\lim_{n \to \infty} S_n = 2 + 7/4 = 15/4$; that is, the new series converges, but its sum is $7/4$ greater than that of the geometric series.

On the other hand, suppose we change the first 100 terms of the harmonic series, which diverges, to 0, but leave the remaining terms unaltered. The new series is

$$\overbrace{0 + 0 + \cdots + 0}^{100 \text{ terms}} + \frac{1}{101} + \frac{1}{102} + \cdots .$$

We know that the sequence of partial sums $\{T_n\}$ of the harmonic series is increasing and unbounded. The nth partial sum S_n of the new series, for $n \geq 100$, is equal to $T_n - k$, where k is the sum of the first 100 terms of the harmonic series. But because $\lim_{n \to \infty} T_n = \infty$, so also must $\lim_{n \to \infty} S_n = \infty$; that is, the new series diverges.

The following theorem indicates that convergent series can be added and subtracted, and multiplied by constants. These properties can be verified using Definition 11.5.

THEOREM 11.11

If series $\sum_{n=1}^{\infty} c_n$ and $\sum_{n=1}^{\infty} d_n$ have sums C and D, then:

(i) $\displaystyle\sum_{n=1}^{\infty} kc_n = kC$ (when k is a constant); (11.35a)

(ii) $\displaystyle\sum_{n=1}^{\infty} (c_n \pm d_n) = C \pm D.$ (11.35b)

Our first convergence test is a corollary to the following theorem.

THEOREM 11.12

If a series $\sum_{n=1}^{\infty} c_n$ converges, then $\lim_{n\to\infty} c_n = 0$.

PROOF If series $\sum c_n$ has a sum S, its sequence of partial sums

$$S_1, S_2, \cdots, S_n, \cdots$$

has limit S. The sequence

$$0, S_1, S_2, \cdots, S_{n-1}, \cdots$$

must also have limit S; it is the sequence of partial sums with an additional term equal to 0 at the beginning. According to Theorem 11.10, if we subtract these two sequences, the resulting sequence must have limit $S - S = 0$; that is,

$$S_1 - 0, S_2 - S_1, \cdots, S_n - S_{n-1}, \cdots \to 0.$$

But $S_1 - 0 = S_1 = c_1$, $S_2 - S_1 = c_2$, etc; that is, we have shown that $\lim_{n\to\infty} c_n = 0$. ∎

Theorem 11.12 states that a necessary condition for a series $\sum c_n$ to converge is that the sequence $\{c_n\}$ of its terms must approach zero. What we really want are sufficient conditions to guarantee convergence or divergence of a series. We can take the contrapositive of the theorem and obtain the following.

COROLLARY (n^{th} **term test**)

If $\lim_{n\to\infty} c_n \neq 0$, or does not exist, then the series $\sum_{n=1}^{\infty} c_n$ diverges.

This is our first convergence test, the nth term test. It is in fact a test for divergence rather than convergence, stating that if $\lim_{n\to\infty} c_n$ exists and is equal to anything but zero, or the limit does not exist, then the series $\sum c_n$ diverges. Note well that the nth *term test never indicates that a series converges.* Even if $\lim_{n\to\infty} c_n = 0$, we can conclude nothing about the convergence or divergence of $\sum c_n$; it may converge or it may diverge. For example, the harmonic series $\sum 1/n$ and the geometric series $\sum 1/2^{n-1}$ both satisfy the condition $\lim_{n\to\infty} c_n = 0$, yet one series diverges and the other converges. The nth term test therefore may indicate that a series diverges, but it never indicates that a series converges.

In particular, both series $\sum (-1)^n |n - 3|$ and $\sum (-1)^{n+1}$ of Example 11.32 diverge by the nth term test.

To understand series, it is crucial to distinguish clearly among three entities: the series itself, its sequence of partial sums, and its sequence of terms. For any series $\sum c_n$, we can form its sequence of terms $\{c_n\}$ and its sequence of partial sums $\{S_n\}$. Each of these sequences may give information about the sum of the series $\sum c_n$, but in very different ways.

The sum of the series is defined by its sequence of partial sums in that $\sum c_n$ has a sum only if $\{S_n\}$ has a limit. The sequence of partial sums therefore tells us definitely whether the series converges or diverges, provided we can evaluate S_n in a simple form.

The sequence of terms $\{c_n\}$, on the other hand, may or may not tell us whether the series diverges. If $\{c_n\}$ has no limit or has a limit other than zero, we know that the series does not have a sum. If $\{c_n\}$ has limit zero, we obtain no information about convergence of the series, and must continue our investigation.

The interval of convergence of a power series (or Taylor series) is all values of x for which the series converges. It is determined, except possibly for end points, by the radius of convergence. As we uncover convergence tests for series of numbers in this and the next three sections, we will be able to discuss end points of intervals of convergence, and consequently be able to convert open intervals of convergence to intervals of convergence. As a first example, we show that the open interval of convergence for the series $\sum_{n=1}^{\infty} x^{2n}/(n\,4^n)$ is its interval of convergence; that is, end points are not included.

EXAMPLE 11.33 ▼

Find the interval of convergence for the power series $\displaystyle\sum_{n=1}^{\infty} \frac{1}{n\,4^n} x^{2n}$.

SOLUTION   If we set $y = x^2$, then

$$\sum_{n=1}^{\infty} \frac{1}{n\,4^n} x^{2n} = \sum_{n=1}^{\infty} \frac{1}{n\,4^n} y^n.$$

According to equation 11.18a, the radius of convergence of this series is

$$R_y = \lim_{n \to \infty} \left| \frac{\dfrac{1}{n\,4^n}}{\dfrac{1}{(n+1)4^{n+1}}} \right| = \lim_{n \to \infty} 4\left(\frac{n+1}{n}\right) = 4.$$

Consequently, the radius of convergence of the power series in x is $R_x = 2$, and the open interval of convergence is $-2 < x < 2$. At $x = \pm 2$, the series becomes

$$\sum_{n=1}^{\infty} \frac{1}{n\,4^n}(\pm 2)^{2n} = \sum_{n=1}^{\infty} \frac{1}{n},$$

the harmonic series, which diverges. The interval of convergence is therefore $-2 < x < 2$.

▲

EXERCISES 11.9

In Exercises 1–10 determine whether the series converges or diverges. Find the sum of each convergent series. To get a feeling for a series, it is helpful to write out its first few terms. Try it.

∗ **1.** $\displaystyle\sum_{n=1}^{\infty} \frac{n+1}{2n}$

∗ **2.** $\displaystyle\sum_{n=1}^{\infty} \frac{2^n}{5^{n+1}}$

∗ **3.** $\displaystyle\sum_{n=1}^{\infty} \cos\left(\frac{n\pi}{2}\right)$

∗ **4.** $\displaystyle\sum_{n=1}^{\infty} \left(\frac{n}{n+1}\right)^n$

∗ **5.** $\displaystyle\sum_{n=1}^{\infty} \frac{7^{2n+3}}{3^{2n-2}}$

∗ **6.** $\displaystyle\sum_{n=1}^{\infty} \frac{7^{n+3}}{3^{2n-2}}$

∗ **7.** $\displaystyle\sum_{n=1}^{\infty} \sqrt{\frac{n^2-1}{n^2+1}}$

∗ **8.** $\displaystyle\sum_{n=1}^{\infty} \frac{\cos(n\pi)}{2^n}$

∗ **9.** $\displaystyle\sum_{n=1}^{\infty} \frac{4^n+3^n}{3^n}$

∗ **10.** $\displaystyle\sum_{n=1}^{\infty} \mathrm{Tan}^{-1} n$

In Exercises 11–14 we have given a repeating decimal. Express the decimal as a geometric series and use formula 11.33b to express it as a rational number.

∗ **11.** $0.666\,666\ldots$

∗ **12.** $0.131\,313\,131\,3\ldots$

∗ **13.** $1.347\,346\,346\,346\ldots$

∗ **14.** $43.020\,502\,050\,205\ldots$

In Exercises 15–17 complete the statement and give a short proof to substantiate your claim.

∗ **15.** If $\sum c_n$ and $\sum d_n$ converge, then $\sum(c_n+d_n)\ldots$

∗ **16.** If $\sum c_n$ converges and $\sum d_n$ diverges, then $\sum(c_n+d_n)\ldots$

∗ **17.** If $\sum c_n$ and $\sum d_n$ diverge, then $\sum(c_n+d_n)\ldots$

In Exercises 18–21 determine whether the series converges or diverges. Find the sum of each convergent series.

∗ **18.** $\displaystyle\sum_{n=1}^{\infty} \frac{2^n+3^n}{4^n}$

∗ **19.** $\displaystyle\sum_{n=1}^{\infty} \frac{3^n-1}{2^n}$

∗ **20.** $\displaystyle\sum_{n=1}^{\infty} \frac{n^2+2^{2n}}{4^n}$

∗ **21.** $\displaystyle\sum_{n=1}^{\infty} \frac{2^n+4^n-8^n}{2^{3n}}$

** **22.** Find the sum of the series

$$\sum_{n=1}^{\infty} \frac{1}{n(n+1)}.$$

Hint: Use partial fractions on the nth term and find the sequence of partial sums.

** **23.** Find the total distance travelled by the superball in Exercise 56 of Section 11.1 (page 728) before it comes to rest.

** **24.** Find the time taken for the superball in Exercise 56 of Section 11.1 to come to rest.

** **25.** What distance does the dog run from the time when it sees the farmer until the farmer reaches the farmhouse in Exercise 57 of Section 11.1?

** **26.** Find a simplified formula for the area A_n in Exercise 61 of Section 11.1 (page 729). What is $\lim_{n \to \infty} A_n$?

** **27.** (a) According to equation 11.33a, the nth partial sum of the geometric series $1 + r + r^2 + r^3 + \cdots$ is $S_n = (1 - r^n)/(1 - r)$. If T_n denotes the nth partial sum of the series

$$1 + 2r + 3r^2 + 4r^3 + \cdots,$$

show that

$$T_n - S_n = r(T_n - nr^{n-1}).$$

Solve this equation for T_n and take the limit as $n \to \infty$ to show that

$$\sum_{n=1}^{\infty} nr^{n-1} = \frac{1}{(1-r)^2}, \quad |r| < 1.$$

(b) Verify this result by setting $S(r) = \sum_{n=1}^{\infty} nr^{n-1}$ and integrating.

In Exercises 28–31 use the result of Exercise 27 to find the sum of the series.

* **28.** $\dfrac{1}{2} + \dfrac{2}{2^2} + \dfrac{3}{2^3} + \dfrac{4}{2^4} + \cdots$ * **29.** $\dfrac{2}{5} + \dfrac{4}{25} + \dfrac{6}{125} + \dfrac{8}{625} + \cdots$

** **30.** $\dfrac{2}{3} + \dfrac{3}{27} + \dfrac{4}{243} + \dfrac{5}{2187} + \cdots$ ** **31.** $\dfrac{12}{5} + \dfrac{48}{25} + \dfrac{192}{125} + \dfrac{768}{625} + \cdots$

** **32.** Two people flip a single coin to see who can first flip a head. Show that the probability that the first person to flip wins the game is represented by the series

$$\frac{1}{2} + \frac{1}{8} + \frac{1}{32} + \cdots + \frac{1}{2^{2n-1}} + \cdots.$$

What is the sum of this series?

** **33.** Two people throw a die to see who can first throw a six. Find the probability that the person who first throws wins the game.

In Exercises 34–37 show that the interval of convergence of the power series is the same as the open interval of convergence.

** **34.** $\displaystyle\sum_{n=0}^{\infty} \frac{1}{2^n} x^n$
      ** **35.** $\displaystyle\sum_{n=1}^{\infty} n^2 3^n x^n$

** **36.** $\displaystyle\sum_{n=2}^{\infty} 2^n \left(\frac{n-1}{n+1}\right)^2 (x-4)^n$
    ** **37.** $\displaystyle\sum_{n=0}^{\infty} (-1)^n x^{3n}$

** **38.** One of *Zeno's paradoxes* describes a race between Achilles and a tortoise. Zeno claims that, if the tortoise is given a head start, then no matter how fast Achilles runs, he can never catch the tortoise. He reasons as follows: In order to catch the tortoise, Achilles must first make up the length of the head start. But while he is running this distance, the tortoise "runs" a further distance. While Achilles makes up this distance, the tortoise covers a further distance, and so on. It follows that Achilles is always making up distance covered by the tortoise, and therefore can never catch the tortoise. If the tortoise is given a head start of length L and Achilles runs $c > 1$ times as fast as the tortoise, use infinite series to show that Achilles does in fact catch the tortoise and that the distance he covers in doing so is $cL(c-1)^{-1}$.

** **39.** Find the time between 1:05 and 1:10 when the minute and hour hands of a clock point in the same direction

      (a) by reasoning with infinite series as in Exercise 38; and

      (b) by finding expressions for the angular displacements of the hands as functions of time.

** **40.** Repeat Exercise 39 for the instant between 10:50 and 10:55 when the hands coincide.

** **41.** A child has a large number of identical cubical blocks, which she stacks as shown in the figure below. The top block protrudes $\frac{1}{2}$ its length over the second block, which protrudes $\frac{1}{4}$ its length over the third block, which protrudes $\frac{1}{6}$ its length over the fourth block, etc. Assuming the centre of mass of each block is at its geometric centre, show that the centre of mass of the top n blocks lies directly over the edge of the $(n+1)^{\text{th}}$ block. Now deduce that if a sufficient number of blocks is piled, the top block can be made to protrude as far over the bottom block as desired without the stack falling.

** **42.** Prove the following result: If $\sum_{n=1}^{\infty} c_n$ converges, then its terms can be grouped in any manner, and the resulting series is convergent with the same sum as the original series.

$*{*}*$ **43.** It is customary to assume that when a drug is administered to the human body it will be eliminated exponentially; that is, if A represents the amount of drug in the body, then

$$A = A_0 e^{-kt},$$

where $k > 0$ is a constant and A_0 is the amount injected at time $t = 0$. Suppose n successive injections of amount A_0 are administered at equally spaced time intervals T, the first injection at time $t = 0$.

(a) Show that the amount of drug in the body at time t between the nth and $(n+1)^{\text{th}}$ injection is given by

$$A_n(t) = A_0 e^{-kt} \left[\frac{1 - e^{knT}}{1 - e^{kT}} \right], \quad (n-1)T < t < nT.$$

(b) Draw graphs of these functions on one set of axes.

(c) What is the amount of drug in the body immediately after the nth injection for very large n; that is, what is

$$\lim_{n \to \infty} A_n((n-1)T)?$$

11.10 Integral Test, Comparison and Limit Comparison Tests

In Section 11.9 we derived the nth term test, a test for divergence of a series of numbers. In order to develop further convergence and divergence tests, we consider two classes of series of numbers:

1. series with terms that are all nonnegative;

2. series with both positive and negative terms.

A series with terms that are all nonpositive is the negative of a series of type 1, and therefore any test applicable to series of type 1 is easily adapted to a series with nonpositive terms.

> **DEFINITION 11.6**
>
> A series $\sum_{n=1}^{\infty} c_n$ is said to be **nonnegative** if each term is nonnegative: $c_n \geq 0$.

For example, the harmonic series $\sum 1/n$ and the geometric series $\sum 1/2^{n-1}$ are both nonnegative series. We have already seen that both series have the property that $\lim_{n \to \infty} c_n = 0$, yet the harmonic series diverges and the geometric series converges. Examination of terms of these series reveals that $1/2^{n-1}$ approaches 0 much more quickly than does $1/n$. In general, whether a nonnegative series does or does not have a sum depends on how quickly its sequence of terms $\{c_n\}$ approaches zero. The study of convergence of nonnegative series is an investigation into the question "How fast must terms of a nonnegative series approach zero in order that the series have a sum?" This is not a simple problem; only a partial answer is provided through the tests in this section and the next. We begin with the integral test.

> **THEOREM 11.13   (Integral test)**
>
> Suppose the terms in a series $\sum_{n=1}^{\infty} c_n$ are denoted by $c_n = f(n)$, and $f(x)$ is a continuous, positive, decreasing function for $x \geq 1$. Then the series converges if and only if the improper integral $\int_1^{\infty} f(x)\, dx$ converges.

PROOF Suppose first that the improper integral converges to value K. We know that this value can be interpreted as the area under the curve $y = f(x)$, above the x-axis, and to the right of the line $x = 1$ in Figure 11.21a. In Figure 11.21b the area of the nth rectangle is $f(n) = c_n$, and clearly it is less than the area under the curve from $x = n - 1$ to $x = n$.

FIGURE 11.21a

FIGURE 11.21b

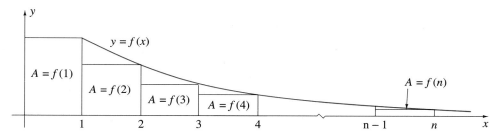

Now, the nth partial sum of the series $\sum c_n$ is

$$
\begin{aligned}
S_n &= c_1 + c_2 + \cdots + c_n \\
&= f(1) + f(2) + \cdots + f(n) \\
&< f(1) + \int_1^2 f(x)\,dx + \int_2^3 f(x)\,dx + \cdots + \int_{n-1}^n f(x)\,dx \\
&= f(1) + \int_1^n f(x)\,dx \\
&< f(1) + \int_1^\infty f(x)\,dx \\
&= f(1) + K.
\end{aligned}
$$

What this shows is that the sequence $\{S_n\}$ is bounded (since $f(1) + K$ is independent of n). Because all terms of the series $\sum c_n$ are positive, the sequence $\{S_n\}$ of partial sums must be increasing. It follows by Theorem 11.7 that the sequence $\{S_n\}$ must have a limit; that is, the series $\sum c_n$ converges.

Conversely, suppose now that the improper integral diverges. This time we draw rectangles to the right of the vertical lines at $n = 1, 2, \ldots$ (Figure 11.22). Then

$$S_n = c_1 + c_2 + \cdots + c_n$$

$$= f(1) + f(2) + \cdots + f(n)$$

$$> \int_1^2 f(x)dx + \int_2^3 f(x)dx + \cdots + \int_n^{n+1} f(x)dx$$

$$= \int_1^{n+1} f(x)dx.$$

Since $\lim_{n \to \infty} \int_1^{n+1} f(x)dx = \infty$, it follows that $\lim_{n \to \infty} S_n = \infty$, and the series therefore diverges.

FIGURE 11.22

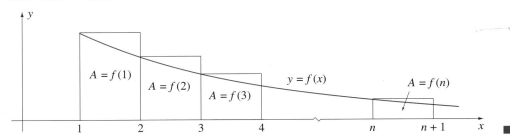

To use the integral test it is necessary to antidifferentiate the function $f(x)$, obtained by replacing n's in c_n with x's. When the antiderivative appears obvious, and other conditions of Theorem 11.13 are met, the integral test may be the easiest way to decide on convergence of the series; when the antiderivative is not obvious, it may be better to try another test.

EXAMPLE 11.34

Determine whether the following series converge or diverge:

$$\text{(a)} \quad \sum_{n=1}^{\infty} \frac{1}{n^2 + 1} \qquad \text{(b)} \quad \sum_{n=2}^{\infty} \frac{1}{n \ln n} \qquad \text{(c)} \quad \sum_{n=1}^{\infty} n e^{-n}$$

SOLUTION

(a) Since $f(x) = 1/(x^2 + 1)$ is continuous, positive, and decreasing for $x \geq 1$, and

$$\int_1^{\infty} \frac{1}{x^2 + 1} dx = \left\{ \text{Tan}^{-1} x \right\}_1^{\infty} = \frac{\pi}{2} - \frac{\pi}{4} = \frac{\pi}{4},$$

it follows that the series $\sum 1/(n^2 + 1)$ converges.

(b) Since this series begins with $n = 2$, we modify the integral test by considering the improper integral of $f(x) = 1/(x \ln x)$ over the interval $x \geq 2$. Since $f(x)$ is positive, continuous, and decreasing for $x \geq 2$, and

$$\int_2^{\infty} \frac{1}{x \ln x} dx = \left\{ \ln (\ln x) \right\}_2^{\infty} = \infty,$$

it follows that $\sum_{n=2}^{\infty} 1/(n \ln n)$ diverges.

(c) Since xe^{-x} is continuous, positive, and decreasing for $x \geq 1$, and

$$\int_1^\infty xe^{-x}dx = \left\{-xe^{-x} - e^{-x}\right\}_1^\infty = \frac{2}{e},$$

the series converges.

We mentioned before that two very important series are geometric series and the harmonic series. Convergence of geometric series was discussed in Section 11.9. The harmonic series belongs to a type of series called *p*-**series** defined as follows

$$\sum_{n=1}^\infty \frac{1}{n^p} = 1 + \frac{1}{2^p} + \frac{1}{3^p} + \cdots . \tag{11.36}$$

When $p = 1$, the *p*-series becomes the harmonic series which we know diverges. The series diverges for $p < 0$ by the nth-term test. Consider the case when $p > 0$, but $p \neq 1$. The function $1/x^p$ is continuous, positive, and decreasing for $x \geq 1$, and

$$\int_1^\infty \frac{1}{x^p}dx = \left\{\frac{1}{-(p-1)x^{p-1}}\right\}_1^\infty = \begin{cases} \infty, & p < 1 \\ \dfrac{1}{p-1}, & p > 1. \end{cases}$$

According to the integral test, the *p*-series converges when $p > 1$ and diverges when $p < 1$. Let us summarize results for geometric series and *p*-series:

Geometric Series   *p*-**series**

$$\sum_{n=1}^\infty ar^{n-1} = \begin{cases} \dfrac{a}{r-1}, & |r| < 1 \\ \text{diverges}, & |r| \geq 1 \end{cases} \qquad \sum_{n=1}^\infty \frac{1}{n^p} = \begin{cases} \text{converges}, & p > 1 \\ \text{diverges}, & p \leq 1 \end{cases} \tag{11.37}$$

It is unfortunate that no general formula can be given for the sum of the *p*-series when $p > 1$. Some interesting cases that arise frequently are cited below:

$$\sum_{n=1}^\infty \frac{1}{n^2} = \frac{\pi^2}{6}, \qquad \sum_{n=1}^\infty \frac{1}{n^3} \approx 1.202\,056\,903\,1, \qquad \sum_{n=1}^\infty \frac{1}{n^4} = \frac{\pi^4}{90}. \tag{11.38}$$

Many series can be shown to converge or diverge by comparing them to known convergent and divergent series. This is the essence of the following two tests.

> **THEOREM 11.14 (Comparison Test)**
>
> If $0 \leq c_n \leq a_n$ for all n and $\sum_{n=1}^{\infty} a_n$ converges, then $\sum_{n=1}^{\infty} c_n$ converges. If $c_n \geq d_n \geq 0$ for all n and $\sum_{n=1}^{\infty} d_n$ diverges, then $\sum_{n=1}^{\infty} c_n$ diverges.

PROOF   Suppose first that $\sum a_n$ converges and $0 \leq c_n \leq a_n$ for all n. Since $a_n \geq 0$ the sequence of partial sums for the series $\sum a_n$ must be nondecreasing. Since the series converges, the sequence of partial sums must also be bounded (corollary to Theorem 11.7). Because $0 \leq c_n \leq a_n$ for all n, it follows that the sequence of partial sums of $\sum c_n$ is nondecreasing and bounded also. This sequence therefore has a limit and series $\sum c_n$ converges. A similar argument can be made for the divergent case. ∎

Theorem 11.14 states that if the terms of a nonnegative series $\sum c_n$ are smaller than those of a known convergent series, then $\sum c_n$ must converge; if they are larger than a known nonnegative divergent series, then $\sum c_n$ must diverge.

In order for the comparison test to be useful, we require a catalogue of known convergent and divergent series with which we may compare other series. The geometric and p-series in equation 11.37 are extremely useful in this respect.

EXAMPLE 11.35

Determine whether the following series converge or diverge:

$$\text{(a)} \quad \sum_{n=2}^{\infty} \frac{\ln n}{n} \qquad \text{(b)} \quad \sum_{n=1}^{\infty} \frac{2n^2 - 1}{15n^4 + 14} \qquad \text{(c)} \quad \sum_{n=1}^{\infty} \frac{2n^2 + 1}{15n^4 - 14}$$

SOLUTION

(a) For this series, we note that when $n \geq 3$,

$$\frac{\ln n}{n} > \frac{1}{n}.$$

Since $\sum_{n=3}^{\infty} 1/n$ diverges (harmonic series with first two terms changed to zero), so does $\sum_{n=3}^{\infty} (\ln n)/n$ by the comparison test. Thus $\sum_{n=2}^{\infty} (\ln n)/n$ diverges. This can also be verified with the integral test.

(b) For the series $\sum (2n^2 - 1)/(15n^4 + 14)$, we note that

$$\frac{2n^2 - 1}{15n^4 + 14} < \frac{2n^2}{15n^4} = \frac{2}{15n^2}.$$

Since $\sum 2/(15n^2) = (2/15) \sum 1/n^2$ converges (p-series with $p = 2$), so does the given series by the comparison test.

(c) For this series we note that the inequality

$$\frac{2n^2 + 1}{15n^4 - 14} \leq \frac{3}{n^2}$$

is valid if and only if

$$45n^4 - 42 \geq 2n^4 + n^2.$$

But this inequality is valid if and only if

$$n^2(43n^2 - 1) \geq 42,$$

which is obviously true for $n \geq 1$. Consequently,

$$\frac{2n^2 + 1}{15n^4 - 14} \leq \frac{3}{n^2},$$

and since $\sum 3/n^2 = 3\sum 1/n^2$ converges, so too does the given series.

Two observations about Example 11.35 are worthwhile:

1. To use the comparison test we must first have a suspicion as to whether the given series converges or diverges in order to discuss the correct inequality; that is, if we suspect that the given series converges, we search for a convergent series for the right side of the inequality $\leq$, and if we suspect that the given series diverges, we search for a divergent series for the right side of the opposite inequality $\geq$.

2. Recall that, when the terms of a nonnegative series approach zero, whether the series has a sum depends on how fast these terms approach zero. The only difference in n^{th} terms of Examples 11.35(b) and (c) is the position of the negative sign. For very large n this difference becomes negligible since each nth term can, for large n, be closely approximated by $2/(15n^2)$. We might then expect similar analyses for these examples, and yet they are quite different. In addition, it is natural to ask where we obtained the factor 3 in part (c). The answer is, "by trial and error."

Each of these observations points out weaknesses in the comparison test, but these problems can be eliminated in many examples with the following test.

> ### THEOREM 11.15   (Limit Comparison Test)
> If $0 \leq c_n$ and $0 < b_n$, and
>
> $$\lim_{n \to \infty} \frac{c_n}{b_n} = \ell, \quad 0 < \ell < \infty, \tag{11.39}$$
>
> then series $\sum_{n=1}^{\infty} c_n$ converges if $\sum_{n=1}^{\infty} b_n$ converges, and diverges if $\sum_{n=1}^{\infty} b_n$ diverges.

PROOF   Suppose that series $\sum b_n$ converges. Since sequence $\{c_n/b_n\}$ converges to ℓ, we can use the result of Exercise 62 in Section 11.8 and say that for all n greater than or equal to some integer N,

$$\frac{c_n}{b_n} < \ell + 1, \qquad \text{or} \qquad c_n < (\ell + 1)b_n.$$

Since the series $\sum_{n=N}^{\infty} (\ell + 1)b_n = (\ell + 1)\sum_{n=N}^{\infty} b_n$ converges, so also must $\sum_{n=N}^{\infty} c_n$ (by the comparison test). Hence, $\sum_{n=1}^{\infty} c_n$ converges. A similar argument can be made for the divergent case.   ∎

If $\lim_{n\to\infty} c_n/b_n = \ell$, then for very large n we can say that $c_n \approx \ell b_n$. It follows that if $\{b_n\}$ approaches zero, $\{c_n\}$ approaches zero $1/\ell$ times as fast as $\{b_n\}$. Theorem 11.15 implies then that, if the sequences of nth terms of two nonnegative series approach zero at proportional rates, the series converge or diverge together.

The limit comparison test avoids inequalities. To use it, we must find a series $\sum b_n$ so that the limit of the ratio c_n/b_n is finite and greater than zero. To obtain this series, it is sufficient in many examples simply to answer the question, "What does the given series really look like for very large n?". In both Examples 11.35(b) and (c) we see that for large n

$$c_n \approx \frac{2n^2}{15n^4} = \frac{2}{15n^2}.$$

Consequently, we calculate in Example 11.35(c) that

$$\ell = \lim_{n\to\infty} \frac{\dfrac{2n^2+1}{15n^4-14}}{\dfrac{2}{15n^2}} = \lim_{n\to\infty} \frac{n^2\left(2+\dfrac{1}{n^2}\right)}{n^4\left(15-\dfrac{14}{n^4}\right)} \cdot \frac{15n^2}{2} = 1.$$

Since $\sum 2/(15n^2)$ converges, so too does $\sum (2n^2+1)/(15n^4-14)$.

EXAMPLE 11.36

Determine whether the following series converge or diverge:

$$\text{(a)} \quad \sum_{n=1}^{\infty} \frac{\sqrt{n^2+2n-1}}{n^{5/2}+15n-3} \qquad \text{(b)} \quad \sum_{n=1}^{\infty} \frac{2^n+1}{3^n+5}$$

SOLUTION

(a) For very large n, the nth term of the series can be approximated by

$$\frac{\sqrt{n^2+2n-1}}{n^{5/2}+15n-3} \approx \frac{n}{n^{5/2}} = \frac{1}{n^{3/2}}.$$

We calculate therefore that

$$\ell = \lim_{n\to\infty} \frac{\dfrac{\sqrt{n^2+2n-1}}{n^{5/2}+15n-3}}{\dfrac{1}{n^{3/2}}} = \lim_{n\to\infty} \frac{n\sqrt{1+\dfrac{2}{n}-\dfrac{1}{n^2}}}{n^{5/2}\left(1+\dfrac{15}{n^{3/2}}-\dfrac{3}{n^{5/2}}\right)} \cdot n^{3/2} = 1.$$

Since $\sum 1/n^{3/2}$ converges (p-series with $p = 3/2$), so too does the given series by the limit comparison test.

(b) For large n,

$$\frac{2^n+1}{3^n+5} \approx \left(\frac{2}{3}\right)^n.$$

We calculate therefore that

$$\ell = \lim_{n\to\infty} \frac{\dfrac{2^n+1}{3^n+5}}{\left(\dfrac{2}{3}\right)^n} = \lim_{n\to\infty} \frac{2^n\left(1+\dfrac{1}{2^n}\right)}{3^n\left(1+\dfrac{5}{3^n}\right)} \cdot \frac{3^n}{2^n} = 1.$$

Since $\sum (2/3)^n$ converges (a geometric series with $r = 2/3$), the given series converges by the limit comparison test.

EXAMPLE 11.37

Find the interval of convergence for the power series $\displaystyle\sum_{n=1}^{\infty} \frac{n}{(n+1)^3 2^n}(x-1)^{2n}$.

SOLUTION   If we set $y = (x-1)^2$, then

$$\sum_{n=1}^{\infty} \frac{n}{(n+1)^3 2^n}(x-1)^{2n} = \sum_{n=1}^{\infty} \frac{n}{(n+1)^3 2^n} y^n.$$

The radius of convergence of the series in y is

$$R_y = \lim_{n\to\infty} \left| \frac{\dfrac{n}{(n+1)^3 2^n}}{\dfrac{n+1}{(n+2)^3 2^{n+1}}} \right| = \lim_{n\to\infty} \frac{2n(n+2)^3}{(n+1)^4} = 2.$$

The radius of convergence of the power series in $x - 1$ is therefore $R_x = \sqrt{2}$, and the open interval of convergence is $1 - \sqrt{2} < x < 1 + \sqrt{2}$. At end points $x = 1 \pm \sqrt{2}$, the power series becomes

$$\sum_{n=1}^{\infty} \frac{n}{(n+1)^3}.$$

Since

$$\ell = \lim_{n\to\infty} \frac{\dfrac{n}{(n+1)^3}}{\dfrac{1}{n^2}} = \lim_{n\to\infty} \frac{n^3}{(n+1)^3} = 1,$$

and $\sum 1/n^2$ converges, so also does $\sum n/(n+1)^3$ (by the limit comparison test). The interval of convergence is therefore $1 - \sqrt{2} \le x \le 1 + \sqrt{2}$.

EXERCISES 11.10

In Exercises 1–22 determine whether the series converges or diverges.

* **1.** $\displaystyle\sum_{n=1}^{\infty} \frac{1}{2n+1}$

* **2.** $\displaystyle\sum_{n=1}^{\infty} \frac{1}{4n-3}$

* **3.** $\displaystyle\sum_{n=1}^{\infty} \frac{1}{2n^2+4}$

* **4.** $\displaystyle\sum_{n=1}^{\infty} \frac{1}{5n^2-3n-1}$

* **5.** $\displaystyle\sum_{n=2}^{\infty} \frac{1}{n^3-1}$

* **6.** $\displaystyle\sum_{n=4}^{\infty} \frac{n^2}{n^4-6n^2+5}$

* **7.** $\displaystyle\sum_{n=1}^{\infty} \frac{1}{(2n-1)(2n+1)}$

* **8.** $\displaystyle\sum_{n=1}^{\infty} \frac{n-5}{n^2+3n-2}$

* **9.** $\displaystyle\sum_{n=2}^{\infty} \frac{1}{\ln n}$

* **10.** $\displaystyle\sum_{n=1}^{\infty} n^2 e^{-2n}$

* **11.** $\displaystyle\sum_{n=2}^{\infty} \frac{\sqrt{n^2+2n-3}}{n^2+5}$

* **12.** $\displaystyle\sum_{n=1}^{\infty} \frac{\sqrt{n+5}}{n^3+3}$

* **13.** $\displaystyle\sum_{n=1}^{\infty} \sqrt{\frac{n^2+2n+3}{2n^4-n}}$

* **14.** $\displaystyle\sum_{n=2}^{\infty} \frac{1}{n^2 \ln n}$

* **15.** $\displaystyle\sum_{n=1}^{\infty} \frac{1}{2^n} \sin\left(\frac{\pi}{n}\right)$

* **16.** $\displaystyle\sum_{n=1}^{\infty} \frac{\sqrt{n^2+1}}{n^3} \mathrm{Tan}^{-1} n$

* **17.** $\displaystyle\sum_{n=1}^{\infty} \frac{2^n+n}{3^n+1}$

* **18.** $\displaystyle\sum_{n=2}^{\infty} \frac{1+\ln^2 n}{n \ln^2 n}$

* **19.** $\displaystyle\sum_{n=1}^{\infty} \frac{1+1/n}{e^n}$

* **20.** $\displaystyle\sum_{n=1}^{\infty} \frac{\ln(n+1)}{n+1}$

* **21.** $\displaystyle\sum_{n=1}^{\infty} n e^{-n^2}$

* **22.** $\displaystyle\sum_{n=2}^{\infty} \frac{1}{n \sqrt[3]{\ln n}}$

In Exercises 23–26 find the interval of convergence of the power series.

** **23.** $\displaystyle\sum_{n=1}^{\infty} \frac{1}{n} x^{2n}$

** **24.** $\displaystyle\sum_{n=1}^{\infty} \frac{1}{n^2} x^{2n}$

** **25.** $\displaystyle\sum_{n=1}^{\infty} \frac{1}{n 2^n} (x-1)^{4n}$

** **26.** $\displaystyle\sum_{n=0}^{\infty} \frac{n 3^n}{(n+1)^3} (x+2)^{2n}$

In Exercises 27–29 find values of p for which the series converges.

** **27.** $\displaystyle\sum_{n=2}^{\infty} \frac{1}{n^p \ln n}$

** **28.** $\displaystyle\sum_{n=2}^{\infty} \frac{1}{n(\ln n)^p}$

*** **29.** $\displaystyle\sum_{n=2}^{\infty} \frac{1}{(\ln n)^p}$

11.11 Limit Ratio and Limit Root Tests

In this section we consider two additional tests to determine whether nonnegative series converge or diverge. The first test, called the limit ratio test, indicates whether a series $\sum c_n$ resembles a geometric series for large n.

THEOREM 11.16   (Limit Ratio Test)

Suppose $c_n > 0$ and

$$\lim_{n \to \infty} \frac{c_{n+1}}{c_n} = L. \tag{11.40}$$

Then,

    (i) $\sum_{n=1}^{\infty} c_n$ converges if $L < 1$;

    (ii) $\sum_{n=1}^{\infty} c_n$ diverges if $L > 1$ (or if $\lim_{n \to \infty} c_{n+1}/c_n = \infty$);

    (iii) $\sum_{n=1}^{\infty} c_n$ may converge or diverge if $L = 1$.

PROOF

(i) By the result of part (a) in Exercise 61 of Section 11.8 (see also Figure 11.23a), we can say that if $\lim_{n \to \infty} c_{n+1}/c_n = L < 1$, there exists an integer N such that for all $n \geq N$,

$$\frac{c_{n+1}}{c_n} < \frac{L + 1}{2}.$$

Consequently,

$$c_{N+1} < \left(\frac{L + 1}{2} \right) c_N;$$

$$c_{N+2} < \left(\frac{L + 1}{2} \right) c_{N+1} < \left(\frac{L + 1}{2} \right)^2 c_N;$$

$$c_{N+3} < \left(\frac{L + 1}{2} \right) c_{N+2} < \left(\frac{L + 1}{2} \right)^3 c_N;$$

and so on. Hence,

$$c_N + c_{N+1} + c_{N+2} + \cdots < c_N + \left(\frac{L + 1}{2} \right) c_N + \left(\frac{L + 1}{2} \right)^2 c_N + \cdots.$$

Since the right side of this inequality is a geometric series with its common ratio $(L + 1)/2 < 1$, it follows by the comparison test that $\sum_{n=N}^{\infty} c_n$ converges. Therefore, $\sum_{n=1}^{\infty} c_n$ converges also.

FIGURE 11.23a

FIGURE 11.23b

(ii) By the result of part (b) in Exercise 61 of Section 11.8 (see also Figure 11.23b), if $\lim_{n\to\infty} c_{n+1}/c_n = L > 1$, there exists an integer N such that for all $n \geq N$,

$$\frac{c_{n+1}}{c_n} > \frac{L+1}{2} > 1.$$

When $\lim_{n\to\infty} c_{n+1}/c_n = \infty$, it is also true that for n greater than or equal to some N, c_{n+1}/c_n must be greater than 1. This implies that for all $n > N$, $c_n > c_N$, and therefore

$$\lim_{n\to\infty} c_n \neq 0.$$

Hence $\sum c_n$ diverges by the nth term test.

(iii) To show that the limit ratio test fails when $L = 1$, consider the two p-series $\sum 1/n$ and $\sum 1/n^2$. For each series, $L = 1$, yet the first series diverges and the second converges. ∎

In a nonrigorous way we can justify the limit ratio test from the following standpoint. If $\lim_{n\to\infty} c_{n+1}/c_n = L$, then for large n, each term of series $\sum c_n$ is essentially L times the term before it; that is, the series resembles a geometric series with common ratio L. We would expect convergence of the series if $L < 1$ and divergence if $L > 1$. We might also anticipate some indecision about the $L = 1$ case, depending on how this limit is reached, since this case corresponds to the common ratio which separates convergent and divergent geometric series.

EXAMPLE 11.38 ▼

Determine whether the following series converge or diverge:

(a) $\displaystyle\sum_{n=1}^{\infty} \frac{2^n}{n^4}$ (b) $\displaystyle\sum_{n=1}^{\infty} \frac{n^{100}}{1 \cdot 3 \cdot 5 \cdot \cdots \cdot (2n-1)}$ (c) $\displaystyle\sum_{n=1}^{\infty} \frac{n^n}{n!}$

SOLUTION

(a) For the series $\sum 2^n/n^4$,

$$L = \lim_{n\to\infty} \frac{\dfrac{2^{n+1}}{(n+1)^4}}{\dfrac{2^n}{n^4}} = \lim_{n\to\infty} 2\left(\frac{n}{n+1}\right)^4 = 2,$$

and the series therefore diverges by the limit ratio test.

(b) For this series,

$$L = \lim_{n\to\infty} \frac{\dfrac{(n+1)^{100}}{1 \cdot 3 \cdot 5 \cdot \cdots \cdot (2n+1)}}{\dfrac{n^{100}}{1 \cdot 3 \cdot 5 \cdot \cdots \cdot (2n-1)}} = \lim_{n\to\infty} \left(\frac{n+1}{n}\right)^{100} \frac{1}{2n+1} = 0,$$

and the series therefore converges by the limit ratio test.

(c) Since

$$L = \lim_{n \to \infty} \frac{\dfrac{(n+1)^{n+1}}{(n+1)!}}{\dfrac{n^n}{n!}} = \lim_{n \to \infty} \left(\frac{n+1}{n} \right)^n = e > 1$$

(see equation 3.33), the series $\sum n^n / n!$ diverges.

We have one last test for nonnegative series.

THEOREM 11.17   (Limit Root Test)

Suppose $c_n \geq 0$ and

$$\lim_{n \to \infty} \sqrt[n]{c_n} = R. \qquad (11.41)$$

Then,

(i) $\sum_{n=1}^{\infty} c_n$ converges if $R < 1$;

(ii) $\sum_{n=1}^{\infty} c_n$ diverges if $R > 1$ (or if $\lim_{n \to \infty} \sqrt[n]{c_n} = \infty$);

(iii) $\sum_{n=1}^{\infty} c_n$ may converge or diverge if $R = 1$.

PROOF

(i) By the result of part (a) in Exercise 61 of Section 11.8, if $\lim_{n \to \infty} \sqrt[n]{c_n} = R < 1$, there exists an integer N such that for all $n \geq N$,

$$\sqrt[n]{c_n} < \frac{R+1}{2}.$$

Consequently,

$$\sqrt[N]{c_N} < \frac{R+1}{2}, \qquad \text{or} \qquad c_N < \left(\frac{R+1}{2} \right)^N;$$

$$\sqrt[N+1]{c_{N+1}} < \frac{R+1}{2}, \qquad \text{or} \qquad c_{N+1} < \left(\frac{R+1}{2} \right)^{N+1};$$

$$\sqrt[N+2]{c_{N+2}} < \frac{R+1}{2}, \qquad \text{or} \qquad c_{N+2} < \left(\frac{R+1}{2} \right)^{N+2};$$

and so on. Hence,

$$c_N + c_{N+1} + c_{N+2} + \cdots < \left(\frac{R+1}{2} \right)^N + \left(\frac{R+1}{2} \right)^{N+1} + \left(\frac{R+1}{2} \right)^{N+2} + \cdots .$$

Since the right side of this inequality is a convergent geometric series, the left side must be a convergent series also by the comparison test. Thus, $\sum_{n=N}^{\infty} c_n$ converges, and so also must $\sum_{n=1}^{\infty} c_n$.

(ii) When $\lim_{n\to\infty} \sqrt[n]{c_n} = R > 1$, there exists an integer N such that for all $n \geq N$,

$$\sqrt[n]{c_n} > \frac{R+1}{2} > 1 \quad \text{or} \quad c_n > 1,$$

as in part (b) of Exercise 61 in Section 11.8. When $\lim_{n\to\infty} \sqrt[n]{c_n} = \infty$, it is also true that for n greater than or equal to some N, $\sqrt[n]{c_n}$ must be greater than 1. But it now follows that $\lim_{n\to\infty} c_n \neq 0$, and the series diverges by the nth term test.

(iii) To show that the test fails when $R = 1$, we show that $R = 1$ for the two p-series $\sum 1/n$ and $\sum 1/n^2$, diverging and converging respectively. For the harmonic series, let $R = \lim_{n\to\infty} (1/n)^{1/n}$. If we take logarithms, then

$$\ln R = \ln \left[\lim_{n\to\infty} \left(\frac{1}{n} \right)^{1/n} \right] = \lim_{n\to\infty} \ln \left(\frac{1}{n} \right)^{1/n} = \lim_{n\to\infty} \frac{1}{n} \ln \left(\frac{1}{n} \right) = -\lim_{n\to\infty} \frac{1}{n} \ln n$$

$$= -\lim_{n\to\infty} \frac{\dfrac{1}{n}}{1} \qquad \text{(by L'Hôpital's rule)}$$

$$= 0.$$

Hence, $R = 1$ for this divergent series. A similar analysis gives $R = 1$ for the convergent series $\sum 1/n^2$. ∎

EXAMPLE 11.39

Determine whether the following series converge or diverge:

$$\text{(a)} \quad \sum_{n=1}^{\infty} \left(\frac{n+1}{n} \right)^{n^2} \qquad \qquad \text{(b)} \quad \sum_{n=1}^{\infty} \frac{n}{(\ln n)^n}$$

SOLUTION

(a) Since

$$R = \lim_{n\to\infty} \left[\left(\frac{n+1}{n} \right)^{n^2} \right]^{1/n} = \lim_{n\to\infty} \left(\frac{n+1}{n} \right)^n = e > 1,$$

the series diverges by the limit root test.

(b) For this series,

$$R = \lim_{n\to\infty} \left[\frac{n}{(\ln n)^n} \right]^{1/n} = \lim_{n\to\infty} \frac{n^{1/n}}{\ln n}.$$

If we set $L = \lim_{n\to\infty} n^{1/n}$, then

$$\ln L = \ln \left(\lim_{n\to\infty} n^{1/n} \right) = \lim_{n\to\infty} \frac{1}{n} \ln n$$

$$= \lim_{n\to\infty} \frac{\dfrac{1}{n}}{1} \qquad \text{(by L'Hôpital's rule)}$$

$$= 0.$$

Thus, $L = 1$, and it follows that

$$R = \lim_{n \to \infty} \frac{n^{1/n}}{\ln n} = 0.$$

The series therefore converges.

We have developed six tests to determine whether series of numbers converge or diverge:

n^{th} term test
integral test
comparison test
limit comparison test
limit ratio test
limit root test

The form of the nth term of a series often suggests which test should be used. Keep the following ideas in mind when choosing a test:

1. The limit ratio test can be effective on factorials, products of the form $1 \cdot 3 \cdot 5 \cdot \; \cdots \; \cdot (2n - 1)$, and constants raised to powers involving n (2^n, 3^{-n}, etc.).

2. The limit root test thrives on functions of n raised to powers involving n (see Example 11.39).

3. The limit comparison test is successful on rational functions of n, and fractional powers as well ($\sqrt{n}$, $\sqrt[3]{n/(n+1)}$, etc.).

4. The integral test can be effective when the nth term is easily integrated. Logarithms often require the integral test.

By definition, a series $\sum c_n$ converges if and only if its sequence of partial sums $\{S_n\}$ converges. The difficulty with using this definition to discuss convergence of a series is that S_n can seldom be evaluated in a simple form, and therefore consideration of $\lim_{n \to \infty} S_n$ is impossible. The above tests have the advantage of avoiding partial sums. On the other hand, they have one disadvantage. Although they may indicate that a series does indeed have a sum, the tests in no way suggest the value of the sum. The problem of calculating the sum often proves more difficult than showing that it exists in the first place. In Section 11.13 we discuss various ways to calculate and approximate sums for known convergent series.

EXERCISES 11.11

In Exercises 1–20 determine whether the series converges or diverges.

* **1.** $\displaystyle\sum_{n=1}^{\infty} \frac{e^n}{n^4}$

* **2.** $\displaystyle\sum_{n=1}^{\infty} \frac{1}{n!}$

* **3.** $\displaystyle\sum_{n=1}^{\infty} \frac{n^3}{2^n}$

* **4.** $\displaystyle\sum_{n=1}^{\infty} \frac{1}{n^n}$

* **5.** $\displaystyle\sum_{n=1}^{\infty} \frac{(n-1)(n-2)}{n^2 2^n}$

* **6.** $\displaystyle\sum_{n=1}^{\infty} \frac{(2n)!}{(n!)^2}$

* **7.** $\displaystyle\sum_{n=1}^{\infty} \frac{\sqrt{n+1}}{n^{n+1/2}}$

* **8.** $\displaystyle\sum_{n=1}^{\infty} \frac{3^{-n} + 2^{-n}}{4^{-n} + 5^{-n}}$

* **9.** $\displaystyle\sum_{n=1}^{\infty} \frac{e^{-n}}{\sqrt{n+\pi}}$

* **10.** $\displaystyle\sum_{n=1}^{\infty} \frac{2 \cdot 4 \cdot \ldots \cdot (2n)}{4 \cdot 7 \cdot \ldots \cdot (3n+1)}$

* **11.** $\displaystyle\sum_{n=1}^{\infty} \frac{n^{n-1}}{3^{n-1}(n-1)!}$

* **12.** $\displaystyle\sum_{n=1}^{\infty} n \left(\frac{3}{4}\right)^n$

* **13.** $\displaystyle\sum_{n=1}^{\infty} \frac{1 + 1/n}{e^n}$

** **14.** $\displaystyle\sum_{n=1}^{\infty} \frac{2 \cdot 4 \cdot \ldots \cdot (2n)}{3 \cdot 5 \cdot \ldots \cdot (2n+1)} \left(\frac{1}{n^2}\right)$

** **15.** $\displaystyle\sum_{n=1}^{\infty} \frac{n^n}{(n+1)^{n+1}}$

** **16.** $\displaystyle\sum_{n=1}^{\infty} \frac{(n+1)^n}{n^{n+1}}$

** **17.** $\displaystyle\sum_{n=1}^{\infty} \frac{n^4 + 3}{5^{n/2}}$

** **18.** $\displaystyle\sum_{n=1}^{\infty} \frac{2^n + n^2 3^n}{4^n}$

** **19.** $\displaystyle\sum_{n=1}^{\infty} \frac{n^2 2^n - n}{n^3 + 1}$

** **20.** $\displaystyle\sum_{n=1}^{\infty} \frac{(2n)!}{(3n)!} 5^{2n}$

** **21.** For what integer values of a is the series

$$\sum_{n=1}^{\infty} \frac{(n!)^2}{(an)!}$$

convergent?

11.12 Absolute and Conditional Convergence, Alternating Series

The convergence tests of Sections 11.10 and 11.11 are applicable to nonnegative series, series whose terms are all nonnegative. Series with infinitely many positive and negative terms are more complicated. It is fortunate, however, that all our tests are still useful in discussing convergence of series with positive and negative terms. What makes this possible is the following definition and Theorem 11.18.

> **DEFINITION 11.7**
>
> A series $\sum_{n=1}^{\infty} c_n$ is said to be **absolutely convergent** if the series of absolute values $\sum_{n=1}^{\infty} |c_n|$ converges.

At first glance it might seem that absolute convergence is a strange concept indeed. What possible good could it do to consider the series of absolute values, which is quite different from the original series? The fact is that when the series of absolute values converges, it automatically follows from the next theorem that the original series converges also. And since the series of absolute values has all nonnegative terms, we can use the comparison, limit comparison, limit ratio, limit root, or integral test to consider its convergence.

> **THEOREM 11.18**
>
> If a series is absolutely convergent, then it is convergent.

PROOF Let $\{S_n\}$ be the sequence of partial sums of the absolutely convergent series $\sum c_n$. Define sequences $\{P_n\}$ and $\{N_n\}$, where P_n is the sum of all positive terms in S_n, and N_n is the sum of the absolute values of all negative terms in S_n. Then

$$S_n = P_n - N_n.$$

The sequence of partial sums for the series of absolute values $\sum |c_n|$ is

$$\{P_n + N_n\},$$

and this sequence must be nondecreasing and bounded. Since each of the sequences $\{P_n\}$ and $\{N_n\}$ is nondecreasing and a part of $\{P_n + N_n\}$, it follows that each is bounded, and therefore has a limit, say, P and N, respectively. As a result, sequence $\{P_n - N_n\} = \{S_n\}$ has limit $P - N$, and series $\sum c_n$ converges to $P - N$. ∎

EXAMPLE 11.40

Show that the following series are absolutely convergent:

(a) $\displaystyle\sum_{n=1}^{\infty} \frac{(-1)^n n}{(n+1)2^n}$

(b) $1 - \dfrac{1}{2^2} - \dfrac{1}{3^3} + \dfrac{1}{4^4} + \dfrac{1}{5^5} + \dfrac{1}{6^6} - \dfrac{1}{7^7} - \dfrac{1}{8^8} - \dfrac{1}{9^9} - \dfrac{1}{10^{10}} + \cdots + \dfrac{1}{15^{15}} - \cdots$

SOLUTION

(a) The series of absolute values is $\sum n/[(n+1)2^n]$. We use the limit comparison test to show that it converges. Since

$$\ell = \lim_{n \to \infty} \frac{\dfrac{n}{(n+1)2^n}}{\dfrac{1}{2^n}} = 1,$$

and $\sum (1/2)^n$ is convergent (a geometric series with $r = 1/2$), it follows that the series of absolute values converges. The given series therefore converges absolutely.

(b) The series of absolute values is $\sum 1/n^n$. We use the comparison test to verify that this series converges. Since

$$\frac{1}{n^n} \leq \frac{1}{n^2}$$

and $\sum 1/n^2$ converges, so does $\sum 1/n^n$. The given series therefore converges absolutely.

In Example 11.40, absolute convergence of the given series implies convergence of the series, but it is customary to omit such a statement. It is important to realize, however, that it is the given series that is being analyzed, and its convergence is guaranteed by Theorem 11.18.

We now ask whether series can converge without converging absolutely. If there are such series, and indeed there are, we must devise new convergence tests. We describe these series as follows.

DEFINITION 11.8

A series that converges but does not converge absolutely is said to **converge conditionally**.

The most important type of series with both positive and negative terms is an alternating series. As the name suggests, an **alternating series** has terms that are alternately positive and negative. For example,

$$\sum_{n=1}^{\infty} \frac{(-1)^{n+1}}{n} = 1 - \frac{1}{2} + \frac{1}{3} - \frac{1}{4} + \frac{1}{5} - \cdots$$

is an alternating series called the **alternating harmonic series**.

Given an alternating series to examine for convergence, we first test for absolute convergence as in Example 11.40(a). Should this fail, we check for conditional convergence with the following test.

THEOREM 11.19   (Alternating Series Test)

An alternating series $\sum_{n=1}^{\infty} c_n$ converges if the sequence of absolute values of the terms $\{|c_n|\}$ is decreasing and has limit zero.

PROOF   If $\{S_n\}$ is the sequence of partial sums of the series, then differences of partial sums are $S_n - S_{n-1} = c_n$. Because the c_n alternate in sign, so also do the differences $S_n - S_{n-1}$. This means that the partial sums $\{S_n\}$ form an oscillating sequence. Since absolute values $|S_n - S_{n-1}| = |c_n|$ are decreasing and approach zero, it follows by Theorem 11.8 of Section 11.8 that the sequence $\{S_n\}$ of partial sums has a limit. ■

| **EXAMPLE 11.41** ▼

Determine whether the following series converge absolutely, converge conditionally, or diverge:

(a) $\displaystyle\sum_{n=1}^{\infty} \frac{(-1)^{n+1}}{n}$ (b) $\displaystyle\sum_{n=1}^{\infty} (-1)^n \frac{\sqrt{n^2 + 5n}}{n^{3/2}}$ (c) $\displaystyle\sum_{n=1}^{\infty} (-1)^{n+1} \frac{4^n}{n^5 3^n}$

SOLUTION

(a) The alternating harmonic series $\sum (-1)^{n+1}/n$ is not absolutely convergent because the series of absolute values $\sum 1/n$ diverges. Since the sequence of absolute values of the terms $\{1/n\}$ is decreasing with limit zero, series $\sum (-1)^{n+1}/n$ converges conditionally.

(b) For this alternating series, we first consider the series of absolute values

$$\sum_{n=1}^{\infty} \frac{\sqrt{n^2 + 5n}}{n^{3/2}}.$$

We use the limit comparison test on this series. Since

$$\ell = \lim_{n \to \infty} \frac{\dfrac{\sqrt{n^2 + 5n}}{n^{3/2}}}{\dfrac{1}{n^{1/2}}} = \lim_{n \to \infty} \frac{n\sqrt{1 + \dfrac{5}{n}}}{n^{3/2}} \cdot n^{1/2} = 1,$$

and $\sum 1/n^{1/2}$ diverges, so does $\sum \sqrt{n^2 + 5n}/n^{3/2}$. The original series does not therefore converge absolutely. We now resort to the alternating series test. The sequence $\{\sqrt{n^2 + 5n}/n^{3/2}\}$ of absolute values of the terms of the series is decreasing if

$$\frac{\sqrt{(n + 1)^2 + 5(n + 1)}}{(n + 1)^{3/2}} < \frac{\sqrt{n^2 + 5n}}{n^{3/2}}.$$

When we square and cross multiply, the inequality becomes

$$n^3(n^2 + 7n + 6) < (n^2 + 5n)(n + 1)^3$$
$$= n^5 + 8n^4 + 18n^3 + 16n^2 + 5n;$$

that is,

$$n^4 + 12n^3 + 16n^2 + 5n > 0,$$

which is obviously valid because $n \geq 1$. Since $\lim_{n \to \infty} \sqrt{n^2 + 5n}/n^{3/2} = 0$, we conclude that the alternating series $\sum (-1)^n(\sqrt{n^2 + 5n}/n^{3/2})$ converges conditionally.

(c) If we apply the limit ratio test to the series of absolute values $\sum 4^n/(n^5 3^n)$, we have

$$L = \lim_{n \to \infty} \frac{\dfrac{4^{n+1}}{(n + 1)^5 3^{n+1}}}{\dfrac{4^n}{n^5 3^n}} = \lim_{n \to \infty} \frac{4}{3}\left(\frac{n}{n + 1}\right)^5 = \frac{4}{3}.$$

Since $L > 1$, the series $\sum 4^n/(n^5 3^n)$ diverges. The original alternating series does not therefore converge absolutely. But $L = 4/3$ implies that for large n, each term in the series of absolute values is approximately $4/3$ times the term that precedes it, and therefore

$$\lim_{n \to \infty} \frac{4^n}{n^5 3^n} = \infty.$$

Consequently,

$$\lim_{n \to \infty} (-1)^{n+1} \frac{4^n}{n^5 3^n}$$

cannot possibly exist, and the given series diverges by the nth term test.

We have noted several times that the essential question for convergence of a nonnegative series is, "Do the terms approach zero quickly enough to guarantee convergence of the series?" With a series that has infinitely many positive and negative terms, this question is inappropriate. Such a series may converge because of a partial cancelling effect; for example, a negative term may offset the effect of a large positive term. This kind of process may produce a convergent series even though the series would be divergent if all terms were replaced by their absolute values. A specific example is the alternating harmonic series which converges (conditionally) because of this cancelling effect, whereas the harmonic series itself, which has no cancellations, diverges.

Absolute and conditional convergence are particularly important when discussing end points of intervals of convergence for power series since one of the end points usually leads to an alternating series. We illustrate in the following examples.

EXAMPLE 11.42

Find the interval of convergence for the power series $\sum_{n=0}^{\infty} \dfrac{1}{(n+1)2^n} x^n$.

SOLUTION   Since the radius of convergence is

$$R = \lim_{n \to \infty} \left| \frac{\dfrac{1}{(n+1)2^n}}{\dfrac{1}{(n+2)2^{n+1}}} \right| = \lim_{n \to \infty} 2\left(\frac{n+2}{n+1} \right) = 2,$$

the open interval of convergence is $-2 < x < 2$. At $x = 2$, the series is $\sum_{n=0}^{\infty} 1/(n+1)$, the harmonic series, which diverges. At $x = -2$, we obtain the alternating harmonic series $\sum_{n=0}^{\infty} (-1)^n/(n+1)$, which converges conditionally. The interval of convergence is therefore $-2 \le x < 2$.

EXAMPLE 11.43

Find the interval of convergence for the power series $\sum_{n=1}^{\infty} \dfrac{n(-1)^n}{(2n+5)^3} (x-3)^n$.

SOLUTION   Since the radius of convergence is

$$R = \lim_{n \to \infty} \left| \frac{\dfrac{n(-1)^n}{(2n+5)^3}}{\dfrac{(n+1)(-1)^{n+1}}{(2n+7)^3}} \right| = \lim_{n \to \infty} \frac{n(2n+7)^3}{(n+1)(2n+5)^3} = 1,$$

the open interval of convergence is $2 < x < 4$. At $x = 2$, the series becomes

$$\sum_{n=1}^{\infty} \frac{n(-1)^n}{(2n+5)^3} (-1)^n = \sum_{n=1}^{\infty} \frac{n}{(2n+5)^3}.$$

Since

$$\ell = \lim_{n \to \infty} \left| \frac{\dfrac{n}{(2n+5)^3}}{\dfrac{1}{8n^2}} \right| = \lim_{n \to \infty} \frac{8n^3}{(2n+5)^3} = 1,$$

and $\sum 1/(8n^2) = (1/8)\sum 1/n^2$ converges ($p = 2$ series), so also does $\sum n/(2n+5)^3$ (by the limit comparison test). At $x = 4$, the power series becomes

$$\sum_{n=1}^{\infty} \frac{n(-1)^n}{(2n+5)^3}.$$

This is an alternating series that converges absolutely, as indicated in the discussion of $x = 2$ on page 824. The interval of convergence is therefore $2 \leq x \leq 4$.

In Sections 11.10–11.12, we have obtained a number of tests for determining whether series of numbers converge or diverge. To test a series for convergence, we suggest the following procedure:

1. Try the nth term test for divergence.
2. If $\{c_n\}$ has limit zero and the series is nonnegative , try the comparison, limit comparison, limit ratio, limit root, or integral test.
3. If $\{c_n\}$ has limit zero and the series contains both positive and negative terms, test for absolute convergence using the tests in 2. If this fails and the series is alternating, test for conditional convergence with the alternating series test.

Each of the comparison, limit comparison, limit ratio, limit root, integral, and alternating series tests requires conditions to be satisfied for all terms of the series. Specifically, the comparison, limit comparison, and limit root tests require $c_n \geq 0$ for all n; the limit ratio test requires $c_n > 0$; the integral test requires $f(n)$ to be positive, continuous, and decreasing; and the alternating series test requires $\{|c_n|\}$ to be decreasing and $\{c_n\}$ to be alternately positive and negative. None of these requirements is essential for all n; in fact, so long as they are satisfied for all terms in the series beyond some point, say for n greater than or equal to some integer N, the particular test may be used on the series $\sum_{n=N}^{\infty} c_n$. The original series $\sum_{n=1}^{\infty} c_n$ then converges if and only if $\sum_{n=N}^{\infty} c_n$ converges.

Before leaving this section, we prove Theorem 11.2. We verify 11.18a using the limit ratio test; verification of 11.18b is similar, using the limit root test.

If the limit ratio test is applied to the series of absolute values $\sum |a_n x^n|$,

$$L = \lim_{n \to \infty} \frac{|a_{n+1} x^{n+1}|}{|a_n x^n|} = |x| \lim_{n \to \infty} \left| \frac{a_{n+1}}{a_n} \right|.$$

Assuming that limit 11.18a exists or is equal to infinity, there are three possibilities:

(i) If $\lim_{n \to \infty} |a_n/a_{n+1}| = 0$, then $\lim_{n \to \infty} |a_{n+1}/a_n| = \infty$. Therefore $L = \infty$, and the power series diverges for all $x \neq 0$. In other words,

$$R = 0 = \lim_{n \to \infty} \left| \frac{a_n}{a_{n+1}} \right|.$$

(ii) If $\lim_{n \to \infty} |a_n/a_{n+1}| = \infty$, then $\lim_{n \to \infty} |a_{n+1}/a_n| = 0$. Therefore $L = 0$, and the power series converges absolutely for all x. Consequently,

$$R = \infty = \lim_{n \to \infty} \left| \frac{a_n}{a_{n+1}} \right|.$$

(iii) If $\lim_{n \to \infty} |a_n/a_{n+1}| = R$, then $\lim_{n \to \infty} |a_{n+1}/a_n| = 1/R$. In this case, $L = |x|/R$. Since the power series converges absolutely for $L < 1$ and diverges for $L > 1$, it follows that absolute convergence occurs for $|x| < R$ and divergence for $|x| > R$. This implies that R is the radius of convergence of the power series.

EXERCISES 11.12

In Exercises 1–14 determine whether the series converges absolutely, converges conditionally, or diverges.

* **1.** $\displaystyle\sum_{n=1}^{\infty} (-1)^n \frac{n}{n^3 + 1}$

* **2.** $\displaystyle\sum_{n=1}^{\infty} (-1)^n \frac{n}{n^2 + 1}$

* **3.** $\displaystyle\sum_{n=1}^{\infty} \frac{\cos(n\pi/2)}{2n^2}$

* **4.** $\displaystyle\sum_{n=1}^{\infty} (-1)^n \frac{n^3}{3^n}$

* **5.** $\displaystyle\sum_{n=1}^{\infty} \frac{(-1)^{n+1}}{\sqrt{n}}$

* **6.** $\displaystyle\sum_{n=1}^{\infty} (-1)^n \frac{3^n}{n^3}$

* **7.** $\displaystyle\sum_{n=1}^{\infty} (-1)^n \frac{n}{n^2 + n + 1}$

* **8.** $\displaystyle\sum_{n=1}^{\infty} \frac{n \sin(n\pi/4)}{2^n}$

* **9.** $\displaystyle\sum_{n=1}^{\infty} (-1)^{n+1} \left(\frac{n}{n+1}\right)$

* **10.** $\displaystyle\sum_{n=1}^{\infty} (-1)^{n+1} \frac{\sqrt{3n-2}}{n}$

** **11.** $\displaystyle\sum_{n=1}^{\infty} (-1)^n \left(\frac{n}{n+1}\right)^n$

** **12.** $\displaystyle\sum_{n=1}^{\infty} (-1)^n \frac{\sqrt{n^2 + 3}}{n^2 + 5}$

** **13.** $\displaystyle\sum_{n=2}^{\infty} (-1)^{n-1} \frac{\ln n}{n}$

** **14.** $\displaystyle\sum_{n=1}^{\infty} \frac{\cos(n\pi/10)\cot^{-1} n}{n^3 + 5n}$

In Exercises 15–22 find the interval of convergence of the power series.

** **15.** $\displaystyle\sum_{n=1}^{\infty} \frac{1}{n} x^n$

** **16.** $\displaystyle\sum_{n=0}^{\infty} \frac{1}{(n+1)^2} x^n$

** **17.** $\displaystyle\sum_{n=1}^{\infty} \frac{1}{n2^n} (x-1)^n$

** **18.** $\displaystyle\sum_{n=1}^{\infty} \frac{1}{\sqrt{n}} (x+2)^n$

** **19.** $\displaystyle\sum_{n=0}^{\infty} \frac{n-1}{n^2 + 1} (2x)^n$

** **20.** $\displaystyle\sum_{n=0}^{\infty} \frac{1}{\sqrt{n+1}} x^{3n+1}$

** **21.** $\displaystyle\sum_{n=2}^{\infty} \frac{1}{\ln n} x^n$

** **22.** $\displaystyle\sum_{n=2}^{\infty} \frac{1}{n^2 \ln n} (x-2)^n$

** **23.** Discuss convergence of the series

$$\sum_{n=1}^{\infty} \frac{\sin(nx)}{n^2}.$$

** **24.** Prove that if $\sum c_n$ converges absolutely, then $\sum c_n^p$ converges absolutely for all integers $p > 1$.

** **25.** Discuss convergence of the series

$$\sum_{n=1}^{\infty} (-1)^n \frac{n^n}{(n+1)^{n+1}}.$$

11.13 Exact and Approximate Values for Sums of Series of Numbers

In Sections 11.9–11.12 we have concentrated on whether series of numbers converge or diverge. But this is only half the problem. The comparison, limit comparison, limit ratio, limit root, integral, and alternating series tests may determine whether a series converges or diverges, but they do not determine the sum of the series in the case of a convergent series. This part of the problem, as suggested before, can sometimes be more complicated.

If the convergent series is a geometric series, no problem exists; we can use formula 11.33b to find its sum. It may also happen that the nth partial sum S_n of the series can be calculated in a simple form, in which case the sum of the series is $\lim_{n \to \infty} S_n$. Cases of this latter type are very rare. By substituting values of x into power series with known sums, we obtain formulas for sums of series of numbers. For instance, in Example 11.8, we verified that the Maclaurin series for e^x is

$$e^x = \sum_{n=0}^{\infty} \frac{1}{n!} x^n, \qquad -\infty < x < \infty.$$

By substituting $x = 1$, we obtain a series which converges to e,

$$e = \sum_{n=0}^{\infty} \frac{1}{n!} = 1 + \frac{1}{1!} + \frac{1}{2!} + \frac{1}{3!} + \cdots .$$

Another illustration is contained in the following example.

EXAMPLE 11.44

Use Example 11.18 to show that the sum of the alternating harmonic series is $\ln 2$.

SOLUTION   According to Example 11.18, the Taylor series about $x = 1$ for $\ln x$ is

$$\ln x = \sum_{n=1}^{\infty} \frac{(-1)^{n+1}}{n} (x-1)^n = (x-1) - \frac{1}{2}(x-1)^2 + \frac{1}{3}(x-1)^3 - \cdots ,$$

with open interval of convergence $0 < x < 2$. At $x = 0$, the series becomes the negative of the harmonic series which diverges, and at $x = 2$, it becomes the conditionally convergent, alternating harmonic series; that is,

$$\ln 2 = \sum_{n=1}^{\infty} \frac{(-1)^{n+1}}{n} = 1 - \frac{1}{2} + \frac{1}{3} - \frac{1}{4} + \cdots .$$

Convergence of the Taylor series at $x = 2$ does not, by itself, imply convergence to $\ln 2$, as we are suggesting. It is, however, true, and this is a direct application of the following theorem.

THEOREM 11.20

If the Taylor series $\sum_{n=0}^{\infty} a_n (x - c)^n$ of a function $f(x)$ converges at the endpoint $x = c + R$ of its interval of convergence, and if $f(x)$ is continuous at $x = c + R$, then the Taylor series evaluated at $c + R$ converges to $f(c + R)$. The same result is valid at the other endpoint $x = c - R$.

This example suggests another possibility for summing convergent series of numbers. Find a power series with known sum that reduces to the given series of numbers upon substitution of a value of x. We illustrate this in the next two examples.

EXAMPLE 11.45

Find $\displaystyle\sum_{n=1}^{\infty} \frac{n}{3^n}$.

SOLUTION This series results if we set $x = 1/3$ in the power series

$$S(x) = \sum_{n=1}^{\infty} n\, x^n.$$

The radius of convergence of this series is

$$R = \lim_{n \to \infty} \left| \frac{n}{n+1} \right| = 1.$$

If we divide both sides by x,

$$\frac{1}{x} S(x) = \sum_{n=1}^{\infty} n x^{n-1},$$

and integrate according to Theorem 11.3,

$$\int \frac{1}{x} S(x)\, dx = \sum_{n=1}^{\infty} x^n.$$

This is a geometric series with sum

$$\int \frac{1}{x} S(x)\, dx = \frac{x}{1-x}, \quad |x| < 1.$$

Differentiation now gives

$$\frac{1}{x} S(x) = \frac{(1-x)(1) - x(-1)}{(1-x)^2} = \frac{1}{(1-x)^2}.$$

Thus,

$$S(x) = \sum_{n=1}^{\infty} n\, x^n = \frac{x}{(1-x)^2}.$$

When we set $x = 1/3$,

$$\sum_{n=1}^{\infty} \frac{n}{3^n} = \frac{1/3}{(1 - 1/3)^2} = \frac{3}{4}.$$

EXAMPLE 11.46

Find the sum of the series $\displaystyle\sum_{n=0}^{\infty} \frac{(-1)^n}{(2n+1)2^n}$.

SOLUTION   There are many power series that reduce to this series upon substitution of a specific value of x. For instance, substitution of $-1/2$, 1, and $1/\sqrt{2}$ into the following power series, respectively, lead to the given series:

$$\sum_{n=0}^{\infty} \frac{1}{2n+1} x^n, \qquad \sum_{n=0}^{\infty} \frac{(-1)^n}{(2n+1)2^n} x^n, \qquad \sum_{n=0}^{\infty} \frac{\sqrt{2}(-1)^n}{2n+1} x^{2n+1}$$

Which should we consider? Although it is not the simplest, the third series looks most promising; the fact that the power on x corresponds to the coefficient in the denominator suggests that we can find the sum of this series. We therefore set

$$S(x) = \sum_{n=0}^{\infty} \frac{\sqrt{2}(-1)^n}{2n+1} x^{2n+1}.$$

To find the radius of convergence of this series, we set $y = x^2$, in which case

$$\sum_{n=0}^{\infty} \frac{\sqrt{2}(-1)^n}{2n+1} x^{2n+1} = x \sum_{n=0}^{\infty} \frac{\sqrt{2}(-1)^n}{2n+1} y^n.$$

The radius of convergence of the y-series is

$$R_y = \lim_{n\to\infty} \left| \frac{\dfrac{\sqrt{2}(-1)^n}{2n+1}}{\dfrac{\sqrt{2}(-1)^{n+1}}{2n+3}} \right| = \lim_{n\to\infty} \left(\frac{2n+3}{2n+1} \right) = 1.$$

The radius of convergence for the power series in x is therefore $R_x = 1$ also. If we differentiate the series with respect to x,

$$S'(x) = \sum_{n=0}^{\infty} \sqrt{2}(-1)^n x^{2n} = \sum_{n=0}^{\infty} \sqrt{2}(-x^2)^n = \frac{\sqrt{2}}{1-(-x^2)} = \frac{\sqrt{2}}{1+x^2}.$$

Antidifferentiation now gives

$$S(x) = \int \frac{\sqrt{2}}{1+x^2}\, dx = \sqrt{2}\,\text{Tan}^{-1} x + C.$$

Since $S(0) = 0$, it follows that $C = 0$, and

$$\sum_{n=0}^{\infty} \frac{\sqrt{2}(-1)^n}{2n+1} x^{2n+1} = \sqrt{2}\,\text{Tan}^{-1} x.$$

If we now set $x = 1/\sqrt{2}$,

$$\sum_{n=0}^{\infty} \frac{\sqrt{2}(-1)^n}{2n+1}\left(\frac{1}{\sqrt{2}}\right)^{2n+1} = \sqrt{2}\,\mathrm{Tan}^{-1}\left(\frac{1}{\sqrt{2}}\right).$$

Consequently,

$$\sum_{n=0}^{\infty} \frac{(-1)^n}{(2n+1)2^n} = \sqrt{2}\,\mathrm{Tan}^{-1}\left(\frac{1}{\sqrt{2}}\right).$$

Approximating the Sum of a Series of Numbers

The above techniques do not find sums for all convergent series of numbers. In fact there are many series for which we would find it impossible to find a sum. But in applications we might be satisfied with a reasonable approximation to the sum of a series, and we therefore turn our attention to the problem of estimating the sum of a convergent series. The easiest method for estimating the sum S of a convergent series $\sum c_n$ is simply to choose the partial sum S_N for some N as an approximation; that is, truncate the series after N terms and choose

$$S \approx S_N = c_1 + c_2 + \cdots + c_N.$$

But an approximation is of value only if we can make some definitive statement about its accuracy. In truncating the series, we have neglected the infinity of terms $\sum_{n=N+1}^{\infty} c_n$, and the accuracy of the approximation is therefore determined by the size of $\sum_{n=N+1}^{\infty} c_n$; the smaller it is, the better the approximation. The problem is that we do not know the exact value of $\sum_{n=N+1}^{\infty} c_n$; if we did, there would be no need to approximate the sum of the original series in the first place. What we must do is estimate the sum $\sum_{n=N+1}^{\infty} c_n$.

When the integral test or the alternating series test is used to prove that a series converges, simple formulas give accuracy estimates on the truncated series. Let us illustrate these first.

Truncating an Alternating Series

It is very simple to obtain an estimate of the accuracy of a truncated alternating series $\sum c_n$ provided the sequence $\{|c_n|\}$ is decreasing with limit zero. For example, suppose $c_1 > 0$ (a similar discussion can be made when $c_1 < 0$). If $\{S_n\}$ is the sequence of partial sums of $\sum c_n$, then even partial sums can be expressed in the form

$$S_{2n} = (c_1 + c_2) + (c_3 + c_4) + \cdots + (c_{2n-1} + c_{2n}).$$

Since $\{|c_n|\}$ is decreasing ($|c_n| > |c_{n+1}|$), each term in the parentheses is positive. Consequently, the subsequence $\{S_{2n}\}$ of even partial sums of $\{S_n\}$ is increasing and approaches the sum of the series $\sum c_n$ from below (Figure 11.24). In a similar way, we can show that the subsequence $\{S_{2n-1}\}$ of odd partial sums is decreasing and approaches the sum of the series from above. It follows that the sum $\sum c_n$ must be between any two terms of the subsequences $\{S_{2n}\}$ and $\{S_{2n-1}\}$. In particular, *the sum of the alternating series must be between any two successive partial sums.* Furthermore, *when the alternating series is truncated, the maximum possible error is the next term.*

FIGURE 11.24

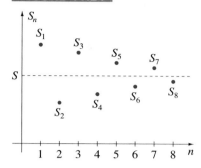

EXAMPLE 11.47

Use the first twenty terms of the series $\sum_{n=1}^{\infty} (-1)^{n+1}/n^3$ to estimate its sum. Obtain an error estimate.

SOLUTION   The sum of the first twenty terms of the series is 0.901 485. Since the series is alternating, and absolute values of terms are decreasing with limit zero, the maximum possible error in this estimate is the twenty-first term, $1/21^3 < 0.000\,108$. Thus,

$$0.901\,485 < \sum_{n=1}^{\infty} \frac{(-1)^{n+1}}{n^3} < 0.901\,593.$$

In practical situations, we often have to decide how many terms of a series to take in order to guarantee a certain degree of accuracy. Once again this is easy for alternating series whose terms satisfy the conditions of the alternating series test.

EXAMPLE 11.48

How many terms in the series $\sum_{n=2}^{\infty} (-1)^{n+1}/(n^3 \ln n)$ ensure a truncation error of less than 10^{-5}?

SOLUTION   Because absolute values of terms are decreasing and have limit zero, the maximum error in truncating this alternating series when $n = N$ is

$$\frac{(-1)^{N+2}}{(N+1)^3 \ln (N+1)}.$$

The absolute value of this error is less than 10^{-5} when

$$\frac{1}{(N+1)^3 \ln (N+1)} < 10^{-5} \qquad \text{or}$$

$$(N+1)^3 \ln (N+1) > 10^5.$$

A calculator quickly reveals that the smallest integer for which this is valid is $N = 30$. Thus, the truncated series has the required accuracy after the 29th term (the first term corresponds to $n = 2$, not $n = 1$).

In Section 11.7 we illustrated the use of Taylor's remainder formula to estimate the error when definite integrals are approximated using Taylor series. Alternating series sometimes provide an easier alternative. We redo the examples of Section 11.7 to demonstrate.

EXAMPLE 11.49

Approximate $\displaystyle\int_0^{1/2} \frac{\sin x}{x}\,dx$ to five decimal places using the Maclaurin series for $\sin x$.

SOLUTION   Using the Maclaurin series for $\sin x$,

$$
\begin{aligned}
\int_0^{1/2} \frac{\sin x}{x}\,dx &= \int_0^{1/2} \frac{1}{x}\left(x - \frac{x^3}{3!} + \frac{x^5}{5!} - \cdots\right)dx \\
&= \int_0^{1/2}\left(1 - \frac{x^2}{3!} + \frac{x^4}{5!} - \cdots\right)dx \\
&= \left\{x - \frac{x^3}{3\cdot 3!} + \frac{x^5}{5\cdot 5!} - \cdots\right\}_0^{1/2} \\
&= \frac{1}{2} - \frac{(1/2)^3}{3\cdot 3!} + \frac{(1/2)^5}{5\cdot 5!} - \cdots .
\end{aligned}
$$

This is a convergent alternating series. If we approximate it after the first three terms, then

$$
\int_0^{1/2} \frac{\sin x}{x}\,dx \approx \frac{1}{2} - \frac{(1/2)^3}{3\cdot 3!} + \frac{(1/2)^5}{5\cdot 5!} = 0.493\,107\,639,
$$

and the maximum error is the next term,

$$
\frac{(1/2)^7}{7\cdot 7!} < 0.000\,000\,222.
$$

Thus,

$$
0.493\,107\,639 - 0.000\,000\,222 < \int_0^{1/2} \frac{\sin x}{x}\,dx < 0.493\,107\,693,
$$

or

$$
0.493\,107\,417 < \int_0^{1/2} \frac{\sin x}{x}\,dx < 0.493\,107\,693.
$$

Consequently, using only three terms of the Maclaurin series for $(\sin x)/x$, and properties of alternating series in place of Taylor remainders, we once again obtain a five-decimal approximation 0.493 11.

EXAMPLE 11.50

Redo Example 11.28 using alternating series to estimate errors.

SOLUTION   If we replace the integrand by its Maclaurin series, we have

$$\frac{\sqrt{\pi}}{2}\text{erf}(1) = \int_0^1 \left[1 - t^2 + \frac{(-t^2)^2}{2!} + \frac{(-t^2)^3}{3!} + \cdots \right] dt$$

$$= \left\{ t - \frac{t^3}{3} + \frac{t^5}{5 \cdot 2!} - \frac{t^7}{7 \cdot 3!} + \cdots \right\}_0^1$$

$$= 1 - \frac{1}{3} + \frac{1}{5 \cdot 2!} - \frac{1}{7 \cdot 3!} + \frac{1}{9 \cdot 4!} - \cdots .$$

This is a convergent alternating series. Since

$$\frac{2}{\sqrt{\pi}} \left(1 - \frac{1}{3} + \frac{1}{5 \cdot 2!} - \frac{1}{7 \cdot 3!} + \frac{1}{9 \cdot 4!} - \frac{1}{11 \cdot 5!} \right) = 0.842\,593\,67$$

and the seventh nonzero term is

$$\frac{2}{\sqrt{\pi}} \frac{1}{13 \cdot 6!} < 0.000\,120\,56,$$

it follows that

$$0.842\,593\,67 < \text{erf}(1) < 0.842\,714\,23,$$

and to three decimal places

$$\text{erf}(1) = 0.843.$$

Truncating a Series Whose Convergence Was Established with the Integral Test

Suppose now that a series $\sum_{n=1}^{\infty} c_n$ has been shown to converge with the integral test; that is, the integral

$$\int_1^{\infty} f(x)\,dx$$

converges where $f(n) = c_n$. If the series is truncated after the Nth term, the error $c_{N+1} + c_{N+2} \cdots$ is shown as the sum of the areas of the rectangles in Figure 11.25. Clearly, the sum of these areas is less than the area under $y = f(x)$ to the right of $x = N$. In other words, the error in truncating the series with the Nth term must be less than

$$\int_N^{\infty} f(x)\,dx. \tag{11.42}$$

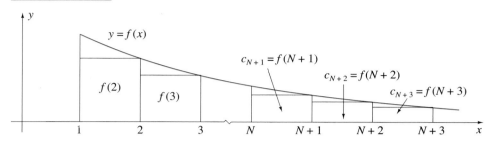

FIGURE 11.25

EXAMPLE 11.51

Obtain an error estimate if the series $\displaystyle\sum_{n=2}^{\infty} \frac{1}{n(\ln n)^4}$ is truncated when $n = 100$.

SOLUTION The error cannot be larger than

$$\int_{100}^{\infty} \frac{1}{x(\ln x)^4} \, dx = \left\{ \frac{-1}{3(\ln x)^3} \right\}_{100}^{\infty} = \frac{1}{3(\ln 100)^3} < 0.0035.$$

When convergence of a series is established by the comparison, limit comparison, limit ratio, or limit root tests, we often estimate the truncation error $\sum_{n=N+1}^{\infty} c_n$ by comparing it to something that is summable. We illustrate this in the following two examples.

EXAMPLE 11.52

In the first paragraph of this section we indicated that e is the sum of the series

$$\sum_{n=0}^{\infty} \frac{1}{n!}.$$

Use the first ten terms to find an approximation for e.

SOLUTION The sum of the first ten terms is

$$\sum_{n=0}^{9} \frac{1}{n!} = 2.718\,281\,526.$$

The truncation error in using this as an approximation for e is

$$\sum_{n=10}^{\infty} \frac{1}{n!} = \frac{1}{10!} + \frac{1}{11!} + \frac{1}{12!} + \frac{1}{13!} + \cdots$$

$$= \frac{1}{10!}\left(1 + \frac{1}{11} + \frac{1}{11 \cdot 12} + \frac{1}{11 \cdot 12 \cdot 13} + \cdots\right)$$

$$< \frac{1}{10!}\left(1 + \frac{1}{11} + \frac{1}{11^2} + \frac{1}{11^3} + \cdots\right) \qquad \text{(a geometric series)}$$

$$= \frac{1}{10!}\frac{1}{1 - 1/11} \qquad \text{(using equation 11.33b)}$$

$$= \frac{11}{10 \cdot 10!}$$

$$< 0.000\,000\,304.$$

We may write, therefore, that

$$2.718\,281\,526 < \sum_{n=0}^{\infty} \frac{1}{n!} < 2.718\,281\,830,$$

and to six decimal places, $e = 2.718\,282$.

EXAMPLE 11.53

How many terms in the convergent series $\sum_{n=1}^{\infty} n/[(n+1)3^n]$ ensure a truncation error of less than 10^{-5}?

SOLUTION If this series is truncated after the Nth term, the error is

$$\sum_{n=N+1}^{\infty} \frac{n}{(n+1)3^n} = \frac{N+1}{(N+2)3^{N+1}} + \frac{N+2}{(N+3)3^{N+2}} + \cdots$$

$$< \frac{1}{3^{N+1}} + \frac{1}{3^{N+2}} + \cdots \qquad \text{(a geometric series)}$$

$$= \frac{\dfrac{1}{3^{N+1}}}{1 - \dfrac{1}{3}} \qquad \text{(using equation 11.33b)}$$

$$= \frac{1}{2 \cdot 3^N}.$$

Consequently, the error is guaranteed to be less than 10^{-5} if N satisfies the inequality

$$\frac{1}{2 \cdot 3^N} < 10^{-5} \qquad \text{or}$$

$$2 \cdot 3^N > 10^5 \qquad \Longrightarrow \qquad N > \frac{\ln(10^5/2)}{\ln 3} = 9.85.$$

Thus, 10 or more terms yield the required accuracy.

EXERCISES 11.13

In Exercises 1–10 verify that the sum of the series is as indicated.

$*$ **1.** $\displaystyle\sum_{n=0}^{\infty} \frac{2^n}{n!} = e^2$
$\qquad$ $*$ **2.** $\displaystyle\sum_{n=0}^{\infty} \frac{(-1)^n}{(2n+1)!} = \sin 1$

$*$ **3.** $\displaystyle\sum_{n=0}^{\infty} \frac{(-1)^n 3^{2n}}{(2n)!} = \cos 3$
$\qquad$ $*$ **4.** $\displaystyle\sum_{n=1}^{\infty} \frac{(-1)^n}{n!} = \frac{1}{e} - 1$

$*$ **5.** $\displaystyle\sum_{n=1}^{\infty} \frac{(-1)^n}{2^{2n}} = -\frac{1}{5}$
$\qquad$ $**$ **6.** $\displaystyle\sum_{n=2}^{\infty} \frac{(-1)^{n+1} 2^{2n+3}}{(2n)!} = -8(1 + \cos 2)$

$**$ **7.** $\displaystyle\sum_{n=1}^{\infty} \frac{2^n}{n 3^n} = \ln 3$
$\qquad$ $**$ **8.** $\displaystyle\sum_{n=1}^{\infty} \frac{1}{n 2^n} = \ln 2$

$**$ **9.** $\displaystyle\sum_{n=1}^{\infty} \frac{(-1)^n}{3^{2n}(2n+1)!} = 3\sin\left(\frac{1}{3}\right) - 1$
$\qquad$ $**$ **10.** $\displaystyle\sum_{n=1}^{\infty} \frac{n}{2^n} = 2$

$**$ **11.** Find the Maclaurin series for $\text{Tan}^{-1} x$ and use it to evaluate

$$\sum_{n=1}^{\infty} \frac{(-1)^n}{2n+1}.$$

$**$ **12.** Find the Maclaurin series for $x/(1+x^2)^2$ and use it to evaluate

$$\sum_{n=1}^{\infty} \frac{n(-1)^n}{3^{2n}}.$$

In Exercises 13 and 14, approximate the sum of the series if it is truncated after the Nth term. Use 11.42 to find an error estimate.

* **13.** $\displaystyle\sum_{n=2}^{\infty} \frac{n^2}{(n^3+1)^4}$, $N = 10$

* **14.** $\displaystyle\sum_{n=1}^{\infty} \frac{n}{e^{3n}}$, $N = 5$

In Exercises 15 and 16, use 11.42 to estimate the error when the series is truncated after the Nth term.

* **15.** $\displaystyle\sum_{n=1}^{\infty} \frac{1}{n^2+1}$, $N = 100$

* **16.** $\displaystyle\sum_{n=1}^{\infty} \frac{1}{n^2} \sin\left(\frac{1}{n}\right)$, $N = 20$

In Exercises 17–24 use the number of terms indicated to find an approximation to the sum of the series. In each case, obtain an estimate of the truncation error.

* **17.** $\displaystyle\sum_{n=2}^{\infty} \frac{(-1)^{n+1}}{n^3 3^n}$ (3 terms)

* **18.** $\displaystyle\sum_{n=1}^{\infty} \frac{(-1)^n}{n^4}$ (20 terms)

** **19.** $\displaystyle\sum_{n=1}^{\infty} \frac{1}{n^n}$ (5 terms)

** **20.** $\displaystyle\sum_{n=1}^{\infty} \frac{1}{n 2^n}$ (10 terms)

** **21.** $\displaystyle\sum_{n=1}^{\infty} \frac{1}{2^n} \sin\left(\frac{\pi}{n}\right)$ (15 terms)

** **22.** $\displaystyle\sum_{n=2}^{\infty} \frac{2^n - 1}{3^n + n}$ (20 terms)

** **23.** $\displaystyle\sum_{n=2}^{\infty} \frac{2^n + 1}{3^n + n}$ (20 terms)

** **24.** $\displaystyle\sum_{n=1}^{\infty} \frac{(-1)^n}{n}$ (100 terms)

In Exercises 25–27 how many terms in the series guarantee an approximation to the sum with a truncation error of less than 10^{-4}?

* **25.** $\displaystyle\sum_{n=1}^{\infty} \frac{(-1)^n}{n^2}$

** **26.** $\displaystyle\sum_{n=1}^{\infty} \frac{1}{n^2 4^n}$

** **27.** $\displaystyle\sum_{n=1}^{\infty} \frac{2^n}{n!}$

** **28.** Suppose the series

$$\sum_{n=1}^{\infty} e^{-n} \sin^2 n$$

is truncated after the 10th term. Obtain an error estimate by (a) using 11.42 (b) using the fact that $e^{-n} \sin^2 n < e^{-n}$. Which gives the better estimate?

In Exercises 29–36, evaluate the integral correct to three decimal places. Compare the work in Exercises 29, 30, 32, and 34 with that in Exercises 11–14 of Section 11.7.

** **29.** $\displaystyle\int_0^1 \frac{\sin x}{x}\,dx$

** **30.** $\displaystyle\int_0^{1/2} \cos(x^2)\,dx$

** **31.** $\displaystyle\int_0^{2/3} \frac{1}{x^4+1}\,dx$

** **32.** $\displaystyle\int_{-1}^1 x^{11}\sin x\,dx$

** **33.** $\displaystyle\int_0^{1/2} \frac{1}{\sqrt{1+x^3}}\,dx$

** **34.** $\displaystyle\int_0^{0.3} e^{-x^2}\,dx$

** **35.** $\displaystyle\int_{-0.1}^0 \frac{1}{x-1}\ln(1-x)\,dx$

** **36.** $\displaystyle\int_0^{1/2} \frac{1}{x^6-3x^3-4}\,dx$

** **37.** In determining the radiated power from a half-wave antenna, it is necessary to evaluate

$$\int_0^{2\pi} \frac{1-\cos\theta}{\theta}\,d\theta.$$

Find a two-decimal approximation for this integral.

** **38.** This exercise shows that we must be very careful in predicting the accuracy of a result. Consider the series

$$S = 3.125\,100\,1 - 0.000\,090\,18\left(1 + \frac{1}{10} + \frac{1}{10^2} + \frac{1}{10^3} + \cdots\right).$$

(a) Show that the sum of this series is exactly $S = 3.124\,999\,9$. To two decimals, then, the value of S is 3.12.

(b) Verify that the first four partial sums of the series are

$$S_1 = 3.125\,100\,1,$$
$$S_2 = 3.125\,009\,92,$$
$$S_3 = 3.125\,000\,902,$$
$$S_4 = 3.125\,000\,000\,2.$$

(c) If $E_n = S_n - S$ are the differences between the sum of the series and its first four partial sums, show that

$$E_1 = 0.000\,100\,2,$$
$$E_2 = 0.000\,010\,02,$$
$$E_3 = 0.000\,001\,002,$$
$$E_4 = 0.000\,000\,100\,2.$$

What can you say about the accuracy of S_1, S_2, S_3, and S_4 as approximations to S?

(d) If S is approximated by any of S_1, S_2, S_3, or S_4 to two decimals, the result is 3.13, not 3.12 as in (a). Thus, in spite of the accuracy predicted in (c), S_1, S_2, S_3, and S_4 do not predict S correctly to two decimals. Do they predict S correctly to three or four decimals?

An infinite sequence of numbers is the assignment of numbers to positive integers. In most applications of sequences, the prime consideration is whether the sequence has a limit. If the sequence has its terms defined explicitly, then our ability to take limits of continuous functions ("limits at infinity" in Chapter 2 and L'Hôpital's rule in Chapter 4) can be very helpful. If the sequence is defined recursively, existence of the limit can sometimes be established by showing that the sequence is monotonic and bounded.

An expression of the form

$$\sum_{n=1}^{\infty} c_n = c_1 + c_2 + \cdots + c_n + \cdots$$

is called an infinite series. We define the sum of this series as the limit of its sequence of partial sums $\{S_n\}$, provided the sequence has a limit. Unfortunately, for most series we cannot find a simple formula for S_n, and therefore analysis of the limit of the sequence $\{S_n\}$ is impossible. To remedy this, we developed various convergence tests that avoided the sequence $\{S_n\}$: n^{th} term, comparison, limit comparison, limit ratio, limit root, integral, and alternating series tests. Note the sequences that are associated with a series $\sum c_n$:

$\{S_n\}$	sequence of partial sums for the definition of a sum;
$\{c_n\}$	sequence of terms for the nth term test;
$\{c_n/b_n\}$	sequence for limit comparison test;
$\{c_{n+1}/c_n\}$	sequence for the limit ratio test;
$\{\sqrt[n]{c_n}\}$	sequence for the limit root test;
$\{|c_n|\}$	sequence for the alternating series test.

Depending on the limits of these sequences—if they exist—we may be able to infer something about convergence of the series.

Infinite sequences and series of functions are important in applications—in particular, power series. (As engineers, you will see other types of series, Fourier series, for example.) We considered situations where a power series was given, and the sum was to be determined. We saw that every power series $\sum a_n(x - c)^n$ has a radius of convergence R and an associated interval of convergence. If $R = 0$, the interval of convergence consists of only one point $x = c$; if $R = \infty$, the power series converges for all x; and if $0 < R < \infty$, the interval of convergence must be one of four possibilities: $c - R < x < c + R$, $c - R \leq x < c + R$, $c - R < x \leq c + R$, or $c - R \leq x \leq c + R$. The radius of convergence is given by $\lim_{n \to \infty} |a_n/a_{n+1}|$ or $\lim_{n \to \infty} |a_n|^{-1/n}$ provided the limits exist or are equal to infinity. If at each point in the interval of convergence of the power series the value of a function $f(x)$ is the same as the sum of the series, we write $f(x) = \sum a_n(x - c)^n$ and call $f(x)$ the sum of the series.

We also considered situations where a function $f(x)$ and a point c are given, and ask whether $f(x)$ has a power series expansion about c. We saw that there can be at most one power series expansion of $f(x)$ about c with a positive radius of convergence, and this series must be its Taylor series. One way to verify that $f(x)$ does indeed have a Taylor series about c and that this series converges to $f(x)$ is to show that the sequence of Taylor's remainders $\{R_n(c, x)\}$ exists and has limit zero. Often, however, it is much easier to find Taylor series by adding, multiplying, differentiating, and integrating known series.

When a Taylor series is truncated, Taylor's remainder $R_n(c, x)$ represents the truncation error and, in spite of the fact that R_n is expressed in terms of some unknown point z_n, it is often possible to calculate a maximum value for the error. Sometimes $R_n(c, x)$ can be avoided altogether. For instance, if the Taylor series is an alternating series, then the maximum possible truncation error is the value of the next term.

Power series are often used in situations that require approximations. Taylor series provide polynomial approximations to complicated functions, and they offer an alternative to the numerical techniques of Section 9.9 in the evaluation of definite integrals. Power series are also useful in situations that do not require approximations. They are sometimes helpful in evaluating limits, and they are the only way to solve many differential equations.

KEY TERMS

In reviewing this chapter, you should be able to define or discuss the following key terms:

Sequence of numbers
Explicit sequence
Recursive sequence
Limit of a sequence
Convergent sequence
Divergent sequence
Increasing sequence
Decreasing sequence
Nonincreasing sequence
Nondecreasing sequence
Monotonic sequence
Oscillating sequence
Truncation error
Power series
Interval of convergence
Radius of convergence
Sum of a power series
Taylor remainder formula
Taylor series
Maclaurin series
Binomial expansion

n^{th} term test
Nonnegative series
Integral test
p-series
Comparison test
Limit comparison test
Limit ratio test
Limit root test
Absolutely convergent series
Conditionally convergent series
Alternating series
Upper bound
Lower bound
Successive approximations
Series of numbers
Geometric series
Harmonic series
Sequence of partial sums
Convergent series
Divergent series

REVIEW EXERCISES

In Exercises 1–4 discuss, with all necessary proofs, whether the sequence is monotonic and has an upper bound, a lower bound, and a limit.

** **1.** $\left\{ \dfrac{n^2 - 5n + 3}{n^2 + 5n + 4} \right\}$

** **2.** $c_1 = 1$, $\quad c_{n+1} = (1/2)\sqrt{c_n^2 + 1}$, $\quad n \geq 1$

** **3.** $\left\{ \dfrac{\text{Tan}^{-1}(1/n)}{n^2 + 1} \right\}$

** **4.** $c_1 = 7$, $\quad c_{n+1} = 15 + \sqrt{c_n - 2}$, $\quad n \geq 1$

* **5.** Use Newton's iterative procedure and the method of successive approximations to approximate the root of the equation

$$x = \left(\frac{x + 5}{x + 4} \right)^2$$

between $x = 1$ and $x = 2$.

* **6.** For what values of k does the sequence

$$c_1 = k, \qquad c_{n+1} = c_n^2, \qquad n \geq 1$$

converge?

* **7.** Find an explicit definition for the sequence

$$c_1 = 1, \qquad c_{n+1} = \sqrt{1 + c_n^2}, \qquad n \geq 1.$$

* **8.** Use the derivative of the function $f(x) = (\ln x)/x$ to prove that the sequence $\{\ln n / n\}$ is decreasing for $n \geq 3$.

In Exercises 9–28 determine whether the series converges or diverges. In the case of a convergent series that has both positive and negative terms, indicate whether it converges absolutely or conditionally.

* **9.** $\displaystyle\sum_{n=1}^{\infty} \frac{n^2 - 3n + 2}{n^3 + 4n}$

* **10.** $\displaystyle\sum_{n=1}^{\infty} \frac{n^2 + 5n + 3}{n^4 - 2n + 5}$

* **11.** $\displaystyle\sum_{n=1}^{\infty} \frac{5^{2n}}{n!}$

* **12.** $\displaystyle\sum_{n=1}^{\infty} \frac{n^2 + 3}{n3^n}$

* **13.** $\displaystyle\sum_{n=1}^{\infty} \frac{(\ln n)^2}{\sqrt{n}}$

* **14.** $\displaystyle\sum_{n=1}^{\infty} (-1)^n \left(\frac{n+1}{n^2}\right)$

* **15.** $\displaystyle\sum_{n=1}^{\infty} (-1)^n \left(\frac{n+1}{n^3}\right)$

* **16.** $\displaystyle\sum_{n=1}^{\infty} \operatorname{Cos}^{-1}\left(\frac{1}{n}\right)$

* **17.** $\displaystyle\sum_{n=1}^{\infty} \frac{1}{n}\operatorname{Cos}^{-1}\left(\frac{1}{n}\right)$

* **18.** $\displaystyle\sum_{n=1}^{\infty} \frac{1}{n^2}\operatorname{Cos}^{-1}\left(\frac{1}{n}\right)$

* **19.** $\displaystyle\sum_{n=1}^{\infty} \frac{2 \cdot 4 \cdot 6 \cdot \ \cdots \ \cdot (2n)}{n!}$

* **20.** $\displaystyle\sum_{n=1}^{\infty} \frac{3 \cdot 6 \cdot 9 \cdot \ \cdots \ \cdot (3n)}{(2n)!}$

* **21.** $\displaystyle\sum_{n=1}^{\infty} \sqrt{\frac{n^2 + 1}{n^2 + 5}}$

* **22.** $\displaystyle\sum_{n=1}^{\infty} (-1)^{n+1}\left(1 + \frac{1}{n}\right)^3$

* **23.** $\displaystyle\sum_{n=1}^{\infty} \frac{1}{n^2}\sin n$

* **24.** $\displaystyle\sum_{n=1}^{\infty} \frac{10^n}{5^{3n+2}}$

* **25.** $\displaystyle\sum_{n=1}^{\infty} (-1)^n \frac{\ln n}{n}$

* **26.** $\displaystyle\sum_{n=1}^{\infty} \frac{1}{e^{n\pi}}$

* **27.** $\displaystyle\sum_{n=1}^{\infty} \frac{2^n + 2^{-n}}{3^n}$

* **28.** $\displaystyle\sum_{n=1}^{\infty} \frac{1}{\sqrt{n}}\cos(n\pi)$

In Exercises 29–36 find the interval of convergence for the power series.

** **29.** $\displaystyle\sum_{n=0}^{\infty} \frac{n+1}{n^2+1} x^n$

** **30.** $\displaystyle\sum_{n=1}^{\infty} \frac{1}{n^2 2^n} x^n$

** **31.** $\displaystyle\sum_{n=0}^{\infty} (n+1)^3 x^n$

** **32.** $\displaystyle\sum_{n=1}^{\infty} \frac{1}{n^n} x^n$

** **33.** $\displaystyle\sum_{n=0}^{\infty} \frac{1}{4^n} (x-2)^n$

** **34.** $\displaystyle\sum_{n=2}^{\infty} \sqrt{\frac{n+1}{n-1}} (x+3)^n$

** **35.** $\displaystyle\sum_{n=1}^{\infty} n 3^n x^{2n}$

** **36.** $\displaystyle\sum_{n=1}^{\infty} \frac{2^n}{n} x^{3n}$

In Exercises 37–45 find the power series expansion of the function about the indicated point.

* **37.** $f(x) = \sqrt{1+x^2}$, about $x=0$

* **38.** $f(x) = e^{x+5}$, about $x=0$

** **39.** $f(x) = \cos(x+\pi/4)$, about $x=0$

** **40.** $f(x) = x \ln(2x+1)$, about $x=0$

** **41.** $f(x) = \sin x$, about $x=\pi/4$

** **42.** $f(x) = x/(x^2+4x+3)$, about $x=0$

* **43.** $f(x) = e^x$, about $x=3$

** **44.** $f(x) = (x+1)\ln(x+1)$, about $x=0$

* **45.** $f(x) = x^3 e^{x^2}$, about $x=0$

** **46.** How many terms in the Maclaurin series for $f(x) = e^{-x^2}$ guarantee a truncation error of less than 10^{-5} for all x in the interval $0 \le x \le 2$?

** **47.** Find a power series solution in powers of x for the differential equation

$$y'' - 4y = 0.$$

** **48.** Find the Maclaurin series for $f(x) = \sqrt{1+\sin x}$ valid for $-\pi/2 \le x \le \pi/2$ by first showing that $f(x)$ can be written in the form

$$f(x) = \sin(x/2) + \cos(x/2).$$

Why is the restriction $-\pi/2 \le x \le \pi/2$ necessary?

** **49.** On a calculator take the cosine of 1 (radian). Take the cosine of this result, and again, and again, and again, ... What happens? Interpret what is going on.

Vectors and Three-Dimensional Analytic Geometry

Chapters 1–11 dealt with single-variable calculus–differentiation and integration of functions *f(x)* of one variable. In Chapters 12–15 we study multivariable calculus. Discussions of three-dimensional analytic geometry and vectors in Sections 1–6 of this chapter prepare the way. In Sections 7–11, we differentiate and integrate vector functions, and apply the results to the geometry of curves in space and the motion of objects.

12.1 Rectangular Coordinates in Space

The coordinate of a point on the real line is its directed distance from the origin. Cartesian coordinates of a point in a plane are its directed distances from the coordinate axes. In space, Cartesian coordinates are directed distances from three fixed planes called the coordinate planes. In particular, we draw through a point O, called the **origin**, three mutually perpendicular lines called the x-, y-, and z-axes (Figure 12.1). Each of the axes is coordinatized with some unit distance (which need not be the same for all three axes). These three coordinate axes determine three planes called **coordinate planes**: The xy-coordinate plane is that plane containing the x- and y-axes, the yz-coordinate plane contains the y- and z-axes, and the xz-coordinate plane contains the x- and z-axes.

FIGURE 12.1

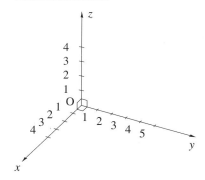

If P is any point in space, we draw lines from P perpendicular to the three coordinate planes (Figure 12.2). The directed distance from the yz-coordinate plane to P is parallel to the x-axis, and is called the x-coordinate of P. Similarly, y- and z-coordinates are defined as directed distances from the xz- and xy-coordinate planes to P. These three coordinates of P, written (x, y, z), are called the **Cartesian** or **rectangular coordinates** of P. Note that if we draw lines through P that are perpendicular to the axes, then the directed distances from O to points of intersection of these perpendiculars with the axes are also the Cartesian coordinates of P (Figure 12.3).

FIGURE 12.2

FIGURE 12.3

By either definition, each point in space has a unique ordered set of Cartesian coordinates (x, y, z); conversely, every ordered triple of real numbers (x, y, z) is the set of coordinates for one and only one point in space. For example, points with coordinates $(1, 1, 1)$, $(2, -3, 4)$, $(3, 4, -1)$, and $(-2, 5, 3)$ are shown in Figure 12.4.

FIGURE 12.4

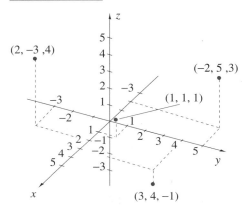

The coordinate systems in Figures 12.1–4 are called **right-handed coordinate systems**, because if we curl the fingers on our right hand from the positive x-direction toward the positive y-direction, then the thumb points in the positive z-direction (Figure 12.5). The coordinate system in Figure 12.6, on the other hand, is a **left-handed coordinate system**, since the thumb of the left hand points in the positive z-direction when the fingers of this hand are curled from the positive x-direction to the positive y-direction. We always use right-handed systems in this book.

FIGURE 12.5

FIGURE 12.6

Suppose we construct for any two points P_1 and P_2 with coordinates (x_1, y_1, z_1) and (x_2, y_2, z_2), respectively, a box with sides parallel to the coordinate planes and with line segment $P_1 P_2$ as diagonal (Figure 12.7). Because triangles $P_1 AB$ and $P_1 BP_2$ are right-angled, we can write

$$\|P_1 P_2\|^2 = \|P_1 B\|^2 + \|B P_2\|^2$$

$$= \|P_1 A\|^2 + \|AB\|^2 + \|B P_2\|^2$$

$$= (x_2 - x_1)^2 + (y_2 - y_1)^2 + (z_2 - z_1)^2.$$

In other words, the length of the line segment joining two points $P_1(x_1, y_1, z_1)$ and $P_2(x_2, y_2, z_2)$ is

$$\|P_1 P_2\| = \sqrt{(x_2 - x_1)^2 + (y_2 - y_1)^2 + (z_2 - z_1)^2}. \tag{12.1}$$

This is the analogue of formula 1.1 for the length of a line segment joining two points in the xy-plane.

FIGURE 12.7

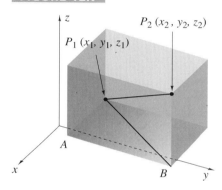

Just as the x- and y-axes divide the xy-plane into four regions called quadrants, the xy-, yz- and xz-coordinate planes divide xyz-space into eight regions called **octants**. The region where x-, y-, and z-coordinates are all positive is called the first octant. There is no commonly accepted way to number the remaining seven octants.

EXERCISES 12.1

∗ **1.** Draw a Cartesian coordinate system and show the points $(1, 2, 1)$, $(-1, 3, 2)$, $(1, -2, 4)$, $(3, 4, -5)$, $(-1, -2, -3)$, $(-2, -5, 4)$, $(8, -3, -6)$, and $(-4, 3, -5)$.

∗ **2.** Find the length of the line segment joining the points $(1, -2, 5)$ and $(-3, 2, 4)$.

∗ **3.** Prove that the triangle with vertices $(2, 0, 4\sqrt{2})$, $(3, -1, 5\sqrt{2})$, and $(4, -2, 4\sqrt{2})$ is right-angled and isosceles.

∗ **4.** A cube has sides of length 2 units. What are coordinates of its corners if one corner is at the origin, three of its faces lie in the coordinate planes, and one corner has all three coordinates positive?

∗ **5.** Show that the (undirected, perpendicular) distances from a point (x, y, z) to the x-, y-, and z-axes are, respectively, $\sqrt{y^2 + z^2}$, $\sqrt{x^2 + z^2}$, and $\sqrt{x^2 + y^2}$.

In Exercises 6–9, find the (undirected) distances from the point to (a) the origin; (b) the x-axis; (c) the y-axis; (d) the z-axis.

∗ **6.** $(2, 3, -4)$

∗ **7.** $(1, -5, -6)$

∗ **8.** $(4, 3, 0)$

∗ **9.** $(-2, 1, -3)$

∗ **10.** Prove that the three points $(1, 3, 5)$, $(-2, 0, 3)$, and $(7, 9, 9)$ are collinear.

∗ **11.** Find that point in the xy-plane that is equidistant from the points $(1, 3, 2)$ and $(2, 4, 5)$ and has a y-coordinate equal to three times its x-coordinate.

∗ **12.** Find an equation describing all points that are equidistant from the points $(-3, 0, 4)$ and $(2, 1, 5)$. What does this equation describe geometrically?

∗ **13.** (a) If $(\sqrt{3} - 3, 2 + 2\sqrt{3}, 2\sqrt{3} - 1)$ and $(2\sqrt{3}, 4, \sqrt{3} - 2)$ are two vertices of an equilateral triangle, and if the third vertex lies on the z-axis, find the third vertex.

(b) Can you find a third vertex on the x-axis?

∗∗ **14.** A birdhouse is built from a box $1/2$ m on each side with a roof as shown in the figure below. If the distance from each corner of the roof to the peak is $3/4$ m, find coordinates of the nine corners of the house. (The sides of the box are parallel to the coordinate planes.)

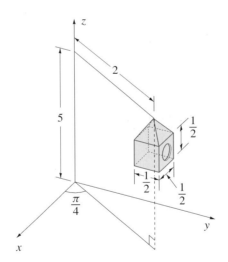

∗∗ **15.** If P and Q in the figure below have coordinates (x_1, y_1, z_1) and (x_2, y_2, z_2), show that coordinates of the point R midway between P and Q are $\left(\dfrac{x_1 + x_2}{2}, \dfrac{y_1 + y_2}{2}, \dfrac{z_1 + z_2}{2} \right)$.

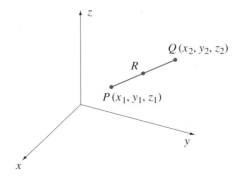

** **16.**   (a) Find the midpoint of the line segment joining the points $P(1, -1, -3)$ and $Q(3, 2, -4)$.

(b) If the line segment joining P and Q is extended its own length beyond Q to a point R, find the coordinates of R.

** **17.** The four-sided object in the figure below is a *tetrahedron*. If the four vertices of the tetrahedron are as shown, prove that the three lines joining the midpoints of opposite edges (one of which is PQ) meet at a point that bisects each of them.

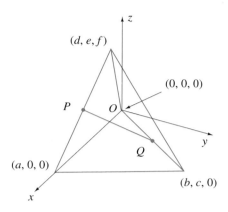

** **18.** Let A, B, C, and D be the vertices of a quadrilateral in space (not necessarily planar). Show that the line segments joining midpoints of opposite sides of the quadrilateral intersect in a point which bisects each.

*** **19.** Generalize the result of Exercise 15 to prove that if a point R divides the length PQ so that $\dfrac{\|PR\|}{\|RQ\|} = \dfrac{r_1}{r_2}$, where r_1 and r_2 are positive integers, then the coordinates of R are

$$x = \frac{r_1 x_2 + r_2 x_1}{r_1 + r_2}, \quad y = \frac{r_1 y_2 + r_2 y_1}{r_1 + r_2}, \quad z = \frac{r_1 z_2 + r_2 z_1}{r_1 + r_2}.$$

*** **20.** A man 2 m tall walks along the edge of a straight road 10 m wide (figure below). On the other edge of the road stands a streetlight 8 m high. A building runs parallel to the road and 1 m from it. If Cartesian coordinates are set up as shown (with x- and y-axes in the plane of the road), find coordinates of the tip of the man's shadow when he is at the position shown.

12.2 Curves and Surfaces

An equation involving the x- and y-coordinates of points in the xy-plane usually specifies a curve. For example, the equation $x^2 + y^2 = 4$ describes a circle of radius 2 centred at the origin (Figure 12.8). We now ask what is defined by an equation involving the Cartesian coordinates (x, y, z) of points in space. For example, the equation $z = 0$ describes all points in the xy-plane since all such points have a z-coordinate equal to zero. Similarly, $y = 2$ describes all points in the plane parallel to and two units to the right of the xz-plane. What does the equation $x^2 + y^2 = 4$ describe? In other words, regarded as a restriction on the x-, y-, and z-coordinates of points in space, rather than a restriction on the x- and y-coordinates of points in the xy-plane, what does it represent? Because the equation says nothing about z, there is no restriction whatsoever on z. In other words, the z-coordinate can take on all possible values, but x- and y-coordinates must be restricted by $x^2 + y^2 = 4$. If we consider those points in the xy-plane $(z = 0)$ that satisfy $x^2 + y^2 = 4$, we obtain the circle in Figure 12.8. In space, each of these points has coordinates $(x, y, 0)$, where x and y still satisfy $x^2 + y^2 = 4$ (Figure 12.9). If we now take any point Q that is either directly above or directly below a point $P(x, y, 0)$ on this circle, it has exactly the same x- and y-coordinates as P; only its z-coordinate differs. Thus the x- and y-coordinates of Q also satisfy $x^2 + y^2 = 4$. Since we can do this for any point P on the circle, it follows that $x^2 + y^2 = 4$ describes the right circular cylinder of radius 2 and infinite extent in Figure 12.9.

FIGURE 12.8

FIGURE 12.9

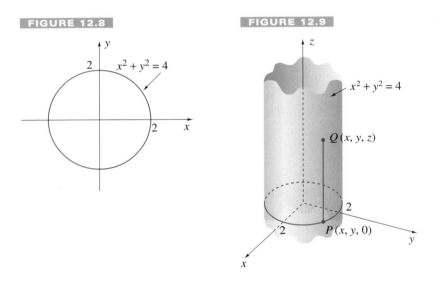

By reasoning similar to that used above, we can show that the equation $2x + y = 2$ describes the plane in Figure 12.10 parallel to the z-axis and standing on the straight line $2x + y = 2$, $z = 0$ in the xy-plane.

FIGURE 12.10

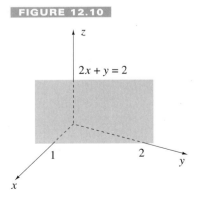

Finally, consider the equation $x^2 + y^2 + z^2 = 9$. Since $\sqrt{x^2 + y^2 + z^2}$ is distance from the origin to a point with coordinates (x, y, z), this equation describes all points that are three units away from the origin. In other words, $x^2 + y^2 + z^2 = 9$ describes points on a sphere of radius 3 centred at the origin.

It appears that one equation in the coordinates (x, y, z) of points in space specifies a surface. The shape of the surface is determined by the form of the equation. If one equation in the coordinates (x, y, z) specifies a surface, it is easy to see what two simultaneous equations specify. For instance, suppose we ask for all points in space whose coordinates satisfy both of the equations

$$x^2 + y^2 = 4, \quad z = 1.$$

By itself, $x^2 + y^2 = 4$ describes the cylinder in Figure 12.9. The equation $z = 1$ describes all points in a plane parallel to the xy-plane and one unit above it. To ask for all points that satisfy $x^2 + y^2 = 4$ and $z = 1$ simultaneously is to ask for all points that lie on both surfaces. Consequently, the equations $x^2 + y^2 = 4$, $z = 1$ describe the curve of intersection of the two surfaces—the circle in Figure 12.11.

FIGURE 12.11

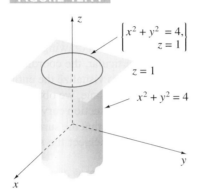

The equation $x = 0$ describes the yz-plane; the equation $y = 0$ describes the xz-plane. If we put the two equations together, $x = 0$ and $y = 0$, we obtain all points that lie on both the yz-plane and the xz-plane, i.e., the z-axis. In other words, equations for the z-axis are $x = 0$, $y = 0$.

Finally, $x^2 + y^2 + z^2 = 9$ is the equation of a sphere of radius 3 centred at the origin, and $y = 2$ is the equation of a plane parallel to the xz-plane and two units to the right. Together, the equations $x^2 + y^2 + z^2 = 9$, $y = 2$ describe the curve of intersection of the two surfaces—the circle in Figure 12.12. Note that by substituting $y = 2$ into the equation of the sphere, we can write alternatively that $x^2 + z^2 = 5$, $y = 2$. This pair of equations is equivalent to the original pair because all points that satisfy $x^2 + y^2 + z^2 = 9$, $y = 2$ also satisfy $x^2 + z^2 = 5$, $y = 2$, and vice versa. This new pair of equations provides an alternative way of visualizing the curve. Again $y = 2$ is the plane of Figure 12.12, but $x^2 + z^2 = 5$ describes a right-circular cylinder of radius $\sqrt{5}$ and infinite extent around the y-axis (Figure 12.13). Our discussion has shown that the cylinder and plane intersect in the same curve as the sphere and plane.

In summary, we have illustrated that one equation in the coordinates (x, y, z) of a point specifies a surface; two simultaneous equations specify a curve, the curve of intersection of the two surfaces (provided, of course, that the surfaces do intersect).

In Chapters 2–10 we learned to appreciate the value of plotting and drawing curves in the xy-plane. Sometimes a plot or drawing serves as a device by which we can interpret algebraic statements geometrically (such as the mean value theorem or the interpretation of a critical point of a function as a point where the tangent line to the graph of the function is horizontal, vertical, or does not exist). Sometimes they play an integral part in the solution of a problem (such as when the definite integral is used to find areas, volumes, etc.). Sometimes a plot or

SOLUTION

(a) If we square the equation, and at the same time complete squares on $-x^2 + 4x$ and $-y^2 + 2y$, we have

$$z^2 = -(x - 2)^2 - (y - 1)^2 + 1,$$

or

$$(x - 2)^2 + (y - 1)^2 + z^2 = 1.$$

Because $\sqrt{(x - 2)^2 + (y - 1)^2 + z^2}$ is the distance from a point (x, y, z) to $(2, 1, 0)$, this equation states that (x, y, z) must always be a unit distance from $(2, 1, 0)$; i.e., the equation $(x - 2)^2 + (y - 1)^2 + z^2 = 1$ defines a sphere of radius 1 centred at $(2, 1, 0)$ (Figure 12.17a). Because the original equation requires z to be nonnegative, the required surface is the upper half of this sphere—the hemisphere in Figure 12.17b.

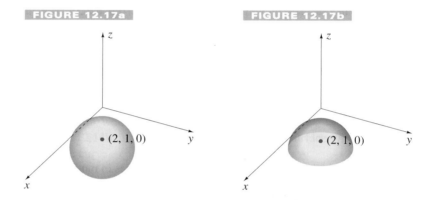

FIGURE 12.17a

FIGURE 12.17b

(b) If we intersect the surface $y = \sqrt{x^2 + z^2}$ with the xy-plane, we obtain the broken (i.e., bent) straight line $y = |x|$, $z = 0$ in Figure 12.18a. Intersections of the surface with planes $y = k$ (k a constant) give

$$y = \sqrt{x^2 + z^2} \qquad \text{or} \qquad x^2 + z^2 = k^2$$
$$y = k \qquad\qquad\qquad\qquad y = k.$$

These define circles of radii k in the planes $y = k$ (Figure 12.18b). Consequently, $y = \sqrt{x^2 + z^2}$ defines the right circular cone in Figure 12.18c. The surface $y = 1 + \sqrt{x^2 + z^2}$ can now be obtained by shifting the cone one unit in the y-direction (Figure 12.18d).

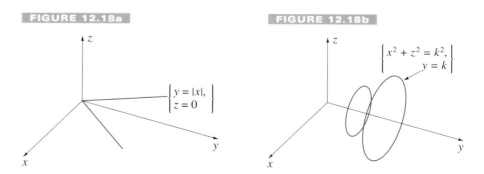

FIGURE 12.18a

FIGURE 12.18b

FIGURE 12.18c

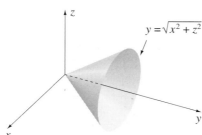

$$y = \sqrt{x^2 + z^2}$$

FIGURE 12.18d

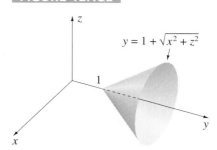

$$y = 1 + \sqrt{x^2 + z^2}$$

Cylinders

When one of the coordinates is missing from the equation of a surface, cross-sections of the surface with planes perpendicular to the axis of the missing variable are all identical. Such a surface is said to be a **cylinder**. For example, z is missing from the equation $x^2 + y^2 = 4$, and we saw in Figure 12.9 that this is the equation of a right circular cylinder around the z-axis. Every cross-section of this surface with a plane $z = k$ is a circle of radius 2 centred on the z-axis. The equation $z = x^2$ is free of y. Each cross-section of this surface with a plane $y = k$ is the parabola $z = x^2$ in the plane $y = k$. Consequently, $z = x^2$ is the equation for the parabolic cylinder in Figure 12.19.

FIGURE 12.19

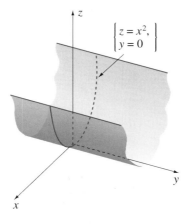

$$\begin{cases} z = x^2, \\ y = 0 \end{cases}$$

Quadric Surfaces

A **quadric surface** is a surface whose equation is quadratic in x, y, and z, the most general such equation being

$$Ax^2 + By^2 + Cz^2 + Dxy + Eyz + Fxz + Gx + Hy + Iz + J = 0. \quad (12.2)$$

For the most part, we encounter quadric surfaces whose equations are of the forms

$$Ax^2 + By^2 + Cz^2 = J \quad \text{or} \quad Ax^2 + By^2 = Iz,$$

or these equations with x, y, and z interchanged. Surfaces with these equations fall into nine major classes depending on whether the constants are positive, negative, or zero. They are illustrated in Figures 12.20–12.28.

FIGURE 12.20

ELLIPTIC CYLINDER

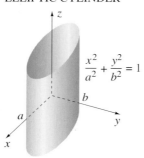

$$\frac{x^2}{a^2} + \frac{y^2}{b^2} = 1$$

FIGURE 12.21

HYPERBOLIC CYLINDER

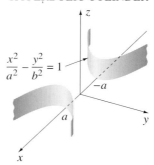

$$\frac{x^2}{a^2} - \frac{y^2}{b^2} = 1$$

FIGURE 12.22

PARABOLIC CYLINDER

$z = Ax^2$

FIGURE 12.23

ELLIPSOID

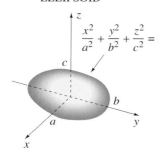

$$\frac{x^2}{a^2} + \frac{y^2}{b^2} + \frac{z^2}{c^2} = 1$$

FIGURE 12.24

ELLIPTIC PARABOLOID

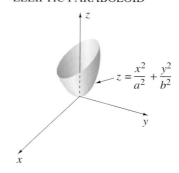

$$z = \frac{x^2}{a^2} + \frac{y^2}{b^2}$$

FIGURE 12.25

ELLIPTIC CONE

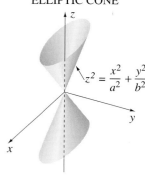

$$z^2 = \frac{x^2}{a^2} + \frac{y^2}{b^2}$$

FIGURE 12.26

HYPERBOLIC PARABOLOID

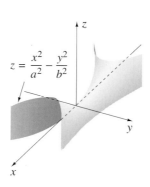

$$z = \frac{x^2}{a^2} - \frac{y^2}{b^2}$$

FIGURE 12.27

ELLIPTIC HYPERBOLOID
of one sheet

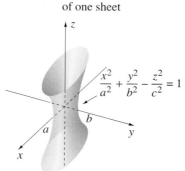

$$\frac{x^2}{a^2} + \frac{y^2}{b^2} - \frac{z^2}{c^2} = 1$$

FIGURE 12.28

ELLIPTIC HYPERBOLOID
of two sheets

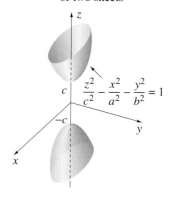

$$\frac{z^2}{c^2} - \frac{x^2}{a^2} - \frac{y^2}{b^2} = 1$$

The names of these surfaces are derived from the fact that their cross-sections are ellipses, hyperbolas, or parabolas. For example, cross-sections of the hyperbolic paraboloid with planes $z = k$ are hyperbolas $x^2/a^2 - y^2/b^2 = k$. cross-sections with planes $x = k$ are parabolas $z = k^2/a^2 - y^2/b^2$, as are cross-sections with planes $y = k$.

In applications of multiple integrals in Chapter 14, it is often necessary to project a space curve into one of the coordinate planes and find equations for the projection. To illustrate, consider the curve of intersection of the cylinder $x^2 + z^2 = 4$ and the plane $2y + z = 4$ (the first octant part of which is shown in Figure 12.29). Since the curve of intersection lies on the cylinder $x^2 + z^2 = 4$, its projection in the xz-plane is the circle $x^2 + z^2 = 4$, $y = 0$. To find its projection in the xy-plane, we eliminate z between the equations $2y + z = 4$ and $x^2 + z^2 = 4$. The result is $x^2 + (4 - 2y)^2 = 4$, or $x^2 + 4(y - 2)^2 = 4$. This shows that the curve of intersection lies on the elliptic cylinder $x^2 + 4(y - 2)^2 = 4$, and therefore it projects onto the ellipse $x^2 + 4(y - 2)^2 = 4$, $z = 0$ in the xy-plane. The projection of the curve in the yz-plane is that part of the line $2y + z = 4$, $x = 0$ between the points $(0, 1, 2)$ and $(0, 3, -2)$.

FIGURE 12.29

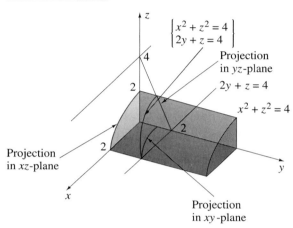

Graphing calculators and computers can plot surfaces provided the equation of the surface is solved for z in terms of x and y. (This may not always be convenient.) We have shown some computer-generated plots in Figures 12.30a–f together with values for x and y specified in generating the plots. With the complexity of the expressions for z, drawing these surfaces by hand would be a formidable task.

FIGURE 12.30a

$z = 3\sqrt{x^2 + y^2} - x^2 - y^2$

$-2 \le x \le 2, -2 \le y \le 2$

FIGURE 12.30b

$z = x\,(y - 1)^2 e^{-(x^2 + y^2)/4}$

$-5 \le x \le 5, -5 \le y \le 5$

FIGURE 12.30c

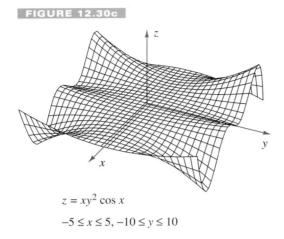

$z = xy^2 \cos x$

$-5 \le x \le 5, -10 \le y \le 10$

FIGURE 12.30d

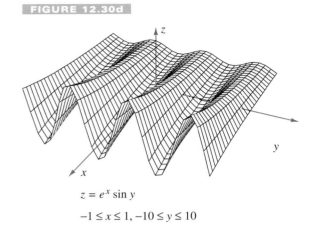

$z = e^x \sin y$

$-1 \le x \le 1, -10 \le y \le 10$

FIGURE 12.30e

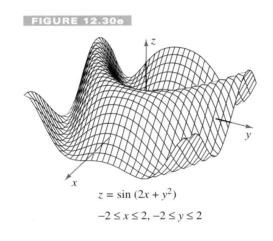

$z = \sin(2x + y^2)$

$-2 \le x \le 2, -2 \le y \le 2$

FIGURE 12.30f

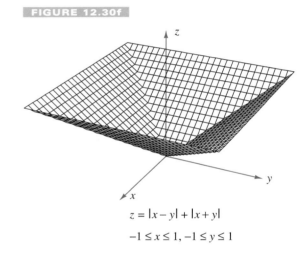

$z = |x - y| + |x + y|$

$-1 \le x \le 1, -1 \le y \le 1$

To appreciate the shape of a plotted surface, it is often necessary to vary the point in space from which the surface is viewed. Computers usually have this ability; graphing calculators may not.

Drawings and plots of curves were invaluable in Chapter 7 when we applied definite integrals to numerous geometric and physical problems. We encounter many of these applications in Chapter 14, but applied to volumes in space rather than areas in the plane. Our ability to visualize and draw surfaces in space proves more indispensable in these problems than plots from computers and graphing calculators. This is especially so when a picture contains a number of intersecting surfaces. For example, the surfaces $z = \sqrt{4 - x^2 - y^2}$ and $z = x^2$ bound volume in the first octant. The drawing in Figure 12.31a gives an excellent visualization of the volume; the plot in Figure 12.31b is not as satisfactory. It is often advantageous to first produce a plot of a surface and use it to render a drawing adding whatever information is important for the application at hand.

FIGURE 12.31a

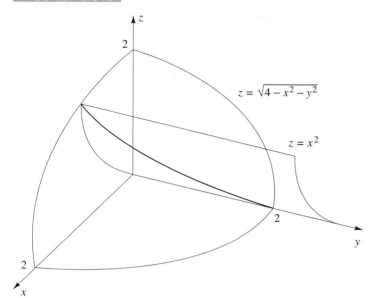

$z = \sqrt{4 - x^2 - y^2}$

$z = x^2$

FIGURE 12.31b

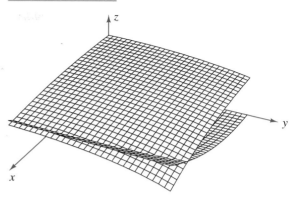

EXERCISES 12.2

In Exercises 1–35 draw the surface defined by the equation. Whenever possible, confirm your drawing with a plot generated by a computer or graphing calculator.

* **1.** $2y + 3z = 6$
* **2.** $2x - 3y = 0$
* **3.** $y = x^2 + 2$
* **4.** $z = x^3$
* **5.** $y^2 + z^2 = 1$
* **6.** $x^2 + y^2 + z^2 = 4$
* **7.** $x^2 + 4y^2 = 1$
* **8.** $y^2 - z^2 = 4$
* **9.** $z = 2(x^2 + y^2)$
* **10.** $x = \sqrt{y^2 + z^2}$
* **11.** $x = \sqrt{1 - y^2}$
* **12.** $z = 2 - x$
* **13.** $x^2 = y^2$
* **14.** $x = z^2 + 2$
* **15.** $z = y + 3$
* **16.** $4z = 3\sqrt{x^2 + y^2}$
* **17.** $x^2 - 2x + z^2 = 0$
* **18.** $yz = 1$
* **19.** $x^2 + y^2 + (z - 1)^2 = 3$
* **20.** $z + 5 = 4(x^2 + y^2)$
* **21.** $x^2 + z^2 = y^2$
* **22.** $x^2 + z^2 = y^2 + 1$
* **23.** $y^2 + z^2 = x$
* **24.** $x^2 + y^2 + 4z^2 = 1$
* **25.** $9z^2 = x^2 + y^2 + 1$
* **26.** $(y^2 + z^2)^2 = x + 1$
* **27.** $z^2 + 4y^2 = 1$
* **28.** $y - z^2 = 0$
* **29.** $x^2 - z^2 = 4$
* **30.** $x^2 + y^2/4 + z^2/9 = 1$

$*$ **31.** $z = x^2/4 + y^2/25$ $\qquad$ $*$ **32.** $x^2 = z^2 + 9y^2$

$*$ **33.** $z = y^2/16 - x^2/4$ $\qquad$ $*$ **34.** $x^2 + y^2/4 - z^2/25 = 1$

$*$ **35.** $z^2 - 9x^2 - 16y^2 = 1$

In Exercises 36–45 draw the curve defined by the equations.

$*$ **36.** $x^2 + y^2 = 2,\ z = 4$ $\qquad$ $*$ **37.** $x + 2y = 6,\ y - 2z = 3$

$*$ **38.** $z = x^2 + y^2,\ x^2 + y^2 = 5$ $\qquad$ $*$ **39.** $x^2 + y^2 = 2,\ x + z = 1$

$*$ **40.** $z = \sqrt{x^2 + y^2},\ y = x$ $\qquad$ $*$ **41.** $z + 2x^2 = 1,\ y = z$

$*$ **42.** $z = \sqrt{4 - x^2 - y^2},\ x^2 + y^2 - 2y = 0$

$*$ **43.** $z = y,\ y = x^2$

$*$ **44.** $x^2 + z^2 = 1,\ y^2 + z^2 = 1$ $\qquad$ $*$ **45.** $z = x^2,\ z = y^2$

In Exercises 46–55 find equations for projections of the curve in the xy-, yz-, and xz-coordinate planes. In each case draw the curve.

$**$ **46.** $x + y = 3,\ 2y + 3z = 4$ $\qquad$ $**$ **47.** $x + y + z = 4,\ 2x - y + z = 6$

$**$ **48.** $x^2 + y^2 = 4,\ z = 4$ $\qquad$ $**$ **49.** $x^2 + y^2 = 4,\ y = x$

$**$ **50.** $x^2 + y^2 = 4,\ x = z$ $\qquad$ $**$ **51.** $x^2 + y^2 = 4,\ x + y + z = 2$

$**$ **52.** $y^2 + z^2 = 3,\ x^2 + z^2 = 3$ $\qquad$ $**$ **53.** $z = x^2 + y^2,\ x + z = 1$

$**$ **54.** $z = \sqrt{x^2 + y^2},\ z = 6 - x^2 - y^2$ $\qquad$ $**$ **55.** $x^2 + y^2 + z^2 = 1,\ y = x$

In Exercises 56–61 find equations for the projection of the curve in the specified plane. Draw each curve.

$**$ **56.** $z = x^2 - y^2,\ z = 2x + 4y$ in the xy-plane

$**$ **57.** $x^2 + y^2 - 4z^2 = 1,\ x + y = 2$ in the xz-plane

$**$ **58.** $y = z + x^2,\ y + z = 1$ in the xy-plane

$**$ **59.** $x = \sqrt{1 + 2y^2 + 4z^2},\ x^2 + 9y^2 + 4z^2 = 36$ in the yz-plane

$**$ **60.** $z = x^2 + y^2,\ z = 4(x - 1)^2 + 4(y - 1)^2$ in the xy-plane

$**$ **61.** $x^2 + y^2 - 2y = 0,\ z^2 = x^2 + y^2$ in the xz-plane

In Exercises 62–71 draw whatever is defined by the equation or equations.

$**$ **62.** $(x - 2)^2 + y^2 + z^2 = 0$ $\qquad$ $**$ **63.** $x = 0,\ y = 5$

$**$ **64.** $\sqrt{x} + \sqrt{y} = 1,\ z = x$ $\qquad$ $**$ **65.** $x + y = 15,\ y - x = 4$

$**$ **66.** $z = 1 - (x^2 + y^2)^{1/3}$ $\qquad$ $**$ **67.** $z = |x|$

$**$ **68.** $z = x^2,\ y = z^2$ $\qquad$ $**$ **69.** $x = \ln (y^2 + z^2)$

$**$ **70.** $z = |x - y|$ $\qquad$ $**$ **71.** $x = 2,\ y = 4,\ z^2 - 1 = 0$

12.3 Vectors

Physical quantities that have associated with them only a magnitude can be represented by real numbers. Some examples are temperature, density, area, moment of inertia, speed, and pressure. They are called **scalars**. There are many quantities, however, that have associated with them both magnitude and direction, and these quantities cannot be described by a single real number. Velocity, acceleration, and force are perhaps the most notable concepts in this category. To represent such quantities mathematically, we introduce **vectors**.

> ### DEFINITION 12.1
> A **vector** is defined as a directed line segment.

To denote a vector we use a letter in boldface type, such as **v**. In Figures 12.32a, b, and c we show two vectors **u** and **v** along a line, three vectors **u**, **v**, and **w** in a plane, and three vectors **u**, **v**, and **w** in space, respectively. It is customary to place an arrowhead on a vector and call this end the **tip** of the vector. The other end is called the **tail** of the vector, and the direction of the vector is from tail to tip. A vector then has both *direction* and *length*.

FIGURE 12.32a **FIGURE 12.32b** **FIGURE 12.32c**

Definition 12.1 for a vector says nothing about its point of application, i.e., where its tail should be placed. This means that we may place the tail anywhere we wish. This suggests the following definition for equality of vectors.

FIGURE 12.33

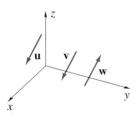

> ### DEFINITION 12.2
> Two vectors are equal if and only if they have the same length and direction. Their points of application are irrelevant.

For example, vectors **u** and **v** in Figure 12.33 have exactly the same length and direction, and are therefore one and the same. Although the vector **w** in the same figure is parallel to **u** and **v** and has the same length, it points in the opposite direction and is not, therefore, the same as **u** and **v**.

Components of Vectors

We realized in Chapter 1 that to solve geometric problems, it is often helpful to represent them algebraically. In fact, our entire development of single-variable calculus has hinged on our ability to represent a curve by an algebraic equation and also to draw the curve described by an equation. We now show that vectors can be represented algebraically. Suppose we denote by **PQ** the vector from point P to point Q in Figure 12.34. If P and Q have coordinates (x_1, y_1, z_1) and (x_2, y_2, z_2) in the coordinate system shown, then the length of **PQ** is

$$\sqrt{(x_2 - x_1)^2 + (y_2 - y_1)^2 + (z_2 - z_1)^2}. \tag{12.3}$$

FIGURE 12.34

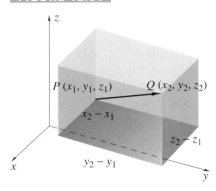

Note also that if we start at point P, proceed $x_2 - x_1$ units in the x-direction, then $y_2 - y_1$ units in the y-direction, and finally $z_2 - z_1$ units in the z-direction, then we arrive at Q. In other words, the three numbers $x_2 - x_1$, $y_2 - y_1$, and $z_2 - z_1$ characterize both the direction and the length of the vector joining P to Q. Because of this we make the following agreement.

DEFINITION 12.3

If the tail of a vector $\mathbf{v}$ is at $P(x_1, y_1, z_1)$ and its tip is at $Q(x_2, y_2, z_2)$, then $\mathbf{v}$ shall be represented by the triple of numbers $x_2 - x_1$, $y_2 - y_1$, $z_2 - z_1$. In such a case we enclose the numbers in parentheses and write

$$\mathbf{v} = (x_2 - x_1, y_2 - y_1, z_2 - z_1). \tag{12.4}$$

The equal sign in 12.4 means "is represented by." The number $x_2 - x_1$ is called the x-**component** of $\mathbf{v}$, $y_2 - y_1$ the y-**component**, and $z_2 - z_1$ the z-**component**. Vectors in the xy-plane have only an x- and a y-component:

$$\mathbf{v} = (x_2 - x_1, y_2 - y_1),$$

where (x_1, y_1) and (x_2, y_2) are the coordinates of the tail and tip of $\mathbf{v}$. Vectors along the x-axis have only an x-component $x_2 - x_1$, where x_1 and x_2 are the coordinates of the tail and tip of $\mathbf{v}$.

We now have an algebraic representation for vectors. Each vector has associated with it a set of components that can be found by subtracting the coordinates of its tail from the coordinates of its tip. Conversely, given a set of real numbers (a, b, c), there is one and only one vector with these numbers as components. We can visualize this vector by placing its tail at the origin and its tip at the point with coordinates (a, b, c) (Figure 12.35). Alternatively, we can place the tail of the vector at any point (x_1, y_1, z_1) and its tip at the point $(x_1 + a, y_1 + b, z_1 + c)$.

FIGURE 12.35

FIGURE 12.36

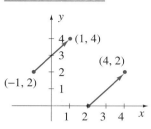

It is worth emphasizing once again that the same components of a vector are obtained for any point of application whatsoever. For example, the two vectors in Figure 12.36 are identical, and in both cases the components $(2, 2)$ are obtained by subtracting the coordinates of the tail from those of the tip. What we are saying is that Definition 12.2 for equality of vectors can be stated algebraically as follows.

THEOREM 12.1

Two vectors are equal if and only if they have the same components.

EXAMPLE 12.2

Find the components of a vector in the xy-plane that has length 5, its tail at the origin, and makes an angle of $\pi/6$ radians with the positive x-axis.

FIGURE 12.37

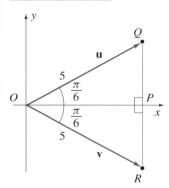

SOLUTION   Figure 12.37 illustrates that there are two such vectors, **u** and **v**. From the triangles shown, it is clear that

$$\|OP\| = 5\cos(\pi/6) = 5\sqrt{3}/2 \quad \text{and} \quad \|PQ\| = \|PR\| = 5\sin(\pi/6) = \frac{5}{2}.$$

Consequently, Q and R have coordinates $Q = (5\sqrt{3}/2, \frac{5}{2})$ and $R = (5\sqrt{3}/2, -\frac{5}{2})$, and

$$\mathbf{u} = \left(\frac{5\sqrt{3}}{2}, \frac{5}{2}\right), \qquad \mathbf{v} = \left(\frac{5\sqrt{3}}{2}, -\frac{5}{2}\right).$$

EXAMPLE 12.3

Find the components of the vector in the xy-plane that has its tail at the point $(4, 5)$, has length 3, and points directly toward the point $(2, -3)$.

FIGURE 12.38

SOLUTION   In Figure 12.38, $\|PQ\| = \sqrt{2^2 + 8^2} = 2\sqrt{17}$. Because of similar triangles, we can write that

$$\frac{\|ST\|}{\|PS\|} = \frac{\|QR\|}{\|PQ\|} \quad \text{or} \quad \|ST\| = \frac{3(2)}{2\sqrt{17}} = \frac{3}{\sqrt{17}}.$$

Similarly,

$$\|PT\| = \|PS\|\frac{\|PR\|}{\|PQ\|} = \frac{3(8)}{2\sqrt{17}} = \frac{12}{\sqrt{17}}.$$

Since $\|ST\|$ and $\|PT\|$ represent differences in the x- and y-coordinates of P and S (except for signs), the components of **PS** are $(-3/\sqrt{17}, -12/\sqrt{17})$.

Unit Vectors and Scalar Multiplication

If the x-, y-, and z-components of a vector $\mathbf{v}$ are (v_x, v_y, v_z), often called the **Cartesian components** of $\mathbf{v}$, then these components represent the differences in the coordinates of its tip and tail. But then, according to equation 12.3, the length of the vector, which we denote by $|\mathbf{v}|$, is

$$|\mathbf{v}| = \sqrt{v_x^2 + v_y^2 + v_z^2}. \tag{12.5}$$

In words, the length of a vector is the square root of the sum of the squares of its components.

DEFINITION 12.4

A vector $\mathbf{v}$ is said to be a **unit vector** if it has length equal to one unit; i.e., $\mathbf{v}$ is a unit vector if

$$v_x^2 + v_y^2 + v_z^2 = 1. \tag{12.6}$$

To indicate that a vector is a unit vector, we place a circumflex $\hat{}$ above it: $\hat{\mathbf{v}}$.

EXAMPLE 12.4

What is the length of the vector from $(1, -1, 0)$ to $(2, -3, -5)$?

SOLUTION   Since the components of the vector are $(1, -2, -5)$, its length is

$$\sqrt{(1)^2 + (-2)^2 + (-5)^2} = \sqrt{30}.$$

We now have vectors, which are directed line segments, and real numbers, which are scalars. We know that scalars can be added, subtracted, multiplied, and divided, but can we do the same with vectors, and can we combine vectors and scalars? In the remainder of this section we show how to add and subtract vectors and multiply vectors by scalars; in Section 12.4 we define two ways to multiply vectors. Each of these operations can be approached either algebraically or geometrically. The geometric approach uses the geometric properties of vectors, namely, length and direction; the algebraic approach uses components of vectors. Neither method is suitable for all situations. Sometimes an idea is more easily introduced with a geometric approach; sometimes an algebraic approach is more suitable. We choose whichever we feel expresses the idea more clearly. But, whenever we take a geometric approach, we are careful to follow it up with the algebraic equivalent; conversely, when an algebraic approach is taken, we always illustrate the geometric significance of the results.

To introduce multiplication of a vector by a scalar, consider the vectors $\mathbf{u}$ and $\mathbf{v}$ in Figure 12.39, both of which have their tails at the origin; $\mathbf{v}$ is in the same direction as $\mathbf{u}$ but is twice as long as $\mathbf{u}$. In such a situation we would like to say that $\mathbf{v}$ is equal to $2\mathbf{u}$ and write $\mathbf{v} = 2\mathbf{u}$. Vector $\mathbf{w}$ is in the opposite direction to $\mathbf{r}$ and is three times as long as $\mathbf{r}$, and we would like to denote this vector by $\mathbf{w} = -3\mathbf{r}$. Both of these situations are realized if we adopt the following definition for multiplication of a vector by a scalar.

FIGURE 12.39

DEFINITION 12.5

If $\lambda > 0$ is a scalar and $\mathbf{v}$ is a vector, then $\lambda\mathbf{v}$ is the vector that is in the same direction as $\mathbf{v}$ and λ times as long as $\mathbf{v}$; if $\lambda < 0$, then $\lambda\mathbf{v}$ is the vector that is in the opposite direction to $\mathbf{v}$ and $|\lambda|$ times as long as $\mathbf{v}$.

This is a geometric definition of scalar multiplication; it describes the length and direction of $\lambda\mathbf{v}$. We now show that the components of $\lambda\mathbf{v}$ are λ times the components of $\mathbf{v}$. In Figure 12.40 we show a box with faces parallel to the coordinate planes and $\lambda\mathbf{v}$ as diagonal, and have given vector $\mathbf{v}$ components (v_x, v_y, v_z). From the pairs of similar triangles OAB and OCD, and OBE and ODF, we can write that

$$\frac{\|OC\|}{v_x} = \frac{\|CD\|}{v_y} = \frac{\|OD\|}{\|OB\|} = \frac{\|DF\|}{v_z} = \frac{|\lambda\mathbf{v}|}{|\mathbf{v}|} = \lambda.$$

Hence

$$\|OC\| = \lambda v_x, \quad \|CD\| = \lambda v_y, \quad \|DF\| = \lambda v_z,$$

where $\|OC\|$, $\|CD\|$, and $\|DF\|$ are the components of $\lambda\mathbf{v}$. In other words, the components of $\lambda\mathbf{v}$ are λ times the components of $\mathbf{v}$:

$$\lambda\mathbf{v} = \lambda(v_x, v_y, v_z) = (\lambda v_x, \lambda v_y, \lambda v_z). \tag{12.7}$$

FIGURE 12.40

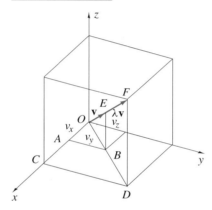

To multiply a vector by a scalar, then, we multiply each component by the scalar.

EXAMPLE 12.5

Find components for the unit vector in the same direction as $\mathbf{v} = (2, -2, 1)$.

SOLUTION   The length of $\mathbf{v}$ is $|\mathbf{v}| = \sqrt{(2)^2 + (-2)^2 + 1^2} = 3$. According to our definition of multiplication of a vector by a scalar, the vector $\frac{1}{3}\mathbf{v}$ must have length 1 ($\frac{1}{3}$ that of $\mathbf{v}$) and the same direction as $\mathbf{v}$. Consequently, a unit vector in the same direction as $\mathbf{v}$ is

$$\hat{\mathbf{v}} = \frac{1}{3}\mathbf{v} = \frac{1}{3}(2, -2, 1) = (\frac{2}{3}, -\frac{2}{3}, \frac{1}{3}).$$

This example illustrates that a unit vector in the same direction as a given vector $\mathbf{v}$ is

$$\hat{\mathbf{v}} = \frac{\mathbf{v}}{|\mathbf{v}|}. \tag{12.8}$$

EXAMPLE 12.6

Find components for the vector of length 4 in the direction opposite that of $\mathbf{v} = (1, 2, -3)$.

SOLUTION   Since $|\mathbf{v}| = \sqrt{1 + 4 + 9} = \sqrt{14}$, a unit vector in the same direction as $\mathbf{v}$ is

$$\hat{\mathbf{v}} = \frac{1}{\sqrt{14}}\mathbf{v}.$$

The vector of length 4 in the opposite direction to $\mathbf{v}$ must therefore be

$$(-4)\hat{\mathbf{v}} = \left(\frac{-4}{\sqrt{14}}\right)\mathbf{v} = \left(\frac{-4}{\sqrt{14}}\right)(1, 2, -3) = \left(-\frac{4}{\sqrt{14}}, -\frac{8}{\sqrt{14}}, \frac{12}{\sqrt{14}}\right).$$

With the operation of scalar multiplication, we can simplify the solution of Example 12.3. The vector that points from $(4, 5)$ to $(2, -3)$ is $\mathbf{v} = (-2, -8)$, and therefore the unit vector in this direction is

$$\hat{\mathbf{v}} = \frac{1}{\sqrt{4 + 64}}(-2, -8) = \frac{1}{2\sqrt{17}}(-2, -8) = \frac{1}{\sqrt{17}}(-1, -4).$$

The required vector of length 3 is

$$3\hat{\mathbf{v}} = \frac{3}{\sqrt{17}}(-1, -4) = \left(-\frac{3}{\sqrt{17}}, -\frac{12}{\sqrt{17}}\right).$$

Addition and Subtraction of Vectors

In Figure 12.41 we show two parallel vectors $\mathbf{u}$ and $\mathbf{v}$ and have placed the tail of $\mathbf{v}$ on the tip of $\mathbf{u}$. It would seem natural to denote the vector that has its tail at the tail of $\mathbf{u}$ and its tip at the tip of $\mathbf{v}$ by $\mathbf{u} + \mathbf{v}$. For instance, if $\mathbf{u}$ and $\mathbf{v}$ were equal, then we would simply be saying that $\mathbf{u} + \mathbf{u} = 2\mathbf{u}$. We use this idea to define addition of vectors even when the vectors are not parallel.

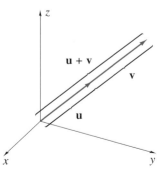

DEFINITION 12.6

The sum of two vectors $\mathbf{u}$ and $\mathbf{v}$, denoted by $\mathbf{u} + \mathbf{v}$, is the vector from the tail of $\mathbf{u}$ to the tip of $\mathbf{v}$ when the tail of $\mathbf{v}$ is placed on the tip of $\mathbf{u}$.

Because the three vectors $\mathbf{u}$, $\mathbf{v}$, and $\mathbf{u} + \mathbf{v}$ then form a triangle (Figure 12.42), we call this **triangular addition** of vectors.

FIGURE 12.42

FIGURE 12.43

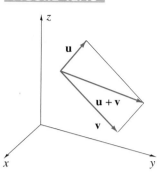

Note that were we to place tails of **u** and **v** both at the same point (Figure 12.43), and complete the parallelogram with **u** and **v** as sides, the diagonal of this parallelogram would also represent the vector **u** + **v**. This is an equivalent method for geometrically finding **u** + **v**, and it is called **parallelogram addition** of vectors.

FIGURE 12.44

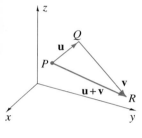

Algebraically, vectors are added component by component; i.e., if $\mathbf{u} = (u_x, u_y, u_z)$ and $\mathbf{v} = (v_x, v_y, v_z)$, then

$$\mathbf{u} + \mathbf{v} = (u_x + v_x, u_y + v_y, u_z + v_z). \tag{12.9}$$

To verify this we simply note that differences in the coordinates of P and Q in Figure 12.44 are (u_x, u_y, u_z), and differences in those of Q and R are (v_x, v_y, v_z). Consequently, differences in the coordinates of P and R must be $(u_x + v_x, u_y + v_y, u_z + v_z)$.

It is not difficult to show (see Exercise 26) that vector addition and scalar multiplication obey the following rules:

$$\mathbf{u} + \mathbf{v} = \mathbf{v} + \mathbf{u}; \tag{12.10a}$$

$$(\mathbf{u} + \mathbf{v}) + \mathbf{w} = \mathbf{u} + (\mathbf{v} + \mathbf{w}); \tag{12.10b}$$

$$\lambda(\mathbf{u} + \mathbf{v}) = \lambda\mathbf{u} + \lambda\mathbf{v}; \tag{12.10c}$$

$$(\lambda + \mu)\mathbf{v} = \lambda\mathbf{v} + \mu\mathbf{v}. \tag{12.10d}$$

If we denote the vector $(-1)\mathbf{v}$ by $-\mathbf{v}$, then the components of $-\mathbf{v}$ are the negatives of those of **v**:

$$-\mathbf{v} = (-1)\mathbf{v} = (-1)(v_x, v_y, v_z) = (-v_x, -v_y, -v_z). \tag{12.11}$$

FIGURE 12.45

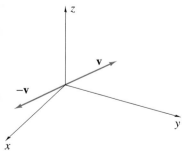

This vector has the same length as **v**, but is opposite in direction to **v** (Figure 12.45). When **v** is added to $-\mathbf{v}$, the resultant vector has components that are all zero:

$$\mathbf{v} + (-\mathbf{v}) = (v_x, v_y, v_z) + (-v_x, -v_y, -v_z) = (0, 0, 0).$$

This vector, called the **zero vector**, is denoted by **0**, and has the property that

$$\mathbf{v} + \mathbf{0} = \mathbf{0} + \mathbf{v} = \mathbf{v} \tag{12.12}$$

for any vector **v** whatsoever.

To subtract a vector $\mathbf{v}$ from $\mathbf{u}$, we add $-\mathbf{v}$ to $\mathbf{u}$.

DEFINITION 12.7

The difference $\mathbf{u} - \mathbf{v}$ between two vectors $\mathbf{u}$ and $\mathbf{v}$ is the vector

$$\mathbf{u} - \mathbf{v} = \mathbf{u} + (-\mathbf{v}). \tag{12.13}$$

In Figure 12.46 $\mathbf{u} - \mathbf{v}$ is determined by a triangle, and in Figure 12.47 by a parallelogram. Alternatively, if we denote by $\mathbf{r}$ the vector joining the tip of $\mathbf{v}$ to the tip of $\mathbf{u}$ in Figure 12.48, then, by triangle addition, we have $\mathbf{v} + \mathbf{r} = \mathbf{u}$. Addition of $-\mathbf{v}$ to each side of this equation gives

$$-\mathbf{v} + \mathbf{v} + \mathbf{r} = -\mathbf{v} + \mathbf{u} \quad \text{or} \quad \mathbf{0} + \mathbf{r} = -\mathbf{v} + \mathbf{u}.$$

Thus $\mathbf{r} = \mathbf{u} - \mathbf{v}$, and $\mathbf{u} - \mathbf{v}$ is the vector joining the tip of $\mathbf{v}$ to the tip of $\mathbf{u}$. Definition 12.7 implies that vectors are subtracted component by component:

$$\mathbf{u} - \mathbf{v} = (u_x, u_y, u_z) - (v_x, v_y, v_z) = (u_x - v_x, u_y - v_y, u_z - v_z). \tag{12.14}$$

FIGURE 12.46	FIGURE 12.47	FIGURE 12.48
		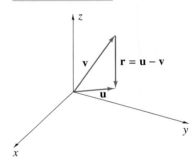

EXAMPLE 12.7

If $\mathbf{u} = (1, 1, 1)$, $\mathbf{v} = (-2, 3, 0)$, and $\mathbf{w} = (-10, 10, -2)$, find:

$$(a)\ \ 3\mathbf{u} + 2\mathbf{v} - \mathbf{w}; \quad (b)\ \ 2\mathbf{u} - 4\mathbf{v} + \mathbf{w}; \quad (c)\ \ |\mathbf{u}|\mathbf{v} + \frac{4}{|\mathbf{v}|}\mathbf{w}.$$

SOLUTION

(a)
$$3\mathbf{u} + 2\mathbf{v} - \mathbf{w} = 3(1, 1, 1) + 2(-2, 3, 0) - (-10, 10, -2)$$
$$= (3, 3, 3) + (-4, 6, 0) + (10, -10, 2)$$
$$= (9, -1, 5)$$

(b)
$$2\mathbf{u} - 4\mathbf{v} + \mathbf{w} = 2(1, 1, 1) - 4(-2, 3, 0) + (-10, 10, -2)$$
$$= (0, 0, 0) = \mathbf{0}$$

(c) Since $|\mathbf{u}| = \sqrt{1^2 + 1^2 + 1^2} = \sqrt{3}$ and $|\mathbf{v}| = \sqrt{(-2)^2 + 3^2} = \sqrt{13}$,

$$|\mathbf{u}|\mathbf{v} + \frac{4}{|\mathbf{v}|}\mathbf{w} = \sqrt{3}(-2, 3, 0) + \frac{4}{\sqrt{13}}(-10, 10, -2)$$

$$= \left(-2\sqrt{3} - \frac{40}{\sqrt{13}}, 3\sqrt{3} + \frac{40}{\sqrt{13}}, -\frac{8}{\sqrt{13}}\right).$$

Forces

We have already mentioned that quantities such as temperature, area, and density have associated with them only a magnitude and are therefore represented by scalars. There are many quantities, however, that have associated with them both magnitude and direction, and these are described by vectors. The most notable of this group are forces. When we speak of a force, we mean a push or pull of some size in some specific direction. For example, when the boy in Figure 12.49a pulls his wagon, he exerts a force in the direction indicated by the handle. Suppose that he pulls with a force of ten newtons (10 N) and that the angle between the handle and the horizontal is $\pi/4$ radians. To represent this force as a vector $\mathbf{F}_1$, we choose the coordinate system in Figure 12.49b, and make the agreement that the *length of $\mathbf{F}_1$ be equal to the magnitude of the force*. Since $\mathbf{F}_1$ represents a force of 10 N, it follows that the length of $\mathbf{F}_1$ is 10 units. Furthermore, because $\mathbf{F}_1$ makes an angle of $\pi/4$ radians with the positive x- and y-axes, the difference in the x-coordinates (and the y-coordinates) of its tip and tail must be $10\cos(\pi/4) = 5\sqrt{2}$. The components of $\mathbf{F}_1$ are therefore $\mathbf{F}_1 = (5\sqrt{2}, 5\sqrt{2})$. If the boy's young sister drags her feet on the ground, then she effectively exerts a force $\mathbf{F}_2$ in the negative x-direction. If the magnitude of this force is 3 N, then its vector representation is $\mathbf{F}_2 = (-3, 0)$. Finally, if the combined weight of the wagon and the girl is 200 N, then the force $\mathbf{F}_3$ of gravity on the wagon and its load is $\mathbf{F}_3 = (0, -200)$. In mechanics we replace the individual forces $\mathbf{F}_1$, $\mathbf{F}_2$, and $\mathbf{F}_3$ by a single force that has the same effect on the wagon as all three forces combined. This force, called the **resultant** of $\mathbf{F}_1$, $\mathbf{F}_2$, and $\mathbf{F}_3$, is represented by the vector $\mathbf{F}$, which is the sum of the vectors $\mathbf{F}_1$, $\mathbf{F}_2$, and $\mathbf{F}_3$:

$$\mathbf{F} = \mathbf{F}_1 + \mathbf{F}_2 + \mathbf{F}_3$$

$$= (5\sqrt{2}, 5\sqrt{2}) + (-3, 0) + (0, -200)$$

$$= (5\sqrt{2} - 3, 5\sqrt{2} - 200).$$

The magnitude of this force corresponds to the length of $\mathbf{F}$,

$$|\mathbf{F}| = \sqrt{(5\sqrt{2} - 3)^2 + (5\sqrt{2} - 200)^2} = 193.0,$$

and must therefore be 193.0 N. Its direction is shown in Figure 12.49c, where

$$\theta = \text{Tan}^{-1}\left(\frac{200 - 5\sqrt{2}}{5\sqrt{2} - 3}\right) = 1.55 \text{ radians.}$$

FIGURE 12.49a

FIGURE 12.49b

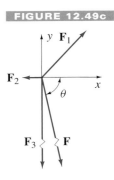

FIGURE 12.49c

EXAMPLE 12.8

If two water skiers S_1 and S_2 exert forces of 500 N and 600 N, respectively, on the boat shown in Figure 12.50a, what is the resultant force on the boat?

FIGURE 12.50a

FIGURE 12.50b

SOLUTION   If we introduce the coordinate system in Figure 12.50b, then lengths of the force vectors $\mathbf{F}_1$ and $\mathbf{F}_2$ exerted by S_1 and S_2 are 500 and 600 units, respectively. The components of $\mathbf{F}_1$ and $\mathbf{F}_2$ are

$$\mathbf{F}_1 = \left(-\frac{500}{\sqrt{2}}, \frac{500}{\sqrt{2}}\right) \qquad \text{and} \qquad \mathbf{F}_2 = \left(-\frac{600\sqrt{3}}{2}, -\frac{600}{2}\right)$$
$$= (-250\sqrt{2}, 250\sqrt{2}) \qquad\qquad\qquad = (-300\sqrt{3}, -300).$$

The resultant force $\mathbf{F}$ on the boat is then the sum of $\mathbf{F}_1$ and $\mathbf{F}_2$:

$$\mathbf{F} = \mathbf{F}_1 + \mathbf{F}_2 = (-250\sqrt{2} - 300\sqrt{3}, 250\sqrt{2} - 300)\text{N}.$$

By the x-, y-, and z-components (v_x, v_y, v_z) of a vector $\mathbf{v}$, we mean that, if we start at a point P (Figure 12.51) and proceed v_x units in the x-direction, v_y units in the y-direction, and v_z units in the z-direction to a point Q, then $\mathbf{v}$ is the directed line segment joining P and Q. To phrase this another way, we introduce three special vectors parallel to the coordinate axes. We define $\hat{\mathbf{i}}$ as a unit vector in the positive x-direction, $\hat{\mathbf{j}}$ as a unit vector in the positive y-direction,

FIGURE 12.51

FIGURE 12.52

and $\hat{\mathbf{k}}$ as a unit vector in the positive z-direction. We have shown these vectors with their tails at the origin in Figure 12.52, and it is clear that their components are

$$\hat{\mathbf{i}} = (1, 0, 0), \quad \hat{\mathbf{j}} = (0, 1, 0), \quad \hat{\mathbf{k}} = (0, 0, 1). \tag{12.15}$$

But note, then, that we can write the vector $\mathbf{v} = (v_x, v_y, v_z)$ in the form

$$\begin{aligned}
\mathbf{v} &= (v_x, 0, 0) + (0, v_y, 0) + (0, 0, v_z) \\
&= v_x(1, 0, 0) + v_y(0, 1, 0) + v_z(0, 0, 1) \\
&= v_x\hat{\mathbf{i}} + v_y\hat{\mathbf{j}} + v_z\hat{\mathbf{k}}.
\end{aligned}$$

In other words, every vector in space can be written as a linear combination of the three vectors $\hat{\mathbf{i}}$, $\hat{\mathbf{j}}$, and $\hat{\mathbf{k}}$ (i.e., as a constant times $\hat{\mathbf{i}}$ plus a constant times $\hat{\mathbf{j}}$ plus a constant times $\hat{\mathbf{k}}$). Furthermore, the constants multiplying $\hat{\mathbf{i}}$, $\hat{\mathbf{j}}$, and $\hat{\mathbf{k}}$ are the Cartesian components of the vector. This result is equally clear geometrically. In Figure 12.51, we have shown the vector $\mathbf{v}$ from P to Q. If we define points A and B as shown in Figure 12.53, then

$$\mathbf{v} = \mathbf{PQ} = \mathbf{PB} + \mathbf{BQ} = \mathbf{PA} + \mathbf{AB} + \mathbf{BQ}.$$

But, because $\mathbf{PA}$ is a vector in the positive x-direction and has length v_x, it follows that $\mathbf{PA} = v_x\hat{\mathbf{i}}$. Similarly, $\mathbf{AB} = v_y\hat{\mathbf{j}}$ and $\mathbf{BQ} = v_z\hat{\mathbf{k}}$, and therefore

$$\mathbf{v} = v_x\hat{\mathbf{i}} + v_y\hat{\mathbf{j}} + v_z\hat{\mathbf{k}}. \tag{12.16}$$

To say then that v_x, v_y, and v_z are the x-, y-, and z-components of a vector $\mathbf{v}$ is to say that $\mathbf{v}$ can be written in form 12.16.

Some authors refer to v_x, v_y, and v_z as the **scalar components** of the vector $\mathbf{v}$, and the vectors $v_x\hat{\mathbf{i}}$, $v_y\hat{\mathbf{j}}$, and $v_z\hat{\mathbf{k}}$ as the **vector components** of $\mathbf{v}$. By "component," we always mean "scalar component."

Vectors in the xy-plane have only an x- and a y-component, and can therefore be written in terms of $\hat{\mathbf{i}}$ and $\hat{\mathbf{j}}$. If $\mathbf{v} = (v_x, v_y)$, then we write equivalently that $\mathbf{v} = v_x\hat{\mathbf{i}} + v_y\hat{\mathbf{j}}$ (Figure 12.54). Vectors along the x-axis have only an x-component and can therefore be written in the form $\mathbf{v} = v_x\hat{\mathbf{i}}$.

FIGURE 12.53

FIGURE 12.54

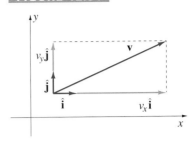

We use this new notation in the following example.

EXAMPLE 12.9

The force $\mathbf{F}$ exerted on a point charge q_1 coulombs by a charge q_2 coulombs is defined by Coulomb's law as

$$\mathbf{F} = \frac{q_1 q_2}{4\pi \epsilon_0 r^2} \hat{\mathbf{r}} \text{ N},$$

FIGURE 12.55

where ϵ_0 is a positive constant, r is the distance in metres between the charges, and $\hat{\mathbf{r}}$ is a unit vector in the direction from q_2 to q_1 (Figure 12.55). When q_1 and q_2 are both positive charges or both negative charges, then $\mathbf{F}$ is repulsive, and when one is positive and the other is negative $\mathbf{F}$ is attractive. In particular, suppose that charges of 2 C and -2 C are placed at $(0, 0, 0)$ and $(3, 0, 0)$, respectively, and a third charge of 1 C is placed at $(1, 1, 1)$. According to Coulomb's law, the 2-C charge will exert a repulsive force on the 1-C charge, and the -2-C charge will exert an attractive force on the 1-C charge. Find the resultant of these two forces on the 1-C charge.

SOLUTION If $\mathbf{F}_1$ is the force exerted on the 1-C charge by the -2-C charge (Figure 12.56), then

$$\mathbf{F}_1 = \frac{(1)(-2)}{4\pi \epsilon_0 r^2} \hat{\mathbf{r}},$$

where the distance between the charges is $r = \sqrt{(-2)^2 + 1^2 + 1^2} = \sqrt{6}$. The vector from $(3, 0, 0)$ to $(1, 1, 1)$ is $(-2, 1, 1)$, and therefore

$$\hat{\mathbf{r}} = \frac{1}{\sqrt{6}}(-2\hat{\mathbf{i}} + \hat{\mathbf{j}} + \hat{\mathbf{k}}).$$

Consequently,

$$\mathbf{F}_1 = \frac{-2}{4\pi \epsilon_0 (6)} \cdot \frac{1}{\sqrt{6}}(-2\hat{\mathbf{i}} + \hat{\mathbf{j}} + \hat{\mathbf{k}}) = \frac{1}{12\sqrt{6}\pi \epsilon_0}(2\hat{\mathbf{i}} - \hat{\mathbf{j}} - \hat{\mathbf{k}}).$$

Similarly, the force $\mathbf{F}_2$ exerted on the 1-C charge by the charge at the origin is

$$\mathbf{F}_2 = \frac{(1)(2)}{4\pi \epsilon_0 (3)} \cdot \frac{1}{\sqrt{3}}(\hat{\mathbf{i}} + \hat{\mathbf{j}} + \hat{\mathbf{k}}) = \frac{1}{6\sqrt{3}\pi \epsilon_0}(\hat{\mathbf{i}} + \hat{\mathbf{j}} + \hat{\mathbf{k}}).$$

The resultant of these forces is

$$\mathbf{F} = \mathbf{F}_1 + \mathbf{F}_2 = \frac{1}{12\sqrt{6}\pi\,\epsilon_0}\{(2 + 2\sqrt{2})\hat{\mathbf{i}} + (2\sqrt{2} - 1)\hat{\mathbf{j}} + (2\sqrt{2} - 1)\hat{\mathbf{k}}\} \text{ N}.$$

FIGURE 12.56

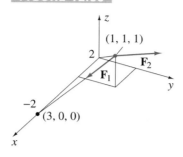

EXERCISES 12.3

If $\mathbf{u} = (1, 3, 6)$, $\mathbf{v} = (-2, 0, 4)$, and $\mathbf{w} = (4, 3, -2)$, find components for the vector in Exercises 1–10.

* **1.** $3\mathbf{u} - 2\mathbf{v}$

* **2.** $2\mathbf{w} + 3\mathbf{v}$

* **3.** $\mathbf{w} - 3\mathbf{u} - 3\mathbf{v}$

* **4.** $\hat{\mathbf{v}}$

* **5.** $2\hat{\mathbf{w}} - 3\mathbf{v}$

* **6.** $|\mathbf{v}|\mathbf{v} - 2|\hat{\mathbf{v}}|\mathbf{w}$

* **7.** $(15 - 2|\mathbf{w}|)(\mathbf{u} + \mathbf{v})$

* **8.** $|3\mathbf{u}|\mathbf{v} - |-2\mathbf{v}|\mathbf{u}$

* **9.** $|2\mathbf{u} + 3\mathbf{v} - \mathbf{w}|\hat{\mathbf{w}}$

* **10.** $\dfrac{\mathbf{v} - \mathbf{w}}{|\mathbf{v} + \mathbf{w}|}$

If $\mathbf{u} = 2\hat{\mathbf{i}} + \hat{\mathbf{j}}$ and $\mathbf{v} = -\hat{\mathbf{i}} + 3\hat{\mathbf{j}}$, find components of the vector in Exercises 11–14 and illustrate the vector geometrically.

* **11.** $\mathbf{u} + \mathbf{v}$

* **12.** $\mathbf{u} - \mathbf{v}$

* **13.** $2\hat{\mathbf{u}}$

* **14.** $\hat{\mathbf{v}} + \hat{\mathbf{u}}$

In Exercises 15–24 find the Cartesian components for the spatial vector described. In each case, draw the vector.

* **15.** From $(1, 3, 2)$ to $(-1, 4, 5)$

* **16.** With length 5 in the positive x-direction

* **17.** With length 2 in the negative z-direction

* **18.** With tail at $(1, 1, 1)$, length 3, and pointing toward the point $(1, 3, 5)$

* **19.** With positive y-component, length 1, and parallel to the line through $(1, 3, 6)$ and $(-2, 1, 4)$

* **20.** In the same direction as the vector from $(1, 0, -1)$ to $(3, 2, -4)$ but only half as long

* **21.** With positive and equal x- and y-components, length 10, and z-component equal to 4

* **22.** Has its tail at the origin, makes angles of $\pi/3$ and $\pi/4$ radians with the positive x- and y-axes respectively, and has length $5/2$

* **23.** From $(1, 3, -2)$ to the midpoint of the line segment joining $(2, 4, -3)$ and $(1, 5, 6)$

* **24.** Has its tail at the origin, makes equal angles with the positive coordinate axes, has all positive components, and has length 2

* **25.** If P, Q, and R are the points with coordinates $(3, 2, -1)$, $(0, 1, 4)$, and $(6, 5, -2)$ respectively, find coordinates of a point S in order that $\mathbf{PQ} = \mathbf{RS}$.

* **26.** Prove that vector addition and scalar multiplication have the properties in equations 12.10.

* **27.** Draw all spatial vectors of length 1 that have equal x- and y-components and tails at the origin.

* **28.** Draw all spatial vectors of length 2 that have their tails at the origin and make an angle of $\pi/4$ with the positive z-axis.

** **29.** If $\mathbf{u} = 3\hat{\mathbf{i}} + 2\hat{\mathbf{j}} - 4\hat{\mathbf{k}}$, and $\mathbf{v} = \hat{\mathbf{i}} + 6\hat{\mathbf{j}} + 5\hat{\mathbf{k}}$, find scalars λ and ρ so that the vector $\mathbf{w} = 5\hat{\mathbf{i}} - 18\hat{\mathbf{j}} - 32\hat{\mathbf{k}}$ can be written in the form $\mathbf{w} = \lambda\mathbf{u} + \rho\mathbf{v}$.

** **30.** Find a vector $\mathbf{T}$ of length 3 along the tangent line to the curve $y = x^2$ at the point $(2, 4)$ (figure below).

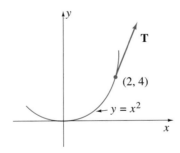

** **31.** Use Coulomb's law (see Example 12.9) to find the force on a charge of 2 coulombs at the origin due to equal charges of 3 coulombs at the points $(1, 1, 2)$ and $(2, -1, -2)$.

** **32.** Newton's universal law of gravitation states that the force of attraction $\mathbf{F}$, in newtons, exerted on a mass of m kilograms by a mass of M kilograms is

$$\mathbf{F} = \frac{GmM}{r^2}\hat{\mathbf{r}},$$

where $G = 6.67 \times 10^{-11}$ is a constant, r is the distance in metres between the masses, and $\hat{\mathbf{r}}$ is a unit vector in the direction from m to M. If point masses, each of 5 kilograms, are situated at $(5, 1, 3)$ and $(-1, 2, 1)$, what is the resultant force on a mass of 10 kilograms at $(2, 2, 2)$?

** **33.** Illustrate geometrically the triangle inequality for vectors $|\mathbf{u} + \mathbf{v}| \leq |\mathbf{u}| + |\mathbf{v}|$. Prove the result algebraically.

Vectors $\mathbf{u}$, $\mathbf{v}$, and $\mathbf{w}$ are said to be linearly dependent if there exist three scalars a, b, and c, not all zero, such that $a\mathbf{u} + b\mathbf{v} + c\mathbf{w} = \mathbf{0}$. If this equation can only be satisfied with $a = b = c = 0$, the vectors are said to be linearly independent. In Exercises 34–37 determine whether the vectors are linearly dependent or linearly independent.

** **34.** $\mathbf{u} = (1, 1, 1)$, $\mathbf{v} = (2, 1, 3)$, $\mathbf{w} = (4, 2, 6)$

** **35.** $\mathbf{u} = (1, 1, 1)$, $\mathbf{v} = (2, 1, 3)$, $\mathbf{w} = (1, 6, 4)$

** **36.** $\mathbf{u} = (-1, 3, -5)$, $\mathbf{v} = (2, 4, -1)$, $\mathbf{w} = (3, 11, -7)$

** **37.** $\mathbf{u} = (4, 2, 6)$, $\mathbf{v} = (1, 3, -2)$, $\mathbf{w} = (7, 1, 4)$

** **38.** Use vectors to show that the line segment joining the midpoints of two sides of a triangle is parallel to the third side and its length is one-half the length of the third side (figure below).

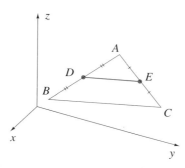

** **39.** Use vectors to show that the medians of a triangle (figure below) all meet in a point with coordinates

$$\left(\frac{x_1 + x_2 + x_3}{3}, \frac{y_1 + y_2 + y_3}{3}, \frac{z_1 + z_2 + z_3}{3} \right).$$

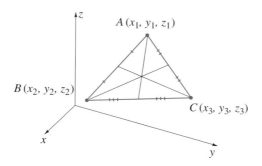

** **40.** If n point masses m_i are located at points (x_i, y_i) in the xy-plane, equations 7.30 and 7.31 define the centre of mass $(\overline{x}, \overline{y})$ of the system. Show that these two scalar equations are represented by the one vector equation

$$M(\overline{x}, \overline{y}) = \sum_{i=1}^{n} m_i \mathbf{r}_i$$

where $\mathbf{r}_i$ is the vector joining the origin to the point (x_i, y_i).

** **41.** If $\mathbf{u} = u_x\hat{\mathbf{i}} + u_y\hat{\mathbf{j}}$ and $\mathbf{v} = v_x\hat{\mathbf{i}} + v_y\hat{\mathbf{j}}$, show that every vector $\mathbf{w}$ in the xy-plane can be written in the form $\mathbf{w} = \lambda\mathbf{u} + \rho\mathbf{v}$ provided $u_xv_y - u_yv_x \neq 0$.

*** **42.** In the figure below, two springs (with constants k_1 and k_2 and unstretched lengths l) are fixed at points A and B. They are joined to a sleeve that slides along the x-axis. Find the resultant force of the springs on the sleeve at any point between O and C.

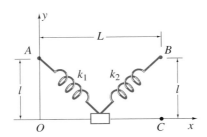

*** **43.** In the left figure below, the xy-plane is the interface between two materials that both transmit light. If a ray of light strikes the surface in a direction defined by the unit vector $\hat{\mathbf{u}} = (u_x, u_y, u_z)$, then some of the light is reflected along a vector $\hat{\mathbf{v}}$, and some is refracted along $\hat{\mathbf{w}}$. The three vectors $\hat{\mathbf{u}}$, $\hat{\mathbf{v}}$, and $\hat{\mathbf{w}}$ all lie in a plane that is perpendicular to the xy-plane.

(a) If the angle of incidence i in the right figure below is equal to the angle of reflection ϕ, find components for $\hat{\mathbf{v}}$ in terms of those of $\hat{\mathbf{u}}$.

(b) If the angle of refraction θ is related to the angle of incidence by $n_1 \sin i = n_2 \sin \theta$, where n_1 and n_2 are the indices of refraction of the two materials, find components of $\hat{\mathbf{w}}$.

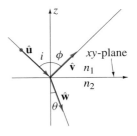

*** **44.** Vectors $\mathbf{v}_1, \mathbf{v}_2, \ldots, \mathbf{v}_n$ are drawn from the centre of a regular n-sided polygon in the plane to each of its vertices. Show that the sum of these vectors is the zero vector.

12.4 The Scalar and Vector Products

The Scalar Product

In Section 12.3 we learned how to multiply a vector by a scalar and how to add and subtract vectors. The next question naturally is, "Can vectors be multiplied?" The answer is yes, and in fact we define two products for vectors, one of which yields a scalar and the other a vector. The first is defined algebraically as follows.

> **DEFINITION 12.8**
>
> The **scalar product** (**dot product** or **inner product**) of two vectors **u** and **v** with Cartesian components (u_x, u_y, u_z) and (v_x, v_y, v_z) is defined as
>
> $$\mathbf{u} \cdot \mathbf{v} = u_x v_x + u_y v_y + u_z v_z. \tag{12.17}$$

If **u** and **v** have only x- and y-components, Definition 12.8 reduces to

$$\mathbf{u} \cdot \mathbf{v} = u_x v_x + u_y v_y, \tag{12.18}$$

and if they have only x-components, it becomes

$$\mathbf{u} \cdot \mathbf{v} = u_x v_x. \tag{12.19}$$

It is straightforward to check that

$$\hat{\mathbf{i}} \cdot \hat{\mathbf{i}} = \hat{\mathbf{j}} \cdot \hat{\mathbf{j}} = \hat{\mathbf{k}} \cdot \hat{\mathbf{k}} = 1, \tag{12.20a}$$

$$\hat{\mathbf{i}} \cdot \hat{\mathbf{j}} = \hat{\mathbf{j}} \cdot \hat{\mathbf{k}} = \hat{\mathbf{i}} \cdot \hat{\mathbf{k}} = 0, \tag{12.20b}$$

and that the scalar product is commutative and distributive:

$$\mathbf{u} \cdot \mathbf{v} = \mathbf{v} \cdot \mathbf{u}, \tag{12.21a}$$

$$\mathbf{u} \cdot (\lambda \mathbf{v} + \rho \mathbf{w}) = \lambda(\mathbf{u} \cdot \mathbf{v}) + \rho(\mathbf{u} \cdot \mathbf{w}). \tag{12.21b}$$

EXAMPLE 12.10

If $\mathbf{u} = (-2, 1, 3)$ and $\mathbf{v} = (3, -2, -1)$, evaluate each of the following:

$$\text{(a) } \mathbf{u} \cdot \mathbf{v}; \quad \text{(b) } 3\mathbf{u} \cdot (2\mathbf{u} - 4\mathbf{v}).$$

SOLUTION

(a) $\mathbf{u} \cdot \mathbf{v} = (-2)(3) + (1)(-2) + (3)(-1) = -11$

(b) $3\mathbf{u} \cdot (2\mathbf{u} - 4\mathbf{v}) = (-6, 3, 9) \cdot (-16, 10, 10)$

$$= (-6)(-16) + (3)(10) + (9)(10)$$

$$= 216.$$

By taking the scalar product of a vector $\mathbf{v} = (v_x, v_y, v_z)$ with itself, we obtain

$$\mathbf{v} \cdot \mathbf{v} = v_x^2 + v_y^2 + v_z^2 = |\mathbf{v}|^2 \iff |\mathbf{v}| = \sqrt{\mathbf{v} \cdot \mathbf{v}}. \tag{12.22}$$

Because Definition 12.8 for the scalar product of two vectors **u** and **v** is phrased in terms of the components of **u** and **v**, and these components depend on the coordinate system used, it follows that this definition also depends on the fact that we have used Cartesian coordinates. Were we to use a different set of coordinates (such as polar coordinates), then the definition of **u** · **v** in terms of components in that coordinate system would be different. For this reason we now find a geometric definition for the scalar product (which is therefore independent of coordinate systems).

THEOREM 12.2

If two nonzero vectors $\mathbf{u}$ and $\mathbf{v}$ are placed tail to tail, and θ is the angle between them $(0 \leq \theta \leq \pi)$, then

$$\mathbf{u} \cdot \mathbf{v} = |\mathbf{u}||\mathbf{v}| \cos \theta. \tag{12.23}$$

FIGURE 12.57

PROOF The cosine law applied to the triangle in Figure 12.57 gives

$$|\mathbf{QR}|^2 = |\mathbf{PQ}|^2 + |\mathbf{PR}|^2 - 2|\mathbf{PQ}||\mathbf{PR}| \cos \theta,$$

or

$$|\mathbf{u} - \mathbf{v}|^2 = |\mathbf{v}|^2 + |\mathbf{u}|^2 - 2|\mathbf{v}||\mathbf{u}| \cos \theta.$$

Consequently,

$$|\mathbf{u}||\mathbf{v}| \cos \theta = \frac{1}{2}\{|\mathbf{u}|^2 + |\mathbf{v}|^2 - |\mathbf{u} - \mathbf{v}|^2\},$$

and if (u_x, u_y, u_z) and (v_x, v_y, v_z) are the Cartesian components of $\mathbf{u}$ and $\mathbf{v}$, then

$$
\begin{aligned}
|\mathbf{u}||\mathbf{v}| \cos \theta &= \frac{1}{2}\{(u_x^2 + u_y^2 + u_z^2) + (v_x^2 + v_y^2 + v_z^2) \\
&\quad - [(u_x - v_x)^2 + (u_y - v_y)^2 + (u_z - v_z)^2]\} \\
&= u_x v_x + u_y v_y + u_z v_z \\
&= \mathbf{u} \cdot \mathbf{v}.
\end{aligned}
$$

An immediate consequence of this result is the following.

COROLLARY

Two nonzero vectors $\mathbf{u}$ and $\mathbf{v}$ are perpendicular if and only if

$$\mathbf{u} \cdot \mathbf{v} = 0. \tag{12.24}$$

For example, the vectors $\mathbf{u} = (1, 2, -1)$ and $\mathbf{v} = (4, 2, 8)$ are perpendicular since $\mathbf{u} \cdot \mathbf{v} = (1)(4) + (2)(2) + (-1)(8) = 0$.

Expression 12.23 doesn't just tell us whether or not the angle between two vectors is $\pi/2$ radians; it can be used to determine the angle between any two nonzero vectors $\mathbf{u}$ and $\mathbf{v}$, simply by solving for θ:

$$\cos \theta = \frac{\mathbf{u} \cdot \mathbf{v}}{|\mathbf{u}||\mathbf{v}|}. \tag{12.25}$$

Since principal values of the inverse cosine function lie between 0 and π, precisely the range for θ, we can write that

$$\theta = \text{Cos}^{-1}\left(\frac{\mathbf{u} \cdot \mathbf{v}}{|\mathbf{u}||\mathbf{v}|}\right). \tag{12.26}$$

EXAMPLE 12.11

Find the angle between the vectors $\mathbf{u} = (2, -3, 1)$ and $\mathbf{v} = (5, 2, 4)$.

SOLUTION   According to formula 12.26,

$$\theta = \text{Cos}^{-1}\left(\frac{\mathbf{u} \cdot \mathbf{v}}{|\mathbf{u}||\mathbf{v}|}\right) = \text{Cos}^{-1}\left(\frac{10 - 6 + 4}{\sqrt{4 + 9 + 1}\sqrt{25 + 4 + 16}}\right)$$

$$= \text{Cos}^{-1}\left(\frac{8}{3\sqrt{70}}\right) = 1.25 \text{ radians.}$$

EXAMPLE 12.12

Find the angle between the lines $x + 2y = 3$ and $4x - 3y = 5$ in the xy-plane.

SOLUTION   Since the slope of $x + 2y = 3$ is $-1/2$, a vector along this line is $\mathbf{u} = (-2, 1)$. Similarly, a vector along $4x - 3y = 5$ is $\mathbf{v} = (3, 4)$. If θ is the angle between these vectors, and therefore between the lines (Figure 12.58), then

$$\theta = \text{Cos}^{-1}\left(\frac{\mathbf{u} \cdot \mathbf{v}}{|\mathbf{u}||\mathbf{v}|}\right) = \text{Cos}^{-1}\left(\frac{-6 + 4}{\sqrt{5}\sqrt{25}}\right) = \text{Cos}^{-1}\left(\frac{-2}{5\sqrt{5}}\right) = 1.75 \text{ radians.}$$

The acute angle between the lines is $\pi - 1.75 = 1.39$ radians. Formula 1.45 gives the same result.

FIGURE 12.58

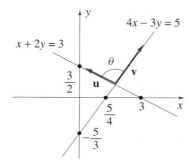

The Vector Product

In many applications we need to find a vector perpendicular to two given vectors. For instance, consider finding a vector $\mathbf{r} = (a, b, c)$ perpendicular to two given vectors $\mathbf{u} = (u_x, u_y, u_z)$ and $\mathbf{v} = (v_x, v_y, v_z)$. The corollary to Theorem 12.2 requires a, b, and c to satisfy

$$0 = \mathbf{r} \cdot \mathbf{u} = au_x + bu_y + cu_z,$$

$$0 = \mathbf{r} \cdot \mathbf{v} = av_x + bv_y + cv_z.$$

When we solve these equations, we find that there is an infinite number of solutions all represented by

$$a = s(u_y v_z - u_z v_y), \quad b = s(u_z v_x - u_x v_z), \quad c = s(u_x v_y - u_y v_x),$$

where s is any real number. In other words, any vector of the form

$$\mathbf{r} = s(u_y v_z - u_z v_y, u_z v_x - u_x v_z, u_x v_y - u_y v_x)$$

is perpendicular to $\mathbf{u} = (u_x, u_y, u_z)$ and $\mathbf{v} = (v_x, v_y, v_z)$. When we choose $s = 1$, the resulting vector is called the vector product of $\mathbf{u}$ and $\mathbf{v}$.

> **DEFINITION 12.9**
>
> The **vector product** (**cross product** or **outer product**) of two vectors $\mathbf{u}$ and $\mathbf{v}$ with Cartesian components (u_x, u_y, u_z) and (v_x, v_y, v_z) is defined as
>
> $$\mathbf{u} \times \mathbf{v} = (u_y v_z - u_z v_y)\hat{\mathbf{i}} + (u_z v_x - u_x v_z)\hat{\mathbf{j}} + (u_x v_y - u_y v_x)\hat{\mathbf{k}}. \qquad (12.27)$$

To eliminate the need for memorizing the exact placing of the six components of $\mathbf{u}$ and $\mathbf{v}$ in this definition, we borrow the notation for determinants from linear algebra. A brief discussion of determinants and their properties is given in Appendix F. We set up a 3×3 determinant with $\hat{\mathbf{i}}, \hat{\mathbf{j}}$, and $\hat{\mathbf{k}}$ across the top row and the components of $\mathbf{u}$ and $\mathbf{v}$ across the second and third rows:

$$\begin{vmatrix} \hat{\mathbf{i}} & \hat{\mathbf{j}} & \hat{\mathbf{k}} \\ u_x & u_y & u_z \\ v_x & v_y & v_z \end{vmatrix}.$$

In actual fact, this is not a determinant, since three entries are vectors and six are scalars. If we ignore this fact, and apply the rules for expansion of a 3×3 determinant along its first row (namely, $\hat{\mathbf{i}}$ times the 2×2 determinant obtained by deleting the row and column containing $\hat{\mathbf{i}}$, minus $\hat{\mathbf{j}}$ times the 2×2 determinant obtained by deleting the row and column containing $\hat{\mathbf{j}}$, plus $\hat{\mathbf{k}}$ times the 2×2 determinant obtained by deleting the row and column containing $\hat{\mathbf{k}}$), we obtain

$$\begin{vmatrix} \hat{\mathbf{i}} & \hat{\mathbf{j}} & \hat{\mathbf{k}} \\ u_x & u_y & u_z \\ v_x & v_y & v_z \end{vmatrix} = \begin{vmatrix} u_y & u_z \\ v_y & v_z \end{vmatrix}\hat{\mathbf{i}} - \begin{vmatrix} u_x & u_z \\ v_x & v_z \end{vmatrix}\hat{\mathbf{j}} + \begin{vmatrix} u_x & u_y \\ v_x & v_y \end{vmatrix}\hat{\mathbf{k}}.$$

But the value of a 2×2 determinant is

$$\begin{vmatrix} a & b \\ c & d \end{vmatrix} = ad - bc.$$

Consequently,

$$
\begin{vmatrix} \hat{\mathbf{i}} & \hat{\mathbf{j}} & \hat{\mathbf{k}} \\ u_x & u_y & u_z \\ v_x & v_y & v_z \end{vmatrix} = (u_y v_z - u_z v_y)\hat{\mathbf{i}} + (u_z v_x - u_x v_z)\hat{\mathbf{j}} + (u_x v_y - u_y v_x)\hat{\mathbf{k}},
$$

and this is the same as the right side of 12.27. We may therefore write, as a memory-saving device, that

$$
\mathbf{u} \times \mathbf{v} = \begin{vmatrix} \hat{\mathbf{i}} & \hat{\mathbf{j}} & \hat{\mathbf{k}} \\ u_x & u_y & u_z \\ v_x & v_y & v_z \end{vmatrix}, \tag{12.28}
$$

so long as we evaluate the right side using the general rules for expansion of a determinant along its first row.

For example, if $\mathbf{u} = (1, -1, 2)$ and $\mathbf{v} = (2, 3, -5)$, then

$$
\mathbf{u} \times \mathbf{v} = \begin{vmatrix} \hat{\mathbf{i}} & \hat{\mathbf{j}} & \hat{\mathbf{k}} \\ 1 & -1 & 2 \\ 2 & 3 & -5 \end{vmatrix} = (5-6)\hat{\mathbf{i}} - (-5-4)\hat{\mathbf{j}} + (3+2)\hat{\mathbf{k}} = -\hat{\mathbf{i}} + 9\hat{\mathbf{j}} + 5\hat{\mathbf{k}}.
$$

It is straightforward to verify that

$$
\hat{\mathbf{i}} \times \hat{\mathbf{i}} = \hat{\mathbf{j}} \times \hat{\mathbf{j}} = \hat{\mathbf{k}} \times \hat{\mathbf{k}} = \mathbf{0} \tag{12.29a}
$$

$$
\hat{\mathbf{i}} \times \hat{\mathbf{j}} = \hat{\mathbf{k}}, \hat{\mathbf{j}} \times \hat{\mathbf{k}} = \hat{\mathbf{i}}, \hat{\mathbf{k}} \times \hat{\mathbf{i}} = \hat{\mathbf{j}}, \tag{12.29b}
$$

and that the cross-product is anticommutative and distributive:

$$
\mathbf{u} \times \mathbf{v} = -\mathbf{v} \times \mathbf{u} \tag{12.30a}
$$

$$
\mathbf{u} \times (\lambda\mathbf{v} + \rho\mathbf{w}) = \lambda(\mathbf{u} \times \mathbf{v}) + \rho(\mathbf{u} \times \mathbf{w}). \tag{12.30b}
$$

Our preliminary analysis indicated that $\mathbf{u} \times \mathbf{v}$ is perpendicular to both $\mathbf{u}$ and $\mathbf{v}$. The following theorem relates the length of $\mathbf{u} \times \mathbf{v}$ to lengths of $\mathbf{u}$ and $\mathbf{v}$.

THEOREM 12.3

If θ is the angle between two vectors $\mathbf{u}$ and $\mathbf{v}$, then

$$
|\mathbf{u} \times \mathbf{v}| = |\mathbf{u}||\mathbf{v}| \sin\theta. \tag{12.31}
$$

PROOF Since θ is an angle between 0 and π, $\sin\theta$ must be positive, and we can write from equation 12.25 that

$$
\sin\theta = \sqrt{1 - \cos^2\theta} = \sqrt{1 - \left(\frac{\mathbf{u} \cdot \mathbf{v}}{|\mathbf{u}||\mathbf{v}|}\right)^2} = \frac{1}{|\mathbf{u}||\mathbf{v}|}\sqrt{|\mathbf{u}|^2|\mathbf{v}|^2 - (\mathbf{u} \cdot \mathbf{v})^2}.
$$

Consequently,

$$
|\mathbf{u}||\mathbf{v}| \sin\theta = \sqrt{|\mathbf{u}|^2|\mathbf{v}|^2 - (\mathbf{u} \cdot \mathbf{v})^2}.
$$

If $\mathbf{u} = (u_x, u_y, u_z)$ and $\mathbf{v} = (v_x, v_y, v_z)$ then

$$
\begin{aligned}
|\mathbf{v}|^2|\mathbf{v}|^2 \sin^2\theta &= (u_x^2 + u_y^2 + u_z^2)(v_x^2 + v_y^2 + v_z^2) - (u_x v_x + u_y v_y + u_z v_z)^2 \\
&= u_x^2(v_x^2 + v_y^2 + v_z^2) + u_y^2(v_x^2 + v_y^2 + v_z^2) \\
&\quad + u_z^2(v_x^2 + v_y^2 + v_z^2) - (u_x^2 v_x^2 + u_y^2 v_y^2 + u_z^2 v_z^2 \\
&\quad + 2u_x v_x u_y v_y + 2u_x v_x u_z v_z + 2u_y v_y u_z v_z) \\
&= (u_y^2 v_z^2 - 2u_y v_y u_z v_z + u_z^2 v_y^2) + (u_x^2 v_x^2 - 2u_x v_x u_z v_z + u_z^2 v_z^2) \\
&\quad + (u_x^2 v_y^2 - 2u_x v_x u_y v_y + u_y^2 v_x^2) \\
&= (u_y v_z - u_z v_y)^2 + (u_x v_z - u_z v_x)^2 + (u_x v_y - u_y v_x)^2 \\
&= |\mathbf{u} \times \mathbf{v}|^2,
\end{aligned}
$$

or

$$
|\mathbf{u} \times \mathbf{v}| = |\mathbf{u}||\mathbf{v}| \sin\theta. \qquad \blacksquare
$$

FIGURE 12.59

We now know that $\mathbf{u} \times \mathbf{v}$ is perpendicular to $\mathbf{u}$ and $\mathbf{v}$, and has length $|\mathbf{u}||\mathbf{v}| \sin\theta$, where θ is the angle between $\mathbf{u}$ and $\mathbf{v}$. Figure 12.59 illustrates that there are only two directions that are perpendicular to $\mathbf{u}$ and $\mathbf{v}$, and one is the negative of the other. Let us denote $\hat{\mathbf{w}}$ the unit vector along that direction which is perpendicular to $\mathbf{u}$ and $\mathbf{v}$ and is determined by the right-hand rule (curl the fingers of the right hand from $\mathbf{u}$ toward $\mathbf{v}$ and the thumb points in direction $\hat{\mathbf{w}}$).

We now show that $\mathbf{u} \times \mathbf{v}$ always points in the direction determined by the right-hand rule (i.e., in direction $\hat{\mathbf{w}}$ rather than $-\hat{\mathbf{w}}$). To see this we place $\mathbf{u}$ and $\mathbf{v}$ tail to tail and establish a coordinate system with this common point as origin and the positive x-axis along $\mathbf{u}$ (Figure 12.60a). Let the plane determined by $\mathbf{u}$ and $\mathbf{v}$ be the xy-plane. In this coordinate system, $\mathbf{u}$ has only an x-component, $\mathbf{u} = u_x\hat{\mathbf{i}}(u_x > 0)$, and $\mathbf{v}$ has only x- and y-components, $\mathbf{v} = v_x\hat{\mathbf{i}} + v_y\hat{\mathbf{j}}$. The cross-product of $\mathbf{u}$ and $\mathbf{v}$ is therefore

$$
\mathbf{u} \times \mathbf{v} = \begin{vmatrix} \hat{\mathbf{i}} & \hat{\mathbf{j}} & \hat{\mathbf{k}} \\ u_x & 0 & 0 \\ v_x & v_y & 0 \end{vmatrix} = u_x v_y \hat{\mathbf{k}}.
$$

For $\mathbf{v}$ in Figure 12.60a, v_y is clearly positive and therefore $u_x v_y$, the component of $\mathbf{u} \times \mathbf{v}$, is also positive. But then $\mathbf{u} \times \mathbf{v}$ is indeed determined by the right-hand rule since $\hat{\mathbf{w}} = \hat{\mathbf{k}}$.

FIGURE 12.60a

FIGURE 12.60b

FIGURE 12.60c

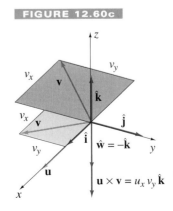

When the tip of **v** lies in the second quadrant of the xy-plane (Figure 12.60b), **u** × **v** is once again in the positive z-direction, the direction of $\hat{\mathbf{w}}$. When the tip of **v** is in the third or fourth quadrant (Figure 12.60c), $v_y < 0$, and **u** × **v** therefore has a negative z-component. But in this case $\hat{\mathbf{w}} = -\hat{\mathbf{k}}$, and the direction of **u** × **v** is once again determined by the right-hand rule.

What we have now established is the following coordinate-free definition for the vector product of two vectors **u** and **v**:

$$\mathbf{u} \times \mathbf{v} = (|\mathbf{u}||\mathbf{v}| \sin\theta)\hat{\mathbf{w}}. \qquad (12.32)$$

The unit vector $\hat{\mathbf{w}}$ defines the direction of **u** × **v** and the factor $|\mathbf{u}||\mathbf{v}| \sin\theta$ is its length.

The fact that the vector product **u** × **v** is perpendicular to both **u** and **v** makes it a powerful tool in many applications. We shall see some of them in Sections 12.5 and 12.6.

EXERCISES 12.4

If $\mathbf{u} = 2\hat{\mathbf{i}} - 3\hat{\mathbf{j}} + \hat{\mathbf{k}}$, $\mathbf{v} = \hat{\mathbf{j}} - \hat{\mathbf{k}}$, and $\mathbf{w} = 6\hat{\mathbf{i}} - 2\hat{\mathbf{j}} + 3\hat{\mathbf{k}}$, evaluate the scalar or find the components of the vector in Exercises 1–10.

* **1.** $\mathbf{u} \cdot \mathbf{v}$
* **2.** $(\mathbf{v} \cdot \mathbf{w})\mathbf{u}$
* **3.** $(2\mathbf{u} - 3\mathbf{v}) \cdot \mathbf{w}$
* **4.** $2\hat{\mathbf{i}} \cdot \hat{\mathbf{u}}$
* **5.** $|2\mathbf{u}|\mathbf{v} \cdot \mathbf{w}$
* **6.** $(3\mathbf{u} - 4\mathbf{w}) \cdot (2\hat{\mathbf{i}} + 3\mathbf{u} - 2\mathbf{v})$
* **7.** $\mathbf{w} \cdot \hat{\mathbf{w}}$
* **8.** $\dfrac{(105\mathbf{u} + 240\mathbf{v}) \cdot (105\mathbf{u} + 240\mathbf{v})}{|105\mathbf{u} + 240\mathbf{v}|^2}$
* **9.** $|\mathbf{u} - \mathbf{v} + \hat{\mathbf{k}}|(\hat{\mathbf{j}} + \mathbf{w}) \cdot \hat{\mathbf{k}}$
* **10.** $\mathbf{u} \cdot \mathbf{v} + \mathbf{v} \cdot \mathbf{w} - (\mathbf{u} + \mathbf{w}) \cdot \mathbf{v}$

If $\mathbf{u} = (3, 1, 4)$, $\mathbf{v} = (-1, 2, 0)$, and $\mathbf{w} = (-2, -3, 5)$, evaluate the scalar or find the components of the vector in Exercises 11–20.

* **11.** $\mathbf{v} \times \mathbf{w}$
* **12.** $(-3\mathbf{u}) \times (2\mathbf{v})$
* **13.** $\mathbf{u} \cdot (\mathbf{v} \times \mathbf{w})$
* **14.** $\hat{\mathbf{u}} \times \hat{\mathbf{w}}$
* **15.** $((3\mathbf{u}) \times \mathbf{w}) + (\mathbf{u} \times \mathbf{v})$
* **16.** $\mathbf{u} \times (3\mathbf{v} - \mathbf{w})$
* **17.** $\dfrac{\mathbf{w} \times \mathbf{u}}{|\mathbf{u} \times \mathbf{v}|}$
* **18.** $(\mathbf{u} \times \mathbf{w}) - (\mathbf{u} \times \mathbf{v}) + (\mathbf{u} \times (2\mathbf{u} + \mathbf{v}))$
* **19.** $(\mathbf{u} \times \mathbf{v}) \times \mathbf{w}$
* **20.** $\mathbf{u} \times (\mathbf{v} \times \mathbf{w})$

In Exercises 21–24 determine whether the vectors are perpendicular.

* **21.** $(1, 2)$, $(3, 5)$
* **22.** $(2, 4)$, $(-8, 4)$
* **23.** $(1, 3, 6)$, $(-2, 1, -4)$
* **24.** $(2, 3, -6)$, $(-6, 6, 1)$

In Exercises 25–30 find the angle between the vectors.

* **25.** $(3, 4)$, $(2, -5)$
* **26.** $(1, 6)$, $(-4, 7)$
* **27.** $(4, 2, 3)$, $(1, 5, 6)$
* **28.** $(3, 1, -1)$, $(-2, 1, 4)$
* **29.** $(2, 0, 5)$, $(0, 3, 0)$
* **30.** $(1, 3, -2)$, $(-2, -6, 4)$

In Exercises 31–33 find components for the vector.

∗ **31.** Perpendicular to the vectors $(1, 3, 5)$ and $(-2, 1, 4)$

∗ **32.** Perpendicular to the y-axis and the vector joining the points $(2, 4, -3)$ and $(1, 5, 6)$

∗ **33.** Perpendicular to the triangle with vertices $(-1, 0, 3)$, $(5, 1, 2)$, and $(-6, 2, 4)$

∗ **34.** Verify the results in equations 12.21.

∗∗ **35.** Use equation 12.22 to prove

$$|\mathbf{u} + \mathbf{v}|^2 + |\mathbf{u} - \mathbf{v}|^2 = 2|\mathbf{u}|^2 + 2|\mathbf{v}|^2.$$

This is often called the parallelogram law. Why?

∗∗ **36.** The angles between a vector $\mathbf{v} = (v_x, v_y, v_z)$ and the vectors $\hat{\mathbf{i}}$, $\hat{\mathbf{j}}$, and $\hat{\mathbf{k}}$, are called direction angles α, β, and γ of $\mathbf{v}$. Show that

$$\alpha = \text{Cos}^{-1}\left(\frac{\mathbf{v} \cdot \hat{\mathbf{i}}}{|\mathbf{v}|}\right) = \text{Cos}^{-1}\left(\frac{v_x}{|\mathbf{v}|}\right);$$

$$\beta = \text{Cos}^{-1}\left(\frac{\mathbf{v} \cdot \hat{\mathbf{j}}}{|\mathbf{v}|}\right) = \text{Cos}^{-1}\left(\frac{v_y}{|\mathbf{v}|}\right);$$

$$\gamma = \text{Cos}^{-1}\left(\frac{\mathbf{v} \cdot \hat{\mathbf{k}}}{|\mathbf{v}|}\right) = \text{Cos}^{-1}\left(\frac{v_z}{|\mathbf{v}|}\right).$$

Find direction angles for the vectors in Exercises 37–40.

∗∗ **37.** $(1, 2, -3)$ ∗∗ **38.** $(0, 1, -3)$

∗∗ **39.** $(-1, -2, 6)$ ∗∗ **40.** $(-2, 3, 4)$

∗∗ **41.** Verify the results in equations 12.35.

∗∗ **42.** Show that the cross-product is not associative; that is, in general $\mathbf{u} \times (\mathbf{v} \times \mathbf{w}) \neq (\mathbf{u} \times \mathbf{v}) \times \mathbf{w}$.

∗∗ **43.** (a) If $\mathbf{u} \neq 0$, show that, if both the conditions $\mathbf{u} \cdot \mathbf{v} = \mathbf{u} \cdot \mathbf{w}$ and $\mathbf{u} \times \mathbf{v} = \mathbf{u} \times \mathbf{w}$ are satisfied, then $\mathbf{v} = \mathbf{w}$.

(b) Show that if one of the conditions in (a) is satisfied, but the other is not, then $\mathbf{v}$ cannot be equal to $\mathbf{w}$.

∗∗ **44.** The scalar $\mathbf{u} \cdot \mathbf{v} \times \mathbf{w}$ is called the *scalar triple product* of $\mathbf{u}$, $\mathbf{v}$, and $\mathbf{w}$.

(a) Find $\mathbf{u} \cdot \mathbf{v} \times \mathbf{w}$ if $\mathbf{u} = (6, -1, 0)$, $\mathbf{v} = (1, 3, 4)$, and $\mathbf{w} = (-2, -1, 4)$.

(b) Prove that $\mathbf{u} \cdot \mathbf{v} \times \mathbf{w} = \mathbf{u} \times \mathbf{v} \cdot \mathbf{w}$.

(c) Show that $|\mathbf{u} \cdot \mathbf{v} \times \mathbf{w}|$ can be interpreted as the volume of the parallelepiped with $\mathbf{u}$, $\mathbf{v}$, and $\mathbf{w}$ as coterminal sides in the figure below.

(d) Verify that three nonzero vectors $\mathbf{u}$, $\mathbf{v}$, and $\mathbf{w}$ all lie in the same plane if and only if $\mathbf{u} \cdot \mathbf{v} \cdot \mathbf{w} = 0$.

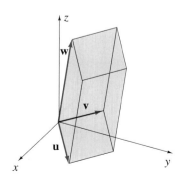

In Exercises 45 and 46 prove the identity.

∗∗ 45. $(\mathbf{u} \times \mathbf{v}) \cdot (\mathbf{w} \times \mathbf{r}) = (\mathbf{u} \cdot \mathbf{w})(\mathbf{v} \cdot \mathbf{r}) - (\mathbf{u} \cdot \mathbf{r})(\mathbf{v} \cdot \mathbf{w})$

∗∗ 46. $\mathbf{u} \times (\mathbf{v} \times \mathbf{w}) = (\mathbf{u} \cdot \mathbf{w})\mathbf{v} - (\mathbf{u} \cdot \mathbf{v})\mathbf{w}$

∗∗ 47. Use vectors to prove the sine law $\dfrac{\sin A}{a} = \dfrac{\sin B}{b} = \dfrac{\sin C}{c}$ for the triangle in the figure below. (Hint: Note that $\mathbf{PQ} + \mathbf{QR} + \mathbf{RP} = \mathbf{0}$. Cross this equation with $\mathbf{PQ}$ and take lengths.)

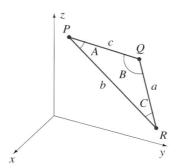

∗∗ 48. Show that the vector $(|\mathbf{v}|\mathbf{u} + |\mathbf{u}|\mathbf{v}) / ||\mathbf{u}|\mathbf{v} + |\mathbf{v}|\mathbf{u}|$ is a unit vector that bisects the angle between $\mathbf{u}$ and $\mathbf{v}$.

12.5 Lines and Planes

Planes

A plane in space can be characterized in various ways: by two intersecting lines, by a line and a point not on the line, or by three noncollinear points. For our present purposes, we use the fact that, given a point $P(x_1, y_1, z_1)$ and a vector (A, B, C) (Figure 12.61), there is one and only one plane through P that is perpendicular to (A, B, C).

FIGURE 12.61

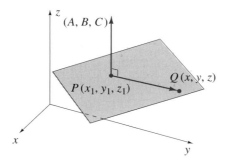

To find the equation of this plane we note that, if $Q(x, y, z)$ is any other point in the plane, then vector $\mathbf{PQ} = (x - x_1, y - y_1, z - z_1)$ lies in the plane. But $\mathbf{PQ}$ must then be perpendicular to (A, B, C); hence, by the corollary to Theorem 12.2,

$$(A, B, C) \cdot (x - x_1, y - y_1, z - z_1) = 0.$$

Because this equation must be satisfied by every point (x, y, z) in the plane (and at the same time is not satisfied by any point not in the plane), it must be the equation of the plane through P and perpendicular to (A, B, C). If we expand the scalar product, we obtain the equation of the plane in the form

$$A(x - x_1) + B(y - y_1) + C(z - z_1) = 0.$$

This result is worth stating as a theorem.

THEOREM 12.4

An equation for the plane through the point (x_1, y_1, z_1) perpendicular to the vector (A, B, C) is

$$A(x - x_1) + B(y - y_1) + C(z - z_1) = 0. \tag{12.33}$$

Equation 12.33 can also be written in the form

$$Ax + By + Cz + D = 0, \tag{12.34}$$

where $D = -(Ax_1 + By_1 + Cz_1)$, and this equation is said to be **linear** in x, y, and z. We have shown then that every plane has a linear equation, and the coefficients A, B, and C of x, y, and z in the equation are the components of a vector (A, B, C) that is perpendicular to the plane. Instead of saying that (A, B, C) is perpendicular to the plane $Ax + By + Cz + D = 0$, we often say that (A, B, C) is **normal** to the plane or that (A, B, C) is a **normal vector** to the plane.

EXAMPLE 12.13

Find an equation for the plane through the point $(4, -3, 5)$ and normal to the vector $(4, -8, 3)$.

SOLUTION   According to 12.33, the equation of the plane is

$$4(x - 4) - 8(y + 3) + 3(z - 5) = 0, \quad \text{or} \quad 4x - 8y + 3z = 55.$$

EXAMPLE 12.14

Determine whether the planes $x + 2y - 4z = 10$ and $2x + 4y - 8z = 11$ are parallel.

SOLUTION   Normal vectors to these planes are $\mathbf{N}_1 = (1, 2, -4)$ and $\mathbf{N}_2 = (2, 4, -8)$. Since $\mathbf{N}_2 = 2\mathbf{N}_1$, the normal vectors are in the same direction, and therefore the planes are parallel.

Lines

In Section 12.2 we indicated that space curves can be described by two simultaneous equations in x, y, and z, and that such a representation describes the curve as the intersection of two surfaces. A straight line results when the surfaces are planes. We shall discuss this further, but we prefer to begin our discussion of straight lines with vectors much as we did for planes. A straight line in space is characterized by a point on it and a vector parallel to it. There is one and only one line ℓ through the point $P_0(x_0, y_0, z_0)$ and in the direction $\mathbf{v} = (v_x, v_y, v_z)$ in Figure 12.62.

FIGURE 12.62

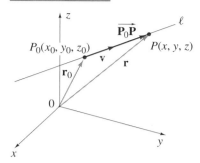

If $P(x, y, z)$ is any point on this line, then the vector **r** joining the origin to P has components $\mathbf{r} = (x, y, z)$. Now this vector can be expressed as the sum of $\mathbf{r}_0$, the vector from O to P_0 and $\mathbf{P_0P}$:

$$\mathbf{r} = \mathbf{r}_0 + \mathbf{P_0P}.$$

But $\mathbf{P_0P}$ is in the same direction as **v**; therefore, it must be some scalar multiple of **v**, i.e., $\mathbf{P_0P} = t\mathbf{v}$. Consequently, the vector joining O to any point on ℓ can be written in the form

$$\mathbf{r} = \mathbf{r}_0 + t\mathbf{v}, \tag{12.35}$$

for an appropriate value of t. Because the components of $\mathbf{r} = (x, y, z)$ are also the coordinates of the point (x, y, z) on ℓ, this equation is called the **vector equation** of the line ℓ through (x_0, y_0, z_0) in direction **v**. If we substitute components for **r**, $\mathbf{r}_0$, and **v**, then

$$(x, y, z) = (x_0, y_0, z_0) + t(v_x, v_y, v_z) = (x_0 + tv_x, y_0 + tv_y, z_0 + tv_z).$$

Since two vectors are equal if and only if corresponding components are identical, we can write that

$$x = x_0 + v_x t, \tag{12.36a}$$

$$y = y_0 + v_y t, \tag{12.36b}$$

$$z = z_0 + v_z t. \tag{12.36c}$$

These three scalar equations are equivalent to vector equation 12.35; they are called **parametric equations** for line ℓ. They illustrate once again that a line in space is characterized by a point (x_0, y_0, z_0) on it and a vector (v_x, v_y, v_z) along it. Each value of t substituted into 12.36 yields a point (x, y, z) on ℓ, and conversely, every point on ℓ is represented by some value of t. For instance, $t = 0$ yields P_0, and $t = 1$ gives the point at the tip of **v** in Figure 12.62.

If none of v_x, v_y, and v_z is equal to zero, we can solve equations 12.36 for t and equate the three expressions to obtain

$$\frac{x - x_0}{v_x} = \frac{y - y_0}{v_y} = \frac{z - z_0}{v_z}. \tag{12.37}$$

These are called **symmetric equations** for the line ℓ through (x_0, y_0, z_0) parallel to $\mathbf{v} = (v_x, v_y, v_z)$. There are only two independent equations in 12.37, which therefore substantiates our previous result that a curve (in this case, a line) can be described by two equations in x, y, and z. We could, for instance, write

$$v_x(y - y_0) = v_y(x - x_0) \quad \text{and} \quad v_z(y - y_0) = v_y(z - z_0),$$

or

$$v_y x - v_x y = v_y x_0 - v_x y_0, \quad v_y z - v_z y = v_y z_0 - v_z y_0.$$

Since the first of these is linear in x and y and the second is linear in y and z, each describes a plane. The line has been described as the curve of intersection of two planes.

EXAMPLE 12.15

Find, if possible, vector, parametric, and symmetric equations for the line through the points $(-1, 2, 1)$ and $(3, -2, 1)$.

SOLUTION A vector along the line is $(3, -2, 1) - (-1, 2, 1) = (4, -4, 0)$, and so too is $(1, -1, 0)$. A vector equation for the line is

$$\mathbf{r} = (-1, 2, 1) + t(1, -1, 0) = (t - 1, -t + 2, 1).$$

Parametric equations are therefore

$$x = t - 1, \quad y = -t + 2, \quad z = 1.$$

Because the z-component of every vector along the line is zero, we cannot write full symmetric equations for the line. By eliminating t between the x- and y-equations, however, we can write

$$x + 1 = \frac{y - 2}{-1}, \quad z = 1.$$

If we set $x + 1 = 2 - y$, $z = 1$, or $x + y = 1$, $z = 1$, we represent the line as the intersection of the planes $x + y = 1$ and $z = 1$ (Figure 12.63).

FIGURE 12.63

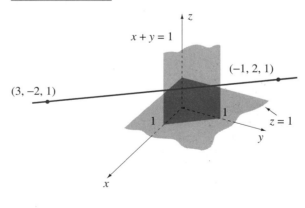

EXAMPLE 12.16

Find the equation of the plane containing the origin and the line $2x + y - z = 4$, $x + z = 5$.

SOLUTION   We can easily find two more points on the plane. For instance, if we set $x = 0$, then the equations of the line require $z = 5$ and $y = 9$; and if we set $z = 0$, then $x = 5$ and $y = -6$. Thus $P(0, 9, 5)$ and $Q(5, -6, 0)$, as well as $O(0, 0, 0)$, are points on the plane. It follows then that $\mathbf{OP} = (0, 9, 5)$ and $\mathbf{OQ} = (5, -6, 0)$ are vectors in the plane, and a vector normal to the plane is

$$\mathbf{OP} \times \mathbf{OQ} = \begin{vmatrix} \hat{\mathbf{i}} & \hat{\mathbf{j}} & \hat{\mathbf{k}} \\ 0 & 9 & 5 \\ 5 & -6 & 0 \end{vmatrix} = (30, 25, -45).$$

The vector $(6, 5, -9)$ is also normal to the plane, and the equation of the plane is

$$0 = (6, 5, -9) \cdot (x - 0, y - 0, z - 0) = 6x + 5y - 9z.$$

EXAMPLE 12.17

Find symmetric equations for the line $x + y - 2z = 6$, $2x - 3y + 4z = 10$.

FIGURE 12.64

$x + y - 2z = 6$

$(2, -3, 4)$

$(1, 1, -2)$

$2x - 3y + 4z = 10$

SOLUTION   To find symmetric equations, we require a vector parallel to the line and a point on it. By setting $x = 0$ and solving $y - 2z = 6$, $-3y + 4z = 10$, we obtain $y = -22$, $z = -14$. Consequently, $(0, -22, -14)$ is a point on the line. To find a vector along the line, we could find another point on the line, say, by setting $z = 0$ and solving $x + y = 6$, $2x - 3y = 10$ for $x = 28/5$, $y = 2/5$. A vector along the line is therefore $(28/5, 2/5, 0) - (0, -22, -14) = (28/5, 112/5, 14)$, and so too is $(5/14)(28/5, 112/5, 14) = (2, 8, 5)$.

   Alternatively, we know that $(1, 1, -2)$ and $(2, -3, 4)$ are vectors that are normal to the planes $x + y - 2z = 6$ and $2x - 3y + 4z = 10$, and a vector along the line of intersection of the planes must be perpendicular to both of these vectors (Figure 12.64). Consequently, a vector along the line of intersection is

$$\begin{vmatrix} \hat{\mathbf{i}} & \hat{\mathbf{j}} & \hat{\mathbf{k}} \\ 1 & 1 & -2 \\ 2 & -3 & 4 \end{vmatrix} = (-2, -8, -5).$$

Symmetric equations for the line are therefore

$$\frac{x}{2} = \frac{y + 22}{8} = \frac{z + 14}{5}.$$

EXERCISES 12.5

In Exercises 1–10 find the equation for the plane.

* **1.** Through the point $(1, -1, 3)$ and normal to the vector $(4, 3, -2)$

* **2.** Through the point $(2, 1, 5)$ and normal to the vector joining $(2, 1, 5)$ and $(4, 2, 3)$

* **3.** Containing the points $(1, 3, 2)$, $(-2, 0, -2)$, $(1, 4, 3)$

* **4.** Containing the point $(2, -4, 3)$ and the line $(x - 1)/3 = (y + 5)/4 = z + 2$

∗ **5.** Containing the lines $x = 2y = (z + 1)/4$ and $x = t$, $y = 2t$, $z = 6t - 1$

∗ **6.** Containing the lines $(x - 1)/6 = y/8 = (z + 2)/2$ and $(x + 1)/3 = (y - 2)/4 = z + 5$

∗ **7.** Containing the line $x - y + 2z = 4$, $2x + y + 3z = 6$ and the point $(1, -2, 4)$

∗ **8.** Containing the lines $x + 2y + 4z = 21$, $x - y + 6z = 13$ and $x = 2 + 3t$, $y = 4$, $z = -3 + 5t$

∗ **9.** Containing the lines $3x + 4y = -6$, $x + 2y + z = 2$ and $2y + 3z = 19$, $3x - 2y - 9z = -58$

∗ **10.** Containing the line $x + y - 4z = 6$, $2x + 3y + 5z = 10$ and

    (a) perpendicular to the xy-plane

    (b) perpendicular to the xz-plane

    (c) perpendicular to the yz-plane

∗ **11.** The acute angle between two intersecting planes is defined as the acute angle between their normals. Find the acute angle between the planes $x - 2y + 4z = 6$ and $2x + y = z + 4$.

In Exercises 12–21 find vector, parametric, and symmetric (if possible) equations for the straight line.

∗ **12.** Through the point $(1, -1, 3)$ and parallel to the vector $(2, 4, -3)$

∗ **13.** Through the point $(-1, 3, 6)$ and parallel to the vector $(2, -3, 0)$

∗ **14.** Through the points $(2, -3, 4)$ and $(5, 2, -1)$

∗ **15.** Through the points $(-2, 3, 3)$ and $(-2, -3, -3)$

∗ **16.** Through the points $(1, 3, 4)$ and $(1, 3, 5)$

∗ **17.** Through the point $(1, -3, 5)$ and parallel to the line

$$\frac{x}{5} = \frac{y - 2}{3} = \frac{z + 4}{-2}$$

∗ **18.** Through the point $(2, 0, 3)$ and parallel to the line $x = 4 + t$, $y = 2$, $z = 6 - 2t$

∗∗ **19.** Through the point of intersection of the lines

$$\frac{x - 1}{2} = \frac{y + 4}{-3} = \frac{z - 2}{5} \quad \text{and} \quad \frac{x - 1}{6} = \frac{y + 4}{3} = \frac{z - 2}{4},$$

and parallel to the line joining the points $(1, 3, -2)$ and $(2, -2, 1)$

∗∗ **20.** $2x - y = 5$, $3x + 4y + z = 10$

∗ **21.** Through the point $(-2, 3, 1)$ and parallel to the line $x + y = 3$, $2x - y + z = -2$

∗∗ **22.** Does the line $(x - 3)/2 = y - 2 = (z + 1)/4$ lie in the plane $x - y + 2z = -1$?

∗∗ **23.** Show that a vector perpendicular to the line $Ax + By + C = 0$ in the xy-plane is (A, B).

∗∗ **24.** Show that if a plane has nonzero intercepts a, b, and c on the x-, y-, and z-axes, then its equation is $x/a + y/b + z/c = 1$.

∗∗ **25.** Find equations for the four faces of the tetrahedron in Exercise 17 of Section 12.1 (page 847).

∗∗ **26.** Find equations for the nine planes forming the sides, bottom, and roof of the birdhouse in Exercises 14 of Section 12.1 (page 846).

∗∗ **27.** Verify that the equation of the plane passing through the three points $P_1(x_1, y_1, z_1)$, $P_2(x_2, y_2, z_2)$, and $P_3(x_3, y_3, z_3)$ can be written in the form $\mathbf{P_1P} \cdot \mathbf{P_1P_2} \times \mathbf{P_1P_3} = 0$, where $P(x, y, z)$ is any point in the plane.

** **28.** The region S_{xy} in the xy-plane bounded by the straight lines $x = 0$, $x = 1$, $y = 0$ and $y = 1$ is a rectangle with unit area.

    (a) Show that the region in the plane $y = z$ that projects onto S_{xy} is also a rectangle, but with area $\sqrt{2}$.

    (b) Show that the region in the plane $x + y - 2z = 0$ that projects onto S_{xy} is a parallelogram with area $\sqrt{6}/2$.

    (c) Generalize the results of (a) and (b) to show that if S is the area in a plane $Ax + By + Cz + D = 0$ $(C \neq 0)$ that projects onto S_{xy}, then the area of S is $\sec \gamma$ where γ is the acute angle between $\hat{\mathbf{k}}$ and the normal to the given plane.

12.6 Applications of Scalar and Vector Products

Components of Vectors in Arbitrary Directions

We have defined what is meant by Cartesian components (v_x, v_y, v_z) of a vector $\mathbf{v}$. They are scalars that multiply $\hat{\mathbf{i}}$, $\hat{\mathbf{j}}$, and $\hat{\mathbf{k}}$ so that $\mathbf{v} = v_x\hat{\mathbf{i}} + v_y\hat{\mathbf{j}} + v_z\hat{\mathbf{k}}$. They can be represented in terms of scalar products as

$$v_x = \mathbf{v} \cdot \hat{\mathbf{i}}, \quad v_y = \mathbf{v} \cdot \hat{\mathbf{j}}, \quad v_z = \mathbf{v} \cdot \hat{\mathbf{k}}. \tag{12.38}$$

Geometrically they can be found by drawing $\mathbf{v}$, $\hat{\mathbf{i}}$, $\hat{\mathbf{j}}$, and $\hat{\mathbf{k}}$ all at the origin and dropping perpendiculars from the tip of $\mathbf{v}$ to the x-, y-, and z-axes (Figure 12.65). We now generalize our definition of a component along an axis to define the component of a vector in any direction whatsoever.

FIGURE 12.65

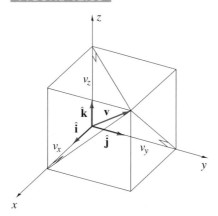

DEFINITION 12.10

To define the component of a vector $\mathbf{v}$ in a direction $\mathbf{u}$, we place $\mathbf{v}$ and $\mathbf{u}$ tail to tail at a point P and draw the perpendicular from the tip of $\mathbf{v}$ to the line containing $\mathbf{u}$ (Figure 12.66a). The directed distance PR is called the component of $\mathbf{v}$ in the direction $\mathbf{u}$. If R is on the same side of P as the tip of $\mathbf{u}$, PR is taken as positive; and if R is on that side of P opposite to the tip of $\mathbf{u}$, PR is negative.

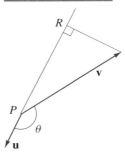

In Figure 12.66a the component of **v** in direction **u** is positive, and in Figure 12.66b the component is negative. Note that the length of **u** is irrelevant; it is only the direction of **u** that determines the component of **v** in the direction **u**. If θ is the angle between **u** and **v**, then

$$PR = |\mathbf{v}| \cos \theta.$$

The right side of this equation looks very much like the scalar product of **v** and a vector that makes an angle θ with **v**. It lacks only the length of this second vector. Clearly, **u** is a vector that makes an angle θ with **v**, but we cannot write $PR = |\mathbf{v}||\mathbf{u}| \cos \theta$, since the length of **u** need not be 1. If, however, $\hat{\mathbf{u}}$ is the unit vector in the same direction as **u**, then we can write

$$PR = |\mathbf{v}||\hat{\mathbf{u}}| \cos \theta \qquad (12.39)$$

In other words, we have the following theorem.

THEOREM 12.5

The component of a vector **v** in a direction **u** is $\mathbf{v} \cdot \hat{\mathbf{u}}$, where $\hat{\mathbf{u}}$ is the unit vector in direction **u**.

This result agrees with equation 12.38 for the x-, y-, and z-components of **v**.

EXAMPLE 12.18 ▼

Find components of $\mathbf{v} = (1, 2, -3)$ in directions specified by the following vectors:

$$(a) \quad \mathbf{u} = (-1, 3, 4) \qquad (b) \quad \mathbf{u} = (1, 3, -2) \qquad (c) \quad \mathbf{u} = (4, 4, 4)$$

SOLUTION   Components of **v** in these directions are

$$(a) \quad \mathbf{v} \cdot \hat{\mathbf{u}} = (1, 2, -3) \cdot \frac{(-1, 3, 4)}{\sqrt{26}} = -\frac{7}{\sqrt{26}}$$

$$(b) \quad \mathbf{v} \cdot \hat{\mathbf{u}} = (1, 2, -3) \cdot \frac{(1, 3, -2)}{\sqrt{14}} = \frac{13}{\sqrt{14}}$$

$$(c) \quad \mathbf{v} \cdot \hat{\mathbf{u}} = (1, 2, -3) \cdot \frac{(4, 4, 4)}{4\sqrt{3}} = 0$$

This last result means that **v** is perpendicular to $(4, 4, 4)$.

Distances between Points, Lines, and Planes

There are three distances between geometric objects in a plane—between points, from a point to a line, and between parallel lines. They were all covered in Section 1.1. Distance formulas 1.1 and 1.7 handle the first two situations. Distance between parallel lines can be calculated by finding a point on one line and using 1.7 for the distance from this point to the other line.

In space, there are six distances of interest—point to point, point to line, point to plane, line to line, line to plane, and plane to plane. Formula 12.1 gives distance between points. We can develop a formula for the distance from a point (x_1, y_1, z_1) to a plane $Ax + By + Cz + D = 0$ very similar to formula 1.7 for distance from a point to a line in the xy-plane. With vectors, the derivation is much simpler than that for 1.7 in Section 1.1. Let R be the foot of the perpendicular from a point $P(x_1, y_1, z_1)$ to the plane in Figure 12.67, and $Q(x, y, z)$ be any other point in the plane. The length of vector **PR** is the distance from P to the plane, and this length is the component of **PQ** in direction **PR**. Hence, $|\mathbf{PR}| = \mathbf{PQ} \cdot \widehat{\mathbf{PR}}$. A normal vector to the plane is (A, B, C), and therefore

$$\widehat{\mathbf{PR}} = \pm \frac{(A, B, C)}{\sqrt{A^2 + B^2 + C^2}}.$$

Thus,

$$|\mathbf{PR}| = \mathbf{PQ} \cdot \left(\frac{\pm(A, B, C)}{\sqrt{A^2 + B^2 + C^2}} \right).$$

We choose the $+$ or $-$ to guarantee a positive result. This can also be accomplished by inserting absolute values,

$$|\mathbf{PR}| = \left| \mathbf{PQ} \cdot \frac{(A, B, C)}{\sqrt{A^2 + B^2 + C^2}} \right|.$$

Since $\mathbf{PQ} = (x - x_1, y - y_1, z - z_1)$,

$$|\mathbf{PR}| = \left| (x - x_1, y - y_1, z - z_1) \cdot \frac{(A, B, C)}{\sqrt{A^2 + B^2 + C^2}} \right|$$

$$= \frac{|Ax + By + Cz - (Ax_1 + By_1 + Cz_1)|}{\sqrt{A^2 + B^2 + C^2}}.$$

Because Q is in the plane, we can replace $Ax + By + Cz$ with $-D$, and the formula then becomes

$$\frac{|Ax_1 + By_1 + Cz_1 + D|}{\sqrt{A^2 + B^2 + C^2}}. \tag{12.40}$$

Notice, as we suggested, the similarity of this formula to 1.7.

FIGURE 12.67

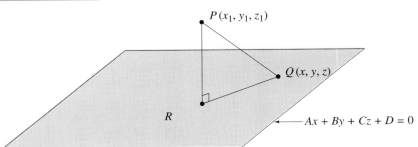

EXAMPLE 12.19

Find the distance from the point $(1, 2, 5)$ to the plane $x + y + 2z = 4$.

SOLUTION   According to 12.40, the distance is

$$\frac{|1 + 2 + 2(5) - 4|}{\sqrt{1 + 1 + 4}} = \frac{9}{\sqrt{6}}.$$

With distance formula 12.40, it is straightforward to find the distance between two parallel planes, or between a plane and a line parallel to the plane.

EXAMPLE 12.20

Show that the line $x = 3 + t$, $y = 1 - 2t$, $z = 4 + 3t$ and the plane $x + 2y + z = 6$ are parallel, and find the distance between them.

SOLUTION   A vector along the line is $\mathbf{v} = (1, -2, 3)$ and a vector normal to the plane is $\mathbf{N} = (1, 2, 1)$. Since $\mathbf{v} \cdot \mathbf{N} = 0$, the vectors are perpendicular, and this confirms that the line is parallel to the plane. (We could also confirm this by substituting from the equation of the line into the plane,

$$6 = (3 + t) + 2(1 - 2t) + (4 + 3t) = 9,$$

a contradiction. The line and plane do not therefore intersect.) To find the distance between the line and plane, we find the distance from $(3, 1, 4)$, a point on the line, to the plane:

$$\frac{|3 + 2(1) + 4 - 6|}{\sqrt{1 + 4 + 1}} = \frac{3}{\sqrt{6}}.$$

For the distance between parallel planes, choose a point on one plane and find the distance to the other plane. There is also a formula (see Exercise 34). This leaves two distances—point to line and line to line. We illustrate each by example.

EXAMPLE 12.21

Find the distance from the point $(1, 3, 6)$ to the line

$$\ell : \quad x - 1 = \frac{y - 2}{2} = \frac{z - 4}{3}.$$

SOLUTION   Clearly $Q(1, 2, 4)$ is a point on the line, and therefore the required distance d is the component of $\mathbf{PQ}$ in the direction $\mathbf{PR}$ (Figure 12.68). By equation 12.39, then, $d = \mathbf{PQ} \cdot \widehat{\mathbf{PR}}$. To find $\widehat{\mathbf{PR}}$ we need a vector in the direction $\mathbf{PR}$. Since $S(0, 0, 1)$ is also a point on ℓ, $\mathbf{QS} = (-1, -2, -3)$ is a vector along ℓ. But then

$$\mathbf{QS} \times \mathbf{PQ} = \begin{vmatrix} \hat{\mathbf{i}} & \hat{\mathbf{j}} & \hat{\mathbf{k}} \\ -1 & -2 & -3 \\ 0 & -1 & -2 \end{vmatrix} = \hat{\mathbf{i}} - 2\hat{\mathbf{j}} + \hat{\mathbf{k}}$$

is perpendicular to both **PQ** and **QS**. It now follows that a vector in direction **PR** is

$$(\mathbf{QS} \times \mathbf{PQ}) \times \mathbf{QS} = \begin{vmatrix} \hat{\mathbf{i}} & \hat{\mathbf{j}} & \hat{\mathbf{k}} \\ 1 & -2 & 1 \\ -1 & -2 & -3 \end{vmatrix} = 8\hat{\mathbf{i}} + 2\hat{\mathbf{j}} - 4\hat{\mathbf{k}}.$$

Finally, then,

$$\widehat{\mathbf{PR}} = \frac{(8, 2, -4)}{\sqrt{64 + 4 + 16}} = \frac{(4, 1, -2)}{\sqrt{21}}$$

and

$$d = (0, -1, -2) \cdot \frac{(4, 1, -2)}{\sqrt{21}} = \frac{\sqrt{21}}{7}.$$

FIGURE 12.68

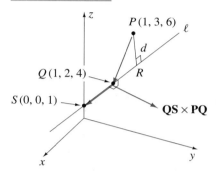

EXAMPLE 12.22

Find the (shortest) distance between the lines

$$\ell_1 : \quad \frac{x - 1}{2} = \frac{y + 3}{3} = z - 4 \text{ and } \ell_2 : \quad x = -1 + t, \ y = 2t, \ z = 3 - 2t.$$

SOLUTION   There will be a point P on ℓ_1 and a point Q on ℓ_2 such that line segment PQ is perpendicular to ℓ_1 and ℓ_2 (Figure 12.69). The length of this line segment is the shortest distance between the lines. There is no problem finding a vector in the same direction as **PQ**, cross vectors $(2, 3, 1)$ and $(1, 2, -2)$ along ℓ_1 and ℓ_2, respectively,

$$\begin{vmatrix} \hat{\mathbf{i}} & \hat{\mathbf{j}} & \hat{\mathbf{k}} \\ 2 & 3 & 1 \\ 1 & 2 & -2 \end{vmatrix} = (-8, 5, 1).$$

FIGURE 12.69

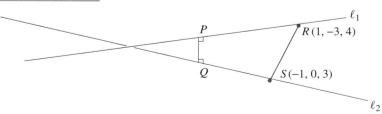

Now comes the hard part to visualize. If we take any point on ℓ_1, say $R(1, -3, 4)$, and any point on ℓ_2, say $S(-1, 0, 3)$, then the component of vector **RS** along **PQ** is the length of **PQ**. (If you are not convinced, consider the following argument. It is easier to see that the component of **PS** along **PQ** is the length of **PQ** so that $|\mathbf{PQ}| = |\mathbf{PS} \cdot \widehat{\mathbf{PQ}}|$. But $\mathbf{PS} = \mathbf{PR} + \mathbf{RS}$, and therefore,

$$|\mathbf{PQ}| = |(\mathbf{PR} + \mathbf{RS}) \cdot \widehat{\mathbf{PQ}}| = |\mathbf{PR} \cdot \widehat{\mathbf{PQ}} + \mathbf{RS} \cdot \widehat{\mathbf{PQ}}|.$$

Since **PR** is perpendicular to $\widehat{\mathbf{PQ}}$, it follows that $\mathbf{PR} \cdot \widehat{\mathbf{PQ}} = 0$, and

$$|\mathbf{PQ}| = |\mathbf{RS} \cdot \widehat{\mathbf{PQ}}|.$$

The right side is the component of **RS** along $\widehat{\mathbf{PQ}}$.) Consequently,

$$|\mathbf{PQ}| = \left| (-2, 3, -1) \cdot \frac{(-8, 5, 1)}{\sqrt{90}} \right| = \sqrt{10}.$$

This procedure fails when the lines are parallel, but in this case, we can pick a point on one line and find the distance to the other line as in Example 12.21.

Work

In Section 7.4, we described work as the product of force and distance. Now that we have represented forces by vectors, we can be more precise. In particular, if a particle moves along the line in Figure 12.70 from P to Q, then vector **PQ** represents its displacement. If a constant force **F** acts on the particle during this motion, then the work done by **F** is defined as

$$W = \mathbf{F} \cdot \mathbf{PQ}. \tag{12.41}$$

It is important to keep in mind exactly when this definition of work can be used: for constant forces acting along straight lines, and by constant **F**, we mean that **F** is constant in both magnitude and direction. Note that when **F** and **PQ** are both in the same direction, then the angle between the vectors is zero. In this case,

$$W = |\mathbf{F}||\mathbf{PQ}| \cos(0) = |\mathbf{F}||\mathbf{PQ}|,$$

and this is essentially the equation dealt with in Section 7.4. When **F** and **PQ** are not in the same direction,

$$W = |\mathbf{F}||\mathbf{PQ}| \cos\theta.$$

Since $|\mathbf{F}| \cos\theta$ is the component of **F** along **PQ**, this equation simply states that when **F** and **PQ** are not in the same direction, $|\mathbf{PQ}|$ should be multiplied by the component of **F** in direction **PQ**.

FIGURE 12.70

EXAMPLE 12.23

If the boy in Figure 12.71 pulls the wagon handle with a force of 10 N at an angle of $\pi/4$ radians with the horizontal, how much work does he do in walking 20 m in a straight line?

SOLUTION   The force **F** exerted by the boy has magnitude $|\mathbf{F}| = 10$, and points in a direction that makes an angle of $\pi/4$ radians with the displacement vector. If **d** is the displacement vector, then

$$W = \mathbf{F} \cdot \mathbf{d} = |\mathbf{F}||\mathbf{d}| \cos\theta = 10(20) \cos\left(\frac{\pi}{4}\right) = \frac{200}{\sqrt{2}} = 100\sqrt{2} \text{ J}.$$

FIGURE 12.71

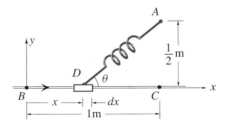

When motion is along a straight line, but **F** is not constant in direction or magnitude or both, we must use integration. The following example illustrates such a situation.

EXAMPLE 12.24

The spring in Figure 12.72 is fixed at A and moves the sleeve frictionlessly along the rod from B to C. If the spring is unstretched when the sleeve is at C, find the work done by the spring.

FIGURE 12.72

SOLUTION   If we set up a coordinate system as shown, then at position D the force **F** exerted by the spring on the sleeve has magnitude

$$|\mathbf{F}| = k \left[\sqrt{(1-x)^2 + \frac{1}{4}} - \frac{1}{2} \right],$$

where k is the spring constant. Since the spring is always stretched during motion, the direction of **F** is along the vector **DA**. Clearly, then, **F** changes in both magnitude and direction as the sleeve moves from B to C. For a small displacement dx at position D, the amount of work done by **F** is (approximately)

$$\mathbf{F} \cdot (dx\hat{\mathbf{i}}) = |\mathbf{F}|dx(\cos\theta) = k\left[\sqrt{(1-x)^2 + \frac{1}{4}} - \frac{1}{2}\right]dx\frac{1-x}{\sqrt{(1-x)^2 + 1/4}}$$

$$= k(1-x)\left[1 - \frac{1/2}{\sqrt{(1-x)^2 + 1/4}}\right]dx.$$

Total work done by the spring as the sleeve moves from B to C must therefore be

$$W = \int_0^1 k(1-x)\left[1 - \frac{1}{2\sqrt{(1-x)^2 + 1/4}}\right]dx$$

$$= k\int_0^1 \left[1 - x - \frac{1-x}{2\sqrt{(1-x)^2 + 1/4}}\right]dx$$

$$= k\left\{x - \frac{x^2}{2} + \frac{1}{2}\sqrt{(1-x)^2 + 1/4}\right\}_0^1$$

$$= \frac{k}{4}(3 - \sqrt{5})\text{ J}.$$

Areas of Triangles and Parallelograms

Vector products provide a simple way to find areas of triangles and parallelograms.

> **THEOREM 12.6**
>
> If A, B, and C are vertices of a triangle, then
>
> $$\text{area of } \triangle ABC = \frac{1}{2}|\mathbf{AB} \times \mathbf{AC}|. \tag{12.42}$$

PROOF   Area of triangle ABC in Figure 12.73 is

$$\frac{1}{2}|\mathbf{AC}||\mathbf{BD}| = \frac{1}{2}|\mathbf{AC}||\mathbf{AB}|\sin\theta = \frac{1}{2}|\mathbf{AB} \times \mathbf{AC}|.$$

For the triangle with vertices $A(1, 1, 1)$, $B(2, -3, 2)$, and $C(4, 1, 5)$ in Figure 12.74,

$$\mathbf{AB} \times \mathbf{AC} = \begin{vmatrix} \hat{\mathbf{i}} & \hat{\mathbf{j}} & \hat{\mathbf{k}} \\ 1 & -4 & 1 \\ 3 & 0 & 4 \end{vmatrix} = (-16, -1, 12).$$

Area of the triangle is therefore

$$\frac{1}{2}|(-16, -1, 12)| = \frac{1}{2}\sqrt{256 + 1 + 144} = \frac{1}{2}\sqrt{401}.$$

FIGURE 12.73

FIGURE 12.74

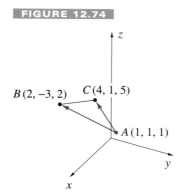

The following corollary is a direct consequence of the previous theorem.

COROLLARY

The area of a parallelogram with coterminal sides **AB** and **AC** (Figure 12.75) is:

$$\text{area} = |\mathbf{AB} \times \mathbf{AC}|. \tag{12.43}$$

FIGURE 12.75

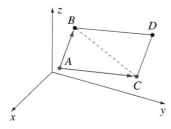

Moments and Torque

Moments play a fundamental role in structural engineering. We encountered them in Section 7.7 in the context of centres of mass, but lacking vectors we could not treat them fully. (There was a hint of vectors when we defined the first moment of an object as its mass multiplied by *directed distance*, directed distances now being associated with vectors.) We discuss first moments of forces here, rather than first moments of masses. Forces were present in Section 7.7, the force of gravity on the masses, but by pushing vectors and forces into the background, we concentrated on the balance concept. Other applications of moments require a vectorial approach.

When a force is to produce a desired effect, where to apply the force may be just as important as the size of the force. For example, it is easier to insert screws using a screwdriver with a large handle than a small one; a long prybar can be more effective in removing spikes than a short one; it is easier to spin a roulette wheel by pulling on its edge rather than somewhere close to its centre. The direction to apply a force is also important. When spinning a roulette wheel, we don't pull away from the centre of the wheel (or toward the centre); we pull in a direction perpendicular to the line joining the centre of the wheel and the point of application of the force. In turning a revolving door, the best direction is perpendicular to the plane of the door. These ideas are captured in the following definition.

> **DEFINITION 12.11**
>
> The **torque** of a force **F**, applied at a point P about a point O (Figure 12.76), is defined as
>
> $$\boldsymbol{\tau} = \mathbf{r} \times \mathbf{F}, \quad \text{where } \mathbf{r} = \overrightarrow{OP}. \tag{12.44}$$

FIGURE 12.76

The magnitude of $\boldsymbol{\tau}$ is

$$|\boldsymbol{\tau}| = |\mathbf{r}||\mathbf{F}| \sin\theta,$$

making it clear that $\boldsymbol{\tau}$ depends on both the point of application (P), and the direction of **F** (angle θ relative to **r**). It is a maximum when **F** is perpendicular to **r**. The direction of $\boldsymbol{\tau}$ is perpendicular to both **r** and **F**. If **r** and **F** are in the plane of the page in Figure 12.76, then $\boldsymbol{\tau}$ points out of the page towards the reader.

Units of torque are newtons multiplied by metres (N·m), the same as for work. For work, a N·m is called a joule. We do not associate joules with torque; torque is a vector, work is a scalar.

EXAMPLE 12.25

What is the torque of the force $\mathbf{F} = 3\hat{\mathbf{i}} - 2\hat{\mathbf{j}} + 8\hat{\mathbf{k}}$ N, applied at the point $(1, 2, -1)$ about the point $(3, 1, 4)$, all distances in metres?

SOLUTION   The torque is

$$\boldsymbol{\tau} = \begin{vmatrix} \hat{\mathbf{i}} & \hat{\mathbf{j}} & \hat{\mathbf{k}} \\ -2 & 1 & -5 \\ 3 & -2 & 8 \end{vmatrix} = -2\hat{\mathbf{i}} + \hat{\mathbf{j}} + \hat{\mathbf{k}} \text{ N·m}.$$

EXAMPLE 12.26

What is the torque of the force $\mathbf{F} = \hat{\mathbf{j}} - \hat{\mathbf{k}}$ N, applied at the point $(3, -2, 3)$ about the point $(3, 1, 0)$, all distances in metres?

SOLUTION   The torque is

$$\boldsymbol{\tau} = \begin{vmatrix} \hat{\mathbf{i}} & \hat{\mathbf{j}} & \hat{\mathbf{k}} \\ 0 & -3 & 3 \\ 0 & 1 & -1 \end{vmatrix} = \mathbf{0}.$$

It is zero because **F** is along the same line as **r**. (This is like trying to close a door by pushing toward the hinges.)

EXERCISES 12.6

In Exercises 1–4 find the area of the triangle with given vertices.

* **1.** $(1, 0)$, $(4, 2)$, $(2, 6)$ * **2.** $(-1, 0, 3)$, $(5, 1, 2)$, $(-6, 2, 4)$

* **3.** $(1, 1, 1)$, $(-3, 4, -2)$, $(-1, -2, 3)$ * **4.** $(1, 2, 3)$, $(3, 5, 10)$, $(-3, -4, -11)$

In Exercises 5 and 6 find the area of the parallelogram with given vertices.

* **5.** $(1, 2, 3)$, $(4, 3, 7)$, $(-1, 3, 6)$, $(2, 4, 10)$

* **6.** $(1, -2, 4)$, $(3, 5, 7)$, $(4, 6, 8)$, $(2, -1, 5)$

In Exercises 7–10 find the component of $\mathbf{v} = (1, -4, 3)$ in the specified direction.

* **7.** $(1, 2, -3)$

* **8.** in the direction from $(-1, 2, 3)$ to $(4, -3, 2)$

* **9.** perpendicular to the plane $x + y + 2z = 4$ in the direction of increasing z

* **10.** along the line $x = 1 - t$, $y = 3 + 2t$, $z = 4 - 3t$ in the direction of increasing x

In Exercises 11–15 calculate the torque of the force about the given point.

* **11.** $\mathbf{F} = 2\hat{\mathbf{i}} + 3\hat{\mathbf{j}} - 4\hat{\mathbf{k}}$ at $(1, 3, 2)$ about $(-1, 4, 2)$

* **12.** $\mathbf{F} = \hat{\mathbf{i}} + 2\hat{\mathbf{j}}$ at $(1, 1, 0)$ about $(2, 1, -5)$

* **13.** $\mathbf{F} = -\hat{\mathbf{i}} + 3\hat{\mathbf{k}}$ at $(0, 0, 0)$ about $(-1, 3, 0)$

* **14.** $\mathbf{F} = 3\hat{\mathbf{i}} - \hat{\mathbf{j}} + 4\hat{\mathbf{k}}$ at $(1, 1, 1)$ about $(2, 2, 2)$

* **15.** $\mathbf{F} = 6\hat{\mathbf{i}}$ at $(0, 1, 3)$ about $(2, 0, 0)$

In Exercises 16–33 find the indicated distance.

* **16.** from the point $(3, 2)$ to the line $2x + 3y = 18$

* **17.** between the lines $x + 4 = y$ and $3x + 7 = 3y$

* **18.** from the point $(1, 3, 4)$ to the plane $x + y - 2z = 0$

* **19.** from the point $(-2, 3, -5)$ to the plane $2x + y + 4z = 6$

* **20.** between the line $x = 1 - t$, $y = 2 + 3t$, $z = 4 - 2t$ and the plane $2x + 4y + 5z = 10$

* **21.** between the line $x + y + z = 6$, $2x - y + z = 3$ and the plane $2x - 7y - z = 5$

* **22.** from the line $x - 1 = 3(y + 4) = -z - 1$ to the plane $2x - 3y + z = 4$

* **23.** from the line $x = 1 - 6t$, $y = 2 + 4t$, $z = -t$ to the plane $x + y - 2z = 1$

* **24.** between the planes $2x + 3y - z = 15$ and $4x + 6y - 2z = 7$

* **25.** between the planes $x - y + 2z = 4$ and $3x - 3y + 6z = 10$

** **26.** from the point $(1, 2, -3)$ to the line $x = 2(y + 1) = (z - 4)/2$

** **27.** from the point $(3, -2, 0)$ to the line $x = t$, $y = 3 - 2t$, $z = 4 + t$

** **28.** from the point $(1, -1, 2)$ to the line $x + y - z = 2$, $2x + 3y + z = 4$

** **29.** from the point $(1, 3, 3)$ to the line $x = 2 - t$, $y = 1 + 2t$, $z = 3$

** **30.** between the lines $(x - 1)/2 = (y + 3)/3 = 4 - z$ and $x = -1 + t$, $y = 2t$, $z = 3 - 2t$

** **31.** between the lines $x = t$, $y = 3t - 1$, $z = 1 + 2t$ and $x = 2t + 1$, $y = 1 - t$, $z = 4 + 2t$

** **32.** between the lines $x + y - z = 4$, $2x - z = 4$ and $x = (y + 1)/2 = (z - 1)/3$

** **33.** between the lines $x + y + z = 2$, $x - y + z = 3$ and $2x - y + z = 3$, $y - 4z = 5$

** **34.** Prove that the (undirected) distance between two parallel planes $Ax + By + Cz + D_1 = 0$ and $Ax + By + Cz + D_2 = 0$ is

$$\frac{|D_1 - D_2|}{\sqrt{A^2 + B^2 + C^2}}.$$

** **35.** Show that the lines joining the midpoints of the sides of any quadrilateral form a parallelo-gram.

** **36.** Show that the three lines $(x - 4)/3 = (y - 8)/4 = (z + 7)/(-4)$; $(x + 5)/3 = (y + 2)/2 = (z + 1)/2$; and $x = 1$, $y = 5 + t$, $z = -6 - 3t$ form a triangle and find its area.

In Exercises 37 and 38 verify that $\hat{\mathbf{v}}$ and $\hat{\mathbf{w}}$ are perpendicular, and then find scalars λ and ρ so that $\mathbf{u} = \lambda\hat{\mathbf{v}} + \rho\hat{\mathbf{w}}$.

** **37.** $\mathbf{u} = (2, 1)$; $\hat{\mathbf{v}} = (1/\sqrt{2}, 1/\sqrt{2})$, $\hat{\mathbf{w}} = (1/\sqrt{2}, -1/\sqrt{2})$

** **38.** $\mathbf{u} = 3\hat{\mathbf{i}} - 2\hat{\mathbf{j}}$; $\hat{\mathbf{v}} = (\hat{\mathbf{i}} - 2\hat{\mathbf{j}})/\sqrt{5}$, $\hat{\mathbf{w}} = (2\hat{\mathbf{i}} + \hat{\mathbf{j}})/\sqrt{5}$

In Exercises 39 and 40 verify that $\hat{\mathbf{u}}$, $\hat{\mathbf{v}}$, and $\hat{\mathbf{w}}$ are mutually perpendicular, and then find scalars λ, ρ, and μ so that $\mathbf{r} = \lambda\hat{\mathbf{u}} + \rho\hat{\mathbf{v}} + \mu\hat{\mathbf{w}}$.

** **39.** $\mathbf{r} = (1, 3, -4)$; $\hat{\mathbf{u}} = (2, 1, 0)/\sqrt{5}$, $\hat{\mathbf{v}} = (-1, 2, 3)/\sqrt{14}$, $\hat{\mathbf{w}} = (3, -6, 5)/\sqrt{70}$

** **40.** $\mathbf{r} = 2\hat{\mathbf{i}} - \hat{\mathbf{k}}$; $\hat{\mathbf{u}} = (\hat{\mathbf{i}} + \hat{\mathbf{j}} + \hat{\mathbf{k}})/\sqrt{3}$, $\hat{\mathbf{v}} = (\hat{\mathbf{i}} + \hat{\mathbf{j}} - 2\hat{\mathbf{k}})/\sqrt{6}$, $\hat{\mathbf{w}} = (\hat{\mathbf{i}} - \hat{\mathbf{j}})/\sqrt{2}$

** **41.** If $\mathbf{u} = (3, 2)$, $\mathbf{v} = (1, -3)$, $\mathbf{w} = (6, 2)$, verify that $\mathbf{v}$ and $\mathbf{w}$ are perpendicular, and find scalars λ and ρ so that $\mathbf{u} = \lambda\mathbf{v} + \rho\mathbf{w}$. Be sure to recognize that $\mathbf{v}$ and $\mathbf{w}$ are not unit vectors.

** **42.** If $\mathbf{u} = (1, 0, 1)$, $\mathbf{v} = (1, 1, -1)$, $\mathbf{w} = (-1, 2, 1)$, and $\mathbf{r} = (-2, -3, 4)$, verify that $\mathbf{u}$, $\mathbf{v}$, and $\mathbf{w}$ are mutually perpendicular, and find scalars λ, ρ, and μ so that $\mathbf{r} = \lambda\mathbf{u} + \rho\mathbf{v} + \mu\mathbf{w}$. As in Exercise 41, take into account that $\mathbf{u}$, $\mathbf{v}$, and $\mathbf{w}$ are not unit vectors.

** **43.** Repeat Example 12.24 (page 895) if the spring has an unstretched length l that is less than the length of AC.

** **44.** Two positive charges q_1 and q_2 are placed in the xy-plane at positions $(5, 5)$ and $(-2, 3)$ respectively. A third positive charge q_3 is moved along the x-axis from $x = 1$ to $x = -1$. Find the total work done by the electrostatic forces of q_1 and q_2 on q_3.

** **45.** When the rocket in the figure below passes close to the spherical asteroid, it is attracted to the asteroid by a gravitational force with magnitude GmM/r^2 where m and M are the masses of the rocket and asteroid, r is the distance from the rocket to the centre of the asteroid, and G is Newton's gravitational constant. Determine the work the rocket must do against this force in order to follow the straight-line path from A to B.

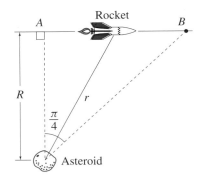

** **46.** If $\mathbf{F}$ is a constant force, show that the work done by $\mathbf{F}$ on an object moving around any closed polygon is zero.

** **47.** Find the equation of a plane normal to $\hat{\mathbf{i}} - 2\hat{\mathbf{j}} + 3\hat{\mathbf{k}}$ and two units from the point $(1, 2, 3)$.

** **48.** If a, b, and c (all positive constants) are the intercepts of a plane with the x-, y-, and z-axes, and p is the length of the perpendicular from the origin to the plane, show that

$$\frac{1}{p^2} = \frac{1}{a^2} + \frac{1}{b^2} + \frac{1}{c^2}.$$

** **49.** Two springs with constants k and $2k$ are joined together at C and have their other ends fixed at points A and B in the figure below. When their join is at C, neither spring is compressed or stretched. If the join is pulled along the straight line CD perpendicular to AB, what work is done?

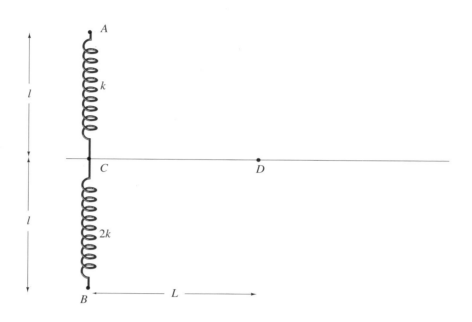

12.7 Differentiation and Integration of Vectors

In Section 12.6, vectors were used to represent forces and moments; they can also be used to describe many other physical quantities, such as position, velocity, acceleration, electric and magnetic fields, and fluid flow. In applications such as these, vectors seldom have constant components; instead they have components that are either functions of position, or functions of some parameter, such as time, or both. For instance, the spring force in Example 12.24 varies in both magnitude and direction as the sleeve moves from B to C. Consequently, components of the vector **F** representing this force are functions of position x between B and C:

$$\mathbf{F} = F_x\hat{\mathbf{i}} + F_y\hat{\mathbf{j}} = F_x(x)\hat{\mathbf{i}} + F_y(x)\hat{\mathbf{j}}.$$

When a particle moves along a curve C in the xy-plane defined parametrically by

$$C: \quad x = x(t), \quad y = y(t), \quad \alpha \leq t \leq \beta,$$

its position (x, y) relative to the origin is represented by the vector $\mathbf{r} = x\hat{\mathbf{i}} + y\hat{\mathbf{j}}$ (Figure 12.77). This vector is called the **position vector** or **displacement vector** of the particle relative to the origin. We will have more to say about it in Section 12.11. Note, however, that if we substitute from the parametric equations for C, we have

$$\mathbf{r} = x(t)\hat{\mathbf{i}} + y(t)\hat{\mathbf{j}},$$

which indicates that the displacement vector has components that are functions of the parameter t.

FIGURE 12.77

In this section we consider the general situation in which the components v_x, v_y, and v_z of a vector $\mathbf{v}$ are functions of some parameter t,

$$\mathbf{v} = v_x(t)\hat{\mathbf{i}} + v_y(t)\hat{\mathbf{j}} + v_z(t)\hat{\mathbf{k}}, \tag{12.45}$$

and show how the operations of differentiation and integration can be applied to such a vector. In Sections 12.9 and 12.10 we will use these results to discuss the geometry of curves, and this will pave the way for an analysis of the motion of particles in Section 12.11.

Because $\mathbf{v}$ in 12.45 has components that are functions of t, we say that $\mathbf{v}$ itself is a function of t, and write $\mathbf{v} = \mathbf{v}(t)$. Each of the component functions $v_x(t)$, $v_y(t)$, and $v_z(t)$ has a domain, and their common domain is called the domain of the vector function $\mathbf{v}(t)$. Given that this domain is some interval $\alpha \leq t \leq \beta$, we express 12.45 more fully in the form

$$\mathbf{v} = \mathbf{v}(t) = v_x(t)\hat{\mathbf{i}} + v_y(t)\hat{\mathbf{j}} + v_z(t)\hat{\mathbf{k}}, \quad \alpha \leq t \leq \beta. \tag{12.46}$$

To differentiate and integrate vector functions, we first require the concept of a limit.

DEFINITION 12.12

If $\mathbf{v}(t) = v_x(t)\hat{\mathbf{i}} + v_y(t)\hat{\mathbf{j}} + v_z(t)\hat{\mathbf{k}}$, then

$$\lim_{t \to t_0} \mathbf{v}(t) = \left[\lim_{t \to t_0} v_x(t)\right]\hat{\mathbf{i}} + \left[\lim_{t \to t_0} v_y(t)\right]\hat{\mathbf{j}} + \left[\lim_{t \to t_0} v_z(t)\right]\hat{\mathbf{k}}, \tag{12.47}$$

provided each of the limits on the right exists.

This definition states that to take the limit of a vector function, we take the limit of each component separately. As an illustration, consider the following example.

EXAMPLE 12.27

If $\mathbf{v} = \mathbf{v}(t) = (t^2 + 1)\hat{\mathbf{i}} + 3t\hat{\mathbf{j}} - (\sin t)\hat{\mathbf{k}}$, calculate $\lim_{t \to 5} \mathbf{v}(t)$.

SOLUTION   According to Definition 12.12,

$$\lim_{t \to 5} \mathbf{v}(t) = \left[\lim_{t \to 5}(t^2 + 1)\right]\hat{\mathbf{i}} + \left[\lim_{t \to 5}(3t)\right]\hat{\mathbf{j}} - \left[\lim_{t \to 5}\sin t\right]\hat{\mathbf{k}}$$

$$= 26\hat{\mathbf{i}} + 15\hat{\mathbf{j}} - (\sin 5)\hat{\mathbf{k}}.$$

In the remainder of this section we define continuity, derivatives, and antiderivatives for vector functions. Each definition is an exact duplicate of the corresponding definition for a scalar function $y(t)$, except that $y(t)$ is replaced by $\mathbf{v}(t)$. We then show that the vector definition can be rephrased in terms of components of the vector. We begin with continuity in the following definition.

> **DEFINITION 12.13**
>
> A vector function $\mathbf{v}(t)$ is said to be **continuous** at a point t_0 if
>
> $$\mathbf{v}(t_0) = \lim_{t \to t_0} \mathbf{v}(t). \tag{12.48}$$

It is a simple matter to prove the next theorem.

> **THEOREM 12.7**
>
> A vector function is continuous at a point if and only if its components are continuous at that point.

PROOF If $\mathbf{v}(t) = v_x(t)\hat{\mathbf{i}} + v_y(t)\hat{\mathbf{j}} + v_z(t)\hat{\mathbf{k}}$, then according to 12.48, $\mathbf{v}(t)$ is continuous at t_0 if and only if

$$v_x(t_0)\hat{\mathbf{i}} + v_y(t_0)\hat{\mathbf{j}} + v_z(t_0)\hat{\mathbf{k}} = \lim_{t \to t_0} \{v_x(t)\hat{\mathbf{i}} + v_y(t)\hat{\mathbf{j}} + v_z(t)\hat{\mathbf{k}}\}.$$

Definition 12.12 implies that we can write this condition in the form

$$v_x(t_0)\hat{\mathbf{i}} + v_y(t_0)\hat{\mathbf{j}} + v_z(t_0)\hat{\mathbf{k}} = \left[\lim_{t \to t_0} v_x(t)\right]\hat{\mathbf{i}} + \left[\lim_{t \to t_0} v_y(t)\right]\hat{\mathbf{j}} + \left[\lim_{t \to t_0} v_z(t)\right]\hat{\mathbf{k}}.$$

But because two vectors are equal if and only if their components are equal, we can say that $\mathbf{v}(t)$ is continuous at t_0 if and only if

$$v_x(t_0) = \lim_{t \to t_0} v_x(t); \quad v_y(t_0) = \lim_{t \to t_0} v_y(t); \quad v_z(t_0) = \lim_{t \to t_0} v_z(t);$$

i.e., $\mathbf{v}(t)$ is continuous at t_0 if and only if its components are continuous at t_0. ∎

For example, the vector function

$$\mathbf{v}(t) = (t - 1)\hat{\mathbf{i}} + (1/t)\hat{\mathbf{j}} + (t^2 - 1)^{-1}\hat{\mathbf{k}}$$

is discontinuous for $t = 0$ (since $v_y(0)$ is not defined) and for $t = \pm 1$ (since $v_z(\pm 1)$ is not defined).

The derivative of a scalar function $y(t)$ is its instantaneous rate of change:

$$\frac{dy}{dt} = \lim_{h \to 0} \frac{y(t + h) - y(t)}{h}.$$

The derivative of a vector function is also a rate of change defined by a similar limit.

DEFINITION 12.14

The derivative of a vector function $\mathbf{v}(t)$ is defined as

$$\frac{d\mathbf{v}}{dt} = \lim_{h \to 0} \frac{\mathbf{v}(t+h) - \mathbf{v}(t)}{h}, \tag{12.49}$$

provided the limit exists.

In practice, we seldom use the definition of a derivative to calculate dy/dt for a scalar function $y(t)$; formulas such as the power, product, quotient, and chain rules are more convenient. It would be helpful to have corresponding formulas for derivatives of vector functions. The following theorem shows that to differentiate a vector function we simply differentiate its Cartesian components.

THEOREM 12.8

If $\mathbf{v}(t) = v_x(t)\hat{\mathbf{i}} + v_y(t)\hat{\mathbf{j}} + v_z(t)\hat{\mathbf{k}}$, then

$$\frac{d\mathbf{v}}{dt} = \frac{dv_x}{dt}\hat{\mathbf{i}} + \frac{dv_y}{dt}\hat{\mathbf{j}} + \frac{dv_z}{dt}\hat{\mathbf{k}}, \tag{12.50}$$

provided the derivatives on the right exist.

PROOF If we substitute the components of $\mathbf{v}(t+h)$ and $\mathbf{v}(t)$ into Definition 12.14, then we have

$$\frac{d\mathbf{v}}{dt} = \lim_{h \to 0} \left\{ \frac{[v_x(t+h)\hat{\mathbf{i}} + v_y(t+h)\hat{\mathbf{j}} + v_z(t+h)\hat{\mathbf{k}}] - [v_x(t)\hat{\mathbf{i}} + v_y(t)\hat{\mathbf{j}} + v_z(t)\hat{\mathbf{k}}]}{h} \right\}$$

$$= \lim_{h \to 0} \left\{ \left[\frac{v_x(t+h) - v_x(t)}{h} \right]\hat{\mathbf{i}} + \left[\frac{v_y(t+h) - v_y(t)}{h} \right]\hat{\mathbf{j}} \right.$$

$$\left. + \left[\frac{v_z(t+h) - v_z(t)}{h} \right]\hat{\mathbf{k}} \right\}$$

(according to equation 12.14), and

$$\frac{d\mathbf{v}}{dt} = \left(\lim_{h \to 0} \frac{v_x(t+h) - v_x(t)}{h} \right)\hat{\mathbf{i}} + \left(\lim_{h \to 0} \frac{v_y(t+h) - v_y(t)}{h} \right)\hat{\mathbf{j}}$$

$$+ \left(\lim_{h \to 0} \frac{v_z(t+h) - v_z(t)}{h} \right)\hat{\mathbf{k}}$$

(according to equation 12.47). Since each of the limits on the right exists, we can write that

$$\frac{d\mathbf{v}}{dt} = \frac{dv_x}{dt}\hat{\mathbf{i}} + \frac{dv_y}{dt}\hat{\mathbf{j}} + \frac{dv_z}{dt}\hat{\mathbf{k}}.$$

Theorem 12.8 gives us a working rule for differentiating vector functions: To differentiate a vector function, we differentiate its Cartesian components.

EXAMPLE 12.28

If $\mathbf{v}(t) = t^2\hat{\mathbf{i}} + (3t^3 - 2t)\hat{\mathbf{j}} + 5\hat{\mathbf{k}}$, find $\mathbf{v}'(3)$.

SOLUTION   According to 12.50,

$$\frac{d\mathbf{v}}{dt} = 2t\hat{\mathbf{i}} + (9t^2 - 2)\hat{\mathbf{j}}.$$

Consequently,

$$\mathbf{v}'(3) = 6\hat{\mathbf{i}} + 79\hat{\mathbf{j}}.$$

The sum rule 3.9 for differentiation of scalar functions has its counterpart in the sum rule for vector functions,

$$\frac{d}{dt}(\mathbf{u} + \mathbf{v}) = \frac{d\mathbf{u}}{dt} + \frac{d\mathbf{v}}{dt} \tag{12.51}$$

(see Exercise 22). There are three types of products associated with vectors: the product of a scalar and a vector, the dot product of two vectors, and the cross product of two vectors. Corresponding to each, we have a product rule for differentiation, but all resemble the product rule for scalar functions.

THEOREM 12.9

If $f(t)$ is a differentiable function and $\mathbf{u}(t)$ and $\mathbf{v}(t)$ are differentiable vector functions, then:

$$\frac{d}{dt}(f\mathbf{v}) = \frac{df}{dt}\mathbf{v} + f\frac{d\mathbf{v}}{dt}, \tag{12.52}$$

$$\frac{d}{dt}(\mathbf{u} \cdot \mathbf{v}) = \mathbf{u} \cdot \frac{d\mathbf{v}}{dt} + \frac{d\mathbf{u}}{dt} \cdot \mathbf{v}, \tag{12.53}$$

$$\frac{d}{dt}(\mathbf{u} \times \mathbf{v}) = \mathbf{u} \times \frac{d\mathbf{v}}{dt} + \frac{d\mathbf{u}}{dt} \times \mathbf{v}. \tag{12.54}$$

For a proof of these results, see Exercise 23.

EXAMPLE 12.29

If $f(t) = t^2 + 2t + 3$, $\mathbf{u}(t) = t\hat{\mathbf{i}} + t^2\hat{\mathbf{j}} - 3\hat{\mathbf{k}}$, and $\mathbf{v}(t) = t(\hat{\mathbf{i}} + \hat{\mathbf{j}} + \hat{\mathbf{k}})$, use (12.52)–(12.54) to evaluate:

(a) $\dfrac{d}{dt}(f\mathbf{u})$;   (b) $\dfrac{d}{dt}(\mathbf{u} \cdot \mathbf{v})$;   (c) $\dfrac{d}{dt}(\mathbf{u} \times \mathbf{v})$.

SOLUTION

(a) With 12.52,

$$\frac{d}{dt}(f\mathbf{u}) = \frac{df}{dt}\mathbf{u} + f\frac{d\mathbf{u}}{dt} = (2t+2)(t\hat{\mathbf{i}} + t^2\hat{\mathbf{j}} - 3\hat{\mathbf{k}}) + (t^2 + 2t + 3)(\hat{\mathbf{i}} + 2t\hat{\mathbf{j}})$$

$$= (3t^2 + 4t + 3)\hat{\mathbf{i}} + (4t^3 + 6t^2 + 6t)\hat{\mathbf{j}} - 6(t+1)\hat{\mathbf{k}}.$$

(b) With 12.53,

$$\frac{d}{dt}(\mathbf{u} \cdot \mathbf{v}) = \mathbf{u} \cdot \frac{d\mathbf{v}}{dt} + \frac{d\mathbf{u}}{dt} \cdot \mathbf{v}$$

$$= (t\hat{\mathbf{i}} + t^2\hat{\mathbf{j}} - 3\hat{\mathbf{k}}) \cdot (\hat{\mathbf{i}} + \hat{\mathbf{j}} + \hat{\mathbf{k}}) + (\hat{\mathbf{i}} + 2t\hat{\mathbf{j}}) \cdot (t\hat{\mathbf{i}} + t\hat{\mathbf{j}} + t\hat{\mathbf{k}})$$

$$= (t + t^2 - 3) + (t + 2t^2) = 3t^2 + 2t - 3.$$

(c) With 12.54,

$$\frac{d}{dt}(\mathbf{u} \times \mathbf{v}) = \mathbf{u} \times \frac{d\mathbf{v}}{dt} + \frac{d\mathbf{u}}{dt} \times \mathbf{v} = (t\hat{\mathbf{i}} + t^2\hat{\mathbf{j}} - 3\hat{\mathbf{k}}) \times (\hat{\mathbf{i}} + \hat{\mathbf{j}} + \hat{\mathbf{k}})$$

$$+ (\hat{\mathbf{i}} + 2t\hat{\mathbf{j}}) \times (t\hat{\mathbf{i}} + t\hat{\mathbf{j}} + t\hat{\mathbf{k}})$$

$$= \begin{vmatrix} \hat{\mathbf{i}} & \hat{\mathbf{j}} & \hat{\mathbf{k}} \\ t & t^2 & -3 \\ 1 & 1 & 1 \end{vmatrix} + \begin{vmatrix} \hat{\mathbf{i}} & \hat{\mathbf{j}} & \hat{\mathbf{k}} \\ 1 & 2t & 0 \\ t & t & t \end{vmatrix}$$

$$= [(t^2 + 3)\hat{\mathbf{i}} - (3 + t)\hat{\mathbf{j}} + (t - t^2)\hat{\mathbf{k}}]$$

$$+ [2t^2\hat{\mathbf{i}} - t\hat{\mathbf{j}} + (t - 2t^2)\hat{\mathbf{k}}]$$

$$= (3t^2 + 3)\hat{\mathbf{i}} - (3 + 2t)\hat{\mathbf{j}} + (2t - 3t^2)\hat{\mathbf{k}}.$$

If vector functions can be differentiated, then they can be antidifferentiated. Formally we make the following statement.

DEFINITION 12.15

A vector function $\mathbf{V}(t)$ is said to be an **antiderivative** or **indefinite integral** of $\mathbf{v}(t)$ on the interval $\alpha < t < \beta$ if

$$\frac{d\mathbf{V}}{dt} = \mathbf{v}(t) \text{ for } \alpha < t < \beta. \tag{12.55}$$

For example, an antiderivative of $\mathbf{v}(t) = 2t\hat{\mathbf{i}} - \hat{\mathbf{j}} + 3t^2\hat{\mathbf{k}}$ is

$$\mathbf{V}(t) = t^2\hat{\mathbf{i}} - t\hat{\mathbf{j}} + t^3\hat{\mathbf{k}}.$$

If we add to $\mathbf{V}(t)$ in 12.55 any vector with constant components, denoted by $\mathbf{C}$, then $\mathbf{V}(t) + \mathbf{C}$ is also an antiderivative of $\mathbf{v}(t)$. We call this vector *the* antiderivative or *the* indefinite integral of $\mathbf{v}(t)$, and write

$$\int \mathbf{v}(t)dt = \mathbf{V}(t) + \mathbf{C}. \tag{12.56}$$

For our example, then,

$$\int (2t\hat{\mathbf{i}} - \hat{\mathbf{j}} + 3t^2\hat{\mathbf{k}})dt = t^2\hat{\mathbf{i}} - t\hat{\mathbf{j}} + t^3\hat{\mathbf{k}} + \mathbf{C}.$$

Because vectors can be differentiated component by component, it follows that they may also be integrated component by component; i.e., if $\mathbf{v}(t) = v_x(t)\hat{\mathbf{i}} + v_y(t)\hat{\mathbf{j}} + v_z(t)\hat{\mathbf{k}}$, then

$$\int \mathbf{v}(t)dt = \left(\int v_x(t)dt\right)\hat{\mathbf{i}} + \left(\int v_y(t)dt\right)\hat{\mathbf{j}} + \left(\int v_z(t)dt\right)\hat{\mathbf{k}}. \tag{12.57}$$

EXAMPLE 12.30

Find the antiderivative of $\mathbf{v}(t) = \sqrt{t-1}\,\hat{\mathbf{i}} + e^t\hat{\mathbf{j}} + 6t^2\hat{\mathbf{k}}$.

SOLUTION   According to equation 12.57,

$$\int \mathbf{v}(t)dt = \left[\frac{2}{3}(t-1)^{3/2} + C_1\right]\hat{\mathbf{i}} + (e^t + C_2)\hat{\mathbf{j}} + (2t^3 + C_3)\hat{\mathbf{k}}$$

$$= \frac{2}{3}(t-1)^{3/2}\hat{\mathbf{i}} + e^t\hat{\mathbf{j}} + 2t^3\hat{\mathbf{k}} + \mathbf{C},$$

where $\mathbf{C} = C_1\hat{\mathbf{i}} + C_2\hat{\mathbf{j}} + C_3\hat{\mathbf{k}}$ is a constant vector.

EXERCISES 12.7

In Exercises 1–5 find the largest possible domain for the vector function.

∗ **1.** $\mathbf{v}(t) = t^2\hat{\mathbf{i}} + \sqrt{t-1}\hat{\mathbf{j}} + \hat{\mathbf{k}}$ ∗ **2.** $\mathbf{v}(t) = (\sin t)\hat{\mathbf{i}} + (\cos t)\hat{\mathbf{j}} - t^3\hat{\mathbf{k}}$

∗ **3.** $\mathbf{v}(t) = (\text{Sin}^{-1} t)\hat{\mathbf{i}} - t^2\hat{\mathbf{j}} + (t+1)\hat{\mathbf{k}}$ ∗ **4.** $\mathbf{v}(t) = \ln(t+4)(\hat{\mathbf{i}} + \hat{\mathbf{j}})$

∗ **5.** $\mathbf{v}(t) = e^t\hat{\mathbf{i}} + (\cos^2 t)\hat{\mathbf{j}} - (e^t \cos t)\hat{\mathbf{k}}$

If $f(t) = t^2 + 3$, $g(t) = 2t^3 - 3t$, $\mathbf{u}(t) = t\hat{\mathbf{i}} - t^2\hat{\mathbf{j}} + 2t\hat{\mathbf{k}}$, and $\mathbf{v}(t) = \hat{\mathbf{i}} - 2t\hat{\mathbf{j}} + 3t^2\hat{\mathbf{k}}$, find the scalar or the components of the vector in Exercises 6–21.

* **6.** $\dfrac{d\mathbf{u}}{dt}$

* **7.** $\dfrac{d}{dt}[f(t)\mathbf{v}(t)]$

* **8.** $\dfrac{d}{dt}[g(t)\mathbf{u}(t)]$

* **9.** $\dfrac{d}{dt}(\mathbf{u} \times \mathbf{v})$

* **10.** $\dfrac{d}{dt}(\mathbf{u} \times t\mathbf{v})$

* **11.** $\dfrac{d}{dt}(2\mathbf{u} \cdot \mathbf{v})$

* **12.** $\dfrac{d}{dt}(3\mathbf{u} + 4\mathbf{v})$

* **13.** $\displaystyle\int \mathbf{u}(t)\, dt$

* **14.** $\dfrac{d}{dt}[f(t)\mathbf{u} + g(t)\mathbf{v}]$

* **15.** $\displaystyle\int 4\mathbf{v}(t)\, dt$

* **16.** $\dfrac{d}{dt}[t(\mathbf{u} \times \mathbf{v})]$

* **17.** $\displaystyle\int [f(t)\mathbf{u}(t)]\, dt$

* **18.** $\displaystyle\int [3g(t)\mathbf{v}(t) + \mathbf{u}(t)]\, dt$

* **19.** $\displaystyle\int [f(t)\mathbf{u} \cdot \mathbf{v}]\, dt$

* **20.** $\mathbf{u} \times \dfrac{d\mathbf{v}}{dt} - f(t)\mathbf{u} \cdot \dfrac{d\mathbf{v}}{dt}\mathbf{v}$

* **21.** $\mathbf{u} \cdot \dfrac{d\mathbf{v}}{dt} - \mathbf{v} \cdot \displaystyle\int \mathbf{u}(t)\, dt$

* **22.** Prove equation 12.51.

* **23.** Verify the results in equations 12.52–12.54.

* **24.** Prove that for differentiable functions $\mathbf{u}(t)$, $\mathbf{v}(t)$, and $\mathbf{w}(t)$,

$$\frac{d}{dt}(\mathbf{u} \cdot \mathbf{v} \times \mathbf{w}) = \frac{d\mathbf{u}}{dt} \cdot \mathbf{v} \times \mathbf{w} + \mathbf{u} \cdot \frac{d\mathbf{v}}{dt} \times \mathbf{w} + \mathbf{u} \cdot \mathbf{v} \times \frac{d\mathbf{w}}{dt}.$$

* **25.** Prove that if a differentiable function $\mathbf{v}(t)$ has constant length, then at any point at which $d\mathbf{v}/dt \neq \mathbf{0}$, the vector $d\mathbf{v}/dt$ is perpendicular to $\mathbf{v}$.

* **26.** If $\mathbf{v} = \mathbf{v}(s)$ is a differentiable vector function and $s = s(t)$ is a differentiable scalar function, prove that

$$\frac{d\mathbf{v}}{dt} = \frac{d\mathbf{v}}{ds}\frac{ds}{dt}.$$

This result is called the *chain rule* for differentiation of vector functions.

* **27.** Show that the following definition for the limit of a vector function is equivalent to Definition 12.12: A vector function $\mathbf{v}(t)$ is said to have limit $\mathbf{V}$ as t approaches t_0 if, given any $\epsilon > 0$, there exists a $\delta > 0$ such that $|\mathbf{v}(t) - \mathbf{V}| < \epsilon$ whenever $0 < |t - t_0| < \delta$.

12.8 Parametric and Vector Representations of Curves

In Section 12.2, we presented curves in space as the intersection of two surfaces. For example, each of the equations

$$x^2 + y^2 + z^2 = 9, \quad y = 2$$

describes a surface (the first is a sphere and the second a plane), and together they describe the curve of intersection of the surfaces—the circle in Figure 12.78.

FIGURE 12.78

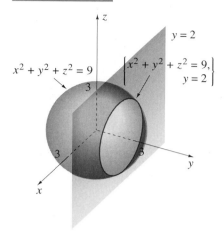

Parametric Representation of Curves

In many applications it is more convenient to have a curve defined *parametrically*.

DEFINITION 12.16

A curve in space is defined parametrically by three functions:

$$C: \quad x = x(t), \quad y = y(t), \quad z = z(t), \quad \alpha \leq t \leq \beta. \tag{12.58}$$

Each value of t in the interval $\alpha \leq t \leq \beta$ is substituted into the three functions, and the triple $(x, y, z) = (x(t), y(t), z(t))$ specifies a point on the curve. Definition 12.16 clearly corresponds to parametric Definition 10.1 for a plane curve.

It is customary to assign a direction to a curve by calling that point on C corresponding to $t = \alpha$ the initial point and that point corresponding to $t = \beta$ the final point, and the direction of C is that direction along C from initial point to final point (Figure 12.79). Note in particular that the direction of a curve always corresponds to the direction in which the parameter *increases* along the curve. Because of this, whenever we describe a curve in nonparametric form but with a specified direction, we must be careful in setting up parametric equations to ensure that the parameter increases in the appropriate direction.

FIGURE 12.79

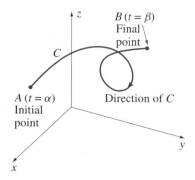

When a curve is described as the curve of intersection of two surfaces, we can obtain parametric equations for the curve by specifying one of x, y, or z as a function of t, and then solving the equations for the other two as functions of t. Considerable ingenuity is sometimes required in arriving at an initial function of t. We illustrate this in the following example.

EXAMPLE 12.31

Find parametric equations for each of the following curves:

    (a) $z - 1 = x^2 + y^2$, $x - y = 0$ directed so that z increases when x and y are positive;

    (b) $x + 2y + z = 4$, $2x + y + 3z = 6$ directed so that y increases along the curve;

    (c) $x^2 + (y - 1)^2 = 4$, $z = x$ directed so that y increases when x is positive.

SOLUTION

    (a) The curve of intersection of the circular paraboloid $z = 1 + x^2 + y^2$ and the plane $y = x$ is shown in Figure 12.80. If we choose x as the parameter along the curve by setting $x = t$, then $y = x$ and $z = 1 + x^2 + y^2$ imply that

$$x = t, \quad y = t, \quad z = 1 + 2t^2.$$

When $t > 0$, so also are x and y, and for these values of t, z increases as t increases. This means that these are acceptable parametric equations for the curve.

FIGURE 12.80

FIGURE 12.81

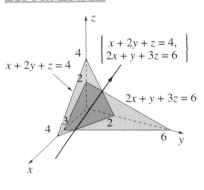

    (b) The straight-line intersection of the two planes is shown in Figure 12.81. If we choose y as the parameter by setting $y = t$ (thus forcing y to increase as t increases), then

$$x + z = 4 - 2t, \quad 2x + 3z = 6 - t.$$

The solution of these equations for x and z in terms of t gives the parametric equations

$$x = 6 - 5t, \quad y = t, \quad z = -2 + 3t.$$

    (c) The curve of intersection of the right circular cylinder $x^2 + (y-1)^2 = 4$ and the plane $z = x$ is shown in Figure 12.82. We saw in Section 10.1 that trigonometric functions are particularly useful for circles. If we set $x = 2\cos t$, then $y = 1 \pm 2\sin t$. A set of parametric equations for the curve is therefore

$$x = 2\cos t, \quad y = 1 + 2\sin t, \quad z = 2\cos t, \quad 0 \le t \le 2\pi.$$

Any range of values of t of length 2π traces the curve exactly once. To check that these equations specify the correct direction along the curve, we note that $t = 0$ gives the point $(2, 1, 2)$ and $t = \pi/2$ gives $(0, 3, 0)$. Since values of t between 0 and $\pi/2$ give one quarter of the curve (rather than three quarters), points are indeed generated in the required direction indicated by the arrowhead in Figure 12.82. Had we chosen the equations

$$x = 2\cos t, \quad y = 1 - 2\sin t, \quad z = 2\cos t, \quad 0 \le t \le 2\pi,$$

we would have generated the same set of points traced in the opposite direction.

FIGURE 12.82

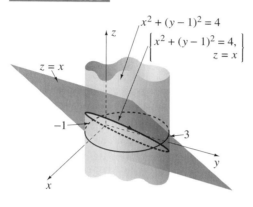

Because computers and graphing calculators produce excellent plots of parametrically-defined, three-dimensional curves, there is incentive to represent curves parametrically rather than as the intersection of two surfaces. Plots of the curves in Example 12.31 are shown in Figure 12.83.

FIGURE 12.83a   **FIGURE 12.83b**   **FIGURE 12.83c**

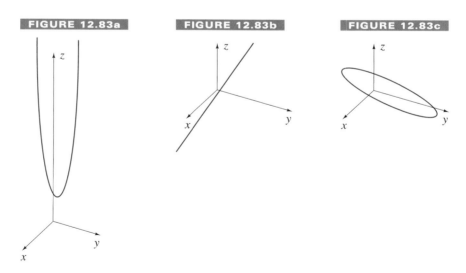

DEFINITION 12.17

A curve C: $x = x(t)$, $y = y(t)$, $z = z(t)$, $\alpha \le t \le \beta$, is said to be **continuous** if each of the functions $x(t)$, $y(t)$, and $z(t)$ is continuous for $\alpha \le t \le \beta$.

Geometrically, this implies that the curve is at no point separated. Each of the curves in Example 12.31 is therefore continuous.

A curve is said to be **closed** if its initial and final points are the same. Circles and ellipses are closed curves. Straight-line segments, parabolas, and hyperbolas are not closed.

Vector Representation of Curves

The position vector or displacement vector of a point $P(x, y, z)$ in space is

$$\mathbf{r} = (x, y, z) = x\hat{\mathbf{i}} + y\hat{\mathbf{j}} + z\hat{\mathbf{k}}.$$

We visualize it as the vector drawn from the origin to P (Figure 12.84). If we consider only points that lie on a curve defined parametrically by 12.58, then for these points we can write that

$$\mathbf{r} = \mathbf{r}(t) = x(t)\hat{\mathbf{i}} + y(t)\hat{\mathbf{j}} + z(t)\hat{\mathbf{k}}, \quad \alpha \leq t \leq \beta. \tag{12.59}$$

As t varies from $t = \alpha$ to $t = \beta$, the tip of this vector traces the curve C from initial point to final point. We call 12.59 the **vector representation** of a curve.

FIGURE 12.84

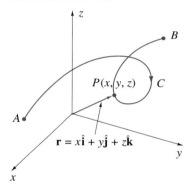

EXAMPLE 12.32

Draw and plot the curve with position vector

$$\mathbf{r} = \mathbf{r}(t) = (2\cos t)\hat{\mathbf{i}} + (3\sin t)\hat{\mathbf{j}} + (t/10)\hat{\mathbf{k}}, \quad 0 \leq t \leq 6\pi.$$

SOLUTION   When we set $x = 2\cos t$, $y = 3\sin t$, and $z = (t/10)$, it is clear that $x^2/4 + y^2/9 = 1$. This means that the curve lines on the elliptic cylinder $x^2/4 + y^2/9 = 1$. As t increases from 0 to 6π, values of z increase linearly from 0 to $3\pi/5$. What we have therefore is three loops that rise around the elliptic cylinder. It is part of what is called an *elliptic helix*. It is drawn in Figure 12.85a and plotted in Figure 12.85b. If the curve is given width, it could represent three coils of a spring or three windings of an inductor.

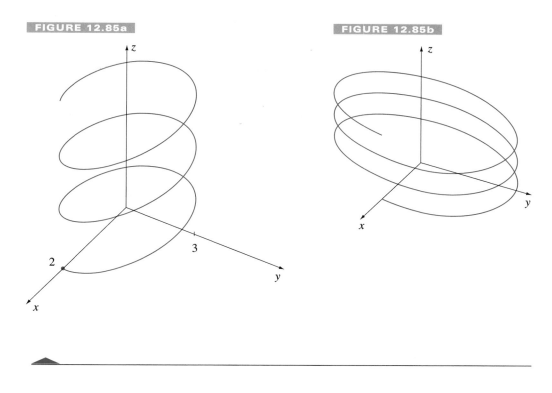

FIGURE 12.85a

FIGURE 12.85b

EXAMPLE 12.33

Draw and plot the curve with position vector

$$\mathbf{r} = \mathbf{r}(t) = t\hat{\mathbf{i}} + t^2\hat{\mathbf{j}} + t\hat{\mathbf{k}}, \quad t \geq 0.$$

SOLUTION   When we set $x = t$, $y = t^2$, and $z = t$, then $z = x$ and $y = x^2$. These imply that $\mathbf{r} = \mathbf{r}(t)$ describes points on the curve of intersection of the surfaces $y = x^2$ and $z = x$ (Figure 12.86a). Because $t \geq 0$, only that half of the curve of intersection in the first octant is defined by $\mathbf{r} = \mathbf{r}(t)$. The curve is plotted in Figure 12.86b.

FIGURE 12.86a

FIGURE 12.86b

In Exercises 1–10 find parametric and vector representations for the curve. Draw and plot each curve.

* **1.** $x + 2y + 3z = 6$, $y - 2z = 3$ directed so that z increases along the curve

* **2.** $x^2 + y^2 = 2$, $z = 4$ directed so that y increases in the first octant

* **3.** $x^2 + y^2 = 2$, $x + y + z = 1$ directed so that y decreases when x is positive

* **4.** $z = x^2 + y^2$, $x^2 + y^2 = 5$ directed clockwise as viewed from the origin

* **5.** $z + 2x^2 = 1$, $y = z$ directed so that x decreases along the curve

* **6.** $z = \sqrt{x^2 + y^2}$, $y = x$ directed so that y increases when x is positive

* **7.** $z = x + y$, $y = x^2$ directed so that x increases along the curve

** **8.** $z = \sqrt{4 - x^2 - y^2}$, $x^2 + y^2 - 2y = 0$ directed so that z decreases when x is positive

** **9.** $x = \sqrt{z}$, $z = y^2$ directed away from the origin in the first octant

** **10.** $z = \sqrt{x^2 + y^2}$, $y = x^2$ directed so that y decreases in the first octant.

In Exercises 11–15 draw and plot the curve with the given position vector.

* **11.** $\mathbf{r}(t) = t\hat{\mathbf{i}} + t\hat{\mathbf{j}} + t^2\hat{\mathbf{k}}$, $t \geq 0$

* **12.** $\mathbf{r}(t) = (2\cos t)\hat{\mathbf{i}} + (2\sin t)\hat{\mathbf{j}} + 3t\hat{\mathbf{k}}$, $0 \leq t \leq 4\pi$

* **13.** $\mathbf{r}(t) = (t - 2)\hat{\mathbf{i}} + (2 - 3t)\hat{\mathbf{j}} + 5t\hat{\mathbf{k}}$

* **14.** $\mathbf{r}(t) = (t^2 - t)\hat{\mathbf{i}} + t\hat{\mathbf{j}} + 5\hat{\mathbf{k}}$

* **15.** $\mathbf{r}(t) = (\cos t)\hat{\mathbf{i}} + (\sin t)\hat{\mathbf{j}} + (\cos t)\hat{\mathbf{k}}$, $0 \leq t \leq \pi$

12.9 Tangent Vectors and Lengths of Curves

If C is a curve in the xy-plane (Figure 12.87), then the tangent line to C at P is defined as the limiting position of the line PQ as Q moves along C toward P (see Section 3.1). We take the same approach in defining tangent vectors to curves in an arbitrary plane or in space. On curve C defined by 12.58 (page 909), we let P and Q be the points corresponding to the parameter values t and $t + h$. Position vectors of P and Q are then

$$\mathbf{r}(t) = x(t)\hat{\mathbf{i}} + y(t)\hat{\mathbf{j}} + z(t)\hat{\mathbf{k}}$$

and

$$\mathbf{r}(t + h) = x(t + h)\hat{\mathbf{i}} + y(t + h)\hat{\mathbf{j}} + z(t + h)\hat{\mathbf{k}}$$

(Figure 12.88), and the vector joining P to Q is

$$\mathbf{PQ} = \mathbf{r}(t + h) - \mathbf{r}(t).$$

FIGURE 12.87

FIGURE 12.87

FIGURE 12.88

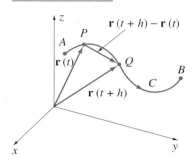

If we let h approach zero, then Q moves along C toward P, and the direction of **PQ** becomes closer to what seems to be a reasonable definition of the tangent direction to C at P. Perhaps then we should define $\lim_{h \to 0}[\mathbf{r}(t+h) - \mathbf{r}(t)]$ as a tangent vector to C at P. Unfortunately, the limit vector has length zero, and therefore

$$\lim_{h \to 0} [\mathbf{r}(t+h) - \mathbf{r}(t)] = \mathbf{0}.$$

If, however, we divide $\mathbf{r}(t+h) - \mathbf{r}(t)$ by h, then the resulting vector

$$\frac{\mathbf{r}(t+h) - \mathbf{r}(t)}{h}$$

is not equal to **PQ**, but it does have the same direction as **PQ**. Consider, then, taking the limit of this vector as h approaches zero:

$$\lim_{h \to 0} \frac{\mathbf{r}(t+h) - \mathbf{r}(t)}{h}.$$

If the limit vector exists, then it will be tangent to C at P. But according to equation 12.49, this limit defines the derivative $d\mathbf{r}/dt$,

$$\frac{d\mathbf{r}}{dt} = \lim_{h \to 0} \frac{\mathbf{r}(t+h) - \mathbf{r}(t)}{h} = \frac{dx}{dt}\hat{\mathbf{i}} + \frac{dy}{dt}\hat{\mathbf{j}} + \frac{dz}{dt}\hat{\mathbf{k}}, \tag{12.60}$$

provided each of the derivatives dx/dt, dy/dt, and dz/dt exists. We have just established the following result.

THEOREM 12.10

If $\mathbf{r} = \mathbf{r}(t) = x(t)\hat{\mathbf{i}} + y(t)\hat{\mathbf{j}} + z(t)\hat{\mathbf{k}}$, $\alpha \leq t \leq \beta$, is the vector representation of a curve C, then at any point on C at which $x'(t)$, $y'(t)$, and $z'(t)$ all exist and do not simultaneously vanish,

$$\mathbf{T} = \frac{d\mathbf{r}}{dt} = \frac{dx}{dt}\hat{\mathbf{i}} + \frac{dy}{dt}\hat{\mathbf{j}} + \frac{dz}{dt}\hat{\mathbf{k}} \tag{12.61}$$

is a tangent vector to C (Figure 12.89).

FIGURE 12.89

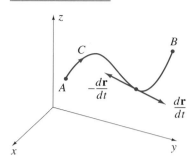

FIGURE 12.90

There are two tangent directions at any point on a curve. One of these has been shown to be $d\mathbf{r}/dt$; the other must be $-d\mathbf{r}/dt$ (Figure 12.90). How can we tell which one is $d\mathbf{r}/dt$? A closer analysis of the limit in 12.60 indicates the following (see Exercise 17).

COROLLARY

The tangent vector $d\mathbf{r}/dt$ to a curve C: $x = x(t)$, $y = y(t)$, $z = z(t)$, $\alpha \leq t \leq \beta$, always points in the direction in which the parameter t increases along C.

DEFINITION 12.18

A curve C: $x = x(t)$, $y = y(t)$, $z = z(t)$, $\alpha \leq t \leq \beta$, is said to be **smooth** if the derivatives $x'(t)$, $y'(t)$, and $z'(t)$ are all continuous for $\alpha < t < \beta$ and do not simultaneously vanish for $\alpha < t < \beta$.

Since $x'(t)$, $y'(t)$, and $z'(t)$ are the components of a tangent vector to C, this definition implies that along a smooth curve, small changes in t produce small changes in the direction of the tangent vector. In other words, the tangent vector turns gradually, or "smoothly." The curve in Figure 12.91 is smooth; that in Figure 12.92 is not because abrupt changes in the direction of the curve occur at P and Q. According to the following definition, this curve is piecewise smooth.

FIGURE 12.91

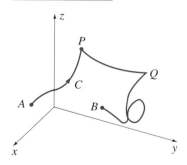

FIGURE 12.92

DEFINITION 12.19

A curve is said to be **piecewise smooth** if it is continuous and can be divided into a finite number of smooth subcurves.

EXAMPLE 12.34

For the curve in Example 12.33, find a tangent vector at the point $(3, 9, 3)$.

SOLUTION   A tangent vector to this curve at any point on the curve is

$$\frac{d\mathbf{r}}{dt} = \frac{dx}{dt}\hat{\mathbf{i}} + \frac{dy}{dt}\hat{\mathbf{j}} + \frac{dz}{dt}\hat{\mathbf{k}} = \hat{\mathbf{i}} + 2t\hat{\mathbf{j}} + \hat{\mathbf{k}}.$$

Since $t = 3$ yields the point $(3, 9, 3)$, a tangent vector at this point is $\mathbf{r}'(3) = \hat{\mathbf{i}} + 6\hat{\mathbf{j}} + \hat{\mathbf{k}}$.

EXAMPLE 12.35

Find a tangent vector at the point $(2, 0, 3)$ to the helix

$$x = 2\cos t, \quad y = 2\sin t, \quad z = \frac{3t}{2\pi}, \quad t \geq 0.$$

Is the helix smooth?

FIGURE 12.93

SOLUTION   A tangent vector to the helix at any point is

$$\frac{d\mathbf{r}}{dt} = \frac{dx}{dt}\hat{\mathbf{i}} + \frac{dy}{dt}\hat{\mathbf{j}} + \frac{dz}{dt}\hat{\mathbf{k}} = (-2\sin t)\hat{\mathbf{i}} + (2\cos t)\hat{\mathbf{j}} + \left(\frac{3}{2\pi}\right)\hat{\mathbf{k}}.$$

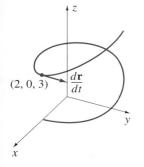

Since $t = 2\pi$ yields the point $(2, 0, 3)$, a tangent vector at this point is $\mathbf{r}'(2\pi) = 2\hat{\mathbf{j}} + (3/(2\pi))\hat{\mathbf{k}}$ (Figure 12.93). Since $x'(t)$, $y'(t)$, and $z'(t)$ are all continuous functions, and they are never simultaneously zero, the helix is indeed smooth.

Unit Tangent Vectors

When a curve in the xy-plane is defined parametrically by

$$C: \quad x = x(t), \quad y = y(t), \quad \alpha \leq t \leq \beta, \tag{12.62}$$

a tangent vector to C is

$$\mathbf{T} = \frac{d\mathbf{r}}{dt} = \frac{dx}{dt}\hat{\mathbf{i}} + \frac{dy}{dt}\hat{\mathbf{j}}, \tag{12.63}$$

and this tangent vector points in the direction in which t increases along C. To produce a unit tangent vector to C at any point, we divide $\mathbf{T}$ by its length:

$$\hat{\mathbf{T}} = \frac{\mathbf{T}}{|\mathbf{T}|} = \frac{d\mathbf{r}/dt}{|d\mathbf{r}/dt|}. \tag{12.64}$$

We now show that if length along C is used as the parameter by which to specify its points, then division by $|\mathbf{T}|$ is unnecessary.

In Section 7.3 we showed that small lengths along a plane curve C can be approximated by straight-line lengths along tangent lines to the curve, and that the total length of a smooth curve from A to B (Figure 12.94) is

$$L = \int_A^B \sqrt{(dx)^2 + (dy)^2}.$$

FIGURE 12.94

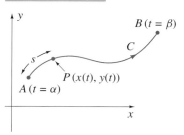

With parametric equations 12.62 we can write this formula as a definite integral with respect to t (see also equation 10.3):

$$L = \int_\alpha^\beta \sqrt{\left[\left(\frac{dx}{dt}\right)^2 + \left(\frac{dy}{dt}\right)^2\right](dt)^2} = \int_\alpha^\beta \sqrt{\left(\frac{dx}{dt}\right)^2 + \left(\frac{dy}{dt}\right)^2}\, dt. \qquad (12.65)$$

Furthermore, if we denote by $s = s(t)$ the length of that part of C from its initial point A (where $t = \alpha$) to any point $P(x(t), y(t))$ on C (Figure 12.94), then $s(t)$ is defined by the integral

$$s(t) = \int_\alpha^t \sqrt{\left(\frac{dx}{dt}\right)^2 + \left(\frac{dy}{dt}\right)^2}\, dt. \qquad (12.66)$$

It follows, then, that the derivative of $s(t)$ is

$$\frac{ds}{dt} = \sqrt{\left(\frac{dx}{dt}\right)^2 + \left(\frac{dy}{dt}\right)^2}. \qquad (12.67)$$

But according to 12.63, $d\mathbf{r}/dt$ is a tangent vector to C with the same length:

$$\left|\frac{d\mathbf{r}}{dt}\right| = \sqrt{\left(\frac{dx}{dt}\right)^2 + \left(\frac{dy}{dt}\right)^2} = \frac{ds}{dt}. \qquad (12.68)$$

When this equation is multiplied by dt, it gives

$$|d\mathbf{r}| = \sqrt{(dx)^2 + (dy)^2} = ds. \qquad (12.69)$$

FIGURE 12.95

This equation states that ds is the length of the tangent vector $d\mathbf{r} = dx\,\hat{\mathbf{i}} + dy\,\hat{\mathbf{j}}$, and therefore ds is a measure of length along the tangent line to C. In spite of this we often think of ds as a measure of small lengths along C itself (Figure 12.95), and that ds is approximated by the tangential straight-line length

$$|d\mathbf{r}| = \sqrt{(dx)^2 + (dy)^2}.$$

Note too that if we use length s along C as a parameter, then the chain rule applied to $\mathbf{r} = \mathbf{r}(s)$, $s = s(t)$ gives

$$\frac{d\mathbf{r}}{dt} = \frac{d\mathbf{r}}{ds}\frac{ds}{dt}. \tag{12.70}$$

(The chain rule is proved in Exercise 26 of Section 12.7.) Consequently, equation 12.68 implies that

$$\frac{d\mathbf{r}}{ds} = \frac{d\mathbf{r}/dt}{ds/dt} = \frac{d\mathbf{r}/dt}{|d\mathbf{r}/dt|}. \tag{12.71}$$

What we have shown, then, is that if we choose length along a curve C as the parameter by which to specify points on the curve (C: $x = x(s)$, $y = y(s)$), then the vector

$$\hat{\mathbf{T}} = \frac{d\mathbf{r}}{ds} = \frac{dx}{ds}\hat{\mathbf{i}} + \frac{dy}{ds}\hat{\mathbf{j}} \tag{12.72}$$

is a unit tangent vector to C. In addition, the corollary to Theorem 12.10 implies that $d\mathbf{r}/ds$ points in the direction in which s increases along C. This suggests perhaps that we should always set up parametric equations for a curve with length along the curve as parameter. Theoretically this is quite acceptable, but practically it is impossible. For most curves we have enough difficulty just finding a set of parametric equations, let alone finding that set with length along the curve as parameter. If we then use a parameter t other than length along the curve, a unit tangent vector is calculated according to 12.64.

These results can be extended to space curves as well. When a smooth curve C has parametric equations 12.58, equation 12.64 still defines a unit tangent vector to C, but because C is a space curve, $d\mathbf{r}/dt$ is calculated according to 12.61.

Corresponding to formula 12.66 for length along a curve in the xy-plane, length along a smooth curve in space is defined by the definite integral:

$$s(t) = \int_{\alpha}^{t} \sqrt{\left(\frac{dx}{dt}\right)^2 + \left(\frac{dy}{dt}\right)^2 + \left(\frac{dz}{dt}\right)^2}\,dt. \tag{12.73}$$

These two results imply that

$$\left|\frac{d\mathbf{r}}{dt}\right| = \sqrt{\left(\frac{dx}{dt}\right)^2 + \left(\frac{dy}{dt}\right)^2 + \left(\frac{dz}{dt}\right)^2} = \frac{ds}{dt}, \tag{12.74}$$

the three-space analogue of 12.68. Once again we are led to the fact that when s is used as parameter along C, then

$$\hat{\mathbf{T}} = \frac{d\mathbf{r}}{ds} = \frac{dx}{ds}\hat{\mathbf{i}} + \frac{dy}{ds}\hat{\mathbf{j}} + \frac{dz}{ds}\hat{\mathbf{k}} \tag{12.75}$$

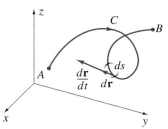

is a unit tangent vector to C. In addition, multiplication of 12.74 by dt yields

$$|d\mathbf{r}| = ds = \sqrt{(dx)^2 + (dy)^2 + (dz)^2}, \tag{12.76}$$

indicating that small lengths ds along C (Figure 12.96) are defined in terms of small lengths $|d\mathbf{r}|$ along the tangent line to C.

EXAMPLE 12.36

Find a unit tangent vector to the curve

$$C : \quad x = \sin t, \quad y = 2\cos t, \quad z = 2t/\pi, \quad t \geq 0$$

at the point $(0, -2, 2)$.

SOLUTION A tangent vector to C at any point is

$$\frac{d\mathbf{r}}{dt} = \frac{dx}{dt}\hat{\mathbf{i}} + \frac{dy}{dt}\hat{\mathbf{j}} + \frac{dz}{dt}\hat{\mathbf{k}} = \cos t\hat{\mathbf{i}} - 2\sin t\hat{\mathbf{j}} + \left(\frac{2}{\pi}\right)\hat{\mathbf{k}}.$$

Since $t = \pi$ yields the point $(0, -2, 2)$, a tangent vector at this point is

$$\mathbf{r}'(\pi) = -\hat{\mathbf{i}} + (2/\pi)\hat{\mathbf{k}}.$$

A unit tangent vector is then

$$\hat{\mathbf{T}} = \frac{-\hat{\mathbf{i}} + (2/\pi)\hat{\mathbf{k}}}{\sqrt{1 + 4/\pi^2}} = \frac{-\pi\hat{\mathbf{i}} + 2\hat{\mathbf{k}}}{\sqrt{4 + \pi^2}}.$$

EXAMPLE 12.37

Find the length of that part of the curve $x = y^{2/3}$, $x = z^{2/3}$ between the points $(0, 0, 0)$ and $(4, 8, 8)$.

SOLUTION If we use $x = t$, $y = t^{3/2}$, $z = t^{3/2}$, $0 \leq t \leq 4$, as parametric equations for the curve, then

$$L = \int_0^4 \sqrt{\left(\frac{dx}{dt}\right)^2 + \left(\frac{dy}{dt}\right)^2 + \left(\frac{dz}{dt}\right)^2} \, dt = \int_0^4 \sqrt{1 + \left(\frac{3}{2}\sqrt{t}\right)^2 + \left(\frac{3}{2}\sqrt{t}\right)^2} \, dt$$

$$= \int_0^4 \sqrt{1 + \frac{9t}{2}} \, dt = \left\{\frac{4}{27}\left(1 + \frac{9t}{2}\right)^{3/2}\right\}_0^4 = \frac{4}{27}(19\sqrt{19} - 1).$$

EXERCISES 12.9

In Exercises 1–5 express the curve in vector form and find the unit tangent vector $\hat{\mathbf{T}}$ at each point on the curve.

* **1.** $x = \sin t$, $y = \cos t$, $z = t$, $-\infty < t < \infty$

* **2.** $x = t$, $y = t^2$, $z = t^3$, $t \geq 1$

* **3.** $x = (t - 1)^2$, $y = (t + 1)^2$, $z = -t$, $-3 \leq t \leq 4$

* **4.** $x + y = 5$, $x^2 - y = z$ from $(5, 0, 25)$ to $(0, 5, -5)$

* **5.** $x + y + z = 4$, $x^2 + y^2 = 4$, $y \geq 0$ from $(2, 0, 2)$ to $(-2, 0, 6)$

In Exercises 6–10 find $\hat{\mathbf{T}}$ at the point.

* **6.** $x = 4\cos t,\ y = 6\sin t,\ z = 2\sin t,\ -\infty < t < \infty;\ (2\sqrt{2}, 3\sqrt{2}, \sqrt{2})$

* **7.** $x = 2 - 5t,\ y = 1 + t,\ z = 6 + 4t,\ -\infty < t < \infty;\ (7, 0, 2)$

* **8.** $x^2 + y^2 + z^2 = 4,\ z = \sqrt{x^2 + y^2}$, directed so that x increases when y is positive; $(1, 1, \sqrt{2})$

* **9.** $x = y^2 + 1,\ z = x + 5$, directed so that y increases along the curve; $(5, 2, 10)$

* **10.** $x^2 + (y - 1)^2 = 4,\ z = x$, directed so that z decreases when y is negative; $(2, 1, 2)$

In Exercises 11–14 find the length of the curve. Draw each curve.

* **11.** $x = 2\cos t,\ y = 2\sin t,\ z = 3t,\ 0 \le t \le 2\pi$

* **12.** $x = 2 - 5t,\ y = 1 + t,\ z = 6 + 4t,\ -1 \le t \le 0$

* **13.** $x = t^3,\ y = t^2,\ z = t^3,\ 0 \le t \le 1$

* **14.** $x = t,\ y = t^{3/2},\ z = 4t^{3/2},\ 1 \le t \le 4$

** **15.** In Definition 12.18 why are the derivatives assumed not to vanish simultaneously? Hint: Consider the curve $x = t^3,\ y = t^2,\ z = 0$.

** **16.** Find a unit tangent vector to the curve $\mathbf{r} = (\cos t + t\sin t)\hat{\mathbf{i}} + (\sin t - t\cos t)\hat{\mathbf{j}},\ 0 \le t \le 2\pi$, called an involute of a circle.

** **17.** Show that the tangent vector $d\mathbf{r}/dt$ to a curve described by equation 12.59 always points in the direction in which t increases along the curve.

** **18.** (a) What happens when equation 12.61 is used to determine a tangent vector to the curve $x = t^2,\ y = t^3,\ z = t^2,\ -\infty < t < \infty$, at the origin?

    (b) Can you devise a way in which to obtain a tangent vector?

12.10 Normal Vectors, Curvature, and Radius of Curvature

In discussing curves we distinguish between two types of properties: intrinsic and not intrinsic. An intrinsic property is one that is independent of the parameter used to specify the curve; a property that is not intrinsic is parameter-dependent. To illustrate, the tangent vector $\mathbf{T} = d\mathbf{r}/dt$ in 12.61 is not intrinsic; a change of parameter results in a change in the length of $\mathbf{T}$. The unit tangent vector $\hat{\mathbf{T}}$, on the other hand, is intrinsic; there is only one unit tangent vector in the direction of the curve. Length of a curve from its initial point to an arbitrary point is an intrinsic property; a change of parameter along the curve does not affect length between points.

Because length along a curve is an intrinsic property, it is customary in theoretical discussions to use it as the parameter by which to specify points on the curve. When C is a smooth curve in the xy-plane, parametric equations for C in terms of length s along C take the form

$$C: \quad x = x(s), \quad y = y(s), \quad 0 \le s \le L. \tag{12.77}$$

Normal Vectors to Curves

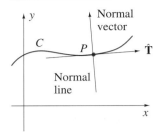

The normal line at a point P on a smooth curve C in the xy-plane is that line which is perpendicular to the tangent line to C at P (Figure 12.97). Any vector along this normal line is said to be a normal vector to the curve at P. Since the unit tangent vector to C at P is

$$\hat{\mathbf{T}} = \frac{d\mathbf{r}}{ds} = \frac{dx}{ds}\hat{\mathbf{i}} + \frac{dy}{ds}\hat{\mathbf{j}},$$

it follows that

$$\hat{\mathbf{N}} = -\frac{dy}{ds}\hat{\mathbf{i}} + \frac{dx}{ds}\hat{\mathbf{j}} \tag{12.78}$$

is a unit normal vector to C at P (note that $\hat{\mathbf{T}} \cdot \hat{\mathbf{N}} = 0$). Because there is only one direction normal to C at P, every normal vector to C at P must be some multiple $\lambda\hat{\mathbf{N}}$ of $\hat{\mathbf{N}}$.

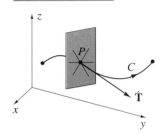

The situation is quite different for space curves (Figure 12.98). If $\hat{\mathbf{T}}$ is the unit tangent vector to a smooth curve C, then there is an entire plane of normal vectors to C at P. In the following discussion, we single out two normal vectors called the **principal normal** and the **binormal**. Suppose that C is defined parametrically by

$$C: \quad x = x(s), \quad y = y(s), \quad z = z(s), \quad 0 \le s \le L, \tag{12.79}$$

and that $\hat{\mathbf{T}}$ is the unit tangent vector to C defined by 12.75. Because $\hat{\mathbf{T}}$ has unit length,

$$1 = \hat{\mathbf{T}} \cdot \hat{\mathbf{T}}.$$

If we use equation 12.53 to differentiate this equation with respect to s, we have

$$0 = \frac{d\hat{\mathbf{T}}}{ds} \cdot \hat{\mathbf{T}} + \hat{\mathbf{T}} \cdot \frac{d\hat{\mathbf{T}}}{ds} = 2\left(\hat{\mathbf{T}} \cdot \frac{d\hat{\mathbf{T}}}{ds}\right).$$

But if neither of the vectors $\hat{\mathbf{T}}$ nor $d\hat{\mathbf{T}}/ds$ is equal to zero, then the fact that their scalar product is equal to zero implies that they are perpendicular. In other words,

$$\mathbf{N} = \frac{d\hat{\mathbf{T}}}{ds} \tag{12.80}$$

is a normal vector to C at any point. The unit normal vector in this direction

$$\hat{\mathbf{N}} = \frac{\mathbf{N}}{|\mathbf{N}|} = \frac{d\hat{\mathbf{T}}/ds}{|d\hat{\mathbf{T}}/ds|} \tag{12.81}$$

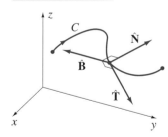

is called the **principal normal** (vector) to C (Figure 12.99).

Because $\hat{\mathbf{N}}$ is defined in terms of intrinsic properties $\hat{\mathbf{T}}$ and s for a curve, it must also be an intrinsic property. It follows, then, that no matter what parameter is used to specify points on a curve, $\hat{\mathbf{N}}$ is always the same. But how do we find $\hat{\mathbf{N}}$ when a curve C is specified in terms of a parameter other than length along C, say, in the form

$$C: \quad x = x(t), \quad y = y(t), \quad z = z(t), \quad \alpha \le t \le \beta? \tag{12.82}$$

If $s = s(t)$ is length along C (measured from $t = \alpha$), then by the chain rule

$$\frac{d\hat{\mathbf{T}}}{ds} = \frac{d\hat{\mathbf{T}}}{dt}\frac{dt}{ds},$$

where dt/ds must be positive since both s and t increase along C. Consequently,

$$\hat{\mathbf{N}} = \frac{d\hat{\mathbf{T}}/ds}{|d\hat{\mathbf{T}}/ds|} = \frac{(d\hat{\mathbf{T}}/dt)(dt/ds)}{|(d\hat{\mathbf{T}}/dt)(dt/ds)|} = \frac{(d\hat{\mathbf{T}}/dt)(dt/ds)}{|d\hat{\mathbf{T}}/dt|dt/ds} = \frac{d\hat{\mathbf{T}}/dt}{|d\hat{\mathbf{T}}/dt|}. \qquad (12.83)$$

In other words, for any parametrization of C whatsoever, the vector $d\hat{\mathbf{T}}/dt$ always points in the direction of the principal normal, and to find $\hat{\mathbf{N}}$, we simply find the unit vector in the direction of $d\hat{\mathbf{T}}/dt$.

In the study of space curves, a second normal vector to C, called the **binormal**, is defined by

$$\hat{\mathbf{B}} = \hat{\mathbf{T}} \times \hat{\mathbf{N}}. \qquad (12.84)$$

Since the cross product of two vectors is always perpendicular to each of the vectors, it follows that the binormal is perpendicular to both $\hat{\mathbf{T}}$ and $\hat{\mathbf{N}}$, and must therefore indeed be a normal vector to C (Figure 12.99).

We have singled out three vectors at each point P on a curve C: a unit tangent vector $\hat{\mathbf{T}}$ and two unit normal vectors $\hat{\mathbf{N}}$ and $\hat{\mathbf{B}}$. As P moves along C, these vectors constantly change direction but always have unit length.

EXAMPLE 12.38

Find $\hat{\mathbf{T}}$, $\hat{\mathbf{N}}$, and $\hat{\mathbf{B}}$ for the curve $x = t$, $y = t^2$, $z = t^2$, $t \geq 0$.

SOLUTION   The unit tangent vector $\hat{\mathbf{T}}$ is defined by

$$\hat{\mathbf{T}} = \frac{d\mathbf{r}/dt}{|d\mathbf{r}/dt|} = \frac{(1, 2t, 2t)}{\sqrt{1 + 4t^2 + 4t^2}} = \frac{(1, 2t, 2t)}{\sqrt{1 + 8t^2}}.$$

The principal normal $\hat{\mathbf{N}}$ lies along the vector $\mathbf{N} = d\hat{\mathbf{T}}/dt$, and according to equation 12.52, we can write that

$$\begin{aligned} \mathbf{N} = \frac{d\hat{\mathbf{T}}}{dt} &= \frac{d}{dt}\left(\frac{1}{\sqrt{1 + 8t^2}}\right)(1, 2t, 2t) + \frac{1}{\sqrt{1 + 8t^2}}\frac{d}{dt}(1, 2t, 2t) \\ &= \frac{-8t}{(1 + 8t^2)^{3/2}}(1, 2t, 2t) + \frac{1}{\sqrt{1 + 8t^2}}(0, 2, 2) \\ &= \frac{1}{(1 + 8t^2)^{3/2}}\{-8t(1, 2t, 2t) + (1 + 8t^2)(0, 2, 2)\} \\ &= \frac{(-8t, 2, 2)}{(1 + 8t^2)^{3/2}}. \end{aligned}$$

The principal normal is therefore

$$\hat{\mathbf{N}} = \frac{\mathbf{N}}{|\mathbf{N}|} = \frac{(-8t, 2, 2)}{\sqrt{64t^2 + 4 + 4}} = \frac{(-4t, 1, 1)}{\sqrt{2 + 16t^2}}.$$

The binormal is

$$\hat{\mathbf{B}} = \hat{\mathbf{T}} \times \hat{\mathbf{N}} = \frac{(1, 2t, 2t)}{\sqrt{1 + 8t^2}} \times \frac{(-4t, 1, 1)}{\sqrt{2 + 16t^2}}$$

$$= \frac{1}{\sqrt{1 + 8t^2}\sqrt{2 + 16t^2}} \begin{vmatrix} \hat{\mathbf{i}} & \hat{\mathbf{j}} & \hat{\mathbf{k}} \\ 1 & 2t & 2t \\ -4t & 1 & 1 \end{vmatrix}$$

$$= \frac{1}{\sqrt{2}\sqrt{1 + 8t^2}\sqrt{1 + 8t^2}}(0, -1 - 8t^2, 1 + 8t^2)$$

$$= \frac{1 + 8t^2}{\sqrt{2}(1 + 8t^2)}(0, -1, 1)$$

$$= \frac{(0, -1, 1)}{\sqrt{2}}.$$

The significance of the fact that the binormal has constant direction can be seen from a drawing of the curve. Because the parametric equations imply that $y = x^2$ and $z = y$, the curve is the curve of intersection of these two surfaces (Figure 12.100). Since the curve lies in the plane $-y + z = 0$, and a normal vector to this plane is $(0, -1, 1)$, it follows that $(0, -1, 1)$ is always normal to the curve. But this is precisely the direction of $\hat{\mathbf{B}}$. In other words, constant $\hat{\mathbf{B}}$ implies that the curve lies in a plane that has $\hat{\mathbf{B}}$ as normal (see Exercise 31).

FIGURE 12.100

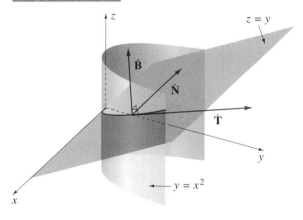

EXAMPLE 12.39

Show that for a smooth curve C: $x = x(t)$, $y = y(t)$, $\alpha \le t \le \beta$ in the xy-plane, the principal normal is

$$\hat{\mathbf{N}} = \text{sgn}\left(\frac{dy}{dt}\frac{d^2x}{dt^2} - \frac{dx}{dt}\frac{d^2y}{dt^2}\right)\frac{(dy/dt, -dx/dt)}{\sqrt{(dx/dt)^2 + (dy/dt)^2}},$$

where the signum function $\text{sgn}(u)$ is defined by

$$\text{sgn}(u) = \begin{cases} 1, & \text{if } u > 0, \\ 0, & \text{if } u = 0, \\ -1, & \text{if } u < 0. \end{cases}$$

SOLUTION   The unit tangent vector to C is

$$\hat{\mathbf{T}} = \frac{(dx/dt, dy/dt)}{\sqrt{(dx/dt)^2 + (dy/dt)^2}}.$$

For simplicity in notation, we use a dot "$\cdot$" above a variable to indicate that the variable is differentiated with respect to t. For example, $\dot{x} = dx/dt$ and $\ddot{x} = d^2x/dt^2$. With this notation,

$$\hat{\mathbf{T}} = \frac{(\dot{x}, \dot{y})}{\sqrt{\dot{x}^2 + \dot{y}^2}}.$$

By equation 12.8, $\hat{\mathbf{N}} = (d\hat{\mathbf{T}}/dt)/|d\hat{\mathbf{T}}/dt|$, where

$$\frac{d\hat{\mathbf{T}}}{dt} = \frac{d}{dt}\frac{(\dot{x}, \dot{y})}{\sqrt{\dot{x}^2 + \dot{y}^2}} = \frac{d}{dt}\left(\frac{1}{\sqrt{\dot{x}^2 + \dot{y}^2}}\right)(\dot{x}\hat{\mathbf{i}} + \dot{y}\hat{\mathbf{j}}) + \frac{1}{\sqrt{\dot{x}^2 + \dot{y}^2}}(\ddot{x}\hat{\mathbf{i}} + \ddot{y}\hat{\mathbf{j}})$$

$$= \left(\frac{-\dot{x}\ddot{x} - \dot{y}\ddot{y}}{(\dot{x}^2 + \dot{y}^2)^{3/2}}\right)(\dot{x}\hat{\mathbf{i}} + \dot{y}\hat{\mathbf{j}}) + \frac{1}{\sqrt{\dot{x}^2 + \dot{y}^2}}(\ddot{x}\hat{\mathbf{i}} + \ddot{y}\hat{\mathbf{j}})$$

$$= \frac{1}{(\dot{x}^2 + \dot{y}^2)^{3/2}}\{-(\dot{x}\ddot{x} + \dot{y}\ddot{y})(\dot{x}\hat{\mathbf{i}} + \dot{y}\hat{\mathbf{j}}) + (\dot{x}^2 + \dot{y}^2)(\ddot{x}\hat{\mathbf{i}} + \ddot{y}\hat{\mathbf{j}})\}$$

$$= \frac{1}{(\dot{x}^2 + \dot{y}^2)^{3/2}}\{(-\dot{x}^2\ddot{x} - \dot{x}\dot{y}\ddot{y} + \dot{x}^2\ddot{x} + \dot{y}^2\ddot{x})\hat{\mathbf{i}}$$

$$+ (-\dot{x}\dot{y}\ddot{x} - \dot{y}^2\ddot{y} + \dot{x}^2\ddot{y} + \dot{y}^2\ddot{y})\hat{\mathbf{j}}\}$$

$$= \frac{1}{(\dot{x}^2 + \dot{y}^2)^{3/2}}\{\dot{y}(\dot{y}\ddot{x} - \dot{x}\ddot{y})\hat{\mathbf{i}} + \dot{x}(\dot{x}\ddot{y} - \dot{y}\ddot{x})\hat{\mathbf{j}}\}$$

$$= \frac{\dot{y}\ddot{x} - \dot{x}\ddot{y}}{(\dot{x}^2 + \dot{y}^2)^{3/2}}(\dot{y}\hat{\mathbf{i}} - \dot{x}\hat{\mathbf{j}}).$$

If $\dot{y}\ddot{x} - \dot{x}\ddot{y}$ is positive, then

$$\hat{\mathbf{N}} = \frac{\dot{y}\hat{\mathbf{i}} - \dot{x}\hat{\mathbf{j}}}{\sqrt{\dot{x}^2 + \dot{y}^2}};$$

whereas if $\dot{y}\ddot{x} - \dot{x}\ddot{y}$ is negative, then

$$\hat{\mathbf{N}} = \frac{-\dot{y}\hat{\mathbf{i}} + \dot{x}\hat{\mathbf{j}}}{\sqrt{\dot{x}^2 + \dot{y}^2}}.$$

In other words,

$$\hat{\mathbf{N}} = \text{sgn}(\dot{y}\ddot{x} - \dot{x}\ddot{y})\left(\frac{\dot{y}\hat{\mathbf{i}} - \dot{x}\hat{\mathbf{j}}}{\sqrt{\dot{x}^2 + \dot{y}^2}}\right).$$

Curvature and Radius of Curvature

FIGURE 12.101

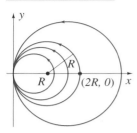

When length s along a smooth curve C is used as the parameter by which to identify points on the curve, the vector $\hat{\mathbf{T}} = d\mathbf{r}/ds$ is a unit tangent vector to C. Suppose we differentiate $\hat{\mathbf{T}}$ with respect to s to form $d\hat{\mathbf{T}}/ds = d^2\mathbf{r}/ds^2$. Since $\hat{\mathbf{T}}$ has constant unit length, only its direction can change; therefore, the derivative $d\hat{\mathbf{T}}/ds$ must be a measure of the rate of change of the direction of $\hat{\mathbf{T}}$. Since $\hat{\mathbf{T}}$ is really our way of specifying the direction of the curve itself, we can also say that $d\hat{\mathbf{T}}/ds$ is a measure of how fast the direction of C changes. But exactly how does a vector $d\hat{\mathbf{T}}/ds$ which has both magnitude and direction measure the rate of change of the direction of C? We illustrate by example that it cannot be the direction of $d\hat{\mathbf{T}}/ds$; it must be its magnitude that measures the rate of change of the direction of C. In Figure 12.101 we show a number of circles in the xy-plane, all of which are tangent to the y-axis at the origin. Parametric equations for the circle with centre $(R, 0)$ and radius R in terms of length s along the circle (as measured from $(2R, 0)$) are:

$$x = R + R\cos(s/R), \quad y = R\sin(s/R), \quad 0 \leq s < 2\pi R.$$

Consequently,

$$\hat{\mathbf{T}} = \frac{d\mathbf{r}}{ds} = -\sin\left(\frac{s}{R}\right)\hat{\mathbf{i}} + \cos\left(\frac{s}{R}\right)\hat{\mathbf{j}}$$

and

$$\frac{d\hat{\mathbf{T}}}{ds} = -\frac{1}{R}\cos\left(\frac{s}{R}\right)\hat{\mathbf{i}} - \frac{1}{R}\sin\left(\frac{s}{R}\right)\hat{\mathbf{j}}.$$

At the origin, $s = \pi R$, and

$$\frac{d\hat{\mathbf{T}}}{ds}_{\,|s=\pi R} = \frac{1}{R}\hat{\mathbf{i}}.$$

Thus, for each of the circles in Figure 12.101, the vector $d\hat{\mathbf{T}}/ds$ has exactly the same direction. Yet the rate of change of the direction of $\hat{\mathbf{T}}$ is not the same for each circle; the direction changes more rapidly as the radius of the circle decreases. We must conclude, therefore, that it cannot be the direction of $d\hat{\mathbf{T}}/ds$ that measures the rate of change of $\hat{\mathbf{T}}$. Since a vector has only length and direction, it must be the length of $d\hat{\mathbf{T}}/ds$ that measures this rate of change. The circles in Figure 12.101 certainly support this claim; the length of $d\hat{\mathbf{T}}/ds$ is $1/R$, and this quantity increases as the radii of the circles decrease. This agrees with the fact that the rate at which $\hat{\mathbf{T}}$ turns increases as R decreases. According to the following definition, we call $|d\hat{\mathbf{T}}/ds|$ **curvature** and $1/|d\hat{\mathbf{T}}/ds|$ **radius of curvature**.

DEFINITION 12.20

If $x = x(s), y = y(s), z = z(s), 0 \leq s \leq L$, are parametric equations for a smooth curve in terms of length s along the curve, we define the **curvature** of the curve at a point as

$$\kappa(s) = \left|\frac{d\hat{\mathbf{T}}}{ds}\right|, \tag{12.85}$$

its **radius of curvature** as

$$\rho(s) = \frac{1}{\kappa(s)}, \tag{12.86}$$

and its **circle of curvature** as that circle in the plane of $\hat{\mathbf{T}}$ and $\hat{\mathbf{N}}$ with centre at $\mathbf{r}(s)+\rho(s)\hat{\mathbf{N}}$ and radius $\rho(s)$.

FIGURE 12.102

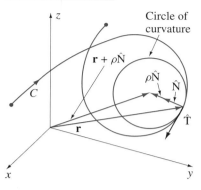

The circle of curvature is illustrated in Figure 12.102.

For the circles in Figure 12.101 we have already shown that $d\hat{\mathbf{T}}/ds = R^{-1}\hat{\mathbf{i}}$. Consequently, for these circles, the curvature is always R^{-1}, and the radius of curvature is R, the radius of the circle. In other words, for a circle, the circle of curvature is the circle itself, its radius of curvature is its radius, and its curvature is the inverse of its radius. For the case when a curve is not a circle, we show in Exercise 27 that at any point on the curve the circle of curvature is in some sense the "best-fitting" circle to the curve at that point.

Because curvature and radius of curvature have been defined in terms of intrinsic properties $\hat{\mathbf{T}}$ and s for a curve, they must also be intrinsic properties. It follows, then, that no matter what parameter is used to specify points on a curve, curvature and radius of curvature are always the same. The following theorem shows how to calculate κ and ρ when the curve is specified in terms of a parameter other than length along the curve.

THEOREM 12.11

When a smooth curve is defined parametrically by

$$C: \quad x = x(t), \quad y = y(t), \quad z = z(t), \quad \alpha \le t \le \beta,$$

its curvature $\kappa(t)$ is given by

$$\kappa(t) = \frac{|\dot{\mathbf{r}} \times \ddot{\mathbf{r}}|}{|\dot{\mathbf{r}}|^3}, \tag{12.87}$$

where $\dot{\mathbf{r}} = d\mathbf{r}/dt$ and $\ddot{\mathbf{r}} = d^2\mathbf{r}/dt^2$.

PROOF   If $s(t)$ is length along C (measured from $t = \alpha$), then by the chain rule

$$\kappa = \left|\frac{d\hat{\mathbf{T}}}{ds}\right| = \left|\frac{d\hat{\mathbf{T}}}{dt}\frac{dt}{ds}\right| = \frac{|d\hat{\mathbf{T}}/dt|}{|ds/dt|} = \frac{|d\hat{\mathbf{T}}/dt|}{ds/dt} = \frac{|d\hat{\mathbf{T}}/dt|}{|\dot{\mathbf{r}}|}$$

(see equation 12.74). Now, we can write $\dot{\mathbf{r}}$ in the form $\dot{\mathbf{r}} = |\dot{\mathbf{r}}|\hat{\mathbf{T}}$; therefore, using 12.52, we have

$$\ddot{\mathbf{r}} = \left(\frac{d}{dt}|\dot{\mathbf{r}}|\right)\hat{\mathbf{T}} + |\dot{\mathbf{r}}|\frac{d\hat{\mathbf{T}}}{dt}$$

$$= \left(\frac{d}{dt}|\dot{\mathbf{r}}|\right)\hat{\mathbf{T}} + (|\dot{\mathbf{r}}||d\hat{\mathbf{T}}/dt|)\frac{d\hat{\mathbf{T}}/dt}{|d\hat{\mathbf{T}}/dt|}$$

$$= \left(\frac{d}{dt}|\dot{\mathbf{r}}|\right)\hat{\mathbf{T}} + \left(|\dot{\mathbf{r}}|\left|\frac{d\hat{\mathbf{T}}}{dt}\right|\right)\hat{\mathbf{N}}.$$

If we take the cross product of this vector with $\dot{\mathbf{r}}$, we get

$$\dot{\mathbf{r}} \times \ddot{\mathbf{r}} = \left(\frac{d}{dt}|\dot{\mathbf{r}}|\right)\dot{\mathbf{r}} \times \hat{\mathbf{T}} + \left(|\dot{\mathbf{r}}|\left|\frac{d\hat{\mathbf{T}}}{dt}\right|\right)\dot{\mathbf{r}} \times \hat{\mathbf{N}}$$

$$= \left(|\dot{\mathbf{r}}|\left|\frac{d\hat{\mathbf{T}}}{dt}\right|\right)\dot{\mathbf{r}} \times \hat{\mathbf{N}}. \quad \text{(since } \dot{\mathbf{r}} \text{ is parallel to } \hat{\mathbf{T}}\text{)}$$

Because $\dot{\mathbf{r}}$ is perpendicular to $\hat{\mathbf{N}}$, it follows that $|\dot{\mathbf{r}} \times \hat{\mathbf{N}}| = |\dot{\mathbf{r}}||\hat{\mathbf{N}}|\sin(\pi/2) = |\dot{\mathbf{r}}|$, and therefore

$$|\dot{\mathbf{r}} \times \ddot{\mathbf{r}}| = \left(|\dot{\mathbf{r}}|\left|\frac{d\hat{\mathbf{T}}}{dt}\right|\right)|\dot{\mathbf{r}}|.$$

Consequently,

$$\left|\frac{d\hat{\mathbf{T}}}{dt}\right| = \frac{|\dot{\mathbf{r}} \times \ddot{\mathbf{r}}|}{|\dot{\mathbf{r}}|^2}$$

and

$$\kappa = \kappa(t) = \frac{|\dot{\mathbf{r}} \times \ddot{\mathbf{r}}|}{|\dot{\mathbf{r}}|^3}.$$ ∎

For the radius of curvature, we have the following.

COROLLARY

When a smooth curve is defined in terms of an arbitrary parameter t,

$$\rho(t) = \frac{|\dot{\mathbf{r}}|^3}{|\dot{\mathbf{r}} \times \ddot{\mathbf{r}}|}.$$ (12.88)

EXAMPLE 12.40

Find curvature and radius of curvature for the curve in Example 12.38.

SOLUTION   According to 12.87,

$$
\begin{aligned}
\kappa(t) &= \frac{|\dot{\mathbf{r}} \times \ddot{\mathbf{r}}|}{|\dot{\mathbf{r}}|^3} = \frac{|(1, 2t, 2t) \times (0, 2, 2)|}{|(1, 2t, 2t)|^3} \\
&= \frac{1}{(1 + 4t^2 + 4t^2)^{3/2}} \begin{vmatrix} \hat{\mathbf{i}} & \hat{\mathbf{j}} & \hat{\mathbf{k}} \\ 1 & 2t & 2t \\ 0 & 2 & 2 \end{vmatrix} = \frac{1}{(1 + 8t^2)^{3/2}} |(0, -2, 2)| \\
&= \frac{2\sqrt{2}}{(1 + 8t^2)^{3/2}}; \\
\rho(t) &= \frac{1}{\kappa(t)} = \frac{(1 + 8t^2)^{3/2}}{2\sqrt{2}}.
\end{aligned}
$$

Note in particular that as t increases, so does ρ, a fact that is certainly supported by Figure 12.100.

EXAMPLE 12.41

Show that for a smooth curve $y = y(x)$ in the xy-plane,

$$\kappa(x) = \frac{|y''|}{[1 + (y')^2]^{3/2}}.$$

SOLUTION   When we use x as parameter along the curve $y = y(x)$, parametric equations are $x = x$, $y = y(x)$. Then,

$$\dot{\mathbf{r}} = (1, y'(x)), \quad \ddot{\mathbf{r}} = (0, y''(x)),$$

and

$$\dot{\mathbf{r}} \times \ddot{\mathbf{r}} = \begin{vmatrix} \hat{\mathbf{i}} & \hat{\mathbf{j}} & \hat{\mathbf{k}} \\ 1 & y' & 0 \\ 0 & y'' & 0 \end{vmatrix} = y''\hat{\mathbf{k}}.$$

Thus

$$\kappa(x) = \frac{|\dot{\mathbf{r}} \times \ddot{\mathbf{r}}|}{|\dot{\mathbf{r}}|^3} = \frac{|y''|}{|(1, y')|^3} = \frac{|y''|}{[1 + (y')^2]^{3/2}}.$$

EXERCISES 12.10

In Exercises 1–5 find $\hat{\mathbf{N}}$ and $\hat{\mathbf{B}}$ at each point on the curve.

* **1.** $x = \sin t$, $y = \cos t$, $z = t$, $-\infty < t < \infty$

* **2.** $x = t$, $y = t^2$, $z = t^3$, $t \geq 1$

* **3.** $x = (t - 1)^2$, $y = (t + 1)^2$, $z = -t$, $-3 \leq t \leq 4$

* **4.** $x + y = 5$, $x^2 - y = z$, from $(5, 0, 25)$ to $(0, 5, -5)$

* **5.** $z = x$, $x^2 + y^2 = 4$, $y \geq 0$, from $(2, 0, 2)$ to $(-2, 0, -2)$

In Exercises 6–10 find $\hat{\mathbf{N}}$ and $\hat{\mathbf{B}}$ at the point.

* **6.** $x = 4 \cos t$, $y = 6 \sin t$, $z = 2 \sin t$, $-\infty < t < \infty$; $(2\sqrt{2}, 3\sqrt{2}, \sqrt{2})$

* **7.** $x = 2 - 5t$, $y = 1 + t$, $z = 6 + 4t^3$, $-\infty < t < \infty$; $(7, 0, 2)$

* **8.** $x^2 + y^2 + z^2 = 4$, $z = \sqrt{x^2 + y^2}$, directed so that x increases when y is positive; $(1, 1, \sqrt{2})$

* **9.** $x = y^2 + 1$, $z = x + 5$, directed so that y increases along the curve; $(5, 2, 10)$

* **10.** $x^2 + (y - 1)^2 = 4$, $x = z$, directed so that z decreases when y is negative; $(2, 1, 2)$

In Exercises 11–18 find the curvature and the radius of curvature of the curve (if they exist). Draw each curve.

* **11.** $(x - h)^2 + (y - k)^2 = R^2$, $z = 0$, directed counterclockwise

* **12.** $x = x_0 + at$, $y = y_0 + bt$, $z = z_0 + ct$, $-\infty < t < \infty$ (x_0, y_0, z_0, a, b, c all constants)

** **13.** $x = t$, $y = t^2$, $z = 0$, $t \geq 0$

** **14.** $x = e^t \cos t$, $y = e^t \sin t$, $z = t$, $-\infty < t < \infty$

** **15.** $x = t$, $y = t^3$, $z = t^2$, $t \geq 0$

** **16.** $x = 2 \cos t$, $y = 2 \sin t$, $z = 2 \sin t$, $0 \leq t < 2\pi$

** **17.** $x = t + 1$, $y = t^2 - 1$, $z = t + 1$, $-\infty < t < \infty$

** **18.** $x = t^2$, $y = t^4$, $z = 2t$, $-1 \leq t \leq 5$

** **19.** At which points on the ellipse $b^2x^2 + a^2y^2 = a^2b^2$ ($a > b$) is the curvature a maximum and at which points is the curvature a minimum?

** **20.** Show that curvature for a smooth curve $x = x(t)$, $y = y(t)$, $\alpha \leq t \leq \beta$, in the xy-plane can be expressed in the form

$$\kappa(t) = \frac{\left| \dfrac{dy}{dt}\dfrac{d^2x}{dt^2} - \dfrac{dx}{dt}\dfrac{d^2y}{dt^2} \right|}{\left[\left(\dfrac{dx}{dt}\right)^2 + \left(\dfrac{dy}{dt}\right)^2 \right]^{3/2}}.$$

** **21.** Show that the only curves for which curvature is identically equal to zero are straight lines.

** **22.** What happens to curvature at a point of inflection on the graph of a function $y = f(x)$?

** **23.** Let C be the curve $x = t$, $y = t^2$ in the xy-plane.
   (a) At each point on C calculate the unit tangent vector $\hat{\mathbf{T}}$ and the principal normal $\hat{\mathbf{N}}$. What is $\hat{\mathbf{B}}$? (See Example 12.39, page 924, for $\hat{\mathbf{N}}$.)
   (b) $\mathbf{F} = t^2\hat{\mathbf{i}} + t^4\hat{\mathbf{j}}$ is a vector that is defined at each point P on C. Denote by F_T and F_N the components of $\mathbf{F}$ in the directions $\hat{\mathbf{T}}$ and $\hat{\mathbf{N}}$ at P. Find F_T and F_N as functions of t.
   (c) Express $\mathbf{F}$ in terms of $\hat{\mathbf{T}}$ and $\hat{\mathbf{N}}$.

** **24.** Repeat Exercise 23 for the curve $C : \; x = 2\cos t$, $y = 2\sin t$, and the vector $\mathbf{F} = x^2\hat{\mathbf{i}} + y^2\hat{\mathbf{j}}$.

** **25.** The vectors $\hat{\mathbf{T}}$, $\hat{\mathbf{N}}$, and $\hat{\mathbf{B}}$ were calculated at each point on the curve $x = t$, $y = t^2$, $z = t^2$ in Example 12.38. If $\mathbf{F} = t^2\hat{\mathbf{i}} + 2t\hat{\mathbf{j}} - 3\hat{\mathbf{k}}$ is a vector defined along C, find the components of $\mathbf{F}$ in the directions $\hat{\mathbf{T}}$, $\hat{\mathbf{N}}$, and $\hat{\mathbf{B}}$. Express $\mathbf{F}$ in terms of $\hat{\mathbf{T}}$, $\hat{\mathbf{N}}$, and $\hat{\mathbf{B}}$.

** **26.** Calculate $\hat{\mathbf{T}}$, $\hat{\mathbf{N}}$, and $\hat{\mathbf{B}}$ for the curve $x = \cos t$, $y = \sin t$, $z = t$. Express the vector $\mathbf{F} = x\hat{\mathbf{i}} + xy^2\hat{\mathbf{j}} + \hat{\mathbf{k}}$ in terms of $\hat{\mathbf{T}}$, $\hat{\mathbf{N}}$, and $\hat{\mathbf{B}}$.

** **27.** In this exercise we discuss our claim that the circle of curvature is the "best-fitting" circle to the curve at a point.
   (a) First, is it true that the circle of curvature at a point on a curve passes through that point?
   (b) Second, show that the circle of curvature and curve share the same tangent line at their common point.
   (c) Finally, verify that the circle of curvature and curve have the same curvature at their common point.

28. If ϕ is the angle between $\hat{\mathbf{i}}$ and $\hat{\mathbf{T}}$ for a curve in the xy-plane (figure below), show that

$$\kappa(s) = \left| \frac{d\phi}{ds} \right|.$$

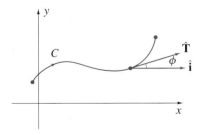

*** **29.** In differential geometry the rates of change $d\hat{\mathbf{T}}/ds$, $d\hat{\mathbf{N}}/ds$, and $d\hat{\mathbf{B}}/ds$ with respect to length along a curve are represented by the Frenet-Serret formulas. We develop these equations in this exercise.

(a) The first Frenet-Serret formula is

$$\frac{d\hat{\mathbf{T}}}{ds} = \kappa \hat{\mathbf{N}}.$$

Verify this result.

(b) The second Frenet-Serret formula is

$$\frac{d\hat{\mathbf{B}}}{ds} = -\tau \hat{\mathbf{N}},$$

where τ is called the torsion of the curve. To verify this result, differentiate the equations $\hat{\mathbf{B}} = \hat{\mathbf{T}} \times \hat{\mathbf{N}}$ and $\hat{\mathbf{B}} \cdot \hat{\mathbf{B}} = 1$ to show that $d\hat{\mathbf{B}}/ds$ is perpendicular to $\hat{\mathbf{T}}$ and $\hat{\mathbf{B}}$ and therefore parallel to $\hat{\mathbf{N}}$.

(c) The third formula is

$$\frac{d\hat{\mathbf{N}}}{ds} = \tau \hat{\mathbf{B}} - \kappa \hat{\mathbf{T}}.$$

Verify this result by showing that $\hat{\mathbf{N}} = \hat{\mathbf{B}} \times \hat{\mathbf{T}}$ and then calculating $d\hat{\mathbf{N}}/ds$.

*** **30.** Show that the torsion in Exercise 29 can be calculated according to

$$\tau(t) = \frac{(\dot{\mathbf{r}} \times \ddot{\mathbf{r}}) \cdot \dddot{\mathbf{r}}}{|\dot{\mathbf{r}} \times \ddot{\mathbf{r}}|^2}.$$

*** **31.** Show that a curve lies in a plane if and only if its torsion vanishes.

12.11 Displacement, Velocity, and Acceleration

In Sections 4.7 and 5.2 we introduced the concepts of displacement, velocity, and acceleration for moving objects, but indicated that our terminology at that time was somewhat loose. In particular, we stated that if $x = x(t)$ represents the position of a particle moving along the x-axis, then the instantaneous velocity of the particle is

$$v = \frac{dx}{dt}, \tag{12.89}$$

provided, of course, that t is time, and the acceleration of the particle is

$$a = \frac{dv}{dt} = \frac{d^2x}{dt^2}. \tag{12.90}$$

We illustrated by examples that, given any one of $x(t)$, $v(t)$, or $a(t)$ and sufficient initial conditions, it is always possible to find the other two. There was nothing wrong with the calculations in the examples—they were correct—but our terminology was not quite correct. We now rectify this situation and give precise definitions of velocity and acceleration.

Suppose a particle moves along some curve C in space (under perhaps the influence of various forces), and that C is defined as a function of time t by the parametric equations

$$C : \quad x = x(t), \quad y = y(t), \quad z = z(t), \quad t \geq 0. \tag{12.91}$$

The position of the particle can then be described as a function of time by its position or displacement vector:

$$\mathbf{r} = \mathbf{r}(t) = x(t)\hat{\mathbf{i}} + y(t)\hat{\mathbf{j}} + z(t)\hat{\mathbf{k}}, \quad t \geq 0. \qquad (12.92)$$

The **velocity v** of the particle at any time t is defined as the time rate of change of its displacement vector:

$$\mathbf{v} = \frac{d\mathbf{r}}{dt}. \qquad (12.93)$$

Velocity, then, is a vector, and because of Theorem 12.8, the components of velocity are the derivatives of the components of displacement:

$$\mathbf{v} = \frac{d\mathbf{r}}{dt} = \frac{dx}{dt}\hat{\mathbf{i}} + \frac{dy}{dt}\hat{\mathbf{j}} + \frac{dz}{dt}\hat{\mathbf{k}}. \qquad (12.94)$$

But according to Theorem 12.10, the vector $d\mathbf{r}/dt$ is tangent to the curve C (Figure 12.103). In other words, if a particle is at position P, and we draw its velocity vector with tail at P, then $\mathbf{v}$ is tangent to the trajectory.

FIGURE 12.103

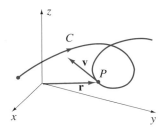

In some applications it is the length or magnitude of velocity that is important, not its direction. This quantity, called **speed**, is therefore defined by

$$|\mathbf{v}| = \sqrt{\left(\frac{dx}{dt}\right)^2 + \left(\frac{dy}{dt}\right)^2 + \left(\frac{dz}{dt}\right)^2}. \qquad (12.95)$$

Equation 12.74 implies that if $s(t)$ is length along the trajectory C (where $s(0) = 0$), then $|\mathbf{v}| = ds/dt$. In other words, speed is the time rate of change of distance travelled along C.

It is important to understand this difference between velocity and speed. Velocity is the time derivative of displacement; speed is the time derivative of distance travelled. Velocity is a vector; speed is a scalar—the magnitude of velocity.

The **acceleration** of the particle as it moves along the curve C in equations 12.91 is defined as the rate of change of velocity with respect to time:

$$\mathbf{a} = \frac{d\mathbf{v}}{dt} = \frac{d^2\mathbf{r}}{dt^2} = \frac{d^2x}{dt^2}\hat{\mathbf{i}} + \frac{d^2y}{dt^2}\hat{\mathbf{j}} + \frac{d^2z}{dt^2}\hat{\mathbf{k}}. \qquad (12.96)$$

Acceleration, then, is also a vector; it is the derivative of velocity, and therefore its components are the derivatives of the components of the velocity vector. Alternatively, it is the second derivative of displacement and has components that are the second derivatives of the components of the displacement vector.

In the special case in which C is a curve in the xy-plane, definitions of displacement, velocity, speed, and acceleration become, respectively,

$$\mathbf{r} = x(t)\hat{\mathbf{i}} + y(t)\hat{\mathbf{j}}, \tag{12.97}$$

$$\mathbf{v} = \frac{d\mathbf{r}}{dt} = \frac{dx}{dt}\hat{\mathbf{i}} + \frac{dy}{dt}\hat{\mathbf{j}}, \tag{12.98}$$

$$|\mathbf{v}| = \sqrt{\left(\frac{dx}{dt}\right)^2 + \left(\frac{dy}{dt}\right)^2}, \tag{12.99}$$

$$\mathbf{a} = \frac{d\mathbf{v}}{dt} = \frac{d^2\mathbf{r}}{dt^2} = \frac{d^2x}{dt^2}\hat{\mathbf{i}} + \frac{d^2y}{dt^2}\hat{\mathbf{j}}. \tag{12.100}$$

For motion along the x-axis,

$$\mathbf{r} = x(t)\hat{\mathbf{i}}, \tag{12.101}$$

$$\mathbf{v} = \frac{d\mathbf{r}}{dt} = \frac{dx}{dt}\hat{\mathbf{i}}, \tag{12.102}$$

$$|\mathbf{v}| = \left|\frac{dx}{dt}\right|, \tag{12.103}$$

$$\mathbf{a} = \frac{d\mathbf{v}}{dt} = \frac{d^2\mathbf{r}}{dt^2} = \frac{d^2x}{dt^2}\hat{\mathbf{i}}. \tag{12.104}$$

If we compare equations 12.102 and 12.104 with equations 12.89 and 12.90, we see that for motion along the x-axis, $x(t)$, $v(t)$, and $a(t)$ are the components of the displacement, velocity, and acceleration vectors, respectively. Because these are the only components of $\mathbf{r}(t)$, $\mathbf{v}(t)$, and $\mathbf{a}(t)$, it follows that consideration of the components of the vectors is equivalent to consideration of the vectors themselves. For one-dimensional motion, then, we can drop the vector notation and work with components (and this is precisely the procedure that we followed in Sections 4.7 and 5.2).

Newton's second law describes the effects of forces on the motion of objects. It states that if an object of mass m is subjected to a force $\mathbf{F}$, then the time rate of change of its momentum $(m\mathbf{v})$ is equal to $\mathbf{F}$:

$$\mathbf{F} = \frac{d}{dt}(m\mathbf{v}). \tag{12.105}$$

In most cases, the mass of the object is constant, and this equation then yields its acceleration:

$$\mathbf{F} = m\frac{d\mathbf{v}}{dt} = m\mathbf{a}. \tag{12.106}$$

If $\mathbf{F}$ is known as a function of time t, $\mathbf{F} = \mathbf{F}(t)$, then 12.106 defines the acceleration of the object as a function of time,

$$\mathbf{a}(t) = \frac{1}{m}\mathbf{F}(t),$$

and integration of this equation leads to expressions for the velocity $\mathbf{v}(t)$ and position $\mathbf{r}(t)$ as functions of time.

EXAMPLE 12.42

Find the velocity and acceleration of a particle and describe its motion, if its position as a function of time is given by

$$x = 2t + \sqrt{2}, \quad y = \sqrt{4t^2 + 4\sqrt{2}t + 1}, \quad z = 0, \quad t \geq 0.$$

SOLUTION   The velocity of the particle is

$$\mathbf{v} = \frac{dx}{dt}\hat{\mathbf{i}} + \frac{dy}{dt}\hat{\mathbf{j}} = 2\hat{\mathbf{i}} + \left(\frac{4t + 2\sqrt{2}}{\sqrt{4t^2 + 4\sqrt{2}t + 1}}\right)\hat{\mathbf{j}},$$

and its acceleration is

$$\mathbf{a} = \frac{d^2x}{dt^2}\hat{\mathbf{i}} + \frac{d^2y}{dt^2}\hat{\mathbf{j}} = \frac{-4}{(4t^2 + 4\sqrt{2}t + 1)^{3/2}}\hat{\mathbf{j}}.$$

Because $t = (x - \sqrt{2})/2$, an explicit definition of the path is

$$y = \sqrt{4\left(\frac{x - \sqrt{2}}{2}\right)^2 + 4\sqrt{2}\left(\frac{x - \sqrt{2}}{2}\right) + 1} = \sqrt{x^2 - 1}.$$

Since $t \geq 0$, it follows that $x \geq \sqrt{2}$, and the path is that part of the hyperbola $x^2 - y^2 = 1$ in Figure 12.104. Note that $dx/dt = 2$, so that the x-component of the velocity is always equal to 2. This is also reflected in the fact that the acceleration has no x-component. Since the y-component of acceleration is always negative, the y-component of velocity is decreasing in time.

FIGURE 12.104

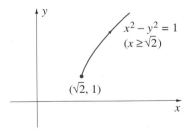

EXAMPLE 12.43

A particle starts at time $t = 0$ from position $(1, 1)$ with speed 2 m/s in the negative y-direction. It is subjected to an acceleration that is given as a function of time by

$$\mathbf{a}(t) = \frac{1}{\sqrt{t + 1}}\hat{\mathbf{i}} + 6t\hat{\mathbf{j}} \ \text{m/s}^2.$$

Find its velocity and position as functions of time.

SOLUTION   If $\mathbf{a} = d\mathbf{v}/dt = (1/\sqrt{t+1})\hat{\mathbf{i}} + 6t\hat{\mathbf{j}}$, then

$$\mathbf{v} = 2\sqrt{t+1}\,\hat{\mathbf{i}} + 3t^2\hat{\mathbf{j}} + \mathbf{C},$$

where $\mathbf{C}$ is some constant vector. Because the initial velocity of the particle is 2 m/s in the negative y-direction, $\mathbf{v}(0) = -2\hat{\mathbf{j}}$. Consequently,

$$-2\hat{\mathbf{j}} = 2\hat{\mathbf{i}} + \mathbf{C},$$

which implies that $\mathbf{C} = -2\hat{\mathbf{i}} - 2\hat{\mathbf{j}}$. The velocity, then, of the particle at any time $t \geq 0$ is

$$\mathbf{v}(t) = (2\sqrt{t+1} - 2)\hat{\mathbf{i}} + (3t^2 - 2)\hat{\mathbf{j}} \text{ m/s}.$$

Because $\mathbf{v} = d\mathbf{r}/dt$, integration gives

$$\mathbf{r} = \left[\frac{4}{3}(t+1)^{3/2} - 2t\right]\hat{\mathbf{i}} + (t^3 - 2t)\hat{\mathbf{j}} + \mathbf{D}.$$

Since the particle starts from position $(1, 1)$, $\mathbf{r}(0) = \hat{\mathbf{i}} + \hat{\mathbf{j}}$, and

$$\hat{\mathbf{i}} + \hat{\mathbf{j}} = \frac{4}{3}\hat{\mathbf{i}} + \mathbf{D}, \text{ or } \mathbf{D} = -\frac{1}{3}\hat{\mathbf{i}} + \hat{\mathbf{j}}.$$

The displacement of the particle is therefore

$$\mathbf{r}(t) = \left[\frac{4}{3}(t+1)^{3/2} - 2t - \frac{1}{3}\right]\hat{\mathbf{i}} + (t^3 - 2t + 1)\hat{\mathbf{j}} \text{ m}.$$

EXAMPLE 12.44

The mass M in Figure 12.105a is dropped from point B. Show that if a mass m is fired from any position A directly at M at the instant M is released, m will always collide with M.

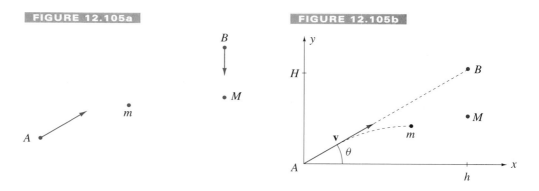

SOLUTION   We choose the coordinate system in Figure 12.105b, and take time $t = 0$ at the instant both masses begin motion. To show that the masses collide, we show that they have the same displacement vector for some time t. The acceleration of each mass is $\mathbf{a} = -9.81\hat{\mathbf{j}}$. Integration of this gives velocities of the masses,

$$\mathbf{v}_m = -9.81t\hat{\mathbf{j}} + \mathbf{C}, \qquad \mathbf{v}_M = -9.81t\hat{\mathbf{j}} + \mathbf{D},$$

Since M is dropped, its initial velocity is zero, $\mathbf{v}_M(0) = \mathbf{O}$. This implies that $\mathbf{D} = \mathbf{O}$. If m is fired at angle θ with speed $v > 0$, then $\mathbf{v}_m(0) = v\cos\theta\hat{\mathbf{i}} + v\sin\theta\hat{\mathbf{j}}$. This implies that $\mathbf{C} = v\cos\theta\hat{\mathbf{i}} + v\sin\theta\hat{\mathbf{j}}$. Integrations of

$$\frac{d\mathbf{r}_m}{dt} = -9.81t\hat{\mathbf{j}} + v\cos\theta\hat{\mathbf{i}} + v\sin\theta\hat{\mathbf{j}} \quad\text{and}\quad \frac{d\mathbf{r}_M}{dt} = -9.81t\hat{\mathbf{j}}$$

give

$$\mathbf{r}_m = v\cos\theta t\hat{\mathbf{i}} + (-4.905t^2 + v\sin\theta t)\hat{\mathbf{j}} + \mathbf{E} \quad\text{and}\quad \mathbf{r}_M = -4.905t^2\hat{\mathbf{j}} + \mathbf{F}.$$

Since $\mathbf{r}_m(0) = \mathbf{O}$ and $\mathbf{r}_M(0) = h\hat{\mathbf{i}} + H\hat{\mathbf{j}}$, we obtain $\mathbf{E} = \mathbf{O}$ and $\mathbf{F} = h\hat{\mathbf{i}} + H\hat{\mathbf{j}}$.

Thus, $\mathbf{r}_m = v\cos\theta t\hat{\mathbf{i}} + (-4.905t^2 + v\sin\theta t)\hat{\mathbf{j}}$ and $\mathbf{r}_M = h\hat{\mathbf{i}} + (H - 4.905t^2)\hat{\mathbf{j}}$.

The masses collide if, and when

$$\mathbf{r}_m = \mathbf{r}_M \Leftrightarrow v\cos\theta t\hat{\mathbf{i}} + (-4.905t^2 + v\sin\theta t)\hat{\mathbf{j}} = h\hat{\mathbf{i}} + (H - 4.905t^2)\hat{\mathbf{j}}.$$

When we equate components,

$$v\cos\theta t = h, \quad -4.905t^2 + v\sin\theta t = H - 4.905t^2,$$

from which

$$t = \frac{h}{v\cos\theta} \quad\text{and}\quad t = \frac{H}{v\sin\theta}.$$

These are compatible since $\tan\theta = H/h$. The time for collision to occur increases as h increases, v decreases, and θ increases.

Tangential and Normal Components of Velocity and Acceleration

For some types of motion it is inconvenient to express velocity and acceleration in terms of Cartesian components; sometimes it is an advantage to resolve these vectors into tangential and normal components. When the trajectory C of a particle is specified as a function of time t by 12.91, its velocity $\mathbf{v} = d\mathbf{r}/dt$ is tangent to C, and we can therefore write

$$\mathbf{v} = |\mathbf{v}|\hat{\mathbf{T}}. \tag{12.107}$$

When we equate components,

$$9.81 \cos \theta = \frac{d}{dt} |\mathbf{v}|, \qquad -N + 9.81m \sin \theta = m|\mathbf{v}| \left| \frac{d\hat{\mathbf{T}}}{dt} \right|.$$

Now, $x = a \cos \theta$ and $y = a \sin \theta$, so that

$$|\mathbf{v}| = \sqrt{ \left(\frac{dx}{dt} \right)^2 + \left(\frac{dy}{dt} \right)^2 } = \sqrt{ a^2 \sin^2 \theta \left(\frac{d\theta}{dt} \right)^2 + a^2 \cos^2 \theta \left(\frac{d\theta}{dt} \right)^2 } = -a \frac{d\theta}{dt}.$$

Hence,

$$9.81 \cos \theta = \frac{d}{dt} \left(-a \frac{d\theta}{dt} \right) = -a \frac{d^2\theta}{dt^2}.$$

Multiplication by $d\theta/dt$ gives

$$-9.81 \cos \theta \frac{d\theta}{dt} = a \frac{d^2\theta}{dt^2} \frac{d\theta}{dt} = \frac{a}{2} \frac{d}{dt} \left(\frac{d\theta}{dt} \right)^2,$$

and we may integrate with respect to t,

$$-9.81 \sin \theta = \frac{a}{2} \left(\frac{d\theta}{dt} \right)^2 + C.$$

Since $d\theta/dt = 0$ when $\theta = \pi/2$, it follows that $C = -9.81$, and

$$-9.81 \sin \theta = \frac{a}{2} \left(\frac{d\theta}{dt} \right)^2 - 9.81 \quad \Rightarrow \quad \left(\frac{d\theta}{dt} \right)^2 = \frac{19.62}{a} (1 - \sin \theta).$$

To tackle the normal components, we first calculate that

$$\mathbf{T} = \frac{d}{dt} (a \cos \theta, \ a \sin \theta) = a(-\sin \theta, \ \cos \theta) \frac{d\theta}{dt} \quad \Rightarrow \quad \hat{\mathbf{T}} = (\sin \theta, \ -\cos \theta).$$

Thus,

$$\frac{d\hat{\mathbf{T}}}{dt} = (\cos \theta, \ \sin \theta) \frac{d\theta}{dt} \quad \Rightarrow \quad \left| \frac{d\hat{\mathbf{T}}}{dt} \right| = -\frac{d\theta}{dt}.$$

Substitution into $-N + 9.81m \sin \theta = m|\mathbf{v}||d\hat{\mathbf{T}}/dt|$ gives

$$-N + 9.81m \sin \theta = m \left(-a \frac{d\theta}{dt} \right) \left(-\frac{d\theta}{dt} \right) = ma \left(\frac{d\theta}{dt} \right)^2.$$

Replacing $(d\theta/dt)^2$ by $(19.62/a)(1 - \sin \theta)$, we obtain

$$-N + 9.81m \sin \theta = 19.62m(1 - \sin \theta) \quad \Rightarrow \quad N = 9.81m(3 \sin \theta - 2).$$

Clearly, $N = 0$ when $\theta = \text{Sin}^{-1}(2/3)$, and this is the angle at which the mass leaves the sphere. Its speed at this point is

$$|\mathbf{v}| = -a\frac{d\theta}{dt} = a\sqrt{\frac{19.62}{a}\left(1 - \frac{2}{3}\right)} = \sqrt{\frac{19.62a}{3}}.$$

EXERCISES 12.11

In Exercises 1–5 find the velocity, speed, and acceleration of a particle if the given equations represent its position as a function of time.

* **1.** $x(t) = \sqrt{t^2 + 1}$, $y(t) = t\sqrt{t^2 + 1}$, $t \geq 0$

* **2.** $x(t) = t + 1/t$, $y(t) = t - 1/t$, $t \geq 1$

* **3.** $x(t) = \sin t$, $y(t) = 3\cos t$, $z(t) = \sin t$, $0 \leq t \leq 10\pi$

* **4.** $x(t) = t^2 + 1$, $y(t) = 2te^t$, $z(t) = 1/t^2$, $1 \leq t \leq 5$

* **5.** $x(t) = e^{-t^2}$, $y(t) = t \ln t$, $z(t) = 5$, $t \geq 1$

In Exercises 6 and 7 a particle at $(1, 2, -1)$ starts from rest at time $t = 0$. Find its position as a function of time if the given function defines its acceleration.

* **6.** $\mathbf{a}(t) = 3t^2\hat{\mathbf{i}} + (t + 1)\hat{\mathbf{j}} - 4t^3\hat{\mathbf{k}}$, $t \geq 0$

* **7.** $\mathbf{a}(t) = 3\hat{\mathbf{i}} + \hat{\mathbf{j}}/(t + 1)^3$, $t \geq 0$

In Exercises 8 and 9 find the tangential and normal components of acceleration for a particle moving with position defined by the given functions (where t is time).

* **8.** $x(t) = t$, $y(t) = t^2 + 1$, $t \geq 0$

* **9.** $x(t) = \cos t$, $y(t) = \sin t$, $z = t$, $t \geq 0$

* **10.** Show that the normal component of acceleration of a particle can be expressed in the form $a_N = |\mathbf{v}|^2/\rho = \kappa|\mathbf{v}|^2$.

* **11.** Find the kinetic energy for each particle in Exercises 1–5 if its mass is 2 g. Assume that x, y, and z are measured in metres and t in seconds.

* **12.** A particle starts at the origin and moves along the curve $4y = x^2$ to the point $(4, 4)$.

    (a) If the y-component of its acceleration is always equal to 2 and the y-component of its velocity is initially zero, find the x-component of its acceleration.

    (b) If the x-component of its acceleration is equal to $24t^2$ (t being time) and the x-component of its velocity is initially zero, find the y-component of its acceleration.

* **13.** A particle moves along the curve $x(t) = t$, $y(t) = t^3 - 3t^2 + 2t$, $0 \leq t \leq 5$ in the xy-plane (where t is time). Is there any point at which its velocity is parallel to its displacement?

* **14.** A particle moves along the curve $y = x^3 - 2x + 3$ so that its x-component of velocity is always equal to 5. Find its acceleration.

* **15.** If a particle starts at time $t = 0$ from rest at position $(3, 4)$ and experiences an acceleration $\mathbf{a} = -5t^4\hat{\mathbf{i}} - (2t^3 + 1)\hat{\mathbf{j}}$, find its speed at $t = 2$.

** **16.** A particle travels counterclockwise around the circle $(x - h)^2 + (y - k)^2 = R^2$ in the figure below. Show that the speed of the particle at any time is $|\mathbf{v}| = \omega R$, where $\omega = d\theta/dt$ is called the *angular speed* of the particle.

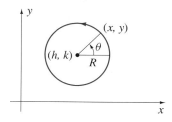

** **17.** A particle travels around the circle $x^2 + y^2 = 4$ counterclockwise at constant speed, making 2 revolutions each second. If x and y are measured in metres, what is the velocity of the particle when it is at the point $(1, -\sqrt{3})$?

** **18.**   (a) Show that, if an object moves with constant speed in a circular path of radius R, the magnitude of its acceleration is $|\mathbf{a}| = |\mathbf{v}|^2/R$.

   (b) If a satellite moves with constant speed in a circular orbit 200 km above the earth's surface, what is its speed? (Hint: Use Newton's universal law of gravitation (see Exercise 32 in Section 12.3) to determine the acceleration $\mathbf{a}$ of the satellite. Assume that the earth is a sphere with radius 6370 km and density 5.52×10^3 kg/m^3.)

** **19.** Two particles move along curves C_1 and C_2 in the figure below. If at some instant of time the particles are at positions P_1 and P_2, then the vector $\mathbf{P_1P_2}$ is the displacement of P_2 with respect to P_1. Clearly, $\mathbf{OP_1} + \mathbf{P_1P_2} = \mathbf{OP_2}$. Show that when this equation is differentiated with respect to time, we have

$$\mathbf{v}_{P_1/O} + \mathbf{v}_{P_2/P_1} = \mathbf{v}_{P_2/O},$$

where $\mathbf{v}_{P_1/O}$ and $\mathbf{v}_{P_2/O}$ are velocities of P_1 and P_2 with respect to the origin, and $\mathbf{v}_{P_2/P_1}$ is the velocity of P_2 with respect to P_1. Can this equation be rewritten in the form

$$\mathbf{v}_{P_1/O} + \mathbf{v}_{O/P_2} = \mathbf{v}_{P_1/P_2}?$$

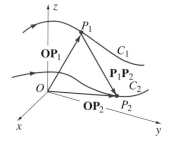

** **20.** A plane flies on a course N30°E with air speed of 650 km/h (i.e., the speed of the plane relative to the air is 650). If the air is moving at 40 km/h due east, find the ground velocity and speed of the plane. (Hint: Use Exercise 19.)

** **21.** A plane flies with speed 600 km/h in still air. The plane is to fly in a straight line from city A to city B where B is 1000 km northwest of A. What should be its bearing if the wind is blowing from the west at 50 km/h? How long will the trip take?

** **22.** A straight river is 200 m wide and the water flows at 3 km/h. If you can paddle your canoe at 4 km/h in still water, in what direction should you paddle if you wish the canoe to go straight across the river? How long will it take to cross?

∗∗ **23.**   (a) In the figure below a cannon is fired up an inclined plane. If the speed at which the ball is ejected from the cannon is S, show that the range R of the ball is given by

$$R = \frac{2S^2 \cos\theta \,\sin(\theta - \alpha)}{g\cos^2\alpha},$$

where g is the acceleration due to gravity.

(b) What angle θ maximizes R?

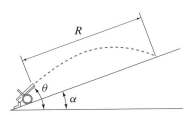

∗∗ **24.** What constant acceleration must a particle experience if it is to travel from $(1, 2, 3)$ to $(4, 5, 7)$ along the straight line joining the points, starting from rest, and covering the distance in 2 units of time?

∗∗ **25.** Calculate the normal component a_N of the acceleration of a particle using equation 12.109 and 12.110 if its position is given by $x = t^2 + 1$, $y = 2t^2 - 1$, $t = t^2 + 5t$, $t \geq 0$ (t being time).

∗∗ **26.** A particle moves along the curve $x(t) = 2 + \sqrt{1 - t^2}$, $y(t) = t$, $0 \leq t \leq 1/2$, where t is time. Is there a time at which its acceleration is perpendicular to its velocity?

∗∗ **27.**   (a) Show that motion along a straight line results in both of the following situations:

(i) The initial velocity is zero, and the acceleration is constant.

(ii) The initial velocity is nonzero, and the acceleration is constant and parallel to the initial velocity.

(b) Can we generalize the results of (a) and state that constant acceleration produces straight-line motion? Illustrate.

∗∗ **28.** A particle starts from position $\mathbf{r}_0 = (x_0, y_0, z_0)$ at time $t = t_0$ with velocity $\mathbf{v}_0$. If it experiences constant acceleration $\mathbf{a}$, show that

$$\mathbf{r} = \mathbf{r}_0 + \mathbf{v}_0(t - t_0) + \frac{1}{2}\mathbf{a}(t - t_0)^2.$$

∗∗ **29.** A ladder 8 m long has its upper end against a vertical wall and its lower end on a horizontal floor. Suppose that the lower end slips away from the wall at constant speed 1 m/s.

(a) Find the velocity and acceleration of the middle point of the ladder when the foot of the ladder is 3 m from the wall.

(b) How fast does the middle point of the ladder strike the floor?

∗∗ **30.** Water issues from the nozzle of a fire hose at speed S (figure at the top of page 945). Show that the maximum height attainable by the water on the building is given by $(S^4 - g^2 d^2)/(2gS^2)$, where g is the acceleration due to gravity.

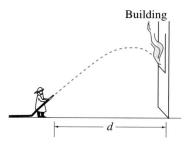

Building

d

** **31.** A boy stands on a cliff 50 m high that overlooks a river 85 m wide (figure below). If he can throw a stone at 25 m/s, can he throw it across the river?

50

85

** **32.** A golfer can drive a maximum of 300 m in the air on a level fairway. From the tee in the figure below, can he expect to clear the stream?

Tee

310 m

Green

20 m

Stream

** **33.** If **r** is the position vector of a particle with mass m moving under the action of a force **F**, the torque of **F** about the origin is $\tau = \mathbf{r} \times \mathbf{F}$. The angular momentum of m about O is defined as $\mathbf{H} = \mathbf{r} \times m\mathbf{v}$. Use Newton's second law in the form 12.105 (page 933) to show that $\tau = d\mathbf{H}/dt$.

** **34.** When a stone is embedded in the tread of a tire and the tire rolls (without slipping) along the x-axis (figure below), the path that it traces is called a cycloid (see Example 10.7 in Section 10.1).

    (a) Verify that if the centre of the tire moves at constant speed S, with $t = 0$ when the stone is at the origin, then $\theta = St/R$.

    (b) Find velocity, speed, and acceleration of the stone at any time.

    (c) What are the normal and tangential components of the stone's acceleration?

y

$2R$ (x, y)

θ

x

** **35.** If the stone in Exercise 34 is embedded in the side of the tire, its path is called a trochoid (see Exercise 48 in Section 10.1).

(a) Find velocity, speed, and acceleration of the stone if the tire rolls so that its centre has constant speed S. (Assume $x = 0$ at time $t = 0$.)

(b) What are normal and tangential components of the stone's acceleration?

** **36.** Circles C_1 and C_2 in the figure below represent cross-sections of two cylinders. The left cylinder remains stationary while the right one rolls (without slipping) around the left one, and the cylinders always remain in contact. If the right cylinder picks up a speck of dirt at point $(R, 0)$, the path that the dirt traces out during one revolution is a cardioid.

(a) Show that parametric equations for the cardioid are

$$x = R(2\cos\theta - \cos 2\theta), \quad y = R(2\sin\theta - \sin 2\theta).$$

(b) Verify that if the point of contact moves at constant speed S, with $t = 0$ when the speck of dirt is picked up, then $\theta = St/R$.

(c) Find velocity, speed, and acceleration of the speck of dirt.

(d) What are normal and tangential components of the dirt's acceleration?

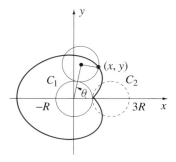

** **37.** Show that the path of a particle lies on a sphere if its displacement and velocity are always perpendicular during its motion.

** **38.** Suppose in Exercise 22 that, because of an injured elbow, you can paddle only at 2 km/h. What should be your heading to travel straight to a point L km downstream on the opposite shore? Are there any restrictions on L?

** **39.** Suppose position vectors of a system of n masses m_i are denoted by $\mathbf{r}_i$ and forces acting on these masses are $\mathbf{F}_i$. Show that if $\mathbf{F} = \sum_{i=1}^{n} \mathbf{F}_i$, then the acceleration $\mathbf{a}$ of the centre of mass of the system (see Section 7.7) is given by $\mathbf{F} = M\mathbf{a}$, where $M = \sum_{i=1}^{n} m_i$.

** **40.** If the force acting on a particle is always tangent to the particle's trajectory, what can you conclude about the trajectory?

** **41.** The second of Kepler's laws states that the line joining the sun to a planet sweeps out equal areas in equal time intervals. To show this, let $A(t)$ be the area swept out by the line beginning at some time t_0 (figure on page 947) and ending at time t.

(a) Verify that

$$A(t) = \int_{\theta_0}^{\theta(t)} \frac{1}{2}r^2 d\theta,$$

and use equation 6.20 (page 427) to prove that

$$\frac{dA}{dt} = \frac{1}{2}r^2\frac{d\theta}{dt}.$$

(b) Show that $\hat{\mathbf{r}} = \cos\theta\,\hat{\mathbf{i}} + \sin\theta\,\hat{\mathbf{j}}$ and $d\hat{\mathbf{r}}/dt$ are perpendicular, and deduce from equation 12.112 that

$$\frac{d\theta}{dt} = \frac{|\mathbf{C}|}{r^2}.$$

(c) Use (a) and (b) to obtain

$$\frac{dA}{dt} = \frac{|\mathbf{C}|}{2} = \text{constant}.$$

Does this verify Kepler's second law?

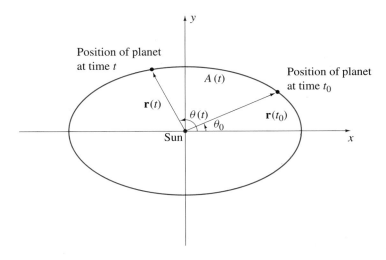

** **42.** Kepler's third law states that the square of the period of revolution of a planet is proportional to the cube of the length of the major axis of its orbit.

(a) Use part (c) of Exercise 41 to show that if P is the time taken for one complete revolution, then

$$P = \frac{2\pi ab}{|\mathbf{C}|}$$

where $2a$ and $2b$ are lengths of the major and minor axes of the elliptic orbit.

(b) Use equation 12.113 (page 939) to show that $b^2/a = |\mathbf{C}|^2/(GM)$, and therefore

$$P^2 = \frac{4\pi^2 a^3}{GM}.$$

SUMMARY

We have now established the groundwork for multivariable calculus. We discussed curves and surfaces in space and introduced vectors. We described points by Cartesian coordinates (x, y, z) and then illustrated that an equation $F(x, y, z) = 0$ in these coordinates usually defines a surface. A second equation $G(x, y, z) = 0$ also defines a surface, and the pair of simultaneous equations $F(x, y, z) = 0$, $G(x, y, z) = 0$ describes the curve of intersection of the two surfaces (if the two surfaces do intersect). It is often more useful to have parametric equations for a curve, and these can be obtained by specifying one of x, y, or z as a function of a parameter t and solving the given equations for the other two in terms of t: $x = x(t)$, $y = y(t)$, $z = z(t)$.

The most common surfaces that we encountered were planes and quadric surfaces. Every plane has an equation of the form $Ax + By + Cz + D = 0$; conversely, every such equation describes a plane (provided A, B, and C are not all zero). A plane is uniquely defined by a vector that is perpendicular to it [and (A, B, C) is one such vector] and a point on it. Quadric surfaces are surfaces whose equations are quadratic in x, y, and z, the most important of which were sketched in Figures 12.20–12.28.

Every straight line in space is characterized by a vector along it and a point on it. (Contrast this with the above characterization of a plane.) If (a, b, c) are the components of a vector along a line and (x_0, y_0, z_0) are the coordinates of a point on it, then vector, symmetric, and parametric equations for the line are, respectively,

$$(x, y, z) = (x_0, y_0, z_0) + t(a, b, c);$$

$$\frac{x - x_0}{a} = \frac{y - y_0}{b} = \frac{z - z_0}{c};$$

$$x = x_0 + at,$$

$$y = y_0 + bt,$$

$$z = z_0 + ct.$$

Geometrically, vectors are defined as directed line segments; algebraically, they are represented by ordered sets of real numbers (v_x, v_y, v_z), called their Cartesian components. Vectors can be added or subtracted geometrically using triangles or parallelograms; algebraically, they are added and subtracted component by component. Vectors can also be multiplied by scalars to give parallel vectors of different lengths.

We defined two products of vectors: the scalar product and the vector product. The scalar product of two vectors $\mathbf{u} = (u_x, u_y, u_z)$ and $\mathbf{v} = (v_x, v_y, v_z)$ is defined as

$$\mathbf{u} \cdot \mathbf{v} = u_x v_x + u_y v_y + u_z v_z = |\mathbf{u}||\mathbf{v}| \cos \theta,$$

where $|\mathbf{u}| = \sqrt{u_x^2 + u_y^2 + u_z^2}$ is the length of $\mathbf{u}$ and θ is the angle between $\mathbf{u}$ and $\mathbf{v}$. If the components of $\mathbf{u}$ and $\mathbf{v}$ are known, this equation can be used to find the angle θ between the vectors. The scalar product has many uses: finding components of vectors in arbitrary directions, calculating distances between geometric objects, and finding mechanical work, among others.

The vector product of two vectors $\mathbf{u}$ and $\mathbf{v}$ is

$$\mathbf{u} \times \mathbf{v} = \begin{vmatrix} \hat{\mathbf{i}} & \hat{\mathbf{j}} & \hat{\mathbf{k}} \\ u_x & u_y & u_z \\ v_x & v_y & v_z \end{vmatrix} = |\mathbf{u}||\mathbf{v}| \sin \theta \, \hat{\mathbf{w}},$$

where $\hat{\mathbf{w}}$ is the unit vector perpendicular to $\mathbf{u}$ and $\mathbf{v}$ determined by the right-hand rule. Because of the perpendicularity property, the vector product is indispensable in finding vectors perpendicular to other vectors. We used this fact when finding a vector along the line of intersection of two planes, a vector perpendicular to the plane containing three given points, and distances between geometric objects. It can also be used to find areas of triangles and parallelograms.

If a curve is represented vectorially in the form $\mathbf{r}(t) = x(t)\hat{\mathbf{i}} + y(t)\hat{\mathbf{j}} + z(t)\hat{\mathbf{k}}$, then a unit vector tangent to the curve at any point is

$$\hat{\mathbf{T}} = \frac{d\mathbf{r}/dt}{|d\mathbf{r}/dt|}.$$

Two unit vectors normal to the curve are the principal normal $\hat{\mathbf{N}}$ and the binormal $\hat{\mathbf{B}}$:

$$\hat{\mathbf{N}} = \frac{d\hat{\mathbf{T}}/dt}{|d\hat{\mathbf{T}}/dt|}; \quad \hat{\mathbf{B}} = \hat{\mathbf{T}} \times \hat{\mathbf{N}}.$$

These three vectors form a moving triad of mutually perpendicular unit vectors along the curve. The curvature of a curve, defined by $\kappa(t) = |\dot{\mathbf{r}} \times \ddot{\mathbf{r}}|/|\dot{\mathbf{r}}|^3$, measures the rate at which the curve changes direction: The larger κ is, the faster the curve turns. The reciprocal of curvature $\rho = \kappa^{-1}$ is called radius of curvature. It is the radius of that circle which best approximates the curve at any point.

If parametric equations for a curve represent the position of a particle and t is time, then the velocity and acceleration of the particle are, respectively,

$$\mathbf{v} = \frac{d\mathbf{r}}{dt}; \quad \mathbf{a} = \frac{d\mathbf{v}}{dt} = \frac{d^2\mathbf{r}}{dt^2};$$

and its speed is the magnitude of velocity, $|\mathbf{v}|$.

Tangential and normal components of velocity and acceleration of the particle are defined by

$$\mathbf{v} = |\mathbf{v}|\hat{\mathbf{T}}; \quad \mathbf{a} = a_T\hat{\mathbf{T}} + a_N\hat{\mathbf{N}};$$

where

$$a_T = \frac{d}{dt}|\mathbf{v}| \text{ and } a_N = |\mathbf{v}|\left|\frac{d\hat{\mathbf{T}}}{dt}\right| = \sqrt{|\mathbf{a}|^2 - a_T^2}.$$

What these results say is that velocity is always tangent to the trajectory of the particle, and its acceleration always lies in the plane of the velocity vector and the principal normal.

KEY TERMS

In reviewing this chapter, you should be able to define or discuss the following key terms:

Coordinate plane	Right(left)-handed coordinate system
Quadric surface	Vector
Tip of a vector	Tail of a vector
Components of a vector	Equality of vectors
Unit vector	Scalar multiplication
Sum of two vectors	Triangular addition of vectors
Parallelogram addition of vectors	Zero vector
Resultant of forces	Scalar product
Equation of a plane	Vector product
Vector equation of a line	Parametric equations of a line
Symmetric equations of a line	Scalar triple product
Continuous vector function	Derivative of a vector function
Antiderivative of a vector function	Parametric definition of a curve
Continuous curve	Closed curve
Displacement vector	Vector representation of a curve
Tangent vector to a curve	Smooth curve
Piecewise smooth curve	Unit tangent vector to a curve
Length of a curve	Principal normal vector
Binormal vector	Curvature
Radius of curvature	Circle of curvature
Velocity	Acceleration
Tangential component of acceleration	Normal component of acceleration

REVIEW
EXERCISES

In Exercises 1–10, find the value of the scalar or the components of the vector if $\mathbf{u} = (1, 3, -2)$, $\mathbf{v} = (2, 4, -1)$, $\mathbf{w} = (0, 2, 1)$, and $\mathbf{r} = (2, 0, -1)$.

* **1.** $2\mathbf{u} - 3\mathbf{w} + \mathbf{r}$
* **2.** $\mathbf{u} \cdot (\mathbf{v} \times \mathbf{w})$
* **3.** $(3\mathbf{u} \times 4\mathbf{v}) - \mathbf{w}$
* **4.** $3\mathbf{u} \times (4\mathbf{v} - \mathbf{w})$
* **5.** $|\mathbf{u}|\mathbf{v} - |\mathbf{v}|\mathbf{r}$
* **6.** $(\mathbf{u} + \mathbf{v}) \cdot (\mathbf{r} - \mathbf{w})$
* **7.** $(\mathbf{u} + \mathbf{v}) \times (\mathbf{r} - \mathbf{w})$
* **8.** $(\mathbf{u} \times \mathbf{v}) \times (\mathbf{r} \times \mathbf{w})$
* **9.** $(\mathbf{u} \cdot \mathbf{v})\mathbf{r} - 3(\mathbf{v} \cdot \mathbf{w})\mathbf{u}$
* **10.** $\dfrac{2\mathbf{r}}{\mathbf{v} \cdot \mathbf{w}} + 3(\mathbf{v} + \mathbf{u})$

In Exercises 11–26 draw whatever the equation, or equations, describe in space.

* **11.** $x - y + 2z = 6$
* **12.** $x^2 + z^2 = 1$
* **13.** $x = \sqrt{y^2 + z^2}$
* **14.** $x - y = 5,\ 2x + y = 6$
* **15.** $x^2 + y^2 + z^2 = 6z + 10$
* **16.** $x^2 + y^2 + z^2 = 6z - 10$
* **17.** $x + y = 5,\ 2x - 3y + 6z = 1,\ y = z$
* **18.** $x = t^2,\ y = t,\ z = t^3$
* **19.** $x = t,\ y = t^3 + 1$
* **20.** $\dfrac{x - 1}{3} = \dfrac{y + 5}{2} = z$
* **21.** $z = 4 - x^2 - 2y^2$
* **22.** $y^2 + z^2 = 1,\ y = z$
* **23.** $y^2 + z^2 = 1,\ x = z$
* **24.** $x^2 + y^2 = z^2 + 1$
* **25.** $x = y^2,\ x = z^2$
* **26.** $z^2 = x^2 - y^2$

In Exercises 27–30 find equations for the line.

* **27.** Through the points $(-2, 3, 0)$ and $(1, -2, 4)$
* **28.** Through $(6, 6, 2)$ and perpendicular to the plane $5x - 2y + z = 4$
* **29.** Parallel to the line $x - y = 5,\ 2x + 3y + 6z = 4$ and through the origin
* **30.** Perpendicular to the line $x = t + 2,\ y = 3 - 2t,\ z = 4 + t$, intersecting this line, and through the point $(1, 3, 2)$

In Exercises 31–34 find the equation for the plane.

* **31.** Through the points $(1, 3, 2)$, $(2, -1, 0)$, and $(6, 1, 3)$
* **32.** Through the point $(1, 2, -1)$ and perpendicular to the line $y = z,\ x + y = 4$
* **33.** Containing the line $x - y + z = 3,\ 3x + 4y = 6$ and the point $(2, 2, 2)$
* **34.** Containing the lines $x = 3t,\ y = 1 + 2t,\ z = 4 - t$, and $x = y = z$

In Exercises 35–39 find the distance.

* **35.** Between the points $(1, 3, -2)$ and $(6, 4, 1)$
* **36.** From the point $(6, 2, 1)$ to the plane $6x + 2y - z = 4$
* **37.** From the line $x - y + z = 2,\ 2x + y + z = 4$ to the plane $x - y = 5$
* **38.** From the line $x - y + z = 2,\ 2x + y + z = 4$ to the plane $3x + 6y = 4$

** **39.** From the point $(6, 2, 3)$ to the line $x - y + z = 6$, $2x + y + 4z = 1$

* **40.** Find the area of the triangle with vertices $(1, 1, 1)$, $(-2, 1, 0)$, and $(6, 3, -2)$.

* **41.** If the points in Exercise 40 are three vertices of a parallelogram, what are possibilities for the fourth vertex? What are areas of these parallelograms?

In Exercises 42 and 43 find the unit tangent vector $\hat{\mathbf{T}}$, the principal normal vector $\hat{\mathbf{N}}$, and the binormal vector $\hat{\mathbf{B}}$ for the curve.

* **42.** $x = 2 \sin t$, $y = 2 \cos t$, $z = t$

* **43.** $x = t^3$, $y = 2t^2$, $z = t + 4$

* **44.** If a particle has a trajectory defined by $x = t$, $y = t^2$, $z = t^2$, where t is time, find its velocity, speed, and acceleration at any time. What are normal and tangential components of its velocity and acceleration?

** **45.** A force $\mathbf{F} = x^{-2}(2\hat{\mathbf{i}} + 3\hat{\mathbf{j}})$ acts on a particle moving from $x = 1$ to $x = 4$ along the x-axis. How much work does it do?

** **46.** A ball rolls off a table 1 m high with speed 0.5 m/s (figure below).

    (a) With what speed does it strike the floor?

    (b) What is its displacement vector relative to the point where it left the table when it strikes the floor?

    (c) If it rebounds in the direction shown but loses 20% of its speed in the bounce, find the position of its second bounce.

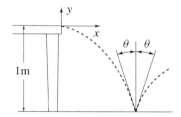

** **47.** In the figure below, a spring (with constant k) is fixed at A and attached to a sleeve at C. The sleeve is free to slide without friction on a vertical rod, and when the spring is horizontal (at B), it is unstretched. If the sleeve is slowly lowered, there is a position at which the vertical component of the spring force on C is balanced by the force of gravity on the sleeve (ignoring the weight of the spring itself). If the mass of the sleeve is m, find an equation determining s in terms of d, m, k, and $g = 9.81$, the acceleration due to gravity.

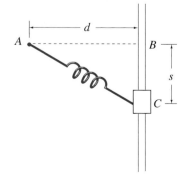

∗∗ **48.** Find Cartesian components of the spring force **F** on the sleeve in Example 12.24 (page 895).

∗∗ **49.** If a toy train travels around the oval track in the figure below with constant speed, show that its acceleration at A (the point at which the circular end meets the straight section) is discontinuous.

Differential Calculus of Multivariable Functions

Few quantities in real life depend on only one variable; most depend on a multitude of interrelated variables. In order to understand such complicated relationships, we initiate discussions in this chapter with derivatives of functions of more than one variable. Much of the theory and many of our examples involve functions of two or three variables, because in these cases we can give geometric as well as analytic explanations. If the situation is completely analogous for functions of more variables, then it is likely that no mention of this fact will be made; on the other hand, if the situation is different for a higher number of variables, we will be careful to point out these differences.

13.1 Multivariable Functions

If a variable T depends on other variables x, y, z, and t, we write $T = f(x, y, z, t)$ and speak of T as a function of x, y, z, and t. For example, T might be temperature, x, y, and z might be the coordinates of points in some region of space, and t might be time. The stopping distance D of a car depends on many factors: the initial speed s, the reaction time t of the driver to move from the accelerator to the brake, the texture T of the road, the moisture level M on the road, etc. We write $D = f(s, t, T, M, \ldots)$ to represent this functional dependence. The function $P = f(I, R) = I^2 R$ represents the power necessary to maintain a current I through a wire with resistance R.

More precisely, a variable z is said to be a function of two independent variables x and y if x and y are not related and each pair of values of x and y determines a unique value of z. We write $z = f(x, y)$ to indicate that z is a function of x and y. Each possible pair of values x and y of the independent variables can be represented geometrically as a point (x, y) in the xy-plane. The totality of all points for which $f(x, y)$ is defined forms a region in the xy-plane called the **domain** of the function. Figure 13.1, for example, illustrates a rectangular domain. If for each point (x, y) in the domain we plot a point $f(x, y)$ units above the xy-plane, we obtain a surface, such as the one in Figure 13.1. Each point on this surface has coordinates (x, y, z) that satisfy the equation

$$z = f(x, y), \tag{13.1}$$

and therefore 13.1 is the equation of the surface. This surface is a pictorial representation of the function.

It is clear that functions of more than two independent variables cannot be represented pictorially as surfaces. For example, if $u = f(x, y, z)$ is a function of three independent variables, values (x, y, z) of these independent variables can be represented geometrically as points in space. To graph $u = f(x, y, z)$ as above would require a u-axis perpendicular to the x-, y-, and z-axes, a somewhat difficult task geometrically. We can certainly think of $u = f(x, y, z)$ as defining a surface in four-dimensional $xyzu$-space, but visually we are stymied.

FIGURE 13.1

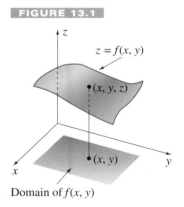

Domain of $f(x, y)$

953

Although every function $f(x, y)$ of two independent variables can be represented geometrically as a surface, not every surface represents a function $f(x, y)$. A given surface does represent a function $f(x, y)$ if and only if every vertical line (in the z-direction) that intersects the surface does so in exactly one point. For example, a sphere such as $x^2 + y^2 + z^2 = 1$ does not determine z as a function of x and y. Most, but not all, vertical lines that intersect the sphere do so in two points. It is also clear algebraically that $x^2 + y^2 + z^2 = 1$ does not define z as a function of x and y. Solving for z gives $z = \pm\sqrt{1 - x^2 - y^2}$, two solutions for each x and y satisfying $x^2 + y^2 < 1$.

EXAMPLE 13.1

Draw the surface defined by the function $f(x, y) = x^2 + 4y^2$.

SOLUTION   To draw the surface $z = x^2 + 4y^2$, we note that if the surface is intersected with a plane $z = k > 0$, then the ellipse $x^2 + 4y^2 = k$, $z = k$ is obtained. As k increases, the ellipse becomes larger. In other words, cross-sections of this surface are ellipses that expand with increasing z. If we intersect the surface with the yz-plane ($x = 0$), we obtain the parabola $z = 4y^2$, $x = 0$. Similarly, intersection of the surface with the xz-plane gives the parabola $z = x^2$, $y = 0$. These facts lead to Figure 13.2a. A computer plot of the surface for $-2 \le x \le 2$ and $-2 \le y \le 2$ is shown in Figure 13.2b.

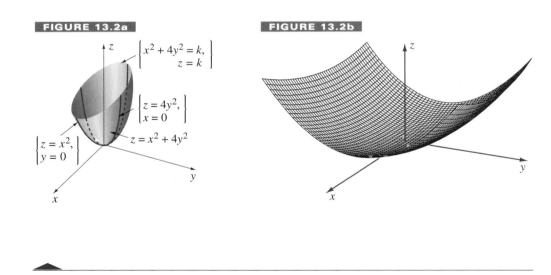

FIGURE 13.2a

FIGURE 13.2b

EXAMPLE 13.2

The ends of a taut string are fixed at $x = 0$ and $x = 2$ on the x-axis. At time $t = 0$, the string is given a displacement in the y-direction of $y = \sin(\pi x/2)$ (Figure 13.3a). If the string is then released, its displacement thereafter is given by

$$y = f(x, t) = \sin(\pi x/2) \cos(8\pi t).$$

FIGURE 13.3a

FIGURE 13.3b

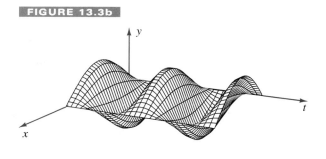

Physically, this function need only be considered for $t \geq 0$ and $0 \leq x \leq 2$, and is plotted for $0 \leq t \leq 1/2$ in Figure 13.3b. Interpret physically the intersections of this surface with planes $t = t_0$ ($t_0 =$ a constant) and $x = x_0$ ($x_0 =$ a constant).

SOLUTION   Grid lines on the surface are curves of intersection of the surface with vertical planes $x = x_0$ and $t = t_0$. The curve of intersection of the surface with a plane $t = t_0$ represents the position of the string at time t_0. For $t = t_0$, the equation of the curve is $y(x, t_0) = \cos(8\pi t_0) \sin(\pi x/2)$. Thus, the string vibrates up and down always in the shape of a sine curve, its amplitude at time t_0 being $|\cos(8\pi t_0)|$.

The curve of intersection of the surface with a plane $x = x_0$, $0 < x_0 < 2$, is a graphical history of the vertical displacement of the particle in the string at position x_0. The equation of the curve is $y(x_0, t) = \sin(\pi x_0/2) \cos(8\pi t)$. The particle at x_0 undergoes simple harmonic motion, the amplitude being $|\sin(\pi x_0/2)|$.

Another way to visualize a function $f(x, y)$ of two independent variables is through level curves. Curves $f(x, y) = C$ are drawn in the xy-plane for various values of C. Effectively, the surface $z = f(x, y)$ is sliced with a plane $z = C$, and the curve of intersection is projected into the xy-plane. Each curve joins all points for which $f(x, y)$ has the same value; or it joins all points which have the same height on the surface $z = f(x, y)$. A few level curves for the surface in Figure 13.2 are shown in Figure 13.4; they are ellipses.

FIGURE 13.4

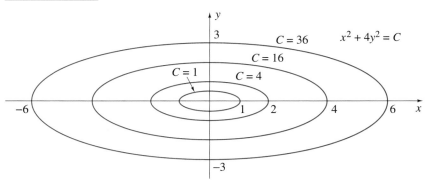

Level curves for the function in Figure 13.5a are shown in Figure 13.5b. This technique is used on topographical maps to indicate land elevation, on marine charts to indicate water depth, and on climatic maps to indicate curves of constant temperature (isotherms) and curves of constant barometric pressure (isobars).

FIGURE 13.5a

FIGURE 13.5b

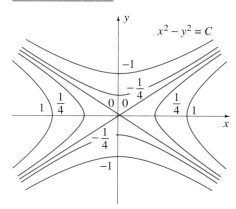

EXERCISES 13.1

* **1.** If $f(x, y) = x^3 y + x \sin y$, evaluate

   (a) $f(1, 2)$

   (b) $f(-2, -2)$

   (c) $f(x^2 + y, x - y^2)$

   (d) $f(x + h, y) - f(x, y)$

* **2.** If $f(x, y, z) = x^2 y^2 - x^4 + 4zx^2$, show that $f(a + b, a - b, ab) = 0$.

In Exercises 3–6 find and illustrate geometrically the largest possible domain for the function.

* **3.** $f(x, y) = \sqrt{4 - x^2 - y^2}$

* **4.** $f(x, y) = \ln(1 - x^2 + y^2)$

* **5.** $f(x, y) = \text{Sin}^{-1}(x^2 y + 1)$

* **6.** $f(x, y, z) = 1/(x^2 + y^2 + z^2)$

* **7.** For what values of x and y is the function $f(x, y) = \dfrac{12xy - x^2 y^2}{2(x + y)}$ equal to zero? Illustrate these values as points in the xy-plane. What is the largest domain of the function?

In Exercises 8–21 draw and plot the surface defined by the function.

* **8.** $f(x, y) = y^2$

* **9.** $f(x, y) = 4 - x - 2y$

* **10.** $f(x, y) = x^2 + y^2$

* **11.** $f(x, y) = \sqrt{x^2 + y^2}$

* **12.** $f(x, y) = y + x$

* **13.** $f(x, y) = 1 - x^3$

* **14.** $f(x, y) = 2(x^2 + y^2)$

* **15.** $f(x, y) = 1 - x^2 - 4y^2$

* **16.** $f(x, y) = xy$

** **17.** $f(x, y) = y - x^2$

** **18.** $f(x, y) = e^{-x^2 - y^2}$

** **19.** $f(x, y) = |x - y|$

** **20.** $f(x, y) = x^2 - y^2$

** **21.** $f(x, y) = \sqrt{1 + x^2 - y^2}$

In Exercises 22–25 draw level curves $f(x, y) = C$ corresponding to the values $C = -2, -1, 0, 1, 2$.

** **22.** $f(x, y) = 4 - \sqrt{4x^2 + y^2}$

** **23.** $f(x, y) = y - x^2$

** **24.** $f(x, y) = \ln(x^2 + y^2)$

** **25.** $f(x, y) = x^2 - y^2$

** **26.** A closed box is to have total surface area 30 m^2. Find a formula for the volume of the box in terms of its length l and width w.

** **27.** (a) A company wishes to construct a storage tank in the form of a rectangular box. If material for sides and top costs $1.25 per square metre and material for the bottom costs $4.75 per square metre, find the cost of building the tank as a function of its length l, width w, and height h.

(b) If the tank must hold 1000 m^3, find the construction cost in terms of l and w.

(c) Repeat (a) and (b) if the 12 edges of the tank must be welded at a cost of $7.50 per metre of weld.

** **28.** A rectangular box is inscribed inside the ellipsoid $x^2/a^2 + y^2/b^2 + z^2/c^2 = 1$ with sides parallel to the coordinate planes and corners on the ellipsoid. Find a formula for the volume of the box in terms of x and y.

** **29.** (a) A silo is to be built in the shape of a right circular cylinder surmounted by a right circular cone. If the radius of each is 6 m, find a formula for the volume V of the silo as a function of the heights H and h of the cylinder and cone.

(b) If the total surface area of the silo must be 200 m^2 (not including the base), find V as a function of h. (The area of the curved surface of a cone is $\pi r \sqrt{r^2 + h^2}$.)

** **30.** A long piece of metal 1 m wide is bent in two places A and B (figure below) to form a channel with three straight sides. Find a formula for the cross-sectional area of the channel in terms of x, θ, and ϕ.

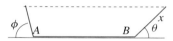

** **31.** (a) A uniform circular rod has flat ends at $x = 0$ and $x = \pi$ on the x-axis. The round side of the rod is insulated and the faces at $x = 0$ and $x = \pi$ are both kept at temperature 0°C for time $t > 0$. If the initial temperature (at time $t = 0$) of the rod is given by $100 \sin x$, $0 \le x \le \pi$, then the temperature thereafter is

$$T = f(x, t) = 100e^{-kt} \sin x \quad (k > 0 \text{ constant}).$$

(b) Draw the surface $T = f(x, t)$.

(c) Interpret physically the curves of intersection of this surface with planes $x = x_0$ and $t = t_0$.

** **32.** A cow's daily diet consists of three foods: hay, grain, and supplements. The animal is always given 11 kg of hay per day, 50% of which is digestive material and 12% of which is protein. Grain is 74% digestive and 8.8% protein. Supplements are 62% digestive material and 34% protein. Hay costs $27.50 for 1000 kg, whereas grain and supplements cost $110 and $175 respectively for 1000 kg. A healthy cow's daily diet must contain between 9.5 kg and 11.5 kg of digestive material and between 1.9 kg and 2.0 kg of protein. Find a formula for the cost per day, C, of feeding a cow in terms of the number of kilograms of grain, G, and supplements, S, fed to the cow daily. What is the domain of this function?

13.2 Limits and Continuity

The concepts of limit and continuity for multivariable functions are exactly the same as for functions of one variable; on the other hand, the work involved with the application of these concepts is much more complicated for multivariable functions.

Intuitively, a function $f(x, y)$ is said to have limit L as x and y approach x_0 and y_0, if $f(x, y)$ gets arbitrarily close to L, and stays close to L, as x and y get arbitrarily close to x_0 and y_0. To say this in a precise mathematical way, it is convenient to represent pairs of independent variables as points (x, y) in the xy-plane. We then have the following definition.

> **DEFINITION 13.1**
>
> A function $f(x, y)$ has limit L as (x, y) approaches (x_0, y_0), written
>
> $$\lim_{(x,y) \to (x_0, y_0)} f(x, y) = L, \tag{13.2}$$
>
> if, given any $\epsilon > 0$, we can find a $\delta > 0$ such that
>
> $$|f(x, y) - L| < \epsilon$$
>
> whenever $0 < \sqrt{(x - x_0)^2 + (y - y_0)^2} < \delta$ and (x, y) is in the domain of $f(x, y)$.

In other words, $f(x, y)$ has limit L as (x, y) approaches (x_0, y_0) if $f(x, y)$ can be made arbitrarily close to L (within ϵ) by choosing points (x, y) sufficiently close to (x_0, y_0) (within a circle of radius δ). Note the similarity of this definition to that for the limit of a function $f(x)$ of one variable in Section 2.5.

It is clear that

$$\lim_{(x,y) \to (2,1)} (x^2 + 2xy - 5) = 3,$$

but the limit

$$\lim_{(x,y) \to (0,0)} \frac{y^2 - x^2}{y^2 + x^2}$$

presents a problem, since both numerator and denominator approach zero as x and y approach zero.

To conclude that $\lim_{x \to a} f(x) = L$, the limit must be L no matter how x approaches a—be it through numbers larger than a, through numbers smaller than a, or through any other approach. For limit 13.2, the limit of $f(x, y)$ must also be L for all possible ways of approaching (x_0, y_0). But in this case there might be a multitude of ways of approaching (x_0, y_0). We might be able to approach (x_0, y_0) along straight lines with various slopes, along parabolas, along cubics, etc. Definition 13.1 implies, then, that the limit exists only if it is independent of the manner of approach. It is assumed, however, that we approach (x_0, y_0) only through points (x, y) that lie in the domain of definition of the function.

For the second example above, suppose we approach the origin along the straight line $y = mx$. Along this line,

$$\lim_{(x,y) \to (0,0)} \frac{y^2 - x^2}{y^2 + x^2} = \lim_{x \to 0} \frac{m^2 x^2 - x^2}{m^2 x^2 + x^2} = \frac{m^2 - 1}{m^2 + 1}.$$

Because this result depends on m, we have shown that as (x, y) approaches $(0, 0)$ along various straight lines, the function $(y^2 - x^2)/(y^2 + x^2)$ approaches different numbers. We conclude, therefore, that the function does not have a limit as (x, y) approaches $(0, 0)$. In Figure 13.6a we show a portion of the surface $z = (y^2 - x^2)/(y^2 + x^2)$ to illustrate our conclusion. The curve of intersection of the surface with the plane $y = 2x$ is shown in Figure 13.6b. Heights of points on the curve represent values of $(y^2 - x^2)/(y^2 + x^2)$ at points on the line $y = 2x$ in the xy-plane; they are always equal to 3/5. For other vertical planes containing the z-axis, curves of intersection are at different heights.

FIGURE 13.6a

FIGURE 13.6b

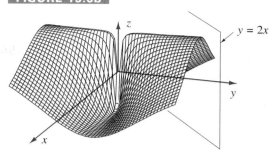

EXAMPLE 13.3

Evaluate

$$\lim_{(x,y)\to(0,0)} \frac{x^2 - y^2}{x + y}$$

if it exists.

SOLUTION   Because points on the line $y = -x$ are not within the domain of definition of the function, we can write that

$$\lim_{(x,y)\to(0,0)} \frac{x^2 - y^2}{x + y} = \lim_{(x,y)\to(0,0)} \frac{(x - y)(x + y)}{x + y} = \lim_{(x,y)\to(0,0)} (x - y) = 0.$$

The concept of continuity for multivariable functions is contained in the following definition.

DEFINITION 13.2

A function $f(x, y)$ is said to be **continuous** at a point (x_0, y_0) if

$$\lim_{(x,y)\to(x_0,y_0)} f(x, y) = f(x_0, y_0). \tag{13.3}$$

Just as Definition 2.1 for continuity of a function of only one variable contains three conditions, so too does Definition 13.2. It demands that:

1. $f(x, y)$ be defined at (x_0, y_0);

2. $\displaystyle\lim_{(x,y)\to(x_0,y_0)} f(x, y)$ exist;

3. the numbers in (1) and (2) be the same.

The function $f(x, y) = (y^2 - x^2)/(y^2 + x^2)$ in Figure 13.6 is discontinuous at $(0, 0)$. The function is undefined at $(0, 0)$ and the limit as $(x, y) \to (0, 0)$ does not exist. Geometrically, a function $f(x, y)$ is continuous at a point (x_0, y_0) if the surface $z = f(x, y)$ is not separated at the point $(x_0, y_0, f(x_0, y_0))$. The function $f(x, y) = 1 - e^{-1/(x^2+y^2)}$ is discontinuous at $(0, 0)$ since it is undefined for $x = 0$ and $y = 0$. The surface has a hole at $(0, 0)$ (Figure 13.7). The function $f(x, y) = \text{sgn}\left[(x - 1)^2 + (y - 2)^2\right]$ (see Exercise 43 in Section 2.4) is discontinuous at $(1, 2)$; it has value 0 at $x = 1$ and $y = 2$, and value 1 for all other values of x and y (Figure 13.8).

FIGURE 13.7

FIGURE 13.8

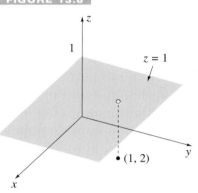

EXERCISES 13.2

In Exercises 1–20 evaluate the limit, if it exists.

* **1.** $\displaystyle \lim_{(x,y)\to(2,-3)} \frac{x^2 - 1}{x + y}$

* **2.** $\displaystyle \lim_{(x,y)\to(1,1)} \frac{x^3 + 2y^3}{x^3 + 4y^3}$

* **3.** $\displaystyle \lim_{(x,y)\to(3,2)} \frac{2x - 3y}{x + y}$

* **4.** $\displaystyle \lim_{(x,y,z)\to(2,3,-1)} \frac{xyz}{x^2 + y^2 + z^2}$

* **5.** $\displaystyle \lim_{(x,y)\to(1,0)} \frac{x}{y}$

* **6.** $\displaystyle \lim_{(x,y,z)\to(0,\pi/2,1)} \text{Tan}^{-1}\left[x/(yz)\right]$

* **7.** $\displaystyle \lim_{(x,y,z)\to(0,\pi/2,1)} \text{Tan}^{-1}\left(yz/x\right)$

* **8.** $\displaystyle \lim_{(x,y,z)\to(0,\pi/2,1)} \text{Tan}^{-1}\left|yz/x\right|$

* **9.** $\displaystyle \lim_{(x,y)\to(3,4)} \frac{|x^2 - y^2|}{x^2 - y^2}$

* **10.** $\displaystyle \lim_{(x,y)\to(3,4)} \frac{|x^2 + y^2|}{x^2 + y^2}$

* **11.** $\displaystyle \lim_{(x,y)\to(2,1)} \frac{x^2 - y^2}{x - y}$

** **12.** $\displaystyle \lim_{(x,y)\to(2,2)} \frac{x^2 - y^2}{x - y}$

** **13.** $\displaystyle \lim_{(x,y)\to(0,0)} \frac{x^2 - y^2}{x - y}$

** **14.** $\displaystyle \lim_{(x,y)\to(0,0)} \frac{x - y}{x + y}$

* **15.** $\displaystyle \lim_{(x,y,z)\to(0,0,0)} \frac{x^2 - y^2}{y^2 + z^2 + 1}$

* **16.** $\displaystyle \lim_{(x,y)\to(2,1)} \frac{(x - 2)^2(y + 1)}{x - 2}$

* **17.** $\displaystyle \lim_{(x,y,z)\to(1,1,1)} |2x - y - z|$

** **18.** $\displaystyle \lim_{(x,y)\to(0,0)} \frac{3x^3 - y^3}{2x^3 + 4y^3}$

** **19.** $\displaystyle \lim_{(x,y)\to(0,0)} \text{Sec}^{-1}\left(\frac{-1}{x^2 + y^2}\right)$

** **20.** $\displaystyle \lim_{(x,y)\to(0,0)} \text{Sec}^{-1}(x^2 + y^2)$

In Exercises 21–26 find all points of discontinuity for the function.

* **21.** $f(x, y) = \dfrac{x^2 - 1}{x + y}$

* **22.** $f(x, y) = \dfrac{xy}{x^2 + y^2}$

* **23.** $f(x, y) = \dfrac{1}{1 - x^2 - y^2}$

* **24.** $f(x, y, z) = \dfrac{1}{xyz}$

* **25.** $f(x, y) = |x - y|$

** **26.** $f(x, y) = \dfrac{x + y}{x^2 y + xy^2}$

** **27.** Evaluate $\lim\limits_{(x,y) \to (a,a)} \left[\cos (x + y) - \sqrt{1 - \sin^2 (x + y)} \right]$ where $0 \le a \le \pi/2$.

In Exercises 28–31 evaluate the limit, if it exists.

** **28.** $\lim\limits_{(x,y) \to (0,0)} \dfrac{x^4 + y^2}{x^4 - y^2}$ Hint: Approach $(0, 0)$ along parabolas.

** **29.** $\lim\limits_{(x,y) \to (1,1)} \dfrac{x^2 - 2x + y^2 + 2y - 2}{x^2 - y^2 - 2x + 2y}$ Hint: Approach $(1, 1)$ along straight lines.

** **30.** $\lim\limits_{(x,y) \to (1,0)} \dfrac{\sqrt{x + y} - \sqrt{x - y}}{y}$

** **31.** $\lim\limits_{(x,y) \to (0,0)} \dfrac{\sin (x^2 + y^2)}{x^2 + y^2}$

** **32.** (a) Does the limit $\lim\limits_{(x,y) \to (1,1)} \dfrac{\sin (x - y)}{x - y}$ exist? Explain. Is the function continuous at $(1, 1)$?

   (b) If we define the function everywhere by giving it the value 1 along the line $y = x$, does the limit of the function exist at $(1, 1)$? Is the function continuous at $(1, 1)$?

** **33.** Give a mathematical definition for $\lim\limits_{(x,y,z) \to (x_0,y_0,z_0)} f(x, y, z) = L$.

** **34.** Is the following statement true or false? If a function $f(x, y)$ is undefined at every point on a curve C, then for any point (x_0, y_0) on C, $\lim\limits_{(x,y) \to (x_0,y_0)} f(x, y)$ does not exist. Explain. Give an example.

** **35.** Let $f(x, y) = \begin{cases} \dfrac{x^2 y^2}{x^4 + y^4} & \text{if } (x, y) \ne (0, 0) \\ 0 & \text{if } (x, y) = (0, 0). \end{cases}$

   (a) Show that $f(x, y)$ is continuous in each variable separately at $(0, 0)$. In other words, show that $f(x, 0)$ and $f(0, y)$ are continuous at $x = 0$ and $y = 0$.

   (b) Show that $f(x, y)$ is not continuous at $(0, 0)$.

*** **36.** Prove that:

   (a) $\lim\limits_{(x,y) \to (0,0)} (xy + 5) = 5$

   (b) $\lim\limits_{(x,y) \to (1,1)} (x^2 + 2xy + 5) = 8$

EXERCISES 13.3

In Exercises 1–20 evaluate $\partial f/\partial x$ and $\partial f/\partial y$.

∗ **1.** $f(x, y) = x^3 y^2 + 2xy$

∗ **2.** $f(x, y) = 3xy - 4x^4 y^4$

∗ **3.** $f(x, y) = x^4/y^3$

∗ **4.** $f(x, y) = x/(x + y) - x/y$

∗ **5.** $f(x, y) = x/(2x^2 + y)$

∗ **6.** $f(x, y) = \sin(xy)$

∗ **7.** $f(x, y) = x \cos(x + y)$

∗ **8.** $f(x, y) = \sqrt{x^2 + y^2}$

∗ **9.** $f(x, y) = x\sqrt{x^2 - y^2}$

∗ **10.** $f(x, y) = \tan(2x^2 + y^2)$

∗ **11.** $f(x, y) = e^{x+y}$

∗ **12.** $f(x, y) = e^{xy}$

∗ **13.** $f(x, y) = xye^{xy}$

∗ **14.** $f(x, y) = \ln(x^2 + y^2)$

∗ **15.** $f(x, y) = (x + 1) \ln(xy)$

∗ **16.** $f(x, y) = \sin(ye^x)$

∗ **17.** $f(x, y) = \text{Tan}^{-1}(x/y)$

∗ **18.** $f(x, y) = \sqrt[3]{1 - \cos^3(x^2 y)}$

∗ **19.** $f(x, y) = \dfrac{\sin x}{\cos y}$

∗ **20.** $f(x, y) = \ln\left(\sec\sqrt{x + y}\right)$

In Exercises 21–30 evaluate the indicated derivative.

∗ **21.** $\partial f/\partial x$ if $f(x, y, z) = xyze^{x^2+y^2}$

∗ **22.** $\partial f/\partial z$ if $f(x, z) = \text{Tan}^{-1}[1/(x^2 + z^2)]$

∗ **23.** $\partial f/\partial y$ at $(1, 1, 0)$ if $f(x, y, z) = xy(x^2 + y^2 + z^2)^{1/3}$

∗ **24.** $\partial f/\partial x$ at $(1, -1, 1, -1)$ if $f(x, y, z, t) = zt/(x^2 + y^2 - t^2)$

∗ **25.** $\partial f/\partial t$ if $f(x, y, t) = x\sqrt{t^2 - y^2}/t^2 + (x/y)\text{Sec}^{-1}(t/3)$

∗ **26.** $\partial f/\partial x$ if $f(x, y, z) = \text{Cot}^{-1}(1 + x + y + z)$

∗ **27.** $\partial f/\partial y$ at $(1, 2, 3)$ if $f(x, y, t) = \text{Sin}^{-1}(xyt)/\text{Cos}^{-1}(xyt)$

∗ **28.** $\partial f/\partial x$ if $f(x, y, z) = x^3/y + x \sin(yz/x)$

∗ **29.** $\partial f/\partial t$ if $f(x, y, z, t) = xyz \ln(x^2 + y^2 + z^2)$

∗ **30.** $\partial f/\partial z$ if $f(x, z) = (z^2/2)\text{Sin}^{-1}(x/z) + (x/2)\sqrt{z^2 - x^2}$

∗∗ **31.** If $f(x, y) = x^3 y/(x - y)$, show that $x\dfrac{\partial f}{\partial x} + y\dfrac{\partial f}{\partial y} = 3f(x, y)$.

∗∗ **32.** If $f(x, y, z) = (x^4 + y^4 + z^4)/(xyz)$, show that $x\dfrac{\partial f}{\partial x} + y\dfrac{\partial f}{\partial y} + z\dfrac{\partial f}{\partial z} = f(x, y, z)$.

∗∗ **33.** If $f(x, y, z) = (x^2 + y^2)\cos[(y + z)/x]$, show that $x\dfrac{\partial f}{\partial x} + y\dfrac{\partial f}{\partial y} + z\dfrac{\partial f}{\partial z} = 2f(x, y, z)$.

∗∗ **34.** To evaluate $\partial f/\partial x$ for $f(x, y)$ at the point $(1, 2)$, state which of the following are acceptable:

(a) Differentiate $f(x, y)$ with respect to x holding y constant, and then set $x = 1$ and $y = 2$.

(b) Set $x = 1$ and $y = 2$, and then differentiate with respect to x.

(c) Set $y = 2$, differentiate with respect to x, and set $x = 1$.

(d) Set $x = 1$, differentiate with respect to x, and set $y = 2$.

∗∗ **35.** Temperature at points (x, y) in a semicircular plate defined by $x^2 + y^2 \leq 4$, $y \geq 0$ is given by $T(x, y) = 16x^2 - 24xy + 40y^2$. Find, if possible:

(a) $T_x(1, 1)$ (b) $T_y(1, 1)$ (c) $T_x(1, 0)$ (d) $T_y(1, 0)$ (e) $T_x(0, 2)$ (f) $T_y(0, 2)$

** **36.** Can you find a function $f(x, y)$ so that $f_x(x, y) = 2x - 3y$ and $f_y(x, y) = 3x + 4y$?

** **37.** Suppose a, b, and c are the lengths of the sides of a triangle and A, B, and C are the opposite angles. Find:

(a) $a_A(b, c, A)$ (b) $A_a(a, b, c)$ (c) $a_b(b, c, A)$ (d) $A_b(a, b, c)$

** **38.** The equation of continuity for three-dimensional unsteady flow of a compressible fluid is

$$\frac{\partial \rho}{\partial t} + \frac{\partial}{\partial x}(\rho u) + \frac{\partial}{\partial y}(\rho v) + \frac{\partial}{\partial z}(\rho w) = 0,$$

where $\rho(x, y, z, t)$ is the density of the fluid, and $u\hat{\mathbf{i}} + v\hat{\mathbf{j}} + w\hat{\mathbf{k}}$ is the velocity of the fluid at position (x, y, z) and time t. Determine whether the continuity equation is satisfied if:

(a) $\rho = $ constant, $u = (2x^2 - xy + z^2)t$, $v = (x^2 - 4xy + y^2)t$,
$w = (-2xy - yz + y^2)t$

(b) $\rho = xy + zt$, $u = x^2y + t$, $v = y^2z - 2t^2$, $w = 5x + 2z$

** **39.** In complex variable theory, two functions $u(x, y)$ and $v(x, y)$ are said to be *harmonic conjugates* in a region R if in R they satisfy the Cauchy-Riemann equations

$$\frac{\partial u}{\partial x} = \frac{\partial v}{\partial y}, \qquad \frac{\partial v}{\partial x} = -\frac{\partial u}{\partial y}.$$

Show that the following pairs of functions are harmonic conjugates:

(a) $u(x, y) = -3xy^2 + y + x^3$, $v(x, y) = 3x^2y - y^3 - x + 5$
(b) $u(x, y) = (x^2 + x + y^2)/(x^2 + y^2)$, $v(x, y) = -y/(x^2 + y^2)$
(c) $u(x, y) = e^x(x \cos y - y \sin y)$, $v(x, y) = e^x(x \sin y + y \cos y)$

** **40.** If r and θ are polar coordinates, then the Cauchy-Riemann equations in Exercise 39 for functions $u(r, \theta)$ and $v(r, \theta)$ take the form

$$\frac{\partial u}{\partial r} = \frac{1}{r}\frac{\partial v}{\partial \theta}, \qquad \frac{1}{r}\frac{\partial u}{\partial \theta} = -\frac{\partial v}{\partial r}, \qquad r \neq 0.$$

Show that the following pairs of functions satisfy these equations:

(a) $u(r, \theta) = (r^2 + r \cos \theta)/(1 + r^2 + 2r \cos \theta)$, $v(r, \theta) = r \sin \theta/(1 + r^2 + 2r \cos \theta)$
(b) $u(r, \theta) = \sqrt{r} \cos(\theta/2)$, $v(r, \theta) = \sqrt{r} \sin(\theta/2)$
(c) $u(r, \theta) = \ln r$, $v(r, \theta) = \theta$

13.4 Gradients

Suppose a function $f(x, y, z)$ is defined at each point in some region of space, and that at each point of the region all three partial derivatives

$$\frac{\partial f}{\partial x}, \quad \frac{\partial f}{\partial y}, \quad \frac{\partial f}{\partial z}$$

exist. For example, if $f(x, y, z)$ represents the present temperature at each point in the room in which you are working, then these derivatives represent rates of change of temperature in directions parallel to the x-, y-, and z-axes, respectively. There is a particular combination of these derivatives that proves very useful in later work. This combination is contained in the following definition.

DEFINITION 13.4

If a function $f(x, y, z)$ has partial derivatives $\partial f/\partial x$, $\partial f/\partial y$, and $\partial f/\partial z$ at each point in some region D of space, then at each point in D we define a vector called the **gradient** of $f(x, y, z)$, written grad f or ∇f, by

$$\text{grad} f = \nabla f = \frac{\partial f}{\partial x}\hat{\mathbf{i}} + \frac{\partial f}{\partial y}\hat{\mathbf{j}} + \frac{\partial f}{\partial z}\hat{\mathbf{k}}. \tag{13.8}$$

For a function $f(x, y)$ of only two independent variables, we have

$$\nabla f = \frac{\partial f}{\partial x}\hat{\mathbf{i}} + \frac{\partial f}{\partial y}\hat{\mathbf{j}}. \tag{13.9}$$

EXAMPLE 13.6

If $f(x, y, z) = x^2 yz - 2x/y$, find ∇f at $(1, -1, 3)$.

SOLUTION   Since $\nabla f = (2xyz - 2/y)\hat{\mathbf{i}} + (x^2 z + 2x/y^2)\hat{\mathbf{j}} + (x^2 y)\hat{\mathbf{k}}$, we have

$$\nabla f|_{(1,-1,3)} = -4\hat{\mathbf{i}} + 5\hat{\mathbf{j}} - \hat{\mathbf{k}}.$$

EXAMPLE 13.7

If $f(x, y, z) = \text{Tan}^{-1}(xy/z)$, what is ∇f?

SOLUTION

$$\nabla f = \left\{\frac{1}{1 + (xy/z)^2}\left(\frac{y}{z}\right)\right\}\hat{\mathbf{i}} + \left\{\frac{1}{1 + (xy/z)^2}\left(\frac{x}{z}\right)\right\}\hat{\mathbf{j}} + \left\{\frac{1}{1 + (xy/z)^2}\left(\frac{-xy}{z^2}\right)\right\}\hat{\mathbf{k}}$$

$$= \frac{yz}{z^2 + x^2y^2}\hat{\mathbf{i}} + \frac{xz}{z^2 + x^2y^2}\hat{\mathbf{j}} - \frac{xy}{z^2 + x^2y^2}\hat{\mathbf{k}}$$

$$= (yz\hat{\mathbf{i}} + xz\hat{\mathbf{j}} - xy\hat{\mathbf{k}})/(z^2 + x^2y^2).$$

Gradients arise in a multitude of applications in applied mathematics—heat conduction, electromagnetic theory, and fluid flow, to name a few—and two of the properties that make them so indispensable are discussed in detail in Sections 13.8 and 13.9. Examples 13.8 and 13.9 suggest these properties, but we make no attempt at a complete discussion here. For the moment we simply want you to be familiar with the definition of gradients and be able to calculate them.

EXAMPLE 13.8

The equation $F(x, y, z) = 0$, where $F(x, y, z) = x^2 + y^2 + z^2 - 4$, defines a sphere of radius 2 centred at the origin. Show that the gradient vector ∇F at any point on the sphere is perpendicular to the sphere.

SOLUTION If $P(x, y, z)$ is any point on the sphere (Figure 13.10), then the position vector $\mathbf{r} = x\hat{\mathbf{i}} + y\hat{\mathbf{j}} + z\hat{\mathbf{k}}$ from the origin to P is clearly perpendicular to the sphere. On the other hand,

$$\nabla F = 2x\hat{\mathbf{i}} + 2y\hat{\mathbf{j}} + 2z\hat{\mathbf{k}} = 2\mathbf{r}.$$

Consequently, at any point P on the sphere, ∇F is also perpendicular to the sphere.

FIGURE 13.10

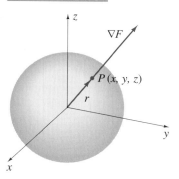

This example suggests that gradients may be useful in finding perpendiculars to surfaces (and, as we will see, perpendiculars to curves).

EXAMPLE 13.9

The function $f(x, y) = 2x^2 - 4x + 3y^2 + 2y + 6$ is defined at every point in the xy-plane. If we start at the origin $(0, 0)$ and move along the positive x-axis, the rate of change of the function is $f_x(0, 0) = -4$; if we move along the y-axis, the rate of change is $f_y(0, 0) = 2$. Calculate the rate of change of $f(x, y)$ at $(0, 0)$ if we move toward the point $(1, 1)$ along the line $y = x$, and show that it is equal to the component of $\nabla f_{|(0,0)}$ in the direction $\mathbf{v} = (1, 1)$.

SOLUTION The difference in values of $f(x, y)$ at any point (x, y) and $(0, 0)$ is $f(x, y) - f(0, 0) = (2x^2 - 4x + 3y^2 + 2y + 6) - (6) = 2x^2 - 4x + 3y^2 + 2y$. If we divide this by the length of the line joining $(0, 0)$ and (x, y), we obtain

$$\frac{f(x, y) - f(0, 0)}{\sqrt{x^2 + y^2}} = \frac{2x^2 - 4x + 3y^2 + 2y}{\sqrt{x^2 + y^2}}.$$

The limit of this quotient as (x, y) approaches $(0, 0)$ along the line $y = x$ should yield the required rate of change. We therefore set $y = x$ and take the limit as x approaches zero through positive numbers:

$$\lim_{x \to 0^+} \frac{2x^2 - 4x + 3x^2 + 2x}{\sqrt{x^2 + x^2}} = \lim_{x \to 0^+} \frac{x(5x - 2)}{\sqrt{2}x} = -\sqrt{2}.$$

This quantity, then, is the rate of change of $f(x, y)$ at $(0, 0)$ along the line $y = x$ toward the point $(1, 1)$.

On the other hand,

$$\nabla f_{|(0,0)} = (4x - 4, 6y + 2)_{|(0,0)} = (-4, 2),$$

and the component of this vector in the direction $\mathbf{v} = (1, 1)$ is

$$\nabla f_{|(0,0)} \cdot \hat{\mathbf{v}} = (-4, 2) \cdot \frac{(1, 1)}{\sqrt{2}} = \frac{-4 + 2}{\sqrt{2}} = -\sqrt{2}.$$

This example indicates that gradients may be useful in calculating rates of change of functions in directions other than those parallel to the coordinate axes.

EXAMPLE 13.10

The electrostatic potential at a point (x, y, z) in space due to a charge q fixed at the origin is given by

$$V = \frac{q}{4\pi \epsilon_0 r},$$

where $r = \sqrt{x^2 + y^2 + z^2}$. If a second charge Q is placed at (x, y, z), it experiences a force $\mathbf{F}$, where

$$\mathbf{F} = \frac{qQ}{4\pi \epsilon_0 r^3} \mathbf{r},$$

where $\mathbf{r} = x\hat{\mathbf{i}} + y\hat{\mathbf{j}} + z\hat{\mathbf{k}}$. Show that $\mathbf{F} = -Q\nabla V$.

SOLUTION   We show that $\nabla V = -\mathbf{F}/Q$,

$$\nabla V = \nabla\left(\frac{q}{4\pi \epsilon_0 r}\right) = \frac{q}{4\pi \epsilon_0}\nabla\left(\frac{1}{\sqrt{x^2 + y^2 + z^2}}\right)$$

$$= \frac{q}{4\pi \epsilon_0}\left\{\frac{-x}{(x^2 + y^2 + z^2)^{3/2}}\hat{\mathbf{i}} + \frac{-y}{(x^2 + y^2 + z^2)^{3/2}}\hat{\mathbf{j}} + \frac{-z}{(x^2 + y^2 + z^2)^{3/2}}\hat{\mathbf{k}}\right\}$$

$$= \frac{-q}{4\pi \epsilon_0 r^3}(x\hat{\mathbf{i}} + y\hat{\mathbf{j}} + z\hat{\mathbf{k}}) = \frac{-q}{4\pi \epsilon_0 r^3}\mathbf{r} = -\frac{\mathbf{F}}{Q}.$$

EXERCISES 13.4

In Exercises 1–10 find the gradient of the function.

* **1.** $f(x, y, z) = x^2y + xz + yz^2$
* **2.** $f(x, y, z) = x^2yz$
* **3.** $f(x, y, z) = x^2y/z - 2xz^6$
* **4.** $f(x, y) = x^2y + xy^2$
* **5.** $f(x, y) = \sin(x + y)$
* **6.** $f(x, y, z) = \text{Tan}^{-1}(xyz)$
* **7.** $f(x, y) = \text{Tan}^{-1}(y/x)$
* **8.** $f(x, y, z) = e^{x+y+z}$
* **9.** $f(x, y) = 1/(x^2 + y^2)$
* **10.** $f(x, y, z) = 1/\sqrt{x^2 + y^2 + z^2}$

In Exercises 11–15 find the gradient of the function at the point.

* **11.** $f(x, y) = xy + x + y$ at $(1, 3)$ * **12.** $f(x, y, z) = \cos(x + y + z)$ at $(-1, 1, 1)$

* **13.** $f(x, y, z) = (x^2 + y^2 + z^2)^2$ at $(0, 3, 6)$

* **14.** $f(x, y) = e^{-x^2 - y^2}$ at $(2, 2)$ * **15.** $f(x, y, z) = xy \ln(x + y)$ at $(4, -2)$

* **16.** The equation $F(x, y, z) = Ax + By + Cz + D = 0$ defines a plane in space. Show that at any point on the plane the vector ∇F is perpendicular to the plane.

* **17.** Use the result of Exercise 16 to illustrate that a vector along the line

$$F(x, y, z) = 2x + 3y - 2z + 4 = 0,$$

$$G(x, y, z) = x - y + 3z + 6 = 0$$

is $\nabla F \times \nabla G$. Find parametric equations for the line.

* **18.** Prove that, if $f(x, y, z)$ and $g(x, y, z)$ both have gradients, then $\nabla(fg) = f\nabla g + g\nabla f$. What does this remind you of?

** **19.** Repeat Example 13.9 for the functions (a) $f(x, y) = x^2 + y^2$ and (b) $f(x, y) = 2x^3 - 3y$.

** **20.** The equation $F(x, y) = x^3 + xy + y^4 - 5 = 0$ implicitly defines a curve in the xy-plane. Show that at any point on the curve ∇F is a normal vector to the curve.

Draw the surface defined by the equations in Exercises 21 and 22. At what points on the surface is ∇F not defined?

** **21.** $F(x, y, z) = z - \sqrt{x^2 + y^2} = 0$ ** **22.** $F(x, y, z) = z - |x - y| = 0$

** **23.** If $f(x, y) = 1 - x^2 - y^2$, find ∇f. Find the point (x, y) at which $\nabla f = \mathbf{0}$, and illustrate graphically the nature of the surface $z = f(x, y)$ at this point.

** **24.** If the gradient of a function $f(x, y)$ is $\nabla f = (2xy - y)\hat{\mathbf{i}} + (x^2 - x)\hat{\mathbf{j}}$, what is $f(x, y)$?

** **25.** Repeat Exercise 24 if $\nabla f = (2x/y + 1)\hat{\mathbf{i}} + (-x^2/y^2 + 2)\hat{\mathbf{j}}$.

** **26.** If the gradient of a function $f(x, y, z)$ is $\nabla f = yz\hat{\mathbf{i}} + (xz + 2yz)\hat{\mathbf{j}} + (xy + y^2)\hat{\mathbf{k}}$, what is $f(x, y, z)$?

** **27.** Repeat Exercise 26 if $\nabla f = (x\hat{\mathbf{i}} + y\hat{\mathbf{j}} + z\hat{\mathbf{k}})/\sqrt{x^2 + y^2 + z^2}$.

** **28.** If $f(x, y)$ and $g(x, y)$ have first partial derivatives in a region R of the xy-plane, and if in R, $\nabla f = \nabla g$, how are $f(x, y)$ and $g(x, y)$ related?

** **29.** If $\nabla f = \mathbf{0}$ for all points in some region R of space, what can we say about $f(x, y, z)$ in R?

*** **30.** Show that if the equation $F(x, y) = 0$ implicitly defines a curve C in the xy-plane, then at any point on C the vector ∇F is perpendicular to C.

13.5 Higher-Order Partial Derivatives

If $f(x, y) = x^3 y^2 + ye^x$, then

$$\frac{\partial f}{\partial x} = 3x^2 y^2 + ye^x \quad \text{and} \quad \frac{\partial f}{\partial y} = 2x^3 y + e^x.$$

Since each of these partial derivatives is a function of x and y, we can take further partial derivatives. The partial derivative of $\partial f/\partial x$ with respect to x is called the second partial derivative of $f(x, y)$ with respect to x, and is written

$$\frac{\partial}{\partial x}\left(\frac{\partial f}{\partial x}\right) = \frac{\partial^2 f}{\partial x^2} = 6xy^2 + ye^x.$$

Similarly, we have three more second partial derivatives:

$$\frac{\partial}{\partial y}\left(\frac{\partial f}{\partial x}\right) = \frac{\partial^2 f}{\partial y \partial x} = 6x^2 y + e^x,$$

$$\frac{\partial}{\partial x}\left(\frac{\partial f}{\partial y}\right) = \frac{\partial^2 f}{\partial x \partial y} = 6x^2 y + e^x,$$

$$\frac{\partial}{\partial y}\left(\frac{\partial f}{\partial y}\right) = \frac{\partial^2 f}{\partial y^2} = 2x^3.$$

Note that the second partial derivatives $\partial^2 f/\partial x \partial y$ and $\partial^2 f/\partial y \partial x$ are identical. This is not a peculiarity of this function; according to the following theorem it is to be expected.

THEOREM 13.1

If $f(x, y)$, $\partial f/\partial x$, $\partial f/\partial y$, $\partial^2 f/\partial x \partial y$, and $\partial^2 f/\partial y \partial x$ are all defined inside a circle centred at a point P, and are continuous at P, then at P

$$\frac{\partial^2 f}{\partial x \partial y} = \frac{\partial^2 f}{\partial y \partial x}. \tag{13.10}$$

Corresponding to the subscript notation $f_x(x, y)$ for $\partial f/\partial x$, we have the following notations for second partial derivatives,

$$\frac{\partial^2 f}{\partial x^2} = f_{xx}, \quad \frac{\partial^2 f}{\partial x \partial y} = f_{yx}, \quad \frac{\partial^2 f}{\partial y \partial x} = f_{xy}, \quad \frac{\partial^2 f}{\partial y^2} = f_{yy}.$$

Notice the reversal in order of x and y in the middle terms. In subscript notation, derivatives are taken in the order in which they appear (left to right, y first, x second, in f_{yx}). In $\partial^2 f/\partial x \partial y$, derivatives are done right to left, y first, x second. Because of Theorem 13.1, the order is usually irrelevant anyway.

Partial derivatives of orders higher than two are also possible. For the above function $f(x, y) = x^3 y^2 + ye^x$, we have:

$$f_{xxy} = \frac{\partial^3 f}{\partial y \partial x^2} = \frac{\partial}{\partial y}\left(\frac{\partial^2 f}{\partial x^2}\right) = \frac{\partial}{\partial y}\left(6xy^2 + ye^x\right) = 12xy + e^x$$

and

$$f_{yxyx} = \frac{\partial}{\partial x}\left[\frac{\partial}{\partial y}\left(\frac{\partial^2 f}{\partial x \partial y}\right)\right] = \frac{\partial}{\partial x}\left[\frac{\partial}{\partial y}\left(6x^2 y + e^x\right)\right] = \frac{\partial}{\partial x}\left[6x^2\right] = 12x.$$

For most functions with which we will be concerned, Theorem 13.1 can be extended to say that a mixed partial derivative may be calculated in any order whatsoever. For example, if we require $\partial^{10} f / \partial x^3 \partial y^7$, where $f(x, y) = \ln(y^y) + x^2 y^{10}$, it is advantageous to reverse the order of differentiation:

$$\frac{\partial^{10} f}{\partial x^3 \partial y^7} = \frac{\partial^{10} f}{\partial y^7 \partial x^3} = \frac{\partial^7}{\partial y^7}\left(\frac{\partial^3 f}{\partial x^3}\right) = 0.$$

EXAMPLE 13.11

Show that $f(x, y) = \mathrm{Tan}^{-1}(2xy/(x^2 - y^2))$ satisfies the equation

$$\frac{\partial^2 f}{\partial x^2} + \frac{\partial^2 f}{\partial y^2} = 0$$

at all points in the xy-plane that are not on the lines $y = \pm x$.

SOLUTION   Since

$$\frac{\partial f}{\partial x} = \frac{1}{1 + \left(\dfrac{2xy}{x^2 - y^2}\right)^2}\left\{\frac{(x^2 - y^2)(2y) - 2xy(2x)}{(x^2 - y^2)^2}\right\}$$

$$= \frac{(x^2 - y^2)^2}{(x^2 - y^2)^2 + 4x^2 y^2}\left\{\frac{2y(x^2 - y^2 - 2x^2)}{(x^2 - y^2)^2}\right\}$$

$$= \frac{-2y(x^2 + y^2)}{x^4 - 2x^2 y^2 + y^4 + 4x^2 y^2} = \frac{-2y(x^2 + y^2)}{(x^2 + y^2)^2} = \frac{-2y}{x^2 + y^2},$$

the second derivative with respect to x is

$$\frac{\partial^2 f}{\partial x^2} = \frac{2y}{(x^2 + y^2)^2}(2x) = \frac{4xy}{(x^2 + y^2)^2}.$$

Similarly,

$$\frac{\partial f}{\partial y} = \frac{1}{1 + \left(\dfrac{2xy}{x^2 - y^2}\right)^2}\left\{\frac{(x^2 - y^2)(2x) - 2xy(-2y)}{(x^2 - y^2)^2}\right\}$$

$$= \frac{(x^2 - y^2)^2}{(x^2 + y^2)^2}\left\{\frac{2x(x^2 - y^2 + 2y^2)}{(x^2 - y^2)^2}\right\} = \frac{2x}{x^2 + y^2}.$$

Consequently,

$$\frac{\partial^2 f}{\partial y^2} = \frac{-2x}{(x^2 + y^2)^2}(2y) = \frac{-4xy}{(x^2 + y^2)^2},$$

and addition of these two expressions for $\partial^2 f / \partial x^2$ and $\partial^2 f / \partial y^2$ completes the proof.

The equation

$$\frac{\partial^2 f}{\partial x^2} + \frac{\partial^2 f}{\partial y^2} = 0 \qquad (13.11)$$

for a function $f(x, y)$ is one of the most important equations in applied mathematics. It is called Laplace's equation in two variables (x and y). Laplace's equation for a function $f(x, y, z)$ of three variables is

$$\frac{\partial^2 f}{\partial x^2} + \frac{\partial^2 f}{\partial y^2} + \frac{\partial^2 f}{\partial z^2} = 0. \qquad (13.12)$$

A function is said to be **harmonic** in a region R if it satisfies Laplace's equation in R and has continuous second partial derivatives in R. In particular, the function $f(x, y)$ in Example 13.11 is harmonic in any region that does not contain points on the lines $y = \pm x$. The next two examples illustrate areas of applied mathematics in which Laplace's equation is prominent. A third is contained in Exercise 30.

EXAMPLE 13.12

Show that the electrostatic potential function of Example 13.10 is harmonic in any region not containing the origin.

SOLUTION With $V = \left[q/(4\pi\epsilon_0)\right](x^2 + y^2 + z^2)^{-1/2}$,

$$\frac{\partial V}{\partial x} = \frac{q}{4\pi\epsilon_0}\left[\frac{-x}{(x^2 + y^2 + z^2)^{3/2}}\right],$$

and

$$\frac{\partial^2 V}{\partial x^2} = \frac{-q}{4\pi\epsilon_0}\left[\frac{1}{(x^2 + y^2 + z^2)^{3/2}} + \frac{-3x^2}{(x^2 + y^2 + z^2)^{5/2}}\right]$$

$$= \frac{q}{4\pi\epsilon_0}\left[\frac{2x^2 - y^2 - z^2}{(x^2 + y^2 + z^2)^{5/2}}\right].$$

Similarly,

$$\frac{\partial^2 V}{\partial y^2} = \frac{q}{4\pi\epsilon_0}\left[\frac{2y^2 - x^2 - z^2}{(x^2 + y^2 + z^2)^{5/2}}\right], \qquad \frac{\partial^2 V}{\partial z^2} = \frac{q}{4\pi\epsilon_0}\left[\frac{2z^2 - x^2 - y^2}{(x^2 + y^2 + z^2)^{5/2}}\right].$$

Addition of these shows that $V_{xx} + V_{yy} + V_{zz} = 0$. Since second partial derivatives are continuous in any region not containing $(0, 0, 0)$, $V(x, y, z)$ is harmonic therein.

EXAMPLE 13.13

Figure 13.11 shows a 1 m by 1 m metal plate that is insulated top and bottom. Temperature along the edges $x = 0$, $x = 1$, and $y = 1$ is held at $0°$C, while that along $y = 0$ is $f(x) = 4\sin\pi x$. Steady-state temperature at points inside the plate is then

$$T(x, y) = \frac{4}{e^\pi - e^{-\pi}}\left[e^{\pi(1-y)} - e^{-\pi(1-y)}\right]\sin\pi x.$$

Show that $T(x, y)$ is harmonic inside the plate.

FIGURE 13.11

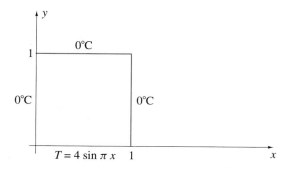

SOLUTION   Partial derivatives of $T(x, y)$ are

$$\frac{\partial T}{\partial x} = \frac{4\pi}{e^\pi - e^{-\pi}} \left[e^{\pi(1-y)} - e^{-\pi(1-y)} \right] \cos \pi x,$$

$$\frac{\partial^2 T}{\partial x^2} = \frac{-4\pi^2}{e^\pi - e^{-\pi}} \left[e^{\pi(1-y)} - e^{-\pi(1-y)} \right] \sin \pi x,$$

$$\frac{\partial T}{\partial y} = \frac{4}{e^\pi - e^{-\pi}} \left[-\pi e^{\pi(1-y)} - \pi e^{-\pi(1-y)} \right] \sin \pi x,$$

$$\frac{\partial^2 T}{\partial y^2} = \frac{4}{e^\pi - e^{-\pi}} \left[\pi^2 e^{\pi(1-y)} - \pi^2 e^{-\pi(1-y)} \right] \sin \pi x.$$

Clearly, $\partial^2 T/\partial x^2 + \partial^2 T/\partial y^2 = 0$. Since second partial derivatives are continuous for $0 < x < 1$ and $0 < y < 1$, the temperature function $T(x, y)$ is harmonic in this region. In practice, we are not given $T(x, y)$; we must find it; that is, we must solve Laplace's equation

$$\frac{\partial^2 T}{\partial x^2} + \frac{\partial^2 T}{\partial y^2} = 0, \qquad 0 < x < 1, \quad 0 < y < 1,$$

subject to the conditions

$$T(0, y) = 0, \qquad 0 < y < 1,$$

$$T(1, y) = 0, \qquad 0 < y < 1,$$

$$T(x, 1) = 0, \qquad 0 < x < 1,$$

$$T(x, 0) = 4 \sin \pi x, \qquad 0 < x < 1.$$

This is called a *boundary-value problem*; it is treated in books dealing with partial differential equations. Laplace's equation is a partial differential equation, that is, a differential equation with partial derivatives.

Longitudinal Vibrations of Bars

Figure 13.12 shows a bar of length L and uniform cross-section. If a longitudinal force (in the x-direction) is applied to each cross-section, the bar stretches and/or compresses. (The bar acts like a very stiff spring.) Suppose we denote displacement of the cross-section normally at position x by $y(x, t)$ where t is time. This allows for the situation when the applied force

$F(x, t)$ varies along the bar and is also a function of time. It can be shown that $y(x, t)$ must satisfy the following partial differential equation, called the one-dimensional wave equation,

$$\frac{\partial^2 y}{\partial t^2} = \frac{E}{\rho} \frac{\partial^2 y}{\partial x^2} + \frac{F(x, t)}{\rho}, \qquad 0 < x < L, \quad t > 0, \qquad (13.13\text{a})$$

where ρ is the density of the material in the bar and E is Young's modulus of elasticity. It is a constant that depends on the material of the bar; the larger the value of E, the more resistant the bar is to stretch or compression. Given $F(x, t)$, ρ, and E, the objective is to solve the wave equation for $y(x, t)$ giving therefore positions of cross-sections of the bar for all time.

FIGURE 13.12

By itself, the wave equation has many solutions; other conditions must be stipulated. Newton's second law governs motion (and it was used in developing the wave equation). Our experience with particle motion suggests that we require two *initial conditions*, one specifying the initial positions of cross-sections of the bar, and a second specifying their velocities. In other words, accompanying 13.13a will be initial conditions of the form

$$y(x, 0) = f(x), \qquad 0 < x < L, \qquad (13.13\text{b})$$

$$y_t(x, 0) = g(x), \qquad 0 < x < L. \qquad (13.13\text{c})$$

There will also be *boundary conditions* specifying what is happening at the ends of the bar. For instance, if end $x = 0$ is clamped in position and, therefore, not allowed to move, $y(x, t)$ satisfies

$$y(0, t) = 0, \qquad t > 0. \qquad (13.13\text{d})$$

If end $x = L$ of the bar is allowed to move freely, then the boundary condition there is

$$y_x(L, t) = 0, \qquad t > 0. \qquad (13.13\text{e})$$

There are many types of boundary conditions that can occur at $x = 0$ and $x = L$. These are two examples.

Let us take a very simple illustration. The problem is much easier if displacement is not a function of time, in which case we solve for what are called *static displacements* of the bar. For this to occur, the applied force must be independent of time. Displacement becomes only a function of position, $y(x)$, initial conditions are dropped, time disappears from the boundary conditions, and the partial differential equation becomes an ordinary differential equation,

$$0 = E\frac{d^2y}{dx^2} + F(x), \qquad 0 < x < L, \tag{13.14a}$$

$$y(0) = 0, \tag{13.14b}$$

$$y'(L) = 0. \tag{13.14c}$$

If, for instance, all cross-sections are subjected to the same force F (a constant), then

$$\frac{d^2y}{dx^2} = -\frac{F}{E} \quad \Rightarrow \quad y(x) = -\frac{Fx^2}{2E} + Cx + D.$$

The boundary conditions require

$$0 = y(0) = D, \quad 0 = y'(L) = -\frac{FL}{E} + C \quad \Rightarrow \quad C = \frac{FL}{E}.$$

Thus, displacement of the cross-section of the bar from its equilibrium position x is

$$y(x) = -\frac{Fx^2}{2E} + \frac{FLx}{E} = \frac{Fx}{2E}(2L - x).$$

Its graph is shown in Figure 13.13. To get an idea of the magnitude of these displacements, we find that for a 1 m steel bar with $E = 2.0 \times 10^{11}$ N/m^2, and an applied force $F = 10^5$ N, the right end has displacement

$$y(1) = \frac{10^5}{2(2 \times 10^{11})}\left[2(1) - 1\right] = 2.5 \times 10^{-5} \text{ m};$$

that is, the bar stretches by only 0.025 mm.

FIGURE 13.13

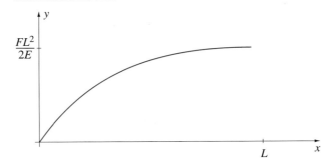

The one-dimensional wave equation 13.13a is satisfied by longitudinal vibrations of bars. It is also satisfied by transverse vibrations of strings and rotational vibrations of bars. These will be discussed in the exercises.

EXERCISES 13.5

In Exercises 1–20 find the derivative.

$*$ **1.** $\partial^2 f/\partial x^2$ if $f(x, y) = x^2 y^2 - 2x^3 y$ $\quad *$ **2.** $\partial^3 f/\partial y^3$ if $f(x, y) = 2x/y + 3x^3 y^4$

$*$ **3.** $\partial^2 f/\partial z^2$ if $f(x, y, z) = \sin(xyz)$ $\quad *$ **4.** $\partial^2 f/\partial y \partial z$ if $f(x, y, z) = xyze^{x+y+z}$

$*$ **5.** $\partial^2 f/\partial y \partial x$ if $f(x, y) = \sqrt{x^2 + y^2}$ $\quad *$ **6.** $\partial^3 f/\partial x^2 \partial y$ if $f(x, y) = e^{x+y} - x^2/y^2$

$*$ **7.** $\partial^3 f/\partial y^3$ at $(1, 3)$ if $f(x, y) = 3x^3 y^3 - 3x/y$

$*$ **8.** $\partial^3 f/\partial x \partial y \partial z$ at $(1, 0, -1)$ if $f(x, y, z) = x^2 y^2 + x^2 z^2 + y^2 z^2$

$*$ **9.** $\partial^2 f/\partial x^2$ if $f(x, y) = \sqrt{1 - x^2 - y^2}$

$*$ **10.** $\partial^2 f/\partial z^2$ if $f(x, y, z) = \ln \sqrt{x^2 + y^2 + z^2}$

$*$ **11.** $\partial^3 f/\partial x^2 \partial y$ if $f(x, y) = x^2 e^y + y^2 e^x$

$*$ **12.** $\partial^2 f/\partial x^2$ if $f(x, y) = \text{Tan}^{-1}(y/x)$

$*$ **13.** $\partial^3 f/\partial x \partial y^2$ if $f(x, y, z) = \cot(x^2 + y^2 + z^2)$

$*$ **14.** $\partial^2 f/\partial x \partial y$ at $(-2, -2)$ if $f(x, y) = \text{Sin}^{-1}(x^2 + y^2)^{-1}$

$*$ **15.** $\partial^{10} f/\partial x^7 \partial y^3$ if $f(x, y) = x^7 e^x y^2 + 1/y^6$

$*$ **16.** $\partial^8 f/\partial x^8$ if $f(x, y, z) = x^8 y^9 z^{10}$

$*$ **17.** $\partial^6 f/\partial x^2 \partial y^2 \partial z^2$ if $f(x, y, z) = 1/x^2 + 1/y^2 + 1/z^2$

$*$ **18.** $\partial^4 f/\partial x^3 \partial y$ if $f(x, y) = \cos(x + y^3)$

$*$ **19.** $\partial^4 f/\partial x \partial y \partial z \partial t$ if $f(x, y, z, t) = \sqrt{x^2 + y^2 + z^2 - t^2}$

$*$ **20.** $\partial^2 f/\partial x \partial y$ if $f(x, y) = \text{Sec}^{-1}(xy)$

$**$ **21.** If $z = x^2 + xy + y^2 \sin(x/y)$, show that

$$x \frac{\partial z}{\partial x} + y \frac{\partial z}{\partial y} = 2z = x^2 \frac{\partial^2 z}{\partial x^2} + 2xy \frac{\partial^2 z}{\partial x \partial y} + y^2 \frac{\partial^2 z}{\partial y^2}.$$

$**$ **22.** If $u = x + y + ze^{y/x}$, show that

$$x^2 \frac{\partial^2 u}{\partial x^2} + y^2 \frac{\partial^2 u}{\partial y^2} + z^2 \frac{\partial^2 u}{\partial z^2} + 2xy \frac{\partial^2 u}{\partial x \partial y} + 2yz \frac{\partial^2 u}{\partial y \partial z} + 2xz \frac{\partial^2 u}{\partial x \partial z} = 0.$$

In Exercises 23–28 find a region (if possible) in which the function is harmonic.

$**$ **23.** $f(x, y) = x^2 - y^2 + 2xy + y$ $\quad **$ **24.** $f(x, y) = \ln(x^2 + y^2)$

$**$ **25.** $f(x, y) = x^3 y^2 - 3xy$ $\quad **$ **26.** $f(x, y, z) = 3x^2 yz - y^3 z + xy$

$**$ **27.** $f(x, y, z) = 1/\sqrt{x^2 + y^2 + z^2}$ $\quad **$ **28.** $f(x, y, z) = x^3 y^3 z^3$

$**$ **29.** If $V(x, y, z)$ represents the electrostatic potential at a point (x, y, z) a system of n point charges at point (x_i, y_i, z_i), does $V(x, y, z)$ satisfy Laplace's equation?

** **30.** The gravitational potential at a point (x, y, z) in space due to a uniform spherical mass distribution (mass M) at the origin is defined as $V = GM/r$ where G is a constant and $r = \sqrt{x^2 + y^2 + z^2}$. Show that $V(x, y, z)$ satisfies Laplace's equation 13.12.

** **31.** The figure below shows a plate bounded by the lines $x = 0$, $y = 0$, $x = 1$, and $y = 1$. Temperature along the first three sides is kept at $0°C$, while that along $y = 1$ varies according to $f(x) = \sin(3\pi x) - 2\sin(4\pi x)$, $0 \le x \le 1$. The temperature at any point interior to the plate is then

$$T(x, y) = C(e^{3\pi y} - e^{-3\pi y}) \sin(3\pi x) + D(e^{4\pi y} - e^{-4\pi y}) \sin(4\pi x),$$

where $C = (e^{3\pi} - e^{-3\pi})^{-1}$ and $D = -2(e^{4\pi} - e^{-4\pi})^{-1}$. Show that $T(x, y)$ is harmonic in the region $0 < x < 1$, $0 < y < 1$, and that it also satisfies the boundary conditions $T(0, y) = 0$, $T(1, y) = 0$, $T(x, 0) = 0$, and $T(x, 1) = f(x)$.

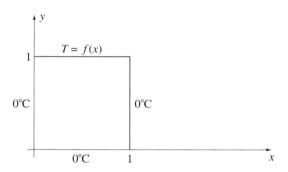

** **32.** When the bar in Figure 13.12 is turned vertically and clamped at $x = 0$ (figure below) static deflections of the bar must satisfy the following problem

$$0 = E\frac{d^2 y}{dx^2} + F(x), \qquad 0 < x < L,$$

$$y(0) = 0, \quad y'(0) = 0,$$

where $F(x)$ is the weight of that part of the bar below the cross-section that would be at position x if the bar were unstretched. Find the length of the bar as it stretches under its own weight.

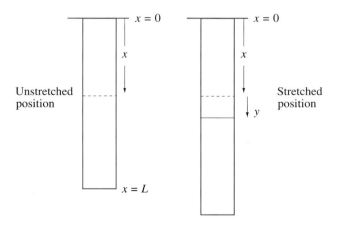

Show that this function satisfies the partial differential equation, boundary conditions, and the second initial condition in Exercise 38. (Assume that differential and summation operations can be interchanged.)

(b) Suppose the first twenty terms of the series are used to approximate $y(x, t)$. Plot the sum of these terms as a function of x for the times $t = 0$, $1/(16c)$, $1/(8c)$, $3/(16/c)$, $1/(4c)$, $5/(16c)$, $3/(8c)$, $7/(16c)$, and $1/(2c)$. Does the string appear to retain its broken line shape?

** **41.** The uniform circular rod in the figure below has flat ends at $x = 0$ and $x = L$. If the round side of the rod is perfectly insulated, heat flows in only the x-direction. When no heat sources exist within the rod, temperature $T(x, t)$ at points in the rod must satisfy the one-dimensional heat conduction equation

$$\frac{\partial T}{\partial t} = k \frac{\partial^2 T}{\partial x^2}, \qquad 0 < x < L, \quad t > 0,$$

where $k > 0$ is a constant called the *thermal diffusivity* of the rod. Accompanying this equation will be two boundary conditions describing what is happening at the ends of the rod. For instance, if the left end is kept at temperature $0°$C and the right end is insulated, $T(x, t)$ must satisfy

$$T(0, t) = 0, \qquad T_x(L, t) = 0 \quad t > 0.$$

In addition, there will be an initial condition describing the temperature distribution in the rod at time $t = 0$,

$$T(x, 0) = f(x), \qquad 0 < x < L.$$

$x = 0$ Perfect insulation L x

(a) Show that the heat conduction equation is satisfied by functions of the form

$$T(x, t) = (A \sin \lambda x + B \cos \lambda x)e^{-k\lambda^2 t}$$

where λ, A, and B are arbitrary constants.

(b) Show that the boundary conditions require $B = 0$ and $\lambda = (2n - 1)\pi/(2L)$, where n is an integer.

(c) When the initial temperature of the rod is $f(x) = x$, so that it increases linearly from $0°$C at its left end to $L°$C at its right end, temperature thereafter is

$$T(x, t) = \frac{8L}{\pi^2} \sum_{n=1}^{\infty} \frac{(-1)^{n+1}}{(2n - 1)^2} e^{-(2n-1)^2 \pi^2 kt/(4L^2)} \sin \frac{(2n - 1)\pi x}{2L}.$$

Show that this function satisfies the partial differential equation and boundary conditions. (Assume that differentiation and summation operations can be interchanged.)

(d) Plot the sum of the first twenty terms of the series as a function of x for the times $t = 0, 10, 100, 1000$, and $10\,000$. Use a 1 m rod and $k = 1.14 \times 10^{-4}$ m²/s (typical for copper). Does the $t = 0$ plot approximate $f(x)$, and do the remaining plots reflect the boundary conditions?

** **42.** If both ends of the rod in Exercise 41 are held at temperature $0°$C, the problem for $T(x, t)$ becomes

$$\frac{\partial T}{\partial t} = k\frac{\partial^2 T}{\partial x^2}, \qquad 0 < x < L, \quad t > 0,$$

$$T(0, t) = T(L, t) = 0, \qquad t > 0,$$

$$T(x, 0) = f(x), \qquad 0 < x < L.$$

(a) Show that for the functions in part (a) of Exercise 41, the boundary conditions require $B = 0$ and $\lambda = n\pi/L$, where n is an integer.

(b) When the initial temperature of the rod is $f(x) = x(L - x)$, temperature thereafter is

$$T(x, t) = \frac{8L^4}{\pi^3}\sum_{n=1}^{\infty}\frac{1}{(2n - 1)^3}e^{-(2n-1)^2\pi^2 kt/L^2}\sin\frac{(2n - 1)\pi x}{L}.$$

Plot the sum of the first twenty terms of this series as a function of x for the times $t = 0, 10, 100, 1000,$ and $10\,000$. Use a 1 m rod and $k = 1.14 \times 10^{-4}$ m^2/s. Does the $t = 0$ plot approximate $f(x)$?

** **43.** If the ends $x = 0$ and $x = L$ of the rod in Exercise 41 are held at temperatures $T_0°$C and $T_L°$C for $t > 0$, temperature $T(x, t)$ must satisfy

$$\frac{\partial T}{\partial t} = k\frac{\partial^2 T}{\partial x^2}, \qquad 0 < x < L, \quad t > 0,$$

$$T(0, t) = T_0, \qquad T(L, t) = T_L, \quad t > 0,$$

$$T(x, 0) = f(x), \qquad 0 < x < L.$$

After a very long time, the temperature at each point in the rod will remain constant, but temperature will vary from point to point. Temperature is said to have reached a steady-state situation. It can be found by removing the term $\partial T/\partial t$ from the heat conduction equation, and the initial condition. Temperature becomes a function of only x, which satisfies

$$0 = \frac{d^2 T}{dx^2}, \qquad 0 < x < L,$$

$$T(0) = T_0, \qquad T(L) = T_L.$$

Find $T(x)$.

** **44.** If heat is added to the rod in Exercise 41, the heat conduction equation takes the form

$$\frac{\partial T}{\partial t} = k\frac{\partial^2 T}{\partial x^2} + F(x, t), \qquad 0 < x < L, \quad t > 0,$$

where $F(x, t)$ is the heat source term. If temperatures of the ends of the rod are held at $T_0°$C and $T_L°$C, then $T(x, t)$ must also satisfy the boundary conditions

$$T(0, t) = T_0, \qquad T(L, t) = T_L, \quad t > 0.$$

If $F(x, t)$ is independent of t, then steady-state temperature of the rod (see Exercise 43) must satisfy

$$0 = k\frac{d^2T}{dx^2} + F(x), \qquad 0 < x < L, \quad t > 0,$$

$$T(0) = T_0, \quad T(L) = T_L.$$

Find $T(x)$ in the case that $F(x)$ is a constant value F.

** **45.** The figure below shows a bar of length L and uniform cross-section. When torque is applied to cross-sections, these cross-sections are forced to rotate. Let x represent distance from the left end of the bar, and $y(x, t)$ the angular displacement of the cross-section at position x and time t from its torque-free position. It can be shown that when $\tau(x, t)$ represents the torque per unit length at position x and time t, and ρ is the (constant) density of the bar, $y(x, t)$ must satisfy the one-dimensional wave equation

$$\frac{\partial^2 y}{\partial t^2} = c^2\frac{\partial^2 y}{\partial x^2} + \frac{\tau(x, t)}{\rho}, \qquad 0 < x < L, \quad t > 0,$$

where $c^2 = E/\rho$ and E is Young's modulus of elasticity of the material in the bar under shear. Accompanying the partial differential equation will be initial conditions 13.13b and c specifying the initial displacements and velocities of cross-sections at time $t = 0$. In addition, there will be two boundary conditions describing end conditions. For example, if the left end is clamped so that no rotation is possible, and the right end is free to move, the boundary conditions will be 13.13d,e. The problem is therefore identical to that for longitudinal vibrations of bars. *Static rotations* occur when $\tau(x, t)$ is independent of t, the time-derivative is removed from the wave equation, and initial conditions are deleted. Displacements $y(x)$ must then satisfy

$$0 = E\frac{d^2y}{dx^2} + \tau(x), \qquad 0 < x < L,$$

$$y(0) = 0, \quad y'(L) = 0,$$

at least for the boundary conditions described above. Solve this problem when $\tau(x) = x$ so that torque increases linearly along the bar.

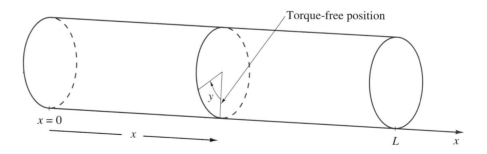

** **46.** Show that when $\tau(x)$ is unspecified static rotations in Exercise 45 can be expressed in the form

$$y(x) = Cx - \frac{1}{E}\int_0^x \int_0^v \tau(u)\,du\,dv,$$

where

$$C = \frac{1}{E}\int_0^L \tau(x)\,dx.$$

** **47.** Two functions $u(x, y)$ and $v(x, y)$ are said to be harmonic conjugates if they satisfy the Cauchy-Riemann equations of Exercise 39 in Section 13.3. Show that if $u(x, y)$ and $v(x, y)$ are harmonic conjugates and have continuous second partial derivatives in a region R, then each is harmonic in R.

** **48.** (a) Show that the function $u(x, y) = x^2 - y^2$ is harmonic in the entire xy-plane.

 (b) Use the Cauchy-Riemann equations in Exercise 39 of Section 13.3 to find a function $v(x, y)$ so that u and v are harmonic conjugates.

** **49.** Repeat Exercise 48 if $u(x, y) = e^x \cos y + x$.

** **50.** For what values of n does the function $(x^2 + y^2 + z^2)^n$ satisfy equation 13.12? In what regions are the functions harmonic?

13.6 Chain Rules for Partial Derivatives

If $y = f(u)$ and $u = g(x)$, the chain rule for the derivative dy/dx of the composite function $f[g(x)]$ is

$$\frac{dy}{dx} = \frac{dy}{du}\frac{du}{dx}. \tag{13.15}$$

Equation 13.15 can be extended in terms of more intermediate variables, say $y = f(u)$, $u = g(s)$, $s = h(x)$, in which case

$$\frac{dy}{dx} = \frac{dy}{du}\frac{du}{ds}\frac{ds}{dx}. \tag{13.16}$$

For multivariable functions, variations in chain rules are countless. We discuss two examples in considerable detail, and then show schematic diagrams that easily lead to chain rules for even the most complicated functional situations.

Suppose z is a function of u and v and each of u and v is a function of x and y,

$$z = f(u, v), \quad u = g(x, y), \quad v = h(x, y). \tag{13.17}$$

By the substitutions

$$z = f[g(x, y), h(x, y)], \tag{13.18}$$

we express z as a function of x and y, and could then calculate the partial derivative $\partial z/\partial x$. However, if the functions in 13.17 are at all complicated, you can imagine how difficult the composite function in 13.18 might be to differentiate. As a result, we search for an alternative procedure for calculating $\partial z/\partial x$, namely, the appropriate chain rule. It is contained in the following theorem.

THEOREM 13.2

Let $u = g(x, y)$ and $v = h(x, y)$ be continuous and have first partial derivatives with respect to x at a point (x, y), and let $z = f(u, v)$ have continuous first partial derivatives inside a circle centred at the point $(u, v) = (g(x, y), h(x, y))$. Then

$$\frac{\partial z}{\partial x} = \frac{\partial z}{\partial u}\frac{\partial u}{\partial x} + \frac{\partial z}{\partial v}\frac{\partial v}{\partial x}. \tag{13.19}$$

PROOF This result can be proved in much the same way as chain rule 3.20a was proved in Section 3.7. By Definition 13.3,

$$\frac{\partial z}{\partial x} = \lim_{\Delta x \to 0} \frac{f[g(x + \Delta x, y), h(x + \Delta x, y)] - f[g(x, y), h(x, y)]}{\Delta x}.$$

Now the increment Δx in x produces changes in u and v, which we denote by

$$\Delta u = g(x + \Delta x, y) - g(x, y), \quad \Delta v = h(x + \Delta x, y) - h(x, y).$$

If we write u and v whenever $g(x, y)$ and $h(x, y)$ are evaluated at (x, y), and substitute for $g(x + \Delta x, y)$ and $h(x + \Delta x, y)$ in the definition for $\partial z/\partial x$, then

$$\frac{\partial z}{\partial x} = \lim_{\Delta x \to 0} \frac{f(u + \Delta u, v + \Delta v) - f(u, v)}{\Delta x}$$

$$= \lim_{\Delta x \to 0} \frac{[f(u + \Delta u, v + \Delta v) - f(u, v + \Delta v)] + [f(u, v + \Delta v) - f(u, v)]}{\Delta x}$$

$$= \lim_{\Delta x \to 0} \left\{ \frac{f(u + \Delta u, v + \Delta v) - f(u, v + \Delta v)}{\Delta x} + \frac{f(u, v + \Delta v) - f(u, v)}{\Delta x} \right\}.$$

We assumed that the derivative

$$\frac{\partial z}{\partial v} = \lim_{\Delta v \to 0} \frac{f(u, v + \Delta v) - f(u, v)}{\Delta v}$$

exists at (u, v). An equivalent way to express the fact that this limit exists is to say that

$$\frac{f(u, v + \Delta v) - f(u, v)}{\Delta v} = \frac{\partial z}{\partial v} + \epsilon_1,$$

where ϵ_1 must satisfy the condition that $\lim_{\Delta v \to 0} \epsilon_1 = 0$. We can write, therefore, that

$$f(u, v + \Delta v) - f(u, v) = [z_v(u, v) + \epsilon_1]\Delta v.$$

Similarly, we can write that

$$f(u + \Delta u, v + \Delta v) - f(u, v + \Delta v) = [z_u(u, v + \Delta v) + \epsilon_2]\Delta u,$$

where $\lim_{\Delta u \to 0} \epsilon_2 = 0$ (provided Δv is sufficiently small). When these expressions are substituted into the limit for $\partial z/\partial x$, we have

$$\frac{\partial z}{\partial x} = \lim_{\Delta x \to 0} \left\{ [z_u(u, v + \Delta v) + \epsilon_2]\frac{\Delta u}{\Delta x} + [z_v(u, v) + \epsilon_1]\frac{\Delta v}{\Delta x} \right\}.$$

We now examine each part of this limit. Clearly,

$$\lim_{\Delta x \to 0} \frac{\Delta u}{\Delta x} = \frac{\partial u}{\partial x} \quad \text{and} \quad \lim_{\Delta x \to 0} \frac{\Delta v}{\Delta x} = \frac{\partial v}{\partial x}.$$

In addition, because $g(x, y)$ and $h(x, y)$ are continuous, $\Delta u \to 0$ and $\Delta v \to 0$ as $\Delta x \to 0$. Consequently,

$$\lim_{\Delta x \to 0} \epsilon_1 = \lim_{\Delta v \to 0} \epsilon_1 = 0 \quad \text{and} \quad \lim_{\Delta x \to 0} \epsilon_2 = \lim_{\Delta v \to 0} \epsilon_2 = 0.$$

Finally, because $\partial z/\partial u$ is continuous,

$$\lim_{\Delta x \to 0} z_u(u, v + \Delta v) = \lim_{\Delta v \to 0} z_u(u, v + \Delta v) = z_u(u, v).$$

When all these results are taken into account, we have

$$\frac{\partial z}{\partial x} = z_u(u, v)\frac{\partial u}{\partial x} + z_v(u, v)\frac{\partial v}{\partial x} = \frac{\partial z}{\partial u}\frac{\partial u}{\partial x} + \frac{\partial z}{\partial v}\frac{\partial v}{\partial x},$$

which completes the proof. ∎

Chain rule 13.19 defines $\partial z/\partial x$ in terms of derivatives of the given functions in 13.17. We could be more explicit by indicating which variable is being held constant in each of the five derivatives:

$$\left.\frac{\partial z}{\partial x}\right)_y = \left.\frac{\partial z}{\partial u}\right)_v \left.\frac{\partial u}{\partial x}\right)_y + \left.\frac{\partial z}{\partial v}\right)_u \left.\frac{\partial v}{\partial x}\right)_y. \tag{13.20}$$

From the point of view of rates of change, this result seems quite reasonable. The left side is the rate of change of z with respect to x holding y constant. The first term $(\partial z/\partial u)(\partial u/\partial x)$ accounts for the rate of change of z with respect to those x's that affect z through u. The second term $(\partial z/\partial v)(\partial v/\partial x)$ accounts for the rate of change of z with respect to those x's that affect z through v. The total rate of change is then the sum of the two parts.

Consider now the functional situation

$$z = f(u, v), \quad u = g(x, y, s), \quad v = h(x, y, s), \quad x = p(t), \quad y = q(t), \quad s = r(t). \tag{13.21}$$

By the substitutions

$$z = f[g(p(t), q(t), r(t)), h(p(t), q(t), r(t))], \tag{13.22}$$

we express z as a function of t alone, and can therefore pose the problem of calculating dz/dt. If we reason as in the last paragraph, the appropriate chain rule for dz/dt must account for all t's affecting z through u and v. We obtain, then,

$$\frac{dz}{dt} = \frac{\partial z}{\partial u}\frac{du}{dt} + \frac{\partial z}{\partial v}\frac{dv}{dt},$$

where we have written du/dt and dv/dt because u and v can be expressed entirely in terms of t:

$$u = g[p(t), q(t), r(t)], \quad v = h[p(t), q(t), r(t)].$$

Chain rules for each of du/dt and dv/dt (similar to 13.19) yield

$$\frac{du}{dt} = \frac{\partial u}{\partial x}\frac{dx}{dt} + \frac{\partial u}{\partial y}\frac{dy}{dt} + \frac{\partial u}{\partial s}\frac{ds}{dt}, \quad \frac{dv}{dt} = \frac{\partial v}{\partial x}\frac{dx}{dt} + \frac{\partial v}{\partial y}\frac{dy}{dt} + \frac{\partial v}{\partial s}\frac{ds}{dt}.$$

Finally, then,

$$\frac{dz}{dt} = \frac{\partial z}{\partial u}\left\{\frac{\partial u}{\partial x}\frac{dx}{dt} + \frac{\partial u}{\partial y}\frac{dy}{dt} + \frac{\partial u}{\partial s}\frac{ds}{dt}\right\} + \frac{\partial z}{\partial v}\left\{\frac{\partial v}{\partial x}\frac{dx}{dt} + \frac{\partial v}{\partial y}\frac{dy}{dt} + \frac{\partial v}{\partial s}\frac{ds}{dt}\right\}, \tag{13.23}$$

which expresses dz/dt in terms of derivatives of the given functions in 13.21.

These two examples suggest the complexities that may be involved in finding chain rules for complicated composite functions. Fortunately, there is an amazingly simple method that gives the correct chain rule in every situation. The method is not designed to help you understand the chain rule, but to find it quickly. We suggest that you test your understanding by developing a few chain rules in the exercises with a discussion such as in the second example above, and then check your result by the quicker method.

In the first example we represent the functional situation described in 13.17 by the schematic diagram below.

At the top of the diagram is the dependent variable z, which we wish to differentiate. In the line below z are the variables u and v in terms of which z is initially defined. In the line below u and v are x's and y's illustrating that each of u and v is defined in terms of x and y. To obtain the partial derivative $\partial z/\partial x$, we take all possible paths in this diagram from z to x. The two paths are from z through u to x, and from z through v to x. For each straight-line segment in a given path, we differentiate the upper variable with respect to the lower variable and multiply together all such derivatives in that path. The products are then added together to form the complete chain rule. In particular, for the path through u we form the product

$$\frac{\partial z}{\partial u}\frac{\partial u}{\partial x},$$

and for the path through v,

$$\frac{\partial z}{\partial v}\frac{\partial v}{\partial x}.$$

The complete chain rule is then the sum of these products,

$$\frac{\partial z}{\partial x} = \frac{\partial z}{\partial u}\frac{\partial u}{\partial x} + \frac{\partial z}{\partial v}\frac{\partial v}{\partial x},$$

and this result agrees with 13.19. The schematic diagram also indicates which variables are to be held constant in the derivatives on the right (as in 13.20). All variables on the same level are held constant.

For the second example in equations 13.21 the schematic diagram is:

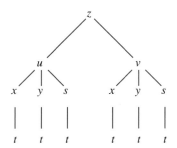

There are six possible paths from z to t so that the chain rule for dz/dt must have six terms. We find

$$\frac{dz}{dt} = \frac{\partial z}{\partial u}\frac{\partial u}{\partial x}\frac{dx}{dt} + \frac{\partial z}{\partial u}\frac{\partial u}{\partial y}\frac{dy}{dt} + \frac{\partial z}{\partial u}\frac{\partial u}{\partial s}\frac{ds}{dt} + \frac{\partial z}{\partial v}\frac{\partial v}{\partial x}\frac{dx}{dt} + \frac{\partial z}{\partial v}\frac{\partial v}{\partial y}\frac{dy}{dt} + \frac{\partial z}{\partial v}\frac{\partial v}{\partial s}\frac{ds}{dt},$$

and this agrees with 13.23. Note too that if, when forming a derivative from the schematic diagram, there are two or more lines emanating from a variable, then we obtain a partial derivative; if there is only one line, then we have an ordinary derivative.

EXAMPLE 13.14

Find chain rules for

$$\left.\frac{\partial z}{\partial x}\right)_y \quad \text{and} \quad \left.\frac{\partial z}{\partial y}\right)_x$$

if

$$z = f(r, s, x), \quad r = g(x, y), \quad s = h(x, y).$$

SOLUTION   From the schematic diagram below,

$$\left.\frac{\partial z}{\partial y}\right)_x = \left.\frac{\partial z}{\partial r}\right)_{s,x}\left.\frac{\partial r}{\partial y}\right)_x + \left.\frac{\partial z}{\partial s}\right)_{r,x}\left.\frac{\partial s}{\partial y}\right)_x,$$

$$\left.\frac{\partial z}{\partial x}\right)_y = \left.\frac{\partial z}{\partial r}\right)_{s,x}\left.\frac{\partial r}{\partial x}\right)_y + \left.\frac{\partial z}{\partial s}\right)_{r,x}\left.\frac{\partial s}{\partial x}\right)_y + \left.\frac{\partial z}{\partial x}\right)_{r,s}.$$

 In the previous example it is essential that we indicate which variables to hold constant in the partial derivatives. If we were to omit these designations, then in the second result we would have a term $\partial z/\partial x$ on both sides of the equation but they would have different meanings. The term $\partial z/\partial x)_y$ indicates the derivative of z with respect to x holding y constant if z were expressed entirely in terms of x and y; the term $\partial z/\partial x)_{r,s}$ indicates the derivative of the given function $f(r, s, x)$ with respect to x holding r and s constant.

EXAMPLE 13.15

Find dz/dt if

$$z = x^3 y^2 + x\sin y + tx, \quad x = 2t + \frac{1}{t}, \quad y = t^2 e^t.$$

SOLUTION   From the schematic diagram below,

$$\frac{dz}{dt} = \frac{\partial z}{\partial x}\frac{dx}{dt} + \frac{\partial z}{\partial y}\frac{dy}{dt} + \frac{\partial z}{\partial t}$$

$$= (3x^2y^2 + \sin y + t)(2 - 1/t^2) + (2x^3y + x\cos y)(2te^t + t^2e^t) + x.$$

When a chain rule is used to calculate a derivative, the result usually involves all intermediate variables. For instance, the derivative dz/dt in Example 13.15 involves not only t, but the intermediate variables x and y as well. Were dz/dt required at $t = 1$, values of x and y for $t = 1$ would be calculated—$x(1) = 3$ and $y(1) = e$—and all three values substituted to obtain

$$\frac{dz}{dt}_{|t=1} = [3(3)^2(e)^2 + \sin(e) + 1](2 - 1)$$

$$+ [2(3)^3e + (3)\cos(e)](2e + e) + 3$$

$$= 1378.6.$$

EXAMPLE 13.16

Find $\partial^2 z/\partial x^2$ if

$$z = s^2t + 2\sin t, \quad s = xy - y, \quad t = x^2 + \frac{y}{x}.$$

SOLUTION   From the schematic diagram below,

$$\frac{\partial z}{\partial x} = \frac{\partial z}{\partial s}\frac{\partial s}{\partial x} + \frac{\partial z}{\partial t}\frac{\partial t}{\partial x}$$

$$= (2st)(y) + (s^2 + 2\cos t)(2x - y/x^2).$$

Now $\partial z/\partial x$ is a function of s, t, x, and y, and therefore in order to find

$$\frac{\partial^2 z}{\partial x^2} = \frac{\partial}{\partial x}\left(\frac{\partial z}{\partial x}\right),$$

we form a schematic diagram for $\partial z/\partial x$:

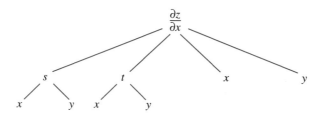

From this schematic diagram, we obtain

$$\frac{\partial^2 z}{\partial x^2} = \frac{\partial}{\partial s}\left(\frac{\partial z}{\partial x}\right)\frac{\partial s}{\partial x} + \frac{\partial}{\partial t}\left(\frac{\partial z}{\partial x}\right)\frac{\partial t}{\partial x} + \frac{\partial}{\partial x}\left(\frac{\partial z}{\partial x}\right)_{s,t,y}$$

$$= [2ty + 2s(2x - y/x^2)](y) + [2sy - 2\sin t(2x - y/x^2)](2x - y/x^2)$$

$$+ (s^2 + 2\cos t)(2 + 2y/x^3).$$

EXAMPLE 13.17

Temperature T at points in the atmosphere depends on both position (x, y, z) and time t: $T = T(x, y, z, t)$. When a weather balloon is released to take temperature readings, it is not free to take readings at just any point, but only at those points along the path that the winds force the balloon to follow. This path is a curve in space represented parametrically by

$$C: \quad x = x(t), \quad y = y(t), \quad z = z(t), \quad t \geq 0,$$

t again being time. If we substitute from the equations for C into the temperature function, then T becomes a function of t alone,

$$T = T[x(t), y(t), z(t), t],$$

and this function of time describes the temperature at points along the path of the balloon. For the derivative of this function with respect to t, the schematic diagram

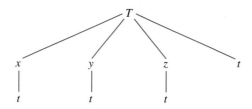

yields

$$\frac{dT}{dt} = \frac{\partial T}{\partial x}\frac{dx}{dt} + \frac{\partial T}{\partial y}\frac{dy}{dt} + \frac{\partial T}{\partial z}\frac{dz}{dt} + \frac{\partial T}{\partial t}.$$

The question we pose is "What is the physical difference between dT/dt and $\partial T/\partial t$?"

SOLUTION   Temperature at a point in space is independent of the observer measuring it; hence $T[x(t), y(t), z(t), t]$ is the temperature at points on C as measured by both the balloon and any observer fixed in the xyz-reference system. If, however, these two observers calculate the rate of change of temperature with respect to time at some point (x, y, z) on C, they calculate different results. The observer fixed in the xyz-reference system (not restricted to move along C) calculates the rate of change of T with respect to t as the derivative of the function $T(x, y, z, t)$ partially with respect to t holding x, y, and z constant; i.e., the fixed observer calculates $\partial T/\partial t$ as the rate of change of temperature in time. The balloon, on the other hand, has no alternative but to take temperature readings as it moves along C; thus its measurement of T as a function of t is

$$T[x(t), y(t), z(t), t].$$

Therefore, when the balloon calculates the time variation of temperature, it is calculating dT/dt. It follows, then, that the terms

$$\frac{\partial T}{\partial x}\frac{dx}{dt} + \frac{\partial T}{\partial y}\frac{dy}{dt} + \frac{\partial T}{\partial z}\frac{dz}{dt}$$

describe that part of dT/dt caused by the motion of the balloon through space.

Discussions like those in Example 13.17 are prominent in the study of fluid motion (gas or liquid). Sometimes rates of change from the point of view of a fixed observer are important; other times, rates of change as measured by an observer moving with the fluid are appropriate.

Many important applications of the chain rule occur in the field of partial differential equations. The following example is an illustration.

EXAMPLE 13.18

The one-dimensional wave equation

$$\frac{\partial^2 y}{\partial t^2} = c^2 \frac{\partial^2 y}{\partial x^2}, \quad c = \text{constant}$$

for functions $y(x, t)$ describes transverse vibrations of taut strings, and longitudinal and rotational vibrations of metal bars. Show that if $f(u)$ and $g(v)$ are twice differentiable functions of u and v, then $y(x, t) = f(x + ct) + g(x - ct)$ satisfies the wave equation.

SOLUTION   The schematic diagram below describes the functional situation

$$y = f(u) + g(v)$$

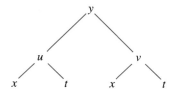

where $u = x + ct$ and $v = x - ct$. The chain rule for $\partial y / \partial t$ is

$$\frac{\partial y}{\partial t} = \frac{\partial y}{\partial u}\frac{\partial u}{\partial t} + \frac{\partial y}{\partial v}\frac{\partial v}{\partial t} = cf'(u) - cg'(v).$$

The schematic diagram for $\partial y / \partial t$ below leads to

$$\frac{\partial^2 y}{\partial t^2} = \frac{\partial}{\partial u}\left(\frac{\partial y}{\partial t}\right)\frac{\partial u}{\partial t} + \frac{\partial}{\partial v}\left(\frac{\partial y}{\partial t}\right)\frac{\partial v}{\partial t}$$

$$= [cf''(u)]c + [-cg''(v)](-c)$$

$$= c^2[f''(u) + g''(v)].$$

A similar calculation gives $\dfrac{\partial^2 y}{\partial x^2} = f''(u) + g''(v)$. Hence $y(x, t)$ does indeed satisfy the wave equation.

We have suggested how important Laplace's equation is to engineering, particularly in electrostatics, heat conduction, fluid flow, and deflection of plates. The two-dimensional Laplace equation in Cartesian coordinates is 13.11. In the next example, we find its form in polar coordinates.

EXAMPLE 13.19

Find Laplace's equation in polar coordinates.

SOLUTION Cartesian coordinates are related to polar coordinates by the equations $x = r\cos\theta$, $y = r\sin\theta$. The inverse transformation is

$$r = \sqrt{x^2 + y^2}, \qquad \theta = \text{Tan}^{-1}\left(\frac{y}{x}\right).$$

The second of these is not always correct; it may need $\pm\pi$ added to it. Since the derivation here requires only derivatives of θ, and not θ itself, the $\pm\pi$ is inconsequential. Suppose $V = f(x, y)$ is a function that satisfies Laplace's equation $\partial^2 V / \partial x^2 + \partial^2 V / \partial y^2 = 0$. We can express V in terms of r and θ by writing $V = F(r, \theta) = f(r\cos\theta, r\sin\theta)$. The left schematic below represents the functional situation where $V = F(r, \theta)$ and r and θ are expressed in terms of x and y.

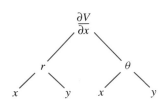

From this schematic,

$$\frac{\partial V}{\partial x} = \frac{\partial V}{\partial r}\frac{\partial r}{\partial x} + \frac{\partial V}{\partial \theta}\frac{\partial \theta}{\partial x},$$

where

$$\frac{\partial r}{\partial x} = \frac{x}{\sqrt{x^2+y^2}} = \frac{r\cos\theta}{r} = \cos\theta,$$

$$\frac{\partial \theta}{\partial x} = \frac{1}{1+(y/x)^2}\left(\frac{-y}{x^2}\right) = \frac{-y}{x^2+y^2} = \frac{-\sin\theta}{r}.$$

Thus,

$$\frac{\partial V}{\partial x} = \cos\theta\,\frac{\partial V}{\partial r} - \frac{\sin\theta}{r}\frac{\partial V}{\partial \theta}.$$

The combination of r and θ derivatives on the right of this equation must be applied to V when it is expressed in terms of r and θ to give the partial derivative with respect to x when V is expressed in terms of x and y. To find $\partial^2 V/\partial x^2$, we use the right schematic above,

$$\frac{\partial^2 V}{\partial x^2} = \frac{\partial}{\partial r}\left(\frac{\partial V}{\partial x}\right)\frac{\partial r}{\partial x} + \frac{\partial}{\partial \theta}\left(\frac{\partial V}{\partial x}\right)\frac{\partial \theta}{\partial x}$$

$$= \left(\cos\theta\,\frac{\partial^2 V}{\partial r^2} + \frac{\sin\theta}{r^2}\frac{\partial V}{\partial \theta} - \frac{\sin\theta}{r}\frac{\partial^2 V}{\partial r\partial\theta}\right)\cos\theta$$

$$+ \left(-\sin\theta\,\frac{\partial V}{\partial r} + \cos\theta\,\frac{\partial^2 V}{\partial\theta\partial r} - \frac{\cos\theta}{r}\frac{\partial V}{\partial \theta} - \frac{\sin\theta}{r}\frac{\partial^2 V}{\partial\theta^2}\right)\left(-\frac{\sin\theta}{r}\right)$$

$$= \cos^2\theta\,\frac{\partial^2 V}{\partial r^2} - \frac{2\sin\theta\cos\theta}{r}\frac{\partial^2 V}{\partial r\partial\theta} + \frac{\sin^2\theta}{r^2}\frac{\partial^2 V}{\partial\theta^2}$$

$$+ \frac{2\sin\theta\cos\theta}{r^2}\frac{\partial V}{\partial \theta} + \frac{\sin^2\theta}{r}\frac{\partial V}{\partial r},$$

where we have assumed that mixed partial derivatives are equal. A similar calculation gives

$$\frac{\partial^2 V}{\partial y^2} = \sin^2\theta\,\frac{\partial^2 V}{\partial r^2} + \frac{2\sin\theta\cos\theta}{r}\frac{\partial^2 V}{\partial r\partial\theta} + \frac{\cos^2\theta}{r^2}\frac{\partial^2 V}{\partial\theta^2} - \frac{2\sin\theta\cos\theta}{r^2}\frac{\partial V}{\partial \theta} + \frac{\cos^2\theta}{r}\frac{\partial V}{\partial r}.$$

When these are added together, the result is Laplace's equation in polar coordinates,

$$\frac{\partial^2 V}{\partial r^2} + \frac{1}{r}\frac{\partial V}{\partial r} + \frac{1}{r^2}\frac{\partial^2 V}{\partial\theta^2} = 0. \tag{13.24}$$

Calculations in this example allow us to emphasize a point that we made in Section 13.3. We said that, although the derivative dy/dx can be considered as a quotient of differentials, we never consider a partial derivative as a quotient. To do so in Example 13.19 would lead to errors. We calculated $\partial r/\partial x = \cos\theta$. Notice that $\partial x/\partial r = \cos\theta$, and therefore to regard $\partial r/\partial x$ as the reciprocal of $\partial x/\partial r$ is incorrect.

Homogeneous Functions

Homogeneous functions arise in numerous areas of applied mathematics. A function $f(x, y, z)$ is said to be **positively homogeneous** of degree n if for every $t > 0$,

$$f(tx, ty, tz) = t^n f(x, y, z). \tag{13.25}$$

For example, the function $f(x, y, z) = x^2 + y^2 + z^2$ is homogeneous of degree 2; the function $f(x, y) = x^3 \cos(y/x) + x^2 y + xy^2$ is homogeneous of degree 3; and $f(x, y, z, t) = \sqrt{x^2 + z^2}(x^2 y + yt^2)$ is homogeneous of degree 4. Partial derivatives of homogeneous functions satisfy many identities. In particular, their first derivatives satisfy Euler's theorem.

THEOREM 13.3   (Euler's Theorem)

If $f(x, y, z)$ is positively homogeneous of degree n, and has continuous first partial derivatives, then

$$x\frac{\partial f}{\partial x} + y\frac{\partial f}{\partial y} + z\frac{\partial f}{\partial z} = nf(x, y, z). \qquad (13.26)$$

PROOF   To verify 13.26 we differentiate 13.25 with respect to t holding x, y, and z constant. For the derivative of the left side we introduce variables $u = tx$, $v = ty$, and $w = tz$, and use the schematic diagram below. The result is

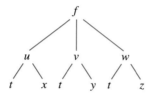

$$\frac{\partial f}{\partial u}\frac{\partial u}{\partial t} + \frac{\partial f}{\partial v}\frac{\partial v}{\partial t} + \frac{\partial f}{\partial w}\frac{\partial w}{\partial t} = nt^{n-1} f(x, y, z),$$

or

$$x\frac{\partial f}{\partial u} + y\frac{\partial f}{\partial v} + z\frac{\partial f}{\partial w} = nt^{n-1} f(x, y, z).$$

When we set $t = 1$, we obtain $u = x$, $v = y$, $w = z$, and the above equation becomes 13.26. ∎

The results of Exercises 31–33 in Section 13.3 are special cases of 13.26.

EXERCISES 13.6

In Exercises 1–10 we have defined a general functional situation and a specific example. Find the chain rule for the indicated derivative in the general situation, and then use that result to calculate the same derivative in the specific example.

* **1.** dz/dt if $z = f(x, t)$, $x = g(t)$; $z = xt^2/(x + t)$, $x = e^{3t}$

* **2.** $\left.\dfrac{\partial z}{\partial t}\right)_s$ if $z = f(x, y)$, $x = g(s, t)$, $y = h(s, t)$; $z = x^2 e^y + y \ln x$, $x = s^2 \cos t$, $y = 4 \operatorname{Sec}^{-1}(t^2 + 2s)$

* **3.** $\left.\dfrac{\partial u}{\partial s}\right)_t$ if $u = f(x, y, z)$, $x = g(s, t)$, $y = h(s, t)$, $z = k(s, t)$; $u = \sqrt{x^2 + y^2 + z^2}$, $x = 2st$, $y = s^2 + t^2$, $z = st$

* **4.** dz/du if $z = f(x, y, v)$, $x = g(u)$, $y = h(u)$, $v = k(u)$; $z = x^2yv^3$,
  $x = u^3 + 2u$, $y = \ln(u^2 + 1)$, $v = ue^u$

* **5.** $\left(\dfrac{\partial u}{\partial r}\right)_t$ if $u = f(x, y, s)$, $x = g(t)$, $y = h(r)$, $s = k(r, t)$; $u = \sqrt{x^2 + y^2 s}$,
  $x = t/(t + 5)$, $y = \mathrm{Sin}^{-1}(r^2 + 5)$, $s = \tan(rt)$

* **6.** $\left(\dfrac{\partial z}{\partial t}\right)_r$ if $z = f(x)$, $x = g(y)$, $y = h(r, t)$; $z = 3^{x+2}$, $x = y^2 + 5$, $y = \csc(r^2 + t)$

* **7.** $\left(\dfrac{\partial u}{\partial x}\right)_y$ if $u = f(x, y, z)$, $z = g(x, y)$; $u = y/\sqrt{x^2 + y^2 + z^2}$, $z = x/y$

* **8.** $\left(\dfrac{\partial x}{\partial y}\right)_z$ if $x = f(r, s, t)$, $r = g(y)$, $s = h(y, z)$, $t = k(y, z)$; $x = s^2r^2t^2$,
  $r = y^{-5}$, $s = 1/(y^2 + z^2)$, $t = 1/y^2 + 1/z^2$

* **9.** $\left(\dfrac{\partial z}{\partial t}\right)_s$ if $z = f(x, y)$, $x = g(r)$, $y = h(r)$, $r = k(s, t)$; $z = e^{x+y}$, $x = 2r + 5$,
  $y = 2r - 5$, $r = t \ln(s^2 + t^2)$

* **10.** dz/dt if $z = f(x, y, u)$, $x = g(v)$, $u = h(x, y)$, $v = k(t)$, $y = p(t)$;
  $z = x^2 + y^2 + u^2$, $x = v^3 - 3v^2$, $u = 1/(x^2 - y^2)$, $v = e^t$, $y = e^{4t}$

In Exercises 11–15 find the derivative.

** **11.** $\left(\dfrac{\partial^2 z}{\partial t^2}\right)_s$ if $z = x^2y^2 + xe^y$, $x = s + t^2$, $y = s - t^2$

** **12.** d^2x/dt^2 if $x = y^2 + yt - t^2$, $y = t^2e^t$

** **13.** $\left(\dfrac{\partial^2 u}{\partial s^2}\right)_t$ if $u = x^2 + y^2 + z^2 + xyz$, $x = s^2 + t^2$, $y = s^2 - t^2$, $z = st$

** **14.** d^2z/dv^2 if $z = \sin(xy)$, $x = 3\cos v$, $y = 4\sin v$

** **15.** $\partial^2 u/\partial x \partial y$ if $u = y/\sqrt{x^2 + y^2 + z^2}$, $z = x/y$

** **16.** Suppose that u is a differentiable function of r and $r = \sqrt{x^2 + y^2 + z^2}$. Show that

$$\left(\frac{\partial u}{\partial x}\right)^2 + \left(\frac{\partial u}{\partial y}\right)^2 + \left(\frac{\partial u}{\partial z}\right)^2 = \left(\frac{du}{dr}\right)^2.$$

** **17.** Consider a gas that is moving through some region D of space. If we follow a particular particle of the gas, it traces out some curved path (figure on page 995)

$$C: \quad x = x(t), \quad y = y(t), \quad z = z(t), \quad t \geq 0.$$

Suppose the density of the gas at any point in the region D at time t is denoted by $\rho(x, y, z, t)$. We can write that along C, $\rho = \rho[x(t), y(t), z(t), t]$.

(a) Obtain the chain rule defining $d\rho/dt$ in terms of $\partial\rho/\partial t$ and derivatives of x, y, and z with respect to t.

(b) Explain the physical difference between $d\rho/dt$ and $\partial\rho/\partial t$.

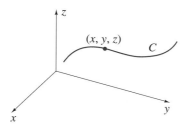

** **18.** The radius and height of a right circular cone are 10 and 20 cm respectively. If the radius is increasing at 1 cm/min and the height is decreasing at 2 cm/min, how fast is the volume changing? Do you need multivariable calculus to solve this problem?

** **19.** If two sides of a triangle have lengths x and y and the angle between them is θ, then the area of the triangle is $A = (1/2)xy\sin\theta$. How fast is the area changing when x is 1 m, y is 2 m, and θ is 1/3 radian, if x and y are each increasing at 1/2 m/s and θ is decreasing at 1/10 radian per second?

** **20.** When a rocket rises from the earth's surface, its mass decreases because fuel is being consumed at the rate of 50 kg/s. Use Newton's universal law of gravitation (see Example 7.32 in Section 7.10) to determine how fast the force of gravity of the earth on the rocket is changing when the rocket is 100 km above the earth's surface and travelling at 2 km/s. Assume that the mass of the rocket at this height is 12×10^6 kg.

** **21.** If $z = f(u, v)$, $u = g(x, y)$, $v = h(x, y)$, find the chain rule for the second derivative $\partial^2 z/\partial x^2$.

** **22.** Determine which of the following functions are positively homogeneous:

(a) $f(x, y) = x^2 + xy + 3y^2$

(b) $f(x, y) = x^2 y + xy - 2xy^2$

(c) $f(x, y, z) = x^2 \sin(y/z) + y^2 + y^3/z$

(d) $f(x, y, z) = xe^{y/z} - xyz$

(e) $f(x, y, z, t) = x^4 + y^4 + z^4 + t^4 - xyzt$

(f) $f(x, y, z, t) = e^{x^2+y^2}(z^2 + t^2)$

(g) $f(x, y, z) = \cos(xy)\sin(yz)$

(h) $f(x, y) = \sqrt{x^2 + xy + y^2}e^{y/x}(2x^2 - 3y^2)$

** **23.** (a) Suppose that the circular plate with radius R in the figure on page 996 has its lower edge held at 0 volts and its upper edge held at 1 volt. Show that the electrostatic potential

$$V(r, \theta) = \frac{1}{2} + \frac{1}{\pi}\text{Tan}^{-1}\left(\frac{2Rr\sin\theta}{R^2 - r^2}\right)$$

satisfies Laplace's equation 13.24 for $r < R$.

(b) The function in (a) is not defined for $r = R$, but it can be shown that as $r \to R$ values of V approach 1 for $0 < \theta < \pi$, and approach 0 for $-\pi < \theta < 0$. Solve the expression in (a) for r in terms of V and θ, and use the result to plot equipotential curves for $V = 1/8, 1/4, 3/8, 1/2, 5/8, 3/4,$ and $7/8$. Set $R = 1$ to do this.

** **24.** Verify that

$$V(r, \theta) = \frac{1}{2}(V_1 + V_2) + \frac{1}{\pi}(V_1 - V_2) \text{Tan}^{-1} \left(\frac{2Rr \sin \theta}{R^2 - r^2} \right)$$

satisfies 13.24 for $r < R$. It represents potential in the circle in the figure above when potential on the upper edge is V_1 and that on the lower edge is V_2.

** **25.** If $f(s)$ and $g(t)$ are differentiable functions, show that $\nabla f(x^2 - y^2) \cdot \nabla g(xy) = 0$.

** **26.** If $f(s)$ is a differentiable function, show that $f(x - y)$ satisfies the equation

$$\frac{\partial f}{\partial y} = -\frac{\partial f}{\partial x}.$$

** **27.** If $f(s)$ is a differentiable function, show that $u(x, y) = f(4x - 3y) + 5(y - x)$ satisfies the equation

$$3\frac{\partial u}{\partial x} + 4\frac{\partial u}{\partial y} = 5.$$

** **28.** If $f(s)$ and $g(t)$ are twice differentiable, show that the function $u(x, y) = xf(x + y) + yg(x + y)$ satisfies

$$\frac{\partial^2 u}{\partial x^2} - 2\frac{\partial^2 u}{\partial x \partial y} + \frac{\partial^2 u}{\partial y^2} = 0.$$

** **29.** If $f(s)$ and $g(t)$ are twice differentiable, show that $f(x - y) + g(x + y)$ satisfies

$$\frac{\partial^2 u}{\partial x^2} - \frac{\partial^2 u}{\partial y^2} = 0.$$

** **30.** Show that if $f(v)$ is differentiable, then $u(x, y) = x^2 f(y/x)$ satisfies

$$x\frac{\partial u}{\partial x} + y\frac{\partial u}{\partial y} = 2u.$$

In Exercises 31–33 suppose that $f(x, y)$ satisfies the first partial differential equation. Show that, with the change of independent variables, function $F(u, v) = f[x(u, v), y(u, v)]$ must satisfy the second partial differential equation.

** **31.** $\left(\frac{\partial f}{\partial x} \right)^2 + \left(\frac{\partial f}{\partial y} \right)^2 = 0$; $u = (x + y)/2$, $v = (x - y)/2$; $\left(\frac{\partial F}{\partial u} \right)^2 + \left(\frac{\partial F}{\partial v} \right)^2 = 0$

** **32.** $\dfrac{\partial^2 f}{\partial x^2} - \dfrac{\partial^2 f}{\partial y^2} = 0$; $u = (x+y)/2$, $v = (x-y)/2$; $\dfrac{\partial^2 F}{\partial u \partial v} = 0$

** **33.** $\left(\dfrac{\partial f}{\partial x}\right)^2 + \left(\dfrac{\partial f}{\partial y}\right)^2 = 0$; $x = u\cos v$, $y = u\sin v$; $\left(\dfrac{\partial F}{\partial u}\right)^2 + \dfrac{1}{u^2}\left(\dfrac{\partial F}{\partial v}\right)^2 = 0$

** **34.** An observer travels along the curve $x = t^2$, $y = 3t^3 + 1$, $z = 2t + 5$, where x, y, and z are in metres and $t \geq 0$ is in seconds. If the density ρ of a gas (in kg/m^3) is given by $\rho = (3x^2 + y^2)/(z^2 + 5)$, find the time rate of change of the density of the gas as measured by the observer when $t = 2$ s.

** **35.** If $f(r)$ is a differentiable function and $r = \sqrt{x^2 + y^2 + z^2}$, show that

$$\nabla f = \frac{f'(r)}{r}(x\hat{\mathbf{i}} + y\hat{\mathbf{j}} + z\hat{\mathbf{k}}).$$

** **36.** If $f(x, y) = 0$ defines y as a function of x, show that

$$\frac{d^2 y}{dx^2} = -\frac{f_{xx}f_y^2 - 2f_{xy}f_x f_y + f_{yy}f_x^2}{f_y^3}.$$

** **37.** If $f(x, y)$ is a harmonic function, show that the function $F(x, y) = f(x^2 - y^2, 2xy)$ is also harmonic.

** **38.** (a) Show that $f(x, y) = \ln(x^2 + y^2)$ satisfies Laplace's equation 13.11.

(b) Transform $f(x, y)$ into polar coordinates and show that the function satisfies 13.24.

** **39.** Find an identity satisfied by the second partial derivatives of a function $f(x, y, z)$ that is positively homogeneous of degree n.

*** **40.** It is postulated in one of the theories of traffic flow that the average speed u at a point x on a straight highway (along the x-axis) is related to the concentration k of traffic by the differential equation

$$u\frac{\partial u}{\partial x} + \frac{\partial u}{\partial t} = -c^2 k^n \frac{\partial k}{\partial x},$$

where t is time, and $c > 0$ and n are constants.

(a) Use chain rules for $\partial u/\partial x$ and $\partial u/\partial t$ in the functional situation $u = f(k)$ and $k = g(x, t)$ to show that

$$\frac{du}{dk}\left(u\frac{\partial k}{\partial x} + \frac{\partial k}{\partial t}\right) + c^2 k^n \frac{\partial k}{\partial x} = 0.$$

(b) The equation of continuity for traffic flow states that

$$\frac{\partial k}{\partial t} + \frac{\partial(ku)}{\partial x} = 0.$$

Use these last two equations to obtain the differential equation relating speed and concentration:

$$\frac{du}{dk} = -ck^{(n-1)/2}.$$

(c) Solve the differential equation in (b) for $u = f(k)$.

*** **41.** A bead slides from rest at the origin on a frictionless wire in a vertical plane to the point (x_0, y_0) under the influence of gravity (figure below). It can be shown that the time for the bead to traverse the path is

$$t = \frac{1}{\sqrt{2g}} \int_0^{x_0} \sqrt{\frac{1 + (y')^2}{y}} \, dx,$$

where g is the acceleration due to gravity and $y' = dy/dx$. The problem of finding that shape of wire which makes t as small as possible is called the *brachistochrone problem*. It is shown in the calculus of variations that $y = f(x)$ must satisfy the equation

$$\frac{d}{dx} \left(\frac{\partial F}{\partial y'} \right) - \frac{\partial F}{\partial y} = 0,$$

where

$$F(y, y') = \sqrt{\frac{1 + (y')^2}{y}}.$$

(a) Show that $f(x)$ must satisfy the differential equation $1 + (y')^2 + 2yy'' = 0$.

(b) Show that the curve that satisfies the equation in (a) is the cycloid defined parametrically by

$$x = a(\theta - \sin \theta), \quad y = a(1 - \cos \theta),$$

where a is a constant.

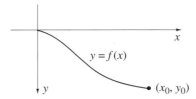

*** **42.** Two equal masses m are connected by springs having equal spring constant k so that the masses are free to slide on a frictionless table (figure on page 999). The walls A and B are fixed.

(a) Use Newton's second law to show that the differential equations for the motions of the masses are

$$m\ddot{x}_1 = k(x_2 - 2x_1), \qquad m\ddot{x}_2 = k(x_1 - 2x_2),$$

where x_1 and x_2 are the displacements of the masses from their equilibrium positions, $\ddot{x}_1 = d^2x_1/dt^2$ and $\ddot{x}_2 = d^2x_2/dt^2$.

(b) The Euler-Lagrange equations from theoretical mechanics for this system are

$$\frac{d}{dt} \left(\frac{\partial L}{\partial \dot{x}_1} \right) - \frac{\partial L}{\partial x_1} = 0, \qquad \frac{d}{dt} \left(\frac{\partial L}{\partial \dot{x}_2} \right) - \frac{\partial L}{\partial x_2} = 0,$$

where L is defined as the kinetic energy of the two masses less the energy stored in the springs. Show that

$$L(x_1, x_2, \dot{x}_1, \dot{x}_2) = \frac{m}{2}(\dot{x}_1^2 + \dot{x}_2^2) - k(x_1^2 + x_2^2 - x_1 x_2).$$

(c) Obtain the equations in (a) from the Euler-Lagrange equations in (b).

*** **43.** Suppose the second-order partial differential equation

$$p\frac{\partial^2 z}{\partial x^2} + q\frac{\partial^2 z}{\partial x \partial y} + r\frac{\partial^2 z}{\partial y^2} = F\left(x, y, z, \frac{\partial z}{\partial x}, \frac{\partial z}{\partial y}\right)$$

(p, q, and r are constants) is subjected to the change of variables

$$s = ax + by, \quad t = cx + dy,$$

where a, b, c, and d are constants. Show that the partial differential equation in s and t is

$$P\frac{\partial^2 z}{\partial s^2} + Q\frac{\partial^2 z}{\partial s \partial t} + R\frac{\partial^2 z}{\partial t^2} = G\left(s, t, z, \frac{\partial z}{\partial s}, \frac{\partial z}{\partial t}\right),$$

where $Q^2 - 4PR = (q^2 - 4pr)(ad - bc)^2$.

44. Show that if a solution $u = f(x, y, z)$ of the three-dimensional Laplace equation 13.12 can be expressed in the form $u = g(r)$, where $r = \sqrt{x^2 + y^2 + z^2}$, then $f(x, y, z)$ must be of the form

$$f(x, y, z) = \frac{C}{\sqrt{x^2 + y^2 + z^2}} + D,$$

where C and D are constants.

13.7 Implicit Differentiation

In Section 3.8 we introduced the technique of implicit differentiation in order to obtain the derivative of a function $y = f(x)$ defined implicitly by an equation

$$F(x, y) = 0. \tag{13.27}$$

Essentially, the technique is to differentiate all terms in the equation with respect to x, considering all the while that y is a function of x. For example, if y is defined implicitly by

$$x^2 y^3 + 3xy = 3x + 2,$$

implicit differentiation gives

$$2xy^3 + 3x^2 y^2\frac{dy}{dx} + 3y + 3x\frac{dy}{dx} = 3.$$

We can now solve to obtain

$$\frac{dy}{dx} = \frac{3 - 2xy^3 - 3y}{3x^2 y^2 + 3x}.$$

With the chain rule we can actually present a formula for dy/dx. Since equation 13.27, when written in the form

$$F[x, f(x)] = 0,$$

must be valid for all x in the domain of the function $f(x)$, we can differentiate it with respect to x. From the schematic diagram below

the derivative of the left side of the equation is

$$\frac{dF}{dx} = \frac{\partial F}{\partial x} + \frac{\partial F}{\partial y}\frac{dy}{dx}.$$

If we equate this to the derivative of the right side of the equation, we find

$$F_x + F_y\frac{dy}{dx} = 0,$$

or

$$\frac{dy}{dx} = -\frac{F_x}{F_y}. \tag{13.28}$$

For the function defined implicitly above by $x^2y^3 + 3xy - 3x - 2 = 0$, equation 13.28 gives

$$\frac{dy}{dx} = -\frac{2xy^3 + 3y - 3}{3x^2y^2 + 3x},$$

and this result is identical to that obtained by implicit differentiation.

Similarly, if the equation

$$F(x, y, z) = 0 \tag{13.29}$$

defines z implicitly as a function of x and y, the schematic diagram below

immediately yields

$$\frac{\partial F}{\partial x} + \frac{\partial F}{\partial z}\frac{\partial z}{\partial x} = 0, \qquad \frac{\partial F}{\partial y} + \frac{\partial F}{\partial z}\frac{\partial z}{\partial y} = 0.$$

From these we obtain the results

$$\frac{\partial z}{\partial x} = -\frac{F_x}{F_z}, \qquad \frac{\partial z}{\partial y} = -\frac{F_y}{F_z}. \tag{13.30}$$

We do not suggest that formulas 13.28 and 13.30 be memorized. On the contrary, we obtain results in this section that include 13.28 and 13.30 as special cases. To develop these results we work with three equations in five variables,

$$F(x, y, u, v, w) = 0,$$
$$G(x, y, u, v, w) = 0,$$
$$H(x, y, u, v, w) = 0.$$

(13.31)

We assume that these equations define u, v, and w as functions of x and y for some domain of values of x and y (and do so implicitly). It might even be possible to solve the system and obtain explicit definitions of the functions

$$u = f(x, y),$$
$$v = g(x, y),$$
$$w = h(x, y).$$

(13.32)

We pose the problem of finding the six first-order partial derivatives of u, v, and w with respect to x and y supposing that it is undesirable or even impossible to obtain the explicit form of the functions. To do this, we note that, were results 13.32 known and substituted into 13.31, then

$$F[x, y, f(x, y), g(x, y), h(x, y)] = 0,$$
$$G[x, y, f(x, y), g(x, y), h(x, y)] = 0,$$
$$H[x, y, f(x, y), g(x, y), h(x, y)] = 0$$

would be identities in x and y. As a result we could differentiate each equation with respect to x, obtaining from the schematic diagram

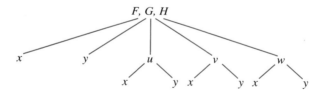

the following:

$$\frac{\partial F}{\partial x} + \frac{\partial F}{\partial u}\frac{\partial u}{\partial x} + \frac{\partial F}{\partial v}\frac{\partial v}{\partial x} + \frac{\partial F}{\partial w}\frac{\partial w}{\partial x} = 0,$$

$$\frac{\partial G}{\partial x} + \frac{\partial G}{\partial u}\frac{\partial u}{\partial x} + \frac{\partial G}{\partial v}\frac{\partial v}{\partial x} + \frac{\partial G}{\partial w}\frac{\partial w}{\partial x} = 0,$$

(13.33)

$$\frac{\partial H}{\partial x} + \frac{\partial H}{\partial u}\frac{\partial u}{\partial x} + \frac{\partial H}{\partial v}\frac{\partial v}{\partial x} + \frac{\partial H}{\partial w}\frac{\partial w}{\partial x} = 0,$$

or

$$F_u\frac{\partial u}{\partial x} + F_v\frac{\partial v}{\partial x} + F_w\frac{\partial w}{\partial x} = -F_x,$$

$$G_u\frac{\partial u}{\partial x} + G_v\frac{\partial v}{\partial x} + G_w\frac{\partial w}{\partial x} = -G_x,$$

(13.34)

$$H_u\frac{\partial u}{\partial x} + H_v\frac{\partial v}{\partial x} + H_w\frac{\partial w}{\partial x} = -H_x.$$

We have in 13.34 three equations in the three unknowns $\partial u/\partial x$, $\partial v/\partial x$, and $\partial w/\partial x$, and, because the equations are linear in the unknowns, solutions can be obtained using Cramer's rule.[†] In particular,

$$\frac{\partial u}{\partial x} = \frac{\begin{vmatrix} -F_x & F_v & F_w \\ -G_x & G_v & G_w \\ -H_x & H_v & H_w \end{vmatrix}}{\begin{vmatrix} F_u & F_v & F_w \\ G_u & G_v & G_w \\ H_u & H_v & H_w \end{vmatrix}} = -\frac{\begin{vmatrix} F_x & F_v & F_w \\ G_x & G_v & G_w \\ H_x & H_v & H_w \end{vmatrix}}{\begin{vmatrix} F_u & F_v & F_w \\ G_u & G_v & G_w \\ H_u & H_v & H_w \end{vmatrix}}. \tag{13.35}$$

The two determinants on the right of 13.35 involve only derivatives of the given functions F, G, and H, and we have therefore obtained a method for finding $\partial u/\partial x$ that avoids solving 13.31 for u, v, and w. We could list similar formulas for the remaining five derivatives, but first we introduce some simplifying notation.

DEFINITION 13.5

The **Jacobian determinant** of functions F, G, and H with respect to variables u, v, and w is denoted by $\dfrac{\partial(F, G, H)}{\partial(u, v, w)}$ and is defined as the determinant

$$\frac{\partial(F, G, H)}{\partial(u, v, w)} = \begin{vmatrix} F_u & F_v & F_w \\ G_u & G_v & G_w \\ H_u & H_v & H_w \end{vmatrix} = \begin{vmatrix} \dfrac{\partial F}{\partial u} & \dfrac{\partial F}{\partial v} & \dfrac{\partial F}{\partial w} \\[2mm] \dfrac{\partial G}{\partial u} & \dfrac{\partial G}{\partial v} & \dfrac{\partial G}{\partial w} \\[2mm] \dfrac{\partial H}{\partial u} & \dfrac{\partial H}{\partial v} & \dfrac{\partial H}{\partial w} \end{vmatrix}. \tag{13.36}$$

With this notation we can write 13.35 in the form

$$\frac{\partial u}{\partial x} = -\frac{\dfrac{\partial(F, G, H)}{\partial(x, v, w)}}{\dfrac{\partial(F, G, H)}{\partial(u, v, w)}}. \tag{13.37}$$

The remaining derivatives of $v = g(x, y)$ and $w = h(x, y)$ with respect to x can also be obtained from equations 13.34 by Cramer's rule:

$$\frac{\partial v}{\partial x} = -\frac{\dfrac{\partial(F, G, H)}{\partial(u, x, w)}}{\dfrac{\partial(F, G, H)}{\partial(u, v, w)}}, \quad \frac{\partial w}{\partial x} = -\frac{\dfrac{\partial(F, G, H)}{\partial(u, v, x)}}{\dfrac{\partial(F, G, H)}{\partial(u, v, w)}}. \tag{13.38}$$

A similar procedure yields

$$\frac{\partial u}{\partial y} = -\frac{\dfrac{\partial(F, G, H)}{\partial(y, v, w)}}{\dfrac{\partial(F, G, H)}{\partial(u, v, w)}}, \quad \frac{\partial v}{\partial y} = -\frac{\dfrac{\partial(F, G, H)}{\partial(u, y, w)}}{\dfrac{\partial(F, G, H)}{\partial(u, v, w)}}, \quad \frac{\partial w}{\partial y} = -\frac{\dfrac{\partial(F, G, H)}{\partial(u, v, y)}}{\dfrac{\partial(F, G, H)}{\partial(u, v, w)}}. \tag{13.39}$$

[†] Cramer's rule is discussed in Appendix B.

Formulas 13.37–13.39 apply only to the situation in which equations 13.31 define u, v, and w as functions of x and y. It is, however, fairly evident how to construct formulas in other situations. Each formula has a Jacobian divided by a Jacobian (and do not forget the negative sign). In the denominator, it is the Jacobian of the functions defining the original equations with respect to the dependent variables. The only difference in the Jacobian in the numerator is that the dependent variable that is being differentiated is replaced by the independent variable with respect to which differentiation is being performed.

The results in equations 13.37–13.39 are valid provided, of course, that the Jacobian

$$\frac{\partial(F, G, H)}{\partial(u, v, w)} \neq 0.$$

In actual fact, it is this condition that guarantees that equations 13.31 do define u, v, and w as functions of x and y in the first place.

As a second example, if

$$F(x, y, s, t) = 0,$$

$$G(x, y, s, t) = 0,$$

define x and y as functions of s and t, then

$$\frac{\partial x}{\partial s} = -\frac{\dfrac{\partial(F, G)}{\partial(s, y)}}{\dfrac{\partial(F, G)}{\partial(x, y)}}, \qquad \frac{\partial x}{\partial t} = -\frac{\dfrac{\partial(F, G)}{\partial(t, y)}}{\dfrac{\partial(F, G)}{\partial(x, y)}},$$

$$\frac{\partial y}{\partial s} = -\frac{\dfrac{\partial(F, G)}{\partial(x, s)}}{\dfrac{\partial(F, G)}{\partial(x, y)}}, \qquad \frac{\partial y}{\partial t} = -\frac{\dfrac{\partial(F, G)}{\partial(x, t)}}{\dfrac{\partial(F, G)}{\partial(x, y)}}.$$

We now show that the results in equations 13.28 and 13.30 are special cases of these considerations. If equation 13.27 defines y implicitly as a function of x, then

$$\frac{dy}{dx} = -\frac{\dfrac{\partial(F)}{\partial(x)}}{\dfrac{\partial(F)}{\partial(y)}} = -\frac{F_x}{F_y}.$$

If equation 13.29 defines z implicitly as a function of x and y, then

$$\frac{\partial z}{\partial x} = -\frac{\dfrac{\partial(F)}{\partial(x)}}{\dfrac{\partial(F)}{\partial(z)}} = -\frac{F_x}{F_z}; \qquad \frac{\partial z}{\partial y} = -\frac{\dfrac{\partial(F)}{\partial(y)}}{\dfrac{\partial(F)}{\partial(z)}} = -\frac{F_y}{F_z}.$$

EXAMPLE 13.20

If $x^2y^2z^3 + zx \sin y = 5$ defines z as a function of x and y, find $\partial z/\partial x$.

SOLUTION If we set $F(x, y, z) = x^2y^2z^3 + zx \sin y - 5 = 0$, then

$$\frac{\partial z}{\partial x} = -\frac{\dfrac{\partial(F)}{\partial(x)}}{\dfrac{\partial(F)}{\partial(z)}} = -\frac{F_x}{F_z} = -\frac{2xy^2z^3 + z \sin y}{3x^2y^2z^2 + x \sin y}.$$

EXAMPLE 13.21

In the theory of thermodynamics, pressure P, temperature T, volume V, and internal energy U are related by two equations of state:

$$F(P, T, V, U) = 0, \quad G(P, T, V, U) = 0.$$

The second law of thermodynamics implies that, if U and P are regarded as functions of T and V, then the functions $U(T, V)$ and $P(T, V)$ must satisfy the equation

$$\frac{\partial U}{\partial V} - T\frac{\partial P}{\partial T} + P = 0.$$

Show that if U and V are chosen as independent variables rather than T and V, then the second law takes the form

$$\frac{\partial T}{\partial V} + T\frac{\partial P}{\partial U} - P\frac{\partial T}{\partial U} = 0.$$

SOLUTION When U and V are taken as independent variables,

$$\frac{\partial T}{\partial V} = -\frac{\dfrac{\partial(F, G)}{\partial(V, P)}}{\dfrac{\partial(F, G)}{\partial(T, P)}},$$

$$\frac{\partial P}{\partial U} = -\frac{\dfrac{\partial(F, G)}{\partial(T, U)}}{\dfrac{\partial(F, G)}{\partial(T, P)}},$$

$$\frac{\partial T}{\partial U} = -\frac{\dfrac{\partial(F, G)}{\partial(U, P)}}{\dfrac{\partial(F, G)}{\partial(T, P)}}.$$

Consequently,

$$\frac{\partial T}{\partial V} + T\frac{\partial P}{\partial U} - P\frac{\partial T}{\partial U} = -\frac{1}{\dfrac{\partial(F, G)}{\partial(T, P)}}\left\{\frac{\partial(F, G)}{\partial(V, P)} + T\frac{\partial(F, G)}{\partial(T, U)} - P\frac{\partial(F, G)}{\partial(U, P)}\right\}.$$

To show that this expression must be equal to zero, we note that, when T and V are independent variables, the second law implies that

$$0 = -\frac{\dfrac{\partial(F, G)}{\partial(V, P)}}{\dfrac{\partial(F, G)}{\partial(U, P)}} + T\frac{\dfrac{\partial(F, G)}{\partial(U, T)}}{\dfrac{\partial(F, G)}{\partial(U, P)}} + P$$

$$= -\frac{1}{\dfrac{\partial(F, G)}{\partial(U, P)}}\left\{\frac{\partial(F, G)}{\partial(V, P)} - T\frac{\partial(F, G)}{\partial(U, T)} - P\frac{\partial(F, G)}{\partial(U, P)}\right\}$$

$$= -\frac{1}{\dfrac{\partial(F, G)}{\partial(U, P)}}\left\{\frac{\partial(F, G)}{\partial(V, P)} + T\frac{\partial(F, G)}{\partial(T, U)} - P\frac{\partial(F, G)}{\partial(U, P)}\right\}.$$

Hence

$$\frac{\partial T}{\partial V} + T\frac{\partial P}{\partial U} - P\frac{\partial T}{\partial U} = 0.$$

EXERCISES 13.7

In Exercises 1–4 y is defined implicitly as a function of x. Find dy/dx.

* **1.** $x^3 y^2 - 2xy + 5 = 0$
* **2.** $(x + y)^2 = 2x$
* **3.** $x(x - y) - 4y^3 = 2e^{xy} + 6$
* **4.** $\sin(x + y) + y^2 = 12x^2 + y$

In Exercises 5–8 z is defined implicitly as a function of x and y. Find $\partial z/\partial x$ and $\partial z/\partial y$.

* **5.** $x^2 \sin z - ye^z = 2x$
* **6.** $x^2 z^2 + yz + 3x = 4$
* **7.** $z \sin^2 y + y \sin^2 x = z^3$
* **8.** $\text{Tan}^{-1}(yz) = xz$

In Exercises 9–13 find the required derivative. Assume that the system of equations does define the function(s) indicated.

* **9.** $\partial u/\partial x$ and $\partial v/\partial y$ if $x^2 - y^2 + u^2 + 2v^2 = 1$, $x^2 + y^2 = 2 + u^2 + v^2$
* **10.** $\partial x/\partial t$ if $\sin(x + t) - \sin(x - t) = z$
* **11.** $(\partial\phi/\partial x)_{y,z}$ if $x = r\sin\phi\cos\theta$, $\quad y = r\sin\phi\sin\theta$, $\quad z = r\cos\phi$
* **12.** dz/dx if $x^2 + y^2 - z^2 + 2xy = 1$, $\quad x^3 + y^3 - 5y = 4$
* **13.** $(\partial u/\partial y)_x$ if $xyu + vw = 4$, $\quad y^2 + u^2 - u^2 v = y$, $\quad yw + xu + v + 4 = 0$
** **14.** Given that the equations

$$x^2 - y\cos(uv) + z^2 = 0,$$

$$x^2 + y^2 - \sin(uv) + 2z^2 = 2,$$

$$xy - \sin u \cos v + z = 0$$

define x, y, and z as functions of u and v, find $\partial x/\partial u)_v$ at the values $x = 1$, $y = 1$, $u = \pi/2$, $v = 0$, and $z = 0$.

** **15.** If the equation $F(x, y, z) = 0$ defines each of x, y, and z as a function of the other two, show that

$$\left(\frac{\partial z}{\partial x}\right)_y \left(\frac{\partial x}{\partial y}\right)_z \left(\frac{\partial y}{\partial z}\right)_x = -1.$$

** **16.** If $z = e^x \cos y$, where x and y are functions of t defined by

$$x^3 + e^x - t^2 - t = 1, \quad yt^2 + y^2t - t + y = 0,$$

find dz/dt.

** **17.** Find $\partial s/\partial u)_v$ if $s = x^2 + y^2$, and x and y are functions of u and v defined by

$$u = x^2 - y^2, \quad v = x^2 - y.$$

** **18.** Find $\partial z/\partial y)_x$ if $z = u^3v + \sin(uv)$, and u and v are functions of x and y defined by

$$x = e^u \cos v, \quad y = e^u \sin v.$$

** **19.** Given that $z^3 - xz - y = 0$ defines z as a function of x and y, show that

$$\frac{\partial^2 z}{\partial x \partial y} = -\frac{3z^2 + x}{(3z^2 - x)^3}.$$

** **20.** If the equations $x = u^2 - v^2$, $y = 2uv$, define u and v as functions of x and y, find $\partial^2 u/\partial x^2$.

** **21.** (a) Given that the equation $z^4x + y^3z + 9x^3 = 2$ defines z as a function of x and y, and x as a function of y and z, are $\partial z/\partial x$ and $\partial x/\partial z$ reciprocals?

(b) Given that the equations $z^4x + y^3z + 9x^3 = 2$, $x^2y + xz = 1$ define z as a function of x, and x as a function of z, are dz/dx and dx/dz reciprocals?

(c) Given that the equations $u^2 - v = 3x + y$, $u - 2v^2 = x - 2y$ define u and v as functions of x and y, and also define x and y as functions of u and v, are $\partial u/\partial x$ and $\partial x/\partial u$ reciprocals?

** **22.** Given that the equations $x^2 - 2y^2s^2t - 2st^2 = 1$, $x^2 + 2y^2s^2t + 5st^2 = 1$ define s and t as functions of x and y, find $\partial^2 t/\partial y^2$.

** **23.** Show that if V and P are chosen as independent variables in Example 13.21, then the second law of thermodynamics takes the form

$$T - P\frac{\partial T}{\partial P} + \frac{\partial(T, U)}{\partial(V, P)} = 0.$$

*** **24.** (a) Suppose the equations $F(u, v, s, t) = 0$, $G(u, v, s, t) = 0$ define u and v as functions of s and t, and the equations $H(s, t, x, y) = 0$, $I(s, t, x, y) = 0$ define s and t as functions of x and y. Show that

$$\frac{\partial(u, v)}{\partial(s, t)} \frac{\partial(s, t)}{\partial(x, y)} = \frac{\partial(u, v)}{\partial(x, y)}.$$

(b) If the equations $F(u, v, x, y) = 0$, $G(u, v, x, y) = 0$ define u and v as functions of x and y, and also define x and y as functions of u and v, show that

$$\frac{\partial(u, v)}{\partial(x, y)} = \frac{1}{\dfrac{\partial(x, y)}{\partial(u, v)}}.$$

*** **25.** Suppose the system of m linear equations in n unknowns $(n > m)$

$$\sum_{j=1}^{n} a_{ij} x_j = c_i, \qquad i = 1, \ldots, m$$

defines $x_1, x_2, \ldots, x_m$ as functions of $x_{m+1}, x_{m+2}, \ldots, x_n$. Show that if $1 \le i \le m$ and $m + 1 \le j \le n$, then

$$\frac{\partial x_i}{\partial x_j} = -\frac{D_{ij}}{D},$$

where $D = |a_{ij}|_{m \times m}$, and D_{ij} is the same as determinant D except that its ith column is replaced by the jth column of $[a_{ij}]_{m \times n}$.

13.8 Directional Derivatives

FIGURE 13.14

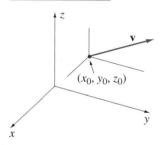

If a function $f(x, y, z)$ is defined throughout some region of space, then at any point (x_0, y_0, z_0) we can calculate its partial derivatives $\partial f / \partial x$, $\partial f / \partial y$, and $\partial f / \partial z$. These derivatives define rates of change of $f(x, y, z)$ at (x_0, y_0, z_0) in directions parallel to the x-, y-, and z-axes. But what if we want the rate of change of $f(x, y, z)$ at (x_0, y_0, z_0) in some arbitrary direction defined by a vector $\mathbf{v}$ (Figure 13.14)? By the rate of change of $f(x, y, z)$ in the direction $\mathbf{v}$, we mean the rate of change with respect to distance as measured along a line through (x_0, y_0, z_0) in direction $\mathbf{v}$. Let us define s as a measure of directed distance along this line, taking $s = 0$ at (x_0, y_0, z_0) and positive s in the direction of $\mathbf{v}$. What we want, then, is the derivative of $f(x, y, z)$ with respect to s at $s = 0$. To express $f(x, y, z)$ in terms of s, we use parametric equations of the line through (x_0, y_0, z_0) along $\mathbf{v}$. If $\hat{\mathbf{v}} = (v_x, v_y, v_z)$ is a unit vector in the direction of $\mathbf{v}$, then parametric equations for this line (see equations 12.36) are

$$x = x_0 + v_x s, \quad y = y_0 + v_y s, \quad z = z_0 + v_z s. \tag{13.40}$$

From the schematic diagram below,

we obtain

$$\frac{df}{ds} = \frac{\partial f}{\partial x}\frac{dx}{ds} + \frac{\partial f}{\partial y}\frac{dy}{ds} + \frac{\partial f}{\partial z}\frac{dz}{ds}$$

$$= \frac{\partial f}{\partial x}v_x + \frac{\partial f}{\partial y}v_y + \frac{\partial f}{\partial z}v_z,$$

where all partial derivatives of $f(x, y, z)$ are to be evaluated at (x_0, y_0, z_0). This derivative df/ds is called the **directional derivative** of $f(x, y, z)$ in the direction $\mathbf{v}$ at the point (x_0, y_0, z_0), and is usually given the alternative notation

$$D_{\mathbf{v}} f = \frac{\partial f}{\partial x} v_x + \frac{\partial f}{\partial y} v_y + \frac{\partial f}{\partial z} v_z. \tag{13.41}$$

Now, v_x, v_y, and v_z are the components of the unit vector $\hat{\mathbf{v}}$ in the direction of $\mathbf{v}$, and $\partial f/\partial x$, $\partial f/\partial y$, and $\partial f/\partial z$ are the components of the gradient of $f(x, y, z)$. We can write, therefore, that

$$D_{\mathbf{v}} f = \nabla f \cdot \hat{\mathbf{v}}. \tag{13.42}$$

Consequently, the derivative (rate of change) of a function in any given direction is the scalar product of the gradient of the function and a unit vector in the required direction. We state this in the following theorem.

> **THEOREM 13.4**
>
> The directional derivative of a function in any direction is the component of the gradient of the function in that direction.

EXAMPLE 13.22

Find $D_{\mathbf{v}} f$ at $(4, 0, 16)$ if $f(x, y, z) = x^3 e^y + xz$ and $\mathbf{v}$ is the vector from $(4, 0, 16)$ to $(-2, 1, 4)$.

SOLUTION Since

$$\nabla f_{|(4,0,16)} = \{(3x^2 e^y + z)\hat{\mathbf{i}} + x^3 e^y \hat{\mathbf{j}} + x\hat{\mathbf{k}}\}_{|(4,0,16)}$$

$$= 64\hat{\mathbf{i}} + 64\hat{\mathbf{j}} + 4\hat{\mathbf{k}}$$

and

$$\hat{\mathbf{v}} = \frac{\mathbf{v}}{|\mathbf{v}|} = \frac{(-6, 1, -12)}{\sqrt{36 + 1 + 144}} = \frac{-1}{\sqrt{181}}(6, -1, 12),$$

we have

$$D_{\mathbf{v}} f = -(64, 64, 4) \cdot \frac{(6, -1, 12)}{\sqrt{181}} = -\frac{368}{\sqrt{181}}.$$

The fact that the derivative is negative means that $f(x, y, z)$ is decreasing in direction $\mathbf{v}$.

The directional derivative gives us insight into some of the properties of the gradient vector. In particular, we have the next theorem.

> **THEOREM 13.5**
>
> The gradient ∇f of a function $f(x, y, z)$ defines the direction in which the function increases most rapidly, and the maximum rate of change is $|\nabla f|$.

PROOF Theorem 13.4 states that the directional derivative of $f(x, y, z)$ in a direction **v** is the component of ∇f in that direction. Figure 13.15, which shows components of ∇f in various directions, makes it clear that $D_\mathbf{v} f$ is greatest when **v** is parallel to ∇f. Alternatively, if θ is the angle between **v** and ∇f, then

$$D_\mathbf{v} f = \nabla f \cdot \hat{\mathbf{v}} = |\nabla f||\hat{\mathbf{v}}| \cos\theta = |\nabla f| \cos\theta.$$

Obviously $D_\mathbf{v} f$ is a maximum when $\cos\theta$ is a maximum—i.e., when $\cos\theta = 1$ or $\theta = 0$—and this occurs when **v** is parallel to ∇f. Finally, when **v** is parallel to ∇f, we have $D_\mathbf{v} f = |\nabla f|$, and this completes the proof. ∎

FIGURE 13.15

Note that for any function $f(x, y, z)$,

$$D_\mathbf{i} f = \frac{\partial f}{\partial x}, \quad D_\mathbf{j} f = \frac{\partial f}{\partial y}, \quad D_\hat{\mathbf{k}} f = \frac{\partial f}{\partial z}.$$

In other words, the partial derivatives of a function are its directional derivatives along the coordinate directions.

EXAMPLE 13.23

Find the direction at the point $(1, 2, -3)$ in which the function $f(x, y, z) = x^2 y + xyz$ increases most rapidly.

SOLUTION   According to Theorem 13.5, $f(x, y, z)$ increases most rapidly in the direction

$$\nabla f_{|(1,2,-3)} = (2xy + yz, x^2 + xz, xy)_{|(1,2,-3)} = (-2, -2, 2).$$

You might feel that, because the definition of the directional derivative $D_\mathbf{v} f$ does not involve a limit process, it is some strange new type of differentiation. To show that this is not the case, let us return to the calculation of the derivative of $f(x, y, z)$ at (x_0, y_0, z_0) in the direction **v** shown in Figure 13.14. With parametric equations 13.40 for the line through (x_0, y_0, z_0) along **v**, the value of $f(x, y, z)$ at any point (x, y, z) along this line is $f(x_0 + v_x s, y_0 + v_y s, z_0 + v_z s)$. If we take the difference between this value and $f(x_0, y_0, z_0)$ and divide by the distance s between (x_0, y_0, z_0) and (x, y, z), then the limit of this expression as $s \to 0^+$ should define the derivative of $f(x, y, z)$ at (x_0, y_0, z_0) in the direction **v**; that is,

$$D_\mathbf{v} f = \lim_{s \to 0^+} \frac{f(x_0 + v_x s, y_0 + v_y s, z_0 + v_z s) - f(x_0, y_0, z_0)}{s}. \tag{13.43}$$

It can be shown that this limit (and this is perhaps the form we might have expected the derivative to take) also leads to the result contained in 13.42 (see Exercise 35).

Consider a curve C in space that is defined parametrically by

$$C: \quad x = x(t), \quad y = y(t), \quad z = z(t), \quad \alpha \le t \le \beta$$

FIGURE 13.16

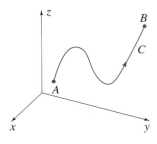

(Figure 13.16). Imagine that C is the path traced out by some particle as it moves through space under the action of some system of forces, and suppose that $f(x, y, z)$ is a function defined along C. Perhaps the particle is a weather balloon and $f(x, y, z)$ is temperature at points along its trajectory C. In such applications we are frequently asked for the rate of change of $f(x, y, z)$ with respect to distance travelled along C. If we use s as a measure of distance along C (taking $s = 0$ at A), then the required rate of change is df/ds. Since the coordinates of points (x, y, z) on C can be regarded as functions of s (although it might be difficult to find these functions explicitly), the chain rule gives

$$\frac{df}{ds} = \frac{\partial f}{\partial x}\frac{dx}{ds} + \frac{\partial f}{\partial y}\frac{dy}{ds} + \frac{\partial f}{\partial z}\frac{dz}{ds}$$

$$= \left(\frac{\partial f}{\partial x}, \frac{\partial f}{\partial y}, \frac{\partial f}{\partial z}\right) \cdot \left(\frac{dx}{ds}, \frac{dy}{ds}, \frac{dz}{ds}\right)$$

$$= \nabla f \cdot \frac{d\mathbf{r}}{ds}.$$

In Section 12.9 we saw that $d\mathbf{r}/ds$ is a unit tangent vector $\hat{\mathbf{T}}$ to C. Consequently,

$$\frac{df}{ds} = \nabla f \cdot \hat{\mathbf{T}}.$$

But this equation states that df/ds is the directional derivative of $f(x, y, z)$ along the tangent direction to the curve C. In other words, to calculate the rate of change of a function $f(x, y, z)$ with respect to distance as measured along a curve C, we calculate the directional derivative of $f(x, y, z)$ in the direction of the tangent vector to C.

EXAMPLE 13.24

Find the rate of change of the function $f(x, y, z) = x^2 y - xz$ along the curve $y = x^2, z = x$ in the direction of decreasing x at the point $(2, 4, 2)$.

SOLUTION   Since parametric equations for the curve are $C: x = -t, y = t^2, z = -t$, a tangent vector to C at any point is $\mathbf{T} = (-1, 2t, -1)$. At $(2, 4, 2)$, $t = -2$, and the tangent vector is $\mathbf{T} = (-1, -4, -1)$. A unit tangent vector to C at $(2, 4, 2)$ in the direction of decreasing x is therefore

$$\hat{\mathbf{T}} = \frac{(-1, -4, -1)}{\sqrt{18}} = \frac{-1}{3\sqrt{2}}(1, 4, 1).$$

The rate of change of $f(x, y, z)$ in this direction is

$$\nabla f \cdot \hat{\mathbf{T}} = (2xy - z, x^2, -x)_{|(2,4,2)} \cdot \frac{(1, 4, 1)}{-3\sqrt{2}}$$

$$= \frac{-1}{3\sqrt{2}}(14, 4, -2) \cdot (1, 4, 1) = -\frac{28}{3\sqrt{2}}.$$

In preparation for maxima and minima of multivariable functions in Section 13.10, we now discuss directional derivatives for a function $f(x, y)$ of two independent variables. Such a function can be represented graphically as a surface $z = f(x, y)$ (Figure 13.17).

FIGURE 13.17

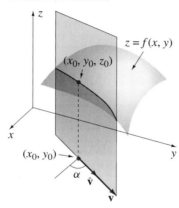

For a direction $\mathbf{v}$ at (x_0, y_0) in the xy-plane,

$$D_{\mathbf{v}}f = \nabla f \cdot \hat{\mathbf{v}},$$

where ∇f is evaluated at (x_0, y_0). Algebraically, this is the rate of change of $f(x, y)$ in direction $\mathbf{v}$. Geometrically, it is the rate of change of the height z of the surface as we move along the curve of intersection of the surface and a vertical plane containing the vector $\mathbf{v}$, or the slope of this curve. Each direction $\mathbf{v}$ at (x_0, y_0) defines an angle α with a line through (x_0, y_0) parallel to the positive x-axis, and for this direction

$$\hat{\mathbf{v}} = \cos\alpha\,\hat{\mathbf{i}} + \sin\alpha\,\hat{\mathbf{j}}.$$

We can write, then,

$$D_{\mathbf{v}}f = \nabla f \cdot \hat{\mathbf{v}} = \left(\frac{\partial f}{\partial x}\hat{\mathbf{i}} + \frac{\partial f}{\partial y}\hat{\mathbf{j}}\right) \cdot (\cos\alpha\,\hat{\mathbf{i}} + \sin\alpha\,\hat{\mathbf{j}})$$

$$= \frac{\partial f}{\partial x}\cos\alpha + \frac{\partial f}{\partial y}\sin\alpha. \tag{13.44}$$

If $D_{\mathbf{v}} f$ represents the slope of the curve of intersection of the surface and the vertical plane through $\mathbf{v}$, then $D_{\mathbf{v}}(D_{\mathbf{v}} f)$ represents the rate of change of this slope. Now

$$D_{\mathbf{v}}(D_{\mathbf{v}} f) = \nabla(D_{\mathbf{v}} f) \cdot \hat{\mathbf{v}},$$

and

$$\nabla(D_{\mathbf{v}} f) = \left(\frac{\partial^2 f}{\partial x^2} \cos\alpha + \frac{\partial^2 f}{\partial x \partial y} \sin\alpha\right)\hat{\mathbf{i}} + \left(\frac{\partial^2 f}{\partial y \partial x} \cos\alpha + \frac{\partial^2 f}{\partial y^2} \sin\alpha\right)\hat{\mathbf{j}}.$$

Thus,

$$D_{\mathbf{v}}(D_{\mathbf{v}} f) = \left(\frac{\partial^2 f}{\partial x^2} \cos\alpha + \frac{\partial^2 f}{\partial x \partial y} \sin\alpha\right) \cos\alpha + \left(\frac{\partial^2 f}{\partial y \partial x} \cos\alpha + \frac{\partial^2 f}{\partial y^2} \sin\alpha\right) \sin\alpha$$

$$= \frac{\partial^2 f}{\partial x^2} \cos^2\alpha + 2\frac{\partial^2 f}{\partial x \partial y} \cos\alpha \sin\alpha + \frac{\partial^2 f}{\partial y^2} \sin^2\alpha. \tag{13.45}$$

We call $D_{\mathbf{v}}(D_{\mathbf{v}} f)$ the second directional derivative of $f(x, y)$ at (x_0, y_0) in direction $\mathbf{v}$. If it is positive, then the curve of intersection is concave upward, whereas if it is negative the curve is concave downward. We will find these results useful in Section 13.10 when we discuss relative extrema of functions of two independent variables.

EXERCISES 13.8

In Exercises 1–8 calculate the directional derivative of the function at the point and in the direction indicated.

* **1.** $f(x, y, z) = 2x^2 - y^2 + z^2$ at $(1, 2, 3)$ in the direction of the vector from $(1, 2, 3)$ to $(3, 5, 0)$

* **2.** $f(x, y, z) = x^2 y + xz$ at $(-1, 1, -1)$ in the direction of the vector that joins $(3, 2, 1)$ to $(3, 1, -1)$

* **3.** $f(x, y) = xe^y + y$ at $(3, 0)$ in the direction of the vector from $(3, 0)$ to $(-2, -4)$

* **4.** $f(x, y, z) = \ln(xy + yz + xz)$ at $(1, 1, 1)$ in the direction from $(1, 1, 1)$ toward the point $(-1, -2, 3)$

* **5.** $f(x, y) = \text{Tan}^{-1}(xy)$ at $(1, 2)$ along the line $y = 2x$ in the direction of increasing x

* **6.** $f(x, y) = \sin(x + y)$ at $(2, -2)$ along the line $3x + 4y = -2$ in the direction of decreasing y

* **7.** $f(x, y, z) = x^3 y \sin z$ at $(3, -1, -2)$ along the line $x = 3 + t$, $y = -1 + 4t$, $z = -2 + 2t$ in the direction of decreasing x

* **8.** $f(x, y, z) = x^2 y + y^2 z + z^2 x$ at $(1, -1, 0)$ along the line $x + 2y + 1 = 0$, $x - y + 2z = 2$ in the direction of decreasing z

In Exercises 9–12 find the rate of change of the function with respect to distance travelled along the curve.

* **9.** $f(x, y) = 2x - 3y$ at $(1, 1)$ along the curve $y = x^2$ in the direction of increasing x

* **10.** $f(x, y) = x^2 + y$ at $(-1, 3)$ along the curve $y = -3x^3$ in the direction of decreasing x

* **11.** $f(x, y, z) = xy + z^2$ at $(1, 0, -2)$ along the curve $y = x^2 - 1$, $z = -2x$ in the direction of increasing x

* **12.** $f(x, y, z) = x^2y + xy^3z$ at $(2, -1, 2)$ along the curve $x^2 - y^2 = 3$, $z = x$ in the direction of increasing x

In Exercises 13–18 find the direction in which the function increases most rapidly at the point. What is the rate of change in that direction?

* **13.** $f(x, y, z) = x^4yz - xy^3 + z$ at $(1, 1, -3)$

* **14.** $f(x, y) = 2xy + \ln(xy)$ at $(2, 1/2)$

* **15.** $f(x, y, z) = 1/\sqrt{x^2 + y^2 + z^2}$ at $(1, -3, 2)$

* **16.** $f(x, y, z) = -1/\sqrt{x^2 + y^2 + z^2}$ at $(1, -3, 2)$

* **17.** $f(x, y, z) = \text{Tan}^{-1}(xyz)$ at $(3, 2, -4)$

* **18.** $f(x, y) = xye^{xy}$ at $(1, 1)$

* **19.** In what direction is the rate of change of $f(x, y, z) = xyz$ smallest at the point $(2, -1, 3)$?

** **20.** In what directions (if any) is the rate of change of the function $f(x, y) = x^2y + y^3$ at the point $(1, -1)$ equal to (a) 0? (b) 1? (c) 20?

** **21.** In what directions (if any) is the rate of change of the function $f(x, y, z) = xy + z$ at the point $(0, 1, -2)$ equal to (a) 0? (b) 1? (c) -20?

** **22.** Must there always be a direction in which the rate of change of a function at a point is equal to (a) 0? (b) 3?

** **23.** In the derivation of 13.41, why was it necessary to use a unit vector $\hat{\mathbf{v}}$ to determine parametric equations for the line through (x_0, y_0, z_0) along $\mathbf{v}$? In other words, why could we not use the components of $\mathbf{v}$ itself to write parametric equations for the line?

** **24.** How fast is the distance to the origin changing with respect to distance travelled along the curve $x = 2\cos t$, $y = 2\sin t$, $z = 3t$ at any point on the curve? What is the rate of change when $t = 0$? Would you expect this?

** **25.** Find points on the curve $C : x = t$, $y = 1 - 2t$, $z = t$ at which the rate of change of $f(x, y, z) = x^2 + xyz$ with respect to distance travelled along the curve vanishes.

** **26.** Repeat Exercise 25 for the curve $C : z = x$, $x = y^2$ and the function $f(x, y, z) = x^2 - y^2 + z^2$.

** **27.** The path of a particle is defined parametrically by $x = (\cos t - t\sin t)\hat{\mathbf{i}} + (\sin t - t\cos t)\hat{\mathbf{j}}$ where t is time. Plot the path called an involute of a circle. Show that, if the particle's speed is constant, then the rate of change of its distance from the origin with respect to distance travelled is also constant. Is the time rate of change of its distance from the origin also constant?

** **28.** If we know the rate of change of a function $f(x, y, z)$ at a point P on a curve C, proceeding in one direction along C, what is the rate of change in the opposite direction along C?

** **29.** What is the rate of change of a function $f(x, y, z)$ in a direction perpendicular to ∇f?

** **30.** The rate of change of a function $f(x, y)$ at a point (x_0, y_0) in direction $\hat{\mathbf{i}} + 2\hat{\mathbf{j}}$ is 3 and the rate of change in direction $-2\hat{\mathbf{i}} - \hat{\mathbf{j}}$ is -1. Find its rate of change in direction $2\hat{\mathbf{i}} + 3\hat{\mathbf{j}}$.

** **31.** Rates of change of a function $f(x, y, z)$ at a point (x_0, y_0, z_0) in directions $\hat{\mathbf{i}} + \hat{\mathbf{j}}$, $2\hat{\mathbf{i}} - \hat{\mathbf{k}}$, and $\hat{\mathbf{i}} - \hat{\mathbf{j}} + \hat{\mathbf{k}}$ are 1, 2, and -3 respectively. What is its partial derivative with respect to z at the point?

** **32.** Find the second directional derivative of the function $f(x, y) = x^3 y^2$ at the point $(1, 1)$ in the direction of the vector $(1, -2)$.

** **33.** Find the second directional derivative of the function $f(x, y, z) = x^2 + 2y^2 + 3z^2$ at the point $(-2, -1, 3)$ in the direction $(1, 1, -1)$.

** **34.** The path followed by a stone embedded in the tread of a tire is a cycloid given parametrically by $x = R(\theta - \sin\theta)$, $y = R(1 - \cos\theta)$, $\theta \geq 0$ (see Example 10.7 in Section 10.1, page 659).

(a) How fast is the distance from the origin changing with respect to distance travelled along the curve at the points corresponding to $\theta = \pi/2$ and $\theta = \pi$?

(b) How fast is the y-coordinate changing at these points?

(c) How fast is the x-coordinate changing at these points?

*** **35.** Verify that expression 13.43 for $D_{\mathbf{v}} f$ leads to formula 13.42.

13.9 Tangent Lines and Tangent Planes

Tangent Lines to Curves

One equation in the coordinates x, y, and z of points in space,

$$F(x, y, z) = 0, \tag{13.46}$$

usually defines a surface. (There are exceptions. The equation $x^2 + y^2 + z^2 = 0$ defines a point, and $x^2 + y^2 + z^2 = -1$ defines nothing.) When each of the equations

$$F(x, y, z) = 0, \quad G(x, y, z) = 0 \tag{13.47}$$

defines a surface, then together they define the curve of intersection of the two surfaces (provided, of course, that the surfaces do intersect). Theoretically, we can find parametric equations for the curve by setting x equal to some function of a parameter t, say $x = x(t)$, and then solving equations 13.47 for y and z in terms of t: $y = y(t)$ and $z = z(t)$. The parametric definition, therefore, takes the form

$$x = x(t), \quad y = y(t), \quad z = z(t), \quad \alpha \leq t \leq \beta, \tag{13.48}$$

where α and β specify the endpoints of the curve. Practical difficulties arise in choosing $x(t)$ and solving for $y(t)$ and $z(t)$. For some examples, it might be more convenient to specify $y(t)$ and solve for $x(t)$ and $z(t)$ or, alternatively, to specify $z(t)$ and solve for $x(t)$ and $y(t)$. We considered examples of such conversions in Section 12.8.

In Section 12.9 we indicated that when a curve C is defined parametrically by 13.48, a tangent vector to C at any point P is

$$\frac{d\mathbf{r}}{dt} = \frac{dx}{dt}\hat{\mathbf{i}} + \frac{dy}{dt}\hat{\mathbf{j}} + \frac{dz}{dt}\hat{\mathbf{k}} \tag{13.49}$$

(Figure 13.18). The tangent line to C at P is defined as the line through P having direction $d\mathbf{r}/dt$. If (x_0, y_0, z_0) are the coordinates of P and t_0 is the value of t yielding P, then the vector equation for the tangent line at P is

$$(x, y, z) = (x_0, y_0, z_0) + u\frac{d\mathbf{r}}{dt}_{|t=t_0} \tag{13.50a}$$

(see equation 12.35). Parametric equations for the tangent line are

$$x = x_0 + x'(t_0)u, \quad y = y_0 + y'(t_0)u, \quad z = z_0 + z'(t_0)u, \tag{13.50b}$$

and in the case where none of $x'(t_0)$, $y'(t_0)$, and $z'(t_0)$ vanishes we can also write symmetric equations for the tangent line:

$$\frac{x - x_0}{x'(t_0)} = \frac{y - y_0}{y'(t_0)} = \frac{z - z_0}{z'(t_0)}. \tag{13.50c}$$

FIGURE 13.18

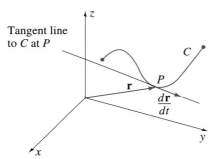

EXAMPLE 13.25

Find equations for the tangent line to the elliptic helix

$$C: \quad x = 2\cos t, \quad y = 4\sin t, \quad z = 2t/\pi$$

at $P(\sqrt{2}, 2\sqrt{2}, \frac{1}{2})$.

SOLUTION   Since $t = \pi/4$ at P, a tangent vector to C at P is

$$\frac{d\mathbf{r}}{dt}_{|t=\pi/4} = (-2\sin t, 4\cos t, 2/\pi)_{|t=\pi/4} = (-\sqrt{2}, 2\sqrt{2}, 2/\pi).$$

FIGURE 13.19

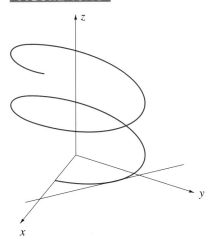

Symmetric equations for the tangent line are therefore

$$\frac{x - \sqrt{2}}{-\sqrt{2}} = \frac{y - 2\sqrt{2}}{2\sqrt{2}} = \frac{z - \frac{1}{2}}{2/\pi}.$$

We have shown the tangent line to the helix in Figure 13.19.

EXAMPLE 13.26

Find equations for the tangent line to the curve

$$C : \quad x = 2t^3 + t^2 + 4t, \quad y = t^2 - 3t + 5, \quad z = t^2 - 1$$

at the point $P(0, 5, -1)$.

SOLUTION   Since $t = 0$ at P, a tangent vector to C at P is

$$\frac{d\mathbf{r}}{dt}\Big|_{t=0} = (6t^2 + 2t + 4, 2t - 3, 2t)_{|t=0} = (4, -3, 0).$$

Because the z-component of this vector is equal to zero, we cannot use the full symmetric equations 13.50c. We can, however, use them for the x- and y- coordinates of points on the line and write

$$\frac{x}{4} = \frac{y - 5}{-3}, \quad \text{or } 3x + 4y = 20.$$

Since the z-component of $d\mathbf{r}/dt$ vanishes, $d\mathbf{r}/dt$ must be parallel to the xy-plane, and so too must the tangent line to C at P. But then every point on the line has the same z-coordinate, and, because the z-coordinate of P is -1, it follows that all points on the tangent line have z-coordinate equal to -1. Equations for the tangent line are therefore $3x + 4y = 20$, $z = -1$. It is shown in Figure 13.20.

FIGURE 13.20

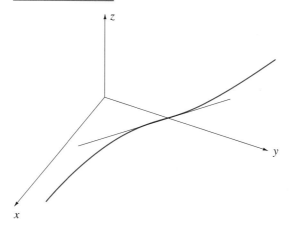

Tangent Planes to Surfaces

We now consider the problem of finding the equation for the tangent plane at a point P on a surface S (Figure 13.21). We define the tangent plane as that plane which contains all tangent lines at P to curves in S through P (provided, of course, that such a plane exists). Suppose that the surface is defined by the equation

$$F(x, y, z) = 0, \tag{13.51}$$

and that

$$C : \quad x = x(t), \quad y = y(t), \quad z = z(t), \quad \alpha \le t \le \beta,$$

is any curve in S through P. Since C is in S, the equation

$$F[x(t), y(t), z(t)] = 0$$

is valid for all t in $\alpha \le t \le \beta$. If $F(x, y, z)$ has continuous first partial derivatives, and $x(t)$, $y(t)$, and $z(t)$ are all differentiable, we may differentiate this equation using the chain rule:

$$\frac{\partial F}{\partial x} \frac{dx}{dt} + \frac{\partial F}{\partial y} \frac{dy}{dt} + \frac{\partial F}{\partial z} \frac{dz}{dt} = 0.$$

This equation, which holds at all points on C, and in particular at P, can be expressed vectorially as

$$0 = \left(\frac{\partial F}{\partial x}, \frac{\partial F}{\partial y}, \frac{\partial F}{\partial z} \right) \cdot \left(\frac{dx}{dt}, \frac{dy}{dt}, \frac{dz}{dt} \right) = \nabla F \cdot \frac{d\mathbf{r}}{dt}.$$

FIGURE 13.23

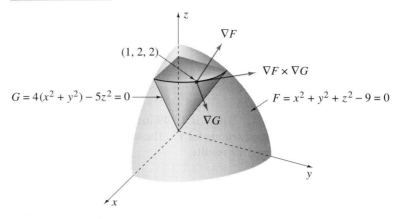

Example 13.28 illustrates that, when a curve is defined as the intersection of two surfaces $F(x, y, z) = 0$, $G(x, y, z) = 0$ (Figure 13.24), then a vector tangent to the curve is

$$\mathbf{T} = \nabla F \times \nabla G. \tag{13.53}$$

Thus to find a tangent vector to a curve we use 13.49 when the curve is defined parametrically. When the curve is defined as the intersection of two surfaces, we can either find parametric equations and use 13.49, or use 13.55. Note, too, that in order to find tangent lines to curves, it is not necessary to have a direction assigned to the curves.

FIGURE 13.24   **FIGURE 13.25**

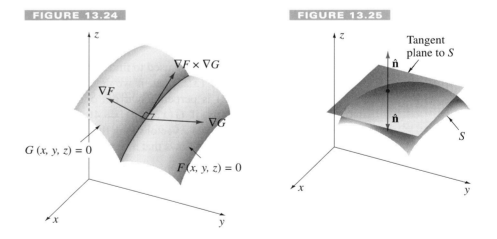

At each point on a surface S at which S has a tangent plane (Figure 13.25), we have defined a normal vector to S as a vector normal to the tangent plane to S. If we denote by $\hat{\mathbf{n}}$ a unit normal vector to S, then the direction of $\hat{\mathbf{n}}$ clearly varies as we move from point to point on S. We say that $\hat{\mathbf{n}}$ is a function of position (x, y, z) on S. Furthermore, at each point at which S has a unit normal vector, it has two such vectors, one in the opposite direction to the other. We say that a surface S is **smooth** if it can be assigned a unit normal $\hat{\mathbf{n}}$ that varies continuously on S. What this means geometrically is that, for small changes in position, the unit normal $\hat{\mathbf{n}}$ will undergo small changes in direction. The sphere in Figure 13.26 is smooth, as is the paraboloid in Figure 13.27.

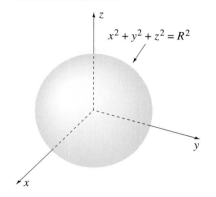

FIGURE 13.26

$$x^2 + y^2 + z^2 = R^2$$

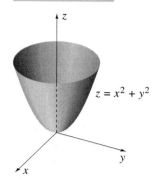

FIGURE 13.27

$$z = x^2 + y^2$$

The surface bounding the cylindrical volume in Figure 13.28 is not smooth; a unit normal that varies continuously over the surface cannot be assigned at points on the circles $x^2 + y^2 = 1$, $z = \pm 1$. This surface can, however, be divided into a finite number of subsurfaces, each of which is smooth. In particular, we choose the three subsurfaces S_1: $z = 1$, $x^2 + y^2 \leq 1$; S_2: $z = -1$, $x^2 + y^2 \leq 1$; S_3: $x^2 + y^2 = 1$, $-1 < z < 1$. Such a surface is said to be **piecewise smooth**.

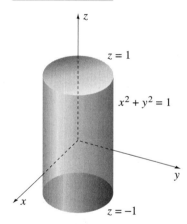

FIGURE 13.28

$$z = 1$$

$$x^2 + y^2 = 1$$

$$z = -1$$

EXERCISES 13.9

In Exercises 1–20 find equations for the tangent line to the curve at the point.

* **1.** $y = x^2$, $z = 0$ at $(-2, 4, 0)$

* **2.** $x = t$, $y = t^2$, $z = t^3$ at $(1, 1, 1)$

* **3.** $x = \cos t$, $y = \sin t$, $z = \cos t$ at $(1, 0, 1)$

* **4.** $y = x^2$, $z = x$ at $(-2, 4, -2)$

* **5.** $x^2 = y$, $z + x = y$ at $(1, 1, 0)$

* **6.** $x = 2 - t^2$, $y = 3 + 2t$, $z = t$ at $(1, 5, 1)$

* **7.** $x = 2\cos t$, $y = 3\sin t$, $z = 5$ at $(\sqrt{2}, -3/\sqrt{2}, 5)$

* **8.** $x^2 y^3 + xy = 68$ at $(1, 4)$

* **9.** $x + y + z = 4$, $x - y = 2$ at $(0, -2, 6)$

* **10.** $x = e^{-t}\cos t$, $y = e^{-t}\sin t$, $z = t$ at $(1, 0, 0)$

* **11.** $x = t^2 + 1$, $y = 2t - 4$, $z = t^3 + 3$ at $(2, -6, 2)$

* **12.** $y^2 + z^2 = 6$, $x + z = 1$ at $(2, -\sqrt{5}, -1)$

* **13.** $x^2 + y^2 + z^2 = 4$, $z^2 = x^2 + y^2$ at $(1, 1, -\sqrt{2})$

* **14.** $x = t$, $y = 1$, $z = \sqrt{1 + t^2}$ at $(4, 1, \sqrt{17})$

* **15.** $x = 1 + \cos t$, $y = 2 - \sin t$, $z = \sqrt{4 + t}$ at $(2, 2, 2)$

* **16.** $x = z^2 + z^3$, $y = z - z^4$ at $(12, -14, 2)$

* **17.** $x = y^2 + 3y^3 - 2y + 5$, $z = 0$ at $(7, 1, 0)$

* **18.** $2x^2 + y^2 + 2y = 3$, $z = x + 1$ at $(0, 1, 1)$

* **19.** $x = t^2$, $y = t$, $z = \sqrt{t + t^4}$ at $(1, 1, \sqrt{2})$

* **20.** $x = t \sin t$, $y = t \cos t$, $z = 2t$ at $(0, 2\pi, 4\pi)$

In Exercises 21–26 find an equation for the tangent plane to the surface at the point.

* **21.** $z = \sqrt{x^2 + y^2}$ at $(1, 1, \sqrt{2})$

* **22.** $x = x^2 - y^3 z$ at $(2, -1, -2)$

* **23.** $x^2 y + y^2 z + z^2 x + 3 = 0$ at $(2, -1, -1)$

* **24.** $x + y + z = 4$ at $(1, 1, 2)$

* **25.** $x = y \sin(\pi z / 2)$ at $(-1, -1, 1)$

* **26.** $x^2 + y^2 + 2y = 1$ at $(1, 0, 3)$

** **27.** Show that the curve $x = 2(t^3 + 2)/3$, $y = 2t^2$, $z = 3t - 2$ intersects the surface $x^2 + 2y^2 + 3z^2 = 15$ at right angles at the point $(2, 2, 1)$.

** **28.** Verify that the curve $x^2 - y^2 + z^2 = 1$, $xy + xz = 2$ is tangent to the surface $xyz - x^2 - 6y + 6 = 0$ at the point $(1, 1, 1)$.

** **29.** Show that the equation of the tangent plane to a surface $S : z = f(x, y)$ at a point (x_0, y_0, z_0) on S can be written in the form

$$z - z_0 = (x - x_0) f_x(x_0, y_0) + (y - y_0) f_y(x_0, y_0).$$

In Exercises 30–32 find the indicated derivative for the function.

** **30.** $f(x, y, z) = 2x^2 + y^2 z^2$ at $(3, 1, 0)$ with respect to distance along the curve $x + y + z = 4$, $x - y + z = 2$ in the direction of increasing x

** **31.** $f(x, y, z) = xyz + xy + xz + yz$ at $(1, -2, 5)$ perpendicular to the surface $z = x^2 + y^2$

** **32.** $f(x, y, z) = x^2 + y^2 - z^2$ at $(3, 4, 5)$ with respect to distance along the curve $x^2 + y^2 - z^2 = 0$, $2x^2 + 2y^2 - z^2 = 25$ in the direction of decreasing x

** **33.** If $F(x, y) = 0$ defines a curve implicitly in the xy-plane, prove that at any point on the curve ∇F is perpendicular to the curve.

** **34.** Find the equation of the tangent plane to the ellipsoid $x^2/a^2 + y^2/b^2 + z^2/c^2 = 1$ at any point (x_0, y_0, z_0) on the surface.

** **35.** Find all points on the surface $z = x^2/4 - y^2/9$ at which the tangent plane is parallel to the plane $x + y + z = 4$.

** **36.** Find all points on the surface $z^2 = 4(x^2 + y^2)$ at which the tangent plane is parallel to the plane $x - y + 2z = 3$.

** **37.** Suppose that the equations $F(x, y, z, t) = 0$, $G(x, y, z, t) = 0$, $H(x, y, z, t) = 0$ implicitly define parametric equations for a curve C (t being the parameter). If $P(x_0, y_0, z_0)$ is a point on C, show that equations for the tangent line to C at P can be written in the form

$$\frac{x - x_0}{\dfrac{\partial(F, G, H)}{\partial(t, y, z)}\Big|_P} = \frac{y - y_0}{\dfrac{\partial(F, G, H)}{\partial(x, t, z)}\Big|_P} = \frac{z - z_0}{\dfrac{\partial(F, G, H)}{\partial(x, y, t)}\Big|_P},$$

provided none of the Jacobians vanish.

** **38.** Find all points on the paraboloid $z = x^2 + y^2 - 1$ at which the normal to the surface coincides with the line joining the origin to the point.

*** **39.** Show that the sum of the intercepts on the x-, y-, and z-axes of the tangent plane to the surface $\sqrt{x} + \sqrt{y} + \sqrt{z} = \sqrt{a}$ at any point is a.

13.10 Relative Maxima and Minima

We now study relative extrema of functions of more than one independent variable. Most of the discussion will be confined to functions $f(x, y)$ of two independent variables because we can discuss the concepts geometrically as well as algebraically. Unfortunately, not all results are easily extended to functions of more than two independent variables, and we will therefore be careful to point out these limitations.

Before beginning the discussion, we briefly review maxima-minima results for functions $f(x)$ of one variable. We do this because maxima-minima theory for multivariable functions is essentially the same as that for single-variable functions. In fact, every definition that we make and every result that we discuss in this section has its counterpart in single-variable theory. Hence, a synopsis of single-variable results is in order. Unfortunately, proving results in the multivariable case is considerably more complicated than in the single-variable case, but if we can keep central ideas foremost in our minds and constantly make comparisons with single-variable calculus, we will find that discussions are not nearly as difficult as they might otherwise be.

Critical points of a function $f(x)$ are points at which $f'(x)$ is either equal to zero or does not exist. Geometrically, this means points at which the graph of $f(x)$ has a horizontal tangent line, a vertical tangent line, or no tangent line at all. Critical points for continuous functions can yield relative maxima, relative minima, horizontal points of inflection, vertical points of inflection, or just corners. There are two tests to determine whether a critical point x_0 yields a relative maximum or a relative minimum. The first-derivative test states that, if $f'(x)$ changes from a positive quantity to a negative quantity as x increases through x_0, then x_0 gives a relative maximum; if $f'(x)$ changes from negative to positive, then a relative minimum is obtained. The second-derivative test indicates the nature of a critical point at which $f'(x_0) = 0$ whenever $f''(x_0) \neq 0$. If $f''(x_0) > 0$, then a relative minimum is obtained, and if $f''(x_0) < 0$, a relative maximum is found.

We begin our study of extrema theory for multivariable functions by defining critical points for functions of two independent variables.

DEFINITION 13.6

A point (x_0, y_0) in the domain of a function $f(x, y)$ is said to be a **critical point** of $f(x, y)$ if

$$\frac{\partial f}{\partial x}\Big|_{(x_0, y_0)} = 0, \qquad \frac{\partial f}{\partial y}\Big|_{(x_0, y_0)} = 0, \tag{13.54}$$

or if one (or both) of these partial derivatives does not exist at (x_0, y_0).

There are two ways to interpret critical points of $f(x, y)$ geometrically. In Section 13.3, we interpreted $\partial f/\partial x$ at (x_0, y_0) as the slope of the tangent line to the curve of intersection of the surface $z = f(x, y)$ and the plane $y = y_0$, and $\partial f/\partial y$ as the slope of the tangent line to the curve of intersection with $x = x_0$. It follows, then, that (x_0, y_0) is critical if both curves have horizontal tangent lines or if either curve has a vertical tangent line or no tangent line at all. Alternatively, recall that the equation of the tangent plane to the surface $z = f(x, y)$ at (x_0, y_0) is

$$z - z_0 = f_x(x_0, y_0)(x - x_0) + f_y(x_0, y_0)(y - y_0)$$

(see Exercise 29 in Section 13.9). If both partial derivatives vanish, then the tangent plane is horizontal with equation $z = z_0$. For example, at each of the critical points in Figures 13.29–13.32, $\partial f/\partial x = \partial f/\partial y = 0$ and the tangent plane is horizontal. The remaining functions in Figures 13.33–13.37 have critical points at which either $\partial f/\partial x$ or $\partial f/\partial y$ or both do not exist. In Figures 13.33–13.36, the surfaces do not have tangent planes at critical points and, in Figure 13.37, the tangent plane is vertical at each critical point. Consequently, (x_0, y_0) is a critical point of a function $f(x, y)$ if at (x_0, y_0) the surface $z = f(x, y)$ has a horizontal tangent plane, a vertical tangent plane, or no tangent plane at all.

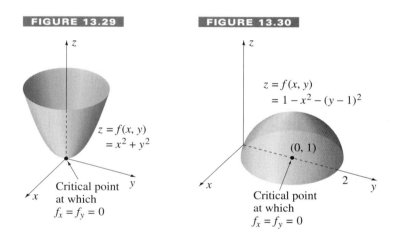

FIGURE 13.29

$z = f(x, y)$
$= x^2 + y^2$

Critical point at which
$f_x = f_y = 0$

FIGURE 13.30

$z = f(x, y)$
$= 1 - x^2 - (y - 1)^2$

$(0, 1)$

Critical point at which
$f_x = f_y = 0$

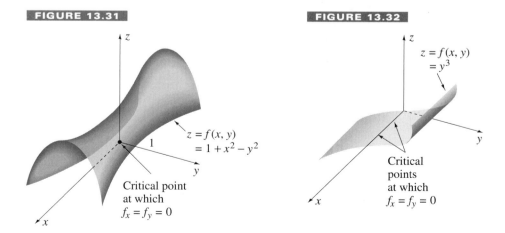

FIGURE 13.31

$z = f(x, y)$
$= 1 + x^2 - y^2$

Critical point at which
$f_x = f_y = 0$

FIGURE 13.32

$z = f(x, y)$
$= y^3$

Critical points at which
$f_x = f_y = 0$

FIGURE 13.33

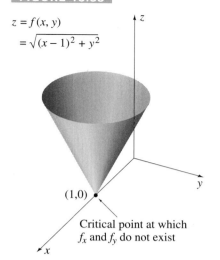

$$z = f(x, y) = \sqrt{(x-1)^2 + y^2}$$

$(1,0)$

Critical point at which f_x and f_y do not exist

FIGURE 13.34

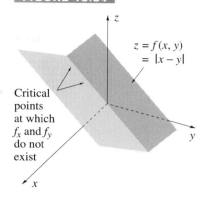

$$z = f(x, y) = |x - y|$$

Critical points at which f_x and f_y do not exist

FIGURE 13.35

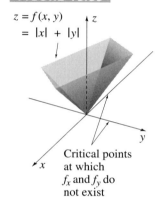

$$z = f(x, y) = |x| + |y|$$

Critical points at which f_x and f_y do not exist

FIGURE 13.36

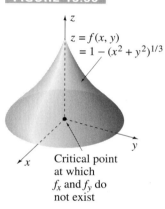

$$z = f(x, y) = 1 - (x^2 + y^2)^{1/3}$$

Critical point at which f_x and f_y do not exist

FIGURE 13.37

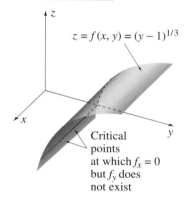

$$z = f(x, y) = (y - 1)^{1/3}$$

Critical points at which $f_x = 0$ but f_y does not exist

EXAMPLE 13.29

Find all critical points for the function

$$f(x, y) = x^2 y - 2xy^2 + 3xy + 4.$$

SOLUTION   For critical points, we first solve

$$0 = \frac{\partial f}{\partial x} = 2xy - 2y^2 + 3y = y(2x - 2y + 3),$$

$$0 = \frac{\partial f}{\partial y} = x^2 - 4xy + 3x = x(x - 4y + 3).$$

To satisfy these two equations simultaneously, there are four possibilities:

1. $x = 0$, $y = 0$, which gives the critical point $(0, 0)$;

2. $y = 0$, $x - 4y + 3 = 0$, which gives the critical point $(-3, 0)$;

3. $x = 0$, $2x - 2y + 3 = 0$, which gives the critical point $(0, \frac{3}{2})$;

4. $2x - 2y + 3 = 0$, $x - 4y + 3 = 0$, which gives the critical point $(-1, \frac{1}{2})$.

Since $\partial f / \partial x$ and $\partial f / \partial y$ are defined for all x and y, these are the only critical points. The plot of the surface $z = f(x, y)$ in Figure 13.38 does not really illustrate the critical points.

FIGURE 13.38

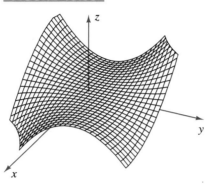

Critical points for functions of more than two independent variables can be defined algebraically, but because we have no geometric representation for such functions, there is no geometric interpretation for their critical points. For example, if $f(x, y, z, t)$ is a function of independent variables x, y, z, and t, then (x_0, y_0, z_0, t_0) is a critical point of $f(x, y, z, t)$ if all four of its first-order partial derivatives vanish at (x_0, y_0, z_0, t_0),

$$\frac{\partial f}{\partial x}\Big|_{(x_0, y_0, z_0, t_0)} = \frac{\partial f}{\partial y}\Big|_{(x_0, y_0, z_0, t_0)} = \frac{\partial f}{\partial z}\Big|_{(x_0, y_0, z_0, t_0)} = \frac{\partial f}{\partial t}\Big|_{(x_0, y_0, z_0, t_0)} = 0, \qquad (13.55)$$

or if at least one of the partial derivatives does not exist at the point. Note that because the partial derivatives of a function are the components of the gradient of the function, we can say that a critical point of a function is a point at which its gradient is either equal to zero or undefined.

EXAMPLE 13.30

Find all critical points for the function

$$f(x, y, z) = xyz\sqrt{x^2 + y^2 + z^2}.$$

SOLUTION   For critical points, we consider the equations

$$0 = \frac{\partial f}{\partial x} = yz\sqrt{x^2 + y^2 + z^2} + \frac{x^2 yz}{\sqrt{x^2 + y^2 + z^2}}$$

$$= \frac{yz}{\sqrt{x^2 + y^2 + z^2}}(2x^2 + y^2 + z^2),$$

$$0 = \frac{\partial f}{\partial y} = xz\sqrt{x^2 + y^2 + z^2} + \frac{xy^2 z}{\sqrt{x^2 + y^2 + z^2}}$$

$$= \frac{xz}{\sqrt{x^2 + y^2 + z^2}}(x^2 + 2y^2 + z^2),$$

$$0 = \frac{\partial f}{\partial z} = xy\sqrt{x^2 + y^2 + z^2} + \frac{xyz^2}{\sqrt{x^2 + y^2 + z^2}}$$

$$= \frac{xy}{\sqrt{x^2 + y^2 + z^2}}(x^2 + y^2 + 2z^2).$$

The partial derivatives are clearly undefined for $x = y = z = 0$, and therefore the origin $(0, 0, 0)$ is a critical point. If x, y, and z are not all zero, then the terms in parentheses cannot vanish, and we must set

$$yz = 0, \quad xz = 0, \quad xy = 0.$$

If any two of x, y, and z vanish, but the third does not, then these equations are satisfied. In other words, every point on the x-axis, every point on the y-axis, and every point on the z-axis is critical.

We now turn our attention to the classification of critical points of a function $f(x, y)$ of two independent variables. Critical points $(0, 1)$ in Figure 13.30 and $(0, 0)$ in Figure 13.36 yield "high" points on the surfaces. We describe this property in the following definition.

DEFINITION 13.7

A function $f(x, y)$ is said to have a **relative maximum** $f(x_0, y_0)$ at a point (x_0, y_0) if there exists a circle in the xy-plane centred at (x_0, y_0) such that for all points (x, y) inside this circle

$$f(x, y) \leq f(x_0, y_0). \tag{13.56}$$

The "low" points on the surfaces at $(0, 0)$ in Figure 13.29, $(1, 0)$ in Figure 13.33, and $(0, 0)$ in Figure 13.35 are relative minima according to the following.

DEFINITION 13.8

A function $f(x, y)$ is said to have a **relative minimum** $f(x_0, y_0)$ at a point (x_0, y_0) if there exists a circle in the xy-plane centred at (x_0, y_0) such that for all points (x, y) inside this circle

$$f(x, y) \geq f(x_0, y_0). \tag{13.57}$$

Every critical point in Figure 13.34 yields a relative minimum of $f(x, x) = 0$.

DEFINITION 13.9

If a critical point of a function $f(x, y)$ at which $\partial f/\partial x = \partial f/\partial y = 0$ yields neither a relative maximum nor a relative minimum, it is said to yield a **saddle point**.

The critical point $(0, 0)$ in Figure 13.31 therefore gives a saddle point, as does each of the critical points in Figure 13.32. Saddle points for surfaces $z = f(x, y)$ are clearly the analogues of horizontal points of inflection for curves $y = f(x)$. In both cases the derivative(s) of the function vanishes but there is neither a relative maximum nor a relative minimum.

The critical points in Figure 13.35 (except $(0, 0)$) are the counterparts of corners for the graph of a function $f(x)$. They are points at which one or both of the partial derivatives of $f(x, y)$ do not exist, but like corners for $f(x)$, they do not necessarily yield relative extrema. Critical points in Figure 13.37 are the analogues of vertical points of inflection for a function $f(x)$.

Our discussion has made it clear that:

   (a) At a relative maximum or minimum of $f(x, y)$, either $\partial f/\partial x$ and $\partial f/\partial y$ both vanish, or one or both of the partial derivatives do not exist;

   (b) Saddle points may also occur where $\partial f/\partial x = \partial f/\partial y = 0$, and points where the derivatives do not exist may fail to yield relative extrema.

In other words, every relative extremum of $f(x, y)$ occurs at a critical point, but critical points do not always give relative extrema.

Given the problem of determining all relative maxima and minima of a function $f(x, y)$, we should first find its critical points. But how do we decide whether these critical points yield relative maxima, relative minima, saddle points, or none of these? We do not have a practical test that is equivalent to the first-derivative test for functions of one variable, but we do have a test that corresponds to the second-derivative test. For functions of two independent variables the situation is more complicated, however, since there are three second-order partial derivatives, but the idea of the test is essentially the same. It determines whether certain curves are concave upward or concave downward at the critical point. The complete result is contained in the following theorem.

THEOREM 13.7

Suppose (x_0, y_0) is a critical point of $f(x, y)$ at which $\partial f/\partial x$ and $\partial f/\partial y$ both vanish. Suppose further that f_x, f_y, f_{xx}, f_{xy}, and f_{yy} are all continuous at (x_0, y_0). Define

$$A = f_{xx}(x_0, y_0), \quad B = f_{xy}(x_0, y_0), \quad C = f_{yy}(x_0, y_0).$$

If

   (i) $B^2 - AC < 0$ and $A < 0$, then $f(x, y)$ has a relative maximum at (x_0, y_0);

   (ii) $B^2 - AC < 0$ and $A > 0$, then $f(x, y)$ has a relative minimum at (x_0, y_0);

   (iii) $B^2 - AC > 0$, then $f(x, y)$ has a saddle point at (x_0, y_0);

   (iv) $B^2 - AC = 0$, then the test fails.

PROOF

(i) Suppose we intersect the surface $z = f(x, y)$ with a plane parallel to the z-axis, through the point $(x_0, y_0, 0)$, and making an angle α with the line through $(x_0, y_0, 0)$ parallel to the positive x-axis (Figure 13.39). The slope of the curve of intersection of these surfaces at the point $(x_0, y_0, f(x_0, y_0))$ is given by the directional derivative

$$D_{\mathbf{v}} f_{|(x_0, y_0)} = \frac{\partial f}{\partial x}_{|(x_0, y_0)} \cos \alpha + \frac{\partial f}{\partial y}_{|(x_0, y_0)} \sin \alpha$$

(see equation 13.44). Since (x_0, y_0) is a critical point at which $\nabla f = \mathbf{0}$, it follows that

$$D_{\mathbf{v}} f_{|(x_0, y_0)} = 0 \text{ for all } \alpha.$$

FIGURE 13.39

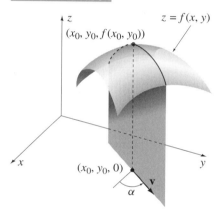

In Figure 13.39 we have illustrated the critical point as a relative maximum. But how do we verify that this is indeed the case? If we can show that each and every curve of intersection of the surface with a vertical plane through $(x_0, y_0, 0)$ is concave downward at $(x_0, y_0, 0)$, then (x_0, y_0) must give a relative maximum. But to discuss concavity of a curve we require the second derivative—in this case, the second directional derivative of $f(x, y)$. According to 13.45, the second directional derivative of $f(x, y)$ at (x_0, y_0) in the direction $\hat{\mathbf{v}} = (\cos \alpha, \sin \alpha)$ is

$$D_{\mathbf{v}}(D_{\mathbf{v}} f) = \frac{\partial^2 f}{\partial x^2}_{|(x_0, y_0)} \cos^2 \alpha + 2 \frac{\partial^2 f}{\partial x \partial y}_{|(x_0, y_0)} \cos \alpha \sin \alpha + \frac{\partial^2 f}{\partial y^2}_{|(x_0, y_0)} \sin^2 \alpha$$

$$= A \cos^2 \alpha + 2B \cos \alpha \sin \alpha + C \sin^2 \alpha,$$

where we understand that here $D_{\mathbf{v}}(D_{\mathbf{v}} f)$ is implicitly suffixed by (x_0, y_0). In order, therefore, to verify that (x_0, y_0) gives a relative maximum, it is sufficient to show that $D_{\mathbf{v}}(D_{\mathbf{v}} f)$ is negative for each value of α in the interval $0 \leq \alpha < 2\pi$. However, because $D_{\mathbf{v}}(D_{\mathbf{v}} f)$ is unchanged if α is replaced by $\alpha + \pi$, it is sufficient to verify that $D_{\mathbf{v}}(D_{\mathbf{v}} f)$ is negative for $0 \leq \alpha < \pi$.

For any of these values of α except $\pi/2$, we can write

$$D_{\mathbf{v}}(D_{\mathbf{v}}f) = \cos^2\alpha(A + 2B\tan\alpha + C\tan^2\alpha),$$

and if we set $u = \tan\alpha$,

$$D_{\mathbf{v}}(D_{\mathbf{v}}f) = \cos^2\alpha(A + 2Bu + Cu^2).$$

It is evident that $D_{\mathbf{v}}(D_{\mathbf{v}}f) < 0$ for all $\alpha \neq \pi/2$ if and only if

$$Q(u) = A + 2Bu + Cu^2 < 0 \text{ for } -\infty < u < \infty.$$

Were we to draw a graph of the quadratic $Q(u)$, we would see that it crosses the u-axis where

$$u = \frac{-2B \pm \sqrt{4B^2 - 4AC}}{2C} = \frac{-B \pm \sqrt{B^2 - AC}}{C}.$$

But because $B^2 - AC < 0$, there are no real solutions of this equation, and therefore $Q(u)$ never crosses the u-axis. Since $Q(0) = A < 0$, it follows that $Q(u) < 0$ for all u. We have shown, then, that

$$D_{\mathbf{v}}(D_{\mathbf{v}}f) < 0 \text{ for all } \alpha \neq \pi/2.$$

When $\alpha = \pi/2$, $D_{\mathbf{v}}(D_{\mathbf{v}}f) = C$. Since $B^2 - AC < 0$ and $A < 0$, it follows that $C < 0$ also. Consequently, if $B^2 - AC < 0$ and $A < 0$, then $D_{\mathbf{v}}(D_{\mathbf{v}}f) < 0$ for all α, and (x_0, y_0) yields a relative maximum.

(ii) If $B^2 - AC < 0$ and $A > 0$, a similar argument leads to the conclusion that (x_0, y_0) yields a relative minimum; the only difference is that inequalities are reversed.

(iii) If $B^2 - AC > 0$, then $Q(u)$ has real distinct zeros, in which case $Q(u)$ is sometimes negative and sometimes positive. This means that the curve of intersection is sometimes concave upward and sometimes concave downward, and the point (x_0, y_0) therefore gives a saddle point.

(iv) If $B^2 - AC = 0$, the classification of the point determined by (x_0, y_0) depends on which of A, B, and C vanish, if any. ∎

To illustrate that we can obtain a relative maximum, a relative minimum, or a saddle point for a critical point at which $B^2 - AC = 0$, consider the three functions $f(x, y) = -y^2$, $f(x, y) = y^2$, and $f(x, y) = y^3$ in Figures 13.40–13.42. The point $(0, 0)$ is a critical point for each function, and at this point $B^2 - AC = 0$. Yet $(0, 0)$ yields a relative maximum for $f(x, y) = -y^2$, a relative minimum for $f(x, y) = y^2$, and a saddle point for $f(x, y) = y^3$. In fact, every point on the x-axis is a relative maximum for $f(x, y) = -y^2$, a relative minimum for $f(x, y) = y^2$, and a saddle point for $f(x, y) = y^3$.

FIGURE 13.40

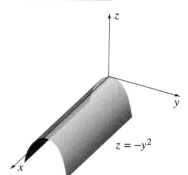

$z = -y^2$

FIGURE 13.41

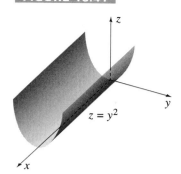

$z = y^2$

FIGURE 13.42

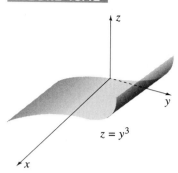

$z = y^3$

EXAMPLE 13.31

Find and classify critical points for each of the following functions as yielding relative maxima, relative minima, saddle points, or none of these:

(a) $f(x, y) = 4xy - x^4 - y^4$

(b) $f(x, y) = y^3 + x^2 - 6xy + 3x + 6y$

(c) $f(x, y) = \dfrac{12xy - x^2y^2}{2(x + y)}$

SOLUTION

(a) Critical points of $f(x, y)$ are given by

$$0 = \frac{\partial f}{\partial x} = 4y - 4x^3, \quad 0 = \frac{\partial f}{\partial y} = 4x - 4y^3.$$

Solutions of these equations are $(0, 0)$, $(1, 1)$, and $(-1, -1)$. We now calculate

$$\frac{\partial^2 f}{\partial x^2} = -12x^2, \quad \frac{\partial^2 f}{\partial x \partial y} = 4, \quad \frac{\partial^2 f}{\partial y^2} = -12y^2.$$

At $(0, 0)$, $B^2 - AC = 16 > 0$, and therefore $(0, 0)$ yields a saddle point. At $(1, 1)$, $B^2 - AC = -128$ and $A = -12$, and therefore $(1, 1)$ gives a relative maximum. At $(-1, -1)$, $B^2 - AC = -128$ and $A = -12$, and $(-1, -1)$ also gives a relative maximum.

(b) Critical points for $f(x, y)$ are given by

$$0 = \frac{\partial f}{\partial x} = 2x - 6y + 3, \quad 0 = \frac{\partial f}{\partial y} = 3y^2 - 6x + 6.$$

Solutions of these equations are $(\frac{27}{2}, 5)$ and $(\frac{3}{2}, 1)$. The second derivatives of $f(x, y)$ are

$$\frac{\partial^2 f}{\partial x^2} = 2, \quad \frac{\partial^2 f}{\partial x \partial y} = -6, \quad \frac{\partial^2 f}{\partial y^2} = 6y.$$

At $(\frac{27}{2}, 5)$, $B^2 - AC = -24$ and $A = 2$, and therefore $(\frac{27}{2}, 5)$ yields a relative minimum. At $(\frac{3}{2}, 1)$, $B^2 - AC = 24$, and therefore $(\frac{3}{2}, 1)$ gives a saddle point.

(c) For critical points, we consider

$$0 = \frac{\partial f}{\partial x} = \frac{y^2(12 - x^2 - 2xy)}{2(x + y)^2}, \quad 0 = \frac{\partial f}{\partial y} = \frac{x^2(12 - y^2 - 2xy)}{2(x + y)^2},$$

or

$$0 = y^2(12 - x^2 - 2xy), \quad 0 = x^2(12 - y^2 - 2xy).$$

Solutions of these equations are $(2, 2)$ and $(-2, -2)$ (the solution $(0, 0)$ has been rejected since $f(x, y)$ is not defined there). To classify these critical points by means of the second-derivative test would lead to some messy calculations; thus we will show that the classification can be achieved without the test. First we note that $f(x, y) = 0$ whenever

$$0 = 12xy - x^2y^2 = xy(12 - xy),$$

i.e., along the x-axis, the y-axis, and the hyperbola $xy = 12$ (except $(0, 0)$; see Figure 13.43). Further, between the axes and that part of the hyperbola in the first quadrant, the function is always positive. Finally, since $(2, 2)$ is the only critical point in this region, it follows that $(2, 2)$ could only yield a relative maximum. A similar argument in the third quadrant indicates that $(-2, -2)$ gives a relative minimum.

FIGURE 13.43

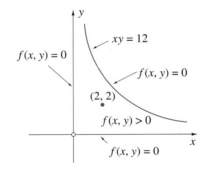

EXAMPLE 13.32

It is straightforward to verify that $(0, 0)$ is the only critical point of the function $f(x, y) = x^2 - 6xy^2 + y^4$ and that the quantity $B^2 - AC$ of Theorem 13.7 is equal to zero at this critical point. Show graphically and algebraically that the critical point gives a saddle point.

SOLUTION   The value of the function at $(0, 0)$ is $f(0, 0) = 0$. There is a relative minimum at $(0, 0)$ if $f(x, y) \geq 0$ in some circle around $(0, 0)$; there is a relative maximum if $f(x, y) \leq 0$ in some such circle; and there is a saddle point if $f(x, y)$ takes on negative and positive values in every circle centred at $(0, 0)$. The plot in Figure 13.44 indicates that the last situation prevails. To show this algebraically we first note that values of $f(x, y)$ are nonnegative along the x-axis and the y-axis. On the parabola $x = y^2$, values of the function are

$$(y^2)^2 - 6(y^2)y^2 + y^4 = -4y^4 \leq 0.$$

Thus, $(0, 0)$ yields a saddle point.

FIGURE 13.44

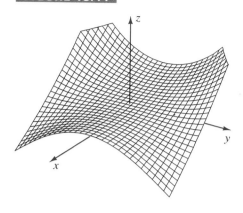

This completes our discussion of relative extrema of functions of two independent variables. Our next step should be to extend the theory to functions of more than two variables. It is a simple matter to give definitions of relative maxima and minima for such functions; they are almost identical to Definitions 13.7 and 13.8 (see Exercise 19). On the other hand, to develop a theorem for functions of more than two independent variables that is analogous to Theorem 13.7 is beyond the scope of this book. We refer the interested reader to more advanced books.

EXERCISES 13.10

In Exercises 1–14 find all critical points for the function and classify each as yielding a relative maximum, a relative minimum, a saddle point, or none of these.

* **1.** $f(x, y) = x^2 + 2xy + 2y^2 - 6y$ * **2.** $f(x, y) = 3xy - x^3 - y^3$

* **3.** $f(x, y) = x^3 - 3x + y^2 + 2y$ * **4.** $f(x, y) = x^2y^2 + 3x$

* **5.** $f(x, y) = xy - x^2 + y^2$ * **6.** $f(x, y) = x \sin y$

* **7.** $f(x, y) = xye^{-(x^2+y^2)}$ * **8.** $f(x, y) = x^2 - 2xy + y^2$

* **9.** $f(x, y) = (x^2 + y^2)^{2/3}$ * **10.** $f(x, y) = x^4y^3$

* **11.** $f(x, y) = 2xy^2 + 3xy + x^2y^3$ ** **12.** $f(x, y) = |x| + y^2$

** **13.** $f(x, y) = (1 - x)(1 - y)(x + y - 1)$

** **14.** $f(x, y) = x^4 + y^4 - x^2 - y^2 + 1$

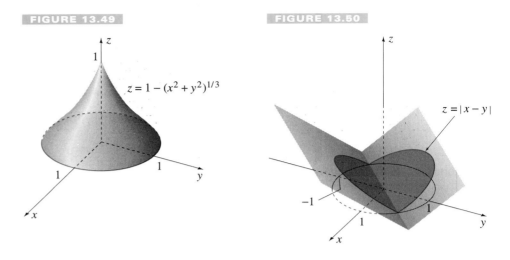

FIGURE 13.49

$z = 1 - (x^2 + y^2)^{1/3}$

FIGURE 13.50

$z = |x - y|$

TABLE 13.1

Function $f(x, y)$	Position of absolute maximum	Value of absolute maximum	Position of absolute minumum	Value absolu minum
$x^2 + y^2$	Every point on $x^2 + y^2 = 1$	1	$(0, 0)$	0
$1 + x^2 - y^2$	$(\pm 1, 0)$	2	$(0, \pm 1)$	0
$\sqrt{(x - 1)^2 + y^2}$	$(-1, 0)$	2	$(1, 0)$	0
$4 - 2x^2 - y^2$	$(0, 0)$	4	$(\pm 1, 0)$	2
$1 - (x^2 + y^2)^{1/3}$	$(0, 0)$	1	Every point on $x^2 + y^2 = 1$	0
$|x - y|$	$(\pm 1/\sqrt{2} \mp 1/\sqrt{2}),$	$\sqrt{2}$	Every point on $y = x,$ $-1/\sqrt{2} \leq x \leq 1/\sqrt{2}$	0

For each of the functions in these figures, *absolute extrema occur at either a critical point or a point on the boundary of R*. This result is true for any *continuous function defined on a finite region that includes all the points on its boundary*. Although this result may seem fairly obvious geometrically, to prove it analytically is very difficult; we will be content to assume its validity and carry on from there.

Sometimes a drawing or plot of the surface defined by a function makes absolute maxima and minima clear. This may not always be the case, however, and we therefore turn our attention to determining absolute extrema algebraically. Suppose, then, that a continuous function $f(x, y)$ is given and we are required to find its absolute extrema on a finite region R (which includes its boundary points). The previous discussion indicated that the extrema must occur either at critical points or on the boundary of R. Consequently, we should first determine all critical points of $f(x, y)$ in R, and evaluate $f(x, y)$ at each of these points. These values should now be compared to the maximum and minimum values of $f(x, y)$ on the boundary of R. But how do we find the maximum and minimum values of $f(x, y)$ on the boundary? If the boundary of R is denoted by C (Figure 13.51), and if C has parametric equations $x = x(t)$, $y = y(t)$, $\alpha \leq t \leq \beta$, then on C we can express $f(x, y)$ in terms of t, and t alone:

$$f[x(t), y(t)], \quad \alpha \leq t \leq \beta.$$

To find the maximum and minimum values of $f(x, y)$ on C is now an absolute extrema problem for a function of one variable. The function $f[x(t), y(t)]$ should therefore be evaluated at each of its critical points and at $t = \alpha$ and $t = \beta$. A plot of $f[x(t), y(t)]$ could be valuable here.

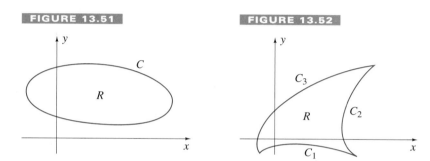

FIGURE 13.51

FIGURE 13.52

If the boundary of R consists of a number of curves (Figure 13.52), then this boundary procedure must be performed for each part. In other words, on each part of the boundary we express $f(x, y)$ as a function of one variable, and then evaluate this function at its critical points and at the ends of that part of the boundary to which it applies.

The absolute maximum of $f(x, y)$ on R is then the largest of all values of $f(x, y)$ evaluated at the critical points inside R, the critical points on the boundary of R, and the endpoints of each part of the boundary. The absolute minimum of $f(x, y)$ on R is the smallest of all these values.

Recall that to find the absolute extrema of a function $f(x)$, continuous on $a \leq x \leq b$, we evaluate $f(x)$ at all critical points and at the boundary points $x = a$ and $x = b$. The procedure that we have established here for $f(x, y)$ is much the same—the difference is that, for $f(x, y)$, the boundary consists not of two points, but of entire curves. Evaluation of $f(x, y)$ on the boundary therefore reduces to one or more extrema problems for functions of one variable. Note too that for $f(x, y)$ (or $f(x)$), it is not necessary to determine the nature of the critical points; it is necessary only to evaluate $f(x, y)$ at these points.

EXAMPLE 13.33

Find the maximum value of the function $z = f(x, y) = 4xy - x^4 - 2y^2$ on the region R: $-2 \leq x \leq 2, -2 \leq y \leq 2$.

SOLUTION   A plot of the function in Figure 13.53 suggests that the maximum value occurs at a critical point in the first and third quadrants, or on the edges of the square. We confirm this with the procedure outlined above.

FIGURE 13.53

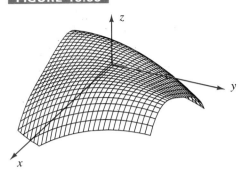

Critical points of $f(x, y)$ are given by

$$0 = \frac{\partial f}{\partial x} = 4y - 4x^3, \quad 0 = \frac{\partial f}{\partial y} = 4x - 4y.$$

Solutions of these equations are $(0, 0)$, $(1, 1)$, and $(-1, -1)$, and the values of $f(x, y)$ at these critical points are

$$f(0, 0) = 0, \quad f(1, 1) = 1, \quad f(-1, -1) = 1.$$

We denote the four parts of the boundary of R by C_1, C_2, C_3, and C_4 (Figure 13.54).

FIGURE 13.54

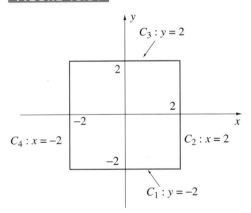

On C_1, $y = -2$, in which case

$$z = -8x - x^4 - 8, \quad -2 \le x \le 2.$$

For critical points of this function, we solve

$$0 = \frac{dz}{dx} = -8 - 4x^3.$$

The only solution is $x = -2^{1/3}$, at which the value of z is:

$$z = 8 \cdot 2^{1/3} - 2^{4/3} - 8 = -0.44.$$

On C_2, $x = 2$, in which case

$$z = 8y - 16 - 2y^2, \quad -2 \leq y \leq 2.$$

Critical points are defined by

$$0 = \frac{dz}{dy} = 8 - 4y.$$

The only solution $y = 2$ defines one of the corners of the square, and at this point

$$z = -8.$$

On C_3, $y = 2$ and

$$z = 8x - x^4 - 8, \quad -2 \leq x \leq 2.$$

For critical points, we solve

$$0 = \frac{dz}{dx} = 8 - 4x^3.$$

At the single point $x = 2^{1/3}$,

$$z = 8 \cdot 2^{1/3} - 2^{4/3} - 8 = -0.44.$$

On the final curve C_4, $x = -2$ and

$$z = -8y - 16 - 2y^2, \quad -2 \leq y \leq 2.$$

Critical points are given by

$$0 = \frac{dz}{dy} = -8 - 4y.$$

The solution $y = -2$ defines another corner of the square at which

$$z = -8.$$

We have now evaluated $f(x, y)$ at all critical points inside R and at all critical points on the four parts of the boundary of R. It remains only to evaluate $f(x, y)$ at the corners of the square. Two corners have already been accounted for; the other two give

$$f(2, -2) = -40, \quad f(-2, 2) = -40.$$

The largest value of $f(x, y)$ produced is 1, and this is therefore the maximum value of $f(x, y)$ on R.

EXAMPLE 13.34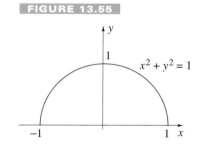

Temperature at each point (x, y) in a semicircular plate defined by $x^2 + y^2 \leq 1$, $y \geq 0$ is given by

$$T(x, y) = 16x^2 - 24xy + 40y^2.$$

Find the hottest and coldest points in the plate.

SOLUTION   For critical points of $T(x, y)$, we solve

$$0 = \frac{\partial T}{\partial x} = 32x - 24y, \quad 0 = \frac{\partial T}{\partial y} = -24x + 80y.$$

The only solution of these equations, $(0, 0)$, is on the boundary. On the upper edge of the plate (Figure 13.55), we set $x = \cos t$, $y = \sin t$, $0 \leq t \leq \pi$, in which case

$$T = 16\cos^2 t - 24\cos t \sin t + 40 \sin^2 t, \quad 0 \leq t \leq \pi.$$

A plot of this function in Figure 13.56 shows the absolute maximum at the relative maximum and the absolute minimum at the relative minimum. To locate them we solve for critical points of this function,

$$0 = \frac{dT}{dt} = -32\cos t \sin t - 24(-\sin^2 t + \cos^2 t) + 80 \sin t \cos t$$

$$= 24(\sin 2t - \cos 2t).$$

If we divide by $\cos 2t$ (since $\cos 2t = 0$ does not lead to a solution of this equation), we have

$$\tan 2t = 1.$$

The only solutions of this equation in the interval $0 \leq t \leq \pi$ are $t = \pi/8$ and $t = 5\pi/8$. When $t = \pi/8$, $T = 11.0$; and when $t = 5\pi/8$, $T = 45.0$. At the ends of this part of the boundary, $t = 0$ and $t = \pi$, and $T(1, 0) = 16$ and $T(-1, 0) = 16$. On the lower edge of the plate, $y = 0$, in which case

$$T = 16x^2, \quad -1 \leq x \leq 1.$$

The only critical point of this function is $x = 0$, at which $T = 0$. The hottest point in the plate is therefore $(\cos(5\pi/8), \sin(5\pi/8)) = (-0.38, 0.92)$, where the temperature is $45°$, and the coldest point is $(0, 0)$ with temperature $0°$.

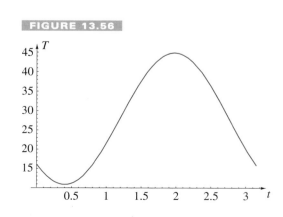

FIGURE 13.55

FIGURE 13.56

EXAMPLE 13.35

Find the point on the first octant part of the plane $6x + 3y + 4z = 6$ closest to the point $(4, 6, 7)$. Assume that lines of intersection of the plane with the coordinate planes are part of the surface.

SOLUTION The distance D from $(4, 6, 7)$ to any point (x, y, z) in space is defined by

$$D^2 = (x - 4)^2 + (y - 6)^2 + (z - 7)^2.$$

We must minimize D, but consider only points (x, y, z) that satisfy the equation of the plane and lie in the first octant (Figure 13.57). At the moment, D^2 is a function of three variables x, y, and z, but they are not all independent because of the planar restriction. If we solve the equation of the plane for z in terms of x and y, and substitute,

$$D^2 = f(x, y) = (x - 4)^2 + (y - 6)^2 + \left(\frac{6 - 6x - 3y}{4} - 7 \right)^2$$

$$= (x - 4)^2 + (y - 6)^2 + \left(\frac{6x + 3y + 22}{4} \right)^2,$$

where x and y are independent variables. Now D is minimized when D^2 is minimized, and we therefore find the point (x, y) that minimizes D^2. The values of x and y that yield points on the first octant part of the plane are those in the triangle of Figure 13.58.

FIGURE 13.57

FIGURE 13.58

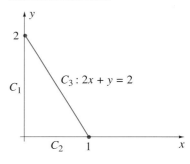

For critical points of D^2, we solve

$$0 = \frac{\partial f}{\partial x} = 2(x - 4) + \frac{3}{4}(6x + 3y + 22),$$

$$0 = \frac{\partial f}{\partial y} = 2(y - 6) + \frac{3}{8}(6x + 3y + 22).$$

✱✱ **22.** Find the maximum and minimum values of $f(x, y) = x^2 - y^2$ on the circle $x^2 + y^2 = 1$.

✱✱ **23.** Find the maximum and minimum values of $f(x, y) = |x - y|$ on the circle $x^2 + y^2 = 1$.

✱✱ **24.** Find the maximum and minimum values of $f(x, y) = x^2 - y^2$ on the curve $|x| + |y| = 1$.

✱✱ **25.** Find the maximum and minimum values of $f(x, y) = |x - 2y|$ on the curve $|x| + |y| = 1$.

✱✱ **26.** If P is the perimeter of a triangle with sides of length x, y, and z, the area of the triangle is

$$A = \sqrt{\frac{P}{2}\left(\frac{P}{2} - x\right)\left(\frac{P}{2} - y\right)\left(\frac{P}{2} - z\right)},$$

where $P = x + y + z$. Show that A is maximized for fixed P when the triangle is equilateral.

✱✱ **27.** Show that for any triangle with interior angles A, B, and C,

$$\sin(A/2) \sin(B/2) \sin(C/2) \leq 1/8.$$

Hint: Find the maximum value of the function $f(A, B, C) = \sin(A/2) \sin(B/2) \sin(C/2)$.

✱✱ **28.** A silo is in the shape of a right circular cylinder surmounted by a right circular cone. If the radius of each is 6 m and the total surface area must be 200 m^2 (not including the base), what heights for the cone and cylinder yield maximum enclosed volume?

✱✱ **29.** What values of x and y maximize the production function $P(x, y) = kx^\alpha y^\beta$ where k, α, and β are positive constants ($\alpha + \beta = 1$) when x and y must satisfy $Ax + By = C$, where A, B, and C are positive constants.

✱✱ **30.** A long piece of metal 1 m wide is bent at A and B, as shown in the figure below, to form a channel with three straight sides. If the bends are equidistant from the ends, where should they be made in order to obtain maximum possible flow of fluid along the channel?

✱✱ **31.** Find maximum and minimum values of the function $f(x, y, z) = xy + xz$ on the region $x^2 + y^2 + z^2 \leq 1$.

✱✱✱ **32.** Find maximum and minimum values of the function $f(x, y, z) = x^2yz$ on the region $x^2 + y^2 \leq 1, 0 \leq z \leq 1$.

✱✱✱ **33.** A cow's daily diet consists of three foods: hay, grain, and supplements. The cow is always given 11 kg of hay per day, 50% of which is digestive material and 12% of which is protein. Grain is 74% digestive material and 8.8% protein, whereas supplements are 62% digestive material and 34% protein. The cost of hay is $27.50 for 1000 kg, and grain and supplements cost $110 and $175 for 1000 kg. A healthy cow's diet must contain between 9.5 and 11.5 kg of digestive material and between 1.9 and 2.0 kg of protein. Determine the daily amounts of grain and supplements that the cow should be fed in order that total food costs be kept to a minimum.

*** **34.** Find the area of the largest triangle that has vertices on the circle $x^2 + y^2 = r^2$.

*** **35.** A rectangle is surmounted by an isosceles triangle as shown in the figure below. Find x, y, and θ in order that the area of the figure will be as large as possible under the restriction that its perimeter must be P.

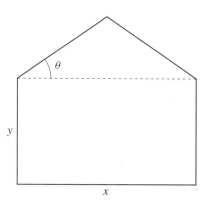

*** **36.** In example 8.12 we considered sesmic prospecting with two media. The figure below shows the situation for three media, the problem being to predict depths d_1 and d_2.

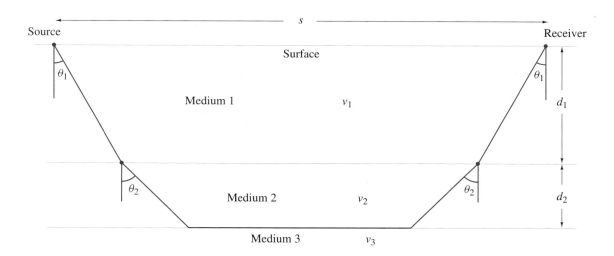

Show that the time for a signal to be emitted by the source, pass through medium 1 with speed v_1, pass through medium 2 with speed v_2, travel along the interface between media 2 and 3 with speed v_3, then through medium 2 and medium 1 to the receiver is given by

$$t = \frac{2d_1 \sec\theta_1}{v_1} + \frac{2d_2 \sec\theta_2}{v_2} + \frac{s - 2d_1 \tan\theta_1 - 2d_2 \tan\theta_2}{v_3}.$$

Verify that t is a minimum when θ_1 and θ_2 are given by

$$\theta_1 = \text{Sin}^{-1}(v_1/v_3), \qquad \theta_2 = \text{Sin}^{-1}(v_2/v_3).$$

13.12 Lagrange Multipliers

Many applied maxima and minima problems result in **constraint problems**. In particular, Examples 13.34 and 13.35 contain such problems. In Example 13.34, to find extreme values of T on the edge of the plate we maximized and minimized $T(x, y) = 16x^2 - 24x + 40y^2$ subject to first the constraint $x^2 + y^2 = 1$, and then the constraint $y = 0$. Our method there was to substitute from the constraint equation into $T(x, y)$ in order to obtain a function of one variable. In Example 13.35, to find the minimum distance from $(4, 6, 7)$ to the plane $6x + 3y + 4z = 6$, we minimized $D^2 = (x - 4)^2 + (y - 6)^2 + (z - 7)^2$ subject to the constraint $6x + 3y + 4z = 6$. Again we substituted from the constraint to obtain D^2 as a function of two independent variables.

A natural question to ask is whether problems of this type can be solved without substituting from the constraint equation, for if the constraint equation is complicated, substitution may be very difficult or even impossible. To show that there is indeed an alternative, consider the situation in which a function $f(x, y, z)$ is to be maximized or minimized subject to two constraints:

$$F(x, y, z) = 0, \qquad (13.60a)$$

$$G(x, y, z) = 0. \qquad (13.60b)$$

Algebraically, we are to find extreme values of $f(x, y, z)$, considering only those values of x, y, and z that satisfy equations 13.60. Geometrically, we can interpret each of these conditions as specifying a surface, so that we are seeking extreme values of $f(x, y, z)$, considering only those points on the curve of intersection C of the surfaces $F(x, y, z) = 0$ and $G(x, y, z) = 0$ (Figure 13.59).

FIGURE 13.59

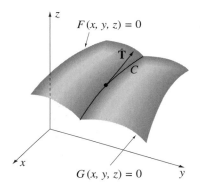

Extreme values of $f(x, y, z)$ along C will occur either at critical points of the function or at the ends of the curve. But what derivative or derivatives of $f(x, y, z)$ are we talking about when we say critical points? Since we are concerned only with values of $f(x, y, z)$ on C, we must mean the derivative of $f(x, y, z)$ along C; i.e., the directional derivative in the tangent direction to C. If **T**, then, is a tangent vector to C, critical points of $f(x, y, z)$ along C are given by

$$0 = D_{\mathbf{T}} f = \nabla f \cdot \hat{\mathbf{T}},$$

or at points where the directional derivative is undefined. According to equation 13.53, a tangent vector to C is $\mathbf{T} = \nabla F \times \nabla G$, and hence a unit tangent vector is

$$\hat{\mathbf{T}} = \frac{\nabla F \times \nabla G}{|\nabla F \times \nabla G|}.$$

The directional derivative of $f(x, y, z)$ at points along C is therefore given by

$$D_{\mathbf{T}} f = \nabla f \cdot \frac{\nabla F \times \nabla G}{|\nabla F \times \nabla G|}.$$

It follows, then, that critical points of $f(x, y, z)$ are points (x, y, z) that satisfy the equation

$$\nabla f \cdot \nabla F \times \nabla G = 0,$$

or points at which the left side is not defined. Now vector $\nabla F \times \nabla G$ is perpendicular to both ∇F and ∇G. Since ∇f is perpendicular to $\nabla F \times \nabla G$ (their dot product is zero), it follows that ∇f must lie in the plane of ∇F and ∇G. Consequently, there exist scalars λ and μ such that

$$\nabla f = (-\lambda)\nabla F + (-\mu)\nabla G,$$

or

$$\nabla f + \lambda \nabla F + \mu \nabla G = \mathbf{0}. \tag{13.61}$$

This vector equation is equivalent to the three scalar equations

$$\frac{\partial f}{\partial x} + \lambda \frac{\partial F}{\partial x} + \mu \frac{\partial G}{\partial x} = 0, \tag{13.62a}$$

$$\frac{\partial f}{\partial y} + \lambda \frac{\partial F}{\partial y} + \mu \frac{\partial G}{\partial y} = 0, \tag{13.62b}$$

$$\frac{\partial f}{\partial z} + \lambda \frac{\partial F}{\partial z} + \mu \frac{\partial G}{\partial z} = 0, \tag{13.62c}$$

and these equations must be satisfied at a critical point at which the directional derivative of $f(x, y, z)$ vanishes. Note too that at a point at which the directional derivative of $f(x, y, z)$ does not exist, one of the partial derivatives in these equations does not exist. In other words, we have shown that critical points of $f(x, y, z)$ are points that satisfy equations 13.62 or points at which the equations are undefined. These equations, however, contain five unknowns: x, y, z, λ, and μ. To complete the system we add equations 13.60 since they must also be satisfied by a critical point. Equations 13.60 and 13.62 therefore yield a system of five equations in the five unknowns x, y, z, λ, and μ; the first three unknowns (x, y, z) define a critical point of $f(x, y, z)$ along C. The advantage of this system of equations lies in the fact that differentiations in 13.62 involve only the given functions (and no substitutions from the constraint equations are necessary). What we have sacrificed is a system of three equations in the three unknowns (x, y, z) for a system of five equations in the five unknowns (x, y, z, λ, μ).

Let us not forget that the original problem was to find extreme values for the function $f(x, y, z)$ subject to constraints 13.60. What we have shown so far is that critical points at which the directional derivative of $f(x, y, z)$ vanishes can be found by solving equations 13.60 and 13.62. In addition, critical points at which the directional derivative of $f(x, y, z)$ does not exist are points at which equations 13.62 are not defined. What remains is to evaluate $f(x, y, z)$ at all critical points and at the ends of C. If C is a *closed curve* (i.e., if C rejoins itself), then $f(x, y, z)$ *needs to be evaluated only at the critical points.*

Through the directional derivative and tangent vectors to curves, we have shown that equations 13.60 and 13.62 define critical points of a function $f(x, y, z)$ that is subject to two constraints: $F(x, y, z) = 0$ and $G(x, y, z) = 0$. But what about other situations? Let us say, for example, that we require extreme values of a function $f(x, y, z, t)$ subject to a single constraint $F(x, y, z, t) = 0$. How shall we find critical points of this function? Fortunately, as we now show, there is a very simple method that yields equations 13.60 and 13.62, and this method generalizes to other situations also.

To find critical points of $f(x, y, z)$ subject to constraints 13.60, we define a function

$$L(x, y, z, \lambda, \mu) = f(x, y, z) + \lambda F(x, y, z) + \mu G(x, y, z),$$

and regard it as a function of five independent variables x, y, z, λ, and μ. To find critical points of this function, we would first solve the equations obtained by setting each of the partial derivatives of L equal to zero:

$$0 = \frac{\partial L}{\partial x} = \frac{\partial f}{\partial x} + \lambda \frac{\partial F}{\partial x} + \mu \frac{\partial G}{\partial x},$$

$$0 = \frac{\partial L}{\partial y} = \frac{\partial f}{\partial y} + \lambda \frac{\partial F}{\partial y} + \mu \frac{\partial G}{\partial y},$$

$$0 = \frac{\partial L}{\partial z} = \frac{\partial f}{\partial z} + \lambda \frac{\partial F}{\partial z} + \mu \frac{\partial G}{\partial z},$$

$$0 = \frac{\partial L}{\partial \lambda} = F(x, y, z),$$

$$0 = \frac{\partial L}{\partial \mu} = G(x, y, z).$$

In addition, we would consider points at which the partial derivatives of L do not exist. Clearly, this means points (x, y, z) at which any of the partial derivatives of $f(x, y, z)$, $F(x, y, z)$, and $G(x, y, z)$ do not exist. But these are equations 13.60 and 13.62. We have shown, then, that finding critical points (x, y, z) of $f(x, y, z)$ subject to $F(x, y, z) = 0$ and $G(x, y, z) = 0$ is equivalent to finding critical points (x, y, z, λ, μ) of $L(x, y, z, \lambda, \mu)$. The two unknowns λ and μ that accompany a critical point (x, y, z) of $f(x, y, z)$ are called **Lagrange multipliers**. They are not a part of the solution (x, y, z) to the original problem, but have been introduced as a convenience by which to arrive at that solution. The function $L(x, y, z, \lambda, \mu)$ is often called the **Lagrangian** of the problem.

The method for other constraint problems should now be evident. Given a function $f(x, y, z, t, \ldots)$ of n variables to maximize or minimize subject to m constraints

$$F_1(x, y, z, t, \ldots) = 0, \quad F_2(x, y, z, t, \ldots) = 0, \ldots, \quad F_m(x, y, z, t, \ldots) = 0, \quad (13.63)$$

we introduce m Lagrange multipliers $\lambda_1, \lambda_2, \ldots, \lambda_m$ into a Lagrangian of $n + m$ independent variables $x, y, z, t, \ldots, \lambda_1, \lambda_2, \ldots, \lambda_m$:

$$L(x, y, z, t, \ldots \lambda_1, \lambda_2, \ldots, \lambda_m) = f(x, y, z, t, \ldots) + \lambda_1 F_1(x, y, z, t, \ldots) + \cdots$$
$$+ \lambda_m F_m(x, y, z, t, \ldots). \quad (13.64)$$

Critical points $(x, y, z, t, \ldots)$ of $f(x, y, z, t, \ldots)$ are then determined by the equations defining critical points of $L(x, y, z, t, \ldots, \lambda_1, \lambda_2, \ldots, \lambda_m)$, namely,

$$0 = \frac{\partial L}{\partial x} = \frac{\partial f}{\partial x} + \lambda_1 \frac{\partial F_1}{\partial x} + \cdots + \lambda_m \frac{\partial F_m}{\partial x}, \quad (13.65a)$$

$$0 = \frac{\partial L}{\partial y} = \frac{\partial f}{\partial y} + \lambda_1 \frac{\partial F_1}{\partial y} + \cdots + \lambda_m \frac{\partial F_m}{\partial y}, \quad (13.65b)$$

$$\vdots \qquad \vdots \qquad \vdots$$

$$0 = \frac{\partial L}{\partial \lambda_1} = F_1(x, y, z, t, \ldots), \tag{13.65c}$$

$$0 = \frac{\partial L}{\partial \lambda_2} = F_2(x, y, z, t, \ldots), \tag{13.65d}$$

$$\vdots \qquad \vdots \qquad \vdots$$

$$0 = \frac{\partial L}{\partial \lambda_m} = F_m(x, y, z, t, \ldots). \tag{13.65e}$$

To use Lagrange multipliers in Example 13.35, where we were minimizing $D^2 = (x - 4)^2 + (y - 6)^2 + (z - 7)^2$ subject to $6x + 3y + 4z = 6$, we define the Lagrangian

$$L(x, y, z, \lambda) = D^2 + \lambda(6x + 3y + 4z - 6)$$

$$= (x - 4)^2 + (y - 6)^2 + (z - 7)^2 + \lambda(6x + 3y + 4z - 6).$$

Critical points of $L(x, y, z, \lambda)$ are defined by

$$0 = \frac{\partial L}{\partial x} = 2(x - 4) + 6\lambda,$$

$$0 = \frac{\partial L}{\partial y} = 2(y - 6) + 3\lambda,$$

$$0 = \frac{\partial L}{\partial z} = 2(z - 7) + 4\lambda,$$

$$0 = \frac{\partial L}{\partial \lambda} = 6x + 3y + 4z - 6.$$

The solution of this linear system is $(x, y, z, \lambda) = (-140/61, 174/61, 171/61, 128/61)$, yielding as before the critical point $(-140/61, 174/61, 171/61)$ of D^2.

EXAMPLE 13.36

Find the maximum and minimum values of the function $f(x, y, z) = xyz$ on the sphere $x^2 + y^2 + z^2 = 1$.

SOLUTION   If we define the Lagrangian

$$L(x, y, z, \lambda) = xyz + \lambda(x^2 + y^2 + z^2 - 1),$$

then critical points of L, and therefore of $f(x, y, z)$, are defined by the equations

$$0 = \frac{\partial L}{\partial x} = yz + 2\lambda x,$$

$$0 = \frac{\partial L}{\partial y} = xz + 2\lambda y,$$

$$0 = \frac{\partial L}{\partial z} = xy + 2\lambda z,$$

$$0 = \frac{\partial L}{\partial \lambda} = x^2 + y^2 + z^2 - 1.$$

If we multiply the first equation by y and the second by x, and equate the resulting expressions for $2\lambda xy$, we have

$$y^2 z = x^2 z.$$

Consequently, either $z = 0$ or $y = \pm x$.

Case I: $\quad z = 0$. In this case the equations reduce to

$$\lambda x = 0, \quad \lambda y = 0, \quad xy = 0, \quad x^2 + y^2 = 1.$$

The first implies that either $x = 0$ or $\lambda = 0$. If $x = 0$, then $y = \pm 1$, and we have two critical points $(0, \pm 1, 0)$. If $\lambda = 0$, then the third equation requires $x = 0$ or $y = 0$. We therefore obtain two additional critical points $(\pm 1, 0, 0)$.

Case II: $\quad y = x$. In this case the equations reduce to

$$xz + 2\lambda x = 0, \quad x^2 + 2\lambda z = 0, \quad 2x^2 + z^2 = 1.$$

The first implies that either $x = 0$ or $z = -2\lambda$. If $x = 0$, then $z = \pm 1$, and we have the two critical points $(0, 0, \pm 1)$. If $z = -2\lambda$, then the last two equations imply that $x = \pm 1/\sqrt{3}$, and we obtain the four critical points

$$(\pm 1/\sqrt{3}, \pm 1/\sqrt{3}, 1/\sqrt{3}) \text{ and } (\pm 1/\sqrt{3}, \pm 1/\sqrt{3}, -1/\sqrt{3}).$$

Case III: $\quad y = -x$. This case is similar to that for $y = x$, and leads to the additional four critical points

$$(\pm 1/\sqrt{3}, \mp 1/\sqrt{3}, 1/\sqrt{3}) \text{ and } (\pm 1/\sqrt{3}, \mp 1/\sqrt{3}, -1/\sqrt{3}).$$

Because $x^2 + y^2 + z^2 = 1$ is a surface without a boundary, we complete the problem by evaluating $f(x, y, z)$ at each of the critical points:

$$f(\pm 1, 0, 0) = f(0, \pm 1, 0) = f(0, 0, \pm 1) = 0,$$

$$f(\pm 1/\sqrt{3}, \pm 1/\sqrt{3}, 1/\sqrt{3}) = f(\pm 1/\sqrt{3}, \mp 1/\sqrt{3}, -1/\sqrt{3}) = \sqrt{3}/9,$$

$$f(\pm 1/\sqrt{3}, \pm 1/\sqrt{3}, -1/\sqrt{3}) = f(\pm 1/\sqrt{3}, \mp 1/\sqrt{3}, 1/\sqrt{3}) = -\sqrt{3}/9.$$

The maximum and minimum values of $f(x, y, z)$ on $x^2 + y^2 + z^2 = 1$ are therefore $\sqrt{3}/9$ and $-\sqrt{3}/9$.

To compare the Lagrangian solution in this example to that without a Lagrange multiplier, see Exercise 21 in Section 13.11.

EXAMPLE 13.37

When a thermonuclear reactor is built in the form of a right circular cylinder, neutron diffusion theory requires its radius and height to satisfy the equation

$$\left(\frac{2.4048}{r}\right)^2 + \left(\frac{\pi}{h}\right)^2 = k,$$

where k is a constant. Find r and h in terms of k if the reactor is to occupy as small a volume as possible.

SOLUTION   The volume of a right circular cylinder is $V = \pi r^2 h$, and were there no constraints on r and h, this function would be considered for all points in the first quadrant of the rh-plane. However, r and h must satisfy a constraint that geometrically can be interpreted as a curve in the rh-plane. What we must do then is minimize $V = \pi r^2 h$, considering only those points (r, h) on the curve defined by the constraint. Clearly there is only one independent variable in the problem—either r or h, but not both. If we choose r as the independent variable, then we note from the constraint that as h becomes very large, r approaches $2.4048/\sqrt{k}$. Since there is no upper bound on r, we can state that the values of r to be considered in the minimization of V are $r > 2.4048/\sqrt{k}$.

To find critical points of V we introduce the Lagrangian

$$L(r, h, \lambda) = \pi r^2 h + \lambda \left\{ \left(\frac{2.4048}{r} \right)^2 + \left(\frac{\pi}{h} \right)^2 - k \right\},$$

and first solve the equations

$$0 = \frac{\partial L}{\partial r} = 2\pi r h + \lambda \left(\frac{-2(2.4048)^2}{r^3} \right),$$

$$0 = \frac{\partial L}{\partial h} = \pi r^2 + \lambda \left(\frac{-2\pi^2}{h^3} \right),$$

$$0 = \frac{\partial L}{\partial \lambda} = \left(\frac{2.4048}{r} \right)^2 + \left(\frac{\pi}{h} \right)^2 - k.$$

If we solve each of the first two equations for λ and equate the resulting expressions, we have

$$\frac{\pi r^4 h}{2.4048^2} = \frac{r^2 h^3}{2\pi}.$$

Since neither r nor h can be zero, we divide by $r^2 h$:

$$\frac{\pi r^2}{2.4048^2} = \frac{h^2}{2\pi},$$

from which

$$r = \frac{2.4048 h}{\sqrt{2\pi}}.$$

Substitution of this result into the constraint equation gives

$$\left(\frac{\sqrt{2\pi}}{h} \right)^2 + \left(\frac{\pi}{h} \right)^2 = k,$$

and this equation can be solved for $h = \pi \sqrt{3/k}$. This gives

$$r = \frac{2.4048 \, \pi \sqrt{3}}{\sqrt{2\pi} \, \sqrt{k}} = 2.4048 \sqrt{3/(2k)}.$$

We have obtained therefore only one critical point (r, h) at which the derivatives of L vanish. The only values of r and h at which the derivatives of L do not exist are $r = 0$ and $h = 0$, but these must be rejected since the constraint requires both r and h to be positive.

To finish the problem we note that

$$\lim_{r \to \infty} V = \infty, \qquad \lim_{r \to 2.4048/\sqrt{k}^{+}} V = \lim_{h \to \infty} V = \infty.$$

It follows, therefore, that the single critical point at which $r = 2.4048\sqrt{3/(2k)}$ and $h = \pi\sqrt{3/k}$ must give the absolute minimum value of $V(r, h)$.

EXERCISES 13.12

In Exercises 1–8 use Lagrange multipliers to find maximum and minimum values of the function subject to the constraints. In each case, interpret the constraints geometrically.

** **1.** $f(x, y) = x^2 + y$ subject to $x^2 + y^2 = 4$

** **2.** $f(x, y, z) = 5x - 2y + 3z + 4$ subject to $x^2 + 2y^2 + 4z^2 = 9$

** **3.** $f(x, y) = x + y$ subject to $(x - 1)^2 + y^2 = 1$

** **4.** $f(x, y, z) = x^3 + y^3 + z^3$ subject to $x^2 + y^2 + z^2 = 9$

** **5.** $f(x, y, z) = xyz$ subject to $x^2 + 2y^2 + 3z^2 = 12$

** **6.** $f(x, y, z) = x^2 y + z$ subject to $x^2 + y^2 = 1$, $z = y$

** **7.** $f(x, y, z) = x^2 + y^2 + z^2$ subject to $x^2 + y^2 + z^2 = 2z$, $x + y + z = 1$

** **8.** $f(x, y, z) = xyz - x^2 z$ subject to $x^2 + y^2 = 1$, $z = \sqrt{x^2 + y^2}$

** **9.** Use Lagrange multipliers to solve Exercise 11 in Section 13.11.

** **10.** Use Lagrange multipliers to solve Exercise 12 in Section 13.11.

** **11.** Use Lagrange multipliers to solve Exercise 20 in Section 13.11.

** **12.** Use Lagrange multipliers to solve Exercise 26 in Section 13.11.

** **13.** Use Lagrange multipliers to solve Exercise 27 in Section 13.11.

** **14.** Use Lagrange multipliers to solve Exercise 28 in Section 13.11.

** **15.** Use Lagrange multipliers to solve Exercise 22 in Section 13.11.

** **16.** Use Lagrange multipliers to solve Exercise 23 in Section 13.11.

** **17.** Use Lagrange multipliers to solve Exercise 24 in Section 13.11.

** **18.** Use Lagrange multipliers to solve Exercise 25 in Section 13.11.

** **19.** Use Lagrange multipliers to solve Exercise 17 in Section 13.11.

** **20.** Suppose that $F(x, y) = 0$ and $G(x, y) = 0$ define two curves C_1 and C_2 in the xy-plane. Let $P(x_0, y_0)$ and $Q(X_0, Y_0)$ be the points on C_1 and C_2 that minimize the distance between C_1 and C_2. If C_1 and C_2 have tangent lines at P and Q, show that the line PQ is perpendicular to these tangent lines.

✱✱ **21.** Find the points on the curve $x^2 + xy + y^2 = 1$ closest to and farthest from the origin.

In Exercises 22–24 use Lagrange multipliers to find maximum and minimum values of the function.

✱✱ **22.** $f(x, y) = 3x^2 + 2xy - y^2 + 5$ for $4x^2 + 9y^2 \le 36$

✱✱ **23.** $f(x, y) = x^2 y + xy^2 + y$ for $-1 \le x \le 1$, $-1 \le y \le 1$

✱✱ **24.** $f(x, y, z) = xy + xz$ for $x^2 + y^2 + z^2 \le 1$

✱✱ **25.** Find the maximum value of $f(x, y, z) = x^2 yz - xzy^2$ subject to constraints $x^2 + y^2 = 1$, $z = \sqrt{x^2 + y^2}$.

✱✱ **26.** The equation $3x^2 + 4xy + 6y^2 = 140$ describes an ellipse which has its centre at the origin, but major and minor axes are not along the x- and y-axes. Find coordinates of the ends of the major and minor axes without rotating axes.

✱✱ **27.** Use Lagrange multipliers to find the point on the first octant part of the plane $Ax + By + Cz = D$ (A, B, C, D all positive constants) which maximizes the function $f(x, y, z) = x^p y^q z^r$ where p, q, and r are positive constants.

✱✱ **28.** The folium of Descartes has parametric equations

$$x = \frac{3at}{1 + t^3}, \qquad y = \frac{3at^2}{1 + t^3}, \qquad (a > 0)$$

(see Exercise 51 in Section 3.8 and Exercise 50 in Section 10.1). Find the point in the first quadrant farthest from the origin in two ways:

(a) Express $D^2 = x^2 + y^2$ in terms of t and maximize this function of one variable.

(b) Show that an implicit equation for the curve is $x^3 + y^3 = 3axy$ and maximize $D^2 = x^2 + y^2$ subject to this constraint.

✱✱✱ **29.** To find the point on the curve $x^2 - xy + y^2 - z^2 = 1$, $x^2 + y^2 = 1$ closest to the origin, we must minimize the function $f(x, y, z) = x^2 + y^2 + z^2$ subject to the constraints defined by the equations of the curve. Show that this can be done by

(a) using two Lagrange multipliers;

(b) expressing $f(x, y, z)$ in terms of x and y alone, $u = 1 - xy$, and minimizing this function subject to $x^2 + y^2 = 1$ (with one Lagrange multiplier);

(c) expressing $f(x, y, z)$ in terms of x alone, $u = 1 \pm x\sqrt{1 - x^2}$, and minimizing these functions on appropriate intervals;

(d) writing $x = \cos t$, $y = \sin t$ along the curve, expressing $f(x, y, z)$ in terms of t, $u = 1 - \sin t \cos t$, and minimizing this function on appropriate intervals.

✱✱✱ **30.** Find the smallest and largest distances from the origin to the curve $x^2 + y^2/4 + z^2/9 = 1$, $x + y + z = 0$.

✱✱✱ **31.** Find the maximum value of $f(x, y, z) = (xy + x^2)/(z^2 + 1)$ subject to the constraint $x^2(4 - x^2) = y^2$.

✱✱✱ **32.** Find the maximum and minimum values of the function $f(x, y, z) = 2x^2 y^2 + 2y^2 z^2 + 3z$ considering only values of x, y, and z satisfying the equations $z = x^2 + y^2$, $x^2 + 3y^2 = 1$. Do this with and without Lagrange multipliers.

13.13 Least-Squares

The basic tool of applied mathematics is the *function*. Given a function such as $f(x) = x^2 - 2x - 3$, and any value of x, say $x = 4$, we calculate the value of the function at this x as $f(4) = 5$. In many problems, the function is not given; it must be found. What we might have is a set of experimental data suggesting that various quantities are related to one another, but the data do not specify exactly how they are related. For example, suppose a variable y is known to depend on a variable x, and an experiment is performed to measure ten values of y corresponding to ten values of x. The results are listed in Table 13.2.

TABLE 13.2

x	1	2	3	4	5	6	7	8	9	10
y	6.05	8.32	10.74	13.43	15.90	18.38	20.93	23.32	24.91	28.36

Consider the problem of finding that function $y = y(x)$ that best describes this data. Two considerations are important—simplicity and accuracy. We would like as simple a function as possible to describe how y depends on x. At the same time, we want the function to be as accurate as possible. For example, we could find a polynomial of degree nine to fit the data exactly; it would give the exact y-value for each value of x. But this would hardly be a simple representation. It would also be unreasonable from the following standpoint. Because y-values are determined experimentally (we assume that x-values are exact), they will be subject to errors both random and systematic. It is pointless to use a ninth-degree polynomial to reproduce the data exactly, since it therefore also reproduces the inherent errors. What we want is a simple function that best fits the *trend* of the data. To discover this *trend*, we plot the data of Table 13.2 as points (x, y) in Figure 13.60.

FIGURE 13.60

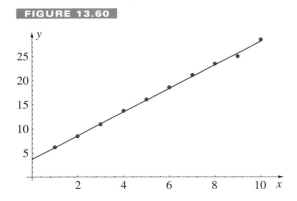

We are immediately impressed by the fact that although the points do not all lie on a straight line, to describe them by a straight line would be a simple representation, and reasonably accurate. We therefore look for a linear function

$$y = y(x) = ax + b \tag{13.66}$$

to describe the data in Table 13.2. Many lines could be drawn to fit the points in Figure 13.60 reasonably accurately, and each line would be characterized by different values of a and b. Our problem then is to find that line (or those values of a and b) that *best* fits the points. Mathematicians have developed a method called **least-squares** to arrive at a best fit.

We denote the x-values in Table 13.2 by $x_i = i$, $i = 1, \ldots, 10$, and corresponding observed values of y by $\overline{y}_i$. The linear function $y(x) = ax + b$ predicts values $y_i = y(x_i) = ax_i + b$ at the x_i. Differences between observed and predicted values of y are

$$y_i - \overline{y}_i = (ax_i + b) - \overline{y}_i.$$

We define a quantity S as the sum of the squares of these differences for all x_i,

$$S = \sum_{i=1}^{10}(y_i - \overline{y}_i)^2 = \sum_{i=1}^{10}(ax_i + b - \overline{y}_i)^2. \tag{13.67}$$

Given any a and b, S is a measure of the degree to which the points vary from that line—the better the fit, the smaller the value of S. The method of least-squares states that one way to approximate the points in Figure 13.60 by a straight line is to choose that line which minimizes the function $S = S(a, b)$. In other words, find values of a and b that minimize $S(a, b)$. For critical points of this function we solve

$$0 = \frac{\partial S}{\partial a} = \sum_{i=1}^{10} 2x_i(ax_i + b - \overline{y}_i),$$

$$0 = \frac{\partial S}{\partial b} = \sum_{i=1}^{10} 2(ax_i + b - \overline{y}_i).$$

If we rewrite these equations in the form

$$\left(\sum_{i=1}^{10} x_i^2\right) a + \left(\sum_{i=1}^{10} x_i\right) b = \sum_{i=1}^{10} x_i \overline{y}_i, \tag{13.68a}$$

$$\left(\sum_{i=1}^{10} x_i\right) a + 10b = \sum_{i=1}^{10} \overline{y}_i, \tag{13.68b}$$

we have a pair of linear equations in a and b. With the data in Table 13.2, we obtain

$$385a + 55b = 1139.27,$$

$$55a + 10b = 170.34,$$

which have solutions $a = 2.45$ and $b = 3.54$. Since there is only one critical point, and we know that $S(a, b)$ must have an absolute minumum for some a and b, it follows that these values must minimize $S(a, b)$. The straight line

$$y = y(x) = 2.45x + 3.54$$

is therefore the best straight line fit (in the least-squares sense) to the data in Table 13.2. This is, in fact, the line in Figure 13.60. It is important to remember that least squares assumes that we have a prior knowledge of the type of function to be determined (in this case a linear function), and the method then proceeds to find the best such function.

Polynomials of higher order (quadratics, cubics, etc.) can be fitted to data points by least-squares (see Exercises 3–5). Other types of functions can also be used, often by reducing the problem to a straight-line situation. Suppose, for example, that we wish to fit a curve to the data in Table 13.3 (which we have plotted in Figure 13.61a).

TABLE 13.3

x	0.5	1.0	1.5	2.0	2.5	3.0	3.5	4.0	4.5	5.0	5.5
y	140	180	230	290	365	455	565	670	785	1000	1230

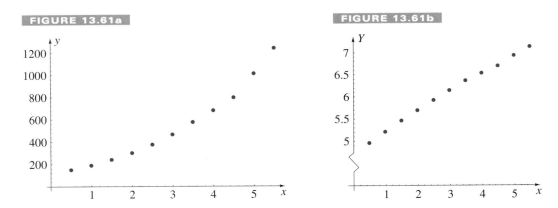

FIGURE 13.61a

FIGURE 13.61b

We could try a quadratic or a cubic but neither would give a satisfactory fit. The fact that y values increase faster and faster as x increases suggests that an exponential function

$$y = y(x) = be^{ax} \qquad (13.69)$$

might best describe the points. To determine whether this is the case, we take logarithms on both sides of 13.69,

$$\ln y = \ln b + ax.$$

If we define $Y = \ln y$ and $B = \ln b$, then

$$Y = ax + B,$$

and this is the equation of a straight line in the xY-plane. In other words, to test whether a set of points can be described by an exponential be^{ax}, we plot $Y = \ln y$ against x, and if these points can be approximated by a straight line, then the original data should be describable by an exponential function. For the points in Table 13.3, we obtain the data in Table 13.4 and the graph in Figure 13.61b, and clearly a close approximation to a straight line.

TABLE 13.4

x	0.5	1.0	1.5	2.0	2.5	3.0	3.5	4.0	4.5	5.0	5.5
y	4.94	5.19	5.44	5.67	5.90	6.12	6.34	6.51	6.67	6.91	7.11

If we fit a straight line $Y = ax + B$ to the data in Table 13.4, the equations corresponding to 13.68 are

$$\left(\sum_{i=1}^{11} x_i^2 \right) a + \left(\sum_{i=1}^{11} x_i \right) B = \sum_{i=1}^{11} x_i \overline{Y}_i, \qquad (13.70a)$$

$$\left(\sum_{i=1}^{11} x_i \right) a + (11)b = \sum_{i=1}^{11} \overline{Y}_i, \qquad (13.70b)$$

from which

$$126.5a + 33B = 212.149,$$

$$33a + 11B = 66.795.$$

The solution of these is $a = 0.43$ and $B = 4.79$. Consequently, the least-squares estimates give

$$\ln y = Y = 4.79 + 0.43x.$$

When we take exponentials,

$$y = e^{4.79+0.43x} = 120.3e^{0.43x},$$

and this exponential approximates the data in Table 13.3, or the points in Figure 13.61a. It is important to notice that we did not apply the method of least-squares directly to the exponential function be^{ax} and the data in Table 13.3. Were we to do so, it would be very difficult to solve the resulting equations for a and b. Try it! Were we successful in doing so, values of a and b would differ from those above, but not significantly.

Representation of data by other types of functions is discussed in the exercises.

EXERCISES 13.13

∗ **1.** The following table shows the ages M (in months) and the average test scores S (obtained on a standard intelligence test) for children between the ages of 9 and 12 years.

Age M (months)	108	112	116	120	124	128	132	136	140	144
Average score S	62	67	72	77	83	87	94	100	105	109

Find least-squares estimates for a linear function to describe $S = S(M)$.

∗ **2.** The table below shows the average systolic blood pressure P of 13 children.

Age A (years)	4	5	6	7	8	9	10	11	12	13	14	15	16
Pressure P	85	87	90	92	95	98	100	105	108	110	112	115	118

Use least-squares to find the equation of a straight line fitting this data.

∗∗ **3.**   (a) Plot the sixteen points in the following table.

x	3.00	3.25	3.50	3.75	4.00	4.25	4.50	4.75	5.00	5.25	5.50	5.75	6.00	6.25	6.50	6.75
y	31.5	30.4	29.2	28.1	26.9	26.4	25.3	25.2	25.1	25.2	25.4	26.3	27.0	28.2	29.3	29.9

Do they seem to follow a parabolic path?

Geometrically, this is the difference in the heights of the surface at the points $(x + dx, y + dy)$ and (x, y). If we draw the tangent plane to the surface at (x, y), then very near (x, y) the height of the tangent plane approximates the height of the surface (Figure 13.63). In particular, the height of the tangent plane at $(x + dx, y + dy)$ for small dx and dy approximates the height of the surface. We define the **differential** dz as the change in z corresponding to the changes dx and dy in x and y if we follow the tangent plane at (x, y) instead of the surface itself. To find dz in terms of dx and dy, we note that the vector joining the points (x, y, z) and $(x + dx, y + dy, z + dz)$ has components (dx, dy, dz), and this vector lies in the tangent plane. Since a normal vector to the tangent plane is

$$\nabla(z - f(x, y)) = (-f_x, -f_y, 1),$$

it follows that the vectors $(-f_x, -f_y, 1)$ and (dx, dy, dz) must be perpendicular. Consequently,

$$0 = (-f_x, -f_y, 1) \cdot (dx, dy, dz) = -\frac{\partial f}{\partial x} dx - \frac{\partial f}{\partial y} dy + dz,$$

and hence

$$dz = \frac{\partial f}{\partial x} dx + \frac{\partial f}{\partial y} dy. \tag{13.71}$$

Note that if y is held constant in the function $f(x, y)$, then $dy = 0$ and 13.71 for dz reduces to the definition of the differential of a function of one variable.

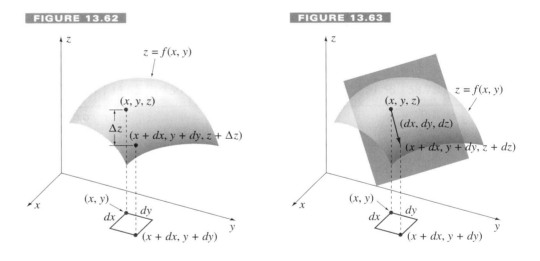

FIGURE 13.62

FIGURE 13.63

In Section 4.11 we indicated that we must be careful in using the differential dy as an approximation for the change Δy in a function $f(x)$. For the same reasons, we must be judicious in our use of dz as an approximation for Δz. Indeed, we have stated that dz is an approximation for Δz for small dx and dy, but the difficulty is deciding how small is small and how good is the approximation. In addition, note that if (x, y) is a critical point of the function $f(x, y)$, then either $dz = 0$ for all dx and dy, or dz is undefined. In other words, dz cannot be used to approximate Δz at a critical point.

EXAMPLE 13.38

If the radius of a right circular cone is changed from 10 cm to 10.1 cm and the height is changed from 1 m to 0.99 m, use differentials to approximate the change in its volume.

SOLUTION   The volume of a cone of radius r and height h is given by the formula $V = \pi r^2 h/3$. The differential of this function is

$$dV = \frac{\partial V}{\partial r}dr + \frac{\partial V}{\partial h}\,dh = \frac{2}{3}\pi r h\, dr + \frac{1}{3}\pi r^2 dh.$$

If $r = 10$, $dr = 0.1$, $h = 100$, and $dh = -1$, then

$$dV = \frac{2}{3}\pi(10)(100)(0.1) + \frac{1}{3}\pi(10)^2(-1) = \frac{100\pi}{3}\ \text{cm}^3.$$

Equation 13.71 suggests the following definition for the differential of a function of more than two independent variables.

DEFINITION 13.11

If $u = f(x, y, z, t, \ldots, w)$, then the **differential** of $f(x, y, z, t, \ldots, w)$ is defined as

$$du = \frac{\partial f}{\partial x}dx + \frac{\partial f}{\partial y}dy + \frac{\partial f}{\partial z}dz + \frac{\partial f}{\partial t}dt + \cdots + \frac{\partial f}{\partial w}dw. \qquad (13.72)$$

EXAMPLE 13.39

The area of the triangle in Figure 13.64 is given by the formula $A = \frac{1}{2}ab \sin\theta$. If when $\theta = \pi/3$, a and b are changed by $\frac{1}{3}\%$ and θ by $\frac{1}{2}\%$, use differentials to find the approximate percentage change in A.

FIGURE 13.64

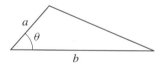

SOLUTION   Since

$$dA = \frac{\partial A}{\partial a}da + \frac{\partial A}{\partial b}db + \frac{\partial A}{\partial \theta}d\theta$$

$$= \frac{1}{2}b\, \sin\theta\, da + \frac{1}{2}a\, \sin\theta\, db + \frac{1}{2}ab\, \cos\theta\, d\theta,$$

the approximate percentage change in A is

$$100\left(\frac{dA}{A}\right) = \frac{100}{A}\left\{\frac{1}{2}b\, \sin\theta\, da + \frac{1}{2}a \sin\theta\, db + \frac{1}{2}ab\, \cos\theta\, d\theta\right\}$$

$$= 100\left\{\frac{da}{a} + \frac{db}{b} + \cot\theta d\theta\right\}.$$

Since a and b are changed by $\frac{1}{3}\%$ and θ by $\frac{1}{2}\%$,

$$100\left(\frac{da}{a}\right) = 100\left(\frac{db}{b}\right) = \frac{1}{3} \text{ and } 100\left(\frac{d\theta}{\theta}\right) = \frac{1}{2}.$$

Thus,

$$100\left(\frac{dA}{A}\right) = \frac{1}{3} + \frac{1}{3} + \frac{\theta}{2}\cot\theta = \frac{2}{3} + \frac{\theta}{2}\cot\theta,$$

and when $\theta = \pi/3$,

$$100\left(\frac{dA}{A}\right) = \frac{2}{3} + \frac{1}{2}\left(\frac{\pi}{3}\right)\left(\frac{1}{\sqrt{3}}\right) = 0.97.$$

The approximate percentage change in A is therefore 1%.

EXERCISES 13.14

In Exercises 1–10 find the differential of the function.

* **1.** $f(x, y) = x^2 y - \sin y$
* **2.** $f(x, y) = \text{Tan}^{-1}(xy)$
* **3.** $f(x, y, z) = xyz - x^3 e^z$
* **4.** $f(x, y, z) = \sin(xyz) - x^2 y^2 z^2$
* **5.** $f(x, y, z) = \ln(x^2 + y^2 + z^2)$
* **6.** $f(x, y) = \text{Sin}^{-1}(xy)$
* **7.** $f(x, y) = \text{Sin}^{-1}(x + y) + \text{Cos}^{-1}(x + y)$
* **8.** $f(x, y, z, t) = xy + yz + zt + xt$
* **9.** $f(x, y, z, w) = xy \tan(zw)$
* **10.** $f(x, y, z, t) = e^{x^2 + y^2 + z^2 - t^2}$
* **11.** A right circular cone has radius 10 cm and height 20 cm. If its radius increases by 0.1 cm and its height decreases by 0.3 cm, use differentials to find the approximate change in its volume. Compare this with the actual change in volume.
* **12.** When the ellipse $b^2 x^2 + a^2 y^2 = a^2 b^2$ is rotated about the x-axis, the volume V of the spheroid is $4\pi ab^2/3$. If a and b are each increased by 1%, use differentials to find the approximate percentage change in V.
* **13.** When two resistors of resistances R_1 and R_2 are connected in parallel, their effective resistance is $R = R_1 R_2/(R_1 + R_2)$. Show that if R_1 and R_2 are both increased by a small percentage c, then the percentage increase of R is also c.

13.15 Taylor Series for Multivariable Functions

Taylor series for functions of one variable can be used to generate Taylor series for multivariable functions. For simplicity, we once again work with functions of two independent variables. Extensions to functions of more than two independent variables will be clear. Suppose that a function $f(x, y)$ has continuous partial derivatives of all orders in some open circle centred at the point (c, d) (Figure 13.65).

FIGURE 13.65

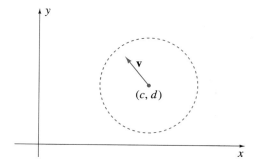

Parametric equations for the line through (c, d) in direction $\mathbf{v} = (v_x, v_y)$ are

$$x = c + v_x t, \quad y = d + v_y t.$$

If we substitute these values into $f(x, y)$, we obtain a function $F(t)$ of one variable,

$$F(t) = f(c + v_x t, d + v_y t),$$

which represents the value of $f(x, y)$ at points along the line through (c, d) in direction $\mathbf{v}$. If we expand this function into its Maclaurin series we obtain

$$F(t) = F(0) + F'(0)t + \frac{F''(0)}{2!}t^2 + \cdots . \tag{13.73}$$

The schematic diagram to the left gives

$$F'(t) = \frac{\partial F}{\partial x}\frac{dx}{dt} + \frac{\partial F}{\partial y}\frac{dy}{dt} = f_x(x, y)v_x + f_y(x, y)v_y,$$

and therefore

$$F'(0) = f_x(c, d)v_x + f_y(c, d)v_y.$$

The second derivative of $F(t)$ is

$$F''(t) = \frac{\partial}{\partial x}[F'(t)]\frac{dx}{dt} + \frac{\partial}{\partial y}[F'(t)]\frac{dy}{dt}$$

$$= \frac{\partial}{\partial x}[f_x(x, y)v_x + f_y(x, y)v_y]v_x$$

$$+ \frac{\partial}{\partial y}[f_x(x, y)v_x + f_y(x, y)v_y]v_y$$

$$= f_{xx}(x, y)v_x^2 + 2f_{xy}(x, y)v_x v_y + f_{yy}(x, y)v_y^2,$$

and therefore

$$F''(0) = f_{xx}(c, d)v_x^2 + 2f_{xy}(c, d)v_x v_y + f_{yy}(c, d)v_y^2.$$

A similar calculation gives

$$F'''(0) = F_{xxx}(c, d)v_x^3 + 3f_{xxy}(c, d)v_x^2 v_y + 3f_{xyy}(c, d)v_x v_y^2 + f_{yyy}(c, d)v_y^3,$$

and the pattern is emerging. When these results are substituted into 13.73,

$$F(t) = f(c + v_x t, d + v_y t)$$

$$= f(c, d) + [f_x(c, d)v_x + f_y(c, d)v_y]t$$

$$+ [f_{xx}(c, d)v_x^2 + 2f_{xy}(c, d)v_x v_y + f_{yy}(c, d)v_y^2]\frac{t^2}{2!} + \cdots . \quad (13.74)$$

We now let **v** be the vector from (c, d) to point (x, y), so that $v_x = x - c$ and $v_y = y - d$, and at the same time set $t = 1$. Then $F(1) = f(c + x - c, d + y - d) = f(x, y)$, and 13.74 becomes

$$f(x, y) = f(c, d) + [f_x(c, d)(x - c) + f_y(c, d)(y - d)] + \frac{1}{2!}[f_{xx}(c, d)(x - c)^2$$

$$+ 2f_{xy}(c, d)(x - c)(y - d) + f_{yy}(c, d)(y - d)^2] + \cdots \quad (13.75)$$

This is the Taylor series for $f(x, y)$ about the point (c, d). It gives the value of the function at the point (x, y) in terms of values of the function and its derivative at the point (c, d).

EXAMPLE 13.40

Find the first six nonzero terms in the Taylor series for $f(x, y) = \sin(2x + 3y)$ about $(0, 0)$.

SOLUTION   We calculate that:

$$f(0, 0) = 0$$

$$f_x(0, 0) = 2\cos(2x + 3y)_{|(0,0)} = 2$$

$$f_y(0, 0) = 3\cos(2x + 3y)_{|(0,0)} = 3$$

$$f_{xx}(0, 0) = -4\sin(2x + 3y)_{|(0,0)} = 0$$

$$f_{xy}(0, 0) = -6\sin(2x + 3y)_{|(0,0)} = 0$$

$$f_{yy}(0, 0) = -9\sin(2x + 3y)_{|(0,0)} = 0$$

$$f_{xxx}(0, 0) = -8\cos(2x + 3y)_{|(0,0)} = -8$$

$$f_{xxy}(0, 0) = -12\cos(2x + 3y)_{|(0,0)} = -12$$

$$f_{xyy}(0, 0) = -18\cos(2x + 3y)_{|(0,0)} = -18$$

$$f_{yyy}(0, 0) = -27\cos(2x + 3y)_{|(0,0)} = -27$$

Formula 13.75 then gives

$$\sin(2x + 3y) = 0 + [2x + 3y] + \frac{1}{2!}[0] + \frac{1}{3!}[-8x^3 - 36x^2y - 54xy^2 - 27y^3] + \cdots$$

$$= (2x + 3y) - \frac{1}{3!}(2x + 3y)^3 + \cdots .$$

This series could also have been obtained by substituting $2x + 3y$ for x in the Maclaurin series for $\sin x$. This is not always an alternative.

EXERCISES 13.15

* **1.** If $f(x, y) = F(x)G(y)$, is the Taylor series of $f(x, y)$ about $(0, 0)$ the product of the Maclaurin series for $F(x)$ and $G(y)$?

* **2.** What are the cubic terms in 13.75?

In Exercises 3–8 find the Taylor series of the function about the point by using Taylor series for functions of one variable.

* **3.** $\cos(xy)$ about $(0, 0)$

* **4.** e^{2x-3y} about $(1, -1)$

* **5.** $x^2 y \sqrt{1 + x}$ about $(0, 0)$

* **6.** $\ln(1 + x^2 + y^2)$ about $(0, 0)$

* **7.** $\dfrac{1}{1 + x + y}$ about $(3, -4)$

* **8.** $\dfrac{xy^2}{1 + y^2}$ about $(-1, 0)$

In Exercises 9–14 find the Taylor series of the function up to and including quadratic terms.

** **9.** $\dfrac{xy}{x^2 + y^2}$ about $(-1, 1)$

** **10.** $\sqrt{1 + xy}$ about $(2, 1)$

** **11.** $e^x \sin(3x - y)$ about $(-1, 0)$

** **12.** $(x + y)^2 \ln(x + y)$ about $(0, 1)$

** **13.** $\mathrm{Tan}^{-1}(3x + 2y)$ about $(1, -1)$

** **14.** $x^8 y^{10}$ about $(0, 0)$

** **15.** What are the terms in the Taylor series for a function $f(x, y, z)$ about the point (c, d, e) corresponding to those in equation 13.75?

** **16.** Express 13.75 in sigma notation. Hint: Think about an operator

$$\left[(x - a)\frac{\partial}{\partial x} + (y - b)\frac{\partial}{\partial y} \right]^n$$

which is expanded as a binomial to operate on functions $f(x, y)$.

SUMMARY

We began the study of multivariable functions in this chapter, concentrating our attention on differentiation and its applications. We introduced two types of derivatives for a multivariable function: partial derivatives and directional derivatives. Partial derivatives are directional derivatives in directions parallel to the coordinate axes. The directional derivative of a function $f(x, y, z)$ in the direction $\mathbf{v}$ is given by the formula $D_{\mathbf{v}} f = \nabla f \cdot \hat{\mathbf{v}}$, where $\hat{\mathbf{v}}$ is the unit vector in the direction of $\mathbf{v}$, and the gradient ∇f is evaluated at the point at which $D_{\mathbf{v}} f$ is required. This formula leads to the fact that the gradient $\nabla f(x, y, z)$ points in the direction in which $f(x, y, z)$ increases most rapidly, and $|\nabla f|$ is the (maximum) rate of change of $f(x, y, z)$ in the direction ∇f. A second property of gradient vectors (which is related to the first) is that if $F(x, y, z) = 0$ is the equation of a surface, then at any point on the surface ∇F is perpendicular to the surface. This property, along with the fact that perpendicularity to a surface is synonymous with perpendicularity to its tangent plane, enables us to find equations for tangent planes to surfaces and tangent lines to curves.

We illustrated various ways to calculate partial derivatives of a multivariable function, depending on whether the function is defined explicitly, implicitly, or as a composite function. Since partial derivatives are ordinary derivatives with other variables held constant, there is no difficulty calculating partial derivatives when the function is defined explicitly; we simply use the rules from single-variable calculus. When the partial derivative of a composite function is required, we use a schematic diagram illustrating functional dependences to develop the appropriate chain rule. Partial derivatives for functions defined implicitly are calculated using Jacobians.

Once antidifferentiation in 14.8 is completed and the limits substituted, the result is a function of y alone. It is then possible to integrate this function with respect to y between any two limits, say from $y = c$ to $y = d$:

$$\int_c^d \left\{ \int_{g(y)}^{h(y)} f(x, y)dx \right\} dy.$$

In practice we omit the braces and simply write

$$\int_c^d \int_{g(y)}^{h(y)} f(x, y)dx \, dy, \tag{14.10}$$

understanding that in the evaluation we proceed from the inner integral to the outer. This is called a **double iterated integral** first with respect to x and then with respect to y (or, more concisely, with respect to x and y). Double iterated integrals with respect to y and x take the form

$$\int_a^b \int_{g(x)}^{h(x)} f(x, y)dy \, dx. \tag{14.11}$$

EXAMPLE 14.1

Evaluate each of the following double iterated integrals:

(a) $\int_0^1 \int_x^{x-1} (x^2 + e^y)dy \, dx$

(b) $\int_{-1}^1 \int_0^y xye^{x^2} dx \, dy$

SOLUTION

(a)
$$\int_0^1 \int_x^{x-1} (x^2 + e^y)dy \, dx = \int_0^1 \{x^2 y + e^y\}_x^{x-1} dx$$

$$= \int_0^1 \{x^2(x - 1) + e^{x-1} - x^3 - e^x\}dx$$

$$= \int_0^1 \{-x^2 + e^{x-1} - e^x\}dx$$

$$= \left\{ -\frac{x^3}{3} + e^{x-1} - e^x \right\}_0^1$$

$$= \left\{ -\frac{1}{3} + 1 - e \right\} - \{e^{-1} - 1\}$$

$$= \frac{5}{3} - e - e^{-1}.$$

(b)
$$\int_{-1}^1 \int_0^y xye^{x^2} dx \, dy = \int_{-1}^1 \left\{ \frac{1}{2} ye^{x^2} \right\}_0^y dy = \frac{1}{2} \int_{-1}^1 (ye^{y^2} - y)dy$$

$$= \frac{1}{2} \left\{ \frac{1}{2} e^{y^2} - \frac{y^2}{2} \right\}_{-1}^1 = \frac{1}{2} \left\{ \frac{e}{2} - \frac{1}{2} - \frac{e}{2} + \frac{1}{2} \right\} = 0.$$

EXERCISES 14.1

In Exercises 1–30 evaluate the double iterated integral.

* **1.** $\displaystyle\int_{-1}^{2}\int_{y}^{y+2} (x^2 - xy)\, dx\, dy$

* **2.** $\displaystyle\int_{-3}^{3}\int_{-\sqrt{18-2y^2}}^{\sqrt{18-2y^2}} x\, dx\, dy$

* **3.** $\displaystyle\int_{0}^{1}\int_{x^2}^{x} (2xy + 3y^2)\, dy\, dx$

* **4.** $\displaystyle\int_{-1}^{0}\int_{y}^{2} (1 + y)^2\, dx\, dy$

* **5.** $\displaystyle\int_{3}^{4}\int_{0}^{\pi/2} x \sin y\, dy\, dx$

* **6.** $\displaystyle\int_{1}^{2}\int_{1}^{y} e^{x+y}\, dx\, dy$

* **7.** $\displaystyle\int_{-1}^{1}\int_{-x}^{5} (x^2 + y^2)\, dy\, dx$

* **8.** $\displaystyle\int_{-1}^{1}\int_{x}^{2x} (xy + x^3 y^3)\, dy\, dx$

* **9.** $\displaystyle\int_{0}^{1}\int_{x}^{1} (x + y)^4\, dy\, dx$

* **10.** $\displaystyle\int_{1}^{2}\int_{x}^{2x} \frac{1}{(x + y)^3}\, dy\, dx$

* **11.** $\displaystyle\int_{0}^{1}\int_{0}^{3x} \sqrt{x + y}\, dy\, dx$

* **12.** $\displaystyle\int_{-1}^{1}\int_{1}^{e} \frac{y}{x}\, dx\, dy$

* **13.** $\displaystyle\int_{1}^{4}\int_{\sqrt{x}}^{x^2} (x^2 + 2xy - 3y^2)\, dy\, dx$

* **14.** $\displaystyle\int_{0}^{2}\int_{x^2}^{2x^2} x \cos y\, dy\, dx$

* **15.** $\displaystyle\int_{0}^{1}\int_{1}^{\tan x} \frac{1}{1 + y^2}\, dy\, dx$

* **16.** $\displaystyle\int_{0}^{1}\int_{0}^{y^3} \frac{1}{1 + y^2}\, dx\, dy$

* **17.** $\displaystyle\int_{2}^{3}\int_{0}^{1} \frac{x}{\sqrt{1 - y^2}}\, dy\, dx$

* **18.** $\displaystyle\int_{0}^{2}\int_{-x}^{x} (8 - 2x^2)^{3/2}\, dy\, dx$

** **19.** $\displaystyle\int_{0}^{1}\int_{0}^{x} \frac{1}{\sqrt{1 - y^2}}\, dy\, dx$

** **20.** $\displaystyle\int_{-9}^{0}\int_{0}^{x^2\sqrt{9+x}} dy\, dx$

** **21.** $\displaystyle\int_{0}^{2}\int_{\sqrt{4-x^2}}^{2} y^2\, dy\, dx$

* **22.** $\displaystyle\int_{-1}^{0}\int_{y}^{0} x\sqrt{x^2 + y^2}\, dx\, dy$

** **23.** $\displaystyle\int_{2}^{3}\int_{1}^{2x} \frac{1}{(xy + x^2)^2}\, dy\, dx$

** **24.** $\displaystyle\int_{0}^{1}\int_{0}^{\cos^{-1}x} x \cos y\, dy\, dx$

** **25.** $\displaystyle\int_{0}^{1}\int_{\sqrt{y^2+y}}^{\sqrt{2y}} x^3\sqrt{x^2 - y^2}\, dx\, dy$

** **26.** $\displaystyle\int_{0}^{1}\int_{\sqrt{2y}}^{\sqrt{y^2+y}} x^3\sqrt{x^2 - y^2}\, dx\, dy$

** **27.** $\displaystyle\int_{-2}^{0}\int_{x^4}^{4x^2} \sqrt{y - x^4}\, dy\, dx$

** **28.** $\displaystyle\int_{-2}^{0}\int_{y}^{0} \frac{x}{\sqrt{x^2 + y^2}}\, dx\, dy$

** **29.** $\displaystyle\int_{-1}^{2}\int_{-1}^{y^3} \sqrt{1 + y}\, dx\, dy$

*** **30.** $\displaystyle\int_{0}^{1}\int_{0}^{x} \sqrt{x^2 + y^2}\, dy\, dx$

$f(x, y)\, dy\, dx,$

and it is these quantities that are added.

14.2 Evaluation of Double Integrals by Double Iterated Integrals

There are two distinct parts to every double integral: first, the function $f(x, y)$ being integrated, which is the integrand; second, the region R over which integration is being performed, and this region determines the limits on the corresponding double iterated integral. Note that we do not use $f(x, y)$ to determine limits on the double iterated integral; the region determines the limits. Conversely, if we are given a double iterated integral, then we know that it represents the double integral of its integrand over some region, and the region is completely defined by the limits on the iterated integral. This point is emphasized in the following example.

EXAMPLE 14.3

Evaluate the double iterated integral

$$\int_0^2 \int_y^2 e^{x^2}\, dx\, dy.$$

SOLUTION   The function e^{x^2} does not have an elementary antiderivative with respect to x, and it is therefore impossible to evaluate the double iterated integral as it now stands. But the double iterated integral represents the double integral of e^{x^2} over some region R in the xy-plane. To find R we note that the inner integral indicates horizontal strips that all start on the line $x = y$ and stop on the line $x = 2$ (Figure 14.9a). The outer limits state that the first and last strips are at $y = 0$ and $y = 2$, respectively. This defines R as the triangle bounded by the straight lines $y = x$, $x = 2$, and $y = 0$ (Figure 14.9b). If we now reverse the order of integration and use vertical strips, we have

$$\int_0^2 \int_y^2 e^{x^2}\, dx\, dy = \iint_R e^{x^2}\, dA = \int_0^2 \int_0^x e^{x^2}\, dy\, dx = \int_0^2 \{ye^{x^2}\}_0^x\, dx$$

$$= \int_0^2 x e^{x^2}\, dx = \left\{\frac{1}{2}e^{x^2}\right\}_0^2 = \frac{e^4 - 1}{2}.$$

FIGURE 14.9a

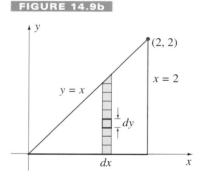

FIGURE 14.9b

This example also points out that an iterated integral in one order may be much easier to evaluate than the corresponding iterated integral in the opposite order.

EXERCISES 14.2

In Exercises 1–12 evaluate the double integral over the region.

* **1.** $\iint_R (x^2 + y^2)\, dA$ where R is bounded by $y = x^2$, $x = y^2$

* **2.** $\iint_R (4 - x^2 - y)\, dA$ where R is bounded by $x = \sqrt{4 - y}$, $x = 0$, $y = 0$

* **3.** $\iint_R (x + y)\, dA$ where R is bounded by $x = y^3 + 2$, $x = 1$, $y = 1$

* **4.** $\iint_R xy^2\, dA$ where R is bounded by $x + y + 1 = 0$, $x + y^2 = 1$

* **5.** $\iint_R xe^y\, dA$ where R is bounded by $y = x$, $y = 0$, $x = 1$

* **6.** $\iint_R (x + y)\, dA$ where R is bounded by $x^2 + y^2 = 9$

* **7.** $\iint_R x^2 y\, dA$ where R is bounded by $y = \sqrt{x + 4}$, $y = 0$, $x + y = 2$

* **8.** $\iint_R (xy + y^2 - 3x^2)\, dA$ where R is bounded by $y = |x|$, $y = 1$, $y = 2$

* **9.** $\iint_R (1 - x)^2\, dA$ where R is bounded by $x + y = 1, x + y = -1,\ x - y = 1,\ y - x = 1$

* **10.** $\iint_R (x + y)\, dA$ where R is bounded by $x = y^2$, $x^2 - y^2 = 12$

* **11.** $\iint_R x\, dA$ where R is bounded by $y = 3x$, $y = x$, $x + y = 4$

* **12.** $\iint_R y^2\, dA$ where R is bounded by $x = 0$, $y = 1$, $y = 1/2$, $x = 1/\sqrt{y^4 + 12y^2}$

In Exercises 13–18 evaluate the double iterated integral by reversing the order of integration.

** **13.** $\displaystyle\int_0^2 \int_0^{\sqrt{4-x^2}} (4 - y^2)^{3/2}\, dy\, dx$

** **14.** $\displaystyle\int_0^1 \int_y^1 \sin(x^2)\, dx\, dy$

** **15.** $\displaystyle\int_{-2}^0 \int_{-y}^2 y(x^2 + y^2)^8\, dx\, dy$

** **16.** $\displaystyle\int_{-2}^0 \int_{-2}^x \frac{x}{\sqrt{x^2 + y^2}}\, dy\, dx$

** **17.** $\displaystyle\int_0^2 \int_0^{x^2/2} \frac{x}{\sqrt{1 + x^2 + y^2}}\, dy\, dx$

** **18.** $\displaystyle\int_0^2 \int_{-x^2/2}^0 \frac{x}{\sqrt{1 + x^2 + y^2}}\, dy\, dx$

** **19.** Verify that if $m \leq f(x, y) \leq M$ for all (x, y) in R, then

$$m(\text{area of } R) \leq \iint_R f(x, y)\, dA \leq M(\text{area of } R).$$

** **20.** Evaluate the double integral of $f(x, y) = 1/\sqrt{2x - x^2}$ over the region in the first quadrant bounded by $y^2 = 4 - 2x$.

In Exercises 21–28 either the integral has value 0 or it can be evaluated by doubling the double integral over half the region. By drawing the region and examining the integrand, determine which situation prevails. Do not evaluate the integral.

* **21.** $\iint_R x^2 y^3\, dA$ where R is bounded by $x = \sqrt{4 - y^2}$, $x = 0$

* **22.** $\iint_R x^2 y^2\, dA$ where R is bounded by $x = \sqrt{4 - y^2}$, $x = 0$

* **23.** $\iint_R (x + y)\, dA$ where R is the square with vertices $(\pm 3, 0)$ and $(0, \pm 3)$

* **24.** $\iint_R x^7 \cos(x^2)\, dA$ where R is bounded by $y = 4 - |x|$, $y = x^2$

* **25.** $\iint_R e^{x^2 + y^2}\, dA$ where R is bounded by $y = 4 - 4x^2$, $y = x^2 - 1$

* **26.** $\iint_R \cos(x^2 y)\, dA$ where R is bounded by $y = 0$, $y = x^3 - x$

* **27.** $\iint_R \sin(x^2 y)\, dA$ where R is bounded by $y = 0$, $y = x^3 - x$

* **28.** $\iint_R (x^2 y^3 + xy^2)\, dA$ where R is bounded by $\sqrt{|x|} + \sqrt{|y|} = 1$

The average value of a function $f(x, y)$ over a region R with area A is defined as

$$\overline{f} = \frac{1}{A} \iint_R f(x, y)\, dA.$$

In Exercises 29–32 find the average value of the function over the region.

* **29.** $f(x, y) = xy$ over the region in the first quadrant bounded by $x = 0$, $y = \sqrt{1 - x^2}$, $y = 0$

* **30.** $f(x, y) = x + y$ over the region bounded by $y = x$, $y = 0$, $y = \sqrt{2 - x}$

** **31.** $f(x, y) = x$ over the region between $y = \sin x$ and $y = 0$ for $0 \leq x \leq 2\pi$

** **32.** $f(x, y) = e^{x+y}$ over the region bounded by $y = x + 1$, $y = x - 1$, $y = 1 - x$, $y = -1 - x$

In Exercises 33–39 evaluate the double integral over the region.

** **33.** $\iint_R x^2\, dA$ where R is bounded by $x^2 + y^2 = 4$

** **34.** $\iint_R (6 - x - 2y)\, dA$ where R is bounded by $x^2 + y^2 = 4$

** **35.** $\displaystyle\iint_R 6x^5\, dA$ where R is the region under $x + 5y = 16$, above $y = x - 4$, and bounded by $x = (y - 2)^2$

** **36.** $\displaystyle\iint_R ye^x\, dA$ where R is bounded by $y = x$, $x + y = 2$, $y = 0$

** **37.** $\displaystyle\iint_R \sqrt{1 + y}\, dA$ where R is bounded by $x = -1$, $y = 2$, $x = y^3$

** **38.** $\displaystyle\iint_R y\sqrt{x^2 + y^2}\, dA$ where R is bounded by $y = x$, $x = -1$, $y = 0$

** **39.** $\displaystyle\iint_R (x^2 + y^2)\, dA$ where R is bounded by $x^2 + y^2 = 9$

** **40.** Evaluate the double iterated integral $\displaystyle\int_0^1 \int_0^1 |x - y|\, dy\, dx$.

** **41.** Evaluate the double integral $\displaystyle\iint_R |y - 2x^2 + 1|\, dA$ where R is the square bounded by $x = \pm 1$, $y = \pm 1$.

14.3 Areas and Volumes of Solids of Revolution

Because equation 14.7 represents the area of a region R as a double integral, and double integrals are evaluated by means of double iterated integrals, it follows that areas can be calculated using double iterated integrals. In particular, to find area of the region in Figure 14.10, we subdivide R into rectangles of dimensions dx and dy and therefore of area $dA = dy\, dx$. Areas of these rectangles are then added in the y-direction to give the area of a vertical strip

$$\int_{g(x)}^{h(x)} dy\, dx,$$

where limits indicate that every vertical strip starts on the curve $y = g(x)$ and ends on the curve $y = h(x)$. Finally, areas of the vertical strips are added together to give the total area:

$$\text{area} = \int_a^b \int_{g(x)}^{h(x)} dy\, dx,$$

where a and b indicate the x-positions of first and last strips.

FIGURE 14.10

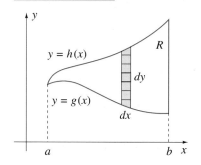

EXAMPLE 14.4

Find the area bounded by the curves $xy = 2$, $x = 2\sqrt{y}$, $y = 4$.

SOLUTION   If we choose horizontal strips for this area (Figure 14.11), we have

$$\text{area} = \int_1^4 \int_{2/y}^{2\sqrt{y}} dx \, dy = \int_1^4 \left\{ 2\sqrt{y} - \frac{2}{y} \right\} dy$$

$$= \left\{ \frac{4}{3} y^{3/2} - 2 \ln |y| \right\}_1^4 = \frac{28}{3} - 2 \ln 4.$$

For vertical strips (Figure 14.12), we require two iterated integrals because, to the left of the line $x = 2$, strips begin on the hyperbola $xy = 2$, whereas, to the right of $x = 2$, they begin on the parabola $x = 2\sqrt{y}$. We obtain

$$\text{area} = \int_{1/2}^2 \int_{2/x}^4 dy \, dx + \int_2^4 \int_{x^2/4}^4 dy \, dx = \int_{1/2}^2 \left\{ 4 - \frac{2}{x} \right\} dx + \int_2^4 \left\{ 4 - \frac{x^2}{4} \right\} dx$$

$$= \{ 4x - 2 \ln |x| \}_{1/2}^2 + \left\{ 4x - \frac{x^3}{12} \right\}_2^4 = \frac{28}{3} - 2 \ln 4.$$

FIGURE 14.11

FIGURE 14.12

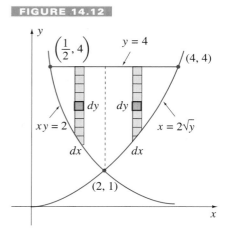

If we compare finding areas by definite integrals (Section 7.1) and finding the same areas by double integrals, it is clear that no great advantage is derived by using double integrals. In fact, it is probably more work because we must perform two, rather than one, integrations, although the first integration is trivial. The advantage of double integrals is therefore not in finding area; it is in finding volumes of solids of revolution, centres of mass, moments of inertia, and fluid forces, among other applications.

Volumes of Solids of Revolution

If the area in Figure 14.13 is rotated around the x-axis, the volume of the resulting solid of revolution can be evaluated by using the washer method introduced in Section 7.2:

$$\text{volume} = \int_a^b \{ \pi [h(x)]^2 - \pi [g(x)]^2 \} dx. \tag{14.17}$$

If this area is rotated around the y-axis, the volume generated is calculated by using the cylindrical shell method:

$$\text{volume} = \int_a^b 2\pi x[h(x) - g(x)]dx. \tag{14.18}$$

Thus, once we have chosen to use vertical rectangles, the axis of revolution determines whether we use either washers or cylindrical shells. We now show that with double integrals one method works for all problems.

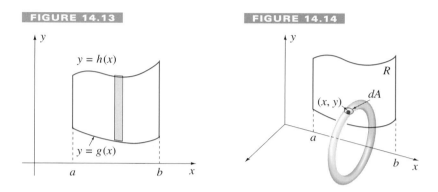

FIGURE 14.13

FIGURE 14.14

To rotate this area around the x-axis we subdivide R into small areas dA (Figure 14.14). If the area dA at a point (x, y) is rotated about the x-axis, it generates a "ring" with cross-sectional area dA. Since (x, y) travels a distance $2\pi y$ in traversing the ring, it follows that the volume in the ring is approximately $2\pi y\, dA$. To find the total volume obtained by rotating R about the x-axis, we add the volumes of all such rings and take the limit as the areas shrink to points. But this is what we mean by the double integral of $2\pi y$ over the region R, and we therefore write

$$\text{volume} = \iint_R 2\pi y\, dA. \tag{14.19}$$

On the other hand, if dA is rotated about the y-axis, it again forms a ring, but with approximate volume $2\pi x\, dA$. The total volume, then, when R is rotated about the y-axis is:

$$\text{volume} = \iint_R 2\pi x\, dA. \tag{14.20}$$

Since double iterated integrals are used to evaluate double integrals, it follows that we can set up double iterated integrals to find volumes represented by equations 14.19 and 14.20. The decision to use a double iterated integral with respect to y and x implies a subdivision of R into rectangles of dimensions dx and dy (Figure 14.15). The volume of the ring formed when this rectangle is rotated around the x-axis is $2\pi y\, dy\, dx$. If we choose to integrate first with respect to y, we are adding over all rectangles in a vertical strip

$$\int_{g(x)}^{h(x)} 2\pi y\, dy\, dx,$$

where limits indicate that all vertical strips start on the curve $y = g(x)$ and end on the curve $y = h(x)$. This integral is the volume generated by rotating the vertical strip around the x-axis. Integration now with respect to x adds over all strips to give the required volume

$$\text{volume} = \int_a^b \int_{g(x)}^{h(x)} 2\pi y\, dy\, dx. \tag{14.21}$$

FIGURE 14.15

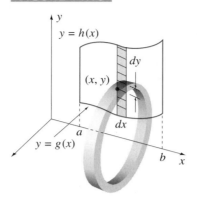

Note that, when we actually do perform the inner integration, we get

$$\text{volume} = \int_a^b \{\pi y^2\}_{g(x)}^{h(x)} dx = \int_a^b \{\pi[h(x)]^2 - \pi[g(x)]^2\}dx,$$

and this is the result contained in equation 14.17.

When R is rotated around the y-axis, the rectangular area $dy\,dx$ generates a ring of volume $2\pi x\,dy\,dx$. Addition over the rectangles in a vertical strip

$$\int_{g(x)}^{h(x)} 2\pi x\,dy\,dx$$

gives the volume generated by rotating the strip about the y-axis. Finally, integration with respect to x adds over all strips to give

$$\text{volume} = \int_a^b \int_{g(x)}^{h(x)} 2\pi x\,dy\,dx. \tag{14.22}$$

This time the inner integration leads to

$$\text{volume} = \int_a^b \{2\pi xy\}_{g(x)}^{h(x)} dx = \int_a^b 2\pi x[h(x) - g(x)]dx,$$

the same result as in equation 14.18.

The advantage, then, in using double integrals to find volumes of solids of revolution is that it requires only one idea, that of rings. The first integration leads to washers or cylindrical shells, but we need never think about this.

EXAMPLE 14.5

Find volumes of the solids of revolution if the area bounded by the curves $y = 2x - x^2$, $y = x^2 - 2x$ is rotated around:

    (a) the y-axis;

    (b) $x = -3$;

    (c) $y = 2$.

SOLUTION

(a) If we use vertical strips (Figure 14.16), then

$$\text{volume} = \int_0^2 \int_{x^2-2x}^{2x-x^2} 2\pi x \, dy \, dx = 2\pi \int_0^2 \{xy\}_{x^2-2x}^{2x-x^2} dx$$

$$= 2\pi \int_0^2 x\{(2x - x^2) - (x^2 - 2x)\}dx$$

$$= 4\pi \int_0^2 (2x^2 - x^3)dx = 4\pi \left\{ \frac{2x^3}{3} - \frac{x^4}{4} \right\}_0^2 = \frac{16\pi}{3}.$$

FIGURE 14.16

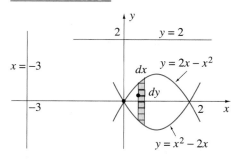

(b) In this case the radius of the ring formed by rotating the rectangle about $x = -3$ is $x + 3$, and therefore

$$\text{volume} = \int_0^2 \int_{x^2-2x}^{2x-x^2} 2\pi (x + 3)dy \, dx = 2\pi \int_0^2 \{y(x + 3)\}_{x^2-2x}^{2x-x^2} dx$$

$$= 2\pi \int_0^2 (x + 3)(4x - 2x^2)dx = 4\pi \int_0^2 (6x - x^2 - x^3)dx$$

$$= 4\pi \left\{ 3x^2 - \frac{x^3}{3} - \frac{x^4}{4} \right\}_0^2 = \frac{64\pi}{3}.$$

(c) When the rectangle is rotated around $y = 2$, the radius of the ring is $2 - y$, and hence

$$\text{volume} = \int_0^2 \int_{x^2-2x}^{2x-x^2} 2\pi (2 - y)dy \, dx = 2\pi \int_0^2 \left\{ -\frac{1}{2}(2 - y)^2 \right\}_{x^2-2x}^{2x-x^2} dx$$

$$= \pi \int_0^2 \{(2 - x^2 + 2x)^2 - (2 - 2x + x^2)^2\}dx = 8\pi \int_0^2 (2x - x^2)dx$$

$$= 8\pi \left\{ x^2 - \frac{x^3}{3} \right\}_0^2 = \frac{32\pi}{3}.$$

In Exercises 1–10 use a double integral to find the area bounded by the curves.

* **1.** $y = 4x^2$, $x = 4y^2$ * **2.** $y = x^2$, $y = 5x + 6$

* **3.** $x = y^2$, $x = 3y - 2$ * **4.** $y = x^3 + 8$, $y = 4x + 8$

* **5.** $y = 4/x^2$, $y = 5 - x^2$ * **6.** $y = xe^{-x}$, $y = x$, $x = 2$

* **7.** $x = 4y - 4y^2$, $y = x - 3$, $y = 1$, $y = 0$

* **8.** $x = y(y - 2)$, $x + y = 12$

* **9.** $y = x^3 - x^2 - 2x + 2$, $y = 2$

* **10.** $x + y = 1$, $x + y = 5$, $y = 2x + 1$, $y = 2x + 6$

In Exercises 11–20 use a double integral to find the volume of the solid of revolution obtained by rotating the area bounded by the curves around the line.

* **11.** $y = -\sqrt{4 - x}$, $x = 0$, $y = 0$ about $y = 0$

* **12.** $4x^2 + 9y^2 = 36$ about $y = 0$

* **13.** $y = (x - 1)^2$, $y = 1$ about $x = 0$

* **14.** $y = x^2 + 4$, $y = 2x^2$ about $y = 0$

* **15.** $x - 1 = y^2$, $x = 5$ about $x = 1$

* **16.** $x = y^3$, $y = \sqrt{2 - x}$, $y = 0$ about $y = 1$

* **17.** $y = 4x^2 - 4x$, $y = x^3$ about $y = -2$

* **18.** $x = 3y - y^2$, $x = y^2 - 3y$ about $y = 4$

* **19.** $x = 2y - y^2 - 2$, $x = -5$ about $x = 1$

* **20.** $x + y = 4$, $y = 2\sqrt{x - 1}$, $y = 0$ about $y = -1$

In Exercises 21–30 use a double integral to find the area bounded by the curves.

** **21.** $y = 2x^3$, $y = 4x + 8$, $y = 0$

** **22.** $y = x/\sqrt{x + 3}$, $x = 1$, $x = 6$, $y = -x^2$

** **23.** $x = y^2 + 2$, $x = -(y - 4)^2$, $y = 4 - x$, $y = 0$

** **24.** $y = x^3 - x$, $x + y + 1 = 0$, $x = \sqrt{y + 1}$

** **25.** $y^2 = x^2(4 - x^2)$

** **26.** $x^2 + y^2 = 4$, $x^2 + y^2 = 4x$ (interior to both)

** **27.** $x = 1/\sqrt{4 - y^2}$, $4x + y^2 = 0$, $y + 1 = 0$, $y - 1 = 0$

** **28.** $y^2 = x^4(9 + x)$

** **29.** $y = (x^2 + 1)/(x + 1)$, $x + 3y = 7$

** **30.** $(x + 2)^2 y = 4 - x$, $x = 0$, $y = 0$ ($x \geq 0$, $y \geq 0$)

In Exercises 31–35 use a double integral to find the volume of the solid of revolution obtained by rotating the area bounded by the curves around the line.

** **31.** $y = 4/(x^2 + 1)^2$, $y = 1$ about $x = 0$

** **32.** $y = x^2 - 2$, $y = 0$ about $y = -1$

** **33.** $y = |x^2 - 1|$, $x = -2$, $x = 2$, $y = -1$ about $y = -2$

** **34.** $x = \sqrt{4 + 12y^2}$, $x - 20y = 24$, $y = 0$ about $y = 0$

** **35.** $y = (x + 1)^{1/4}$, $y = -(x + 1)^2$, $x = 0$ about $x = 0$

** **36.** Find the area common to the two circles $x^2 + y^2 = 4$ and $x^2 + y^2 = 6x$.

In Exercises 37–40 find the volume of the solid of revolution obtained by rotating the area bounded by the curves around the line. (Hint: Use distance formula 1.7.)

** **37.** $x = 1$, $y = 1$, $x = 0$, $y = 0$ about $x + y = 2$

** **38.** $y = x^2$, $y = 2x + 3$ about $y = 2x + 3$

** **39.** $y = \sqrt{x}$, $y = 0$, $x = 1$ about $y = 3x + 2$

** **40.** $x = 2y$, $y = x - 1$, $y = 0$ about $x + y + 1 = 0$

*** **41.** Prove that the area above the line $y = h$ and under the circle $x^2 + y^2 = r^2$ $(r > h)$ is given by

$$A = \pi r^2/2 - h\sqrt{r^2 - h^2} - r^2 \mathrm{Sin}^{-1}(h/r).$$

14.4 Fluid Pressure

In Section 7.6 we defined pressure at a point in a fluid as the magnitude of the force per unit area that would act on any surface placed at that point. We discovered that at a depth $d > 0$ below the surface of a fluid, pressure is given by

$$P = 9.81\rho d, \tag{14.23}$$

where ρ is the density of the fluid. With these ideas and the definite integral, we were able to calculate fluid forces on flat surfaces in the fluid. In particular, the magnitude of the total force on each side of the vertical surface in Figure 14.17 is given by the definite integral:

$$\text{force} = \int_a^b -9.81\rho y[h(y) - g(y)]dy. \tag{14.24}$$

Although horizontal rectangles are convenient for this problem, it is clear that they are not reasonable for the surface in Figure 14.18.

FIGURE 14.17

FIGURE 14.18

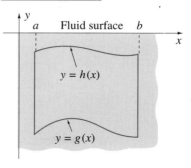

Double integrals, on the other hand, can be applied with equal ease to both surfaces. To see this we consider, first, force on each side of the surface in Figure 14.19. If the surface is divided into small areas dA, then the force on dA is $P\,dA$, where P is pressure at that depth. Total force on R is the sum of the forces on all such areas in R as areas dA shrink to a point. But once again this is the double integral, and we therefore write

$$\text{force} = \iint_R P\,dA. \tag{14.25}$$

To set up a double iterated integral in order to evaluate this double integral, we use our interpretation of the double iterated integral as a limit of a sum in which the areas dA have been chosen as rectangles. In particular, for the surface in Figure 14.17 we draw rectangles of dimensions dx and dy, as shown in Figure 14.20. The force on this rectangle is its area $dx\,dy$ multiplied by pressure $-9.81\rho y$ at that depth, $-9.81\rho y\,dx\,dy$. Addition of these quantities over all rectangles in a horizontal strip gives the force on the strip,

$$\int_{g(y)}^{h(y)} -9.81\rho y\,dx\,dy,$$

where the limits indicate that all horizontal strips start on the curve $x = g(y)$ and end on the curve $x = h(y)$. Integration with respect to y now adds over all horizontal strips to give the total force on the surface:

$$\text{force} = \int_a^b \int_{g(y)}^{h(y)} -9.81\rho y\,dx\,dy. \tag{14.26}$$

When we perform the inner integration, we obtain

$$\text{force} = \int_a^b \{-9.81\rho y x\}_{g(y)}^{h(y)}\,dy = \int_a^b -9.81\rho y[h(y) - g(y)]\,dy,$$

and this is the result contained in equation 14.24.

FIGURE 14.19

FIGURE 14.20

FIGURE 14.21

For the surface in Figure 14.18 we again draw rectangles of dimensions dx and dy and calculate the force on such a rectangle, $-9.81\rho y\,dy\,dx$. In this case it is more convenient to add over rectangles in a vertical strip to give the force on the strip (Figure 14.21):

$$\int_{g(x)}^{h(x)} -9.81\rho y\,dy\,dx.$$

Force on the entire surface can now be found by adding over all vertical strips:

$$\text{force} = \int_a^b \int_{g(x)}^{h(x)} -9.81\rho y \, dy \, dx. \tag{14.27}$$

EXAMPLE 14.6 ▼

The face of a dam is parabolic with breadth 100 m and height 50 m. Find the magnitude of the total force due to fluid pressure on the face.

SOLUTION   If we use the coordinate system in Figure 14.22, then the edge of the dam has an equation of the form $y = kx^2$. Since (50, 50) is a point on this curve, it follows that $k = \frac{1}{50}$. Because force on the left half of the dam is the same as that on the right half, we can integrate for the right half and double the result; that is,

$$\text{force} = 2 \int_0^{50} \int_{x^2/50}^{50} 9.81(1000)(50 - y) \, dy \, dx$$

$$= 19\,620 \int_0^{50} \left\{ 50y - \frac{y^2}{2} \right\}_{x^2/50}^{50} dx = 19\,620 \int_0^{50} \left\{ 1250 - x^2 + \frac{x^4}{5000} \right\} dx$$

$$= 19\,620 \left\{ 1250x - \frac{x^3}{3} + \frac{x^5}{25\,000} \right\}_0^{50} = 6.54 \times 10^8 \text{N}.$$

FIGURE 14.22

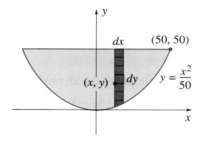

EXAMPLE 14.7 ▼

A tank in the form of a right circular cylinder of radius $\frac{1}{2}$ m and length 10 m has its axis horizontal. If it is full of water, find the force due to water pressure on each end of the tank.

SOLUTION   Since the force on that part of the end to the left of the y-axis (Figure 14.23) is identical to the force on that part to the right, we double the force on the right half, i.e.,

$$\text{force} = 2 \int_{-1/2}^{1/2} \int_0^{\sqrt{1/4-y^2}} 9.81(1000)(\tfrac{1}{2} - y) \, dx \, dy$$

$$= 9810 \int_{-1/2}^{1/2} \int_0^{\sqrt{1/4-y^2}} dx \, dy - 19\,620 \int_{-1/2}^{1/2} \int_0^{\sqrt{1/4-y^2}} y \, dx \, dy.$$

To calculate moments of inertia (second moments) of thin plates about lines parallel to the x- and y-axes is as easy as calculating first moments if we use double integrals. In particular, the mass in area dA in Figure 14.24 is $\rho\,dA$; thus its moments of inertia about the x- and y-axes are, respectively, $y^2\rho\,dA$ and $x^2\rho\,dA$. Moments of inertia of the entire plate about the x- and y-axes are therefore given by the double integrals:

$$I_x = \iint_R y^2\rho\,dA \quad \text{and} \quad I_y = \iint_R x^2\rho\,dA. \tag{14.34}$$

For a plate such as that shown in Figure 14.28, we evaluate these double integrals by means of double iterated integrals with respect to x and y:

$$I_x = \int_a^b \int_{g(y)}^{h(y)} y^2\rho\,dx\,dy, \tag{14.35}$$

$$I_y = \int_a^b \int_{g(y)}^{h(y)} x^2\rho\,dx\,dy. \tag{14.36}$$

FIGURE 14.28

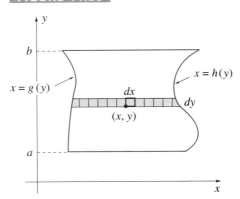

EXAMPLE 14.10

Find the moment of inertia about the x-axis of a thin plate with constant mass per unit area ρ if its edges are defined by the curves $y = x^3$, $y = \sqrt{2 - x}$, and $x = 0$.

SOLUTION   With vertical strips as shown in Figure 14.29, the moment of inertia about the x-axis is

$$\int_0^1 \int_{x^3}^{\sqrt{2-x}} y^2\rho\,dy\,dx = \rho\int_0^1 \left\{\frac{y^3}{3}\right\}_{x^3}^{\sqrt{2-x}} dx = \frac{\rho}{3}\int_0^1 \{(2-x)^{3/2} - x^9\}dx$$

$$= \frac{\rho}{3}\left\{-\frac{2}{5}(2-x)^{5/2} - \frac{x^{10}}{10}\right\}_0^1 = \frac{16\sqrt{2}-5}{30}\rho.$$

FIGURE 14.29

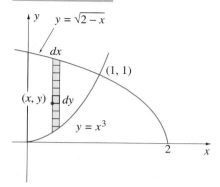

EXAMPLE 14.11

Find the second moment of area about the line $y = -1$ of the area bounded by the curves $x = y^2$ and $x = 2y$.

SOLUTION   The second moment of area about the line $y = -1$ (Figure 14.30) is

$$\int_0^2 \int_{y^2}^{2y} (y+1)^2 dx\, dy = \int_0^2 \{x(y+1)^2\}_{y^2}^{2y} dy = \int_0^2 (y+1)^2 (2y - y^2) dy$$

$$= \int_0^2 (-y^4 + 3y^2 + 2y) dy = \left\{ -\frac{y^5}{5} + y^3 + y^2 \right\}_0^2$$

$$= \frac{28}{5}.$$

FIGURE 14.30

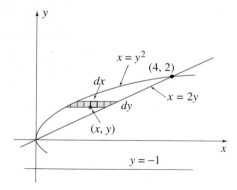

In Exercises 1–10 find the centroid of the area bounded by the curves.

* **1.** $x = y + 2$, $x = y^2$

* **2.** $y = 8 - 2x^2$, $y + x^2 = 4$

* **3.** $y = x^2 - 1$, $y + (x + 1)^2 = 0$

* **4.** $x + y = 5$, $xy = 4$

* **5.** $y = e^x$, $y = 0$, $x = 0$, $x = 1$

* **6.** $y = \sqrt{4 - x^2}$, $y = x$, $x = 0$

* **7.** $y = 1/(x - 1)$, $y = 1$, $y = 2$, $x = 0$

* **8.** $x = 4y - 4y^2$, $x = y + 3$, $y = 1$, $y = 0$

* **9.** $y = |x^2 - 1|$, $y = 2$

* **10.** $y = x$, $y = 2x$, $2y = x + 3$

In Exercises 11–15 find the second moment of area of the region bounded by the curves about the line.

* **11.** $y = x^2$, $y = x^3$ about the y-axis

* **12.** $y = x$, $y = 2x + 4$, $y = 0$ about the x-axis

* **13.** $y = x^2$, $2y = x^2 + 4$ about $y = 0$

* **14.** $y = x^2 - 4$, $y = 2x - x^2$ about $x = -2$

* **15.** $x = 1/\sqrt{y^4 + 12y^2}$, $x = 0$, $y = 1/2$, $y = 1$ about $y = 0$

* **16.** Find the first moment about the line $y = -2$ of a thin plate of constant mass per unit area ρ if its edges are defined by the curves $y = 2 - 2x^2$ and $y = x^2 - 1$.

In Exercises 17–23 find the centroid of the area bounded by the curves.

** **17.** $x = \sqrt{y + 2}$, $y = x$, $y = 0$

** **18.** $y + x^2 = 0$, $x = y + 2$, $x + y + 2 = 0$, $y = 2$ (above $y + x^2 = 0$)

** **19.** $y^2 = x^4(1 - x^2)$ (right loop)

** **20.** $3x^2 + 4y^2 = 48$, $(x - 2)^2 + y^2 = 1$

** **21.** $y = \ln x$, $y + \sqrt{x - 1} = 0$, $x = 2$

** **22.** $y = \sqrt{2 - x}$, $15y = x^2 - 4$

** **23.** $y = x\sqrt{1 - x^2}$, $x \geq 0$ and the x-axis

** **24.** Find the moment of inertia of a uniform rectangular plate a units long and b units wide about a line through the centre of the plate and perpendicular to the plate.

In Exercises 25–27 find the second moment of area of the region bounded by the curves about the line.

** **25.** $4x^2 + 9y^2 = 36$ about $y = -2$

** **26.** $x = y^2$, $x + y = 2$ about $x = -1$

** **27.** $y = \sqrt{a^2 - x^2}$, $y = a$, $x = a$ $(a > 0)$ about the x-axis

** **28.** Find the first moment of area about the line $x + y = 1$ for the area bounded by the curves $x = y^2 - 2$ and $y = x$.

** **29.** Prove the *theorem of Pappus*: If a plane area is revolved about a coplanar axis not crossing the area, the volume generated is equal to the product of the area and the circumference of the circle described by the centroid of the area.

** **30.** A thin flat plate of area A is immersed vertically in a fluid with density ρ. Show that the total force (due to fluid pressure) on each side of the plate is equal to the product of 9.81, A, ρ, and the depth of the centroid of the plate below the surface of the fluid. Use this result to find the forces in some of the problems in Exercises 14.4, say 1, 3, 4, 5, 7, 11, 12, 13, 16, and 17 (page 1090). For those problems involving triangles recall the result of Exercise 36 in Section 7.7 or Exercise 39 in Section 12.3.

** **31.** Prove the *parallel axis theorem*: The moment of inertia of a thin plate (with constant mass per unit area) with respect to any coplanar line is equal to the moment of inertia with respect to the parallel line through the centre of mass plus the mass multiplied by the square of the distance between the lines.

14.6 Surface Area

To find the length of a curve in Section 7.3 we approximated the curve by tangent line segments. To find the area of a surface we follow a similar procedure by approximating the surface with tangential planes. In particular, consider finding the area of a smooth surface S given that every vertical line that intersects the surface does so in exactly one point (Figure 14.31). If S_{xy} is the area in the xy-plane onto which S projects, we divide S_{xy} into n sub-areas ΔA_i in any fashion whatsoever, and choose a point (x_i, y_i) in each ΔA_i. At the point (x_i, y_i, z_i) on the surface S that projects onto (x_i, y_i), we draw the tangent plane to S. Suppose we now project ΔA_i upward onto S and onto the tangent plane at (x_i, y_i, z_i) and denote these projected areas by ΔS_i and ΔS_{Ti}, respectively.

FIGURE 14.31

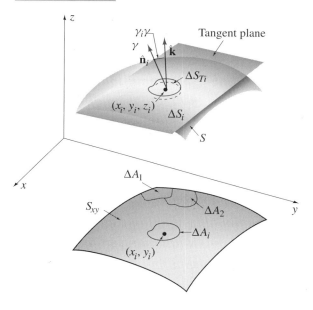

Now ΔS_{Ti} is an approximation to ΔS_i and, as long as ΔA_i is small, a reasonably good approximation. In fact, the smaller ΔA_i, the better the approximation. We therefore define the area of S as

$$\text{area of } S = \lim_{n\to\infty} \sum_{i=1}^{n} \Delta S_{Ti}, \qquad (14.37)$$

where in taking the limit we demand that each ΔA_i shrink to a point. We have therefore defined area on a curved surface in terms of flat areas on tangent planes to the surface. The advantage of this definition is that we can calculate ΔS_{Ti} in terms of ΔA_i. To see how, we denote by $\hat{\mathbf{n}}_i$

the unit normal vector to S at (x_i, y_i, z_i) with positive z-component, and by γ_i the acute angle between $\hat{\mathbf{n}}_i$ and $\hat{\mathbf{k}}$. Now ΔS_{Ti} projects onto ΔA_i, and γ_i is the acute angle between the planes containing ΔA_i and ΔS_{Ti} (Figure 14.32). It follows that ΔA_i and ΔS_{Ti} are related by

$$\Delta A_i = \cos \gamma_i \Delta S_{Ti} \tag{14.38}$$

(see Exercise 28 in Section 12.5). Note that, if ΔS_{Ti} is horizontal, then $\gamma_i = 0$ and $\Delta S_{Ti} = \Delta A_i$; and if ΔS_{Ti} tends toward the vertical ($\gamma_i \to \pi/2$), then ΔS_{Ti} becomes very large for fixed ΔA_i.

FIGURE 14.32

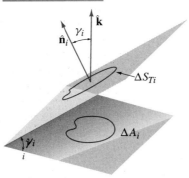

Because the surface projects in a one-to-one fashion onto the area S_{xy} in the xy-plane, we can take the equation for S in the form $z = f(x, y)$. A vector normal to S at any point is therefore

$$\nabla(z - f(x, y)) = \left(-\frac{\partial f}{\partial x}, -\frac{\partial f}{\partial y}, 1\right);$$

hence

$$\hat{\mathbf{n}} = \frac{\left(-\dfrac{\partial f}{\partial x}, -\dfrac{\partial f}{\partial y}, 1\right)}{\sqrt{1 + \left(\dfrac{\partial f}{\partial x}\right)^2 + \left(\dfrac{\partial f}{\partial y}\right)^2}} = \frac{\left(-\dfrac{\partial z}{\partial x}, -\dfrac{\partial z}{\partial y}, 1\right)}{\sqrt{1 + \left(\dfrac{\partial z}{\partial x}\right)^2 + \left(\dfrac{\partial z}{\partial y}\right)^2}}.$$

Since $\hat{\mathbf{n}}_i \cdot \hat{\mathbf{k}} = |\hat{\mathbf{n}}_i||\hat{\mathbf{k}}| \cos \gamma_i = \cos \gamma_i$, it follows that

$$\cos \gamma_i = \hat{\mathbf{n}}_i \cdot \hat{\mathbf{k}} = \frac{1}{\sqrt{1 + z_x^2(x_i, y_i) + z_y^2(x_i, y_i)}}.$$

When we substitute this expression into 14.38, we obtain the result that area ΔS_{Ti} on the tangent plane to $z = f(x, y)$ at the point (x_i, y_i, z_i) is related to its projection ΔA_i in the xy-plane according to

$$\Delta S_{Ti} = \sqrt{1 + z_x^2(x_i, y_i) + z_y^2(x_i, y_i)}\, \Delta A_i. \tag{14.39}$$

Definition 14.37 for the area of S can now be written in the form

$$\text{area of } S = \lim_{\|\Delta A_i\| \to 0} \sum_{i=1}^{n} \sqrt{1 + z_x^2(x_i, y_i) + z_y^2(x_i, y_i)}\, \Delta A_i, \tag{14.40}$$

where the summation is carried out over all areas ΔA_i in S_{xy} as each ΔA_i shrinks to a point. But this is the definition of the double integral of the function $\sqrt{1 + (\partial z/\partial x)^2 + (\partial z/\partial y)^2}$ over the region S_{xy}. In other words, on the basis of definition 14.37, areas on surfaces can be calculated according to

$$\text{area of } S = \iint_{S_{xy}} \sqrt{1 + \left(\frac{\partial z}{\partial x}\right)^2 + \left(\frac{\partial z}{\partial y}\right)^2}\, dA. \tag{14.41}$$

Note the analogy between equations 14.41 and 7.15. In equation 7.15 we think of $\sqrt{1 + (dy/dx)^2}\,dx$ as a small length along a curve $C: y = f(x)$ that projects onto the length dx along the x-axis. In fact, $\sqrt{1 + (dy/dx)^2}\,dx$ is along the tangent line to C, but we think of it as along C itself (see Section 12.9). The total length of C is then found by adding over all projections of C from $x = a$ to $x = b$. Similarly, in equation 14.41 we think of $\sqrt{1 + (\partial z/\partial x)^2 + (\partial z/\partial y)^2}\,dA$ as a small area on a surface $S: z = f(x, y)$ that projects onto the area dA in the xy-plane. It is, in fact, a small area on the tangent plane to S, but it is usually easier to think of it as being on S itself. The total area of S is then the addition over all projections S_{xy} of S.

Note too that when S is smooth, $\partial z/\partial x$ and $\partial z/\partial y$ are continuous, thus guaranteeing existence of double integral 14.41. If S is piecewise smooth, we divide it into smooth subsurfaces and integrate over each piece separately.

This discussion has been based on the assumption that S projects one-to-one onto some region S_{xy} in the xy-plane. If this condition is not met, one possibility is to subdivide S into parts, each of which projects one-to-one onto the xy-plane. The total area of S is then the sum of the areas of its parts. For instance, to find the area of the surface S in Figure 14.33, we could subdivide S into S_1 and S_2 along the curve C. The area of S is then the sum of the areas of S_1 and S_2, each of which projects one-to-one onto the xy-plane.

FIGURE 14.33

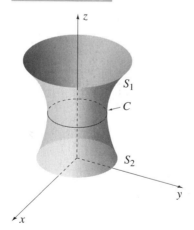

Alternatively, we could note that there is nothing sacred about projecting S onto the xy-plane. We could develop similar results if S projects one-to-one onto either the yz- or the xz-plane. If S_{yz} and S_{xz} represent these projections, then

$$\text{area of } S = \iint_{S_{yz}} \sqrt{1 + \left(\frac{\partial x}{\partial y}\right)^2 + \left(\frac{\partial x}{\partial z}\right)^2}\, dA, \tag{14.42a}$$

$$\text{area of } S = \iint_{S_{xz}} \sqrt{1 + \left(\frac{\partial y}{\partial x}\right)^2 + \left(\frac{\partial y}{\partial z}\right)^2}\, dA. \tag{14.42b}$$

EXAMPLE 14.12

Find the area of that part of the plane $x + 2y + 3z = 6$ in the first octant.

SOLUTION   This area projects one-to-one onto the triangular area S_{xy} in the xy-plane bounded by the lines $x = 0$, $y = 0$, $x + 2y = 6$ (Figure 14.34). Since $z = 2 - x/3 - 2y/3$,

$$\text{area} = \iint_{S_{xy}} \sqrt{1 + \left(\frac{\partial z}{\partial x}\right)^2 + \left(\frac{\partial z}{\partial y}\right)^2} \, dA = \iint_{S_{xy}} \sqrt{1 + \left(-\frac{1}{3}\right)^2 + \left(-\frac{2}{3}\right)^2} \, dA$$

$$= \frac{\sqrt{14}}{3} \iint_{S_{xy}} dA = \frac{\sqrt{14}}{3} \, (\text{area of } S_{xy}) = \frac{\sqrt{14}}{3} \left[\frac{1}{2}(3)(6)\right] = 3\sqrt{14}.$$

FIGURE 14.34

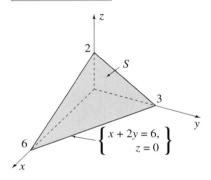

$$\left\{ \begin{matrix} x + 2y = 6, \\ z = 0 \end{matrix} \right\}$$

We also could use formula 12.42.

EXAMPLE 14.13

Find the area of the surface $z = x^{3/2}$ that projects onto the rectangle in the xy-plane bounded by the straight lines $x = 0$, $x = 2$, $y = 1$, and $y = 3$.

SOLUTION   Since the surface projects one-to-one onto the rectangle (Figure 14.35), we find that

$$\text{area} = \iint_{S_{xy}} \sqrt{1 + \left(\frac{\partial z}{\partial x}\right)^2 + \left(\frac{\partial z}{\partial y}\right)^2} \, dA = \iint_{S_{xy}} \sqrt{1 + \left(\frac{3}{2} x^{1/2}\right)^2} \, dA$$

$$= \frac{1}{2} \int_0^2 \int_1^3 \sqrt{4 + 9x} \, dy \, dx = \frac{1}{2} \int_0^2 \{y\sqrt{4 + 9x}\}_1^3 dx = \int_0^2 \sqrt{4 + 9x} \, dx$$

$$= \left\{\frac{2}{27}(4 + 9x)^{3/2}\right\}_0^2 = \frac{2}{27}(22\sqrt{22} - 8).$$

FIGURE 14.35

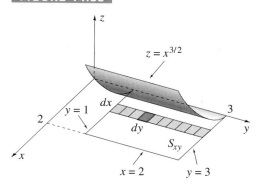

EXAMPLE 14.14

Find the area of the cone $y = \sqrt{x^2 + z^2}$ to the left of the plane $y = 1$.

SOLUTION   **Method 1**   The surface projects one-to-one onto the interior of the circle $x^2 + z^2 = 1$ in the xz-plane (Figure 14.36). Since the area of the surface is four times that in the first octant, if we let S_{xz} be the quarter-circle $x^2 + z^2 \leq 1$, $x \geq 0$, $z \geq 0$, then

$$
\begin{aligned}
\text{area} &= 4 \iint_{S_{xz}} \sqrt{1 + \left(\frac{\partial y}{\partial x}\right)^2 + \left(\frac{\partial y}{\partial z}\right)^2} \, dA \\
&= 4 \iint_{S_{xz}} \sqrt{1 + \left(\frac{x}{\sqrt{x^2 + z^2}}\right)^2 + \left(\frac{z}{\sqrt{x^2 + z^2}}\right)^2} \, dA \\
&= 4 \iint_{S_{xz}} \sqrt{1 + \frac{x^2}{x^2 + z^2} + \frac{z^2}{x^2 + z^2}} \, dA = 4 \iint_{S_{xz}} \sqrt{2} \, dA \\
&= 4\sqrt{2}(\text{area of } S_{xz}) = 4\sqrt{2}\{\tfrac{1}{4}\pi(1)^2\} = \sqrt{2}\pi.
\end{aligned}
$$

FIGURE 14.36

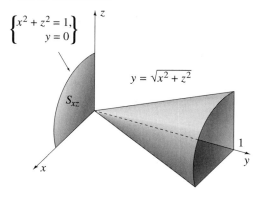

Method 2   Suppose instead that we project that part of the surface in the first octant onto the triangle S_{xy} in Figure 14.37. We write the equation of this part of the surface in the form $z = \sqrt{y^2 - x^2}$ and calculate

$$
\begin{aligned}
\text{area} &= 4 \iint_{S_{xy}} \sqrt{1 + \left(\frac{\partial z}{\partial x}\right)^2 + \left(\frac{\partial z}{\partial y}\right)^2}\, dA \\
&= 4 \iint_{S_{xy}} \sqrt{1 + \left(\frac{-x}{\sqrt{y^2 - x^2}}\right)^2 + \left(\frac{y}{\sqrt{y^2 - x^2}}\right)^2}\, dA \\
&= 4 \iint_{S_{xy}} \sqrt{1 + \frac{x^2}{y^2 - x^2} + \frac{y^2}{y^2 - x^2}}\, dA = 4\sqrt{2} \iint_{S_{xy}} \frac{y}{\sqrt{y^2 - x^2}}\, dA.
\end{aligned}
$$

To evaluate this double integral, it is advantageous to integrate first with respect to y:

$$
\begin{aligned}
\text{area} &= 4\sqrt{2} \int_0^1 \int_x^1 \frac{y}{\sqrt{y^2 - x^2}}\, dy\, dx = 4\sqrt{2} \int_0^1 \left\{\sqrt{y^2 - x^2}\right\}_x^1 dx \\
&= 4\sqrt{2} \int_0^1 \sqrt{1 - x^2}\, dx.
\end{aligned}
$$

If we now set $x = \sin\theta$, then $dx = \cos\theta\, d\theta$, and

$$
\begin{aligned}
\text{area} &= 4\sqrt{2} \int_0^{\pi/2} \cos\theta \cos\theta\, d\theta = 4\sqrt{2} \int_0^{\pi/2} \left\{\frac{1 + \cos 2\theta}{2}\right\} d\theta \\
&= 2\sqrt{2} \left\{\theta + \frac{1}{2}\sin 2\theta\right\}_0^{\pi/2} = \sqrt{2}\pi.
\end{aligned}
$$

FIGURE 14.37

FIGURE 14.42

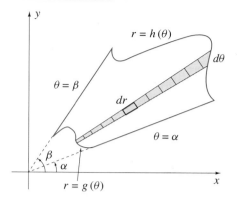

r = h(θ)

dθ

θ = β

dr

θ = α

β

α

r = g(θ)

x

y

EXAMPLE 14.15

Find the area inside the circle $x^2 + y^2 = 4x$ and outside the circle $x^2 + y^2 = 1$.

SOLUTION   If R is the region bounded by these circles and above the x-axis (Figure 14.43), then the required area is

$$2 \iint_R dA.$$

Since the curves intersect in the first quadrant at a point where $\theta = \overline{\theta} = \text{Cos}^{-1}(\frac{1}{4})$, then

$$\text{area} = 2 \int_0^{\overline{\theta}} \int_1^{4\cos\theta} r \, dr \, d\theta = 2 \int_0^{\overline{\theta}} \left\{ \frac{r^2}{2} \right\}_1^{4\cos\theta} d\theta = \int_0^{\overline{\theta}} (16\cos^2\theta - 1) d\theta$$

$$= \int_0^{\overline{\theta}} \left\{ 16\left(\frac{1 + \cos 2\theta}{2} \right) - 1 \right\} d\theta = \int_0^{\overline{\theta}} (7 + 8\cos 2\theta) d\theta = \{7\theta + 4\sin 2\theta\}_0^{\overline{\theta}}$$

$$= 7\overline{\theta} + 4\sin 2\overline{\theta} = 7\text{Cos}^{-1}(\tfrac{1}{4}) + 8\cos\overline{\theta}\sin\overline{\theta}$$

$$= 7\text{Cos}^{-1}(\tfrac{1}{4}) + 8(\tfrac{1}{4})\sqrt{1 - \tfrac{1}{16}} = 7\text{Cos}^{-1}(\tfrac{1}{4}) + \sqrt{15}/2.$$

FIGURE 14.43

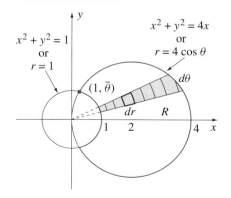

$x^2 + y^2 = 1$
or
$r = 1$

$x^2 + y^2 = 4x$
or
$r = 4\cos\theta$

$(1, \overline{\theta})$

dθ

dr   R

1   2   4   x

y

EXAMPLE 14.16

Evaluate the double iterated integral

$$\int_0^1 \int_0^{\sqrt{-x^2+x}} y^2 \, dy \, dx.$$

SOLUTION   The limits identify the region of integration as the interior of the semicircle in Figure 14.44. The integrand suggests an interpretation of the integral as the second moment of area of this semicircle about the x-axis. Since the semicircle R in Figure 14.45 has exactly the same second moment about the x-axis, we can state that

$$\int_0^1 \int_0^{\sqrt{-x^2+x}} y^2 \, dy \, dx = \iint_R y^2 \, dA = \int_0^\pi \int_0^{1/2} (r^2 \sin^2 \theta) r \, dr \, d\theta$$

$$= \int_0^\pi \left\{ \frac{r^4}{4} \sin^2 \theta \right\}_0^{1/2} d\theta = \frac{1}{64} \int_0^\pi \sin^2 \theta \, d\theta$$

$$= \frac{1}{128} \int_0^\pi (1 - \cos 2\theta) d\theta = \frac{1}{128} \left\{ \theta - \frac{\sin 2\theta}{2} \right\}_0^\pi = \frac{\pi}{128}.$$

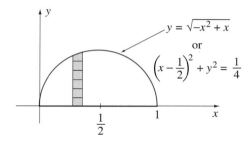

FIGURE 14.44

$y = \sqrt{-x^2 + x}$
or
$\left(x - \dfrac{1}{2}\right)^2 + y^2 = \dfrac{1}{4}$

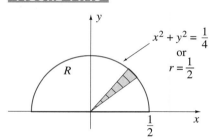

FIGURE 14.45

$x^2 + y^2 = \dfrac{1}{4}$
or
$r = \dfrac{1}{2}$

EXAMPLE 14.17

Find the centroid of the region R in Figure 14.46.

SOLUTION   Evidently, $\overline{x} = 0$, and the area of the region is $A = (\pi b^2 - \pi a^2)/2$. Since

$$A\overline{y} = \iint_R y \, dA = \int_0^\pi \int_a^b (r \sin \theta) r \, dr \, d\theta = \int_0^\pi \left\{ \frac{r^3}{3} \sin \theta \right\}_a^b d\theta$$

$$= \frac{1}{3}(b^3 - a^3) \int_0^\pi \sin \theta \, d\theta$$

$$= \frac{1}{3}(b^3 - a^3)\{-\cos \theta\}_0^\pi = \frac{2}{3}(b^3 - a^3),$$

it follows that

$$\overline{y} = \frac{2}{3}(b^3 - a^3)\frac{2}{\pi(b^2 - a^2)} = \frac{4}{3\pi} \frac{b^2 + ab + a^2}{a + b}.$$

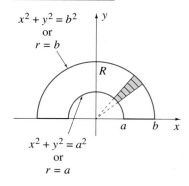

FIGURE 14.46

$x^2 + y^2 = b^2$
or
$r = b$

R

a b x

$x^2 + y^2 = a^2$
or
$r = a$

EXAMPLE 14.18

Find the area of that portion of the sphere $x^2 + y^2 + z^2 = 2$ inside the cone $z = \sqrt{x^2 + y^2}$.

SOLUTION   If S is that portion of the sphere that is inside the cone and also in the first octant (Figure 14.47), then the required area is four times that of S; i.e.,

$$\text{area} = 4 \iint_{S_{xy}} \sqrt{1 + \left(\frac{\partial z}{\partial x}\right)^2 + \left(\frac{\partial z}{\partial y}\right)^2} \, dA,$$

where S_{xy} is the projection of S on the xy-plane. The curve of intersection of the cone and the sphere has equations

$$x^2 + y^2 + z^2 = 2, \qquad \text{or equivalently} \qquad x^2 + y^2 = 1,$$

$$z = \sqrt{x^2 + y^2}, \qquad\qquad\qquad\qquad z = 1.$$

Consequently, S_{xy} is the interior of the quarter-circle $x^2 + y^2 \leq 1$, $x \geq 0$, $y \geq 0$. On S,

$$\frac{\partial z}{\partial x} = -\frac{x}{z} \quad \text{and} \quad \frac{\partial z}{\partial y} = -\frac{y}{z},$$

so that

$$\text{area} = 4 \iint_{S_{xy}} \sqrt{1 + \frac{x^2}{z^2} + \frac{y^2}{z^2}} \, dA = 4 \iint_{S_{xy}} \sqrt{\frac{x^2 + y^2 + z^2}{z^2}} \, dA$$

$$= 4 \iint_{S_{xy}} \sqrt{\frac{2}{z^2}} \, dA = 4\sqrt{2} \iint_{S_{xy}} \frac{1}{\sqrt{2 - x^2 - y^2}} \, dA.$$

If we now use polar coordinates to evaluate this double integral, we have

$$\text{area} = 4\sqrt{2} \int_0^{\pi/2} \int_0^1 \frac{1}{\sqrt{2 - r^2}} r \, dr \, d\theta = 4\sqrt{2} \int_0^{\pi/2} \{-\sqrt{2 - r^2}\}_0^1 d\theta$$

$$= 4\sqrt{2}(\sqrt{2} - 1) \int_0^{\pi/2} d\theta = 4\sqrt{2}(\sqrt{2} - 1)\frac{\pi}{2} = 2\sqrt{2}\pi(\sqrt{2} - 1).$$

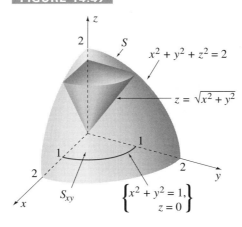

FIGURE 14.47

EXERCISES 14.7

In Exercises 1–5 evaluate the double integral of the function over the region R.

* **1.** $f(x, y) = e^{x^2+y^2}$ where R is bounded by $x^2 + y^2 = a^2$

* **2.** $f(x, y) = x$ where R is bounded by $x = \sqrt{2y - y^2}$, $x = 0$

* **3.** $f(x, y) = \sqrt{x^2 + y^2}$ where R is bounded by $y = \sqrt{9 - x^2}$, $y = x$, $x = 0$

* **4.** $f(x, y) = 1/\sqrt{x^2 + y^2}$ where R is the region outside $x^2 + y^2 = 4$ and inside $x^2 + y^2 = 4x$

* **5.** $f(x, y) = \sqrt{1 + 2x^2 + 2y^2}$ where R is bounded by $x^2 + y^2 = 1$, $x^2 + y^2 = 4$

Evaluate the double iterated integral in Exercises 6 and 7.

* **6.** $\displaystyle\int_0^1 \int_0^{\sqrt{1-x^2}} \sqrt{x^2 + y^2}\, dy\, dx$ * **7.** $\displaystyle\int_{-\sqrt{2}}^0 \int_{-y}^{\sqrt{4-y^2}} x^2\, dx\, dy$

In Exercises 8–12 find the area bounded by the curves.

* **8.** Outside $x^2 + y^2 = 9$ and inside $x^2 + y^2 = 2\sqrt{3}y$

* **9.** $r = 9(1 + \cos\theta)$

* **10.** $r = \cos 3\theta$

** **11.** Common to $r = 2$ and $r^2 = 9\cos 2\theta$

** **12.** Common to $r = 1 + \sin\theta$ and $r = 2 - 2\sin\theta$

** **13.** Find the centroid of the area bounded by the curves $y = x$, $y = -x$, $x = \sqrt{2 - y^2}$

** **14.** Find the second moment of area for a circular plate of radius R about any diameter.

** **15.** A water tank in the form of a right circular cylinder with radius R and length h has its axis horizontal. If it is full, what is the force due to water pressure on each end?

In Exercises 16–18 find the area of the surface.

** **16.** The area of $z = x^2 + y^2$ below $z = 4$

** **17.** The area of $x^2 + y^2 + z^2 = 4$ inside $x^2 + y^2 = 1$

** **18.** The area of $z = xy$ inside $x^2 + y^2 = 9$

** **19.** Prove that the area of a sphere of radius R is $4\pi R^2$.

** **20.** Find the area of the hyperbolic paraboloid $z = x^2 - y^2$ between the cylinders $x^2 + y^2 = 1$ and $x^2 + y^2 = 4$

In Exercises 21 and 22 find the volume of the solid of revolution obtained by rotating the area bounded by the curve about the line.

** **21.** $r = \cos^2\theta$ about the x-axis

** **22.** $r = 1 + \sin\theta$ about the y-axis

** **23.** Find the area bounded by the curve $(x^2 + y^2)^3 = 4a^2x^2y^2$.

** **24.** Find the volume of the solid of revolution when a circle of radius R is rotated about a tangent line.

** **25.** A circular plate of radius R (figure below) has a uniform charge distribution of ρ coulombs per square metre. If P is a point directly above the centre of the plate and dA is a small area on the plate, then the potential at P due to dA is given by

$$\frac{1}{4\pi\epsilon_0}\frac{\rho\,dA}{s},$$

where s is the distance from P to dA.

(a) Show that, in terms of polar coordinates, the potential V at P due to the entire plate is

$$V = \frac{\rho}{4\pi\epsilon_0}\int_{-\pi}^{\pi}\int_0^R \frac{r}{\sqrt{r^2 + d^2}}\,dr\,d\theta,$$

where d is the distance from P to the centre of the plate.

(b) Evaluate the double iterated integral to find V.

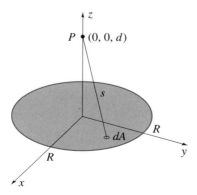

** **26.** Use Coulomb's law (see Example 12.9 in Section 12.3) to find the force on a charge q at point P due to the charge on the plate in Exercise 25. What happens to this force as the radius of the plate gets very large?

In Exercises 27–29 find the area bounded by the curves.

** **27.** $(x^2 + y^2)^2 = 2xy$

** **28.** Inside both $r = 6\cos\theta$ and $r = 4 - 2\cos\theta$

** **29.** $r = \cos^2\theta \, \sin\theta$

** **30.** Find the centroid of the cardioid $r = 1 + \cos\theta$.

** **31.** Find the second moment of area about the x-axis for the area bounded by $r^2 = 9\cos 2\theta$.

** **32.** Evaluate the double integral of

$$f(x, y) = \sqrt{\frac{1 - x^2 - y^2}{1 + x^2 + y^2}}$$

over the area inside the circle $x^2 + y^2 = 1$.

** **33.** The figure below illustrates a piece of an artery or vein with circular cross-section (radius R). The speed of blood flowing through the blood vessel is not uniform because of the viscosity of the blood and friction at the walls. Poiseuille's law states that, for laminar blood flow, the speed v of blood at a distance r from the centre of the vessel is given by

$$v = \frac{P}{4nL}(R^2 - r^2),$$

where P is the pressure difference between the ends of the vessel, L is the length of the vessel, and n is the viscosity of the blood. Find the amount of blood flowing over a cross-section of the blood vessel per unit time.

*** **34.** Find the area of that part of the sphere $x^2 + y^2 + z^2 = a^2$ inside $(x^2 + y^2)^2 = a^2(x^2 - y^2)$.

*** **35.** Find the area of that portion of the surface $x^2 + z^2 = a^2$ cut out by $x^2 + y^2 = a^2$.

*** **36.** A very important integral in statistics is

$$I = \int_0^\infty e^{-x^2} \, dx.$$

To evaluate the integral we set

$$I = \int_0^\infty e^{-y^2} \, dy,$$

and then multiply these two equations. Do this to prove that $I = \sqrt{\pi}/2$.

*** **37.** Use the result of Exercise 36 to evaluate the gamma function

$$\Gamma(n) = \int_0^\infty x^{n-1} e^{-x} \, dx$$

at $n = 1/2$.

14.8 Triple Integrals and Triple Iterated Integrals

Triple integrals are defined in much the same way as double integrals. Suppose $f(x, y, z)$ is a function defined in some region V of space that has finite volume (Figure 14.48). We divide V into n subregions of volumes $\Delta V_1, \Delta V_2, \ldots, \Delta V_n$ in any manner whatsoever, and in each subregion $\Delta V_i (i = 1, \ldots, n)$ we choose an arbitrary point (x_i^*, y_i^*, z_i^*). We then form the sum

$$f(x_1^*, y_1^*, z_1^*)\Delta V_1 + f(x_2^*, y_2^*, z_2^*)\Delta V_2 + \cdots + f(x_n^*, y_n^*, z_n^*)\Delta V_n$$

$$= \sum_{i=1}^{n} f(x_i^*, y_i^*, z_i^*)\Delta V_i. \qquad (14.47)$$

If this sum approaches a limit as the number of subregions becomes increasingly large and every subregion shrinks to a point, we call the limit the triple integral of $f(x, y, z)$ over the region V and denote it by

$$\iiint_V f(x, y, z)dV = \lim_{\|\Delta V_i\| \to 0} \sum_{i=1}^{n} f(x_i^*, y_i^*, z_i^*)\Delta V_i. \qquad (14.48)$$

FIGURE 14.48

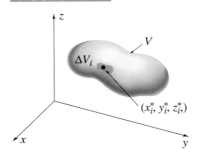

As in the case of double integrals, we require that this limit be independent of the manner of subdivision of V and the choice of star-points in the subregions. This is guaranteed for continuous functions by the following theorem.

THEOREM 14.2

Let S be a piecewise-smooth surface that encloses a region V with finite volume. If $f(x, y, z)$ is continuous inside and on S, then the triple integral of $f(x, y, z)$ over V exists.

Properties analogous to those in equations 14.4–14.7 hold for triple integrals, although we will not list the first three here. Corresponding to equation 14.7, the volume of a region V is given by the triple integral

$$\text{volume of } V = \iiint_V dV. \qquad (14.49)$$

We evaluate triple integrals with triple iterated integrals. If we use Cartesian coordinates there are six possible triple iterated integrals of a function $f(x, y, z)$, corresponding to the six permutations of the product of the differentials dx, dy, and dz:

$$dz\, dy\, dx, \quad dz\, dx\, dy, \quad dx\, dz\, dy, \quad dx\, dy\, dz, \quad dy\, dx\, dz, \quad dy\, dz\, dx.$$

The general triple iterated integral of $f(x, y, z)$ with respect to z, y, and x is of the form

$$\int_a^b \int_{g_1(x)}^{g_2(x)} \int_{h_1(x,y)}^{h_2(x,y)} f(x, y, z)dz\, dy\, dx. \tag{14.50}$$

Because the first integration with respect to z holds x and y constant, the limits on z may therefore depend on x and y. Similarly, the second integration with respect to y holds x constant, and the limits may be functions of x.

EXAMPLE 14.19

Evaluate the triple iterated integral

$$\int_0^1 \int_0^{x^2} \int_{xy}^{x+y} xyz\, dz\, dy\, dx.$$

SOLUTION

$$\int_0^1 \int_0^{x^2} \int_{xy}^{x+y} xyz\, dz\, dy\, dx = \int_0^1 \int_0^{x^2} \left\{ \frac{xyz^2}{2} \right\}_{xy}^{x+y} dy\, dx$$

$$= \frac{1}{2} \int_0^1 \int_0^{x^2} \{xy(x+y)^2 - xy(xy)^2\}dy\, dx$$

$$= \frac{1}{2} \int_0^1 \int_0^{x^2} (x^3y + 2x^2y^2 + xy^3 - x^3y^3)dy\, dx$$

$$= \frac{1}{2} \int_0^1 \left\{ \frac{x^3y^2}{2} + \frac{2x^2y^3}{3} + \frac{xy^4}{4} - \frac{x^3y^4}{4} \right\}_0^{x^2} dx$$

$$= \frac{1}{24} \int_0^1 (6x^7 + 8x^8 + 3x^9 - 3x^{11})dx$$

$$= \frac{1}{24} \left\{ \frac{3x^8}{4} + \frac{8x^9}{9} + \frac{3x^{10}}{10} - \frac{x^{12}}{4} \right\}_0^1 = \frac{19}{270}.$$

Because of the analogy between double and triple integrals, we accept without proof that triple integrals can be evaluated with triple iterated integrals. We must, however, examine how triple iterated integrals bring about the summations represented by triple integrals, for it is only by thoroughly understanding this process that we can obtain limits for triple iterated integrals.

In Section 14.3 we discussed in considerable detail evaluation of double integrals by means of double iterated integrals. In particular, we showed that double iterated integrals in Cartesian coordinates represent the subdivision of an area into small rectangles by coordinate lines $x =$

constant and $y =$ constant. The first integration creates a summation over rectangles in a strip, and the second integration adds over all strips. It is fairly straightforward to generalize these ideas to triple integrals. Consider evaluating the triple integral

$$\iiint_V f(x, y, z)dV$$

over the region V in Figure 14.49 bounded above by the surface $z = h(x, y)$, below by the area R in the xy-plane, and on the sides by a cylindrical wall standing on the curve bounding R.

FIGURE 14.49

The choice of a triple iterated integral in Cartesian coordinates to evaluate this triple integral implies a subdivision of V into small rectangular parallelepipeds (boxes for short) by means of coordinate planes $x =$ constant, $y =$ constant, and $z =$ constant. The dimensions of a representative box at position (x, y, z) in V are denoted by dx, dy, and dz, with resulting volume $dx\, dy\, dz$. If we decide on a triple iterated integral with respect to z, y, and x, then the first integration on z holds x and y constant. This integration therefore adds the quantities

$$f(x, y, z)dz\, dy\, dx$$

over boxes in a vertical column of cross-sectional dimensions dx and dy. Lower and upper limits on z identify where each and every column starts and stops, and must consequently be 0 and $h(x, y)$:

$$\int_0^{h(x,y)} f(x, y, z)dz\, dy\, dx.$$

Since this integration produces a function of x and y alone, the remaining integration with respect to y and x is essentially a double iterated integral in the xy-plane. These integrations must account for all columns in V and therefore the area in the xy-plane over which this double iterated integral is performed is the area R upon which all columns in V stand. Since the y-integration adds inside a strip in the y-direction and limits identify where all strips start and stop, they must therefore be $g_1(x)$ and $g_2(x)$. We have now

$$\int_{g_1(x)}^{g_2(x)} \int_0^{h(x,y)} f(x, y, z) dz\, dy\, dx.$$

Finally, the x-integration adds over all strips and the limits are $x = a$ and $x = b$:

$$\iiint_V f(x, y, z) dV = \int_a^b \int_{g_1(x)}^{g_2(x)} \int_0^{h(x,y)} f(x, y, z) dz\, dy\, dx.$$

Suppose now that V is the region bounded above by the surface $z = h_2(x, y)$ and below by $z = h_1(x, y)$ (Figure 14.50). In this case, the limits on the first integration with respect to z are $h_1(x, y)$ and $h_2(x, y)$ since every column starts on the surface $z = h_1(x, y)$ and ends on the surface $z = h_2(x, y)$:

$$\int_{h_1(x,y)}^{h_2(x,y)} f(x, y, z) dz\, dy\, dx.$$

For the volume of Figure 14.49 we interpreted the final two integrations as a double iterated integral in the xy-plane over the area R from which all columns emanated. For the present volume we interpret R as the area in the xy-plane onto which all columns project. We obtain then

$$\iiint_V f(x, y, z) dV = \int_a^b \int_{g_1(x)}^{g_2(x)} \int_{h_1(x,y)}^{h_2(x,y)} f(x, y, z) dz\, dy\, dx.$$

FIGURE 14.50

Schematically we have obtained the following interpretation for the limits of triple iterated integrals in Cartesian coordinates:

$$\int_{\substack{\text{position of} \\ \text{first strip}}}^{\substack{\text{position of} \\ \text{last strip}}} \int_{\substack{\text{where every} \\ \text{strip starts}}}^{\substack{\text{where every} \\ \text{strip stops}}} \int_{\substack{\text{where every} \\ \text{column starts}}}^{\substack{\text{where every} \\ \text{column stops}}} f(x, y, z) \begin{Bmatrix} dz\,dy\,dx \\ dz\,dx\,dy \\ dx\,dz\,dy \\ dx\,dy\,dz \\ dy\,dx\,dz \\ dy\,dz\,dx \end{Bmatrix}.$$

EXAMPLE 14.20

Set up the six triple iterated integrals in Cartesian coordinates for the triple integral of a function $f(x, y, z)$ over the region V in the first octant bounded by the surfaces

$$y^2 + z^2 = 1, \quad y = x, \quad z = 0, \quad x = 0.$$

SOLUTION   The triple integral of $f(x, y, z)$ over V is given by each of the following triple iterated integrals (see Figure 14.51):

$$\int_0^1 \int_x^1 \int_0^{\sqrt{1-y^2}} f(x, y, z)\,dz\,dy\,dx,$$

$$\int_0^1 \int_0^y \int_0^{\sqrt{1-y^2}} f(x, y, z)\,dz\,dx\,dy,$$

$$\int_0^1 \int_0^{\sqrt{1-z^2}} \int_0^y f(x, y, z)\,dx\,dy\,dz,$$

$$\int_0^1 \int_0^{\sqrt{1-y^2}} \int_0^y f(x, y, z)\,dx\,dz\,dy,$$

$$\int_0^1 \int_0^{\sqrt{1-x^2}} \int_x^{\sqrt{1-z^2}} f(x, y, z)\,dy\,dz\,dx,$$

$$\int_0^1 \int_0^{\sqrt{1-z^2}} \int_x^{\sqrt{1-z^2}} f(x, y, z)\,dy\,dx\,dz.$$

FIGURE 14.51

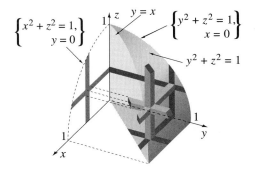

$$\begin{Bmatrix} x^2 + z^2 = 1, \\ y = 0 \end{Bmatrix}$$

$$\begin{Bmatrix} y^2 + z^2 = 1, \\ x = 0 \end{Bmatrix}$$

$y = x$

$y^2 + z^2 = 1$

EXAMPLE 14.21

Set up triple iterated integrals with respect to z, y, and x for the triple integral of the function $f(x, y, z) = x^2 y \sin z$ over the region V bounded by the surfaces

$$z = \sqrt{y}, \quad y + z = 2, \quad x = 0, \quad z = 0, \quad x = 2.$$

SOLUTION   The problem requires triple iterated integrals first with respect to z, then with respect to y and x. Since some columns end on the parabolic cylinder $z = \sqrt{y}$ (Figure 14.52) and others on the plane $y + z = 2$, we require two iterated integrals:

$$\iiint_V x^2 y \sin z \, dV = \int_0^2 \int_0^1 \int_0^{\sqrt{y}} x^2 y \sin z \, dz \, dy \, dx$$

$$+ \int_0^2 \int_1^2 \int_0^{2-y} x^2 y \sin z \, dz \, dy \, dx.$$

FIGURE 14.52

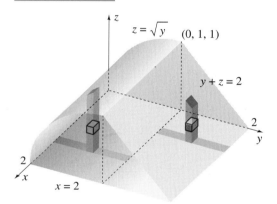

EXERCISES 14.8

In Exercises 1–12 evaluate the triple integral over the region.

* **1.** $\iiint_V (x^2 z + y e^x) \, dV$ where V is bounded by $x = 0$, $x = 1$, $y = 1$, $y = 2$, $z = 0$, $z = 1$

* **2.** $\iiint_V x \, dV$ where V is bounded by $x = 0$, $y = 0$, $z = 0$, $x + y + z = 4$

* **3.** $\iiint_V \sin(y+z)\,dV$ where V is bounded by $z = 0,\ y = 2x,\ y = 0,\ x = 1, z = x+2y$

* **4.** $\iiint_V xy\,dV$ where V is enclosed by $z = \sqrt{1 - x^2 - y^2},\ z = 0$

* **5.** $\iiint_V dV$ where V is bounded by $x^2 + y^2 = 1,\ x^2 + z^2 = 1$

* **6.** $\iiint_V (x^2 + 2z)\,dV$ where V is bounded by $z = 0,\ y + z = 4,\ y = x^2$

* **7.** $\iiint_V x^2 y^2 z^2\,dV$ where V is bounded by $z = 1+y,\ y+z = 1,\ x = 1,\ x = 0,\ z = 0$

* **8.** $\iiint_V xyz\,dV$ where V is the first octant volume cut out by $z = x^2 + y^2,\ z = \sqrt{x^2 + y^2}$

* **9.** $\iiint_V dV$ where V is bounded by $z = x^2,\ y + z = 4,\ y = 0$

* **10.** $\iiint_V (x + y + z)\,dV$ where V is bounded by $x = 0,\ x = 1, z = 0,\ y+z = 2,\ y = z$

* **11.** $\iiint_V xyz\,dV$ where V is bounded by $z = 1,\ z = x^2/4 + y^2/9$

* **12.** $\iiint_V x^2 y\,dV$ where V is the first octant volume bounded by $z = 1,\ z = x^2/4 + y^2/9$

** **13.** Set up the six triple iterated integrals in Cartesian coordinates for the triple integral of a function $f(x, y, z)$ over the volume enclosed by the surfaces $y = 1 - x^2, z = 0$, and $y = z$.

In Exercises 14–17 set up, but do not evaluate, a triple iterated integral for the triple integral.

** **14.** $\iiint_V (x^2 + y^2 + z^2)\,dV$ where V is bounded by $z = \sqrt{1 - x^2 - y^2},\ z = x^2$

** **15.** $\iiint_V xz \sin(x + y)\,dV$ where V is bounded by $y^2 = 1 + 4x^2 + 4z^2,\ y = \sqrt{4 + x^2}$

** **16.** $\iiint_V xyz\,dV$ where V is bounded by $z = x^2 + 4y^2,\ 2x + 8y + z = 4$

** **17.** $\iiint_V x^2 y^2 z^2\,dV$ where V is bounded by $x = y^2 + z^2,\ x + 1 = (y^2 + z^2)^2$

In Exercises 18–23 evaluate the triple integral over the region.

** **18.** $\iiint_V (y + x^2)\,dV$ where V is bounded by $x + z^2 = 1,\ z = x + 1,\ y = 1,\ y = -1$

** **19.** $\iiint_V (xy + z)\,dV$ where V is bounded by $y + z = 1,\ z = 2y,\ z = y,\ x = 0,\ x = 3$

** **20.** $\iiint_V dV$ where V is bounded by $z = 0$, $x^2 + y^2 = 1$, $x + y + z = 2$

** **21.** $\iiint_V dV$ where V is bounded by $z = x^2 + y^2$, $z = 4 - x^2 - y^2$

** **22.** $\iiint_V (x + y + z) \, dV$ where V is bounded by $2z = y^2 - x^2$, $z = 1 - x^2$

** **23.** $\iiint_V |yz| \, dV$ where V is bounded by $z^2 = 1 + x^2 + y^2$, $z = \sqrt{4 - x^2 - y^2}$

** **24.** Set up, but do not evaluate, triple iterated integrals to evaluate the triple integral of the function $f(x, y, z) = x^2 + y^2 + z^2$ over the volume bounded by the surfaces $x^2 + y^2 = z^2 + 1$, $2z = \sqrt{x^2 + y^2}$, $z = 0$.

14.9 Volumes

Because the volume of a region V is represented by triple integral 14.49 and triple integrals are evaluated by means of triple iterated integrals, it follows that volumes can be evaluated with triple iterated integrals. For example, to evaluate the volume of the region in Figure 14.50 (page 1118) using a triple iterated integral in x, y, and z, we subdivide V into boxes of dimensions dx, dy, and dz and therefore of volume

$$dz \, dy \, dx.$$

Integration with respect to z adds these volumes in the z-direction to give the volume of a vertical column (Figure 14.53):

$$\int_{h_1(x,y)}^{h_2(x,y)} dz \, dy \, dx.$$

Limits indicate that all columns start on the surface $z = h_1(x, y)$ and end on the surface $z = h_2(x, y)$. Integration with respect to y now adds volumes of columns that project onto a strip in the y-direction:

$$\int_{g_1(x)}^{g_2(x)} \int_{h_1(x,y)}^{h_2(x,y)} dz \, dy \, dx,$$

where the limits indicate that all strips start on the curve $y = g_1(x)$ and end on the curve $y = g_2(x)$. Evidently this integration yields the volume of a slab as shown in Figure 14.53. Finally, integration with respect to x adds the volumes of all such slabs in V:

$$\int_a^b \int_{g_1(x)}^{g_2(x)} \int_{h_1(x,y)}^{h_2(x,y)} dz \, dy \, dx,$$

where a and b designate positions of first and last strips.

FIGURE 14.53

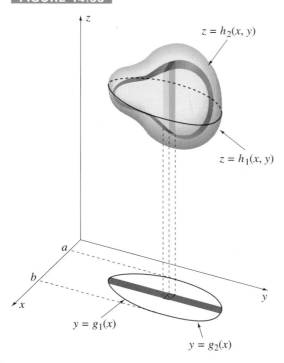

EXAMPLE 14.22

Find the volume bounded by the planes $z = x + y$, $y = 2x$, $z = 0$, $x = 0$, $y = 2$.

SOLUTION   If we use vertical columns (Figure 14.54), we have

$$\text{volume} = \int_0^1 \int_{2x}^2 \int_0^{x+y} dz\, dy\, dx = \int_0^1 \int_{2x}^2 (x+y) dy\, dx = \int_0^1 \left\{ xy + \frac{y^2}{2} \right\}_{2x}^2 dx$$

$$= 2 \int_0^1 (1 + x - 2x^2) dx = 2 \left\{ x + \frac{x^2}{2} - \frac{2x^3}{3} \right\}_0^1 = \frac{5}{3}.$$

FIGURE 14.54

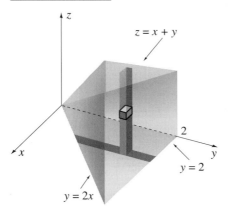

EXAMPLE 14.23

Find the volume in the first octant cut from the cyclinder $x^2 + z^2 = 4$ by the plane $y + z = 6$.

SOLUTION With columns in the y-direction (Figure 14.55), we have

$$
\text{volume} = \int_0^2 \int_0^{\sqrt{4-x^2}} \int_0^{6-z} dy\, dz\, dx = \int_0^2 \int_0^{\sqrt{4-x^2}} (6-z)dz\, dx
$$

$$
= \int_0^2 \left\{ 6z - \frac{z^2}{2} \right\}_0^{\sqrt{4-x^2}} dx
$$

$$
= \int_0^2 \left\{ 6\sqrt{4-x^2} - \frac{1}{2}(4-x^2) \right\} dx.
$$

In the first term we set $x = 2\sin\theta$, from which we get $dx = 2\cos\theta\, d\theta$, and

$$
\text{volume} = 6 \int_0^{\pi/2} (2\cos\theta)2\cos\theta\, d\theta + \left\{ -2x + \frac{x^3}{6} \right\}_0^2
$$

$$
= 12 \int_0^{\pi/2} (1 + \cos 2\theta)d\theta - \frac{8}{3}
$$

$$
= 12 \left\{ \theta + \frac{1}{2}\sin 2\theta \right\}_0^{\pi/2} - \frac{8}{3} = 6\pi - \frac{8}{3}.
$$

FIGURE 14.55

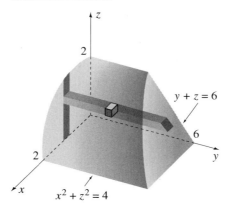

Had we used iterated integrals with respect to z, y, and x in this example, we would have had two integrals

$$
\text{volume} = \int_0^2 \int_0^{6-\sqrt{4-x^2}} \int_0^{\sqrt{4-x^2}} dz\, dy\, dx + \int_0^2 \int_{6-\sqrt{4-x^2}}^6 \int_0^{6-y} dz\, dy\, dx.
$$

EXERCISES 14.9

In Exercises 1–17 find the volume of the region bounded by the surfaces.

* **1.** $y = x^2$, $y = 1$, $z = 0$, $z = 4$

* **2.** $x = z^2$, $z = x^2$, $y = 0$, $y = 2$

* **3.** $x = 3z$, $z = 3x$, $y = 1$, $y = 0$, $x = 2$

* **4.** $x + y + z = 6$, $y = 4 - x^2$, $z = 0$, $y = 0$

* **5.** $z = x^2 + y^2$, $y = x^2$, $y = 4$, $z = 0$

* **6.** $x + y + z = 4$, $y = 3z$, $x = 0$, $y = 0$

* **7.** $x^2 + y^2 = 4$, $y^2 + z^2 = 4$

* **8.** $y = x^2 - 1$, $y = 1 - x^2$, $x + z = 1$, $z = 0$

* **9.** $z = 16 - x^2 - 4y^2$, $x + y = 1$, $z = 16$, $x = 0$, $y = 0$ (in the first octant)

* **10.** $z = x^2 + y^2$, $x = 1$, $z = 0$, $x = y$, $x = 2y$

* **11.** $z = 1 - x^2 - y^2$, $z = 0$

* **12.** $x - z = 0$, $x + z = 3$, $y + z = 1$, $z = y + 1$, $z = 0$

* **13.** $x + y + z = 2$, $x^2 + y^2 = 1$, $z = 0$

* **14.** $y + z = 1$, $z = 2y$, $z = y$, $x = 0$, $x + y + z = 4$

* **15.** $x^2 + 4y^2 = z$, $x^2 + 4y^2 = 12 - 2z$

** **16.** $y = 1 - z^2$, $y = z^2 - 1$, $x = 1 - z^2$, $x = z^2 - 1$

** **17.** $x + 3y + 2z = 6$, $z = 0$, $y = x$, $y = 2x$

** **18.** Find the volume in the first octant bounded by the plane $2x + y + z = 2$ and inside the cylinder $y^2 + z^2 = 1$.

** **19.** A pyramid has a square base with side length b and has height h at its centre.

   (a) Find its volume by taking cross-sections parallel to the base (see Section 7.9).

   (b) Find its volume using triple integrals.

The average value of a function $f(x, y, z)$ over a region with volume V is defined as

$$\overline{f} = \frac{1}{V} \iiint_V f(x, y, z) \, dV.$$

In Exercises 20–22 find the average value of the function over the region.

* **20.** $f(x, y, z) = xy$ over the region bounded by the surfaces $x = 0$, $y = 0$, $z = 0$, $x + y + z = 1$

* **21.** $f(x, y, z) = x + y + z$ over the region in the first octant bounded by the surfaces $z = 9 - x^2 - y^2$, $z = 0$ and for which $0 \le x \le 1$, $0 \le y \le 1$

* **22.** $f(x, y, z) = x^2 + y^2 + z^2$ over the region bounded by the surfaces $x = 0$, $x = 1$, $y + z = 2$, $y = 2$, $z = 2$

** **23.** Find the volume bounded by the surfaces $z = x^2 - y^2$ and $z = 4 - 2(x^2 + y^2)$.

** **24.** Verify that the surfaces $z = x^2 - y^2$ and $z = 4 - x^2 - y^2$ do not bound a finite volume.

** **25.** Find the volume bounded by the surfaces $x + z = 2$, $z = 0$, $4y^2 = x(2 - z)$.

** **26.** Find the volume bounded by the surfaces $z = (x - 1)^2 + y^2$, $2x + z = 2$.

** **27.** Find the volume inside the ellipsoid $x^2/a^2 + y^2/b^2 + z^2/c^2 = 1$.

*** **28.** The bottom and sides of a boat are defined by the surface equation $x = 10(1 - y^2 - z^2)$, $0 \le x \le 10$, where all dimensions are in metres.

  (a) Find the volume of water displaced by the boat when the water level on the side of the boat is d metres below the top of the boat.

  (b) Archimedes' principle states that the buoyant force on an object when immersed or partially immersed in a fluid is equal to the weight of the fluid displaced by the object. Find the maximum weight of the boat and contents just before sinking.

*** **29.** Find the volume inside all three surfaces $x^2 + y^2 = a^2$, $x^2 + z^2 = a^2$, $y^2 + z^2 = a^2$.

14.10 Centres of Mass and Moments of Inertia

In this section we discuss centres of mass and moments of inertia for three-dimensional objects of density $\rho(x, y, z)$ (mass per unit volume). If we divide the object occupying region V in Figure 14.56 into small volumes dV, then the amount of mass in dV is $\rho \, dV$. The triple integral

$$M = \iiint_V \rho \, dV \qquad (14.51)$$

adds the masses of all such volumes (of ever-decreasing size) to produce the total mass M of the object.

FIGURE 14.56

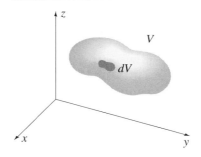

Corresponding to equations 14.29 and 14.30 for first moments of planar masses about the coordinate axes are the following formulas for first moments of the object about the coordinate planes:

$$\text{First moment of object about } yz\text{-plane} = \iiint_V x\rho \, dV, \qquad (14.52)$$

$$\text{First moment of object about } xz\text{-plane} = \iiint_V y\rho \, dV, \qquad (14.53)$$

$$\text{First moment of object about } xy\text{-plane} = \iiint_V z\rho \, dV. \qquad (14.54)$$

The centre of mass of the object is defined as that point $(\overline{x}, \overline{y}, \overline{z})$ at which a particle of mass M would have the same first moments about the coordinate planes as the object itself. Since the first moments of M at $(\overline{x}, \overline{y}, \overline{z})$ about the coordinate planes are $M\overline{x}$, $M\overline{y}$, and $M\overline{z}$, it follows that we can use the equations

$$M\overline{x} = \iiint_V x\rho \, dV, \tag{14.55}$$

$$M\overline{y} = \iiint_V y\rho \, dV, \tag{14.56}$$

$$M\overline{z} = \iiint_V z\rho \, dV, \tag{14.57}$$

to solve for $(\overline{x}, \overline{y}, \overline{z})$ once M and the integrals on the right have been evaluated.

If we use a triple iterated integral with respect to z, y, and x to evaluate 14.57, say, for the object in Figure 14.53, then

$$\iiint_V z\rho \, dV = \int_a^b \int_{g_1(x)}^{g_2(x)} \int_{h_1(x,y)}^{h_2(x,y)} z\rho \, dz \, dy \, dx.$$

Quantity $z\rho \, dz \, dy \, dx$ is the first moment about the xy-plane of the mass in an elemental box of dimensions dx, dy, and dz. The z-integration then adds these moments over boxes in the z-direction to give the first moment about the xy-plane of the mass in a vertical column:

$$\int_{h_1(x,y)}^{h_2(x,y)} z\rho \, dz \, dy \, dx.$$

The y-integration then adds the first moments of columns that project onto a strip in the y-direction:

$$\int_{g_1(x)}^{g_2(x)} \int_{h_1(x,y)}^{h_2(x,y)} z\rho \, dz \, dy \, dx.$$

This quantity therefore represents the first moment about the xy-plane of the slab in Figure 14.53. Finally, the x-integration adds first moments of all such slabs to give the total first moment of V about the xy-plane:

$$\int_a^b \int_{g_1(x)}^{g_2(x)} \int_{h_1(x,y)}^{h_2(x,y)} z\rho \, dz \, dy \, dx.$$

EXAMPLE 14.24   ◤

Find the centre of mass of an object of constant density ρ if it is bounded by the surfaces $z = 1 - y^2$, $x = 0$, $z = 0$, $x = 2$.

SOLUTION   From the symmetry of the object (Figure 14.57), we see $\overline{x} = 1$ and $\overline{y} = 0$. Now

$$M = 2\int_0^2 \int_0^1 \int_0^{1-y^2} \rho \, dz \, dy \, dx = 2\rho \int_0^2 \int_0^1 (1 - y^2) \, dy \, dx$$

$$= 2\rho \int_0^2 \left\{ y - \frac{y^3}{3} \right\}_0^1 dx = \frac{4\rho}{3} \int_0^2 dx = \frac{8\rho}{3};$$

and

$$M\bar{z} = 2\int_0^2\int_0^1\int_0^{1-y^2} z\rho\,dz\,dy\,dx = 2\rho\int_0^2\int_0^1\left\{\frac{z^2}{2}\right\}_0^{1-y^2}dy\,dx$$

$$= \rho\int_0^2\int_0^1(1 - 2y^2 + y^4)dy\,dx$$

$$= \rho\int_0^2\left\{y - \frac{2y^3}{3} + \frac{y^5}{5}\right\}_0^1 dx = \frac{8\rho}{15}\int_0^2 dx = \frac{16\rho}{15}.$$

Thus,

$$\bar{z} = \frac{16\rho}{15}\cdot\frac{3}{8\rho} = \frac{2}{5}.$$

FIGURE 14.57

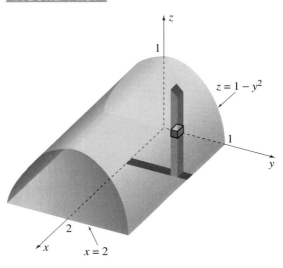

$z = 1 - y^2$

In view of our discussion on first moments in this section and on moments of inertia of thin plates in Section 14.5, it should not be necessary to give a full treatment of moments of inertia of three-dimensional objects. Instead, we simply note that the distances from a point (x, y, z) to the x-, y-, and z-axes are, respectively, $\sqrt{y^2 + z^2}$, $\sqrt{x^2 + z^2}$, and $\sqrt{x^2 + y^2}$. If follows, then, that if an object of density $\rho(x, y, z)$ occupies a region V of space:

$$\text{Moment of inertia of object about } x\text{-axis} = \iiint_V (y^2 + z^2)\rho\,dV, \qquad (14.58)$$

$$\text{Moment of inertia of object about } y\text{-axis} = \iiint_V (x^2 + z^2)\rho\,dV, \qquad (14.59)$$

$$\text{Moment of inertia of object about } z\text{-axis} = \iiint_V (x^2 + y^2)\rho\,dV. \qquad (14.60)$$

EXAMPLE 14.25

Find the moment of inertia of a right circular cylinder of constant density ρ about its axis.

SOLUTION   Let the length and radius of the cylinder be h and r and choose the coordinate system in Figure 14.58. The required moment of inertia about the z-axis is four times the moment of inertia of that part of the cylinder in the first octant. Hence,

$$I = 4 \int_0^r \int_0^{\sqrt{r^2-x^2}} \int_0^h (x^2 + y^2)\rho \, dz \, dy \, dx = 4\rho h \int_0^r \int_0^{\sqrt{r^2-x^2}} (x^2 + y^2) dy \, dx$$

$$= 4\rho h \int_0^r \left\{ x^2 y + \frac{y^3}{3} \right\}_0^{\sqrt{r^2-x^2}} dx = 4\rho h \int_0^r \left\{ x^2\sqrt{r^2 - x^2} + \frac{1}{3}(r^2 - x^2)^{3/2} \right\} dx$$

$$= \frac{4\rho h}{3} \int_0^r \{ r^2\sqrt{r^2 - x^2} + 2x^2\sqrt{r^2 - x^2} \} dx.$$

To evaluate this definite integral we set $x = r \sin\theta$, which implies that $dx = r \cos\theta \, d\theta$, and

$$I = \frac{4\rho h}{3} \int_0^{\pi/2} \{ r^2 r \cos\theta + 2r^2 \sin^2\theta r \cos\theta \} r \cos\theta \, d\theta$$

$$= \frac{4\rho h r^4}{3} \int_0^{\pi/2} \{ \cos^2\theta + 2\sin^2\theta \cos^2\theta \} d\theta$$

$$= \frac{4\rho h r^4}{3} \int_0^{\pi/2} \left\{ \frac{1 + \cos 2\theta}{2} + \frac{1 - \cos 4\theta}{4} \right\} d\theta$$

$$= \frac{4\rho h r^4}{3} \left\{ \frac{3\theta}{4} + \frac{\sin 2\theta}{4} - \frac{\sin 4\theta}{16} \right\}_0^{\pi/2} = \frac{\rho h r^4 \pi}{2}.$$

FIGURE 14.58

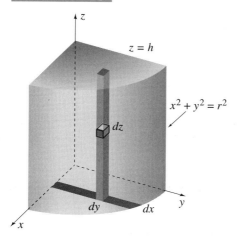

EXERCISES 14.10

In Exercises 1–5 the surfaces bound a solid object of constant density. Find its centre of mass.

∗ **1.** $z = x^2 + y^2$, $z = 0$, $x = 0$, $y = 0$, $x = 1$, $y = 1$

∗ **2.** $x + y + z = 1$, $x = 0$, $y = 0$, $z = 0$

∗ **3.** $z = x^2$, $y + z = 4$, $y = 0$

∗ **4.** $y = 4 - x^2$, $z = 0$, $y = z$

∗ **5.** $x + y + z = 4$, $y = 3z$, $x = 0$, $y = 0$

In Exercises 6–10 the surfaces bound a solid object of constant density ρ. Find its moment of inertia about the line.

∗ **6.** $x = 0$, $y = 0$, $z = 0$, $x = 1$, $y = 1$, $z = 1$ about the x-axis

∗ **7.** $z = 2x$, $z = 0$, $y = 0$, $y = 2$, $x = 3$ about the y-axis

∗ **8.** $x + y + z = 2$, $y = 0$, $x = 0$, $0 \leq z \leq \sqrt{1 - y}$ about the x-axis

∗ **9.** $z = xy$, $x^2 + y^2 = 1$, $z = 0$ (first octant) about the z-axis

∗ **10.** $y + z = 2$, $x + z = 2$, $x = 0$, $y = 0$, $z = 0$ about the z-axis

∗ **11.** Find the first moment about the xy-plane of a solid of constant density ρ if it is bounded by the surfaces $x = z$, $x + z = 0$, $z = 2$, $y = 0$, $y = 2$.

In Exercises 12–14 the surfaces bound a solid object of constant density. Find its centre of mass.

∗∗ **12.** $y = x^3$, $x = y^2$, $z = 1 + x^2 + y^2$, $z = -x^2 - y^2$

∗∗ **13.** $z = x^2$, $x + z = 2$, $z = y$, $y = 0$

∗∗ **14.** $y + z = 0$, $y - z = 0$, $x + z = 0$, $x - z = 0$, $z = 2$

In Exercises 15–17 the surfaces bound a solid object of constant density ρ. Find its moment of inertia about the line.

∗∗ **15.** $z = 4 - x^2$, $x + z + 2 = 0$, $y = 0$, $y = 2$ about the y-axis

∗∗ **16.** $x^2 + z^2 = a^2$, $x^2 + y^2 = a^2$ about the x-axis

∗∗ **17.** $x + y - z = 0$, $x = 3y$, $3y = 2x$, $x = 3$, $z = 0$ about the z-axis

∗∗ **18.** Find the first moment about the plane $x + y + z = 1$ of a solid object of constant density ρ if it is bounded by the surfaces $x + 2y + 4z = 12$, $x = 0$, $y = 0$, $z = 0$.

∗∗ **19.** Prove the parallel axis theorem for solid objects: The moment of inertia of a uniform solid about a line is equal to the moment of inertia about a parallel line through the centre of mass of the solid plus the mass multiplied by the square of the distance between the lines.

∗∗ **20.** Find the centre of mass of a uniform solid in the first octant bounded by the ellipsoid $x^2/a^2 + y^2/b^2 + z^2/c^2 = 1$.

*** **21.** Let $\hat{\mathbf{v}} = (v_x, v_y, v_z)$ be a unit vector with its tail at the origin. Show that the moment of inertia I of any solid object occupying volume V about the line containing $\hat{\mathbf{v}}$ can be expressed in the form

$$I = v_x^2 I_x + v_y^2 I_y + v_z^2 I_z - 2v_x v_y I_{xy} - 2v_y v_z I_{yz} - 2v_z v_x I_{xz}$$

where I_x, I_y, and I_z are moments of inertia about the x-, y-, and z-axes, and

$$I_{xy} = \iiint_V xy\rho \, dV, \quad I_{yz} = \iiint_V yz\rho \, dV,$$

$$I_{xz} = \iiint_V xz\rho \, dV.$$

*** **22.** Find the moment of inertia of a uniform solid sphere of radius R about any tangent line.

14.11 Triple Iterated Integrals in Cylindrical Coordinates

In Section 14.7 we saw that polar coordinates are sometimes more convenient than Cartesian coordinates in evaluating double integrals. It should come as no surprise, then, that other co-ordinate systems can simplify the evaluation of triple integrals. Two of the most common are cylindrical and spherical coordinates. Cylindrical coordinates are useful in problems involving an axis of symmetry. They are based on a Cartesian coordinate along the axis of symmetry and polar coordinates in a plane perpendicular to the axis of symmetry. If the z-axis is the axis of symmetry and polar coordinates are defined in the xy-plane with the origin as pole and the positive x-axis as polar axis, then cylindrical coordinates and Cartesian coordinates are related by the equations

$$x = r\cos\theta, \quad y = r\sin\theta, \quad z = z \tag{14.61}$$

(see Figure 14.59). Recall that r can be expressed in terms of x and y by

$$r = \sqrt{x^2 + y^2}, \tag{14.62a}$$

and θ is defined implicitly by the equations

$$\cos\theta = \frac{x}{\sqrt{x^2 + y^2}}, \quad \sin\theta = \frac{y}{\sqrt{x^2 + y^2}}. \tag{14.62b}$$

FIGURE 14.59

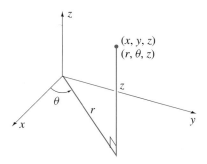

To use cylindrical coordinates in the evaluation of triple integrals, we must express equations of surfaces in terms of these coordinates. But this is very simple, for, if $F(x, y, z) = 0$ is the equation of a surface in Cartesian coordinates, then to express this equation in cylindrical coordinates we substitute from equations 14.61: $F(r\cos\theta, r\sin\theta, z) = 0$. For example, the right circular cylinder $x^2 + y^2 = 9$, which has the z-axis as its axis of symmetry, has the very simple equation $r = 3$ in cylindrical coordinates. The right circular cone $z = \sqrt{x^2 + y^2}$ also has the z-axis as its axis of symmetry, and its equation in cylindrical coordinates takes the simple form $z = r$.

Suppose that we are to evaluate the triple integral of a continuous function $f(x, y, z)$,

$$\iiint_V f(x, y, z)\, dV,$$

over some region V of space. The choice of a triple iterated integral in cylindrical coordinates implies a subdivision of V into small volumes dV by means of coordinate surfaces $r = $ constant, $\theta = $ constant, and $z = $ constant (Figure 14.60). Surfaces $r = $ constant are right circular cylinders coaxial with the z-axis; surfaces $\theta = $ constant are planes containing the z-axis and therefore perpendicular to the xy-plane; and surfaces $z = $ constant are planes parallel to the xy-plane.

FIGURE 14.60

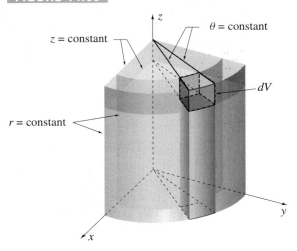

If we denote small variations in r, θ, and z for the element dV by dr, $d\theta$, and dz (Figure 14.61), then volume of the element is approximately $(r\, dr\, d\theta)dz$, where $r\, dr\, d\theta$ is the polar cross-sectional area parallel to the xy-plane. Hence, in cylindrical coordinates we set

$$dV = r\, dz\, dr\, d\theta. \tag{14.63}$$

FIGURE 14.61

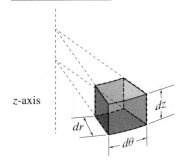

The integrand $f(x, y, z)$ is expressed in cylindrical coordinates as $f(r\cos\theta, r\sin\theta, z)$. It remains only to affix appropriate limits to the triple iterated integral, and these will, of course, depend on which of the six possible iterated integrals in cylindrical coordinates we choose. The most commonly used triple iterated integral is with respect to z, r, and θ, and in this case the z-integration adds the quantities

$$f(r\cos\theta, r\sin\theta, z)r\, dz\, dr\, d\theta$$

in a vertical column, where r and θ are constant. The limits therefore identify surfaces on which each and every column starts and stops, and generally depend on r and θ:

$$\int_{h_1(r,\theta)}^{h_2(r,\theta)} f(r\cos\theta, r\sin\theta, z)r\,dz\,dr\,d\theta.$$

The remaining integrations with respect to r and θ perform additions over the area in the xy-plane onto which all vertical columns project. Since r and θ are simply polar coordinates, the r-integration adds over small areas in a wedge and the θ-integration adds over all wedges. The triple iterated integral with respect to z, r, and θ therefore has the form

$$\int_a^b \int_{g_1(\theta)}^{g_2(\theta)} \int_{h_1(r,\theta)}^{h_2(r,\theta)} f(r\cos\theta, r\sin\theta, z)r\,dz\,dr\,d\theta.$$

We comment on the geometric aspects of these additions more fully in the following examples.

EXAMPLE 14.26

Find the volume inside both the sphere $x^2 + y^2 + z^2 = 2$ and the cylinder $x^2 + y^2 = 1$.

SOLUTION   The required volume is eight times the first octant volume shown in Figure 14.62. If we use cylindrical coordinates, volume of an elemental piece is $r\,dz\,dr\,d\theta$. A z-integration adds these pieces to give volume in a vertical column:

$$\int_0^{\sqrt{2-r^2}} r\,dz\,dr\,d\theta,$$

where limits indicate that for volume in the first octant all columns start on the xy-plane (where $z = 0$) and end on the sphere (where $r^2 + z^2 = 2$). An r-integration now adds volumes of all columns that stand on a wedge:

$$\int_0^1 \int_0^{\sqrt{2-r^2}} r\,dz\,dr\,d\theta,$$

where limits indicate that all wedges start at the origin (where $r = 0$) and end on the curve $x^2 + y^2 = 1$ (or $r = 1$) in the xy-plane. This integration therefore yields the volume of a slice (Figure 14.62). Finally, the θ-integration adds volumes of all such slices

$$\int_0^{\pi/2} \int_0^1 \int_0^{\sqrt{2-r^2}} r\,dz\,dr\,d\theta,$$

where limits 0 and $\pi/2$ identify positions of first and last wedges, respectively, in the first quadrant. We obtain the required volume, then, as

$$8\int_0^{\pi/2} \int_0^1 \int_0^{\sqrt{2-r^2}} r\,dz\,dr\,d\theta = 8\int_0^{\pi/2} \int_0^1 r\sqrt{2-r^2}\,dr\,d\theta$$

$$= 8\int_0^{\pi/2} \left\{ -\frac{1}{3}(2-r^2)^{3/2} \right\}_0^1 d\theta$$

$$= \frac{8}{3}(2\sqrt{2}-1) \int_0^{\pi/2} d\theta = \frac{4\pi}{3}(2\sqrt{2}-1).$$

FIGURE 14.62

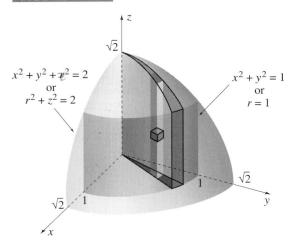

Had we used a triple iterated integral with respect to z, y, and x in Example 14.26, we would have obtained

$$\text{volume} = 8 \int_0^1 \int_0^{\sqrt{1-x^2}} \int_0^{\sqrt{2-x^2-y^2}} dz\, dy\, dx.$$

To appreciate the value of cylindrical coordinates, try to evaluate this triple iterated integral.

   As further evidence of the value of cylindrical coordinates, repeat Examples 14.23 and 14.25 using cylindrical coordinates.

EXAMPLE 14.27

Evaluate the triple iterated integral

$$I = \int_0^1 \int_0^{\sqrt{1-y^2}} \int_0^{x^2+y^2} y^2\, dz\, dx\, dy.$$

SOLUTION   The first two integrations with respect to z and x are quite straightforward, giving

$$I = \frac{1}{3} \int_0^1 y^2(1+2y^2)\sqrt{1-y^2}dy.$$

To evaluate this definite integral we could make a trigonometric substitution $y = \sin\theta$. (Try it.) Alternatively, we could consider using cylindrical coordinates on the original triple iterated integral. We first obtain the region over which integration is being performed. Limits on z indicate that all columns begin on the xy-plane and end on the paraboloid $z = x^2 + y^2$. Limits on x indicate that in the xy-plane all strips begin on the y-axis ($x = 0$) and end on the curve $x = \sqrt{1-y^2}$ (i.e., the circle $x^2 + y^2 = 1$). First and last strips are at $y = 0$ and $y = 1$, respectively. These facts determine the region of integration as that region in the first octant bounded by the paraboloid $z = x^2 + y^2$ and the right circular cylinder $x^2 + y^2 = 1$ (Figure 14.63a). If we use a triple iterated integral with respect to z, r, and θ (Figure 14.63b), then

$$I = \int_0^{\pi/2} \int_0^1 \int_0^{r^2} r^2 \sin^2 \theta \, r \, dz \, dr \, d\theta = \int_0^{\pi/2} \int_0^1 r^5 \sin^2 \theta \, dr \, d\theta = \int_0^{\pi/2} \frac{1}{6} \sin^2 \theta \, d\theta$$

$$= \frac{1}{6} \int_0^{\pi/2} \left\{ \frac{1 - \cos 2\theta}{2} \right\} d\theta = \frac{1}{12} \left\{ \theta - \frac{1}{2} \sin 2\theta \right\}_0^{\pi/2} = \frac{\pi}{24}.$$

FIGURE 14.63a

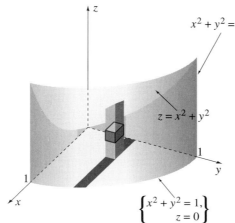

$$\left\{ \begin{array}{l} x^2 + y^2 = 1, \\ z = 0 \end{array} \right\}$$

FIGURE 14.63b

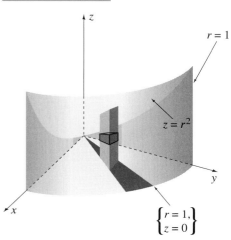

$$\left\{ \begin{array}{l} r = 1, \\ z = 0 \end{array} \right\}$$

EXERCISES 14.11

In Exercises 1–10 find the equation for the surface in cylindrical coordinates. Draw each surface and indicate whether it is symmetric about the z-axis.

∗ **1.** $x^2 + y^2 + z^2 = 4$ ∗ **2.** $x^2 + y^2 = 1$

∗ **3.** $y^2 + z^2 = 6$ ∗ **4.** $x + y = 5$

∗ **5.** $z = 2\sqrt{x^2 + y^2}$ ∗ **6.** $z = x^2$

∗ **7.** $x^2 + 4y^2 = 4$ ∗ **8.** $4z = x^2 + y^2$

∗ **9.** $y = x$ ∗ **10.** $x^2 + y^2 = 1 + z^2$

In Exercises 11–15 find the volume bounded by the surfaces.

∗ **11.** $z = \sqrt{x^2 + y^2}$, $x^2 + y^2 = 4$, $z = 0$ ∗ **12.** $z = \sqrt{2 - x^2 - y^2}$, $z = x^2 + y^2$

∗ **13.** $z = xy$, $x^2 + y^2 = 1$, $z = 0$ ∗ **14.** $z = x^2 + y^2$, $z = 4 - x^2 - y^2$

∗ **15.** $x + y + z = 2$, $x^2 + y^2 = 1$, $z = 0$

∗∗ **16.** Find the volume inside the sphere $x^2 + y^2 + z^2 = 4$ but outside the cylinder $x^2 + y^2 = 1$.

∗∗ **17.** Find the centre of mass of a uniform hemispherical solid.

** **18.** Set up the six triple iterated integrals in cylindrical coordinates for the triple integral of a function $f(x, y, z)$ over the volume bounded by surfaces $z = 1 + x^2 + y^2$, $x^2 + y^2 = 9$, $z = 0$.

** **19.** Find the moment of inertia of a uniform right circular cylinder of radius R and height h (a) about its axis; (b) about a line through the centre of its base and perpendicular to its axis.

** **20.** Find the moment of inertia of a uniform sphere of radius R about any line through its centre.

In Exercises 21–24 evaluate the triple iterated integral.

** **21.** $\displaystyle\int_0^3 \int_0^{\sqrt{9-x^2}} \int_0^{\sqrt{x^2+y^2}} dz\, dy\, dx$

** **22.** $\displaystyle\int_0^9 \int_0^{\sqrt{81-y^2}} \int_0^{\sqrt{81-x^2-y^2}} \frac{1}{\sqrt{x^2+y^2}} dz\, dx\, dy$

** **23.** $\displaystyle\int_0^4 \int_0^{\sqrt{4y-y^2}} \int_0^{y+x^2} dz\, dx\, dy$

** **24.** $\displaystyle\int_0^{\sqrt{3}/2} \int_{5-\sqrt{21-y^2}}^{\sqrt{1-y^2}} \int_0^{x^2+y^2} y\, dz\, dx\, dy$

** **25.** Find the centre of mass for the uniform solid bounded by the surfaces $x^2 + y^2 = 2x$, $z = \sqrt{x^2 + y^2}$, $z = 0$.

** **26.** Find the moment of inertia of a uniform right circular cone of radius R and height h about its axis.

** **27.** A casting is in the form of a sphere of radius b with two cylindrical holes of radius $a < b$ such that the axes of the holes pass through the centre of the sphere and intersect at right angles. What volume of metal is required for the casting?

In Exercises 28–38 find the volume described.

** **28.** Bounded by $x^2 + y^2 - z^2 = 1$, $4z^2 = x^2 + y^2$

** **29.** Bounded by $z = x^2 + y^2$, $z = 0$, $(x^2 + y^2)^2 = x^2 - y^2$

** **30.** Bounded by $x^2 + y^2 + z^2 = 4$, $x^2 + y^2 + z^2 = 16$, $z = \sqrt{x^2 + y^2}$ (smaller piece)

** **31.** Bounded by $x^2 + y^2 + z^2 = 1$, $y = x$, $x = 2y$, $z = 0$ (in the first octant)

** **32.** Inside both $2x^2 + 2y^2 + z^2 = 8$ and $x^2 + y^2 = 1$

** **33.** Bounded by $z^2 = (x^2 + y^2)^2$, $x^2 + y^2 = 2y$

** **34.** Inside $x^2 + y^2 + z^2 = a^2$ but outside $x^2 + y^2 = ay$

** **35.** Bounded by $z = 0$, $x^2 + y^2 = 1$, $z = e^{-x^2-y^2}$

** **36.** Inside $x^2 + y^2 + z^2 = 9$ but outside $x^2 + y^2 = 1 + z^2$

** **37.** Cut off from $z = x^2 + y^2$ by $z = x + y$

** **38.** Inside $x^2 + y^2 + z^2 = 4$ and below $3z = x^2 + y^2$

** **39.** Evaluate the triple integral of $\sqrt{x^2 + y^2 + z^2}$ over the volume bounded by $z = 3$ and $z = \sqrt{x^2 + y^2}$.

** **40.** Evaluate the triple integral of $f(y, z) = |yz|$ over the volume bounded by $z^2 = 1 + x^2 + y^2$ and $z = \sqrt{4 - x^2 - y^2}$.

*** **41.** A tumbler in the form of a right circular cylinder of radius R and height h is full of water. As the axis of the tumbler is tilted from the vertical, water pours over the side. Find the volume of water remaining in the tumbler as a function of the angle between the vertical and the axis of the tumbler.

*** **42.** Use cylindrical coordinates to find the volume of the torus $(\sqrt{x^2 + y^2} - a)^2 + z^2 = b^2$, $b < a$.

14.12 Triple Iterated Integrals in Spherical Coordinates

Spherical coordinates are useful in solving problems where figures are symmetric about a point. If the origin is that point, then spherical coordinates $(\Re, \theta, \phi)$ in Figure 14.64 are related to Cartesian coordinates (x, y, z) by the equations

$$x = \Re \sin \phi \cos \theta,$$
$$y = \Re \sin \phi \sin \theta, \qquad (14.64)$$
$$z = \Re \cos \phi.$$

As is the case for polar and cylindrical coordinates, without restrictions on $\Re$, θ, and ϕ, each point in space has many sets of spherical coordinates. The positive value of its spherical coordinate $\Re$ is given by

$$\Re = \sqrt{x^2 + y^2 + z^2}. \qquad (14.65)$$

The θ-coordinate in cylindrical and spherical coordinates is identical, so that no simple formula for θ in terms of x, y, and z exists. The ϕ-coordinate is the angle between the positive z-axis and the line joining the origin to the point (x, y, z), and that value of ϕ in the range $0 \leq \phi \leq \pi$ is determined by the formula

$$\phi = \text{Cos}^{-1}\left(\frac{z}{\sqrt{x^2 + y^2 + z^2}}\right) \qquad (14.66)$$

FIGURE 14.64

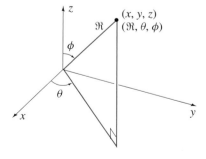

To transform equations $F(x, y, z) = 0$ of surfaces from Cartesian to spherical coordinates, we substitute from equations 14.64:

$$F(\Re \sin \phi \cos \theta, \Re \sin \phi \sin \theta, \Re \cos \phi) = 0.$$

For example, the sphere $x^2 + y^2 + z^2 = 4$ is symmetric about its centre, and its equation in spherical coordinates is simply $\Re = 2$. For the right circular cone $z = \sqrt{x^2 + y^2}$, we write

$$\Re \cos \phi = \sqrt{\Re^2 \sin^2 \phi \cos^2 \theta + \Re^2 \sin^2 \phi \sin^2 \theta} = \Re \sin \phi.$$

Consequently, $\tan \phi = 1$ or $\phi = \pi/4$; i.e., $\phi = \pi/4$ is the equation of the cone in spherical coordinates.

Suppose that we are to evaluate the triple integral of a function $f(x, y, z)$,

$$\iiint_V f(x, y, z) dV,$$

over some region V of space. The choice of a triple iterated integral in spherical coordinates implies a subdivision of V into small volumes by means of coordinate surfaces $\Re = $ constant, $\theta = $ constant, and $\phi = $ constant (Figure 14.65). Surfaces $\Re = $ constant are spheres centred at the origin; surfaces $\theta = $ constant are planes containing the z-axis; and surfaces $\phi = $ constant are right circular cones symmetric about the z-axis with the origin as apex.

FIGURE 14.65

FIGURE 14.66

If we denote small variations in $\Re, \theta$, and ϕ for the element dV by $d\Re, d\theta$, and $d\phi$ (Figure 14.66) and approximate dV by a rectangular parallelepiped with dimensions $d\Re, \Re \, d\phi$, and $\Re \sin \phi \, d\theta$, then

$$dV = (\Re \sin \phi \, d\theta)(\Re \, d\phi) d\Re = \Re^2 \sin \phi \, d\Re \, d\theta \, d\phi. \tag{14.67}$$

The integrand $f(x, y, z)$ is expressed in spherical coordinates as

$$f(\Re \sin \phi \cos \theta, \Re \sin \phi \sin \theta, \Re \cos \phi).$$

It remains only to affix appropriate limits to the triple iterated integral, and these limits depend on which of the six possible triple iterated integrals in spherical coordinates we choose. If we use a triple iterated integral with respect to $\Re$, ϕ, and θ, then it is of the form

$$\iiint_V f(x, y, z)\,dV$$
$$= \int_a^b \int_{g_1(\theta)}^{g_2(\theta)} \int_{h_1(\theta,\phi)}^{h_2(\theta,\phi)} f(\Re \sin\phi \cos\theta, \Re \sin\phi \sin\theta, \Re \cos\phi)\Re^2 \sin\phi\, d\Re\, d\phi\, d\theta.$$

The geometric interpretations of the additions represented by these integrations with respect to $\Re$, ϕ, and θ are left to the examples.

EXAMPLE 14.28 ▼

Find the volume of a sphere.

SOLUTION   The equation of a sphere of radius R centred at the origin is $x^2 + y^2 + z^2 = R^2$ or, in spherical coordinates, $\Re = R$. The volume of this sphere is eight times the first octant volume shown in Figure 14.67. If we use spherical coordinates, the volume of an elemental piece is

$$\Re^2 \sin\phi\, d\Re\, d\phi\, d\theta.$$

An $\Re$-integration adds these volumes for constant ϕ and θ to give the volume in a "spike,"

$$\int_0^R \Re^2 \sin\phi\, d\Re\, d\phi\, d\theta,$$

where limits indicate that all spikes start at the origin (where $\Re = 0$) and end on the sphere (where $\Re = R$). A ϕ-integration now adds the volumes of spikes for constant θ. This yields the volume of a slice

$$\int_0^{\pi/2} \int_0^R \Re^2 \sin\phi\, d\Re\, d\phi\, d\theta,$$

where limits indicate that all slices in the first octant start on the z-axis (where $\phi = 0$) and end on the xy-plane (where $\phi = \pi/2$). Finally, the θ-integration adds the volumes of all such slices

$$\int_0^{\pi/2} \int_0^{\pi/2} \int_0^R \Re^2 \sin\phi\, d\Re\, d\phi\, d\theta,$$

where limits 0 and $\pi/2$ identify positions of first and last slices, respectively, in the first octant. The volume of the sphere is therefore

$$8 \int_0^{\pi/2} \int_0^{\pi/2} \int_0^R \Re^2 \sin\phi\, d\Re\, d\phi\, d\theta = 8 \int_0^{\pi/2} \int_0^{\pi/2} \left\{ \frac{\Re^3}{3} \sin\phi \right\}_0^R d\phi\, d\theta$$
$$= \frac{8R^3}{3} \int_0^{\pi/2} \{-\cos\phi\}_0^{\pi/2} d\theta$$
$$= \frac{8R^3}{3} \{\theta\}_0^{\pi/2} = \frac{4}{3}\pi R^3.$$

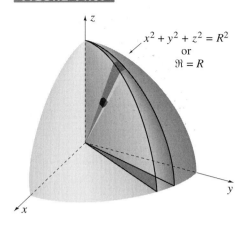

FIGURE 14.67

$$x^2 + y^2 + z^2 = R^2$$
or
$$\Re = R$$

EXAMPLE 14.29

Find the centre of mass of a solid object of constant density if it is in the shape of a right circular cone.

SOLUTION   Let the cone have altitude h and base radius R. Then its mass is

$$M = \frac{1}{3}\pi R^2 h\rho,$$

where ρ is the density of the object. If we place axes as shown in Figure 14.68, then $\overline{x} = \overline{y} = 0$; i.e., the centre of mass is on the axis of symmetry of the cone. To find $\overline{z}$ we offer three solutions.

Method 1   If we use Cartesian coordinates, the equation of the surface of the cone is of the form $z = k\sqrt{x^2 + y^2}$. Since $(0, R, h)$ is a point on the cone, $h = kR$, and therefore $k = h/R$. Now

$$M\overline{z} = 4\int_0^R \int_0^{\sqrt{R^2-x^2}} \int_{k\sqrt{x^2+y^2}}^h z\rho\,dz\,dy\,dx$$

$$= 4\rho\int_0^R \int_0^{\sqrt{R^2-x^2}} \left\{\frac{z^2}{2}\right\}_{k\sqrt{x^2+y^2}}^h dy\,dx$$

$$= 2\rho\int_0^R \int_0^{\sqrt{R^2-x^2}} \{h^2 - k^2(x^2 + y^2)\}dy\,dx$$

$$= 2\rho\int_0^R \left\{h^2 y - k^2 x^2 y - \frac{k^2 y^3}{3}\right\}_0^{\sqrt{R^2-x^2}} dx$$

$$= 2\rho\int_0^R \left\{h^2\sqrt{R^2 - x^2} - k^2 x^2\sqrt{R^2 - x^2} - \frac{k^2}{3}(R^2 - x^2)^{3/2}\right\} dx.$$

FIGURE 14.68

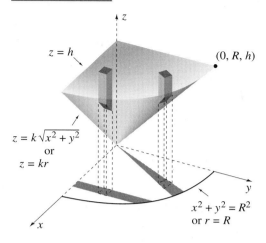

If we set $x = R\sin\theta$, then $dx = R\cos\theta\,d\theta$, and

$$M\overline{z} = 2\rho \int_0^{\pi/2} \left\{ h^2 R\cos\theta - k^2(R^2\sin^2\theta)R\cos\theta - \frac{k^2}{3}R^3\cos^3\theta \right\} R\cos\theta\,d\theta$$

$$= 2\rho R^2 \int_0^{\pi/2} \left\{ \frac{h^2}{2}(1 + \cos 2\theta) - \frac{k^2 R^2}{8}(1 - \cos 4\theta) \right.$$

$$\left. - \frac{k^2 R^2}{12}\left(1 + 2\cos 2\theta + \frac{1 + \cos 4\theta}{2}\right) \right\} d\theta$$

$$= \rho R^2 \left\{ h^2\left(\theta + \frac{\sin 2\theta}{2}\right) - \frac{k^2 R^2}{4}\left(\theta - \frac{\sin 4\theta}{4}\right) \right.$$

$$\left. - \frac{k^2 R^2}{6}\left(\frac{3\theta}{2} + \sin 2\theta + \frac{\sin 4\theta}{8}\right) \right\}_0^{\pi/2}$$

$$= \rho R^2 \left\{ \frac{\pi h^2}{2} - \frac{k^2 R^2 \pi}{8} - \frac{k^2 R^2 \pi}{8} \right\} = \frac{\pi R^2 h^2 \rho}{4}.$$

Thus,

$$\overline{z} = \frac{\pi R^2 h^2 \rho}{4} \cdot \frac{3}{\pi R^2 h\rho} = \frac{3h}{4}.$$

Method 2   If we use cylindrical coordinates, the equation of the surface of the cone is $z = kr$. From Figure 14.68, we see that

$$M\overline{z} = 4\int_0^{\pi/2}\int_0^R\int_{kr}^h z\rho r\,dz\,dr\,d\theta = 4\rho\int_0^{\pi/2}\int_0^R \left\{\frac{rz^2}{2}\right\}_{kr}^h dr\,d\theta$$

$$= 2\rho\int_0^{\pi/2}\int_0^R r(h^2 - k^2 r^2)dr\,d\theta = 2\rho\int_0^{\pi/2}\left\{\frac{r^2 h^2}{2} - \frac{k^2 r^4}{4}\right\}_0^R d\theta$$

$$= \rho\left(R^2 h^2 - \frac{k^2 R^4}{2}\right)\{\theta\}_0^{\pi/2} = \rho\left(R^2 h^2 - \frac{k^2 R^4}{2}\right)\frac{\pi}{2} = \frac{\pi R^2 h^2 \rho}{4}.$$

Again, then, $\overline{z} = 3h/4$.

Method 3 If we use spherical coordinates, then the equation of the surface of the cone is

$$\phi = \text{Cos}^{-1}\frac{h}{\sqrt{h^2 + R^2}} = \phi_1.$$

From Figure 14.69, we see that

$$M\bar{z} = 4\int_0^{\pi/2}\int_0^{\phi_1}\int_0^{h\sec\phi}(\Re\cos\phi)\rho\Re^2\sin\phi\,d\Re\,d\phi\,d\theta$$

$$= 4\rho\int_0^{\pi/2}\int_0^{\phi_1}\left\{\frac{\Re^4}{4}\sin\phi\cos\phi\right\}_0^{h\sec\phi}d\phi\,d\theta$$

$$= \rho h^4\int_0^{\pi/2}\int_0^{\phi_1}\frac{\sin\phi}{\cos^3\phi}d\phi\,d\theta = \rho h^4\int_0^{\pi/2}\left\{\frac{1}{2\cos^2\phi}\right\}_0^{\phi_1}d\theta$$

$$= \frac{\rho h^4}{2}\left\{\frac{1}{\cos^2\phi_1} - 1\right\}\{\theta\}_0^{\pi/2} = \frac{\pi\rho h^4}{4}\left\{\frac{h^2 + R^2}{h^2} - 1\right\} = \frac{\pi h^2 R^2\rho}{4}.$$

Again, $\bar{z} = 3h/4$.

FIGURE 14.69

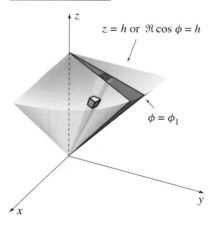

$z = h$ or $\Re\cos\phi = h$

$\phi = \phi_1$

EXERCISES 14.12

In Exercises 1–7 find the equation of the surface in spherical coordinates. Draw the surface.

∗ **1.** $x^2 + y^2 + z^2 = 4$ ∗ **2.** $x^2 + y^2 = 1$

∗ **3.** $3z = \sqrt{x^2 + y^2}$ ∗ **4.** $4z = x^2 + y^2$

∗ **5.** $y = x$ ∗ **6.** $x^2 + y^2 = 1 + z^2$

∗ **7.** $z = -2\sqrt{x^2 + y^2}$

In Exercises 8–12 find the volume described.

* **8.** Bounded by $z = \sqrt{x^2 + y^2}$, $z = \sqrt{1 - x^2 - y^2}$

* **9.** Bounded by $z = 1$, $z = \sqrt{4 - x^2 - y^2}$

* **10.** Bounded by $x^2 + y^2 + z^2 = 1$, $y = x$, $y = 2x$, $z = 0$ (in the first octant)

* **11.** Inside $x^2 + y^2 + z^2 = 2$ but outside $x^2 + y^2 = 1$

* **12.** Bounded by $z = 2\sqrt{x^2 + y^2}$, $x^2 + y^2 = 4$, $z = 0$

** **13.** Find the centre of mass of a uniform hemispherical solid.

** **14.** Find the moment of inertia of a uniform solid sphere of radius R about any line through its centre.

** **15.** Find the first moment about the yz-plane of the uniform solid in the first octant bounded by the surfaces $x^2 + y^2 + z^2 = 4$, $x^2 + y^2 + z^2 = 9$, $y = 0$, and $y = \sqrt{3}x$.

** **16.** A solid sphere of radius R and centre at the origin has a continuous charge distribution throughout. If the density of the charge is $\rho(x, y, z) = k\sqrt{x^2 + y^2 + z^2}$ coulombs per cubic metre, find the total charge in the sphere.

** **17.** Set up the six triple iterated integrals in spherical coordinates for the triple integral of a function $f(x, y, z)$ over the volume in the first octant under the sphere $x^2 + y^2 + z^2 = 2$ and inside the cylinder $x^2 + y^2 = 1$.

Evaluate the triple iterated integral in Exercises 18 and 19.

** **18.** $\displaystyle\int_0^9 \int_0^{\sqrt{81-y^2}} \int_0^{\sqrt{81-x^2-y^2}} \frac{1}{x^2 + y^2 + z^2} dz\, dx\, dy$

** **19.** $\displaystyle\int_0^1 \int_0^{\sqrt{1-x^2}} \int_{\sqrt{x^2+y^2}}^{\sqrt{2-x^2-y^2}} dz\, dy\, dx$

** **20.** Find a formula for the smaller volume bounded by $x^2 + y^2 + z^2 = R^2$ and $z = k\sqrt{x^2 + y^2}$ $(k > 0)$.

** **21.** Find the volume bounded by $(x^2 + y^2 + z^2)^2 = x$.

** **22.**   (a) Use Archimedes' principle to determine the density of a spherical ball if it floats half submerged in water.

   (b) What force is required to keep the ball with its centre at a depth of one-half the radius of the ball?

** **23.** Find the volume bounded by the surface $(x^2 + y^2 + z^2)^2 = 2z(x^2 + y^2)$.

** **24.** A sphere of radius R carries a uniform charge distribution of ρ coulombs per cubic metre (figure below). If P is a point on the z-axis (distance $d > R$ from the centre of the sphere) and dV is a small element of volume of the sphere, then potential at P due to dV is given by

$$\frac{1}{4\pi \epsilon_0} \frac{\rho dV}{s},$$

where s is distance from P to dV.

   (a) Show that in terms of spherical coordinates potential V at P due to the entire sphere is

$$V = \frac{\rho}{4\pi \epsilon_0} \int_{-\pi}^{\pi} \int_0^{\pi} \int_0^R \frac{\mathfrak{R}^2 \sin\phi}{\sqrt{\mathfrak{R}^2 + d^2 - 2\mathfrak{R}d \cos\phi}} d\mathfrak{R}\, d\phi\, d\theta.$$

(b) Because this iterated integral is very difficult to evaluate, we replace ϕ with the variable $s = \sqrt{\Re^2 + d^2 - 2\Re d \cos \phi}$. Show that with this change

$$V = \frac{\rho}{4\pi \epsilon_0 d} \int_{-\pi}^{\pi} \int_0^R \int_{d-\Re}^{d+\Re} \Re \, ds \, d\Re \, d\theta.$$

(c) Evaluate the integral in (b) above to verify that $V = Q/(4\pi \epsilon_0 d)$, where Q is total charge on the sphere.

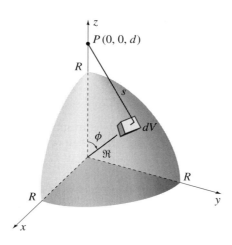

** 25. A sphere of constant density ρ and radius R is located at the origin (figure on page 1145). If a mass m is situated at a point P on the z-axis (distance $d > R$ from the centre of the sphere) and dV is a small element of volume of the sphere, then according to Newton's universal law of gravitation, the z-component of the force on m due to the mass in dV is given by

$$-\frac{Gm\rho \, dV \cos \psi}{s^2},$$

where G is a constant and s is distance between P and dV.

(a) Show that in spherical coordinates total force on m due to the entire sphere has z-component

$$F_z = -\frac{Gm\rho}{2d} \int_{-\pi}^{\pi} \int_0^{\pi} \int_0^R \left(\frac{s^2 + d^2 - \Re^2}{s^3}\right) \Re^2 \sin \phi \, d\Re \, d\phi \, d\theta.$$

(b) Use the transformation in Exercise 24(b) to write F_z in the form

$$F_z = -\frac{Gm\rho}{2d^2} \int_{-\pi}^{\pi} \int_0^R \int_{d-\Re}^{d+\Re} \Re \left(\frac{s^2 + d^2 - \Re^2}{s^2}\right) ds \, d\Re \, d\theta,$$

and show that $F_z = -GmM/d^2$ where M is the total mass of the sphere.

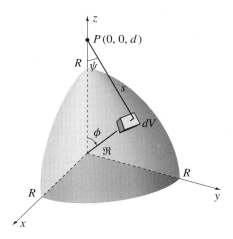

$***$ **26.** A homogeneous solid is bounded by two concentric spheres of radii a and b ($a < b$). Verify that the force that this layer exerts on a point mass at any point interior to the shell vanishes.

14.13 Derivatives of Definite Integrals

If a function $f(x, y)$ of two independent variables is integrated with respect to y from $y = a$ to $y = b$, the result depends on x. Suppose we denote this function by $F(x)$:

$$F(x) = \int_a^b f(x, y)dy. \tag{14.68}$$

To calculate the derivative $F'(x)$ of this function, we should first integrate with respect to y and then differentiate with respect to x. The following theorem indicates that differentiation can be done first and integration later.

> **THEOREM 14.3**
>
> If the partial derivative $\partial f/\partial x$ of $f(x, y)$ is continuous on a rectangle $a \le y \le b$, $c \le x \le d$, then for $c < x < d$
>
> $$\frac{d}{dx} \int_a^b f(x, y)dy = \int_a^b \frac{\partial f(x, y)}{\partial x}dy. \tag{14.69}$$

PROOF If we define

$$g(x) = \int_a^b \frac{\partial f(x, y)}{\partial x}dy$$

as the right-hand side of equation 14.69, then this function is defined for $c \le x \le d$. In fact, because $\partial f/\partial x$ is continuous, $g(x)$ is also continuous. We can therefore integrate $g(x)$ with respect to x from $x = c$ to any value of x in the interval $c \le x \le d$:

$$\int_c^x g(x)dx = \int_c^x \int_a^b \frac{\partial f(x, y)}{\partial x}dy\,dx.$$

This double iterated integral represents the double integral of $\partial f/\partial x$ over the rectangle in Figure 14.70, and if we reverse the order of integration, we have

$$\int_c^x g(x)dx = \int_a^b \int_c^x \frac{\partial f(x,y)}{\partial x}dx\, dy = \int_a^b \{f(x,y)\}_c^x dy$$

$$= \int_a^b \{f(x,y) - f(c,y)\}dy = \int_a^b f(x,y)dy - \int_a^b f(c,y)dy.$$

Because the second integral on the right is independent of x, if we differentiate this equation with respect to x, we get

$$\frac{d}{dx}\int_c^x g(x)dx = \frac{d}{dx}\int_a^b f(x,y)dy.$$

But Theorem 6.7 gives

$$g(x) = \frac{d}{dx}\int_a^b f(x,y)dy.$$

This completes the proof.

FIGURE 14.70

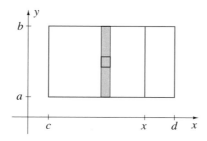

The limits on the integral in 14.68 need not be numerical constants; so far as the integration with respect to y is concerned, x is constant, and therefore a and b could be functions of x:

$$F(x) = \int_{a(x)}^{b(x)} f(x,y)dy.$$

Indeed, this is precisely what does occur in the evaluation of double integrals by means of double iterated integrals. To differentiate $F(x)$ now is more complicated than in Theorem 14.3 because we must also account for the x's in $a(x)$ and $b(x)$. The chain rule can be used to develop a formula for $F'(x)$.

THEOREM 14.4 (Leibnitz's rule)

If the partial derivative $\partial f/\partial x$ is continuous on the area bounded by the curves $y = a(x)$, $y = b(x)$, $x = c$, and $x = d$, then

$$\frac{d}{dx}\int_{a(x)}^{b(x)} f(x,y)dy = \int_{a(x)}^{b(x)} \frac{\partial f(x,y)}{\partial x}dy + f[x,b(x)]\frac{db}{dx} - f[x,a(x)]\frac{da}{dx}.$$

$$\text{(14.70)}$$

PROOF   If $F(x) = \int_{a(x)}^{b(x)} f(x, y)dy$, the schematic diagram

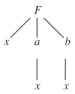

gives the chain rule

$$F'(x) = \left.\frac{\partial F}{\partial x}\right)_{a,b} + \left.\frac{\partial F}{\partial b}\right)_{a,x} \frac{db}{dx} + \left.\frac{\partial F}{\partial a}\right)_{b,x} \frac{da}{dx}.$$

The first term is the situation covered in Theorem 14.3, and therefore

$$\left.\frac{\partial F}{\partial x}\right)_{a,b} = \int_a^b \frac{\partial f(x, y)}{\partial x} dy.$$

Since Theorem 6.7 indicates that

$$\frac{d}{db} \int_a^b f(y)dy = f(b),$$

it follows that

$$\frac{\partial}{\partial b} \int_a^b f(x, y)dy = f(x, b).$$

In other words,

$$\left.\frac{\partial F}{\partial b}\right)_{a,x} = f(x, b).$$

Furthermore,

$$\left.\frac{\partial F}{\partial a}\right)_{b,x} = \frac{\partial}{\partial a} \int_a^b f(x, y)dy = -\frac{\partial}{\partial a} \int_a^b f(x, y)dy = -f(x, a).$$

Substitution of these facts into the chain rule now gives Leibnitz's rule. ∎

Our derivation of Leibnitz's rule for the differentiation of a definite integral that depends on a parameter (x in this case) shows that the first term accounts for those x's in the integrand, and the second and third terms for the x's in the upper and lower limits. The following geometric interpretation of Leibnitz's rule emphasizes this same point.

Suppose the function $f(x, y)$ has only positive values so that the surface $z = f(x, y)$ lies completely above the xy-plane (Figure 14.71). Equations $y = a(x)$ and $y = b(x)$ describe cylindrical walls standing on the curves $y = a(x)$, $z = 0$ and $y = b(x)$, $z = 0$ in the xy-plane. Were we to slice through the surface $z = f(x, y)$ with a plane $x = $ constant, an area would be defined in this plane bounded on the top by $z = f(x, y)$, on the sides by $y = a(x)$ and $y = b(x)$, and on the bottom by the xy-plane. This area is clearly defined by the definite integral

$$\int_{a(x)}^{b(x)} f(x, y)dy$$

in Leibnitz's rule, and as x varies so too does the area. Note, in particular, that as the plane varies, area changes, not only because height of the surface $z = f(x, y)$ changes but also because width of the area varies (i.e., the two cylindrical walls are not a constant distance apart). The first term in Leibnitz's rule accounts for vertical variation, whereas the remaining two terms represent variations due to fluctuating width.

FIGURE 14.71

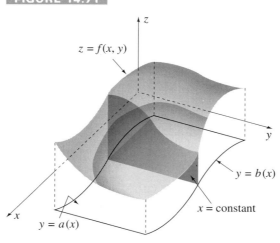

EXAMPLE 14.30

If $F(x) = \int_x^{2x} (y^3 \ln y + x^3 e^y)dy$, find $F'(x)$.

SOLUTION   With Leibnitz's rule, we have

$$F'(x) = \int_x^{2x} (3x^2 e^y)dy + \{(2x)^3 \ln (2x) + x^3 e^{2x}\}(2) - \{x^3 \ln x + x^3 e^x\}(1)$$

$$= \{3x^2 e^y\}_x^{2x} + 16x^3(\ln 2 + \ln x) + 2x^3 e^{2x} - x^3 \ln x - x^3 e^x$$

$$= 3x^2 e^{2x} - 3x^2 e^x + (16 \ln 2)x^3 + 15x^3 \ln x + 2x^3 e^{2x} - x^3 e^x$$

$$= 3x^2 e^x(e^x - 1) + x^3(16 \ln 2 + 15 \ln x + 2e^{2x} - e^x).$$

In Exercise 6 you are asked to evaluate the integral defining $F(x)$ in this example and then to differentiate the resulting function with respect to x. It will be clear, then, that for this example Leibnitz's rule simplifies the calculations considerably.

EXAMPLE 14.31

Evaluate

$$\int_0^1 \frac{y^x - 1}{\ln y} dy \quad \text{for } x > -1.$$

SOLUTION We use Leibnitz's rule in this example to avoid finding an antiderivative for $(y^x - 1)/\ln y$. If we set

$$F(x) = \int_0^1 \frac{y^x - 1}{\ln y} dy$$

and use Leibnitz's rule, we have

$$F'(x) = \int_0^1 \frac{\partial}{\partial x}\left\{\frac{y^x - 1}{\ln y}\right\} dy = \int_0^1 \frac{y^x \ln y}{\ln y} dy = \int_0^1 y^x dy = \left\{\frac{y^{x+1}}{x + 1}\right\}_0^1 = \frac{1}{x + 1}.$$

It follows, therefore, that $F(x)$ must be of the form

$$F(x) = \ln(x + 1) + C.$$

But, from the definition of $F(x)$ as an integral, it is clear that $F(0) = 0$, and hence $C = 0$. Thus,

$$\int_0^1 \frac{y^x - 1}{\ln y} dy = \ln(x + 1).$$

Note that this problem originally had nothing whatsoever to do with Leibnitz's rule. But by introducing the rule, we were able to find a simple formula for $F'(x)$, and this immediately led to $F(x)$. This can be a very useful technique for evaluating definite integrals that depend on a parameter.

EXAMPLE 14.32

FIGURE 14.72

When a shell is fired from the artillery gun in Figure 14.72, the barrel recoils along a well-lubricated guide and its motion is braked by a battery of heavy springs. We set up a coordinate system where $x = 0$ represents the firing position of the gun when no stretch or compression exists in the springs. Suppose that, when the gun is fired, the horizontal component of the force causing recoil is $g(t)$ ($t = $ time). If the mass of the gun is m and the effective spring constant for the battery of springs is k, then Newton's second law states that

$$m\frac{d^2x}{dt^2} = g(t) - kx,$$

or

$$m\frac{d^2x}{dt^2} + kx = g(t).$$

This differential equation can be solved by a method called **variation of parameters**, and the solution is

$$x(t) = A\cos\sqrt{\frac{k}{m}}t + B\sin\sqrt{\frac{k}{m}}t + \frac{1}{\sqrt{mk}}\int_0^t g(u)\sin\left(\sqrt{\frac{k}{m}}(t - u)\right) du,$$

where A and B are arbitrary constants. We do not discuss differential equations until Chapter 16, but with Leibnitz's rule it is possible to verify that this function is indeed a solution. Do so.

SOLUTION   To differentiate the integral we rewrite Leibnitz's rule in terms of the variables of this problem:

$$\frac{d}{dt}\int_{a(t)}^{b(t)} f(t,u)du = \int_{a(t)}^{b(t)} \frac{\partial f(t,u)}{\partial t}du + f[t,b(t)]\frac{db}{dt} - f[t,a(t)]\frac{da}{dt}.$$

If we now apply this formula to the definite integral in $x(t)$ where

$$f(t,u) = g(u)\sin\left[\sqrt{\frac{k}{m}}(t-u)\right],$$

then

$$\frac{dx}{dt} = -\sqrt{\frac{k}{m}}A\sin\sqrt{\frac{k}{m}}t + \sqrt{\frac{k}{m}}B\cos\sqrt{\frac{k}{m}}t$$

$$+ \frac{1}{\sqrt{mk}}\left\{\int_0^t g(u)\sqrt{\frac{k}{m}}\cos\left(\sqrt{\frac{k}{m}}(t-u)\right)du + g(t)\sin\left(\sqrt{\frac{k}{m}}(t-t)\right)\right\}$$

$$= \sqrt{\frac{k}{m}}\left\{-A\sin\sqrt{\frac{k}{m}}t + B\cos\sqrt{\frac{k}{m}}t\right\} + \frac{1}{m}\int_0^t g(u)\cos\left(\sqrt{\frac{k}{m}}(t-u)\right)du.$$

We now use Leibnitz's rule once more to find d^2x/dt^2:

$$\frac{d^2x}{dt^2} = \sqrt{\frac{k}{m}}\left\{-\sqrt{\frac{k}{m}}A\cos\sqrt{\frac{k}{m}}t - \sqrt{\frac{k}{m}}B\sin\sqrt{\frac{k}{m}}t\right\}$$

$$+ \frac{1}{m}\left\{\int_0^t -g(u)\sqrt{\frac{k}{m}}\sin\left(\sqrt{\frac{k}{m}}(t-u)\right)du + g(t)\cos\left(\sqrt{\frac{k}{m}}(t-t)\right)\right\}$$

$$= -\frac{k}{m}\left\{A\cos\sqrt{\frac{k}{m}}t + B\sin\sqrt{\frac{k}{m}}t\right\}$$

$$- \sqrt{\frac{k}{m^3}}\int_0^t g(u)\sin\left(\sqrt{\frac{k}{m}}(t-u)\right)du + \frac{1}{m}g(t)$$

$$= -\frac{k}{m}\left\{A\cos\sqrt{\frac{k}{m}}t + B\sin\sqrt{\frac{k}{m}}t + \frac{1}{\sqrt{mk}}\int_0^t g(u)\sin\left(\sqrt{\frac{k}{m}}(t-u)\right)du\right\}$$

$$+ \frac{1}{m}g(t)$$

$$= -\frac{k}{m}x(t) + \frac{1}{m}g(t).$$

In other words, the function $x(t)$ satisfies the differential equation

$$m\frac{d^2x}{dt^2} = -kx(t) + g(t),$$

and the proof is complete.

This example illustrates that Leibnitz's rule is essential to the manipulation of solutions of differential equations that are represented as definite integrals. Here is another example, but with a partial differential equation instead of an ordinary differential equation.

| **EXAMPLE 14.33**

The one-dimensional heat conduction equation

$$\frac{\partial T}{\partial t} = k\frac{\partial^2 T}{\partial x^2},$$

where $k > 0$ is a constant, describes temperature in a rod (see Exercise 41 in Section 13.5). Show that the function defined by the definite integral

$$T(x, t) = \frac{2T_0}{\sqrt{\pi}}\int_0^{x/(2\sqrt{kt})} e^{-v^2}\, dv$$

satisfies the heat conduction equation.

SOLUTION   We use Leibnitz's rule 14.70 to obtain $\partial T/\partial t$ and $\partial T/\partial x$:

$$\frac{\partial T}{\partial t} = \frac{2T_0}{\sqrt{\pi}}e^{-[x/(2\sqrt{kt})]^2}\left(\frac{-x}{4\sqrt{k}t^{3/2}}\right) = \frac{-xT_0}{2\sqrt{k\pi}\,t^{3/2}}e^{-x^2/(4kt)},$$

$$\frac{\partial T}{\partial x} = \frac{2T_0}{\sqrt{\pi}}e^{-[x/(2\sqrt{kt})]^2}\left(\frac{1}{2\sqrt{kt}}\right) = \frac{T_0}{\sqrt{k\pi t}}e^{-x^2/(4kt)}.$$

It follows that

$$\frac{\partial^2 T}{\partial x^2} = \frac{T_0}{\sqrt{k\pi t}}e^{-x^2/(4kt)}\left(\frac{-2x}{4kt}\right) = \frac{1}{k}\left[\frac{-xT_0}{2\sqrt{k\pi}\,t^{3/2}}e^{-x^2/(4kt)}\right] = \frac{1}{k}\frac{\partial T}{\partial t}.$$

In Exercises 1–5 use Leibnitz's rule to find the derivative of $F(x)$. Check your result by evaluating the integral and then differentiating.

* **1.** $F(x) = \displaystyle\int_0^3 (x^2 y^2 + 3xy)\,dy$ 　　　　 * **2.** $F(x) = \displaystyle\int_1^x (x^2/y^2 + e^y)\,dy$

* **3.** $F(x) = \displaystyle\int_{x-1}^{x^2} (x^3 y + y^2 + 1)\,dy$ 　　 * **4.** $F(x) = \displaystyle\int_{x^2}^{x^3-1} (x + y \ln y)\,dy$

* **5.** $F(x) = \displaystyle\int_0^x \dfrac{y - x}{y + x}\,dy$

* **6.** Find $F'(x)$ in Example 14.30 (page 1148) by first evaluating the definite integral.

* **7.** Use the result of Example 14.31 (page 1149) to prove that

$$\int_0^1 \frac{x^p - x^q}{\ln x}\,dx = \ln\left(\frac{p+1}{q+1}\right),$$

provided $p > -1$ and $q > -1$.

* **8.** Find $F'(x)$ if $F(x) = \displaystyle\int_{\sin x}^{e^x} \sqrt{1 + y^3}\,dy$.

In Exercises 9–11 use Leibnitz's rule to verify that the function $y(x)$ satisfies the differential equation.

** **9.** $y(x) = \dfrac{1}{x^2} \displaystyle\int_0^x t^2 f(t)\,dt; \quad x\dfrac{dy}{dx} + 2y = xf(x)$

** **10.** $y(x) = \dfrac{1}{2} \displaystyle\int_0^x f(t)(e^{x-t} - e^{t-x})\,dt; \quad \dfrac{d^2 y}{dx^2} - y = f(x)$

** **11.** $y(x) = \dfrac{1}{\sqrt{2}} \displaystyle\int_0^x e^{2(t-x)} \sin[\sqrt{2}(x - t)]f(t)\,dt; \quad \dfrac{d^2 y}{dx^2} + 4\dfrac{dy}{dx} + 6y = f(x)$

** **12.** Given that

$$\int_0^b \frac{1}{1 + ax}\,dx = \frac{1}{a}\ln(1 + ab),$$

find a formula for $\displaystyle\int_0^b \frac{x}{(1 + ax)^2}\,dx$.

** **13.** Given that

$$\int \frac{1}{\sqrt{a^2 - x^2}}\,dx = \operatorname{Sin}^{-1}\left(\frac{x}{a}\right) + C,$$

find a formula for $\displaystyle\int \frac{1}{(a^2 - x^2)^{3/2}}\,dx$.

** **14.** Given that

$$\int \frac{1}{a^2 + x^2} dx = \frac{1}{a} \operatorname{Tan}^{-1}\left(\frac{x}{a}\right) + C,$$

find a formula for $\int \frac{1}{(a^2 + x^2)^3} dx$.

** **15.** Use the result that

$$\int_0^{\pi/2} \frac{1}{a^2 \cos^2 x + b^2 \sin^2 x} dx = \frac{\pi}{2|ab|}$$

to find a formula for $\int_0^{\pi/2} \frac{1}{(a^2 \cos^2 x + b^2 \sin^2 x)^2} dx$.

In Exercises 16 and 17 use Leibnitz's rule to evaluate the integral.

** **16.** $\int_0^{\pi} \frac{\ln(1 + a \cos x)}{\cos x} dx$ where $|a| < 1$

** **17.** $\int_0^{\infty} \frac{\operatorname{Tan}^{-1}(ax)}{x(1 + x^2)} dx$ where $a > 0$

** **18.** Verify that the function

$$T(x, t) = \int_0^{(1-x)/(2\sqrt{t})} e^{-v^2} dv + \int_0^{(1+x)/(2\sqrt{t})} e^{-v^2} dv$$

satisfies the partial differential equation

$$\frac{\partial T}{\partial t} = \frac{\partial^2 T}{\partial x^2}.$$

** **19.** (a) What is the domain of the function $F(x) = \int_0^9 \ln(1 - x^2 y^2) \, dy$? What is $F(0)$?

(b) Find $F'(x)$ by Leibnitz's rule. What is $F'(0)$?

(c) Show that the graph of the function $F(x)$ is concave downward for all x in its domain of definition.

*** **20.** Laplace's equation for a function $u(r, \theta)$ in polar coordinates is

$$\frac{\partial^2 u}{\partial r^2} + \frac{1}{r} \frac{\partial u}{\partial r} + \frac{1}{r^2} \frac{\partial^2 u}{\partial \theta^2} = 0.$$

(see Example 13.19 in Section 13.6). If the values of $u(r, \theta)$ are specified on the circle $r = R$ as $u(R, \phi)$, $-\pi < \phi \leq \pi$, then Poisson's integral formula states that the value of $u(r, \theta)$ interior to this circle is defined by

$$u(r, \theta) = \frac{R^2 - r^2}{2\pi} \int_{-\pi}^{\pi} \frac{u(R, \phi)}{R^2 + r^2 - 2rR \cos(\theta - \phi)} d\phi.$$

Show that this function does indeed satisfy Laplace's equation.

∗∗∗ **21.** If the function $u(r, \theta)$ in Exercise 20 must satisfy the condition

$$\frac{\partial u(R, \theta)}{\partial r} = f(\theta), \quad -\pi < \theta \le \pi,$$

on the circle $r = R$ (rather than have its values prescribed on the circle), then values of $u(r, \theta)$ interior to the circle are given by

$$u(r, \theta) = C - \frac{R}{2\pi} \int_{-\pi}^{\pi} f(u) \ln \left[R^2 + r^2 - 2rR \cos(\theta - u) \right] du$$

where C is an arbitrary constant. This is called *Dini's integral*. Verify that it satisfies Laplace's equation in polar coordinates.

SUMMARY

The definite integral of a function $f(x)$ from $x = a$ to $x = b$ is a limit of a sum

$$\int_a^b f(x)dx = \lim_{\|\Delta x_i\| \to 0} \sum_{i=1}^n f(x_i^*)\Delta x_i.$$

In this chapter we extended this idea to define double integrals of functions $f(x, y)$ over regions in the xy-plane and triple integrals of functions $f(x, y, z)$ over regions of space. Each is once again the limit of a sum:

$$\iint_R f(x, y)dA = \lim_{\|\Delta A_i\| \to 0} \sum_{i=1}^n f(x_i^*, y_i^*)\Delta A_i,$$

$$\iiint_V f(x, y, z)dV = \lim_{\|\Delta V_i\| \to 0} \sum_{i=1}^n f(x_i^*, y_i^*, z_i^*)\Delta V_i.$$

To evaluate double integrals we use double iterated integrals in Cartesian or polar coordinates. Which is the more useful in a given problem depends on the shape of the region R and the form of the function $f(x, y)$. For instance, circles centred at the origin and straight lines through the origin are represented very simply in polar coordinates, and therefore a region R with these curves as boundaries immediately suggests the use of polar coordinates. On the other hand, curves that can be described in the form $y = f(x)$, where $f(x)$ is a polynomial, a rational function, or a transcendental function, often suggest using double iterated integrals in Cartesian coordinates. Each integration in a double iterated integral can be interpreted geometrically, and through these interpretations it is a simple matter to find appropriate limits for the integrals. In particular, for a double iterated integral in Cartesian coordinates, the inner integration is over the rectangles in a strip (horizontal or vertical), and the outer integration adds over all strips. In polar coordinates, the inner integral is inside either a wedge or a ring, and the outer integral adds over all wedges or rings.

To evaluate triple integrals we use triple iterated integrals in Cartesian, cylindrical, or spherical coordinates. Once again the limits on these integrals can be determined by interpreting the summations geometrically. For example, the first integration in a triple iterated integral in Cartesian coordinates is always over the boxes in a column, the second over the rectangles inside a strip, and the third over all strips.

We used double integrals to find plane areas, volumes of solids of revolution, centroids, moments of inertia, fluid forces, and areas of surfaces. We dealt with these same applications (with the exception of surface area) in Chapter 7 using the definite integral, but with some difficulty: Volumes required two methods, shells and washers; centroids required an averaging formula for the first moment of a rectangle that has its length perpendicular to the axis about which a moment is required; moments of inertia needed a "one-third cubed formula" for rectangles with lengths perpendicular to the axis about which the moment of inertia is required; and fluid forces required horizontal rectangles. On the other hand, double integrals eliminate these difficulties but, more importantly, provide a unified approach to all applications.

In Section 14.13 we used double integrals to verify Leibnitz's rule for differentiating definite integrals that depend on a parameter.

KEY TERMS

In reviewing this chapter, you should be able to define or discuss the following key terms:

Double integral	Double iterated integral
Area	Volume of a solid of revolution
Fluid pressure	Centre of mass
Moment of inertia	Surface area
Double integrals in polar coordinates	Triple integral
Triple iterated integral	Volume
Cylindrical coordinates	Triple integrals in cylindrical coordinates
Spherical coordinates	Triple integrals in spherical coordinates
Leibnitz's rule	

REVIEW EXERCISES

In Exercises 1–21 evaluate the integral over the region.

* **1.** $\iint_R (2x + y)\, dA$ where R is bounded by $y = x$, $y = 0$, $x = 2$

* **2.** $\iiint_V xyz\, dV$ where V is bounded by $y = z$, $x = 0$, $y = 3$, $x = 1$, $z = 0$

* **3.** $\iint_R x^3 y^2\, dA$ where R is bounded by $x = 1$, $x = -1$, $y = 2$, $y = -2$

* **4.** $\iiint_V (x^2 - y^3)\, dV$ where V is bounded by $z = xy$, $z = 0$, $x = 1$, $y = 1$

* **5.** $\iiint_V (x^2 - y^2)\, dV$ where V is the region in Exercise 4

* **6.** $\iint_R y\, dA$ where R is bounded by $y = (x - 1)^2$, $y = x + 1$

* **7.** $\iint_R xy^2\, dA$ where R is bounded by $x = 2 - 2y^2$, $x = -y^2$

* **8.** $\iint_R x^2 y\, dA$ where R is the region of Exercise 7

* **9.** $\displaystyle\iiint_V (x^2 + y^2 + z^2)\, dV$ where V is bounded by $z = x$, $z = -x$, $y = 0$, $y = 1$, $z = 2$

* **10.** $\displaystyle\iint_R (xy - x^2 y^2)\, dA$ where R is bounded by $y = 2x^2$, $y = 4 - 2x^2$

* **11.** $\displaystyle\iint x \sin y\, dA$ where R is bounded by $x = \sqrt{1 - y}$, $x = 0$, $y = 0$

is known to be the gradient of some function $f(x, y, z)$, then

$$\frac{\partial f}{\partial x}\hat{\mathbf{i}} + \frac{\partial f}{\partial y}\hat{\mathbf{j}} + \frac{\partial f}{\partial z}\hat{\mathbf{k}} = (3x^2 yz + z^2)\hat{\mathbf{i}} + (x^3 z + 2y)\hat{\mathbf{j}} + (x^3 y + 2xz + 1)\hat{\mathbf{k}}.$$

Since two vectors are equal if and only if they have identical components, we can say that

$$\frac{\partial f}{\partial x} = 3x^2 yz + z^2, \quad \frac{\partial f}{\partial y} = x^3 z + 2y, \quad \frac{\partial f}{\partial z} = x^3 y + 2xz + 1.$$

Integration of the first of these with respect to x, holding y and z constant, implies that $f(x, y, z)$ must be of the form

$$f(x, y, z) = x^3 yz + xz^2 + v(y, z)$$

for some function $v(y, z)$. To determine $v(y, z)$, we substitute this $f(x, y, z)$ into the second equation,

$$x^3 z + \frac{\partial v}{\partial y} = x^3 z + 2y,$$

and this implies that

$$\frac{\partial v}{\partial y} = 2y.$$

Consequently,

$$v(y, z) = y^2 + w(z)$$

for some $w(z)$, and therefore

$$f(x, y, z) = x^3 yz + xz^2 + y^2 + w(z).$$

We now know both the x- and y-dependence of $f(x, y, z)$. To find $w(z)$ we substitute into the equation for $\partial f / \partial z$:

$$x^3 y + 2xz + \frac{dw}{dz} = x^3 y + 2xz + 1.$$

This equation requires

$$\frac{dw}{dz} = 1,$$

from which we have

$$w(z) = z + C, \quad C = \text{a constant.}$$

Thus,

$$f(x, y, z) = x^3 yz + xz^2 + y^2 + z + C,$$

and this represents all functions that have a gradient equal to the given vector $\mathbf{F}$.

    A more difficult question is to determine whether a given vector field $\mathbf{F}(x, y, z)$ is the gradient of some scalar function $f(x, y, z)$. In many examples, we can say that if the above procedure fails, then the answer is no. For instance, if $\mathbf{F} = x^2 y\hat{\mathbf{i}} + xy\hat{\mathbf{j}} + z\hat{\mathbf{k}}$, and we attempt to find a function $f(x, y, z)$ so that $\nabla f = \mathbf{F}$, then

$$\frac{\partial f}{\partial x} = x^2 y, \quad \frac{\partial f}{\partial y} = xy, \quad \frac{\partial f}{\partial z} = z.$$

The first implies that

$$f(x, y, z) = \frac{x^3 y}{3} + v(y, z),$$

and when this is substituted into the second, we get

$$\frac{x^3}{3} + \frac{\partial v}{\partial y} = xy, \quad \text{or} \quad \frac{\partial v}{\partial y} = xy - \frac{x^3}{3}.$$

But this is an impossible situation since v is to be a function of y and z only. How then could its derivative depend on x? Although this type of argument will suffice in most examples, it is really not a satisfactory mathematical answer. The following theorem gives a test by which to determine whether a given vector function is the gradient of some scalar function.

> ### THEOREM 15.1
>
> Suppose the components $P(x, y, z)$, $Q(x, y, z)$, and $R(x, y, z)$ of $\mathbf{F} = P\hat{\mathbf{i}} + Q\hat{\mathbf{j}} + R\hat{\mathbf{k}}$ have continuous first partial derivatives in a domain D. If there exists a function $f(x, y, z)$ defined in D such that $\nabla f = \mathbf{F}$, then $\nabla \times \mathbf{F} = \mathbf{0}$. Conversely, if D is simply-connected, and $\nabla \times \mathbf{F} = \mathbf{0}$ in D, then there exists a function $f(x, y, z)$ such that $\nabla f = \mathbf{F}$ in D.

It is obvious that if $\mathbf{F} = \nabla f$ then $\nabla \times \mathbf{F} = \mathbf{0}$, for this is the result of equation 15.14. To prove the converse result requires Stokes's theorem from Section 15.10, and a proof is therefore delayed until that time. Notice that in the converse result the domain (open, connected set) must be simply-connected. This is our first encounter with a situation where the nature of a region is important to the result.

In the special case in which $\mathbf{F}$ is a vector field in the xy-plane, the equation $\nabla \times \mathbf{F} = \mathbf{0}$ is still the condition for existence of a function $f(x, y)$ such that $\nabla f = \mathbf{F} = P\hat{\mathbf{i}} + Q\hat{\mathbf{j}}$, but the condition reduces to

$$\mathbf{0} = \begin{vmatrix} \hat{\mathbf{i}} & \hat{\mathbf{j}} & \hat{\mathbf{k}} \\ \partial/\partial x & \partial/\partial y & \partial/\partial z \\ P & Q & 0 \end{vmatrix} = \left(\frac{\partial Q}{\partial x} - \frac{\partial P}{\partial y} \right) \hat{\mathbf{k}},$$

or

$$\frac{\partial Q}{\partial x} = \frac{\partial P}{\partial y}. \tag{15.16}$$

EXAMPLE 15.3

Find, if possible, a function $f(x, y)$ such that

$$\nabla f = \left(\frac{x^3 - 2y^2}{x^3 y} \right) \hat{\mathbf{i}} + \left(\frac{y^2 - x^3}{x^2 y^2} \right) \hat{\mathbf{j}}.$$

SOLUTION We first note that

$$\frac{\partial}{\partial x} \left(\frac{y^2 - x^3}{x^2 y^2} \right) = \frac{\partial}{\partial x} \left(\frac{1}{x^2} - \frac{x}{y^2} \right) = \frac{-2}{x^3} - \frac{1}{y^2}$$

and

$$\frac{\partial}{\partial y} \left(\frac{x^3 - 2y^2}{x^3 y} \right) = \frac{\partial}{\partial y} \left(\frac{1}{y} - \frac{2y}{x^3} \right) = \frac{-1}{y^2} + \frac{-2}{x^3}.$$

Since the components of **F** are undefined whenever $x = 0$ or $y = 0$, we can state that in any simply-connected domain that does not contain points on either of the axes there is a function $f(x, y)$ such that $\nabla f = \mathbf{F}$. To find $f(x, y)$, we set

$$\frac{\partial f}{\partial x} = \frac{1}{y} - \frac{2y}{x^3}, \quad \frac{\partial f}{\partial y} = \frac{1}{x^2} - \frac{x}{y^2}.$$

From the first equation, we have

$$f(x, y) = \frac{x}{y} + \frac{y}{x^2} + v(y),$$

which substituted into the second equation gives us

$$-\frac{x}{y^2} + \frac{1}{x^2} + \frac{dv}{dy} = \frac{1}{x^2} - \frac{x}{y^2}.$$

Consequently,

$$\frac{dv}{dy} = 0 \quad \text{or} \quad v(y) = C,$$

and therefore

$$f(x, y) = \frac{x}{y} + \frac{y}{x^2} + C.$$

EXERCISES 15.1

In Exercises 1–9 determine whether the set of points in the xy-plane is open, closed, connected, a domain, and/or a simply-connected domain.

∗ **1.** $x^2 + (y + 1)^2 < 4$ ∗ **2.** $x^2 + (y - 3)^2 \leq 4$

∗ **3.** $0 < x^2 + (y - 1)^2 < 16$ ∗ **4.** $1 < (x - 4)^2 + (y + 1)^2 \leq 9$

∗ **5.** $x > 3$ ∗ **6.** $y \leq -2$

∗ **7.** $2(x - 1)^2 - (y + 2)^2 < 16$

∗ **8.** All points satisfying $x^2 + y^2 < 1$ or $(x - 2)^2 + y^2 < 1$

∗ **9.** $4(x + 1)^2 + 9(y - 2)^2 > 20$

∗∗ **10.** Give definitions of the following for sets of points in xyz-space: interior, exterior, and boundary points; open, closed, and connected sets; domain and simply-connected domain.

In Exercises 11–19 determine whether the set of points in space is open, closed, connected, a domain, and/or a simply-connected domain.

∗ **11.** $x^2 + y^2/4 + z^2/9 < 1$ ∗ **12.** $z \geq x^2 + y^2$

∗ **13.** $x^2 + y^2 + z^2 > 0$

∗ **14.** $1 < x^2 + y^2 + z^2 < 4$, $x \geq 0$, $y \geq 0$, $z \geq 0$

∗ **15.** $x^2 + y^2 > 0$ ∗ **16.** $|z| > 0$

∗ **17.** $z^2 > x^2 + y^2$ ∗ **18.** $z^2 > x^2 + y^2 - 1$

∗ **19.** $z^2 < x^2 + y^2 - 1$

∗∗ **20.** Prove that the only nonempty set in the xy-plane that is both open and closed is the whole plane.

In Exercises 21–40 calculate the required quantity.

* **21.** ∇f if $f(x, y, z) = 3x^2y - y^3z^2$

* **22.** ∇f if $f(x, y, z) = (x^2 + y^2 + z^2)^{-1/2}$

* **23.** ∇f if $f(x, y) = \text{Tan}^{-1}(y/x)$

* **24.** ∇f at $(1, 2)$ if $f(x, y) = x^3y - 2x \cos y$

* **25.** ∇f at $(1, -1, 4)$ if $f(x, y, z) = e^{xyz}$

* **26.** $\nabla \cdot \mathbf{F}$ if $\mathbf{F}(x, y, z) = 2xe^y\hat{\mathbf{i}} + 3x^2z\hat{\mathbf{j}} - 2x^2yz\hat{\mathbf{k}}$

* **27.** $\nabla \cdot \mathbf{F}$ if $\mathbf{F}(x, y) = x \ln y\hat{\mathbf{i}} - y^3e^x\hat{\mathbf{j}}$

* **28.** $\nabla \cdot \mathbf{F}$ if $\mathbf{F}(x, y, z) = \sin(x^2 + y^2 + z^2)\hat{\mathbf{i}} + \cos(y + z)\hat{\mathbf{j}}$

* **29.** $\nabla \cdot \mathbf{F}$ if $\mathbf{F}(x, y) = e^x\hat{\mathbf{i}} + e^y\hat{\mathbf{j}}$

* **30.** $\nabla \cdot \mathbf{F}$ at $(1, 1, 1)$ if $\mathbf{F}(x, y, z) = x^2y^3\hat{\mathbf{i}} - 3xy\hat{\mathbf{j}} + z^2\hat{\mathbf{k}}$

* **31.** $\nabla \cdot \mathbf{F}$ at $(-1, 3)$ if $\mathbf{F}(x, y) = (x + y)^2(\hat{\mathbf{i}} + \hat{\mathbf{j}})$

* **32.** $\nabla \cdot \mathbf{F}$ if $\mathbf{F}(x, y, z) = (x\hat{\mathbf{i}} + y\hat{\mathbf{j}} + z\hat{\mathbf{k}})/\sqrt{x^2 + y^2 + z^2}$

* **33.** $\nabla \cdot \mathbf{F}$ if $\mathbf{F}(x, y) = \text{Cot}^{-1}(xy)\hat{\mathbf{i}} + \text{Tan}^{-1}(xy)\hat{\mathbf{j}}$

* **34.** $\nabla \times \mathbf{F}$ if $\mathbf{F}(x, y, z) = x^2z\hat{\mathbf{i}} + 12xyz\hat{\mathbf{j}} + 32y^2z^4\hat{\mathbf{k}}$

* **35.** $\nabla \times \mathbf{F}$ if $\mathbf{F}(x, y) = xe^y\hat{\mathbf{i}} - 2xy^2\hat{\mathbf{j}}$

* **36.** $\nabla \times \mathbf{F}$ if $\mathbf{F}(x, y, z) = x^2\hat{\mathbf{i}} + y^2\hat{\mathbf{j}} + z^2\hat{\mathbf{k}}$

* **37.** $\nabla \times \mathbf{F}$ at $(1, -1, 1)$ if $\mathbf{F}(x, y, z) = xz^3\hat{\mathbf{i}} - 2x^2yz\hat{\mathbf{j}} + 2yz^4\hat{\mathbf{k}}$

* **38.** $\nabla \times \mathbf{F}$ at $(2, 0)$ if $\mathbf{F}(x, y) = y\hat{\mathbf{i}} - x\hat{\mathbf{j}}$

* **39.** $\nabla \times \mathbf{F}$ if $\mathbf{F}(x, y, z) = \ln(x + y + z)(\hat{\mathbf{i}} + \hat{\mathbf{j}} + \hat{\mathbf{k}})$

* **40.** $\nabla \times \mathbf{F}$ if $\mathbf{F}(x, y) = \text{Sec}^{-1}(x + y)\hat{\mathbf{i}} + \text{Csc}^{-1}(y + x)\hat{\mathbf{j}}$

* **41.** If $\mathbf{F} = x^2y\hat{\mathbf{i}} - 2xz\hat{\mathbf{j}} + 2yz\hat{\mathbf{k}}$, find $\nabla \times (\nabla \times \mathbf{F})$.

* **42.**   (a) Verify that Laplace's equation 13.12 can be expressed in the form $\nabla \cdot \nabla f = 0$.
   (b) Show that $f(x, y, z) = (x^2 + y^2 + z^2)^{-1/2}$ satisfies Laplace's equation.

** **43.** Prove properties 15.7–15.12, 15.14, and 15.15.

** **44.** A gas is moving through some region D of space. If we follow a particular particle of the gas, it traces out some curved path $C : x = x(t)$, $y = y(t)$, $z = z(t)$, where t is time. If $\rho(x, y, z, t)$ is the density of gas at any point in D at time t, then along C we can express density in terms of t only, $\rho = \rho[x(t), y(t), z(t), t]$. Show that along C,

$$\frac{d\rho}{dt} = \frac{\partial \rho}{\partial t} + \nabla \rho \cdot \frac{d\mathbf{r}}{dt},$$

where $\mathbf{r} = x\hat{\mathbf{i}} + y\hat{\mathbf{j}} + z\hat{\mathbf{k}}$.

In Exercises 45–49 find all functions $f(x, y)$ such that $\nabla f = \mathbf{F}$.

* **45.** $\mathbf{F}(x, y) = 2xy\hat{\mathbf{i}} + x^2\hat{\mathbf{j}}$ * **46.** $\mathbf{F}(x, y) = (3x^2y^2 + 3)\hat{\mathbf{i}} + (2x^3y + 2)\hat{\mathbf{j}}$

* **47.** $\mathbf{F}(x, y) = e^y\hat{\mathbf{i}} + (xe^y + 4y^2)\hat{\mathbf{j}}$ * **48.** $\mathbf{F}(x, y) = (x + y)^{-1}(\hat{\mathbf{i}} + \hat{\mathbf{j}})$

** **49.** $\mathbf{F}(x, y) = -xy(1 - x^2y^2)^{-1/2}(y\hat{\mathbf{i}} + x\hat{\mathbf{j}})$

In Exercises 50–55 find all functions $f(x, y, z)$ such $\nabla f = \mathbf{F}$.

$*$ **50.** $\mathbf{F}(x, y, z) = x\hat{\mathbf{i}} + y\hat{\mathbf{j}} + z\hat{\mathbf{k}}$

$*$ **51.** $\mathbf{F}(x, y, z) = yz\hat{\mathbf{i}} + xz\hat{\mathbf{j}} + (yx - 3)\hat{\mathbf{k}}$

$*$ **52.** $\mathbf{F}(x, y, z) = (1 + x + y + z)^{-1}(\hat{\mathbf{i}} + \hat{\mathbf{j}} + \hat{\mathbf{k}})$

$*$ **53.** $\mathbf{F}(x, y, z) = (2x/y^2 + 1)\hat{\mathbf{i}} - (2x^2/y^3)\hat{\mathbf{j}} - 2z\hat{\mathbf{k}}$

$*$ **54.** $\mathbf{F}(x, y, z) = (1 + x^2y^2)^{-1}(y\hat{\mathbf{i}} + x\hat{\mathbf{j}}) + z\hat{\mathbf{k}}$

$*$ **55.** $\mathbf{F}(x, y, z) = (3x^2y + yz + 2xz^2)\hat{\mathbf{i}} + (xz + x^3 + 3z^2 - 6y^2z)\hat{\mathbf{j}} + (2x^2z + 6yz - 2y^3 + xy)\hat{\mathbf{k}}$

$**$ **56.**   (a) Find constants a, b, and c in order that the vector field

$$\mathbf{F} = (x^2 + 2y + az)\hat{\mathbf{i}} + (bx - 3y - z)\hat{\mathbf{j}} + (4x + cy + 2z)\hat{\mathbf{k}}$$

will be irrotational.

(b) If $\mathbf{F}$ is irrotational, find a scalar function $f(x, y, z)$ such that $\nabla f = \mathbf{F}$.

$**$ **57.** A vector field $\mathbf{F}(x, y, z)$ is said to be *solenoidal* if $\nabla \cdot \mathbf{F} = 0$.

(a) Are either of $\mathbf{F} = (2x^2 + 8xy^2z)\hat{\mathbf{i}} + (3x^3y - 3xy)\hat{\mathbf{j}} - (4y^2z^2 + 2x^3z)\hat{\mathbf{k}}$ or $xyz^2\mathbf{F}$ solenoidal?

(b) Show that $\nabla f \times \nabla g$ is solenoidal for arbitrary functions $f(x, y, z)$ and $g(x, y, z)$ which have continuous second partial derivatives.

$**$ **58.** Associated with every electric field is a scalar function $V(x, y, z)$ called potential. It is defined by $\mathbf{E} = -\nabla V$, where $\mathbf{E}$ is a vector field called the electric field intensity. In addition, if a point charge Q is placed at a point (x, y, z) in the electric field, then the force $\mathbf{F}$ on Q is $\mathbf{F} = Q\mathbf{E}$.

(a) If the force on Q due to a charge q at the origin is

$$\mathbf{F} = \frac{qQ}{4\pi\epsilon_0|\mathbf{r}|^3}\mathbf{r},$$

where $\mathbf{r} = x\hat{\mathbf{i}} + y\hat{\mathbf{j}} + z\hat{\mathbf{k}}$ and ϵ_0 is a constant, find $V(x, y, z)$ for the field due to q.

(b) If the entire xy-plane is given a uniform charge density σ units of charge per unit area, it is found that the force on a charge Q placed z units above the plane is $\mathbf{F} = [Q\sigma/(2\epsilon_0)]\hat{\mathbf{k}}$. Find the potential V for the electric field due to this charge distribution.

$**$ **59.** Show that if $\mathbf{v} = \boldsymbol{\omega} \times \mathbf{r}$, where $\boldsymbol{\omega}$ is a constant vector, and $\mathbf{r} = x\hat{\mathbf{i}} + y\hat{\mathbf{j}} + z\hat{\mathbf{k}}$, then $\boldsymbol{\omega} = (1/2)(\nabla \times \mathbf{v})$.

$**$ **60.** Show that if a function $f(x, y, z)$ satisfies Laplace's equation 13.12, then its gradient is both irrotational and solenoidal.

$**$ **61.** Theorem 15.1 indicates that a vector field $\mathbf{F}$ is the gradient of some scalar field if $\nabla \times \mathbf{F} = \mathbf{0}$. Sometimes a given vector field $\mathbf{F}$ is the curl of another field $\mathbf{v}$, that is, $\mathbf{F} = \nabla \times \mathbf{v}$. The following theorem indicates when this is the case: Let D be the interior of a sphere in which the components of a vector field $\mathbf{F}$ have continuous first partial derivatives. Then there exists a vector field $\mathbf{v}$ defined in D such that $\mathbf{F} = \nabla \times \mathbf{v}$ if and only if $\nabla \cdot \mathbf{F} = 0$ in D. In other words, $\mathbf{F}$ is the curl of a vector field if and only if $\mathbf{F}$ is solenoidal.

(a) Show that if $\mathbf{F} = P\hat{\mathbf{i}} + Q\hat{\mathbf{j}} + R\hat{\mathbf{k}}$ is solenoidal, then the components of $\mathbf{v} = L\hat{\mathbf{i}} + M\hat{\mathbf{j}} + N\hat{\mathbf{k}}$ would have to satisfy the equations

$$P = \frac{\partial N}{\partial y} - \frac{\partial M}{\partial z}, \quad Q = \frac{\partial L}{\partial z} - \frac{\partial N}{\partial x}, \quad R = \frac{\partial M}{\partial x} - \frac{\partial L}{\partial y}.$$

(b) Show that the vector field $\mathbf{v}(x, y, z)$ defined by

$$\mathbf{v}(x, y, z) = \int_0^1 t\mathbf{F}(tx, ty, tz) \times (x, y, z)\, dt$$

satisfies these equations. In other words, this formula defines a possible vector $\mathbf{v}$. Is it unique?

(c) Show that if $\mathbf{F}$ satisfies the property that $\mathbf{F}(tx, ty, tz) = t^n \mathbf{F}(x, y, z)$, then

$$\mathbf{v}(x, y, z) = \frac{1}{n+2} \mathbf{F} \times \mathbf{r}, \quad \mathbf{r} = x\hat{\mathbf{i}} + y\hat{\mathbf{j}} + z\hat{\mathbf{k}}.$$

In Exercises 62–64 verify that the vector field is solenoidal, and then use the formulas in Exercise 61 to find a vector field $\mathbf{v}$ such that $\mathbf{F} = \nabla \times \mathbf{v}$.

** **62.** $\mathbf{F} = x\hat{\mathbf{i}} + y\hat{\mathbf{j}} - 2z\hat{\mathbf{k}}$

** **63.** $\mathbf{F} = (1+x)\hat{\mathbf{i}} - (x+z)\hat{\mathbf{k}}$

** **64.** $\mathbf{F} = 2x^2\hat{\mathbf{i}} - y^2\hat{\mathbf{j}} + (2yz - 4xz)\hat{\mathbf{k}}$

15.2 Line Integrals

Just as definite integrals, double integrals, and triple integrals are defined as limits of sums, so too are line and surface integrals. The only difference is that line integrals are applied to functions defined along curves, and surface integrals involve functions defined on surfaces.

A curve C in space is defined parametrically by three functions

$$C : \begin{array}{l} x = x(t), \\ y = y(t), \quad \alpha \le t \le \beta \\ z = z(t), \end{array} \tag{15.17}$$

where α and β specify initial and final points A and B of the curve, respectively (Figure 15.9). The direction of a curve is from initial to final point, and in Section 12.7 we agreed to parametrize a curve using parameters that increase in the direction of the curve.

FIGURE 15.9

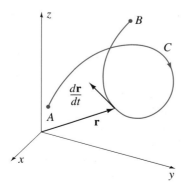

The curve is said to be continuous if each of the functions $x(t)$, $y(t)$, and $z(t)$ is continuous (implying that C is at no point separated). It is said to be smooth if each of these functions has a continuous first derivative; geometrically, this means that the tangent vector $d\mathbf{r}/dt = x'(t)\hat{\mathbf{i}} + y'(t)\hat{\mathbf{j}} + z'(t)\hat{\mathbf{k}}$ turns gradually or smoothly along C. A continuous curve that is not smooth but can be divided into a finite number of smooth subcurves is said to be piecewise smooth.

Suppose a function $f(x, y, z)$ is defined along a curve C joining A to B (Figure 15.10). We divide C into n subcurves of lengths Δs_1, Δs_2, ..., Δs_n by any $n - 1$ consecutive points $A = P_0$, P_1, P_2, ..., P_{n-1}, $P_n = B$, whatsoever. On each subcurve of length Δs_i $(i = 1, \ldots, n)$ we choose an arbitrary point $P_i^*(x_i^*, y_i^*, z_i^*)$. We then form the sum

$$f(x_1^*, y_1^*, z_1^*)\Delta s_1 + f(x_2^*, y_2^*, z_2^*)\Delta s_2 + \cdots + f(x_n^*, y_n^*, z_n^*)\Delta s_n$$

$$= \sum_{i=1}^{n} f(x_i^*, y_i^*, z_i^*)\Delta s_i. \tag{15.18}$$

FIGURE 15.10

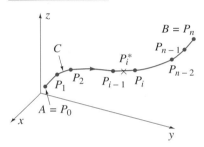

If this sum approaches a limit as the number of subcurves becomes increasingly large and the length of every subcurve approaches zero, we call the limit the line integral of $f(x, y, z)$ along the curve C, and denote it by

$$\int_C f(x, y, z)\,ds = \lim_{\|\Delta s_i\| \to 0} \sum_{i=1}^{n} f(x_i^*, y_i^*, z_i^*)\Delta s_i. \tag{15.19}$$

A more appropriate name might be curvilinear integral, rather than line integral, but line integral is the accepted terminology. We regard the word "line" as meaning "curved line" rather than "straight line."

For definition 15.19 to be useful we demand that the limit be independent of the manner of subdivision of C and choice of star-points on the subcurves. Theorem 15.2 indicates that for continuous functions defined on smooth curves, this requirement is always satisfied.

THEOREM 15.2

Let $f(x, y, z)$ be continuous on a smooth curve C of finite length, $C : x = x(t)$, $y = y(t)$, $z = z(t)$, $\alpha \leq t \leq \beta$. Then the line integral of $f(x, y, z)$ along C exists and can be evaluated by means of the following definite integral:

$$\int_C f(x, y, z)\,ds = \int_{\alpha}^{\beta} f[x(t), y(t), z(t)]\sqrt{\left(\frac{dx}{dt}\right)^2 + \left(\frac{dy}{dt}\right)^2 + \left(\frac{dz}{dt}\right)^2}\,dt. \tag{15.20}$$

It is not necessary to memorize 15.20 as a formula. Simply put, the right side is obtained by expressing x, y, z, and ds in terms of t and interpreting the result as a definite integral with respect to t. To be more explicit, recall from equation 12.76 that, when length along a curve is measured from its initial point, then an infinitesimal length ds along C corresponding to an increment dt in t is given by

$$ds = \sqrt{(dx)^2 + (dy)^2 + (dz)^2} = \sqrt{\left\{ \left(\frac{dx}{dt} \right)^2 + \left(\frac{dy}{dt} \right)^2 + \left(\frac{dz}{dt} \right)^2 \right\} (dt)^2}$$

$$= \sqrt{\left(\frac{dx}{dt} \right)^2 + \left(\frac{dy}{dt} \right)^2 + \left(\frac{dz}{dt} \right)^2} \, dt.$$

If we substitute this into the left side of 15.20 and at the same time use the equations for C to express $f(x, y, z)$ in terms of t, then

$$\int_C f(x, y, z) ds = \int_C f[x(t), y(t), z(t)] \sqrt{\left(\frac{dx}{dt} \right)^2 + \left(\frac{dy}{dt} \right)^2 + \left(\frac{dz}{dt} \right)^2} \, dt.$$

But if we now interpret the right side of this equation as the definite integral of $f[x(t), y(t), z(t)]$ $\sqrt{(dx/dt)^2 + (dy/dt)^2 + (dz/dt)^2}$ with respect to t and affix limits $t = \alpha$ and $t = \beta$ that identify end points of C, we obtain 15.20.

To evaluate a line integral, then, we express $f(x, y, z)$ and ds in terms of some parameter along C and evaluate the resulting definite integral. If equations for C are given in the form $C : y = y(x)$, $z = z(x)$, $x_A \le x \le x_B$, then x is a convenient parameter, and equation 15.20 takes the form

$$\int_C f(x, y, z) ds = \int_{x_A}^{x_B} f[x, y(x), z(x)] \sqrt{1 + \left(\frac{dy}{dx} \right)^2 + \left(\frac{dz}{dx} \right)^2} \, dx. \qquad (15.21)$$

Similar expressions exist if either y or z is a convenient parameter.

When C is piecewise smooth rather than smooth, we evaluate line integrals along each smooth subcurve and add results.

| **EXAMPLE 15.4** ◀

Evaluate the line integral of $f(x, y, z) = 8x + 6xy + 30z$ from $A(0, 0, 0)$ to $B(1, 1, 1)$

(a) along the straight line joining A to B with parametrization

$$C_1 : x = t, \quad y = t, \quad z = t, \quad 0 \le t \le 1.$$

(b) along the straight line in (a) with parametrization

$$C_1 : x = -1 + \frac{t}{2}, \quad y = -1 + \frac{t}{2}, \quad z = -1 + \frac{t}{2}, \quad 2 \le t \le 4.$$

(c) along the curve (Figure 15.11)

$$C_2 : x = t, \quad y = t^2, \quad z = t^3, \quad 0 \le t \le 1.$$

FIGURE 15.11

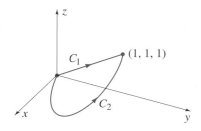

SOLUTION

(a)

$$\int_{C_1} (8x + 6xy + 30z)ds = \int_0^1 (8t + 6t^2 + 30t)\sqrt{(1)^2 + (1)^2 + (1)^2}dt$$

$$= \sqrt{3} \int_0^1 (38t + 6t^2)dt = \sqrt{3}\{19t^2 + 2t^3\}_0^1$$

$$= 21\sqrt{3}.$$

(b)

$$\int_{C_1} (8x + 6xy + 30z)ds = \int_2^4 \{8(-1 + t/2) + 6(-1 + t/2)^2 + 30(-1 + t/2)\}$$

$$\times \sqrt{(1/2)^2 + (1/2)^2 + (1/2)^2}dt$$

$$= \frac{\sqrt{3}}{2} \int_2^4 \{38(-1 + t/2) + 6(-1 + t/2)^2\}dt$$

$$= \frac{\sqrt{3}}{2}\{38(-1 + t/2)^2 + 4(-1 + t/2)^3\}_2^4$$

$$= \frac{\sqrt{3}}{2}\{38 + 4\} = 21\sqrt{3}.$$

(c)

$$\int_{C_2} (8x + 6xy + 30z)ds = \int_0^1 (8t + 6t^3 + 30t^3)\sqrt{(1)^2 + (2t)^2 + (3t^2)^2}dt$$

$$= \int_0^1 (8t + 36t^3)\sqrt{1 + 4t^2 + 9t^4}dt$$

$$= \{\tfrac{2}{3}(1 + 4t^2 + 9t^4)^{3/2}\}_0^1 = \tfrac{2}{3}(14\sqrt{14} - 1).$$

Parts (a) and (b) of this example suggest that the value of a line integral does not depend on the particular parametrization of the curve used in its evaluation. This is indeed true, and should perhaps be expected since definition 15.20 makes no reference whatsoever to parametrization of the curve. For a proof of this fact see Exercise 37. Different parameters normally lead to different definite integrals, but they all give exactly the same value for the line integral. Parts (a) and (c) illustrate that the line integral does depend on the curve joining points A and B; i.e., the value of the line integral changed when the curve C joining A and B changed.

EXAMPLE 15.5

Evaluate the line integral of $f(x, y) = x^2 + y^2$ once clockwise around the circle $x^2 + y^2 = 4$, $z = 0$.

SOLUTION   If we use the parametrization

$$C : x = 2\cos t, \quad y = -2\sin t, \quad z = 0, \quad 0 \le t \le 2\pi$$

(Figure 15.12), then

$$\int_C (x^2 + y^2)\,ds = \int_0^{2\pi} (4)\sqrt{\left(\frac{dx}{dt}\right)^2 + \left(\frac{dy}{dt}\right)^2}\,dt$$

$$= 4\int_0^{2\pi} \sqrt{(-2\sin t)^2 + (-2\cos t)^2}\,dt = 8\int_0^{2\pi} dt = 16\pi.$$

FIGURE 15.12

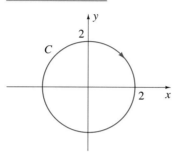

The value of this line integral, as well as many other line integrals in the xy-plane, can be given a geometric interpretation. Suppose a function $f(x, y)$ is positive along a curve C in the xy-plane. If at each point of C we draw a vertical line of height $z = f(x, y)$, then a vertical wall is constructed as shown in Figure 15.13. Since ds is an elemental piece of length along C, the quantity $f(x, y)\,ds$ can be interpreted as approximately the area of the vertical wall projecting onto ds. Because the line integral

$$\int_C f(x, y)\,ds,$$

like all integrals, is a limit-summation process, we interpret the value of this line integral as the total area of the vertical wall. Correct as this interpretation is, it really is of little use in the evaluation of line integrals and, in addition, the interpretation is valid only if the curve C along which the line integral is performed is contained in a plane.

FIGURE 15.13

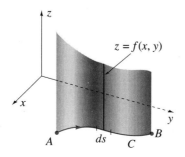

Because a line integral is a limit-summation, it should be obvious that the line integral

$$\int_C ds \tag{15.22}$$

represents the length of the curve C. If C is a curve in the xy-plane, we substitute $ds = \sqrt{(dx)^2 + (dy)^2}$, and if C is a curve in space, then $ds = \sqrt{(dx)^2 + (dy)^2 + (dz)^2}$. This agrees with the results of equations 12.69 and 12.76.

We make one last point about notation. To indicate that a line integral is being evaluated around a closed curve, we usually draw a circle on the integral sign, as follows:

$$\oint_C f(x, y, z)ds.$$

Such would be the case for the curve of intersection of the cylinder $x^2 + y^2 = 1$ and the plane $x + z = 1$ (Figure 15.14).

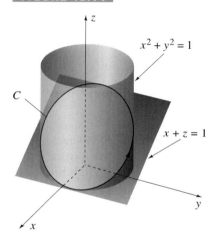

When C is a closed curve in the xy-plane that does not cross itself, we indicate the direction along C by an arrowhead on the circle. For the curves shown in Figure 15.15 we write

$$\oint_{C_1} f(x, y)ds \quad \text{and} \quad \oint_{C_2} f(x, y)ds.$$

In Exercises 30–33 use the fact that in polar coordinates small lengths along a curve can be expressed in the form $ds = \sqrt{r^2 + (dr/d\theta)^2}\, d\theta$ (see formula 10.14) to evaluate the line integral.

** **30.** $\displaystyle\int_C \frac{x}{\sqrt{x^2 + y^2}}\, ds$ where C is the first quadrant part of the limacon $r = 2 - \sin\theta$ starting from the point on the x-axis

** **31.** $\displaystyle\oint_C (x^2 + y^2)\, ds$ where C is the cardioid $r = 1 + \cos\theta$

** **32.** $\displaystyle\int_C xy\, ds$ where C is the spiral $r = e^\theta$ from $\theta = 0$ to $\theta = 2\pi$

** **33.** $\displaystyle\oint_C \cos^3 2\theta\, ds$ around the first quadrant loop of the lemniscate $r^2 = \sin 2\theta$

In Exercises 34 and 35 find a definite integral which can be used to evaluate the line integral. Use Simpson's rule with 10 equal subdivisions to approximate the definite integral.

** **34.** $\displaystyle\int_C (x^2 y + z)\, ds$ where C is the curve $z = x^2 + y^2$, $y + x = 1$ from $(-1, 2, 5)$ to $(1, 0, 1)$

** **35.** $\displaystyle\oint_C x^2 y^2\, ds$ where C is the ellipse $4x^2 + 9y^2 = 36$, $z = 0$

*** **36.** Find the surface area of the torus obtained by rotating, around the y-axis, the circle $(x - a)^2 + y^2 = b^2$ $(a > b)$.

*** **37.** Show that the value of a line integral is independent of the parameter used to specify the curve.

15.3 Line Integrals Involving Vector Functions

There are many ways in which $f(x, y, z)$ in the line integral

$$\int_C f(x, y, z)\, ds \tag{15.23}$$

can arise. According to equation 15.22 we choose $f(x, y, z) = 1$ in order to find the length of the curve C; Exercise 17 in Section 15.2 indicates that for areas of the surfaces traced out when a curve in the xy-plane is rotated around the y- and x-axes, we choose $f(x, y)$ equal to $2\pi x$ and $2\pi y$, respectively.

The most important and common type of line integral occurs when $f(x, y, z)$ is specified as the tangential component of some given vector field $\mathbf{F}(x, y, z)$ defined along C; i.e., $f(x, y, z)$ itself is not given, but $\mathbf{F}$ is, and to find $f(x, y, z)$ we must calculate the tangential component of $\mathbf{F}$ along C. By the tangential component of $\mathbf{F}(x, y, z)$ along C we mean the component of $\mathbf{F}$ along that tangent vector to C which points in the same direction as C (Figure 15.16).

FIGURE 15.16

FIGURE 15.16

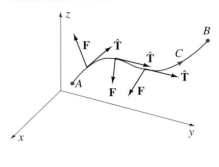

FIGURE 15.17

In Section 12.9 we saw that, if s is a measure of length along a curve C from A to B (Figure 15.17), and if s is chosen equal to zero at A, then a unit tangent vector pointing in the direction of motion along C is

$$\hat{\mathbf{T}} = \frac{d\mathbf{r}}{ds}. \tag{15.24}$$

Consequently, if $f(x, y, z)$ is the tangential component of $\mathbf{F}(x, y, z)$ along C, then

$$f(x, y, z) = \mathbf{F} \cdot \hat{\mathbf{T}} = \mathbf{F} \cdot \frac{d\mathbf{r}}{ds}, \tag{15.25}$$

and we can write that

$$\int_C f(x, y, z)\,ds = \int_C \mathbf{F} \cdot \hat{\mathbf{T}}\,ds = \int_C \mathbf{F} \cdot \frac{d\mathbf{r}}{ds}\,ds = \int_C \mathbf{F} \cdot d\mathbf{r}. \tag{15.26}$$

If the components of the vector field $\mathbf{F}(x, y, z)$ are

$$\mathbf{F}(x, y, z) = P(x, y, z)\hat{\mathbf{i}} + Q(x, y, z)\hat{\mathbf{j}} + R(x, y, z)\hat{\mathbf{k}}, \tag{15.27}$$

then

$$\int_C \mathbf{F} \cdot d\mathbf{r} = \int_C (P\,dx + Q\,dy + R\,dz),$$

and if parentheses are omitted, we have

$$\int_C \mathbf{F} \cdot d\mathbf{r} = \int_C P\,dx + Q\,dy + R\,dz. \tag{15.28}$$

This discussion has shown that when the integrand $f(x, y, z)$ of a line integral is specified as the tangential component of $\mathbf{F} = P\hat{\mathbf{i}} + Q\hat{\mathbf{j}} + R\hat{\mathbf{k}}$ along C, the product $f(x, y, z)\,ds$ can be replaced by the sum of products $P\,dx + Q\,dy + R\,dz$:

$$\int_C f(x, y, z)\,ds = \int_C \mathbf{F} \cdot d\mathbf{r} = \int_C P\,dx + Q\,dy + R\,dz. \tag{15.29}$$

According to the results of Section 15.2, evaluation of this line integral can be accomplished by expressing $P\,dx + Q\,dy + R\,dz$ in terms of any parametric representation of C and evaluating the resulting definite integral.

EXAMPLE 15.6

Evaluate

$$\int_C \frac{z}{y}dx + (x^2 + y^2 + z^2)dz,$$

where C is the first octant intersection of $x^2 + y^2 = 1$ and $z = 2x + 4$ joining $(1, 0, 6)$ to $(0, 1, 4)$.

SOLUTION   If we choose the parametrization

$$x = \cos t, \quad y = \sin t, \quad z = 2\cos t + 4, \quad 0 \le t \le \pi/2,$$

for C (Figure 15.18), then

$$\int_C \frac{z}{y}dx + (x^2 + y^2 + z^2)dz$$

$$= \int_0^{\pi/2} \left\{ \left(\frac{2\cos t + 4}{\sin t} \right)(-\sin t\, dt) \right.$$

$$\left. + (\cos^2 t + \sin^2 t + 4\cos^2 t + 16\cos t + 16)(-2\sin t\, dt) \right\}$$

$$= -2\int_0^{\pi/2} \{\cos t + 2 + 17\sin t + 4\cos^2 t \sin t + 16\cos t \sin t\}dt$$

$$= -2\left\{ \sin t + 2t - 17\cos t - \frac{4\cos^3 t}{3} + 8\sin^2 t \right\}_0^{\pi/2} = -2\pi - \frac{164}{3}.$$

FIGURE 15.18

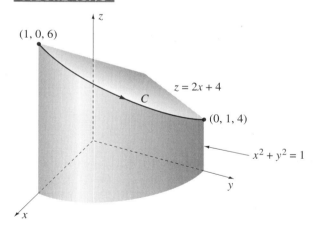

$(1, 0, 6)$

$z = 2x + 4$

C

$(0, 1, 4)$

$x^2 + y^2 = 1$

EXAMPLE 15.7

Evaluate

$$\oint_C y^2 dx + x^2 dy,$$

where C is the closed curve in Figure 15.19.

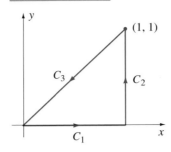

FIGURE 15.19

SOLUTION   If we start at the origin and denote the three straight-line paths by

$$C_1 : \quad y = 0, \quad 0 \le x \le 1,$$

$$C_2 : \quad x = 1, \quad 0 \le y \le 1,$$

$$C_3 : \quad y = x = 1 - t, \quad 0 \le t \le 1,$$

then

$$\oint_C y^2 dx + x^2 dy = \int_{C_1} y^2 dx + x^2 dy + \int_{C_2} y^2 dx + x^2 dy + \int_{C_3} y^2 dx + x^2 dy$$

$$= \int_0^1 0\, dx + x^2 0 + \int_0^1 y^2 0 + 1\, dy$$

$$+ \int_0^1 (1 - t)^2 (-dt) + (1 - t)^2 (-dt)$$

$$= \{y\}_0^1 + \left\{ \frac{2}{3}(1 - t)^3 \right\}_0^1 = 1 - \frac{2}{3} = \frac{1}{3}.$$

EXAMPLE 15.8

If a curve C has initial and final points A and B, then the curve that traces the same points but has initial and final points B and A is denoted by $-C$. Show that

$$\int_{-C} \mathbf{F} \cdot d\mathbf{r} = - \int_C \mathbf{F} \cdot d\mathbf{r}. \tag{15.30}$$

SOLUTION   When the functions in 15.17 are parametric equations for C, then parametric equations for $-C$ are

$$-C : \quad x = x(t), \quad y = y(t), \quad z = z(t), \quad \beta \ge t \ge \alpha.$$

To obtain an increasing parameter along $-C$, we set $u = -t$, in which case

$$-C : \quad x = x(-u), \quad y = y(-u), \quad z = z(-u), \quad -\beta \le u \le -\alpha.$$

If $\mathbf{F} = P\hat{\mathbf{i}} + Q\hat{\mathbf{j}} + R\hat{\mathbf{k}}$, then the value of the line integral along $-C$ can be expressed as a definite integral with respect to u:

$$
\int_{-C} \mathbf{F} \cdot d\mathbf{r} = \int_{-C} P\,dx + Q\,dy + R\,dz
$$

$$
= \int_{-\beta}^{-\alpha} \left\{ P[x(-u), y(-u), z(-u)]\frac{dx}{du} \right.
$$

$$
\left. + Q[x(-u), y(-u), z(-u)]\frac{dy}{du} + R[x(-u), y(-u), z(-u)]\frac{dz}{du} \right\} du.
$$

If we now change variables of integration by setting $t = -u$,

$$
\frac{dx}{du} = \frac{dx}{dt}\frac{dt}{du} = -\frac{dx}{dt},
$$

and similarly for dy/du and dz/du. Consequently,

$$
\int_{-C} \mathbf{F} \cdot d\mathbf{r} = \int_{\beta}^{\alpha} \left\{ P[x(t), y(t), z(t)]\left(-\frac{dx}{dt} \right) + Q[x(t), y(t), z(t)]\left(-\frac{dy}{dt} \right) \right.
$$

$$
\left. + R[x(t), y(t), z(t)]\left(-\frac{dz}{dt} \right) \right\}(-dt)
$$

$$
= -\int_{\alpha}^{\beta} \left\{ P[x(t), y(t), z(t)]\frac{dx}{dt} + Q[x(t), y(t), z(t)]\frac{dy}{dt} \right.
$$

$$
\left. + R[x(t), y(t), z(t)]\frac{dz}{dt} \right\} dt
$$

$$
= -\int_{C} \mathbf{F} \cdot d\mathbf{r}.
$$

According to this example, when the direction along a curve is reversed, the value of a line integral of form 15.28 along the new curve is the negative of its value along the original curve. This is because the signs of dx, dy, and dz are reversed when the direction along C is reversed. This is not the case for line integral 15.20; ds does not change sign when the direction along C is reversed.

Line integrals of form 15.28 are singled out for special consideration because they arise in so many physical problems. For example, suppose $\mathbf{F}$ represents a force, and we consider work done by this force as a particle moves along a curve C from A to B. We begin by dividing C into n subcurves of lengths Δs_i as shown in Figure 15.20.

FIGURE 15.20

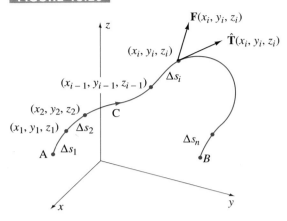

If $\mathbf{F}(x, y, z)$ is continuous along C, then along any given Δs_i, $\mathbf{F}(x, y, z)$ does not vary greatly (provided, of course, that Δs_i is small). If we approximate $\mathbf{F}(x, y, z)$ along Δs_i by its value $\mathbf{F}(x_i, y_i, z_i)$ at the final point (x_i, y_i, z_i) of Δs_i, then an approximation to the work done by $\mathbf{F}$ along Δs_i is $\mathbf{F}(x_i, y_i, z_i) \cdot \hat{\mathbf{T}}(x_i, y_i, z_i) \Delta s_i$. An approximation to the total work done by $\mathbf{F}$ along C is therefore

$$\sum_{i=1}^{n} \mathbf{F}(x_i, y_i, z_i) \cdot \hat{\mathbf{T}}(x_i, y_i, z_i) \Delta s_i.$$

To obtain the work done by $\mathbf{F}$ along C we take the limit of this sum as the number of subdivisions becomes larger and larger and each Δs_i approaches zero. But this is the definition of the line integral of $\mathbf{F} \cdot \hat{\mathbf{T}}$, and we therefore write

$$W = \int_C \mathbf{F} \cdot \hat{\mathbf{T}} \, ds = \int_C \mathbf{F} \cdot d\mathbf{r}. \tag{15.31}$$

EXAMPLE 15.9

The force of repulsion between two positive point charges, one of size q and the other of size unity, has magnitude $q/(4\pi\epsilon_0 r^2)$, where ϵ_0 is a constant and r is the distance between the charges. The potential V at any point P due to charge q is defined as the work required to bring the unit charge to P from an infinite distance along the straight line joining q and P. Find V.

SOLUTION   If that part of the line joining q and P from infinity to P is denoted by C (Figure 15.21), then

$$V = \int_C \mathbf{F} \cdot d\mathbf{r} = -\int_{-C} \mathbf{F} \cdot d\mathbf{r}$$

(see equation 15.30), where $\mathbf{F}$, the force necessary to overcome the electrostatic repulsion, is given by

$$\mathbf{F} = \frac{-q}{4\pi\epsilon_0 x^2} \hat{\mathbf{i}}.$$

Along $-C$, $d\mathbf{r} = dx\,\hat{\mathbf{i}}$, and therefore V can be evaluated by the (improper) definite integral

$$V = -\int_r^{\infty} \frac{-q}{4\pi\epsilon_0 x^2} dx = -\left\{ \frac{q}{4\pi\epsilon_0 x} \right\}_r^{\infty} = \frac{q}{4\pi\epsilon_0 r}.$$

FIGURE 15.21

If the vector field $\mathbf{F}$ in equation 15.27 has an x-component that is only a function of x, $P = P(x)$, and if the curve C is a portion of the x-axis from $x = a$ to $x = b$, then

$$\int_C \mathbf{F} \cdot d\mathbf{r} = \int_a^b P(x)\,dx.$$

This equation indicates that definite integrals with respect to x can be regarded as line integrals along the x-axis.

EXERCISES 15.3

In Exercises 1–10 evaluate the line integral.

* **1.** $\displaystyle\int_C x\,dx + x^2 y\,dy$ where C is the curve $y = x^3$, $z = 0$ from $(-1, -1, 0)$ to $(2, 8, 0)$

* **2.** $\displaystyle\int_C x\,dx + yz\,dy + x^2\,dz$ where C is the curve $y = x$, $z = x^2$ from $(-1, -1, 1)$ to $(2, 2, 4)$

* **3.** $\displaystyle\int_C x\,dx + (x + y)\,dy$ where C is the curve $x = 1 + y^2$ from $(2, 1)$ to $(2, -1)$

* **4.** $\displaystyle\int_C x^2\,dx + y^2\,dy + z^2\,dz$ where C is the curve $x + y = 1$, $x + z = 1$ from $(-2, 3, 3)$ to $(1, 0, 0)$

* **5.** $\displaystyle\int_C (y + 2x^2 z)\,dx$ where C is the curve $x = y^2$, $z = x^2$ from $(4, -2, 16)$ to $(1, 1, 1)$

* **6.** $\displaystyle\oint_C x^2 y\,dx + (x - y)\,dy$ once counterclockwise around the curve bounding the area described by the curves $x = 1 - y^2$, $y = x + 1$

* **7.** $\displaystyle\int_C y^2\,dx + x^2\,dy$ where C is the semicircle $x = \sqrt{1 - y^2}$ from $(0, 1)$ to $(0, -1)$

* **8.** $\displaystyle\int_C y\,dx + x\,dy + z\,dz$ where C is the curve $z = x^2 + y^2$, $x + y = 1$ from $(1, 0, 1)$ to $(-1, 2, 5)$

* **9.** $\displaystyle\oint_C x^2 y\,dy + z\,dx$ where C is the curve $x^2 + y^2 = 1$, $x + y + z = 1$ directed so that x decreases when y is positive

* **10.** $\displaystyle\oint_C y^2\,dx + x^2\,dy$ once clockwise around the curve $|x| + |y| = 1$

* **11.** Find the work done by the force $\mathbf{F} = x^2 y\hat{\mathbf{i}} + x\hat{\mathbf{j}}$ as a particle moves from $(1, 0)$ to $(6, 5)$ along the straight line joining these points.

* **12.** Consider the line integral $\displaystyle\int_C xy\,dx + x^2\,dy$, where C is the quarter-circle $x^2 + y^2 = 9$ from $(3, 0)$ to $(0, 3)$. Show that for each of the following parametrizations of C the value of the line integral is the same:

   (a) $x = 3\cos t$, $y = 3\sin t$, $0 \le t \le \pi/2$

   (b) $x = \sqrt{9 - y^2}$, $0 \le y \le 3$

* **13.** Evaluate the line integral $\displaystyle\int_C xy\,dx + x\,dy$ from $(-5, 3, 0)$ to $(4, 0, 0)$ along each of the following curves:

   (a) the straight line joining the points

   (b) $x = 4 - y^2$, $z = 0$

   (c) $3y = x^2 - 16$, $z = 0$

* **14.** Find the work done by a force $\mathbf{F} = x\hat{\mathbf{i}} + y\hat{\mathbf{j}}$ on a particle as it moves once counterclockwise around the ellipse $b^2 x^2 + a^2 y^2 = a^2 b^2$, $z = 0$.

In Exercises 15–19 evaluate the line integral.

** **15.** $\displaystyle\int_C \frac{1}{yz}\,dx$ where C is the curve $z = \sqrt{1 - x^2}$, $y = \sqrt{1 - x^2}$ from $(1/\sqrt{2}, 1/\sqrt{2}, 1/\sqrt{2})$ to $(-1\sqrt{2}, 1/\sqrt{2}, 1/\sqrt{2})$

** **16.** $\displaystyle\oint_C (x^2 + 2y^2)\,dy$ twice clockwise around the circle $(x - 2)^2 + y^2 = 1$, $z = 0$

** **17.** $\displaystyle\int_C y\,dx - y(x - 1)\,dy + y^2 z\,dz$ where C is the first octant intersection of $x^2 + y^2 + z^2 = 4$ and $(x - 1)^2 + y^2 = 1$ from $(2, 0, 0)$ to $(0, 0, 2)$

** **18.** $\displaystyle\int_C x^2 y\,dx + y\,dy + \sqrt{1 - x^2}\,dz$ where C is the curve $y - 2z^2 = 1$, $z = x + 1$ from $(0, 3, 1)$ to $(1, 9, 2)$

** **19.** $\displaystyle\int_C x\,dx + xy\,dy + 2\,dz$ where C is the curve $x + 2y + z = 4$, $4x + 3y + 2z = 13$ from $(2, -1, 4)$ to $(3, 1, -1)$

** **20.** Evaluate the line integral $\displaystyle\int_C \frac{x^3}{(1 + x^4)^3}\,dx + y^2 e^y\,dy + \frac{z}{\sqrt{1 + z^2}}\,dz$ where C consists of line segments joining successively the points $(0, -1, 1)$, $(1, -1, 1)$, $(1, 0, 1)$, and $(1, 0, 2)$.

** **21.** One end of a spring (with constant k) is fixed at point D in the figure below. The other end is moved along the x-axis from A to B. If the spring is stretched an amount l at A, find the work done against the spring.

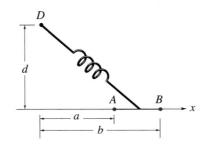

** **22.** Two positive charges q_1 and q_2 are placed at positions $(5, 5)$ and $(-2, 3)$ respectively, in the xy-plane. A third positive charge q_3 is moved along the x-axis from $x = 1$ to $x = -1$. Find the work done by the electrostatic forces of q_1 and q_2 on q_3.

In Exercises 23–26 set up a definite integral to evaluate the line integral. Use Simpson's rule with 10 equal subdivisions to approximate the definite integral.

** **23.** $\int_C xy\,dx + xy^2\,dy$ where C is the curve $z = 0$, $y = 1/\sqrt{1 + x^3}$, from $(0, 1, 0)$ to $(2, 1/3, 0)$

** **24.** $\int_C xz\,dx + \tan x\,dy + e^{xy}\,dz$ where C is the curve $x = y^2$, $z = y^3$ from $(1, -1, -1)$ to $(1, 1, 1)$

** **25.** $\int_C \sqrt{1 + y^2}\,dz + zy\,dy$ where C is curve $y = \cos^3 t$, $z = \sin^3 t$, $x = 0$, $0 \le t \le \pi/2$

** **26.** $\int_C xyz\,dy$ where C is the curve $x = (1 - t^2)/(1 + t^2)$, $y = t(1 - t^2)/(1 + t^2)$, $z = t$, $-1 \le t \le 1$

In Exercises 27 and 28 evaluate the line integral along the polar coordinate curve.

** **27.** $\oint_C y\,dx$ where C is the cardioid $r = 1 - \cos\theta$

** **28.** $\int_C y\,dx + x\,dy$ where C is the curve $r = \theta$, $0 \le \theta \le \pi$

** **29.** Suppose a gas flows through a region D of space. At each point $P(x, y, z)$ in D and time t, the gas has velocity $\mathbf{v}(x, y, z, t)$. If C is a closed curve in D, the line integral

$$\Gamma = \oint_C \mathbf{v} \cdot d\mathbf{r}$$

is called the circulation of the flow for the curve C. If C is the circle $x^2 + y^2 = r^2$, $z = 1$ (directed clockwise as viewed from the origin), calculate Γ for the following flow vectors:

(a) $\mathbf{v}(x, y, z) = (x\hat{\mathbf{i}} + y\hat{\mathbf{j}} + z\hat{\mathbf{k}})/(x^2 + y^2 + z^2)^{3/2}$

(b) $\mathbf{v}(x, y, z) = -y\hat{\mathbf{i}} + x\hat{\mathbf{j}}$

** **30.** We have shown that given a line integral 15.28, it is always possible to write it uniquely in form 15.23, where $f = \mathbf{F} \cdot \hat{\mathbf{T}}$. Show that the converse is not true; that is, given $f(x, y, z)$, there does not exist a unique $\mathbf{F}(x, y, z)$ such that $\mathbf{F} \cdot d\mathbf{r} = f(x, y, z)\,ds$.

** **31.** Explain why the line integral $\oint_C f(x)\,dx + g(y)\,dy + h(z)\,dz$ must have value zero when $f(x)$, $g(y)$, and $h(z)$ are continuous functions in some domain containing C.

** **32.** The cycloid $x = R(\theta - \sin\theta)$, $y = R(1 - \cos\theta)$ is the curve traced out by a fixed point on the circumference of a circle of radius R rolling along the x-axis (see Example 10.7 in Section 10.1). Suppose the point is acted on by a force of unit magnitude directed toward the centre of the rolling circle.

(a) Find the work done by the force as the point moves from $\theta = 0$ to $\theta = \pi$.

(b) How much of the work in (a) is done by the vertical component of the force?

15.4 Independence of Path

In Sections 15.2 and 15.3 we illustrated that the value of a line integral joining two points usually depends on the curve joining the points. In this section we show that certain line integrals have the same value for all curves joining the same two points. We formalize this idea in the following definition.

DEFINITION 15.4

A line integral $\int \mathbf{F} \cdot d\mathbf{r}$ is said to be **independent of path** in a domain D if for each pair of points A and B in D, the value of the line integral

$$\int_C \mathbf{F} \cdot d\mathbf{r}$$

is the same for all piecewise-smooth paths C in D from A to B.

The value of such a line integral for given $\mathbf{F}$ will then depend only on the end points A and B. Note that we speak of independence of path only for the special class of line integrals of the form $\int \mathbf{F} \cdot d\mathbf{r}$. The question we must now ask is "How do we determine whether a given line integral is independent of path?" One answer is contained in the following theorem.

THEOREM 15.3

Suppose $P(x, y, z)$, $Q(x, y, z)$, and $R(x, y, z)$ are continuous functions in some domain D. The line integral

$$\int \mathbf{F} \cdot d\mathbf{r} = \int P\,dx + Q\,dy + R\,dz$$

is independent of path in D if and only if there exists a function $\phi(x, y, z)$ defined in D such that

$$\nabla \phi = \mathbf{F} = P\hat{\mathbf{i}} + Q\hat{\mathbf{j}} + R\hat{\mathbf{k}}. \tag{15.32}$$

Essentially, then, a line integral is independent of path if $\mathbf{F}$ is the gradient of some scalar function.

PROOF Suppose first of all that in D there exists a function $\phi(x, y, z)$ such that $\nabla \phi = P\hat{\mathbf{i}} + Q\hat{\mathbf{j}} + R\hat{\mathbf{k}}$. If

$$C: \quad x = x(t), \quad y = y(t), \quad z = z(t), \quad \alpha \leq t \leq \beta$$

is any smooth curve in D from A to B, then

$$\int_C \mathbf{F} \cdot d\mathbf{r} = \int_C P\,dx + Q\,dy + R\,dz = \int_\alpha^\beta \left\{ \frac{\partial \phi}{\partial x}\frac{dx}{dt} + \frac{\partial \phi}{\partial y}\frac{dy}{dt} + \frac{\partial \phi}{\partial z}\frac{dz}{dt} \right\} dt.$$

The term in braces is the chain rule for the derivative of composite function $\phi[x(t), y(t), z(t)]$, and we can therefore write that

$$\int_C \mathbf{F} \cdot d\mathbf{r} = \int_\alpha^\beta \frac{d\phi}{dt} dt = \{\phi[x(t), y(t), z(t)]\}_\alpha^\beta$$

$$= \phi[x(\beta), y(\beta), z(\beta)] - \phi[x(\alpha), y(\alpha), z(\alpha)]$$

$$= \phi(x_B, y_B, z_B) - \phi(x_A, y_A, z_A).$$

(The same result is obtained even when C is piecewise smooth rather than smooth.) Because this last expression does not depend on the curve C taken from A to B, it follows that the line integral is independent of path in D.

Conversely, suppose now that the line integral

$$\int \mathbf{F} \cdot d\mathbf{r} = \int P \, dx + Q \, dy + R \, dz$$

is independent of path in D, and A is chosen as some fixed point in D. If $P(x, y, z)$ is any other point in D (Figure 15.22), and C is a piecewise-smooth curve in D from A to P, then the line integral

$$\phi(x, y, z) = \int_C \mathbf{F} \cdot d\mathbf{r}$$

defines a single-valued function $\phi(x, y, z)$ in D, and the value of $\phi(x, y, z)$ is the same for all piecewise-smooth curves from A to P.

FIGURE 15.22

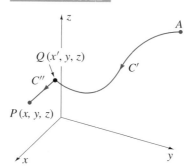

Consider a curve C composed of two parts: a straight-line portion C'' parallel to the x-axis from a fixed point $Q(x', y, z)$ to $P(x, y, z)$, and any other piecewise-smooth curve C' in D from A to Q. Then

$$\phi(x, y, z) = \int_{C'} \mathbf{F} \cdot d\mathbf{r} + \int_{C''} \mathbf{F} \cdot d\mathbf{r}.$$

Now along C'', y and z are both constant, and therefore

$$\phi(x, y, z) = \int_{C'} \mathbf{F} \cdot d\mathbf{r} + \int_{x'}^x P(t, y, z) dt.$$

The partial derivative of this function with respect to x is

$$\frac{\partial \phi}{\partial x} = \frac{\partial}{\partial x} \int_{C'} \mathbf{F} \cdot d\mathbf{r} + \frac{\partial}{\partial x} \int_{x'}^{x} P(t, y, z)dt,$$

but because $Q(x', y, z)$ is fixed,

$$\frac{\partial}{\partial x} \int_{C'} \mathbf{F} \cdot d\mathbf{r} = 0.$$

Consequently,

$$\frac{\partial \phi}{\partial x} = \frac{\partial}{\partial x} \int_{x'}^{x} P(t, y, z)dt = P(x, y, z).$$

By choosing other curves with straight-line portions parallel to the y- and z-axes, we can also show that $\partial \phi / \partial y = Q$ and $\partial \phi / \partial z = R$. Thus, $\mathbf{F} = \nabla \phi$, and this completes the proof. ∎

This theorem points out that it is very simple to evaluate a line integral that is independent of path. We state this in the following corollary.

COROLLARY 1

When a line integral is independent of path in a domain D, and A and B are points in D, then

$$\int_{C} \mathbf{F} \cdot d\mathbf{r} = \phi(x_B, y_B, z_B) - \phi(x_A, y_A, z_A),$$

where $\nabla \phi = \mathbf{F}$, for every piecewise smooth curve C in D from A to B.

EXAMPLE 15.10

Evaluate $\int_C 2xy\, dx + x^2 dy + 2z\, dz$, where C is the first octant intersection of $x^2 + y^2 = 1$ and $z = 2x + 4$ from $(0, 1, 4)$ to $(1, 0, 6)$.

SOLUTION Since $\nabla(x^2 y + z^2) = 2xy\hat{\mathbf{i}} + x^2\hat{\mathbf{j}} + 2z\hat{\mathbf{k}}$, the line integral is independent of path everywhere, and

$$\int_C 2xy\, dx + x^2 dy + 2z\, dz = \left\{ x^2 y + z^2 \right\}_{(0,1,4)}^{(1,0,6)} = 36 - 16 = 20.$$

The following corollary is also an immediate consequence of Theorem 15.3.

COROLLARY 2

The line integral $\int \mathbf{F} \cdot d\mathbf{r}$ is independent of path in a domain D if and only if

$$\oint_C \mathbf{F} \cdot d\mathbf{r} = 0 \tag{15.33}$$

for every closed path in D.

Theorem 15.3 states that a necessary and sufficient condition for line integral 15.28 to be independent of path is the existence of a function $\phi(x, y, z)$ such that $\nabla \phi = \mathbf{F}$. For most problems it is obvious whether such a function $\phi(x, y, z)$ exists; but for a few, it is not. Since much time could be wasted searching for $\phi(x, y, z)$ (when in fact it does not exist), it would be helpful to have a test that states *a priori* whether $\phi(x, y, z)$ does indeed exist. Such a test is contained in Theorem 15.1. It states that $\mathbf{F}$ is the gradient of some scalar function $\phi(x, y, z)$ if $\nabla \times \mathbf{F} = \mathbf{0}$. When this result is combined with Theorem 15.3, we obtain this important theorem.

THEOREM 15.4

Let D be a domain in which $P(x, y, z)$, $Q(x, y, z)$, and $R(x, y, z)$ have continuous first derivatives. If the line integral $\int \mathbf{F} \cdot d\mathbf{r} = \int P\,dx + Q\,dy + R\,dz$ is independent of path in D, then $\nabla \times \mathbf{F} = \mathbf{0}$ in D. Conversely, if D is simply-connected, and $\nabla \times \mathbf{F} = \mathbf{0}$ in D, then the line integral is independent of path in D.

We have in Theorem 15.4 a simple test to determine whether a given line integral is independent of path: We see whether the curl of $\mathbf{F}$ is zero. Evaluation of a line integral that is independent of path still requires the function $\phi(x, y, z)$, but it is at least nice to know that ϕ exists before searching for it.

Theorems 15.1, 15.3, and 15.4 have identified an important equivalence, at least in simply-connected domains:

1. $\int \mathbf{F} \cdot d\mathbf{r}$ is independent of path in D;

2. $\mathbf{F} = \nabla \phi$ for some function $\phi(x, y, z)$ defined in D;

3. $\nabla \times \mathbf{F} = \mathbf{0}$ in D.

Theorem 15.1 states that (2) and (3) are equivalent; Theorem 15.3 verifies the equivalence of (1) and (2); and these two imply the equivalence of (1) and (3) (Theorem 15.4).

For line integrals in the xy-plane, this equivalence is still valid except that $\nabla \times \mathbf{F} = \mathbf{0}$ can be stated more simply as

$$\frac{\partial Q}{\partial x} = \frac{\partial P}{\partial y} \tag{15.34}$$

(see equation 15.16).

EXAMPLE 15.11

Evaluate $\int_C 2xye^z\,dx + (x^2e^z + y)\,dy + (x^2ye^z - z)\,dz$ along the straight line C from $(0, 1, 2)$ to $(2, 1, -8)$.

SOLUTION   **Method 1**   Parametric equations for the straight line are

$$C: \quad x = 2t, \quad y = 1, \quad z = 2 - 10t, \quad 0 \le t \le 1.$$

If I is the value of the line integral, then

$$
\begin{aligned}
I &= \int_0^1 \{2(2t)(1)e^{2-10t}(2\,dt) + [(2t)^2 e^{2-10t} + 1](0) \\
&\quad + [(2t)^2(1)e^{2-10t} - 2 + 10t](-10\,dt)\} \\
&= \int_0^1 \{8e^2(t - 5t^2)e^{-10t} + 20 - 100t\}dt \\
&= 8e^2 \left\{ (t - 5t^2)\frac{e^{-10t}}{-10} \right\}_0^1 + \frac{4e^2}{5} \int_0^1 (1 - 10t)e^{-10t}\,dt + 10\{2t - 5t^2\}_0^1 \\
&= 8e^2 \left\{ \frac{2}{5}e^{-10} \right\} - 30 + \frac{4}{5}e^2 \left\{ (1 - 10t)\frac{e^{-10t}}{-10} \right\}_0^1 + \frac{2e^2}{25} \int_0^1 -10e^{-10t}\,dt \\
&= \frac{16}{5}e^{-8} - 30 + \frac{4}{5}e^2 \left\{ \frac{9}{10}e^{-10} + \frac{1}{10} \right\} + \frac{2e^2}{25} \left\{ e^{-10t} \right\}_0^1 = 4e^{-8} - 30.
\end{aligned}
$$

Method 2   It is evident that

$$\nabla\left(x^2 y e^z + \frac{y^2}{2} - \frac{z^2}{2} \right) = 2xye^z \hat{\mathbf{i}} + (x^2 e^z + y)\hat{\mathbf{j}} + (x^2 y e^z - z)\hat{\mathbf{k}},$$

and hence the line integral is independent of path. Its value is therefore

$$I = \left\{ x^2 y e^z + \frac{y^2}{2} - \frac{z^2}{2} \right\}_{(0,1,2)}^{(2,1,-8)} = \left\{ 4e^{-8} + \frac{1}{2} - 32 \right\} - \left\{ \frac{1}{2} - 2 \right\} = 4e^{-8} - 30.$$

Method 3   Since

$$
\nabla \times (2xye^z \hat{\mathbf{i}} + (x^2 e^z + y)\hat{\mathbf{j}} + (x^2 y e^z - z)\hat{\mathbf{k}}) = \begin{vmatrix} \hat{\mathbf{i}} & \hat{\mathbf{j}} & \hat{\mathbf{k}} \\ \partial/\partial x & \partial/\partial y & \partial/\partial z \\ 2xye^z & x^2 e^z + y & x^2 y e^z - z \end{vmatrix}
$$

$$
\begin{aligned}
&= (x^2 e^z - x^2 e^z)\hat{\mathbf{i}} \\
&\quad + (2xye^z - 2xye^z)\hat{\mathbf{j}} \\
&\quad + (2xe^z - 2xe^z)\hat{\mathbf{k}} = \mathbf{0},
\end{aligned}
$$

the line integral is independent of path. Thus, there exists a function $\phi(x, y, z)$ such that

$$\nabla\phi = 2xye^z \hat{\mathbf{i}} + (x^2 e^z + y)\hat{\mathbf{j}} + (x^2 y e^z - z)\hat{\mathbf{k}}$$

or

$$\frac{\partial\phi}{\partial x} = 2xye^z, \quad \frac{\partial\phi}{\partial y} = x^2 e^z + y, \quad \frac{\partial\phi}{\partial z} = x^2 y e^z - z.$$

Integration of the first of these equations yields

$$\phi(x, y, z) = x^2 y e^z + K(y, z).$$

Substitution of this function into the left side of the second equation gives

$$x^2 e^z + \frac{\partial K}{\partial y} = x^2 e^z + y,$$

which implies that

$$\frac{\partial K}{\partial y} = y.$$

Consequently,

$$K(y, z) = \frac{y^2}{2} + L(z),$$

and we know both the x- and y-dependence of ϕ:

$$\phi(x, y, z) = x^2 y e^z + \frac{y^2}{2} + L(z).$$

To obtain the z-dependence contained in $L(z)$, we substitute into the left side of the third equation,

$$x^2 y e^z + \frac{dL}{dz} = x^2 y e^z - z,$$

from which we have

$$\frac{dL}{dz} = -z.$$

Hence, $L(z) = -z^2/2 + C$ (C a constant), and

$$\phi(x, y, z) = x^2 y e^z + \frac{y^2}{2} - \frac{z^2}{2} + C.$$

Finally, then, we have

$$I = \left\{ x^2 y e^z + \frac{y^2}{2} - \frac{z^2}{2} \right\}_{(0,1,2)}^{(2,1,-8)} = 4e^{-8} - 30.$$

Method 1 is one of "brute force." The function $x^2 y e^z + y^2/2 - z^2/2$ in Method 2 was obtained by observation. Method 3 is the systematic procedure suggested in Section 15.1 for finding the function $\phi(x, y, z)$.

EXAMPLE 15.12

Evaluate

$$I = \int_C \left(\frac{x^3 - 2y^2}{x^3 y} \right) dx + \left(\frac{y^2 - x^3}{x^2 y^2} \right) dy + 2z^2 dz,$$

where C is the curve $y = x^2$, $z = x - 1$ from $(1, 1, 0)$ to $(2, 4, 1)$.

SOLUTION   If we set

$$\mathbf{F} = \left(\frac{x^3 - 2y^2}{x^3 y} \right) \hat{\mathbf{i}} + \left(\frac{y^2 - x^3}{x^2 y^2} \right) \hat{\mathbf{j}} + 2z^2 \hat{\mathbf{k}},$$

$$= \left(\frac{1}{y} - \frac{2y}{x^3} \right) \hat{\mathbf{i}} + \left(\frac{1}{x^2} - \frac{x}{y^2} \right) \hat{\mathbf{j}} + 2z^2 \hat{\mathbf{k}},$$

it is evident that $\mathbf{F} = \nabla\phi$ if

$$\phi(x, y, z) = \frac{x}{y} + \frac{y}{x^2} + \frac{2z^3}{3},$$

and this is valid in any domain that does not contain points on the yz-plane ($x = 0$) or xz-plane ($y = 0$). Since C does not pass through either of these planes, then

$$I = \left\{ \frac{x}{y} + \frac{y}{x^2} + \frac{2z^3}{3} \right\}_{(1,1,0)}^{(2,4,1)} = \left\{ \frac{1}{2} + 1 + \frac{2}{3} \right\} - \left\{ 1 + 1 \right\} = \frac{1}{6}.$$

EXAMPLE 15.13

In thermodynamics the state of a gas is described by four variables—pressure P, temperature T, internal energy U, and volume V. These variables are related by two equations of state,

$$F(P, T, U, V) = 0 \quad \text{and} \quad G(P, T, U, V) = 0,$$

so that two of the variables are independent and two are dependent. If U and V are chosen as independent variables, then $T = T(U, V)$ and $P = P(U, V)$. An experimental law called the second law of thermodynamics states that the line integral

$$\int_C \frac{1}{T} dU + \frac{P}{T} dV$$

is independent of path in the UV-plane. Show that the second law can be expressed in the differential form

$$T \frac{\partial P}{\partial U} - P \frac{\partial T}{\partial U} + \frac{\partial T}{\partial V} = 0.$$

SOLUTION   According to equation 15.34, the line integral is independent of path if and only if

$$\frac{\partial}{\partial U}\left(\frac{P}{T}\right) = \frac{\partial}{\partial V}\left(\frac{1}{T}\right),$$

or

$$0 = \frac{T\dfrac{\partial P}{\partial U} - P\dfrac{\partial T}{\partial U}}{T^2} + \frac{1}{T^2}\frac{\partial T}{\partial V};$$

that is,

$$0 = T\frac{\partial P}{\partial U} - P\frac{\partial T}{\partial U} + \frac{\partial T}{\partial V},$$

and the proof is complete.

Since the above line integral is independent of path, there exists a function $S(U, V)$ such that

$$\frac{\partial S}{\partial U} = \frac{1}{T}, \qquad \frac{\partial S}{\partial V} = \frac{P}{T},$$

and the value of the line integral is given by

$$\int_C \frac{1}{T}dU + \frac{P}{T}dV = S(B) - S(A),$$

where C joins A and B. This function, called **entropy**, plays a key role in thermodynamics.

EXERCISES 15.4

In Exercises 1–10 show that the line integral is independent of path, and evaluate it.

* **1.** $\displaystyle\int_C xy^2\, dx + x^2 y\, dy$ where C is the curve $y = x^2$, $z = 0$ from $(0, 0, 0)$ to $(1, 1, 0)$

* **2.** $\displaystyle\int_C (3x^2 + y)\, dx + x\, dy$ where C is the straight line from $(2, 1, 5)$ to $(-3, 2, 4)$

* **3.** $\displaystyle\int_C 2xe^y\, dx + (x^2 e^y + 3)\, dy$ where C is the curve $y = \sqrt{1 - x^2}$, $z = 0$ from $(1, 0, 0)$ to $(-1, 0, 0)$

* **4.** $\displaystyle\int_C 3x^2 yz\, dx + x^3 z\, dy + (x^3 y - 4z)\, dz$ where C is the curve $x^2 + y^2 + z^2 = 3$, $y = x$ from $(-1, -1, 1)$ to $(1, 1, -1)$

* **5.** $\displaystyle\int_C -\frac{y}{z}\sin x\, dx + \frac{1}{z}\cos x\, dy - \frac{y}{z^2}\cos x\, dz$ where C is the helix $x = 2\cos t$, $y = 2\sin t$, $z = t$ from $(2, 0, 2\pi)$ to $(2, 0, 4\pi)$

* **6.** $\displaystyle\oint_C y\cos x\, dx + \sin x\, dy$ once clockwise around the circle $x^2 + y^2 - 2x + 4y = 7$, $z = 0$

* **7.** $\displaystyle\int_C x^2\, dx + y^2\, dy + z^2\, dz$ where C is the curve $x + y = 1$, $x + z = 1$ from $(-2, 3, 3)$ to $(1, 0, 0)$

* 8. $\displaystyle\int_C y\,dx + x\,dy + z\,dz$ where C is the curve $z = x^2 + y^2$, $x + y = 1$ from $(1, 0, 1)$ to $(-1, 2, 5)$

* 9. $\displaystyle\int_C \frac{1}{y}\,dx - \frac{x}{y^2}\,dy + dz$ where C is the curve $y = x^2 + 1$, $x + y + z = 2$ from $(0, 1, 1)$ to $(3, 10, -11)$

* 10. $\displaystyle\int_C 3x^2y^3\,dx + 3x^3y^2\,dy$ where C is the curve $y = e^x$ from $(0, 1)$ to $(1, e)$

* 11. Show that if $f(x)$, $g(y)$, and $h(z)$ have continuous first derivatives, then the line integral

$$\int_C f(x)\,dx + g(y)\,dy + h(z)\,dz$$

is independent of path.

** 12. If $\nabla \times \mathbf{F} = \mathbf{0}$ in a domain D which is not simply-connected, can you conclude that the line integral $\displaystyle\int_C \mathbf{F} \cdot d\mathbf{r}$ is not independent of path in D? Explain.

In Exercises 13–18 evaluate the line integral.

** 13. $\displaystyle\int_C zye^{xy}\,dx + zxe^{xy}\,dy + (e^{xy} - 1)\,dz$ where C is the curve $y = x^2$, $z = x^3$ from $(1, 1, 1)$ to $(2, 4, 8)$

** 14. $\displaystyle\oint_C y(\tan x + x\sec^2 x)\,dx + x\tan x\,dy + dz$ once around the circle $x^2 + y^2 = 1$, $z = 0$

** 15. $\displaystyle\int_C \left(\frac{1 + y^2}{x^3}\right)dx - \left(\frac{y + x^2y}{x^2}\right)dy + z\,dz$ where C is the broken line joining successively $(1, 0, 0)$, $(25, 2, 3)$, and $(5, 2, 1)$

** 16. $\displaystyle\oint_C \frac{zy\,dx - xz\,dy + xy\,dz}{y^2}$ where C is the curve $x^2 + z^2 = 1$, $y + z = 2$

** 17. $\displaystyle\int_C -\frac{1}{x^2}\mathrm{Tan}^{-1}y\,dx + \frac{1}{x + xy^2}\,dy$ where C is the curve $x = y^2 + 1$ from $(2, -1)$ to $(10, 3)$

** 18. $\displaystyle\int_C \frac{1}{(x-3)^2(y+5)}\,dx + \frac{1}{(x-3)(y+5)^2}\,dy + \frac{1}{z+4}\,dz$ where C is the curve $x = y = z$ from $(0, 0, 0)$ to $(2, 2, 2)$

** 19. Evaluate $\displaystyle\oint_C \frac{-y\,dx + x\,dy}{x^2 + y^2}$

(a) once counterclockwise around the circle $x^2 + y^2 = 1$, $z = 0$

(b) once counterclockwise around the circle $(x - 2)^2 + y^2 = 1$, $z = 0$

** 20. Evaluate $\displaystyle\int_C \frac{y}{x^2 + y^2}\,dx - \frac{x}{x^2 + y^2}\,dy$ where C consists of line segments joining successively the points $(1, 0)$, $(1, 1)$, $(-1, 1)$, and $(-1, 0)$.

** **21.** Is the line integral

$$\int_C \frac{x}{\sqrt{x^2 + y^2}} dx + \frac{y}{\sqrt{x^2 + y^2}} dy$$

independent of path in the domain consisting of the xy-plane with the origin removed? Is the line integral

$$\int_C \frac{x}{\sqrt{x^2 + y^2 + z^2}} dx + \frac{y}{\sqrt{x^2 + y^2 + z^2}} dy + \frac{z}{\sqrt{x^2 + y^2 + z^2}} dz$$

independent of path in the domain consisting of xyz-space with the origin removed?

** **22.** In which of the following domains is the line integral $\displaystyle\int_C \frac{y \, dx - x \, dy}{x^2 + y^2}$ independent of path:

(a) $x > 0$, (b) $x < 0$, (c) $y > 0$, (d) $y < 0$, (e) $x^2 + y^2 > 0$.

** **23.** The second law of thermodynamics states that the line integral $I = \displaystyle\int T^{-1}(dU + P \, dV)$ is independent of path in the UV-plane (see Example 15.13).

(a) The equations of state for an ideal gas are $PV = nRT$, $U = f(T)$, where n and R are constants and $f(T)$ is some given function. Because of these, it is more convenient to choose T and V as independent variables and to express P and U in terms of T and V. If this is done, show that

$$I = \int kT^{-1} \, dT + nRV^{-1} \, dV \quad \text{where } k = dU/dT.$$

(b) Since the line integral is independent of path, there exists a function $S(T, V)$, called entropy, such that

$$\int_C \frac{k}{T} dT + \frac{nR}{V} dV = S(B) - S(A),$$

where C is any curve joining points A and B. Show in the case that k is constant that $S = k \ln T + nR \ln V + S_0$, where S_0 is a constant.

** **24.** Evaluate the line integral $\displaystyle\oint_C (2xye^{x^2 y} + x^2 y) \, dx + x^2 e^{x^2 y} \, dy$ once clockwise around the ellipse $x^2 + 4y^2 = 4$, $z = 0$.

** **25.** A spring has one end fixed at point P. The other end is moved along the curve $y = f(x)$ from $(x_0, f(x_0))$ to $(x_1, f(x_1))$. If initial and final stretches in the spring are a and b ($b > a$), what work is done against the spring?

*** **26.** Electrostatic forces due to point charges and gravitational forces due to point masses are examples of inverse square force fields—force fields of the form $\mathbf{F} = k\hat{\mathbf{r}}/|\mathbf{r}|^2$, where k is a constant and $\mathbf{r} = x\hat{\mathbf{i}} + y\hat{\mathbf{j}} + z\hat{\mathbf{k}}$.

(a) Is the line integral representing work done by such a force field independent of path?

(b) What is the work done by $\mathbf{F}$ in moving a particle from (x_1, y_1, z_1) to (x_2, y_2, z_2)?

15.5 Energy and Conservative Force Fields

In Section 7.5 we introduced the idea of potential energy associated with forces and the law of conservation of energy. We now complete this discussion by showing to which forces can be attributed potential energy.

> **DEFINITION 15.5**
>
> A force field $\mathbf{F}(x, y, z)$, defined in a domain D of space, is said to be **conservative** in D if the line integral $\int \mathbf{F} \cdot d\mathbf{r}$ is independent of path in D.

Since $\int_C \mathbf{F} \cdot d\mathbf{r}$ can be interpreted as the work done by $\mathbf{F}$ along C, a force field is conservative if work that it does is independent of path taken from one point to another. According to the results of Section 15.4, we can also state that a force field $\mathbf{F}$ is conservative if and only if there exists a function $\phi(x, y, z)$ such that $\mathbf{F} = \nabla\phi$.

It is customary to associate a potential energy function $U(x, y, z)$ with a conservative force field $\mathbf{F}$. This function assigns a potential energy to each point (x, y, z) in such a way that if a particle moves from point A to point B, then the difference in potential energy $U(A) - U(B)$ is the work done by $\mathbf{F}$; i.e., if C is any curve joining A and B, then

$$U(A) - U(B) = \int_C \mathbf{F} \cdot d\mathbf{r}. \tag{15.35}$$

If $\int \mathbf{F} \cdot d\mathbf{r} > 0$, then potential energy at A is greater than potential energy at B; if $\int \mathbf{F} \cdot d\mathbf{r} < 0$, then potential energy at B is greater than that at A. To find $U(x, y, z)$, we use the fact that, because $\mathbf{F}$ is conservative, there exists a function $\phi(x, y, z)$ such that $\mathbf{F} = \nabla\phi$, and

$$\int_C \mathbf{F} \cdot d\mathbf{r} = \phi(B) - \phi(A).$$

If follows, then, that $U(x, y, z)$ must satisfy the equation

$$U(A) - U(B) = \phi(B) - \phi(A),$$

or

$$U(A) + \phi(A) = U(B) + \phi(B).$$

Since A and B are arbitrary points in D, this last equation states that the value of the function $U(x, y, z) + \phi(x, y, z)$ is the same at every point in the force field,

$$U(x, y, z) + \phi(x, y, z) = C,$$

where C is a constant. Thus,

$$U(x, y, z) = -\phi(x, y, z) + C. \tag{15.36}$$

Equation 15.36 shows that the force field $\mathbf{F}$ defines a potential energy function $U(x, y, z)$ up to an additive constant. (This seems reasonable in that ϕ itself is defined only to an additive constant.) Because $U = -\phi + C$ and $\mathbf{F} = \nabla\phi$, we can also regard U as being defined by the equation

$$\mathbf{F} = -\nabla U. \tag{15.37}$$

The advantage of this equation is that it defines U directly, not through the function ϕ.

For a conservative force field $\mathbf{F}$, then, we define a potential energy function $U(x, y, z)$ by equation 15.37. If a particle moves from A to B, then the work done by $\mathbf{F}$ is

$$W = U(A) - U(B); \tag{15.38}$$

in other words, work done by a conservative force field is equal to loss in potential energy.

On the other hand, if a particle moves under the action of a force $\mathbf{F}$ (and only $\mathbf{F}$), be it conservative or nonconservative, then it does so according to Newton's second law,

$$\mathbf{F} = \frac{d}{dt}(m\mathbf{v}) = m\frac{d\mathbf{v}}{dt},$$

where m is the mass of the particle (assumed constant), $\mathbf{v}$ is its velocity, and t is time. The action of $\mathbf{F}$ produces motion along some curve C, and the work done by $\mathbf{F}$ along this curve from A to B is

$$W = \int_C \mathbf{F} \cdot d\mathbf{r} = \int_\alpha^\beta m\frac{d\mathbf{v}}{dt} \cdot \frac{d\mathbf{r}}{dt}dt = \int_\alpha^\beta m\frac{d\mathbf{v}}{dt} \cdot \mathbf{v}\, dt = \int_\alpha^\beta \frac{d}{dt}\left\{\frac{1}{2}m\mathbf{v} \cdot \mathbf{v}\right\} dt$$

$$= \left\{\frac{1}{2}m\mathbf{v} \cdot \mathbf{v}\right\}_\alpha^\beta = \left\{\frac{1}{2}m|\mathbf{v}|^2\right\}_\alpha^\beta.$$

Thus, if $K(x, y, z) = \frac{1}{2}m|\mathbf{v}|^2$ represents kinetic energy of the particle, the work done by $\mathbf{F}$ is equal to the gain in kinetic energy of the particle,

$$W = K(B) - K(A) \tag{15.39}$$

(and this is true for any force $\mathbf{F}$ as long as $\mathbf{F}$ is the total resultant force producing motion).

If the total force producing motion is a conservative force field $\mathbf{F}$ we have two expressions 15.38 and 15.39 for the work done as a particle moves from one point to another under the action of $\mathbf{F}$. If we equate them, we have

$$U(A) - U(B) = K(B) - K(A),$$

or

$$U(A) + K(A) = U(B) + K(B). \tag{15.40}$$

We have shown then that, if a particle moves under the action of a conservative force field *only*, the sum of the kinetic and potential energies at B must be the same as the sum of the kinetic and potential energies at A. In other words, if E is the total energy of the particle, kinetic plus potential, then

$$E(A) = E(B). \tag{15.41}$$

Since B can be any point along the path of the particle, it follows that when a particle moves under the action of a conservative force field, and only a conservative force field, then at every point along its trajectory

$$E = \text{a constant.} \tag{15.42}$$

This is the **law of conservation of energy** for a conservative force field.

EXAMPLE 15.14

Show that the electrostatic force due to a point charge is conservative, and determine a potential energy function for the field.

FIGURE 15.23

SOLUTION   The electrostatic force on a charge Q due to a charge q is

$$\mathbf{F} = \frac{qQ}{4\pi\epsilon_0 |\mathbf{r}|^3}\mathbf{r},$$

where $\mathbf{r}$ is the vector from q to Q. If we choose a Cartesian coordinate system with q at the origin (Figure 15.23), then

$$\mathbf{F} = \frac{qQ}{4\pi\epsilon_0 (x^2 + y^2 + z^2)^{3/2}}(x\hat{\mathbf{i}} + y\hat{\mathbf{j}} + z\hat{\mathbf{k}}).$$

Since

$$\nabla\left\{\frac{-qQ}{4\pi\epsilon_0 (x^2 + y^2 + z^2)^{1/2}}\right\} = \mathbf{F},$$

the force field is conservative, and possible potential energy functions are

$$U(x, y, z) = \frac{qQ}{4\pi\epsilon_0 (x^2 + y^2 + z^2)^{1/2}} + C = \frac{qQ}{4\pi\epsilon_0 r} + C,$$

where $r = |\mathbf{r}|$. In electrostatics it is customary to choose $U(x, y, z)$ so that $\lim_{r\to\infty} U = 0$, in which case $C = 0$, and

$$U(x, y, z) = \frac{qQ}{4\pi\epsilon_0 r}.$$

In addition, if V is defined as the potential energy per unit test charge Q, then

$$V(x, y, z) = \frac{U}{Q} = \frac{q}{4\pi\epsilon_0 r}.$$

This result agrees with that in Example 15.9.

EXERCISES 15.5

In Exercises 1–5 determine whether the force field is conservative. Identify each conservative force field, and find a potential energy function.

* **1.** $\mathbf{F}(x, y, z) = \dfrac{q_1 q_2}{4\pi\epsilon_0}\dfrac{x\hat{\mathbf{i}} + y\hat{\mathbf{j}} + z\hat{\mathbf{k}}}{(x^2 + y^2 + z^2)^{3/2}}$

* **2.** $\mathbf{F}(x, y) = mx\hat{\mathbf{i}} + xy\hat{\mathbf{j}}$, m is a constant

* **3.** $\mathbf{F}(x) = -kx\hat{\mathbf{i}}$, k is a constant

* **4.** $\mathbf{F}(x, y, z) = -mg\hat{\mathbf{k}}$, m and g are constants

* **5.** $\mathbf{F}(x, y, z) = GMm\dfrac{x\hat{\mathbf{i}} + y\hat{\mathbf{j}} + z\hat{\mathbf{k}}}{(x^2 + y^2 + z^2)^{3/2}}$, G, M, and m are constants

* **6.** Suppose that $\mathbf{F}(x, y, z)$ is a conservative force field in some domain D, and $U(x, y, z)$ is a potential energy function associated with $\mathbf{F}$. The surfaces $U(x, y, z) = C$, where C is a constant, are called *equipotential surfaces*. Through each point P in D there is one and only one such equipotential surface for $\mathbf{F}$. Show that at P the force $\mathbf{F}$ is normal to the equipotential surface through P.

* **7.** Draw equipotential surfaces for the forces in Exercises 1, 4, and 5.

** **8.** One end of a spring with unstretched length L is fixed at the origin in space. If the other end is at point (x, y, z) (all coordinates in metres), what is the force exerted by the spring? Is this force conservative?

** **9.** Explain why friction is not conservative.

** **10.**

(a) When students in Universityland leave their houses, a supernatural power attracts them to university in such a way that the magnitude of the force at any point is inversely proportional to the square of the distance from university. This force acts until they are 100 m from university and then it disappears. Is this force conservative?

(b) If someone diverts the power so that the force attracts students to the local donut shop, is this force conservative?

*** **11.** A force field $\mathbf{F}(x, y, z)$ is said to be radially symmetric about the origin if it can be written in the form

$$\mathbf{F}(x, y, z) = f\left(\sqrt{x^2 + y^2 + z^2}\right)\mathbf{r}, \qquad \mathbf{r} = x\hat{\mathbf{i}} + y\hat{\mathbf{j}} + z\hat{\mathbf{k}},$$

for some function f. We often write in such a case that

$$\mathbf{F}(x, y, z) = f(r)\mathbf{r}, \quad \text{where } r = |\mathbf{r}| = \sqrt{x^2 + y^2 + z^2}.$$

(a) Use Theorem 15.4 to show that such a force is conservative in suitably defined domains (provided $f(r)$ has a continuous first derivative).

(b) If A and B are the points in the figure below joined by the curve C, show that

$$\int_C \mathbf{F} \cdot d\mathbf{r} = \int_a^b r f(r) \, dr,$$

where the limits a and b are the distances from the origin to A and B.

(c) Have we discussed any radially symmetric force fields in this chapter?

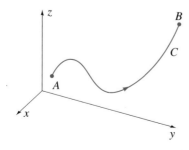

15.6 Green's Theorem

Line integrals in the xy-plane are of the form $\int_C f(x, y)ds$, and in the special case that $f(x, y)$ is the tangential component of some vector field $\mathbf{F}(x, y) = P(x, y)\hat{\mathbf{i}} + Q(x, y)\hat{\mathbf{j}}$ along C, they take the form

$$\int_C \mathbf{F} \cdot d\mathbf{r} = \int_C P(x, y)dx + Q(x, y)dy. \qquad (15.43)$$

We now show that when C is a closed curve, line integral 15.43 can usually be replaced by a double integral. The precise result is contained in the following theorem.

THEOREM 15.5   (Green's theorem)

Let C be a piecewise-smooth, closed curve in the xy-plane that does not intersect itself and that encloses a region R (Figure 15.24). If $P(x, y)$ and $Q(x, y)$ have continuous first partial derivatives in a domain D containing C and R, then

$$\oint_C P\,dx + Q\,dy = \iint_R \left(\frac{\partial Q}{\partial x} - \frac{\partial P}{\partial y} \right) dA. \qquad (15.44)$$

FIGURE 15.24

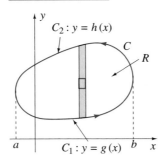

PROOF   First consider a simple region R for which every line parallel to the x- and y-axes that intersects C does so in at most two points (Figure 15.24). Then C can be subdivided into an upper and a lower part,

$$C_2: \quad y = h(x) \quad \text{and} \quad C_1: \quad y = g(x).$$

If we consider the second term on the right of equation 15.44, we have

$$\iint_R -\frac{\partial P}{\partial y}dA = \int_a^b \int_{g(x)}^{h(x)} -\frac{\partial P}{\partial y}dy\,dx = \int_a^b \left\{ -P \right\}_{g(x)}^{h(x)} dx$$

$$= \int_a^b \{P[x, g(x)] - P[x, h(x)]\}dx.$$

On the other hand, the first term on the left of 15.44 is

$$\oint_C P\,dx = \int_{C_1} P\,dx + \int_{C_2} P\,dx = \int_{C_1} P\,dx - \int_{-C_2} P\,dx,$$

and if we use x as a parameter along C_1 and $-C_2$, then

$$\oint_C P\,dx = \int_a^b P[x, g(x)]dx - \int_a^b P[x, h(x)]dx$$

$$= \int_a^b \{P[x, g(x)] - P[x, h(x)]\}dx.$$

We have shown therefore that

$$\oint_C P\,dx = \iint_R -\frac{\partial P}{\partial y}dA.$$

By subdividing C into two parts of the type $x = g(y)$ and $x = h(y)$ (where $g(y) \leq h(y)$), we can also show that

$$\oint_C Q\,dy = \iint_R \frac{\partial Q}{\partial x}dA.$$

Addition of these results gives Green's theorem for this C and R.

Now consider a more general region R such as that in Figure 15.25; it can be decomposed into n subregions R_i, each of which satisfies the condition that lines parallel to the coordinate axes intersect its boundary in at most two points. For each subregion R_i, Green's theorem gives

$$\oint_{C_i} P\,dx + Q\,dy = \iint_{R_i} \left(\frac{\partial Q}{\partial x} - \frac{\partial P}{\partial y}\right)dA.$$

If these results are added, we get

$$\sum_{i=1}^n \oint_{C_i} P\,dx + Q\,dy = \sum_{i=1}^n \iint_{R_i} \left(\frac{\partial Q}{\partial x} - \frac{\partial P}{\partial y}\right)dA.$$

Now, R is composed of the subregions R_i; thus the right side of this equation is the double integral over R. Figure 15.25 illustrates that when line integrals over the C_i are added, contributions from ancillary (interior) curves cancel in pairs, leaving the line integral around C. This completes the proof.

FIGURE 15.25

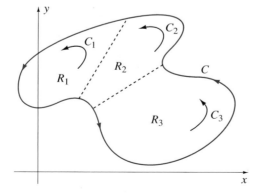

We omit a proof for even more general regions that cannot be divided into a finite number of these subregions. The interested reader should consult more advanced books.

EXAMPLE 15.15

Evaluate the line integral of Example 15.7.

SOLUTION   By Green's theorem (see Figure 15.26), we have

$$\oint_C y^2 dx + x^2 dy = \iint_R (2x - 2y) dA = 2 \int_0^1 \int_0^x (x - y) dy\, dx$$

$$= 2 \int_0^1 \left\{ xy - \frac{y^2}{2} \right\}_0^x dx = \int_0^1 x^2 dx = \left\{ \frac{x^3}{3} \right\}_0^1 = \frac{1}{3}.$$

FIGURE 15.26

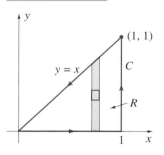

EXAMPLE 15.16

Evaluate

$$\oint_C (5x^2 + ye^x + y) dx + (e^x + e^y) dy,$$

where C is the circle $(x - 1)^2 + y^2 = 1$.

SOLUTION   By Green's theorem (see Figure 15.27), we have

$$\oint_C (5x^2 + ye^x + y) dx + (e^x + e^y) dy = \iint_R (e^x - e^x - 1) dA = - \iint_R dA$$

$$= -(\text{area of } R) = -\pi(1)^2 = -\pi.$$

FIGURE 15.27

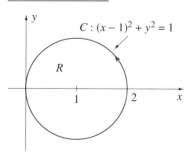

EXAMPLE 15.17

Show that the area of a region R is defined by each of the line integrals

$$\oint_C x\,dy = \oint_C -y\,dx = \frac{1}{2}\oint_C x\,dy - y\,dx,$$

where C is the boundary of R.

SOLUTION   By Green's theorem, we have

$$\oint_C x\,dy = \iint_R 1\,dA = \text{area of } R$$

and

$$\oint_C -y\,dx = \iint_R 1\,dA = \text{area of } R.$$

The third expression for area is the average of these two equations. These formulas are of particular value when the curve C is defined parametrically.

Green's theorem cannot be used to evaluate a line integral around a closed curve that contains a point at which either $P(x, y)$ or $Q(x, y)$ fails to have continuous first partial derivatives. For example, it cannot be used to evaluate the line integral in Exercise 19(a) of Section 15.4 because neither P nor Q is defined at the origin. Try it and compare the result to the correct answer (2π). A generalization of Green's theorem that can be useful in such situations is discussed in Exercise 30.

EXERCISES 15.6

In Exercises 1–11 use Green's theorem (if possible) to evaluate the line integral.

* **1.** $\oint_C y^2\,dx + x^2\,dy$ where C is the circle $x^2 + y^2 = 1$

* **2.** $\oint_C (x^2 + 2y^2)\,dy$ where C is the curve $(x - 2)^2 + y^2 = 1$

* **3.** $\oint_C x^2 e^y\,dx + (x + y)\,dy$ where C is the square with vertices $(\pm 1, 1)$ and $(\pm 1, -1)$

* **4.** $\oint_C xy^3\,dx + x^2\,dy$ where C is the curve enclosing the area bounded by $x = \sqrt{1 + y^2}$, $x = 2$

* **5.** $\oint_C (x^3 + y^3)\,dx + (x^3 - y^3)\,dy$ where C is the curve enclosing the area bounded by $x = y^2 - 1$, $x = 1 - y^2$

* **6.** $\oint_C 2 \operatorname{Tan}^{-1}(y/x)\, dx + \ln(x^2 + y^2)\, dy$ where C is the circle $(x - 4)^2 + (y - 1)^2 = 2$

* **7.** $\oint_C (3x^2 y^3 + y)\, dx + (3x^3 y^2 + 2x)\, dy$ where C is the boundary of the area enclosed by $x + y = 1$, $x = -1$, $y = -1$

* **8.** $\oint_C (x^3 + y^3)\, dx + (x^3 - y^3)\, dy$ where C is the curve $2|x| + |y| = 1$

* **9.** $\oint_C (x^2 y^2 + 3x)\, dx + (2xy - y)\, dy$ where C is the boundary of the area enclosed by $x = 1 - y^2 \ (x \geq 0)$, $y = x + 1$, $y + x + 1 = 0$

* **10.** $\oint_C (xy^2 + 2x)\, dx + (x^2 y + y + x^2)\, dy$ where C is the boundary of the area enclosed by $y^2 - x^2 = 4$, $x = 0$, $x = 3$

* **11.** $\oint_C \dfrac{-y\, dx + x\, dy}{x^2 + y^2}$ where C is the circle $x^2 + y^2 = 1$

* **12.** Show that Green's theorem can be expressed vectorially in the form

$$\oint_C \mathbf{F} \cdot d\mathbf{r} = \iint_R (\nabla \times \mathbf{F}) \cdot \hat{\mathbf{k}}\, dA.$$

** **13.** If a curve C is traced out in the direction defined by Green's theorem, it can be shown that a normal vector to C that always points to the outside of C is $\mathbf{n} = (dy, -dx)$. Show that Green's theorem can be written vectorially in the form

$$\oint_C \mathbf{F} \cdot \hat{\mathbf{n}}\, ds = \iint_R \nabla \cdot \mathbf{F}\, dA.$$

In Exercises 14–19 use the results of Example 15.17 to find the area enclosed by the curve.

** **14.** $x^2/a^2 + y^2/b^2 = 1$

** **15.** The strophoid $x = (1 - t^2)/(1 + t^2)$, $y = (t - t^3)/(1 + t^2)$ (see Example 10.6)

** **16.** The astroid $x = \cos^3 \theta$, $y = \sin^3 \theta$ (see Exercise 51 in Section 10.1)

** **17.** The right loop of the curve of Lissajous (see Exercise 49 in Section 10.1)

** **18.** The deltoid $x = 2\cos t + \cos 2t$, $y = 2\sin t - \sin 2t$

** **19.** The droplet $x = 2\cos t - \sin 2t$, $y = \sin t$

** **20.** (a) If C is the straight-line segment joining points $P_1(x_1, y_1)$ and $P_2(x_2, y_2)$ in the xy-plane, show that

$$\int_C x\, dy - y\, dx = x_1 y_2 - x_2 y_1.$$

   (b) Let $P_1(x_1, y_1)$, $P_2(x_2, y_2)$, $\cdots$, $P_n(x_n, y_n)$ be the coordinates of a polygon labelled in the counterclockwise direction (figure on page 1207). Use the result in (a) and Example 15.17 to derive the following formula for the area A of the polygon:

$$A = \frac{1}{2}\Big[(x_1 y_2 - x_2 y_1) + (x_2 y_3 - x_3 y_2) + \cdots$$
$$+ (x_{n-1} y_n - x_n y_{n-1}) + (x_n y_1 - x_1 y_n)\Big].$$

(c) Show that, if the coordinates of the vertices are arranged in a vertical column with (x_1, y_1) repeated at the bottom (figure below), then

$$A = \frac{1}{2}\Big\{[\text{sum of downward products to the right } (\searrow)]$$

$$- [\text{sum of downward products to the left } (\swarrow)]\Big\}.$$

In Exercises 21–24 use the result of Exercise 20 to find the area of the polygon with the points as successive vertices.

∗ **21.** $(1, 2)$, $(-3, 2)$, $(4, 1)$

∗ **22.** $(2, -2)$, $(1, -3)$, $(-2, 1)$, $(5, 6)$

∗ **23.** $(3, 0)$, $(1, 1)$, $(2, 5)$, $(-4, -4)$

∗ **24.** $(0, 4)$, $(-1, 0)$, $(-2, 0)$, $(-3, -4)$, $(0, -5)$, $(6, -2)$, $(3, 0)$, $(2, 2)$

In Exercises 25–29 evaluate the line integral.

∗∗ **25.** $\oint_C (2xye^{x^2y} + 3x^2y)\,dx + x^2e^{x^2y}\,dy$ where C is the ellipse $x^2 + 4y^2 = 4$

∗∗ **26.** $\oint_C (3x^2y^3 - x^2y)\,dx + (xy^2 + 3x^3y^2)\,dy$ where C is the circle $x^2 + y^2 = 9$

∗∗ **27.** $\oint_C -x^3y^2\,dx + x^2y^3\,dy$ where C is the right loop of $(x^2 + y^2)^{3/2} = x^2 - y^2$

∗∗ **28.** $\int_C (x - y)(dx + dy)$ where C is the semicircular part of $x^2 + y^2 = 4$ above $y = x$ from $(-\sqrt{2}, -\sqrt{2})$ to $(\sqrt{2}, \sqrt{2})$

∗∗ **29.** $\int_C (e^y - y\sin x)\,dx + (\cos x + xe^y)\,dy$ where C is the curve $x = 1 - y^2$ from $(0, -1)$ to $(0, 1)$

✱✱ **30.** The result of this exercise is useful when the curve C in Green's theorem contains a point (or points) at which either P or Q fails to have continuous first partial derivatives (see Exercises 31–35).

(a) Suppose a piecewise-smooth curve C (figure below) contains in its interior another piecewise smooth curve C', and that $P(x, y)$ and $Q(x, y)$ have continuous first partial derivatives in a domain containing C and C' and the area R between them. Prove that

$$\oint_C P\,dx + Q\,dy + \oint_{C'} P\,dx + Q\,dy = \iint_R \left(\frac{\partial Q}{\partial x} - \frac{\partial P}{\partial y} \right) dA.$$

Hint: Join C and C' by two curves such as those in the figure.

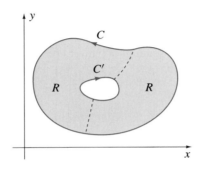

(b) Extend this result to show that when C' is replaced by n distinct curves C_i (figure below), and P and Q have continuous first partial derivatives in a domain containing C and the C_i and the area R between them,

$$\oint_C P\,dx + Q\,dy + \oint_{C_1} P\,dx + Q\,dy + \cdots + \oint_{C_n} P\,dx + Q\,dy$$

$$= \iint_R \left(\frac{\partial Q}{\partial x} - \frac{\partial P}{\partial y} \right) dA.$$

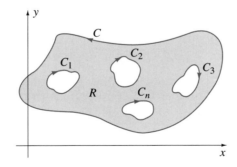

(c) What can we conclude in (a) and (b) if $\partial Q/\partial x = \partial P/\partial y$ in R?

In Exercises 31–33 use the result of Exercise 30(a) to evaluate the line integral.

✱✱ **31.** $\displaystyle\oint_C \frac{y\,dx - (x - 1)\,dy}{(x - 1)^2 + y^2}$ where C is the circle $x^2 + y^2 = 4$

** **32.** $\displaystyle\oint_C \frac{-x^2 y\,dx + x^3\,dy}{(x^2 + y^2)^2}$ where C is the ellipse $4x^2 + y^2 = 1$

** **33.** $\displaystyle\oint_C \frac{-y\,dx + x\,dy}{x^2 + y^2}$ where C is the square with vertices $(\pm 2, 0)$ and $(0, \pm 2)$

** **34.** Show that the line integral of Exercise 19 in Section 4 (page 1196) has value $\pm 2\pi$ for every piecewise-smooth, closed curve enclosing the origin that does not intersect itself.

** **35.** (a) In what domains is the line integral $\displaystyle\int_C \frac{x\,dx + y\,dy}{x^2 + y^2}$ independent of path?

  (b) Evaluate the integral clockwise around the curve $x^2 + y^2 - 2y = 1$.

In Exercises 36–38 assume that $P(x, y)$ and $Q(x, y)$ have continuous second partial derivatives in a domain containing R and C (figure below). Let $\hat{\mathbf{n}} = (dy/ds, -dx/ds)$ be the outward pointing normal to C, and let

$$\frac{\partial P}{\partial n} = \nabla P \cdot \hat{\mathbf{n}}, \qquad \frac{\partial Q}{\partial n} = \nabla Q \cdot \hat{\mathbf{n}}$$

be the directional derivatives of P and Q in the direction $\hat{\mathbf{n}}$.

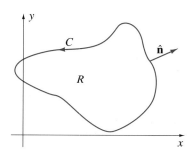

** **36.** Show that

$$\oint_C \frac{\partial P}{\partial n}\,ds = \iint_R \nabla^2 P\,dA$$

where $\nabla^2 P = \dfrac{\partial^2 P}{\partial x^2} + \dfrac{\partial^2 P}{\partial y^2}$. Hint: See Exercise 13. What can we conclude if $P(x, y)$ satisfies Laplace's equation in R?

** **37.** Show that

$$\oint_C P\frac{\partial Q}{\partial n}\,ds = \iint_R P\nabla^2 Q\,dA + \iint_R \nabla P \cdot \nabla Q\,dA.$$

Hint: Use identity 15.11. This result is often called Green's first identity (in the plane).

** **38.** Prove that

$$\oint_C \left(P\frac{\partial Q}{\partial n} - Q\frac{\partial P}{\partial n} \right) ds = \iint_R (P\nabla^2 Q - Q\nabla^2 P)\,dA.$$

This is often called Green's second identity (in the plane).

*** **39.** Find all possible values for the line integral $\displaystyle\oint_C \frac{-y\,dx + x\,dy}{x^2 + y^2}$ for curves in the xy-plane not passing through the origin.

15.7 Surface Integrals

Consider a function $f(x, y, z)$ defined on some surface S (Figure 15.28). We divide S into n subsurfaces of areas $\Delta S_1, \Delta S_2, \ldots, \Delta S_n$ in any manner whatsoever. On each subsurface $\Delta S_i (i = 1, \ldots, n)$ we choose an arbitrary point (x_i^*, y_i^*, z_i^*), and form the sum

$$f(x_1^*, y_1^*, z_1^*)\Delta S_1 + f(x_2^*, y_2^*, z_2^*)\Delta S_2 + \cdots + f(x_n^*, y_n^*, z_n^*)\Delta S_n$$

$$= \sum_{i=1}^{n} f(x_i^*, y_i^*, z_i^*)\Delta S_i. \qquad (15.45)$$

If this sum approaches a limit as the number of subsurfaces becomes increasingly large and every subsurface shrinks to a point, we call the limit the surface integral of $f(x, y, z)$ over the surface S, and denote it by

$$\iint_S f(x, y, z)\,dS = \lim_{\|\Delta S_i\| \to 0} \sum_{i=1}^{n} f(x_i^*, y_i^*, z_i^*)\Delta S_i. \qquad (15.46)$$

FIGURE 15.28

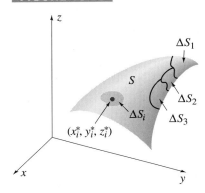

Surface integrals, like all integrals, are limit-summations. We think of dS as a small piece of area on S, and each dS is multiplied by the value of $f(x, y, z)$ for that area. All such products are then added together and the limit taken as the pieces of area shrink to points.

The following theorem guarantees existence of surface integrals of continuous functions over smooth surfaces.

THEOREM 15.6

If a function $f(x, y, z)$ is continuous on a smooth surface S of finite area, then the surface integral of $f(x, y, z)$ over S exists. If S projects in a one-to-one fashion onto a region S_{xy} in the xy-plane, then the surface integral of $f(x, y, z)$ over S can be evaluated by means of the following double integral:

$$\iint_S f(x, y, z)\,dS = \iint_{S_{xy}} f[x, y, g(x, y)]\sqrt{1 + \left(\frac{\partial z}{\partial x}\right)^2 + \left(\frac{\partial z}{\partial y}\right)^2}\,dA, \qquad (15.47)$$

where $z = g(x, y)$ is the equation of S.

It is not necessary to memorize 15.47 as a formula. The right side is obtained by expressing z and dS in terms of x and y and interpreting the result as a double integral over the projection of S in the xy-plane. To be more explicit, recall from Section 14.7 that, when a surface S can be represented in the form $z = g(x, y)$, a small area dS on S is related to its projection dA in the xy-plane according to the formula

$$dS = \sqrt{1 + \left(\frac{\partial z}{\partial x}\right)^2 + \left(\frac{\partial z}{\partial y}\right)^2}\, dA. \tag{15.48}$$

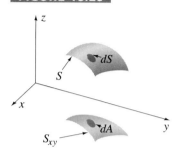

(Figure 15.29). If we substitute this into the left side of 15.47 and at the same time use $z = g(x, y)$ to express $f(x, y, z)$ in terms of x and y, then

$$\iint_S f(x, y, z)\,dS = \iint_S f[x, y, g(x, y)]\sqrt{1 + \left(\frac{\partial z}{\partial x}\right)^2 + \left(\frac{\partial z}{\partial y}\right)^2}\, dA.$$

But if we now interpret the right side of this equation as the double integral of $f[x, y, g(x, y)]$ $\sqrt{1 + (\partial z/\partial x)^2 + (\partial z/\partial y)^2}$ over the projection S_{xy} of S onto the xy-plane, we obtain 15.47.

Note the analogy between equations 15.47 and 15.20. Equation 15.20 states that the line integral on the left can be evaluated by means of the definite integral on the right. Equation 15.47 states that the surface integral on the left can be evaluated by means of the double integral on the right.

If a surface does not project one-to-one onto an area in the xy-plane, then one possibility is to subdivide it into parts, each of which projects one-to-one onto the xy-plane. The total surface integral over the surface is then the sum of the surface integrals over the parts. For example, if we require the surface integral of a function $f(x, y, z)$ over the sphere $S: x^2 + y^2 + z^2 = 1$ (Figure 15.30), we could subdivide S into two hemispheres,

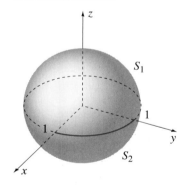

$$S_1: \quad z = \sqrt{1 - x^2 - y^2},$$
$$S_2: \quad z = -\sqrt{1 - x^2 - y^2},$$

each of which projects onto

$$S_{xy}: \quad x^2 + y^2 \leq 1.$$

Then,

$$\iint_S f(x, y, z)\,dS = \iint_{S_1} f(x, y, z)\,dS + \iint_{S_2} f(x, y, z)\,dS.$$

A second possibility is to project surfaces onto either the xz-plane or the yz-plane. If a surface projects one-to-one onto areas S_{xz} and S_{yz} in these planes, then

$$\iint_S f(x, y, z)\,dS = \iint_{S_{xz}} f[x, g(x, z), z]\sqrt{1 + \left(\frac{\partial y}{\partial x}\right)^2 + \left(\frac{\partial y}{\partial z}\right)^2}\, dA \tag{15.49}$$

and

$$\iint_S f(x, y, z)\,dS = \iint_{S_{yz}} f[g(y, z), y, z]\sqrt{1 + \left(\frac{\partial x}{\partial y}\right)^2 + \left(\frac{\partial x}{\partial z}\right)^2}\, dA. \tag{15.50}$$

We should also note that the area of a surface S is defined by the surface integral

$$\iint_S dS, \tag{15.51}$$

and if S projects one-to-one onto S_{xy}, then

$$\text{area of } S = \iint_{S_{xy}} \sqrt{1 + \left(\frac{\partial z}{\partial x}\right)^2 + \left(\frac{\partial z}{\partial y}\right)^2}\, dA. \tag{15.52}$$

This is formula 14.41.

The results in equations 15.47–15.52 were based on a smooth surface S. If S is piecewise smooth, rather than smooth, we subdivide S into smooth parts and apply each of these results to the smooth parts. The surface integral over S is then the summation of the surface integrals over its parts.

| **EXAMPLE 15.18** ▼

Evaluate $\iint_S (x + y + z)\, dS$, where S is that part of the plane $x + 2y + 4z = 8$ in the first octant.

SOLUTION The surface S projects one-to-one onto the triangle S_{xy} in the xy-plane in Figure 15.31. Since $z = (8 - x - 2y)/4$ on S,

$$\iint_S (x + y + z)\, dS = \iint_{S_{xy}} \left(x + y + 2 - \frac{x}{4} - \frac{y}{2}\right) \sqrt{1 + \left(\frac{\partial z}{\partial x}\right)^2 + \left(\frac{\partial z}{\partial y}\right)^2}\, dA$$

$$= \frac{1}{4} \iint_{S_{xy}} (3x + 2y + 8) \sqrt{1 + \left(-\frac{1}{4}\right)^2 + \left(-\frac{1}{2}\right)^2}\, dA$$

$$= \frac{\sqrt{21}}{16} \int_0^4 \int_0^{8-2y} (3x + 2y + 8)\, dx\, dy$$

$$= \frac{\sqrt{21}}{16} \int_0^4 \left\{ \frac{3x^2}{2} + (2y + 8)x \right\}_0^{8-2y} dy$$

$$= \frac{\sqrt{21}}{8} \int_0^4 (80 - 24y + y^2)\, dy$$

$$= \frac{\sqrt{21}}{8} \left\{ 80y - 12y^2 + \frac{y^3}{3} \right\}_0^4 = \frac{56\sqrt{21}}{3}.$$

FIGURE 15.31

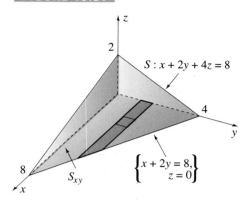

$$S : x + 2y + 4z = 8$$

$$\begin{Bmatrix} x + 2y = 8, \\ z = 0 \end{Bmatrix}$$

S_{xy}

EXAMPLE 15.19

Evaluate $\iint_S z^2 \, dS$ where S is the sphere $x^2 + y^2 + z^2 = 4$.

FIGURE 15.32

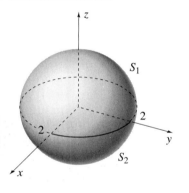

SOLUTION   We divide S into two hemispheres (Figure 15.32),

$$S_1 : z = \sqrt{4 - x^2 - y^2}, \quad S_2 : z = -\sqrt{4 - x^2 - y^2}$$

each of which projects one-to-one onto the circle $S_{xy} : x^2 + y^2 \le 4$, $z = 0$ in the xy-plane. For each hemisphere,

$$dS = \sqrt{1 + \left(\frac{\partial z}{\partial x}\right)^2 + \left(\frac{\partial z}{\partial y}\right)^2} \, dA$$

$$= \sqrt{1 + \frac{x^2}{4 - x^2 - y^2} + \frac{y^2}{4 - x^2 - y^2}} \, dA = \frac{2}{\sqrt{4 - x^2 - y^2}} \, dA,$$

and therefore

$$\iint_S z^2 \, dS = \iint_{S_1} z^2 \, dS + \iint_{S_2} z^2 \, dS$$

$$= \iint_{S_{xy}} (4 - x^2 - y^2) \frac{2}{\sqrt{4 - x^2 - y^2}} \, dA$$

$$+ \iint_{S_{xy}} (4 - x^2 - y^2) \frac{2}{\sqrt{4 - x^2 - y^2}} \, dA$$

$$= 4 \iint_{S_{xy}} \sqrt{4 - x^2 - y^2} \, dA.$$

If we use polar coordinates to evaluate this integral over S_{xy}, then

$$\iint_S z^2 dS = 4 \int_0^{2\pi} \int_0^2 \sqrt{4 - r^2}\, r\, dr\, d\theta$$

$$= 4 \int_0^{2\pi} \left\{ -\frac{1}{3}(4 - r^2)^{3/2} \right\}_0^2 d\theta = \frac{32}{3} \{\theta\}_0^{2\pi} = \frac{64\pi}{3}.$$

If parameters can be found to describe a surface, it may not be necessary to project the surface into one of the coordinate planes. Such is the case for a sphere centred at the origin. Formula 14.67 for the volume element in spherical coordinates indicates that an area element on the surface of a sphere of radius R can be expressed in terms of angles θ and ϕ (Figure 15.33) as

$$dS = R^2 \sin \phi\, d\phi\, d\theta. \tag{15.53}$$

With this choice of area element, the surface integral in Example 15.19 is evaluated as follows:

$$\iint_S z^2 dS = \int_0^{2\pi} \int_0^{\pi} (2\cos\phi)^2 4 \sin\phi\, d\phi\, d\theta = 16 \int_0^{2\pi} \int_0^{\pi} \cos^2\phi \sin\phi\, d\phi\, d\theta$$

$$= 16 \int_0^{2\pi} \left\{ -\frac{1}{3}\cos^3\phi \right\}_0^{\pi} d\theta = \frac{32}{3}\{\theta\}_0^{2\pi} = \frac{64\pi}{3}.$$

FIGURE 15.33

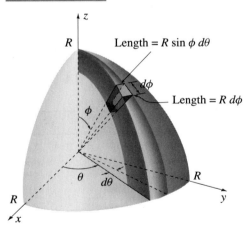

In Exercises 1–8 evaluate the surface integral.

EXERCISES 15.7

* **1.** $\displaystyle\iint_S (x^2 y + z)\, dS$ where S is the first octant part of $2x + 3y + z = 6$

* **2.** $\displaystyle\iint_S (x^2 + y^2) z\, dS$ where S is that part of $z = x + y$ cut out by $x = 0$, $y = 0$, $x + y = 1$

* **3.** $\iint_S xyz \, dS$ where S is the surface of the cube $0 \leq x \leq 1$, $0 \leq y \leq 1$, $0 \leq z \leq 1$

* **4.** $\iint_S xy \, dS$ where S is the first octant part of $z = \sqrt{x^2 + y^2}$ cut out by $x^2 + y^2 = 1$

* **5.** $\iint_S \dfrac{1}{\sqrt{z - y + 1}} \, dS$ where S is the surface defined by $2z = x^2 + 2y$, $0 \leq x \leq 1$, $0 \leq y \leq 1$

* **6.** $\iint_S \sqrt{4y + 1} \, dS$ where S is the first octant part of $y = x^2$ cut out by $2x + y + z = 1$

* **7.** $\iint_S x^2 z \, dS$ where S is the surface $x^2 + y^2 = 1$ for $0 \leq z \leq 1$

* **8.** $\iint_S (x + y) \, dS$ where S is the surface bounding the volume enclosed by $x = 0$, $y = 0$, $z = 0$, $6x - 3y + 2z = 6$

** **9.** Set up double iterated integrals for the surface integral of a function $f(x, y, z)$ over the surface defined by $z = 4 - x^2 - 4y^2$ ($x, y, z \geq 0$) if the surface is projected onto the xy-, the xz-, and the yz-planes.

** **10.** Use a surface integral to find the area of the curved portion of a right circular cone of radius R and height h.

In Exercises 11–17 evaluate the surface integral.

** **11.** $\iint_S xyz^3 \, dS$ where S is the surface defined by $x = y^2$, $0 \leq x \leq 4$, $0 \leq z \leq 1$

** **12.** $\iint_S xyz \, dS$ where S is the surface defined by $2y = \sqrt{9 - x}$, $x \geq 0$, $0 \leq z \leq 3$

** **13.** $\iint_S \dfrac{1}{\sqrt{2az - z^2}} \, dS$ where S is that part of $x^2 + y^2 + (z - a)^2 = a^2$ inside the cylinder $x^2 + y^2 = ay$, underneath the plane $z = a$, and in the first octant

** **14.** $\iint_S z \, dS$ where S is that part of the surface $x^2 + y^2 - z^2 = 1$ between the planes $z = 0$ and $z = 1$

** **15.** $\iint_S x^2 y^2 \, dS$ where S is that part of $z = x^2 + y^2$ inside $x^2 + y^2 + z^2 = 2$

** **16.** $\iint_S x^2 \, dS$ where S is that part of $z = xy$ inside $x^2 + y^2 = 4$

** **17.** $\iint_S z(y + x^2) \, dS$ where S is that part of $y = 1 - x^2$ bounded by $z = 0$, $z = 2$, and $y = 0$

In Exercises 18–22 evaluate the surface integral by projecting the surface into one of the coordinate planes and also by using area element 15.53.

** **18.** $\iint_S dS$ where S is the sphere $x^2 + y^2 + z^2 = R^2$. Is this the formula for the area of a sphere?

** **19.** $\iint_S x^2 z^2 \, dS$ where S is the sphere $x^2 + y^2 + z^2 = 1$

** **20.** $\iint_S (x^2 - y^2) \, dS$ where S is the hemisphere $z = \sqrt{9 - x^2 - y^2}$

** **21.** $\iint_S (x^2 + y^2) \, dS$ where S is the sphere $x^2 + y^2 + z^2 = R^2$

** **22.** $\iint_S \dfrac{1}{x^2 + y^2} \, dS$ where S is that part of the sphere $x^2 + y^2 + z^2 = 4R^2$ between the planes $z = 0$ and $z = R$

** **23.** A viscous material is allowed to drip onto the sphere $x^2 + y^2 + z^2 = 1$ at the point $(0, 0, 1)$ (all dimensions in metres). The material spreads out evenly in all directions and runs down the sphere, becoming more and more viscous as it does so. The thickness of the material increases linearly with respect to angle ϕ (Figure 15.33) from 0.001 m at $(0, 0, 1)$ to 0.005 m at $(0, 0, -1)$. Find the volume of material on the sphere.

** **24.** Show that if a surface S defined implicitly by the equation $F(x, y, z) = 0$ projects one-to-one onto the area S_{xy} in the xy-plane, then

$$ \iint_S f(x, y, z) \, dS = \iint_{S_{xy}} f[x, y, g(x, y)] \frac{|\nabla F|}{|\partial F / \partial z|} \, dA. $$

** **25.**   (a) Find the area cut from the cones $z^2 = x^2 + y^2$ by the cylinder $x^2 + y^2 = 2x$.

(b) Find the area cut from the cylinder $x^2 + y^2 = 2x$ by the cones $z^2 = x^2 + y^2$.

15.8 Surface Integrals Involving Vector Fields

The most important and common type of surface integral occurs when $f(x, y, z)$ in 15.46 is specified as the normal component of some given vector field $\mathbf{F}(x, y, z)$ defined on S. In other words, $f(x, y, z)$ itself is not given, but $\mathbf{F}$ is, and to find $f(x, y, z)$ we must calculate the component of $\mathbf{F}$ normal to S.

This presupposes that surfaces are two-sided and that a normal vector to a surface can be assigned in an unambiguous way. When this is possible, the surface is said to be **orientable**. All surfaces in this book are orientable, with the exception of the Möbius strip mentioned below.

Take a thin rectangular strip of paper and label its corners A, B, C, and D (Figure 15.34a). Give the strip a half twist and join A and C, and B and D (Figure 15.34b).

FIGURE 15.34a

FIGURE 15.34b

This surface, called a Möbius strip, cannot be assigned a unique normal vector which varies continuously over the surface. To illustrate, suppose at point P in Figure 15.34b, we assign a unit normal vector $\hat{\mathbf{n}}$ as shown. By moving once around the strip, we can vary the direction of $\hat{\mathbf{n}}$ continuously and arrive back at P with $\hat{\mathbf{n}}$ pointing in the opposite direction. This surface is said to have only one side, or to be nonorientable.

We consider only surfaces that are orientable, or have two sides, and can therefore be assigned a unit normal vector in an unambiguous way.

Suppose again that $\mathbf{F}(x, y, z)$ is a vector field defined on an (orientable) surface S and $\hat{\mathbf{n}}$ is the unit normal on one side of S. If $f(x, y, z)$ is the component of $\mathbf{F}$ in the direction $\hat{\mathbf{n}}$, then $f(x, y, z) = \mathbf{F} \cdot \hat{\mathbf{n}}$, and the surface integral of $f(x, y, z)$ over S can be expressed in the form

$$\iint_S f(x, y, z)\, dS = \iint_S \mathbf{F} \cdot \hat{\mathbf{n}}\, dS. \tag{15.54}$$

EXAMPLE 15.20

Evaluate $\iint_S \mathbf{F} \cdot \hat{\mathbf{n}}\, dS$, where $\mathbf{F} = x^2 y \hat{\mathbf{i}} + xz \hat{\mathbf{j}}$ and $\hat{\mathbf{n}}$ is the upper normal to the surface $S : z = 4 - x^2 - y^2, z \geq 0$.

SOLUTION   Since a normal vector to S is

$$\nabla(z - 4 + x^2 + y^2) = (2x, 2y, 1),$$

it follows that

$$\hat{\mathbf{n}} = \frac{(2x, 2y, 1)}{\sqrt{4x^2 + 4y^2 + 1}}.$$

Thus,

$$\mathbf{F} \cdot \hat{\mathbf{n}} = \frac{2x^3 y + 2xyz}{\sqrt{4x^2 + 4y^2 + 1}}.$$

If we project S onto $S_{xy} : x^2 + y^2 \leq 4$ in the xy-plane (Figure 15.35), then

$$\begin{aligned}
\iint_S \mathbf{F} \cdot \hat{\mathbf{n}}\, dS &= \iint_{S_{xy}} \frac{2x^3 y + 2xyz}{\sqrt{4x^2 + 4y^2 + 1}} \sqrt{1 + \left(\frac{\partial z}{\partial x}\right)^2 + \left(\frac{\partial z}{\partial y}\right)^2}\, dA \\
&= \iint_{S_{xy}} \frac{2x^3 y + 2xy(4 - x^2 - y^2)}{\sqrt{4x^2 + 4y^2 + 1}} \sqrt{1 + (-2x)^2 + (-2y)^2}\, dA \\
&= 2 \iint_{S_{xy}} (4xy - xy^3)\, dA \\
&= 2 \int_{-2}^{2} \int_{-\sqrt{4-x^2}}^{\sqrt{4-x^2}} (4xy - xy^3)\, dy\, dx \\
&= 2 \int_{-2}^{2} \left\{ 2xy^2 - \frac{xy^4}{4} \right\}_{-\sqrt{4-x^2}}^{\sqrt{4-x^2}}\, dx \\
&= 0.
\end{aligned}$$

FIGURE 15.35

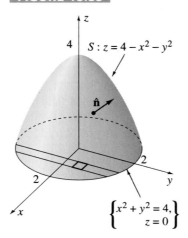

EXAMPLE 15.21

Evaluate

$$\oiint_S \mathbf{F} \cdot \hat{\mathbf{n}} \, dS,$$

where $\mathbf{F} = x\hat{\mathbf{i}} + y\hat{\mathbf{j}} + z\hat{\mathbf{k}}$ and $\hat{\mathbf{n}}$ is the unit outward-pointing normal to the surface S enclosing the volume bounded by $x^2 + y^2 = 4$, $z = 0$, $z = 2$.

SOLUTION   We divide S into four parts (Figure 15.36):

$$S_1 : \quad z = 0, \quad x^2 + y^2 \le 4;$$

$$S_2 : \quad z = 2, \quad x^2 + y^2 \le 4;$$

$$S_3 : \quad y = \sqrt{4 - x^2}, \quad 0 \le z \le 2;$$

$$S_4 : \quad y = -\sqrt{4 - x^2}, \quad 0 \le z \le 2.$$

FIGURE 15.36

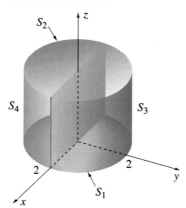

On S_1, $\hat{\mathbf{n}} = -\hat{\mathbf{k}}$; on S_2, $\hat{\mathbf{n}} = \hat{\mathbf{k}}$; and on S_3 and S_4,

$$\hat{\mathbf{n}} = \frac{\nabla(x^2 + y^2 - 4)}{|\nabla(x^2 + y^2 - 4)|} = \frac{(2x, 2y, 0)}{\sqrt{4x^2 + 4y^2}} = \frac{(x, y, 0)}{2}.$$

We know that S_1 and S_2 project onto $S_{xy} : x^2 + y^2 \leq 4$ in the xy-plane, and S_3 and S_4 project onto the rectangle $S_{xz} : -2 \leq x \leq 2, 0 \leq z \leq 2$ in the xz-plane. Consequently,

$$\oiint_S \mathbf{F} \cdot \hat{\mathbf{n}} \, dS = \iint_{S_1} -z \, dS + \iint_{S_2} z \, dS + \iint_{S_3} \left(\frac{x^2 + y^2}{2}\right) dS$$

$$+ \iint_{S_4} \left(\frac{x^2 + y^2}{2}\right) dS$$

$$= 0 + \iint_{S_{xy}} 2\sqrt{1} \, dA$$

$$+ \frac{1}{2} \iint_{S_{xz}} (x^2 + 4 - x^2)\sqrt{1 + \left(\frac{-x}{\sqrt{4 - x^2}}\right)^2} \, dA$$

$$+ \frac{1}{2} \iint_{S_{xz}} (x^2 + 4 - x^2)\sqrt{1 + \left(\frac{x}{\sqrt{4 - x^2}}\right)^2} \, dA$$

$$= 2 \iint_{S_{xy}} dA + 4 \iint_{S_{xz}} \frac{2}{\sqrt{4 - x^2}} \, dA$$

$$= 2(\text{area of } S_{xy}) + 8 \int_{-2}^{2} \int_{0}^{2} \frac{1}{\sqrt{4 - x^2}} \, dz \, dx$$

$$= 2(4\pi) + 16 \int_{-2}^{2} \frac{1}{\sqrt{4 - x^2}} \, dx.$$

If we set $x = 2 \sin \theta$, then $dx = 2 \cos \theta \, d\theta$ and

$$\oiint_S \mathbf{F} \cdot \hat{\mathbf{n}} \, dS = 8\pi + 16 \int_{-\pi/2}^{\pi/2} \frac{1}{2 \cos \theta} 2 \cos \theta \, d\theta = 8\pi + 16 \left\{\theta\right\}_{-\pi/2}^{\pi/2} = 24\pi.$$

Note that the notation $\oiint$ for the surface integral over a closed surface (a closed surface being one that encloses a volume) is similar to that for a line integral around a closed curve. In Exercise 16 of Section 7.6 we discussed Archimedes' principle for a right circular cylinder. In the next example we discuss it for a sphere. It will be verified for arbitrary shapes in Section 15.9.

EXAMPLE 15.22

If a spherical object is submerged in a fluid with density ρ, it experiences a buoyant force due to fluid pressure. Show that the magnitude of this force is exactly the weight of fluid displaced by the object.

SOLUTION   Suppose the object is represented by the sphere $x^2 + y^2 + z^2 = R^2$ and the surface of the fluid by the plane $z = h$ $(h > R)$ (Figure 15.37). If dS is a small area on the surface, then the force due to fluid pressure on dS has magnitude $P\,dS$, where P is pressure, and this force acts normal to the surface of the sphere. If $\hat{\mathbf{n}}$ is the unit inward-pointing normal to the sphere, then the force on dS is

$$(P\,dS)\hat{\mathbf{n}}.$$

Clearly, the resultant force on the sphere will be in the z-direction, the x- and y-components cancelling because of the symmetry of the sphere. The z-component of the force on dS is $(P\,dS)\hat{\mathbf{n}} \cdot \hat{\mathbf{k}}$; thus the magnitude of the resultant force on the sphere is

$$\iint_S P\hat{\mathbf{k}} \cdot \hat{\mathbf{n}}\,dS.$$

A normal to the surface is $\nabla(x^2 + y^2 + z^2 - R^2) = (2x, 2y, 2z)$, and therefore the unit inward-pointing normal is

$$\hat{\mathbf{n}} = \frac{-(x, y, z)}{\sqrt{x^2 + y^2 + z^2}} = -\frac{1}{R}(x, y, z).$$

With formula 15.53 for an area element on the sphere, and $P = 9.81\rho(h - z)$,

$$\iint_S P\hat{\mathbf{k}} \cdot \hat{\mathbf{n}}\,dS = \iint_S 9.81\rho(h - z)\left(-\frac{z}{R}\right) dS$$

$$= -\frac{9.81\rho}{R} \int_0^{2\pi} \int_0^{\pi} (h - R\cos\phi)(R\cos\phi)R^2 \sin\phi\,d\phi\,d\theta$$

$$= -9.81\rho R^2 \int_0^{2\pi} \left\{-\frac{h}{2}\cos^2\phi + \frac{R}{3}\cos^3\phi\right\}_0^{\pi} d\theta$$

$$= \frac{2(9.81)\rho R^3}{3}\{\theta\}_0^{2\pi}$$

$$= \frac{4}{3}\pi R^3(9.81\rho).$$

FIGURE 15.37

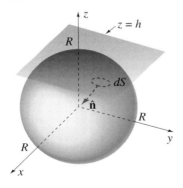

EXERCISES 15.8

In Exercises 1–16 evaluate the surface integral.

* **1.** $\iint_S (x\hat{\mathbf{i}} + z\hat{\mathbf{k}}) \cdot \hat{\mathbf{n}}\,dS$ where S is the first octant part of $x + y + z = 3$, and $\hat{\mathbf{n}}$ is the unit normal to S with positive z-component

* **2.** $\iint_S (yz^2\hat{\mathbf{i}} + ye^x\hat{\mathbf{j}} + x\hat{\mathbf{k}}) \cdot \hat{\mathbf{n}}\,dS$ where S is defined by $y = x^2$, $0 \le y \le 4$, $0 \le z \le 1$, and $\hat{\mathbf{n}}$ is the unit normal to S with positive y-component

* **3.** $\iint_S (x\hat{\mathbf{i}} + y\hat{\mathbf{j}} + z\hat{\mathbf{k}}) \cdot \hat{\mathbf{n}}\,dS$ where S is the hemisphere $z = \sqrt{1 - x^2 - y^2}$, and $\hat{\mathbf{n}}$ is its upper normal

* **4.** $\iint_S (yz\hat{\mathbf{i}} + zx\hat{\mathbf{j}} + xy\hat{\mathbf{k}}) \cdot \hat{\mathbf{n}}\,dS$ where S is that part of the surface $z = x^2 + y^2$ cut out by the planes $x = 1$, $x = -1$, $y = 1$, $y = -1$, and $\hat{\mathbf{n}}$ is the unit lower normal to S

* **5.** $\iint_S (z\hat{\mathbf{i}} - x\hat{\mathbf{j}} + y\hat{\mathbf{k}}) \cdot \hat{\mathbf{n}}\,dS$ where S is the surface that encloses the volume defined by $z = \sqrt{4 - x^2 - y^2}$, $z = 0$, and $\hat{\mathbf{n}}$ is the unit outer normal to S

** **6.** $\iint_S (x\hat{\mathbf{i}} + y\hat{\mathbf{j}}) \cdot \hat{\mathbf{n}}\,dS$ where S is that part of the surface $z = \sqrt{x^2 + y^2}$ below $z = 1$, and $\hat{\mathbf{n}}$ is the unit normal to S with negative z-component

** **7.** $\iint_S (xyz\hat{\mathbf{i}} - x\hat{\mathbf{j}} + z\hat{\mathbf{k}}) \cdot \hat{\mathbf{n}}\,dS$ where S is the smaller part of $x^2 + y^2 = 9$ cut out by $z = 0$, $z = 2$, $y = |x|$, and $\hat{\mathbf{n}}$ is the unit normal to S with positive y-component

** **8.** $\iint_S (x^2y\hat{\mathbf{i}} + xy\hat{\mathbf{j}} + z\hat{\mathbf{k}}) \cdot \hat{\mathbf{n}}\,dS$ where S is defined by $z = 2 - x^2 - y^2$, $z \ge 0$, and $\hat{\mathbf{n}}$ is the unit normal to S with negative z-component

** **9.** $\iint_S (yz\hat{\mathbf{i}} + xz\hat{\mathbf{j}} + xy\hat{\mathbf{k}}) \cdot \hat{\mathbf{n}}\,dS$ where S is the surface enclosing the volume defined by $x = 0$, $x = 2$, $z = 0$, $z = y$, $y + z = 2$, and $\hat{\mathbf{n}}$ is the unit outer normal to S

** **10.** $\iint_S (x\hat{\mathbf{i}} + y\hat{\mathbf{j}}) \cdot \hat{\mathbf{n}}\,dS$ where S is the surface $x^2 + y^2 + z^2 = 4$, $z \ge 1$, and $\hat{\mathbf{n}}$ is the unit upper normal to S

** **11.** $\iint_S (x^2\hat{\mathbf{i}} + y^2\hat{\mathbf{j}} + z^2\hat{\mathbf{k}}) \cdot \hat{\mathbf{n}}\,dS$ where S is the sphere $x^2 + y^2 + z^2 = a^2$, and $\hat{\mathbf{n}}$ is the unit outer normal to S

** **12.** $\iint_S (y\hat{\mathbf{i}} - x\hat{\mathbf{j}} + \hat{\mathbf{k}}) \cdot \hat{\mathbf{n}}\,dS$ where S is the smaller surface cut, by the plane $y + z = 1$, from the sphere $x^2 + y^2 + z^2 = 1$, and $\hat{\mathbf{n}}$ is the unit upper normal to S

** **13.** $\iint_S \mathbf{F} \cdot \hat{\mathbf{n}}\,dS$ where $\mathbf{F} = (z^2 - x)\hat{\mathbf{i}} - xy\hat{\mathbf{j}} + 3z\hat{\mathbf{k}}$, S is the surface enclosing the volume defined by $z = 4 - y^2$, $x = 0$, $x = 3$, $z = 0$, and $\hat{\mathbf{n}}$ is the unit outer normal to S

** **14.** $\iint_S (x^2\hat{\mathbf{i}} + xy\hat{\mathbf{j}} + xz\hat{\mathbf{k}}) \cdot \hat{\mathbf{n}}\, dS$ where S is that part of the surface $z = \sqrt{4 + y^2 - x^2}$ in the first octant cut out by the planes $y = 0$, $y = 1$, $x = 0$, $z = 0$, and $\hat{\mathbf{n}}$ is the unit normal to S with positive z-component

** **15.** $\iint_S (x^2\hat{\mathbf{i}} + yz\hat{\mathbf{j}} - x\hat{\mathbf{k}}) \cdot \hat{\mathbf{n}}\, dS$ where S is that part of the surface $x = yz$ in the first octant cut out by $y^2 + z^2 = 1$, and $\hat{\mathbf{n}}$ is the unit normal to S with positive x-component

** **16.** $\oiint_S (yx\hat{\mathbf{i}} + y^2\hat{\mathbf{j}} + yz\hat{\mathbf{k}}) \cdot \hat{\mathbf{n}}\, dS$ where S is the ellipsoid $x^2 + y^2/4 + z^2 = 1$, and $\hat{\mathbf{n}}$ is the unit outer normal to S

** **17.** Show that if a surface S projects one-to-one onto an area S_{xy} in the xy-plane, then

$$\iint_S (P\hat{\mathbf{i}} + Q\hat{\mathbf{j}} + R\hat{\mathbf{k}}) \cdot \hat{\mathbf{n}}\, dS = \pm \iint_{S_{xy}} \left(-P\frac{\partial z}{\partial x} - Q\frac{\partial z}{\partial y} + R \right) dA,$$

the $\pm$ depending on whether $\hat{\mathbf{n}}$ is the upper or lower normal to S. What are corresponding formulas when S projects one-to-one onto areas S_{yz} and S_{xz} in the yz- and xz-coordinate planes?

** **18.** Evaluate $\iint_S (y\hat{\mathbf{i}} - x\hat{\mathbf{j}} + z\hat{\mathbf{k}}) \cdot \hat{\mathbf{n}}\, dS$ where

  (a) S is that part of $z = 9 - x^2 - y^2$ cut out by $z = 2y$,
  (b) S is that part of $z = 2y$ cut out by $z = 9 - x^2 - y^2$, and $\hat{\mathbf{n}}$ is the unit upper normal to S in each case. Hint: Use polar coordinates with pole at $(0, -1)$.

** **19.** Show that, if a surface S, defined implicitly by the equation $G(x, y, z) = 0$, projects one-to-one onto the area S_{xy} in the xy-plane, then

$$\iint_S \mathbf{F} \cdot \hat{\mathbf{n}}\, dS = \pm \iint_{S_{xy}} \frac{\mathbf{F} \cdot \nabla G}{|\partial G/\partial z|}\, dA.$$

** **20.** A circular tube $S : x^2 + z^2 = 1$, $0 \le y \le 2$ is a model for a part of an artery. Blood flows through the artery and the force per unit area at any point on the arterial wall is given by

$$\mathbf{F} = e^{-y}\hat{\mathbf{n}} + \frac{1}{y^2 + 1}\hat{\mathbf{j}},$$

where $\hat{\mathbf{n}}$ is the unit outer normal to the arterial wall. Blood diffuses through the wall in such a way that if dS is a small area on S, the amount of diffusion through dS in one second is $\mathbf{F} \cdot \hat{\mathbf{n}}\, dS$. Find the total amount of blood leaving the entire wall per second.

** **21.** A beam of light travelling in the positive y-direction has circular cross-section $x^2 + z^2 \le a^2$. It strikes a surface $S : x^2 + y^2 + z^2 = a^2$, $y \ge a/2$. The intensity of the beam is given by

$$\mathbf{I} = \frac{e^{-t}}{y^2}\hat{\mathbf{j}}, \qquad \text{where } t \text{ is time.}$$

The absorption of light by a small area dS on S in time dt is $\mathbf{I} \cdot \hat{\mathbf{n}}\, dS\, dt$ where $\hat{\mathbf{n}}$ is the unit normal to S at dS.

  (a) Find the total absorption over S in time dt.
  (b) Find the total absorption over S from time $t = 0$ to time $t = 5$.

15.9 The Divergence Theorem

In this section and in Section 15.10 we show that relationships may exist between line integrals and surface integrals and between surface integrals and triple integrals. In Section 15.11 we indicate how useful these relationships can be in the fields of electromagnetic theory and fluid dynamics.

The divergence theorem relates certain surface integrals over surfaces that enclose volumes to triple integrals over the enclosed volume. More precisely, we have the following.

> ### THEOREM 15.7 (Divergence theorem)
> Let S be a piecewise-smooth surface enclosing a region V (Figure 15.38). Let $\mathbf{F}(x, y, z) = L(x, y, z)\hat{\mathbf{i}} + M(x, y, z)\hat{\mathbf{j}} + N(x, y, z)\hat{\mathbf{k}}$ be a vector field whose components L, M, and N have continuous first partial derivatives in a domain containing S and V. If $\hat{\mathbf{n}}$ is the unit outer normal to S, then
>
> $$\oiint_S \mathbf{F} \cdot \hat{\mathbf{n}}\, dS = \iiint_V \nabla \cdot \mathbf{F}\, dV, \tag{15.55a}$$
>
> or
>
> $$\oiint_S (L\hat{\mathbf{i}} + M\hat{\mathbf{j}} + N\hat{\mathbf{k}}) \cdot \hat{\mathbf{n}}\, dS = \iiint_V \left(\frac{\partial L}{\partial x} + \frac{\partial M}{\partial y} + \frac{\partial N}{\partial z}\right) dV. \tag{15.55b}$$

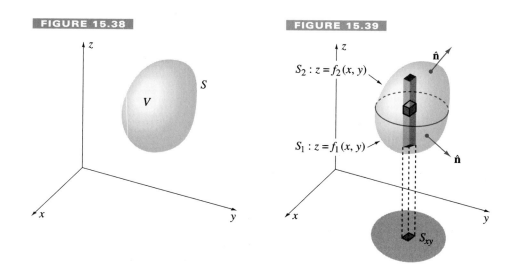

FIGURE 15.38

FIGURE 15.39

PROOF We consider first of all a surface S for which any line parallel to any coordinate axis intersects S in at most two points (Figure 15.39). We can then divide S into an upper and a lower portion, $S_2 : z = f_2(x, y)$ and $S_1 : z = f_1(x, y)$, both of which have the same projection S_{xy} in the xy-plane. We consider the third term in the surface integral on the left side of equation (15.55b):

$$\oiint_S N\hat{\mathbf{k}} \cdot \hat{\mathbf{n}}\, dS = \iint_{S_1} N\hat{\mathbf{k}} \cdot \hat{\mathbf{n}}\, dS + \iint_{S_2} N\hat{\mathbf{k}} \cdot \hat{\mathbf{n}}\, dS.$$

On S_1,

$$\hat{\mathbf{n}} = \frac{\left(\dfrac{\partial f_1}{\partial x}, \dfrac{\partial f_1}{\partial y}, -1\right)}{\sqrt{1 + \left(\dfrac{\partial f_1}{\partial x}\right)^2 + \left(\dfrac{\partial f_1}{\partial y}\right)^2}},$$

and on S_2,

$$\hat{\mathbf{n}} = \frac{-\left(\dfrac{\partial f_2}{\partial x}, \dfrac{\partial f_2}{\partial y}, -1\right)}{\sqrt{1 + \left(\dfrac{\partial f_2}{\partial x}\right)^2 + \left(\dfrac{\partial f_2}{\partial y}\right)^2}}.$$

Consequently,

$$
\begin{aligned}
\oiint_S N\hat{\mathbf{k}} \cdot \hat{\mathbf{n}}\, dS &= \iint_{S_1} \frac{-N}{\sqrt{1 + \left(\dfrac{\partial f_1}{\partial x}\right)^2 + \left(\dfrac{\partial f_1}{\partial y}\right)^2}}\, dS \\
&\quad + \iint_{S_2} \frac{N}{\sqrt{1 + \left(\dfrac{\partial f_2}{\partial x}\right)^2 + \left(\dfrac{\partial f_2}{\partial y}\right)^2}}\, dS \\
&= \iint_{S_{xy}} \frac{-N[x, y, f_1(x, y)]}{\sqrt{1 + \left(\dfrac{\partial f_1}{\partial x}\right)^2 + \left(\dfrac{\partial f_1}{\partial y}\right)^2}} \sqrt{1 + \left(\dfrac{\partial f_1}{\partial x}\right)^2 + \left(\dfrac{\partial f_1}{\partial y}\right)^2}\, dA \\
&\quad + \iint_{S_{xy}} \frac{N[x, y, f_2(x, y)]}{\sqrt{1 + \left(\dfrac{\partial f_2}{\partial x}\right)^2 + \left(\dfrac{\partial f_2}{\partial y}\right)^2}} \sqrt{1 + \left(\dfrac{\partial f_2}{\partial x}\right)^2 + \left(\dfrac{\partial f_2}{\partial y}\right)^2}\, dA \\
&= \iint_{S_{xy}} \{N[x, y, f_2(x, y)] - N[x, y, f_1(x, y)]\}\, dA.
\end{aligned}
$$

On the other hand,

$$
\begin{aligned}
\iiint_V \frac{\partial N}{\partial z}\, dV &= \iint_{S_{xy}} \left\{ \int_{f_1(x,y)}^{f_2(x,y)} \frac{\partial N}{\partial z}\, dz \right\} dA = \iint_{S_{xy}} \left\{ N \right\}_{f_1(x,y)}^{f_2(x,y)}\, dA \\
&= \iint_{S_{xy}} \{N[x, y, f_2(x, y)] - N[x, y, f_1(x, y)]\}\, dA.
\end{aligned}
$$

We have shown then that

$$\oiint_S N\hat{\mathbf{k}} \cdot \hat{\mathbf{n}}\, dS = \iiint_V \frac{\partial N}{\partial z}\, dV.$$

FIGURE 15.42

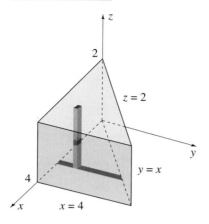

EXAMPLE 15.26

Prove Archimedes' principle, which states that, when an object is submerged in a fluid, it experiences a buoyant force equal to the weight of the fluid displaced.

SOLUTION   Suppose the surface of the fluid is taken as the xy-plane ($z = 0$), and the object occupies a region V with bounding surface S (Figure 15.43). The force due to fluid pressure P on a small area dS on S is $(P\,dS)\hat{\mathbf{n}}$, where $\hat{\mathbf{n}}$ is the unit inner normal to S at dS. If ρ is the density of the fluid, then $P = -9.81\rho z$, and the force on dS is

$$(-9.81\rho z\,dS)\hat{\mathbf{n}}.$$

The resultant buoyant force is in the positive z-direction (the x- and y-components cancelling), so that we require only the z-component of this force,

$$(-9.81\rho z\,dS)\hat{\mathbf{n}} \cdot \hat{\mathbf{k}}.$$

The total buoyant force must therefore have z-component

$$\iint_S (-9.81\rho z)\hat{\mathbf{n}} \cdot \hat{\mathbf{k}}\,dS = \iint_S (-9.81\rho z\hat{\mathbf{k}}) \cdot \hat{\mathbf{n}}\,dS = \iint_S (9.81\rho z\hat{\mathbf{k}}) \cdot (-\hat{\mathbf{n}})dS,$$

where $-\hat{\mathbf{n}}$ is the unit outer normal to S. If we now use the divergence theorem, we have

$$\iint_S (-9.81\rho z)\hat{\mathbf{n}} \cdot \hat{\mathbf{k}}\,dS = \iiint_V \nabla \cdot (9.81\rho z\hat{\mathbf{k}})dV$$

$$= \iiint_V 9.81\rho\,dV = 9.81\rho \text{ (Volume of } V),$$

and this is the weight of the fluid displaced by the object.

FIGURE 15.43

In Exercises 1–12 use the divergence theorem to evaluate the surface integral.

* **1.** $\iint_S (x\hat{\mathbf{i}} + y\hat{\mathbf{j}} - 2z\hat{\mathbf{k}}) \cdot \hat{\mathbf{n}}\,dS$ where S is the surface bounding the volume defined by the surfaces $z = 2x^2 + y^2$, $x^2 + y^2 = 3$, $z = 0$, and $\hat{\mathbf{n}}$ is the unit outer normal to S

* **2.** $\iint_S (x^2\hat{\mathbf{i}} + y^2\hat{\mathbf{j}} + z^2\hat{\mathbf{k}}) \cdot \hat{\mathbf{n}}\,dS$ where S is the sphere $x^2 + y^2 + z^2 = a^2$, and $\hat{\mathbf{n}}$ is the unit outer normal to S

* **3.** $\iint_S (yz\hat{\mathbf{i}} + xz\hat{\mathbf{j}} + xy\hat{\mathbf{k}}) \cdot \hat{\mathbf{n}}\,dS$ where S is the surface enclosing the volume defined by $x = 0$, $x = 2$, $z = 0$, $z = y$, $y + z = 2$, and $\hat{\mathbf{n}}$ is the unit outer normal to S

* **4.** $\iint_S [(z^2 - x)\hat{\mathbf{i}} - xy\hat{\mathbf{j}} + 3z\hat{\mathbf{k}}] \cdot \hat{\mathbf{n}}\,dS$ where S is the surface enclosing the volume defined by $z = 4 - y^2$, $x = 0$, $x = 3$, $z = 0$, and $\hat{\mathbf{n}}$ is the unit outer normal to S

* **5.** $\iint_S \mathbf{F} \cdot \hat{\mathbf{n}}\,dS$ where $\mathbf{F} = (x^2y\hat{\mathbf{i}} + y^2z\hat{\mathbf{j}} + z^2x\hat{\mathbf{k}})/2$, S is the surface bounding the volume in the first octant defined by $x = 0$, $y = 0$, $z = 1$, $z = 0$, $x^2 + y^2 = 1$, and $\hat{\mathbf{n}}$ is the unit inner normal to S

** **6.** $\iint_S (x\hat{\mathbf{i}} + y\hat{\mathbf{j}} + 2z\hat{\mathbf{k}}) \cdot \hat{\mathbf{n}}\,dS$ where S is the surface bounding the volume defined by the surfaces $z = 2x^2 + y^2$, $x^2 + y^2 = 3$, $z = 0$, and $\hat{\mathbf{n}}$ is the unit outer normal to S

** **7.** $\iint_S (z\hat{\mathbf{i}} - x\hat{\mathbf{j}} + y\hat{\mathbf{k}}) \cdot \hat{\mathbf{n}}\,dS$ where S is the surface enclosing the volume defined by the surfaces $z = \sqrt{4 - x^2 - y^2}$, $z = 0$, and $\hat{\mathbf{n}}$ is the unit outer normal to S

** **8.** $\iint_S (2x^2y\hat{\mathbf{i}} - y^2\hat{\mathbf{j}} + 4xz^2\hat{\mathbf{k}}) \cdot \hat{\mathbf{n}}\,dS$ where S is the surface enclosing the volume in the first octant defined by $y^2 + z^2 = 9$, $x = 2$, and $\hat{\mathbf{n}}$ is the unit outer normal to S

** **9.** $\iint_S (yx\hat{\mathbf{i}} + y^2\hat{\mathbf{j}} + yz\hat{\mathbf{k}}) \cdot \hat{\mathbf{n}}\, dS$ where S is the ellipsoid $x^2 + y^2/4 + z^2 = 1$, and $\hat{\mathbf{n}}$ is the unit outer normal to S

** **10.** $\iint_S (x^3\hat{\mathbf{i}} + y^3\hat{\mathbf{j}} - z^3\hat{\mathbf{k}}) \cdot \hat{\mathbf{n}}\, dS$ where S is the surface enclosing the volume defined by $z = 6 - x^2 - y^2$, $z = \sqrt{x^2 + y^2}$, and $\hat{\mathbf{n}}$ is the unit outer normal to S

** **11.** $\iint_S (y\hat{\mathbf{i}} - xy\hat{\mathbf{j}} + zy^2\hat{\mathbf{k}}) \cdot \hat{\mathbf{n}}\, dS$ where S is the surface enclosing the volume defined by $y^2 - x^2 - z^2 = 4$, $y = 4$, and $\hat{\mathbf{n}}$ is the unit inner normal to S

** **12.** $\iint_S (xy\hat{\mathbf{i}} + z^2\hat{\mathbf{k}}) \cdot \hat{\mathbf{n}}\, dS$ where S is the surface enclosing the volume in the first octant bounded by the planes $z = 0$, $y = x$, $y = 2x$, $x + y + z = 6$, and $\hat{\mathbf{n}}$ is the unit outer normal to S

In Exercises 13–15 use the divergence theorem to evaluate the surface integral. In each case an additional surface must be introduced in order to enclose a volume.

** **13.** $\iint_S (x\hat{\mathbf{i}} + y\hat{\mathbf{j}} + z\hat{\mathbf{k}}) \cdot \hat{\mathbf{n}}\, dS$ where S is the top half of the ellipsoid $x^2 + 4y^2 + 9z^2 = 36$, and $\hat{\mathbf{n}}$ is the unit upper normal to S

** **14.** $\iint_S (xy\hat{\mathbf{i}} - yz\hat{\mathbf{j}} + x^2z\hat{\mathbf{k}}) \cdot \hat{\mathbf{n}}\, dS$ where S is that part of the cone $z = \sqrt{x^2 + y^2}$ below $z = 2$, and $\hat{\mathbf{n}}$ is the unit normal to S with positive z-component

** **15.** $\iint_S (y^2e^z\hat{\mathbf{i}} - xy\hat{\mathbf{j}} + z\hat{\mathbf{k}}) \cdot \hat{\mathbf{n}}\, dS$ where S is that part of $z = 4 - x^2 - y^2$ cut out by $z = 2y$, and $\hat{\mathbf{n}}$ is the unit upper normal to S

** **16.** Show that if $\hat{\mathbf{n}}$ is the unit outer normal to a surface S, then the region enclosed by S has volume

$$V = \frac{1}{3} \iint_S \mathbf{r} \cdot \hat{\mathbf{n}}\, dS, \qquad \mathbf{r} = x\hat{\mathbf{i}} + y\hat{\mathbf{j}} + z\hat{\mathbf{k}}.$$

** **17.** If $\hat{\mathbf{n}}$ is the unit outer normal to a surface S that encloses a region V, show that the area of S can be expressed in the form

$$\text{area}(S) = \iiint_V \nabla \cdot \hat{\mathbf{n}}\, dV.$$

** **18.** How would you prove Archimedes' principle in the case that an object is only partially submerged? (See Example 15.26.)

In Exercises 19–21 evaluate the surface integral.

** **19.** $\iint_S [(x + y)\hat{\mathbf{i}} + y^3\hat{\mathbf{j}} + x^2z\hat{\mathbf{k}}] \cdot \hat{\mathbf{n}}\, dS$ where S is the surface enclosing the volume defined by $x^2 + y^2 - z^2 = 1$, $2z^2 = x^2 + y^2$, and $\hat{\mathbf{n}}$ is the unit outer normal to S

** **20.** $\iint_S [(x + y)^2\hat{\mathbf{i}} + x^2y\hat{\mathbf{j}} - x^2z\hat{\mathbf{k}}] \cdot \hat{\mathbf{n}}\, dS$ where $\hat{\mathbf{n}}$ is the unit inner normal to the surface S enclosing the volume defined by $z^2 = (1 - x^2 - 2y^2)^2$

** **21.** $\iint_S [(y^3 + x^2y)\hat{\mathbf{i}} + (x^3 - xy^2)\hat{\mathbf{j}} + z\hat{\mathbf{k}}] \cdot \hat{\mathbf{n}}\, dS$ where $\hat{\mathbf{n}}$ is the unit upper normal to the surface $S : z = \sqrt{1 - x^2 - y^2}$

∗∗ **22.** If V is a region bounded by a closed surface S, and $\mathbf{B} = \nabla \times \mathbf{A}$, show that

$$\iint_S \mathbf{B} \cdot \hat{\mathbf{n}}\, dS = 0.$$

∗∗ **23.** Is Green's theorem related to the divergence theorem? (See Exercise 13 in Section 15.6.)

In Exercises 24–26 assume that $P(x, y, z)$ and $Q(x, y, z)$ have continuous first and second partial derivatives in a domain containing a closed surface S and its interior V. Let $\hat{\mathbf{n}}$ be the unit outer normal to S.

∗∗ **24.** Show that

$$\iint_S \nabla P \cdot \hat{\mathbf{n}}\, dS = \iiint_V \nabla^2 P\, dV,$$

where $\nabla^2 P = \dfrac{\partial^2 P}{\partial x^2} + \dfrac{\partial^2 P}{\partial y^2} + \dfrac{\partial^2 P}{\partial z^2}$. What can we conclude if $P(x, y, z)$ satisfies Laplace's equation in V?

∗∗ **25.** Show that

$$\iint_S P\nabla Q \cdot \hat{\mathbf{n}}\, dS = \iiint_V (P\nabla^2 Q + \nabla P \cdot \nabla Q)\, dV.$$

This result is called *Green's first identity*.

∗∗ **26.** Prove that

$$\iint_S (P\nabla Q - Q\nabla P) \cdot \hat{\mathbf{n}}\, dS = \iiint_V (P\nabla^2 Q - Q\nabla^2 P)\, dV.$$

This result is called *Green's second identity*.

∗∗ **27.** Compare Exercises 24–26 with Exercises 36–38 in Section 15.6.

∗∗∗ **28.** Let S be a closed surface, and let $\hat{\mathbf{n}}$ be the unit outer normal to S. If $\mathbf{r}_0 = x_0\hat{\mathbf{i}} + y_0\hat{\mathbf{j}} + z_0\hat{\mathbf{k}}$ is the position vector of some fixed point P_0, show that

$$\iint_S \frac{\mathbf{r} - \mathbf{r}_0}{|\mathbf{r} - \mathbf{r}_0|^3} \cdot \hat{\mathbf{n}}\, dS = \begin{cases} 0 & \text{if } S \text{ does not enclose } P_0 \\ 4\pi & \text{if } S \text{ does enclose } P_0. \end{cases}$$

∗∗ **29.** In this problem we use the result of Exercise 28 to prove Gauss's law for electrostatic fields. Let S be a closed surface containing n point charges q_i $(i = 1, \ldots, n)$ at points $\mathbf{r}_i = (x_i, y_i, z_i)$ in its interior. According to Coulomb's law, the electric field $\mathbf{E}$ at a point $\mathbf{r} = (x, y, z)$ due to this charge distribution is defined by $\mathbf{E} = -\nabla V$, where V, the potential at $\mathbf{r}$, is

$$V(x, y, z) = \sum_{i=1}^{n} \frac{q_i}{4\pi \epsilon_0 |\mathbf{r} - \mathbf{r}_i|}.$$

(a) Verify that

$$\mathbf{E} = \sum_{i=1}^{n} \frac{q_i(\mathbf{r} - \mathbf{r}_i)}{4\pi \epsilon_0 |\mathbf{r} - \mathbf{r}_i|^3}.$$

(b) Now use the result of Exercise 28 to show that if $\hat{\mathbf{n}}$ is the unit outer normal to S, then

$$\iint_S \mathbf{E} \cdot \hat{\mathbf{n}}\, dS = \frac{Q}{\epsilon_0},$$

where Q is the total charge inside S. This result is Gauss's law.

15.10 Stokes's Theorem

Stokes's theorem relates certain line integrals around closed curves to surface integrals over surfaces that have the curves as boundaries.

THEOREM 15.8 (Stokes's theorem)

Let C be a closed, piecewise-smooth curve that does not intersect itself and let S be a piecewise-smooth, orientable surface with C as boundary (Figure 15.44). Let $\mathbf{F}(x, y, z) = P(x, y, z)\hat{\mathbf{i}} + Q(x, y, z)\hat{\mathbf{j}} + R(x, y, z)\hat{\mathbf{k}}$ be a vector field whose components P, Q, and R have continuous first partial derivatives in a domain that contains S and C. Then

$$\oint_C \mathbf{F} \cdot d\mathbf{r} = \iint_S (\nabla \times \mathbf{F}) \cdot \hat{\mathbf{n}} \, dS, \tag{15.56a}$$

or

$$\oint_C P \, dx + Q \, dy + R \, dz = \iint_S \left\{ \left(\frac{\partial R}{\partial y} - \frac{\partial Q}{\partial z} \right) \hat{\mathbf{i}} \right.$$
$$\left. + \left(\frac{\partial P}{\partial z} - \frac{\partial R}{\partial x} \right) \hat{\mathbf{j}} + \left(\frac{\partial Q}{\partial x} - \frac{\partial P}{\partial y} \right) \hat{\mathbf{k}} \right\} \cdot \hat{\mathbf{n}} \, dS, \tag{15.56b}$$

where $\hat{\mathbf{n}}$ is the unit normal to S chosen in the following way: If when moving along C the surface S is on the left, then $\hat{\mathbf{n}}$ must be chosen as the unit normal on that side of S. On the other hand if, when moving along C, the surface is on the right, then $\hat{\mathbf{n}}$ must be chosen on the opposite side of S.

FIGURE 15.44

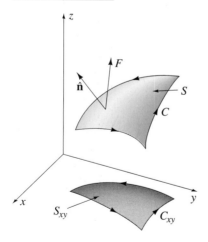

PROOF   We first consider a surface S that projects in a one-to-one fashion onto each of the three coordinate planes. Because S projects one-to-one onto some region S_{xy} in the xy-plane, we can take the equation for S in the form $z = f(x, y)$. If the direction along C is as indicated in Figure 15.44, and C_{xy} is the projection of C on the xy-plane, then

$$\oint_C P \, dx = \oint_{C_{xy}} P[x, y, f(x, y)] \, dx.$$

If we use Green's theorem on this line integral around C_{xy}, we have

$$\oint_C P\,dx = \iint_{S_{xy}} -\left\{\frac{\partial P}{\partial y} + \frac{\partial P}{\partial z}\frac{\partial z}{\partial y}\right\} dA.$$

On the other hand, since a unit normal to S is

$$\hat{\mathbf{n}} = \frac{\left(-\dfrac{\partial z}{\partial x}, -\dfrac{\partial z}{\partial y}, 1\right)}{\sqrt{1 + \left(\dfrac{\partial z}{\partial x}\right)^2 + \left(\dfrac{\partial z}{\partial y}\right)^2}},$$

it also follows that

$$\iint_S \left(\frac{\partial P}{\partial z}\hat{\mathbf{j}} - \frac{\partial P}{\partial y}\hat{\mathbf{k}}\right) \cdot \hat{\mathbf{n}}\,dS = \iint_S \frac{-\dfrac{\partial P}{\partial z}\dfrac{\partial z}{\partial y} - \dfrac{\partial P}{\partial y}}{\sqrt{1 + \left(\dfrac{\partial z}{\partial x}\right)^2 + \left(\dfrac{\partial z}{\partial y}\right)^2}}\,dS$$

$$= \iint_{S_{xy}} \frac{-\dfrac{\partial P}{\partial z}\dfrac{\partial z}{\partial y} - \dfrac{\partial P}{\partial y}}{\sqrt{1 + \left(\dfrac{\partial z}{\partial x}\right)^2 + \left(\dfrac{\partial z}{\partial y}\right)^2}} \times$$

$$\sqrt{1 + \left(\frac{\partial z}{\partial x}\right)^2 + \left(\frac{\partial z}{\partial y}\right)^2}\,dA$$

$$= \iint_{S_{xy}} -\left\{\frac{\partial P}{\partial y} + \frac{\partial P}{\partial z}\frac{\partial z}{\partial y}\right\} dA.$$

We have shown, then, that

$$\oint_C P\,dx = \iint_S \left(\frac{\partial P}{\partial z}\hat{\mathbf{j}} - \frac{\partial P}{\partial y}\hat{\mathbf{k}}\right) \cdot \hat{\mathbf{n}}\,dS.$$

By projecting C and S onto the xz- and yz-planes, we can show similarly that

$$\oint_C R\,dz = \iint_S \left(\frac{\partial R}{\partial y}\hat{\mathbf{i}} - \frac{\partial R}{\partial x}\hat{\mathbf{j}}\right) \cdot \hat{\mathbf{n}}\,dS$$

and

$$\oint_C Q\,dy = \iint_S \left(\frac{\partial Q}{\partial x}\hat{\mathbf{k}} - \frac{\partial Q}{\partial z}\hat{\mathbf{i}}\right) \cdot \hat{\mathbf{n}}\,dS.$$

Addition of these three results gives Stokes's theorem for $\mathbf{F}$ and S.

The proof can be extended to more general curves and surfaces that do not project in a one-to-one fashion onto all three coordinate planes. Most surfaces S with bounding curves C can be subdivided into n subsurfaces S_i with bounding curves C_i that do satisfy this condition (Figure 15.45). For each such subsurface, Stokes's theorem applies

$$\oint_{C_i} \mathbf{F} \cdot d\mathbf{r} = \iint_{S_i} (\nabla \times \mathbf{F}) \cdot \hat{\mathbf{n}}\,dS, \quad i = 1, \ldots, n.$$

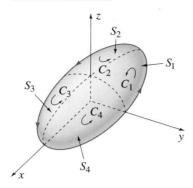

FIGURE 15.45

If these n equations are now added together, we have

$$\sum_{i=1}^{n} \oint_{C_i} \mathbf{F} \cdot d\mathbf{r} = \sum_{i=1}^{n} \iint_{S_i} (\nabla \times \mathbf{F}) \cdot \hat{\mathbf{n}} \, dS.$$

Since the S_i constitute S, the right side of this equation is the surface integral of $(\nabla \times \mathbf{F}) \cdot \hat{\mathbf{n}}$ over S. Figure 15.45 illustrates that when line integrals over the C_i are added, contributions from auxiliary (interior) curves cancel in pairs, and the remaining line integrals give the line integral of $\mathbf{F} \cdot d\mathbf{r}$ along C. For general surfaces that cannot be divided into a finite number of subsurfaces of this type, the reader should consult a more advanced book. ∎

Green's theorem is a special case of Stokes's theorem. For if $\mathbf{F} = P(x, y)\hat{\mathbf{i}} + Q(x, y)\hat{\mathbf{j}}$, and C is a closed curve in the xy-plane, then by Stokes's theorem

$$\oint_C P \, dx + Q \, dy = \iint_S \left(\frac{\partial Q}{\partial x} - \frac{\partial P}{\partial y} \right) \hat{\mathbf{k}} \cdot \hat{\mathbf{n}} \, dS,$$

where S is any surface for which C is the boundary. If we choose S as that part of the xy-plane bounded by C, then $\hat{\mathbf{n}} = \hat{\mathbf{k}}$ and

$$\oint_C P \, dx + Q \, dy = \iint_S \left(\frac{\partial Q}{\partial x} - \frac{\partial P}{\partial y} \right) dA.$$

With Stokes's theorem, it is straightforward to verify the sufficiency half of Theorem 15.1. Suppose that the curl of a vector field $\mathbf{F}$ vanishes in a simply-connected domain D. If C is any piecewise-smooth, closed curve in D, then there exists a piecewise-smooth surface S in D with C as boundary. By Stokes's theorem,

$$\oint_C \mathbf{F} \cdot d\mathbf{r} = \iint_S (\nabla \times \mathbf{F}) \cdot \hat{\mathbf{n}} \, dS = 0.$$

According to Corollary 2 of Theorem 15.3, the line integral is independent of path in D, and the theorem itself implies the existence of a function $f(x, y, z)$ such that $\nabla f = \mathbf{F}$.

EXAMPLE 15.27

Verify Stokes's theorem if $\mathbf{F} = x^2\hat{\mathbf{i}} + x\hat{\mathbf{j}} + xyz\hat{\mathbf{k}}$, and S is that part of the sphere $x^2 + y^2 + z^2 = 4$ above the plane $z = 1$.

SOLUTION   If we choose $\hat{\mathbf{n}}$ as the upper normal to S, then C, the boundary of S, must be traversed in the direction shown in Figure 15.46. (If $\hat{\mathbf{n}}$ is chosen as the lower normal, then C must be traversed in the opposite direction.) Since parametric equations for C are

$$x = \sqrt{3}\cos t, \quad y = \sqrt{3}\sin t, \quad z = 1, \quad 0 \le t \le 2\pi,$$

$$
\begin{aligned}
\oint_C \mathbf{F} \cdot d\mathbf{r} &= \oint_C x^2\,dx + x\,dy + xyz\,dz \\
&= \int_0^{2\pi} \{3\cos^2 t(-\sqrt{3}\sin t\,dt) + \sqrt{3}\cos t(\sqrt{3}\cos t\,dt)\} \\
&= \int_0^{2\pi} \left\{ -3\sqrt{3}\cos^2 t\sin t + \frac{3}{2}(1 + \cos 2t) \right\} dt \\
&= \left\{ \sqrt{3}\cos^3 t + \frac{3t}{2} + \frac{3\sin 2t}{4} \right\}_0^{2\pi} \\
&= 3\pi.
\end{aligned}
$$

FIGURE 15.46

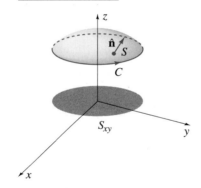

Alternatively,

$$
\begin{aligned}
\iint_S (\nabla \times \mathbf{F}) \cdot \hat{\mathbf{n}}\,dS &= \iint_S (xz\hat{\mathbf{i}} - yz\hat{\mathbf{j}} + \hat{\mathbf{k}}) \cdot \frac{(2x, 2y, 2z)}{\sqrt{4x^2 + 4y^2 + 4z^2}}\,dS \\
&= \iint_S \frac{x^2 z - y^2 z + z}{\sqrt{x^2 + y^2 + z^2}}\,dS \\
&= \iint_{S_{xy}} \frac{z(x^2 - y^2 + 1)}{2}\sqrt{1 + \left(\frac{\partial z}{\partial x}\right)^2 + \left(\frac{\partial z}{\partial y}\right)^2}\,dA \\
&= \frac{1}{2}\iint_{S_{xy}} z(x^2 - y^2 + 1)\sqrt{1 + (-x/z)^2 + (-y/z)^2}\,dA \\
&= \frac{1}{2}\iint_{S_{xy}} z(x^2 - y^2 + 1)\frac{\sqrt{x^2 + y^2 + z^2}}{z}\,dA \\
&= \iint_{S_{xy}} (x^2 - y^2 + 1)\,dA.
\end{aligned}
$$

If we use polar coordinates to evaluate this double integral over $S_{xy} : x^2 + y^2 \leq 3$, we have

$$\iint_S (\nabla \times \mathbf{F}) \cdot \hat{\mathbf{n}} \, dS = \int_{-\pi}^{\pi} \int_0^{\sqrt{3}} (r^2 \cos^2 \theta - r^2 \sin^2 \theta + 1) r \, dr \, d\theta$$

$$= \int_{-\pi}^{\pi} \left\{ \frac{r^4}{4} (\cos^2 \theta - \sin^2 \theta) + \frac{r^2}{2} \right\}_0^{\sqrt{3}} d\theta$$

$$= \int_{-\pi}^{\pi} \left\{ \frac{9}{4} \cos 2\theta + \frac{3}{2} \right\} d\theta$$

$$= \left\{ \frac{9}{8} \sin 2\theta + \frac{3\theta}{2} \right\}_{-\pi}^{\pi} = 3\pi.$$

EXAMPLE 15.28

Evaluate

$$\oint_C 2xy^3 \, dx + 3x^2 y^2 \, dy + (2z + x) dz,$$

where C consists of line segments joining $A(2, 0, 0)$ to $B(0, 1, 0)$ to $D(0, 0, 1)$ to A.

SOLUTION   By Stokes's theorem,

$$\oint_C 2xy^3 \, dx + 3x^2 y^2 \, dy + (2z + x) dz = \iint_S \nabla \times (2xy^3, 3x^2 y^2, 2z + x) \cdot \hat{\mathbf{n}} \, dS,$$

where S is any surface with C as boundary. If we choose S as the flat triangle bounded by C (Figure 15.47), then a normal vector to S is

$$\mathbf{BD} \times \mathbf{BA} = \begin{vmatrix} \hat{\mathbf{i}} & \hat{\mathbf{j}} & \hat{\mathbf{k}} \\ 0 & -1 & 1 \\ 2 & -1 & 0 \end{vmatrix} = (1, 2, 2),$$

and therefore

$$\hat{\mathbf{n}} = \frac{(1, 2, 2)}{3}.$$

Since

$$\nabla \times (2xy^3, 3x^2 y^2, 2z + x) = \begin{vmatrix} \hat{\mathbf{i}} & \hat{\mathbf{j}} & \hat{\mathbf{k}} \\ \partial/\partial x & \partial/\partial y & \partial/\partial z \\ 2xy^3 & 3x^2 y^2 & 2z + x \end{vmatrix} = -\hat{\mathbf{j}},$$

it follows that

$$\nabla \times (2xy^3, 3x^2 y^2, 2z + x) \cdot \hat{\mathbf{n}} = -\hat{\mathbf{j}} \cdot \frac{(1, 2, 2)}{3} = -\frac{2}{3}$$

and

$$\oint_C 2xy^3 \, dx + 3x^2 y^2 \, dy + (2z + x) dz = \iint_S -\frac{2}{3} dS = -\frac{2}{3} (\text{area of } S).$$

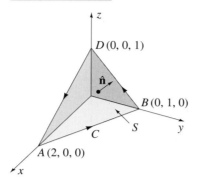

FIGURE 15.47

But from equation (12.42), the area of triangle S is

$$\frac{1}{2}|\mathbf{BD} \times \mathbf{BA}| = \frac{1}{2}|(1, 2, 2)| = \frac{3}{2}.$$

Finally, then,

$$\oint_C 2xy^3\, dx + 3x^2 y^2\, dy + (2z + x)dz = -\frac{2}{3}\left(\frac{3}{2}\right) = -1.$$

In Exercises 1–14 use Stokes's theorem to evaluate the line integral.

* **1.** $\displaystyle\oint_C x^2 y\, dx + y^2 z\, dy + z^2 x\, dz$ where C is the curve $z = x^2 + y^2$, $x^2 + y^2 = 4$, directed counterclockwise as viewed from the origin

* **2.** $\displaystyle\oint_C y^2\, dx + xy\, dy + xz\, dz$ where C is the curve $x^2 + y^2 = 2y$, $y = z$, directed so that y increases when x is positive

* **3.** $\displaystyle\oint_C (xyz + 2yz)\, dx + xz\, dy + 2xy\, dz$ where C is the curve $z = 1$, $x^2 + y^2 + z^2 = 4$, directed clockwise as viewed from the origin

* **4.** $\displaystyle\oint_C (2xy + y)\, dx + (x^2 + xy - 3y)\, dy + 2xz\, dz$ where C is the curve $z = \sqrt{x^2 + y^2}$, $z = 4$

* **5.** $\displaystyle\oint_C x^2\, dx + y^2\, dy + (x^2 + y^2)\, dz$ where C is the boundary of the first octant part of the plane $x + y + z = 1$, directed counterclockwise as viewed from the origin

* **6.** $\displaystyle\oint_C y\, dx + x\, dy + (x^2 + y^2 + z^2)\, dz$ where C is the curve $x^2 + y^2 = 1$, $z = xy$, directed clockwise as viewed from the point $(0, 0, 1)$

* **7.** $\oint_C zy^2\,dx + xy\,dy + (y^2 + z^2)\,dz$ where C is the curve $x^2 + z^2 = 9$, $y = \sqrt{x^2 + z^2}$, directed counterclockwise as viewed from the origin

* **8.** $\oint_C y\,dx + z\,dy + x\,dz$ where C is the curve $x + y = 2b$, $x^2 + y^2 + z^2 = 2b(x + y)$, directed clockwise as viewed from the origin

* **9.** $\oint_C y^2\,dx + (x + y)\,dy + yz\,dz$ where C is the curve $x^2 + y^2 = 2$, $x + y + z = 2$, directed clockwise as viewed from the origin

** **10.** $\oint_C (x + y)^2\,dx + (x + y)^2\,dy + yz^3\,dz$ where C is the curve with equations $z = \sqrt{x^2 + y^2}$, $(x - 1)^2 + y^2 = 1$

** **11.** $\oint_C xy\,dx - zx\,dy + yz\,dz$ where C is the boundary of that part of $z = x + y$ in the first octant cut off by $x + y = 1$, directed counterclockwise as viewed from the point $(0, 0, 1)$

** **12.** $\oint_C y^3\,dx - x^3\,dy + xyz\,dz$ where C is the curve $x^2 + y^2 = z^2 + 3$, $z = 3 - \sqrt{x^2 + y^2}$, directed clockwise as viewed from the origin

** **13.** $\oint_C z(x + y)^2\,dx + (y - x)^2\,dy + z^2\,dz$ where C is the smooth curve of intersection of the surfaces $x^2 + z^2 = a^2$, $y^2 + z^2 = a^2$ which has a portion in the first octant, directed so that z decreases in the first octant

** **14.** $\oint_C -2y^3x^2\,dx + x^3y^2\,dy + z\,dz$ where C is the curve $x^2 + y^2 + z^2 = 4$, $x^2 + 4y^2 = 4$, directed so that x decreases along that part of the curve in the first octant

** **15.** Evaluate the line integral $\oint_C 2x^2y\,dx - yz\,dy + xz\,dz$ where C is the curve $x^2 + y^2 + z^2 = 4$, $z = \sqrt{3}(x^2 + y^2)$, directed clockwise as viewed from the origin, in four ways:

    (a) directly as a line integral

    (b) using Stokes's theorem with S as that part of $z = \sqrt{4 - x^2 - y^2}$ bounded by C

    (c) using Stokes's theorem with S as that part of $z = \sqrt{3}(x^2 + y^2)$ bounded by C

    (d) using Stokes's theorem with S as that part of $z = \sqrt{3}$ bounded by C

** **16.** Let S_1 be that part of $x^2 + y^2 + z^2 = 1$ above the xy-plane and S_2 be that part of $z = 1 - x^2 - y^2$ above the xy-plane. Show that if $\hat{\mathbf{n}}_1$ and $\hat{\mathbf{n}}_2$ are the unit upper normals to these surfaces, and $\mathbf{F}$ is a vector field defined on both S_1 and S_2, then

$$\iint_{S_1} (\nabla \times \mathbf{F}) \cdot \hat{\mathbf{n}}_1\,dS = \iint_{S_2} (\nabla \times \mathbf{F}) \cdot \hat{\mathbf{n}}_2\,dS.$$

15.11 Flux and Circulation

In many branches of engineering and physics we encounter the concepts of **flux** and **circulation**. In this section we discuss the relationships between flux and divergence and between circulation and curl. We find a physical setting mentioned in Section 15.1 most useful in developing these ideas.

Fluid Flow

Consider a gas flowing through a region D of space. At time t and point $P(x, y, z)$ in D, gas flows through P with velocity $\mathbf{v}(x, y, z, t)$. If A is a unit area around P perpendicular to $\mathbf{v}$ (Figure 15.48) and $\rho(x, y, z, t)$ is density of the gas at P, then the amount of gas crossing A per unit time is $\rho\mathbf{v}$. At every point P in D, then, the vector $\rho\mathbf{v}$ is such that its direction $\mathbf{v}$ gives velocity of gas flow, and its magnitude $\rho|\mathbf{v}|$ describes the mass of gas flowing in that direction per unit time.

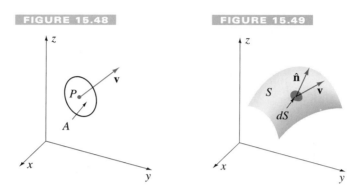

FIGURE 15.48

FIGURE 15.49

Consider some surface S in D (Figure 15.49). If $\hat{\mathbf{n}}$ is a unit normal to S, then $\rho\mathbf{v} \cdot \hat{\mathbf{n}}$ is the component of $\rho\mathbf{v}$ normal to the surface S. If dS is an element of area on S, then $\rho\mathbf{v} \cdot \hat{\mathbf{n}}\,dS$ describes the mass of gas flowing through dS per unit time. Consequently,

$$\iint_S \rho\mathbf{v} \cdot \hat{\mathbf{n}}\,dS$$

is the mass of gas flowing through S per unit time. This quantity is called **flux** for surface S.

If S is a closed surface (Figure 15.50) and $\hat{\mathbf{n}}$ is the unit outer normal to S, then

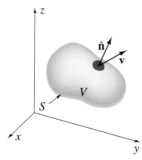

FIGURE 15.50

$$\oiint_S \rho\mathbf{v} \cdot \hat{\mathbf{n}}\,dS$$

is the mass of gas flowing out of surface S per unit time. If this flux (for closed S) is positive, then there is a net outward flow of gas through S (i.e., more gas is leaving the volume bounded by S than is entering); if the flux is negative, the net flow is inward.

If we apply the divergence theorem to the flux integral over the closed surface S, we have

$$\oiint_S \rho\mathbf{v} \cdot \hat{\mathbf{n}}\,dS = \iiint_V \nabla \cdot (\rho\mathbf{v})dV. \tag{15.57}$$

Now the flux (on the left side of this equation) is the mass of gas per unit time leaving S. In order for the right side to represent the same quantity, $\nabla \cdot (\rho\mathbf{v})$ must be interpreted as the mass of gas leaving unit volume per unit time, because then $\nabla \cdot (\rho\mathbf{v})dV$ represents the mass per unit time leaving dV, and the triple integral is the mass per unit time leaving V.

We have obtained, therefore, an interpretation of the divergence of $\rho\mathbf{v}$. The divergence of $\rho\mathbf{v}$ is the flux per unit volume per unit time at a point: the mass of gas leaving unit volume in unit time. We can use this idea of flux to derive the "equation of continuity" for fluid flow. The triple integral

$$\iiint_V \frac{\partial\rho}{\partial t}\,dV = \frac{\partial}{\partial t}\iiint_V \rho\,dV$$

measures the time rate of change of mass in a volume V. If this triple integral is positive, then there is a net inward flow of mass; if it is negative, the net flow is outward. We conclude, therefore, that this triple integral must be the negative of the flux for the volume V; i.e.,

$$\iiint_V \frac{\partial\rho}{\partial t}\,dV = -\iiint_V \nabla\cdot(\rho\mathbf{v})\,dV,$$

or

$$\iiint_V \left(\nabla\cdot(\rho\mathbf{v}) + \frac{\partial\rho}{\partial t}\right)\,dV = 0.$$

If $\nabla\cdot(\rho\mathbf{v})$ and $\partial\rho/\partial t$ are continuous functions, then this equation can hold for arbitrary volume V only if

$$\nabla\cdot(\rho\mathbf{v}) + \frac{\partial\rho}{\partial t} = 0. \tag{15.58}$$

This equation, called the **equation of continuity**, expresses conservation of mass; it is basic to all fluid flow.

If C is a closed curve in the flow region D, then the **circulation** of the flow for the curve C is defined by

$$\Gamma = \oint_C \mathbf{v}\cdot d\mathbf{r}. \tag{15.59}$$

To obtain an intuitive feeling for Γ, we consider two very simple two-dimensional flows. First, suppose $\mathbf{v} = x\hat{\mathbf{i}} + y\hat{\mathbf{j}}$, so that all particles of gas flow along radial lines directed away from the origin (Figure 15.51). In this case, the line integral defining Γ is independent of path and $\Gamma = 0$ for any curve whatsoever.

Second, suppose $\mathbf{v} = -y\hat{\mathbf{i}} + x\hat{\mathbf{j}}$, so that all particles of the gas flow counterclockwise around circles centred at the origin (Figure 15.52). In this case Γ does not generally vanish. In particular, if C is the circle $x^2 + y^2 = r^2$, then $\mathbf{v}$ and $d\mathbf{r}$ are parallel, and

$$\Gamma = \oint_C \mathbf{v}\cdot d\mathbf{r} = \oint_C |\mathbf{v}|ds = \oint_C \sqrt{y^2 + x^2}ds = \oint_C r\,ds = 2\pi r^2.$$

FIGURE 15.51

FIGURE 15.52

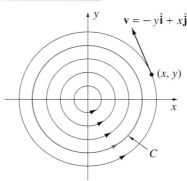

Differential Equations

Differential equations serve as models for many problems in engineering and physics. In this chapter we discuss some of the methods for solving first-order and simple second-order equations. We also give a fairly thorough treatment of linear differential equations. We include a wide variety of applications to illustrate the relevance of differential equations in applied mathematics.

16.1 Introduction

A **differential equation** is an equation that must be solved for an unknown function. What distinguishes a differential equation from other equations is the fact that it contains at least one derivative of the unknown function. For example, each of the following equations is a differential equation in y as a function of x:

$$\frac{dy}{dx} + \frac{2}{3}y = 9.81, \tag{16.1}$$

$$\frac{d^2y}{dx^2} = k\sqrt{1 + \left(\frac{dy}{dx}\right)^2}, \tag{16.2}$$

$$x\frac{d^2y}{dx^2} + \frac{dy}{dx} + xy = 0, \tag{16.3}$$

$$\frac{d^4y}{dx^4} - k^4y = 0. \tag{16.4}$$

Equation 16.1 can be used to determine the position of a paratrooper who falls under the influences of gravity, and of air resistance which is proportional to velocity (see Example 16.9); equation 16.2 describes the shape of a hanging cable (Exercise 13 in Section 16.4); equation 16.3, called Bessel's differential equation of order zero, is found in heat flow and vibration problems; and equation 16.4 is used to determine the deflection of beams.

DEFINITION 16.1

The **order** of a differential equation is the order of the highest derivative in the equation.

Of the four differential equations 16.1–16.4, the first is first order, the second and third are second order, and the last is fourth order.

We have considered quite a number of differential equations in Chapters 3, 5, 8, and 9. In Section 5.4 we concentrated on describing physical situations by means of differential equations; in Chapters 3 and 8 we verified that particular combinations of transcendental functions satisfied certain differential equations; and in Chapter 9 we used our integration techniques to solve various equations. Almost all of these differential equations were based on applications, most from physics and engineering, but also some from geometry and other fields such as ecology, chemistry, and psychology. In applications, differential equations are almost always accompanied by subsidiary conditions called **initial** or **boundary conditions**. For example, if a mass m

falls from rest under gravity and is acted on by a force due to air resistance that is proportional to its instantaneous velocity, then the differential equation that describes its velocity $v(t)$ as a function of time t is

$$m\frac{dv}{dt} = -kv + mg, \quad \text{where } k \text{ is a constant.} \tag{16.5}$$

$y = 0,$
$t = 0,$
$v = 0$

y

This is simply a statement of Newton's second law, where dv/dt is the vertical component of the acceleration of m, and $-kv + mg$ is the vertical component of the total force on m due to gravity (mg) and air resistance ($-kv$). If distance y is chosen as positive downward (Figure 16.1), then $g = 9.81$, and k is positive. Furthermore, if we choose time $t = 0$ at the instant m is dropped, then because m falls from rest, the condition $v(0) = 0$ must be added to the differential equation. In other words, the real problem is to find the solution of the differential equation that also satisfies the initial condition:

$$m\frac{dv}{dt} = -kv + mg, \quad v(0) = 0. \tag{16.6}$$

This is the form in which applied mathematicians use differential equations—the differential equation is accompanied by subsidiary conditions that express extra requirements of the solution. It is not difficult to show that the solution of equation 16.6 is

$$v(t) = \frac{mg}{k} - \frac{mg}{k}e^{-kt/m}. \tag{16.7}$$

(All we need do to verify this is substitute the function into the differential equation to see that it does indeed satisfy the equation. It is clear that it does satisfy the initial condition.)

If we change the initial condition to $v(0) = v_0$, so that the initial velocity of m has vertical component v_0, then the solution becomes

$$v(t) = \frac{mg}{k} - \left(\frac{mg}{k} - v_0\right)e^{-kt/m}. \tag{16.8}$$

In other words, every solution of differential equation 16.5 can be written in the form

$$v(t) = \frac{mg}{k} + Ce^{-kt/m}, \tag{16.9}$$

and when we impose the initial condition $v(0) = v_0$, then $C = v_0 - mg/k$.

In a similar way, equation 16.4 for beam deflections is normally accompanied by four boundary conditions that specify the types of supports at the ends of the beam (see Section 5.1). For simple supports at both $x = 0$ and $x = L$, $y(x)$ must satisfy

$$y(0) = y(L) = 0;$$

$$y''(0) = y''(L) = 0.^{\dagger}$$

---

† In this chapter it is frequently convenient to use the notation $y', y'', y''', \ldots$ to represent dy/dx, d^2y/dx^2, d^3y/dx^3, etc. In this notation $y''(a)$ is the second derivative of y evaluated at $x = a$. In addition, we denote the solution of a differential equation in y as a function of x by $y(x)$.

** **24.** (a) Verify that $y(x) = Ce^{2x}$ is a one-parameter family of solutions for the differential equation

$$\frac{dy}{dx} = 2y.$$

   (b) Draw the one-parameter family of curves defined by this general solution.

   (c) Show that there is a particular solution that passes through any given point (x_0, y_0), and that this solution can be obtained by choosing C appropriately.

** **25.** (a) Verify that a one-parameter family of solutions for the differential equation

$$2x\frac{dy}{dx} = y$$

   is defined implicitly by the equation $y^2 = Cx$.

   (b) Draw the one-parameter family of curves defined by this equation.

   (c) Show that, with the exception of points on the y-axis, there is a particular solution that passes through any given point (x_0, y_0), and that this solution can be obtained by specifying C appropriately.

** **26.** (a) Draw the one-parameter family of curves defined by the solution in Exercise 22.

   (b) Show that, with the exception of points on the x-axis, there is a solution passing through any given point in the xy-plane.

** **27.** Consider the differential equation

$$\frac{dy}{dx} = \frac{1}{x^2}.$$

   (a) Find a solution that satisfies the condition $y(1) = 1$.

   (b) Find a solution that satisfies the condition $y(-1) = 2$.

   (c) Find a solution that satisfies the conditions in both (a) and (b).

16.2 Separable Differential Equations

In this section and the next, we consider first-order differential equations that can be written in the form

$$\frac{dy}{dx} = F(x, y). \tag{16.16}$$

Since any function $F(x, y)$ can always be considered as the quotient of two other functions,

$$F(x, y) = \frac{M(x, y)}{N(x, y)},$$

equation 16.16 can also be written in the equivalent form

$$N(x, y)dy = M(x, y)dx. \tag{16.17}$$

Depending on the form of $F(x, y)$ (or $M(x, y)$ and $N(x, y)$) various methods can be used to obtain the unknown function $y(x)$. Two of the more important techniques are considered here and in Section 16.3; others are discussed in the exercises.

Differential equation 16.16 is said to be separable if it can be expressed in the form

$$\frac{dy}{dx} = \frac{M(x)}{N(y)},$$
(16.18)

that is, if dy/dx is equal to a function of x divided by a function of y. Equivalently, a differential equation is said to be separable if it can be written in the form

$$N(y)dy = M(x)dx.$$
(16.19)

When a differential equation is written in this way, it is said to be separated—separated in the sense that x- and y-variables appear on opposite sides of the equation. For a separated equation we can write therefore that

$$N(y)\frac{dy}{dx} = M(x),$$
(16.20)

and if we integrate both sides with respect to x, we have

$$\int N(y)\frac{dy}{dx}dx = \int M(x)dx + C.$$
(16.21)

Cancellation of differentials on the left leads to the solution

$$\int N(y)dy = \int M(x)dx + C.$$
(16.22)

What we mean by saying that 16.21 and 16.22 represent one-parameter families of solutions for 16.19 is that any function defined *implicitly* by 16.21 or 16.22 is a solution of 16.19. For example, the differential equation

$$\frac{dy}{dx} = \frac{3x^2}{1 - y}$$

is separable. Multiplying by $1 - y$ leads to

$$(1 - y)dy = 3x^2\,dx.$$

According to 16.21 we should divide by dx and integrate both sides with respect to x:

$$\int (1 - y)\frac{dy}{dx}dx = \int 3x^2\,dx = x^3 + C.$$

Antidifferentiation on the left must be interpreted as "implicit antidifferentiation," asking for that function which when differentiated with respect to x gives $(1 - y)dy/dx$. Since an antiderivative is $y - y^2/2$, a one-parameter family of solutions for the differential equation is

$$y - \frac{y^2}{2} = x^3 + C.$$

Were we to use 16.22 after separation (instead of 16.21), we would write

$$\int (1 - y)dy = \int 3x^2\, dx + C,$$

and integrate for

$$y - \frac{y^2}{2} = x^3 + C.$$

By saying that $y - y^2/2 = x^3 + C$ represents a one-parameter family of solutions of the original differential equation, we mean that any function defined implicitly by this equation is a solution. We can find explicit solutions by solving the equation for y in terms of x. Multiplying by -2 expresses the equation as a quadratic in y,

$$y^2 - 2y + 2(x^3 + C) = 0,$$

and therefore

$$y = \frac{2 \pm \sqrt{4 - 8(x^3 + C)}}{2} = 1 \pm \sqrt{1 - 2(x^3 + C)}.$$

Explicit solutions of the differential equation are therefore

$$y(x) = 1 + \sqrt{1 - 2(x^3 + C)} \quad \text{and} \quad y(x) = 1 - \sqrt{1 - 2(x^3 + C)},$$

provided expressions on the right are indeed functions of x. Once C is determined, this will be true only for certain values of x. For example, suppose we require the solution of the differential equation that satisfies the initial condition $y(0) = 3$. The second function $y(x) = 1 - \sqrt{1 - 2(x^3 + C)}$ cannot satisfy this condition because y cannot be greater than 1. If we substitute $x = 0$ and $y = 3$ into the other function,

$$3 = 1 + \sqrt{1 - 2C},$$

and this requires $C = -3/2$. Thus, the solution of the differential equation for which $y(0) = 3$ is

$$y(x) = 1 + \sqrt{1 - 2(x^3 - 3/2)} = 1 + \sqrt{4 - 2x^3}.$$

Since $4 - 2x^3$ must be nonnegative for this function to be defined, the solution is valid only on the interval $x \leq 2^{1/3}$.

EXAMPLE 16.3

Find one-parameter families of solutions for the following differential equations:

(a) $2x^3 y^2\, dx = xy^3\, dy$

(b) $\dfrac{dy}{dx} = \dfrac{y \sin x + y^3 \sin x}{(1 + y^2)^2}$

SOLUTION

(a) If we divide the differential equation by xy^2, we obtain

$$y\,dy = 2x^2\,dx,$$

which is separated. A one-parameter family of solutions is therefore defined implicitly by

$$\int y\,dy = \int 2x^2\,dx + C,$$

or

$$\frac{y^2}{2} = \frac{2x^3}{3} + C.$$

Note that $y(x) = 0$ is also a solution, but it cannot be obtained by specifying C. In other words, $y(x) = 0$ is a singular solution. We removed this solution when we divided the original equation by xy^2.

(b) Since

$$\frac{dy}{dx} = \frac{y\sin x(1 + y^2)}{(1 + y^2)^2} = \frac{y\sin x}{1 + y^2},$$

the differential equation can be separated:

$$\sin x\,dx = \frac{1 + y^2}{y}dy = \left(\frac{1}{y} + y\right)dy.$$

A one-parameter family of solutions is therefore defined implicitly by

$$\int \sin x\,dx + C = \int \left(\frac{1}{y} + y\right)dy,$$

or

$$-\cos x + C = \ln|y| + \frac{y^2}{2}.$$

EXAMPLE 16.4

Find the function (in explicit form) that satisfies the differential equation

$$e^{2x+y}\,dx - 2e^{x-y}\,dy = 0, \quad y(0) = 1.$$

SOLUTION   If we divide the equation by e^{x+y}, we have

$$e^x\,dx = 2e^{-2y}\,dy.$$

A one-parameter family of solutions is defined implicitly by

$$e^x = -e^{-2y} + C.$$

The condition $y(0) = 1$ requires $e^0 = -e^{-2} + C$, which implies that $C = 1 + e^{-2}$. The required solution is therefore defined implicitly by

$$e^x = -e^{-2y} + e^{-2} + 1,$$

and from this equation, we have

$$e^{-2y} = e^{-2} + 1 - e^x \quad \text{or} \quad y(x) = -\tfrac{1}{2} \ln\left(e^{-2} + 1 - e^x\right).$$

EXAMPLE 16.5

According to the modified Torricelli law 5.12 in Section 5.4, when water leaks from the tank in the form of a right circular cone in Figure 16.2, it does so with speed $v = c\sqrt{2gy}$, where $g = 9.81$, $0 < c < 1$ is the discharge coefficient, and y is depth of water in the tank. Determine how long it takes for a full tank to empty.

FIGURE 16.2

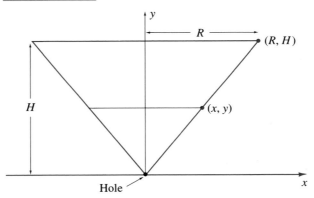

SOLUTION   The volume of water in the tank at any given time is $V = \pi x^2 y / 3$, and since $y = Hx/R$,

$$V = \frac{\pi}{3}\left(\frac{Ry}{H}\right)^2 y = \frac{\pi R^2 y^3}{3H^2}.$$

Differentiation of this equation with respect to time t gives

$$\frac{dV}{dt} = \frac{\pi R^2 y^2}{H^2}\frac{dy}{dt},$$

and both derivatives are negative. On the other hand, the rate at which water exits through the hole is the area A of the hole multiplied by the speed of the water throught the hole. Hence, we may set

$$-\frac{\pi R^2 y^2}{H^2}\frac{dy}{dt} = Av = Ac\sqrt{2gy}.$$

This is a separable differential equation,

$$y^{3/2}dy = \frac{AcH^2\sqrt{2g}}{\pi R^2}dt,$$

and, therefore, a one-parameter family of solutions is defined implicitly by

$$\frac{2}{5}y^{5/2} = -\frac{AcH^2\sqrt{2g}}{\pi R^2}t + D.$$

If we set $y = H$ when $t = 0$, then $D = (2/5)H^{5/2}$, and

$$\frac{2}{5}y^{5/2} = \frac{2}{5}H^{5/2} - \frac{AcH^2\sqrt{2g}}{\pi R^2}t.$$

The tank empties when $y = 0$, at which time

$$0 = \frac{2}{5}H^{5/2} - \frac{AcH^2\sqrt{2g}}{\pi R^2}t.$$

This can be solved for

$$t = \frac{\pi R^2\sqrt{2H}}{5Ac\sqrt{g}}.$$

EXERCISES 16.2

In Exercises 1–10 find a one-parameter family of solutions for the differential equation.

∗ **1.** $y^2\,dx - x^2\,dy = 0$

∗ **2.** $\dfrac{dy}{dx} + 2xy = 4x$

∗ **3.** $2xy\,dx + (x^2 + 1)\,dy = 0$

∗ **4.** $\dfrac{dy}{dx} = 3y + 2$

∗ **5.** $3(y^2 + 2)\,dx = 4y(x - 1)\,dy$

∗ **6.** $(x^2y + x^2)\,dx + (xy^2 - y^2)\,dy = 0$

∗ **7.** $\dfrac{dy}{dx} = -\dfrac{\cos y}{\sin x}$

∗ **8.** $(x^2ye^x - y)\,dx + xy^3\,dy = 0$

∗ **9.** $(x^2y^2\sec x\,\tan x + xy^2\sec x)\,dx + xy^3\,dy = 0$

∗ **10.** $\dfrac{dy}{dx} = \dfrac{1 + y^2}{1 + x^2}$

In Exercises 11–15 find a solution of the differential equation that also satisfies the given condition.

∗ **11.** $2y\,dx + (x + 1)\,dy = 0,\ \ y(1) = 2$

∗ **12.** $(xy + y)\,dx - (xy - x)\,dy = 0,\ \ y(1) = 2$

∗ **13.** $\dfrac{dy}{dx} = e^{x+y},\ \ y(0) = 0$

∗ **14.** $\dfrac{dy}{dx} = 2x(1 + y^2),\ \ y(2) = 4$

∗ **15.** $\dfrac{dy}{dx} = \dfrac{\sin^2 y}{\cos^2 x},\ \ y(0) = \pi/2$

✷✷ **16.** A girl lives 6 km from school. She decides to travel to school so that her speed is always proportional to the square of her distance from the school.

(a) Find her distance from school at any time.

(b) When does she reach school?

✷✷ **17.** Find a one-parameter family of solutions for the differential equation

$$\frac{dy}{dx} = -\frac{1 + y^3}{xy^2 + x^3y^2}.$$

✷✷ **18.** When a container of water at temperature $80°$C is placed in a room at temperature $20°$C, Newton's law of cooling states that the time rate of change of the temperature of the water is proportional to the difference between the temperature of the water and room temperature. If the water cools to $60°$C in 2 min, find a formula for its temperature as a function of time.

✷✷ **19.** A thermometer reading $23°$C is taken outside where the temperature is $-20°$C. If the reading drops to $0°$C in 4 min, when will it read $-19°$C?

✷✷ **20.** The amount of a drug such as penicillin injected into the body is used up at a rate proportional to the amount still present. If a dose decreases by 5% in the first hour, when does it decrease to one-half its original amount?

✷✷ **21.** When a deep-sea diver inhales air, his body tissues absorb extra amounts of nitrogen. Suppose the diver enters the water at time $t = 0$, drops very quickly to depth d, and remains at this depth for a very long time. The amount N of nitrogen in his body tissues increases as he remains at this depth until a maximum amount $\overline{N}$ is reached. The time rate of change of N is proportional to the difference $\overline{N} - N$. Show that, if N_0 is the amount of nitrogen in his body tissues when he enters the water, then

$$N = N_0 e^{-kt} + \overline{N}(1 - e^{-kt}),$$

where $k > 0$ is a constant.

✷✷ **22.** When a substance such as glucose is administered intravenously into the bloodstream, it is used up by the body at a rate proportional to the amount present at that time. If it is added at a constant rate of R units per unit time, and A_0 is the amount present in the bloodstream when the intravenous feeding begins, find a formula for the amount in the bloodstream at any time.

✷✷ **23.** Find the equations for all curves that satisfy the condition that the normal at any point on the curve, and the line joining the point to the origin, form an isosceles triangle with the x-axis as base.

✷✷ **24.** When two substances A and B are brought together in one solution, they react to form a third substance C in such a way that one gram of A reacts with one gram of B to produce two grams of C. The rate at which C is formed is proportional to the product of the amounts of A and B still present in the solution. If 10 g of A and 15 g of B are originally brought together, find a formula for the amount of C present in the mixture at any time.

✷✷ **25.** What is the solution to Exercise 24 when the initial amounts of A and B are both 10 g?

✷✷ **26.** A first-order differential equation in $y(x)$ is said to be *homogeneous* if it can be written in the form

$$\frac{dy}{dx} = f\left(\frac{y}{x}\right).$$

Show that the change of dependent variable $v = y/x$ yields a differential equation in $v(x)$ that is always separable.

✷✷ **27.** State a condition on the functions M and N in equation 16.17 in order that the differential equation be homogeneous in the sense of Exercise 26.

In Exercises 28–33 show that the differential equation is homogeneous, and use the change of variable in Exercise 26 to find a one-parameter family of solutions.

** **28.** $(y^2 - x^2)\, dx + xy\, dy = 0$ ** **29.** $2x\, dy - 2y\, dx = \sqrt{x^2 + 4y^2}\, dx$

** **30.** $\dfrac{dy}{dx} = \dfrac{y + x}{y - x}$ ** **31.** $x\, dy - y\, dx = x \cos(y/x)\, dx$

** **32.** $\dfrac{dy}{dx} = \dfrac{x^2 e^{-y/x} + y^2}{xy}$ ** **33.** $(x^2 y + y^3)\, dx + x^3\, dy = 0$

** **34.** If a curve passes through the point $(1, 2)$ and is such that the length of that part of the tangent line at (x, y) from (x, y) to the y-axis is equal to the y-intercept of the tangent line, find the equation of the curve.

** **35.** Find in explicit form a function that satisfies the differential equation $dy/dx = \csc y$ and the following conditions: (a) $y(0) = \pi/4$ and (b) $y(0) = 7\pi/4$

** **36.** In a chemical reaction, one molecule of trypsinogen yields one molecule of trypsin. In order for the reaction to take place, an initial amount of trypsin must be present. Suppose that the initial amount is y_0. Thereafter, the rate at which trypsinogen is changed into trypsin is proportional to the product of the amounts of each chemical in the reaction. Find a formula for the amount of trypsin if the initial amount of trypsinogen is A.

** **37.** If a first-order differential equation can be written in the form

$$\frac{dy}{dx} = f(ax + by),$$

where a and b are constants, show that the change of dependent variable $v = ax + by$ always gives a differential equation in $v(x)$ that is separable.

Use the method of Exercise 37 to find a one-parameter family of solutions for the differential equation in Exercises 38–41.

** **38.** $\dfrac{dy}{dx} = x + y$ ** **39.** $\dfrac{dy}{dx} = (x + y)^2$

** **40.** $\dfrac{dy}{dx} = \dfrac{1}{2x + 3y}$ ** **41.** $\dfrac{dy}{dx} = \sin^2(x - y)$

** **42.** A certain chemical dissolves in water at a rate proportional to the product of the amount of undissolved chemical and the difference between concentrations in a saturated solution and the existing concentration in the solution. A saturated solution contains 25 g of chemical in 100 mL of solution.

     (a) If 50 g of chemical are added to 200 mL of water, find a formula for the amount of chemical dissolved as a function of time. Draw its graph.

     (b) Repeat (a) if the 50 g of chemical are added to 100 mL of water.

     (c) Repeat (a) if 10 g of chemical are added to 100 mL of water.

** **43.** Snow has been falling for some time when a snowplow starts plowing the highway. The plow begins at 12:00 and travels 2 km during the first hour and 1 km during the second hour. Make reasonable assumptions to find out when the snow started falling.

** **44.** In order to perform a one-hour operation on a dog a veterinarian anesthetizes the dog with sodium pentobarbital. During the operation, the dog's body breaks down the drug at a rate proportional to the amount still present, and only half an original dose remains after 5 h. If the dog has mass 20 kg, and 20 mg of sodium pentobarbital per kilogram of body mass are required to maintain surgical anesthesia, what original dose is required?

** **45.** Explain what the cancellation of differentials in proceeding from 16.21 to 16.22 really means.

** **46.** Solve the differential equation

$$(x^3 y^4 + 2xy^4)\, dx + (x - xy^6)\, dy = 0, \quad y(1) = 1.$$

** **47.** Two substances A and B react to form a third substance C in such a way that 2 g of A react with 1 g of B to produce 3 g of C. The rate at which C is formed is proportional to the amounts of A and B still present in the mixture. Find the amount of C present in the mixture as a function of time when the original amounts of A and B brought together at time $t = 0$ are as follows. Draw graphs of all three functions on the same axes.

    (a) 20 g and 10 g

    (b) 20 g and 5 g

    (c) 20 g and 20 g

*** **48.** A bird is due east of its nest a distance L away and at the same height above the ground as the nest. Wind is blowing due north at speed v. If the bird flies horizontally with constant speed V always pointing straight at its nest, what is the equation of the curve that it follows. Take the nest at the origin and the x- and y-directions as east and north.

*** **49.** Find the equation of the curve that passes through $(1, 1)$ and is such that the tangent and normal lines at any point (x, y) make with the x-axis a triangle whose area is equal to the slope of the tangent line at (x, y).

16.3 Linear First-Order Differential Equations

A first-order differential equation that can be written in the form

$$\frac{dy}{dx} + P(x)y = Q(x) \tag{16.23}$$

is said to be linear. We will explain the significance of the adjective "linear" in Section 16.6. To illustrate how to solve such differential equations, consider the equation

$$\frac{dy}{dx} + \frac{1}{x}y = 1.$$

If we multiply both sides by x, we have

$$x\frac{dy}{dx} + y = x.$$

But note now that the left side is the derivative of the product xy; i.e.,

$$\frac{d}{dx}(xy) = x\frac{dy}{dx} + y.$$

In other words, we can write the differential equation in the form

$$\frac{d}{dx}(xy) = x,$$

and integration immediately gives a one-parameter family of solutions

$$xy = \frac{x^2}{2} + C.$$

This is the principle behind all linear first-order equations: Multiply the equation by a function of x in order that the left side be expressible as the derivative of a product. To show that this is always possible, we turn now to the general equation 16.23. If the equation is multiplied by a function $\mu(x)$,

$$\mu\frac{dy}{dx} + \mu P(x)y = \mu Q(x). \qquad (16.24)$$

This equation is equivalent to 16.23 in the sense that $y(x)$ is a solution of 16.23 if and only if it is a solution of 16.24. We ask whether it is possible to find μ so that the left side of 16.24 can be written as the derivative of the product μy; that is, can we find $\mu(x)$ so that

$$\mu\frac{dy}{dx} + \mu P(x)y = \frac{d}{dx}(\mu y)?$$

If we expand the right side, μ must satisfy

$$\mu\frac{dy}{dx} + \mu P(x)y = \mu\frac{dy}{dx} + y\frac{d\mu}{dx},$$

from which we get

$$\mu P = \frac{d\mu}{dx} \quad \text{or} \quad \frac{d\mu}{\mu} = P\,dx.$$

Thus μ must satisfy a separated differential equation, one solution of which is

$$\ln|\mu| = \int P(x)dx \quad \text{or} \quad \mu = \pm e^{\int P(x)dx}.$$

We have shown then that, if 16.23 is multiplied by the factor

$$e^{\int P(x)dx} \quad (\text{or by} -e^{\int P(x)dx}),$$

then the differential equation becomes

$$e^{\int P(x)dx}\frac{dy}{dx} + P(x)ye^{\int P(x)dx} = Q(x)e^{\int P(x)dx},$$

and the left side can be expressed as the derivative of a product:

$$\frac{d}{dx}\{ye^{\int P(x)dx}\} = Q(x)e^{\int P(x)dx}.$$

Integration now gives a one-parameter family of solutions

$$ye^{\int P(x)dx} = \int Q(x)e^{\int P(x)dx}dx + C. \qquad (16.25)$$

The quantity $e^{\int P(x)dx}$ is called an **integrating factor** for equation 16.23 because when the equation is multiplied by this factor, it becomes immediately integrable.

In summary, if linear differential equation 16.23 is multiplied by the function $e^{\int P(x)dx}$, then the left side of the equation becomes the derivative of the product of y and this function. In this form the equation can immediately be integrated. Given a specific linear equation to solve, we have three choices on how to proceed:

1. Memorize 16.25 as a formula.
2. Memorize that an integrating factor is $e^{\int P(x)dx}$.
3. Develop the integrating factor in every example.

We feel that (2) is a reasonable compromise in that we must memorize something, but not too much, and we must do some calculations, but not too many.

EXAMPLE 16.6

Find one-parameter families of solutions for the following differential equations:

(a) $\dfrac{dy}{dx} + xy = x$

(b) $(y - x \sin x)dx + x\, dy = 0$

(c) $\cos x \dfrac{dy}{dx} + y \sin x = 1$

SOLUTION

(a) An integrating factor for this linear equation is

$$e^{\int x\, dx} = e^{x^2/2}.$$

If we multiply the equation by this integrating factor, we have

$$e^{x^2/2}\frac{dy}{dx} + yxe^{x^2/2} = xe^{x^2/2},$$

or

$$\frac{d}{dx}\{ye^{x^2/2}\} = xe^{x^2/2}.$$

Integration yields

$$ye^{x^2/2} = \int xe^{x^2/2}dx = e^{x^2/2} + C,$$

or

$$y(x) = 1 + Ce^{-x^2/2}.$$

(b) If we write the differential equation in the form

$$\frac{dy}{dx} + \frac{y}{x} = \sin x,$$

then we see that it is linear first order. An integrating factor is therefore

$$e^{\int 1/x\, dx} = e^{\ln|x|} = |x|.$$

If we multiply the differential equation by this factor, we get

$$|x|\frac{dy}{dx} + \frac{|x|}{x}y = |x|\sin x.$$

If $x > 0$, then we write

$$x\frac{dy}{dx} + \frac{x}{x}y = x \sin x,$$

whereas if $x < 0$,

$$-x\frac{dy}{dx} - \frac{x}{x}y = -x\sin x.$$

In either case, however, the equation simplifies to

$$x\frac{dy}{dx} + y = x\sin x,$$

or

$$\frac{d}{dx}(xy) = x\sin x.$$

Integration now gives

$$xy = \int x\sin x\,dx = -x\cos x + \sin x + C.$$

Finally, then,

$$y(x) = -\cos x + \frac{\sin x}{x} + \frac{C}{x}.$$

(c) An integrating factor for the linear equation

$$\frac{dy}{dx} + y\tan x = \frac{1}{\cos x}$$

is

$$e^{\int \tan x\,dx} = e^{\ln|\sec x|} = |\sec x|.$$

For either $\sec x < 0$ or $\sec x > 0$, we obtain

$$\frac{1}{\cos x}\frac{dy}{dx} + y\frac{\sin x}{\cos^2 x} = \frac{1}{\cos^2 x},$$

or

$$\frac{d}{dx}\left\{\frac{y}{\cos x}\right\} = \sec^2 x.$$

Integration now yields

$$\frac{y}{\cos x} = \tan x + C,$$

and hence

$$y(x) = \sin x + C\cos x.$$

A one-paramater family of solutions of a first-order differential equation is a general solution of the differential equation if and only if it contains all solutions of the differential equation. If the family has singular solutions, it cannot be a general solution. In Section 16.8, we indicate that a one-parameter family of solutions of a linear first-order differential equation cannot have singular solutions. In other words, a one-parameter family of solutions for a linear first-order differential equation is always a general solution.

In Section 3.10 we introduced differential equations that arise when energy balance is applied to a system such as that in Figure 16.3. We repeat the discussion here but in a more general way. Liquid at temperature $T_0\,^\circ$C enters the tank, the liquid is heated, and it then leaves at higher temperature $T\,^\circ$C. The tank is perfectly insulated so that no heat can escape from its sides, and therefore all heat supplied by the heater raises the temperature of the liquid. When the tank is full, and we assume that this is always the case, the mass of liquid is M kg. Liquid enters the tank at a rate denoted by $\dot{m}$ kg/s and leaves at the same rate. The heater adds energy at the rate of q J/s (where J is joules). In reality, temperature of the liquid in the tank depends on both time and position in the tank. The mixer is added to remove spatial dependence. We assume that the mixer is so efficient that temperature of the liquid is the same at every point in the tank. Temperature then depends only on time t, denoted by $T(t)$, and this is also the temperature at which liquid leaves the tank.

FIGURE 16.3

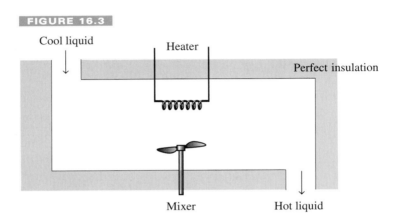

Energy balance for this system states that

$$\left\{\begin{array}{c}\text{Rate at which}\\\text{energy enters tank}\\\text{in cool liquid}\end{array}\right\} + \left\{\begin{array}{c}\text{Rate at which}\\\text{energy is added}\\\text{by heater}\end{array}\right\} = \left\{\begin{array}{c}\text{Rate at which}\\\text{energy leaves tank}\\\text{in hot liquid}\end{array}\right\} + \left\{\begin{array}{c}\text{Rate at which}\\\text{energy is used to raise}\\\text{temperature of}\\\text{liquid in tank}\end{array}\right\}.$$

(For simplicity, we have ignored energy associated with the mixer.) The first three rates are known; the fourth can be expressed in terms of dT/dt, the time rate of change of temperature. The second term on the left is q. For the remaining terms, we must introduce the *specific heat c_p* of the liquid. It is the energy required to raise the temperature of 1 kg of the liquid by 1°C. (For example, the specific heat of water is $c_p = 4190$ J/kg·°C. It requires 4190 joules of energy to raise 1 kg of water 1°C. To raise 5 kg of water 10° requires $(5)(10)(4190) = 209\,500$ joules.) Since $\dot{m}$ kg of liquid at temperature $T_0\,^\circ$C enter the tank each second, the rate at which energy enters the tank due to this liquid is $\dot{m}c_p T_0$ J/s. (This is the energy required to raise $\dot{m}$ kg of liquid from 0°C to $T_0\,^\circ$C.) In a similar way, the rate at which energy leaves the tank in the liquid at temperature T is $\dot{m}c_p T$ J/s. This leaves only the last term on the right side of the energy balance equation. The rate of change of temperature of the M kg of liquid in the tank is dT/dt; that is, the temperature changes dT/dt degrees each second. It follows that the rate at which energy is used to raise the temperature of this liquid is $Mc_p(dT/dt)$.

When these rates are substituted into the energy balance equation, the result is

$$\dot{m}c_p T_0 + q = \dot{m}c_p T + Mc_p\frac{dT}{dt}.$$

Division of Mc_p gives

$$\frac{dT}{dt} + \frac{\dot{m}}{M}T = \frac{\dot{m}}{M}T_0 + \frac{q}{Mc_p}.$$

Of the quantities $\dot{m}$, c_p, T_0, q, and M in this equation, only c_p and M must be constant; $\dot{m}$, T_0, and q could be functions of time t. In an actual problem, all five quantities are given, and the objective is to solve this linear first-order differential equation for $T(t)$. Multiplication of the differential equation by the integrating factor

$$e^{\int (\dot{m}/M)dt}$$

leads to

$$\frac{d}{dt}\left[T e^{\int (\dot{m}/M)dt} \right] = \left(\frac{\dot{m}}{M} T_0 + \frac{q}{Mc_p} \right) e^{\int (\dot{m}/M)dt}.$$

Integration gives

$$T e^{\int (\dot{m}/M)dt} = \int \left(\frac{\dot{m}}{M} T_0 + \frac{q}{Mc_p} \right) e^{\int (\dot{m}/M)dt}\, dt + F$$

or

$$T(t) = F e^{-\int (\dot{m}/M)dt} + e^{-\int (\dot{m}/M)dt} \int \left(\frac{\dot{m}}{M} T_0 + \frac{q}{Mc_p} \right) e^{\int (\dot{m}/M)dt}\, dt,$$

where F is the constant of integration. Once $\dot{m}$, c_p, T_0, q, and M are specified, integrations can be performed to yield $T(t)$. Let us consider the special case in which all five quantities are constants (independent of time). Then

$$T(t) = F e^{-\dot{m}t/M} + e^{-\dot{m}t/M} \int \left(\frac{\dot{m}}{M} T_0 + \frac{q}{Mc_p} \right) e^{\dot{m}t/M}\, dt$$

$$= F e^{-\dot{m}t/M} + e^{-\dot{m}t/M} \left(\frac{\dot{m}}{M} T_0 + \frac{q}{Mc_p} \right) \left(\frac{M}{\dot{m}} \right) e^{\dot{m}t/M}$$

$$= T_0 + \frac{q}{\dot{m}c_p} + F e^{-\dot{m}t/M}.$$

If we assume that T_0 is not only the temperature at which liquid enters the tank, but also the temperature of the water at time $t = 0$, then $F = -q/(\dot{m}c_p)$, and

$$T(t) = T_0 + \frac{q}{\dot{m}c_p}\left(1 - e^{-\dot{m}t/M} \right).$$

A graph of this function is shown in Figure 16.4.

FIGURE 16.4

In Exercises 1–12 find a general solution for the differential equation.

* **1.** $\dfrac{dy}{dx} + 2xy = 4x$

* **2.** $\dfrac{dy}{dx} + \dfrac{2}{x}y = 6x^3$

* **3.** $(2y - x)\,dx + dy = 0$

* **4.** $\dfrac{dy}{dx} + y\cot x = 5e^{\cos x}$

* **5.** $(x^2 + 2xy)\,dx + (x^2 + 1)\,dy = 0$

* **6.** $(x + 1)\dfrac{dy}{dx} - 2y = 2(x + 1)$

* **7.** $\dfrac{1}{x}\dfrac{dy}{dx} - \dfrac{y}{x^2} = \dfrac{1}{x^3}$

* **8.** $(y + e^{2x})\,dx = dy$

* **9.** $\dfrac{dy}{dx} + y = 2\cos x$

* **10.** $x^3\dfrac{dy}{dx} + (2 - 3x^2)y = x^3$

* **11.** $\dfrac{dy}{dx} + \dfrac{y}{x\ln x} = x^2$

* **12.** $(-2y\cot 2x - 1 + 2x\cot 2x + 2\csc 2x)\,dx + dy = 0$

In Exercises 13–15 solve the differential equation.

* **13.** $\dfrac{dy}{dx} + 3x^2 y = x^2, \;\; y(1) = 2$

* **14.** $(-e^x \sin x + y)\,dx + dy = 0, \;\; y(0) = -1$

* **15.** $\dfrac{dy}{dx} + \dfrac{x^3 y}{x^4 + 1} = x^7, \;\; y(0) = 1$

** **16.** Find a general solution for the differential equation $(y^3 - x)\,dy = y\,dx$.

** **17.** A differential equation of the form

$$\frac{dy}{dx} + P(x)y = y^n Q(x)$$

is called a *Bernoulli equation*. Show that the change of dependent variable $z = y^{1-n}$ gives

$$\frac{dz}{dx} + (1 - n)Pz = (1 - n)Q,$$

a linear first-order equation in $z(x)$.

In Exercises 18–22 use the change of variable in Exercise 17 to find a general solution for the differential equation.

** **18.** $\dfrac{dy}{dx} + y = y^2 e^x$

** **19.** $\dfrac{dy}{dx} + \dfrac{y}{x} = \dfrac{y^2}{x^2}$

** **20.** $\dfrac{dy}{dx} - y + (x^2 + 2x)y^2 = 0$

** **21.** $x\,dy + y\,dx = x^3 y^5\,dx$

** **22.** $\dfrac{dy}{dx} + y\tan x = y^4 \sin x$

** **23.** Repeat Exercise 22 in Section 16.2 if the glucose is added at a rate $R(t)$ that is a function of time t.

** **24.** A tank has 100 L of solution containing 4 kg of sugar. A mixture with 10 g of sugar per litre of solution is added at a rate of 200 mL per minute. At the same time, 100 mL of well-stirred mixture are removed each minute. Find the amount of sugar in the tank as a function of time.

** **25.** Repeat Exercise 24 if 300 mL of mixture are removed each minute.

** **26.** A tank originally contains 1000 L of water in which 5 kg of salt have been dissolved. A mixture containing 2 kg of salt for each 100 L of solution is added to the tank at 10 mL/s. At the same time, the well-stirred mixture in the tank is removed at the rate of 5 mL/s. Find the amount of salt in the tank as a function of time. Draw a graph of the function.

** **27.** Repeat Exercise 26 if the mixture is removed at 10 mL/s. What is the limit of the amount of salt in the tank for long periods of time?

** **28.** Repeat Exercise 26 if the mixture is removed at 20 mL/s.

** **29.** A room with volume 100 m³ initially contains 0.1% carbon dioxide. Beginning at time $t = 0$, fresher air containing 0.05% carbon dioxide flows into the room at 5 m³/min. The well-mixed air in the room flows out at the same rate. Find the amount of carbon dioxide in the room as a function of time. What is the limit of the function as $t \to \infty$?

** **30.** Water at temperature 10°C enters the tank in Figure 16.3 at a rate of 0.03 kg/s. At time $t = 0$, the tank reaches capacity (100 kg), and water then leaves at the same rate. The heater is turned on at $t = 0$ adding energy to the water at 2000 J/s. Find the temperature of the water in the tank as a function of time t, and plot its graph.

** **31.** Repeat Exercise 30 if the rate at which water enters the tank is $100/(t + 1)$ kg/s.

** **32.** Repeat Exercise 30 if the amount of energy supplied by the heater for the first ten minutes is $q = 20t$ J/s. Solve only for $0 \leq t \leq 600$.

** **33.** Repeat Exercise 30 if the temperature of the incoming water is $T_0(t) = 10e^{-t}$, $t \geq 0$.

** **34.** (a) The current I in the LR-circuit on page 1270 must satisfy the differential equation

$$L\frac{dI}{dt} + RI = E.$$

If $E(t) = E_0 \sin(\omega t)$, $t \geq 0$, where E_0 and ω are constants, solve this differential equation for $I(t)$, and show that the solution can be written in the form

$$I(t) = Ae^{-Rt/L} + \frac{E_0}{Z}\sin(\omega t - \phi),$$

where A is an arbitrary constant and

$$Z = \sqrt{R^2 + \omega^2 L^2}, \qquad \phi = \text{Tan}^{-1}\left(\frac{\omega L}{R}\right).$$

(b) What is the value of A if the current in the circuit at time $t = 0$ when $E(t)$ is connected is I_0?

** **35.** (a) The current I in the RC-circuit on page 1270 must satisfy the differential equation

$$R\frac{dI}{dt} + \frac{I}{C} = \frac{dE}{dt}.$$

If $E(t)$ is as in Exercise 34, show that the solution can be written in the form

$$I(t) = Ae^{-t/(RC)} + \frac{E_0}{Z} \sin(\omega t - \phi),$$

where A is an arbitrary constant and

$$Z = \sqrt{R^2 + \frac{1}{\omega^2 C^2}}, \qquad \phi = \mathrm{Tan}^{-1}\left(-\frac{1}{\omega C R}\right).$$

(b) What is the value of A if the current in the circuit at time $t = 0$ is I_0 when $E(t)$ is connected?

*** **36.** Repeat Exercise 21 in Section 16.2 given that the diver descends slowly to the bottom. Assume that his descent is at a constant rate over a time interval of length T. Assume also that maximum pressure $\overline{N}$ is proportional to depth below the surface.

16.4 Second-Order Equations Reducible to Two First-Order Equations

Second-order differential equations in y as a function of x can be expressed symbolically in the form

$$F(x, y, y', y'') = 0. \tag{16.26}$$

When F is independent of x or y, the second-order equation can be replaced by two first-order equations.

Type I: Dependent Variable Missing

If a second-order differential equation in $y(x)$ is explicitly independent of y, then it is of the form

$$F(x, y', y'') = 0. \tag{16.27}$$

In such a case we set

$$v = \frac{dy}{dx} \quad \text{and} \quad \frac{dv}{dx} = \frac{d^2y}{dx^2}. \tag{16.28}$$

If we substitute these into 16.27, we obtain

$$F(x, v, v') = 0,$$

a first-order differential equation in $v(x)$. If we can solve this equation for $v(x)$, we can then integrate $dy/dx = v(x)$ for $y(x)$.

EXAMPLE 16.7

Find two-parameter families of solutions for the following differential equations:
(a) $xy'' - y' = 0$
(b) $y'' = (y')^2$

SOLUTION

(a) Since y is explicitly missing, we use substitutions 16.28:

$$x\frac{dv}{dx} - v = 0.$$

Variables are now separable,

$$\frac{dv}{v} = \frac{dx}{x},$$

and a solution for $v(x)$ is

$$\ln|v| = \ln|x| + C \quad \text{or} \quad v = Dx \quad (D = \pm e^{C}).$$

Because $v = dy/dx$,

$$\frac{dy}{dx} = Dx,$$

and we can integrate for

$$y(x) = \frac{D}{2}x^2 + E$$
$$= Fx^2 + E \quad (F = D/2).$$

(b) Since y is again missing, we substitute from 16.28 to get

$$\frac{dv}{dx} = v^2.$$

Variables are again separable,

$$\frac{dv}{v^2} = dx,$$

and a solution for $v(x)$ is

$$-\frac{1}{v} = x + C.$$

Consequently,

$$v = \frac{dy}{dx} = -\frac{1}{x + C}.$$

Integration now yields

$$y(x) = -\ln|x + C| + D.$$

Type II: Independent Variable Missing

If a second-order differential equation in $y(x)$ is explicitly independent of x, then it is of the form

$$F(y, y', y'') = 0. \qquad (16.29)$$

In this case we set

$$\frac{dy}{dx} = v \qquad \text{and} \qquad \frac{d^2y}{dx^2} = \frac{dv}{dx} = \frac{dv}{dy}\frac{dy}{dx} = v\frac{dv}{dy}. \qquad (16.30)$$

When we substitute these into 16.29, we obtain

$$F\left(y, v, v\frac{dv}{dy}\right) = 0,$$

a first-order differential equation in $v(y)$. If we can solve this equation for $v(y)$, we can separate $dy/dx = v(y)$ and integrate for $y(x)$.

EXAMPLE 16.8

Find two-parameter families of solutions for the following differential equations:

(a) $yy'' + (y')^2 = 1$

(b) $y'' = (y')^2$

SOLUTION

(a) Since x is explicitly missing, we substitute from 16.30:

$$yv\frac{dv}{dy} + v^2 = 1.$$

If variables are separated, we have

$$\frac{v\,dv}{v^2 - 1} = -\frac{dy}{y},$$

and integration gives

$$\tfrac{1}{2}\ln|v^2 - 1| = -\ln|y| + C.$$

Thus,

$$|v^2 - 1| = \frac{e^{2C}}{y^2},$$

from which we have

$$\frac{dy}{dx} = v = \frac{\pm\sqrt{D + y^2}}{y} \qquad (D = \pm e^{2C}).$$

We separate variables again to get

$$\frac{y\,dy}{\sqrt{y^2 + D}} = \pm dx,$$

and obtain an implicit definition of the solution $y(x)$,

$$\sqrt{y^2 + D} = \pm x + E.$$

(b) Since x is explicitly missing, we again substitute from 16.30 to obtain

$$v \frac{dv}{dy} = v^2.$$

If variables are separated,

$$\frac{dv}{v} = dy,$$

and a solution for $v(y)$ is

$$\ln |v| = y + C.$$

Thus,

$$|v| = e^{y+C}$$

and

$$\frac{dy}{dx} = v = De^y \quad (D = \pm e^C).$$

We separate variables again,

$$e^{-y} dy = D dx,$$

and find an implicit definition of the solution $y(x)$,

$$-e^{-y} = Dx + E.$$

In each of Examples 16.7 and 16.8 we solved the differential equation $y'' = (y')^2$ since both the independent variable x and the dependent variable y are missing. Although the solutions appear different, each is easily derivable from the other.

Let us summarize the results of this section. Substitutions 16.28 for differential equation 16.27 with the dependent variable missing replace the second-order differential equation with two first-order equations: a first-order equation in $v(x)$, followed by a first-order equation in $y(x)$. Contrast this with the method for equation 16.29 with the independent variable missing. Substitutions 16.30 again replace the second-order equation with two first-order equations. However, the first first-order equation is in $v(y)$, so that for this equation y is the independent variable rather than the dependent variable. The second first-order equation is again one for $y(x)$.

is described by

$$\mathbf{F} = m\mathbf{a}. \tag{16.31}$$

When **F** is given, this is an algebraic equation giving acceleration **a**. If we substitute $\mathbf{a} = d\mathbf{v}/dt$, we obtain a first-order differential equation

$$\mathbf{F} = m\frac{d\mathbf{v}}{dt} \tag{16.32}$$

for velocity **v** as a function of time t. If we substitute $\mathbf{a} = d^2\mathbf{r}/dt^2$, we obtain a second-order differential equation for position **r** as a function of time:

$$\mathbf{F} = m\frac{d^2\mathbf{r}}{dt^2}. \tag{16.33}$$

In practice, it is seldom this simple. Often **F** is not given as a function of time, but as a function of position, or velocity, or some other variable. In such cases we may have to change the dependent or independent variable, or both, in order to solve the differential equation.

In this section we are concerned with motion in one direction only. If $r(t)$, $v(t)$, $a(t)$, and F are the components of $\mathbf{r}(t)$, $\mathbf{v}(t)$, $\mathbf{a}(t)$, and **F** in this direction, then equations 16.31–16.33 take the forms

$$F = ma, \tag{16.34}$$

$$F = m\frac{dv}{dt}, \tag{16.35}$$

$$F = m\frac{d^2r}{dt^2}, \tag{16.36}$$

respectively.

EXAMPLE 16.9

A paratrooper and his parachute have mass 100 kg. At the instant his parachute opens, the paratrooper is travelling vertically downward at 10 m/s. Air resistance on the chute varies directly as the instantaneous velocity and is 800 N when the paratrooper's velocity is 12 m/s. Find the position and velocity of the paratrooper as functions of time.

SOLUTION   Let us measure y as positive in the downward direction, taking $y = 0$ and time $t = 0$ at the instant the paratrooper opens his parachute (Figure 16.5). If $F_\mathbf{a}$ is the vertical component of the force of air resistance, then

$$F_\mathbf{a} = -kv.$$

FIGURE 16.5

$t = 0,$
$y = 0,$
$v = 10$

y

Since $F_\mathbf{a} = -800$ when $v = 12$, it follows that $-800 = -12k$ or $k = 200/3$. Since the total force on the paratrooper during the fall has component $100g - 200v/3$ ($g = 9.81$), Newton's second law gives

$$100\frac{dv}{dt} = 100g - \frac{200}{3}v, \quad t \geq 0.$$

Note that the force on the right is not a function of t so that the equation is not immediately integrable. It is however separable,

$$\frac{dv}{2v - 3g} = -\frac{dt}{3},$$

and solutions are defined implicitly by

$$\frac{1}{2} \ln |2v - 3g| = -\frac{t}{3} + C.$$

An explicit solution is found by solving this equation for v:

$$v(t) = \frac{3g}{2} + De^{-2t/3} \quad (D = \pm e^{2C}/2).$$

Since $v = 10$ when $t = 0$,

$$10 = \frac{3g}{2} + D \quad \text{or} \quad D = 10 - \frac{3g}{2}.$$

The velocity of the paratrooper as a function of time is therefore

$$v(t) = \frac{3g}{2} + \left(10 - \frac{3g}{2}\right)e^{-2t/3}.$$

Because $v(t) = dy/dt$, we set

$$\frac{dy}{dt} = \frac{3g}{2} + \left(10 - \frac{3g}{2}\right)e^{-2t/3},$$

and this equation is immediately integrable to

$$y(t) = \frac{3g}{2}t - \frac{3}{2}\left(10 - \frac{3g}{2}\right)e^{-2t/3} + E.$$

Since $y(0) = 0$,

$$0 = -\frac{3}{2}\left(10 - \frac{3g}{2}\right) + E \quad \text{or} \quad E = 15 - \frac{9g}{4}.$$

Consequently, the distance fallen by the paratrooper as a function of time is given by

$$y(t) = \frac{3g}{2}t - \left(15 - \frac{9g}{4}\right)e^{-2t/3} + \left(15 - \frac{9g}{4}\right) = 14.715t + 7.0725(e^{-2t/3} - 1).$$

Note that the velocity of the paratrooper does not increase indefinitely. In fact, as time passes a limiting velocity is approached:

$$\lim_{t \to \infty} v(t) = \lim_{t \to \infty} \left\{\frac{3g}{2} + \left(10 - \frac{3g}{2}\right)e^{-2t/3}\right\} = \frac{3g}{2} = 14.715 \text{ m/s}.$$

This is called the terminal velocity of the paratrooper; it is a direct result of the assumption that air resistance is proportional to instantaneous velocity.

Whenever an object such as the block in Figure 16.6 is moving over a surface, there is resistance to the motion. This resistance, called **friction**, is due to the fact that the interface between the block and the surface is not smooth; each surface is inherently rough, and this roughness retards the motion of one surface over the other. In effect, a force slowing the motion of the block is created, and this force is called the **force of friction**. Many experiments have been performed to obtain a functional representation for this force. It turns out that, when the block in Figure 16.6 slides along a horizontal surface, the magnitude of the force of friction opposing the motion is given by

$$|\mathbf{F}| = \mu mg, \tag{16.37}$$

where m is the mass of the block, g is the acceleration due to gravity, and μ is a constant called **the coefficient of kinetic friction**. In other words, the force of friction is directly proportional to the weight mg of the block. We caution the reader that this result is valid for the situation shown in Figure 16.6, but it may not be valid for other configurations (say, perhaps, for an inclined plane). Furthermore, there is a coefficient of static friction that is used in place of μ if the block is being accelerated from rest.

FIGURE 16.6

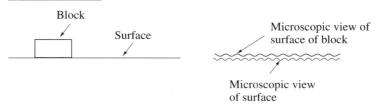

Block

Surface

Microscopic view of surface of block

Microscopic view of surface

EXAMPLE 16.10

A block of mass 2 kg is given initial speed 5 m/s along a horizontal surface. If the coefficient of kinetic friction between the block and surface is $\mu = 0.25$, how far does the block slide before stopping?

FIGURE 16.7

$t = 0$

$x = 0$

x

SOLUTION   Let us measure x as positive in the direction of motion (Figure 16.7), taking $x = 0$ and $t = 0$ at the instant the block is released. The x-component of the force of friction on the block is

$$F = -0.25(2)g = -\frac{g}{2} \quad (g = 9.81).$$

According to Newton's second law,

$$2\frac{dv}{dt} = -\frac{g}{2},$$

from which we get

$$v(t) = -\frac{g}{4}t + C.$$

Since $v(0) = 5$, it follows that $C = 5$, and

$$v(t) = -\frac{g}{4}t + 5.$$

But $v = dx/dt$, and hence

$$\frac{dx}{dt} = -\frac{g}{4}t + 5.$$

Integration gives

$$x(t) = -\frac{g}{8}t^2 + 5t + D.$$

Because we chose $x = 0$ at time $t = 0$, D must also be zero, and

$$x(t) = -\frac{g}{8}t^2 + 5t.$$

The block comes to rest when $v = 0$; i.e., when

$$-\frac{g}{4}t + 5 = 0, \text{ or } t = \frac{20}{g}.$$

The position of the block at this time is

$$x = -\frac{g}{8}\left(\frac{20}{g}\right)^2 + 5\left(\frac{20}{g}\right) = 5.10.$$

The block therefore slides 5.10 m before stopping.

EXERCISES 16.5

∗ **1.** A car of mass 1500 kg starts from rest at an intersection. The engine exerts a constant force of 3000 N, and air friction causes a resistive force whose magnitude in newtons is equal to the speed of the car in metres per second. Find the speed of the car and its distance from the intersection after 10 s.

∗∗ **2.** You are called on as an expert to testify in a traffic accident hearing. The question concerns the speed of a car that made an emergency stop with brakes locked and wheels sliding. The skid mark on the road measured 9 m. If you assume that the coefficient of kinetic friction between the tires and road was less than one, what can you say about the speed of the car before the brakes were applied? Are you testifying for the prosecution or the defence?

∗∗ **3.** A boat and its contents have mass 250 kg. Water exerts a resistive force on the motion of the boat that is proportional to the instantaneous speed of the boat and is 200 N when the speed is 30 km/h.

    (a) If the boat starts from rest and the engine exerts a constant force of 250 N in the direction of motion, find the speed of the boat as a function of time.

    (b) What is the limiting speed of the boat?

∗∗ **4.**   (a) In Example 16.9 it was necessary to solve the equation

$$\frac{1}{2}\ln|2v - 3g| = -\frac{t}{3} + C$$

    for $v = v(t)$ and use the condition $v(0) = 10$ to evaluate C. We chose first to solve for $v(t)$ and then to evaluate the constant. Show the details of this analysis.

    (b) Instead, first use the condition $v(0) = 10$ to evaluate C, and then solve the equation for $v(t)$.

∗∗ **5.** A body of weight W falls from rest under gravity. It is acted on by a resistive force that is proportional to the square root of the speed at any instant. If the magnitude of this resistive force is F^* when the speed is v^*, find an expression for its terminal speed.

∗∗ **6.** A spring (with constant k) is attached on one end to a wall and on the other end to a mass M (figure on page 1280). The mass is set into motion along the x-axis by pulling it a distance x_0 to the right of the position it would occupy were the spring unstretched and given speed v_0 to the left. During the subsequent motion, there is a frictional force between M and the horizontal surface with coefficient of kinetic friction equal to μ.

(a) Show that the differential equation describing the motion of M is

$$M\frac{d^2x}{dt^2} = -kx + \mu Mg, \quad x(0) = x_0, \quad v(0) = -v_0,$$

if we take $t = 0$ at the instant that motion is initiated. When is this equation valid?

(b) Since t is explicitly missing from the equation in (a), show that it can be rewritten in the form

$$Mv\frac{dv}{dx} = -kx + \mu Mg, \quad v(x_0) = -v_0,$$

and that therefore

$$\frac{k}{2}(x_0^2 - x^2) = \frac{M}{2}(v^2 - v_0^2) + \mu Mg(x_0 - x).$$

Interpret each of the terms in this equation physically.

(c) If x^* represents the position at which M comes to rest for the first time, use the equation in (b) to determine x^* as a function of μ, M, g, k, x_0, and v_0. Discuss the possibilities of x^* being positive, negative, and zero.

** **7.** When a body falls in air, it is acted on by gravity and also by a force due to air resistance that is proportional to its instantaneous speed.

(a) If the body is initially projected upward, find its velocity as a function of time, and draw a graph of the function.

(b) Repeat (a) if the body is projected downward with velocity less than its terminal velocity.

(c) Repeat (b) if the initial velocity is greater than its terminal velocity.

** **8.** A 1-kg rock is thrown vertically upward with speed 20 m/s. Air resistance to its motion when measured in newtons has magnitude equal to one-tenth its speed in metres per second. Find the maximum height attained by the rock.

** **9.** A stone of mass 100 g is thrown vertically upward with speed 20 m/s. Air exerts a resistive force on the stone proportional to its instantaneous speed, and has magnitude $1/10$ N when the speed of the stone is 10 m/s. Find an equation defining the time when the stone returns to its projection point. Solve this equation numerically, and compare the result with the time taken if air resistance is neglected.

** **10.**   (a) A 1-kg mass falls under the influence of gravity. It is also acted on by air resistance proportional to the square of its velocity and is 5 N when its velocity is 50 m/s. If the velocity of the mass has magnitude 20 m/s at time $t = 0$, find a formula for its velocity as a function of time.

(b) Plot a graph of the velocity function. Is there a terminal velocity?

** **11.** Repeat Exercise 10 if the initial velocity has magnitude 100 m/s.

** **12.** For how long does the mass in Exercise 10 rise if it is thrown upward with velocity 20 m/s?

** **13.** A mass m falls under the influence of gravity and air resistance which is proportional to the square of velocity. If the speed of the mass is v_0 at time $t = 0$, find a formula for its speed as a function of time.

** **14.** (a) If the mass in Exercise 13 is thrown upward with speed v_0, find a formula for speed on its ascent. Is this formula also valid for its speed when it begins to fall?

(b) Find a formula for its height. How high does it rise?

** **15.** A mass m slides from rest down a frictionless plane inclined at angle α to the horizontal. Find a formula for the time taken to travel a distance D down the plane. What is its speed at this time?

** **16.** Find formulas for speed and distance travelled for the mass in Exercise 15 if air resistance proportional to velocity also acts on the mass.

** **17.** In Example 7.35 of Section 7.10 we derived the escape velocity of a projectile from the earth's surface based on energy principles. In this exercise we obtain the same result using differential equations. When a projectile of mass m, fired from the earth's surface, is a distance r from the centre of the earth, the magnitude of the force of attraction on it is given by Newton's universal law of gravitation, $F = GMm/r^2$, where M is the mass of the earth and G is a constant. Use Newton's second law and a substitution corresponding to 16.30 to find the velocity of the projectile as a function of r. What minimum initial velocity guarantees that the projectile escapes the gravitational field of the earth?

*** **18.** A huge cannon fires a projectile with initial velocity v_0 directly toward the moon (figure below). When the projectile is a distance r above the earth's surface, the force of attraction of the earth on the projectile has magnitude

$$\frac{GMm}{(r + R)^2},$$

where G is a constant, R is the radius of the earth, M is the mass of the earth, and m is the mass of the projectile. At this point the moon's gravitational attraction has magnitude

$$\frac{GM^*m}{(a + R^* - r)^2},$$

where M^* is the mass of the moon.

(a) Show that if only the above two forces are considered to act on the projectile, then the differential equation describing its motion is

$$\frac{d^2r}{dt^2} = -\frac{gR^2}{(r + R)^2} + \frac{g^*R^{*2}}{(a + R^* - r)^2},$$

where g and g^* are gravitational accelerations on the surfaces of the earth and the moon.

(b) Prove that the velocity of the projectile at a distance r above the surface of the earth is defined by

$$v^2 = \frac{2gR^2}{r + R} + \frac{2g^*R^{*2}}{a + R^* - r} + v_0^2 - 2gR - \frac{2g^*R^{*2}}{a + R^*}.$$

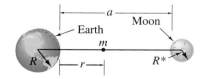

✳✳✳ **19.** Newton's second law states that if an object of variable mass $m(t)$ is subjected to a force $\mathbf{F}(t)$, then

$$\frac{d}{dt}(m\mathbf{v}) = \mathbf{F},$$

where t is time and $\mathbf{v}$ is the velocity of the object. A uniform chain of length 3 m and mass 6 kg is held by one end so that the other end just touches the floor. If the chain is released, find the velocity of the falling chain as a function of the length of chain still falling. How fast does the end hit the floor?

16.6 Linear Differential Equations

Throughout Chapters 5, 8, and 9 and Sections 16.1–16.5, we have stressed the use of differential equations in solving applied problems. We have considered examples from such diverse areas as engineering, geometry, ecology, and psychology, hoping thereby to illustrate how valuable differential equations can be in modelling situations mathematically. Perhaps the most important type of differential equation is the **linear differential equation**.

> **DEFINITION 16.4**
>
> A differential equation of the form
>
> $$a_0(x)\frac{d^n y}{dx^n} + a_1(x)\frac{d^{n-1}y}{dx^{n-1}} + a_2(x)\frac{d^{n-2}y}{dx^{n-2}} + \cdots + a_{n-1}(x)\frac{dy}{dx} + a_n(x)y = F(x)$$
>
> is called a **linear differential equation**. It is n^{th}-order when $a_0(x) \not\equiv 0$.　　(16.38)

Note in particular that none of the derivatives of $y(x)$ are multiplied together, nor are they squared or cubed or taken to any other power, nor do they appear as the argument of any transcendental function. All we see is a function of x multiplying y, plus a function of x multiplying the first derivative of y, plus a function of x multiplying the second derivative of y, and so on.

If $n = 1$, equation 16.38 reduces to

$$a_0(x)\frac{dy}{dx} + a_1(x)y = F(x),$$

and at any point at which $a_0(x) \neq 0$, we can divide to obtain

$$\frac{dy}{dx} + \frac{a_1(x)}{a_0(x)}y = \frac{F(x)}{a_0(x)}.$$

If we set $P(x) = a_1(x)/a_0(x)$ and $Q(x) = F(x)/a_0(x)$, we have

$$\frac{dy}{dx} + P(x)y = Q(x);$$

i.e., every linear first-order differential equation can be expressed in this form. We discussed equations of this type in Section 16.3, where it was shown that such equations have a general solution

$$y(x) = e^{-\int P(x)dx}\left\{\int Q(x)e^{\int P(x)dx}dx + C\right\}.$$

In other words, we already know how to solve linear first-order differential equations, and therefore our discussion in the next five sections is directed primarily at second- and higher-order equations. Keep in mind, however, that all results are also valid for first-order linear equations.

Because equation 16.38 is so cumbersome, we introduce notation to simplify its representation. In particular, if we use the notation $D = d/dx$, $D^2 = d^2/dx^2$, etc., we can write

$$a_0(x)D^n y + a_1(x)D^{n-1}y + a_2(x)D^{n-2}y + \cdots + a_{n-1}(x)Dy + a_n(x)y = F(x), \quad (16.39)$$

or

$$\{a_0(x)D^n + a_1(x)D^{n-1} + a_2(x)D^{n-2} + \cdots + a_{n-1}(x)D + a_n(x)\}y = F(x). \quad (16.40)$$

The quantity in braces is called a **differential operator**; it operates on whatever follows it—in this case, y. It is a "differential" operator because it operates by taking derivatives. Because the operator involves only x's and D's, we denote it by

$$\phi(x, D) = a_0(x)D^n + a_1(x)D^{n-1} + a_2(x)D^{n-2} + \cdots + a_{n-1}(x)D + a_n(x). \quad (16.41)$$

The general linear n^{th}-order differential equation can then be represented very simply by

$$\phi(x, D)y = F(x). \quad (16.42)$$

For example, differential equation 16.3,

$$xy'' + y' + xy = 0$$

is called Bessel's differential equation of order zero. In operator notation we write

$$(xD^2 + D + x)y = 0,$$

or

$$\phi(x, D)y = 0,$$

where $\phi(x, D) = xD^2 + D + x$.

For the differential equation

$$y'' + 2y' - 3y = e^{-x},$$

we write

$$\phi(D)y = e^{-x},$$

where $\phi(D) = D^2 + 2D - 3$.

We now indicate the meaning of the term "linear." Suppose that L is an operator that operates on each function $y(x)$ in some set S. For example, L might be the operation that multiplies each function by 5, or perhaps squares each function, or perhaps differentiates each function. It is said to be a linear operator if it satisfies the following definition.

DEFINITION 16.5

An operator L is said to be **linear** on a set of functions S if for any two functions $y_1(x)$ and $y_2(x)$ in S, and any constant c,

$$L(y_1 + y_2) = Ly_1 + Ly_2, \tag{16.43a}$$

$$L(cy_1) = c(Ly_1). \tag{16.43b}$$

Many of the operations in calculus are therefore linear. For instance, taking limits is a linear operation, as is differentiation, antidifferentiation, and taking definite integrals. On the other hand, taking the square root of a positive function is not a linear operation, since

$$L(y_1 + y_2) = \sqrt{y_1 + y_2} \neq L(y_1) + L(y_2).$$

It is not difficult to show that the differential operator $\phi(x, D)$ in 16.42 is linear; that is,

$$\phi(x, D)(y_1 + y_2) = \phi(x, D)y_1 + \phi(x, D)y_2,$$

$$\phi(x, D)(cy_1) = c[\phi(x, D)y_1],$$

and because of this, differential equation 16.42 is also said to be linear. The following two differential equations are not linear. We say that they are nonlinear.

$$\frac{d^2y}{dx^2} + y^2 = x, \qquad \frac{d^2y}{dx^2}\frac{dy}{dx} = e^x.$$

In particular, if we substitute $y_1(x) + y_2(x)$ into the left side of the first equation, we obtain

$$\frac{d^2}{dx^2}(y_1 + y_2) + (y_1 + y_2)^2 = \frac{d^2y_1}{dx^2} + \frac{d^2y_2}{dx^2} + y_1^2 + 2y_1y_2 + y_2^2.$$

If we substitute $y_1(x)$, then $y_2(x)$, and then add the results, we find a different expression:

$$\frac{d^2y_1}{dx^2} + y_1^2 + \frac{d^2y_2}{dx^2} + y_2^2.$$

Unless otherwise indicated, we assume that coefficient functions $a_0(x), a_1(x), \ldots, a_n(x)$ and the function $F(x)$ in linear differential equation 16.38 are all continuous on some open interval I. We also assume that $a_0(x) \neq 0$ at any point I. Solutions of the differential equation must necessarily have derivatives of orders up to and including n at each pont in I. But existence of the nth derivative implies continuity of all lower order derivatives. Furthermore, we can write 16.38 in the form

$$\frac{d^ny}{dx^n} = \frac{1}{a_0(x)}\left[F(x) - a_1(x)\frac{d^{n-1}y}{dx^{n-1}} - \cdots - a_n(x)y\right],$$

where all functions on the right are continuous and $a_0(x) \neq 0$. Hence, when derivatives of orders up to and including $n - 1$ are continuous on I, so also is the nth derivative d^ny/dx^n. Thus, solutions must have continuous derivates of orders up to and including n.

EXERCISES 16.6

In Exercises 1–10 prove either that the operator L is linear or that it is not linear. In each case assume that the set S of functions on which L operates is the set of all functions on which L can operate. For instance, in Exercise 6, assume that S is the set of all functions $y(x)$ that have a first derivative dy/dx.

* **1.** L multiplies functions $y(x)$ by 5

* **2.** L multiplies functions $y(x)$ by $15x$

* **3.** L adds the fixed function $z(x)$ to functions $y(x)$

* **4.** L takes the limit of functions $y(x)$ as x approaches 3

* **5.** L takes the limit of functions $y(x)$ as x approaches infinity

* **6.** L takes the first derivative of functions $y(x)$ with respect to x

* **7.** L takes the third derivative of functions $y(x)$ with respect to x

* **8.** L takes the antiderivative of functions $y(x)$ with respect to x

* **9.** L takes the definite integral of functions $y(x)$ with respect to x from $x = -1$ to $x = 4$

* **10.** L takes the cube root of functions $y(x)$

In Exercises 11–20 determine whether the differential equation is linear or nonlinear. Write those equations that are linear in operator notation 16.42.

* **11.** $2x\dfrac{d^2y}{dx^2} + x^3y = x^2 + 5$ * **12.** $2x\dfrac{d^2y}{dx^2} + x^3y = x^2 + 5y$

* **13.** $2x\dfrac{d^2y}{dx^2} + x^3y = x^2 + 5y^2$ * **14.** $x\dfrac{d^3y}{dx^3} + 3x\dfrac{d^2y}{dx^2} - 2\dfrac{dy}{dx} + y = 10\sin x$

* **15.** $x\dfrac{d^3y}{dx^3} + 3x\dfrac{d^2y}{dx^2} - 2\dfrac{dy}{dx} + y^2 = 10\sin x$

* **16.** $y\dfrac{d^3y}{dx^3} + 3x\dfrac{d^2y}{dx^2} - 2\dfrac{dy}{dx} + y = 10\sin x$

* **17.** $y'' - 3y' - 2y = 9\sec^2 x$ * **18.** $yy'' + 3y' - 2y = e^x$

* **19.** $\sqrt{1 + y'} + x^2 = 4$ * **20.** $y'''' + y'' - y = \ln x$

** **21.** The Laplace transform of a function $y(t)$ is defined as the function $L(y) = \displaystyle\int_0^\infty e^{-st}y(t)\,dt$, provided the improper integral converges. Show that if S is the set of all functions that have a Laplace transform, then the operation of taking the Laplace transform is linear on S.

** **22.** The finite Fourier cosine transform of a function $y(x)$ is defined as $L(y) = \displaystyle\int_0^{2\pi} y(x)$ $\cos nx\,dx$, where n is a nonnegative integer, provided the definite integral exists. Show that if S is the set of all functions that have a finite Fourier cosine transform, then the operation of taking the transform is linear on S.

16.7 Homogeneous Linear Differential Equations

Two classes of linear differential equations present themselves: those for which $F(x) \equiv 0$, and those for which $F(x) \not\equiv 0$. In this section and the next we consider equations for which $F(x) \equiv 0$; we discuss the more difficult class, in which $F(x) \not\equiv 0$, in Section 16.9. First let us name each of these classes of linear differential equations.

DEFINITION 16.6
A linear differential equation $\phi(x, D)y = F(x)$ is said to be **homogeneous** if $F(x) \equiv 0$, and **nonhomogeneous** otherwise.

The meaning of "homogeneous" to describe a property of linear differential equation in this definition is totally different from the meaning in Exercise 16.2:26.

The fundamental idea behind the solution of all linear differential equations is the following theorem.

THEOREM 16.1 (Superposition principle)
If $y_1(x), y_2(x), \ldots, y_m(x)$ are solutions of a homogeneous linear differential equation

$$\phi(x, D)y = 0,$$

on an interval I, then so too is any linear combination thereof,

$$C_1 y_1(x) + C_2 y_2(x) + \cdots + C_m y_m(x)$$

(for arbitrary constants $C_1, C_2, \ldots, C_m$).

PROOF The proof requires only linearity of the operator $\phi(x, D)$, for

$$\phi(x, D)[C_1 y_1(x) + \cdots + C_m y_m(x)] = C_1[\phi(x, D)y_1] + \cdots + C_m[\phi(x, D)y_m]$$
$$= 0 + 0 + \cdots + 0 = 0. \qquad \blacksquare$$

Solutions of a linear differential equation that are linearly combined to produce other solutions are said to be superposed—consequently, the name "superposition principle" for Theorem 16.1. For example, it is straightforward to verify that $y_1(x) = e^{4x}$ and $y_2(x) = e^{-3x}$ are solutions of the homogeneous equation $y'' - y' + 12y = 0$. The superposition principle then states that for any constants C_1 and C_2, the function $y(x) = C_1 y_1(x) + C_2 y_2(x) = C_1 e^{4x} + C_2 e^{-3x}$ must also be a solution. It is a two-parameter family of solutions.

Similarly, superposition of the three solutions e^x, xe^x, and $x^2 e^x$ of the linear differential equation $y''' - 3y'' + 3y' - y = 0$ gives a three-parameter family of solutions $y(x) = C_1 e^x + C_2 xe^x + C_3 x^2 e^x$.

Now we begin to see the importance of the superposition principle. If we can find n solutions $y_1(x), y_2(x), \ldots, y_n(x)$ of an nth-order homogeneous linear differential equation, then an n-parameter family of solutions is

$$y(x) = C_1 y_1(x) + C_2 y_2(x) + \cdots + C_n y_n(x).$$

In other words, all that we need do is find n solutions; the superposition principle will do the rest. There is a problem, however, if we take things a little too literally. For instance, $y_1(x) = e^{4x}$ and $y_2(x) = 10e^{4x}$ are both solutions of $y'' - y' + 12y = 0$. By superposition, so too then is $y(x) = C_1 y_1 + C_2 y_2 = C_1 e^{4x} + 10 C_2 e^{4x}$. But is it a two-parameter family of solutions? The answer is no, because we could write $y(x) = (C_1 + 10 C_2)e^{4x}$, and by setting $C_3 = C_1 + 10 C_2$, we have $y(x) = C_3 e^{4x}$. Superposition of the solutions e^{4x} and $10e^{4x}$ has not therefore led to a two-parameter family of solutions, and the reason is that they are essentially the same solution: $y_2(x)$ is $y_1(x)$ multiplied by a constant. Superposition does not therefore lead to a solution with two arbitrary constants.

In a similar way, $y_1(x) = e^x$, $y_2(x) = xe^x$, and $y_3(x) = 2e^x - 3xe^x$ are all solutions of $y''' - 3y'' + 3y' - y = 0$, and therefore so is $y(x) = C_1 y_1 + C_2 y_2 + C_3 y_3$. But because we can write

$$y(x) = C_1 e^x + C_2 x e^x + C_3 (2e^x - 3xe^x)$$

$$= (C_1 + 2C_3)e^x + (C_2 - 3C_3)xe^x$$

$$= C_4 e^x + C_5 x e^x,$$

$y(x)$ is not a three-parameter family of solutions; it is only a two-parameter family. This is a direct result of the fact that $y_3(x)$ is a linear combination of the solutions $y_1(x)$ and $y_2(x)$; it is twice $y_1(x)$ minus three times $y_2(x)$.

Our problem seems to come down to this: If we have n solutions of an nth-order homogeneous linear differential equation, how can we determine whether superposition leads to a solution that contains n arbitrary constants? Our examples have suggested that if any one of the solutions is a linear combination of the others, then an n-parameter family of solutions is not obtained, and this is indeed true. If one of the solutions is a linear combination of the others, we say that the n solutions are **linearly dependent**; if no solution is a linear combination of the others, we say that the n solutions are **linearly independent**. We summarize these results in the following theorem.

THEOREM 16.2

If $y_1(x)$, $y_2(x)$, ..., $y_n(x)$ are n linearly independent solutions of an nth-order homogeneous linear differential equation on an interval I, then $y(x) = C_1 y_1(x) + C_2 y_2(x) + \cdots + C_n y_n(x)$ is an n-parameter family of solutions of the differential equation on I.

What we should now do is devise a test to determine whether a set of n solutions is linearly independent or linearly dependent. For most examples, no test is really necessary; it is obvious whether one of the solutions can be written as a linear combination of the others. For those rare occasions when it is not obvious, Exercise 10 describes a test that can be used to determine whether functions are linearly independent.

We pointed out in Section 16.1 that an n-parameter family of solutions for an nth-order differential equation might not contain all solutions of the equation, and might not therefore be a general solution of the differential equation. This is not the case for linear n^{th}-order differential equations. It can be shown that, if $y(x)$ is an n-parameter family of solutions for a linear n^{th}-order differential equation on an interval I, then every solution of the differential equation on I can be obtained by specifying particular values for the arbitrary constants in $y(x)$. Here then is a very important class of differential equations for which an n-parameter family of solutions is always a general solution. Let us state this as a corollary to Theorem 16.2.

COROLLARY

If $y_1(x)$, $y_2(x)$, ..., $y_n(x)$ are n linearly independent solutions of an nth-order homogeneous linear differential equation on an interval I, then $y(x) = C_1 y_1(x) + C_2 y_2(x) + \cdots + C_n y_n(x)$ is a general solution of the differential equation on I.

In summary, the superposition principle states that solutions of a homogeneous linear differential equation can be superposed to produce other solutions. If n linearly independent solutions of an nth-order equation are superposed, a general solution is obtained. This is the importance of the superposition principle. We need not devise a method that takes us directly to a general solution; we need a method for finding n linearly independent solutions—superposition does the rest. Unfortunately, for completely general coefficients $a_i(x)$ in $\phi(x, D)$ (see equation 16.41), it is impossible to give a method that always yields n linearly independent solutions of a homogeneous equation. There is, however, one special case of great practical importance in which it is always possible to produce n linearly independent solutions in a very simple way. This special case occurs when the coefficients $a_i(x)$ are all constants a_i, and this is the subject of Section 16.8.

EXAMPLE 16.11

If $y_1(x) = \cos 3x$ and $y_2(x) = \sin 3x$ are solutions of the differential equation $y'' + 9y = 0$, find a general solution.

SOLUTION   Since $y_1(x)$ and $y_2(x)$ are linearly independent solutions (one is not a constant times the other), a general solution can be obtained by superposition:

$$y(x) = C_1 \cos 3x + C_2 \sin 3x.$$

EXAMPLE 16.12

Given that $y_1(x) = e^{2x} \cos x$ and $y_2(x) = e^{2x} \sin x$ are solutions of the homogeneous linear differential equation $y'' - 4y' + 5y = 0$, find that solution which satisfies the conditions $y(\pi/4) = 1$, $y(\pi/3) = 2$.

SOLUTION   By superposition, a general solution of the differential equation is

$$y(x) = C_1 e^{2x} \cos x + C_2 e^{2x} \sin x = e^{2x}(C_1 \cos x + C_2 \sin x).$$

To satisfy the conditions $y(\pi/4) = 1$ and $y(\pi/3) = 2$, we have

$$1 = e^{\pi/2}(C_1/\sqrt{2} + C_2/\sqrt{2}), \quad 2 = e^{2\pi/3}(C_1/2 + \sqrt{3}C_2/2).$$

Thus C_1 and C_2 are defined by the pair of equations

$$C_1 + C_2 = \sqrt{2}e^{-\pi/2}, \quad C_1 + \sqrt{3}C_2 = 4e^{-2\pi/3},$$

the solution of which is

$$C_2 = \frac{4e^{-2\pi/3} - \sqrt{2}e^{-\pi/2}}{\sqrt{3} - 1}, \quad C_1 = \sqrt{2}e^{-\pi/2} - C_2.$$

The required solution is therefore $y(x) = e^{2x}(C_1 \cos x + C_2 \sin x)$ with these values for C_1 and C_2.

In Exercises 1–8 show that the functions are solutions of the differential equation. Check that the differential equation is linear and homogeneous, and then find a general solution.

* **1.** $y'' + y' - 6y = 0$; $y_1(x) = e^{2x}$, $y_2(x) = e^{-3x}$

* **2.** $y' + y \tan x = 0$; $y_1(x) = \cos x$

* **3.** $y'''' + 5y'' + 4y = 0$; $y_1(x) = \cos 2x$, $y_2(x) = \sin 2x$, $y_3(x) = \cos x$, $y_4(x) = \sin x$

* **4.** $2y'' - 16y' + 32y = 0$; $y_1(x) = 3e^{4x}$, $y_2(x) = -2xe^{4x}$

* **5.** $y''' - 3y'' + 2y' = 0$; $y_1(x) = 10$, $y_2(x) = 3e^x$, $y_3(x) = 4e^{2x}$

* **6.** $2y'' - 8y' + 9y = 0$; $y_1(x) = e^{2x} \cos(x/\sqrt{2})$, $y_2(x) = e^{2x} \sin(x/\sqrt{2})$

* **7.** $x^2 y'' + xy' + (x^2 - 1/4)y = 0$; $y_1(x) = (\sin x)/\sqrt{x}$, $y_2(x) = (\cos x)/\sqrt{x}$

* **8.** $x^2 y'' + xy' + 4y = 0$; $y_1(x) = \cos(2 \ln x)$, $y_2(x) = \sin(2 \ln x)$

* **9.** Show that $y_1(x) = -2/(x+1)$ and $y_2(x) = -2/(x+2)$ are both solutions of the differential equation $y'' = yy'$. Is $y(x) = y_1(x) + y_2(x)$ a solution? Explain.

** **10.** We stated in this section that n functions $y_1(x), \ldots, y_n(x)$ are linearly dependent if at least one of the functions can be expressed as a linear combination of the others; they are linearly independent if none of the functions is a linear combination of the others. Another way of saying this is as follows: Functions $y_1(x), \ldots, y_n(x)$ are linearly dependent on an interval I if there exist constants $C_1, \ldots, C_n$, not all zero, such that on I

$$C_1 y_1(x) + \cdots + C_n y_n(x) \equiv 0.$$

If this equation can be satisfied only with $C_1 = C_2 = \cdots = C_n = 0$, the functions are linearly independent. In this exercise we give a test to determine whether functions are linearly independent or dependent. If $y_1(x), \ldots, y_n(x)$ have derivatives up to and including order $n - 1$ on the interval I, we define the Wronskian of the functions as the $n \times n$ determinant:

$$W(y_1, \ldots, y_n) = \begin{vmatrix} y_1 & y_2 & \cdots & y_n \\ y_1' & y_2' & \cdots & y_n' \\ y_1'' & y_2'' & \cdots & y_n'' \\ \vdots & \vdots & \ddots & \vdots \\ y_1^{(n-1)} & y_2^{(n-1)} & \cdots & y_n^{(n-1)} \end{vmatrix}.$$

Show that if $y_1, \ldots, y_n$ are linearly dependent on I, then $W(y_1, \ldots, y_n) \equiv 0$ on I. It follows then that if there exists at least one point in I at which $W(y_1, \ldots, y_n) \neq 0$, the functions $y_1, \ldots, y_n$ are linearly independent on I.

In Exercises 11–15 use the method of Exercise 10 to determine whether the functions are linearly dependent or independent on the interval.

* **11.** $\{1, x, x^2\}$ on $-\infty < x < \infty$

* **12.** $\{x, 2x - 3x^2, x^2\}$ on $-\infty < x < \infty$

* **13.** $\{\sin x, \cos x\}$ on $0 \le x \le 2\pi$

* **14.** $\{x, xe^x, x^2 e^x\}$ on $0 \le x \le 1$

* **15.** $\{x \sin x, e^{2x}\}$ on $-\infty < x < \infty$

16.8 Homogeneous Linear Differential Equations with Constant Coefficients

We now consider homogeneous linear differential equations

$$a_0 D^n y + a_1 D^{n-1} y + a_2 D^{n-2} y + \cdots + a_{n-1} Dy + a_n y = 0, \tag{16.44}$$

where the coefficients $a_0, a_1, \ldots, a_n$ are all constants. In operator notation we write

$$\phi(D)y = 0, \tag{16.45a}$$

where

$$\phi(D) = a_0 D^n + a_1 D^{n-1} + \cdots + a_{n-1} D + a_n. \tag{16.45b}$$

The superposition principle states that a general solution of equation 16.44 is $y(x) = C_1 y_1(x) + \cdots + C_n y_n(x)$, provided $y_1(x), \ldots, y_n(x)$ are any n linearly independent solutions of the equation. Our problem then is to devise a technique for finding n linearly independent solutions; to illustrate a possible procedure, we first consider three second-order equations. The first is

$$y'' + 2y' - 3y = 0.$$

It is not unreasonable to expect that, for some value of m, the function $y(x) = e^{mx}$ might be a solution of this equation. After all, the equation says that the second derivative of the function must be equal to three times the function minus twice its first derivative. Since the exponential function reproduces itself when differentiated, perhaps m can be chosen to produce this combination. To see whether this is possible, we substitute $y = e^{mx}$ into the differential equation, and find that if $y = e^{mx}$ is to be a solution, then

$$m^2 e^{mx} + 2m e^{mx} - 3e^{mx} = 0;$$

this implies that

$$0 = m^2 + 2m - 3 = (m + 3)(m - 1).$$

Thus $y = e^{mx}$ is a solution if m is chosen as either 1 or -3; i.e., $y_1 = e^x$ and $y_2 = e^{-3x}$ are solutions of the differential equation. Since they are linearly independent, a general solution is $y(x) = C_1 e^x + C_2 e^{-3x}$.

For our second example, we take

$$y'' + 2y' + y = 0.$$

Since exponentials worked in the first example, we once again try a solution of the form $y(x) = e^{mx}$. If we substitute into the differential equation, we obtain

$$m^2 e^{mx} + 2m e^{mx} + e^{mx} = 0,$$

which implies that

$$0 = m^2 + 2m + 1 = (m + 1)^2.$$

Thus $y_1 = e^{-x}$ is a solution, but unfortunately it is the only solution that we obtain as a result of our guess. We need a second linearly independent solution $y_2(x)$ in order to obtain a general solution. Clearly, no other exponential will work. Perhaps if we multiplied e^{-x} by another function, we might find a second solution; in other words, perhaps there is a solution of the form $y(x) = v(x)e^{-x}$ for some $v(x)$. To see, we again substitute into the differential equation,

$$0 = \{v''e^{-x} - 2v'e^{-x} + ve^{-x}\} + 2\{v'e^{-x} - ve^{-x}\} + ve^{-x} = v''e^{-x}.$$

Consequently, $v'' = 0$, and this implies that $v(x) = Ax + B$, for any constants A and B. In particular, if $A = 1$ and $B = 0$, $v(x) = x$, and $y_2(x) = v(x)e^{-x} = xe^{-x}$ is also a solution of the differential equation. By superposition, then, we find that a general solution is $y(x) = C_1e^{-x} + C_2xe^{-x} = (C_1 + C_2x)e^{-x}$. Note that if we had set $A = 0$ and $B = 1$, then $v(x) = 1$, and the solution $y(x) = v(x)e^{-x}$ would have been $y_1(x)$. Further, if we had simply set $y(x) = v(x)e^{-x} = (Ax + B)e^{-x}$, we would have the general solution.

Our third example is

$$y'' + 2y' + 10y = 0.$$

As in the previous two examples, if we assume a solution $y = e^{mx}$, then

$$m^2e^{mx} + 2me^{mx} + 10e^{mx} = 0,$$

or

$$0 = m^2 + 2m + 10.$$

The solutions of this quadratic equation are the complex numbers

$$m = \frac{-2 \pm \sqrt{4 - 40}}{2} = -1 \pm 3i.$$

(A brief introduction to complex numbers can be found in Appendix C.) Because m is complex, no real exponential $y = e^{mx}$ satisfies the differential equation. If, however, we form complex exponentials $y_1(x) = e^{(-1+3i)x}$ and $y_2(x) = e^{(-1-3i)x}$, and superpose these solutions, then

$$y(x) = Ae^{(-1+3i)x} + Be^{(-1-3i)x},$$

and this must also be a solution. When we use Euler's identity for complex exponentials, $e^{i\theta} = \cos\theta + i\sin\theta$, we can write $y(x)$ in the form

$$y(x) = Ae^{-x}e^{3xi} + Be^{-x}e^{-3xi}$$

$$= Ae^{-x}(\cos 3x + i\sin 3x) + Be^{-x}(\cos 3x - i\sin 3x)$$

$$= e^{-x}[(A + B)\cos 3x + i(A - B)\sin 3x]$$

$$= e^{-x}(C_1\cos 3x + C_2\sin 3x),$$

where $C_1 = A+B$ and $C_2 = i(A-B)$. In other words, the function $y(x) = e^{-x}(C_1\cos 3x + C_2\sin 3x)$ is a general solution of the differential equation, and it has been derived from the complex roots $m = -1 \pm 3i$ of the equation $m^2 + 2m + 10 = 0$. Note that what multiplies x in the exponential is the real part of these complex numbers, and what multiplies x in the trigonometric functions is the imaginary part. In Exercise 17 we show that this solution can also be derived without complex numbers, but we feel that in general the use of complex numbers is the best method.

In each of these examples we guessed $y = e^{mx}$ as a possible solution. We then substituted into the differential equation to obtain an algebraic equation for m. In each case the equation for m was

$$\phi(m) = 0;$$

that is, take the operator $\phi(D)$, replace D by m, and set the polynomial equal to zero. This is not a peculiarity of these examples for it is straightforward to show that for any homogeneous linear equation 16.45, if we assume a solution of the form $y = e^{mx}$, then m must satisfy the equation $\phi(m) = 0$. We name this equation in the following definition.

DEFINITION 16.7

With every linear differential equation that has constant coefficients $\phi(D)y = F(x)$, we associate an equation

$$\phi(m) = 0 \tag{16.46}$$

called the **auxiliary equation**.

To summarize, in each of the examples we assumed a solution $y = e^{mx}$ and found that m had to satisfy the auxiliary equation $\phi(m) = 0$. From the roots of the auxiliary equation we obtained solutions of the differential equation, and superposition then led to a general solution. This procedure works on every homogeneous linear differential equation with constant coefficients. But if the procedure is the same in every case, surely we can set down rules that eliminate the necessity of tediously repeating these steps in every example. This we do in the following theorem.

THEOREM 16.3

If $\phi(m) = 0$ is the auxiliary equation associated with the homogeneous linear differential equation $\phi(D)y = 0$, then there are two possibilities:

(i) $\phi(m) = 0$ has a real root m of multiplicity k. Then a solution of the differential equation is

$$(C_1 + C_2 x + \cdots + C_k x^{k-1})e^{mx}. \tag{16.47a}$$

(ii) $\phi(m) = 0$ has a pair of complex conjugate roots $a \pm bi$ each of multiplicity k. Then a solution of the differential equation is

$$e^{ax}[(C_1 + C_2 x + \cdots + C_k x^{k-1})\cos bx$$
$$+ (D_1 + D_2 x + \cdots + D_k x^{k-1})\sin bx]. \tag{16.47b}$$

A general solution of the differential equation is obtained by superposing all solutions in (i) and (ii).

For a proof of this theorem see Exercise 20. Let us now apply the theorem to our previous examples. The auxiliary equation for $y'' + 2y' - 3y = 0$ is

$$0 = m^2 + 2m - 3 = (m + 3)(m - 1)$$

with solutions $m = 1$ and $m = -3$. If we now use part (i) of Theorem 16.3 with two real roots, each of multiplicity 1, a general solution of the differential equation is

$$y(x) = C_1 e^x + C_2 e^{-3x}.$$

The auxiliary equation for $y'' + 2y' + y = 0$ is

$$0 = m^2 + 2m + 1 = (m + 1)^2$$

with solutions $m = -1$ and $m = 1$. Part (i) of Theorem 16.3 with a single real root of multiplicity 2 gives the general solution

$$y(x) = (C_1 + C_2 x)e^{-x}.$$

The auxiliary equation for $y'' + 2y' + 10y = 0$ is

$$0 = m^2 + 2m + 10$$

with solutions $m = -1 \pm 3i$. Part (ii) of Theorem 16.3 with a pair of complex conjugate roots, each of multiplicity 1, gives the general solution

$$y(x) = e^{-x}(C_1 \cos 3x + C_2 \sin 3x).$$

EXAMPLE 16.13

Find a general solution for the differential equation

$$y''' - 3y'' - 4y' + 6y = 0.$$

SOLUTION   The auxiliary equation is

$$0 = m^3 - 3m^2 - 4m + 6 = (m - 1)(m^2 - 2m - 6)$$

with solutions $m = 1$ and $m = 1 \pm \sqrt{7}$. A general solution of the differential equation is therefore

$$y(x) = C_1 e^x + C_2 e^{(1+\sqrt{7})x} + C_3 e^{(1-\sqrt{7})x}.$$

EXAMPLE 16.14

Find a general solution for $y''' - y = 0$.

SOLUTION   The auxiliary equation is

$$0 = m^3 - 1 = (m - 1)(m^2 + m + 1)$$

with solutions $m = 1$ and $m = -(1/2) \pm (\sqrt{3}/2)i$. A general solution of the differential equation is therefore

$$y(x) = C_1 e^x + e^{-x/2}[C_2 \cos(\sqrt{3}x/2) + C_3 \sin(\sqrt{3}x/2)].$$

EXAMPLE 16.15

If the roots of the auxiliary equation $\phi(m) = 0$ are

$$3, \ 3, \ 3, \ \pm 2i, \ -2, \ 1 \pm \sqrt{3}, \ -4 \pm i, \ -4 \pm i,$$

find a general solution of the differential equation $\phi(D)y = 0$.

SOLUTION A general solution is

$$y(x) = (C_1 + C_2 x + C_3 x^2)e^{3x} + C_4 \cos 2x + C_5 \sin 2x + C_6 e^{-2x}$$

$$+ C_7 e^{(1+\sqrt{3})x} + C_8 e^{(1-\sqrt{3})x} + e^{-4x}[(C_9 + C_{10}x) \cos x$$

$$+ (C_{11} + C_{12}x) \sin x].$$

EXERCISES 16.8

In Exercises 1–12 find a general solution for the homogeneous differential equation.

* **1.** $y'' + y' - 6y = 0$ * **2.** $2y'' - 16y' + 32y = 0$

* **3.** $2y'' + 16y' + 82y = 0$ * **4.** $y'' + 2y' - 2y = 0$

* **5.** $y'' - 4y' + 5y = 0$ * **6.** $y''' - 3y'' + y' - 3y = 0$

* **7.** $y'''' + 2y'' + y = 0$ * **8.** $y''' - 6y'' + 12y' - 8y = 0$

* **9.** $3y''' - 12y'' + 18y' - 12y = 0$ * **10.** $y'''' + 5y'' + 4y = 0$

* **11.** $y''' - 3y'' + 2y' = 0$ ** **12.** $y'''' + 16y = 0$

In Exercises 13–16 find a homogeneous linear differential equation that has the function as general solution.

** **13.** $y(x) = C_1 e^x + (C_2 + C_3 x)e^{-4x}$

** **14.** $y(x) = e^{-2x}(C_1 \cos 4x + C_2 \sin 4x)$

** **15.** $y(x) = C_1 + C_2 e^{\sqrt{3}x} + C_3 e^{-\sqrt{3}x}$

** **16.** $y(x) = e^x(C_1 + C_2 x) \cos \sqrt{2}x + e^x(C_3 + C_4 x) \sin \sqrt{2}x$

** **17.** Show that if we assume that $y(x) = e^{ax} \sin bx$ is a solution of the differential equation $y'' + 2y' + 10y = 0$, then a and b must be equal to -1 and ± 3 respectively. Verify that, for this a and b, $y(x) = e^{ax} \cos bx$ is also a solution, and therefore a general solution is $y(x) = e^{-x}(C_1 \cos 3x + C_2 \sin 3x)$.

** **18.** The equation $y''' + ay'' + by' + cy = 0$, where a, b, and c are constants, has solution $y(x) = C_1 e^{-x} + e^{-2x}(C_2 \sin 4x + C_3 \cos 4x)$. Find a, b, and c.

*** **19.** Show that if p is constant and $f(x)$ is differentiable, then

$$D\{e^{px} f(x)\} = e^{px}\{(D + p)f(x)\}.$$

Now use mathematical induction to prove that if $f(x)$ is k times differentiable, then

$$D^k\{e^{px} f(x)\} = e^{px}\{(D + p)^k f(x)\}.$$

Finally, verify that

$$\phi(D)\{e^{px}f(x)\} = e^{px}\{\phi(D+p)f(x)\},$$

a result called the *operator shift theorem*.

$***$ **20.**   (a) If m_0 is a real root of multiplicity k for the auxiliary equation $\phi(m) = 0$, show that the operator $\phi(D)$ can be expressed in the form

$$\phi(D) = (D - m_0)^k \psi(D),$$

where $\psi(D)$ is a polynomial in D. Now use the operator shift theorem of Exercise 19 to verify that $(C_1 + C_2 x + \cdots + C_k x^{k-1})e^{m_0 x}$ is a solution of $\phi(D)y = 0$.

(b) If $a \pm bi$ are complex conjugate roots each of multiplicity k for the auxiliary equation $\phi(m) = 0$, show that $\phi(D)$ can be expressed in the form

$$\phi(D) = (D - a - bi)^k (D - a + bi)^k \psi(D),$$

where $\psi(D)$ is a polynomial in D. Now use the operator shift theorem of Exercise 19 to verify that 16.47b is a solution of $\phi(D)y = 0$.

$***$ **21.** If M, β, and k are all positive constants, find a general solution for the linear differential equation

$$M\frac{d^2 x}{dt^2} + \beta \frac{dx}{dt} + kx = 0.$$

Discuss all possibilities.

16.9 Nonhomogeneous Linear Differential Equations with Constant Coefficients

The general nonhomogeneous linear differential equation with constant coefficients is

$$\phi(D)y = F(x), \tag{16.48a}$$

where

$$\phi(D) = a_0 D^n + a_1 D^{n-1} + \cdots + a_{n-1}D + a_n. \tag{16.48b}$$

It is natural to ask whether we can use the results of Section 16.8 concerning homogeneous equations with constant coefficients to solve nonhomogeneous problems. Fortunately the answer is yes, as shown by the following definition and theorem.

> **DEFINITION 16.8**
>
> With every nonhomogeneous linear differential equation with constant coefficients
>
> $$\phi(D)y = F(x),$$
>
> we associate a homogeneous equation
>
> $$\phi(D)y = 0, \tag{16.49}$$
>
> called the **homogeneous (reduced**, or **complementary)** equation associated with $\phi(D)y = F(x)$.

We now prove the following theorem.

> ### THEOREM 16.4
>
> A general solution of the linear differential equation $\phi(D)y = F(x)$ is $y(x) = y_h(x) +$ $y_p(x)$, where $y_h(x)$ is a general solution of the associated homogeneous equation, and $y_p(x)$ is any particular solution of the given equation.

PROOF   Since $\phi(D)$ is a linear operator,

$$\phi(D)(y_h + y_p) = \phi(D)y_h + \phi(D)y_p = 0 + F(x) = F(x),$$

so that $y_h + y_p$ is indeed a solution of the given differential equation. Because $y_h(x)$ is a general solution of the associated homogeneous equation, it contains the requisite number of arbitrary constants for $y(x)$ to be an n-parameter family of solutions of $\phi(D)y = F(x)$. What remains is to show that any (and every) solution $y_1(x)$ of $\phi(D)y = F(x)$ can be expressed in the form $y_h(x) + y_p(x)$ for some choice of the constants in $y_h(x)$. Consider the function $y(x)$ defined as the difference between $y_1(x)$ and $y_p(x)$, that is, $y(x) = y_1(x) - y_p(x)$. Because of linearity,

$$\phi(D)y = \phi(D)(y_1 - y_p) = \phi(D)y_1 - \phi(D)y_p = F(x) - F(x) = 0.$$

But every solution of the homogeneous equation $\phi(D)y = 0$ can be expressed in the form $y_h(x)$ for some choice of the constants; that is $y(x) = y_1(x) - y_p(x)$ can be expressed in the form

$$y_1(x) - y_p(x) = y_h(x)$$

for some choice of constants in $y_h(x)$. This completes the proof that $y_1(x)$ can be expressed in the form $y_h(x) + y_p(x)$, and therefore $y_h(x) + y_p(x)$ is a general solution of 16.48a.   ∎

We note in passing that Theorem 16.4 is also valid for linear differential equations with variable coefficients.

Theorem 16.4 indicates that discussions of nonhomogeneous differential equations can be divided into two parts. First, find a general solution $y_h(x)$ of the associated homogeneous equation 16.49, and this can be done using the results of Section 16.8. To this, add any particular solution $y_p(x)$ of 16.48. We present two methods for finding a particular solution: (1) the method of undetermined coefficients, and (2) the method of operators. Both methods apply in general only to differential equations in which $F(x)$ is a power (x^n, n a nonnegative integer), an exponential (e^{px}), a sine ($\sin px$), a cosine ($\cos px$), and/or any sums or products thereof.

Method of Undetermined Coefficients for a Particular Solution

As stated above, the method of undetermined coefficients is to be used only when $F(x)$ in equation 16.48 is of the form x^n, e^{px}, $\sin px$, $\cos px$, and/or sums or products thereof. For example, if

$$y'' + y' - 6y = e^{4x},$$

the method essentially says that Ae^{4x} is the simplest function that could conceivably yield e^{4x} when substituted into the left side of the differential equation. Consequently, it is natural to assume $y_p = Ae^{4x}$ and attempt to determine the unknown coefficient A. Substitution of this function into the differential equation gives

$$16Ae^{4x} + 4Ae^{4x} - 6Ae^{4x} = e^{4x}.$$

If we divide by e^{4x}, then

$$14A = 1 \quad \text{and} \quad A = \tfrac{1}{14}.$$

A particular solution is therefore $y_p = e^{4x}/14$.

Before stating a general rule, we illustrate a few more possibilities in the following example.

EXAMPLE 16.16

Find a particular solution of $y'' + y' - 6y = F(x)$ in each case.

(a) $F(x) = 6x^2 + 2x + 3$

(b) $F(x) = 2 \sin 2x$

(c) $F(x) = xe^{-x} - e^{-x}$

SOLUTION

(a) Since terms in x^2, x, and constants yield terms in x^2, x, and constants when substituted into the left side of the differential equation, we attempt to find a particular solution of the form $y_p = Ax^2 + Bx + C$. Substitution into the differential equation gives

$$(2A) + (2Ax + B) - 6(Ax^2 + Bx + C) = 6x^2 + 2x + 3,$$

or

$$(-6A)x^2 + (2A - 6B)x + (2A + B - 6C) = 6x^2 + 2x + 3.$$

But this equation can hold for all values of x only if coefficients of corresponding powers of x are identical (see Exercise 35 in Section 3.8). Equating coefficients then gives

$$-6A = 6,$$
$$2A - 6B = 2,$$
$$2A + B - 6C = 3.$$

These imply that $A = -1$, $B = -2/3$, $C = -17/18$, and

$$y_p = -x^2 - \frac{2x}{3} - \frac{17}{18}.$$

(b) Since terms in $\sin 2x$ and $\cos 2x$ yield terms in $\sin 2x$ when substituted into the left side of the differential equation, we assume that $y_p = A \sin 2x + B \cos 2x$. Substitution into the differential equation gives

$$(-4A \sin 2x - 4B \cos 2x) + (2A \cos 2x - 2B \sin 2x)$$
$$- 6(A \sin 2x + B \cos 2x) = 2 \sin 2x,$$

or

$$(-10A - 2B) \sin 2x + (2A - 10B) \cos 2x = 2 \sin 2x.$$

Equating coefficients of $\sin 2x$ and $\cos 2x$ gives

$$-10A - 2B = 2,$$
$$2A - 10B = 0.$$

These imply that $A = -5/26$, $B = -1/26$, and hence

$$y_p = -\frac{1}{26}(5\sin 2x + \cos 2x).$$

(c) Since terms in xe^{-x} and e^{-x} yield terms in xe^{-x} and e^{-x} when substituted into the left side of the differential equation, we assume that $y_p = Axe^{-x} + Be^{-x}$. Substitution into the differential equation gives

$$(Axe^{-x} - 2Ae^{-x} + Be^{-x}) + (-Axe^{-x} + Ae^{-x} - Be^{-x})$$

$$- 6(Axe^{-x} + Be^{-x}) = xe^{-x} - e^{-x},$$

or

$$(-6A)xe^{-x} + (-A - 6B)e^{-x} = xe^{-x} - e^{-x}.$$

Equating coefficients of e^{-x} and xe^{-x} yields

$$-6A = 1,$$

$$-A - 6B = -1.$$

These imply that $A = -1/6$, $B = 7/36$, and hence

$$y_p = -(1/6)xe^{-x} + (7/36)e^{-x}.$$

The following rule encompasses each part of this example.

RULE 1

If a term of $F(x)$ consists of a power (x^n), an exponential (e^{px}), a sine $(\sin px)$, a cosine $(\cos px)$, or any product thereof, assume as a part of y_p a constant multiplied by that term plus a constant multiplied by any linearly independent function arising from it by differentiation.

For Example 16.16(a), since $F(x)$ contains the term $6x^2$, we assume y_p contains Ax^2. Differentiation of Ax^2 yields a term in x and a constant so that we form $y_p = Ax^2 + Bx + C$. No new terms for y_p are obtained from the terms $2x$ and 3 in $F(x)$.

For Example 16.16(b), we assume that y_p contains $A\sin 2x$ to account for the term $2\sin 2x$ in $F(x)$. Differentiation of $A\sin 2x$ gives a linearly independent term in $\cos 2x$ so that we form $y_p = A\sin 2x + B\cos 2x$.

For Example 16.16(c), since $F(x)$ contains the term xe^{-x}, we assume that y_p contains Axe^{-x}. Differentiation of Axe^{-x} yields a term in e^{-x} so that we form $y_p = Axe^{-x} + Be^{-x}$. No new terms for y_p are obtained from the term $-e^{-x}$ in $F(x)$.

EXAMPLE 16.17

What is the form of the particular solution predicted by Rule 1 for the differential equation

$$y'' + 15y' - 6y = x^2 e^{4x} + x + x\cos x?$$

SOLUTION   Rule 1 suggests that

$$y_p = Ax^2 e^{4x} + Bxe^{4x} + Ce^{4x} + Dx + E + Fx \cos x + Gx \sin x$$
$$+ H \cos x + I \sin x.$$

Unfortunately, exceptions to Rule 1 do occur. For the differential equation $y'' + y = \cos x$, Rule 1 would predict $y_p = A \cos x + B \sin x$. If we substitute this into the differential equation we obtain the absurd identity $0 = \cos x$, and certainly no equations to solve for A and B. This result could have been predicted had we first calculated $y_h(x)$. The auxiliary equation $m^2 + 1 = 0$ has solutions $m = \pm i$ so that $y_h(x) = C_1 \cos x + C_2 \sin x$. Since y_p as suggested by Rule 1 is precisely y_h with different names for the constants, then certainly $y_p'' + y_p = 0$. Suppose as an alternative we multiply this y_p by x, and assume that $y_p = Ax \cos x + Bx \sin x$. Substitution into the differential equation now gives

$$-2A \sin x + 2B \cos x = \cos x.$$

Identification of coefficients requires $A = 0$, $B = 1/2$, and $y_p = (1/2)x \sin x$.

This example suggests that if y_p predicted by Rule 1 is already contained in y_h, then a modification of y_p is necessary. A precise statement of the situation is given in the following rule.

> **RULE 2**
>
> Suppose that a term in $F(x)$ is of the form $x^n f(x)$ (n a nonnegative integer). Suppose further that $f(x)$ can be obtained from $y_h(x)$ by specifying values for the arbitrary constants. If this term in y_h results from a root of the auxiliary equation of multiplicity k, then corresponding to $x^n f(x)$, assume as a part of y_p the term $Ax^k(x^n f(x)) = Ax^{n+k} f(x)$, plus a constant multiplied by any linearly independent function arising from it by differentiation.

To use this rule we first require $y_h(x)$. Then, and only then, can we decide on the form of $y_p(x)$. As an illustration, consider the following example.

EXAMPLE 16.18

Find a general solution for $y''' - y = x^3 e^x$.

SOLUTION   In Example 16.14 we solved the auxiliary equation to obtain $m = 1$ and $m = -1/2 \pm (\sqrt{3}/2)i$, from which we formed

$$y_h(x) = C_1 e^x + e^{-x/2}[C_2 \cos(\sqrt{3}x/2) + C_3 \sin(\sqrt{3}x/2)].$$

Now $x^3 e^x$ is x^3 times e^x, and e^x can be obtained from y_h by specifying $C_1 = 1$ and $C_2 = C_3 = 0$. Since this term results from the root $m = 1$ of multiplicity 1, we assume that y_p contains $Ax(x^3 e^x) = Ax^4 e^x$. Differentiation of this function gives terms in $x^3 e^x$, $x^2 e^x$, xe^x, and e^x. We therefore take

$$y_p = Ax^4 e^x + Bx^3 e^x + Cx^2 e^x + Dxe^x.$$

(We do not include a term in e^x since it is already in y_h.) Substitution into the differential equation and simplification gives

$$(12A)x^3e^x + (36A + 9B)x^2e^x + (24A + 18B + 6C)xe^x$$
$$+ (6B + 6C + 3D)e^x = x^3e^x.$$

Equating coefficients gives

$$12A = 1,$$
$$36A + 9B = 0,$$
$$24A + 18B + 6C = 0,$$
$$6B + 6C + 3D = 0.$$

These imply that $A = 1/12$, $B = -1/3$, $C = 2/3$, $D = -2/3$, and

$$y_p = \tfrac{1}{12}x^4e^x - \tfrac{1}{3}x^3e^x + \tfrac{2}{3}x^2e^x - \tfrac{2}{3}xe^x.$$

A general solution of the differential equation is therefore

$$y(x) = C_1e^x + e^{-x/2}[C_2\cos(\sqrt{3}x/2) + C_3\sin(\sqrt{3}x/2)]$$
$$+ \frac{x^4e^x}{12} - \frac{x^3e^x}{3} + \frac{2x^2e^x}{3} - \frac{2xe^x}{3}.$$

EXAMPLE 16.19

If the roots of the auxiliary equation $\phi(m) = 0$ for the differential equation $\phi(D)y = x^2 - 2\sin x + xe^{-2x}$ are $\pm i$, -2, -2, -2, 4, and 4, what is the form of y_p predicted by the method of undetermined coefficients?

SOLUTION   From the roots of the auxiliary equation, we can form

$$y_h(x) = C_1\cos x + C_2\sin x + (C_3 + C_4x + C_5x^2)e^{-2x} + (C_6 + C_7x)e^{4x}.$$

Corresponding to the term x^2 in $F(x)$, Rule 1 requires that y_p contain $Ax^2 + Bx + C$. Because $-2\sin x$ can be obtained from $y_h(x)$ by specifying $C_2 = -2$, $C_1 = C_3 = C_4 = C_5 = C_6 = C_7 = 0$, and this term results from the roots $m = \pm i$, each of multiplicity 1, Rule 2 suggests that y_p contain $Dx\sin x + Ex\cos x$. (We do not include terms in $\sin x$ and $\cos x$ since they are already in y_h.) Finally, xe^{-2x} is x times e^{-2x}, and this function can be obtained from $y_h(x)$ by setting $C_3 = 1$, and $C_1 = C_2 = C_4 = C_5 = C_6 = C_7 = 0$. Because this term results from the root $m = -2$ of multiplicity 3, y_p must contain $Fx^4e^{-2x} + Gx^3e^{-2x}$ (but not terms in x^2e^{-2x}, xe^{-2x}, and e^{-2x} since they are in y_h). The total particular solution is therefore

$$y_p = Ax^2 + Bx + C + Dx\sin x + Ex\cos x + Fx^4e^{-2x} + Gx^3e^{-2x}.$$

Operator Method for a Particular Solution

This method, like that of undetermined coefficients, is designed only for functions $F(x)$ in equation 16.48 of the form x^n, e^{px}, $\sin px$, $\cos px$, and/or sums or products thereof. Essentially, the operator method says that if

$$\phi(D)y = F(x), \tag{16.50}$$

then

$$y = \frac{1}{\phi(D)}F(x). \tag{16.51}$$

But there is a problem. What does it mean to say that

$$\frac{1}{\phi(D)} = \frac{1}{a_0 D^n + a_1 D^{n-1} + \cdots + a_{n-1}D + a_n} \tag{16.52}$$

operates on $F(x)$? The operator method then depends on our explaining how $1/\phi(D)$ operates on $F(x)$. The simplest $\phi(D)$ is $\phi(D) = D$. In this case the differential equation is

$$Dy = F(x),$$

and the solution is

$$y(x) = \int F(x)dx.$$

If the solution is also to be represented in operator notation by

$$y(x) = \frac{1}{D}F(x),$$

then we must define the operator $1/D$ by

$$\frac{1}{D}F(x) = \int F(x)dx. \tag{16.53}$$

If $1/D$ means integrate, then $1/D^2$ must mean integrate twice, $1/D^3$ integrate three times, and so on.

We have stated that the method of operators is applicable in general only when $F(x)$ consists of powers, exponentials, and/or sines and cosines. Consider first the case in which $F(x)$ is a power x^n, in which case equations 16.50 and 16.51 become

$$\phi(D)y = x^n \tag{16.54}$$

and

$$y = \frac{1}{\phi(D)}x^n. \tag{16.55}$$

If we forget for the moment that D is a differential operator, and simply regard $1/\phi(D)$ as a rational function of a variable D, then we can express $1/\phi(D)$ as an infinite series of the form

$$\frac{1}{\phi(D)} = \frac{1}{D^k}\{b_0 + b_1 D + b_2 D^2 + \cdots\} \tag{16.56}$$

For all other cases of $F(x)$, we make use of a theorem called the **inverse operator shift theorem**. This theorem states that

$$\frac{1}{\phi(D)}\{e^{px}f(x)\} = e^{px}\frac{1}{\phi(D+p)}f(x) \qquad (16.59)$$

(see Exercise 16). The theorem enables us to shift the exponential e^{px} past the operator $1/\phi(D)$, but to do so, the operator must be modified to $1/\phi(D+p)$. Equation 16.59 immediately yields $y_p(x)$ whenever $f(x)$ is a power x^n, n a nonnegative integer (or a polynomial). This is illustrated in the following example.

| EXAMPLE 16.22

Find particular solutions for the following differential equations:

(a) $y'' + 3y' + 10y = x^2 e^{-x}$

(b) $y'' - 4y' + 4y = e^{2x}$

(c) $y''' + 2y'' + 3y' - y = e^{2x}$

SOLUTION

(a) Equation 16.59 gives

$$y_p = \frac{1}{D^2 + 3D + 10}x^2 e^{-x} = e^{-x}\frac{1}{(D-1)^2 + 3(D-1) + 10}x^2$$

$$= e^{-x}\frac{1}{D^2 + D + 8}x^2,$$

and we now proceed as in Example 16.20:

$$y_p = e^{-x}\frac{1}{8\left(1 + \dfrac{D+D^2}{8}\right)}x^2$$

$$= \frac{e^{-x}}{8}\left\{1 - \left(\frac{D+D^2}{8}\right) + \left(\frac{D+D^2}{8}\right)^2 - \cdots\right\}x^2$$

$$= \frac{e^{-x}}{8}\left\{1 - \frac{D}{8} - \frac{7D^2}{64} + \cdots\right\}x^2 = \frac{e^{-x}}{8}\left\{x^2 - \frac{x}{4} - \frac{7}{32}\right\}.$$

(b) Once again we use the inverse operator shift theorem to get

$$y_p = \frac{1}{D^2 - 4D + 4}e^{2x} = e^{2x}\frac{1}{(D+2)^2 - 4(D+2) + 4}(1)$$

$$= e^{2x}\frac{1}{D^2}(1) = \frac{x^2}{2}e^{2x},$$

since $1/D^2$ means to integrate twice.

(c) For this differential equation,

$$y_p = \frac{1}{D^3 + 2D^2 + 3D - 1}e^{2x} = e^{2x}\frac{1}{(D+2)^3 + 2(D+2)^2 + 3(D+2) - 1} \quad (1)$$

$$= e^{2x}\frac{1}{D^3 + 8D^2 + 23D + 21}(1) = e^{2x}\frac{1}{21\left(1 + \dfrac{23D + 8D^2 + D^3}{21}\right)}(1)$$

$$= \frac{e^{2x}}{21}\{1 + \cdots\}(1) = \frac{e^{2x}}{21}.$$

EXAMPLE 16.23

Find a general solution for $y''' - 3y'' + 3y' - y = 2x^2e^x$.

SOLUTION   The auxiliary equation is

$$0 = m^3 - 3m^2 + 3m - 1 = (m - 1)^3$$

with solutions 1, 1, and 1. Thus,

$$y_h(x) = (C_1 + C_2x + C_3x^2)e^x.$$

A particular solution is

$$y_p(x) = \frac{2}{(D - 1)^3}x^2e^x = 2e^x\frac{1}{(D + 1 - 1)^3}x^2 = 2e^x\frac{1}{D^3}x^2 = \frac{x^5}{30}e^x.$$

A general solution of the differential equation is therefore

$$y(x) = (C_1 + C_2x + C_3x^2)e^x + \frac{x^5}{30}e^x.$$

EXAMPLE 16.24

Find a particular solution for the differential equation

$$y''' + 3y'' - 4y = xe^{-2x} + x^2.$$

SOLUTION

$$
y_p = \frac{1}{D^3 + 3D^2 - 4}(xe^{-2x} + x^2)
$$

$$
= e^{-2x} \frac{1}{(D-2)^3 + 3(D-2)^2 - 4}x + \frac{1}{D^3 + 3D^2 - 4}x^2
$$

$$
= e^{-2x} \frac{1}{D^3 - 3D^2}x - \frac{1}{4\left(1 - \dfrac{3D^2 + D^3}{4}\right)}x^2
$$

$$
= e^{-2x} \frac{1}{D^2}\frac{1}{D-3}x - \frac{1}{4}\left\{1 + \left(\frac{3D^2 + D^3}{4}\right) + \cdots\right\}x^2
$$

$$
= e^{-2x} \frac{1}{D^2}\frac{1}{-3\left(1 - \dfrac{D}{3}\right)}x - \frac{1}{4}\left(x^2 + \frac{3}{2}\right)
$$

$$
= \frac{e^{-2x}}{-3}\frac{1}{D^2}\left\{1 + \frac{D}{3} + \cdots\right\}x - \frac{x^2}{4} - \frac{3}{8}
$$

$$
= \frac{e^{-2x}}{-3}\frac{1}{D^2}\left\{x + \frac{1}{3}\right\} - \frac{x^2}{4} - \frac{3}{8}
$$

$$
= \frac{e^{-2x}}{-3}\left\{\frac{x^3}{6} + \frac{x^2}{6}\right\} - \frac{x^2}{4} - \frac{3}{8}.
$$

When $F(x)$ in 16.51 is of the form $x^n \sin px$ or $x^n \cos px$, n a nonnegative integer and p a constant, we introduce complex exponentials. Specifically, because

$$
x^n e^{ipx} = x^n(\cos px + i \sin px),
$$

we can write

$$
x^n \cos px = \text{Real part of } x^n e^{ipx} = \text{Re}\{x^n e^{ipx}\}, \tag{16.60a}
$$

$$
x^n \sin px = \text{Imaginary part of } x^n e^{ipx} = \text{Im}\{x^n e^{ipx}\}. \tag{16.60b}
$$

To operate on either of these functions by $1/\phi(D)$, we interchange the operations of $1/\phi(D)$ and take real and imaginary parts:

$$
\frac{1}{\phi(D)}\{x^n \cos px\} = \frac{1}{\phi(D)}\text{Re}\{x^n e^{ipx}\} = \text{Re}\left\{\frac{1}{\phi(D)}(x^n e^{ipx})\right\}, \tag{16.61a}
$$

$$
\frac{1}{\phi(D)}\{x^n \sin px\} = \frac{1}{\phi(D)}\text{Im}\{x^n e^{ipx}\} = \text{Im}\left\{\frac{1}{\phi(D)}(x^n e^{ipx})\right\}. \tag{16.61b}
$$

We can now proceed by using inverse operator shift theorem 16.59.

EXAMPLE 16.25

Find particular solutions for the following differential equations:

(a) $y'' + y = \sin 2x$;

(b) $y'' + 4y = x^2 \cos x$;

(c) $y'' + 9y = \sin 3x$;

(d) $y'' + 4y = x \sin 2x$;

(e) $y'' + 2y' + 4y = e^{-x} \sin \sqrt{3}x$.

SOLUTION

(a) Equation 16.61b gives

$$y_p = \frac{1}{D^2 + 1} \sin 2x = \frac{1}{D^2 + 1}\{\text{Im}(e^{2ix})\} = \text{Im}\left\{\frac{1}{D^2 + 1} e^{2ix}\right\}.$$

We now use inverse operator shift theorem 16.59,

$$y_p = \text{Im}\left\{e^{2ix} \frac{1}{(D + 2i)^2 + 1}(1)\right\}$$

$$= \text{Im}\left\{e^{2ix} \frac{1}{D^2 + 4iD - 3}(1)\right\} = \text{Im}\left\{e^{2ix} \frac{1}{-3\left(1 - \frac{4iD + D^2}{3}\right)}(1)\right\}$$

$$= \text{Im}\left\{\frac{e^{2ix}}{-3}\right\} = -\frac{1}{3}\sin 2x.$$

(b)

$$y_p = \frac{1}{D^2 + 4} x^2 \cos x = \frac{1}{D^2 + 4}\text{Re}(x^2 e^{ix}) = \text{Re}\left\{\frac{1}{D^2 + 4} x^2 e^{ix}\right\}$$

$$= \text{Re}\left\{e^{ix} \frac{1}{(D + i)^2 + 4} x^2\right\} = \text{Re}\left\{e^{ix} \frac{1}{D^2 + 2iD + 3} x^2\right\}$$

$$= \text{Re}\left\{e^{ix} \frac{1}{3\left(1 + \frac{2iD + D^2}{3}\right)} x^2\right\}$$

$$= \text{Re}\left\{\frac{e^{ix}}{3}\left[1 - \left(\frac{2iD + D^2}{3}\right) + \left(\frac{2iD + D^2}{3}\right)^2 - \cdots\right] x^2\right\}$$

$$= \frac{1}{3}\text{Re}\left\{e^{ix}\left[1 - \frac{2iD}{3} - \frac{7D^2}{9} + \cdots\right] x^2\right\} = \frac{1}{3}\text{Re}\left\{e^{ix}\left[x^2 - \frac{4ix}{3} - \frac{14}{9}\right]\right\}$$

$$= \frac{1}{3}\left\{x^2 \cos x + \frac{4x}{3}\sin x - \frac{14}{9}\cos x\right\}.$$

(c) For the differential equation $y'' + 9y = \sin 3x$, we have

$$y_p = \frac{1}{D^2 + 9} \sin 3x = \frac{1}{D^2 + 9} \text{Im}(e^{3ix}) = \text{Im}\left\{ \frac{1}{D^2 + 9} e^{3ix} \right\}$$

$$= \text{Im}\left\{ e^{3ix} \frac{1}{(D + 3i)^2 + 9}(1) \right\} = \text{Im}\left\{ e^{3ix} \frac{1}{D^2 + 6iD}(1) \right\}$$

$$= \text{Im}\left\{ e^{3ix} \frac{1}{D} \frac{1}{D + 6i}(1) \right\} = \text{Im}\left\{ \frac{e^{3ix}}{6i} \frac{1}{D} \frac{1}{1 + D/6i}(1) \right\}$$

$$= \text{Im}\left\{ -\frac{i}{6} e^{3ix} \frac{1}{D}\left(1 - \frac{D}{6i} + \cdots\right)(1) \right\} = \text{Im}\left\{ -\frac{i}{6} e^{3ix} \frac{1}{D}(1) \right\}$$

$$= \text{Im}\left\{ -\frac{i}{6} e^{3ix} x \right\} = -\frac{x}{6} \cos 3x.$$

(d) For the differential equation $y'' + 4y = x \sin 2x$, we get

$$y_p = \frac{1}{D^2 + 4} x \sin 2x = \frac{1}{D^2 + 4} \text{Im}(xe^{2ix}) = \text{Im}\left\{ \frac{1}{D^2 + 4} xe^{2ix} \right\}$$

$$= \text{Im}\left\{ e^{2ix} \frac{1}{(D + 2i)^2 + 4} x \right\} = \text{Im}\left\{ e^{2ix} \frac{1}{D^2 + 4iD} x \right\}$$

$$= \text{Im}\left\{ e^{2ix} \frac{1}{D} \frac{1}{D + 4i} x \right\} = \text{Im}\left\{ \frac{1}{4i} e^{2ix} \frac{1}{D}\left(1 - \frac{D}{4i} + \cdots\right) x \right\}$$

$$= \text{Im}\left\{ -\frac{i}{4} e^{2ix} \frac{1}{D}\left(x - \frac{1}{4i}\right) \right\} = \text{Im}\left\{ -\frac{i}{4} e^{2ix}\left(\frac{x^2}{2} + \frac{ix}{4}\right) \right\}$$

$$= -\frac{x^2}{8} \cos 2x + \frac{x}{16} \sin 2x.$$

(e) For the differential equation $y'' + 2y' + 4y = e^{-x} \sin \sqrt{3}x$, we have

$$y_p = \frac{1}{D^2 + 2D + 4} e^{-x} \sin \sqrt{3}x = \frac{1}{D^2 + 2D + 4} \text{Im}(e^{-x} e^{\sqrt{3}ix})$$

$$= \text{Im}\left\{ \frac{1}{D^2 + 2D + 4} e^{(-1+\sqrt{3}i)x} \right\}$$

$$= \text{Im}\left\{ e^{(-1+\sqrt{3}i)x} \frac{1}{(D - 1 + \sqrt{3}i)^2 + 2(D - 1 + \sqrt{3}i) + 4}(1) \right\}$$

$$= \text{Im}\left\{ e^{(-1+\sqrt{3}i)x} \frac{1}{D^2 + 2\sqrt{3}iD}(1) \right\} = \text{Im}\left\{ e^{(-1+\sqrt{3}i)x} \frac{1}{D} \frac{1}{D + 2\sqrt{3}i}(1) \right\}$$

$$= \text{Im}\left\{ e^{(-1+\sqrt{3}i)x} \frac{1}{D} \frac{1}{2\sqrt{3}i} \right\} = \text{Im}\left\{ -\frac{i}{2\sqrt{3}} xe^{(-1+\sqrt{3}i)x} \right\}$$

$$= -\frac{xe^{-x}}{2\sqrt{3}} \cos \sqrt{3}x.$$

EXAMPLE 16.26

Find a general solution of $y'' + 6y' + y = \sin 3x$.

SOLUTION   The auxiliary equation is $m^2 + 6m + 1 = 0$ with solutions $m = -3 \pm 2\sqrt{2}$. Consequently,

$$y_h(x) = C_1 e^{(-3+2\sqrt{2})x} + C_2 e^{(-3-2\sqrt{2})x}.$$

A particular solution is

$$y_p = \frac{1}{D^2 + 6D + 1} \operatorname{Im}(e^{3ix}) = \operatorname{Im}\left\{ \frac{1}{D^2 + 6D + 1} e^{3ix} \right\}$$

$$= \operatorname{Im}\left\{ e^{3ix} \frac{1}{(D + 3i)^2 + 6(D + 3i) + 1} (1) \right\}$$

$$= \operatorname{Im}\left\{ e^{3ix} \frac{1}{D^2 + (6 + 6i)D + (-8 + 18i)} (1) \right\} = \operatorname{Im}\left\{ e^{3ix} \frac{1}{-8 + 18i} \right\}$$

$$= \operatorname{Im}\left\{ e^{3ix} \frac{1}{-8 + 18i} \frac{-8 - 18i}{-8 - 18i} \right\} = -\frac{1}{194} (4 \sin 3x + 9 \cos 3x).$$

Finally, then, a general solution of the differential equation is

$$y(x) = C_1 e^{(-3+2\sqrt{2})x} + C_2 e^{(-3-2\sqrt{2})x} - \frac{1}{194} (4 \sin 3x + 9 \cos 3x).$$

Examples 16.20–16.26 have illustrated that, with identity 16.59 and the concept of series, we can obtain a particular solution for 16.50 whenever $F(x)$ is x^n, e^{px}, $\sin px$, $\cos px$, and/or any sums or products thereof. In summary:

When $F(x) = x^n$, use series 16.56 for $1/\phi(D)$.

In any other situation, use 16.59, and then series 16.56 for $1/\phi(D + p)$.

We make one final comment. We introduced complex exponentials e^{ipx} to handle terms involving $\sin px$ and $\cos px$. This is not the only way to treat trigonometric functions, since there do exist other methods that completely avoid complex numbers. Unfortunately, these methods require memorization of somewhat involved identities. Because of this, we prefer the use of complex numbers.

EXERCISES 16.9

In Exercises 1–11 find a particular solution for the differential equation, by both the method of operators and the method of undetermined coefficients. Find a general solution of the equation.

In order to describe the position of M as a function of time t, we must choose a vertical coordinate system. There are two natural places to choose the origin $y = 0$, one being the position of M when the spring is unstretched. Suppose we do this and choose y as positive upward. When M is a distance y away from the origin, the restoring force of the spring has y-component $-ky$ $(k > 0)$. In addition, if $g = 9.81$ is the acceleration due to gravity, then the force of gravity on M has y-component $-Mg$. Finally, suppose oscillations take place in a medium that exerts a damping force proportional to the instantaneous velocity of M. This damping force must therefore have a y-component of the form $-\beta(dy/dt)$, where β is a positive constant. The total force on M therefore has y-component

$$-ky - Mg - \beta\frac{dy}{dt},$$

and Newton's second law states that the acceleration d^2y/dt^2 of M must satisfy the equation

$$-ky - Mg - \beta\frac{dy}{dt} = M\frac{d^2y}{dt^2}.$$

Consequently, the differential equation that determines the position $y(t)$ of M relative to the unstretched position of the spring is

$$M\frac{d^2y}{dt^2} + \beta\frac{dy}{dt} + ky = -Mg. \tag{16.62}$$

The alternative possibility for describing oscillations is to attach M to the spring and slowly lower M until it reaches an equilibrium position. At this position, the restoring force of the spring is exactly equal to the force of gravity on the mass, and the mass, left by itself, will remain motionless. If s is the amount of stretch in the spring at equilibrium, then at equilibrium

$$ks - Mg = 0, \quad \text{where } s > 0. \tag{16.63}$$

Suppose we take the equilibrium position as $x = 0$ and x as positive upward (Figure 16.9). When M is a distance x away from its equilibrium position, the restoring force on M has x-component $k(s - x)$. The x-component of the force of gravity remains as $-Mg$, and that of the damping force is $-\beta(dx/dt)$. Newton's second law therefore implies that

$$M\frac{d^2x}{dt^2} = k(s - x) - Mg - \beta\frac{dx}{dt},$$

or

$$M\frac{d^2x}{dt^2} + \beta\frac{dx}{dt} + kx = -Mg + ks.$$

But according to 16.63, $-Mg + ks = 0$, and hence

$$M\frac{d^2x}{dt^2} + \beta\frac{dx}{dt} + kx = 0. \tag{16.64}$$

This is the differential equation describing the displacement $x(t)$ of M relative to the equilibrium position of M.

FIGURE 16.9

x

M $x = 0$ at
equilibrium

Note that both equations 16.62 and 16.64 are linear second-order differential equations with constant coefficients. The advantage of 16.64 is that it is homogeneous as well, and this is simply due to a convenient choice of dependent variable (x as opposed to y). Physically, we are saying that there are two parts to the spring force $k(s - x)$: a part ks and a part $-kx$. Gravity is always acting on M, and that part ks of the spring force is counteracting it in an attempt to restore the spring to its unstretched position. Because these forces always cancel, we might just as well eliminate both of them from our discussion. This would leave us $-kx$, and we therefore interpret $-kx$ as the *spring force attempting to restore the mass to its equilibrium position.*

If we choose equation 16.64 to describe the motion of M (and this equation is usually chosen over equation 16.62), we must remember three things: x is measured from equilibrium, $-kx$ is the spring force attempting to restore M to its equilibrium position, and gravity has been taken into account.

There are three basic ways to initiate the motion. First, we can move the mass away from its equilibrium position and then release it, giving it an initial displacement but no initial velocity. Second, we can strike the mass at the equilibrium position, imparting an initial velocity but no initial displacement. And finally, we can give the mass both an initial displacement and an initial velocity. Each of these methods adds two initial conditions to the differential equation.

To be complete we note that, when (in addition to the forces already mentioned) there is an externally applied force acting on the mass that is represented as a function of time by $F(t)$, then equation 16.64 is modified to

$$M\frac{d^2x}{dt^2} + \beta\frac{dx}{dt} + kx = F(t). \qquad (16.65)$$

Perhaps, for example, M contains some iron, and $F(t)$ is due to a magnet directly below M that exerts a time dependent attractive force on M.

EXAMPLE 16.27

A 2-kg mass is suspended vertically from a spring with constant 16 N/m. The mass is raised 10 cm above its equilibrium position and then released. If damping is ignored, find the amplitude, period, and frequency of the motion.

SOLUTION   If we choose $x = 0$ at the equilibrium position of the mass and x positive upward (Figure 16.9), differential equation 16.64 for motion $x(t)$ of the mass is

$$2\frac{d^2x}{dt^2} + 16x = 0,$$

or

$$\frac{d^2x}{dt^2} + 8x = 0,$$

along with the initial conditions

$$x(0) = 1/10, \quad x'(0) = 0.$$

The auxiliary equation is $m^2 + 8 = 0$ with solutions $m = \pm 2\sqrt{2}i$. Consequently,

$$x(t) = C_1 \cos(2\sqrt{2}t) + C_2 \sin(2\sqrt{2}t).$$

The initial conditions require

$$1/10 = C_1, \quad 0 = 2\sqrt{2}C_2.$$

Thus,

$$x(t) = \frac{1}{10}\cos(2\sqrt{2}t).$$

The amplitude of the oscillations is 1/10 m, the period is $2\pi/(2\sqrt{2}) = \pi/\sqrt{2}$ s, and the frequency is $\sqrt{2}/\pi \ s^{-1}$. A graph of this function (Figure 16.10) illustrates the oscillations of the mass about its equilibrium position. This is an example of **simple harmonic motion**.

FIGURE 16.10

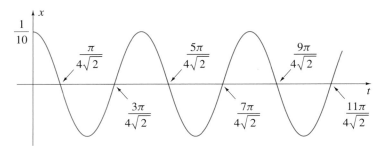

EXAMPLE 16.28

A 100-g mass is suspended vertically from a spring with constant 5 N/m. The mass is pulled 5 cm below its equilibrium position and given velocity 2 m/s upward. If, during the motion, the mass is acted on by a damping force in newtons numerically equal to one-twentieth the instantaneous velocity in metres per second, find the position of the mass at any time.

SOLUTION   If we choose $x = 0$ at the equilibrium position of the mass and x positive upward (Figure 16.9), then differential equation 16.64 for motion $x(t)$ of the mass is

$$\frac{1}{10}\frac{d^2x}{dt^2} + \frac{1}{20}\frac{dx}{dt} + 5x = 0,$$

or

$$2\frac{d^2x}{dt^2} + \frac{dx}{dt} + 100x = 0,$$

along with the initial conditions

$$x(0) = -1/20, \quad x'(0) = 2.$$

The auxiliary equation is $2m^2 + m + 100 = 0$ with solutions

$$m = \frac{-1 \pm \sqrt{1 - 800}}{4} = \frac{-1 \pm \sqrt{799}i}{4}.$$

Consequently,

$$x(t) = e^{-t/4}[C_1 \cos(\sqrt{799}t/4) + C_2 \sin(\sqrt{799}t/4)].$$

The initial conditions require

$$-1/20 = C_1, \quad 2 = -C_1/4 + \sqrt{799}C_2/4,$$

from which we get

$$C_2 = \frac{159\sqrt{799}}{15\,980}.$$

Finally, then,

$$x(t) = e^{-t/4}\left[-\frac{1}{20}\cos\left(\frac{\sqrt{799}t}{4}\right) + \frac{159\sqrt{799}}{15\,980}\sin\left(\frac{\sqrt{799}t}{4}\right)\right].$$

The graph of this function in Figure 16.11 clearly indicates how the amplitude of the oscillations decreases in time.

FIGURE 16.11

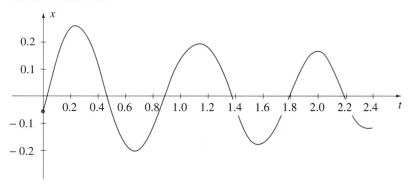

LCR Circuits

If a resistance R, an inductance L, and a capacitance C are connected in series with an electromotive force $E(t)$ (Figure 16.12) and the switch is closed, current flows in the circuit and charge builds up in the capacitor. If at any time t, Q is the charge on the capacitor and I is the current in the loop, then Kirchhoff's law states that

$$L\frac{dI}{dt} + RI + \frac{Q}{C} = E(t), \tag{16.66}$$

where $L\, dI/dt$, RI, and Q/C represent the voltage drops across the inductor, the resistor, and the capacitor, respectively. If we substitute $I = dQ/dt$, then

$$L\frac{d^2Q}{dt^2} + R\frac{dQ}{dt} + \frac{1}{C}Q = E(t),$$ (16.67)

a second-order linear differential equation for $Q(t)$.

Alternatively, if we differentiate this equation, we obtain

$$L\frac{d^2I}{dt^2} + R\frac{dI}{dt} + \frac{1}{C}I = E'(t),$$ (16.68)

a second-order linear differential equation for $I(t)$.

FIGURE 16.12

EXAMPLE 16.29

At time $t = 0$, a 25-Ω resistor, a 2-H inductor, and a 0.01-F capacitor are connected in series with a generator producing an alternating voltage of $10\sin(5t)$, $t \geq 0$ (Figure 16.13). Find the charge on the capacitor and the current in the circuit if the capacitor is uncharged when the circuit is closed.

FIGURE 16.13

SOLUTION   Differential equation 16.67 for the charge Q on the capacitor is

$$2\frac{d^2Q}{dt^2} + 25\frac{dQ}{dt} + 100Q = 10\sin(5t),$$

to which we add the initial conditions

$$Q(0) = 0, \quad Q'(0) = I(0) = 0.$$

The auxiliary equation is $2m^2 + 25m + 100 = 0$ with solutions

$$m = \frac{-25 \pm \sqrt{625 - 800}}{4} = \frac{-25 \pm 5\sqrt{7}i}{4}.$$

Consequently, a general solution of the homogeneous equation is .

$$Q_h(t) = e^{-25t/4}[C_1 \cos(5\sqrt{7}t/4) + C_2 \sin(5\sqrt{7}t/4)].$$

To find a particular solution of the nonhomogeneous equation by undetermined coefficients, we set

$$Q_p(t) = A \sin(5t) + B \cos(5t).$$

Substitution of this function into the differential equation gives

$$2\{-25A \sin(5t) - 25B \cos(5t)\} + 25\{5A \cos(5t) - 5B \sin(5t)\}$$
$$+ 100\{A \sin(5t) + B \cos(5t)\} = 10 \sin(5t).$$

This equation requires A and B to satisfy

$$50A - 125B = 10, \quad 125A + 50B = 0,$$

the solution of which is $A = 4/145$, $B = -10/145$. A particular solution is therefore

$$Q_p(t) = \frac{2}{145}\left[-5\cos(5t) + 2\sin(5t)\right],$$

and a general solution is

$$Q(t) = Q_h(t) + Q_p(t)$$
$$= e^{-25t/4}\left[C_1 \cos\left(\frac{5\sqrt{7}t}{4}\right) + C_2 \sin\left(\frac{5\sqrt{7}t}{4}\right)\right]$$
$$+ \frac{2}{145}[2\sin(5t) - 5\cos(5t)].$$

The initial conditions require

$$0 = C_1 - \frac{10}{145}, \quad 0 = -\frac{25}{4}C_1 + \frac{5\sqrt{7}}{4}C_2 + \frac{20}{145},$$

and these imply that $C_1 = 10/145$, $C_2 = 34/(145\sqrt{7})$. Consequently,

$$Q(t) = \frac{e^{-25t/4}}{145\sqrt{7}}\left[10\sqrt{7}\cos\left(\frac{5\sqrt{7}t}{4}\right) + 34\sin\left(\frac{5\sqrt{7}t}{4}\right)\right]$$
$$+ \frac{2}{145}[2\sin(5t) - 5\cos(5t)].$$

The current in the circuit is

$$I(t) = \frac{dQ}{dt} = \left(-\frac{25}{4}\right)\frac{e^{-25t/4}}{145\sqrt{7}}\left[10\sqrt{7}\cos\left(\frac{5\sqrt{7}t}{4}\right) + 34\sin\left(\frac{5\sqrt{7}t}{4}\right)\right]$$

$$+ \frac{e^{-25t/4}}{145\sqrt{7}}\left[-\frac{175}{2}\sin\left(\frac{5\sqrt{7}t}{4}\right) + \frac{85\sqrt{7}}{2}\cos\left(\frac{5\sqrt{7}t}{4}\right)\right]$$

$$+ \frac{2}{145}[10\cos(5t) + 25\sin(5t)]$$

$$= -\frac{e^{-25t/4}}{29\sqrt{7}}\left[4\sqrt{7}\cos\left(\frac{5\sqrt{7}t}{4}\right) + 60\sin\left(\frac{5\sqrt{7}t}{4}\right)\right]$$

$$+ \frac{2}{29}[2\cos(5t) + 5\sin(5t)].$$

The solution $Q(t)$ contains two parts. The first two terms (containing the exponential $e^{-25t/4}$) are $Q_h(t)$ with the constants C_1 and C_2 determined by the initial conditions; the last two terms are $Q_p(t)$. We point this out because the two parts display completely different characteristics. Their graphs are shown in Figure 16.14. For small t, both parts of $Q(t)$ are present and contribute significantly, but for large t, the first two terms become negligible. In other words, after a long time, the charge $Q(t)$ on the capacitor is defined essentially by $Q_p(t)$. We call $Q_p(t)$ the steady-state part of the solution, and the two other terms in $Q(t)$ are called the transient part of the solution. Similarly, the first two terms in $I(t)$ are called the transient part of the current and the last two terms the steady-state part of the current.

Finally, note that the frequency of the steady-state part of either $Q(t)$ or $I(t)$ is exactly that of the forcing voltage $E(t)$.

FIGURE 16.14a

FIGURE 16.14b

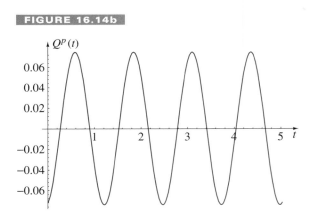

The similarity between differential equations 16.65 and 16.67 cannot go unmentioned:

$$M\frac{d^2x}{dt^2} + \beta\frac{dx}{dt} + kx = F(t),$$

$$L\frac{d^2Q}{dt^2} + R\frac{dQ}{dt} + \frac{1}{C}Q = E(t).$$

Each of the coefficients M, β, and k for the mechanical system has its analogue L, R, and $1/C$ in the electrical system.

EXERCISES 16.10

** **1.** A 1-kg mass is suspended vertically from a spring with constant 16 N/m. The mass is pulled 10 cm below its equilibrium position, and then released. Find the position of the mass, relative to its equilibrium position, at any time if

    (a) damping is ignored.

    (b) a damping force in newtons equal to one-tenth the instantaneous velocity in metres per second acts on the mass.

    (c) a damping force in newtons equal to ten times the instantaneous velocity in metres per second acts on the mass.

** **2.** A 200-g mass suspended vertically from a spring with constant 10 N/m is set into vibration by an external force in newtons given by $4 \sin 10t$, $t \geq 0$. During the motion a damping force in newtons equal to $3/2$ the velocity of the mass in metres per second acts on the mass. Find the position of the mass as a function of time t.

** **3.** A 0.001-F capacitor and a 2-H inductor are connected in series with a 20-V battery. If there is no charge on the capacitor before the battery is connected, find the current in the circuit as a function of time.

** **4.** At time $t = 0$, a 0.02-F capacitor, a 100-Ω resistor, and 1-H inductor are connected in series. If the charge on the capacitor is initially 5 C, find its charge as a function of time.

** **5.** A 5-H inductor and 20-Ω resistor are connected in series with a generator supplying an oscillating voltage of $10 \sin 2t$, $t \geq 0$. What are the transient and steady-state currents in the circuit?

** **6.** At time $t = 0$, a mass M is attached to the end of a hanging spring with constant k, and then released. Assuming that damping is negligible, find the subsequent displacement of the mass as a function of time.

** **7.** A 0.5-kg mass sits on a table attached to a spring with constant 18 N/m (figure below). The mass is pulled so as to stretch the spring 5 cm and is then released.

    (a) If friction between the mass and the table creates a force of 0.5 N that opposes motion, show that the differential equation determining motion is

$$\frac{d^2x}{dt^2} + 36x = 1, \quad x(0) = 0.05, \quad x'(0) = 0.$$

    (b) Find where the mass comes to rest for the first time. Will it move from this position?

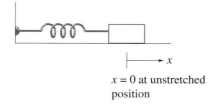

$x = 0$ at unstretched
position

** **8.** Repeat Exercise 7 given that the mass is pulled 25 cm to the right.

** **9.** At time $t = 0$ an uncharged 0.1-F capacitor is connected in series with 0.5-H inductor and a 3-Ω resistor. If the current in the circuit at this instant is 1 A, find the maximum charge that the capacitor stores.

** **10.** Differential equation 16.64 describes the motion of a mass M at the end of a spring, taking damping proportional to velocity into account. Show each of the following:

(a) If $\beta = 0$,

$$x(t) = C_1 \cos(\sqrt{k/M}\,t) + C_2 \sin(\sqrt{k/M}\,t),$$

called *simple harmonic motion*.

(b) If $\beta \neq 0$ and $\beta^2 - 4kM < 0$,

$$x(t) = e^{-\beta t/(2M)}(C_1 \cos \omega t + C_2 \sin \omega t), \quad \text{where } \omega = \frac{\sqrt{4kM - \beta^2}}{2M},$$

called *damped oscillatory motion*.

(c) If $\beta \neq 0$ and $\beta^2 - 4kM > 0$,

$$x(t) = e^{-\beta t/(2M)}(C_2 e^{\omega t} + C_2 e^{-\omega t}), \quad \text{where } \omega = \frac{\sqrt{\beta^2 - 4kM}}{2M},$$

called *overdamped motion*.

(d) If $\beta \neq 0$ and $\beta^2 - 4kM = 0$,

$$x(t) = (C_1 + C_2 t)e^{-\beta t/(2M)},$$

called *critically damped motion*.

** **11.** A 100-g mass is suspended from a spring with constant 4000 N/m. At its equilibrium position, it is suddenly (time $t = 0$) given an upward velocity of 10 m/s. If an external force $3 \cos 200t$, $t \geq 0$, acts on the mass, find its displacement as a function of time. Do you note anything strange?

** **12.** A vertical spring having constant 64 N/m has a 1-kg mass attached to it. An external force $F(t) = 2 \sin 8t$, $t \geq 0$ is applied to the mass. If the mass is at rest at its equilibrium position at time $t = 0$, and damping is negligible, find the position of the mass as a function of time. What happens to the oscillations as time progresses? This phenomenon is called *resonance*; it occurs when the frequency of the forcing function is identical to the natural frequency of the undamped system.

** **13.** A mass M is suspended from a vertical spring with constant k. If an external force $F(t) = A \cos \omega t$, $t \geq 0$, is applied to the mass, find the value of ω that causes resonance.

** **14.** A $(25/9)$-H inductor, a 0.04-F capacitor, and a generator with voltage $15 \cos 3t$ are connected in series at time $t = 0$. Find the current in the circuit as a function of time. Does resonance occur?

** **15.** A mass M is suspended from a vertical spring with constant k. When an external force $F(t) = A \cos \omega t$, $t \geq 0$ is applied to the mass, and a damping force proportional to velocity acts on the mass during its subsequent oscillations, the differential equation governing the motion of M is

$$M\frac{d^2x}{dt^2} + \beta\frac{dx}{dt} + kx = A \cos \omega t.$$

(a) Show that steady-state oscillations of M are defined by

$$x(t) = \frac{A}{(k - M\omega^2)^2 + \beta^2\omega^2}\left[(k - M\omega^2)\cos \omega t + \beta\omega \sin \omega t\right].$$

(b) Verify that this function can be expressed in the form

$$x(t) = \frac{A}{\sqrt{(k - M\omega^2)^2 + \beta^2\omega^2}} \sin(\omega t + \phi),$$

where

$$\sin\phi = \frac{k - M\omega^2}{\sqrt{(k - M\omega^2)^2 + \beta^2\omega^2}},$$

$$\cos\phi = \frac{\beta\omega}{\sqrt{(k - M\omega^2)^2 + \beta^2\omega^2}}.$$

(c) If resonance is said to occur when the amplitude of the steady-state oscillations is a maximum, what value of ω yields resonance? What is the maximum amplitude?

** **16.** (a) A cube L m on each side and with mass M kg floats half submerged in water. If it is pushed down slightly and then released, oscillations take place. Use Archimedes' principle to find the differential equation governing these oscillations. Assume no damping forces due to the viscosity of the water.

(b) What is the frequency of the oscillations?

** **17.** A cylindrical buoy 20 cm in diameter floats partially submerged with its axis vertical. When it is depressed slightly and released, its oscillations have a period equal to 4 s. What is the mass of the buoy?

*** **18.** A sphere of radius R floats half submerged in water. It is set into vibration by pushing it down slightly and then releasing it. If y denotes the instantaneous distance of its centre below the surface, show that

$$\frac{d^2y}{dt^2} = \frac{-3g}{2R^3}\left(R^2y - \frac{y^3}{3}\right),$$

where g is the acceleration due to gravity.

*** **19.** A cable hangs over a peg, 10 m on one side and 15 m on the other. Find the time for it to slide off the peg

(a) if friction at the peg is negligible.

(b) if friction at the peg is equal to the weight of 1 m of cable.

SUMMARY

A differential equation is an equation that contains an unknown function and some of its derivatives, and the equation must be solved for this function. Depending on the form of the equation, various techniques may be used to find the solution. From this point of view, solving differential equations is much like evaluating antiderivatives: We must first recognize the technique appropriate to the particular problem at hand, and then proceed through the mechanics of the technique. It is important, then, to recognize immediately the type of differential equation under consideration.

Broadly speaking, we divided the differential equations we considered into two main groups: first-order equations together with simple second-order equations, and linear differential equations. A first-order differential equation in $y(x)$ is said to be separable if it can be written in the form $N(y)dy = M(x)dx$, and a general solution for such an equation is defined implicitly by

$$\int N(y)dy = \int M(x)dx + C.$$

A differential equation of the form $dy/dx + P(x)y = Q(x)$ is said to be linear first order. If this equation is multiplied by the integrating factor $e^{\int P(x)dx}$, then the left side becomes the derivative of $ye^{\int P(x)dx}$, and the equation is immediately integrable.

If a second-order differential equation in $y(x)$ has either the dependent variable or the independent variable explicitly missing, it can be reduced to a pair of first-order equations. This is accomplished in the former case by setting $v = y'$ and $v' = y''$, and in the latter case by setting $v = y'$ and $y'' = v\, dv/dy$.

A differential equation in $y(x)$ is said to be linear if it is of the form

$$a_0 \frac{d^n y}{dx^n} + a_1 \frac{d^{n-1} y}{dx^{n-1}} + \cdots + a_{n-1} \frac{dy}{dx} + a_n y = F(x).$$

The general solution of such an equation is composed of two parts: $y(x) = y_h(x) + y_p(x)$. The function $y_h(x)$ is a general solution of the associated homogeneous equation obtained by replacing $F(x)$ by 0; $y_p(x)$ is any particular solution of the given equation whatsoever. In the special case that the a_i are constants, it is always possible to find $y_h(x)$. This is done by calculating all solutions of the auxiliary equation

$$a_0 m^n + a_1 m^{n-1} + \cdots + a_{n-1} m + a_n = 0,$$

a polynomial equation in m, and then using the rules of Theorem 16.3.

When the a_i are constants and $F(x)$ is a polynomial, an exponential, a sine, a cosine, or any sums or products thereof, we can find $y_p(x)$ either by undetermined coefficients or by operators. The method of undetermined coefficients is simply an intelligent way of guessing at $y_p(x)$ on the basis of $F(x)$ and $y_h(x)$. Note once again that $y_h(x)$ must be calculated before using the method of undetermined coefficients. The method of operators, on the other hand, is based on formal algebraic manipulations and the inverse operator shift theorem. It does not require prior calculation of $y_h(x)$ and, in its simplest form, uses complex numbers.

KEY TERMS

In reviewing this chapter, you should be able to define or discuss the following key terms:

Differential equation	Order of a differential equation
Initial conditions	Boundary conditions
General solution	Particular solution
Singular solution	Separable differential equation
Homogeneous differential equation	Linear differential equation
Integrating factor	Bernoulli equation
Friction	Force of friction
Coefficient of kinetic friction	Differential operator
Linear operator	Homogeneous linear differential equation
Superposition principle	Linearly independent solutions
Auxiliary equation	Method of undetermined coefficients
Operator method	Operator shift theorem
Inverse operator shift theorem	Cauchy-Euler equation
Simple harmonic motion	Resonance

**REVIEW
EXERCISES**

In Exercises 1–20 find a general solution for the differential equation.

* **1.** $x^2\,dy - y\,dx = 0$

* **2.** $(x+1)\,dx - xy\,dy = 0$

* **3.** $\dfrac{dy}{dx} + 3xy = 2x$

* **4.** $\dfrac{dy}{dx} + 4y = x^2$

* **5.** $\dfrac{d^2y}{dx^2} + 4\dfrac{dy}{dx} + 3y = 2$

* **6.** $\dfrac{d^2y}{dx^2} + 3\dfrac{dy}{dx} + 4y = 2$

* **7.** $yy' = \sqrt{1+y^2}$

* **8.** $\dfrac{d^2y}{dx^2} + \dfrac{1}{x}\dfrac{dy}{dx} = x$

* **9.** $y'' + 6y' + 3y = xe^x$

* **10.** $\dfrac{dy}{dx} + 2xy = 2x^3$

* **11.** $y^2 y'' = y'$

* **12.** $y'' - 4y' + 4y = \sin x$

* **13.** $y'' - 4y' + 4y = x^2 e^{2x}$

* **14.** $y'' + 4y = \sin 2x$

* **15.** $y'' + 4y' = x^2$

* **16.** $2xy^2\dfrac{dy}{dx} + (x+1)^2 y^3 = 0$

* **17.** $\dfrac{d^3y}{dx^3} + 3\dfrac{d^2y}{dx^2} + 3\dfrac{dy}{dx} + y = 2e^{-x}$

* **18.** $y'' + 2y' + 4y = e^{-x}\cos\sqrt{3}x$

* **19.** $\dfrac{dy}{dx} = y\tan x + \cos x$

* **20.** $(2y^2 + 3x)\,dy + dx = 0$

In Exercises 21–24 find the solution of the differential equation satisfying the given condition(s).

* **21.** $y^2\,dx + (x+1)\,dy = 0,\ y(0) = 3$

* **22.** $y'' - 8y' - 9y = 2x + 4,\ y(0) = 3,\ y'(0) = 7$

* **23.** $y'' + 9y = e^x,\ y(0) = 0,\ y(\pi/2) = 4$

** **24.** $y' + \dfrac{2}{x}y = \sin x,\ y(1) = 1$

** **25.** The quantity of radioactive material present in a sample decays at a rate proportional to the amount of the material in the sample (see Section 9.2). If one-quarter of a sample decays in 5 years, how long does it take for 90% of the sample to decay?

** **26.** (a) A piece of wood rises from the bottom of a container of oil 1 m deep. If the wood has mass 0.5 g and volume 1 cm^3, and the density of the oil is 0.9 g/cm^3, show that Archimedes' principle predicts a buoyant force due to fluid pressure of 8.829×10^{-3} N. What is the force on the piece of wood due to gravity and fluid pressure?

(b) The viscosity of the oil opposes motion by exerting a force equal (in newtons) to twice its velocity (in metres per second). Find the distance travelled by the wood as a function of time, assuming that it starts from rest on the bottom.

✱✱ **27.** A 100-g mass is suspended vertically from a spring with constant 1 N/m. The mass is pulled 4 cm above its equilibrium position and then released. Find the position of the mass at any time if

(a) damping is ignored.

(b) a damping force in newtons equal to $1/5$ the instantaneous velocity of the mass in metres per second acts on the mass.

(c) a damping force in newtons equal to $\sqrt{2/5}$ the instantaneous velocity of the mass in metres per second acts on the mass.

✱✱ **28.** A 10-g stone is dropped over the side of a bridge 50 m above a river 10 m deep. Air resistance is negligible, but as the stone sinks in the water, its motion is retarded by a force in newtons equal to one-fifth the speed of the stone in metres per second. If the stone loses 10% of its speed in penetrating the surface of the water, find

(a) its velocity as a function of time.

(b) its position relative to the drop point as a function of time.

(c) the time it takes to reach the bottom of the river from the point at which it was dropped.

Assume that the water is stationary and that buoyancy due to Archimedes' principle may be neglected.

✱✱ **29.** Repeat Exercise 28 but assume that, for the purpose of Archimedes' principle, the volume of the stone is 3 cm^3.

APPENDIX A

Mathematical Induction

Mathematical induction is a type of proof used to verify propositions that involve integers. It is best introduced through a simple example.

Suppose we are required to verify that the sum of the first n positive integers is

$$1 + 2 + 3 + \cdots + n = \frac{n(n + 1)}{2}. \tag{A.1}$$

It is a simple matter to check that this result is valid for the first few integers:

For $n = 1$, the sum is 1, and the formula gives $1(1 + 1)/2 = 1$.
For $n = 2$, the sum is $1 + 2 = 3$; the formula gives $2(2 + 1)/2 = 3$.
For $n = 3$, the sum is $1 + 2 + 3 = 6$; the formula gives $3(3 + 1)/2 = 6$.

We shall see later that to use induction, it is necessary to verify the result at this stage only for $n = 1$.

We now suppose that k is some integer for which the result is valid; that is, we suppose that

$$1 + 2 + 3 + \cdots + k = \frac{k(k + 1)}{2}. \tag{A.2}$$

On the basis of this supposition, the sum of the first $k + 1$ integers is

$$
\begin{aligned}
1 + 2 + \cdots + (k + 1) &= (1 + 2 + \cdots + k) + (k + 1) \\
&= \frac{k(k + 1)}{2} + (k + 1) \qquad \text{(by assumption A.2)} \\
&= \frac{1}{2}[k(k + 1) + 2(k + 1)] \\
&= \frac{(k + 1)(k + 2)}{2}.
\end{aligned}
$$

But this is equation A.1 with n replaced by $k + 1$. We have shown that if A.1 is valid for some integer k, then it must also be valid for the next integer $k + 1$.

We now put the above facts together to verify that A.1 is valid for all $n \geq 1$. We have proved that it is valid for $n = 1, 2,$ and 3. But if it is true for 3, then it must be true for 4. If it is true for 4, then it is true for 5. If it is true for 5, it is true for 6; and so on, and so on. Consequently, A.1 must be valid for every positive integer n.

This example contains the two essential parts of every proof by what we call mathematical induction:

1. Verification of the result for the smallest integer in question;

2. Verification of the fact that if the result is valid for some integer k, then it must be valid for the next integer $k + 1$.

Once these two facts are established, it follows by the same argument as above that the result is valid for the given set of integers. This is the principle of mathematical induction, which we state formally as follows.

THEOREM B.5 (Cramer's rule)

A system of n linear equations in n unknowns $x_1, x_2, \ldots, x_n$,

$$
\begin{aligned}
a_{11}x_1 + a_{12}x_2 + \cdots + a_{1n}x_n &= c_1, \\
a_{21}x_1 + a_{22}x_2 + \cdots + a_{2n}x_n &= c_2, \\
&\ \ \vdots \qquad\quad \vdots \\
a_{n1}x_1 + a_{n2}x_2 + \cdots + a_{nn}x_n &= c_n,
\end{aligned}
\tag{B.6a}
$$

can be represented compactly in the form

$$
\sum_{j=1}^{n} a_{ij}x_j = c_i, \qquad i = 1, \ldots, n.
\tag{B.6b}
$$

From the determinant $D = \left| a_{ij} \right|_{n \times n}$ of the system, we define n other determinants D_k by replacing the kth column of D by the column of constants $c_1, c_2, \ldots, c_n$. The solution of equations B.6 is then

$$
x_k = \frac{D_k}{D}, \qquad k = 1, \ldots, n,
\tag{B.7}
$$

provided $D \neq 0$.

PROOF We multiply the first equation in B.6 by the cofactor A_{1k}, the second equation by A_{2k}, and so on, until the last equation is multiplied by A_{nk}. Symbolically, this is represented by multiplying the ith equation by A_{ik} to get

$$
\sum_{j=1}^{n} a_{ij} A_{ik} x_j = c_i A_{ik}, \qquad i = 1, \ldots, n.
$$

We now add all these equations together:

$$
\sum_{i=1}^{n} \sum_{j=1}^{n} a_{ij} A_{ik} x_j = \sum_{i=1}^{n} c_i A_{ik}.
$$

The right side of this equation is the expansion of D_k along its kth column. Consequently,

$$
D_k = \sum_{j=1}^{n} \left(\sum_{i=1}^{n} a_{ij} A_{ik} \right) x_j.
$$

But according to Theorem B.4, the summation in parentheses is zero unless $j = k$; and when $j = k$, the result is

$$
D_k = \left(\sum_{i=1}^{n} a_{ik} A_{ik} \right) x_k = D x_k, \quad \text{or} \quad x_k = \frac{D_k}{D}.
$$

EXAMPLE B.5

Use Cramer's rule to solve

$$3x - 2y = -1,$$
$$x + 4y - 2z = 6,$$
$$3y + 4z = 7.$$

SOLUTION We have

$$D = \begin{vmatrix} 3 & -2 & 0 \\ 1 & 4 & -2 \\ 0 & 3 & 4 \end{vmatrix} = 3(22) + 2(4) = 74;$$

$$D_x = \begin{vmatrix} -1 & -2 & 0 \\ 6 & 4 & -2 \\ 7 & 3 & 4 \end{vmatrix} = -1(22) + 2(38) = 54;$$

$$D_y = \begin{vmatrix} 3 & -1 & 0 \\ 1 & 6 & -2 \\ 0 & 7 & 4 \end{vmatrix} = 3(38) + 1(4) = 118;$$

$$D_z = \begin{vmatrix} 3 & -2 & -1 \\ 1 & 4 & 6 \\ 0 & 3 & 7 \end{vmatrix} = 3(10) - 1(-11) = 41.$$

Hence, $x = 54/74 = 27/37$, $y = 118/74 = 59/37$, $z = 41/74$.

EXERCISES

In Exercises 1–9 evaluate the determinant.

∗ **1.** $\begin{vmatrix} 3 & 2 \\ -1 & 4 \end{vmatrix}$

∗ **2.** $\begin{vmatrix} 1 & 0 \\ 0 & 1 \end{vmatrix}$

∗ **3.** $\begin{vmatrix} -2 & -4 \\ -6 & -8 \end{vmatrix}$

∗ **4.** $\begin{vmatrix} 1 & 2 & 3 \\ 2 & 4 & 6 \\ -1 & 3 & 0 \end{vmatrix}$

∗ **5.** $\begin{vmatrix} -2 & 0 & 5 \\ 1 & 3 & 6 \\ -7 & 8 & 10 \end{vmatrix}$

∗ **6.** $\begin{vmatrix} 10 & 20 & 30 \\ 16 & 32 & 64 \\ -1 & 2 & -3 \end{vmatrix}$

∗∗ **7.** $\begin{vmatrix} 1 & 1 & 1 & -3 \\ 2 & 1 & 3 & 6 \\ 7 & -8 & 9 & 10 \\ 3 & 4 & -2 & 1 \end{vmatrix}$

∗∗ **8.** $\begin{vmatrix} 3 & -2 & 1 & 6 \\ 4 & 5 & 2 & -1 \\ 0 & 0 & 3 & 2 \\ -1 & 3 & 4 & -1 \end{vmatrix}$

∗∗ **9.** $\begin{vmatrix} 0 & 1 & 2 & 3 & 4 \\ -1 & 0 & 5 & 6 & 7 \\ -2 & -5 & 0 & 8 & 9 \\ -3 & -6 & -8 & 0 & 10 \\ -4 & -7 & -9 & -10 & 0 \end{vmatrix}$

In Exercises 10–15 use Cramer's rule to solve the system of equations.

∗ **10.** $-3x + 4y = 2$, $x - 2y = 6$

∗ **11.** $2x - 3y = -10$, $x + y = 0$

∗ **12.** $4r - 2s = 6$, $3r - s = -1$

∗ **13.** $2x - 3y + z = 2$, $6x - y + 2z = 4$, $x - y = 1$

∗ **14.** $3z - 2y + x = 6$, $z + y + 4x = 2$, $-z - x + 2y = -1$

∗∗ **15.** $2x - 3y + 4z + w = 1$, $x - 3y + 2w = 6$, $3y + 4z - w = 2$, $3x - y + z = 0$

∗∗ **16.** (a) Use mathematical induction to verify that when two parallel lines of a determinant with value D are interchanged, the value of the new determinant is $-D$.

(b) Use part (a) to prove that a determinant with two identical parallel lines has value zero.

∗∗ **17.** A determinant is said to be *skew-symmetric* if its elements satisfy the property $a_{ij} + a_{ji} = 0$. (The determinant in Exercise 9 is skew-symmetric.) Show that a skew-symmetric determinant of odd order has value zero.

∗∗ **18.** Solve the system of equations

$$2x + 3y - 4z + w = 0,$$

$$x + y - 2z + 3w = 0,$$

$$2x - 3y + z - 2w = 0,$$

$$x + y - z + w = 0.$$

∗∗ **19.** Can we use Cramer's rule to solve the system

$$2x + 3y - 4z + w = 0,$$

$$x + y - 2z + 3w = 0,$$

$$2x - 3y + z - 2w = 0,$$

$$5x + y - 5z + 2w = 0?$$

∗∗ **20.** (a) Show that the equation of the straight line in the xy-plane through the two points (x_1, y_1) and (x_2, y_2) can be expressed in the form

$$\begin{vmatrix} x & y & 1 \\ x_1 & y_1 & 1 \\ x_2 & y_2 & 1 \end{vmatrix} = 0.$$

(b) Use the result in (a) to find the equation of the line through $(1, 1)$ and $(-2, 3)$.

∗∗ **21.** (a) Show that the equation of the circle in the xy-plane through the three points (x_1, y_1), (x_2, y_2), and (x_3, y_3) can be expressed in the form

$$\begin{vmatrix} x^2 + y^2 & x & y & 1 \\ x_1^2 + y_1^2 & x_1 & y_1 & 1 \\ x_2^2 + y_2^2 & x_2 & y_2 & 1 \\ x_3^2 + y_3^2 & x_3 & y_3 & 1 \end{vmatrix} = 0.$$

(b) Use the result in (a) to find the circle through $(2, 1)$, $(-3, -3)$, and $(7, -5)$.

** **22.** A parabola of the form $y = ax^2 + bx + c$ is to pass through the three points $(1, 0)$, $(2, 11)$, and $(-2, 1)$. Find a determinant that implicitly defines its equation.

** **23.** Find a condition in the form of a determinant that serves as a test to determine whether a circle can be drawn through four given points, no three of which are collinear. (Hint: See Exercise 21.)

** **24.** Evaluate
$$\begin{vmatrix} a+b & a & a & \cdots & a \\ a & a+b & a & \cdots & a \\ a & a & a+b & \cdots & a \\ \vdots & \vdots & \vdots & \ddots & \vdots \\ a & a & a & \cdots & a+b \end{vmatrix}_{n \times n}.$$

** **25.**

(a) Evaluate
$$\begin{vmatrix} a & b & b & b & \cdots & b & b \\ b & a & b & b & \cdots & b & b \\ b & b & a & b & \cdots & b & b \\ b & b & b & a & \cdots & b & b \\ \vdots & \vdots & \vdots & \vdots & \ddots & \vdots & \vdots \\ b & b & b & b & \cdots & a & b \\ b & b & b & b & \cdots & b & a \end{vmatrix}_{n \times n}.$$

(b) Use the result in (a) to solve the system of equations

$$\begin{aligned} ax_1 + bx_2 + bx_3 + \cdots + bx_n &= 1, \\ bx_1 + ax_2 + bx_3 + \cdots + bx_n &= 1, \\ bx_1 + bx_2 + ax_3 + \cdots + bx_n &= 1, \\ \vdots \qquad\qquad \vdots \qquad\quad \vdots \\ bx_1 + bx_2 + bx_3 + \cdots + ax_n &= 1. \end{aligned}$$

Complex Numbers

The fundamental complex number is i, a number whose square is -1; that is, i is defined as a number satisfying $i^2 = -1$. The **complex number system** is all numbers of the form

$$z = x + yi \tag{C.1}$$

where x and y are real. The number x is called the **real part** of z, and y is called the **imaginary part** of z. For example, real and imaginary parts of $6 - 2i$ are 6 and -2. Both real and imaginary parts of a complex number are themselves real numbers. The real number system is a subset of the complex number system obtained when $y = 0$. We call $x + yi$ the Cartesian form for a complex number.

Complex numbers can be visualized geometrically as points in the complex (Argand) plane. Some fixed point O is chosen to represent the complex number $0 + 0i$. Through O are drawn two mutually perpendicular axes (Figure C.1), one called the real axis, and the other called the imaginary axis. The complex number $x + yi$ is then represented by the point x units in the real direction and y units in the imaginary direction. For example, the complex numbers $1 + 2i$, $-1 - i$, $4 - 3i$, and $-2 + 2i$ are shown in Figure C.2. The real number system is represented by points on the real axis.

FIGURE C.1

FIGURE C.2

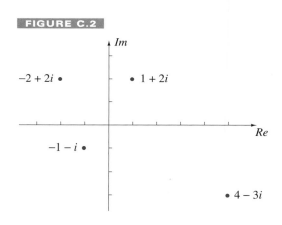

Two complex numbers $x + yi$ and $a + bi$ are said to be equal if their real and imaginary parts are identical; that is,

$$x + yi = a + bi \qquad \Leftrightarrow \qquad x = a \quad \text{and} \quad y = b. \tag{C.2}$$

Geometrically, two complex numbers are equal if they correspond to the same point in the complex plane.

We add and subtract complex numbers $z_1 = x + yi$ and $z_2 = a + bi$ as follows:

$$z_1 + z_2 = (x + a) + (y + b)i, \tag{C.3a}$$

$$z_1 - z_2 = (x - a) + (y - b)i. \tag{C.3b}$$

In words, complex numbers are added and subtracted by adding and subtracting their real and imaginary parts. For example,

$$(3 - 2i) + (6 + i) = (3 + 6) + (-2 + 1)i = 9 - i,$$

$$(3 - 2i) - (6 + i) = (3 - 6) + (-2 - 1)i = -3 - 3i.$$

Complex numbers are multiplied according to the following definition. If $z_1 = x + yi$ and $z_2 = a + bi$, then

$$z_1 z_2 = (x + yi)(a + bi) = (xa - yb) + (xb + ya)i. \tag{C.4}$$

For example,

$$(3 - 2i)(6 + i) = \left[(3)(6) - (-2)(1)\right] + \left[(3)(1) + (-2)(6)\right]i = 20 - 9i.$$

It is not necessary to memorize C.4 when we note that this definition is precisely what we would expect if the usual laws for multiplying binomials were applied, together with the fact that $i^2 = -1$:

$$(3 - 2i)(6 + i) = (3)(6) + (3)(i) + (-2i)(6) + (-2i)(i)$$

$$= 18 + 3i - 12i - 2i^2$$

$$= 18 - 9i - 2(-1)$$

$$= 20 - 9i.$$

With addition, subtraction, and multiplication taken care of, it is natural to turn to division of complex numbers. If we accept that division of any complex number by itself should be equal to 1, and ordinary rules of algebra should prevail, a definition of division of complex numbers is not necessary; it follows from C.4. When $z_1 = x + yi$ and $z_2 = a + bi$, we calculate

$$\frac{z_1}{z_2} = \frac{x + yi}{a + bi}$$

by multiplying numerator and denominator by $a - bi$. This results in

$$\frac{z_1}{z_2} = \frac{x + yi}{a + bi} = \frac{(x + yi)(a - bi)}{(a + bi)(a - bi)}$$

$$= \frac{(xa + yb) + (-xb + ya)i}{a^2 + b^2} \qquad \text{(using C.4)}$$

$$= \left(\frac{xa + yb}{a^2 + b^2}\right) + \left(\frac{ya - xb}{a^2 + b^2}\right)i.$$

For example,

$$\frac{3 - 2i}{6 + i} = \frac{(3 - 2i)(6 - i)}{(6 + i)(6 - i)} = \frac{16 - 15i}{37} = \frac{16}{37} - \frac{15}{37}i.$$

In summary, addition, subtraction, multiplication, and division of complex numbers are performed using ordinary rules of algebra with the extra condition that i^2 is always replaced by -1.

EXAMPLE C.1

Write the following complex numbers in Cartesian form:

$$\text{(a)} \quad (3+i)(2-i)^2 - i \qquad \text{(b)} \quad \frac{i^3}{2+i} \qquad \text{(c)} \quad \frac{4-3i^2+2i}{(2-2i^3)^2}$$

SOLUTION

(a) $\quad (3+i)(2-i)^2 - i = (3+i)(3-4i) - i = (13-9i) - i = 13 - 10i$

(b) $\qquad \dfrac{i^3}{2+i} = \dfrac{-i(2-i)}{(2+i)(2-i)} = \dfrac{-1-2i}{5} = -\dfrac{1}{5} - \dfrac{2}{5}i$

(c) $\dfrac{4-3i^2+2i}{(2-2i^3)^2} = \dfrac{4+3+2i}{(2+2i)^2} = \dfrac{7+2i}{8i} = \dfrac{(7+2i)(-i)}{(8i)(-i)} = \dfrac{2-7i}{8} = \dfrac{1}{4} - \dfrac{7}{8}i$

Notice in part (c) of this example that we multiplied numerator and denominator by $-i$ rather than $-8i$; the result is the same in either case. Both lead to a real denominator.

The complex conjugate $\bar{z}$ of a complex number $z = x + yi$ is

$$\bar{z} = x - yi. \qquad (C.5)$$

Geometrically, $\bar{z}$ is the reflection of z in the real axis (Figure C.3).

FIGURE C.3

The procedure by which two complex numbers are divided can be stated as follows. To divide z_1 by z_2, multiply z_1 and z_2 by $\overline{z_2}$,

$$\frac{z_1}{z_2} = \frac{z_1\overline{z_2}}{z_2\overline{z_2}}.$$

The denominator will be real, and the Cartesian form is immediate.

With complex numbers in place, quadratic equations can be discussed fully. When the discriminant of a quadratic equation is positive, the equation has two real solutions. For example, the discriminant of

$$x^2 + 4x - 2 = 0$$

is $16 + 8 = 24$, and solutions of the equation are

$$x = \frac{-4 \pm \sqrt{16 + 8}}{2} = -2 \pm \sqrt{6}.$$

When the discriminant is zero, we regard the quadratic as having two real solutions which are identical. For example, the discriminant of

$$x^2 + 4x + 4 = 0$$

is zero. The left side may be factored in the form

$$(x + 2)^2 = 0.$$

We say that -2 is a double root of the equation.

For quadratics with negative discriminants, we first consider the equation

$$x^2 + 1 = 0.$$

The complex number i is a solution, by so also is $-i$ since $(-i)^2 + 1 = -1 + 1 = 0$. The quadratic equation

$$x^2 + 16 = 0$$

has two solutions $x = \pm 4i$. If we apply the quadratic formula to the equation

$$x^2 + 2x + 5 = 0,$$

the result is

$$x = \frac{-2 \pm \sqrt{4 - 20}}{2} = \frac{-2 \pm \sqrt{-16}}{2}.$$

By $\sqrt{-16}$ we would seem to mean the number that multiplied by itself is -16. But there are two such numbers, namely $\pm 4i$. Let us make the agreement that $\sqrt{-16}$ shall denote that complex number whose square is -16, and which has a positive imaginary part. By this agreement,

$$\sqrt{-16} = 4i \quad \text{and} \quad -\sqrt{-16} = -4i.$$

The quadratic formula applied to $x^2 + 2x + 5 = 0$ therefore gives two complex numbers

$$x = \frac{-2 \pm 4i}{2} = -1 \pm 2i.$$

It is straightforward to verify that these two complex conjugates actually satisfy $x^2 + 2x + 5 = 0$.

The agreement made in this last example is worth reiterating as a general principle: When $a > 0$ (is a real number),

$$\sqrt{-a} = \sqrt{a}\,i. \tag{C.6}$$

We call $\sqrt{a}\,i$ the **principal square root** of $-a$; the other square root is $-\sqrt{a}\,i$.

The above examples lead us to state that every real quadratic equation

$$ax^2 + bx + c = 0 \tag{C.7a}$$

has two solutions

$$x = \frac{-b \pm \sqrt{b^2 - 4ac}}{2a}. \tag{C.7b}$$

When $b^2 - 4ac > 0$, roots are real and distinct; when $b^2 - 4ac = 0$, roots are real and equal; and when $b^2 - 4ac < 0$, roots are complex conjugates. Verification of this is a matter of substituting C.7b into C.7a.

The only other aspect of complex numbers that we use in the context of differential equations in Chapter 16 is Euler's identity. This is usually developed in courses in complex function theory. We can get to it in a formal way using Maclaurin series from Chapter 11. The real exponential function e^x can be defined by its Maclaurin series (see Example 11.8 in Section 11.3). Consider whether meaning can be given to e raised to a purely imaginary number θi where θ is real; that is, can we give meaning to $e^{\theta i}$? One way would be to replace x in the Maclaurin series for e^x by θi, and call the result $e^{\theta i}$,

$$e^{\theta i} = \sum_{n=0}^{\infty} = 1 + \theta i + \frac{(\theta i)}{2!} + \frac{(\theta i)^3}{3!} + \frac{(\theta i)}{4!} + \cdots.$$

If we gather even and odd terms in separate groups,

$$e^{\theta i} = \left(1 - \frac{\theta^2}{2!} + \frac{\theta^4}{4!} - \frac{\theta^6}{6!} + \cdots\right) + \left(\theta - \frac{\theta^3}{3!} + \frac{\theta^5}{5!} - \cdots\right)i.$$

But the two series on the right are Maclaurin series for $\cos\theta$ and $\sin\theta$; that is, we may write that

$$e^{\theta i} = \cos\theta + \sin\theta i.$$

In other words, if $e^{\theta i}$ is to have a value and this value is to be consistent with our theory on infinite series, then it must be defined as

$$e^{\theta i} = \cos\theta + \sin\theta i. \tag{C.8}$$

This is called Euler's identity. For instance, e^{2i} must be the complex number

$$e^{2i} = \cos 2 + \sin 2i.$$

It follows immediately that

$$e^{\pi i/2} = i, \quad e^{\pi i} = -1, \quad e^{3\pi i/2} = -i, \quad e^{2\pi i} = 1.$$

Eulers identity C.8 implies that

$$e^{-\theta i} = \cos(-\theta) + \sin(-\theta)i = \cos\theta - \sin\theta i. \tag{C.9}$$

When C.8 is added to this equation, we obtain

$$e^{\theta i} + e^{-\theta i} = 2\cos\theta \qquad \text{or} \qquad \cos\theta = \frac{e^{\theta i} + e^{-\theta i}}{2}.$$

This expresses the real cosine function in terms of two complex exponentials. By subtracting rather than adding, we can express $\sin \theta$ in terms of complex exponentials

$$\sin \theta = \frac{e^{\theta i} - e^{-\theta i}}{2i}.$$

EXERCISES

* **1.** Show each of the following complex numbers in the complex plane: $2 - i$, $3 + 4i$, $-1 - 5i$, $-3 + 2i$, $5i$, $2(1 + i)$

In Exercises 2–26 write the complex expression in Cartesian form.

* **2.** $(2 + 4i) - (3 - 2i)$

* **3.** $(1 + 2i)^2$

* **4.** $(-2 + i)(3 - 4i)$

* **5.** $3i(4i - 1)^2$

* **6.** $i^3 - 3i^2 + 2i + 4$

* **7.** $(1 + i)^6$

* **8.** $\dfrac{1 - i}{3 + 2i}$

* **9.** $\dfrac{(3 + i)^2}{2 - i}$

* **10.** $i^{24} - 3i^{13} + 4$

* **11.** $(i - 2)\left[(2 + i)(1 - i) + 3i - 2\right]$

* **12.** $6i\left(\dfrac{1 + i}{2 - i}\right) + 3\left(\dfrac{i - 4}{2i + 1}\right)$

* **13.** $\overline{2 + i} - (3 + \overline{4i})$

* **14.** $\overline{1 + i}^2 + \overline{(1 + i)^2}$

* **15.** $\left(\dfrac{1}{2} - \dfrac{\sqrt{3}}{2}i\right)^3$

* **16.** $(1 + \overline{2 - i})^2$

* **17.** $\dfrac{(2i + 3)(4 - i)}{(3 + i)(-6 + 2i)}$

* **18.** $\dfrac{(1 + i)^2(2 - i)}{(3 + 2i)^2}$

* **19.** $(1 - i)^{12}(2i + 3)$

** **20.** $\overline{(4 - i)^2}$

** **21.** $\dfrac{1}{2i}\left(\dfrac{1}{\sqrt{2}} + \dfrac{i}{\sqrt{2}}\right)^{18}$

** **22.** $\dfrac{1}{1 + \frac{1}{1 + 2i}}$

** **23.** $\dfrac{\frac{i}{3 + i}}{(2 - i)^2}$

** **24.** $\dfrac{(1 + i^3)^2(2 - i)}{4 - 5i}$

** **25.** $\left(\dfrac{1}{\sqrt{2}} + \dfrac{i}{\sqrt{2}}\right)^4 \left(\dfrac{1}{\sqrt{2}} - \dfrac{i}{\sqrt{2}}\right)^4$

** **26.** $\left(\dfrac{2i}{1 + i}\dfrac{3 - 4i}{3 + 4i}\right)^2$

In Exercises 27–36 find all solutions of the equation.

* **27.** $x^2 + 5x + 3 = 0$

* **28.** $x^2 + 3x + 5 = 0$

* **29.** $x^2 + 8x + 16 = 0$

* **30.** $x^2 + 2x - 7 = 0$

* **31.** $x^2 + 2x + 7 = 0$

* **32.** $4x^2 - 2x + 5 = 0$

** **33.** $\sqrt{3}x^2 + 5x + \sqrt{15} = 0$

** **34.** $x^4 + 4x^2 - 5 = 0$

** **35.** $x^4 + 4x^2 + 3 = 0$

** **36.** $x^4 + 6x^2 + 3 = 0$

** **37.** Verify the following properties for the complex conjugation operation:

(a) $\overline{z_1 + z_2} = \overline{z_1} + \overline{z_2}$

(b) $\overline{z_1 - z_2} = \overline{z_1} - \overline{z_2}$

(c) $\overline{z_1 z_2} = \overline{z_1}\,\overline{z_2}$

(d) $\overline{\left(\dfrac{z_1}{z_2}\right)} = \dfrac{\overline{z_1}}{\overline{z_2}}$

(e) $\overline{z^n} = \overline{z}^n$, n a positive integer.

** **38.** Verify that all complex numbers z satisfying the equation $z\overline{z} = r^2$, $r > 0$ a real constant, lie on a circle. What is its centre and radius?

** **39.** Prove that if $z_1 z_2 = 0$, then at least one of z_1 and z_2 must be zero.

** **40.** We have made the agreement that when $a > 0$ is a real number, $\sqrt{-a}$ denotes a complex number with positive imaginary part. Show that with this agreement, $\sqrt{z_1 z_2}$ is not always equal to $\sqrt{z_1}\sqrt{z_2}$.

** **41.** Explain the fallacy in

$$-1 = \sqrt{-1}\sqrt{-1} = \sqrt{(-1)(-1)} = \sqrt{1} = 1.$$

** **42.** Find two numbers whose sum is 6 and whose product is 10.

** **43.** Verify that the values of x in C.7b satisfy C.7a.

** **44.** To find the square roots of a complex number, say i, we could set $(x + yi)^2 = i$, and solve the equation for x and y. Do this by using C.2 for equality of complex numbers.

** **45.** Use the technique of Exercise 44 to find square roots for (a) $-7 - 24i$ (b) $2 + i$.

Answers to Even-Numbered Exercises

Chapter 1

Exercises 1.1

2. $3\sqrt{5}$

4. $\sqrt{58}$

6. $(1, -1/2)$

8. $(-1/2, 1/2)$

10.

12.

14.

16.

18.

20. Parallel **22.** Perpendicular **24.** Neither

26. Neither **28.** $(1, 2)$ **30.** $(-10, -14)$

32. $(37/73, 153/146)$ **34.** $8/\sqrt{5}$ **36.** $2\sqrt{2}$

38. $67/\sqrt{229}$ **40.** $x + y = 2$ **42.** $3y = 2x$

44. $5x + 3y = 30$ **46.** $y = -1$ **50.** $(-11/10, 19/10)$

Exercises 1.2A

2.

4.

6.

8.

10.

12.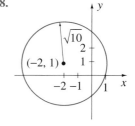

14. $y = x^2/2 + 1$ **16.** $y = (x^2 - 2x - 3)/3$

18. $(0, 1), (-1, 0)$ **20.** No points

22. $(0, 1), (-3, -2)$ **24.** $\pi/4$ radians

26. $(1, 1), (4, 4)$ **28.** $y = (x^2 - 2x + 5)/2$

Exercises 1.2B

2. $(x + 5)^2 + (y - 2)^2 = 6$

4. $x^2 + y^2 - 4y + 1 = 0$

6. $2x^2 + 2y^2 + 6x = 25$

8. $x^2 + 4x + y^2 - 2y = 5$

10. No points satisfy the equation **12.** $(x - 1)^2 + y^2 = 1$

14. $(x - 3/2)^2 + (y + 3/2)^2 = 9/2$

68.

$$A = \begin{cases} t^2/4, & 0 \le t \le 4\sqrt{2} \\ 4\sqrt{2}\,t - t^2/4 - 16, & 4\sqrt{2} \le t \le 8\sqrt{2} \end{cases}$$

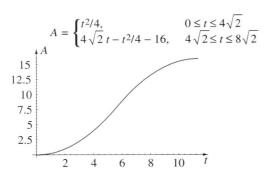

70.

$$y = \begin{cases} 0, & 5 \le x \le 5\sqrt{7} \\ -16 + \sqrt{x^2 + 81}, & 5\sqrt{7} \le x \le 4\sqrt{34} \end{cases}$$

72.   **74.**

76.

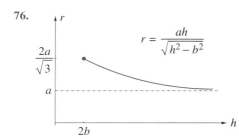

$$r = \frac{ah}{\sqrt{h^2 - b^2}}$$

Exercises 1.4

2. (a) left, right (b) right (c) 6.2 (d) 5.2 (e) left, right (f) three
   (g) 9/2

4. 5/3   **6.** 12

8. 9.5 m

10. (2.2, 2.5)

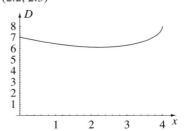

12. (a) 1 C (b) Positive, negative, positive (c) 7 s (d) 2 s, 5 s, 10 s
   (e) 3.5 ≤ t ≤ 7.5 s For what times is the current in the
   circuit counterclockwise? (f) Clockwise, counterclockwise
   (g) 2 ≤ t ≤ 5 s (h) Decreasing

Exercises 1.5

2. $\pi/3$ rad   **4.** $-\pi/2$ rad   **6.** $17\pi/4$ rad

8. $32\pi/45$ rad   **10.** $-213\pi/180$ rad   **12.** $-225°$

14. $1440°$ **16.** $180/\pi°$ **18.** $450/\pi°$

20. $1980/\pi°$

22.

	$\sin x$	$\cos x$	$\tan x$	$\csc x$	$\sec x$	$\cot x$
0	0	1	0		1	
$\pi/6$	1/2	$\sqrt{3}/2$	$1/\sqrt{3}$	2	$2/\sqrt{3}$	$\sqrt{3}$
$\pi/4$	$1/\sqrt{2}$	$1/\sqrt{2}$	1	$\sqrt{2}$	$\sqrt{2}$	1
$\pi/3$	$\sqrt{3}/2$	1/2	$\sqrt{3}$	$2/\sqrt{3}$	2	$1/\sqrt{3}$
$\pi/2$	1	0		1		0

24. 0.927 rad, 0.644 rad, $\pi/2$ rad

26. 2.57, 2.19, $19\pi/30$ rad

32. 2.68 rad   **34.** 1.25 rad   **36.** 0 rad

38. Parallel   **40.** Perpendicular   **42.** 1.37 rad

44. 1.11 rad

46.   **48.**

$y = \sin 2x$ $y = \sin\left(x + \dfrac{\pi}{4}\right)$

50.

$$y = \sin\left(4x + \frac{\pi}{4}\right)$$

52.

$$y = 4\cos\left(x/3\right)$$

54.

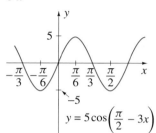

$$y = 5\cos\left(\frac{\pi}{2} - 3x\right)$$

56.

$$y = \tan\left(3x\right)$$

58.

$$y = \cot\left(x + \frac{\pi}{4}\right)$$

60.

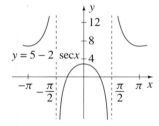

$$y = \sqrt{1 - \cos^2 x}$$

62.

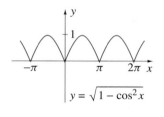

$$y = 5 - 2\,\sec x$$

64.

$$y = \tan|x|$$

66.

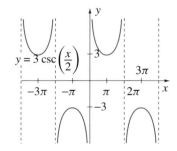

$$y = 3\csc\left(\frac{x}{2}\right)$$

68. No solutions   **70.** $\pm 0.955 + n\pi$ rad, n an integer

72. $-0.392 + 2n\pi$ rad, $2.75 + 2n\pi$ rad, n an integer

74. $2n\pi$, $(4n + 1)\pi/2$, n an integer

76. $\sqrt{17}\sin\left(2x + 0.245\right)$; Amplitude $= \sqrt{17}$, Period $= \pi$, Phase shift $= -0.1225$

78. $(1/2)\sin 2x$; Amplitude $= 1/2$, Period $= \pi$, Phase shift $= 0$

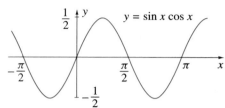

86. $(2n + 1)\pi/4$, $\pm\pi/6 + (4n + 1)\pi/4$, n an integer
88. $(2n + 1)\pi/6$, $(12n + 1)\pi/6$, n an integer

Exercises 1.6
2. $(1/3)\log_{10} 5$ **4.** No solution   **6.** No solution

8. No solution   **10.** 5   **12.** $-1 \pm \sqrt{1 + a^2}$

14.

$$y = e^{-x^2}$$

16.

$$y = \log_a|x^2 - 1|$$

18.

$$y = \log_{10}(4x)$$

20.

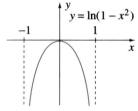

$$y = \ln(1 - x^2)$$

22.

$$y = 10^{x+2}$$

24.

$$y = e^{x^2}$$

26.

$$y = e^{-x^2}\cos x$$

28. Yes. The domain of $f(x)$ is $x \neq 0$, but that of $g(x)$ is $x > 0$.

30. $20\,000(3/4)^t$ **34.** $x_1 > 0$ and $x_2 > 0$

36. (b) 9 years   **38.** 0.694 **40.** $\sqrt{2}$

42. $\{20\,000[1 - (3/4)^t] + 250[(6/5)^t - 1]\}/t$

44. $(1/2)\ln(y + \sqrt{1 + y^2})$ **46.** $(1/2)\ln[(1 + y)/(1 - y)]$

Exercises 1.7

2. $-1.561\,553,\ 2.561\,553$ **4.** $3.044\,723$

6. $0.754\,878$ **8.** $2.485\,584$

10. $\pm 0.795\,324$

12. $-2.931\,137,\ -2.467\,518,\ -1.555\,365,\ -0.787\,653,$
     $0.056\,258,\ 0.642\,581$

14. $4.188\,760$ **16.** $0.852\,606$

18. $0,\ -1.2483$ **20.** $-1.892,\ -0.173,\ 3.064$

22. ± 1.0986 **24.** 0.3212

26. $(-0.1725, 0.6848),\ (-1.8920, 0.7956)$

28. $(-1.3532, 3.8312)$

30. (a) 3.83 s, (b) 4.08 s   **32.** $0.000\,289\,7$

Review Exercises

2. $5\sqrt{2},\ (-1/2, -3/2)$ **4.** $y = 2x + 4$

6. $x + y = 2$ **8.** $|x| \geq \sqrt{5}$

10. $x \neq 0, -1$ **12.** $x \geq 0$

14. $|x + 1| > \sqrt{14}/2$ **16.** $-1 \leq x < 0,\ x \geq 1$

18. Parabola   **20.** Ellipse

22. None of these   **24.** Parabola

26. Circle   **28.** Hyperbola

30. $21/\sqrt{10}$

32.

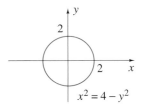

$x^2 = 4 - y^2$

34.

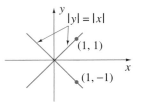

$|y| = |x|$
$(1, 1)$
$(1, -1)$

36.

$x^2 + 3y^2 = 6$

38.

$y = x + 1$
$(1, 2 + \sqrt{2})$
$(1, 2)$
$(1, 2 - \sqrt{2})$
$y = 3 - x$
$x^2 - 2x - y^2 + 4y = 1$

40.

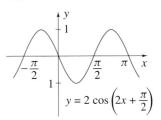

$y = 2\cos\left(2x + \dfrac{\pi}{2}\right)$

42.

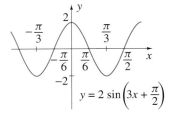

$y = 2\sin\left(3x + \dfrac{\pi}{2}\right)$

44.

$y = \sqrt{-x^2 + 4x + 4}$

46.

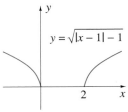

$y = \sqrt{|x - 1|} - 1$

48.

$2\ln 4$
$y = 2\ln(3x + 4)$

50.

$y = \sin|x|$

52.

$y = \sqrt{\sin 2x}$

54. 0.833 rad   **56.** $\sqrt{(x + 1)(2 - x)}$

58. $1 + \sqrt{-x}$ **60.** 2π

62.

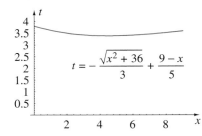

$t = -\dfrac{\sqrt{x^2 + 36}}{3} + \dfrac{9 - x}{5}$

64. $-0.637,\ 1.410$

66. $-4.9385,\ -3.6980,\ 0.0416,\ 2.8422$

Chapter 2

Exercises 2.1

2. 0 **4.** $-3/2$ **6.** 0

8. -12 **10.** 6 **12.** -4

14. 0 **16.** -1 **18.** 0

20. $-7/4$ **22.** does not exist   **24.** 2

26. $1/\sqrt{2}$ **28.** $2\sqrt{2}$ **30.** 1

32. does not exist   **34.** $-1/4$ **36.** 1

38. 4   **40.** -2 **42.** 1

44. False   **46.** $1/(2\sqrt{x})$ **50.** L

52. Cannot find   **54.** L

56. 0 if $F = 0$; does not exist if $F \neq 0$

Exercises 2.2

2. $-\infty$ **4.** ∞ **6.** ∞

8. 13/2   **10.** does not exist   **12.** ∞

14. 1   **16.** ∞ **18.** does not exist

20. ∞ **22.** 0   **24.** ∞

26. ∞

Exercises 2.3

2. $-1/2$ **4.** ∞ **6.** 1/3

8. $-\infty$ **10.** 3/2   **12.** 0

14. 1   **16.** ∞ **18.** ∞

20. $3/4^{1/3}$ **22.** 0   **24.** 0

26. does not exist   **28.** 0   **30.** ∞

32. 1   **34.** -1 **36.** 0

38. ∞

Exercises 2.4

2.

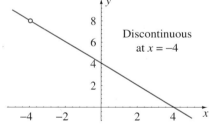

Discontinuous at $x = -4$

4.

6.

8.

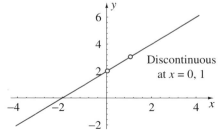

Discontinuous at $x = 0, 1$

10.

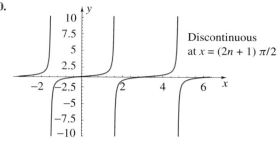

Discontinuous at $x = (2n + 1)\,\pi/2$

12. See Figure 2.8(b). Discontinuous at $x = 0$.

14.

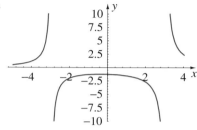

Discontinuous at $x = \pm 3$

16.

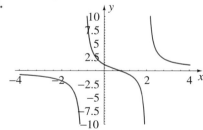

Discontinuous at $x = 3$

18.

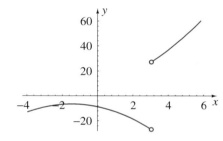

Discontinuous at $x = -1, 2$

20.

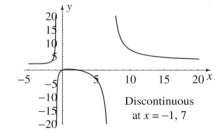

Discontinuous at $x = -1, 7$

22.

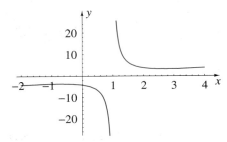

Discontinuous at $x = 1$

24.

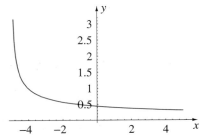

Discontinuous at $x = -5$

26.

Discontinuous at $x = -5, 0$

28.

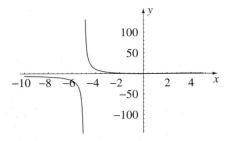

Discontinuous at $x = -5$

30.

Discontinuous at $x = -2$

32. $1 - x^2 + (2x^2 - 1)H(x)$

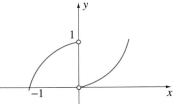

34. $(x - 2)H(x + 1) - 2H(x - 3) - (x - 4)H(x - 5)$

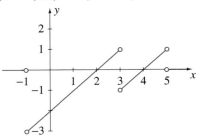

36. $\sin x[H(x) - 3H(x - \pi) + 5H(x - 2\pi) - 7H(x - 3\pi) + 4H(x - 4\pi)]$

38. No

40.

42.

44.

$f(x) = \text{sgn}(x + 1)$
$-\text{sgn}(x - 1)$

Discontinuous at $x = \pm 1$

46.

Discontinuous
at $x = \dfrac{n}{10}$,
n an integer

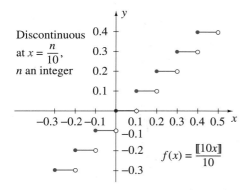

$f(x) = \dfrac{[\![10x]\!]}{10}$

48.

Discontinuous at
$x = n + \dfrac{1}{2}$, n an integer

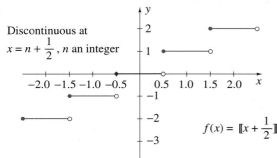

$f(x) = \left[\!\!\left[x + \dfrac{1}{2} \right]\!\!\right]$

50.

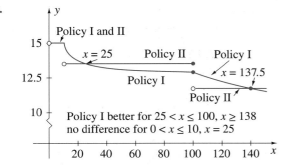

Policy I and II

$x = 25$ Policy II   Policy I

Policy I $x = 137.5$

Policy II

Policy I better for $25 < x \le 100$, $x \ge 138$
no difference for $0 < x \le 10$, $x = 25$

Exercises 2.5

10. $\lim_{x \to a^+} f(x) = L$ if given any $\epsilon > 0$, there exists a $\delta > 0$ such that $|f(x) - L| < \epsilon$ whenever $0 < x - a < \delta$.

12. $\lim_{x \to \infty} f(x) = L$ if given any $\epsilon > 0$, there exists an X such that $|f(x) - L| < \epsilon$ whenever $x > X$.

14. $\lim_{x \to a} f(x) = \infty$ if given any $M > 0$, there exists a $\delta > 0$ such that $f(x) > M$ whenever $0 < |x - a| < \delta$.

16. $\lim_{x \to \infty} f(x) = \infty$ if given any $M > 0$, there exists an X such that $f(x) > M$ whenever $x > X$.

18. $\lim_{x \to -\infty} f(x) = \infty$ if given any $M > 0$, there exists an X such that $f(x) > M$ whenever $x < X$.

30. No

Review Exercises

2. -2 **4.** 1 **6.** $-1/4$

8. 0 **10.** 0 **12.** ∞

14. $1/2$ **16.** 0 **18.** $-\sqrt{3}/2$

20. $-\infty$

22.

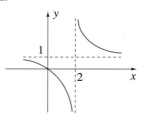

$f(x) = \dfrac{x}{x - 2}$ Discontinuous at $x = 2$

24.

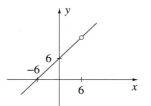

$f(x) = \dfrac{x^2 - 36}{x - 6}$ Discontinuous at $x = 6$

26.

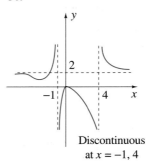

$f(x) = \left| \dfrac{x + 1}{x - 1} \right|$ Discontinuous at $x = 1$

28.

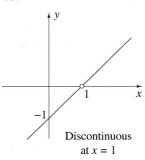

$f(x) = \dfrac{x + 1}{|x - 1|}$ Discontinuous at $x = 1$

30.

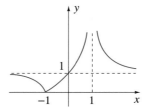

Discontinuous at $x = -1, 4$

$f(x) = \dfrac{2x^2}{x^2 - 3x - 4}$

32.

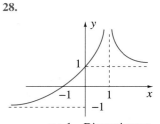

Discontinuous at $x = 1$

$f(x) = \dfrac{x^3 - 3x^2 + 3x - 1}{x^2 - 2x + 1}$

Chapter 3

Exercises 3.1

2. $6x$ **4.** $3x^2 + 4x$ **6.** $3x^2 + 4x$

8. $-9/(x - 5)^2$ **10.** $(2x^2 - 2)/(x^2 + x + 1)^2$

12. $2\pi r$ **14.** $4\pi r^2$ **16.** m

18. 0 **20.** $y = 2x + 2$ **22.** $x + 4y = 3$

24. 1.107 **26.** 2.897 **28.** $8x^7$

30. $-4/(x - 2)^5$ **32.** $(3x + 2)/(2\sqrt{x + 1})$

34. $\sqrt{A}/(4\sqrt{\pi})$ **36.** $|x|/x$

38.

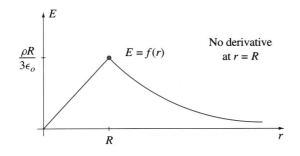

$\dfrac{\rho R}{3\epsilon_o}$ $E = f(r)$ No derivative at $r = R$

Exercises 3.2

2. $9x^2 + 4$ **4.** $20x^4 - 30x^2 + 3$ **6.** $-6/x^4$

8. $1/x^3 - 12/x^5$ **10.** $20x^3 - 5/(4x^6)$ **12.** $1/(2\sqrt{x})$

14. $-(3/2)/x^{5/2} + (3/2)\sqrt{x}$

16. $\pi^2 x^{\pi-1}$ **18.** $4 + 3/x^4$ **20.** $6(2x + 5)^2$

22. $4y = x + 24,\ 4x + y = 23$

24. $y = 6x - 10,\ x + 6y = 51$

26. $75 - 7x/50,\ 50 - x/15 - x^2/20\,000,\ 25 - 11x/150 + x^2/20\,000$

28. $(2, -12)$ **32.** $n|x|^n/|x|$

Exercises 3.3

2. right; no left; no derivative   **4.** no right; no left; no derivative

6. right; no left; no derivative   **8.** right; no left; no derivative

10. True   **12.** True

14.   **16.**

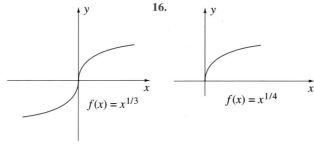

$f(x) = x^{1/3}$ $f(x) = x^{1/4}$

18. 1.23 rad at $(\pm 1/\sqrt{2},\ 1/2)$ **24.** 2/3

26. Yes   **28.** Not necessarily

30. Yes; derivative is zero

Exercises 3.4

2. $4(2 - 3x^2 - x^3)$ **4.** $-10x/(4x^2 - 5)^2$

6. $x^2(4x^2 + 3)/(4x^2 + 1)^2$ **8.** $(2 - 3x)/[2\sqrt{x}(3x + 2)^2]$

10. $-(2x^2 + 20x + 1)/(2x^2 - 1)^2$

12. $(17 - 4x - 7x^2)/(x^2 - 5x + 1)^2$

14. $-27/(x + 1)^4$

---

16. $(x - 8\sqrt{x} - 2)/[\sqrt{x}(\sqrt{x} - 4)^2]$

18. (b) $x(x - 4)/(x^2 + x - 2)^2$

20. $f'(x)g(x)h(x) + f(x)g'(x)h(x) + f(x)g(x)h'(x)$

22. 0.668 at $(\sqrt{2},\ 2\sqrt{2} + 2)$; 0.415 at $(-\sqrt{2},\ 2 - 2\sqrt{2})$

24. $(5 + \sqrt{55},\ -\sqrt{55}/(11 + \sqrt{55})),\ (5 - \sqrt{55},\ \sqrt{55}/(11 - \sqrt{55}))$

26. (a)

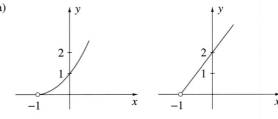

$H(x + 1)(x + 1)^2$ $2H(x + 1)(x + 1)$

(b) Yes, $f'(-1) = 0$

Exercises 3.5

2. 6   **4.** 18   **6.** $6 + 60/t^6$

8. $(3u^2 - 6u - 1)/[4u^{3/2}(u + 1)^3]$

10. $-(\sqrt{x} + 3)/[4\sqrt{x}(\sqrt{x} + 1)^3]$

12. (b) $(bT_b - aT_a)/(b - a),\ ab(T_a - T_b)/(b - a)$

14. $-1, 3$

16. (a) $f'''g + 3f''g' + 3f'g'' + fg'''$
(b) $f''''g + 4f'''g' + 6f''g'' + 4f'g''' + fg''''$

Exercises 3.6

2. (a) Left, right (b) Left, right (c) Three (d) 7/2 (e) 9/2

4. (a) Right, left (b) Right, left (c) Two (d) 9/2 (e) 9/2

6. $v = 2t - 7$ m/s, $a = 2$ m/s^2

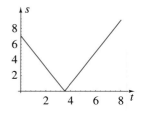

8. $v = -6t^2 + 4t + 16$ m/s, $a = -12t + 4$ m/s²

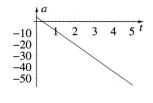

10. $v = -12 \sin 4t$ m/s, $a = -48 \cos 4t$ m/s²

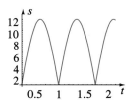

12. $v = 1 - 4/t^2$ m/s, $a = 8/t^3$ m/s²

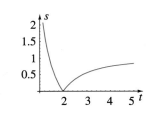

14. $v = (5t^2 - 6t + 1)/(2\sqrt{t})$ m/s,
    $a = (15t^2 - 6t - 1)/(4t^{3/2})$ m/s²

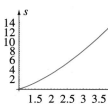

16. (a) -11 m, -12 m/s, 12 m/s, 0 m/s² (b) $t = 1$ s, $t = 5$ s
    (c) $t = 3$ s (d) $0 \le t < 1$, $t > 5$; $1 < t \le 5$
    (e) $(9 \pm \sqrt{39})/3$ s (f) Never (g) $(9 + \sqrt{106})/3$ s
    (h) $(9 + \sqrt{106})/3$ s

18. $a(t_1 + t_2)/2 + b$ (b) $(t_1 + t_2)/2$, closer to x_1

Exercises 3.7

2. $1/[2\sqrt{x}(u + 1)^2]$ 4. $-2(x - 1)(s^2 + 2)/(s^2 - 2)^2$

6. $-21/[(t - 4)^2(x + 1)^2]$

8. $(4u + 5u^{3/2})(x^2 + 2x - 1)/[2(x - x^2)^2]$

10. $(3x + 2)/[2\sqrt{x + 1}]$ 12. $(x + 1)/(2x + 1)^{3/2}$

14. $(2x - 1)(6x + 23)/(3x + 5)^2$

16. $x^2(18 - 55x^2)/[3(2 - 5x^2)^{2/3}]$

18. $(x + 1)(15x + 11)(3x + 1)^2$

20. $(9x - 8)/[2x^3\sqrt{2 - 3x}]$ 22. $-1/[(2 + x)^{5/4}(2 - x)^{3/4}]$

24. $(x + 5)^3(11x^3 + 15x^2 + 8)/[2\sqrt{1 + x^3}]$

26. $(x + 5)^3(13x^4 + 25x^3 + 10x + 10)/[2\sqrt{1 + x^3}]$

28. $(7x^2 + 6x + 4\sqrt{1 + x})/[4\sqrt{1 + x}\sqrt{1 + x\sqrt{1 + x}}]$

30. $(4/3)x(s - 1)/[(x^2 + 5)^2(2s - s^2)^{2/3}]$

32. $(5/2)(2 - x)/[x^2\sqrt{x - 1}(u + 5)^2]$

34. $[1 + (2\sqrt{t} + 1)/(4\sqrt{t}\sqrt{t + \sqrt{t}})][-4x/(x^2 - 1)^2]$

36. $(1 - k - 3k^2)(x^2 + 5)^4(11x^2 + 5)/[2\sqrt{k}(1 + k + k^2)^2]$

38. -0.067 N/s

44. $(6u^4 + 6u^3 - 2)(2\sqrt{x + 1} + 1)^2/[4u^3(x + 1)] -$
    $[3(u^2 + u)^2 + 1]/[4u^2(x + 1)^{3/2}]$

46. $-3[2\sqrt{x} + (s + 6)(1 + 3x + 4\sqrt{x})]/[2x^{3/2}(1 + \sqrt{x})^4(s + 6)^3]$

48. $-8f(3 - 4x)f'(3 - 4x)$; $8(4x - 3)(16x^2 - 24x + 7)$
    $(48x^2 - 72x + 25)$

50. $(1 - 1/x^2)f'(x + 1/x)$; $(x^2 - 1)(3x^4 + 6x^2 + 1)/x^4$

52. $12f(1 - 3x)f'(1 - 3x)/\sqrt{3 - 4[f(1 - 3x)]^2}$;
$12(1 - 3x)(9x^2 - 6x - 1)(27x^2 - 18x + 1)/$
$\sqrt{3 - 4(1 - 3x)^2(9x^2 - 6x - 1)^2}$

54. $f'(x - f(x))[1 - f'(x)]$; $3(1 - x^2)(3x^6 - 18x^4 + 27x^2 - 2)$

56.   **58.**

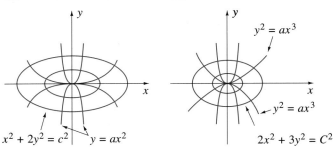

$x^2 + 2y^2 = c^2$ $y = ax^2$ $2x^2 + 3y^2 = C^2$

60. 444   **62.** $-x(x + 1)^2/\sqrt{1 - x^2}$

64. $(4.05, -0.164), (-4.51, -0.521)$

66. $(-4^{1/3}, \pm(4^{1/3} - 1)^{3/2})$

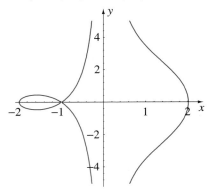

Exercises 3.8

2. $-4x^3/(2y + 3y^2)$

4. $(6x^2 - 3y^4 + 5y)/(12xy^3 - 5x)$

6. $(1 - x - y)/(x + y)$

8. $(48x\sqrt{x + y} - 1)/[1 + (4y - 2)\sqrt{x + y}]$

10. $-(6x + 5y)/(5x + 2y)$

12. $-[2(3y^2 + 1)^2 + 24x^2y]/(3y^2 + 1)^3$

14. $-(25/4)/(y + 1)^3$ **16.** -1

18. $-(2xy^3 + 2)/(3x^2y^2 + 4)$, $[-2y^3(3x^2y^2 + 4)^2 + 12xy^2$
$(2xy^3 + 2)(3x^2y^2 + 4) - 6x^2y(2xy^3 + 2)^2]/(3x^2y^2 + 4)^3$

20. $\sqrt{1 - y^2}/(1 - 2y^2)$ **22.** $-3/4$

24. $0, -1/2$ **26.** $(1, \pm1), (-1, \pm1)$

28. (a) $(x^2 + 4x + 2)/(x^2 + 3x + 2)$ (b) $(2y + 1)(3 - y)/(7y)$

30. $(1, 2)$ **34.** $(3r^3 - 6ar^2 + x)/(3r^2 - 6ar)$

36. (a) -1 **42.** (a) $-\sqrt{y/x}$

44. $2xu/[(u^2 - 1)(1 + 2x^2\sqrt{u^2 - 1})]$

Exercises 3.9

2. $-\sin x - 20\cos 5x$ **4.** $-9\sec^2 3x/\tan^4 3x$

6. $2\csc(4 - 2x)\cot(4 - 2x)$ **8.** $\cot x^2 - 2x^2\csc^2 x^2$

10. $[(x^2 + x)\cos x + \sin x]/(x + 1)^2$

12. $2\cos 4x$

14. $(3/4)\tan^2 x\sec^2 x/(1 + \tan^3 x)^{3/4}$

16. $(\cos y + y\sin x)/(\cos x + x\sin y)$

18. $\sec^2(x + y)/[1 - \sec^2(x + y)]$

20. $3(1 - x^2y)/(x^3 + 2\tan y\sec^2 y)$

22. $-\sec^2 x\sin(\tan x)$ **24.** $4x\sin(2x^2)$

26. $-\sec v\tan v\sec^2 \sqrt{x}/[4\sqrt{x}\sqrt{3 - \sec v}]$

28. $-x\sin x^2\sec^3 u\tan u/[\sqrt{1 + \cos x^2}(1 + \sec^3 u)^{2/3}]$

30. $[18x^2\tan^2(3x^2 - 4)\sec^2(3x^2 - 4) - (2 + x\cot x)$
$(1 + \tan^3(3x^2 - 4))]/(x^3\sin x)$

32. $2(y + 1)[\sec^2 y - x - (y + 1)\sec^2 y\tan y]/(\sec^2 y - x)^3$

34. 0 **36.** 2

38. Does not exist   **40.** No

44. $\cos x|\sin x|/\sin x$, $x \neq n\pi$

Exercises 3.10

2. $6x/(3x^2 + 1)$ **4.** $-2e^{1-2x}$

6. $\ln x + 1$ **8.** $-4\log_{10} e/(3 - 4x)$

10. $-\tan x$ **12.** $2x + (4x^3 + 3x^2)e^{4x}$

14. $2e^{2x}\cos(e^{2x})$ **16.** $e^{-2x}(3\cos 3x - 2\sin 3x)$

18. $4(e^x + e^{-x})^{-2}$ **20.** $2\sin(\ln x)$

22. $-\tan v\sin 2x$

24. $-(y\cos x + 2x\ln y + e^y)/(xe^y + \sin x + x^2/y)$

26. $2x^2/\sqrt{x^2 + 1}$ **28.** $(1 + 5e^{4x})^{-1}$

30. $(y^2/x^2)(e^{1/x} - 1)/(1 - e^{1/y})$

32. (b) $c = [T_b - T_a + k(a - b)]/\ln(b/a)$,
$d = [T_a\ln b - T_b\ln a + k(b\ln a - a\ln b)]/\ln(b/a)$

34. (a) $dT/dt + 3T/10\,000 = (20t + 1257)/419\,000$
(b)

38.

40. (b)

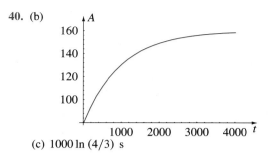

(c) $1000 \ln (4/3)$ s

42.

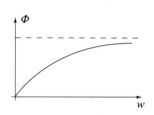

44. $1/\sqrt{x^2 + 4}$

Exercises 3.11

2. $4x^{4\cos x}(\cos x/x - \sin x \ln x)$

4. $(\sin x)^x[x \cot x + \ln (\sin x)]$

6. $x(1 + 1/x)^{x^2}[2 \ln (1 + 1/x) - 1/(1 + x)]$

8. $(3/x^2)(2/x)^{3/x}(\ln x - \ln 2 - 1)$

10. $(1/x)(\ln x)^{\ln x}[\ln (\ln x) + 1]$

12. $(1 + 9x^2 + 6x^4)/[2\sqrt{x}(1 + x^2)^{3/2}]$

14. $(x^2 + 3x)^2(x^2 + 5)^3(14x^3 + 33x^2 + 30x + 45)$

16. $\sqrt{x}e^{-2x}(3 - 4x)/2$

18. $e^x[(x - 1) \ln (x - 1) - 1]/\{(x - 1)[\ln (x - 1)]^2\}$

20. $(1 - 6x^2 - 3x^4)/[2\sqrt{x}(1 + x^2)^{3/2}]$

22. $x^3(13x^4 - 44x^3 + 10x^2 - 32)/(2\sqrt{1 + x^3})$

24. $\sin 2x \sec 5x[2 \cot 2x + 5 \tan 5x - 6 \csc^2 x/(1 - 2 \cot x)]/$
$(1 - 2 \cot x)^3$

26. (b) $(a - br)^{-1}$

Exercises 3.12

2. 2

4. Cannot be applied since $f'(0)$ does not exist

6. $(-2 \pm \sqrt{19})/3$ **8.** $1 + \sqrt{3}$ **10.** $-3 + \sqrt{6}$

12. $(4 - \ln 5 + \ln 2)/[2(\ln 5 - \ln 2)]$

14. Cannot be applied since $f(\pi/2)$ is not defined

16. Cannot be applied since $g'(0) = 0$

18. $(1 + \sqrt{6})/(1 - \sqrt{6})$

Review Exercises

2. $6x + 2 - 1/x^2$ **4.** $(1/3)x^{-2/3} - (10/9)x^{2/3}$

6. $x(x^2 + 2)(x^3 - 3)^2(13x^3 + 18x - 12)$

8. $17/(x + 5)^2$ **10.** $-4(x^2 + 2)/(x^2 + 5x - 2)^2$

12. $2x(y - 1)/(2y - x^2)$

14. $-(4xy\sqrt{1 + x} + y)/[2\sqrt{1 + x}(x^2 + \sqrt{1 + x})]$

16. $-8 \sec^2 (1 - 4x) \tan (1 - 4x)$

18. $2 \sec^2 2x \sec (\tan 2x) \tan (\tan 2x)$

20. $(1/2) \sin 4x$ **22.** $(1 - 2 \sin 2t)(1 + 2 \sin 2x)$

24. $(-x/v)(\cos^2 v - v \sin 2v)$ **26.** $(1/2)e^{-2y}$

28. $2x(x^2 + 1)/(x^2 - 1) + 2x \ln (x^2 - 1)$

30. 1 **32.** $16x/(x^2 + 1)^2$

34. $x(20 - 23x - 3x^2)/[2\sqrt{1 - x}(x + 5)^2]$

36. $(x - y)/(x + y)$

38. $(y^3 + 4y + 6)^2/(-y^4 + 4y^3 + 4y^2 + 12y - 12)$

40. $1/(2\sqrt{x})$ **42.** $(\cos x)^x[\ln (\cos x) - x \tan x]$

44. $(\log_{10} e)^2/(x \log_{10} x)$ **46.** $1/(e^y - e^{-y})$

48. $[(x + y)(1 - 2xy) - 1]/[x^2(x + y) + 1]$

50. $x + 25y = 5$, $125x - 5y + 1 = 0$

52. $23x + 9y = 11$, $27x - 69y = 119$

54. $-42(x^2 - 5xy + y^2)/(5x - 2y)^3$

56. No

58. $(-3, 1)$; finding the shortest distance from $(-6, 7)$ to the curve

60. $\sqrt{3}\ell/2$

62. $(\sqrt{6}/4, \pm\sqrt{2}/4)$, $(-\sqrt{6}/4, \pm\sqrt{2}/4)$

64. $(\sqrt{93} - 9)/6$

Chapter 4

Exercises 4.1

2. $-1.561\,553$, $2.561\,553$ **4.** $3.044\,723$

6. $0.754\,878$ **8.** $2.485\,584$

10. $\pm 0.795\,324$

12. $-2.931\,137$, $-2.467\,518$, $-1.555\,365$, $-0.787\,653$,
$0.056\,258$, $0.642\,851$

14. $4.188\,760$ **16.** $0.852\,606$

18. 0, -1.2483 **20.** -1.892, -0.173, 3.064

22. ± 1.0986 **24.** 0.3212

26. $(-0.1725, 0.6848)$, $(-1.8920, 0.7956)$, $(3.0644, 16.5196)$

28. $(-1.3532, 3.8312)$ **30. (a)** 3.83 s **(b)** 4.08 s

Exercises 4.2

2. Decreasing for all x

4. Increasing for $x \leq 5/4$; decreasing for $x \geq 5/4$

6. Increasing for $x \leq 1/4$; decreasing for $x \geq 1/4$

8. Increasing for all x

10. Decreasing for all x

12. Decreasing for $x \leq 1$; increasing for $x \geq 1$

14. Decreasing for $-1 \leq x \leq 1$; increasing for $x \leq -1$ and $x \geq 1$

16. Decreasing for $x \leq -1$ and $0 < x \leq 1$; increasing for $-1 \leq x < 0$ and $x \geq 1$

18. Decreasing for $0 \leq x < 1$ and $x > 1$; increasing for $x < -1$ and $-1 < x \leq 0$

20. Decreasing for $x \leq -1$ and $0 \leq x \leq 1$; increasing for $-1 \leq x \leq 0$ and $x \geq 1$

22. Decreasing for $x \leq 0$ and $x \geq 2$; increasing for $0 \leq x \leq 2$

24. Decreasing for $0 < x \leq 1/e$; increasing for $x \geq 1/e$

26. Decreasing for $x < -2$, $-2 < x < 1$, $x > 1$

28.
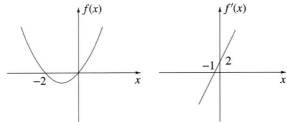
$$f(x) = x^2 + 2x \qquad f'(x) = 2x + 2$$

30.
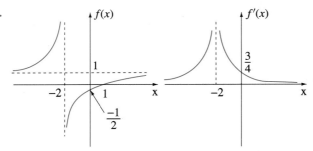
$$f(x) = \frac{x-1}{x+2} \qquad f'(x) = \frac{3}{(x+2)^2}$$

32.
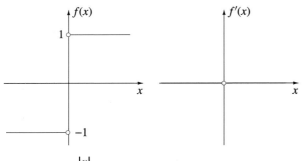
$$f(x) = \frac{|x|}{x} \qquad f'(x) = 0, \ x \neq 0$$

34.
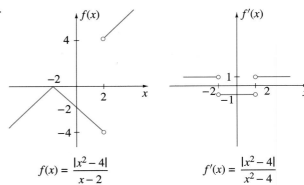
$$f(x) = \frac{|x^2 - 4|}{x - 2} \qquad f'(x) = \frac{|x^2 - 4|}{x^2 - 4}$$

36. Decreasing for $x \leq -0.8612$; increasing for $x \geq 0.8612$

38. Decreasing for $-\pi \leq x \leq -2.2889$ and $2.2889 \leq x \leq \pi$; increasing for $-2.2889 \leq x \leq 2.2889$

46. Yes

48. One at most

Exercises 4.3

2. Relative minimum at $x = -1$; relative maximum at $x = -4$

4. $x = 1$ critical, but does not give a relative extremum

6. Relative maximum at $x = 1 - \sqrt{2}$; relative minimum at $x = 1 + \sqrt{2}$

8. $x = 2$ critical, but does not give a relative extremum

10. $x = 0$ critical, but does not give a relative extremum

12. Relative minimum at $x = -3^{1/4}$; relative maximum at $x = 3^{1/4}$; $x = 0$ critical, but does not give a relative extremum

14. Relative maximum at $x = -1$; relative minimum at $x = 1$

16. $x = 0$ critical, but does not give a relative extremum

18. Relative maximum at $x = 7$; $x = 1$ critical, but does not give a relative extremum

20. Relative minima at $x = 2(3n + 2)\pi/3$; relative maxima at $x = 2(3n + 1)\pi/3$

22. Relative minimum at $x = 2 + 50^{1/3}$

24. Relative maximum at $x = 1$

26. Relative maximum at $x = 0$

28. Relative maximum at $x = -2$; relative minimum at $x = 2$

30. Relative maxima at $x = (8n + 1)\pi/4$; relative minima at $x = (8n + 5)\pi/4$

32. Relative minimum at $x = 0.464$

34. $x = 3\pi/4$ and $7\pi/4$ are critical, but do not give relative extrema

36. Relative minimum at $x = 1/e$

38. Relative maximum at $x = 1/2$

40. Relative maximum at $x = -2/3$; relative minimum at $x = 0$

42. No   44. False   46. True

48. False   50. False

52. (a)

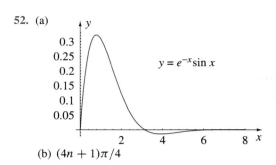

$y = e^{-x}\sin x$

(b) $(4n + 1)\pi/4$

54. relative minima at $x = 2.3301$ and $x = -2.1284$; relative maximum at $x = -0.2016$

56. relative maximum at $x = -4.1072$ **60.** 1.6299

62. (c) No   **64.** (c) Relative maximum

Exercises 4.4

2. Concave upward on $x \le -2/3$ and $x \ge 0$; concave downward on $-2/3 \le x \le 0$; points of inflection at $(0, 2)$ and $(-2/3, 470/27)$

4. Concave upward on $x < -1$ and $x > 1$; concave downward on $-1 < x < 1$

6. Concave upward for all x

8. Concave upward on $-2\pi < x \le -5\pi/6$, $-\pi/6 \le x \le 7\pi/6$, $11\pi/6 \le x < 2\pi$; concave downward on $-5\pi/6 \le x \le -\pi/6$, $7\pi/6 \le x \le 11\pi/6$; points of inflection at $(-5\pi/6, 2 + 25\pi^2/36)$, $(-\pi/6, 2 + \pi^2/36)$, $(7\pi/6, 2 + 49\pi^2/36)$, $(11\pi/6, 2 + 121\pi^2/36)$

10. Concave downward on $0 < x \le e^{-3/2}$; concave upward on $x \ge e^{-3/2}$; point of inflection at $(e^{-3/2}, -3/(2e^3))$

12. Concave downward on $x \le 1$; concave upward on $x \ge 1$; point of inflection at $(1, e^{-2})$

14. Concave downward on $x \le -\ln 2$; concave upward on $x \ge -\ln 2$; point of inflection at $(-\ln 2, (\ln 2)^2 - 2)$

16. Relative minimum at $x = 1$; relative maximum at $x = -1$

18. Relative minumum at $x = 1/5$

20. Relative minimum at $x = 1/\sqrt{e}$

22. Relative minimum at $x = 0$; relative maximum at $x = 1$

26.

28.

30.

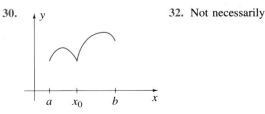

32. Not necessarily

---

38. (a)

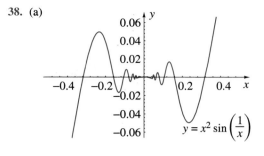

$y = x^2 \sin\left(\dfrac{1}{x}\right)$

Exercises 4.5

2.

$y = x^3 - 6x^2 + 12x + 9$

4.

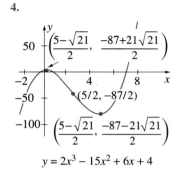

$\left(\dfrac{5 - \sqrt{21}}{2}, \dfrac{-87 + 21\sqrt{21}}{2}\right)$

$(5/2, -87/2)$

$\left(\dfrac{5 - \sqrt{21}}{2}, \dfrac{-87 - 21\sqrt{21}}{2}\right)$

$y = 2x^3 - 15x^2 + 6x + 4$

6.

$\left(-2\sqrt{3}, -\sqrt{3}/8\right)$ $\left(2, \dfrac{1}{4}\right)$

$\left(-2, -\dfrac{1}{4}\right)$ $\left(2\sqrt{3}, \sqrt{3}/8\right)$

$y = \dfrac{x}{x^2 + 4}$

8.

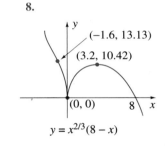

$(-1.6, 13.13)$

$(3.2, 10.42)$

$(0, 0)$

$y = x^{2/3}(8 - x)$

10.

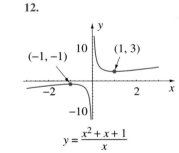

$(2, 16\sqrt{2} - 18)$

$(1, 5)$ $(4, 4)$

$y = 2x^{3/2} - 9x + 12x^{1/2}$

12.

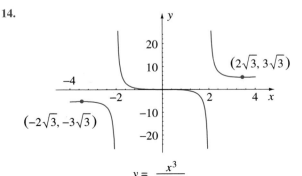

$(-1, -1)$ $(1, 3)$

$y = \dfrac{x^2 + x + 1}{x}$

14.

$(2\sqrt{3}, 3\sqrt{3})$

$(-2\sqrt{3}, -3\sqrt{3})$

$y = \dfrac{x^3}{x^2 - 4}$

16.

$$y = \frac{x^2 + 1}{x^2 - 1}$$

18.

$$y = (x-1)^{2/3}$$

20.

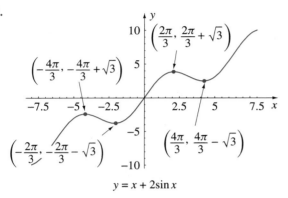

$$y = x + 2\sin x$$

22.

$$P = \frac{MI}{I + K} - R$$

24.

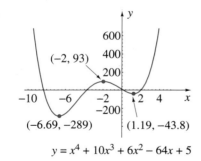

$$y = x^4 + 10x^3 + 6x^2 - 64x + 5$$

26.

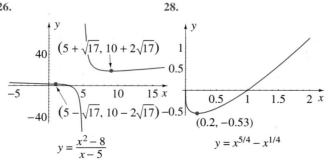

$$y = \frac{x^2 - 8}{x - 5}$$

28.

$$y = x^{5/4} - x^{1/4}$$

30.

$$y = \frac{(2x - 1)(x - 8)}{(x - 1)(x - 4)}$$

32.

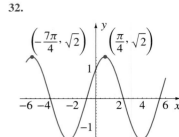

$$y = \sin x + \cos x$$

34.

$$y = \left(\frac{x + 8}{x}\right)\sqrt{x^2 + 100}$$

36.

$$P = \left(\frac{4r^2}{a^2}\right)e^{-2r/a}$$

38.

40. (a) (3.336, 0.0416)

42.

$$y = \frac{1 + x + x^2 + x^3}{1 + x^3}$$

Exercises 4.6

2. $-4, 6/11$

4. $\pi/3 - \sqrt{3}, 11\pi/3 + \sqrt{3}$

6. no absolute minimum, 12

8. no absolute minimum, $\sqrt{3}/6$

12. $720\sqrt{10}$ m

14. $12^{1/3} \times 12^{1/3} \times 12^{1/3}/2$ m

16. $90 \times 180/\pi$ m

18. $ab/2$

20. $a/\sqrt{3}$

22. 15 cm $\times$ 22.5 cm

24. $(-1 - \sqrt{6}/2, 5/2 + \sqrt{6})$

26. smallest area is 20

28. 32

30. (a) 7/4 km from P (b) directly to Q

32. lengths in x- and y-directions are $\sqrt{2}a$ and $\sqrt{2}b$

34. $\sqrt{2/e}$

36. $4\pi hr^2/27$

38. 11.8 m

40. $[p\beta/(r\alpha - r)]^{1/\alpha}$

42. $dI_1^{1/3}/(I_1^{1/3} + I_2^{1/3})$ from I_1 source

44. (a) width $= 5/3^{1/3}$ m, height $= 4(3^{2/3})$ m
(b) width $= 4.19$ m, height $= 5.70$ m

48. $\ell = 1/3$ m, $w = h = 2/3$ m; 1/3 m apart

50. $2\pi(1 - \sqrt{6}/3)$

54. $(2, 2\sqrt{5}/3)$

56. $x = -(A + bB)/(2aB)$

Exercises 4.7

2. $v = 2t - 7$ m/s, $a = 2$ m/s^2

4. $v = -6t^2 + 4t + 6$ m/s, $a = -12t + 4$ m/s^2

6. $v = -12 \sin 4t$ m/s, $a = -48 \cos 4t$ m/s^2

8. $v = 1 - 4/t^2$ m/s, $a = 8/t^3$ m/s^2

10. $v = (5t^2 - 6t + 1)/(2\sqrt{t})$ m/s,
$a = (15t^2 - 6t - 1)/(4t^{3/2})$ m/s^2

12. (a) Neither (b) 12 m/s, -15 m/s (c) 12 m/s, 0 m/s (d) 18 m/s^2, -18 m/s^2 (e) Never

14. (a) $0 \le t \le (4 - \sqrt{7})/3, (4 + \sqrt{7})/3 \le t \le 4$;
$(4 - \sqrt{7})/3 \le t \le (4 + \sqrt{7})/3$ (b) $0 \le t \le (4 - \sqrt{7})/3$,
$1 \le t \le (4 + \sqrt{7})/3, 3 \le t \le 4; (4 - \sqrt{7})/3 \le t \le 1$,
$(4 + \sqrt{7})/3 \le t \le 3$ (c) $t = 4$ s, $t = (4 + \sqrt{7})/3$ s
(d) $t = 4$ s, $t = 0, 1, 3$ s (e) 34 m (f) 30 m

16. Horizontal point of inflection **18.** 144 m/s

20. Not always

22. Not always

Exercises 4.8

2. $2\sqrt{13}/3$ m/s

4. $13/(5\pi)$ cm/s

6. decreasing at 20 m^2/s

8. -2.5 N/m^2/s

10. $7/\sqrt{5}$ m/s

12. (a) $2/(25\pi)$ m/min (b) $2/(25\pi)$ m/min

14. (a) $10(50 - x)/y$ m/s (b) $10/\sqrt{3}$ m/s (c) Car 1

16. -27.14 cm/s **18.** 594.5 cm/s **20.** $-31\sqrt{15}/10$

22. 102 m/s **24.** $6/\sqrt{13}$ m/s

26. (a) $1/(50\pi)$ m/min (b) $\sqrt{13}/(300\pi)$ m/min

28. $4\pi x\ell(k - x) \sin\theta/[y(\ell \cos\theta - x)]$ m/s

30. (a) $\pm 11\pi/360$ rad/min (b) $-11\pi/60$ cm/min (c) 0.23 cm/min

Exercises 4.9

2. (b) $-V/R$

Exercises 4.10

2. 6

4. 0

6. ∞

8. 0

10. Does not exist

12. ∞

14. $1/\sqrt{5}$

16. na^{n-1}

18. 1

20. 0

22. $1/12$

24. 2^{15}

26. 0

28. 4

30. 0

32. 1

34. e

36. 1

38. Does not exist

40. 2

42. $-1/3$

44.

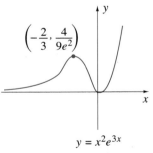

$\left(-\dfrac{2}{3}, \dfrac{4}{9e^2}\right)$

$y = x^2 e^{3x}$

46.

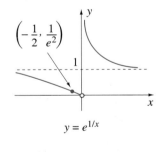

$\left(-\dfrac{1}{2}, \dfrac{1}{e^2}\right)$

$y = e^{1/x}$

48.

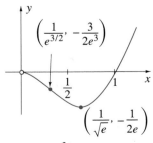

$\left(\dfrac{1}{e^{3/2}}, -\dfrac{3}{2e^3}\right)$

$\left(\dfrac{1}{\sqrt{e}}, -\dfrac{1}{2e}\right)$

$y = x^2 \ln x$

50.

$(\sqrt{e}, 2e)$

$y = \dfrac{x^2}{\ln x}$

52.

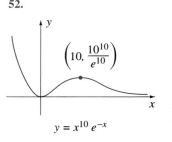

$$\left(10, \frac{10^{10}}{e^{10}}\right)$$

$$y = x^{10} e^{-x}$$

54.

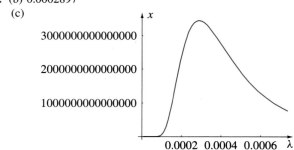

$$\left(\frac{\pi}{3}, \sqrt{3}\right)$$

$$y = 2\csc x - \cot x$$

56. ∞

58. (b) 0.0002897

(c)

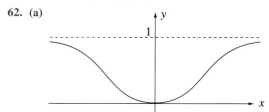

60. $a = b = \pm 7$, c arbitrary

62. (a)

Exercises 4.11

2. $-2dx/(x-1)^2$

4. $[2x\cos(x^2+2)+\sin x]dx$

6. $x^2(9-16x^2)dx/\sqrt{3-4x^2}$

8. $[1-12/(x-1)^3]dx$

10. $2(x^2+5x-5)dx/(x^2+5)^2$ **12.** -4%

14. 3250 m

Exercises 4.12

2. $P_0(x) = 1$, $P_1(x) = 1$, $P_2(x) = 1 - 9x^2/2$,
$P_3(x) = 1 - 9x^2/2$, $P_4(x) = 1 - 9x^2/2 + 27x^4/80$

4. $P_0(x) = 1$, $P_1(x) = 1$, $P_2(x) = 1 - (x-\pi/2)^2/2$,
$P_3(x) = 1 - (x-\pi/2)^2/2$,
$P_4(x) = 1 - (x-\pi/2)^2/2 + (x-\pi/2)^4/24$

6. $P_0(x) = 1$, $P_1(x) = 1+(x-1)$, $P_2(x) = 1+(x-1)+(x-1)^2$,
$P_3(x) = 1 + (x-1) + (x-1)^2 + (x-1)^3$,
$P_4(x) = 1 + (x-1) + (x-1)^2 + (x-1)^3 + (x-1)^4$

8. $P_0(x) = 0$, $P_1(x) = x$, $P_2(x) = x + x^2/2$,
$P_3(x) = x + x^2/2 + x^3/6$, $P_4(x) = x + x^2/2 + x^3/6 + x^4/24$

10. $P_0(x) = 1$, $P_1(x) = 1 + 3x/2$, $P_2(x) = 1 + 3x/2 - 9x^2/8$,
$P_3(x) = 1 + 3x/2 - 9x^2/8 + 27x^3/16$,
$P_4(x) = 1 + 3x/2 - 9x^2/8 + 27x^3/16 - 405x^4/128$

12. $P_0(x) = 6^{-1/3}$, $P_1(x) = 6^{-1/3}[1 - (x-2)/18]$,
$P_2(x) = 6^{-1/3}[1 - (x-2)/18 + (x-2)^2/162]$, $P_3(x) = 6^{-1/3}[1 - (x-2)/18 + (x-2)^2/162 - 7(x-2)^3/8748]$,
$P_4(x) = 6^{-1/3}[1 - (x-2)/18 + (x-2)^2/162 - 7(x-2)^3/8748 + 35(x-2)^4/3888]$

14. $10^{-3}(0.9)^{-4}$

Review Exercises

2. (a)

$$\left(-\frac{1}{12}, \frac{253}{216}\right)$$
$$\left(-\frac{1}{2}, \frac{7}{4}\right)$$
$$\left(\frac{1}{3}, \frac{16}{27}\right)$$

$$y = 4x^3 + x^2 - 2x + 1$$

(b)

$$y = \frac{x^2 - 2x + 4}{x^2 - 2x + 1}$$

4. No solution   **6.** No solution   **8.** $-3/2$

10. 8   **12.** 0   **14.** 0

16. 0   **18.** e^2

20. (a) 0.312 908 (b) 1.051 888   **24.** $2\ell^3/27$

26. (a)

$$\left(-\sqrt{3}, \frac{3\sqrt{3}}{2}\right)$$
$$(0, 0)$$
$$\left(\sqrt{3}, \frac{3\sqrt{3}}{2}\right)$$

$$y = \frac{x^3}{x^2 - 1}$$

(b)

$$(0, 0)$$

$$y = x^2 \sin^2 x$$

28. 67 ¢

30. $4\sqrt{5}$ m/s

32. $x = 100bq/(ap + bq)$ hectares, $y = 100ap/(ap + bq)$ hectares

Chapter 5

Exercises 5.1

2. $x^5/5 + x^3 + 5x^2/2 + C$ **4.** $-\cos x + C$

6. $(2/3)x^{3/2} + C$ **8.** $-1/x + 2/(3x^3) + C$

10. $-1/x + \sqrt{x} + C$ **12.** $1/(2x) + (3/4)x^4 + C$

14. $(4/3)x^{3/2} + (6/5)x^{5/2} - (10/7)x^{7/2} + C$

16. $(2/5)x^{5/2} + (2/3)x^{3/2} + C$

18. $x^3/3 + 2x^5/5 + x^7/7 + C$

20. $(2/5)x^{5/2} - (4/3)x^{3/2} + 2\sqrt{x} + C$

22. $y = x^4/2 + 2x^2 + 5$ **24.** $y = 2x - 2x^2 + x^8$

26. $f(x) = -5x^3/6 + 10x - 31/3$

28. $(2/3)(x + 2)^{3/2} + C$

30. $-(2/3)(2 - x)^{3/2} + C$

32. $(1/5)(2x - 3)^{5/2} + C$

34. $-(1/16)(1 - 2x)^8 + C$

36. $-(1/15)/(1 + 3x)^5 + C$

38. $(1/72)(2 + 3x^3)^8 + C$

40. $(1/2) \sin 2x + C$

42. $(3/4) \sin^2 2x + C$

44. $-(1/4) \cot 4x + C$

46. $-(1/2)e^{-x^2} + C$

48. $(1/4)e^{4x-3} + C$

50. $-(2/5) \ln |7 - 5x| + C$

52. $-(1/4) \ln |1 - 4x^3| + C$

54. $(1/2)3^{2x} \log_3 e + C$

56. $-(1/5)(1 + \cos x)^5 + C$

58. $(1/8)(1 + e^{2x})^4 + C$

60. $y = x^4/4 + 1/x + C$

62. $-(2/3)/(3x + 5)^{1/2} + C$

64. $-(1/12)/(2 + 3x^4) + C$

66. $y = -3 - 1/x, \ x < 0; \ y = 2 - 1/x, \ x > 0$

68. $-9.81m(x^4 - 2Lx^3 + L^3x)/(24EIL)$

70. $-9.81m(x^4 - 4Lx^3 + 6L^2x^2)/(24EIL)$

72. $(-4.0875x^4 + 245.25x^3 - 4291.875x^2)/(EI), \ 0 < x \le 5;$
$(-8.175x^4 + 327x^3 - 4909x^2 + 510.9375x - 2554.6875)/(EI),$
$5 < x \le 10;$ larger

Exercises 5.2

2. $6t - t^2 + 5, \ 3t^2 - t^3/3 + 5t$ 4. $60t^2 - 4t^3, \ 20t^3 - t^4 + 4$

6. $t^3/3 + 5t^2/2 + 4t - 2, \ t^4/12 + 5t^3/6 + 2t^2 - 2t - 3$

8. $4 - 3 \cos t, \ 4 + 4t - 3 \sin t$

10. (a) $t^3 - t^2 - 3t + 1$ (b) $(\sqrt{10} + 1)/3$

12. (a) 200/3 m (b) 500/3 m, 20 s

14. 5.1 m

16. $|v| \ge 19.8$ m/s

18.

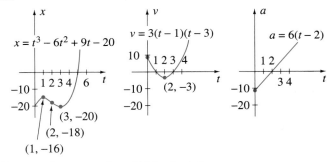

20. Less than or equal to 47.8 km/h; defense

22. (a) Yes (b) No

Exercises 5.3

2. $-(1/3)(1 - 2x)^{3/2} + C$ 4. $-(10/63)(5 - 42x)^{3/4} + C$

6. $-(1/2)/(x^2 + 4) + C$

8. $-1/(x - 2) - 2/(x - 2)^2 - (4/3)/(x - 2)^3 + C$

10. $(1/6)(2x + 3)^{3/2} - (3/2)\sqrt{2x + 3} + C$

12. $(1/5)(s^2 + 5)^{5/2} - (5/3)(s^2 + 5)^{3/2} + C$

14. $(2/3)(1 - \cos x)^{3/2} + C$

16. $(2/7)(y - 4)^{7/2} + (16/5)(y - 4)^{5/2} + (32/3)(y - 4)^{3/2} + C$

18. $(1/729)(3x^3-5)^9+(5/324)(3x^3-5)^8+(25/567)(3x^3-5)^7+C$

20. $(3/4)(x^2 + 2x + 2)^{2/3} + C$

22. $(7/192)/(3 - 4 \sin x)^3 + (3/64)/(3 - 4 \sin x)^2 - (1/64)/$
$(3 - 4 \sin x) + C$

24. $(4/5)(1 + \sqrt{x})^{5/2} - (4/3)(1 + \sqrt{x})^{3/2} + C$

26. $(1/2) \tan^2 x + C$ 28. $(\ln x)^2/2 + C$

30. $-(1/2)/ \ln (x^2 + 1) + C$

32. $(2/3)(1+x)^{3/2} - 2\sqrt{1 + x} + C$ if $x \ge 0$; $-(2/3)(1+x)^{3/2} +$
$2\sqrt{1 + x} + C$ if $-1 < x \le 0$

34. $-(2/5)(1/x - 1)^{5/2} - (2/3)(1/x - 1)^{3/2} + C$

36. $-2/[\sqrt{(1 - x)/(1 + x)} - 3] + C$

Exercises 5.4

2. (a) $(24EI)^{-1}gx^2[4M(3L - x) + m(6L^2 - 4Lx + x^2)]$
   (b) $(24EI)^{-1}gL^3(8M + 3mL)$

4. $x^2 + (y - 2)^2 = 13$ 6. $\pi R^2 \sqrt{2H}/(5ca\sqrt{g})$

8. $-\sqrt{2GMh/(R^2 + Rh)}, \ \sqrt{2GM/R}$

10. $y = Cx^k$

12. (a) $(192EI)^{-1}[8mgx^4 + 32Ax^3 - (8AL^2 + mgL^3)x]$
    (b) $-3mgL/16$

14. $(1 + e^{-t/20})/20$ m³, 1/20 m³

Review Exercises

2. $-(1/4)/x^4 + x^2 + (1/2)/x^2 + C$

4. $-1/x - (4/3)x^{3/2} + C$ 6. $(1/30)(1 + 3x^2)^5 + C$

8. $(2/5)x^{5/2} + 10\sqrt{x} + C$ 10. $-2/\sqrt{x} - 30\sqrt{x} + C$

12. $-(1/3)(1 - x^2)^{3/2} + C$

14. $x^3/3 - (4/5)x^5 + (4/7)x^7 + C$

16. $(2/3)(2 - x)^{3/2} - 4\sqrt{2 - x} + C$

18. $4x + (8/3)x^{3/2} + x^2/2 + C$

20. $(1/5) \sin^5 x + C$ 22. $-(1/8)e^{-4x^2} + C$

24. $(1/5) \ln | \ln x| + C$ 26. $y = x^4 + 7x/2 - 1/2$

28. $[(\sqrt{3} - \sqrt{6})t + \sqrt{6}]^2$ km, $\sqrt{6}/(\sqrt{6} - \sqrt{3})$ hours

30. 11.9 m/s   32. $(2/3)(1 + x)^{3/2} - x + C$

34. $-(3/56)(3 - 2x^3)^7 + (1/32)(3 - 2x^3)^8 - (1/216)(3 - 2x^3)^9 + C$

36. $(1/4) \sin^4 x - (1/6) \sin^6 x + C$

38. $- \ln | \cos x| + C$ 40. $y = 2x/(1 + x)$

42. $9.81M(-150x^2 + 20x^3 - x^4)/(120EI), \ 0 < x \le 5;$
$49.05M(5 - 4x)/(4EI), \ 5 < x \le 10;$ yes

44. $9.81M(x^3 - 15x^2)/(6EI), \ 0 < x \le 5/2;$
$9.81M(-x^4 + 30x^3 - 675x^2/2 + 125x/2 - 625/16)/(120EI),$
$5/2 < x \le 15/2; \ 6131.25M(2 - x)/(48EI),$
$15/2 < x \le 10.$ No and yes

Chapter 6

Exercises 6.1

2. $\sum_{k=1}^{10} k/2^k$

4. $\sum_{k=1}^{14\,641} \sqrt{k}$

6. $\sum_{k=1}^{1019} (-1)^k (k+1)$

8. $\sum_{k=1}^{225} (\tan k)/(1+k^2)$

10. $\sum_{k=1}^{9} (10^k - 1)/10^k$

16. 258

18. $4n(4n^2 - 1)/3$

20. 5350

24. $f(n) - f(0)$

26. No

28. $(1 - 2^{-18})/4$

30. $3960[1 - (0.99)^{15}]$

Exercises 6.3

2. 6

4. 4

6. 3/2

8. 0

10. (c) $\log_2 e$

Exercises 6.4

2. 20/3

4. 2/3

6. 1

8. $-7/2$

10. 3/4

12. 0

14. $5/16 - 1/(\pi + 1)$

16. 125/12

18. $-65/4$

20. $3/\sqrt{2}$

22. $-1/2$

24. 0

26. $(e^3 - 1)/e$

28. $\ln(2/3)$

30. $20\log_3 e$

32. 88/3

34. $-1/6$

36. $-475, 475$

38. 1/6, 1/6

40. 9/4, 11/4

42. $\pi/2, 0$ 44. $4\pi/(\pi^2 + 160), 0$ 46. $2\sqrt{31}, 2\sqrt{5}$

Exercises 6.5

2. $1/\sqrt{x^2 + 1}$

4. $-x^3 \cos x$

6. $2\sqrt{2x + 1}$

8. $5\sqrt{(5x + 4)^3 + 1}$

10. $256(3x^2 + 3x + 1) + 2/\sqrt{x} - 2/\sqrt{x + 1}$

12. $2x[-2 \sec(1 + 2x^2) + \sec(1 + x^2)]$

14. $\cos x/\sqrt{\sin x + 1} + \sin x/\sqrt{\cos x + 1}$

16. $(2\sqrt{2} - 1)/(2x^{1/4})$

18. $-\ln(x^2 + 1)$

20. $3e^{-36x^2} + 2e^{-16x^2}$

Exercises 6.6

2. 0

4. 31/5

6. $2\sqrt{2}/3$

8. 11/5

10. $2/\pi$

12. 1

14. 23/9

16. 1/2

18. 0

20. 9/7

22. $2cR^2/3$

24. 0

26. $\pi/2$

28. 0.6032

30. 1.400

Exercises 6.7

2. 4/15

4. 7/384

6. Does not exist

8. $\sqrt{3/2} - 2\sqrt{2}/3$

10. $3(5^{2/3} - 2^{2/3})/4$

12. $\sqrt{2}(7\sqrt{7} - 8)/18$

14. 52/5

16. $(16 - 9\sqrt{2})/6$

18. 5/18

20. $(1/3)(\ln 2)^3$

22. $2(\sqrt{7} - 1)/3$

24. $1/\sqrt{3}$

26. $-8\sqrt{5}/45$

Review Exercises

2. 4/15

4. $-4/3$

6. 1/6

8. 0

10. 66/5

12. $16\sqrt{6}$

14. $-4/15$

16. $(2112\sqrt{6} - 704)/105$

18. 120 852.5

20. 20

24. 2

26. $4/(15\pi)$

28. $-x^2(x + 1)^3$

30. $-4x \cos 2x$

32. $4x \sin^2 x^2$

34. 0, 1/2

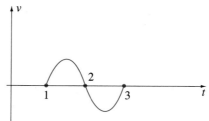

36. $(33\sqrt{6} - 56)/60$

38. $(24\sqrt{3} - 14\sqrt{2} - 8\sqrt{5})/3$

40. $24\sqrt{5} - 8$

42. $(3\sqrt{3} - 2\sqrt{2} - 1)/3$

Chapter 7

Exercises 7.1

2. 8   **4.** $343/6$ **6.** $(e^6 - e^3)/3 + 3/2$

8. $8/3$ **10.** 2   **12.** $20/3$

14. $4\sqrt{2}/3 - e\ln 2$ **16.** $10/3$

18. (a) 22 (b) 37, 15 (c) 29, 7 (d) $203/8$, $27/8$

20. $4\int_1^2 \sqrt{4 - x^2}\, dx$

22. $2\int_0^{\sqrt{(\sqrt{65}-1)/2}} \left(\sqrt{16 - y^2} - y^2\right) dy$

24. $235/3$ **26.** $7/6$ **28.** $1 - \ln 2$

30. $2\sqrt{e} - 8/3$ **32.** $32/3$ **34.** $14/\sqrt{21}$

36. 2.182   **38.** 8.436   **40.** 0.952

42. 7.177   **44.** $0 < m < 1$, $(m - 1 - \ln m)/3$

Exercises 7.2

2. $80\sqrt{5}\pi/3$ **4.** $5888\pi/15$ **6.** π

8. $1088\pi/15$ **10.** $32\pi/3$ **12.** $\pi(e^2 - 4e + 5)/2$

14. 8π **16.** $14\pi/3$ **18.** $40\pi/3$

20. $\pi/3$ **22.** $344\pi/3$ **24.** $1472\pi/15$

26. $775\pi/6$ **28.** $272\pi/15$ **30.** $68\pi/9$

32. $16\pi(13 + 2^{17/4} - 3^{9/4})/45$

34. 111.303   **36.** 11.761   **38.** $4\pi/3$

40. (a) $x^2(6 - x)/16$ (b) $1161\pi/10$ m^3

Exercises 7.3

2. $17/3$ **4.** $(85^{3/2} - 13^{3/2})/27$

6. $(e - e^{-1})/2$ **8.** $23/18$

10. $3011/480$ **12.** $\int_1^2 \sqrt{36x^2 - 48x + 17}\, dx$

14. $\int_{-\sqrt{3}}^{2\sqrt{2}} \sqrt{(1 + 2y^2)/(1 + y^2)}\, dy$

16. $\int_0^{\pi/4} \sec x\, dx$

18. $\sqrt{2}\int_0^1 (3 - 2x^2)/\sqrt{1 - x^2}\, dx$

20. $2\int_0^2 \sqrt{(16 + 5x^2)/(4 - x^2)}\, dx$

22. 6

24. $(b^{2n+1} - a^{2n+1})/[4(2n - 1)] + (1/a^{2n-1} - 1/b^{2n-1})/(2n + 1)$

Exercises 7.4

2. $8/15$ J   **4.** 8829 J   **6.** 98.1 J

8. 184 J   **10.** 1.78×10^7 J   **12.** 114 450 J

14. 7.60×10^6 J   **16.** $q_3(q_1 - 8q_2)/(48\pi\epsilon_0)$

18. kb^2 J   **20.** 3.22×10^5 J

22. (b) 9.8087×10^5 J (c) 9.82×10^5 J

24. $C\ln 2$

Exercises 7.5

2. (a) $kx^2 + mv^2 = kx_0^2 + mv_0^2$ (b) $\sqrt{x_0^2 + mv_0^2/k}$
(c) $\sqrt{v_0^2 + kx_0^2/m}$

4. $qQ/(4\pi\epsilon_0 r)$

6. (a) $9.81m/k$ (b) $(9.81m + \sqrt{9.81^2 m^2 + kmv_0^2})/k$

8. (a) 9.81×10^5 J (b) 4.43×10^2 m/s

Exercises 7.6

2. 6.78×10^{10} N   **4.** 9.25×10^3 N

6. 7.60×10^5 N, 4.41×10^5 N, 4.91×10^4 N

12. 2.943×10^6 N, 2.969×10^6 N   **14.** 1.60 m $\times$ 1.60 m

20. (a) 10.2 cm   **22.** 0.52

Exercises 7.7

2. $(3a/8, 3h/5)$ **4.** $(4r/(3\pi), 4r/(3\pi))$

6. $(-1/2, -1/2)$ **8.** $(0, 0)$

10. $(33/5, 1/2)$ **12.** $(177/85, 9/17)$

14. $25\rho/3$

16. $(M + 2\rho L)^{-1}(\sum_{i=1}^6 m_i x_i + 2\rho L^2)$

18. $(0, 192/205)$

20. On the axis of symmetry, $1637/350$ from smaller end

24. $(12/5, -3/5)$ **26.** $\overline{x} = -2/(8\sqrt{3} - 1)$, $\overline{y} = 0$

28. $(0, 1.944)$ **30.** $(0.510, 0.238)$

32. $2\pi^2 ab^2$

Exercises 7.8

2. $32\rho/3$ **4.** $603\rho/10$ **6.** $2\rho/15$

8. $344\rho/105$ **10.** $1143\rho/20$

12. $\rho h(y_2^3 + y_2^2 y_1 - y_1^2 y_2 - y_1^3)/4$; no

16. $\tilde{x} = \overline{x}$ **18.** 0.650   **20.** 0.438

Exercises 7.9

2. $\pi r\sqrt{r^2 + h^2}$ **4.** $b^2 h/3$ **6.** (a) 20 000 (b) 7000

8. $31\,250/\sqrt{3}$ cm^3 **10.** $\pi/2$ **12.** $r^3(3\pi - 4)/6$

14. $AEL/(9.81M + AE)$

16. $L + FL/(AE) + 9.81\rho L^2/(2E)$

18. Discontinuity at $y = L$

20. $L + 9.81\rho L^2(2a^3 - 3a^2 b + b^3)/[6Eb(a - b)^2]$

22. $L + 9.81\rho L^2(2a^3 - 3a^2 b + b^3)/[6Eb(a - b)^2] + 9.81ML/(abE)$

24. 15   **26.** 16.5   **28.** 140 m^3

Exercises 7.10

2. Diverges

4. $-1/16\,464$

6. Diverges

8. 2

10. $\sqrt{21}$

12. Diverges

14. Diverges

16. Diverges

18. $17\sqrt{5}/3$

20. (a) No (b) No

22. (a) 1/2 (b) $\pi/5$

28. No

30. Diverges

32. Converges

Review Exercises

2. (a) 5/4 (b) $46\pi/21$, $14\pi/15$ (c) $(28/75, 92/105)$ (d) 73/60, 1/4

4. (a) 1/2 (b) $\pi/3$, π (c) $(1, 1/3)$ (d) 1/12, 7/12

6. $10\pi/3$

8. $16\pi/105$

10. $43\pi/12$

12. $-5/6$, 3/5

14. $-28/3$, 14

16. 1.04×10^5 J

18. $4\sqrt{5}\pi$

22. $2\sqrt{3}$

24. 0

26. Diverges

28. Diverges

30. 6.58×10^4 J

Chapter 8

Exercises 8.1

2.

$f^{-1}(x) = x^2 - 1 (x \geq 0)$

4.
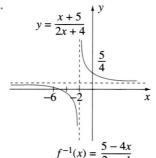
$y = \dfrac{x + 5}{2x + 4}$
$f^{-1}(x) = \dfrac{5 - 4x}{2x - 1}$

6.
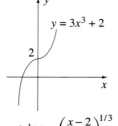
$y = 3x^3 + 2$
$f^{-1}(x) = \left(\dfrac{x - 2}{3}\right)^{1/3}$

8.
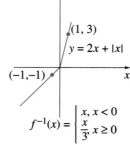
$y = 2x + |x|$
$f^{-1}(x) = \begin{cases} x, & x < 0 \\ \dfrac{x}{3}, & x \geq 0 \end{cases}$

10.
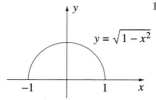
$y = \sqrt{1 - x^2}$
No inverse function

12.

$y = x^2 - 2x + 4$
$f^{-1}(x) = 1 + \sqrt{x - 3}$

14.

$y = \dfrac{x}{3 + x^2}$
No inverse function

16. $-x^{-1/4}$, $x < 0$; $x^{-1/4}$, $x > 0$

18. $-\sqrt{\sqrt{2 + x} - 2}$, $x < 0$; $\sqrt{\sqrt{2 + x} - 2}$, $x \geq 0$

20. $-\sqrt{(x + \sqrt{x^2 + 16x})/2}$, $x < 0$; $\sqrt{(x + \sqrt{x^2 + 16x})/2}$, $x \geq 0$

22.

$r = f^{-1}(x)$

Exercises 8.2

2. 0.253

4. $-2\pi/3$

6. Does not exist

8. $-\pi/4$

10. Does not exist

12. 0.164

14. $\pi/4$

16. $-\pi/4$

18. $-0.876 + n\pi$

20. No solutions

22. No solutions

24.

$y = \sqrt{\text{Tan}^{-1}x} + \sqrt{\text{Sec}^{-1}x}$

26.

$y = \text{Sin}^{-1}x + \text{Csc}^{-1}x$

28. $(3n \pm 1)\pi/3$

30. No solutions

32. $\text{Tan}^{-1}(2n\pi \pm 3\pi/4) + m\pi$

34. (a)

$y = \sin(\text{Sin}^{-1}x)$

(b)
$y = \text{Sin}^{-1}(\sin x)$

36. $\sqrt{17} \sin(2x + 0.24)$

38. $2\sqrt{5} \sin(2x + 2.03)$

44. $E_0/\sqrt{R^2 + [\omega L - 1/(\omega C)^2]^2} \cos(\omega t + \phi)$ where $\phi = \text{Tan}^{-1}\{[\omega L - 1/(\omega C)]/R\}$

52. Decreasing on $0 \leq x \leq 0.208$; increasing on $0.208 \leq x \leq 5$

Chapter 9

Exercises 8.3

2. $-2x/[1 + (x^2 + 2)^2]$ **4.** $-2x/[1 + (2 - x^2)^2]$

6. $\text{Csc}^{-1}(x^2 + 5) - 2x^2/[(x^2 + 5)\sqrt{(x^2 + 5)^2 - 1}]$

8. $1/[2(x + 3)\sqrt{x + 2}]$ **10.** $-1/(x\sqrt{x^2 - 1})$

12. $2x\text{Sec}^{-1}x + x/\sqrt{x^2 - 1}$ **14.** $-1/(1 + x^2)$

16. $-1/[\sqrt{x}(x + 1)]$

18. $x(t - \sqrt{1 - t^2}\text{Cos}^{-1}t)/[\sqrt{1 - x^2}\sqrt{1 - t^2}]$

20. $(5\sqrt{1 - x^2y^2} - y)/(x - 2\sqrt{1 - x^2y^2})$

22. $-18/(x^3\sqrt{x^2 - 9})$

24. $-(1/x^2)\text{Csc}^{-1}(3x) - 2/(x^2\sqrt{9x^2 - 1})$

26. $(\text{Cos}^{-1}x)^2 - 2 + 2x(1 - \text{Cos}^{-1}x)/\sqrt{1 - x^2}$

28. $2\sqrt{4x - x^2}$

30. $\sqrt{2}(1 - x^2)/[(1 + x^2)\sqrt{1 + x^4}]$

32.

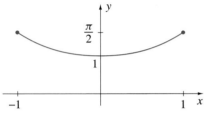

$$y = x\,\text{Sin}^{-1}x + \sqrt{1 - x^2}$$

34. 1.23   **36.** $\text{Tan}^{-1}\mu$

38. (a) 2/15 rad/s

Exercises 8.4

2. 2.30   **4.** 9.08×10^{-5}

6. 1.00   **8.** 0.266

10. 5.07   **12.** $\sinh(x/2) + (x/2)\cosh(x/2)$

14. $(1/x)\,\text{sech}^2(\ln x)$

16. $2\sqrt{1 + y}\,\text{csch}^2 x/(2\sqrt{1 + y} - 1)$

18. $-\sinh t\,\text{sech}^2 x\,\sec^2(\cosh t)\sin(\tanh x)$

20. $2\,\text{csch}\,4x$ **26.** (a) 189.3 N (b) 17.0 m

Review Exercises

2. $(4n + 1)\pi/4 \pm 0.659$ **4.** -1.79

6. $(4n + 1)\pi/2 \pm 0.604$

8. $\ln[(4n + 1)\pi/2 \pm 0.84] - 2,\ n \geq 0$

10. 2.09   **12.** $6x\cosh x^2$

14. $(x^2 - 1)/(x^4 + 3x^2 + 1)$ **16.** $\cos x\,\text{sech}\,y$

18. $\text{Csc}^{-1}(1/x^2) + 2x^2/\sqrt{1 - x^4}$ **20.** -1

Exercises 9.1

2. $-(1/4)e^{-2x^2} + C$ **4.** $\ln(1 + e^x) + C$

6. $-(2/27)(1 - 3x^3)^{3/2} + C$ **8.** $-(1/6)(1 + x^3)^{-2} + C$

10. $2\sqrt{x} - x + C$ **12.** $\text{Tan}^{-1}x + x + C$

14. $(1/6)(2x + 4)^{3/2} + \sqrt{2x + 4} + C$

16. $(1/8)\sin^4 2x + C$

18. $(1/2)(x + 5)^2 - 15x + 75\ln|x + 5| + 125/(x + 5) + C$

20. $(1/3)\ln|\sec 3x| + C$

22. $(4/5)x^{5/4} - x + (4/3)x^{3/4} - 2\sqrt{x} + 4x^{1/4} - 4\ln(x^{1/4} + 1) + C$

24. $(35 - 12\ln 6)/6$

28. (a) $1962(1 - e^{-t/200})$ m/s (b) $1962t + 392\,400(e^{-t/200} - 1)$ m

30. $(1/(3n))\ln|x^n/(3 + 2x^n)| + C$

Exercises 9.2

2. 4.75 h   **4.** 19.74 s   **6.** $k\ln(S/S_0)$

8. 13.51 h   **10.** (a) Finitely long (b) Infinitely long

12. 33 574 years ago

14.

$$C = \frac{R}{k}(1 - e^{-kt}) + C_0 e^{-kt}$$

16. $20 + 70e^{-0.01399t}$

18.   **20.**

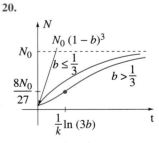

$$N = Ae^{-be^{-kt}}$$ $$N = N_0(1 - be^{-kt})^3$$

22.

$$s = 1 - (1 - e^{-kx})^n$$

24. 3:25 A.M.

Exercises 9.3

2. $(x^2/2 - x/2 + 1/4)e^{2x} + C$

4. $(2/3)x^{3/2} \ln (2x) - (4/9)x^{3/2} + C$

6. $-(2/5)(2 + x)(3 - x)^{3/2} + C$

8. $(2x^2/3)(x+5)^{3/2} - (8x/15)(x+5)^{5/2} + (16/105)(x+5)^{7/2} + C$

10. $2x^2\sqrt{x+2} - (8x/3)(x+2)^{3/2} + (16/15)(x+2)^{5/2} + C$

12. $(1/3)(x-1)^3 \ln x - x^3/9 + x^2/2 - x + (1/3)\ln x + C$

14. $x\,\mathrm{Tan}^{-1} x - (1/2)\ln (1 + x^2) + C$

16. $e^{2x}(2\cos 3x + 3\sin 3x)/13 + C$

18. $x\ln (x^2 + 4) - 2x + 4\mathrm{Tan}^{-1}(x/2) + C$

20. 5.92×10^9 N   **22.** $-2L^2/(n\pi),\ 0$

24. $4L^2/(n\pi),\ 0$ **28.** $2/s^3$

30. $1/(s + 1)^2$ **32.** $i(e^{-i\omega L/2} - e^{i\omega L/2})/\omega$

34. $2a/(\omega^2 + a^2)$

36. $x\,\mathrm{Tan}^{-1}\sqrt{x} - \sqrt{x} + \mathrm{Tan}^{-1}\sqrt{x} + C$

38. $(x/2)e^x(\sin x - \cos x) + (1/2)e^x \cos x + C$

Exercises 9.4

2. $-(1/2)\sin^{-2} x + C$ **4.** $-(1/3)\csc^3 x + C$

6. $(2/3)\tan^{3/2} x + (2/7)\tan^{7/2} x + C$

8. $(1/18)\sec^6 3x + C$ **10.** $(1/4)\sec^4 x + C$

12. $\tan\theta + C$ **14.** $2\sqrt{1 + \tan x} + C$

16. $7\tan x - 3x + C$

18. $(1/32)(12x - 8\sin 2x + \sin 4x) + C$

20. $(1/2)\tan^2 x + \ln|\tan x| + C$

22. $\pi(1 - \pi/4 + \ln 2)$

24. $-(1/3)\cot^3 z + \cot z + z + C$

26. $\theta/2 + (1/4)\sin 2\theta + (1/3)\cos^3\theta + C$

28. $(1/8)\sin 4x + (1/16)\sin 8x + C$

30. $\sec x + \ln|\csc x - \cot x| + C$

34. A **36.** $1/2$

Exercises 9.5

2. $(1/\sqrt{5})\mathrm{Sin}^{-1}(\sqrt{5}x/3) + C$

4. $-\sqrt{4 - x^2}/(4x) + C$

6. $(1/15)(5x^2 + 3)^{3/2} + C$

8. $(1/2)\ln|(1 + x)/(1 - x)| + C$

10. $(1/20)\ln (5x^2 + 1) + (\sqrt{5}/2)\mathrm{Tan}^{-1}(\sqrt{5}x) + C$

12. $\sqrt{4 - x^2} + 2\ln|(2 - \sqrt{4 - x^2})/x| + C$

14. $\ln|x + \sqrt{x^2 - 16}| - \sqrt{x^2 - 16}/x + C$

16. $(1/16)\mathrm{Sec}^{-1}(x/2) + \sqrt{x^2 - 4}/(8x^2) + C$

18. $(1/3)(y^2 + 4)^{3/2} - 4\sqrt{y^2 + 4} + C$

20. $2\ln|x| - (1/2)\ln (1 + x^2) + C$

24. $L\ln[(L + \sqrt{L^2 - x^2})/x] - \sqrt{L^2 - x^2}$

26. 1.053   **28.** $0.404b$ **30.** $9.81\pi\rho r^3$

32. $(147/8)\mathrm{Sin}^{-1}(x/\sqrt{7}) + (x/8)(35 - 2x^2)\sqrt{7 - x^2} + C$

34. $-6\mathrm{Sec}^{-1}(2x) - (12x^2 - 1)/(2x^2\sqrt{4x^2 - 1}) + C$

36. $(1/2)[x\sqrt{1 + 3x^2} + (1/\sqrt{3})\ln|\sqrt{3}x + \sqrt{1 + 3x^2}|] + C$

38. $\sqrt{2\pi}$ **40.** $r/2$

Exercises 9.6

2. $\ln|\sqrt{x^2 + 2x + 2} + x + 1| + C$

4. $-(2/\sqrt{7})\mathrm{Tan}^{-1}[(2x - 3)/\sqrt{7}] + C$

6. $(x/2)/\sqrt{4x - x^2} + C$

8. $\ln (x^2 + 6x + 13) - (9/2)\mathrm{Tan}^{-1}[(x + 3)/2] + C$

10. $\ln|\sqrt{6 + 4\ln x + (\ln x)^2} + \ln x + 2| + C$

12. $39\,240\pi$ N

14. $(1/2)(x - 1)\sqrt{x^2 - 2x - 3} - 2$
$\ln|x - 1 + \sqrt{x^2 - 2x - 3}| + C$

Exercises 9.7

2. $-(1/2)(y + 1)^{-2} + C$

4. $x + (10/3)\ln|x - 4| + (2/3)\ln|x + 2| + C$

6. $-(1/6)\ln|y| - (2/15)\ln|y + 3| + (3/10)\ln|y - 2| + C$

8. $(1/2)\ln (x^2 + 2) + (x^2 + 2)^{-1} + C$

10. $y - 4\ln|y + 2| + \ln|y + 1| + C$

12. $\ln (y^2 + 1) + \mathrm{Tan}^{-1} y - \ln (y^2 + 4) + C$

14. $-[1/(3\sqrt{2})]\mathrm{Tan}^{-1}(x/\sqrt{2}) + (2/3)\ln|(x - 1)/(x + 1)| + C$

16. $-(5/27)\ln|x - 1| - (7/9)(x - 1)^{-1} + (32/27)\ln|x + 2| + (2/9)(x + 2)^{-1} + C$

14.

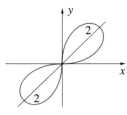

$$(x^2 + y^2)^2 = 8xy$$

16.

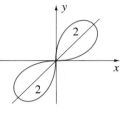

$$(x^2 + y^2)^{3/2} = 4xy$$

18.

$$x^2 + y^2 = 3\sqrt{x^2 + y^2} - 4x$$

20.

$$y^3 = x^2$$

22. $(2, \pm\pi/6)$, $(2, \pm 5\pi/6)$ **24.** $(0, \theta)$, $(4/3, \pm 1.23)$

26. $\sqrt{3}/7$ **28.** $3/(5\sqrt{2} + 3)$

30. $\sqrt{3}$

32.

$$r = \sin 3\theta$$

34.

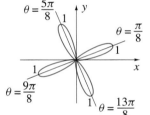

$$r = \sin 4\theta$$

36.

$$r = e^\theta$$

38.

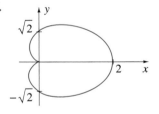

$$r = -2\cos(\theta/2)$$

40. $8a$

42.

$$(x^2 + y^2)^2 = x^2 y$$

44.

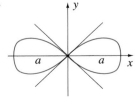

$$(x^2 + y^2)^2 = a^2(x^2 - y^2)$$

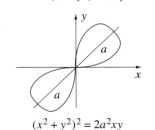

$$(x^2 + y^2)^2 = 2a^2 xy$$

Exercises 10.4

2. 9π **4.** 2

6. 6π **8.** 18π

10. 2π **12.** $(5\pi - 8)/4$

14. $6\pi - 16$ **16.** $(7\pi - 12\sqrt{3})/12$

18. $33[\pi + 2\operatorname{Sin}^{-1}(2/3)]/4 + 7\sqrt{5}$

20. $\pi/32$

Exercises 10.5

2. Circle   **4.** None of these

6. Hyperbola   **8.** Parabola

10. None of these   **12.** Ellipse

14. Straight line

16.

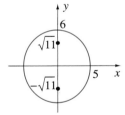

$$\frac{x^2}{25} + \frac{y^2}{36} = 1$$

18.

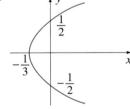

$$3x = 4y^2 - 1$$

20.

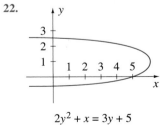

$$7x^2 + 3y^2 = 16$$

22.

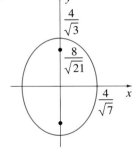

$$2y^2 + x = 3y + 5$$

24.

$$y^2 = 10(2 - x^2)$$

26.

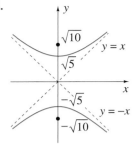

$$y^2 - x^2 = 5$$

28.

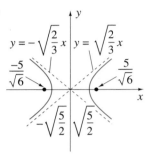

$$2x^2 - 3y^2 = 5$$

30.

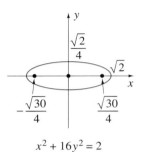

$$x^2 + 16y^2 = 2$$

32.

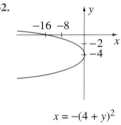

$$x = -(4 + y)^2$$

34.

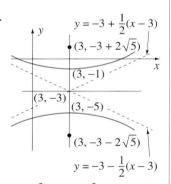

$$y = -3 + \tfrac{1}{2}(x - 3)$$
$$(3, -3 + 2\sqrt{5})$$
$$(3, -1)$$
$$(3, -3)$$
$$(3, -5)$$
$$(3, -3 - 2\sqrt{5})$$
$$y = -3 - \tfrac{1}{2}(x - 3)$$
$$x^2 - 6x - 4y^2 - 24y = 11$$

36.

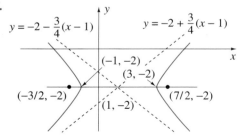

$$y = -2 - \tfrac{3}{4}(x - 1) \qquad y = -2 + \tfrac{3}{4}(x - 1)$$
$$(-1, -2)$$
$$(3, -2)$$
$$(-3/2, -2) \qquad (7/2, -2)$$
$$(1, -2)$$
$$9x^2 - 16y^2 - 18x - 64y = 91$$

38. $3x^2 + y^2 = 28$ **40.** $25/4$

42. $9x^2 + 25y^2 = 225$ **48.** πab

50. $4\pi ab^2/3,\ 4\pi a^2 b/3$

54. In Exercise 15, focus is $(0, -7/8)$, directrix is $y = -9/8$; in Exercise 18, focus is $(-7/48, 0)$, directrix is $x = -25/48$; in Exercise 21, focus is $(3/4, 0)$, directrix is $x = 5/4$; in Exercise 22, focus is $(6, 3/4)$, directrix is $x = 25/4$; in Exercise 27, focus is $(3, -1/4)$, directrix is $y = 1/4$; in Exercise 32, focus is $(-1/4, -4)$, directrix is $x = 1/4$.

Exercises 10.6

2.

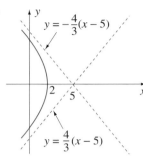

$$y = -\tfrac{4}{3}(x - 5)$$
$$y = \tfrac{4}{3}(x - 5)$$
$$r = \frac{16}{3 + 5\cos\theta}$$

4.

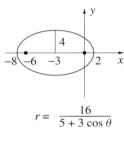

$$r = \frac{16}{5 + 3\cos\theta}$$

6.

$$r = \frac{4}{4 - 3\sin\theta}$$

8.

$$r = \frac{1}{2 + \sin\theta}$$

10.

$$r = \frac{4\csc\theta}{7\csc\theta - 2}$$

12. $8x^2 + 2x + 9y^2 = 1$ **14.** $8x^2 + 12x - y^2 + 4 = 0$

16. $40x + 25y^2 = 16$

18. With $x = \sqrt{5} + r\cos\theta$, $y = r\sin\theta$, the equation becomes $r = 4/(3 + \sqrt{5}\cos\theta)$.

20. With $x = \sqrt{10} + r\cos\theta$, $y = 2 + r\sin\theta$, the equation becomes $r = 1/(3 - \sqrt{10}\cos\theta)$.

22. $\epsilon \to 0$

Exercises 10.7

2.

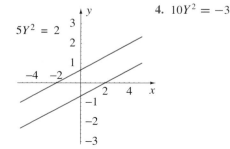

$$5Y^2 = 2$$

4. $10Y^2 = -3$

6.

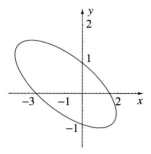

$$11\sqrt{10}X^2 + 2X + \sqrt{10}Y^2 - 6Y = 8\sqrt{10}$$

8. $11\sqrt{10}X^2 + \sqrt{10}Y^2 = 6X + 2Y - 20\sqrt{10}$

10.

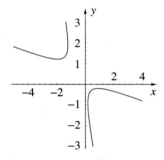

$$3X^2 - Y^2 + \sqrt{2}Y = -1$$

12.

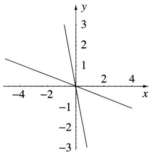

$$3X^2 - Y^2 = 0$$

14.

$$5Y^2 = 8$$

16.

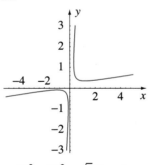

$$6Y^2 - 4X^2 + \sqrt{2}X = -1$$

18.

22.

Review Exercises

2.

$$r = -\sin\theta$$

4.

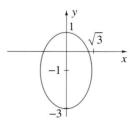

$$r = \frac{3}{2 + \sin\theta}$$

6.

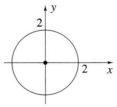

$$r = \frac{3}{2 - 2\cos\theta}$$

8.

$$r^2 = 4\cos\theta$$

10.

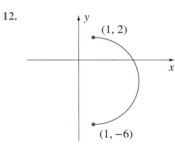

$$x = 4 + t$$
$$y = 5 - 3t^2$$

12.

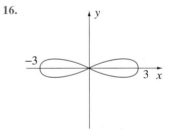

$$x = 1 + 4\sin t \quad,\quad 0 \le t \le \pi$$
$$y = -2 + 4\cos t$$

14.

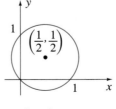

$$x^2 + y^2 = 2\sqrt{x^2 + y^2}$$

16.

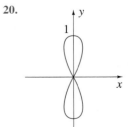

$$r = 3\cos 2\theta$$

18.

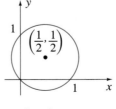

$$x^2 + y^2 = x + y$$

20.

$$r = \sin^2\theta - \cos^2\theta$$

22.

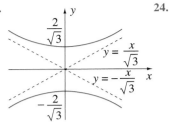

$$x^2 - 3y^2 + 4 = 0$$

24.

(4, −1)

$$y^2 + x + 2y = 3$$

26.

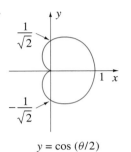

$$y = \cos(\theta/2)$$

28.

$y = x - \frac{1}{2}$

$-\frac{1}{2}$

$y = -x - \frac{1}{2}$

$$x^2 = y^2 + y$$

30.

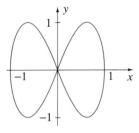

$$x = \sin\theta$$
$$y = \sin 2\theta$$

32. 12π

34. $3\pi/8$

36. $(3 - 3t^2)/(2 + 3t^2)$, $-30t/(2 + 3t^2)^3$

38. 16

40. $\sqrt{2}(e^{\pi/2} - 1)$

42. $X^2 - Y^2 - \sqrt{2}X - 3\sqrt{2}Y + 6 = 0$, hyperbola

44. $10\sqrt{10}X^2 + X + 3Y = 2\sqrt{10}$, parabola

Chapter 11

Exercises 11.1

2. Divergent

4. Convergent, 0

6. Convergent, 0

8. Convergent, 0

10. Convergent, $\pi/2$

12. Divergent

14. Convergent, 0

16. Convergent, 0

18. Convergent, 0

20. Convergent, $\pi/2$

22. $(3n + 1)/n^2$

24. $[1 + (-1)^{n+1}]/2$

26. 0

28. 4

32. $-0.381\,966\,0$

34. x_2 is not defined

36. $3.044\,723\,1$

38. $1.214\,648\,0$

40. $0.334\,734\,1$

42. $-1.388\,792\,0$

44. 2.4142

46. 0.1667

48. 0.5453

50. 0.2648

52. 0.7391

54. 0.2402

56. (a) $40(0.99)^n$ m (b) $4(0.981)^{-1/2}(0.99)^{n/2}$ s

58. $4P/3$, $16P/9$, $(4/3)^n P$; does not exist

60. 0.0625, 0.1125

Exercises 11.2

2.

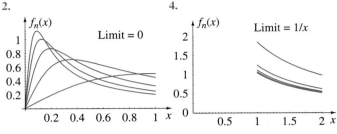

Limit = 0

4.

Limit = 1/x

6.

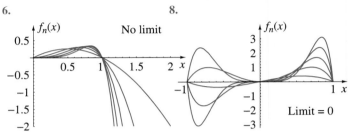

No limit

8.

Limit = 0

10.

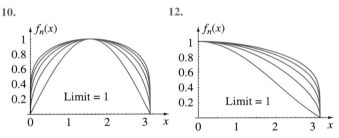

Limit = 1

12.

Limit = 1

14.

Limit = 0

Exercises 11.3

2. $\sum_{n=0}^{\infty} (5^n/n!)x^n$, $-\infty < x < \infty$

4. $(1/\sqrt{2})[1 + (x - \pi/4) - (x - \pi/4)^2/2! - (x - \pi/4)^3/$
$3! + (x - \pi/4)^4/4! + \cdots]$, $-\infty < x < \infty$

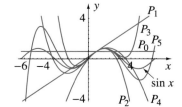

6. $\sum_{n=0}^{\infty} (2^n/n!)x^n$, $-\infty < x < \infty$

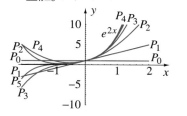

8. $\sum_{n=0}^{\infty} x^n$, $-1 < x < 1$

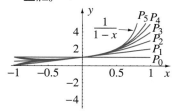

10. $\sum_{n=1}^{\infty} (-1)^{n+1} 2^{n-1} x^n$, $-1/2 < x < 1/2$

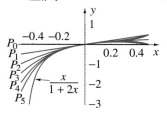

12. $\sum_{n=0}^{\infty} (n+1)(-1)^n 3^n x^n$, $-1/3 < x < 1/3$

14. $1 + 3x/2 + \sum_{n=2}^{\infty} \{[(-1)^{n+1} 3^n (2n-3)!]/[2^{2n-2} n! (n-2)!]\} x^n$, $-1/3 \le x \le 1/3$

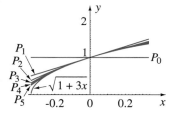

Exercises 11.4

2. $-1 < x < 1$ **4.** $-1/3 < x < 1/3$ **6.** $-4 < x < -2$

8. $7/2 < x < 9/2$ **10.** $-1 < x < 1$ **12.** $-1 < x < 1$

14. $-1/e < x < 1/e$ **16.** $x = 0$ **18.** $-1/3 < x < 1/3$

20. $-5^{1/3} < x < 5^{1/3}$ **22.** $-1 < x < 1$ **24.** $-1 < x < 1$

26. $4/(4 - x^3)$, $-4^{1/3} < x < 4^{1/3}$

28. $(x-1)/(10-x)$, $-8 < x < 10$

30. $\cos(x^2)$, $-\infty < x < \infty$

---

32. $x \sin(x/3)$, $-\infty < x < \infty$

34. $e^{-x} - 1$, $-\infty < x < \infty$

36. e^{2x}, $-\infty < x < \infty$

38. (a) $1 - x^2/2^2 + x^4/[2^4(2!)^2] - x^6/[2^6(3!)^2] + x^8/[2^8(4!)^2] - \cdots$; $x/2 - x^3/(2^3 2!) + x^5/(2^5 2! \, 3!) - x^7/(2^7 3! \, 4!) + x^9/(2^9 4! \, 5!) + \cdots$; $x^m/(2^m m!) - x^{m+2}/[2^{m+2}(m+1)!] + x^{m+4}/[2^{m+4} 2!(m+2)!] - x^{m+6}/[2^{m+6} 3!(m+3)!] + x^{m+8}/[2^{m+8} 4!(m+4)!] - \cdots$ (b) $-\infty < x < \infty$

Exercises 11.5

2. $\sum_{n=0}^{\infty} [(-1)^n/4^{n+1}] x^{2n}$, $-2 < x < 2$

4. $\sum_{n=0}^{\infty} (5^n/n!) x^n$, $-\infty < x < \infty$

6. $\sum_{n=0}^{\infty} [e(-2)^n/n!] x^n$, $-\infty < x < \infty$

8. $\sum_{n=0}^{\infty} [1/(2n)!] x^{2n}$, $-\infty < x < \infty$

10. $x^4 + 3x^2 - 2x + 1$, $-\infty < x < \infty$

12. $\sum_{n=0}^{\infty} [(-1)^n/5^{n+1}](x-2)^n$, $-3 < x < 7$

14. $-4/5 - 4(x-2)/25 + \sum_{n=2}^{\infty} \{9(-1)^{n+1} 4^n/[80(5^n)]\}(x-2)^n$, $6/5 < x < 16/5$

16. $\sum_{n=1}^{\infty} [(-1)^{n+2} 2^n/n] x^n$, $-1/2 < x < 1/2$

18. $\ln 2 + \sum_{n=1}^{\infty} [(-1)^{n+1}/(n 2^n)](x-2)^n$, $0 < x < 4$

20. $\sum_{n=0}^{\infty} [(-1)^n/4^{n+1}](x-4)^n$, $0 < x < 8$

22. $\sum_{n=0}^{\infty} (-1)^n (n+1)(x-3)^n$, $2 < x < 4$

24. $\sum_{n=0}^{\infty} [(-1)^n (1/3^{n+1} - 1/5^{n+1})/2] x^n$, $-3 < x < 3$

26. $\sqrt{3}(1 + x/6) + \sum_{n=2}^{\infty} \{4\sqrt{3}(-1)^{n-1}(2n-3)!/[12^n n! (n-2)!]\} x^n$, $-3 < x < 3$

28. $-1 - 2(x-1)/3 + \sum_{n=2}^{\infty} \{(-1)^n 2^n [2 \cdot 5 \cdot 8 \cdots (3n-4)]/(3^n n!)\}(x-1)^n$, $1/2 \le x \le 3/2$

30. $x - x^2/3 - \sum_{n=3}^{\infty} \{2 \cdot 5 \cdot 8 \cdots (3n-7)/[3^{n-1}(n-1)!]\} x^n$, $-1 \le x \le 1$

32. $1 + x^2/2 + 5x^4/24 + 61x^6/720$

34. $1 + \sum_{n=1}^{\infty} [(-1)^n 2^{2n-1}/(2n)!] x^{2n}$, $-\infty < x < \infty$

36. $x^2 + \sum_{n=1}^{\infty} \{(2n)!/[(2n+1) 2^{2n}(n!)^2]\} x^{4n+2}$, $-1 < x < 1$

38. $\sqrt{2} \sum_{n=0}^{\infty} \{1/[2^n(2n+1)]\} x^{2n+1}$, $-\sqrt{2} < x < \sqrt{2}$

42. (a) $1/p$ (b) 6

44. $\sum_{n=0}^{\infty} \{(-1)^n \pi^{2n}/[(4n+1) 2^{2n}(2n)!]\} x^{4n+1}$, $-\infty < x < \infty$; $\sum_{n=0}^{\infty} \{(-1)^n \pi^{2n+1}/[(4n+3) 2^{2n+1}(2n+1)!]\} x^{4n+3}$, $-\infty < x < \infty$

46. $n(-2)^{n-1}$

48. $(-1)^{n+1}(n-2)/e^2$

54. (a) $-1/2$, $1/6$, 0, $-1/30$, 0

Exercises 11.6

2. $2/(1-x)^3$, $-1 < x < 1$

4. $(x+1)/(1-x)^3$, $-1 < x < 1$

6. $-(1/x)\ln(1-x)$, $-1 < x < 1$

8. $-\ln(1+x^2)$, $-1 < x < 1$

10. $1/(x-x^2) - (1/x^2)\ln(1-x)$, $-1 < x < 1$

Exercises 11.7

2. 4.2×10^{-10} 4. 4.2×10^{-10} 6. 1.4×10^{-9}

8. 0.115 10. 0.00625 12. 0.497

14. 0.291 16. $1/2$ 18. $1/2$

20. 0 22. $1 - x^3/2 + 3x^6/8 - 5x^9/16$

24. $1 - x^2$

26. $a_0 + a_1 \sum_{n=1}^{\infty} [(-1)^{n+1}/n!]x^n = C + De^{-x}$

28. $a_0 \sum_{n=0}^{\infty} [(-1)^n/(2n)!]x^n$

30. $a_1 \sum_{n=1}^{\infty} \{(-1)^{n+1}/[n!\,(n-1)!]\}x^n$

Exercises 11.8

2. False, $\{n\}$ 4. False, $\{-n\}$ 6. True

8. True   10. True   12. False, $\{(-1)^n/n\}$

14. False, $\{(-1)^n/n\}$ 16. False, $\{(-1)^n\}$ 18. False, $\{-3^n\}$

20. False, $\{-n\}$ 22. True

24. False, $\{[1 + (-1)^{n+1}]/(2n)\}$

26. Increasing, $U = 1$, $V = 0$, $L = 0.419\,241$

28. Increasing, $U = 5$, $V = 0$, $L = (\sqrt{21} + 1)/2$

30. Increasing, $U = 10$, $V = 3$, $L = (3 + \sqrt{29})/2$

32. Decreasing, $U = 4$, $V = 0$, $L = (7 - \sqrt{5})/2$

34. Decreasing, $U = 1$, $V = 0$, $L = (2 - \sqrt{2})/2$

36. Increasing, $U = 1$, $V = 0$, $L = 1/2$

38. Decreasing, $U = 4$, $V = 0$, $L = 1$

40. Decreasing, $U = 3/2$, $V = 0$, $L = 1$

42. $0.273\,89$ 44. $4/3$ 46. $(\sqrt{5} - 1)/2$

48. $(\sqrt{13} - 1)/2$ 50. $(\sqrt{21} - 1)/5$ 52. $(\sqrt{89} - 3)/2$

56. 2 60. Increasing, $U = 30$, $V = -30$, $L = 30$

64. $c_n = d$

Exercises 11.9

2. $2/15$ 4. Diverges   6. $10\,804.5$

8. $-1/3$ 10. Diverges   12. $13/99$

14. $430\,162/9999$ 16. Diverges   18. 4

20. Diverges   22. 1 24. 804 s

26. $(\sqrt{3}P^2/180)[8 - 3(4/9)^n]$, $2\sqrt{3}P^2/45$

28. 2 30. $51/64$ 32. $2/3$

40. $600/11$ minutes after 10:00

Exercises 11.10

2. Diverges   4. Converges   6. Converges

8. Diverges   10. Converges   12. Converges

14. Converges   16. Converges   18. Diverges

20. Diverges   22. Diverges   24. $-1 \le x \le 1$

26. $-7/3 \le x \le -5/3$

28. Converges for $p > 1$, diverges for $p \le 1$

Exercises 11.11

2. Converges   4. Converges   6. Diverges

8. Diverges   10. Converges   12. Converges

14. Converges   16. Diverges   18. Converges

20. Converges

Exercises 11.12

2. Converges conditionally   4. Converges absolutely

6. Diverges   8. Converges absolutely

10. Converges conditionally   12. Converges conditionally

14. Converges absolutely   16. $-1 \le x \le 1$

18. $-3 \le x < -1$ 20. $-1 \le x < 1$

22. $-1 \le x \le 3$

Exercises 11.13

12. $-9/100$

14. $0.055\,140\,9$, error $< 6 \times 10^{-7}$

16. error < 0.00125

18. -0.9470, error $< 5.2 \times 10^{-6}$

20. $0.693\,064\,9$, error $< 1/(11 \cdot 2^{10})$

22. 1.07, error $< 4.1 \times 10^{-4}$ 24. -0.7, error $< 1/101$

26. 4

28. (a) Error $< 3.6 \times 10^{-5}$ (b) Error $< 2.7 \times 10^{-5}$

30. 0.497 32. 0.133

34. 0.291 36. -0.122

Review Exercises

2. Decreasing, $U = 1$, $V = 0$, $L = 1/\sqrt{3}$

4. Increasing, $U = 100$, $V = 7$, $L = (31 + \sqrt{53})/2$

6. $|k| \le 1$ 10. Converges

12. Converges   14. Converges conditionally

16. Diverges   18. Converges

20. Converges   22. Diverges

24. Converges   26. Converges

28. Converges conditionally   30. $-2 \le x \le 2$

32. $-\infty < x < \infty$ 34. $-4 < x < -2$

36. $-2^{-1/3} \le x < 2^{-1/3}$

38. $e^5 \sum_{n=0}^{\infty} (1/n!)x^n$, $-\infty < x < \infty$

40. $\sum_{n=2}^{\infty} [(-1)^n 2^{n-1}/(n-1)]x^n$, $-1/2 < x \le 1/2$

42. $(1/2) \sum_{n=0}^{\infty} (-1)^{n+1}(1 - 1/3^n)x^n$, $|x| < 1$

44. $x + \sum_{n=2}^{\infty} \{(-1)^n/[n(n-1)]\}x^n$, $-1 < x \le 1$

46. $\sum_{n=0}^{18} [(-1)^n/n!]x^{2n}$

48. $\sum_{n=0}^{\infty} \{(-1)^n/[2^{2n}(2n)!]\}x^{2n} + \sum_{n=0}^{\infty}\{(-1)^n/$
$[2^{2n+1}(2n+1)!]\}x^{2n+1}$

Chapter 12

Exercises 12.1

2. $\sqrt{33}$

4. $(0, 0, 0)$, $(2, 0, 0)$, $(0, 2, 0)$, $(0, 0, 2)$, $(2, 2, 0)$, $(2, 0, 2)$, $(0, 2, 2)$, $(2, 2, 2)$

6. $\sqrt{29}$, 5, $2\sqrt{5}$, $\sqrt{13}$

8. 5, 3, 4, 5

12. $10x + 2y + 2z = 5$, plane

14. $(\sqrt{2}, \sqrt{2}, 5)$, $(\sqrt{2} \pm 1/4, \sqrt{2} \pm 1/4, 5 - \sqrt{7}/4)$, $(\sqrt{2} \pm 1/4, \sqrt{2} \pm 1/4, 9/2 - \sqrt{7}/4)$

16. (a) $(2, 1/2, -7/2)$ (b) $(5, 5, -5)$

Exercises 12.2

2.
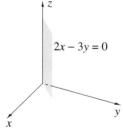
$2x - 3y = 0$

4.

$z = x^2$

6.
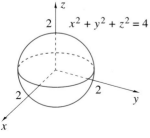
$x^2 + y^2 + z^2 = 4$

8.
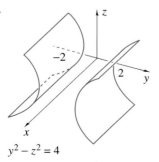
$y^2 - z^2 = 4$

10.

$x = \sqrt{y^2 + z^2}$

12.
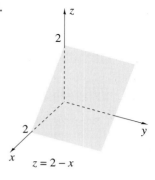
$z = 2 - x$

14.
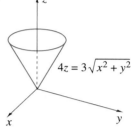
$x = z^2 + 2$

16.

$4z = 3\sqrt{x^2 + y^2}$

18.

$yz = 1$

20.

$z + 5 = 4(x^2 + y^2)$

22.
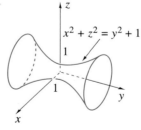
$x^2 + z^2 = y^2 + 1$

24.
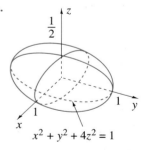
$x^2 + y^2 + 4z^2 = 1$

26.
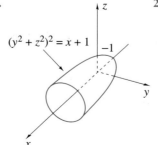
$(y^2 + z^2)^2 = x + 1$

28.

$y - z^2 = 0$

30.

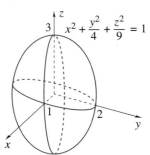

$$x^2 + \frac{y^2}{4} + \frac{z^2}{9} = 1$$

32.

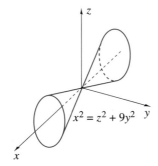

$$x^2 = z^2 + 9y^2$$

46.

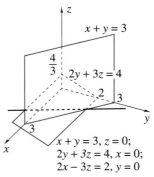

$x + y = 3$
$2y + 3z = 4$

$x + y = 3, z = 0;$
$2y + 3z = 4, x = 0;$
$2x - 3z = 2, y = 0$

48.

$z = 4$
$x^2 + y^2 = 4$

$x^2 + y^2 = 4, z = 0;$
$z = 4, x = 0, |y| \le 2;$
$z = 4, y = 0, |x| \le 2$

34.

$$x^2 + \frac{y^2}{4} - \frac{z^2}{25} = 1$$

36.

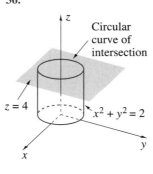

Circular curve of intersection

$z = 4$
$x^2 + y^2 = 2$

50.

$x^2 + y^2 = 4$
$x = z$

$y^2 + z^2 = 4, x = 0;$
$x^2 + y^2 = 4, z = 0;$
$z = x, y = 0, |x| \le 2$

52.

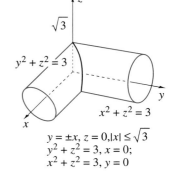

$\sqrt{3}$
$y^2 + z^2 = 3$
$x^2 + z^2 = 3$

$y = \pm x, z = 0, |x| \le \sqrt{3};$
$y^2 + z^2 = 3, x = 0;$
$x^2 + z^2 = 3, y = 0$

38.

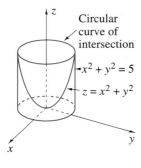

Circular curve of intersection

$x^2 + y^2 = 5$
$z = x^2 + y^2$

40.

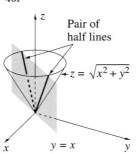

Pair of half lines

$z = \sqrt{x^2 + y^2}$

$y = x$

54.

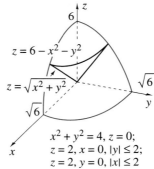

$z = 6 - x^2 - y^2$
$z = \sqrt{x^2 + y^2}$
$\sqrt{6}$
$\sqrt{6}$

$x^2 + y^2 = 4, z = 0;$
$z = 2, x = 0, |y| \le 2;$
$z = 2, y = 0, |x| \le 2$

56.

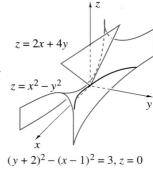

$z = 2x + 4y$
$z = x^2 - y^2$

$(y + 2)^2 - (x - 1)^2 = 3, z = 0$

42.

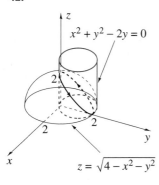

$x^2 + y^2 - 2y = 0$

$z = \sqrt{4 - x^2 - y^2}$

44.

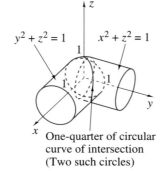

$y^2 + z^2 = 1$ $x^2 + z^2 = 1$

One-quarter of circular curve of intersection (Two such circles)

58.

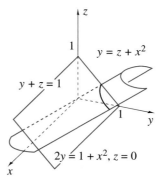

$y = z + x^2$
$y + z = 1$

$2y = 1 + x^2, z = 0$

60.

$z = x^2 + y^2$

$(1, 1)$

$z = 4(x - 1)^2 + 4(y - 1)^2$

$$\left(x - \frac{4}{3}\right)^2 + \left(y - \frac{4}{3}\right)^2 = \frac{8}{9}, z = 0$$

62.

64.

66.

68.

70.

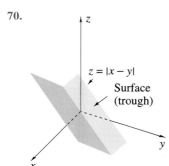

Exercises 12.3

2. $(2, 6, 8)$ **4.** $(-1/\sqrt{5}, 0, 2/\sqrt{5})$

6. $(-4\sqrt{5} - 8, -6, 8\sqrt{5} + 4)$

8. $(-6\sqrt{46} - 4\sqrt{5}, -12\sqrt{5}, 12\sqrt{46} - 24\sqrt{5})$

10. $(-6/\sqrt{17}, -3/\sqrt{17}, 6/\sqrt{17})$ **12.** $3\hat{\mathbf{i}} - 2\hat{\mathbf{j}}$

14. $(2/\sqrt{5} - 1/\sqrt{10})\hat{\mathbf{i}} + (1/\sqrt{5} + 3/\sqrt{10})\hat{\mathbf{j}}$

16. $(5, 0, 0)$ **18.** $(0, 3/\sqrt{5}, 6/\sqrt{5})$

20. $(1, 1, -3/2)$ **22.** $(5/4, 5\sqrt{2}/4, 5/4)$

24. $(2/\sqrt{3}, 2/\sqrt{3}, 2/\sqrt{3})$ **30.** $(3/\sqrt{17}, 12/\sqrt{17})$

32. $5G(-3/\sqrt{10} + 30\sqrt{11}/121, -10\sqrt{11}/121, -1/\sqrt{10} + 10\sqrt{11}/121)$ N

34. Linearly dependent   **36.** Linearly dependent

Exercises 12.4

2. $(-10, 15, -5)$ **4.** $4/\sqrt{14}$ **6.** -178

8. 1   **10.** 0   **12.** $(48, 24, -42)$

14. $(2\sqrt{247})^{-1}(17, -23, -7)$

16. $(-41, 11, 28)$ **18.** $(17, -23, -7)$ **20.** $(-10, 19, 5)$

22. Yes   **24.** Yes   **26.** 0.684 rad

28. 2.20 rad   **30.** π rad   **32.** $\lambda(9, 0, 1)$

38. $\pi/2$, 1.25, 2.82   **40.** 1.95, 0.980, 0.734

44. (a) 101

Exercises 12.5

2. $2x + y - 2z + 5 = 0$ **4.** $19x - 14y - z = 91$

6. $2x - y - 2z = 6$ **8.** $10x + 89y - 6z = 394$

10. (a) $13x + 17y = 70$ (b) $x - 17z = 8$ (c) $y + 13z = -2$

12. $(x, y, z) = (1, -1, 3) + t(2, 4, -3)$; $x = 1 + 2t$, $y = -1 + 4t$, $z = 3 - 3t$; $(x - 1)/2 = (y + 1)/4 = (z - 3)/(-3)$

14. $(x, y, z) = (2, -3, 4) + t(3, 5, -5)$; $x = 2 + 3t$, $y = -3 + 5t$, $z = 4 - 5t$; $(x - 2)/3 = (y + 3)/5 = (z - 4)/(-5)$

16. $(x, y, z) = (1, 3, 4) + t(0, 0, 1)$; $x = 1$, $y = 3$, $z = 4 + t$

18. $(x, y, z) = (2, 0, 3) + t(1, 0, -2)$; $x = 2 + t$, $y = 0$, $z = 3 - 2t$; $x - 2 = (z - 3)/(-2)$, $y = 0$

20. $(x, y, z) = (0, -5, 30) + t(1, 2, -11)$; $x = t$, $y = -5 + 2t$, $z = 30 - 11t$; $x = (y + 5)/2 = (z - 30)/(-11)$

22. No

26. $x = \sqrt{2} \pm 1/4$, $y = \sqrt{2} \pm 1/4$, $z = 9/2 - \sqrt{7}/4$, $\sqrt{7}x - z = \sqrt{14} - 5$, $\sqrt{7}x + z = \sqrt{14} + 5$, $\sqrt{7}y - z = \sqrt{14} - 5$, $\sqrt{7}y + z = \sqrt{14} + 5$

Exercises 12.6

2. $\sqrt{299}/2$ **4.** 0   **6.** $\sqrt{42}$

8. $20/\sqrt{35}$ **10.** $18/\sqrt{14}$ **12.** $(-10, 5, -2)$

14. $(-5, 1, 4)$ **16.** $6/\sqrt{13}$ **18.** $4/\sqrt{6}$

20. $4\sqrt{5}/3$ **22.** $9/\sqrt{14}$ **24.** $23\sqrt{14}/28$

26. $\sqrt{710}$ **28.** $155/\sqrt{4030}$ **30.** $16/\sqrt{26}$

32. $6/\sqrt{3}$ **36.** $\sqrt{154}$ **38.** $7/\sqrt{5}$, $4/\sqrt{5}$

40. $1/\sqrt{3}$, $4/\sqrt{6}$, $\sqrt{2}$ **42.** 1, -3, 0

44. $q_3(4\pi\epsilon_0)^{-1}[q_1(1/\sqrt{41} - 1/\sqrt{61}) + q_2(\sqrt{2}/6 - 1/\sqrt{10})]$

Exercises 12.7

2. $-\infty < t < \infty$ **4.** $t > -4$ **6.** $\hat{\mathbf{i}} - 2t\hat{\mathbf{j}} + 2\hat{\mathbf{k}}$

8. $2t(4t^2 - 3)\hat{\mathbf{i}} + t^2(9 - 10t^2)\hat{\mathbf{j}} + 4t(4t^2 - 3)\hat{\mathbf{k}}$

10. $3t^2(4 - 5t^2)\hat{\mathbf{i}} + 4t(1 - 3t^2)\hat{\mathbf{j}} - 3t^2\hat{\mathbf{k}}$

12. $3\hat{\mathbf{i}} - 2(3t + 4)\hat{\mathbf{j}} + 6(1 + 4t)\hat{\mathbf{k}}$

14. $9t^2\hat{\mathbf{i}} + (6t - 20t^3)\hat{\mathbf{j}} + (30t^4 - 21t^2 + 6)\hat{\mathbf{k}}$

16. $(-15t^4 + 12t^2)\hat{\mathbf{i}} + (4t - 12t^3)\hat{\mathbf{j}} - 3t^2\hat{\mathbf{k}}$

18. $(3t^4/2 - 4t^2)\hat{\mathbf{i}} + (17t^3/3 - 12t^5/5)\hat{\mathbf{j}} + (3t^6 - 27t^4/4 + t^2)\hat{\mathbf{k}} + C$

20. $(-14t^4 - 6t^3 - 42t^2 + 4t)\hat{\mathbf{i}} + (28t^5 + 84t^3 - 6t^2)\hat{\mathbf{j}} - (126t^4 + 42t^6 + 2t)\hat{\mathbf{k}}$

Exercises 12.8

2.

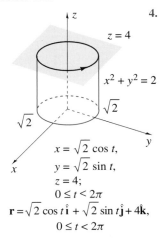

$x = \sqrt{2}\cos t,$
$y = \sqrt{2}\sin t,$
$z = 4;$
$0 \le t < 2\pi$
$\mathbf{r} = \sqrt{2}\cos t\,\hat{\mathbf{i}} + \sqrt{2}\sin t\,\hat{\mathbf{j}} + 4\hat{\mathbf{k}},$
$0 \le t < 2\pi$

4.

$x = \sqrt{5}\cos t,$
$y = \sqrt{5}\sin t,$
$z = 5;$
$0 \le t < 2\pi$
$\mathbf{r} = \sqrt{5}\cos t\,\hat{\mathbf{i}} + \sqrt{5}\sin t\,\hat{\mathbf{j}} + 5\hat{\mathbf{k}},$
$0 \le t < 2\pi$

6.

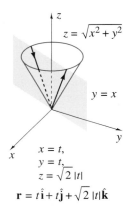

$x = t,$
$y = t,$
$z = \sqrt{2}\,|t|$
$\mathbf{r} = t\hat{\mathbf{i}} + t\hat{\mathbf{j}} + \sqrt{2}\,|t|\hat{\mathbf{k}}$

8.

$x = \cos t,$
$y = 1 + \sin t,$ $\Bigg\} \; 0 \le t < 2\pi$
$z = \sqrt{2 - 2\sin t}$
$\mathbf{r} = \cos t\hat{\mathbf{i}} + (1 + \sin t)\hat{\mathbf{j}}$
$+ \sqrt{2 - 2\sin t}\,t\hat{\mathbf{k}}, \; 0 \le t < 2\pi$

10.

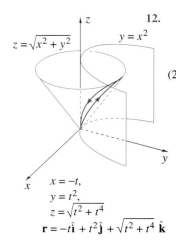

$x = -t,$
$y = t^2,$
$z = \sqrt{t^2 + t^4}$
$\mathbf{r} = -t\hat{\mathbf{i}} + t^2\hat{\mathbf{j}} + \sqrt{t^2 + t^4}\,\hat{\mathbf{k}}$

12.

14.

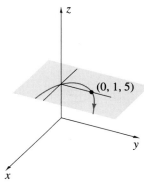

Exercises 12.9

2. $\mathbf{r} = t\hat{\mathbf{i}} + t^2\hat{\mathbf{j}} + t^3\hat{\mathbf{k}}; \; \hat{\mathbf{T}} = (\hat{\mathbf{i}} + 2t\hat{\mathbf{j}} + 3t^3\hat{\mathbf{k}})/\sqrt{1 + 4t^2 + 9t^4}$

4. $\mathbf{r} = -t\hat{\mathbf{i}} + (5 + t)\hat{\mathbf{j}} + (t^2 - t - 5)\hat{\mathbf{k}}, \; -5 \le t \le 0;$
$\hat{\mathbf{T}} = [-\hat{\mathbf{i}} + \hat{\mathbf{j}} + (2t - 1)\hat{\mathbf{k}}]/\sqrt{4t^2 - 4t + 3}$

6. $(-2\hat{\mathbf{i}} + 3\hat{\mathbf{j}} + \hat{\mathbf{k}})/\sqrt{14}$ **8.** $(\hat{\mathbf{i}} - \hat{\mathbf{j}})/\sqrt{2}$

10. $-\hat{\mathbf{j}}$ **12.** $\sqrt{42}$

14. $(616\sqrt{616} - 157\sqrt{157})/459$ **16.** $\cos t\,\hat{\mathbf{i}} + \sin t\,\hat{\mathbf{j}}$

18. (a) $(0, 0, 0)$ (b) $2\hat{\mathbf{i}} + 2\hat{\mathbf{k}}$

Exercises 12.10

2. $[-(2t + 9t^3)\hat{\mathbf{i}} + (1 - 9t^4)\hat{\mathbf{j}} + (3t + 6t^3)\hat{\mathbf{k}}]/\sqrt{1 + 13t^2 + 54t^4 + 117t^6 + 81t^8};$
$(3t^2\hat{\mathbf{i}} - 3t\hat{\mathbf{j}} + \hat{\mathbf{k}})/\sqrt{1 + 9t^2 + 9t^4}$

4. $[(2t - 1)\hat{\mathbf{i}} + (1 - 2t)\hat{\mathbf{j}} + 2\hat{\mathbf{k}}]/\sqrt{8t^2 - 8t + 6}; \; (\hat{\mathbf{i}} + \hat{\mathbf{j}})/\sqrt{2}$

6. $-(5\hat{\mathbf{i}} + 3\hat{\mathbf{j}} + \hat{\mathbf{k}})/\sqrt{35}; \; (-\hat{\mathbf{j}} + 3\hat{\mathbf{k}})/\sqrt{10}$

8. $-(\hat{\mathbf{i}} + \hat{\mathbf{j}})/\sqrt{2}; \; -\hat{\mathbf{k}}$

10. $-(\hat{\mathbf{i}} + \hat{\mathbf{k}})/\sqrt{2}; \; (\hat{\mathbf{i}} - \hat{\mathbf{k}})/\sqrt{2}$

12. $\kappa = 0, \; \rho$ undefined

14. $\kappa = 2e^t\sqrt{1 + e^{2t}}/(1 + 2e^{2t})^{3/2},$
$\rho = (1 + 2e^{2t})^{3/2}/(2e^t\sqrt{1 + e^{2t}})$

16. $\kappa = 1/[\sqrt{2}(1 + \cos^2 t)^{3/2}], \; \rho = \sqrt{2}(1 + \cos^2 t)^{3/2}$

18. $\kappa = \sqrt{1 + 36t^4 + 16t^6}/[2(1 + t^2 + 4t^6)^{3/2}],$
$\rho = 2(1 + t^2 + 4t^6)^{3/2}/\sqrt{1 + 36t^4 + 16t^6}$

22. 0

24. (a) $-\sin t\,\hat{\mathbf{i}} + \cos t\,\hat{\mathbf{j}}; \; -(\cos t\,\hat{\mathbf{i}} + \sin t\,\hat{\mathbf{j}}); \; \hat{\mathbf{k}}$
(b) $2\sin 2t(\sin t - \cos t), -4(\cos^3 t + \sin^3 t);$
(c) $2\sin 2t(\sin t - \cos t)\hat{\mathbf{T}} - 4(\cos^3 t + \sin^3 t)\hat{\mathbf{N}}$

26. $(-\sin t\,\hat{\mathbf{i}} + \cos t\,\hat{\mathbf{j}} + \hat{\mathbf{k}})/\sqrt{2};$
$-(\cos t\,\hat{\mathbf{i}} + \sin t\,\hat{\mathbf{j}}); \; (\sin t\,\hat{\mathbf{i}} - \cos t\,\hat{\mathbf{j}} + \hat{\mathbf{k}})/\sqrt{2};$
$(1 - \cos t\,\sin t + \cos^2 t\,\sin^2 t)\hat{\mathbf{T}}/\sqrt{2} - \cos t(\cos t + \sin^3 t)\hat{\mathbf{N}} +$
$(-\cos^2 t\,\sin^2 t + \cos\,\sin t + 1)\hat{\mathbf{B}}/\sqrt{2}$

Exercises 12.11

2. $[(t^2 - 1)\hat{\mathbf{i}} + (t^2 + 1)\hat{\mathbf{j}}]/t^2;\ \sqrt{2 + 2t^4}/t^2;\ 2(\hat{\mathbf{i}} - \hat{\mathbf{j}})/t^3$

4. $2t\hat{\mathbf{i}} + 2e^t(t + 1)\hat{\mathbf{j}} - 2\hat{\mathbf{k}}/t^3;\ 2\sqrt{t^2 + e^{2t}(t + 1)^2 + 1/t^6};$
$2\hat{\mathbf{i}} + 2e^t(t + 2)\hat{\mathbf{j}} + 6\hat{\mathbf{k}}/t^4$

6. $(t^4/4 + 1)\hat{\mathbf{i}} + (t^3 + 3t^2 + 12)\hat{\mathbf{j}}/6 - (t^5/5 + 1)\hat{\mathbf{k}}$

8. $4t/\sqrt{1 + 4t^2};\ 2/\sqrt{1 + 4t^2}$

12. (a) 0 (b) $56t^6$ **14.** $150x\hat{\mathbf{j}}$ **18.** (b) 7.79 km/s

20. $365\hat{\mathbf{i}} + 325\sqrt{3}\hat{\mathbf{j}}$ km/h; 670.9 km/h

22. $\text{Sin}^{-1}(3/4)$ radians upstream, $1/(5\sqrt{7})$ h

24. $(3/2, 3/2, 2)$ **26.** $(3, 0)$ **32.** Yes

34. (b) $S[(1 - \cos\theta)\hat{\mathbf{i}} + \sin\theta\,\hat{\mathbf{j}}]$,
$S\sqrt{2 - 2\cos\theta},\ (S^2/R)(\sin\theta\,\hat{\mathbf{i}} + \cos\theta\,\hat{\mathbf{j}})$
(c) $(S^2/R)\sqrt{(1 - \cos\theta)/2},\ (S^2/R)\sin\theta/\sqrt{2 - 2\cos\theta}$

36. (c) $2S[(\sin 2\theta - \sin\theta)\hat{\mathbf{i}} + (\cos\theta - \cos 2\theta)\hat{\mathbf{j}}],\ 2\sqrt{2}S\sqrt{1 - \cos\theta},$
$(2S^2/R)[(2\cos 2\theta - \cos\theta)\hat{\mathbf{i}} + (2\sin 2\theta - \sin\theta)\hat{\mathbf{j}}]$
(d) $(3\sqrt{2}S^2/R)\sqrt{1 - \cos\theta},\ (\sqrt{2}S^2/R)\sin\theta/\sqrt{1 - \cos\theta}$

38. At an angle $\text{Sin}^{-1}[(15L + \sqrt{100L^2 - 5})/(50L^2 + 2)]$ downstream. $L > 5/\sqrt{10}$ km

40. Straight line path

Review Exercises

2. -8 **4.** $(39, -33, -30)$

6. -2 **8.** $(-16, -24, -4)$

10. $(67/7, 21, -65/7)$

12.   **14.**

16. No points   **18.**

20.

22.

24.   **26.**

28. $x = 6 + 5t,\ y = 6 - 2t,\ z = 2 + t$

30. $x = 1 + u,\ y = 3 + 2u,\ z = 2 + 3u$

32. $x = y + z$ **34.** $3x - 4y + z = 0$

36. $35/\sqrt{41}$ **38.** $2\sqrt{5}/15$

40. $\sqrt{59}$

42. $(2\cos t\,\hat{\mathbf{i}} - 2\sin t\,\hat{\mathbf{j}} + \hat{\mathbf{k}})/\sqrt{5};\ -\sin t\,\hat{\mathbf{i}} - \cos t\,\hat{\mathbf{j}};$
$(\cos t\,\hat{\mathbf{i}} - \sin t\,\hat{\mathbf{j}} - 2\hat{\mathbf{k}})/\sqrt{5}$

44. $\hat{\mathbf{i}} + 2t\hat{\mathbf{j}} + 2t\hat{\mathbf{k}};\ \sqrt{1 + 8t^2};\ 2(\hat{\mathbf{j}} + \hat{\mathbf{k}});\ 0,\ \sqrt{1 + 8t^2};$
$2\sqrt{2}/\sqrt{1 + 8t^2},\ 8t/\sqrt{1 + 8t^2}$

46. (a) 4.46 m/s (b) $0.226\hat{\mathbf{i}} - \hat{\mathbf{j}}$ m (c) 0.255 m from point on floor directly below point it left table

48. $k(1 - x)[1 - 1/\sqrt{1 + 4(1 - x)^2}]\hat{\mathbf{i}} + (k/2)[1 - 1/\sqrt{1 + 4(1 - x)^2}]\hat{\mathbf{j}}$ N

Chapter 13

Exercises 13.1

4. All points between the branches of the hyperbola $x^2 - y^2 = 1$

6. All points in space except $(0, 0, 0)$

8.

10.

12.

14.

16.

18.

20.

22.

24.
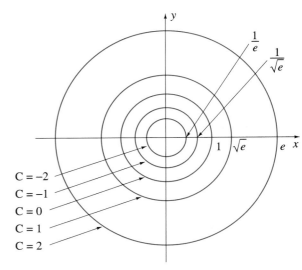

26. $lw(15 - lw)/(w + l)$

28. $8c|xy|\sqrt{1 - x^2/a^2 - y^2/b^2}$

30. $x \sin \theta (1 - x - x \sin \theta \, \csc \phi) + (x^2/2) \sin \theta (\cos \theta + \sin \theta \, \cot \phi)$

32. $30.25 + 11G + 17.5S$ cents. The domain consists of all points (G, S) satisfying the inequalities $200 \leq 37G + 31S \leq 300$, $145 \leq 22G + 85S \leq 170$.

Exercises 13.2

2. $3/5$ **4.** $-3/7$ **6.** 0

8. $\pi/2$ **10.** 1 **12.** 4

14. Does not exist   **16.** 0 **18.** Does not exist

20. Does not exist   **22.** $(0, 0)$

24. $x = 0$; $y = 0$; $z = 0$

26. $x = 0$; $y = 0$; $y = -x$

28. Does not exist   **30.** 1 **32.** (a) 1, no (b) 1, yes

34. False

Exercises 13.3

2. $3y - 16x^3 y^4$, $3x - 16x^4 y^3$

4. $-x(x + 2y)/[y(x + y)^2]$, $x^2(x + 2y)/[y^2(x + y)^2]$

6. $y \cos(xy)$, $x \cos(xy)$

8. $x/\sqrt{x^2 + y^2}$, $y/\sqrt{x^2 + y^2}$

10. $4x \sec^2(2x^2 + y^2)$, $2y \sec^2(2x^2 + y^2)$

12. ye^{xy}, xe^{xy}

14. $2x/(x^2 + y^2)$, $2y/(x^2 + y^2)$

16. $ye^x \cos(ye^x)$, $e^x \cos(ye^x)$

18. $2xy \cos^2(x^2 y) \sin(x^2 y)/[1 - \cos^3(x^2 y)]^{2/3}$, $x^2 \cos^2(x^2 y) \sin(x^2 y)/[1 - \cos^3(x^2 y)]^{2/3}$

20. $\tan\sqrt{x + y}/(2\sqrt{x + y})$, $\tan\sqrt{x + y}/(2\sqrt{x + y})$

22. $-2z/[1 + (x^2 + z^2)^2]$ **24.** 2

26. $-1/[1 + (1 + x + y + z)^2]$

28. $3x^2/y + \sin(yz/x) - (yz/x)\cos(yz/x)$

30. $z\,\text{Sin}^{-1}(x/z)$ if $z > 0$; $z\,\text{Sin}^{-1}(x/z) + xz/\sqrt{z^2 - x^2}$ if $z < 0$

34. (a) Yes (b) No (c) Yes (d) No

36. No   **38.** (a) Yes (b) No

Exercises 13.4

2. $2xyz\hat{\mathbf{i}} + x^2z\hat{\mathbf{j}} + x^2y\hat{\mathbf{k}}$ **4.** $(2xy + y^2)\hat{\mathbf{i}} + (x^2 + 2xy)\hat{\mathbf{j}}$

6. $(1 + x^2y^2z^2)^{-1}(yz\hat{\mathbf{i}} + xz\hat{\mathbf{j}} + xy\hat{\mathbf{k}}$

8. $e^{x+y+z}(\hat{\mathbf{i}} + \hat{\mathbf{j}} + \hat{\mathbf{k}})$

10. $-(x^2 + y^2 + z^2)^{-3/2}(x\hat{\mathbf{i}} + y\hat{\mathbf{j}} + z\hat{\mathbf{k}})$

12. $-\sin 1(\hat{\mathbf{i}} + \hat{\mathbf{j}} + \hat{\mathbf{k}})$ **14.** $-4e^{-8}(\hat{\mathbf{i}} + \hat{\mathbf{j}})$

22. ∇F not defined along the line $y = x$

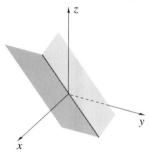

24. $x^2y - xy + C$ **26.** $xyz + y^2z + C$

28. $f(x, y) = g(x, y) + C$

Exercises 13.5

2. $-12x/y^4 + 72x^3y$ **4.** $x(1 + z + y + yz)e^{x+y+z}$

6. $e^{x+y} + 4/y^3$ **8.** 0

10. $(x^2 + y^2 - z^2)/(x^2 + y^2 + z^2)^2$

12. $2xy/(x^2 + y^2)^2$ **14.** $127/(756\sqrt{7})$

16. $8!y^9z^{10}$ **18.** $3y^2\cos(x + y^3)$

20. $-xy/(x^2y^2 - 1)^{3/2}$

24. Entire plane with $(0, 0)$ deleted

26. All space   **28.** Not harmonic

32. $L + 9.81L^2/(2c^2)$ **34.** (c) 0 (d) 0

38. (b)

(c)

---

40. (b)

42. (b)

44. $T_0 + (T_L - T_0)x/L + Fx(L - x)/(2k)$

48. (b) $2xy + C$

50. $n = 0$, all space; $n = -1/2$, $x^2 + y^2 + z^2 > 0$

Exercises 13.6

2. $(2xe^y + y/x)(-s^2\sin t) + (x^2e^y + \ln x)$
$\left[8t/\left((t^2 + 2s)\sqrt{(t^2 + 2s)^2 - 1}\right)\right]$

4. $xv^2[2yv(3u^2 + 2) + 2xuv/(u^2 + 1) + 3xye^u(u + 1)]$

6. $-2(\ln 3)y3^{x+2}\csc(r^2 + t)\cot(r^2 + t)$

8. $-2rst[5st/y^6 + 2yrt/(y^2 + z^2)^2 + 2rs/y^3]$

10. $6xve^t(v - 2)[1 - 2u/(x^2 - y^2)^2] + 8ye^{4t}[1 + 2u/(x^2 - y^2)^2]$

12. $2(t^2 + 2t)^2e^{2t} + (t^3 + 6t^2 + 6t + 2yt^2 + 4y + 8yt)e^t - 2$

14. $-(3y\sin v - 4x\cos v)^2\sin(xy) - (24\sin v\cos v + 3y\cos v + 4x\sin v)\cos(xy)$

18. $200\pi/3$ cc/min; no

20. -7.11×10^4 N/s

22. (a) Yes (b) No (c) Yes (d) No (e) Yes (f) No (g) No (h) Yes

34. 18.77 kg/m^3/s

Exercises 13.7

2. $1/(x + y) - 1$

4. $[24x - \cos(x + y)]/[2y - 1 + \cos(x + y)]$

6. $-(2xz^2 + 3)/(2x^2z + y)$, $-z/(2x^2z + y)$

8. $z(1 + y^2z^2)/(y - x - xy^2z^2)$, $z/(x + xy^2z^2 - y)$

10. $[\cos(x + t) + \cos(x - t)]/[\cos(x - t) - \cos(x + t)]$

12. $(x + y)(3y^2 - 3x^2 - 5)/(3y^2z - 5z)$

14. 0

16. $e^x[(2t + 1)\cos y/(3x^2 + e^x) + (y^2 + 2yt - 1)\sin y/(t^2 + 2yt + 1)]$

18. $[u^3 + u\cos(uv)]e^{-u}\cos v + [3u^2v + v\cos(uv)]e^{-u}\sin v$

20. $u(3v^2 - u^2)/[4(u^2 + v^2)^3]$

22. $-2t/(9y^2)$

Exercises 13.8

2. $1/\sqrt{5}$ **4.** $-2/\sqrt{17}$ **6.** $1/5$

8. $-13/\sqrt{29}$ **10.** $11/\sqrt{82}$ **12.** $-40/\sqrt{6}$

14. $(1, 4)$ **16.** $(1, -3, 2)$ **18.** $(1, 1)$

20. (a) $\pm(2, 1)$ (b) $(-1 \pm 2\sqrt{19}, 2 \pm \sqrt{19})$ (c) No direction

22. (a) Yes (b) No

24. $9t/[\sqrt{13}\sqrt{4 + 9t^2}]$, 0

26. $(0, 0, 0)$, $(1/4, \pm 1/2, 1/4)$

28. $-k$ **30.** $\sqrt{65}/3$ **32.** -2

34. (a) $\pi/(\sqrt{2}\sqrt{8 - 4\pi + \pi^2})$, $\pi/\sqrt{4 + \pi^2}$ (b) $1/\sqrt{2}$, 0 (c) $1/\sqrt{2}$, 1

Exercises 13.9

2. $x = 1 + u$, $y = 1 + 2u$, $z = 1 + 3u$

4. $x = -2 + t$, $y = 4 - 4t$, $z = -2 + t$

6. $x = 1 - 2u$, $y = 5 + 2u$, $z = 1 + u$

8. $132x + 49y = 328$, $z = 0$

10. $x = 1 - u$, $y = u$, $z = u$

12. $x = 2 + t$, $y = -\sqrt{5} + (1/\sqrt{5})t$, $z = -1 - t$

14. $x = 4 + \sqrt{17}u$, $y = 1$, $z = \sqrt{17} + 4u$

16. $x = 12 + 16t$, $y = -14 - 31t$, $z = 2 + t$

18. $x = t$, $y = 1$, $z = 1 + t$

20. $x = 2\pi u$, $y = 2\pi + u$, $z = 4\pi + 2u$

22. $3x + 6y + z + 2 = 0$ **24.** $x + y + z = 4$

26. $x + y = 1$ **30.** $6\sqrt{2}$

32. 0

34. $x_0 x/a^2 + y_0 y/b^2 + z_0 z/c^2 = 1$

36. No points

38. $(0, 0, -1)$ and $x^2 + y^2 = 1/2$, $z = -1/2$

Exercises 13.10

2. $(0, 0)$ saddle point; $(1, 1)$ relative maximum

4. None

6. $(0, n\pi)$ saddle points

8. (x, x) relative minima

10. $(x, 0)$ saddle points; $(0, y)$ $y > 0$ relative minima; $(0, y)$ $y < 0$ relative maxima

12. $(0, 0)$ relative minimum; $(0, y)$ $y \neq 0$ none of these

14. $(0, 0)$ relative maximum; $(0, \pm 1/\sqrt{2})$, $(\pm 1/\sqrt{2}, 0)$ saddle points; $(1/\sqrt{2}, \pm 1/\sqrt{2})$, $(-1/\sqrt{2}, \pm 1/\sqrt{2})$ relative minima

16. None

18. All points on the coordinate axes

22. $(0, 0)$ relative minimum; $(-1/3, \pm 2/9)$ saddle points

Exercises 13.11

2. $4, -1/3$ **4.** $3, -3$ **6.** $3, -2$

8. $(2\sqrt{2} + 5)/\sqrt{2}$, $(2\sqrt{2} - 5)/\sqrt{2}$

10. $(1, 1, -2)$ **12.** $(1/2, 1/2, 1/2)$ **14.** $2, -32$

16. $100(2/3)^{1/3} \times 100(2/3)^{1/3} \times 100(3/2)^{2/3}$ cm

20. $2a/\sqrt{3} \times 2b/\sqrt{3} \times 2c/\sqrt{3}$

22. $1, -1$ **24.** $1, -1$

28. $12/\sqrt{5}$ m, $(50\sqrt{5} - 27\pi)/(3\sqrt{5}\pi)$ m

30. $\|AB\| = 1/3$ m, $\theta = \pi/3$

Exercises 13.12

2. $(8 \pm 9\sqrt{13})/2$ **4.** ± 27 **6.** $\pm\sqrt{32/27}$

8. $(\pm\sqrt{2} - 1)/2$ **22.** $33.12, -0.12$ **24.** $\pm 1/\sqrt{2}$

26. $(\pm 2\sqrt{14}, \mp\sqrt{14})$, $(\pm 2, \pm 4)$

28. $(3a/2, 3a/2)$

Exercises 13.13

2. $P = 2.8187A + 72.967$

4. (a) $y = 5.921x^2 - 5.553x - 5.368$ (b) 1.394

6. $N = 6.308 \times 10^{-21} e^{0.0268}$ (millions)

8. (a) $\left(\sum_{i=1}^{n} \bar{y}_i/x_i^2\right) / \left(\sum_{i=1}^{n} 1/x_i^4\right)$ (b) 0.5590

10. $y = 1/(0.04274x + 0.4968)$

Exercises 13.14

2. $y(1 + x^2 y^2)^{-1} dx + x(1 + x^2 y^2)^{-1} dy$

4. $[yz \cos(xyz) - 2xy^2 z^2] dx + [xz \cos(xyz) - 2x^2 yz^2] dy + [xy \cos(xyz) - 2x^2 y^2 z] dz$

6. $(1 - x^2 y^2)^{-1/2}(y\, dx + x\, dy)$

8. $(y + t)\, dx + (x + z)\, dy + (y + t)\, dz + (z + x)\, dt$

10. $2e^{x^2 + y^2 + z^2 - t^2}(x\, dx + y\, dy + z\, dz - t\, dt)$

12. 3%

Exercises 13.15

2. $\frac{1}{3!}[f_{xxx}(c, d)(x - c)^3 + 3f_{xxy}(c, d)(x - c)^2(y - d) + 3f_{xyy}(c, d)(x - c)(y - d)^2 + f_{yyy}(c, d)(y - d)^3]$

4. $\displaystyle\sum_{n=0}^{\infty}\sum_{r=0}^{n} \frac{e^5(-1)^{n-r} 2^r 3^{n-r}}{(n - r)!\, r!}(x - 1)^r (y + 1)^{n-r}$

6. $\displaystyle\sum_{n=1}^{\infty}\sum_{r=0}^{n} \frac{(-1)^{n+1}(n - 1)!}{(n - r)!\, r!} x^{2r} y^{2n-2r}$

8. $[(x + 1) - 1] \displaystyle\sum_{n=0}^{\infty} (-1)^n y^{2n+2}$

10. $[72 + 12(x - 2) + 24(y - 1) - (x - 2)^2 + 8(x - 2)(y - 1) - 4(y - 1)^2]/(24\sqrt{3})$

12. $[2x + 2(y - 1) + 3x^2 + 6x(y - 1) + 3(y - 1)^2]/2$

14. 0

16. $\displaystyle\sum_{n=0}^{\infty} \frac{1}{n!}\left[(x - c)\frac{\partial}{\partial x} + (y - d)\frac{\partial}{\partial y}\right]^n f(c, d)$

Review Exercises

2. $2(x^2 - y^2 + z^2)/(x^2 + y^2 + z^2)^2$

4. $(3 - z^2)(1 + z^2)/(1 + 2xz + 2xz^3)$

6. $3(2x - ye^{xy})(t^2 + 1) + (2y - xe^{xy})(\ln t + 1)$

8. $(y + 3yv^2 - 2xuv)/(2u + 6uv^2 + 3v + vx^2)$

10. $e^t(t + 1)(y - 2x) + e^{-t}(1 - t)(x - 2y)$

12. $\cos\theta$

14. $(2x - 3x^2y^2)\cos\theta - 2yx^3\sin\theta$

16. $6(6x^4t^6z + 9x^5 + x^4t^9z^2 + 2xt^9z^5 - 9t^{12}z^5 - 6xt^6z^4 - 4x^5t^3z - 2x^2t^3z^4)/(x^4t^8z^4)$

18. $xz[(1 - 3xz^2)(1 - 6xz^2 + 6x^2z^4) + (1 - 2xz^2)(1 - 3xz^2 + 6x^2z^4)]/(x - 3x^2z^2)^3$

20. $2y(v^2t - 1)(t^2 - 3/\sqrt{1 - x^2y^2}) + xv\sec^2 t\, (t^2 - 3/\sqrt{1 - x^2y^2}) + 2xyt$

26. $\sqrt{6}$ 28. $9/\sqrt{14}$ 30. 2

32. $x + 3y - 6z + 4 = 0$

34. $x = 2 + u, \ y = u, \ z = 6 + 4u$

36. $x = 1 + t, \ y = 1 - t, \ z = 1$

38. None

40. $(0, 0)$ relative maximum; all points on $x^2 + y^2 = 1$ relative minima

42. $1/2, -1/2$

44. $\left(\pm\sqrt{(\sqrt{5} - 1)/2}, 0\right), \ (0, \pm1)$

46. $x = 100bcqr/(acpr + abpq + bcqr)$, $y = 100acpr/(acpr + abpq + bcqr)$, $z = 100abpq/(acpr + abpq + bcqr)$

48. $1/\sqrt{} + \sqrt{2}(\pi + 6)(x - 1)/4 + (y - \pi/4)/\sqrt{2} + \sqrt{2}(24 + 14\pi - \pi^2)(x - 1)^2/16 + \sqrt{2}(10 - \pi)(x - 1)(y - \pi/4)/4 - \sqrt{2}(y - \pi/4)^2/4 + \cdots$

Chapter 14

Exercises 14.1

2. 0 4. $3/4$ 6. $e^2(1 - e)^2/2$

8. 0 10. $5/144$ 12. 0

14. -0.54 16. $(1 - \ln 2)/2$ 18. $128\sqrt{2}/5$

20. $11\,664/35$ 22. $(1 - 2\sqrt{2})/12$ 24. $1/3$

26. $8/189$ 28. $2(1 - \sqrt{2})$

Exercises 14.2

2. $128/15$ 4. $-621/140$ 6. 0

8. 0 10. $304/15$ 12. $\sqrt{13} - 7/2$

14. $(1 - \cos 1)/2$ 16. $2(1 - \sqrt{2})$ 18. $(5/4)\ln 5 - 1$

20. 4 22. Double   24. 0

26. Double   28. 0 30. $101/70$

32. $(e^2 - 1)/(2e)$ 34. 24π 36. $e^2 - 2e - 1$

38. $(1 - 2\sqrt{2})/12$ 40. $1/3$

Exercises 14.3

2. $343/6$ 4. 8 6. $1 + 3e^{-2}$

8. $343/6$ 10. $20/3$ 12. 16π

14. $1024\pi/15$ 16. $26\pi/15$ 18. 45π

20. $34\pi/3$ 22. $235/3$ 24. $7/6$

26. $8\pi/3 - 2\sqrt{3}$ 28. $23\,328/35$ 30. $2 - \ln 3$

32. $16\pi(1 + \sqrt{2})/15$

34. $68\pi/9$ 36. 5.38 38. $512\pi/(15\sqrt{5})$

40. $7\sqrt{2}\pi/6$

Exercises 14.4

2. $256\rho g/5$ 4. $48\rho g$ 6. $370\rho g/3$

8. $250\rho g/3$ 10. 7.63×10^6 N   12. $5\sqrt{29}\rho g$

14. $10\rho g$ 16. $\pi\rho g r^3$ 18. 9.00×10^4 N

Exercises 14.5

2. $(0, 24/5)$ 4. $\bar{x} = \bar{y} = 9/(15 - 16\ln 2)$

6. $\left(8(2 - \sqrt{2})/(3\pi), 8\sqrt{2}/(3\pi)\right)$

8. $(177/85, 9/17)$ 10. $(4/3, 5/3)$ 12. $32/3$

14. $603/10$ 16. $48\rho/5$

18. $(0, 192/205)$

20. $(-2/(8\sqrt{3} - 1), 0)$

22. $(-61/28, 807/700)$

24. $\rho ab(a^2 + b^2)/12$

26. $4761/140$ 28. $81\sqrt{2}/40$

Exercises 14.6

2. $4\sqrt{14}/3$ 4. $4(5\sqrt{3} - 2\sqrt{6} - 3 + \sqrt{2})/3$

6. $(247\sqrt{13} + 64)/1215$

8. $2\int_0^4 \int_0^{\sqrt{16 - y^2}} \sqrt{4 + y^2 + z^2}\, dz\, dy$

10. $4\int_0^{\sqrt{2}} \int_0^{\sqrt{2 - x^2}} \sqrt{1 + 16(x^2 + y^2)^3}\, dy\, dx$

12. $\int_0^1 \int_0^{1 - x^2} \sqrt{1 + 2/(1 + x + y)^2}\, dy\, dx$

16. $\sqrt{2}\int_{-1/\sqrt{2}}^{1/\sqrt{2}}\int_{x}^{\sqrt{1-x^2}}\sqrt{2+1/(y-x)}\,dy\,dx$

18. $\int_{0}^{1}\int_{1-x}^{2-x}\sqrt{1+9x^4+9y^4}\,dy\,dx+$
$\int_{1}^{2}\int_{0}^{2-x}\sqrt{1+9x^4+9y^4}\,dy\,dx$

20. $(37\sqrt{74}-5\sqrt{10})/24$

Exercises 14.7

2. $2/3$ 4. $4(3\sqrt{3}-\pi)/3$

6. $\pi/6$ 8. $(3\sqrt{3}-\pi)/2$

10. $\pi/4$

12. $15\pi/4-19\sqrt{2}/3-(9/2)\mathrm{Sin}^{-1}(1/3)$

14. $\pi R^4/4$ 16. $\pi(17\sqrt{17}-1)/6$

18. $2\pi(10\sqrt{10}-1)/3$ 20. $(17\sqrt{17}-5\sqrt{5})\pi/6$

22. $8\pi/3$ 24. $2\pi^2 R^3$

26. $[q\rho/(2\epsilon_0)](1-d/\sqrt{R^2+d^2})$, $q\rho/(2\epsilon_0)$

28. $9\pi-12\sqrt{3}$ 30. $(5/6,0)$

32. $\pi(\pi-2)/2$

Exercises 14.8

2. $32/3$ 4. 0 6. $1024/21$

8. $1/96$ 10. $11/6$ 12. $48/35$

14. $4\int_{0}^{\sqrt{(\sqrt{5}-1)/2}}\int_{0}^{\sqrt{1-x^2-x^4}}\int_{x^2}^{\sqrt{1-x^2-y^2}}(x^2+y^2+z^2)\,dz\,dy\,dx$

16. $\int_{-5/2}^{1/2}\int_{-1-\sqrt{9-4(y+1)^2}}^{-1+\sqrt{9-4(y+1)^2}}\int_{x^2+4y^2}^{4-2x-8y}xyz\,dz\,dx\,dy$

18. $729/70$ 20. 2π 22. $\pi/3$

24. $4\int_{0}^{1}\int_{0}^{\sqrt{1-x^2}}\int_{0}^{\sqrt{x^2+y^2}/2}(x^2+y^2+z^2)\,dz\,dy\,dx+$
$4\int_{0}^{1}\int_{\sqrt{1-x^2}}^{\sqrt{4/3-x^2}}\int_{\sqrt{x^2+y^2-1}}^{\sqrt{x^2+y^2}/2}(x^2+y^2+z^2)\,dz\,dy\,dx+$
$4\int_{1}^{2/\sqrt{3}}\int_{0}^{\sqrt{4/3-x^2}}\int_{\sqrt{x^2+y^2-1}}^{\sqrt{x^2+y^2}/2}(x^2+y^2+z^2)\,dz\,dy\,dx$

Exercises 14.9

2. $2/3$ 4. $704/15$ 6. 8

8. $8/3$ 10. $19/96$ 12. $7/3$

14. $5/18$ 16. $64/15$ 18. $\pi/4-1/3$

20. $1/20$ 22. $13/3$ 26. $\pi/2$

Exercises 14.10

2. $(1/4,1/4,1/4)$ 4. $(0,16/7,8/7)$ 6. $2\rho/3$

8. $773\rho/2520$ 10. $64\rho/15$

12. $(6772/11\,847, 7300/14\,001, 1/2)$

14. $(0,0,3/2)$ 16. $128\rho a^5/45$ 18. $51\sqrt{3}\rho$

20. $(3a/8,3b/8,3c/8)$

Exercises 14.11

2.

4.

6.

8.

10.

12. $(8\sqrt{2}-7)\pi/6$ 14. 4π 16. $4\sqrt{3}\pi$

18. $\int_{0}^{2\pi}\int_{0}^{3}\int_{0}^{1+r^2}f(r\cos\theta,r\sin\theta,z)\,r\,dz\,dr\,d\theta$,
$\int_{0}^{3}\int_{0}^{2\pi}\int_{0}^{1+r^2}f(r\cos\theta,r\sin\theta,z)\,r\,dz\,d\theta\,dr$,
$\int_{0}^{3}\int_{0}^{1+r^2}\int_{0}^{2\pi}f(r\cos\theta,r\sin\theta,z)\,r\,d\theta\,dz\,dr$,
$\int_{0}^{1}\int_{0}^{3}\int_{0}^{2\pi}f(r\cos\theta,r\sin\theta,z)\,r\,d\theta\,dr\,dz+$
$\int_{1}^{10}\int_{\sqrt{z-1}}^{3}\int_{0}^{2\pi}f(r\cos\theta,r\sin\theta,z)\,r\,d\theta\,dr\,dz$,
$\int_{0}^{2\pi}\int_{0}^{1}\int_{0}^{3}f(r\cos\theta,r\sin\theta,z)\,r\,dr\,dz\,d\theta+$
$\int_{0}^{2\pi}\int_{1}^{10}\int_{\sqrt{z-1}}^{3}f(r\cos\theta,r\sin\theta,z)\,r\,dr\,dz\,d\theta$,
$\int_{0}^{1}\int_{0}^{2\pi}\int_{0}^{3}f(r\cos\theta,r\sin\theta,z)\,r\,dr\,d\theta\,dz+$
$\int_{1}^{10}\int_{0}^{2\pi}\int_{\sqrt{z-1}}^{3}f(r\cos\theta,r\sin\theta,z)\,r\,dr\,d\theta\,dz$

20. $8\pi\rho R^5/15$ 22. $81\pi^2/8$ 24. 0.084

26. $\pi h\rho R^4/10$ 28. $4\sqrt{3}\pi/9$ 30. $56\pi(2-\sqrt{2})/3$

32. $4\pi(8\sqrt{2}-3\sqrt{6})/3$ 34. $2a^3(3\pi+4)/9$ 36. $64\pi/3$

38. $15\pi/2$ 40. $3\sqrt{6}/5$

Exercises 14.12

2. $\Re\sin\phi=1$ (see figure for exercise 14.11–2)

4. $\Re=4\cot\phi\csc\phi$ (see figure for exercise 14.11–8)

6. $\Re^2=-\sec 2\phi$ (see figure for exercise 14.11–10)

8. $(2 - \sqrt{2})\pi/3$

10. $[4\text{Tan}^{-1} 2 - \pi]/12$

12. $32\pi/3$

14. $8\pi\rho R^5/15$

16. $k\pi R^4$ coulombs

18. $9\pi/2$

20. $(2\pi R^3/3)(1 - k/\sqrt{1 + k^2})$

22. (a) $\rho_b = \rho_w/2$ (b) $11\pi\rho_w g R^3/24$

Exercises 14.13

2. $2x - 1 + e^x$

4. $4x^3 - 3x^2 - 1 + 3x^2(x^3 - 1)\ln(x^3 - 1) - 2x^3\ln(x^2)$

8. $e^x\sqrt{1 + e^{3x}} - \cos x\sqrt{1 + \sin^3 x}$

12. $(1/a^2)\ln(1 + ab) - b/(a + a^2 b)$

14. $[3/(8a^5)]\text{Tan}^{-1}(x/a) + x(3x^2 + 5a^2)/[8a^4(a^2 + x^2)^2] + C$

16. $\pi\text{Sin}^{-1} a$

Review Exercises

2. $81/16$

4. $1/40$

6. $36/5$

8. 0

10. $-4544/945$

12. $\pi/6$

14. $32\pi/3$

16. $2\pi/5$

18. $3\pi/2$

20. $(\pi \ln 2)/4$

22. (a) $\iint_R dA$ (b) $\iint_R 2\pi(2 - x)\, dA$, $\iint_R 2\pi(y + 4)\, dA$
(c) $\iint_R (x - 1)\, dA$, $\iint_R (y + 1)\, dA$ (d) $\iint_R (x + 1)^2\, dA$,
$\iint_R (y - 4)^2\, dA$ (e) $\iint_R \sigma(x, y)\, dA$ (f) $\iint_R \rho(x, y)\, dA$
(g) $\iint_R P(x, y)\, dA$

24. $32/3$

26. $16/15$

28. $\pi(17\sqrt{17} - 1)/96$

30. $128\pi/15, 4\pi^2$

32. $1/4$

34. 2

36. $14\rho\pi/15$

38. $(0, 1/8)$

40. $(9\pi/64, 9\pi/64, 3/8)$

42. $16\pi/15$

Chapter 15

Exercises 15.1

2. Closed, simply-connected

4. Connected

6. Closed, simply-connected

8. Open

10. For interior, exterior, and boundary points replace circle with sphere in planar definitions. Open, closed, connected, and domain definitions are identical. A domain is simply-connected if every closed curve in the domain is the boundary of a surface that contains only points of the domain.

12. Closed, simply-connected

14. Connected

16. Open

18. Open, connected, simply-connected domain

22. $-(x^2 + y^2 + z^2)^{-3/2}(x\hat{\mathbf{i}} + y\hat{\mathbf{j}} + z\hat{\mathbf{k}})$

24. $(6 - 2\cos 2)\hat{\mathbf{i}} + (1 + 2\sin 2)\hat{\mathbf{j}}$ **26.** $2(e^y - x^2 y)$

28. $2x\cos(x^2 + y^2 + z^2) - \sin(y + z)$

30. 1

32. $2/\sqrt{x^2 + y^2 + z^2}$

34. $4y(16z^4 - 3x)\hat{\mathbf{i}} + x^2\hat{\mathbf{j}} + 12yz\hat{\mathbf{k}}$

36. 0

38. $-2\hat{\mathbf{k}}$

40. $-2\hat{\mathbf{k}}/[(x + y)\sqrt{(x + y)^2 - 1}]$ **46.** $x^3 y^2 + 3x + 2y + C$

48. $\ln|x + y| + C$

50. $(x^2 + y^2 + z^2)/2 + C$

52. $\ln|1 + x + y + z| + C$

54. $\text{Tan}^{-1}(xy) + z^2/2 + C$

56. (a) $4, 2, -1$ (b) $x^3/3 + 2xy + 4xz - 3y^2/2 - yz + z^2 + C$

58. (a) $q/(4\pi\epsilon_0|\mathbf{r}|) + C$ (b) $-\sigma z/(2\epsilon_0) + C$

62. $yz\hat{\mathbf{i}} - xz\hat{\mathbf{j}}$

64. $(1/4)[(4xyz - 3y^2 z)\hat{\mathbf{i}} + (2xyz - 6x^2 z)\hat{\mathbf{j}} + (2x^2 y + xy^2)\hat{\mathbf{k}}]$

Exercises 15.2

2. $32/3$

4. $50\sqrt{10}/3$

6. $37\sqrt{2}/3$

8. $9\sqrt{1 + 16\pi^2}$ cm **10.** $37/80$

12. 0.78

14. 0

16. $(1 - 161\sqrt{161})/6$

18. $\pi(145\sqrt{145} - 10\sqrt{10})/27$

20. $\pi/2$

22. 0.007

24. 2

26. 0.242

28. π

30. $(5\sqrt{5} - 1)/6$

32. $\sqrt{2}(1 - e^{6\pi})/13$

34. 17.08

Exercises 15.3

2. $51/4$

4. -15

6. $-99/140$

8. 10

10. 0

12. 9

14. 0

16. -8π

18. $(768 + 5\pi)/20$

20. $67/32 + \sqrt{5} - \sqrt{2} - 5/e$

22. $(4\pi\epsilon_0)^{-1}[q_1 q_3(1/\sqrt{41} - 1/\sqrt{61}) + q_2 q_3(1/\sqrt{18} - 1/\sqrt{10})]$

24. 3.719

26. -4.26×10^{-4}

28. 0

32. (a) $2a$ (b) 0

Exercises 15.4

2. -43

4. -2

6. 0

8. 10

10. e^3

12. No

14. 0

16. 0

18. $8/105 + \ln(3/2)$

20. $-\pi$

22. (a) Yes (b) Yes (c) Yes (d) Yes (e) No

24. 2π

Exercises 15.5

2. Not conservative

4. mgz

8. $-k(\sqrt{x^2 + y^2 + z^2} - L)(x\hat{\mathbf{i}} + y\hat{\mathbf{j}} + z\hat{\mathbf{k}})/\sqrt{x^2 + y^2 + z^2}$, yes

10. (a) Yes (b) Yes

Exercises 15.6

2. 4π **4.** $2\sqrt{3}/5$ **6.** 0

8. $-3/8$ **10.** $4(13\sqrt{13}-8)/3$ **14.** πab

16. $3\pi/8$ **18.** 2π **22.** 24

24. $77/2$ **26.** $81\pi/2$ **28.** -4π

32. π

Exercises 15.7

2. $2\sqrt{3}/15$ **4.** $\sqrt{2}/8$

6. $(-61+44\sqrt{2})/5$ **8.** -3

10. $\pi R\sqrt{R^2+h^2}$ **12.** $3(145^{5/2}-361)/5120$

14. $\pi(3\sqrt{3}-1)/3$ **16.** $(50\sqrt{5}+2)\pi/15$

18. $4\pi R^2$ **20.** 0

22. $\pi\ln 3$

Exercises 15.8

2. $2e^2-10e^{-2}$ **4.** 0

6. $2\pi/3$ **8.** -2π

10. $10\pi/3$ **12.** $\sqrt{2}\pi/4$

14. $2\sqrt{5}+8\ln[(\sqrt{5}+1)/2]$ **16.** 0

18. (a) 30π (b) -20π **20.** $2\pi(1-e^{-2})$

Exercises 15.9

2. 0 **4.** 16 **6.** 27π

8. 180 **10.** $-1328\pi/5$ **12.** $57/2$

14. $52\pi/5$ **20.** 0

Exercises 15.10

2. 0 **4.** $\pm 16\pi$ **6.** 0

8. $-2\sqrt{2}\pi b^2$ **10.** $\pm 2\pi$ **12.** -24π

14. 3π

Review Exercises

2. $3x^2y+x^2/y^2$ **4.** 0

6. $ye^x+ze^y+xe^z$ **8.** $(\hat{\mathbf{i}}+\hat{\mathbf{j}})/\sqrt{1-(x+y)^2}$

10. $x(-y\hat{\mathbf{j}}+z\hat{\mathbf{k}})/(1+x^2y^2z^2)$ **12.** $2\sqrt{3}$

14. π **16.** $\sqrt{2}/3$

18. 2π **20.** $8\pi(18-7\sqrt{6})/3$

22. 0 **24.** 0

26. $-\pi/120$ **28.** $(25\sqrt{5}-11)/120$

30. $8\pi/3$

Chapter 16

Exercises 16.1

12. $-3\sin 3x+2\cos 3x$

14. $[(2\cos 3-\cos 6)\sin 3x+(\sin 6-2\sin 3)\cos 3x]/\sin 3$

16. $(1/3)\mathrm{Tan}^{-1}(x/3)+C$

18. $(x^3/6)\ln x-5x^3/36+C_1x+C_2$

20. (b) $L\ln x-\sqrt{L^2-x^2}-L\ln(L-\sqrt{L^2-x^2})$

22. $y=0$

24. (b)

26. (a)

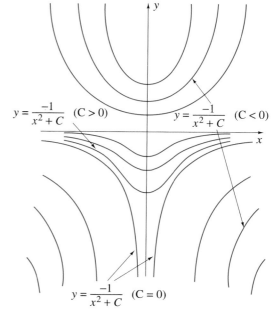

Exercises 16.2

2. $y(x)=2+Ce^{-x^2}$ **4.** $y(x)=Ce^{3x}-2/3$

6. $y^2+x^2+2(x-y)+2\ln|y+1|+2\ln|x-1|=C$

8. $y(x) = [C + 3 \ln |x| + 3e^x (1 - x)]^{1/3}$

10. $y(x) = (x + C)/(1 - Cx)$ **12.** $xy = 2e^{y-x-1}$

14. $y(x) = [\tan (x^2 - 4) + 4]/[1 - 4 \tan (x^2 - 4)]$

16. (a) $6(1 - 6kt)^{-1}$ km (b) never

18. $20 + 60e^{-0.203t}$ **20.** 13.51 h

22. $(R/k)(1 - e^{-kt}) + A_0 e^{-kt}$ **24.** $60(1 - e^{kt})/(2 - 3e^{kt})$ gm

28. $x^2(x^2 - 2y^2) = C$ **30.** $x^2 + 2xy - y^2 = C$

32. $(y - x)e^{y/x} = x \ln |x| + Cx$ **34.** $x^2 + y^2 = 5x$

36. $y_0(A + y_0)/[y_0 + Ae^{-k(A+y_0)t}]$

38. $y = Ce^x - x - 1$

40. $6y - 3 \ln |4x + 6y + 3| = C$

42. (a) (b)

(c)

44. 459.5 milligrams **46.** $y^3(x^3 + 6x - 5) = 1 + y^6$

Exercises 16.3

2. $y(x) = x^4 + C/x^2$ **4.** $y \sin x = C - 5e^{\cos x}$

6. $y(x) = C(x + 1)^2 - 2x - 2$ **8.** $y(x) = Ce^x + e^{2x}$

10. $y(x) = x^3/2 + Cx^3 e^{1/x^2}$

12. $y(x) = x + \cos 2x + C \sin 2x$

14. $y(x) = e^x(2 \sin x - \cos x)/5 - 4e^{-x}/5$

16. $xy = C + y^4/4$ **18.** $y(x) = e^{-x}/(C - x)$

20. $x^2 y + Cye^{-x} = 1$

22. $y(x) = [(3/4) \cos x + C \sec^3 x]^{-1/3}$

24. $1000 + t + 3 \times 10^6/(1000 + t)$ gm

26.

$$S = \frac{10^6 + 5t}{50}$$

$$S = \frac{10^6 + 5t}{50} - \frac{15 \times 10^9}{10^6 + 5t}$$

28.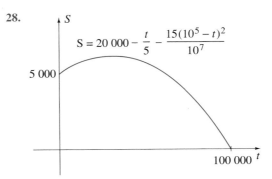

$$S = 20\,000 - \frac{t}{5} - \frac{15(10^5 - t)^2}{10^7}$$

30. $32\,570/1257 - (20\,000/1257)e^{-3t/10\,000}$

32. $(1/1257)(20t - 7430 + 20\,000e^{-3t/10\,000})$

34. (b) $I_0 + \omega L E_0/(R^2 + \omega^2 L^2)$

Exercises 16.4

2. $2\sqrt{Cy - 1} = \pm Cx + D$

4. $y(x) = Cx - C^2 \ln |x + C| + D$

6. $(x + C)^2 + (y + D)^2 = 1$

8. $D + x/2 = \{-1/y; \text{ or } C\text{Tan}^{-1}(Cy);$
$\text{or } [1/(2C)] \ln |(y - C)/(y + C)|\}$

10. $y(x) = (1/2)(Ce^x + C^{-1}e^{-x}) + D$

12. $kr + \ln r[(T_b - T_a) + k(a-b)]/\ln(b/a) + [(T_a - T_b)\ln b + k(b\ln a - a\ln b)]/\ln(b/a)$

Exercises 16.5

2. Less than 47.8 km/h; defence

6. (c) $x^* = (\mu Mg \pm \sqrt{\mu^2 M^2 g^2 + Mkv_0^2 + k^2 x_0^2 - 2k\mu M x_0})/k$

8. 18.0 m

10. (a) $70.0(1 - 0.556e^{-0.280t})/(1 + 0.556e^{-0.280t})$ m/s
    (b)

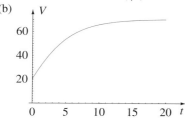

12. 1.99 s

14. (a) $v_T \tan[\text{Tan}^{-1}(v_0/v_T) - kv_T t/m]$ where $v_T = \sqrt{9.81m/k}$ (b) $(m/k)\ln|\cos[\text{Tan}^{-1}(v_0/v_T) - kv_T t/m]| + (m/k)\ln(\sqrt{v_0^2 + v_T^2}/v_T)$, $(m/k)\ln(\sqrt{v_0^2 + v_T^2}/v_T)$

16. $(9.81m\sin\alpha/k)(1 - e^{-kt/m})$, $(9.81m^2\sin\alpha/k^2)(kt/m + e^{-kt/m} - 1)$

Exercises 16.6

2. Linear   4. Linear   6. Linear

8. Linear   10. Not linear   12. Linear

14. Linear   16. Not linear   18. Not linear

20. Linear

Exercises 16.7

2. $C\cos x$ 4. $C_1 e^{4x} + C_2 x e^{4x}$

6. $e^{2x}[C_1\cos(x/\sqrt{2}) + C_2\sin(x/\sqrt{2})]$

8. $C_1\cos(2\ln x) + C_2\sin(2\ln x)$

12. Dependent   14. Independent

Exercises 16.8

2. $(C_1 + C_2 x)e^{4x}$ 4. $C_1 e^{(-1+\sqrt{3})x} + C_2 e^{(-1-\sqrt{3})x}$

6. $C_1 e^{3x} + C_2\cos x + C_3\sin x$ 8. $(C_1 + C_2 x + C_3 x^2)e^{2x}$

10. $C_1\cos x + C_2\sin x + C_3\cos 2x + C_4\sin 2x$

12. $e^{\sqrt{2}x}[C_1\cos(\sqrt{2}x) + C_2\sin(\sqrt{2}x)] + e^{-\sqrt{2}x}[C_1\cos(\sqrt{2}x) + C_4\sin(\sqrt{2}x)]$

14. $y'' + 4y' + 20y = 0$

16. $y'''' - 4y''' + 10y'' - 12y' + 9y = 0$

18. 5, 24, 20

Exercises 16.9

2. $-(x^2/3 + 2/9)e^{-x}$,
   $C_1 e^{(-1+\sqrt{3})x} + C_2 e^{-(1+\sqrt{3})x} - (x^2/3 + 2/9)e^{-x}$

4. $(1/9)\cos 2x$, $(C_1 + C_2 x)\cos x + (C_3 + C_4 x)\sin x + (1/9)\cos 2x$

6. $e^{-2x}/40$, $C_1\cos x + C_2\sin x + C_3\cos 2x + C_4\sin 2x + e^{-2x}/40$

8. $e^{2x}(\cos x - 5\sin x)/312$, $e^{-4x}(C_1\cos 5x + C_2\sin 5x) + e^{2x}(\cos x - 5\sin x)/312$

10. $[(2x+1)\cos x - 2(x+1)\sin x]/16$, $e^{2x}(C_1\cos x + C_2\sin x) + [(2x+1)\cos x - 2(x+1)\sin x]/16$

12. $Axe^{3x} + Be^{3x} + Cx\cos x + Dx\sin x + E\cos x + F\sin x$

14. $Ax^2 e^x + Bxe^x + Cx^3 + Dx^2 + Ex + F + G\cos x + H\sin x$

18. $e^x[C_1\cos(x/\sqrt{2}) + C_2\sin(x/\sqrt{2})] + (4\cos 3x - 5\sin 3x)/246 + (\sin x + 4\cos x)/34$

20. $-(1/108)\sin 3x + (x^2/12)(\sin 3x - \cos 3x) + (x/36)(\sin 3x + \cos 3x)$

24. $C_1\cos(2\ln x) + C_2\sin(2\ln x) + 1/4$

Exercises 16.10

2. $e^{-15t/4}[(12/65)\cos(5\sqrt{23}t/4) + (20/(13\sqrt{23}))\sin(5\sqrt{23}t/4)] - (4/65)[3\cos(10t) + 2\sin(10t)]$ m

4. $-0.0253e^{-99.50t} + 5.03e^{-0.50t}$ C

6. $(M|g|/k)\cos(\sqrt{k/M}t)$ from equilibrium

8. (b) $x = -7/36$ m, yes

12. $(1/64)\sin 8t - (t/8)\cos 8t$ m

14. $(9/10)[3t\cos 3t - \sin 3t]A$, yes

16. (b) $0.705/\sqrt{L}$

Review Exercises

2. $y = \pm\sqrt{2(x + \ln|x|) + C}$

4. $y = Ce^{-4x} + (8x^2 - 4x + 1)/32$

6. $y = e^{-3x/2}[C_1\cos(\sqrt{7}x/2) + C_2\sin(\sqrt{7}x/2)] + 1/2$

8. $y = x^3/9 + C_1\ln|x| + C_2$

10. $y = Ce^{-x^2} + x^2 - 1$

12. $y = (C_1 + C_2 x)e^{2x} + (3\sin x + 4\cos x)/25$

14. $y = C_1\cos 2x + C_2\sin 2x - (x/4)\cos 2x$

16. $xy^2 = Ce^{-2x - x^2/2}$

18. $e^{-x}[C_1\cos(\sqrt{3}x) + C_2\sin(\sqrt{3}x)] + (\sqrt{3}x/6)e^{-x}\sin(\sqrt{3}x)$

20. $x = Ce^{-3y} + (2/27)(-9y^2 + 6y - 2)$

22. $y = (424/405)e^{9x} + (11/5)e^{-x} - 2x/9 - 20/81$

24. $y = -\cos x + (2\sin x)/x + (2\cos x + 1 - \cos 1 - 2\sin 1)/x^2$

26. (a) 3.924×10^{-3} N (b) $1.962 \times 10^{-3}t +$
$4.905 \times 10^{-7}(e^{-4000t} - 1)$ m

28. (a) In air, $v = 9.81t$; in water $v = 0.34335 + 1.502 \times 10^{29}e^{-20t}$
(b) In air, $y = 4.905t^2$; in water,
$y = 50.30 + 0.34435t - 7.51 \times 10^{27}e^{-20t}$ (c) 28.25 s

Appendix B

Exercises

2. 1

4. 0

6. −640

8. 68

10. $x = -14, y = -10$

12. $r = -4, s = -11$

14. $x = -5/18, y = 11/18, z = 5/2$

18. $x = y = z = w = 0$

20. (b) $2x + 3y = 5$

22.
$$\begin{vmatrix} y & x^2 & x & 1 \\ 0 & 1 & 1 & 1 \\ 11 & 4 & 2 & 1 \\ 1 & 4 & -2 & 1 \end{vmatrix} = 0$$

24. $b^{n-1}(an + b)$

Comprehensive Review Problems and Applications

When you solve the exercises in a particular section of this text, you already know the nature of the problem and the techniques that should be used. For example, a problem in Exercises 3.8 will use implicit differentiation, a problem in Exercises 9.5 will use a trigonometric substitution, etc. To ensure that you understand all parts of a chapter, we recommended the review exercises at the end of each chapter. They require you to first identify the particular section that is useful in solving a problem, and then to follow through with its application. When you have finished your calculus studies, you should test your understanding of the entire subject. The following exercises, which review all aspects of the course, will do this. They test your ability to recognize the essential nature of a problem—be it a maximum-minimum problem, a related rate problem, an application of definite integrals, a problem in infinite series, etc.—and they tax your organizational, analytical, and interpretive skills.

The problems are divided into two groups, problems on single-variable calculus (Chapters 1–11) and problems on multivariable calculus (Chapters 11–16), and each group is graded from straightforward to thought-provoking. We have included Problems for Chapter 11 in both groups to accommodate schools that treat infinite series early in calculus and those that treat the topic later. Set your sights high; attempt as many problems as you can. Answers to selected exercises are found at the end of each group.

1. Find all points (x, y) satisfying $y + |y| = x + |x|$.

2. (a) Draw the graph of an even, 2π-periodic function $f(x)$ for which

$$f(x) = x^3 - \pi x, \quad 0 < x < \pi.$$

Can $f(x)$ be made continuous?

(b) Draw the graph of an odd, 2π-periodic function $f(x)$ for which

$$f(x) = x^3 - \pi x, \quad 0 < x < \pi.$$

Can it be made continuous?

In Exercises 3–18 evaluate the limit if it exists. If a limit does not exist, explain why not.

3. $\lim\limits_{x \to 0} \dfrac{x^2 - x}{2x + x^2}$

4. $\lim\limits_{x \to -4^-} \dfrac{x^2 - 16}{x + 4}$

5. $\lim\limits_{x \to 0} \dfrac{\sqrt{2 + x} - \sqrt{2 - x}}{x}$

6. $\lim\limits_{x \to 5} \dfrac{|x^3 - 125|}{x - 5}$

7. $\lim\limits_{x \to 3} \dfrac{x - 3}{x^2 - 6x + 9}$

8. $\lim\limits_{x \to 0^-} \dfrac{3x}{2 - \sqrt{x^2 + 4}}$

9. $\lim\limits_{x \to \infty} \dfrac{2 - x^5}{3 + 4x^5}$

10. $\lim\limits_{x \to -\infty} \tan x$

11. $\lim\limits_{x \to 1^+} \left[\left(1 - \dfrac{1}{x}\right)^3 + x - \dfrac{2}{x^2} \right]$

12. $\lim\limits_{x \to 0} \dfrac{\sin 4x^{12}}{3x^{12}}$

13. $\lim\limits_{x \to \infty} \left(\sqrt{x^2 + 3} - \sqrt{x^2 + 1} \right)$

14. $\lim\limits_{x \to -\infty} \left(\sqrt{x^2 + x} - \sqrt{x^2 + 2x} \right)$

15. $\lim\limits_{x \to -\pi/2^+} \tan x$

16. $\lim\limits_{x \to 1} e^{\frac{1}{x-1}}$

17. $\lim\limits_{x \to 2} \dfrac{\sqrt{x^2 - 4x + 4}}{x - 2}$

18. $\lim\limits_{x \to 1} \dfrac{x + 2\sqrt{x} - 3}{x^2 - 1}$

In Exercises 19–69 assume that y is defined as a function of x by the given equation(s), and find dy/dx.

19. $y = x^2(1 - 3x)^4$

20. $y = \dfrac{x^2}{\sqrt{1 - 3x^2}}$

21. $y = \dfrac{\sin 2x}{\cos 3x}$

22. $y = (2x + 1)^3 \ln(x + 1)$

23. $y = (x^2 + 2x - 3)e^{3x}$

24. $y = \sqrt{\ln(e^{2x} + 1)}$

25. $y = \text{Sin}^{-1}\left(\sqrt{1 - 4x}\right)$

26. $y = x^2 \csc^2(x^2 - 1)$

27. $y = \dfrac{x}{1 + \sec(1 - 2x^3)}$

28. $y = \sqrt[3]{1 + \sqrt{2 - x}}$

29. $x^2 y + y^3 x - 2x = 5$

30. $\sin(x + y) = xy$

31. $x^3 e^y + x - 2y = 3x^2$

32. $x\mathrm{Tan}^{-1}(xy) = y + 3$

33. $y = \cos(v^3 + 2), \quad v = \dfrac{x}{x+1}$

34. $y = \mathrm{Sec}^{-1}(1 - 2u), \quad u = \tan x + 1$

35. $y = \sqrt{v - \sqrt{v}}, \quad v = \sqrt{x + \sqrt{x}}$

36. $y = t^2 e^{-3t}, \quad t = x + \ln x$

37. $x = t + \dfrac{1}{t}, \quad y = 2t - \dfrac{3}{t^2}$

38. $x = 4 + 3\cos t, \quad y = 2 - 4\sin t$

39. $x = \dfrac{\theta}{1 + \theta^2}, \quad y = \dfrac{\theta^2}{1 + \theta^2}$

40. $x = e^{3v} - v, \quad y = e^{-3v} - v^2$

41. $y = 3 + 2x - 6\ln(3 + 2x) - \dfrac{9}{3 + 2x}$

42. $y = \ln\left(\dfrac{4 - x}{x}\right)$

43. $y = \ln\left[\dfrac{(2 + x)^3}{8 + x^3}\right] + 2\sqrt{3}\,\mathrm{Tan}^{-1}\left(\dfrac{x - 1}{\sqrt{3}}\right)$

44. $y = \ln\left(\dfrac{2x^2 - 2x + 1}{2x^2 + 2x + 1}\right) + 2\mathrm{Tan}^{-1}\left(\dfrac{2x}{1 - 2x^2}\right)$

45. $y = (8 - 12x + 27x^2)\sqrt{1 + 3x}$

46. $y = \ln\left(x + \sqrt{x^2 + 9}\right)$

47. $y = \sqrt{x^2 - 4} - 2\mathrm{Sec}^{-1}(x/2)$

48. $y = -\dfrac{\sqrt{7x^2 + 49}}{x^2} + \ln\left(\dfrac{\sqrt{7} + \sqrt{x^2 + 7}}{x}\right)$

49. $y = 2x(3 - x^2)^{3/2} + 9x\sqrt{3 - x^2} + 27\mathrm{Sin}^{-1}\left(\dfrac{x}{\sqrt{3}}\right)$

50. $y = (3x^2 + 2)(1 - x^2)^{3/2}$

51. $y = x\sqrt{16 - x^2} + 16\cos(x/4)$

52. $y = \dfrac{5}{\sqrt{25 - x^2}} - \ln\left(\dfrac{5 + \sqrt{25 - x^2}}{x}\right)$

53. $y = \ln\left(\dfrac{3\sqrt{16 - x^2} + \sqrt{7}x}{\sqrt{9 - x^2}}\right)$

54. $y = \dfrac{3x + 12}{\sqrt{6 + 3x + 4x^2}}$

55. $y = \mathrm{Cos}^{-1}\left(1 - \dfrac{x}{4}\right)$

56. $y = \dfrac{x - 7}{\sqrt{14x - x^2}}$

57. $y = \ln\left(x + 3 + \sqrt{6x + x^2}\right)$

58. $y = \mathrm{Csc}^{-1}\sqrt{\dfrac{3x^5}{2}}$

59. $y = \mathrm{Tan}^{-1}\left(\dfrac{4\tan(x/2) + 3}{\sqrt{7}}\right)$

60. $y = \mathrm{Tan}^{-1}\left(\dfrac{3\tan x}{2}\right)$

61. $y = 3\cot\left(\dfrac{3x}{2}\right) - \cot^3\left(\dfrac{3x}{2}\right)$

62. $y = (3x^2 - 6)\cos x + (x^3 - 6x)\sin x$

63. $y = \mathrm{Sin}^{-1}\left(\dfrac{3\cos 2x}{\sqrt{10}}\right)$

64. $y = \ln\left(7\sin 3x + \sqrt{1 + 49\sin^2 3x}\right)$

65. $y = (1 + x^2)\mathrm{Cot}^{-1}x + x$

66. $y = \dfrac{\sqrt{4x^2 - 1}}{x} - \dfrac{\mathrm{Sec}^{-1}(2x)}{x}$

67. $y = e^{3x}(3\sin 2x - 2\cos 2x)$

68. $y = 2x - \ln\left(5 + 3e^{2x}\right)$

69. $y = 13xe^{2x}(2\sin 3x - 3\cos 3x) + e^{2x}(5\sin 3x + 12\cos 3x)$

70. For what value of k are the curves $y = 1 - x^2$ and $y = x^2 - 4x + k + 4$ tangent? What is the point of tangency?

71. Find the equation of the tangent line to the curve $y = (x + 9)/(x + 5)$ that passes through the origin.

72. Find the first and second derivatives of the function $y = f(x)$ defined by the equation $y^3 + (x - 1)y + 16 = 3x$ when $x = 2$.

73. Verify that when n is a positive integer

$$\frac{d}{dx}(\sin^n x \cos nx) = n\sin^{n-1}x \cos(n + 1)x.$$

74. Show that $f(x) = \ln\left(x + \sqrt{x^2 + 1}\right)$ is an odd function.

75. A chord of the parabola $y = x^2 - 2x + 5$ joins the points with abscissas $x = 1$ and $x = 3$. Find the equation of the tangent line to the parabola parallel to the chord.

76. Find the second derivative of the function $y = f(x)$ defined by the equation $xy^2 + x^2y = 16$ at the point where its first derivative is equal to 0.

77. Show that the function defined by $y^3 + x^4y = 3$ is increasing when $x > 0$.

In Exercises 78–103 evaluate the indefinite integral.

78. $\displaystyle\int \dfrac{1}{3 + 4e^{2x}}\,dx$

79. $\displaystyle\int \dfrac{1}{4e^{3x} - 5e^{-3x}}\,dx$

80. $\displaystyle\int (x - 1)^2 \ln x\,dx$

81. $\displaystyle\int \ln(x^2 + 9)\,dx$

82. $\displaystyle\int \sin^2 3x\,dx$

83. $\displaystyle\int \cos^3(5x + 1)\,dx$

84. $\displaystyle\int \sin 3x \cos^2 4x\,dx$

85. $\displaystyle\int \dfrac{1}{\sqrt{x^2 + 5}}\,dx$

86. $\displaystyle\int \dfrac{x + 5}{\sqrt{x^2 - 3}}\,dx$

87. $\displaystyle\int \dfrac{\sqrt{x^2 + 1}}{x^4}\,dx$

88. $\displaystyle\int \dfrac{1}{x^2\sqrt{7 - x^2}}\,dx$

89. $\displaystyle\int \dfrac{1}{(x^2 + 2x + 5)^{5/2}}\,dx$

90. $\displaystyle\int \dfrac{x}{\sqrt{2x - x^2}}\,dx$

91. $\displaystyle\int \dfrac{2 - 5x - x^2}{x^3 + x^2 + 5x + 5}\,dx$

92. $\displaystyle\int \dfrac{x^3 + 5x^2 + 12x + 12}{x^3 + 2x^2 + 4x}\,dx$

93. $\int \csc^3 2x \, dx$

94. $\int \csc^4 x \, dx$

95. $\int \dfrac{7x^4 - 6x^3 + 16x^2 - 6x + 9}{2x^5 - 4x^4 + 5x^3 - 6x^2 + 3x} \, dx$

96. $\int \dfrac{e^{2x} + 1}{(2e^x + 5)^3} \, dx$

97. $\int \dfrac{x + 2}{\sqrt{1 - x^2}} \, dx$

98. $\int \dfrac{x^2}{(5 - x^2)^{3/2}} \, dx$

99. $\int \dfrac{1}{\sqrt{-2x^2 + 4x + 3}} \, dx$

100. $\int (6x - 2) \ln (x^2 - x + 5) \, dx$

101. $\int \cot^4 \left(\dfrac{x}{2} \right) dx$

102. $\int \dfrac{\tan^3 x}{\sec x} \, dx$

103. $\int \dfrac{1}{\sec x - \tan x} \, dx$

In Exercises 104–123 determine whether the series converges or diverges. In the case of a series with both positive and negative terms, determine whether convergence is absolute or conditional.

104. $\displaystyle\sum_{n=1}^{\infty} \dfrac{n^2 + 1}{3n^3 + 5}$

105. $\displaystyle\sum_{n=1}^{\infty} \dfrac{(n^3 + 4n)^{2/3}}{n^4 - 3n^3 + 1}$

106. $\displaystyle\sum_{n=1}^{\infty} \dfrac{(-1)^{n+1} n^2}{2^n}$

107. $\displaystyle\sum_{n=1}^{\infty} \dfrac{(-1)^n}{\sqrt{n} + 1}$

108. $\displaystyle\sum_{n=2}^{\infty} \dfrac{1}{n \ln n}$

109. $\displaystyle\sum_{n=1}^{\infty} n^{1/n}$

110. $\displaystyle\sum_{n=1}^{\infty} \dfrac{\sin (n + 1)}{n^2 + 4}$

111. $\displaystyle\sum_{n=1}^{\infty} \dfrac{1 + 2^{-n}}{1 + 3^{-n}}$

112. $\displaystyle\sum_{n=1}^{\infty} \sqrt{\dfrac{n^2 + n - 1}{2n^4 + n^3 + 5}}$

113. $\displaystyle\sum_{n=1}^{\infty} \left(\dfrac{n^2 - 2}{2n^2 + 5} \right)^n$

114. $\displaystyle\sum_{n=1}^{\infty} \dfrac{n^2}{n^{3n}}$

115. $\displaystyle\sum_{n=1}^{\infty} \dfrac{\mathrm{Tan}^{-1} (n^2 + 5)}{2n^2 + 4}$

116. $\displaystyle\sum_{n=1}^{\infty} \dfrac{1 \cdot 3 \cdot 5 \cdots (2n + 1)}{2 \cdot 4 \cdot 6 \cdots (2n)} \left(\dfrac{1}{n} \right)$

117. $\displaystyle\sum_{n=1}^{\infty} \dfrac{(n!)^2}{(2n + 1)!}$

118. $\displaystyle\sum_{n=1}^{\infty} \dfrac{\sqrt{n + \sqrt{1 + n^2}}}{n^2}$

119. $\displaystyle\sum_{n=1}^{\infty} \left(\dfrac{1}{2} - \dfrac{1}{n + 1} \right)^{10}$

120. $\displaystyle\sum_{n=1}^{\infty} \dfrac{3}{n(1 + \ln n)^3}$

121. $\displaystyle\sum_{n=1}^{\infty} \dfrac{3^{-2n} + 2^{-2n}}{4^{-n} + 5^{-n}}$

122. $\displaystyle\sum_{n=1}^{\infty} \dfrac{(n - 1)^n}{n^{n-1}}$

123. $\displaystyle\sum_{n=1}^{\infty} \mathrm{Sin}^{-1} \left(\dfrac{1}{n^2} \right)$

124. Find point(s) on the curve $y = 3x^2/(x + 1)$ where the inclination of the tangent line is equal to 1.

125. Can a continuously differentiable function defined by the equation $\sin (x + y) = 2x + 3$ have a critical point?

126. When a projectile (such as an arrow) is fired up the inclined plane in the figure below, the distance that it travels before striking the plane is called its range R. It can be shown that

$$R = \dfrac{v^2 \cos \theta \, \sin (\theta - \alpha)}{4.905 \cos^2 \alpha},$$

where v is the initial speed of the projectile, α is the angle of inclination of the plane, and θ is the angle of projection with the horizontal. Find the angle θ for which R is greatest.

127. Evaluate

$$\lim_{x \to 0} \dfrac{\sqrt{1 + \sin x} - \sqrt{1 - \sin x}}{\tan x}.$$

128. A differential equation in $y = f(x)$ is said to be homogeneous, linear, second order if it is of the form

$$a(x) \dfrac{d^2 y}{dx^2} + b(x) \dfrac{dy}{dx} + c(x) y = 0,$$

where $a(x)$, $b(x)$, and $c(x)$ are functions of x. Show that if $f_1(x)$ and $f_2(x)$ are two solutions of this equation, so also is $y = p_1 f_1(x) + p_2 f_2(x)$ for any constants p_1 and p_2.

In Exercises 129–132 find the area bounded by the curves.

129. $y = 2/\sqrt{x - 1}, \ x + 3y = 8$

130. $y = 8 - 3x^2, \ y = 3x^2/(x^2 + 1)$

131. $x + y = 2e, \ x = y \ln y, \ x = 0$

132. $x = (2/\pi)\mathrm{Sin}^{-1} y, \ y = 2 - x^2, \ y = 0$

In Exercises 133–135 find the volume of the solid of revolution obtained by rotating the area bounded by the curves around the line indicated.

133. $x = y(y - 1), \ x = 4y(y - 1)$ around $x = 0$

134. $y = (x^2 + x)/(x - 2), \ y = 0$ around $x = 1$

135. $x = e^{2y}, \ x = 2, \ y = 0$ around $y = 2$

136. Find the length of $y = \ln (1 - x^2)$ between the lines $x = 0$ and $x = 1/2$.

137. A thin uniform rod of length L and mass M spins in a plane about one of its ends with angular velocity ω. What is its kinetic energy?

138. The length of the hypotenuse of a right-angled triangle is 1 metre, and the lengths of the other two sides are a and b. When the triangle is submerged vertically in a fluid so that the edge with length b is in the surface of the fluid, the fluid force on each side of the triangle is twice that when the side of length a is in the surface. What are a and b?

139. Find the x-coordinate of the centroid of the area bounded by the curves $y = (x^2 + 2x - 3)/(x^2 + 1)$ and $y = 0$.

140. Find moments of inertia of an elliptic plate of uniform mass per unit area about its major and minor axes.

141. If $f(x)$ satisfies the condition $f(f(x)) = x$ for all x in its domain, what is $f^{-1}(x)$? Give an example.

142. In Kintsch's model of choice behaviour, he states that

$$\sum_{n=2}^{\infty} (1 - b)^{n-2} b s^n = \frac{bs^2}{1 - (1 - b)s}.$$

Is this true, and would you place any restrictions on b and s?

143. Let P be a point on the circumference of a circle. Where should a chord QR be drawn parallel to the tangent at P in order that triangle PQR have maximum area?

144. In the figure below, a 5 m rod CD is such that its end C can only move horizontally along the line AGB and end D can only move horizontally along EF. When C is 3 m to the left of G, it is moving toward A at 4 m/s. In what direction and how fast is D moving at this instant?

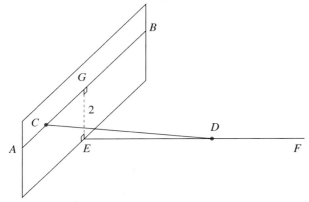

145. Evaluate $\lim_{n \to \infty} \frac{1}{n^2} (1 + 2 + 3 + \cdots + n)$.

146. Find d^2y/dx^2 if $y = \int_0^{x^2} t \cos t \, dt$.

147. Are either of the following valid?

$$\int_0^a f(x) \, dx = \int_0^a f(a - x) \, dx;$$

$$\int_0^a f(x) \, dx = \int_0^a f(b - x) \, dx.$$

148. Find the volume of the solid of revolution obtained by rotating the area bounded by $(x - 4)y^2 = x(x - 3)$ and the x-axis, about the x-axis.

149. Find the Maclaurin series for $f(x) = 1/(x^4 - x^2 + 2)$.

150. If $f(x)$ is a polynomial of degree 4, and it is known that $f(1) = 1$, $f'(1) = -2$, $f''(1) = 0$, $f'''(1) = 3$, and $f''''(1) = -5$, what is $f(-2)$?

151.   (a) A function $f(x)$ is said to be odd-harmonic if it satisfies $f(L + x) = -f(x)$ for all x. Show that $f(x)$ must be $2L$-periodic.

   (b) A function $f(x)$ is said to be odd and odd-harmonic if it satisfies $f(-x) = -f(x)$ and $f(L - x) = f(L + x)$. Show that $f(x)$ must be $4L$-periodic.

   (c) Draw the graph of the extension of $f(x) = x$, $0 < x < L$, to a continuous, odd, and odd-harmonic function.

   (d) A function $f(x)$ is said to be even and odd-harmonic if it satisfies $f(-x) = f(x)$ and $f(L + x) = -f(L - x)$. Show that $f(x)$ must be $4L$-periodic.

   (e) Draw the graph of the extension of $f(x) = x$, $0 < x < L$, to an even and odd-harmonic function.

152. Find the distance from the point $(1, 2)$ to the tangent line to the curve $y = x^2 \ln x$ at the point with abscissa equal to 2.

153. Show that a function defined by the equation $xye^{xy} = 1$ cannot have a first derivative equal to 1.

154. Find all critical points of the function $f(x) = x^{ax}$, $x > 0$ where $a > 0$ is a constant.

155. Show that the equation $x^n + px + q = 0$ cannot have more than two real roots when n is an even integer, and not more than three real roots when n is an odd integer.

156. The reproduction rate of Drosophila melanogaster supposedly decreases with population. Show that this is the case if the progeny per female per day, p, when x denotes the number of flies per bottle is

$$p = 34.53 x^{-0.658} e^{-0.018x}, \qquad x \geq 1.$$

157. Can a continuous function $y = f(x)$ with a first derivative satisfying

$$\frac{dy}{dx} + \sin y = 2 + x^2$$

have relative extrema?

158. For what value of k does the function $f(x) = k \cos x + 3x \sin x$ have a relative extremum at $x = \pi/3$? Is it a relative maximum or minimum?

159. Show that the angle of inclination of a curve has relative extrema at points of inflection.

160. Is the curve

$$x = t^2 - \frac{1}{t}, \qquad y = t^3 + 3t, \qquad t \geq 1$$

concave upward or concave downward?

161. Thurstone's learning curve is

$$y = \frac{Lx + Lc}{x + c + a}, \qquad x \geq 0$$

where $L > 0$ is the limit of practice, x is the amount of formal practice, c is the equivalent previous practice, and a is the rate of learning. Draw a graph of the learning curve.

162. One end of a ladder 20 m long slips down the side of a house at 1 m/s. What is the acceleration of the foot of the ladder when it is 15 m from the house?

163. If a point moves around the ellipse $b^2 x^2 + a^2 y^2 = a^2 b^2$, at what points are the rates of change of its abscissa and ordinate equal?

164. If the velocity of a particle moving along the x-axis is given by $A \cos(\omega t + \phi)$, and its position at time t_1 is x_1, what is its position at time t_2?

165. If $|f(x) - g(x)| \le \epsilon$ for $a \le x \le b$, what is the maximum value of
$$\left| \int_a^b [f(x) - g(x)] \, dx \right|?$$

166. Find the length of the loop of the curve $9ay^2 = x(x - 3a)^2$ where $a > 0$ is a constant.

167. A block of wood (density 500 kg/m^3) is cubical with side lengths L m. If it floats in water, what work is required to lift it out of the water?

168. The lengths of the major and minor axes of an elliptic plate are $2a$ and $2b$. Let F_a denote the force on each side of the plate when it is half submerged in a fluid with major axis in the surface, and let F_b denote the force when the minor axis is in the fluid. What is the ratio F_a/F_b?

169. A thin plate with constant mass per unit area has the shape of a sector of a circle of radius r and angle α where $0 < \alpha < \pi/2$. How far from the vertex of the sector is the centre of mass?

170. Find the inverse function for $f(x) = \dfrac{e^x - e^{-x}}{e^x + e^{-x}} + 1$.

171. Approximate
$$\int_0^{1/2} \frac{1}{1 + x^4} \, dx$$

   (a) using Simpson's rule with 10 subdivisions;
   (b) using the first four terms of the Maclaurin series of the integrand.

   Discuss the relative merits of these techniques for this example.

172. Find the centroid of the area bounded by the x-axis and the arch of the cycloid described by
$$x = R(\theta - \sin \theta), \quad y = R(1 - \cos \theta), \quad 0 \le \theta \le 2\pi.$$

173. Find the length of the curve $r = \sin^3\left(\dfrac{\theta}{3}\right)$.

174. Find the Maclaurin series for $f(x) = \ln[(1 + x)^{1+x}] + \ln[(1 - x)^{1-x}]$. What is the interval of convergence of the series?

175. What are the first three terms in the Maclaurin series for the function defined by the equation $y + \ln(1 + xy) + x = 5$?

176. Evaluate $\lim\limits_{x \to 0} \dfrac{6x - x^3 - 6\sin x}{x^5}$.

177. Find constants b_n in order that the function $y = f(x) = \sum_{n=1}^{\infty} b_n \sin nx$ will satisfy the differential equation
$$\frac{d^2 y}{dx^2} - 3y = \sum_{n=1}^{\infty} \frac{1}{n^2} \sin nx.$$

   (Assume that the series for $f(x)$ can be differentiated term-by-term.)

178. Find the distance from the point of intersection of the curves $y = 1/(x - 1)$ and $4y = x^2$ to the normal line to the curve $x^3 y + xy^2 = 6$ at the point with ordinate $y = -1$.

179. Verify that for $0 < a < b$,
$$\frac{b - a}{b} < \ln\left(\frac{b}{a}\right) < \frac{b - a}{a}.$$

180. It is known that a continuous function $y = f(x)$ always has a positive derivative which satisfies the equation
$$\frac{dy}{dx} = y^3 - 3y.$$

   Show that its graph can have a point of inflection only at a point where $y = 1$ or $y = -1$.

181. The circles in the figure below represent ferris wheels with radii 20 m and 10 m. Their centres are displaced horizontally by 50 m and vertically by 30 m. The larger rotates half as fast as the small, making one revolution every 10 seconds. How fast is the distance changing between riders at positions P and Q?

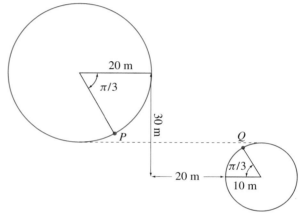

182. At time $t = 0$, a body is moving along the x-axis with velocity $v_0 > 0$. If the only force acting on it is resistance, and this force is proportional to the square of the velocity of the body, where and when does the body stop?

183. Find the length of the curve
$$y = L \ln\left(\frac{L + \sqrt{L^2 - x^2}}{x}\right) - \sqrt{L^2 - x^2}$$

   from $(L, 0)$ to any other point (x, y) on the curve (see Exercise 24 in Section 9.5).

184. Find the centroid of the area bounded by the curve $|y| = x\sqrt{1 - x^2}$.

185. If the number of subdivisions in the approximation of a definite integral by Simpson's rule is doubled, does the error decrease by a factor of 16? Explain.

186. Show that a function $f(x)$ and the nth degree polynomial
$$P_n(x) = f(c) + f'(c)(x - c) + \cdots + \frac{f^{(n)}(c)}{n!}(x - c)^n$$

   have the same first n derivatives at $x = c$.

187. Find a cubic polynomial approximating the function
$$f(x) = \int_0^x e^{-t^2} \, dt.$$

   How accurate is it on the interval $-0.2 \le x \le 0.2$?

188. Show that if a continuous function $y = f(x)$, which satisfies the equation

$$\frac{dy}{dx} - y^3 = y,$$

has a point of inflection on its graph, it must be a horizontal point of inflection.

189. Find circle(s) that have centres on the line $2x + 3y = 6$ and intersect the circle $x^2 + y^2 = 1$ at only one point.

190. One hundred dollars is invested at 10% compounded quarterly. Draw a graph of the accumulated value of the investment as a function of time t (in months) taking $t = 0$ at the time it is initially invested. Express this function in terms of the greatest integer function.

191. Evaluate $\displaystyle\lim_{y \to 0} \frac{(x + y)^{1/3} - x^{1/3}}{y}$.

192. Evaluate $\displaystyle\lim_{x \to \infty} x \left(\sqrt{x^2 + \sqrt{x^4 + 6}} - \sqrt{2}x \right)$.

193. If $f(x)$ is such that $|f(x)| \leq x^2$ for all x, what is $f'(0)$?

194. Show that the angle of inclination of the tangent line to the curve $y = \text{Tan}^{-1}(2x^3)$ always lies in the interval $0 \leq \phi \leq \text{Tan}^{-1} 2$.

195. Calculate $\dfrac{d^{20}}{dx^{20}}[(x^2 + 1)\sin x]$.

196. Show that any tangent to the curve $y = (1/2)\sqrt{x - 4x^2}$ intersects the y-axis at a point equidistant from the point of tangency and the origin.

197. Let P be any point on the hyperbola $b^2x^2 - a^2y^2 = a^2b^2$. If the tangent line at P cuts the asymptotes of the hyperbola at points Q and R, show that P is the midpoint of the line segment joining Q and R.

198. Show that the x-intercept of the tangent line to the curve $a/x^2 + b/y^2 = 1$ at any point is proportional to the cube of the abscissa of the point of tangency.

199. (a) Show that when the area of a circle of radius r is approximated by an inscribed polygon with n equal sides, the area of the polygon is

$$A_n = \frac{nr^2}{2} \sin\left(\frac{2\pi}{n}\right).$$

(b) Verify that $\lim_{n \to \infty} A_n = \pi r^2$.

200. Prove that the function

$$f(x) = \frac{x^2}{2} + \frac{1}{2}x\sqrt{x^2 + 1} + \ln\sqrt{x + \sqrt{x^2 + 1}}$$

satisfies the differential equation

$$x\frac{dy}{dx} + \ln\left(\frac{dy}{dx}\right) = 2y.$$

201. Verify that if a polynomial of even degree ever takes on a value opposite in sign to the coefficient of its highest-degree term, it must have at least two real zeros.

202. For what values of a and b does the function $f(x) = ax\ln x + bx^2 + x$ have relative extrema at $x = 1$ and $x = 2$? Are they relative maxima or minima?

203. A man's cottage is situated on a lake that is more or less circular with radius 2 km. Diametrically opposite is a store where he buys groceries. If the man can paddle his canoe at v km/h and can walk along the shore at V km/h, what combination of paddling and walking produces the quickest trip from cottage to store?

204. A fish swims upstream at a constant speed v relative to the water. The water itself has speed v_1 relative to the ground. The fish intends to reach a point at distance s upstream. The energy required is determined by the friction in water and by the time t taken. Experiments have shown that this energy is $E = cv^kt$ where $c > 0$ and $k > 2$ are constants (k depending on the shape of the fish). Given that v_1 is constant, what speed v minimizes E?

205. (a) One end of a ladder slips away from a wall at a uniform rate. Show that the speed of the other end down the wall is always increasing.

(b) Show that when the upper end falls at a constant rate, the speed of the lower end decreases.

206. An eagle is gliding horizontally 100 m above the ground at 15 m/s. It is 200 m south of a tree and is heading due north. A rabbit is 50 m east of the tree hopping along at 2 m/s toward the tree. How fast is the distance between the eagle and rabbit changing at this instant?

207. Show that $\displaystyle\lim_{x \to \infty} \frac{x - \sin x}{x + \sin x}$ exists, but that it cannot be evaluated with L'Hôpital's rule.

208. A particle moves along the x-axis with position (in metres) given by $x(t) = t^3 - 9t^2 + 24t + 3$, $t \geq 0$, where t is time (in seconds). Determine:

(a) the speed of the particle when it is at position $x = 39$;

(b) when the particle is speeding up;

(c) the closest it gets to the point $x = -1$ during the time interval $1 \leq t \leq 3$;

(d) its maximum speed when it is between $x = 20$ and $x = 21$.

209. A man eats a diet of 2500 calories per day; 1200 of them go to basic metabolism (i.e., get used up automatically). He also spends 16 calories per day for each kilogram of body mass in exercise. Assume that the storage of calories as fat is 100% efficient and that 1 kilogram of fat contains 10 000 calories. Find his mass as a function of time, and sketch this function, if initially his mass is (a) 100 kg (b) 70 kg. Is he on a diet?

210. Evaluate $\displaystyle\lim_{n \to \infty} \frac{1 - 2 + 3 - 4 + \cdots - 2n}{(3 + 8n^3)^{1/3}}$.

211. Show that the definite integral from $x = 0$ to $x = 1$ with respect to x does not exist for the function

$$f(x) = \begin{cases} 1 & x \text{ is a rational number} \\ 0 & x \text{ is an irrational number.} \end{cases}$$

212. Without evaluating the integral $\displaystyle\int_0^{10} \frac{x}{x^3 + 16}\,dx$, show that it must be less than $5/6$.

213. Assuming that y is defined implicitly as a function of x by

$$\int_0^y \frac{1}{1+t^3}\,dt + \int_0^x \cos t^2\,dt = 5,$$

find dy/dx.

214. Evaluate $\int_{-a}^a \sin^2 x\,\ln\left(\dfrac{1+x}{1-x}\right)dx$ when $0 < a < 1$.

215. (a) What is the area bounded by the line $y = 2x + 1$ and the parabola $y = x^2$?

(b) Find the line parallel to $y = 2x + 1$ which along with $y = x^2$ bounds an area equal to half that in (a).

216. Find the volume of the solid of revolution obtained by rotating the area bounded by $x^4 + y^4 = x^3$ about the x-axis.

217. A tank is in the form of a frustrum of a right-circular cone (figure below). If the tank is full of oil with density ρ, what work is required to empty the tank over its top edge?

218. The lower half of a cubical tank 2 m on each side is occupied by water, and the upper half by oil (density 0.90 grams per cubic centimetre).

(a) What is the force on each side of the tank due to the pressure of the water and the oil?

(b) If the oil and water are stirred to create a uniform mixture, does the force on each side change? If not, explain why not. If so, by how much does it increase or decrease?

219. Find the second moment of area of the larger loop of $y^2 = (x+1)^2(4-x^2)$ about the y-axis.

220. Find the volume cut from a right circular cylinder of radius r by a plane passing through its diameter (figure below).

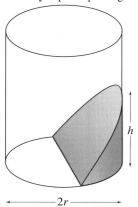

221. A thin rod of length L has σ coulombs of positive charge uniformly distributed over each metre of length (figure below). Suppose a positive charge q coulombs is placed at point P, a distance of a units along the right-bisector of the rod. According to Coulomb's law, the magnitude of the force of repulsion on q by a small amount of charge Q at point R is $qQ/(kr^2)$ where k is a constant and $r = \|PR\|$. A similar charge equally distant from the other end of the rod also repels q but in a different direction. The combined repulsive force on q by these two charges is perpendicular to the rod and has magnitude

$$\frac{2qQ\cos\theta}{kr^2},$$

where θ is the angle shown.

(a) Find the total force on q.

(b) What is the force when the rod is infinitely long?

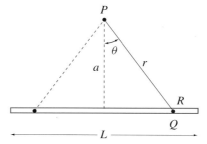

222. (a) Your friend says that he wants to paint the area bounded by the curves

$$y = \frac{1}{x}, \quad x = 1, \quad y = 0.$$

Why would you advise him to forget it?

(b) Your friend now says that he is going to make a funnel in the shape of the surface of revolution obtained by rotating that part of $y = 1/x$ to the right of $x = 1$ about the x-axis. He intends pouring paint into the funnel until it is full. Can he do this with a finite amount of paint? If he can, what do you reply when he points out that he has now painted both sides of the area in (a)?

(c) Would you advise him to attempt to paint the outside of the funnel? How about the inside?

223. Show that the function $f(x) = 4 + 3\sin\left(\dfrac{x-2}{x+2}\right)$ has an inverse function on the interval $x \geq 0$, and find it.

224. Prove that $\sin x$ is a transcendental function. Hint: Assume

$$P_0(x)\sin^n x + P_1(x)\sin^{n-1} x$$

$$+ \cdots + P_{n-1}(x)\sin x + P_n(x) = 0$$

and evaluate at $x = n\pi$.

225. Show that the function

$$f(x) = \mathrm{Cos}^{-1}\left(\frac{a\cos x + b}{a + b\cos x}\right)$$

$$- 2\,\mathrm{Tan}^{-1}\left[\sqrt{\frac{a-b}{a+b}}\,\tan\left(\frac{x}{2}\right)\right],$$

where $0 < b \leq a$, is constant for $0 \leq x \leq \pi$. Find the value of the constant.

226. Evaluate

$$\int_0^{\pi/2} \frac{\sin x \, \cos x}{a^2 \cos^2 x + b^2 \sin^2 x} \, dx$$

where $ab \neq 0$.

227. Find the area bounded by $y = \ln x$ and $y = (\ln x)^2$.

228. Evaluate $\int_0^{\pi} \sqrt{1 - \cos 3x} \, dx$.

In Exercises 229–230 evaluate the indefinite integral.

229. $\displaystyle\int \frac{1}{x^3 \sqrt{x^2 + 5}} \, dx$ **230.** $\displaystyle\int (4 - x^2)^{3/2} \, dx$

231. Find the area bounded by the curve $x^4 - ax^3 + a^2 y^2 = 0$ where a is a constant.

232. Find the area bounded the the curve $(y - \text{Sin}^{-1} x)^2 = x - x^2$.

In Exercises 233–235 evaluate the indefinite integral.

233. $\displaystyle\int \frac{1}{x^2 (9 - x^2)^{3/2}} \, dx$ **234.** $\displaystyle\int \sqrt{\frac{3 + x}{3 - x}} \, dx$

235. $\displaystyle\int \frac{15 + 38x - 14x^2}{15 - 14x - 11x^2 + 4x^3} \, dx$

236. In a chemical reaction, one molecule of trypsinogen yields one molecule of trypsin. In order for the reaction to take place, an initial amount of trypsin must be present. Suppose that the initial amount is y_0. Thereafter, the rate at which trypsinogen is changed into trypsin is proportional to the product of the amounts of each chemical in the reaction. Find a formula for the amount of trypsin if the initial amount of trypsinogen is A.

237. A very long, thin thread is wrapped clockwise around the circle $x^2 + y^2 = a^2$ so that one end of the string is at the point $(a, 0)$. If the thread is unwound in such a way that it is always kept taut, the path followed by the end of the string is called an involute of the circle.

    (a) Find equation(s) for the involute.

    (b) Find a formula for the slope of the involute.

    (c) What is the limit of the slope as the point $(a, 0)$ is approached along the curve?

    (d) What is the slope of the curve when it crosses the positive x-axis for the first time?

238. Find the area enclosed by the astroid

$$x = a \cos^3 t, \quad y = a \sin^3 t,$$

a a constant.

239. Show that the curves

$$3(x^2 + y^2)^2 = 25(x^2 - y^2), \qquad 22y = 26 - x^2,$$

are tangent at their points of intersection.

240. Find the area between the loops of $r = a \sin^3 (\theta / 3)$.

241. Find the area bounded by the hyperbola $b^2 x^2 - a^2 y^2 = a^2 b^2$ and the line through its focus parallel to the y-axis.

242. A model of learning proposed by Bush and Mosteller assumes that on the nth trial of a sequence of trials, the probability p_n of obtaining a certain response is related to the probability on the immediately preceding trial by $p_n = a + (1 - a - b)p_{n-1}$ where a and b are constants which depend on the amounts of reward and inhibition. Prove that an explicit formula for p_n is

$$p_n = p_0(1 - a - b)^n + \frac{a}{a + b}[1 - (1 - a - b)^n]$$

where p_0 is the initial probability of response.

243. In Iraq an epidemic of methylmercury poisoning killed 459 people in 1972. Human beings became exposed to the poison when they ate homemade bread accidentally prepared from seed-wheat treated with a methylmercurial fungicide. Symptoms appeared only after weeks of exposure. Assume for simplicity that a person takes a constant daily dose d of the poison and that a certain percentage p of the accumulated poison is excreted each day. Find a formula which relates the amount of poison stored in the body to the number of days.

244. Evaluate $\displaystyle\lim_{n \to \infty} \frac{\frac{1}{4} + \frac{1}{8} + \frac{1}{16} + \cdots + \frac{1}{2^n}}{1 + \frac{2}{5} + \frac{4}{25} + \frac{8}{125} + \cdots + \left(\frac{2}{5}\right)^{n-1}}$.

245. Show that if $\sum_{n=1}^{\infty} a_n^2$ and $\sum_{n=1}^{\infty} b_n^2$ are convergent series, then $\sum_{n=1}^{\infty} a_n b_n$ is absolutely convergent.

246. Find $f^{(6)}(0)$ if $f(x) = x^2 / \sqrt{1 - x^2}$.

247. Find the first three nonzero terms in the Maclaurin series for $\ln(\cos x)$.

248. Find the Maclaurin series for $\cos^4 x$.

249. Find the Maclaurin series for $x(2 + 3x)^{3/2}$.

250. Find $\sqrt{4.000\,000\,000\,01}$ accurate to 20 decimal places.

251. Prove that if $f(x)$ is continuous at $x = a$ and $f(a) > 0$, then there exists an open interval I containing a such that $f(x) > 0$ for all x in I.

252. (a) Show that the tangent line to the curve $y = x^3$ at any point P (other than the origin) always intersects the curve again.

    (b) Verify that the slope of the curve at the second point in (a) is four times the slope at P.

253. Prove that when $f'(a)$ exists

$$\lim_{x \to a} \frac{xf(a) - af(x)}{x - a} = f(a) - af'(a).$$

254. (a) Show that if $y = (1 - x)^{-a} e^{-ax}$, then

$$(1 - x) \frac{dy}{dx} = axy.$$

    (b) Now show that

$$(1 - x) \frac{d^{n+1} y}{dx^{n+1}} - (n + ax) \frac{d^n y}{dx^n} - an \frac{d^{n-1} y}{dx^{n-1}} = 0.$$

255. Find all points on the curve $x^2 y^2 + xy = 2$ where the slope is -1.

256. Prove that the segment of the tangent line to the tractrix

$$y = \frac{a}{2} \ln\left(\frac{a + \sqrt{a^2 - x^2}}{a - \sqrt{a^2 - x^2}}\right) - \sqrt{a^2 - x^2}$$

between the point of tangency and the y-axis always has length a.

257. Find d^2y/dx^2 if $y = |\sin x|^x$, $x > 0$.

258. A manufacturer making $144\,000$ units of an item during a year has a choice of different production schedules. He can make the entire lot early in the year in one production run. By doing so, he holds production costs down; but he must then carry much of the production run in inventory, and inventory costs might overbalance savings from mass production economies. Suppose that it costs $500 to make the factory ready for a production run, and $5 for each unit made after the factory has been readied. Assume that after a batch is made, the units are placed in inventory and then used up at a uniform rate such that inventory is zero when the next batch appears. If annual cost of the inventory is the cost of the average inventory, what batch size should be run to minimize annual production and inventory costs?

259. A log 20 m long has the form of a truncated cone with bases of diameters 1 m and 2 m. A beam with square cross-section is to be cut from the log so that axes of log and beam coincide. Find the dimensions of the beam with maximum volume.

260. A street light is 25 m high. A ball is thrown vertically upward from a point 10 m away from the lightpost. If the speed of the ball is 20 m/s when it is 15 m above the ground, what is the acceleration of the shadow of the ball along the ground at this instant?

261. (a) Show that the volume of the container in the figure below is $abh/3$.

(b) If water is poured into the container at a rate of k cubic units per unit time, find a formula for the rate at which the depth of water is increasing when the container is half full by volume.

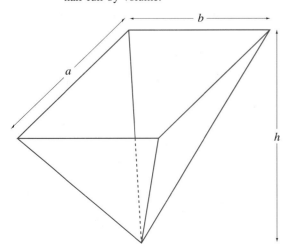

262. If a particle moves along the x-axis so that its acceleration is always proportional to its velocity, show that the percentage change in its velocity in every time interval of length T is always the same.

263. Prove that

$$\int_0^{\pi/2} f(\cos x)\, dx = \int_0^{\pi/2} f(\sin x)\, dx.$$

264. Find the volume of the barrel in the following figure. The cross-section is elliptic.

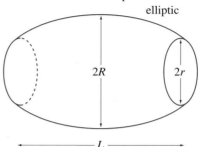

265. A piece of wood with density ρ is in the shape of a pyramid; its base is a square of side length b and its vertex is a distance a from the base. If it floats in a liquid of density $\overline{\rho}$ with its base horizontal and above the surface, find the work required to lift the block from the liquid.

266. (a) A sphere of radius r and density $\overline{\rho}$ floats partially submerged in a liquid with density ρ. Show that the depth d below the surface of the lowest point of the sphere must satisfy the equation

$$\rho(d^3 - 3rd^2) + 4r^3\overline{\rho} = 0.$$

(b) What is d if $\overline{\rho} = \rho/10$?

267. Newton's universal law of gravitation states that the force of attraction between two point masses m and M at distance r apart has magnitude $F = GmM/r^2$, where $G = 6.67 \times 10^{-11}$ m^3/(kg·s^2). A thin uniform rod AB of length L and mass M attracts a point mass m at point C which lies at a distance d from the end closest to B on the extension of AB. What point mass placed at A has the same force of attraction on m?

268. Under what condition will the function $f(x) = (ax + b)/(cx + d)$, where $ad - bc \neq 0$ be equal to its inverse?

269. Evaluate $\displaystyle\int_0^x \frac{1}{1 + t^2}\, dt + \int_0^{1/x} \frac{1}{1 + t^2}\, dt.$

In Exercises 270–278 evaluate the indefinite integral.

270. $\displaystyle\int (\text{Sin}^{-1} x)^2\, dx$

271. $\displaystyle\int \frac{1}{x^2} \text{Cos}^{-1} x\, dx$

272. $\displaystyle\int \frac{1}{\sin x \cos^2 x}\, dx$

273. $\displaystyle\int \frac{1}{(x - 4)\sqrt{x^2 - 16}}\, dx$

274. $\displaystyle\int x^2 \text{Tan}^{-1} 2x\, dx$

275. $\displaystyle\int e^{2x+3} \cos(3x - 1)\, dx$

276. $\displaystyle\int \frac{x + \sin x}{1 + \cos x}\, dx$

277. $\displaystyle\int \sin(2 \ln x)\, dx$

278. $\displaystyle\int \frac{x}{\cos^2 3x}\, dx$

279. The *curvature* of a curve $y = f(x)$ at a point on the curve is defined as

$$\kappa(x) = \frac{|f''(x)|}{\{1 + [f'(x)]^2\}^{3/2}}.$$

(a) Find the curvature of the parabola $y = x^2 - 2x - 3$. Where is it a maximum?

(b) Show that the only curves for which curvature is identically equal to zero are straight lines.

(c) What happens to curvature at a point of inflection on the graph of a function?

280. Show that when a curve is defined parametrically, its curvature (Exercise 279) can be expressed in the form

$$\kappa(t) = \frac{\left| \dfrac{dy}{dt}\dfrac{d^2x}{dt^2} - \dfrac{dx}{dt}\dfrac{d^2y}{dt^2} \right|}{\left[\left(\dfrac{dx}{dt}\right)^2 + \left(\dfrac{dy}{dt}\right)^2 \right]^{3/2}}.$$

281. The *radius of curvature* of a curve at a point is defined as $\rho = 1/\kappa$ where κ is curvature at the point (see Exercises 279 and 280).

(a) Show that the radius of curvature of a circle is constant and equal to the radius of the circle.

(b) At what points on the ellipse $b^2 x^2 + a^2 y^2 = a^2 b^2$, $(a > b)$, is the radius of curvature a maximum and a minimum?

282. Suppose a curve $y = f(x)$ is concave upward at a point (x_0, y_0). The *circle of curvature* at (x_0, y_0) is defined as the circle above the curve with radius equal to ρ (the radius of curvature in Exercise 281), and centre along the normal to $y = f(x)$ at (x_0, y_0) at a distance ρ from (x_0, y_0). If the curve is concave downward, the circle is drawn below the curve.

(a) Find the circle of curvature to the parabola $y = x^2 - 2x - 3$ at $(1, -4)$.

(b) Is it true that the circle of curvature at a point on a curve passes through that point? Draw a picture.

(c) Show that the circle of curvature and curve share the same tangent line at their common point.

283. Find the equation of the tangent line to the curve

$$x = 4(2\cos\theta - \cos 2\theta), \quad y = 4(2\sin\theta - \sin 2\theta),$$

$$0 \le \theta < 2\pi,$$

at the point $(4\sqrt{3} - 2, 4 - 2\sqrt{3})$.

284. The curve $r = Aq^\theta$, where $A > 0$ and $q > 1$ is called a logarithmic spiral. Show that its slope at every point at which it intersects the radial line $\theta = k$ is the same.

285. Dropping leaflets by air is a means sometimes used in advertising to reach a large group of customers. Suppose it is assumed that a constant proportion λ of the leaflets actually survive a given time period T, while the remainder are lost, destroyed, or otherwise rendered unreadable. It is further assumed that each leaflet that survives until the ith time period will reach β people, on the average, during that period. Given that N leaflets are dropped, how many people are reached by the leaflets?

286. Find the Taylor series about $x = 1$ for the function $\sqrt[3]{7 + x}$.

287. Use the Maclaurin series for $f(x) = (1 + x)e^{-x} - (1 - x)e^x$ to find the sum of the series $\sum_{n=1}^{\infty} n/(2n + 1)!$.

288. Find point(s) on the curve $y = \ln(x^2 + 1)$ where the tangent line passes through the point $(1, -2 + \ln 2)$.

289. What point on the ellipse $x^2 + 3y^2 = 9$ is furthest from $(5, 7)$?

290. Let P be any point on the hyperbola $x^2 - y^2 = a^2$. If the tangent line at P cuts the asymptotes of the hyperbola at points Q and R, show that the area of triangle OQR is a^2.

291. Show that if $y = f(x)$ is concave upward on an interval I, then

$$\frac{f(x_1) + f(x_2)}{2} \ge f\left(\frac{x_1 + x_2}{2}\right)$$

for all x_1 and x_2 in I.

292. The rate of flow of water through a circular pipe is directly proportional to its hydraulic radius R defined as $R = A/s$. Quantity A is the cross-sectional area of the water flow in the pipe, and s is the length of the circular arc of this cross-sectional area. The degree to which the pipe is filled is measured by the angle θ from the centre of the pipe to the horizontal surface of the water. For what degree of filling is R a maximum?

293. Sand from a conveyor belt drops at the rate of 10 cubic centimetres per second. It creates a pile on a level floor in the form of a right-circular cone. The height h of the pile increases at the same rate as the diameter D of the circular area A on the floor covered by the sand. How fast is A changing when D is 0.5 metres?

294. Find the volume of the solid of revolution obtained by rotating the area bounded by the curves

$$y = x\sin x, \quad y = 0, \quad 0 \le x \le \pi,$$

around the x-axis.

295. A triangular plate has one vertex in the surface of a fluid. The opposite side has length a, is parallel to the surface of the fluid, and is b units below the surface. What is the force on each side of the plate?

296. If $y = f(x)$ has inverse function $x = f^{-1}(y)$, find a formula for d^2x/dy^2 in terms of derivatives of $f(x)$. Illustrate with an example.

297. Verify that

$$\int (\ln x)^n \, dx = (-1)^n n! x \sum_{r=0}^{n} \frac{(-\ln x)^r}{r!} + C.$$

298. Find the volume of the solid of revolution when the area under one arch of the cycloid of Example 10.7 in Section 10.1 is rotated about (a) the x-axis (b) the y-axis (c) the line $x = \pi R$.

299. Find the area of the surface of revolution when the lemniscate $(x^2 + y^2)^2 = a^2(x^2 - y^2)$ is rotated about the x-axis.

300. Find the area of the surface of revolution when the cardioid $r = a(1 - \cos\theta)$ is rotated about the x-axis.

301. Inscribed in a circle of radius R is a square. A circle is then inscribed in the square, a square in the circle, and so on and so on. Find the limit of the sum of the areas of all the circles and the limit of the sum of the areas of all the rectangles.

302. Are there angles at which the sine of the angle changes twice as fast as the cosine?

303. Find a quadratic polynomial in x which approximates the function defined implicitly by the equation $y^5 + xy + 1 = 0$.

304. How many terms in the Maclaurin series for the integrand guarantee a value for

$$\int_1^2 \frac{\sin 2x}{x} \, dx$$

accurate to 10^{-3}?

305. Evaluate $\lim\limits_{x \to 0} \dfrac{\sin (x^n)}{(\sin x)^m}$ where n and m are positive integers.

306. Show that the segment of the tangent line to the astroid $x^{2/3} + y^{2/3} = a^{2/3}$ contained between the coordinate axes always has length a.

307. Show that the points of inflection on the curve $y = (x + 1)/(x^2 + 1)$ are collinear.

308. At what angle(s) does the sine of the angle change half as fast as the angle itself?

309. A rectangular plate with sides of length L and ℓ ($L > \ell$) is submerged in a liquid of density ρ. Its longer sides are parallel to the surface of the liquid, the plate makes an angle α with the surface, and the edge closest to the surface is at depth d. What is the force on each side of the plate?

310. The gamma function $\Gamma(\nu)$ for $\nu > 0$ is defined by

$$\Gamma(\nu) = \int_0^\infty x^{\nu-1} e^{-x} \, dx.$$

Show that $\Gamma(\nu + 1) = \nu \Gamma(\nu)$.

311. At what point(s) on the curve $x = R(1 - \cos\theta)$, $y = R(\theta - \sin\theta)$ is the slope of the tangent line equal to 1?

312. Find a polynomial approximating the indefinite integral

$$\int \frac{1}{\sqrt[4]{1 + x^4}} \, dx$$

to within 10^{-6} on the interval $0 \le x \le 0.8$.

313. Show that points of inflection of the curve $y = (\sin x)/x$ lie on the curve $y^2(4 + x^4) = 4$.

314. (a) A spherical shell with inner and outer radii 1 m and 2 m is made from styrofoam with mass 2 kg/m^3. The shell is cut into two equal pieces by a plane, and the flat edge of one piece is carefully placed on the surface of a large container of water. How far will the styrofoam sink?

   (b) If a small hole is drilled in the top of the floating shell, will the styrofoam sink further? If so, to what depth?

315. Two substances A and B react to form a third substance C in such a way that 2 gm of A react with 1 gm of B to produce 3 gm of C. The rate at which C is formed is equal to the product of the amounts of A and B present in the mixture. Find the amount of C present in the mixture as a function of time t when the original amounts of A and B brought together at time $t = 0$ are as follows.

   (a) 20 gm and 10 gm

   (b) 20 gm and 5 gm

   (c) 20 gm and 20 gm

316. Find a polynomial in x approximating $f(x) = \cos^3 x$ on $0 \le x \le 0.5$ with maximum error 10^{-6}.

317. For what values of c is the range of the function

$$f(x) = \frac{x^2 + 2x + c}{x^2 + 4x + 3c}$$

equal to $-\infty < x < \infty$?

318. Tangents at points A and B on a circle intersect at point F. The tangent at the midpoint C of the smaller arc of the circle joining A and B intersects AF and BF in D and E respectively. What is the limit of the ratio of the areas of triangles ABF and DEF as A approaches B?

319. Use the definition of a limit to prove that $\lim_{x \to 2} (x^2 + 3)/(x - 1) = 7$.

320. Find the equation of the smaller circle with centre on the line $2y + \sqrt{3}x = 4\sqrt{3} + \sqrt{7}$ which is tangent to the ellipse $9x^2 + 4y^2 = 36$.

321. (a) Find point(s) where the tangent line to the piriform $y^2 = x^3(a - x)$ is horizontal.

   (b) Where is the point of inflection on the top half of this curve?

322. The first derivative test for a relative minimum states that: If $f'(x)$ changes from a negative quantity to a positive quantity as x increases through a critical point, then that critical point yields a relative minimum. Is the converse statement true? To answer, consider the function

$$f(x) = \begin{cases} \dfrac{x^2}{2}\left[1 + \sin\left(\dfrac{1}{x}\right)\right], & x \ne 0 \\ 0, & x = 0. \end{cases}$$

323. Find that point P on the ellipse $b^2x^2 + a^2y^2 = a^2b^2$ so that the triangle with vertices P, $Q(0, b)$ and $R(a, 0)$ has maximum possible area.

324. Two corridors, one 3 m wide and the other 6 m wide, meet at right angles. Find the length of the longest square beam, 1/3 m by 1/3 m, that can be transported horizontally from one corridor to the other.

325. A hemispherical tank with radius r has its flat face horizontal and uppermost. It is originally (time $t = 0$) full of water, and for $t > 0$, water leaks out through a hole in the bottom. Use the modified Torricelli law to determine the time for the tank to empty.

326. Evaluate

$$\lim_{n\to\infty} \frac{1}{n^2}\left[\sin\left(\frac{1}{n^2}\right) + 2\sin\left(\frac{4}{n^2}\right) + 3\sin\left(\frac{9}{n^2}\right)\right.$$

$$\left. + \cdots + n\sin 1\right].$$

327. Show that when $f'(x)$ and $f''(x)$ are positive on an interval $a \le x \le b$, then

$$(b-a)f(a) < \int_a^b f(x)\,dx < \frac{1}{2}(b-a)[f(a) + f(b)].$$

328. Show that the function

$$y = \frac{1}{3}\int_0^x f(t)e^{t-x}\sin 3(x-t)\,dt$$

satisfies the differential equation

$$\frac{d^2y}{dx^2} + 2\frac{dy}{dx} + 10y = f(x).$$

329. Find a in order that the area in the figure below be as large as possible.

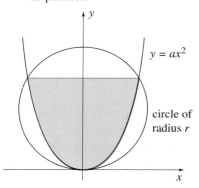

$y = ax^2$

circle of radius r

330. A square is rotated about a line which lies in its plane and passes through one of its vertices. What position of the line yields maximum volume for the resulting solid of revolution?

331. The two people in the figure below each pull in the cable at 1 m/s. The curved part of the cable always has shape $y = k^{-1}\cosh kx + C$, where k and C, although independent of x, depend on the height of the minimum point above the x-axis. Find how fast this minimum point is rising when it is 5 m above the x-axis.

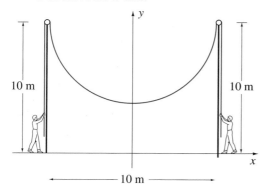

10 m

10 m

10 m

332. A hole is drilled through a long right-circular cylinder of radius R. If the radius of the hole is $r < R$, and axes of the cylinder and hole meet at right angles, set up a definite integral for the volume removed from the cylinder by the hole.

333. For what values of k do the following integrals converge:

(a) $\displaystyle\int_2^\infty \frac{1}{x^k \ln x}\,dx$ (b) $\displaystyle\int_2^\infty \frac{1}{x(\ln x)^k}\,dx$

(c) $\displaystyle\int_2^\infty \frac{1}{(\ln x)^k}\,dx$

334. Prove that e^x is a tanscendental function. Hint: Assume

$$P_0(x)e^{nx} + P_1(x)e^{(n-1)x} + \cdots + P_{n-1}(x)e^x + P_n(x) = 0$$

and take limits as $x \to -\infty$.

335. Let P be any point on the right half of the curve $y = k\cosh(x/k)$, and Q be the foot of the perpendicular from P to the x-axis. Draw a semicircle to the left of PQ with PQ as diameter. Let R be the point on the semicircle at a distance k from Q. Show that PR is tangent to the curve.

336. Prove that when $P(x)$ is a polynomial of degree n,

$$\int e^{mx} P(x)\,dx = e^{mx}\sum_{r=0}^n \frac{(-1)^r}{m^{r+1}} P^{(r)}(x) + C$$

where $P^{(r)}(x)$ is the rth derivative of $P(x)$ and $P^{(0)}(x) = P(x)$.

337. Prove that for constant a and $n \ge 1$ an integer:

(a) $\displaystyle\int \sin^{2n} ax\,dx = -\frac{1}{a}\cos ax \sum_{r=0}^{n-1} \frac{(2n)!(r!)^2}{2^{2n-2r}(2r+1)!(n!)^2} \times$

$$\sin^{2r+1} ax + \frac{(2n)!}{2^{2n}(n!)^2}x + C$$

(b) $\displaystyle\int \sin^{2n+1} ax\,dx = -\frac{1}{a}\cos ax \sum_{r=0}^n \frac{2^{2n-2r}(n!)^2(2r)!}{(2n+1)!(r!)^2} \times$

$$\sin^{2r} ax + C$$

338. Hecht assumes that the amount S of a retinal substance which remains when exposed to light of intensity I is characterized by the equation

$$\frac{dS}{dt} = bI(a - S) - kS^2$$

where a denotes the amount of S prior to stimulation, and b and k are positive constants. Show that

$$S(t) = \frac{-bI}{2k} + \frac{B(1 + De^{-2kBt})}{1 - De^{-2kBt}},$$

where $B^2 = 4b^2I^2/k^2 + abI/k$ and $D = [2k(a - B) + bI]/[2k(a + B) + bI]$.

339. Circles C_1 and C_2 in figure (a) below have radius R m and circle C_1 rolls around C_2 without slipping.

(a) Show that the curve traced out by the point $P(x, y)$ on C_1 which is initially at the origin has equations

$$x = 2R \cos\theta(1 - \cos\theta), \qquad y = 2R \sin\theta(1 - \cos\theta)$$

(see figure (b)).

(b) Show that if ϕ is the angle between the positive x-axis and the line joining $P(x, y)$ to the origin, then $\phi = \theta$. Hence, show that the polar equation of the curve is $r = 2R(1 - \cos\theta)$. Draw the curve.

(c) If C_1 rotates twice around C_2 every second, what is the speed of P when it is at the points with Cartesian coordinates $(-4R, 0)$ and $(0, 0)$?

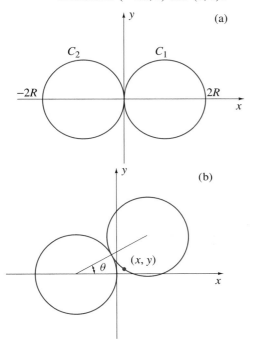

340. A piece of sheet metal $1/6$ cm thick is stored as a coil. Inner and outer radii of the coil are 40 cm and 80 cm. How long is the piece of metal?

341. Find the area bounded by the curve $x^4 + y^4 = x^2 + y^2$.

342. Suppose functions $f(x)$ and $g(x)$ are positive and their graphs are concave upward on the interval $a < x < b$. Show that if they have a relative minimum at the same value of x in the interval, the graph of $f(x)g(x)$ is concave upward on $a < x < b$.

343. Show that the function

$$f(x) = \begin{cases} x^4 \left[2 + \sin\left(\dfrac{1}{x}\right)\right], & x \neq 0 \\ 0, & x = 0 \end{cases}$$

has a relative minimum at $x = 0$ but this cannot be verified with the first derivative test.

344. The line $y = x + 1$ intersects the ellipse $4x^2 + 9y^2 = 36$ in two points P and Q. Find the point R on the ellipse in order that triangle PQR have maximum possible area.

345. Tangent lines to the curve $y = \sqrt{1 - x}$ make triangles with the positive x- and y-axes.

(a) Find the triangle with smallest area.

(b) Find the triangle with smallest perimeter.

346. (a) Show that

$$\left[\int_a^b f(x)g(x)\,dx\right]^2 \leq \int_a^b [f(x)]^2\,dx \int_a^b [g(x)]^2\,dx.$$

Hint: Consider the following function of t,

$$\int_a^b [tf(x) + g(x)]^2\,dx.$$

(b) Verify that

$$\left[\int_a^b f(x)\,dx\right]^2 \leq (b - a)\int_a^b [f(x)]^2\,dx.$$

347. Show that if $f(x)$ is a continuous periodic function with period p, then

$$\int_a^{a+p} f(x)\,dx = \int_0^p f(x)\,dx$$

for any a.

348. Use the result of Exercise 347 to show that when $f(x)$ is an odd, p-periodic continuous function, then the function

$$\int_a^x f(t)\,dt$$

is also p-periodic for any a. Is it odd?

349. A spy satellite is in circular orbit 1500 km above the earth's surface. The satellite's camera returns to earth a picture of every point on earth in direct "line of sight" from the satellite.

(a) At any instant, how much area does the satellite spy on?

(b) What is the total area under the satellite's surveillance in one complete revolution around the earth? Assume the earth is a perfect sphere with radius 6370 km.

350. Find the area bounded by the curve $x^2 y^2 = 4(x - 1)$ and the line through its points of inflection.

351. A bead slides from rest at the origin on a frictionless wire in a vertical plane to the point (x_0, y_0) under the influence of gravity (figure below). It can be shown that the time for the bead to traverse the path is

$$t = \frac{1}{\sqrt{2g}} \int_0^{x_0} \sqrt{\frac{1 + (y')^2}{y}}\,dx$$

where $g = 9.81$ and $y' = dy/dx$. The problem of finding that shape $y = f(x)$ of wire that makes t as small as possible is often called the brachistochrone problem. In the area of mathematics called "calculus of variations", it is shown that $y = f(x)$ must satisfy the differential equation

$$2y\frac{d^2y}{dx^2} + \left(\frac{dy}{dx}\right)^2 + 1 = 0.$$

(a) Show that the cycloid defined parametrically by

$$x = a(\theta - \sin\theta), \qquad y = a(1 - \cos\theta),$$

where a is a constant satisfies the differential equation.

(b) Show that if $2x_0 = \pi y_0$, then the value of a that makes the cycloid pass through (x_0, y_0) is $y_0/2$.

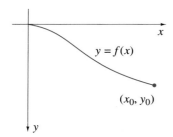

352. Find the volume of the solid of revolution when the area inside the lemniscate $(x^2 + y^2)^2 = a^2(x^2 - y^2)$ is rotated about the x-axis.

353. Find the volume of the smallest cone that can be circumscribed about a sphere of radius R.

Evaluate the integrals in Exercises 354 and 355.

354. $\displaystyle\int (x^2 - 9)^{3/2}\,dx$

355. $\displaystyle\int_{\pi/6}^{\pi/3} \ln(\tan x)\,dx$

356. Two identical rods of length L and mass M lie on the same straight line a distance d apart. Use Newton's universal law of gravitation (Exercise 267) to find the force of attraction between the rods.

357. A number x_0 is called a fixed point of a function $f(x)$ if $f(x_0) = x_0$. How many fixed points can a function have if $f'(x) < 1$ for all x?

Comprehensive Review Problems and Applications for Chapters 11–16

In Exercises 358–377 determine whether the series converges or diverges. In the case of a series with both positive and negative terms, determine whether convergence is absolute or conditional.

358. $\displaystyle\sum_{n=1}^{\infty} \frac{n^2 + 1}{3n^3 + 5}$

359. $\displaystyle\sum_{n=1}^{\infty} \frac{(n^3 + 4n)^{2/3}}{n^4 - 3n^3 + 1}$

360. $\displaystyle\sum_{n=1}^{\infty} \frac{(-1)^{n+1} n^2}{2^n}$

361. $\displaystyle\sum_{n=1}^{\infty} \frac{(-1)^n}{\sqrt{n} + 1}$

362. $\displaystyle\sum_{n=2}^{\infty} \frac{1}{n \ln n}$

363. $\displaystyle\sum_{n=1}^{\infty} n^{1/n}$

364. $\displaystyle\sum_{n=1}^{\infty} \frac{\sin(n+1)}{n^2 + 4}$

365. $\displaystyle\sum_{n=1}^{\infty} \frac{1 + 2^{-n}}{1 + 3^{-n}}$

366. $\displaystyle\sum_{n=1}^{\infty} \sqrt{\frac{n^2 + n - 1}{2n^4 + n^3 + 5}}$

367. $\displaystyle\sum_{n=1}^{\infty} \left(\frac{n^2 - 2}{2n^2 + 5}\right)^n$

368. $\displaystyle\sum_{n=1}^{\infty} \frac{n^2}{n^{3n}}$

369. $\displaystyle\sum_{n=1}^{\infty} \frac{\mathrm{Tan}^{-1}(n^2 + 5)}{2n^2 + 4}$

370. $\displaystyle\sum_{n=1}^{\infty} \frac{1 \cdot 3 \cdot 5 \cdots (2n+1)}{2 \cdot 4 \cdot 6 \cdots (2n)}\left(\frac{1}{n}\right)$

371. $\displaystyle\sum_{n=1}^{\infty} \frac{(n!)^2}{(2n+1)!}$

372. $\displaystyle\sum_{n=1}^{\infty} \frac{\sqrt{n + \sqrt{1 + n^2}}}{n^2}$

373. $\displaystyle\sum_{n=1}^{\infty} \left(\frac{1}{2} - \frac{1}{n+1}\right)^{10}$

374. $\displaystyle\sum_{n=1}^{\infty} \frac{3}{n(1 + \ln n)^3}$

375. $\displaystyle\sum_{n=1}^{\infty} \frac{3^{-2n} + 2^{-2n}}{4^{-n} + 5^{-n}}$

376. $\displaystyle\sum_{n=1}^{\infty} \frac{(n-1)^n}{n^{n-1}}$

377. $\displaystyle\sum_{n=1}^{\infty} \mathrm{Sin}^{-1}\left(\frac{1}{n^2}\right)$

In Exercises 378–403 find a general solution for the differential equation.

378. $\displaystyle\frac{dy}{dx} = \frac{1 - 2x}{y^2}$

379. $\displaystyle\frac{dy}{dx} = -\sqrt{\frac{1 - y^2}{1 - x^2}}$

380. $\displaystyle\frac{dy}{dx} = \frac{1}{x + y^2}$

381. $(1 + e^x)\dfrac{dy}{dx} + e^x y = \sin x$

382. $x\dfrac{d^2y}{dx^2} = \dfrac{dy}{dx} + \left(\dfrac{dy}{dx}\right)^3$

383. $\dfrac{d^2y}{dx^2} + \left(\dfrac{dy}{dx}\right)^2 + 1 = 0$

384. $y'' - 3y' - 4y = x^2 + e^{2x}$

385. $y'' + 4y = 3\cos 2x$

386. $y'' + 6y' + 9y = \sin x$

387. $y''' + 5y'' + 5y' + y = 2\sin 3x$

388. $\dfrac{dy}{dx} + \sin\left(\dfrac{x+y}{2}\right) = \sin\left(\dfrac{x-y}{2}\right)$

389. $x^2\dfrac{dy}{dx} + (2x + x^2)y = e^x$

390. $\dfrac{d^2y}{dx^2} + \dfrac{dy}{dx} = \sin x$

391. $\dfrac{d^2y}{dx^2} + 12\dfrac{dy}{dx} + 2y = e^{-6x}$

392. $y''' - 3y'' + 4y = e^{-x} + 3x + 2$

393. $(x^2 + 1)\dfrac{d^2y}{dx^2} + 2x\dfrac{dy}{dx} + x^2 = 0$

394. $\cos^2 x \dfrac{dy}{dx} + y = 1$

395. $xe^{2y} \cos x \, dx - y^2 \, dy = 0$

396. $y''' + 3y'' + 3y' + y = 2xe^{-x}$

397. $y''' + 3y'' + 3y' + y = e^{2x}$

398. $\sqrt{y} \dfrac{d^2 y}{dx^2} = \dfrac{dy}{dx}$

399. $y'' + 6y' + 9y = x^2 e^{-3x}$

400. $\dfrac{d^2 y}{dx^2} + 4y = 3 \cos x + \sin x$

401. $y'' + 4y = 3x \cos 2x$

402. $\dfrac{d^2 y}{dx^2} + 6 \dfrac{dy}{dx} + 9y = x \sin x - x$

403. $x^2 \dfrac{d^2 y}{dx^2} - 2x \dfrac{dy}{dx} - 4y = 0$

404. Find the components of a vector with length 5 which makes angles of 1 and 0.8 radians with the positive x- and y-axes.

405. Is there a direction in which the vector $2\hat{\mathbf{i}} - 3\hat{\mathbf{j}} + 4\hat{\mathbf{k}}$ has a component of length 6?

406. If the components of a vector along the vectors $\mathbf{v} = 2\hat{\mathbf{i}} - 3\hat{\mathbf{j}}$ and $\mathbf{w} = \hat{\mathbf{i}} + \hat{\mathbf{j}}$ are 2 and -3, what are its Cartesian components?

In Exercises 407–410 find the equation of the plane.

407. containing the point $(1, 2, -3)$ and the line $3x + 4y - 2z = 5$, $x + y + 2z = 6$

408. containing the point $(2, -1, 3)$ and perpendicular to the line $x = 2y = z + 2$

409. containing the points $(1, 4, 0)$, $(2, -2, 1)$, and $(0, 3, 2)$

410. through the point of intersection of the plane $x - 2y + 3z = 4$ and the line $x = 1 - t$, $y = 2 + 3t$, $z = 2(1 + t)$, and containing the line $(3x + 1)/2 = (1 - y)/3 = z$

In Exercises 411 and 412 find equations for the line.

411. through the point $(0, -1, -2)$ and perpendicular to the plane $2x - 3y + 5z = 6$

412. through the midpoint of the line segment joining $(-1, 2, 0)$ and $(3, 2, 2)$ and perpendicular to the plane containing the aforementioned line segment and the line $x = 3 + 3t$, $y = 2 - t$, $z = 2 + 4t$

413. Find the length of the curve $3x = y^2$, $9z = 2xy$ from $(0, 0, 0)$ to $(3, 3, 2)$.

414. Find a unit tangent vector to the curve $x = t^2 \cos 2t$, $y = t^2 \sin 2t$, $z = 1 - t$ at the point $(16\pi, 0, 1 - 4\pi)$.

415. Find a tangent vector at each point on the curve $x + y + 2z = 4$, $z = x^2 + y^2$ directed clockwise as viewed from the origin.

416. Find the unit vector which is normal to the curve $x = t^2$, $y = t^3 + t$, $z = 2t - 1$ at the point $(4, 10, 3)$ and has an x-component equal to the sum of its y- and z-components.

417. Show that the ratio of curvature to torsion for the curve $x = e^t \cos t$, $y = e^t \sin t$, $z = e^t$ is always $\sqrt{2}$.

418. Find the curvature of the curve $x = \ln(\cos t)$, $y = \ln(\sin t)$, $z = \sqrt{2}t$, $0 < t < \pi/2$.

419. Evaluate the following limits if they exist:

(a) $\lim_{(x,y) \to (0,0)} \dfrac{x^2 + y^2}{3x^2 - 4y^2}$

(b) $\lim_{(x,y) \to (1,3)} \dfrac{xy^2 + 2xy + x - y^2 - 2y - 1}{xy - 2x - y + 2}$

420. In Kintsch's model of choice behaviour, he states that

$$\sum_{n=2}^{\infty} (1 - b)^{n-2} bs^n = \dfrac{bs^2}{1 - (1 - b)s}.$$

Is this true, and would you place any restrictions on b and s?

421. Along what straight line(s) approaching $(0, 0)$ is the limit of the function $f(x, y) = (2x - 3y)/(4x + 2y)$ equal to 5?

In Exercises 422–426 find the indicated derivative.

422. $\partial z/\partial y$ if $z = \ln(y + \sqrt{x^2 + y^2})$

423. $\partial z/\partial x$ and $\partial z/\partial y$ if $z = \operatorname{Tan}^{-1} \sqrt{x^y}$

424. $\partial z/\partial u$ if $z = \operatorname{Sin}^{-1}\left(\dfrac{\sqrt{u^2 - v^2}}{\sqrt{u^2 + v^2}}\right)$

425. $\partial z/\partial x$ if $z = u^2 + u \sin v$, $u = \operatorname{Tan}^{-1}(x + y)$, $v = \ln(e^x + e^y)$

426. $\partial u/\partial x)_y$ if $u^2 \sin x - vy = 3x + 2$, $u^3 \cos x + v^3 y - v = 3xy - 4yu$

427. Find all points at which the gradient of the function $f(x, y) = x^2 y + 3x$ is equal to (a) $2\hat{\mathbf{i}} + 3\hat{\mathbf{j}}$ (b) $2\hat{\mathbf{i}} - 3\hat{\mathbf{j}}$.

428. If u and v are functions of x, y, and z defined by $uv = 3x - 2y + z$, $v^2 = x^2 + y^2 + z^2$, show that $x \dfrac{\partial u}{\partial x} + y \dfrac{\partial u}{\partial y} + z \dfrac{\partial u}{\partial z} = 0$.

429. Find the following rates of change of the function $f(x, y, z) = x^2 yz - xy^3 z$:

(a) $\partial f/\partial x$

(b) $\partial f/\partial y$

(c) $\partial f/\partial z$

(d) $\partial^2 f/\partial x^2$ at $(1, -2, 3)$

(e) at $(1, -2, 3)$ toward the point $(3, 2, 4)$

(f) with respect to t along the curve $x = t^3 + 3t + 1$, $y = \sin t$, $z = 2 \cos t$

(g) at $(1, 0, 2)$ with respect to distance travelled along the curve in (f)

(h) at $(2, 2, 2)$ with respect to distance travelled along the curve $x^3 + y^3 + z^3 = 24$, $y = x$ directed so that z increases in the first octant

(i) at $(1, 1, 3)$ normal to the surface $x^2 + y + z = 5$

(j) at $(1, 2, -1/6)$ in any direction tangent to the surface $x^2 yz - xy^3 z = 1$

430. Find the tangent plane to the surface $x^2y + xyz + z^2 = 5$ at the point $(2, 2, -1)$.

431. Find the point(s) where the curve $x + y + z = 1$, $y = x^2 - 4$ intersects the tangent plane to the surface $x^3 + y^3 + z^3 = 10$ at the point $(1, 1, 2)$.

432. Show that the tangent line to the curve $x = 18t - 3$, $y = 3t^2 + 1/t$, $z = t + 1/t^2$ at the point $(15, 4, 2)$ and the tangent plane to the surface $x^2 + y^2 - z^2 = 3x + 1$ at $(1, 2, -1)$ do not intersect.

In Exercises 433–435 find and classify the critical points of the function as yielding relative maxima, relative minima, or saddle points.

433. $f(x, y) = y\sqrt{1 + x} + x\sqrt{1 + y}$

434. $f(x, y) = xy(a - bx - cy)$ where a, b, and c are nonzero constants

435. $f(x, y) = (2x - x^2)(2y - y^2)$

436. Find the Maclaurin series for $f(x) = 1/(x^4 - x^2 + 2)$.

437. If $f(x)$ is a polynomial of degree 4, and it is known that $f(1) = 1$, $f'(1) = -2$, $f''(1) = 0$, $f'''(1) = 3$, and $f''''(1) = -5$, what is $f(-2)$?

438. Find the point in the xy-plane for which the sum of the squares of the distances from n fixed points $P_i(x_i, y_i)$, $(i = 1, \ldots, n)$ is a minimum.

439. Find the point in the xy-plane which minimizes the sum of the squares of the distances from the x-axis, the y-axis, and the line $Ax + By + C = 0$.

440. Show that a positively homogeneous function $f(x, y)$ of degree k can be expressed in the form $x^k F(y/x)$ when $x > 0$. What is the extension of this result for a positively homogeneous function $f(x, y, z)$?

441. Find the Maclaurin series for $f(x) = \ln[(1 + x)^{1+x}] + \ln[(1 - x)^{1-x}]$. What is the interval of convergence of the series?

In Exercises 442–488 evaluate the integral.

442. $\displaystyle\int_{-1}^{0} \int_{x^2}^{3x+1} (x + y)^2 \, dy \, dx$

443. $\displaystyle\int_{0}^{2} \int_{0}^{y} x \sin y \, dx \, dy$

444. $\displaystyle\int_{1}^{2} \int_{x}^{1} \frac{1}{(2x - y)^3} \, dy \, dx$

445. $\displaystyle\int_{0}^{1} \int_{0}^{y} x^2 e^{y^2} \, dx \, dy$

446. $\displaystyle\int_{0}^{2} \int_{0}^{1} |2x - y| \, dy \, dx$

447. $\displaystyle\int_{0}^{2} \int_{y}^{2} \frac{y}{\sqrt{x^2 + y^2}} \, dx \, dy$

448. $\displaystyle\int_{0}^{2} \int_{0}^{\sqrt{4-x^2}} \sqrt{x^2 + y^2} \, dy \, dx$

449. $\displaystyle\iint_{R} \frac{1}{(16 - x^2 - y^2)^{3/2}} \, dA$ where R is the area in the first quadrant bounded by the curves $y = x$, $y = \sqrt{9 - x^2}$, $y = 0$

450. $\displaystyle\int_{-1}^{1} \int_{x}^{x^2} \int_{z+x}^{2} (x + y) \, dy \, dz \, dx$

451. $\displaystyle\int_{0}^{2} \int_{0}^{1} \int_{x}^{1} z \sin y^2 \, dy \, dx \, dz$

452. $\displaystyle\int_{0}^{2} \int_{0}^{\sqrt{4-x^2}} \int_{0}^{4-\sqrt{x^2+y^2}} (x + y)^2 \, dz \, dy \, dx$

453. $\displaystyle\int_{0}^{2} \int_{0}^{\sqrt{4-x^2}} \int_{-\sqrt{4-x^2-y^2}}^{\sqrt{4-x^2-y^2}} x^2z^2 \, dz \, dy \, dx$

454. $\displaystyle\int_{C} (x^2y - z) \, ds$ where C is the straight line from $(0, 1, 2)$ to $(-1, 2, -3)$

455. $\displaystyle\int_{C} (x^2 + y^2) \, dx + (y^2 + z^2) \, dy + (z^2 + x^2) \, dz$ where C is the curve $y = 2x^2$, $z + y = 0$ from $(1, 2, -2)$ to $(-2, 8, -8)$

456. $\displaystyle\int_{C} \frac{2xy}{z} \, dx + \frac{x^2}{z} \, dy - \frac{x^2y}{z^2} \, dz$ where C is the shorter part of the curve $x^2 + y^2 = 8$, $z = y$ from $(2, 2, 2)$ to $(-1, \sqrt{7}, \sqrt{7})$

457. $\displaystyle\oint_{C} x^2y \, dx - y^3x \, dy$ where C is the boundary of the area defined by the lines $x = 1$, $y = 1$, $x + 2y = 5$

458. $\displaystyle\iint_{S} (x + y)^2 \, dS$ where S is that part of $z = x^2 + y^2$ inside $x^2 + y^2 = 4$

459. $\displaystyle\iint_{S} (x^2 + y^2 + z^2) \, dS$ where S is the sphere $x^2 + y^2 + z^2 = a^2$

460. $\displaystyle\iint_{S} (x\hat{\mathbf{i}} + y\hat{\mathbf{j}} + z\hat{\mathbf{k}}) \cdot \hat{\mathbf{n}} \, dS$ where S is that part of $x + 2y + 3z = 6$ in the first octant and $\hat{\mathbf{n}}$ is the unit normal to S with positive y-component

461. $\displaystyle\iint_{S} (x\hat{\mathbf{i}} + y^2\hat{\mathbf{j}} - z^3\hat{\mathbf{k}}) \cdot \hat{\mathbf{n}} \, dS$ where S is the surface bounding the volume $0 \le x \le 1$, $0 \le y \le 2$, $0 \le z \le 3$ and $\hat{\mathbf{n}}$ is the unit outer normal

462. $\displaystyle\oint_{C} x^2z \, dx - yz \, dy + z \, dz$ where C is the curve $x^2 + y^2 = 4$, $y = z$ directed so that x decreases in the first octant

463. $\displaystyle\int_{0}^{\sqrt{3}} \int_{-\sqrt{4-y^2}}^{-y/\sqrt{3}} (x^2 - y^2) \, dx \, dy$

464. $\displaystyle\iiint_{V} xye^z \, dV$ where V is bounded by $z = x^2 + y^2$, $z = 0$, $y = 0$, $x = 1$, $y = x$

465. $\displaystyle\iiint_{V} \sqrt{x^2 + y^2} \, dV$ where V is the volume bounded by $z = \sqrt{x^2 + y^2}$, $z = 2 - x^2 - y^2$

466. $\iiint_V (x^2 + y^2 + z^2)^2 \, dV$ where V is the volume between the spheres $x^2 + y^2 + z^2 = a^2$ and $x^2 + y^2 + z^2 = b^2$ $(a > b)$

467. $\int_C (x - y) \, ds$ where C is the curve $z = x^2 + y^2$, $y + x = 1$ from $(1, 0, 1)$ to $(3, -2, 13)$

468. $\oint_C y \, dx + x^2 \, dy$ where C is the ellipse $x^2 + 4y^2 = 4$

469. $\int_C e^y \sin 2y \, dx + xe^y (\sin 2y + 2\cos 2y) \, dy$ where C is the curve $x = t^3 + 3t$, $y = 1 + t^2$ from $t = -1$ to $t = 0$

470. $\iint_S (x^2 z^3 + y) \, dS$ where S is that part of $3x + 6y + 2z = 6$ cut out by $x = 0$, $z = 0$, $2x + z = 2$

471. $\oint_C e^{xy}(1 + xy) \, dx + (x^2 e^{xy} + xy^2) \, dy$ around the ellipse $x^2 + 4y^2 = 4$

472. $\iint_S (x^2 \hat{\mathbf{i}} + y^2 \hat{\mathbf{j}} + z^2 \hat{\mathbf{k}}) \cdot \hat{\mathbf{n}} \, dS$ where S is that part of $z = 4 - x^2 - y^2$ above the xy-plane, and $\hat{\mathbf{n}}$ is the unit upper normal to S

473. $\iint_S (x^3 \hat{\mathbf{i}} + y^3 \hat{\mathbf{j}} + z^3 \hat{\mathbf{k}}) \cdot \hat{\mathbf{n}} \, dS$ where S is the surface enclosing the volume $x^2 + y^2 \leq 1$, $0 \leq z \leq 1$ and $\hat{\mathbf{n}}$ is the unit outer normal to S

474. $\oint_C yz^2 \, dx + x^2 z \, dy + xy^2 \, dz$ where C is the smooth curve of intersection of the surfaces $x^2 + z^2 = a^2$, $y^2 + z^2 = a^2$ $(a > 0$ constant) which has a portion in the first octant (directed so that z increases in the first octant)

475. $\iint_R \frac{y}{x} \, dA$ where R is the area inside $x^2 + y^2 = 2x$ and above the x-axis

476. $\iiint_V yz \, dV$ where V is the volume in the first octant bounded by $z = 2x + 3y$, $z = 0$, $y = 0$, $x = 1$, $y = x^2 + 1$, $x = 0$

477. $\iiint_V x^2 y^2 z \, dV$ where V is the volume bounded by $x^2 + y^2 = z^2 + 1$, $z = 1$, $z = 0$

478. $\iiint_V (x^2 + y^2) \, dV$ where V is the volume inside the sphere $x^2 + y^2 + z^2 = 1$ and outside the cones $z^2 = x^2 + y^2$

479. $\int_C (x^2 + y^2)^{3/2} ds$ where C is the curve $x = y$, $z = x^2$ from $(-1, -1, 1)$ to $(1, 1, 1)$

480. $\oint_C x \, dy$ once around the polar coordinate curve $r = 4 + 3\sin\theta$

481. $\iint_S y^2(\hat{\mathbf{i}} + 2\hat{\mathbf{j}} + 3\hat{\mathbf{k}}) \cdot \hat{\mathbf{n}} \, dS$ where S is the surface $x^2 + y^2 = 1$, $0 \leq z \leq 1$ and $\hat{\mathbf{n}}$ is the unit normal pointing away from the z-axis

482. $\iint_S (xz^2 \hat{\mathbf{i}} + x^2 y \hat{\mathbf{j}} + z\hat{\mathbf{k}}) \cdot \hat{\mathbf{n}} \, dS$ where S is the surface enclosing the volume bounded by $x^2 + z^2 = 4 + y^2$, $x^2 + z^2 = 9$ and $\hat{\mathbf{n}}$ is the unit outer normal to S

483. $\oint_C (x^2 + 3) \, dx + 2xyz \, dy - x^2 y^2 \, dz$ where C is the curve $x^2 + 4z^2 = 4$, $y = x^2$ directed clockwise as viewed from the negative y-axis

484. $\int_C \left(\frac{y}{x^2 + y^2} + z^2 \right) dx - \frac{x}{x^2 + y^2} \, dy + 2xz \, dz$ where C is the upper half of the curve $y = 2x^2 - 1$, $x^2 + z^2 = 1$ from $(-1, 1, 0)$ to $(1, 1, 0)$

485. $\oint_C \sin y \, dx - \cos x \, dy$ where C is the curve $|x| + |y| = 1$

486. $\iint_S (y\hat{\mathbf{i}} + x\hat{\mathbf{j}} - \hat{\mathbf{k}}) \cdot \hat{\mathbf{n}} \, dS$ where S is the first octant part of $x^2 + y^2 + z^2 = 4$ above $z = y$ and $\hat{\mathbf{n}}$ is the unit lower normal to S

487. $\iint_S (x^3 \hat{\mathbf{i}} + 3xz\hat{\mathbf{j}} - e^{x+y}\hat{\mathbf{k}}) \cdot \hat{\mathbf{n}} \, dS$ where S is the surface enclosing the volume bounded by $z^2 + 9y^2 = 16$, $x = 0$, $x = 4$ and $\hat{\mathbf{n}}$ is the unit inner normal to S

488. $\int_C \sin(x + y + z) \, dx$ where C is the boundary of that part of $2x + 3y + 6z = 6$ in the first octant directed counterclockwise as viewed from the origin

489. What are the first three terms in the Maclaurin series for the function defined by the equation $y + \ln(1 + xy) + x = 5$.

490. At what points is the function $f(x, y) = 1/(\sin^2 \pi x + \sin^2 \pi y)$ discontinuous?

491. Show that the function $f(x, y) = \text{Tan}^{-1}\left(\frac{x + y}{x - y} \right)$ satisfies

$$\frac{\partial f}{\partial x} + \frac{\partial f}{\partial y} = \frac{x - y}{x^2 + y^2}.$$

492. Show that $f(x, y) = \ln(e^x + e^y)$ satisfies

$$\frac{\partial^2 f}{\partial x^2} \frac{\partial^2 f}{\partial y^2} - \left(\frac{\partial^2 f}{\partial x \partial y} \right)^2 = 0.$$

493. Show that $f(x, y) = \sin x + F(\sin y - \sin x)$, where F is a differentiable function, satisfies

$$\frac{\partial f}{\partial y} \cos x + \frac{\partial f}{\partial x} \cos y = \cos x \cos y.$$

494. In what direction(s) from the point $(2, 3)$ does the function $f(x, y) = x^2 y - xy$ have a rate of change equal to 4?

495. Show that the tangent plane to the surface $z = xf(y/x)$, where f is a differentiable function, always passes through the origin.

496. Find the point in space which minimizes the sum of the squares of the distances from the coordinate planes and the plane $Ax + By + Cz + D = 0$.

497. Of all right-angled triangles with perimeter P find the one with largest area.

498. Show that for small $|x|$ and $|y|$,

$$\frac{\cos x}{\cos y} \approx 1 - \frac{1}{2}(x^2 - y^2).$$

499. Use double integrals to find the volumes of the solids of revolution when the area bounded by the curves $y = x^2 + 1$, $y = 2x + 1$ is rotated about the lines (a) $x = 0$ (b) $y = 0$ (c) $x = 4$ (d) $y = -2$ (e) $x + y = 1$

500. Repeat Exercise 499 for the area bounded by the curves $x = y(y + 1)$, $x = -2y(y + 1)$.

501. Find the centroid of the area bounded by the curves $x = 0$, $y = 0$, $y = 1$, $x = \sqrt{1 + y^2}$.

502. A flat vertical plate is immersed vertically in oil with density 900 kg/m³. It is a parallelogram with sides of lengths 2 m and 1 m with the longer sides parallel to the surface of the oil. If one of the longer sides is $1/2$ m below the surface and the parallel side is 1 m below the surface, use a double integral to find the force on each side of the plate due to the surrounding oil.

503. What is the force on the plate in Exercise 502 if the longer diagonal is vertical, and the uppermost vertex is in the surface of the oil?

504. Find the centre of mass of a uniform plate whose edges are defined by the curves $y = \sqrt{4 - x^2}$, $4y = 4 - x^2$.

505. Find constants b_n in order that the function $y = f(x) = \sum_{n=1}^{\infty} b_n \sin nx$ will satisfy the differential equation

$$\frac{d^2 y}{dx^2} - 3y = \sum_{n=1}^{\infty} \frac{1}{n^2} \sin nx.$$

(Assume that the series for $f(x)$ can be differentiated term-by-term.)

In Exercises 506–509 set up, but do not evaluate, double iterated integrals for the area described.

506. The area of $x^3 + y^3 + z^2 = 1$ in the first octant

507. The area of $y = x + z^2$ inside the cylinder $x^2 + z^2 = 4$

508. The area of $x^2 + y^2 - z^2 = 9$ between $z = 1$ and $z = 2$

509. The area of $z = 1 - x^4 - y^4$ above the xy-plane

510. Find the volume of the solid of revolution when the area inside $r^2 = 4|\cos \theta|$ is rotated around the x-axis.

511. Show that a function $f(x)$ and the nth degree polynomial

$$P_n(x) = f(c) + f'(c)(x - c) + \cdots + \frac{f^{(n)}(c)}{n!}(x - c)^n$$

have the same first n derivatives at $x = c$.

512. If A, B, and C are positive, and $D < 0$, find the volume of the tetrahedron formed by the plane $Ax + By + Cz + D = 0$ and the coordinate planes.

513. Show that the tangent plane to the surface $xyz = a$ ($a > 0$ constant) cuts a constant volume from the first octant.

514. A solid object of uniform density ρ is bounded by the surfaces $z = 4x^2$, $z = 4$, $y = -1$, $y = 1$. Find its centre of mass and its moments of inertia about the coordinate axes.

515. Find the volume bounded by $z = x^2 + y^2 - \sqrt{x^2 + y^2}$ and the xy-plane.

516. Find the volume of the region inside the sphere $x^2 + y^2 + z^2 = a^2$ and outside the cones $z^2 = b^2(x^2 + y^2)$.

517. Given that $f(x, y, z) = xyz$ and $\mathbf{F}(x, y, z) = x^2\hat{\mathbf{i}} - yz\hat{\mathbf{j}} + xz\hat{\mathbf{k}}$, find (a) ∇f (b) $\nabla \cdot \mathbf{F}$ (c) $\nabla \times \mathbf{F}$ (d) $\mathbf{F} \times (\nabla \cdot \mathbf{F})\nabla f$ (e) $\nabla f \cdot \nabla \times (x^2\mathbf{F})$ (f) $\nabla(f^2) \times (\nabla \times \nabla f)$ (g) $\nabla \cdot \mathbf{F}(\nabla f \times y\mathbf{G})$

518. Show that the gradient of the function $\text{Tan}^{-1}(y/x)$ is irrotational and solenoidal in any domain that does not contain points on the y-axis.

519. Calculate $\dfrac{d^{20}}{dx^{20}}[(x^2 + 1)\sin x]$.

520. Show that vector fields of the form $\mathbf{F} = f(\sqrt{x^2 + y^2 + z^2})(x\hat{\mathbf{i}} + y\hat{\mathbf{j}} + z\hat{\mathbf{k}})$ are irrotational.

521. Show that the function $y = \dfrac{1}{3} \displaystyle\int_0^x f(t)e^{t-x} \sin 3(x - t)\, dt$ satisfies the differential equation

$$\frac{d^2 y}{dx^2} + 2\frac{dy}{dx} + 10y = f(x).$$

522. Can the function $f(x, y) = xy/(x^2 + y^2)$, $(x, y) \neq (0, 0)$ be made continuous at $(0, 0)$?

523. Find all points at which the sum of the first partial derivatives of the function $f(x, y, z) = x^3 + y^3 - z^3$ is equal to (a) 3 (b) -3 (c) 0.

524. Verify that $f(x, y) = e^x(x \cos y - y \sin y)$ is harmonic.

525. Suppose that $f(x, y) = xF(ax + by) + yG(ax + by)$ where a and b are constants and F and G are arbitrary twice differentiable functions. Show that $\dfrac{\partial^2 f}{\partial x^2} - 2\dfrac{\partial^2 f}{\partial x \partial y} + \dfrac{\partial^2 f}{\partial y^2} = 0$ if and only if $a = b$.

526. In what direction(s) from the point $(1, 2, 3)$ does the function $f(x, y, z) = xyz + x^2 + y^2 + z^2$ have a rate of change equal to -2?

527. Find all tangent planes to the ellipsoid $x^2/a^2 + y^2/b^2 + z^2/c^2 = 1$ which have equal intercepts on the coordinate axes.

528. Radioactive substances decay at a rate proportional to the amount of radioactive material present at any time. If 10% of a radioactive sample decays in 1 year, how long does it take for 50% to decay (called the half-life of the material)?

529. Solve the equation $\dfrac{dy}{dx} = y^2 - 4$ subject to (a) $y(0) = 1$ (b) $y(0) = -2$.

530. Verify that the length of the curve $4y = (z + x)^2$, $4y^2 + 3z^2 = 3x^2$ from $(0, 0, 0)$ to any point (x, y, z) where $x > 0$ is $\sqrt{2}x$. Hint: Regard x as the parameter and use implicit differentiation to find dy/dx and dz/dx.

531. Find a continuous function satisfying

$$\frac{dy}{dx} + y = f(x), \quad y(0) = 1,$$

where
$$f(x) = \begin{cases} 3, & 0 \le x \le 2 \\ 0, & x > 2 \end{cases}.$$

532. Find all points at which the gradient of the function $f(x, y, z) = x^2 y^2 z^2 + xyz$ is equal to $2\hat{\mathbf{i}} + 3\hat{\mathbf{j}} - \hat{\mathbf{k}}$.

533. If $f(x, y) = [F(ax - by) + G(ax + by)]/x$, where a and b are constants and F and G are twice differentiable functions, show that $b^2 \dfrac{\partial}{\partial x}\left(x^2 \dfrac{\partial f}{\partial x}\right) = a^2 x^2 \dfrac{\partial^2 f}{\partial y^2}$.

534. If y is defined as a function of x by $y = f(x, t)$, $F(x, y, t) = 0$, show that $\dfrac{dy}{dx} = \dfrac{f_x F_t - f_t F_x}{f_t F_y + F_t}$.

535. Find point(s) on the curve $x^2 + y^2 + z^2 = 1$, $y = z$ at which the rate of change of the function $f(x, y, z) = 3x + 2y - 2z$ with respect to length along the curve is equal to 1.

536. A model of learning proposed by Bush and Mosteller assumes that on the nth trial of a sequence of trials, the probability p_n of obtaining a certain response is related to the probability on the immediately preceding trial by

$$p_n = a + (1 - a - b)p_{n-1}$$

where a and b are constants which depend on the amounts of reward and inhibition. Prove that an explicit formula for p_n is

$$p_n = p_0(1 - a - b)^n + \frac{a}{a + b}[1 - (1 - a - b)^n]$$

where p_0 is the initial probability of response.

537. Find the second moment of area of the region bounded by the curves $y = 1 - x^2$, $x + y + 1 = 0$ about the lines (a) $y = 0$ (b) $x = 0$ (c) $y = 1$.

538. The area bounded by the curves $y = x^2$ and $y = 1$ (x and y in metres) is to be lowered into water until the force on each side is 10^5 N. Find the position of the horizontal side relative to the surface.

539. Repeat Exercise 538 if the force must be 1000 N.

540. Find the area of $z/3 = 1 - |x| - |y|$ above the xy-plane.

541. Find the area of that part of $y = 1 + 2\sqrt{x^2 + z^2}$ to the left of the plane $y = 2$.

542. Find the area inside each petal of the rose $r = a \sin n\theta$.

543. Find the volume bounded by the surfaces $3y + z = 6$, $3y + 2z = 12$, $x + y + z = 6$, $x = 0$, $y = 0$, $z = 0$.

544. Find the centre of mass of the uniform solid bounded by $4x^2 + y^2 = 4$, $z = 0$, $y + z = 2$.

545. A man eats a diet of 2 500 calories per day; 1 200 of them go to basic metabolism (i.e., get used up automatically). He also spends 16 calories per day for each kilogram of body mass in exercise. Assume that the storage of calories as fat is 100% efficient and that 1 kilogram of fat contains 10 000 calories. Find his mass as a function of time, and sketch this function if initially his mass is (a) 100 kg (b) 70 kg. Is he on a diet?

546. Prove that the function

$$f(x) = \frac{x^2}{2} + \frac{1}{2}x\sqrt{x^2 + 1} + \ln\sqrt{x + \sqrt{x^2 + 1}}$$

satisfies the differential equation $x\dfrac{dy}{dx} + \ln\left(\dfrac{dy}{dx}\right) = 2y$.

547. A 100 gm mass falls from rest under gravity. It is also acted on by air resistance which is 2 N when the speed of the mass is 10 m/s. Find the velocity of the mass when the resistive force is proportional to (a) velocity (b) the square of velocity (c) the square root of velocity. Find distance fallen in (a) also.

548. Evaluate $\displaystyle \lim_{n \to \infty} \frac{\dfrac{1}{4} + \dfrac{1}{8} + \dfrac{1}{16} + \cdots + \dfrac{1}{2^n}}{1 + \dfrac{2}{5} + \dfrac{4}{25} + \dfrac{8}{125} + \cdots + \left(\dfrac{2}{5}\right)^{n-1}}.$

549. Prove that if the amount of radioactive material in a sample is A_1 at time t_1 and the amount is A_2 at time $t_2 > t_1$, then the half-life of the material is $(t_2 - t_1)\ln 2 / \ln(A_1/A_2)$. (See Exercise 528 for the rate of decay of radioactive materials.)

550. In the figure below the four squares each have area one. A rectangle with one vertex at the origin, and two sides along the x- and y-axes has sides of length x and y. Find a formula for the amount of shaded area enclosed by this rectangle as a function of x and y.

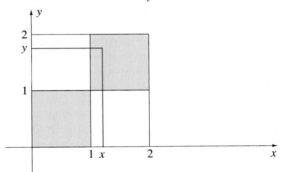

551. (a) The logistic model in population dynamics for the size $N(t)$ of a population is

$$\frac{dN}{dt} = kN\left(1 - \frac{N}{C}\right),$$

where k is a constant (the difference between the natural birth rate and the natural death rate), and C is a constant called the *carrying capacity* of the environment. Solve this differential equation subject to the condition $N(0) = N_0$.

(b) Solve the differential equation in (a) if the term $1 - N/C$ is replaced by $1 - \ln N / \ln C$.

(c) Does the solution in (a) or (b) approach C more rapidly?

552. Find a formula for $d^2 y/dx^2$ in terms of derivatives of $f(x, y)$ when $f(x, y) = 0$ defines y as a function of x.

553. A function $f(x, y)$ satisfies Laplace's equation 13.11. It is also known to be radially symmetric; that is, when expressed in polar coordinates r and θ, it is independent of θ. Find the form of $f(x, y)$. (Hint: see equation 13.24)

554. The area of a triangle with sides of length a, b, and c is $A = \sqrt{s(s-a)(s-b)(s-c)}$ where s is half the perimeter of the triangle. If the length of each side is increased by 1%, what is the approximate percentage increase in its area?

555. Suppose the line $Ax + By + C = 0$ has nonzero x- and y-intercepts. Find the point in the xy-plane which minimizes the sum of the distances from the x-axis, the y-axis, and the line.

556. Beer containing 6% alcohol is pumped into a vat containing 10 000 L of beer with 4% alcohol at 100 L per minute. If well-mixed beer is removed at the same rate, what is the percentage of alcohol in the vat as a function of time t? When is the beer 5% alcohol?

557. Show that if $\sum_{n=1}^{\infty} a_n^2$ and $\sum_{n=1}^{\infty} b_n^2$ are convergent series, then $\sum_{n=1}^{\infty} a_n b_n$ is absolutely convergent.

558. Find $f^{(6)}(0)$ if $f(x) = x^2/\sqrt{1-x^2}$.

559. Find the first three nonzero terms in the Maclaurin series for $\ln(\cos x)$.

560. Find the Maclaurin series for $\cos^4 x$.

561. Find the Maclaurin series for $x(2+3x)^{3/2}$.

562. Of all right-angled triangles with area A, find the one with smallest perimeter.

563. (a) Show that $y_1(x) = (\sin x)/\sqrt{x}$ is a solution of the differential equation
$$x^2 \frac{d^2 y}{dx^2} + x \frac{dy}{dx} + \left(x^2 - \frac{1}{4}\right) y = 0.$$

(b) Show that a second solution $y_2(x) = (\cos x)/\sqrt{x}$ can be obtained by setting $y_2(x) = v(x) y_1(x)$.

(c) What is a general solution of the differential equation?

564. A pyramid with rectangular horizontal base is inscribed in a sphere of radius r (figure below). Find a formula for the volume in the pyramid in terms of x and y.

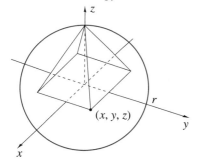

565. Of all planes passing through the point (a, b, c) in the first octant, find the one for which the tetrahedron formed by the plane and the three coordinate planes has the least possible volume.

566. Find $\sqrt{4.000\,000\,000\,01}$ accurate to 20 decimal places.

567. Find the Taylor series about $x = 1$ for the function $\sqrt[3]{7+x}$.

568. Use the Maclaurin series for $f(x) = (1+x)e^{-x} - (1-x)e^x$ to find the sum of the series $\sum_{n=1}^{\infty} n/(2n+1)!$.

569. Find the first moment of area of the region bounded by the curves $y = x^2$, $y = 2 - x^2$ about the line $x + y + 1 = 0$.

570. Find the area of $x^2 + y^2 + z^2 = 4$ between $z = \sqrt{x^2 + y^2}$ and $z = 2\sqrt{x^2 + y^2}$.

571. Find the first moment of area of the bifolium $r = \cos^2 \theta \sin \theta$ about the x-axis.

572. A 100 gm mass is attached to a vertical spring with constant 1000 N/m, and slowly lowered to its equilibrium position. If it is given an initial speed of 4 m/s downward, find its position as a function of time if a damping force proportional to velocity acts on the mass and has magnitude (a) 20 N (b) 40 N (c) 80 N when speed is 2 m/s.

573. (a) Show that if $y = (1-x)^{-a} e^{-ax}$, then
$$(1-x)\frac{dy}{dx} = axy.$$

(b) Now show that
$$(1-x)\frac{d^{n+1}y}{dx^{n+1}} - (n+ax)\frac{d^n y}{dx^n} - an\frac{d^{n-1}y}{dx^{n-1}} = 0.$$

574. Inscribed in a circle of radius R is a square. A circle is then inscribed in the square, a square in the circle, and so on and so on. Find the limit of the sum of the areas of all the circles and the limit of the sum of the areas of all the rectangles.

575. Show that when a surface $z = f(r, \theta)$ in cylindrical coordinates projects one-to-one onto an area S_{xy} in the xy-plane, its area is given by
$$\iint_{S_{xy}} \sqrt{1 + \left(\frac{\partial z}{\partial r}\right)^2 + \frac{1}{r^2}\left(\frac{\partial z}{\partial \theta}\right)^2}\, r\, dr\, d\theta.$$

576. Use the result of Exercise 575 to find the area of the sphere $x^2 + y^2 + z^2 = 4$ inside the cylinder $x^2 + y^2 = 1$.

577. When chemicals A, B, and C are brought together, 1 gm of A reacts with 2 gm of B and 3 gm of C to form 6 gm of substance D. Initially there are 10 gm of A, 20 gm of B, and 30 gm of C. Find the amount of D as a function of time under each of the following conditions:

(a) the rate at which D is formed is proportional to the amounts of A, B, and C present in the reaction

(b) the rate at which D is formed is proportional to the amounts of B and C present in the reaction.

578. Find a polynomial approximating the indefinite integral
$$\int \frac{1}{\sqrt[4]{1+x^4}}\, dx$$
to within 10^{-6} on the interval $0 \leq x \leq 0.8$.

579. Find the volumes bounded by the following surfaces:

(a) $4x^2 + y^2 = 4$, $4x + 3y + z = 12$, $z = 1$

(b) $4x^2 + y^2 = 4$, $4x + 3y + z = 12$, $z = 24$

580. Find a polynomial in x approximating $f(x) = \cos^3 x$ on $0 \leq x \leq 0.5$ with maximum error 10^{-6}.

581. Suppose the initial amounts of A, B, and C are all 10 gm in Exercise 577. When is the amount of D equal to 15 gm if there is 10 gm in 5 minutes?

582. What is the minimum value of $f(x, y) = x^3 + y^3$ on that part of the line $Ax + By = C$ (where $A \geq 0$, $B \geq 0$, and $C \geq 0$) for which $x \geq 0$ and $y \geq 0$.

583. Find the moment of inertia about the line $y = 2x$ of a plate with uniform mass per unit area ρ if its edges are defined by $y = -x$, $y = x$, $x = 1$.

584. If $f(x, y)$ and its first partial derivatives are continuous in a simpy-connected domain containing a non-selfintersecting, piecewise smooth, closed curve C, and $\partial f/\partial y = \partial f/\partial x$ inside C, what is the value of

$$\oint_C f(x, y)(dx + dy)?$$

585. Show that if $f(x, y)$ is harmonic in a simply-connected domain D and C is a closed, piecewise smooth curve in D, then

$$\oint_C \frac{\partial f}{\partial y} dx - \frac{\partial f}{\partial x} dy = 0.$$

586. If $r = \sqrt{x^2 + y^2 + z^2}$, and $f(r)$ is a continuous function, show that

$$\iint_S f(r)\, dS = 4\pi a^2 f(a)$$

when S is the sphere $x^2 + y^2 + z^2 = a^2$.

587. Show that if $y_1(x)$ is a solution of the homogeneous, linear differential equation $y'' + P(x)y' + Q(x)y = 0$, then a second solution is

$$y_2(x) = y_1(x) \int \frac{e^{\int P(x)\,dx}}{[y_1(x)]^2} dx.$$

Hint: Let $y_2(x) = v(x)y_1(x)$.

588. (a) Use the transformation of Exercise 22 in Section 16.9 to solve $x^2 y'' - 4xy' + 4y = 0$.

   (b) Solve the differential equation by setting $y = x^m$ and finding appropriate values for m.

589. Use the techniques in Exercise 588 to solve $x^2 y'' + 3xy' + 4y = 0$.

590. Use the methods of Exercise 588(b) and Exercise 587 to solve $4x^2 y'' + 8xy' + y = 0$.

591. Find the points on the ellipsoid $x^2 + 96y^2 + 96z^2 = 96$ closest to and farthest from the plane $3x + 4y + 12z = 288$.

592. Assuming that the ellipsoid $x^2/a^2 + y^2/b^2 + z^2/c^2 = 1$ and the plane $Ax + By + Cz + D = 0$ do not intersect, find the points on the ellipsoid closest to and farthest from the plane.

Answers

1. $(0, 0)$, all points on $y = x$ in the first quadrant, all points on the negative x-axis, all points on the negative y-axis, and all points in the third quadrant

2. (a)

(b)

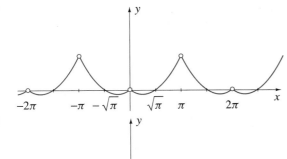

3. $-1/2$

4. -8

5. $1/\sqrt{2}$

6. does not exist

7. does not exist

8. ∞

9. $-1/4$

10. does not exist

11. -1

12. $4/3$

13. 0

14. $1/2$

15. $-\infty$

16. does not exist

17. does not exist

18. 1

19. $2x(1 - 9x)(1 - 3x)^3$

20. $x(2 - 3x^2)/(1 - 3x^2)^{3/2}$

21. $(2\cos 3x \cos 2x - 3\sin 3x \sin 2x)/\cos^2 3x$

22. $6(2x + 1)^2 \ln (x + 1) + (2x + 1)^3/(x + 1)$

23. $(3x^2 + 8x - 7)e^{3x}$

24. $e^{2x}/[(e^{2x} + 1)\sqrt{\ln (e^{2x} + 1)}]$

25. $-1/\sqrt{x - 4x^2}$

26. $2x \csc^2(x^2 - 1)[1 - 2x^2 \cot (x^2 - 1)]$

27. $\{1 + \sec (1 - 2x^3)[1 + 6x^3 \tan (1 - 2x^3)]\}/[1 + \sec (1 - 2x^3)]^2$

28. $-1/[6\sqrt{2 - x}(1 + \sqrt{2 - x})^{2/3}]$

29. $(2 - 2xy - y^3)/(x^2 + 3xy^2)$

30. $[\cos (x + y) - y]/[x - \cos (x + y)]$

31. $(6x - 1 - 3x^2 e^y)/(x^3 e^y - 2)$

32. $[xy + (1 + x^2 y^2)\text{Tan}^{-1} (xy)]/(1 + x^2 y^2 - x^2)$

33. $-3v^2 \sin (v^3 + 2)/(x + 1)^2$

34. $-\sec^2 x/[(1 - 2u)\sqrt{u^2 - u}]$

35. $(2\sqrt{v} - 1)(2\sqrt{x} + 1)/[16\sqrt{vx}\sqrt{v - \sqrt{v}}\sqrt{x + \sqrt{x}}]$

36. $te^{-3t}(2 - 3t)(x + 1)/x$

37. $(2t^3 + 6)/(t^3 - t)$

38. $(4/3)\cot t$

39. $2\theta/(1 - \theta^2)$

40. $(3e^{-3v} + 2v)/(1 - 3e^{3v})$

41. $8x^2/(3 + 2x)^2$

42. $4/(x^2 - 4x)$

43. $24/(8 + x^3)$

44. $16x^2/(1 + 4x^4)$

45. $405x^2/[2\sqrt{1 + 3x}]$

46. $1/\sqrt{x^2 + 9}$

47. $\sqrt{x^2 - 4}/x$

48. $98/(x^3\sqrt{7x^2 + 49})$

49. $8(3 - x^2)^{3/2}$

50. $-15x^3\sqrt{1 - x^2}$

51. $(16 - 2x^2)/\sqrt{16 - x^2} - 4\sin(x/4)$

52. $125/[x(25 - x^2)^{3/2}]$

53. $3\sqrt{7}/[(9 - x^2)\sqrt{16 - x^2}]$

54. $-87x/[2(6 + 3x + 4x^2)^{3/2}]$

55. $1/\sqrt{8x - x^2}$

56. $49/(14x - x^2)^{3/2}$

57. $1/\sqrt{6x + x^2}$

58. $-5/[\sqrt{2}x\sqrt{3x^5 - 2}]$

59. $\sqrt{7}/(8 + 6\sin x)$

60. $6/(4\cos^2 x + 9\sin^2 x)$

61. $18\cos 3x/(1 - \cos 3x)^2$

62. $x^3\cos x$

63. $-6\sin 2x\sqrt{1 + 9\sin^2 2x}$

64. $21\cos 3x/\sqrt{1 + 49\sin^2 3x}$

65. $2x\mathrm{Cot}^{-1} x$

66. $(1/x^2)\mathrm{Sec}^{-1}(2x)$

67. $13e^{3x}\sin 2x$

68. $10/(5 + 3e^{2x})$

69. $169xe^{2x}\sin 3x$

70. -1, $(1, 0)$

71. $x + y = 0$, $x + 25y = 0$

72. $5/13$, $170/2197$

75. $y = 2x + 1$

76. $-2/3$

78. $x/3 - (1/6)\ln(3 + 4e^{2x}) + C$

79. $[1/(6\sqrt{5})]\mathrm{Tan}^{-1}(2e^{3x}/\sqrt{5}) + C$

80. $(1/3)(x - 1)^3\ln x - x^3/9 + x^2/2 - x + (1/3)\ln x + C$

81. $x\ln(x^2 + 9) - 2x + 6\mathrm{Tan}^{-1}(x/3) + C$

82. $x/2 - (1/12)\sin 6x + C$

83. $(1/5)\sin(5x + 1) - (1/15)\sin^3(5x + 1) + C$

84. $(1/20)\cos 5x - (1/6)\cos 3x - (1/44)\cos 11x + C$

85. $\ln\left(\sqrt{x^2 + 5} + x\right) + C$

86. $\sqrt{x^2 - 3} + 5\ln|x + \sqrt{x^2 - 3}| + C$

87. $-(x^2 + 1)^{3/2}/(3x^3) + C$

88. $-\sqrt{7 - x^2}/(7x) + C$

89. $(x+1)/\left(16\sqrt{x^2 + 2x + 5}\right) - (x+1)^3/[48(x^2+2x+5)^{3/2}] + C$

90. $\mathrm{Sin}^{-1}(x - 1) - \sqrt{2x - x^2} + C$

91. $\ln|x + 1| - \ln(x^2 + 5) - (3/\sqrt{5})\mathrm{Tan}^{-1}(x/\sqrt{5}) + C$

92. $x + 3\ln|x| + (2/\sqrt{3})\mathrm{Tan}^{-1}[(x + 1)/\sqrt{3}] + C$

93. $(1/4)\ln|\csc 2x - \cot 2x| - (1/4)\csc 2x\cot 2x + C$

94. $-\cot x - (1/3)\cot^3 x + C$

95. $(1/4)\ln(2x^2 + 3) + 3\ln|x| - 4/(x - 1) + C$

96. $x/125 - (1/125)\ln(2e^x + 5) - (21/100)/(2e^x + 5) +$ $(29/40)/(2e^x + 5)^2 + C$

97. $-\sqrt{1 - x^2} + 2\mathrm{Sin}^{-1} x + C$

98. $x/\sqrt{5 - x^2} - \mathrm{Sin}^{-1}(x/\sqrt{5}) + C$

99. $(1/\sqrt{2})\mathrm{Sin}^{-1}[\sqrt{2}(x - 1)/\sqrt{5}] + C$

100. $3(x^2 - x + 5)\ln(x^2 - x + 5) - 3x^2 + 3x + C$

101. $x + 2\cot(x/2) - (2/3)\cot^3(x/2) + C$

102. $\sec x + \cos x + C$

103. $-\ln(1 - \sin x) + C$

104. Divergent

105. Absolutely convergent

106. Absolutely convergent

107. Conditionally convergent

108. Divergent

109. Divergent

110. Absolutely convergent

111. Divergent

112. Divergent

113. Absolutely convergent

114. Convergent

115. Convergent

116. Divergent

117. Convergent

118. Convergent

119. Divergent

120. Convergent

121. Divergent

122. Divergent

123. Convergent

124. $(0.442, 0.407)$, $(-2.442, -12.407)$

125. No

126. $\pi/4 + \alpha/2$

127. 1

129. $1/2$

130. $6(\sqrt{2} + \mathrm{Tan}^{-1}\sqrt{2})$

131. $(3e^2 + 1)/4$

132. $2/\pi + (4\sqrt{2} - 5)/3$

133. $\pi/2$

134. $2\pi[6\ln(3/2) - 2/3]$

135. $\pi[10\ln 2 - (\ln 2)^2 - 5]/2$

136. $-1/2 + \ln 3$

137. $M\omega^2 L^2/6$

138. $2/\sqrt{5}$ m, $1/\sqrt{5}$ m

139. -0.548

140. $\pi ab^3\rho/4$, $\pi a^3b\rho/4$

141. $f(x)$, $f(x) = 1/x$

142. $|(1 - b)s| < 1$

143. $3/4$ of length of diameter from P

144. $2\sqrt{3}$ m/s toward E

145. $1/2$

146. $6x^2\cos x^2 - 4x^4\sin x^2$

147. Yes, no

148. $(15 - 8\ln 4)\pi/2$

149. $(1/3)\sum_{n=0}^{\infty}[(-1)^n - 1/2^n]x^{2n}$, $|x| < 1$

150. $-187/8$

151. (c)

(d)

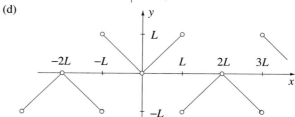

152. $4/\sqrt{5 + 16\ln 2 + 16(\ln 2)^2}$

154. $1/e$ **157.** No

158. $3 + \pi/\sqrt{3}$, relative maximum

160. Concave upward

161.

162. $-16/135$ m/s^2

163. $\left(\pm a^2/\sqrt{a^2 + b^2}, \mp b^2/\sqrt{a^2 + b^2}\right)$

164. $x_1 + (A/\omega)[\sin(\omega t_2 + \phi) - \sin(\omega t_1 + \phi)]$

165. $\epsilon(b - a)$ **166.** $4\sqrt{3}a$

167. $1.23 \times 10^3 L^4$ J   **168.** b/a

169. $2r(1 - \cos\alpha)\sqrt{1 + 4\cot^2\alpha}/(3\alpha)$

170. $(1/2)\ln[x/(2 - x)]$ **171.** (a) 0.4939579 (b) 0.4939576

172. $(\pi R, 5R/6)$ **173.** $3\pi/2$

174. $\sum_{n=1}^{\infty} x^{2n}/[n(2n - 1)]$, $-1 \le x \le 1$

175. $5 - 6x + 37x^2/2$ **176.** $-1/20$

177. $-1/[n^2(n^2 + 3)]$ **178.** $22/\sqrt{137}$

181. -20.2 m/s   **182.** It doesn't

183. $L\ln(x/L)$ **184.** $(3\pi/16, 0)$

185. Not necessarily   **187.** $x - x^3/3$, 3.2×10^{-5}

189. $(x - 12/13)^2 + (y - 18/13)^2 = 49/13 \pm 12/\sqrt{13}$

190. $100(1.025)^{\llbracket t/3 \rrbracket}$ **191.** $(1/3)x^{-2/3}$

192. 0   **193.** 0

195. $(x^2 - 379)\sin x - 40x\cos x$

202. $1/(\ln 2 - 1)$, $\ln 2/(2 - 2\ln 2)$, relative maximum at $x = 1$, relative minimum at $x = 2$

203. Walk all the way if $V > \pi v/2$; paddle all the way if $V < \pi v/2$; walk all the way or paddle all the way if $V = \pi v/2$

204. $kv_1/(k - 1)$ **206.** decreasing at $62/\sqrt{21}$ m/s

207. 1

208. (a) 24 m/s (b) $2 < t < 3$, $t > 4$ (c) 19 m (d) 7.592 m/s

209. (a) $(25/4)(13 + 3e^{-t/625})$ (b) $(5/4)(65 - 9e^{-t/625})$

210. $-1/2$ **213.** $-(1 + y^3)\cos x^2$

214. 0

215. (a) $8\sqrt{2}/3$ (b) $y = 2x + 2^{1/3} - 1$

216. $\pi^2/16$

217. $\rho g\pi H^2(R^2 + 2rR + 3r^2)/12$

218. (a) 36297 N (b) increases by 981 N

219. $(80\pi + 117\sqrt{3})/30$ **220.** $2r^2h/3$

221. (a) $2q\sigma L/[ka\sqrt{L^2 + 4a^2}]$ (b) $2q\sigma/(ka)$

222. (a) No (b) Yes (c) No, Yes

223. $2\{1 + \text{Sin}^{-1}[(x - 4)/3]\}/\{1 - \text{Sin}^{-1}[(x - 4)/3]\}$

225. 0

226. $(b^2 - a^2)^{-1}\ln(b/a)$ if $b^2 \ne a^2$; $1/(2a^2)$ if $b^2 = a^2$

227. $3 - e$ **228.** $2\sqrt{2}$

229. $-\sqrt{x^2 + 5}/(10x^2) - [1/(10\sqrt{5})]\ln|\sqrt{x^2 + 5} - \sqrt{5})/x| + C$

230. $6\text{Sin}^{-1}(x/2) + 2x\sqrt{4 - x^2} + (x/4)(2 - x^2)\sqrt{4 - x^2} + C$

231. $a^2\pi/8$ **232.** $\pi/4$

233. $(2x^2 - 9)/[81x\sqrt{9 - x^2}] + C$

234. $3\text{Sin}^{-1}(x/3) - \sqrt{9 - x^2} + C$

235. $-(3/2)\ln|3 - 4x| + (\sqrt{6}/4 - 1)\ln|1 + \sqrt{6} - x| - (\sqrt{6}/4 + 1)\ln|x - 1 + \sqrt{6}| + C$

236. $y_0(A + y_0)/[y_0 + Ae^{-k(A+y_0)t}]$

237. (a) $x = a(\theta\sin\theta + \cos\theta)$, $y = a(\sin\theta - \theta\cos\theta)$ (b) $\tan\theta$ (c) 0 (d) 7.73

238. $3\pi a^2/8$ **240.** $(5\pi + 18\sqrt{3})a^2/32$

241. $b^2\sqrt{a^2 + b^2}/a - ab\ln[(b + \sqrt{a^2 + b^2})/a]$

243. $d(q^n - 1)/(q - 1)$, where $q = 1 - p/100$

244. 3/10   **246.** 270

247. $-x^2/2 - x^4/12 - x^6/45$

248. $1 + \sum_{n=1}^{\infty}(-1)^n(2^{2n-1} + 2^{4n-3})x^{2n}/(2n)!$

249. $2\sqrt{2}[x + 9x^2/4 + 27x^3/32 + \sum_{n=4}^{\infty}(-1)^{n+1}3^n \times [1 \cdot 3 \cdots (2n - 7)]x^n/[2^{2n-2}(n - 1)!]]$, $|x| \le 2/3$

250. 2.000 000 000 002 5   **255.** $(\pm 1, \pm 1)$

257. $|\sin x|^x[(x\cot x + \ln|\sin x|)^2 + 2\cot x - x\csc^2 x]$

258. 5367

259. length $= 40/3$ m, width $= 2\sqrt{2}/3$ m

260. 175.475 m/s^2 **261.** (b) $2^{2/3}k/(ab)$

264. $\pi L(2R^2 + r^2)/3$ **265.** $9.81\rho a^2 b^2(\rho/\overline{\rho})^{1/3}/4$

266. (b) $d = 0.3916r$ **267.** $M(L + d)/d$

268. $f(x) = x$ or $f(x) = (ax + b)/(cx - a)$

269. $\pi|x|/(2x)$

270. $x(\text{Sin}^{-1}x)^2 + 2\sqrt{1 - x^2}\text{Sin}^{-1}x - 2x + C$

271. $-(1/x)\text{Cos}^{-1}x - \ln|(1 - \sqrt{1 - x^2})/x| + C$

272. $\ln|\csc x - \cot x| + \sec x + C$

273. $-(1/4)(x + 4)/\sqrt{x^2 - 16} + C$

274. $(x^3/3)\text{Tan}^{-1}(2x) + (1/48)\ln(1 + 4x^2) - x^2/12 + C$

275. $(1/13)e^{2x+3}[3\sin(3x - 1) + 2\cos(3x - 1)] + C$

276. $-x\cot(x/2) + C$

277. $(x/5)[\sin(2\ln x) - 2\cos(2\ln x)] + C$

278. $(x/3)\tan 3x - (1/9)\ln|\sec 3x| + C$

279. (a) $2/(4x^2 - 8x + 5)^{3/2}$, $x = 1$ (c) equal to zero

281. (b) $(0, \pm b)$, $(\pm a, 0)$

282. (a) $(x - 1)^2 + (y + 7/2)^2 = 1/4$

283. $y = x + 6 - 6\sqrt{3}$ **285.** $\beta\lambda N/(1 - \lambda)$

286. $2 + (x - 1)/12 + 2\sum_{n=2}^{\infty}(-1)^{n+1} \times$
        $[2 \cdot 5 \cdot 8 \cdots (3n - 4)](x - 1)^n/(24^n n!)$, $-7 \le x \le 9$

287. $1/(2e)$ **288.** $(-1, \ln 2)$

289. $(-2.570, -0.8933)$ **292.** 4.4934

293. $2/5$ cm^2/s   **294.** $\pi^2(2\pi^2 - 3)/12$

295. $9.81\rho ab^2/3$ **296.** $-f''(x)/[f'(x)]^3$

298. (a) $5\pi^2 R^3$ (b) $6\pi^3 R^3$ (c) $\pi R^3(9\pi^2 - 16)/6$

299. $2\pi a^2(2 - \sqrt{2})$ **300.** $32\pi a^2/5$

301. $2\pi R^2$, $4R^2$ **302.** $\mathrm{Tan}^{-1}(-1/2) + n\pi$

303. $-1 + x/5 + x^2/25$ **304.** 7

305. 0 if $n > m$, 1 if $n = m$, does not exist if $n < m$

308. $(2n \pm 1)\pi/3$ **309.** $19.62\rho\ell L(2d + \ell\sin\alpha)$

311. $(R, (4n + 1)\pi R/2 - R)$, n an integer

312. $x + \sum_{n=1}^{8} \{(-1)^n[1 \cdot 5 \cdots (4n - 3)]/[4^n n!(4n + 1)]\}x^{4n+1}$

314. (a) 2.33 mm (b) 3.11 mm

315. (a) $600t/(20t + 3)$ (b) $30(1 - e^{-10t/3})/(2 - e^{-10t/3})$
        (c) $60(1 - e^{-20t/3})/(2 - e^{-20t/3})$

316. $1 - 3x^2/2 + 7x^4/8 - 61x^6/240 + 547x^8/13440$
        $- 703x^{10}/172800$

317. $0 \le c \le 1$ **318.** 4

320. $(x - 1 - \sqrt{3}/\sqrt{7})^2 + (y - 3\sqrt{3}/2 - 2/\sqrt{7})^2 = 1$

321. (a) $(0, 0)$, $(3a/4, 27a^{7/2}/128)$
        (b) $((3 - \sqrt{3})a/4, \sqrt{6\sqrt{3} - 9a^2}/8)$

322. No   **323.** $(-a/\sqrt{2}, -b/\sqrt{2})$

324. 11.8 m   **325.** $14\pi r^{5/2}/(15\sqrt{2}gac)$

326. $(1 - \cos 1)/2$ **329.** $2/r$

330. Perpendicular to the diagonal through the given vertex

331. 1.265 m/s   **332.** $8\int_0^r \sqrt{R^2 - z^2}\sqrt{r^2 - z^2}\,dz$

333. (a) $k > 1$ (b) $k > 1$ (c) Diverges for all k

339. (c) $16\pi R$ m/s, 0   **340.** 9.0×10^4 cm

341. $\sqrt{2}\pi$ **344.** $(9/\sqrt{13}, -4/\sqrt{13})$

345. (a) Smallest area is 4/3 (b) Smallest perimeter is 4.25

348. Not necessarily

349. (a) 4.86×10^3 km^2 (b) 5.10×10^8 km^2

350. $8\left[\sqrt{1 + 2/\sqrt{3}} - \mathrm{Tan}^{-1}\sqrt{1 + 2/\sqrt{3}}\right]$

352. $\pi a^3[3\sqrt{2}\ln(3 + 2\sqrt{2}) - 4]/24$

353. $8\pi R^3/3$

354. $(x^3/4)\sqrt{x^2 - 9} - (45x/8)\sqrt{x^2 - 9} + (243/8)$
        $\ln|x + \sqrt{x^2 - 9}| + C$

355. 0

356. $(GM^2/L^2)\ln[(L + d)^2/(2Ld + d^2)]$

357. One at most

Answers to Comprehensive Review Problems and Applications for Chapters 11–16

358. Divergent   **359.** Absolutely convergent

360. Absolutely convergent   **361.** Conditionally convergent

362. Divergent   **363.** Divergent

364. Absolutely convergent   **365.** Divergent

366. Divergent   **367.** Absolutely convergent

368. Convergent   **369.** Convergent

370. Divergent   **371.** Convergent

372. Convergent   **373.** Divergent

374. Convergent   **375.** Divergent

376. Divergent   **377.** Convergent

378. $(C + 3x - 3x^2)^{1/3}$

379. $x\sqrt{1 - y^2} + y\sqrt{1 - x^2} = C$

380. $x = -y^2 - 2y - 2 + Ce^y$

381. $y = (C - \cos x)/(1 + e^x)$

382. $y = \pm\sqrt{1 - Dx^2}/\sqrt{D} + C$

383. $y = \ln|\cos(C - x)| + D$

384. $y = C_1 e^{4x} + C_2 e^{-x} - x^2/4 + 3x/8 - 13/32 - (1/6)e^{2x}$

385. $y = C_1\cos 2x + (C_2 + 3x/4)\sin 2x$

386. $y = (C_1 + C_2 x)e^{-3x} + (2/25)\sin x - (3/50)\cos x$

387. $y = C_1 e^{-x} + C_2 e^{(\sqrt{3}-2)x} + C_3 e^{-(\sqrt{3}+2)x} +$
        $(3\cos 3x - 11\sin 3x)/260$

388. $\ln|\tan(y/4)| = -2\sin(x/2) + C$

389. $y = (e^x + Ce^{-x})/(2x^2)$

390. $y = Ce^{-x} + D - (1/2)(\cos x + \sin x)$

391. $y = C_1 e^{(\sqrt{34}-6)x} + C_2 e^{-(\sqrt{34}+6)x} - (1/34)e^{-6x}$

392. $y = (C_1 + C_2 x)e^{2x} + (C_3 + x/9)e^{-x} + 3x/4 + 1/2$

393. $y = C\mathrm{Tan}^{-1} x - x^2/6 + (1/6)\ln(x^2 + 1) + D$

394. $y = 1 + Ce^{-\tan x}$

395. $4(x\sin x + \cos x) + e^{-2y}(2y^2 + 2y + 1) = C$

396. $y = (C_1 + C_2 x + C_3 x^2 + x^4/12)e^{-x}$

397. $y = (C_1 + C_2 x + C_3 x^2)e^{-x} + (1/27)e^{2x}$

398. $y = C$ or $x + C = \sqrt{y} - (D/2)\ln|2\sqrt{y} + D|$

399. $y = (C_1 + C_2 x + x^4/12)e^{-3x}$

400. $y = C_1 \cos 2x + C_2 \sin 2x + \cos x + (1/3)\sin x$

401. $y = (C_1 + 3x/16)\cos 2x + (C_2 + 3x^2/8)\sin 2x$

402. $y = (C_1 + C_2 x)e^{-3x} + (2x/25)\sin x - (3x/50)\cos x +$ $(7/1200)\sin x + (97/1600)\cos x - x/9 + 2/27$

403. $y = C_1 x^4 + C_2/x$

404. $(2.70, 3.48, \pm 5.57)$

405. No

406. $((2\sqrt{3} - 9\sqrt{2})/5, -(2\sqrt{3} + 6\sqrt{2})/5)$

407. $13x + 16y + 2z = 39$

408. $2x + y + 2z = 9$

409. $11x + 3y + 7z = 23$

410. $6x + 7y - 25z = 5$

411. $x/2 = -(y+1)/3 = (z+2)/5$

412. $1 - x = (y-2)/5 = (z-1)/2$

413. 5

414. $(8\pi, 32\pi^2, -1)/\sqrt{1024\pi^4 + 64\pi^2 + 1}$

415. $-(1 + 4y)\hat{\mathbf{i}} + (4x + 1)\hat{\mathbf{j}} + 2(y - x)\hat{\mathbf{k}}$

416. $\pm(0.521, 0.284, -0.805)$

418. $(1/\sqrt{2})\sin 2t$

419. (a) Does not exist (b) 16

420. $|(1 - b)s| < 1$

421. $18x + 13y = 0$

422. $1/\sqrt{x^2 + y^2}$

423. $y\sqrt{x^y}/[2x(1 + x^y)]$, $\sqrt{x^y}\ln x/(2 + 2x^y)$

424. $\sqrt{2}u|v|/[(u^2 + v^2)\sqrt{u^2 - v^2}]$

425. $(2u + \sin v)/[1 + (x + y)^2] + ue^x \cos v/(e^x + e^y)$

426. $[y(u^3 \sin x + 3y) + (1 - 3v^2 y)(u^2 \cos x - 3)]/$ $[2u \sin x(3v^2 y - 1) + y(3u^2 \cos x + 4y)]$

427. (a) $(\pm\sqrt{3}, \mp\sqrt{3}/6)$ (b) Not possible

429. (a) $2xyz - y^3 z$ (b) $x^2 z - 3xy^2 z$ (c) $x^2 y - xy^3$ (d) -12 (e) $-102/\sqrt{21}$ (f) $(2xyz - y^3 z)(3t^2 + 3) + (x^2 z - 3xy^2 z)(\cos t) - 2(x^2 y - xy^3)\sin t$ (g) $2/\sqrt{10}$ (h) $34/\sqrt{6}$ (i) 0 (j) 0

430. $3x + y + z = 7$

431. $(-2, 0, 3)$, $(1, -3, 3)$

433. $(-2/3, -2/3)$, saddle point

434. $(0, 0)$, $(0, a/c)$, $(a/b, 0)$ give saddle points; $(a/(3b), a/(3c))$ gives a relative maximum if $abc > 0$ and a relative minimum if $abc < 0$

435. $(0, 0)$, $(2, 0)$, $(0, 2)$, $(2, 2)$ give saddle points; $(1, 1)$ gives a relative maximum

436. $(1/3)\sum_{n=0}^{\infty}[(-1)^n - 1/2^n]x^{2n}$, $|x| < 1$

437. $-187/8$

438. $\left((1/n)\sum_{i=1}^{n} x_i, (1/n)\sum_{i=1}^{n} y_i\right)$

439. $(-AC/(2A^2 + 2B^2), -BC/(2A^2 + 2B^2))$

440. $x^k F(y/x, z/x)$

441. $\sum_{n=1}^{\infty} x^{2n}/(2n^2 - n)$, $|x| \leq 1$

442. $-233/140$

443. $2\sin 2 - \cos 2 - 1$

444. $-1/12$

445. $1/6$

446. $19/6$

447. $2(\sqrt{2} - 1)$

448. $4\pi/3$

449. $(4 - \sqrt{7})\pi/(16\sqrt{7})$

450. $-68/105$

451. $1 - \cos 1$

452. $24(\pi + 2)/5$

453. $128\pi/105$

454. $13\sqrt{3}/4$

455. $618/5$

456. -3

457. $-28/5$

458. $(136\sqrt{17} + 1)\pi/60$

459. $4\pi a^4$

460. 18

461. -36

462. -4π

463. $\sqrt{3}$

464. $(e^2 - 2e)/8$

465. $13\pi/30$

466. $4\pi(a^7 - b^7)/7$

467. $\sqrt{2}(51\sqrt{51} - 3\sqrt{3})/12$

468. 2π

469. $4e^2 \sin 4$

470. $409/540$

471. $-\pi/2$

472. $64\pi/3$

473. $5\pi/2$

474. $\pi a^4/2$

475. 1

476. $787/84$

477. $5\pi/64$

478. $\pi(\pi + 2)/8$

479. $4(6\sqrt{3} + 1)/15$

480. $41\pi/2$

481. 0

482. $160\pi\sqrt{5}/3$

483. 16π

484. $\pi/2$

485. $4(\cos 1 - 1)$

486. $\pi/\sqrt{2} - 4(2 - \sqrt{2})/3$

487. $1024\pi/3$

488. $(3/2)(2\cos 2 - \cos 1 - \cos 3)$

489. $5 - 6x + 37x^2/2$

490. (x, y) where x and y are integers

494. $(36 \pm 2\sqrt{69}, 8 \mp 9\sqrt{69})$

496. $x = -AD/[2(A^2 + B^2 + C^2)]$, $y = -BD/[2(A^2 + B^2 + C^2)]$, $z = -CD/[2(A^2 + B^2 + C^2)]$

497. Sides of lengths $(\sqrt{2} - 1)P/2$, $(\sqrt{2} - 1)P/2$, $\sqrt{3 - 2\sqrt{2}}P$

499. (a) $8\pi/3$ (b) $104\pi/15$ (c) 8π (d) $368\pi/30$ (e) $52\sqrt{2}\pi/15$

500. (a) $2\pi/15$ (b) $\pi/2$ (c) $39\pi/10$ (d) $3\pi/2$ (e) $7\sqrt{2}\pi/10$

501. $(0.581, 0.531)$

502. 6.62×10^3 N

503. 1.28×10^4 N

504. $\bar{x} = 0$, $\bar{y} = 32/(15\pi - 20)$

505. $-1/(n^4 + 3n^2)$

506. $(1/2)\int_0^1 \int_0^{(1-x^3)^{1/3}}$ $\sqrt{(4 - 4x^3 - 4y^3 + 9x^4 + 9y^4)/(1 - x^3 - y^3)}\, dy\, dx$

507. $4\sqrt{2}\int_0^2\int_0^{\sqrt{4-x^2}}\sqrt{1+2z^2}\,dz\,dx$

508. $4\int_0^{\pi/2}\int_{\sqrt{10}}^{\sqrt{13}}r\sqrt{(2r^2-9)/(r^2-9)}\,dr\,d\theta$

509. $4\int_0^1\int_0^{(1-x^4)^{1/4}}\sqrt{1+16x^6+16y^6}\,dy\,dx$

510. $64\pi/15$ **512.** $-D^3/(6ABC)$

514. $\bar{x}=0,\ \bar{y}=0,\ \bar{z}=12/5,\ I_x=4832\rho/63,\ I_y=7904\rho/105,\ I_z=256\rho/45$

515. $\pi/6$ **516.** $(4/3)\pi a^3 b/\sqrt{1+b^2}$

517. (a) $yx\hat{\mathbf{i}}+xz\hat{\mathbf{j}}+xy\hat{\mathbf{k}}$ (b) $3x-z$ (c) $y\hat{\mathbf{i}}-z\hat{\mathbf{j}}$
(d) $(3x-z)[-(xy^2z+x^2z^2)\hat{\mathbf{i}}+xy(z^2-x^2)\hat{\mathbf{j}}+z(x^3+y^2z)\hat{\mathbf{k}}]$
(e) $-x^2z(3xz+y^2)$ (f) $\mathbf{0}$ (g) $(3x-z)y[xz(y^2+xz)\hat{\mathbf{i}}+xy(x^2-z^2)\hat{\mathbf{j}}-z(x^3+y^2z)\hat{\mathbf{k}}]$

519. $(x^2-379)\sin x-40x\cos x$

522. No

523. (a) $x^2+y^2=z^2+1$ (b) $x^2+y^2+1=z^2$ (c) $z^2=x^2+y^2$

526. (a,b,c) where $8a+7b+8c=-2$

527. $x+y+z=\pm\sqrt{a^2+b^2+c^2}$

528. 6.58 years

529. (a) $y=2(3-e^{4x})/(3+e^{4x})$ (b) $y=-2$

531. $y=3-2e^{-x}$ for $0\le x\le 2$, $(3e^2-2)e^{-x}$ for $x>2$

532. $(0.9848, 0.6565, -1.9696)$

535. $(2\sqrt{2}/3,\pm\sqrt{2}/6,\pm\sqrt{2}/6)$, $(-2\sqrt{2}/3,\pm\sqrt{2}/6,\pm\sqrt{2}/6)$

537. (a) $729/140$ (b) $63/20$ (c) $423/28$

538. 7.25 m below   **539.** 0.484 m above

540. $2\sqrt{19}$ **541.** $\sqrt{5}\pi/4$

542. $\pi a^2/(4n)$ **543.** 12

544. $\bar{x}=0,\ \bar{y}=-1/2,\ \bar{z}=5/4$

545. (a) $(25/4)(13+3e^{-t/625})$ (b) $(5/4)(65-9e^{-t/625})$

547. (a) $(g/2)(1-e^{-2t}),\ gt/2+(g/4)(e^{-2t}-1)$
(b) $\sqrt{5g}(1-e^{-2\sqrt{g/5}t})/(1+e^{-2\sqrt{g/5}t})$
(c) $t-(g/20)\ln g=\sqrt{v/10}-(g/20)\ln(g-2\sqrt{1})$

548. $3/10$

550. xy for $0\le x\le 1,\ 0\le y\le 1$; x for $0\le x\le 1,\ 1\le y$ and $1\le x\le 2,\ 2\le y$; y for $1\le x,\ 0\le y\le 1$ and $2\le x,\ 1\le y\le 2$; $xy-x-y+2$ for $1\le x\le 2,\ 1\le y\le 2$; 2 for $x\ge 2,\ y\ge 2$

551. (a) $CN_0/[N_0+(C-N_0)e^{-kt}]$ (b) $Ce^{\ln(N_0/C)e^{-kt/\ln C}}$
(c) Function in (a)

552. $[2f_{xy}f_xf_y-f_{xx}(f_y)^2-f_{yy}(f_x)^2]/(f_y)^3$

553. $C\ln(x^2+y^2)+D$

554. 2

555. $(0,0)$ **556.** 69.3 minutes

558. 270   **559.** $-x^2/2-x^4/12-x^6/45$

560. $1+\sum_{n=1}^\infty(-1)^n(2^{2n-1}+2^{4n-3})x^{2n}/(2n)!$

561. $2\sqrt{2}[x+9x^2/4+27x^3/32+\sum_{n=4}^\infty(-1)^{n+1}3^n\times[1\cdot 3\cdots(2n-7)]x^n/[2^{2n-2}(n-1)!]],\ |x|\le 2/3$

562. Sides of lengths $\sqrt{2A},\ \sqrt{2A},\ 2\sqrt{A}$

563. (c) $(C_1\sin x+C_2\cos x)/\sqrt{x}$

564. $(4/3)|xy|(r\pm\sqrt{r^2-x^2-y^2})$

565. $x/a+y/b+z/c=3$ **566.** $2.000\,000\,000\,002\,5$

567. $2+(x-1)/12+2\sum_{n=2}^\infty(-1)^{n+1}\times[2\cdot 5\cdots(3n-4)](x-1)^n/(24^n n!),\ -7\le x\le 9$

568. $1/(2e^2)$ **569.** $8\sqrt{2}/3$

570. $4\pi(4-\sqrt{10})/\sqrt{5}$ **571.** $\pi/256$

572. (a) $-(2\sqrt{3}/75)e^{-50t}\sin(50\sqrt{3}t)$ (b) $-4te^{-100t}$
(c) $(\sqrt{3}/150)[e^{-(200+100\sqrt{3})t}-e^{(-200+100\sqrt{3})t}]$

573. (a) $e^{-50t}[(1/40560)\cos(50\sqrt{3}t)-(6763\sqrt{3}/253500)\sin(50\sqrt{3}t)+(3/33800)\sin(20t)-(1/40560)\cos(20t)$
(b) $(1/29280-9763t/2440)e^{-1000t}+(3/48800)\sin(20t)-(1/29280)\cos(20t)$

574. $2\pi R^2,\ 4R^2$ **576.** $8\pi(2-\sqrt{3})$

577. (a) $60-60/\sqrt{1+7200kt}$ (b) $60-60/(1+60kt)$

578. $x+\sum_{n=1}^8\{(-1)^n[1\cdot 5\cdots(4n-3)]/[4^n n!(4n+1)]\}x^{4n+1}$

579. (a) 22π (b) 24π

580. $1-3x^2/2+7x^4/8-61x^6/240+547x^8/13440-703x^{10}/172800$

581. (a) 13.08 minutes (b) 12.05 minutes

582. $C^3/(A^{3/2}+B^{3/2})^2$ **583.** $13\sqrt{5}\rho/30$

584. 0   **588.** $C_1x+C_2x^4$

589. $[C_1\cos(\sqrt{3}\ln x)+C_2\sin(\sqrt{3}\ln x)]/x$

590. $(C_1+C_2\ln x)/\sqrt{x}$

591. $(9,1/8,3/8),\ (-9,-1/8,-3/8)$

592. $(\pm a^2A/\sqrt{a^2A^2+b^2B^2+c^2C^2},\ \pm b^2B/\sqrt{a^2A^2+b^2B^2+c^2C^2},\ \pm c^2C/\sqrt{a^2A^2+b^2B^2+c^2C^2})$

Index

M

N

O

Derivative Formulas

$$\frac{d}{du}(u+v) = \frac{du}{dx} + \frac{dv}{dx}$$

$$\frac{d}{dx}(u^n) = nu^{n-1}\frac{du}{dx}$$

$$\frac{d}{dx}(uv) = u\frac{dv}{dx} + v\frac{du}{dx}$$

$$\frac{d}{dx}\left(\frac{u}{v}\right) = \frac{v\dfrac{du}{dx} - u\dfrac{dv}{dx}}{v^2}$$

$$\frac{d}{dx}\sin u = \cos u \frac{du}{dx}$$

$$\frac{d}{dx}\cos u = -\sin u \frac{du}{dx}$$

$$\frac{d}{dx}\tan u = \sec^2 u \frac{du}{dx}$$

$$\frac{d}{dx}\cot u = -\csc^2 u \frac{du}{dx}$$

$$\frac{d}{dx}\sec u = \sec u \tan u \frac{du}{dx}$$

$$\frac{d}{dx}\csc u = -\csc u \cot u \frac{du}{dx}$$

$$\frac{d}{dx}\text{Sin}^{-1} u = \frac{1}{\sqrt{1-u^2}}\frac{du}{dx}$$

$$\frac{d}{dx}\text{Cos}^{-1} u = \frac{-1}{\sqrt{1-u^2}}\frac{du}{dx}$$

$$\frac{d}{dx}\text{Tan}^{-1} u = \frac{1}{1+u^2}\frac{du}{dx}$$

$$\frac{d}{dx}\text{Cot}^{-1} u = \frac{-1}{1+u^2}\frac{du}{dx}$$

$$\frac{d}{dx}\text{Sec}^{-1} u = \frac{1}{u\sqrt{u^2-1}}\frac{du}{dx}$$

$$\frac{d}{dx}\text{Csc}^{-1} u = \frac{-1}{u\sqrt{u^2-1}}\frac{du}{dx}$$

$$\frac{d}{dx}a^u = a^u \ln a \frac{du}{dx}$$

$$\frac{d}{dx}e^u = e^u \frac{du}{dx}$$

$$\frac{d}{dx}\log_a u = \frac{1}{u}\log_a e \frac{du}{dx}$$

$$\frac{d}{dx}\ln u = \frac{1}{u}\frac{du}{dx}$$

$$\frac{d}{dx}\sinh u = \cosh u \frac{du}{dx}$$

$$\frac{d}{dx}\cosh u = \sinh u \frac{du}{dx}$$

$$\frac{d}{dx}\tanh u = \text{sech}^2 u \frac{du}{dx}$$

$$\frac{d}{dx}\coth u = -\text{csch}^2 u \frac{du}{dx}$$

$$\frac{d}{dx}\text{sech}\, u = -\text{sech}\, u \tanh u \frac{du}{dx}$$

$$\frac{d}{dx}\text{csch}\, u = -\text{csch}\, u \coth u \frac{du}{dx}$$

$$\frac{d}{dx}\text{Sinh}^{-1} u = \frac{1}{\sqrt{u^2+1}}\frac{du}{dx}$$

$$\frac{d}{dx}\text{Cosh}^{-1} u = \frac{1}{\sqrt{u^2-1}}\frac{du}{dx}$$

$$\frac{d}{dx}\text{Tanh}^{-1} u = \frac{1}{1-u^2}\frac{du}{dx}$$

$$\frac{d}{dx}\text{Coth}^{-1} u = \frac{1}{1-u^2}\frac{du}{dx}$$

$$\frac{d}{dx}\text{Sech}^{-1} u = \frac{-1}{u\sqrt{1-u^2}}\frac{du}{dx}$$

$$\frac{d}{dx}\text{Csch}^{-1} u = \frac{-1}{|u|\sqrt{1+u^2}}\frac{du}{dx}$$